The Design of Building Structures

Wolfgang Schueller

Professor of Architecture
University of Florida

PRENTICE HALL, Upper Saddle River, New Jersey 07458

Library of Congress Cataloging-in-Publication Data
Schueller, Wolfgang,
   The design of building structures : Wolfgang Schueller.
      p.   cm. -- (Prentice-Hall international series in civil
engineering and engineering mechanics)
   Includes bibliographical references and index.
   ISBN 0-13-346560-8
   1. Structural design.   2. Structural analysis (Engineering)
I. Title. II. Series.
TA658.S15   1995
624/1'77--dc20   95-13790
                                                      CIP

Acquisitions editor: William Stenquist
Editorial/production supervision: TKM Productions
Cover design: Wendy Alling Judy
Manufacturing buyer: Donna Sullivan

©1996 by Prentice-Hall, Inc.
Simon & Schuster / A Viacom Company
Upper Saddle River, New Jersey 07458

Printed in the United States of America
10 9 8 7 6 5 4 3 2

ISBN 0-13-346560-8

Prentice-Hall International (UK) Limited, *London*
Prentice-Hall of Australia Pty. Limited, *Sydney*
Prentice-Hall Canada Inc., *Toronto*
Prentice-Hall Hispanoamericana, S.A., *Mexico*
Prentice-Hall of India Private *Limited, New Delhi*
Prentice-Hall of Japan, Inc., *Tokyo*
Simon & Schuster Asia Pte. Ltd., *Singapore*
Editora Prentice-Hall do Brasil, Ltda., *Rio de Janeiro*

Contents

Preface   viii

Acknowledgments   xii

chapter 1 **Introduction to Building Structures   1**

1.1   **General Design Determinants   6**
*Environmental Context   6 / Introduction to Architectural Design
Considerations   7 / Building Functions   7 / Fire Safety   8 /
Construction and Economy   12*

1.2   **The Early Beginnings of the Science of Mechanics   15**

1.3   **Structure as Geometry   22**

1.4   **The Support Structure   29**
*The Building as a Whole   31 / Structure Systems   32 /
Force Flow   42 / The Effect of Scale and Rules of Thumb   45*

1.5   **Basic Structure Components   49**
*Beams   49 / Floor and Low-slope Roof Structures   50 /
Columns   53 / Walls   55 / Joints   62*

chapter 2 **Basic Structure Concepts   66**

2.1   **Building Loads   69**
*Dead Loads   72 / Live Loads   75 / Wind and Earthquake Loads   82 /
Water and Earth Pressure Loads   84 / Loads Due to Restrained Volume
Change: Hidden Loads   84 / Dynamic Loads   86 /
Abnormal Loads   88 / Load Combinations   88*

2.2   **Statics   89**

Properties of Forces   89 / Force Systems   91 /

2.3   Internal Forces in Beams and Columns   102
*The Relationship Among Load, Shear, and Moment   103 /
Beam Types: The Effect of Boundary Conditions   105 /
Load Types and Load Arrangements   108*

2.4   Properties of Sections   113
Centroids   113 / The Second Moment of Area or Moment of Inertia   116

2.5   General Material *Properties*   120
*Mechanical Material Properties   120 / Building Materials   124 /
Comparison of Materials   128*

2.6   Stress and Deformation   130
*Simple Stresses   130 / Stresses in Beams   132 / Elastic Deflections of
Shallow Beams   138 / General Notes on Bending Members   143*

2.7   Torsion   148
*Circular Sections   150 / Noncircular Closed Sections   151 /
Open Sections   152*

2.8   Prestressing   153
Application of the Prestress Principle to Various Structure Systems   156 /

2.9   Soil and Foundations   161
*Soil Properties   163 / Foundation Systems   165 / Sizing Shallow
Footings   167*

Problems   172

**chapter 3   Approximate Structural Design of Common
Member Types   175**

3.1   Materials   176
*Structural Steel   176 / Structural Wood   179 /
Reinforced Concrete   186*

3.2   Design of Beams   191
*Steel Beams   193 / Steel Plate Girders   206 / Wood Beams   209 /
Built-up Wood Beams   216 / Reinforced Concrete Beams   224 /
Prestressed Concrete Beams   239 / Composite Steel Concrete
Beams   246*

3.3   Design of Tension Members   248
*Single Structural Steel Shapes and Built-up Steel Members   248 /
Round and Square Steel Rods and Flat Steel Bars   253 / Steel
Cables   254 / Wood Tension Members   254*

3.4   Design of Columns and Beam Columns   255
*Steel Columns   261 / Wood Columns   270 / Reinforced Concrete
Columns   275*

3.5   Design of Connections   284
*Steel Connections   286 / Wood Connections   299 / Reinforced Concrete
Connections   301*

**3.6   Design of Simple Reinforced Concrete Column and
Wall Footings   302**

Problems   309

chapter 4   **The Lateral Stability of Buildings   316**

          4.1   **Lateral Load Action   318**
                  Wind Loads   319 / Seismic Loads   328 / Load Combinations   345 /

          4.2   **The Response of Buildings to Lateral Force Action   347**
                  Diaphragm Action of Horizontal or Sloped Building Planes   350
                  Lateral Building Deflection   355 /
                  The Distribution of Lateral Forces to the Vertical Lateral Force-resisting
                  Structures   365 / Overturning   378

          4.3   **Water and Earth Pressure Loads: Basement Walls**
                    and Retaining Walls   380
                  Basement Walls   382 / Retaining Walls   385

                  Problems   391

chapter 5   **Frames, Arches, and Trusses   393**

          5.1   **Introduction to Frames   394**
                  General Design Considerations   396

          5.2   **Rectangular Frames   403**
                  One-story, Single-bay Frames   403 / One-story, Multibay
                  Frames   430 / Multistory Rigid Frames   441 /
                  The Vierendeel Truss   444

          5.3   **Pitched Frames   446**
                  Introduction to General Structural Concepts   448 / Pitched Roof
                  Structures in Residential Construction   452 / Gable Frame
                  Structures   460

          5.4   **Arches   467**
                  Historical Development   468 / Response of the Arch to Loading   473 /
                  Preliminary Design of Common Arches   485 / Minor Masonry
                  Arches   489

          5.5   **Trusses   491**
                  Basic Truss Characteristics   492 / Preliminary Design of Ordinary
                  Trusses   500

          5.6   **Long-span Skeleton Structures   508**
                  Cantilever Structures   508 / Beam Buildings   514

                  Problems   515

chapter 6   **Space Frames   521**

          6.1   **The Development of Space Frames   522**

          6.2   **Simple Single-layer Space Frames   524**
                  Introduction to Space Statics   527

          6.3   **Multilayer Space Frames   535**
                  Design Considerations   535 / Flat Space Frame Roofs   548 /
                  Approximate Design of Flat Double-layer Space Frames   551 /
                  Other Double-layer Space Frames   560

                  Problems   562

chapter 7   **Folded Plate Structures   564**

7.1   **Folded Plate Structure Types   565**

7.2   **Structural Design of Ordinary Folded Plate Roof Structures in Concrete, Steel, and Wood   569**

          Problems   581

chapter 8   **Shell Structures   584**

8.1   **Introduction to Thin-Shell and Skeletal-Shell Structures   585**
          Bent Surface Structures in Nature   585 / The Development of Bent Surface Structures in Architecture   587 / Surface Classification   598 / Membrane Forces   603 / Shell Structure Characteristics   604 / Shell Material   607

8.2   **Cylindrical Shells   617**
          Cylindrical Shell Types   618 / Membrane Forces in Circular Cylindrical Shells   622 / Approximate Design of Barrel Shells   624 / Cylindrical Grid Structures   635 / Asymmetrical Shell Beams   637

8.3   **Thin-shell Domes and Skeletal Domes   637**
          Dome Types   638 / Membrane Forces in Spherical Dome Shells   647 / Structural Behavior of Dome Shells   650 / Approximate Design of Concrete Dome Shells   654 / Approximate Design of Skeletal Domes   658 / Approximate Design of Other Dome Shapes   663

8.4   **The Hyperbolic Paraboloid   666**
          Hypar Types   666 / Membrane Analysis   672 / Supporting Structure Systems   676
          Structural Behavior and Approximate Design of Hypar Shells   678

8.5   **Other Shell Forms   686**

          Problems   692

chapter 9   **Tension Roof Structures   695**

9.1   **General Principles   697**
          Stability Considerations   698 / Anchorage of Tensile Forces   700 / Materials   702 / Loads   707

9.2   **The Single Cable   708**
          Cable Action under Transverse Loads   710 / Cable Action under Radial Loads   725 / The Prestretched Cable   729 / Dynamic Behavior   731 /

9.3   **Cable Beams and Cable Trusses   732**

9.4   **Cable-supported Roof Structures   736**
          Single-strut and Multistrut Cable-supported Beams   737 / Cable-stayed Bridges   738 / Cable-stayed Roof Structures   741

9.5   **Simply Suspended Roofs   751**
          Double-layer Simply Suspended Roofs   755

9.6   **Prestressed Membranes and Cable Nets   756**
          Edge-supported Saddle Roof Structures   757 / Arch- supported Saddle Surface Structures and Mast-supported Conical Surface Structures   759 / Approximate Design of Anticlastic Prestressed Membranes   763 /

Approximate Design of Edge Members   770

9.7   Hybrid Tensile Surface Structures   775

9.8   Pneumatic Structures   779
*Air-supported Structures   780 / Air-inflated Structures   800 /
Some Other Soft Shell Structures   803*

9.9   Typical Membrane Roof Details   803

9.10   Tensile Foundations   805

Problems   808

chapter 10   High-rise Building Structures   813

10.1   General Introduction to High-rise Structure Systems   817
Considerations of Efficiency   821

10.2   Force Flow in High-rise Building Structures   822

**10.3   Introduction to Basic Behavior of High-rise
Building Structures   825**

10.4   Brief Investigation of Common High-rise Structures   827
The New Generation of Tall Structures   837

Appendix A: Tables for Structural Design   842

Appendix B: Answers to Selected Problems   855

Bibliography and References   857

Index   863

Preface

This book grew out of a need to make the field of structures a part of the domain of building and accessible to building designers in general, rather than relying on separate literature in the fields of structural engineering, architecture, construction, and history. Structural engineering books usually treat the subjects of analysis and design of structures separately and place their emphasis on solving given problems where the physical reality often is lost in mathematical abstractions, so that the field of structures becomes nearly inaccessible to other building professionals. Similarly, architects use broad conceptual interpretations developed during the 1960s as based on functional considerations that result in generalizations only partially relevant today and that do not respond to the explosion of knowledge that has taken place since then. Neither architects nor engineers reinforce the discovery of form finding in structures because often the conceptual design determinants are neglected to a large degree.

Furthermore, there is a lack of communication on the academic level among structure, construction, and building design. In courses on structures, the structural element, together with its loading, is usually already isolated from its context so that primary emphasis can be placed on various analytical methods of solution. In contrast, this book emphasizes the setting up of problems based on actual buildings; loads, member shapes, and boundary conditions are derived from the layout of a structure, from construction methods, and from detailing and appearance requirements. The student may have to resolve a complex continuous system into several basic ones so that the structural elements can be evaluated and proportioned quickly by possibly using deformed shape and load path approximations for the quick preliminary investigation of indeterminate structures.

By deriving the mathematical and behavioral concepts in the field of mechanics from an actual building, the student can better understand the purpose and logic of structure and be helped to perceive how it functions and affects the building form. But, most importantly, this experience teaches students to see mathematics as an efficient means of communication, rather than an end in itself. Through this approach, the student will develop a sense for the behavior of structures, the magnitude of stresses, the structure in its deformed state, and the structure as a collapse mechanism, therefore gaining confidence in dealing with the concept of structure, not only to analyze and design structures but also to develop a basic feel for structural planning.

To develop an understanding for the building structure as a system that supports and as a pattern that orders space, it must be seen in the context of the building's anatomy in addition to other design determinants. Therefore, buildings are also studied from a geometrical, aesthetic, historical, functional, environmental, and construction point of view. Through this approach, the various design specialists can learn to appreciate each other's concerns and thus develop a basis for successful teamwork. At the same time, the treatment of structures is broadened and enriched by relating it to the traditionally separate fields of construction, structural analysis and design, materials, history, geometry, and graphics. Nearly equal emphasis has been placed on the descriptive, analytical, and graphical investigations of the topics. The visual portion of the book is essential to the education of the building designer; it leads to structural engineering intuition, which, to a large extent, is founded in visual experiences; the graphical analysis of hundreds of images attempts to support this point.

Not all the illustrations are diagrams of analytical nature or precisely explained in accordance with engineering tradition. Some of the drawings are not simple and clearly defined; on the contrary, they are complex and presented similar to a collage to deliberately be open to interpretation and thus assist in the discovery of architecture. They tell a story, forcing the reader to participate in the design through the drawing; there is no intention to define the limits (i.e., to solve the problem) by explaining exactly what is portrayed in the figures. These types of presentations stand by themselves, only connected with the respective text in its overall spirit. They project beyond the scope of the written descriptions to transfer a feeling to the designer that there is much more than the limits set by the statements. This approach relates to visual thinking, a form of art, rather than to graphical analysis—the illustrations are supposed to initiate curiosity with respect to the sources of building design, in addition to providing esthetic pleasure.

Descriptions of the numerous building cases should help students to expand their ability to communicate about structure as well as to relate abstract principles to physical reality; they should understand that structure does not just happen on a mathematical level. A wealth of building structures is presented in an ordered manner and comparative fashion according to form, and structural behavior. The structures range from small- to long-span structures, from single-bay to multibay to high-rise structures, from rigid to flexible construction, and from skeleton to surface structures. The building cases were selected solely to exemplify structural concepts and to develop a feeling for structure and form, rather than to support specific architectural styles or structural acrobatics. The presentation of actual buildings will help designers to explore the three-dimensional order of structure, as well as to visualize the force flow and spatial interaction of structural elements. Architects will be able to perceive the purpose of structure and will learn to control the interplay of material and nonmaterial spaces during the design stage. Similarly, structural engineers will be able to translate the real structure into an idealized structure, that is, the structural behavior of building elements into abstract images with which they are already familiar.

The writing of this book was also motivated by current methods of teaching the field of structure in colleges. Architectural students in the design studios are primarily concerned with presentation and appearance, independent of the building as an organism and how it is made. They are familiarized with technology in lecture courses that are separate from design. These courses introduce information about basic concepts only, and they often do not keep up with the explosion of technical knowledge that has occurred in recent years. On the other hand, graduating civil engineers are familiar and comfortable with computers—they know how to input information and to get solutions. But they do not understand the overall structural behavior of the building, and thus may use an abstract model that imperfectly simulates the reality of its structure.

Young engineers have not learned to draw or how a building is physically constructed, so naturally they tend to overanalyze and underdetail the structure.

Today, computer-aided structural engineering is commonplace; sophisticated software and fast hardware can solve virtually any problem. The engineer does not have to spend most of his or her time anymore going through laborious hand calculations to solve indeterminate structures or writing computer programs. Because of the power of computers and because building structures have become much more complex, computer users should be experienced engineers, which, unfortunately, is not always the case. Now, young engineers with little training can achieve almost at once what took many years of experience in the past. Some young engineers may be more intrigued by rapid solutions and addicted to standardized techniques as ends in themselves without doubting the results, rather than developing an understanding for the behavior of structure and the reality of construction. In other words, they use computers to input data and collect output, lacking the ability to evaluate whether the solutions are correct. They often have obtained their training from computer outputs and have not developed the feel for numbers and their order of magnitude that is conceptual design and therefore do not have the background to verify computer answers with hand calculations. This qualitative feel for structural behavior is the very essence of a good structural engineer.

The goal of computer-aided structural design cannot just be to obtain quick results; it is the responsibility of the engineer to spend the necessary time to check the output. Increased productivity cannot come at the expense of safety. Misuse of computers through incorrect input, wrong structural model (i.e., discrepancy between idealized structure and real structure), not knowing the capabilities and limitations of program, defective software, and so on, can lead to catastrophic building failure.

One of the most important purposes of this book is to interpret the structural behavior of the elements within various building forms by using a minimum of mathematics while estimating the preliminary member sizes with reasonable accuracy. The mathematics is deliberately kept at a basic level so that the primary emphasis on behavioral aspects is not hidden behind complex analytical processes. Throughout the book, however, the reasons for any of the simplifications are explained—the process of design is never reduced to merely plugging data into ready-made formulas. The discipline of thought, as established by the engineering sciences over a long period of time, is an important aspect of the designer's education.

Although there are many handbooks and much computer software available for the sizing of beams and columns, it is our intention here to develop a feeling for structural behavior and self-confidence in being able to quickly proportion members, thus being in full control of the design without having to depend completely on ready-made solutions. Simple equations have been developed to make the quick estimation of member proportions possible. This is especially important at various times during design and construction. Examples of their use would be when a sense for the order of magnitude of forces and stresses must be developed at an early design stage, when different structure schemes are compared and evaluated, when member sizes must be known for esthetic purposes during the developmental stage of architectural design, when a fast judgment of sizes is needed for checking drawings and verifying computer outputs or for the designer to sense on the construction site when member sizes don't seem right, or, finally, when first trial sections may be needed for the solution of indeterminate structures. Approximations for analysis of redundant structures may be based on estimating the location of inflection points in deformed structures. They also may be based on choosing a load path that yields a determinate structure (e.g., truss).

The student of structures will never truly comprehend the complexity of structural behavior just by reading descriptive material and/or listening to fascinating lectures. He or she must actually solve problems in order to really learn the subject matter and to discover what is not understood. To support this goal, emphasis is placed on practice-oriented problems, which emphasize not only the concept of equilibrium and structural design, but also the visualization of the deformed structure and building collapse mechanisms. Eventually, the student will have developed that certain feeling which brings the building structure alive, because suddenly he perceives himself as being the structure and thus experiencing the pain of stress concentrations and distortions of his bones.

This book is organized as follows: In Chapter 1, basic concepts of design are introduced to provide the reader with a general understanding of structures and of structures as part of buildings. General design determinants are discussed, such as building context, architectural considerations, building functions, safety considerations, and construction. The historical development of the science of mechanics is presented. General principles as related to structure as a system of organization and as support, referring to the building as a whole as well as to its components, are described to lay the foundation for the following chapters.

The complexity of load actions due to gravity, wind, earthquakes, and hidden and dynamic loads are discussed in Chapter 2, as are the basic principles of statics and strength of materials and the concepts of prestressing, and foundation systems.

Chapter 3 deals with approximate design methods for steel, wood, reinforced concrete, and prestressed concrete members. Particularly, the simplified approach to the design of reinforced concrete members should be helpful to the student of architecture and the practicing building designer. The distribution of lateral forces to the resisting vertical building structures, including lateral building stability, is studied in Chapter 4. Also presented in this chapter is the preliminary design of basement walls and retaining walls. In Chapter 5, planar rigid enclosure systems, such as portal frames, A-frames, gable frames, arches, and trusses, are approximately designed. Space frames, folded plate structures, and shell structures are introduced in Chapters 6, 7, and 8. In Chapter 9, general principles related to tension structures are discussed, including prestressed membranes, pneumatic structures and, the new breed of hybrid tensile surface structures. Finally, in Chapter 10, high-rise building structures are briefly introduced.

This book can be used not only as a text for courses in building structures, construction, and design engineering, but also as a reference for design studios. The book will be extremely helpful to the young engineer or architect who is faced in practice with the reality of a building for the first time. The comparative presentation of the many building cases, often given in historical context, together with the references, should be an asset to the practicing architectural and structural designers during the preliminary design stage.

When this book is used as an introductory text to structures in architecture or building construction in a two-semester course, then Chapters 1, 2, 3, and part of 4 should be covered. This portion provides the necessary reference material for the building structure portion of the Architect Registration Examination.

Should this book be used for advanced structure courses in architecture, building construction, or design engineering (i.e., civil or architectural engineering), two courses could be taught: one on *Skeleton Structures* (Chapters 4, 5, 6, and 10) and one on *Surface Structures* (Chapters 7, 8, and 9). Obviously, all kinds of course organizations are possible, depending on the teacher's educational goals.

It should be helpful for the student to select as a project one or more building cases to study graphically and analytically the layout and behavior of a particular structure in more detail. Most of the figure references, which can be obtained from the author (see Acknowledgments), will provide the necessary background information for a conceptual investigation.

It can only be hoped that the approach used in this book will aid the building designer to further develop a sense of structural behavior and knowledge about structures, as well as develop critical thinking, initiate enough curiosity for further studies, and strengthen the creative response to the design and construction of buildings. This book attempts to overcome the current tendency of architects and engineers to be isolated from each other. It establishes a position of structures in architecture, which constitutes a complementary role to the structural engineer's traditional view. This book should provide another bridge for communication and understanding of the various professionals involved with the making of buildings.

ACKNOWLEDGMENTS

My sincere gratitude to the dedicated group of students of the Department of Architecture at Virginia Polytechnic Institute and State University, who have, under my guidance, diligently developed in my studio many of the illustrations in this book. The students who have been involved more substantially in recent years and to whom I am deeply indebted are: Vernon Abelsen, Jaime Bustamante, Nikola Doichev, Benedict Dubbs, Jr., Barry Light, Vincent Marquardt, Edmond Rahme, and Todd Shoaf.

Greatly appreciated is the help of my assistants at Virginia Tech: Todd Shoaf did several outstanding drawings, it was a pleasure to work with him. I thank Benedict Dubbs, Jr., William Gray, Erika Markussen, and Sharon Pitt for their commitment in preparing many of the graphical presentations. The positive attitude of all my assistants has been of stimulating support to me.

Since this book incorporates part of the material of my previous book, *Horizontal-Span Building Structures*, John Wiley & Sons (1983), I must thank again the students of the School of Architecture at Syracuse University for preparing some of the graphical work.

I also wish to thank those other students and assistants whose names I have not mentioned, but who have also been involved in the preparation of drawings and who have supported me through their critical and constructive thinking in the writing of this book.

My sincere appreciation to Trevia Moses for her patience, cooperation, and dedication in preparing the entire manuscript using GML script word processing on Virginia Tech's mainframe.

This book would not have been possible without the contributions of the many architects and engineers whose design of buildings or whose mathematical interpretation of structural behavior has provided a basis for this text. These individuals are too numerous to identify here, but they are given credit in the references and the List of Buildings in Figures, which can be obtained from the author (Prof. W. Schueller, Department of Architecture, University of Florida, Gainesville, FL 32611; Tel. (904) 392-0215, Fax (904) 392-7266).

Finally, I wish to thank Bill Stenquist, the publisher's executive editor, and Ralph Pescatore of TKM Productions, who was responsible for the editorial and production supervision of the book, for their sincere support.

chapter 1

Introduction to Building Structures

Structure is a necessary part of life—it occurs at any level, ranging from the molecular structure of material to the laws of the universe. As order, it relates all the parts of a whole reflecting some pattern of organization. Everything has structure, even if we have not yet recognized it. Societies are structured to properly function—language has structure; the interrelationship of plants and animals with their environment (ecology) represents equilibrium in nature.

The purpose of structure in buildings may be threefold. As an ordering system, it functions as a spatial and dimensional organizer, besides identifying construction systems. As a form giver, it defines the spatial configuration and reflects other meanings and is part of esthetics. As support structure, it holds the building up so that it does not collapse or deform excessively; it makes the building and spaces within the building possible. Structure gives support to the material and therefore is necessary. Building and structure are inseparable and intimately related to each other. Support structures have the necessary *strength* and *stiffness* to resist the vertical loads as caused by gravity, and horizontal loads due to wind and earthquakes to transmit them safely to the ground. In addition to strength and stiffness, *stability* is a basic requirement for structures to maintain their shape.

The richness of structures can only be suggested by the wealth of *building structure types,* ranging from the long-span stadium to the massive building block to the slender tower, from structures above or below ground or in water to structures in outer space. They range from simple symmetrical to complex asymmetrical forms, from boxes to terraced and inverted stepped buildings, from low-rise to high-rise buildings, from horizontal-span to vertical-span buildings, from ordinary bearing wall, skeleton, and core constructions to bridge buildings, cellular clusters, tubes, superframes, suspension buildings, soft and rigid shells, tensegrity structures, to the new breed of compound hybrid forms.

The configuration of buildings may be controlled by large (open or closed) single volumes with long spans, or it may be controlled by multiple (horizontal or vertical) subdivisions with relatively short spans as based on functional requirements. Building shapes range from the small scale of dwellings, such as single and cluster houses, or free-standing and merging houses, to the large scale of skyscrapers. There is almost no limit to the shapes of buildings, as indicated in Fig. 1.1, ranging from boxy to compound hybrid to organic and crystalline shapes. Most conventional building shapes are prisms derived from the rectangle, square, triangle, circle, trapezoid, cruciform, pinwheel, letter shapes, and other linked figures usually composed of rectangles. Odd-shaped buildings may have irregular plans that may change with height so that the floors are not repetitive anymore. From the point of view of overall proportions, buildings can be organized as

— horizontal slabs   —vertical slabs   —massive blocks   —tower shapes

Buildings may be arranged typically in a linear (one-dimensional), gridded or netlike (two-dimensional), or nuclear (centralized) fashion.

Although structure is a necessary part of the building, it is not a necessary part of *architecture.* The relationship of structure to architecture or the interdependence of architectural form and structure is most critical for the broader understanding of structure and the design of buildings in general. On the one hand, the support structure may be exposed to be part of architecture and thus be subject to the laws of esthetics; on the other hand, the structure may be hidden by being disregarded in the form-giving process, as is the case for many postmodern buildings.

In the extreme, the positions of structuralism and symbolism, as basic models of architectural design, may be contrasted. Here *structuralism* addresses the syntactic dimension of the building and reveals meaning through order or the process of establishing order. *Symbolism,* in contrast, addresses the semantic dimensions of the building by being concerned with meaning through image as a motivating force of form giving.

Whatever the designer's position, architecture is science as well as art. It translates abstract ideas into physical *form* consisting of space and the material defining space. Architecture depends on the situation, but also derives its order from culture, which requires an understanding of the unity of the past and the present, possibly the balance between tradition and innovation. One side of architecture is objective and intellectual, while the other side is subjective, irrational, and emotional. The architecture, as an object, may be measured. However, architecture, as an experience, depends on the rather unpredictable response of the human being out of her or his needs and views, which cannot be easily quantified or may be immeasurable. The infinite richness of building shapes and forms can only be suggested by the collage-type visual study in Fig. 1.1. Traditional architecture shapes are derived from the basic geometrical solids: the prism, pyramid, cylinder, cone, and sphere (Fig. 1.1, bottom). The architecture derived from the classical world is very much concerned with esthetics (Fig. 1.1, middle) by which number, proportion, and harmonies are intimately related to each other to define beauty. Design is controlled by the symbolism of geometry and well-balanced composition of form. The unit of composition is achieved, for instance, through proportion, scale, and rhythm. Essential are the proportioning systems as the determinant of beauty and order in architecture.

In contrast to the centripetal space of classical architecture is the centrifugal space of modern architecture, where the shaping of space proceeds primarily from the inside outward (Fig. 1.1, part of right side). Rather than letting the facade wall appear as a closed and purely decorative element, the modern movement led to the resolution of the solidity of the exterior wall through the glass curtain by allowing the zoning of the building organism to expose itself to the outside. The modernists abolished the

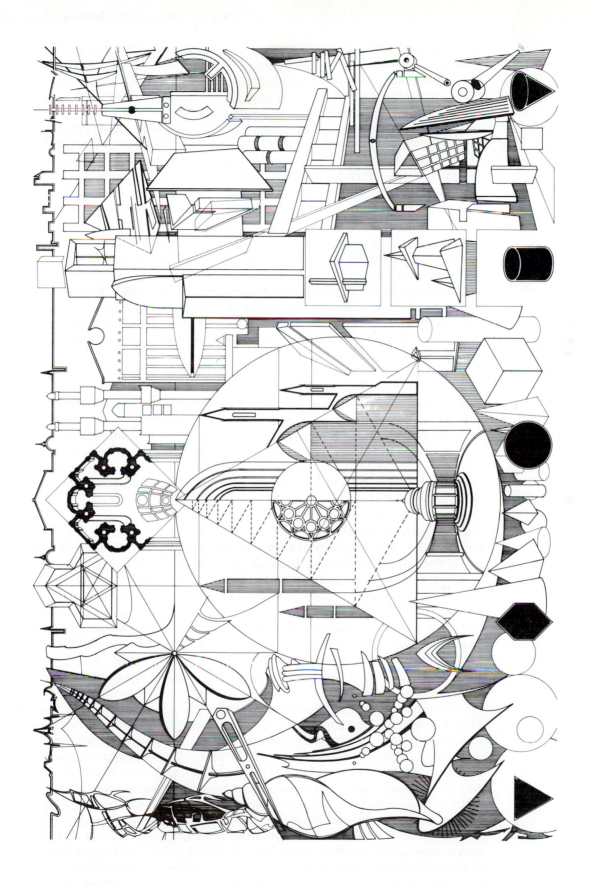

Figure 1.1 Building shapes and forms.

3

facade wall by lifting the building on columns and freeing it from the ground. They invented an almost inexhaustible number of new building shapes through transformation and arrangement of basic shapes, through analogies with biology, the human body, crystallography, machines, tinkertoys, flowforms (forms born out of movement), and so on, or through other personal approaches.

Especially, organic analogies and natural structures have been a rich source of inspiration (Fig. 1.1, left). As a model of design they occur, for example, as shell and cable structures and tensile membranes together with space frames symbolizing minimization of material. The right angle and straight line (in a way representing antinature architecture) are replaced by the curve and the continuous spatial shape. The study of structure and form (referred to as morphology) in nature has been most influential for the development of building structures, as is discussed in other parts of this book.

Currently, deconstructive architecture celibrates order in visual disorder in response to corruption, violence, and irrationality in life. It lets architectural form wildly spin out of control to violate perfection and cause torture and pain in reaction to the traditional values of architecture (Fig. 1.1, part of right side). It has its source in philosophical skepticism and chaos theory in science, which is based on the randomness and uncertainty that occur in catastrophes, failures, instabilities, accidental impacts, turbulences, and so on, in contrast to the linear models and predictability of the deterministic world.

A position of structure to architecture is identified by the buildings in Fig. 1.2 addressing horizontal-span conditions. Here, structure is a generator of form or an essential part of it. As exposed structures, they may express the minimal character of support, the tectonic play of force and form, sculptural qualities, or complex forms composed of several layers of meaning. In the case of the roof structure for the Stansted Airport Terminal in London, U.K. (1991, Fig. 1.2a, Fig. 4.1, left bottom), designed by Foster Associates and the structural engineers Ove Arup & Partners, it has a dominant position. The roof consists of 59×59-ft square lattice domes (tied together horizontally) that span between 69-ft-tall prestressed trees 118 ft on center. A typical tree consists of the trunk composed of four tubular steel columns (acting as vertical Vierendeel cantilever beams) with a four-member pyramid on top. To the trunk columns are hinged inclined slender tubular branches forming an inverted truncated pyramid, which is stabilized and stiffened internally by posttensioned diagonal rods anchored to the pyramid on top of the trunk.

Fumihiko Maki's Fujisawa Gymnasium (1984, Fig. 1.2b), on the other hand, transmits a strong symbolic meaning reminding one of a Japanese warrior helmet. Although its form is obviously not derived from structural minimalism, engineering is wonderfully integrated in the design of the support structure. The roof of the main arena building is resting on two giant 11.5-ft deep trussed arches of triangular cross section spanning 263 ft. The space perpendicular to them is bridged by latticed steel arches forming vaults.

At present, structures resist loads in a passive manner by being unable to adjust to various loading conditions. However, some of the building structures of the future may be of the active type, having their own intelligence. They will have the ability to determine the nature and magnitude of forces through laser sensors, which in turn will feed the information into computers that activate control devices (e.g., prestress devices: hydraulic jacks) to counteract stressing and deformation. In other words, these intelligent structures will self-control geometry, compensate for displacement, control vibrations, maintain a constant stress level, adjust material properties, and so on. As early as the 1960s, Eugene Freyssinet of France and Lev Zetlin of the United States, proposed to control the lateral deflection of tall buildings using the variable prestress concept by placing vertical stressed tendons within the structure near the facade to generate an opposing deformation. Powered passive tuned mass dampers have been used for the first time on a large scale for Boston's John Hancock Tower (1977) and

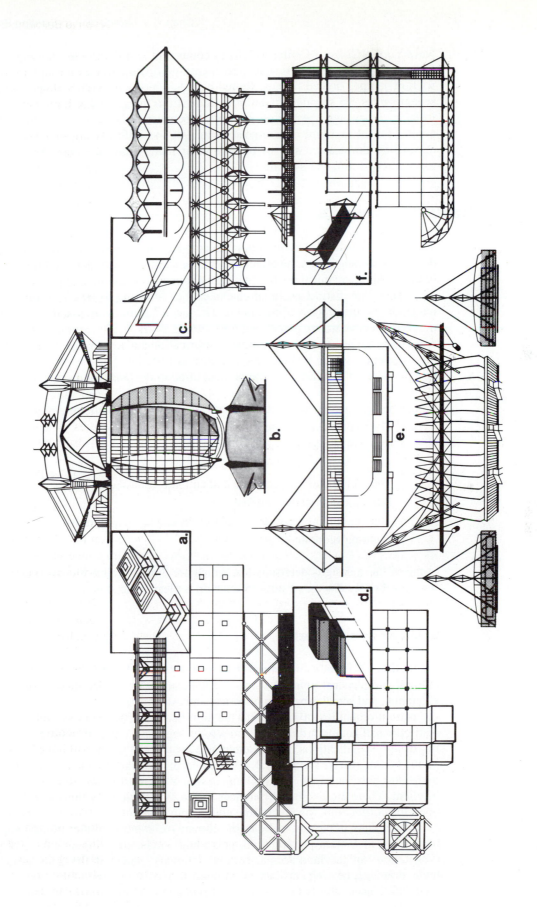

Figure 1.2 Some roof support structures.

New York's Citicorp Center (1978) to counteract and damp the buildings' motion. Examples of passive damping such as energy-dissipating devices (e.g., base isolators, friction damped braced frames, slotted bolted connections) have become common in seismic areas. To separate an entire building from the ground and to place it on base isolators to change the period of motion between ground and structure has been applied for the first time on a large scale in 1986 in California. Although these energy-dissipating devices are still of passive nature, they may be considered a first step toward active structure systems

1.1 GENERAL DESIGN DETERMINANTS

The design of buildings evolves out of a complex interactive process. The many form determinants range from the effect of environmental context, be it cultural or physical, to the building organism itself, which must properly function.

There are distinct building characteristics referring to building form, function, material, and the process of making it. The factors defining architecture range from a purely subjective nature perceiving the building as art or as an idea, to rational considerations based on an organized body of information and knowledge treating the building as science, which includes the building as material. The various determinants of design address the following criteria in addition to *the building as support structure:*

- Building context (including legal restrictions)
- The building as architecture
- The building as an organization of spaces
- The building as an activity and functioning system
- The building as an enclosure and climatic control
- Building safety (fire and health)
- The building as a construction process and as an assembly
- Building economy

Some of these designs determinants, which are directly related to the building structure, are briefly discussed now to develop a clear understanding of the building as structure.

Environmental Context

The building's form, as influenced by its context, may be broadly approached from a cultural and physical point of view. The wide scope of *culture* ranges from political, economic, and social aspects to historical and esthetic ones. Included in this group are the natural amenities of views and the significance of the site, as well as the legal restrictions of zoning ordinances, codes (city, regional, state, and national), and building design specifications, which attempt to protect the user and public at large (i.e., public health, safety, and general welfare); especially, the fire codes play a major role in the design of buildings. The building design is obviously also influenced by the type of client. There are distinct differences in the expectations of the speculative developer, the corporation, the institution, and the government.

The *physical context* includes the climate (orientation of the site, sun angles and intensities, maximum precipitation, prevailing winds, etc.), topography and geology (vegetation, unique land forms, contour intervals, potential flooding, groundwater table, drainage, bearing capacity of ground, etc.), existing urban fabric (land use, services, etc.), accessibility (access to pedestrian and vehicular traffic, parking), capacity and location of existing utilities (sewer, water, electricity, gas, etc.), and ecology (effect of the building on its environment).

On the tight sites of high-density urban areas, the building shape is controlled very much by the lot size, neighboring buildings, and zoning regulations, while the orientation and shapes of high-rise buildings on larger, more open land present many more options, as influenced by the climate, sun, topography, view, and some of the other criteria previously mentioned.

From an economic point of view, location of the building on the site may be dependent on the contour lines, which indicate where to excavate and fill so that the least amount of earth has to be moved a minimum distance. The site geology, that is, the bearing capacity of the soil, may also determine the building location and plan shape.

Introduction to Architectural Design Considerations

When a building has meaning by expressing an idea or by being a special kind of place, it is called *architecture.* Unfortunately, architecture in our culture simply cannot be defined by an all-encompassing universal theory or by formulas, as in the sciences, because architecture is an art as well as a science; it translates abstract ideas into form. One side of architecture is objective and intellectual; the other side is subjective, irrational, and emotional. The architecture, as an object, may be measured. However, the architecture as an experience depends on the rather unpredictable response of the human being, out of his or her needs and views, which cannot be easily quantified or may be immeasurable. Furthermore, external effects influence this experience of architecture; architecture depends on the situation (i.e., values and attitudes), but at the same time derives its order from culture, which requires an understanding of the unity of the past and the present or, in other words, the balance between tradition and innovation.

Early architectural theories by the Roman Vitruvius, and later in the Renaissance by Alberti and Palladio, can only be considered general in the context of their time and culture; in earlier periods, styles lasted for a prolonged length of time, reflecting the fact that architectural concepts were unified. In contrast, there are a multitude of architectural theories, expressed in a diversity of styles, in the twentieth century, partially because of the democratic structure of society in many countries and also because new building materials and construction methods make any form possible. For example, contemporary theories may be based primarily on functionalism, science, constructivism, deconstruction, nature, economy, behavioralism, philosophy (e.g., symbolism), and on art history. These individualistic and pluralistic theories of architecture may contradict each other, thereby reflecting the fact that life is dominated by conflicts or an apparent disorder (i.e., entropy); this disorder, however, is the source of the continuous search for truth, which in turn is order. But all these theories more or less include Vitruvius's basic principles of *commodity* or functional serviceability, *firmness* or strength, safety and durability of materials to assure permanence, *economy,* and *delight* or appealing appearance. The classic concept of beauty, as expressed by a well-balanced compositional equilibrium and visual integrity of the whole, has been transmitted through the ages from Vitruvius, Palladio, and Ledoux to Mies van der Rohe and Le Corbusier.

Building Function

Architectural space is not only static and fixed, as characterized by material space, but as enclosure space, it is alive and dynamic, with people and life-supporting energy moving horizontally as well as vertically from one location to another. There is circulatory space for service and movement, space that is served, as well as the critical space that connects. The circulation zones the building volume and establishes a hierarchy among the various spaces. From a functional point of view, the building volume can be grouped into activity zones or served spaces, the circulation network or service spaces, and access or connective spaces.

The activity zones are clearly defined by the closed spaces of certain building types, such as housing, dormitories, and hotels, but they may also be open office landscapes or open-plan flexible layouts, possibly with movable partitions, as may be the case for exhibition spaces. Often composite floor plans with both open and closed spaces offer a dynamic exchange between privacy and openness. The type of activity depends on the use or occupancy of the building. It may be organized as follows:

- *Residential*
  Low-rise buildings (walkups): low density (e.g., detached housing) and high density (e.g., semidetached housing such as duplex and triplex and row housing)
  Mid-rise buildings (say four to six stories with or without elevators): party-wall housing, block housing, dormitories
  High-rise buildings: block housing, terrace housing, slab blocks, towers, clusters, etc.
  Combinations of the above: urban housing
- *Commercial:* offices, retail stores, shopping centers, hotels, restaurants, etc.
- *Industrial:* light and heavy manufacturing, warehouses, etc.
- *Institutional:* schools, hospitals, prisons, churches, museums, government buildings
- *Special:* towers, sports complexes, exhibition halls, convention centers, bridges, airports, offshore structures, parking facilities, etc.
- *Mixed occupancy* (multiuse): Urban high-rise buildings have been combining living, working, and servicing activities. For example, a mixed-use 70-story skyscraper may consist of 3 underground levels of basement parking and service areas, 8 floors of retail, 20 floors of office space, 2 mechanical levels, 18 floors of hotel rooms, 2 mechanical levels, 20 floors of condominium apartments, and the mechanical penthouse.
- *Megastructure:* The large-scale urban building, as realized by some of the new towns in Europe, such as around Paris, includes functions for living, working, education, and recreation.

Building function clearly influences the selection of structure systems. For example, the typical cellular layout for residential buildings with short spans is different from the longer spans of office buildings with open spaces in the 30- to 40-ft range or from the continuous open spaces of parking garages. Similarly, the long-span structures to cover stadiums cannot be the same as for high school gymnasiums or atria.

Fire Safety

The protection of life during a fire is one of the primary and most complex design considerations. Not only the integrity of the sturcture, as expressed in its fire resistance, but also the escape routes for the occupants and the safety of the fire-fighting teams must be considered, which includes proper access to the building on the site. The fire safety in high-rise buildings and its potential failure is much more critical than for low buildings, which are accessible from the outside with aerial ladder equipment up to 10 floors in some cities. It must also be kept in mind that fires in high-rise buildings tend to spread much faster than in low-rise buildings, due to the stack effect of the temperature difference between inside and ouside. In tall structures, rescue efforts must be fought from the inside; ground-based fire services cannot be provided. Only people in the lower floors can be evacuated, while all the other occupants must be moved to safe refuge areas within the building, thus requiring the vertical and horizontal compartmentalization of a building. These compartments form firetight cells that consist of a continuous fire barrier membrane of wall and floor/ceiling surfaces with special, tightly closed doors that form an envelope with a thermal resistance capable of containing the fire for a certain time period, ideally surviving a burn-out of the contents without failure of the barrier.

The layout of the building must provide smoke-free (vented) and fire-protected horizontal and vertical enclosures as escape routes and adequate exits, as well as access paths for fire fighters; in this context, vestibules may act as smoke barriers between shafts and floors. The minimum dimensions for escape routes to accommodate the occupants in case of an emergency as well as the maximum travel distance along the corridor to the exits or emergency staircases (and elevators, which may become unsafe, since the shafts act as smoke routes) are given by the codes. Actually, the length of the travel paths should be directly related to the combustible contents. It should also not be overlooked that the use of a fire wall may eliminate the need for a staircase. Most people get trapped by smoke and are injured or killed by toxic gases. Thus, the control of smoke infiltration to protect the egress routes is critical. It is obvious that the early detection of fire by some alarm system is essential.

A basic consideration for fire safety and smoke control at the design stage is the restricton of the combustible content or the fire loading; fire loads are generally given in terms of the heat produced (in Btu/ft^2). The designer must be aware that plastic fuels cause much more intense fires than cellulosic fuels and a much higher production of smoke and toxic gas. The use of synthetic insulation, finishes, and carpeting must be restricted. Whereas the fixed fire loads of the exposed structural and nonstructural materials can be controlled by avoiding a highly combustible finishing, this is not the case for the movable fire loads of the furnishing materials or contents, over which the designer has no control. The duration and severity of a fire depend, among other criteria, on the type of combustible contents, which in turn is directly related to the occupancy of the enclosure! Therefore, codes require minimum times of fire resistance for the supporting structures, such as 1½ hours (hr) for apartment buildings, 2 hr for office and hotel buildings, and 3 hr for industrial and mercantile structures.

A building structure must be designed with the necessary *fire endurance* to guarantee stability and to guard against the spread of fire beyond the compartment of origin so that people's lives are protected by making an escape route possible. Building codes protect structural integrity by regulating the fire resistance of the horizontal and vertical building planes that form fire-tight cells, on the basis not only of occupancy, but also of the type of construction, which is determined by the building size (height and floor area), structural materials, building location, sprinkler protection, and other criteria. In addition, insurance rates reflect the degree of fire resistance of structures.

It is apparent that the structural members must be able to withstand, for a certain time period, the stresses due to fully developed fires, together with the other load actions, without collapse. Therefore, fire resistance is rated in terms of the number of hours that the building element must resist the exposure. However, this does not mean that for a given occupancy rate all building components must have the same rating. In this case, the more critical columns and bearing walls may have the maximum rating of a 4-hr fire resistance while the floor assembly is rated at only 3 hr.

Building codes control the height of buildings and floor areas as based on their use and construction types. Exceptions are permitted, for example, when automatic sprinkler systems are used and where fire walls divide a building into separate parts. Construction types may be identified according to the required fire resistance of the floor structures as follows:

- 3-Hr noncombustible construction
- 2-Hr noncombustible construction
- 1½-Hr noncombustible construction
- 1-Hr noncombustible or combustible construction (e.g., light framing of wood or light gauge steel framing, heavy timber construction)

A high fire endurance is required for tall buildings to ensure fire integrity for the main structural members. Typical code requirements are the following:

- Floor construction   2 to 3 hr
- Frames, including columns and interior bearing walls   3 to 4 hr
- Shaft enclosures   2 hr
- Roofs   1½ to 2 hr

A low fire endurance of only 1-hr rating may be required for ordinary low-rise buildings such as for residential use, but this obviously cannot apply to health care facilities, assembly occupancies, or buildings with hazardous contents. The fire resistance rating for various structural assemblies, such as floors, ceilings, walls, columns, beams, and so on, are based on fire tests according to ASTM E119, such as those conducted by Underwriters' Laboratories, Inc., and as accepted by the building codes. A list of fire resistance ratings is available from various agencies, such as Underwriters' Laboratories, Inc. (UL), American Insurance Association (AIA), and the National Bureau of Standards (NBS).

For example, the minimum concrete slab thickness for a 2-hr fire resistance rating is from 3.5 to 5 in., depending on the type of aggregate. Typical for office buildings is the 2- or 3-in. steel deck with 2½-in. normal-weight concrete with spray-on fireproofing, or with 3¼-in. lightweight concrete (115 lb/ft^3) without fireproofing, to provide the required 2-hr fire rating. Concrete columns of 8×8 in. have a fire rating of 1½ hr, while 12×12 in. concrete columns may have roughly a 3-hr rating. For preliminary design purposes, it may be assumed that 2 in. of fire insulation around steel I-sections will provide a 4-hr rating.

Ordinary building fires may reach typical temperatures on the order of about 1300° to 1650°F, keeping in mind that the temperature rises rapidly when a fire starts, possibly reaching 900°F in about 5 minutes. It is apparent that some of the structural materials must be protected. Of the major ones, wood, steel, concrete, and masonry, only wood is combustible, but only concrete and masonry are fire resistant. Although wood burns, heavy members will retain strength under fire longer than some unprotected metals. To achieve the necessary fire rating, the wood members should be oversized, allowing the outer layer of char to reduce further burning, but steel connections must be hidden behind insulation. Wood ignites at about 480°F but has already started charring at approximately 300°F. Wood is assumed to char at 0.025-in./min, so that after half an hour, only 3/4 in. of a wood beam is damaged. This self-insulating character of the char causes the cross section to be weakened much more slowly. Should fire-retardant treated wood be used, some codes allow wood buildings up to five stories if sprinklered, rather than the usual three- or four-story buildings if fully sprinklered.

Most aluminum alloys start losing strength immediately as the temperature is raised and melt at 900° to 1200°F. Masonry materials have been used for a long time as fire protection for buildings. The compressive strength of masonry under high temperatures follows a pattern similar to concrete. Concrete is one of the most highly fire resistant materials. The strength of normal-weight concrete remains relatively stable up to about 900°F, while lightweight concretes perform much better. To control the temperature in the reinforcing, the concrete cover below and above the bars must be thick enough. Usually, the minimum thickness of the concrete cover for the reinforce-ment, as specified by the ACI code, satisfies the requirements based on the fire-resis-tance rating.

In contrast to concrete, when a steel member is exposed to the high temperatures of a fire for long enough, its strength will decrease substantially in a short time This loss of strength, together with excessive deflections and distortions that cause addi-tional stresses in continuous structures, can lead to failure of the steel structure. At roughly 600°F, the capacity of steel rapidly decreases, and at about 1000°F it is almost equal to its allowable stress; steel melts at 2400° to 2750°F. Steel does not burn and contribute fuel to feed a fire, but it is an excellent heat conductor and has a low thermal capacity, so uniform critical temperatures will be quickly achieved throughout the member early in the fire. It is apparent that structural steel members must be ade-quately fire protected for some anticipated fire intensity and duration so that the aver-age steel temperature does not exceed the critical temperature of approximately 1000°F.

Most codes allow rational methods of fire engineering calculations to determine the fire resistance as an alternative to conventional hourly fire rating. In other words, a high-temperature structural analysis is performed to determine the thermal loads for various members due to different fire conditions so that the maximum steel tempera-ture is below the accepted temperature of 1000°F. Typical fire loads are in the range of 5 to 10 psf depending on the combustible material.

Fire protection of steel members in buildings can be achieved either by using a *membrane* as a fire-resistant barrier, such as a wall or a ceiling, to protect the floor framing and ductwork, or it can be achieved by direct *contact,* that is, by protecting the individual steel member with insulation to keep the heat away or to absorb the heat with special coatings. It must be emphasized that, in the membrane and compartment approach, proper fire stops along the edges (such as ceilings and walls) must be pro-vided, similar to fire breaks between floors and exterior cladding.

Typical methods of *heat-resistive insulation* for steel members are based on solid encasement, a boxlike membrane or assembly enclosure of the dry or wet type, a con-tour protection of the spray-on type, or fire-retardant paints. *Intumescent materials* (e.g., coatings or sheets) are only activated as protection under high temperatures. Usu-ally, intumescent mastic coatings are painted or sprayed on steel. Under elevated heat exposure, the coating chemically reacts by expanding into a thick thermal barrier, thus forming an insulation blanket that protects the steel member. *Subliming materials* (e.g., coatings or lightweight foam) do not insulate a steel member from the fire's heat, but absorb heat and act in a manner similar to a coolant system. High temperatures cause sublimers to turn from a solid to a gas. Since this change of state requires a large amount of heat, they are effective as a coolant in removing heat, that is, as a heat absorber.

While the methods of fireproofing steel inside the building are standardized, this is not necessarily true for the exterior structure, where visual considerations become a very important criterion. Should the structure be hidden behind a facade wall, then conventional methods of fire insulation can be used. However, if the designer wants to express the structure, special techniques must be developed. In the traditional approach, the load-bearing frame is enclosed by a steel cladding; in other words, the exterior structure is insulated with fire-resistant material and then wrapped in steel covers to simulate the structural shapes, but thereby altering the appearance by distort-ing the true proportions of the frame.

To eliminate the need for cladding and to expose the true structure, a high-tem-perature structural analysis must be performed and special construction techniques are used to keep the exposed members to an acceptable maximum-temperature level. These methods are based on concepts of *isolation, separation* or *shielding,* and *cooling.*

1. *Air separation principle.* An exposed exterior steel structure does not need to be fireproofed against the flames inside the building as long as it is located sufficiently far away from the glass line or curtain wall.

2. *Flame-impingement shields.* Shielding, for example, can be achieved by providing a fireproof wall barrier behind the exposed columns, together with a protective ceiling to prevent the flames from reaching the exterior structure. In the case of glass walls or windows, a fire-resistive deflector may be employed to move the flames away from the exposed structure. The layout (i.e., geometry) of the assembly may be designed such that the structural steel is kept away from the window, the primary source of heat and flames.

3. *Liquid-filled column systems.* In this system the exposed exterior steel columns consist of hollow sections that are filled with water and possibly with antifreeze and corrosion-inhibiting additives to act as a heat sink. Therefore, the steel does not reach critical temperature as long as an uninterrupted water supply is available to carry the heat away. As the member heats up, the heat is absorbed by the liquid while it rises by convection to storage tanks and is replaced by cooler water from below, thus causing circulating currents from cold to warm between the members and storage tank. At severe fire exposures, steam is generated, which is vented from the top of the storage tank to avoid pressure buildup in the members and tank.

The Pompidou Center (1977) in Paris (see Fig. 2.1) uses virtually all types of fire protection. The exterior columns are water filled, the cantilever brackets are shielded by fire-resistant panels in the facade, the outside tension rods are 25 ft away from the windows, sprinklers are used on the external wall, and the lattice trusses and floor beams are encased to provide heat insulation.

In buildings with low fire exposure, the steel should not reach critical temperatures, as may be the case for open-deck, multilevel parking garages, the roof framing of single-story industrial buildings, and large-volume horizontal-span enclosures where the steel is protected by its height above the floor of more than 25 ft. The roof structures of open, one-story buildings are usually exposed if enough escape routes are available and adequate redundancy of the structure allows local failure. However, for exhibition spaces with potentially flammable displays, the steel roof framing may still have to be partially protected, for instance, with intumescent paint and a sprinkler system.

Construction and Economy

Designs are often only concerned with the final building performance; they may not place much emphasis on the constructibility, that is, the actual building of structures with the interacting flow of activities of the various trades. The construction of buildings involves a sophisticated process, not just in terms of fabricating, shipping, storing, and assembling the various building components, but also in organizing and scheduling this flow of labor and equipment. Besides construction time, the availability of materials, skilled labor, and equipment for a selected structure system must be taken into account. One must realize that axial structure systems may be more economical from a material point of view than bending systems, but they also may require higher labor costs for assembly.

The designer often lacks the field experience to visualize how the contractor is going to build it. The designer should take basic concepts and methods of construction into account during the design stage, when he or she lays out and dimensions the building and responds to the economy of construction, possibly including the contractor in the design team, and during the later quality control stage when the designer makes sure that the building is constructed according to the specifications and drawings. By communicating and working together with the fabricator and erector, the design team will be familiar with the flow of component fabrication, transportation from the shop to the site, accessibility of site, the handling of materials, the size and location of the storage space on the site, the energy supply sources, the process of assembly, including the allowable geometric tolerances and how to allow compensations, as well as the capacity and position of the erection equipment and the availability of local materials and construction expertise. All this should make the designer appreciate the complexity of the construction process and give her or him a sense of control so that complete specifications and clear detailed drawings can be produced to facilitate the building process and thus prevent future confusions and misunderstandings. When the architect uses the result-oriented *performance specifications* rather than the common *descriptive specifications,* which describe in detail the method of construction in a cookbook fashion, then he or she can take advantage of the rapid development of new technologies in certain fields such as curtain wall design, although the designer must deal with elaborate documents that specify design criteria and testing procedures.

The greatest economy and efficiency of construction are achieved by using a *minimum number of operations* on site, which includes minimizing the *number of different components,* which should be assembled in a *repetitive, continuous* process, *simplifying* the field connections and minimizing the *start–stop* of any activity that is, each trade should be in control of its own activity without interference. Hence, a proper construction sequence of the different trades is required, which (in turn) has an effect on the speed of construction. A *dimensional organization* of the building is necessary that reflects *simplicity* and *modular coordination* of standardized elements, including duplication of parts and methods of linkage, and thus includes a means of the control of accuracy, tolerances, and fit. For a regular building, this means uniform bay sizes, symmetry, and alignment of columns, but even a nonregular structure of complex configuration should be the result of a simple *unit addition.* This repetition allows less piecing together, less waste and less time; that is, a smooth flow of construction where machines and other equipment can be effectively utilized. Considering the high cost of borrowing money to finance construction—time is money. Hence, moving personnel and materials onto the site as rapidly as possible is a basic requirement. Traffic time on the site can be kept to a minimum by a computer system that schedules and monitors the time frame of deliveries through specific gates and for specific hoists. Rationalization and automation of the building process will eventually make construction robots feasible, as most manufacturing industries are already in the process of robotizing their production activities.

The weight, size, and shape of the prefabricated units are limited by shipping clearances, available storage space, especially for city sites, and the handling capacity of trucks and hoisting equipment. For example, for truck equipment the common maximum length is 60 ft and the typical maximum height and width is 14 ft.

The typical construction methods are the following:

- Conventional construction
- Industrialized construction
- Special construction techniques

In *conventional construction* methods, nearly all structural components are custom designed for a specific building from standardized rolled sections and/or standard formwork. They use rationalized traditional methods of construction with a certain degree of industrialization. Conventional construction methods are based on materials such as the following:

- Steel construction
- Cast-in-place concrete construction
- Composite steel–concrete construction
- Precast concrete construction
- Masonry construction
- Wood construction
- Cast metal construction
- Fabric construction
- Mixed construction

In *industrialized construction,* manufacturers have mass produced more finished component assemblies. Industrialization in the United States is currently mainly associated with small-scale, low-rise commercial and industrial metal construction, as well as residential construction of factory-made houses and mobile homes.

The concept of mass producing building components and assembling them into a building unit was introduced for the first time on a large scale by Joseph Paxton for the Crystal Palace in London (1851). By the middle of the nineteenth century, U. S. trade catalogs were offering standardized mass-produced building components. Also, the development of the balloon frame at about the same period reflects the transition from craftsmanship by less skilled workers in a relatively short time to the trade approach, emphasizing the assemblage of standardized components. Walter Gropius talked, as early as 1910, about the industrialization of housing through machine-produced standardized building parts. HUD's Operation Breakthrough of 1970 intended to establish a self-sustaining mechanism for rapid volume production of marketable housing at progressively lower costs for people of all income levels. The demand for prefabricated building systems for housing in the United States and western Europe subsided in the 1970s, which, however, is not the case for eastern Europe. Among the pioneers important for the development of industrialization in architecture besides Joseph Paxton and Water Gropius are Jean Prouvé of France, and Konrad Wachsmann and Ezra Ehrenkrantz of the United States.

Industrialized construction may be organized based on the degree of completion of the building assembly, ranging from the completely finished product of the mobile home to prefabricated structure systems.

- Complete building product:
    Mobile home manufacturers
- Partial building product:
    Home manufacturers (single section houses)
    Sectionalized home manufacturers (multisection houses)
    Metal building manufacturers for nonresidential construction
- Structure systems: skeleton systems, panel-frame systems, wall-panel systems, box systems, mixed systems
- Subsystems: utility modules, ceiling systems, curtain systems, partition systems, etc.

In high-rise construction, industrialization consists primarily of the addition of structure assemblies, mechanical systems, and architectural systems (partitions, curtains, ceilings, etc.), rather than the finished product of the home manufacturers. Generally, these structures do not represent closed finished systems, but rather unfinished, more open ones; in other words, buildings delivered to the site may only be about 40% complete. We must keep in mind that, because of all the layers of specialization, quality control becomes difficult and the resulting building may occasionally be costly and may not always be efficient.

Besides conventional and industrialized construction, there exist some *special construction techniques.* For example, it may be economical to construct entire building portions at the ground level and then raise them into their final position by either lifting them in tension or pushing them up in compression. These packages can also be assembled at other building levels, so that the building is erected from the top down rather than from the bottom up. Some of the more common special construction systems are as follows:

- Lift-slab construction
- Push-up construction
- Suspension erection method
- Balanced cantilever erection method
- Tilt-up construction

The selection of a construction system from an economy point of view not only includes the initial *construction costs* of materials, structure system, labor, equipment, and construction time (i.e., cost of financing), but also the future maintenance and operating costs, or capital investment and maintenance replacement costs, in other words, the *life-cycle costs.* Usually, the speculative developer is concerned only with the short-range construction costs, whereas corporate and institutional clients must consider the total project development costs.

1.2 THE EARLY BEGINNINGS OF THE SCIENCE OF MECHANICS

Today, we do not question the fact that engineers design buildings so that they safely stand up and perform properly using mathematical models based on scientific theories and observation of prototype structures or scale models. These engineered buildings, however, are an invention of the nineteenth century in response to the development of numerous new building types, including the emergence of the materials iron, steel, and reinforced concrete and the occurrence of many new structure systems. We must keep in mind that before this time buildings were designed as based on past practical experience derived from construction and failures, by developing structures and construction methods for various building types.

Before briefly introducing the early historical development of structural analysis consisting of mechanics and aided by mathematics, mechanics must be first defined.

- Mechanics of rigid bodies
    Statics
    Dynamics: kinematics, kinetics
- Mechanics of deformable rigid bodies (mechanics of materials)
- Mechanics of fluids

Mechanics of rigid bodies is subdivided into statics and dynamics. *Statics* addresses the study of bodies at rest, that is, equilibrium of forces. *Dynamics* deals with bodies in motion; the study of the geometry of motion is called *kinematics,* and the study of unbalanced forces and the motion they produce is called *kinetics.* The basic concepts of the mechanics of rigid bodies are *force, mass, space,* and *time.*

Mechanics of deformable rigid bodies, also called *mechanics of materials,* deals with the behavior of basic members (beams and columns) under force action, which involves not only the principles of statics (equilibrium of external and internal forces), but also the mechanical properties of materials. The basic concepts of the mechanics of materials are *stress, strength, stiffness* (deformation characteristics), and *stability*.

The application of the rigid-body assumption in statics and dynamics is an approximation, but is generally acceptable since most members (bodies) are subject to only *small deformations*. In other words, even though forces are obtained as based on the rigid-body assumption, they generally can be used to determine the deformation of members.

The study of mechanics, which were first simple concepts of statics and kinematics, can be traced back to the ancient Greeks, who developed a good understanding of levers, pulleys, inclined planes (ramps), simple machines, and floating bodies. The understanding of motion, however, was not clear. Greek mathematics was almost entirely limited to geometry as demonstrated by the work of Euclid, the most famous master of geometry (ca. 300 B.C.). Aristotle (384–322 B.C.), one of the great thinkers of all time, included in his philosophy the sciences, which consisted primarily of mechanics and the arts and humanities—they all were one thing. In contrast to many of his contemporaries, he dealt with mechanics—not on a mathematical but on an abstract philosophical level. Through observation he drew his conclusions about the motion of terrestrial and celestial bodies. He considered bodies at rest as the normal state unless motion was caused by forces; in other words, forces were needed to keep a body moving. The concept of acceleration was not known. Bodies were thought to fall at the same rate with constant speeds proportional to their weight only, without any consideration to gravity. Motion had to be caused by forces, with the exception of celestial bodies following a natural, continuous, never-ending uniform circular motion. It was inconceivable that there was a force that could move the Earth; therefore, the Earth had to be at rest and at the center of the universe. Aristotelian philosophy has influenced western thought for nearly 2000 years.

Often the great mathematician, natural scientist, and engineer Archimedes (287–212 B.C.), who lived in Syracuse, Sicily, is considered the father of mechanics. In contrast to Aristotle, he is not a philosopher. He may be the first who actually applied his knowledge of mechanics by developing the principle of the lever or seesaw balance, which was applied to building ingenious war machines. He also investigated how to find the center of gravity of a body. His discoveries communicate concepts of equilibrium of forces as well as moments. Archimedes is probably best known for the law of buoyancy of floating bodies which states that *any solid body submerged in a liquid loses the weight of the liquid displaced by it.*

Archimedes had very little effect on building construction since Greek architecture was concerned with perfect forms expressed in rules of proportion, so there was no need to develop a structural theory. During the next approximately 2000 years, between the period of Archimedes and Galilei, almost no development of mechanics took place.

Roman buildings were based on Greek orders, with the exception of the arch, which was added and sized as derived from empirical knowledge accumulated over time from material characteristics, construction processes, correction of failures, geometric symbolism, and so on. The rules of proportion were based on geometry and were closely related to esthetics, not structural behavior or mechanics. The principles of design were described by Vitruvius (first century B.C.) in his *Ten Books of Architecture.* Keep in mind, however, that Roman engineers developed impressive new construction systems, as demonstrated by their long-span, usually semicircular arch structures and vaulting systems, such as the 143-ft-span Pantheon in Rome (ca. A.D.123), which was only surpassed in span in the second half of the nineteenth century. Some Roman structures are discussed elsewhere in this book.

The great medieval cathedrals of the Gothic period, admired for their seemingly weightless interior spaces, were made possible by an intricate, delicate skeletal construction, which, in turn, could not have been realized without a profound intuitive understanding of force action, although it was never formulated. The master builders overcame the limits of stone through inventing form-resistant structures so that material was stressed primarily in compression. They equated structure with geometry and developed a great skill with sophisticated geometric techniques; their proportional systems were based on geometry. The geometry as the basis of architecture was achieved through polygonal patterns (tesselations, tilings) and transformations of basic geometric units (e.g., translation, rotation, reflections). Through trial and error methods, the building crafts accumulated an empirical knowledge that became rules of thumb and part of the tradition.

The beginning of the Renaissance, the rediscovery of the classical heritage, is often associated with the work of Filippo Brunelleschi (1377–1446) in Florence and the completion of the enormous 138-ft-span dome of the cathedral, the first modern large-scale structure after the 112-ft-span Hagia Sophia (A.D. 537) in Constantinople, nearly 900 years earlier. This achievement clearly reflects the emergence of a new spirit and a new era in Florence, which called itself the "new Athens" during the first half of the fifteenth century. Brunelleschi's approach to constructing the dome was quite different from the traditional medieval practice. He used a double-layer vault with radial ribs where the lateral thrust is resisted by hidden tension rings (see Chapter 8). Furthermore, he did not use any central shoring during construction to support the dome but employed horizontal sandstone rings to prevent the arches from tilting inward.

During the Renaissance, building technology was primarily concerned with the construction of arches, domes, and bridges. The interest in bridges had its origin probably with the shallow Ponte Vecchio in Florence (1367) that had a depth-to-span ratio of only 1:6.5. The engineers and builders developed their own rules of thumb for the construction of those structures.

Building designers, on the other hand, were deeply involved with the geometry and composition of form in general. They tried to reflect the harmony of nature in architectural proportions by reducing it to the abstract world of mathematical concepts. There were many proportioning systems, some quite complex, but all were based on mathematics as the determinant of beauty and order in architecture. These universal rules of harmony defining esthetic principles are, for example, described by the architectural theories of Leone Battista Alberti (1404–1472) during the early Renaissance period and later by Andrea Palladio (1518–1580).

Although total dedication to the classical past did not permit any innovations in the sciences in general, the new spirit allowed certain individuums to search for an order in the universe separate from medieval restraint of religion, which is the classical perception of the world, thereby developing the basis for the modern sciences. It is the genius of Leonardo da Vinci (1452–1519) more than 100 years after Brunellesci at the beginning of the High Renaissance, architect, engineer, scientist, mathematician and artist, all in one person, that best reflects this new era. He is the first since Archimedes, after more than 1700 years, to investigate concepts of mechanics using a scientific approach. Although he recognized many of the basic laws of statics, dynamics, and strength of materials, he never considered to formulate them as a comprehensive theory because the time was just not ready for that. In his sketches he showed a perfect understanding of the nature of forces and rotational equilibrium. He experimented with the parallelogram of forces and the lateral thrust of arches, as well as investigating the strength of beams, trusses, and columns and testing the tensile strength of wires. Leonardo da Vinci described his discoveries in his notebooks, which, however, were not published during his lifetime; therefore, his influence on the subsequent development of the physical sciences was minimal.

About 100 years later, in 1586, the Dutch engineer and mathematician Simon Stevin (1548–1620) published his book on statics. He established the science of statics as it is known today, especially by his formulation of the parallelogram of forces and the addition of forces. He also laid the basis for the development of graphic statics during the nineteenth century. Naturally, Newton's first and third laws (1687) give a much broader base to the field of statics. Stevin continued the work on water pressure that Archimedes had started with the discovery of the principle of buoyancy almost 1900 years earlier. He is considered the founder of hydrostatics. But it was the French mathematician and philosopher Blaise Pascal (1623–1662) who gave hydrostatics its present form with his treatise on the *Equilibrium of Liquids and the Weight of the Mass of Air,* published in 1663.

However, it was about 100 years after Leonardo da Vinci that his thinking led to the great physicist and astronomer Galileo Galilei (1564–1642) and the rebirth of scientific thinking during the beginning of the Baroque. Galileí challenged the orthodox doctrine of medieval thinkers, which had been based on Aristotelian philosophy. He gave credence to the astronomer Nicolaus Copernicus, who had formulated a new theory of the universe with the sun at its center and the earth moving around it (1543), through observations with the telescope, which had just been invented. Galilei redefined science by using experimentations and developing an abstract language based on mathematics to explain the physical world; in other words, he replaced the Aristotelian natural philosophy as the definition of science.

Galilei is considered the father of modern mechanics by greatly contributing to the formulation of basic principles of dynamics and mechanics of deformable bodies. Among his numerous discoveries through experiments (rather than discussions and arguments), he found that free-fall objects move faster and faster (i.e., accelerate). He concluded that the velocity of that motion must increase in simple proportion to time, thereby establishing the first idea of acceleration [i.e., the rate at which velocity changes with time, V/time or $(m/s)/s = m/s^2$]. He demonstrated that a stone twice as heavy as another one does not fall twice as fast, unlike Aristotelian theory, which considered objects to fall at the same rate. Galilei also found that bodies fall with equal acceleration and hence sensed the concept of gravity.

Galilei was the first to present the basic principles of dynamics, although in primitive form, which were then expanded and formulated by Isaac Newton (1642–1727) using mathematics as the essential means for communication. Newton is considered the father of classical mechanics and represents the culmination of the scientific revolution. He presented in his *Principia* (1687) the famous three fundamental laws of motion, which became the basis of most ordinary engineering applications of mechanics. The power of Newton's three laws lies in their universal application, ranging from the motions of planets and motions on Earth to the design of buildings. He used science as a powerful tool to explain the world around us (positivism).

> **First law**, also called the *law of inertia:* Every body continues in its state of rest, or of uniform motion in a straight line, unless it is compelled to change that state by forces impressed upon it. In other words, a body on which no forces act has zero acceleration.
>
> **Second law**, also called the *law of dynamics:* The acceleration of a body, **a,** is directly proportional to the net force, **F,** acting on the body and inversely proportional to the mass, M, of the body.

$$\mathbf{F} = M\mathbf{a} \qquad\qquad (1.1)$$

> Here the symbols **F** and **a** in boldface type represent vectors, in other words, quantities that have both magnitude and direction.
>
> **Third law**, also called the *law of equilibrium:* To every action there is always opposed an equal reaction.

Newton's first and third laws are the basis of the theory of statics, although the first law, in general, is classified as kinematics. The second law is part of kinetics, which becomes an essential consideration of structural design when buildings are subject to vibrational loading such as caused by wind or earthquakes.

Newton showed that the mass of a falling object is independent of the Earth's acceleration, g, as established earlier by Galilei through his demonstrations of falling objects from the leaning tower of Pisa. Hence, according to Eq. (1.1), the mass of the object is attracted toward the earth by a force due to gravity, which is the weight W of the object, of

$$W = Mg \qquad (1.2)$$

The acceleration due to gravity near Earth's surface is generally taken as

$$1.0g = W/M = 32.2 \text{ ft/s}^2 = 9.81 \text{ m/s}^2 \cong 10 \text{ m/s}^2 \qquad (1.3)$$

According to Newton's second law (Eq. 1.1), for every object the force F is directly proportional to the acceleration a. In other words, the larger the force, the more acceleration it produces, but the more massive the body, the less its acceleration. The constant of proportionality is a property of each object; it is the quantity of matter in a body and is called mass, M.

$$M = \mathbf{F}/\mathbf{a}$$

Here the mass per unit volume is defined as *density* (e.g., kg/m^3). The object has the tendency to resist any change in motion according to Newton's first law, which is called *inertia* as measured by mass. The greater the mass of a body, the greater its inertia. Visualize a body of mass M on which a force F acts so that it experiences an acceleration a. Applying an additional fictitious force, $F = Ma$, exactly opposite would cause the body to be in equilibrium. This fictitious force or force of inactivity is called the *inertia force*.

The kilogram (kg) is the standard unit of mass in the SI system (International System of Units). The unit of force is the newton (N), which, according to Eq. (1.2), is equal to

$$1N = 1 \text{ kg} \cdot \text{m/s}^2 \qquad (1.4)$$

In other words, Eq. (1.2) states that 1 kg will exert a force of 9.81 N downward toward the center of Earth, or 1 kg weighs 9.81 N. Other common units for force are kilonewton, kN = 1000 N, and meganewton, MN = 1000 kN. In the U.S. system of measurements, the unit of mass is the pound-mass (lbm) or the slug (sl) and the unit of force is the pound-force (lbf), which, according to Eq. (1.2), is defined as

$$1 \text{ lbf} = 1 \text{ lbm} \cdot \text{ft/s}^2 = 1 \text{ sl} \cdot \text{ft/s}^2 \qquad (1.5)$$

Unfortunately, the terms of "weight" and "mass" are generally confused in the U.S. system of measurements by using the pound (lb) both as a unit of force (lbf) and a unit of mass (lbm).

The concepts of Newton's dynamics were applied primarily to astronomy. It was only at the turn of this century that they were used in machine design, and, in the 1950s, began to be applied to buildings that had become flexible and structurally more daring with respect to minimum weight construction and, therefore, had become vulnerable to dynamic loading.

Galilei is also considered the founder of *mechanics of materials* since he was the first to attempt to explain on a rational basis the behavior of some members under load. In his classic experiment of a cantilever beam of length L and rectangular cross section bd, supporting a single load P at the end, he tried to predict the bending strength at the support (1638). Although he used the proper concept of rotational equilibrium, that is, the external moment to be balanced by an internal force couple, $M = PL = Td/2$, his assumption of stress distribution was wrong. He assumed uniform tensile resistance T of the fibers with the point of rotation (i.e., neutral axis) at the bottom of the beam resulting in the following stress, f

$$M = fbd\,(d/2), \quad \text{or} \quad f = M/(bd^2/2) \tag{1.6}$$

The real maximum stress, however, is $f = M/S = M/(bd^2/6)$, assuming Hooke's law applicable up to failure. Hence he underestimated the stress, which actually is three times higher.

After taking more than 100 years, the French physicist and engineer Charles Augustin Coulomb (1736–1806) finally formulated in 1776 the true flexural stress distribution with the neutral axis at mid-depth of a rectangular section, as based on work by Robert Hooke (1635–1703) and the Swiss Jakob I. Bernoulli (1654–1705), among others. It was Hooke in 1678 who found that materials deform under loading and that they are elastic, in other words, recover their original shape when the load is removed. He established that the deflection of structures is proportional to the applied loads, which is generally referred to as *Hooke's law*.

Finally, in 1826 Louis Marie Navier (1785–1836), a professor at the Ecole des Ponts et Chaussées, published the first comprehensive treatise on mechanics of materials as it is known today. It should be mentioned that the split between architecture and engineering is often associated with the founding of the École des Ponts et Chaussées in 1747.

The first column formula was presented in 1729 by the Dutch mathematician Pieter van Musschenbroek. His formula was an empirical expression for estimating the strength of rectangular columns. But it was the famous Swiss mathematician Leonhard Euler (1707–1783) who was the first to recognize buckling of columns in 1757 as a mode of instability. He derived the critical buckling load using differential and integral calculus. Unfortunately, Euler's simple column formula was only applicable to slender columns failing in elastic buckling and hence did not predict the strength of ordinary building columns, and therefore was not of practical use. It was more than 100 years later that empirical formulas were presented taking into account material failure together with instability (e.g., inelastic buckling of steel columns).

In conclusion, it may be said that a basic understanding of the fundamental principles of statics and mechanics of materials based on linear elastic analysis, as we know them today, was reached at the beginning of the nineteenth century. However, these principles were rapidly developed further, requiring specialization and a breakdown into several structural engineering subjects as reflected by the introduction of engineering schools at universities.

The appearance of the metal skeleton replacing the traditional masonry and wood construction is closely related to the explosion of nineteenth-century engineering. The *iron frame* is often treated as a symbol of the Industrial Revolution, which is accompanied by urbanization and rapid population growth. During this period, new building types were born, ranging from the *long-span* structure for the great exhibition halls and train sheds, the multibay framing for mills, factories, and warehouses, and the numerous bridge types, to high-rise skeleton construction.

These new building types challenged engineers to develop structural theories to predict the behavior (e.g., strength, deflection, solution of indeterminate structures) of new structure systems (e.g., continuous beams, rigid frames, arches, trusses, suspen-

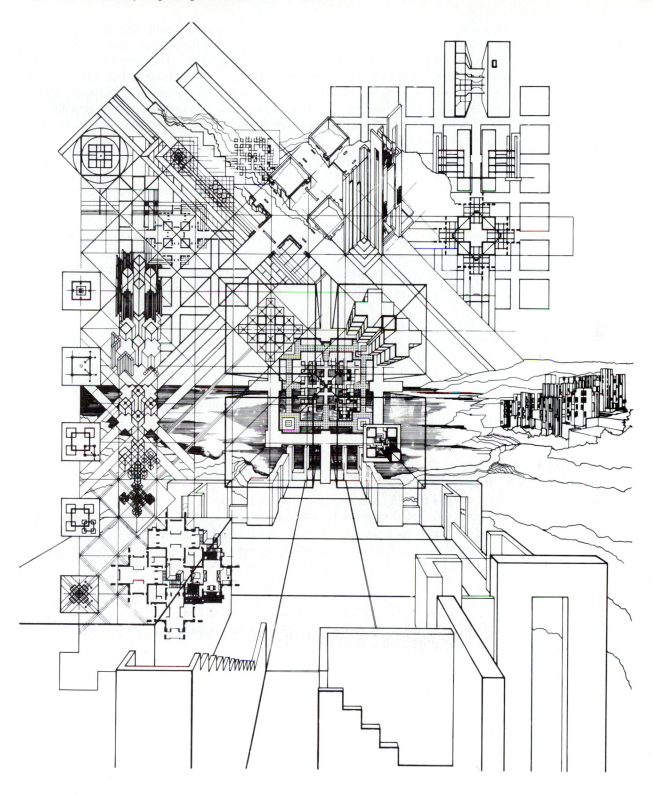

Figure 1.3 La Muralla Roja, Sitges, Spain

sion systems, braced domes, shells, foundations, retaining walls). The development of graphic statics during the nineteenth century eliminated the previous lengthy analytical approach; Carl Culman was among the first to formulate it in the late 1850s. Engineers also had to develop a clear understanding of materials including its use, first for iron and steel and later reinforced concrete (e.g., creep, fatigue), and the various loading conditions as well as loading types (e.g., vibrations). It is quite clear from this discussion that mechanics had become an integral part of the technological process and the making of buildings.

Among the architectural engineering landmarks of the nineteenth century reflecting the unbelievable progress in engineering science, as demonstrated by incredible records with respect to enormous volume, long span, and height, are the following:

- *Crystal Palace* (1851) in London, a lightweight iron-glass structure covering a ground area of 18 acres, which was constructed in only 6 months
- *Brooklyn Bridge* (1883) in New York with a span of 1595 ft
- *Eiffel Tower* (1889) in Paris reaching almost a height of 1000 ft

Further historical development of engineering is discussed in context, under various topics in this book. The classical methods of linear elastic analysis, which have been the basis of structural engineering education for a long time, are currently being challenged by collapse load design methods. In reinforced concrete, the strength design method has replaced to a large extent during the 1970s the working stress approach, and the load and resistance factor method is considered presently for the design of other materials.

1.3 STRUCTURE AS GEOMETRY

The form of a building is in response to a rather complex interplay of design generators caused by functional, behavioral, political, economic, and esthetic forces. Its geometry is defined by its enclosed spaces (e.g., rooms) and by the shapes of its solid components (walls, floors, etc.), all in an ordered relationship to each other. This interdependence and **dimensional coordination** of solid and void are described in planar views by horizontal and vertical sections, each with its own formal organization. The resulting patterns establish a visual order in themselves. It is the recognition, or rather appreciation, of this visual order that is so important for the further discussion in this book. Often it is not possible to develop an objective understanding of this organization, but only a certain sensitivity, since its planar dimensions can only describe the geometrical result, while the process of creation and change, that is, the higher dimensions of the building as an idea, can only be interpreted and seen in the general design context. Some of the complexity of hidden geometry is unveiled in the visual study of Ricardo Bofill's La Muralla Roja in Sitges, Spain (Fig. 1.3). The building appears like stepped towers, almost like a fortified medieval city. The geometrical language is derived from a system of Greek crosses.

There is no set limit to the many patterns possible as is suggested in Fig. 1.4. They range, at one end, from the regular surface subdivisions that occur on tiled floors and wallpaper, which have been systematically explored by Islamic art, to the other end of a complex, ever-changing organism, the city.

The regular linear grids can be replaced by curvilinear ones, as expressed so powerfully by the late Baroque architects or as revived by some modern architects, such as Paolo Portoghesi and Vittorio Gigliotti, with their overlapping circles as space organizers. The nature of curvilinear patterns is further investigated in Fig. 1.5, where

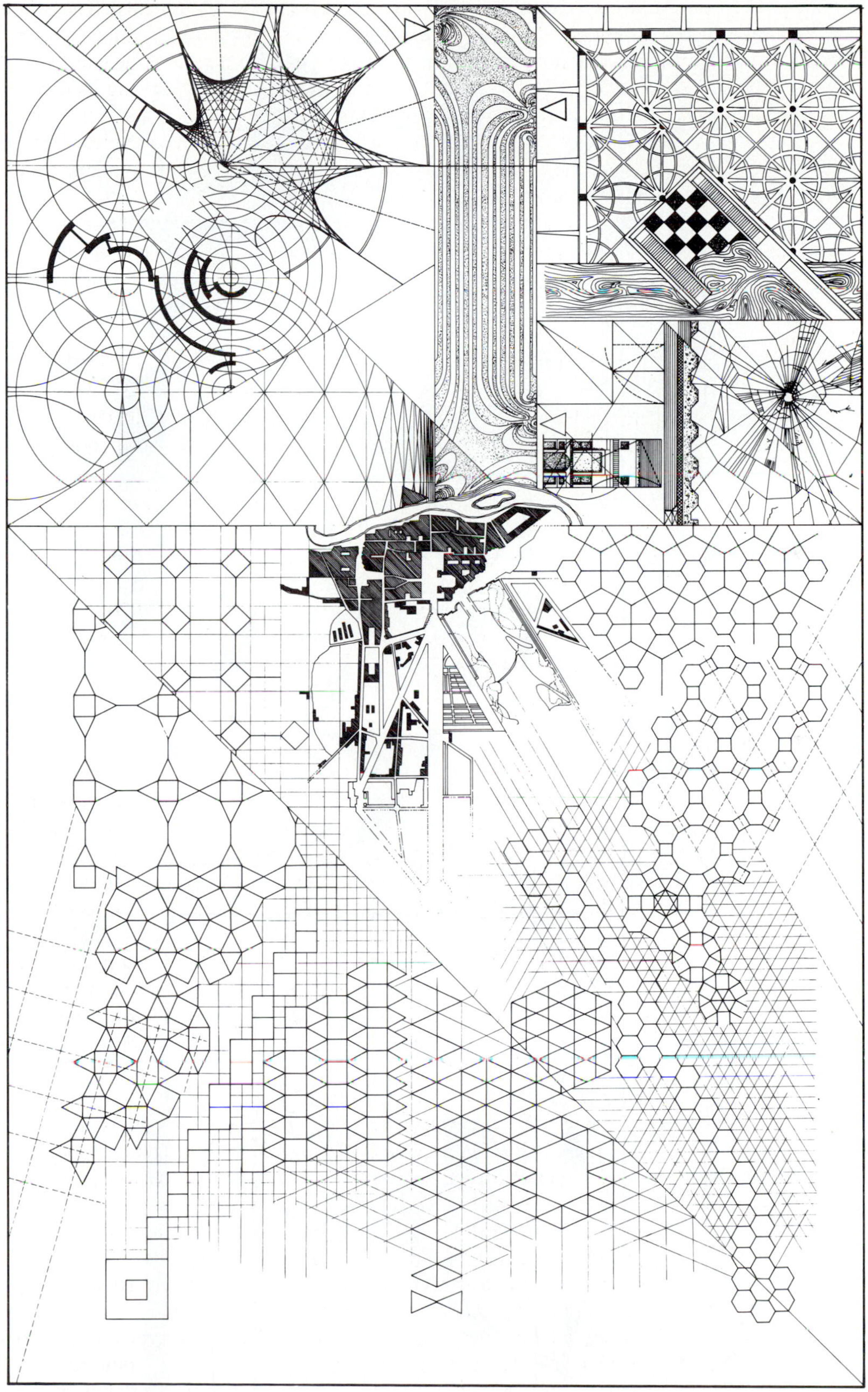

Figure 1.4 Geometric patterns.

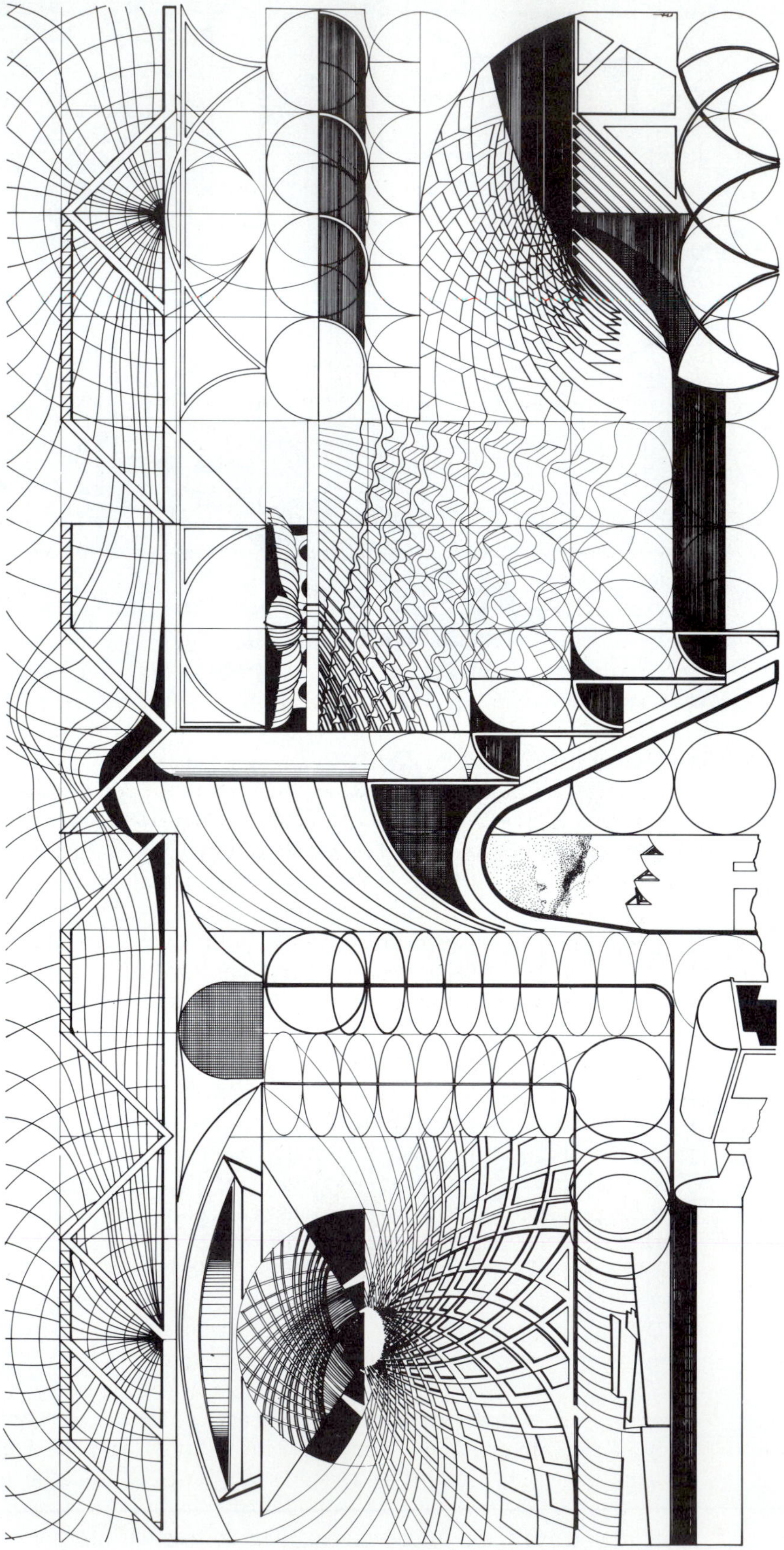

Figure 1.5 Curvilinear patterns.

the geometries of various buildings are superimposed in such a manner as to balance the lateral thrust of arched forms and to express the effect of weightlessness, as well as the dynamics of continuity.

This dynamics of movement is experienced in the structure of the wood grain as its growth responds to stress flow and is affected by the interference of knots and notches (Fig. 1.4), resembling the contours of the earth or the wind pressure distribution along a building surface. Also, the photoelastic diagram of the simple beam reacting to two single loads exposes similar phenomena of stress concentrations at the loads, while the equally spaced lines in the center portion indicate the linear stress distribution. Similar *behavioral patterns* are experienced by the effects of an island in a river, which interferes with the natural flow of the current by pushing aside the stream lines and causing turbulences, or as experienced around a hole in a member that is under tension.

The emergence of form out of force flow is expressed by the slab rib pattern following the isostatic lines of the principal bending moments as based on isotropic plate behavior. This interplay of force and form is often called *tectonics* and is experienced so convincingly in the architecture of the Gothic cathedrals.

Finally, the pattern that is derived from a proportional system in search of absolute harmony and a universal eternal order seems to be in exact opposition to the sudden random pattern caused by a gunshot wound or explosion, as seen in the crack pattern of window glass.

A building is a cellular aggregate of spaces that must be dimensionally coordinated so that it can be constructed. This dimensional network forms patterns of a certain order. The key to understanding that order lies in the nature of mathematical figures, which shall be reviewed briefly.

Most interesting to the field of construction are the *regular polygons* with identical edge lengths and the same angles, reflecting basic considerations of economy. There exists an infinite number of regular polygons, ranging from the simple ones to the more complex star-shaped and other compound forms. Within this context, the simple regular polygons are of more interest. They may be organized according to the number of sides they have as the equilateral triangle, the square, the pentagon, the hexagon, the heptagon, the octagon, the nonagon, the decagon, and finally approaching the circle with the increase in the number of edges. This transition from the regular triangle to the circle also reflects the transition from the longest perimeter necessary to enclose a given area to the least perimeter in the case of the circle.

The regular polygons are symmetrical with respect to the folding line (mirror or bilateral symmetry) and rotation (point symmetry). A square, for example, has four lines of symmetry; that is, when folded about any one of these lines, both portions match exactly. The circle is the most symmetrical of all geometrical figures; every corner of regular polygons lies on the circle.

The perfect regularity of the simple polygons can be relaxed by either allowing unequal angles with still equal sides, called *semiregular polygons* (e.g., rhombus) or allowing only equal angles (e.g., rectangle).

The simplest planar networks (grids, lattices, tiles, nets, mosaics, tessellations, etc.) are derived from the regular polygons. There are only three regular polygons that can fill a surface just by themselves. They are the *triangle, hexagon,* and *square;* the grids that they form are called the three *regular equipartitions* (Fig. 1.4 left).

The *eight semiregular equipartitions* (Fig. 1.4) are made up of two and three regular polygons, with the same ones meeting at each vertex; however, the different polygons do not form equal angles any more. In addition to the three regular surface grids, the semiregular polygons of the rhombus and pentagon can each also fill a surface.

Relaxing the regularity of lattices results in an infinite number of patterns; random arrangements can be developed with the regular polygons. Furthermore, there exists an infinite number of grids that are comprised of identical polygons of unequal-sided figures (e.g., rectangular, rectangular triangle, parallelogram, trapezoid). Although the patterns of grids tend to be repetitive, lattice irregularities occur due to holes, setbacks, corners, discontinuities, defects, or other disturbances and imperfections.

The layout of a building is coordinated with a *dimensional grid* that can be derived from its modularity (Fig. 1.6). The basic building module is a frame of reference, a unit measurement that attempts to encompass, in a purely geometrical fashion, all the building modules, that is, all the requirements of its components (production, transportation, size, position, performance, etc.) and construction process (planning, installation, communication, tolerances, etc.). In *systems building* the degree of complexity of the modular coordination is quite high, since it not only responds to the standardization necessary for mass production, but also provides the rules and framework for the interaction of design, manufacturing, and construction as represented by the various building professionals.

The layout of most buildings is coordinated by a repetitive rectangular network in horizontal and vertical directions, as explained visually in Fig. 1.6; obviously, any other dimensional lattices can be used if the arrangement of the building components so requires. Visualize the typical building to be subdivided by a three-dimensional spatial network of reference lines as derived from the international basic module of $M = 100$ mm $= 10$ cm $(\cong 4$ in.$)$. The typical grids of organization for a building are the following:

- The basic three-dimensional modular grid or the multimodule as a multiple of the basic module $n \cdot M$
- The planning grid
- The structural grid, shown as a dash–dotted line

The spatial modular grid provides the reference for all building components. The structural grid coordinates the location and dimensions of every structural element in the building (Fig. 1.7). The regularity of the grid allows for a standardized construction process and makes it easier to control and reduce sources of errors. We must distinguish between the structural grids defining the floor and roof framing of the horizontal building planes, and the walls, frames, and stairs of the vertical support structure. An example of a modular grid is a multimodule of 2 or 4 ft as usually used in home building and developed around the manufacturing of building materials, in particular the standard panel size of 4×8 ft. This modular grid is a multiple of the reference grid of 4 in. Similarly, the Japanese employed for the design of their traditional houses the Tatami, a straw mat of 3- $\times$ 6-ft size, as the module from which the building grew.

From a mathematical point of view, these dimensions for proportioning are usually drawn from arithmetic or geometric series. An *arithmetic sequence* is built by always adding the same number to grow at a constant rate such as the following number sequences:

$$1, 2, 3, 4, \ldots \quad \text{or} \quad 3, 7, 11, 15, \ldots$$

In *geometric sequences,* growth is at an increasing rate. Each term here is found by multiplying the previous number by a given number, *a,* such as

$$1, 2, 4, 8, 16, \ldots \quad \text{or} \quad 5, 10, 20, 40, \ldots$$

Each building element is a rectangle with each of its dimensions being a multiple of the basic module of 100 mm (or 4 in.). The resulting proportions of the rectangles consist

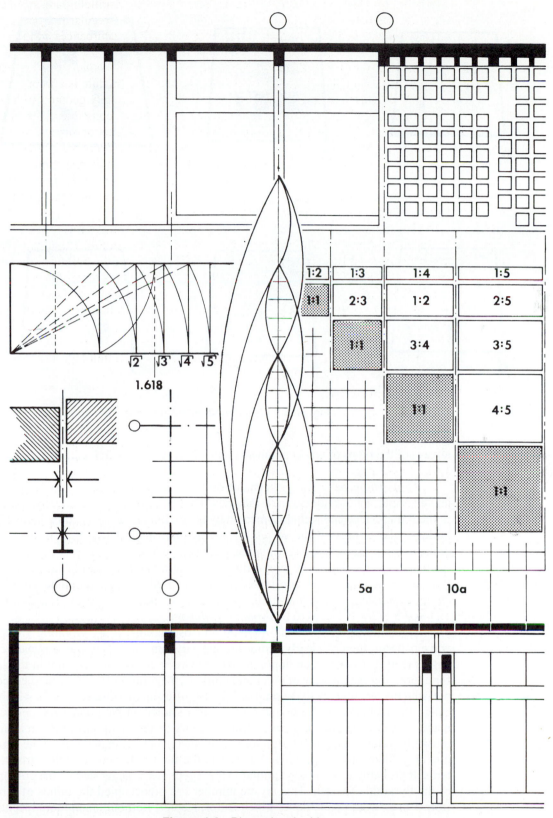

Figure 1.6 Dimensional grids.

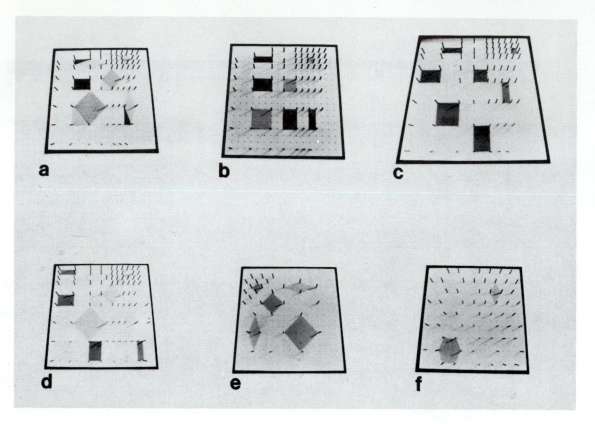

Figure 1.7 Support structure patterns.

always of the ratios of whole numbers (1:1, 1:2, 1:3, 1:4, 2:3, 3:4, etc., as shown in Fig. 1.6).

Proportions have always been one of the most essential elements of visual order and an important consideration of esthetics. They have been systematically researched in the Renaissance and well documented by the architects Alberti and Palladio. Gothic builders also expressed the secrets of geometrical interrelationships in their cathedrals. Earlier the Greek mathematicians developed rational models as seen in the mathematical regularities of proportions to explain harmony and beauty as an absolute, universal order. They considered the ratios based on rational numbers to reflect a *static symmetry,* while the ratio of irrational numbers (e.g., $1:\sqrt{2}$, $1:\sqrt{3}$, $1:(1 + \sqrt{5})/2 = 1:1.618$) to be *dynamic symmetry.*

Dynamic or *harmonic series* represent oscillations (e.g., vibrating strings) or wave patterns, thereby clearly demonstrating a dynamic act. The root or dynamic rectangles (Fig. 1. 6) played an important part in Greek art and architecture in the search for expressing the mystery of order and unity. Pythagoras and Euclid found especially pleasing the rectangle with the ratio of 1/1.618 (often approximated as 3/5), which they called the rectangle of the divine section, also known as the *golden rectangle.* This magic ratio of the *golden mean* not only occurs in the creation of art, architecture, music, literature, and other design forms, but also in nature, in the form of the logarithmic or golden spiral, especially abundant in the field of botany, defining growth patterns. It probably was already familiar to the Egyptians as suggested by the proportions of the Pyramid of Gizeh. This mystic number has preoccupied the minds of scholars from Vitruvius to Leonardo da Vinci, Dürer, Kepler, Bernoulli, and more recently Le Corbusier.

The golden mean may also be derived from the *Fibonacci series,* which had so much influence upon modern architecture. Leonardo Fibonacci of Pisa, Italy, invented his famous sequence of numbers in the early thirteenth century. In this series, each number is equal to the sum of the two preceding numbers, starting with a pair of initially chosen numbers, for example, 3 and 5, resulting in

$$3, 5, 8, 13, 21, 34, 55, 89, \ldots$$

Dividing any of the Fibonacci numbers by the next higher number yields the golden mean, although the early small numbers in the sequence only approximate the magic ratio.

Le Corbusier derived his proportional system, the *Modulor,* for the dimensional coordination of the building by subdividing the human body, assumed to be 2.26 m to the finger tips of the raised arm, first into two basic parts of 0.86 m from the feet to the hanging arm and $2.26 - 0.86 = 1.40$ m for the upper portion as based on the golden section. With these consecutive numbers of 0.86 and 1.40, he constructed his *blue series* upward and downward according to the Fibonacci series.

Whatever the basis is for the derivation of the modular network, be it from the simple arithmetic and geometric series or the more complex harmonic series, the dimensional coordination is essential for the design, fabrication, and construction process, although it may not necessarily be expressed by the appearance of the building. On the other hand, it can be the design-generating determinant, as exemplified by Walter Netsch, of Skidmore, Owings and Merrill (SOM) in his *field theory,* where interlocking lattices reflect the dynamic interplay of various behavioral systems; he applied his theory for the first time in 1965 for the student center at Grinnel College, Iowa. Or, going even one step further, the building module may be a prefabricated room unit. In this case, the cells are assembled and stacked adjacent to and on top of each other, as in Moshe Safdie's Habitat in Montreal (1967).

1.4 THE SUPPORT STRUCTURE

Buildings basically consist of the support structure, the exterior envelope, the ceilings, and the partitions. Structure make spaces within a building possible—it gives support to the material. Whereas the structure holds the building up, the exterior envelope provides a protective shield against the outside environment, and the partitions form interior space dividers.

Most buildings consist of horizontal planes (floor and roof structures), the supporting vertical planes (walls, frames, etc.), and the foundations. The horizontal planes tie the vertical planes together to achieve somewhat of a box effect, and the foundations make the transition from the building to the ground possible. Keep in mind, however, that structure not only occurs on the large scale of the building, which tends to be more of organizational nature, but also on the small scale of the detail, which is more figurative and tactile and on a more human scale.

Although the structure's primary responsibility is that of support to transfer loads to the ground, it also functions as a spatial and dimensional organizer. Should the designer decide to expose the structure rather than hide it behind skin in order to articulate its purpose, then the structure may also enrich the quality of space. The designer may treat the structure not just in the minimal sense as support, but superimpose other layers of meaning to enrich its expression.

The structure resists the vertical action of the gravity loads that is its own weight, as well as the nonpermanent live or occupancy loads. It also resists the horizontal force action of wind and earthquakes; in other other words, it must guarantee lateral stability of the building. Gravity, however, can also generate lateral thrust due to the spatial geometry of structure, as demonstrated in Fig. 1.8. Structural designers predict the response of a structure and evaluate whether the loads can be safely withstood by con-

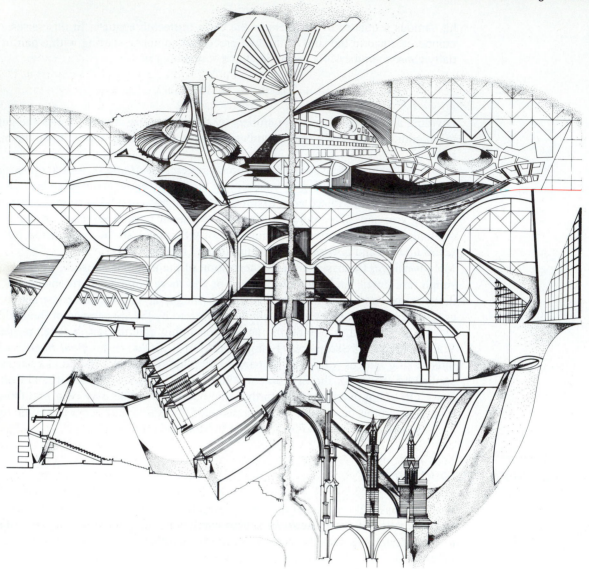

Figure 1.8 Lateral thrust.

structing ideal mathematical models to replace the real building structure, and they often verify the results by physically testing scale models. In general, however, numbers are calculated from ideal abstract models, rather than measured with instruments from physical models.

The primary purpose of the structure is to stand up and maintain its shape under load action by considering the three fundamental principles of **stability**, **strength**, and **stiffness**. For a building not to fail, members must be properly connected and not form a collapse mechanism; member strength must be adequate to resist loads, which includes considerations of member stability. The stiffness of buildings must be controlled so that it does not distort and deflect enough to cause excessive eccentric action of gravity loads (P–Δ effect) or to stress curtains, partitions and mechanical systems. Furthermore, excessive lateral sway of tall buildings together with oscillations due to gusty winds may not be acceptable for human comfort. Examples on the local scale include vibrational problems and large vertical deflections of flexible long-span floor structures that must be controlled.

Whereas the structural design of ordinary buildings is generally *stress governed* that of large-scale buildings, that is, very tall buildings and long-span structures, is *stiffness governed,* keeping in mind that stability must always be considered.

Every building consists of the *load-bearing structure* and the nonload-bearing portion. The main load-bearing structure, in turn, is subdivided into *gravity structure* and the *lateral force-resisting structure.* For the condition where the lateral bracing only resists horizontal forces but does not carry gravity loads, with the exception of its own weight, it is considered a *secondary structure.* Whereas the tower shape may also represent the structure shape, a massive building block only needs some stabilizing vertical elements to give lateral support to the entire building. For ordinary low-rise buildings, only some stand-alone systems are required to provide lateral stability. In other words, large buildings to a high degree are dependent on the support structure, whereas for small buildings the designer has more freedom of form giving.

Loads are resisted on a large scale by the building structure as a whole and on a small scale, locally, by the individual structure components such as beams, columns, slabs, walls, and foundations, as well as by the connections.

The Building as a Whole

Structures range from the horizontal slab and massive gravity block to the slender tower (Fig. 1.9). Whereas the gravity block is kept stable by its own mass, the free-standing slender TV tower seems to provide a bare minimum of structure and one is afraid it may tilt over and fall as the wind pushes and twists it. Buildings range from low-rise structures up to about five stories, to medium-rise structures below about twenty stories, to high-rise structures. On a large scale, structures range from long-span, column-free enclosed spaces covering 10 acres or so, as for superdomes, to the cellular subdivision of multiuse skyscrapers more than 1400 ft high. Whereas in tall buildings, the vertical planes (i.e., walls, frames) control the design of the structure and the horizontal planes (floor structures) are of secondary importance, in long-span structures the horizontal planes (roof or floor framing) are clearly the controlling design element.

From the point of view of overall building proportions, rectangular building shapes may be organized, as indicated in Fig. 1.9 as follows:

—horizontal slabs   —massive blocks   —vertical slabs
—gravity towers   —cantilever towers

While the slender tower cantilevers out of the ground and uses all its energy to resist the lateral forces, the flat horizontal slab is spread out on the ground. It hardly provides any resistance to wind, and the path of the vertical gravity flow is short; in this case, the structural behavior is primarily based on *horizontal gravity flow,* that is *gravity bending* for normal-weight structures (i.e., not lightweight structures). The massive building block, in contrast, is controlled by *axial gravity flow.* Not gravity, but the lateral loads due to wind and seismic action become dominant design determinants as the building increases in height. For the typical, slender, slab-type building, axial gravity, together with lateral force action, must be considered. As the slenderness of towers increases from 5:1 to 8:1 for buildings (12:1 is considered as the upper limit for buildings where gravity towers change into cantilever towers) to 30:1 for TV towers, the effect of wind, together with oscillations and the flexibility of the structure, becomes extremely critical. For the purpose of simplicity, it has been assumed in our discussion that the building shape is equal to the shape of the lateral force-resisting structure, which obviously does not have to be true.

To demonstrate the scale of horizontal-span and high-rise structures, the dimensions of some historically important buildings are given for the purpose of comparison in Fig. 1.9.

- 4625-ft-span Humber Bridge, U.K. 1981
- 680-ft-span, 273-ft-high Louisiana Superdome, New Orleans, 1975
- 1454-ft-high Sears Tower, Chicago, 1974
- 1815-ft CN Tower, Toronto, Canada, 1976
- 2108-ft guyed mast, Warsaw, Poland, 1974

Structure Systems

Structure systems are only introduced briefly since they are discussed in more detail throughout this text. They are classified as follows:

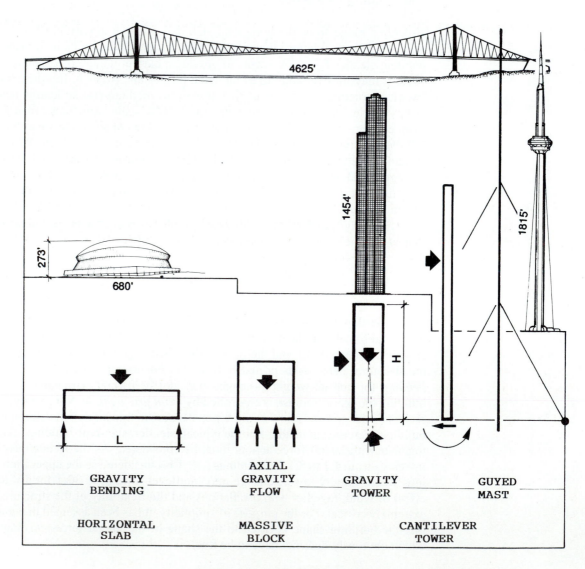

Figure 1.9 From the horizontal slab to the slender tower.

- Horizontal-span structure systems
    Floor and roof structures
    Single-story enclosure systems
- Vertical building structure systems: wall, frame, core structures, etc.

Although buildings are three-dimensional, their vertical support structures or horizontal-span enclosure structures can often be treated as an assembly of two-dimensional structures. In other words, structures can usually be subdivided into a few simpler assemblies unless they are of truly three dimensional nature.

A structure represents an assembly system that consists of components and their linkages. The components or members may be rigid or flexible, linear, planar, or spatial—they may be folded or curved to form a two-dimensional or three-dimensional enclosure structure. The shape of the components may be of constant or variable cross section; in other words, it may be tapered, curved, or only haunched at the supports. The member structure may be solid (homogeneous or composite), trussed, framed (e.g., Vierendeel type), or composite (e.g., cable-supported strutted beams or columns, cable beams). The member weight may be minimalized and the shape optimized to yield a funicular shape (e.g., curved prestressed cable–strut system). In other words, the spatial configuration of a structure and the arrangement of its members can represent an equilibrium form, where the form of the structure makes a natural equilibrium of external forces possible. These forms are derived from structural behavior and not from mathematical shapes. Typical equilibrium-form systems are the following:

- Funicular systems or form-resistant structures
- Self-stressed systems where the induced internal forces are in equilibrium(e.g., fabric structures, tensegrity systems)
- Intelligent systems responding actively to loads through material and geometry adjustment

The structures of ordinary small-scale buildings, such as for low-rise detached and row housing, are usually governed by local construction methods. Other structures and especially large-scale construction, such as for high-rise and long-span buildings, however, are controlled by engineering considerations.

Horizontal-span Structure Systems.   From a geometrical point of view, horizontal-span structures may consist of linear, planar, or spatial elements. Two- and three-dimensional assemblies may be composed of linear or surface elements. Two-dimensional (planar) assemblies may act as one- or two-way systems. For example, one-way floor or planar roof structures (or bridges) typically consist of linear elements spanning in one direction where the loads are transferred from slab to secondary beams to primary beams. Two-way systems, on the other hand, carry loads to the supports along different paths, that is, in more than one direction; here members interact and share the load resistance (e.g., two-way ribbed slabs, space frames).

Single-story building enclosures may be two-dimensional assemblies of linear members (e.g., frames, arches), or they may be three-dimensional assemblies of linear or surface elements. Whereas two-dimensional enclosure systems may resist forces in bending and/or axial action, three-dimensional systems may be form-resistant structures that use their profile to support loads primarily in axial action (i.e. spatial axial systems). Spatial structures are obviously more efficient regarding material (i.e., require less weight) than flexural planar structures. From a structural point of view, horizontal-span structures may be organized as follows:

- **Flexural systems** of the one- or two-way type: e.g., beams, girders, trusses, Vierendeel trusses, prestressed concrete beams, floor grids, flat space frames
- **Two-dimensional flexural and/or axial systems:** e.g., frames, arches, suspension cables
- **Form-resistant structures** (rigid or flexible): e.g., folded plates, shells, tensile membranes

Some common rigid horizontal-span structure systems are shown in Fig. 1.10; they may be organized as follows:

- *Straight or inclined line elements:* beams, columns, struts, hangers, etc.
- *Folded or curved line elements*

    Horizontal-span structure: one- or two-way beam grids

    Enclosure structure (e.g., frames, arches, suspended cables)

    Two-dimensional structure (e.g., parallel arrangement)

    Three-dimensional structure (e.g., spatial arrangement)

- *Straight or folded surface elements*

    Horizontal-span structure: one- or two-way slabs or plates, one- two-way, or three-way folded plate beams or continuous folded plates

    Enclosure structure

    Two-dimensional structure (e.g., folded plate frames or arches)

    Three-dimensional structure (e.g., folded plate dome)

- *Curved surface elements*

    Horizontal span structure: single- or double-curvature shell beams, shallow domes

    Enclosure structure: shells are generally double-curvature systems and form by their very nature three-dimensional structures (e.g., domes, saddle shells, groined shells)

Keep in mind that the structure components in Fig. 1.10 are treated as lines or solid surfaces. In reality they may represent linear truss beams, trussed portal frames or trussed arches, and space frame surface structures. Critical with respect to horizontal-span structure systems is the length of the span. They range from the short span of floor joists to long-span buildings such as exhibition halls, convention centers, stadiums, and airplane hangars. The definition of long span is not always that clear. What may be considered a long span for a wood beam may not be true for a steel beam. A 60-ft span in a high-rise building is considered a long span, although that is not the case for a gymnasium.

Critical with respect to long-span structures is their degree of *redundancy*. Failure of a beam in a multibay structure is contained locally and may hardly affect the behavior of the overall structure; in other words, failure is not synonymous with collapse. In contrast, failure of a structure portion in a long-span structure without interior supports (especially for the condition of a minimum number of supports and for one-way systems) may cause collapse of the entire building, as happened in the Hartford Civic Center Coliseum in Hartford, Connecticut, in 1978. Also critical for long-span systems are the great changes in length of roof structures due to temperature changes. We must keep in mind that for long-span structures not necessarily least-weight systems, yield the most economical solution as considerations of constructability must be taken into account. Usually for spans larger than about 300 ft, some in-place assembly is needed possibly requiring shoring towers and tower cranes with a high lifting capacity, which in turn significantly influence the overall costs. From a material efficiency

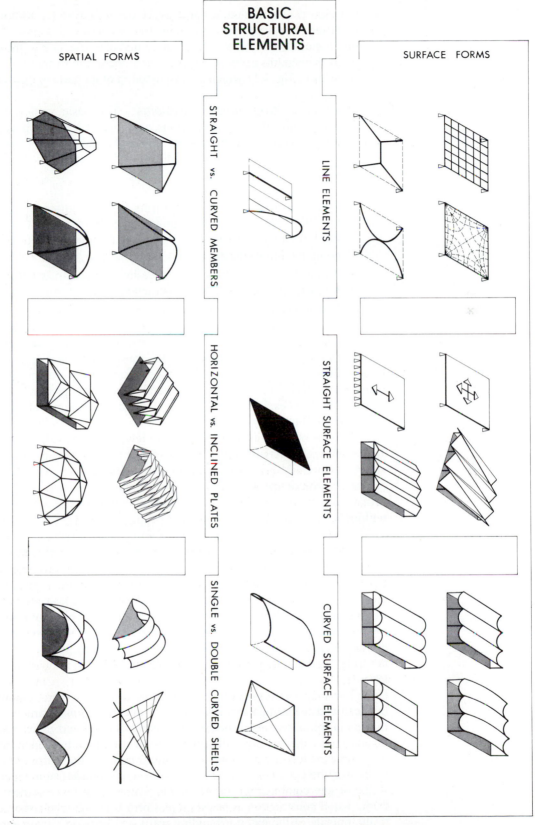

Figure 1.10 Horizontal-span rigid structure systems.

point of view (i.e., minimum weight), every structure type has its upper limit; in other words, different scales require different structure systems, as is discussed in the next section on the effect of scale. It is apparent that classification of the structure systems presented as diagrams cannot express the architecture in its true richness, as for example the cases in Fig. 1.11 which are discussed in other parts of this book.

Vertical Building Structure Systems.   The basic vertical building structure types are bearing wall systems, skeletons, and core structures. They are laterally stabilized by shear walls, moment-resisting frames, braced frames, or any combination of these. High-rise structure systems are discussed in Chapter 10 (Figs. 10.2 and 10.3) as based on the following classification:

- *Two-dimensional structures:* bearing wall structures, light framing construction (e.g., wood platform framing), moment-resisting frame structures (e.g., rigid frames), flat slab structures, braced-frame structures, core structures, combinations of these
- *Three-dimensional structures:* core-interaction structures, tubular structures, megastructures, form-resistant structures, hybrid structures
- Combinations of the preceding

To gain a clearer understanding of vertical structure systems, they must be seen within building space; hence their location must be known. For this reason, solid surface elements have been placed into the uniform beam–column grid of the various plans in Fig. 1.12. They represent the lateral force-resisting structure systems of walls, cores, frames, tubes, and so on. They may form either planar or spatial assemblies or any other combination.

From a material efficiency point of view, each vertical structure system is only applicable within certain height limits (refer to the next section for a discussion of the effect of scale). To demonstrate that structural concepts expressed in diagrams cannot encompass the richness and complexity of architecture, structure systems are briefly investigated now in the context of small residential houses and single-story, multibay buildings.

Structure Systems in House Construction.   Among the earliest European methods of house construction were the polebeam, the braced pole, and the log cabin in timber, as well as stone and brick wall structures. In the Middle Ages, half-timber construction developed in central Europe with each region establishing its own style of framing pattern. The wood framework, which is visible from the outside, is tied together and stiffened with horizontal struts and diagonal braces and is filled in with plaited reeds and clay or brickwork or stone. The many styles of half-timber framing are truly amazing, they range from simple structures to networks of quite decorative layouts.

The small houses in ordinary residential construction are usually not engineered, but designed according to accepted practice of conventional framing techniques. Since the spatial requirements of homes are clearly defined and rather standardized in the housing field, the bearing wall principle is most frequently used because the walls not only carry the loads, but also act as partitions between the various spaces. The bearing walls may be concrete, masonry (brick, concrete block), stone, metal or wood stud walls, or any combination. In the United States, most low-rise homes employ light-frame wood construction as based on platform framing, where the joist floors extend to the outside of the exterior building walls and provide a platform upon which the one-story high stud walls are placed (Fig. 1.21). The traditional balloon framing is still to be found in use for split-level houses, where the floor joists are nailed directly to the

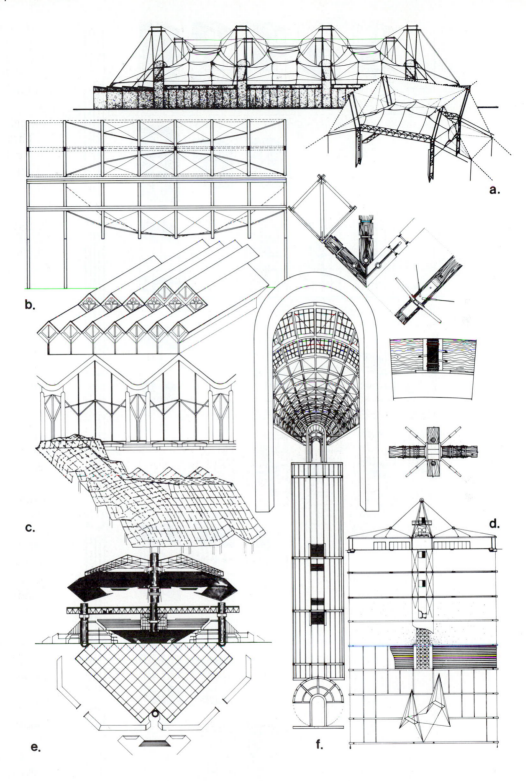

Figure 1.11 Examples of horizontal-span roof structure systems.

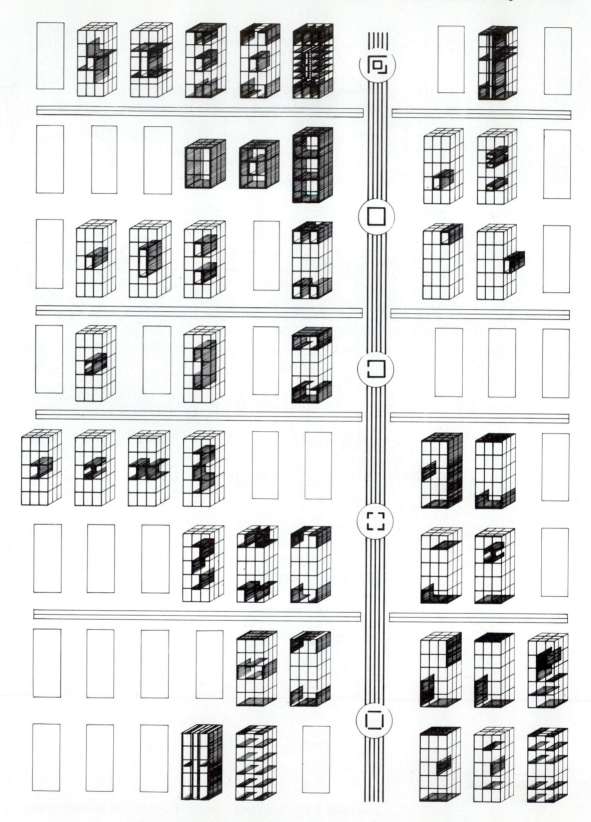

Figure 1.12 Possible location of lateral force-resisting structures within the building. Reproduced with permission from *The Vertical Building Structure,* Wolfgang Schueller, copyright © 1990 by Van Nostrand Reinhold.

sides of the wall studs, which are continuous over the full building height. It is custom-ary to space the studs of bearing walls (normally 2 × 4's for single-story houses) at 16 in. on center.

Nearly all the cases in Fig. 1.13 are custom designs and not typical examples for home construction; however, they exemplify the organization of structure systems. The various arrangements of bearing walls indicate distinctly different features. The wall may form a protective enclosure such as case *a* (Smith Residence by Richard Meier, 1967). The walls may be arranged in a parallel manner and tied together by the roof platform, allowing nature to enter the building, as in case *b* (Hilton Residence by Arthur Erickson, 1975). In contrast is the geometrical, more organic wall pattern of case *c* (Duncan House by Bruce Goff, 1965), which resembles the continuous mean-dering flow around cores in response to topography.

The lightness and strength of the steel frame seem to be so perfectly expressed in Mies van der Rohe's famous Farnsworth house (case *d*, 1950). Rather than emphasiz-ing the beam–column principle, Mies articulated the cantilevering roof and floor planes by letting them float effortlessly between the eight exterior columns located on the long faces. This dynamic effect is amplified by the bare minimum amount of steel material needed for vertical support and by the glass skin between the horizontal slabs.

The modular skeleton structure of case *e* (Kohler House by Booth and Nagle, 1974) seems to be of exact opposite spirit, and not only because of its different planar order. Here the post–beam idea articulates an organic quality of stability by letting the flow of gravity continue directly to the ground in a very static manner. The hinged tim-ber frame is laterally braced by the solid in-fill panels.

Pole construction offers an advantage for difficult sloping sides or where the first floor must be raised because of flooding or other functional requirements. The poles of the building in case *f* (Villa Coupe by K. Yashida, 1972) are tied together by the floor platforms and the walls. In case *i*, independent lightweight steel modules, as derived from assembly line procedures of the mobile home manufacturers, are stacked in alter-nate directions around a central utility shaft. In case *g* (Capsule House K by Kisho Kurokawa, 1972), the steel capsules are simply cantilevered out from the concrete core.

The structural principle of case *h* (Dissentshik House by Zvi Hecker, 1972) is based on stacking and clustering self, all-space filling truncated octahedra. Often space frames or paperboard structures are associated with polyhedral forms. The strength is derived from the continuous double curvature of the folded surface, which allows a thin shell of a low modulus of elasticity, such as corrugated paperboard coated on both sides with glass-reinforced polyester resin, to be used.

The ferrocement shell in case *j* (Lubetkin House by Ant Farm/Jost, Lord and Michels, 1973), familiar from boat construction, consists of a 6-in. network of $\frac{3}{8}$-in. steel rods to which are applied four layers of chicken wire on each side and three layers of cement mortar.

Single-story Multibay Structures.   The various buildings in Fig. 1.14 suggest a wide range of support structure types ranging from simple beams, cantilever beams, stayed beams, flat slabs to shells, using simple or treelike columns. The selection of a structure is related to the span range; long-span structures naturally require special considerations. Typical multibay structures for commercial or industrial buildings are not shown; they usually employ frame construction on multibay column grids (see Fig. 5.18).

For the Labor Palace in Turin, Italy, 1961 (Fig. 5.14 a), Pier Luigi Nervi used independent 82-ft-high umbrella units consisting of tapering cruciform concrete col-umns and radially arranged welded steel girders that carry a 125-ft square concrete roof. The 6.5-ft space between the independent mushroom units is covered by skylight ribbons.

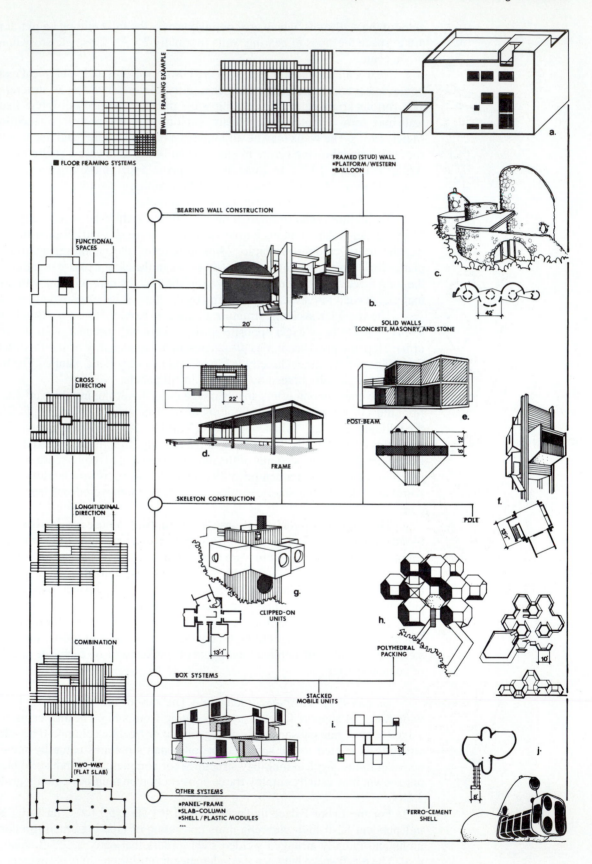

Figure 1.13 Structure systems in residential construction.

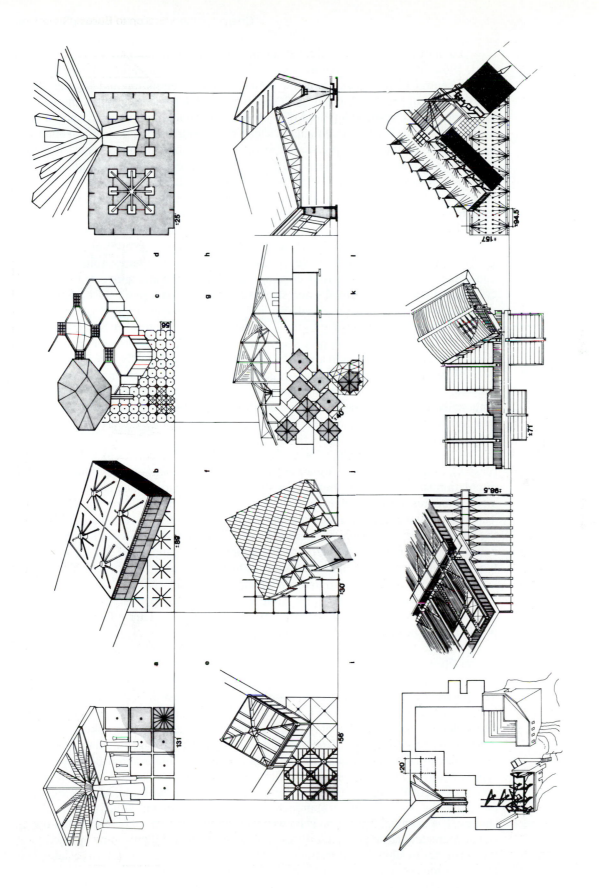

Figure 1.14 Multibay, single-story structure systems.

41

Each of the self-contained 89-ft square units of a printing plant in Tapiola, Finland, 1966, by Aarno Ruusuvuori (case *b*), is suspended with eight concrete hangers from a central 10-ft-diameter concrete core containing ventilation and heating ducts. For the Olivetti plant in Harrisburg, Pennsylvania, 1970 (case *c*), Louis Kahn and the engineer August Komendant employed inverted concrete umbrella shells based on a 56-ft square grid. Each of the prismatic shells with prestressed upper rim is supported by a central column (with prestressed capital panel) containing drainage. The free space between the shells is covered by fiber-glass-reinforced plastic skylights.

The concrete waffle slab of St. John's University Library, Collegeville, Minnesota, 1968 (case *d*), by Marcel Breuer, is supported by perimeter walls and massive interior concrete trees with eight large and four small arms. The concrete columns of the Tokyo YMCA, 1976 (case *i*), by Shozo Uchii, flare out to form abstract trees.

The complex roof of the library at Wells College, Aurora, New York, 1968 (case *g*), by Walter Netsch of SOM, is developed from a series of squares, rotated squares, and octagonal modules. The roof units are radially framed and supported by central columns or clusters of inclined wood struts sitting on masonry piers. Kisho Kurokawa used for Nitto Food Company Cannery in Sagae City, Japan, 1964 (case *e*), tubular steel trusses to span diagonally to the compound corner columns of X-section pipes. The two diagonal trusses support secondary trusses. Skylight ribbons are provided along the diagonal trusses and a ventilation shaft is located at the center intersection.

Helmut Jahn of C. F. Murphy organized the Public Library in Michigan City, Indiana, 1978 (case *f*), on a rectangular grid, while the sawtooth roof framing is placed diagonally to it to enable better lighting conditions. The two-span roof structure of case *h* is composed of single-span steel trusses resting on perimeter frames and central V-shaped concrete columns.

The major spaces of the Osaka Prefectural Sports Center, Osaka, Japan, 1972 (case *k*), by Fumihiko Maki, are arranged in a staggered fashion along both sides of a circulation spine. The main supporting structure for a typical space is a giant tubular steel beam resting on cylindrical concrete columns at the building facade. Kenzo Tange applied the stayed bridge principle to the Printing Plant at Haramachi, Japan, in the late 1960s (case *j*). Along the central longitudinal spine, two rows of columns carry a large concrete box girder which, in turn, supports the spatial steel trusses as well as the suspended steel tubes that give support to the cantilevering trusses. The cable-stayed principle is also applied to the West Japan General Exhibition Center, Fukuoka Prefecture, Japan, 1977 (case *l*), by Arata Isozaki. The rectangular roof modules are composed of four primary beams, which are supported by diagonal cables that are suspended from the tubular steel masts at the periphery.

Force Flow

The horizontal and vertical structural building planes must disperse the external and internal loads to the ground. The load path may be short and direct or long and indirect and suddenly interrupted, causing a detour. The paths the loads may take along horizontal and vertical building planes depend on the structure layout, which must respond to the functional organization of the building where the columns and walls may help to separate and reinforce the spaces to allow for different activities. Various arrangements of vertical support structures for a typical rectangular building unit are compared in Fig. 1.15; a similar approach could be taken for buildings of other basic shapes. The essential distinction is between *multibay structures* of one- and two-directional character and single unit *long-span structures* where columns only appear along the perimeter to allow for unobstructed interior space. The typical patterns established by the various column layouts are apparent in Fig. 1.15.

A similar study is presented in Fig. 1.16 for a multicell rectangular high-rise building as derived from a basic grid by using different column–wall arrangements.

Here the formal transition from a column system laterally stabilized by three walls (*a*) to a pure wall structure (*p*) is presented, clearly identifying the increase in *structural plan density*. The wall components have various shapes, ranging from linear and open compound forms (e.g., *l, j, h,* and *t*) to closed cores. They may be combined and arranged to result in the patterns of Fig. 1.16.

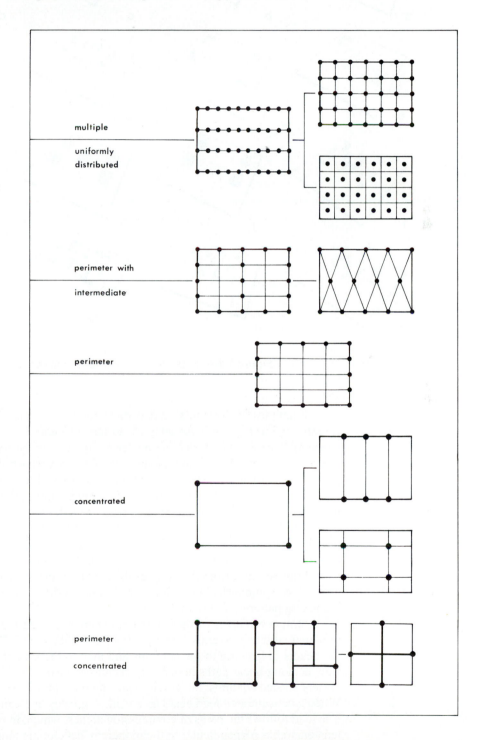

Figure 1.15 Typical locations of support structure.

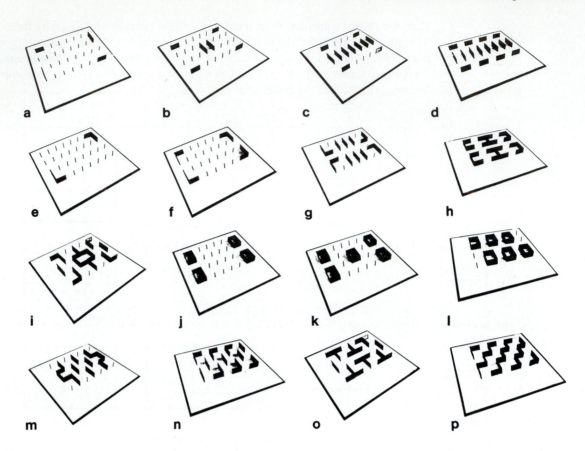

Figure 1.16 From skeleton to wall structure.'

Examples of horizontal and vertical force flow due to gravity load action are investigated in Fig. 1.17; for lateral load transmission, refer to Chapter 4. A concentrated load acting upon a slab is transferred by the floor or roof framing in bending to the vertical structural building planes, which usually transmit the load axially straight down to the ground. The type and pattern of horizontal force flow naturally depend on the arrangement of the vertical structural members. In Fig. 1.17b, only four corner columns are used to demonstrate the diversity of the force flow.

The horizontal load path follows the member layout from the secondary to the primary elements and then vertically downward; the stiffest member attracts most of the loads. Force flow patterns range from one- to two-directional systems, from linear to curvilinear ones, from two-way grids parallel to the perimeter to diagonal two-way grids, from symmetrical to asymmetrical radial systems, from direct to more complex branching patterns, and so on.

Simple examples of vertical load flow are shown in Fig. 1.17a, for various types of planar structure systems (see Fig. 10.4 for further study). First the load is carried by the beams in bending to the columns and then to the ground. While the vertical load flow is along linear paths in ordinary skeleton construction, in tubular structures with closely spaced columns, the gravity loads tend to spread out similar to a bearing wall. When a concentrated load bears on a wall, it spreads out, causing a complex flow net. The load follows the paths of compression arches, which, in turn, must be balanced by tension arches perpendicular to them, thereby developing tensile stresses in the wall.

Should there be an opening in the frame, as required for an atrium, then the force flow is not continuous and straight, but interrupted, and must be transferred to another

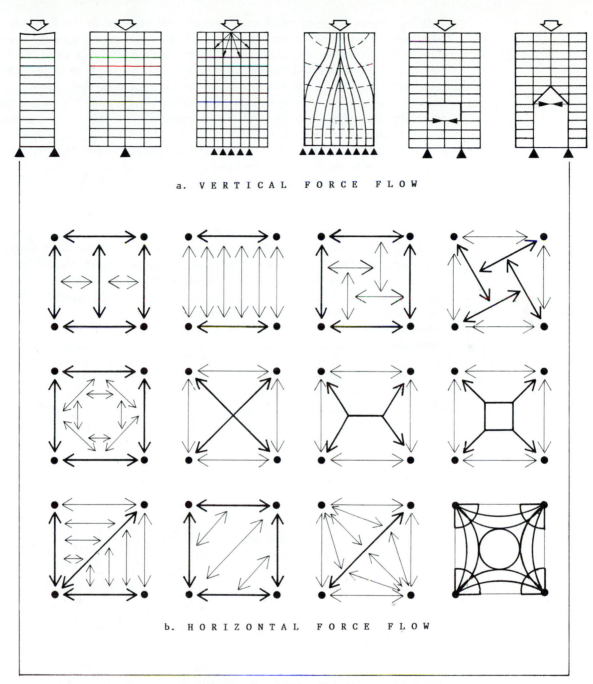

a. VERTICAL FORCE FLOW

b. HORIZONTAL FORCE FLOW

Figure 1.17 Gravity force flow in planar structures.

column line, which may be achieved through frame action or inclined columns (Fig. 1.17a). This action causes lateral thrust; in a symmetrical structure the thrust due to the dead load will self-balance.

The Effect of Scale and Rules of Thumb

With increase in building size, not only the magnitude of forces increases and becomes a dominant design consideration, but also new types of forces appear. Small buildings do not weigh much in contrast to large buildings and therefore generate much smaller inertial forces during an earthquake, indicating also that the effect of irregularity of building shapes becomes much more critical with increase in building size. Large buildings are dependent on support structure to a large degree since only a restricted number of alternatives is available. In small structures, on the other hand, the designer has much more choice and freedom of form giving. The design of large-scale buildings, that is, long-span structures and high-rise buildings, is not only controlled by the support structure, but also by construction methods and the life-supporting systems, including fire safety. It is apparent that, because of their higher complexity, large buildings are more vulnerable to failure.

The effect of scale in structures can be demonstrated with the simple example of the increase of beam span. Here, the bending moment increases with the square of the span and the deflection much more rapidly with the span to the fourth power. Hence, the beam depth must increase not in proportion to the span, but the square of the span for the condition where the bending stresses control the design and cannot be increased. Clearly, at a certain stage the material, structure system, and/or beam shape must change to overcome the effect of increase in dead weight; in other words, bending systems may have to be replaced by axial systems or form-resistant structures.

The effect of scale is known from nature, where animal skeletons become much bulkier with increase of size, as reflected by the change from the tiny ant to the delicate gazelle and finally to the massive rhinoceros and elephant. While an ant can support a multiple of its own weight, it could not even carry itself if its size were proportionally increased to the size of an elephant, since the weight increases with the cube, while the supporting area only increases with the square as the linear dimensions are enlarged. Thus, the dimensions are not in linear relationship to each other; the weight increases much faster than the corresponding cross-sectional area. Hence, either the proportions of the ant's skeleton would have to be changed, or the material made lighter, or the strength and stiffness of the bones increased. It is also interesting to note that the bones of a mouse make up only approximately 8% of the total mass in contrast to about 18% for the human body.

We may conclude that structure proportions in nature are derived from behavioral considerations and cannot remain constant. Galileo Galilei (1638) was the first to demonstrate the effect of scale in nature, particularly the relationships of surface to volume and weight. Later many other scientists, like D'Arcy Thompson (1917), became fascinated by the same phenomenon. The impact of scale on structure and form is apparent from nature not only with respect to animals but also plants. For instance, the slenderness of the wheat stalk (height to diameter) is around 500, while it decreases to 133 for bamboo and to about 36 for a giant redwood tree, clearly illustrating again that proportions are not constant but change.

This phenomenon of scale is taken into account by the various building structure systems relating to horizontal span and vertical span (height), as have been introduced previously. With increase of span or height, material, member proportions, member structure, and structure layout must be altered and optimized to achieve higher strength and stiffness with less weight. The primary design characteristics to be considered with respect to the span of the support structure are *stability, strength,* and *stiffness.*

Stability is taken care of by controlling the member proportions, such as the *depth-to-thickness ratio* of a member or element of a member and the *slenderness ratio* for the overall behavior of a compression member. The efficient use of strength is considered by the *strength-to-weight ratio* (i.e., specific strength). Stiffness is controlled by deflection limitations, which are given in terms of some fraction of the span or in terms of *the member span-to-member depth (L/t) ratio.* Its effect is evaluated by the *elastic modulus-to-weight ratio* (i.e., specific elasticity). Strength and stiffness are naturally affected also by the live- to dead-load ratio. The design characteristics are discussed further in Sections 2.5 and 3.2 and elsewhere in the book.

For preliminary design purposes, rules of thumb have been developed over time from experience, keeping in mind that member proportions may not be controlled by structural requirements, but by dimensional (clearance, structural detail, story height, formwork, etc.), environmental, and esthetic considerations. Typical empirical design aids as expressed in span-to-depth ratios for horizontal-span roof structures (or for lightly loaded floor structures such as for residential usage) are presented in Fig. 1.18. Typical average L/t ratios for the various structure systems are as follows:

- Reinforced concrete shells · $L/t \cong 400$
- Shell beams · $L/t \cong 100$
- Vaults and arches · $L/t \cong 60$
- Decking: e.g., continuous one-way slabs · · · · · · · · · · · · $L/t \cong 36$
  or $t \cong L/3$ (t in inches, L in feet)
- Shallow beams e.g., average floor framing · · · · · · · · · · · $L/t \cong 24$
  or $t \cong L/2$ (t in inches, L in feet)
- Deep beams: e.g., trusses, girders · · · · · · · · · · · · · · · · · · $L/t \cong 12$
  or $t \cong L$ (t in inches, L in feet)
- Envelope systems and pitched trusses: · · · · · · · · · · · · · · $L/h \cong 6$
  refer to the respective sections in this book for L/h rules of thumb

The thickness, $t,$ of shells is by far less than that of the other systems since they resist loads through geometry as membranes in axial and shear action (i.e., strength through form), in contrast to other structures, which are bending systems.

The systems in Fig. 1.18 are rigid and gain weight rapidly as the span increases, so they may have to be replaced at a certain point by flexible, lightweight cable structures. The characteristics of tension systems are discussed in Chapter 9.

The effect of scale is demonstrated by the increase of member thickness t or depth h as the members become larger, that is, change from decking to shallow beams to deep beams to envelope systems. Each system in Fig. 1.18 is applicable for a certain scale range only. This change of structure systems with increase of span can also be seen, for example, in bridge design, where beam bridges (rolled beams to plate girders) change to box girder bridges to truss bridges to arch bridges, cable-stayed bridges, and finally suspension bridges.

Keep in mind, however, that it is impossible to clearly identify in which range, exactly, specific structure systems constitute an optimum solution, since several concepts are competing with each other and may all be efficient, and the selection may not necessarily be based on structural considerations of efficiency either.

The spans given in Fig. 1.18 for the various one- and two-way structure systems represent typical values, although each structure has a practical span range as, for example, for the following *long-span systems* starting at approximately 40- to 50-ft span and ranging usually to roughly the following:

- Flat wood truss · 120 ft
- Flat steel truss · 300 ft
- Timber frames and arches · 250 ft
- Folded plates · 120 ft
- Cylindrical shells · 180 ft
- Thin shell domes · 250 ft
- Space frames · 400 ft
- Skeletal domes · 400 ft
- Two-way trussed box mega-arches · · · · · · · · · · · · · · · · · · 400 ft
- Two-way cable supported strutted mega-arches · · · · · · · · 500 ft
- Composite tensegrity fabric structures · · · · · · · · · · · · · · · 800 ft

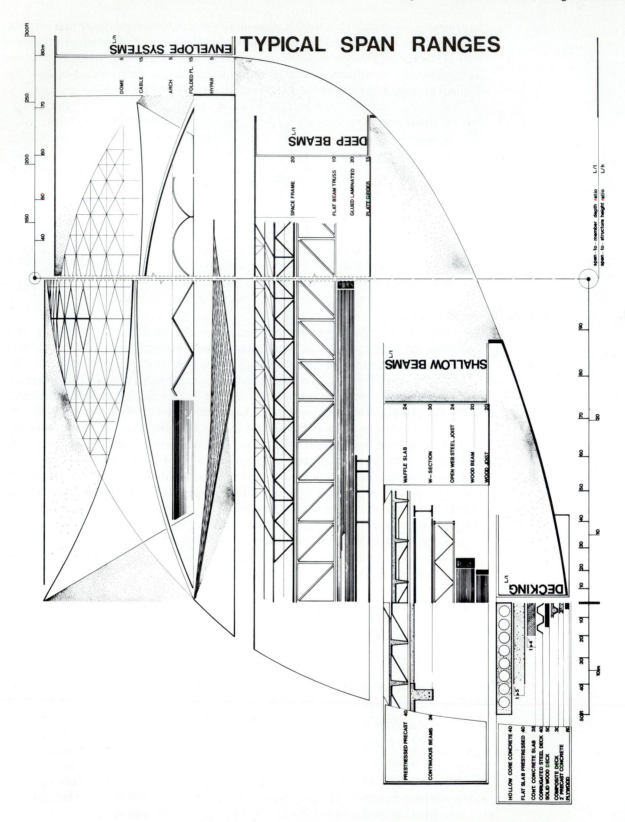

Figure 1.18   Typical span-to-depth ratios for horizontal-span members.

More precise information about the various characteristics of the structure types is presented elsewhere in this book.

The rules of thumb in Fig. 1.18 are given only for preliminary design purposes. The designer in the early planning stage must make structural decisions, possibly without the help of an engineer, and must determine the layout of the physical building elements defining the spaces, proportion the members, and thereby possibly influence the esthetics of the building. Even if the visual image of the building is derived from other proportional orders or subjective opinions, that is, design theories that do not include structures in their vocabulary, the architect who intends to build must have a sense of control by understanding the scale and purpose of the structure, that is, the physical reality of the building independent of stylistic trends. It is impossible to build a body without the knowledge of the anatomy of the bone structure supporting its organs—the understanding of structure is a necessity! To give a design idea a sense of physical reality in the conceptual design stage, empirical structural design aids are necessary so that member sizes can be estimated without the help of rigorous mathematical investigations.

1.5 BASIC STRUCTURE COMPONENTS

The building's superstructure and substructure are defined by geometry, that is, points, lines, surfaces, spaces, and bodies (solids). The basic structure components in an ordinary building superstructure are the nodes or the *point elements* of joints (connections), the *linear members* of beams and columns, the *surface elements* of slabs and walls, and *spatial elements* (e.g., cores). Planar elements can be solid surfaces or can be composed of linear elements (e.g., floor and roof framing, trusses). Other building elements such as infill panels and cladding are treated as nonstructural and, therefore, are ignored for the structural analysis of the building to simplify the calculations.

Typical beams are linear horizontal elements supporting transverse loads and possibly end bending moments. They may also carry axial forces, which, however, are small in comparison to column loads. Typical building columns are vertical and support axial loads and possibly significant bending moments. Whereas slabs are primarily bending systems, walls are essentially axial compressive systems.

Only linear and planar rigid members, as well as joints, of the superstructure of ordinary buildings are briefly introduced in the following sections, but they are discussed further in other parts of the book. Foundations or substructure are introduced in Section 2.9, flexible members in Chapter 9, and three-dimensional elements in Chapters 6 to 9.

Beams

Beams cannot transfer loads directly to the boundaries as axial members do; they must bend in order to transmit external forces to the supports. The deflected member shape is caused by the bending moments. Beams are distinguished in *shape* (e.g., straight, tapered, curved), *cross section* (e.g., rectangular, round, T-, or I-sections), material (e.g., homogeneous, mixed, composite), and *support conditions* (simple, continuous, fixed, etc.). Beams may be part of a *repetitive grid* (e.g., parallel or two-way joist system) or may represent individual members—they may support ordinary floor and roof structures or span a stadium—they may form a stair, a bridge, or an entire building. In other words, there is no limit to the application of the beam principle.

Depending on their span-to-depth ratio (L/t), beams are organized as *shallow beams* with $L/t > 5$ (e.g., rectangular solid, box or flanged sections), *deep beams* (e.g., girders, trusses), and *wall beams* (e.g., walls, trusses, frames). They may not only be the common planar beams, but also spatial beams such as folded plate and shell beams (i.e., corrugated sections) or space trusses.

Furthermore, beams are not always straight as, for example, the typical floor beams and inclined roof joists; they may also be curved (e.g., arches) and folded (e.g., stairs, frames). Besides the common solid and trussed beams, some other types are identified in Fig. 1.19. Shown are a prestressed concrete Vierendeel truss (a), a tree-supported truss (b), and a braced truss (d). New beam types are developed through intermediate supports using tensile members (e.g., rods, cables) from below or above. For example, the overlapping single-strut cable-supported beam (c) represents a sub-tensioned beam (i.e., underslung beam), whereas the beam supported by stayed cables (e) or by suspended cable (f) are supported from above.

The longitudinal profile of beams may be shaped as a funicular form in response to a particular force action, which is usually uniform gravity loading; that is, the beam shape matches the shape of the bending moment diagram to achieve constant maximum stresses. The cases in Fig. 1.19g to l, demonstrate the principle; here the funicular shape of the members is resolved efficiently by trusses or trusslike guyed systems. For example, the bowstring truss (g) consists of a funicular arch with a constant-stress bottom chord. The web members are zero and only activated under asymmetrical loading. Similarly, the fish-belly truss or underslung beam (i) has a funicular bottom chord and a constant-stress top chord. The lens-shaped, cable-supported strutted arch, which is a prestressed cable beam or truss (h), consists of funicular top and bottom chords. The gabled truss (j) has a funicular bottom chord and constant-stress top chord. This shape can be derived by simply folding the moment diagram of the horizontal-span beam (i) to form a gable; see also Fig. 3.10a for M/S diagram. Robert Maillart applied this principle in 1925 to the cantilevered gabled roof truss of the warehouse for the S. A. Magazzini Generali in Chiasso, Switzerland.

The cable–arch configuration in case (k) follows the moment diagram of a portal frame, as does the cable configuration of case (l). Also mentioned should be the shape of the Firth of Forth Bridge in Scotland (1890), which was developed in direct response to force flow intensity for the first time on a large scale similar to (Fig. 5.7 bottom); it became a source of inspiration for numerous designers. For further discussion of beams, refer to Chapters 2 and 3 and other parts of this book.

Floor and Low-slope Roof Structures

Floor and roof structures consist of the skin or slab and the framing. Occasionally, the slab is not supported by beams, as in flat plate construction. The basic floor or flat roof framing systems, as derived from the direction of their beam layout, are identified in Fig. 3.2. They are arranged in parallel, radially, or diagonally, in one, two, or multiple directions. Since the roof live loads are generally low, the roof framing should only be equated to the floor framework supporting light loads, as for residential usage.

The type of framing depends on the building shape, the type as well as location of support structure, the scale of span and loading, disturbances (e.g., openings), and other functional and possibly esthetical considerations. There is no magic formula for choosing a floor structure. Each building has its own unique conditions relating to site, functional needs, support of equipment, openings, labor practices, fireproofing, underfloor ductwork, electrical conduit distribution, sprinkler systems, suspended ceiling, light fixtures, adaptability to future changes, the consideration of less weight for deeper floor framing versus shallower but heavier floors yielding less overall building height, the designer's experience and prejudice, etc.

The **roof structure** must not only provide the necessary strength to support itself and any superimposed loads, but must also control the climate, that is, the flow of heat, sound, water, air, and water vapor; in addition, it must have a certain degree of fire resistance. As for the floor structure, it may have to accommodate electrical, plumbing,

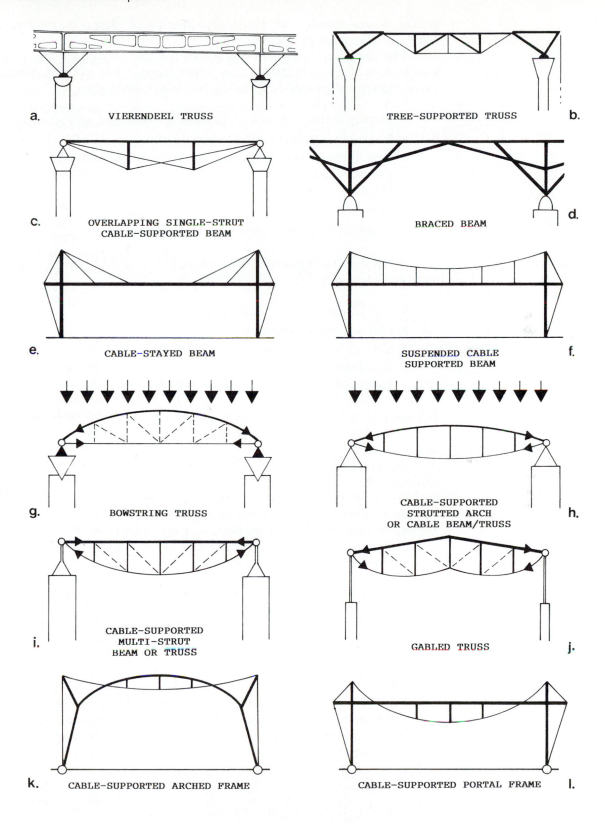

Figure 1.19 Other beam types.

heating, air-conditioning, and sprinkler systems; it must adapt to the attachment of the ceiling or other finishing materials.

Flat roofs include any roof with a slope less than 1 in./ft (5°); they should have a slight slope of at least 0.25/12 to have proper drainage. To prevent the penetration of water, most low-slope roofs may use the following roofing types:

- Built-up roofing (i.e., layers of roofing felt combined with bitumen)
- Single-ply roofing (i.e., single-sheet membrane)
- Liquid applied roofing (e.g., foamed insulation such as urethane covered with a water barrier)
- Metal roofs (e.g., steel, copper, aluminum):
- Architectural metal roofs (metal applied to deck)
- Structural metal roof (metal spans roof framing)

Steep slope roofs generally have a slope of 1/12 to 4/12 (5 to 18°) and are organized according to roofing types as follows:

- Asphalt roofs (e.g., roll roofing, shingles)
- Clay and concrete tile roofs
- Slate roofs
- Wood shingle roofs
- Metal roofs

The roof may have to be insulated against heat loss and gain (e.g., blanket, rigid board, foamed-in-place); a vapor barrier may have to be provided at the underside of the insulation, possibly together with vents.

The surface skin (i.e., deck or slab) may be a one- or two-way structure. It is either directly supported on the primary structural members (walls, columns, primary beams) or it rests on secondary filler beams or joists (i.e., bay subframing). The skin may just be sitting on the framework or it may be a composite part of it, such as cast-in-place concrete, stressed-skin panelized systems, or composite construction between concrete slab, corrugated steel deck, and steel beams. Some of the familiar structural skins are the following:

- Structural decks on joists or beams (plywood sheathing, corrugated steel deck, precast concrete planks, gypsum deck, solid wood deck, etc.)
- One-way solid concrete slab on primary beams or walls (4 to 7 in. thick, 10 to 20 ft typical span range)
- Precast hollow-core concrete slabs (4 to 12 in. thick, 15 to 38 ft typical span range)
- Composite 1 1/2-in. steel deck concrete slab (12 to 15 ft typical span range)
- Concrete joist slabs (6 to 24 in. thick, 35 to 40 ft typical span range)
- Precast, prestressed concrete single and double tees (12 to 36 in. deep, 30 to 100 ft typical span range)
- Two-way concrete flat plate (20 to 25 ft typical span range)
- Two-way concrete flat slabs (25 to 30 ft typical span range)
- Two-way concrete waffle slab (35 to 40 ft typical span range)

The span range of concrete slabs can be increased by about 30% to 40% when post-tensioning is used.

Columns

Columns support loads in compression, whereas hangers, ties, and thin braces do so in tension. The capacity of *slender* columns under concentric compression, such as struts, is controlled by instability, that is, a sudden lateral bending called elastic buckling. Elastic buckling, in turn, depends on the member length and shape, its cross-sectional area, and material properties (see Section 3.4). Stocky columns and hangers (or stays), in contrast, are controlled by the strength of the material. It is apparent that tensile members are very thin in comparison to typical columns, especially when high-strength cable material is used. Whereas struts are predominately compression members, columns carry bending and axial compression where the bending can be more important than the axial load.

Columns are the primary components of skeleton structures, but they are also found as masts, posts, struts, pillars, pilasters, pylons, buttresses, caissons, and piers; as monumental concrete pilotis, they may carry an entire building. Columns can be short or long, slender or stocky, ground supported or flying in the air (i.e., flying guyed masts). They range from single members, such as rectangular or round concrete columns, pipe or rolled W-sections, to multiple-member built-up sections composed, for instance, of angles, tees, channels, or W-sections (e.g., cruciform or box sections). Multiple-member systems may represent frames (e.g., ladder columns), trusses, or cable-braced columns. Columns may represent simple shafts, shafts with capitals, or sculptured two- or three-forked tree columns. They may have heads, rather than capitals, which are formed by cones, blocks, plates, mushrooms, umbrellas, bracing, trees, and so on. The cross section of columns may be varied (sloped, curved, twisted, stepped, etc.) and may have its thinnest part at mid-height or top and/or bottom in x or y directions. On the other hand, the column cross section may be enlarged at mid-height to strengthen its buckling resistance at this critical location.

Columns range from one-story to multistory, from single columns to column groups (e.g., clustered pipe columns), from exterior to interior, from vertical to inclined columns, and so on. Columns may be separated from beams so that mainly axial forces are transferred, or they may be continuous with beams to form beam columns. Several types of columns are shown in Fig. 1.20. Columns usually are in a vertical position supporting roof or floor loads in groups and possibly also resisting wind and seismic loads. However, their primary responsibility may also be to resist lateral thrust as caused by spatial horizontal-span structures such as frames and arches. For this situation the columns may be inclined and free standing, acting alone or together with others (e.g., V-columns) as abutments such as the flying buttresses and the pier buttresses with pinnacles of Gothic cathedrals and as demonstrated by the other cases in Fig. 1.8.

Columns normally may not solely serve their function as support, according to engineering requirements, but also may define space and may provide visual order to a wall. They may also help to refine proportion and compositional considerations, or they may be more ornamental, expressing other layers of meaning than just function. On the other hand, in the current vogue of postmodernism or period of contradictions, the column may not express its purpose as support at all—structural honesty may be replaced by artificial mannered styles. Decorative layers conceal the physical support of the building and its organization. There are no limits set to the form giving of these types of columns, as is suggested by some of the cases in Fig. 1.19. The shapes may be derived from classic order or mannerism. They may articulate conflict and ambiguity with respect to force flow and support. For example, columns may be huge and massive but only support themselves, or at the point of support, where the beam meets the column and where all the accumulated loads must be transferred, the least amount of material is provided, as demonstrated by long slender necks, pencil columns, or trees,

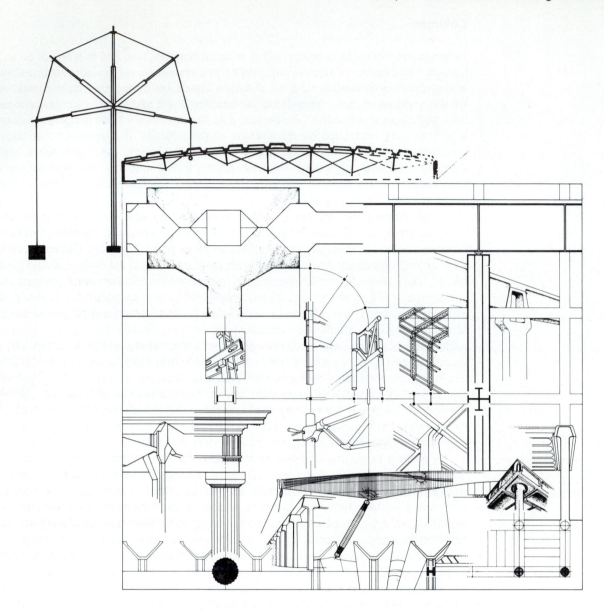

Figure 1.20 Beams and columns.

which, in turn, sit on heavy piers or overdesigned columns with capitals, thereby giving the illusion that the force of gravity does not exist.

Various column types expressing structural honesty are identified in Fig. 1.20. It is important to realize, however, that the column as a structure often can only be understood in the immediate context of the beam that it supports, as well as in the larger context of the building. Therefore, the discussion of the columns as isolated can only be taken with caution.

The column is an essential element of Greek architecture, particularly the Greek temple. It is clearly defined by capital, shaft, and base in the Ionic order; in the Doric order, the shaft sits directly on the stylobate. The column not only reflects the play between load and support, but also other symbolic meanings. The capital represents the beginning of the column or the transition between beam and column, and the base articulates the end of the column, the separation from and at the same time the joining to the ground. The column shaft is fluted to transmit a feeling of height and energy; it is not straight but curved to form a bulge called *entasis,* to give a sense of elasticity and compression.

This classical tradition of the column has been completely disregarded by most of the modernists. Here, the capital and base are eliminated or reduced to a minimum of drop panels and base plates or a hinged base, according to engineering requirements. In other words, the traditional meaning of the column was replaced with a new set of values, these of structural behaviorism and function.

In the truly functional architecture, capital and base disappear, as exemplified in Le Corbusier's flat plate concrete construction for the Dom-ino houses of 1914 (Fig. 2.1) which are treated like objects related to mass production. In the Johnson Wax Company Administration Building at Racine, Wisconsin (1939), Frank Lloyd Wright used double-cantilever, slender, mushroom-headed, tapered columns almost hinged at their base to give support to each other as based on the principle of the three-hinged frame. Columns may be reduced to a minimum, as for Mies van der Rohe's New National Gallery in Berlin (1968, Fig. 1.20, top right). In the Farnsworth house (1950, Fig. 1.13d), Mies van der Rohe supports the slabs between the slender exterior columns rather than on top, thereby generating the impression that the columns are attached to the slabs, or that they are free of them, which further supports the illusion of the almost floating cantilevering floors.

Other designers were influenced by structural expressionism and nature, such as Marcel Breuer's heavy, almost brutal concrete tree columns for St. John's University Library, Collegeville, Minnesota (1968, Fig. 1.14d). Quite different in spirit are Christoph Langhof's slender and minimal, abstract planar, treelike, 100-ft-high masts for the Horst Korber Sports Center in Berlin (1990, Fig. 1.20, top left), with their five branches linked by cables from which the light cable roof trusses are hung.

The influence of the interplay of force and form or *tectonics* on column design is articulated by Pier Luigi Nervi's large, tapered concrete columns capped with steel umbrellas for the Palace of Labor, Turin, Italy (1961, Fig. 1.14a). The large, slender steel trees of the new Stuttgart airport terminal in Germany (1991, Fig. 2.1, top left), designed by von Gerkan, Marg Associates, with their spatial network of branches, give a continuous arched support to the roof structure, thereby almost eliminating the separation between columns and slab.

In many of Calatrava's buildings the bony-shaped members respond to the intensity of force flow—columns and beams are an integral part of the total structure and bring it alive like an organism reflecting an exciting inventiveness, with a rich formal expression articulating the principles of statics in addition to other layers of meaning (e.g., Figs. 2.1 and 5.1).

Other designers were influenced by the process of assembly, or the joining of beam to column. Angelo Mangiarotti's precast concrete hammerhead columns for the ELMAG factory building near Monza, Italy (1964, Fig. 1.20, bottom) have interlocking joints to allow efficient connection to the beams. The reader may want to study the other column types in Fig. 1.20 to sense some of the possible richness of expression.

Walls

Traditional bearing-wall construction is directly associated with brick or stone masonry, which can be traced back thousands of years. The nature of bearing is probably most convincingly demonstrated by the sculptured stone walls of the Incas in the Peruvian Andes, where each block was shaped and fitted with incredible precision. The daring use of stone is proudly expressed by the tall, medieval cathedrals of Europe. Also inspiring are the Gothic brick buildings of the Hanseatic towns along the Baltic Sea.

The exterior walls are most challenging to the designer, since not only must they provide functional integrity, but at the same time they must represent an architectural expression. The term *facade wall* expresses a contrast between this and the other exterior walls of a building, which may not give the primary identity to the building, but just provide protection, although an isolated glass-wall building revealing its interior may not be perceived as having facade walls.

The term *wall* reminds us of solidity, as expressed in nature by a rock formation; it exposes weight and strength, as developed over time from materials like stone, masonry, and concrete; it demonstrates a continuous surface character. The spirit of the wall as an integral part of the building structure is contrasted to the envelope that forms an attached skin, which typically consists of vision and spandrel panels. In other words, exterior walls may function merely as envelopes or skins to protect the interior environment from the outside. They can be thin and nearly weightless, like a membrane, and hang like a curtain from the building structure. In contrast are the heavy, load-bearing walls that must also support the floor loads and must carry these forces directly down to their own foundations. Although a masonry or concrete wall with punched openings for the placement of the window units may appear as a bearing wall, it may still function only as a curtain that supports the wind loads like a vertical slab; in other words, it may just be a veneer hanging onto a concealed frame in the back.

Wall structures are of many various types and materials (Fig. 1.21); they may be organized as *exterior* and *interior,* as *load bearing* or *nonload bearing,* and as prefabricated or built-in-place walls. The interior walls may be nonload-bearing partitions that subdivide the functional space (i.e., space dividers), or they also may be load-bearing walls that must carry the floor loads and act as *shear walls* to resist lateral forces; thus these load-bearing walls represent the building structure. The bearing walls may form the exterior perimeter and may be part of the long-wall or cellular building structure systems; they may be solid, perforated, or coupled walls. The structural walls may just be isolated, stand-alone elements, or they may be part of a spatial all-bearing wall building. On the other hand, exterior walls may only have to provide the protective environmental shield and serve as envelopes or nonload-bearing curtain walls.

Walls may form either single- or multiple-wythe wall systems. Single-shell walls can be either solid, hollow, ribbed, or curved. Typical multiple-shell systems are cavity walls, veneer walls, or composite walls. The exterior support structure may be exposed or it may be cladded. Typical exposed building structures are masonry and concrete walls. Cladding may be provided through layered construction of curtain walls acting as envelopes independent of the building support structure, or it may be provided through monolithic construction by being an integral part of the building structure, such as cast-in-place spandrel panels; occasionally, the exterior envelope may act compositely with the building structure (i.e., composite construction of layers) as in stressed-skin construction.

Curtain Walls.   Curtain walls as exterior envelopes must satisfy three primary criteria. From a *functional* point of view they must provide an environmental shield by controlling thermal movement and noise by keeping air and water out but letting light in. From an *esthetical* point of view they assist in giving expression to the building. From a *structural* point of view they must resist the primary loads due to wind, earthquake, and gravity.

Curtain walls are nonload bearing. They carry their own weight and act as vertical slabs to transfer lateral loads due to wind and seismic action to the building support structure. Keep in mind that local wind pressure will be affected by corners and edges and other peculiar building geometries. In other words, high local pressure and especially suction may develop at critical locations.

Typical curtain walls consist of the exterior cladding material, support framing, insulation, waterproofing, joint treatment, and internal drainage. The cladding systems

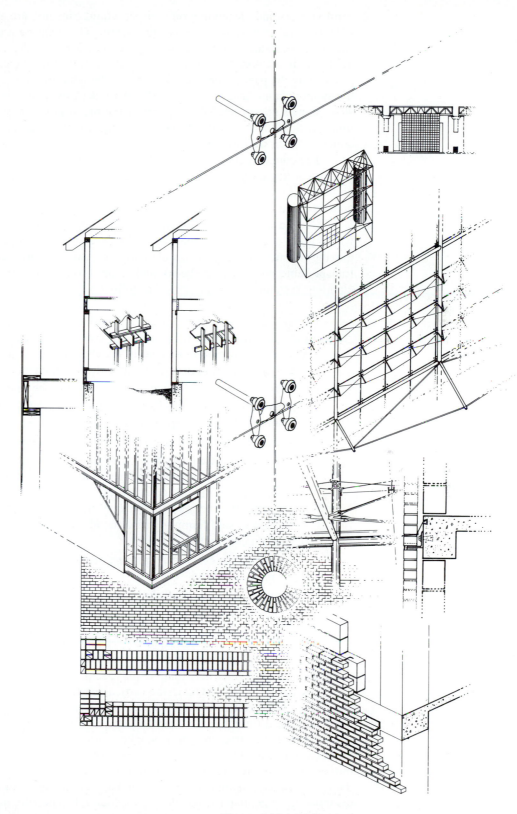

Figure 1.21 Walls.

form veneers and, depending on their structural capacity, are adhered or anchored to the structural or nonstructural backup system. The *adhered veneers,* such as ceramic tiles or thin, natural-stone plates, are directly affixed to the backing with mortar or adhesive. In *anchored veneers,* metal, glass, or thin stone panels are inserted in metal grids. Here the panels must be able to resist the wind in bending by spanning between the support frame grid. On the other hand, *thick slabs* such as brick veneers and heavy stone panels do not need the support of the metal grid—they are supported at floor lines on shelf angles attached to the building structure and anchored, for example, to concrete block wall backup.

The common cladding materials are metal (e.g., aluminum, stainless steel), glass (e.g., clear glass, tinted or heat-absorbing glazing, reflective or coated glass, selective nonreflective coated glazing), stone (e.g, granite, marble, limestone, sandstone, slate), concrete, and masonry, as well as ceramics.

Curtain walls may be organized according to cladding type as metal curtain walls, glass curtain walls (conventional glass infill panels versus structural silicone glazing), stone cladding, canvas cladding as used for fabric-covered structures (e.g., transportable prefabricated sheds), and so on. From a support framing point of view, cladding systems may be organized as follows:

- *Framing systems* use exposed or hidden mullion framing to support metal, glass, or thin stone infill panels. The components can be field assembled (e.g., stick system) or prefabricated in the shop and installed as large framed units (e.g., unit system such as aluminum stick-framed panels).
- *Column and spandrel cover systems* express the layout of the building columns and spandrel beams by covering them directly with cladding material.
- Cast-in-place concrete spandrel-panel construction.
- *Infill systems* are installed within the building frame.
- *Panel systems* form internal units of more surface action in contrast to framed unit systems. They are preassembled units made from sheet metal or formed as castings (e.g., aluminum, concrete, plastics) with a minimum of integral joints. Their support structures can be steel-truss framed panels, diaphragm panels such as steel stud framing with a metal deck, precast concrete, or glass-fiber-reinforced concrete panels.
- *Conventionally set stone or masonry veneer* anchored to concrete block or stud walls used as backup.
- *Composite skins* (stressed skins) are an integral part of the building structure, as in tubular construction.

The nature of curtain walls and the type of support framing depend on the position of the building columns, which may be located inside or outside the building or along the wall line; thus, they may be fully, semi-, face-, or unexposed.

The curtain wall must resist the vertical loads due to its own weight, lateral loads, and hidden loads caused by movement. The movement of the individual wall components and the differential movements between wall components, as well as between wall and building structure, must be taken into account. The various types of movement (which should not interfere with each other) are story drift, including torsional movement, building structure shortening for tall buildings, spandrel deflection, foundation settlement, thermal movement, and any combination of these. The facade panels must, in some manner, be allowed to float independently of each other so that the hidden loads due to prevention of differential movement will not be absorbed by the panels themselves. Since a primary reason for the failure of cladding is a result of overstressing caused by additional loads being transferred from the building, we must consider the compatibility of the exterior wall and the building structure, which includes the movement of the curtain itself and that of the supporting structure, that is, the relationship of the building flexibility to the panel rigidity. Therefore, curtain walls consist of individual panels that are separated by horizontal and vertical soft joints so that they can behave as independent elements.

Other important considerations for the design of curtain walls are air infiltration, water penetration, condensation, thermal resistance, fire resistance, noise and light control and maintenance and repair. For further discussion of curtain walls, refer to Schueller, 1990.

Only briefly introduced here are the **new composite tensile cladding systems of glass and stainless steel** as have been pioneered recently. They form large, suspended, all-glass surfaces (in contrast to ground-based single-glazed systems of much less height) with continuous glazing; the glass plates are sealed together by structural silicone joints and supported by lightweight structures in the back. The goal of these new types of glass skins is maximum transparency, the celebration of lightness and light, and the dematerialization of space with a minimum of visual interference from the support structure, for instance, suggesting guyed glass structures where the rods prestress the glass plates in compression. Keep in mind that the glass as structural material is fragile to shock loads but also strong in direct pressure.

Examples of minimal support structures are vertical suspended glass mullions (fins) along the vertical joints to give linear support and the necessary stiffness to the glass sheets with respect to wind pressure, which are held together at their corners by metal fittings. Cable trusses, on the other hand, possibly using glass rods as compression struts, give point supports. Here, the glass panels are connected at the corners to point fittings which are bolted to the trusses.

The structural and thermal movements in the glass wall are taken up by the resiliency of the glass-to-glass silicone joints and, for example, by ball-jointed metal links at the glass-to-truss connections, thereby preventing stress concentrations and bending of the glass at the corners. The lateral wind pressure is carried by the glass panels in bending to the suspended vertical support structures (which act as beams) or occasionally to horizontal trusses. The tensile trusses are laterally stabilized by the glass or braced by crossed stainless steel rods. The size and thickness of the glass panels are determined by the magnitude of the wind loads.

The wall dead loads are usually transferred from the glass panels to vertical tension rods, or each panel is hung directly from the next panel above; in other words, the upper panels carry the deadweight of the lower glass panels in tension. The height of the suspended glass wall is a function of the weight of the total assembly.

The development of *suspended glass skins* has been significantly influenced by the glass walls for the three monumental greenhouse structures attached to the south side of the Museum of Science and Technology, Parc de la Villette in Paris (1986), designed by the renowned structural engineer Peter Rice. The towerlike structures are approximately 107 ft wide by 107 ft high and 27 ft deep. They capture and store heat to be distributed according to the needs of the museum (Fig. 1.21, top).

The glass wall is subdivided into 16 approximately 27-ft square modules, which also form the basis for the primary stainless steel tubular frame, which is laterally supported against wind action by cable trusses. Each of the 27-ft square modules, in turn, consists of sixteen 6.66-ft square glass sheets laterally supported by a secondary system of horizontal cable beams (tension mullions), which, in turn, are stabilized by the glass. The glass panels are suspended from the main frame. They are attached to each other with a clear silicone sealant and are joined at their corners by a molded steel fixing that allows movement in any direction and reduces stress concentrations to a minimum. The glass weight is transferred in tension from the lower panels to the upper ones and at the top of the four glass sheets is hung from the main frame beam by prestressed spring devices that act as shock absorbers and allow flexibility and readjustment for extraordinary loading conditions (including breakage of the panels) and for the forces to be evenly distributed.

Partitions.   Partitions are nonload bearing. They are interior walls that divide space to provide a visual, acoustic, and, possibly, fire barrier. They only carry their own weight and must resist a minimum lateral load of 5 psf, but they do not support floor loads. Some fixed partitions are designed as shear walls to provide additional lateral stiffness to the building. Partitions are classified as (1) movable partitions and (2) fixed partitions.

Permanent partitions are either constructed of masonry (e.g., brick, lightweight concrete, or gypsum block) or they are framed partitions finished on each side with a single layer or multilayer skin system, such as plaster and/or gypsum board(s). Demountable partitions consist of prefabricated units; they are usually patented and hence much more expensive. They are organized according to their support structure as frame plus infill panels, frame plus overlay panels, or panel systems.

The material for the frames is either metal or wood. In contrast to masonry partitions, framed partitions are of lightweight construction and are easier to assemble. Typical framed partitions consist of stud drywall construction where fire- and sound-resistant sheet materials, possibly in several layers, are attached to the studs, and sound insulation blankets are installed in the cavities between the studs. For example, a movable partition may consist of gypsum panels, steel H-studs, the floor runner, and the flanged top rail.

Generally, partitions, in a manner similar to glass, must float in a frame opening so that they will not be stressed by the vertical beam deflection and crack; also, loads must not be transferred through lateral frame wracking due to wind and earthquake action or by vertical frame wracking in the upper floors due to temperature differences. Special joints are required along the sides and top of the wall to allow for building movement. More sophisticated joints, like slip joints, are required for seismic design.

Bearing Walls.   Bearing walls carry roof and floor loads as well as the weight of the walls above; as exterior walls, they also must perform as curtain walls to resist lateral forces, besides having to act as shear walls. Lateral force-resisting wall structures, shear walls, and retaining walls are discussed in Chapter 4. As the height of a plain bearing wall increases, it may be necessary to stiffen it with piers or buttresses or by using spatial geometries such as cellular, corrugated, or undulating forms. Common bearing wall structures are the following:

- Masonry walls
- Cast-in-place concrete walls
- Precast concrete walls
- Framed walls (e.g., stud walls, tubular structures)
- Trussed walls (e.g., traditional half-timber construction in Europe, braced framed tubes, latticed tubes)

The following are the most common wall structures: masonry walls, concrete walls, and stud walls.

A. Masonry Walls: Masonry and unbaked earth construction have served people for thousands of years. While stone is the oldest raw building material, sun-dried or hard burnt bricks are the oldest man-made ones. Precast concrete blocks, on the other hand, are a more recent development; by the 1870s they had become quite popular in Chicago as so-called artificial stones. Inexpensive *adobe* bricks, also often named sun-dried mud bricks, are used in many regions. In this case, earth is mixed with water and vegetable fibers, such as chopped straw, shaped in molds, and then dried in the sun. In the *pisé de terre* method, hard earth walls are constructed by ramming earth within formwork. Rather than fitting bricks or blocks precisely with dry joints, as the Greeks did for their stone temples and the Incas for their stone walls, it is surely easier to use them together with mortar joints. The Egyptians used gypsum as mortar. Lime mortar was already used by the Minoans on Crete in 2000 B.C., and the Greeks discovered pozzolan cement around 600 B.C., which was further perfected by the Romans.

In high-rise building construction, the massive walls of the past, where the weight had to suppress the tensile stresses due to lateral load action, have been replaced by the engineered thin-wall construction of the present using brick or concrete block. The compressive strength of masonry is controlled by the strength of the mortar and the strength of the masonry units, as well as by the quality of workmanship. Masonry walls can be load-bearing and nonload-bearing veneers or partitions; they form either single- or multiple-wythe wall systems. The basic wall types are as follows:

- *Veneer walls* (e.g., brick facing with stud wall backing)
- *Single-wythe walls* (e.g., solid, hollow, grouted, or reinforced hollow masonry)
- *Cavity walls* (e.g., brick facing with concrete block backup)
- *Composite walls* (e.g., brick-block walls)
- *Reinforced masonry walls* (e.g., reinforced brick walls in double-wythe construction, reinforced concrete block walls)

For further discussion of masonry wall types, refer to Schueller, 1990.

B. Concrete Walls: There is a substantial difference between the monolithic, cast-in-place concrete structure and the precast concrete structure. The monolithic structure allows a continuous interaction between the vertical and horizontal planes, thereby providing more stiffness and redundancy, in comparison to the structure built from prefabricated elements, such as large panels or boxes, which depends on its connections for integrity and thus may not provide much reserve strength. Often the degree of continuity is difficult to evaluate, similar to the partial restraint between floors and masonry walls. For further discussion of concrete walls, refer to Schueller, 1990.

C. Stud Walls: Light framing construction is typical for residential and small-scale commercial buildings. It consists of the closely spaced light frame members (i.e., wall studs and floor/roof joists) made of dimension lumber or light-gage metal members, spaced generally not farther than 2 ft and tied together with sheathing to form the support structure.

Light wood framing is most common today in residential construction with the typical 2- × 4-in. studs extending 8 ft and spaced 16 in. on center for one-story buildings. In steel-framed houses, screws are used instead of nails and the typical 2- × 4-in. wood studs are replaced by light-gage, C-shaped galvanized steel members. Stud walls can be designed as load-bearing shear walls (Fig. 4.21) and nonload bearing. The framing system was introduced in 1832 in Chicago by the engineer and contractor George Washington Snow for the construction of wood houses. It was called balloon frame because its wood members were so thin; they were simply nailed together, in contrast to the heavy timber used in traditional post and beam framing with its complex mortise and tenon joinery. The development of plywood sheathing in the early 1900s made a much stiffer building possible and allowed the transformation of balloon framing into western or platform framing.

Platform framing is the most common method of construction today. Here, one-story stud walls are placed on the floor platform and in turn, support the next floor

(Fig. 1.21, left); floors and walls are assembled independently of each other. The light frame walls consist of the one-story studs attached at the top and bottom to horizontal plates (i.e., single bottom plate and double top plates) and the sheathing material. The horizontal wall plates provide the necessary bearing area for the distribution of the vertical loads. In the rarely used *balloon construction,* the studs are continuous over the full building height, thereby evading the shrinkage due to the compression perpendicular to the grain as caused by the studs on the horizontal plates. The second floor joists rest on a continuous 1- × 4-in. ribband recessed into the studs and are also nailed into the sides of the joists (Fig. 1.21, left).

Platform frame construction in wood or steel can be used usually up to four stories, but diagonal bracing is required to resist wracking due to lateral force action. Stud wall openings are framed with headers acting as lintels (e.g., two 2 × 4s for small openings up to 3 ft) supported on top of studs adjacent to the wall studs.

Joints

At the intersection of every building component occurs a joint. Joints range from the large scale of a building joining the ground to the beam–column connection and finally to the small scale of mortar joints bonding brick together in certain patterns in a masonry wall. With respect to the large scale of a whole building, the intersections of wall and roof and floors, or of wall and base, or of wall and wall (e.g., corner), or wall opening and window framing and light all constitute primary joints. From a visual point of view, joints become an abstract issue, for example, in how a building meets the ground and sky or how the entrance connects the inside to the outside of a building, or how materials join each other, particularly at wall penetrations.

Because of all the layers of specialization in building construction, the traditional monolithic construction has been replaced to a large extent by layered construction (e.g., veneered building) where more automated operations together with assembly of various component systems have placed an increasing importance on building joints and jointing. In other words, there are numerous joints for nonstructural elements and their framing, as for ceilings, partitions, and exterior envelopes.

The type and importance of joints are very much dependent on how structure is related to building form. In contrast to the cladded buildings of *postmodernism,* where structural materials are covered with finish materials and sometimes with extensive ornamentation, *modern architecture* often makes no distinction between structure and finish materials—it has been expressing honestly structure and materials. It has been exposing structure and especially celebrates the spirit of the joints, as the beam–column connection examples in Fig. 1.20 clearly demonstrate, to articulate tectonic qualities, tactile experiences, construction process, and other meanings. Contrasted are pin-jointed assemblies to continuous ones, where, for example, stresses are concentrated at knees, requiring the joints to be strengthened so that they can resist lateral forces.

Mies van der Rohe considered the design of joints an opportunity by claiming: *God is in the details.* He was deeply concerned with the clarity of visual expression in exposed structure connections and the balance of elements and forces joining each other. As Mies van der Rohe was dedicated to the detail in steel, Angelo Mangiarotti was greatly influenced by structural and constructive considerations in concrete. For Carlo Scarpa, the detail was the theme of design, it was the building itself—architectural unity evolved out of the detail.

Although joint details are small in comparison to members and are disregarded in conceptual investigations and therefore seem to be subordinate to the whole, they are usually a cause of headache for the designer since the whole can only work if the details are solved. This fact is reflected by the often quoted statement, *the devil is in the detail.* The success of a building from any point of view, ranging from economical to esthetical considerations, depends very much on the success of the joining or mating of the various elements. Some designers even claim that a building is the sum of its details.

The joint type is dependent on the location and position of adjacent members. The components may be closely fitted or a gap may be left between them deliberately. This gap may have to be sealed by familiar materials, such as bedding, caulking and glazing compounds, putties, mastics, or gaskets. The members at a joint may either stay clear of each other or they may be interlocked in some fashion. Similarly, on a large scale, joints may separate or partially separate entire building blocks that have different mass and stiffness characteristics (i.e., allowing certain movements); here the joints are usually formed between double members (e.g., beams and columns) or at the ends of cantilevered members. A joint may have to satisfy any of the following performance requirements:

- *Environmental control:* sealing (air moisture, wind, dirt, water, sound, fire, insects)
- *Dimensional control:* (component and construction tolerances)
- *Functional control:* (maintenance, replacement, assembly, fit, etc.)
- *Movement control:* (sliding, rotation due to temperature, shrinkage, creep, settlement, elastic deformations, etc.)
- *Structural control:* load-bearing connections (see Section 3.5), energy dissipating connections
- *Esthetics:* for exposed joints

The main joint types are *structural joints,* and *movement joints* which include *soft joints* (e.g., expansion joints, isolation joints) and *control joints* (e.g., slip joints, shrinkage strips). *Construction joints* may be required when, for instance, the process of concreting is interrupted so that the concrete pour is no longer continuous. And keep in mind that fabrication and erection tolerances have to be taken into account in the design of connections (e.g., slotted holes).

Joints may be designed as *open joints* to allow for more movement and less precision in the fitting of members (i.e., larger dimensional tolerances). In *closed joints* the gap between the components is simply closed by a weatherproofing seal of mastics or gaskets. Whatever the joint type, the width of the joint and the sealant must respond to the performance criteria; the sealants must be sufficiently elastic to permit the movement between the elements.

Structural connections may not always be fixed in place by transferring loads without changing position. They may also allow free movement (e.g., elastomeric bearings, rollers, sliding supports) or may act as artificial damping devices similar to shock absorbers in cars. Familiar are the energy-dissipating connections used in seismic areas or where oscillations due to gusty winds are critical. Examples include the slotted, bolted connections dissipating energy by means of friction between the sliding surfaces and the base isolators (e.g., viscoelastic dampers) that change the period of motion between ground and structure; in other words, they are horizontally flexible systems that lengthen the period of vibration to reduce the response of the structure.

It is not the purpose of this discussion to study and classify the seemingly endless number of joint systems that have been established by the various traditional building trades or that are developing presently from the new technologies, nor to teach joint detailing, but rather to identify movement joints and to examine the sources and nature of movement for the building components as exemplified by the examples in Fig. 1.22. For a discussion on the structural performance of joints the reader may refer to the section on connections in Section 3.5.

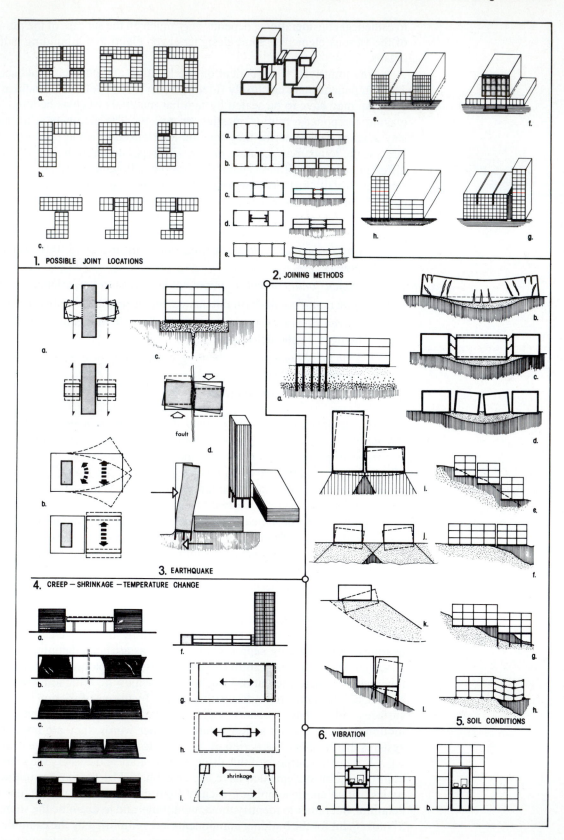

Figure 1.22 Building joints.

The principal causes for the movement of individual components or entire building sections are *changes in material volume, earth settlement, seismic action* and *direct force action.* Whatever the source for the displacement of the members, if they should be held back from free movement, additional forces will be induced into the component itself as well as the adjacent members, preventing the free deformation.

Whatever the main purpose of an expansion joint, be it for the control of temperature and moisture, creep and shrinkage, settlements, seismic or dynamic loading, or any other reason, the following criteria should be considered:

- Long buildings should be subdivided into units, realizing that the joints are spaced closer for stiffer than for flexible buildings.
- Buildings with compound or irregular plan shapes should be separated into units as indicated in Fig. 1.22 (la) to (1c).
- Buildings consisting of different blocks of high- and low-rise sections, each with its own structural system, should be separated at their junction.

chapter 2

Basic Structure Concepts

Basic concepts must be known to perform the structural analysis and design of buildings. The external forces that act on buildings cause internal forces within buildings. The forces flow along the structure members to the ground, requiring foundations as transition structures to the comparatively weak soil. The members must be strong and stiff enough to resist the internal forces. Forces can also be introduced deliberately in members, as when due to prestressing.

The general properties of forces and the responses of the structure members, including the soil, to these forces are investigated. To evaluate the stressing of the members by the forces, in turn, requires an understanding of the geometric properties of member cross sections, as well as of certain material properties. The visual study in Fig. 2.1 attempts to reflect some of the spirit of this chapter and is now briefly discussed.

Although the support structure is necessary for every building, it is not necessarily articulated as an architectural expression. For example, Les Espaces d'Abraxas housing complex, Marne-la-Vallée near Paris, France (1983, Fig. 2.1, top, middle, and right) by Ricardo Bofill, represents a monumentlike classical composition rooted in the past. The appearance of mass and solidity is achieved not through the actual support structure, but by thin veneers of prefabricated concrete panels mirroring the qualities of stone; in other words, the building appearance does not reflect the building organism. In contrast, the other cases in Fig. 2.1 express the purpose of structure and possibly of how they were built. They range from the small scale of Helmut Jahn's beautifully crafted capital for the clustered battened pipe columns (forming a single Vierendeel column) of the United Airlines terminal at O'Hare Airport in Chicago (1987, bottom right), to the large scale of La Grande Arche in Paris (1989, bottom left) by Johan Otto von Spreckelsen with its free-standing, nearly 300-ft high, cable-braced steel lattice elevator tower anchored laterally to the arch with horizontal guyed columns.

66

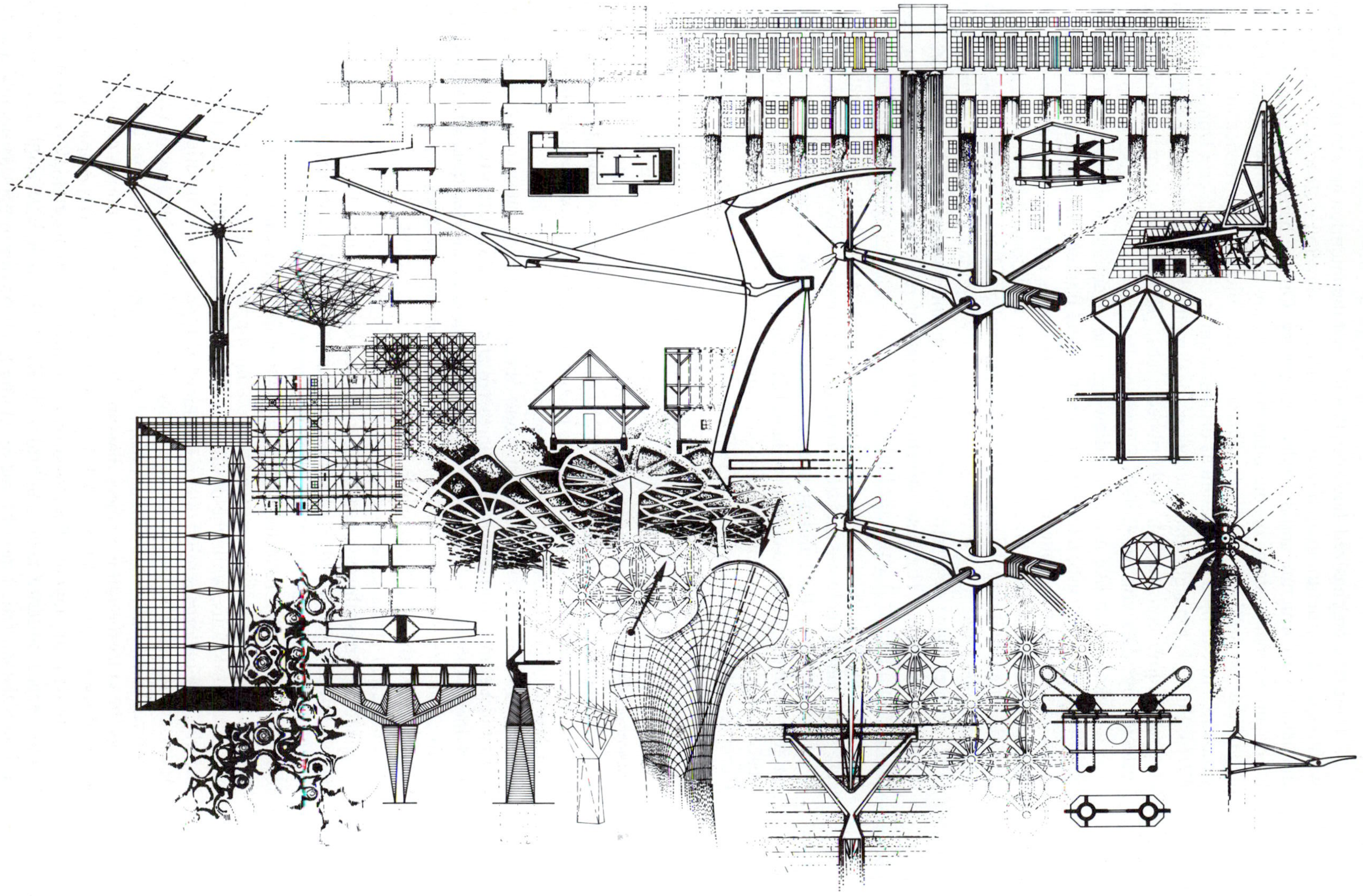

Figure 2.1 Basic structure concepts.

It must be emphasized, however, that it is not the intention here to take a position with respect to structure in architecture but rather to develop a clear understanding of the behavior of structure and its underlying theory. On one hand, structure represents an organizational system; on the other hand, it may be integrating and celebrating details that do not have to be subordinated or are solely in the service of the whole structure. An example of structure as organization is Le Corbusier's Dom-ino system (1914, Fig. 2.1, top right) proposed as an ideal method of mass housing consisting of concrete frame construction with ribbed slabs. The discipline of composition is reflected by Mies von der Rohe's Barcelona Pavilion (1929, Fig. 2.1, top middle), which looks in plan like a De Stijl drawing. The structural walls are used as shear-resisting elements to give lateral stability to the steel frames under loads from all directions. The symmetrical pitched roof building below is supported by simple tree columns, which form knee-braced portal frames in traditional timber construction to provide lateral stability. In contrast to conventional rigid construction is the tension-braced, hinged-assembly structure of the six-story Pompidou Center in Paris, France (1977, center right) designed by architects Piano and Rogers and the engineers Ove Arup & Associates. The basic structure consists of parallel, 8-ft-deep Warren truss beams spanning 147 ft across the building to rest on small cantilever beams called *gerberettes*. These gerberettes are small cast-steel beams pin connected to water-filled, cast-steel, tubular columns and tied down along the exterior by vertical tension rods.

In contrast to the building structure as an overall ordering system is the *detail,* ranging from the joint and column to James Stirling's decorative and playful asymmetrical colored metal entrance canopy structure of the new Staatsgalerie in Stuttgart, Germany (1984, top right). The MERO joint (middle right) reflects adaptibility to various situations and mass production. The typical node is spherical with a series of flat faces and 18 tapped holes (as indicated by the adjacent corresponding space frame with 18 squares and 8 equilateral triangles) to which are connected tubular members with cone-shaped steel ends.

To appreciate some of the quality of a detail, it is helpful to look at the force flow that it must transmit as, for example, occurs at the head of a human thighbone in response to applied loads (Fig. 2.1, center bottom; reference: K. Wunderlich and W. Gloede, 1977). The effect of the notch and cantilever portion causes stress concentrations and a complex interplay of compressive and tensile stress trajectories. Here, the solid lines indicate the direction of principal compression and the dotted lines tension; the force lines cross each other at right angles. Nature responds to stress intensity by strengthening the bone exactly as is required.

Principal stress patterns can also be investigated by photoelasticity by letting polarized light pass through certain plastic-based materials under stress, as indicated at bottom left of Fig. 2.1. From a behavioral point of view, flat slabs are highly complex structures. The intricacy of the force flow along an isotropic plate in response to uniform gravity action is reflected by the principal moment contours in Fig. 2.1, center. In this case, the main moments around the column support are negative and have circular and radial directions, while the positive field moments basically connect the columns linearly. The patterns remind one of organic structures, such as the branching grids of leaves, the delicate network of insect wings, radial spider webs, and the contour lines of conical tents, realizing a similar relationship between cable response and loading, as well as the corresponding moment diagram. Pier Luigi Nervi, for the Gatti Wool Factory in Rome, Italy (1953, Fig. 2.1, center), actually followed the principal bending moments with the layout of the floor ribs. Similarly, Michael Hopkins for the new Schlumberger building near Cambridge, U.K. (1992, below Nervi's slab) visually expresses the isostatic lines of force reflecting the behavior of the slab under loads.

The quality of the detail is articulated by some of the columns in Fig. 2.1. The columns range from ordinary straight ones with or without capitals, Y-shaped columns such as Giovanni Mechelucci's steel column of 1962 (bottom center), or Marcel

Breuer's three-forked spreading concrete column gathering the forces from the mullions above (1960, bottom left) to Jochen Brandi's bent steel pipes bundled together to form tree columns (1985, center right). The huge steel trees of the new Stuttgart airport terminal in Germany (1991, top left), designed by von Gerkan, Marg Associates, with their spatial networks of branches, give a continuous arched support to the roof structure, thereby almost eliminating the separation between columns and slab.

In conclusion, Santiago Calatrava's organic structures must be mentioned, which seem to articulate force diagrams and transmit a feeling of balance or completeness. The bony-shaped members follow effortlessly the force action, and their sizes respond to the intensity of stress flow, thereby bringing the structure alive like an organism. The interaction of the elements expresses a natural equilibrium and is in control of movement that is instability, as Calatrava convincingly demonstrates with two structures in Lucerne, Switzerland.

The steel and glass canopy of the Lucerne postal service center (1984, bottom right) is hung from the existing building. It consists of the frontal cantilevered box beam of wing-shaped cross sections attached to a longitudinal torsion pipe, which, in turn, is supported by bearing and suspension struts forming the transparent section in the back. The entrance hall to Lucerne railway station (1985, center) consists of the metal and glass roof suspended from the existing station building and the entrance arcade, which is formed by F-shaped-like, prefabricated concrete columns, in turn stabilized by slender pendulum columns.

2.1 BUILDING LOADS

A structure must be strong and stiff enough to resist the many types of physical forces imposed on it. The magnitude and direction of these forces vary with the material, type of structural system, purpose of building, and locality. The most obvious loads are due to gravity action, as caused by the self-weight of the building, snow, and occupancy. Lateral forces are exerted on the structure by wind and earthquakes, as well as by earth and hydrostatic pressure. The lateral forces tend to slide and rotate the building block, and the wind attempts to lift up the roof; gravity, in contrast, will counteract and stabilize the structure. While weight and lateral pressure induce a direct force action, movement or deformation generate an indirect action. Building loads may represent *applied* (i.e., contact) forces such as due to wind or snow, or they may be *nonapplied* forces such as due to the weight of building components (gravitational forces), inertia forces (e.g., seismic forces), and magnetic forces. There must also be a distinction between the forces acting on the overall building and those acting locally on the individual framing elements. Loads may be distributed as point, line, or surface loads.

Among the many examples of lateral-force action are the large lateral pressures and impact forces generated by a crane. A traveling elevator causes pumping action, particularly in a single-elevator shaft, together with pressures from the wind entering through the vent shafts; the shaft walls must be designed to resist these forces. Similarly, the walls of the stair shaft for a tall building must resist the lateral air pressure due to pressurization in case of fire, especially close to the fan location. A fire in an interior bay of a multistory building will cause an expansion of the concrete floor, with corresponding horizontal forces. This thermal thrust must be resisted by the cooler adjacent frames and/or shear walls; the thrust acts similarly to an eccentric external prestressing force that increases the moment capacity. Other examples of lateral-force action include possibly excessive stresses in columns, resulting from the internal hydrostatic pressure of liquid used for fireproofing due to the height of the columns. Should a large panel structure be designed to avoid a progressive collapse, then the building must withstand an internal blast pressure of 5 psi.

Loads may also be distinguished according to their variability with respect to location and time. They may be *permanent,* such as the dead load due to the structure itself, or they may be *variable,* as is the case for live loads (such as occupancy loads or wind). Variable loads may be of short duration (such as due to people) or of long duration (such as caused by movable partition walls and furnishings). While these loads are fixed in place, car loads, in contrast, are free in location. The duration of the live load is also of importance for deflection considerations (e.g., creep). Live loads may be *static* or they may be *dynamic,* as when they cause vibration of the structure. While the ever-changing occupancy loads are generally static, since they do not change rapidly, gusty winds (depending on the stiffness and mass of the building) may have to be considered dynamic. The dynamic loads may be *cyclic,* as due to vibrations caused by a machine, or *random,* such as the impact loads due to a collision; the dynamic sea wave action is both regular and random. But there are not only the constant or repetitive load actions, but also accidental loads, as due to explosions.

Vibrational loads may be transmitted to the occupied portions from the mechanical equipment rooms, which may be located at any level of a high-rise building. Machinery such as fans, pumps, and chillers must be isolated by pad materials (e.g., neoprene, cork, fiberglass), steel spring isolation material, or floating concrete bases. In this case, the isolators must resonate at a much lower frequency than that of the machinery, requiring large deflections of the isolators to eliminate potential vibration problems.

Forces may be induced deliberately, as in prestressing, or involuntarily, such as residual stresses due to the production and fabrication process. They also may be locked into members when the material is prevented from responding to changes in temperature and humidity and when the material cannot creep and displace, as caused by constant loading or support movement, thus causing reactive loading. Some loads are time independent (such as dead load), while others are time dependent (for instance, the shrinkage of concrete occurs at a decreasing rate in its early stage of hardening). Most loads, whether geophysical or man-made (Fig. 2.2) are extremely complex, and care must be taken by the designer to properly predict their action.

The following discussion of the various loads should be considered introductory, as needed for preliminary design purposes. For a precise description of loads, the appropriate state or local construction codes, as well as one of the major model building codes listed next, should be consulted:

- Standard Building Code (SBCCI), widely used in the Southeast
- BOCA National Building Code, widely used in the East and Midwest
- Uniform Building Code (UBC), widely used in the West

The model codes become law when adopted by a state or city. In the absence of any governing code, the best reference for building loads is the *Minimum Design Loads for Buildings and Other Structures* (ASCE 7-93), American Society of Civil Engineers. The designer must always keep in mind that the information given in codes is only for minimum loading and not necessarily sufficient; codes may be inadequate for special loading conditions. The lateral-force action due to wind, earthquakes, and water and earth pressure loads is discussed in Chapter 4 dealing with the lateral stability of buildings. The loading of flexible structures is introduced in Chapter 9.

Buildings are classified with respect to determining snow, wind, and earthquake loads as follows:

    I.   Standard occupancy structures except the ones listed below.
   II. Buildings where the primary occupancy is for more than 300 people in one area.

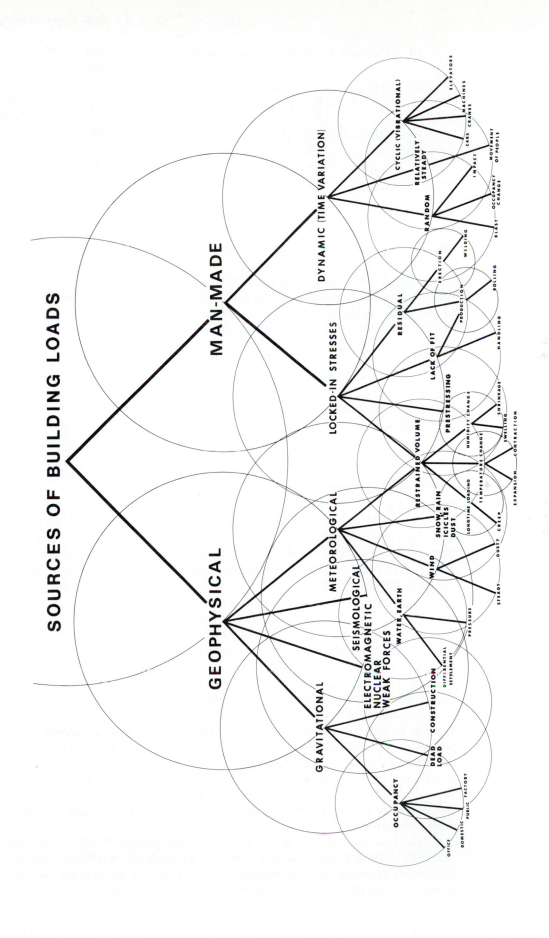

Figure 2.2 Sources of building loads.

III.   Essential facilities, such as hospitals; fire, rescue, or police stations; communication centers; and shelters for emergency situations.

IV.   Buildings that are not hazardous to human life in case of failure, such as agricultural buildings, minor storage facilities, and temporary facilities.

It must be emphasized that it is not the intention here to introduce and discuss precisely code requirements with respect to loading, but to develop an overall understanding and feeling for building loads, as well as an appreciation for the spirit and intentions of codes.

Dead Loads

The loads caused by the weight of the building, called dead loads, and its occupancy, called live loads, represent gravity loads. In contrast to dead loads, which are static (since they do not change), live loads are dynamic and not constant, although they are usually treated as static because they are slowly applied. Dead loads include the weight of the load-bearing structure, ceilings, flooring, partitions, curtains, storage tanks, mechanical and electrical distribution systems, and so on. It may be important to consider the portion of the dead load that is superimposed in a manner similar to the portion of the live load that is sustained. The gravity loads that are not part of the dead load must be considered under live loads. In this century, especially since the 1960s, buildings have become much lighter, so the effect of live load in comparison to dead load has become much more significant.

Although it appears to be a simple matter to determine the weights of materials or the dead load of a structure, it may be impossible at an early design stage to accurately predict the weight of materials not yet selected. Specific nonstructural materials to be chosen include prefabricated facade panels, light fixtures, ceiling systems, pipes, ducts, electrical lines, and other interior components. The weight of stiffening elements and joinery systems for steel structures is estimated only on a percentage basis. The unit weights of materials given by producers or codes are not always consistent with those of the manufactured product. The nominal sizes of building elements differ from the actual sizes; the formwork for poured-in-place concrete may have inaccuracies of $\frac{1}{2}$ in. It must also be kept in mind, in order to accommodate the future computerization of offices, that heavy-duty floor loads have to be planned with high electrical capacity and provisions for raised floors and an uninterruptable power supply. These few examples indicate that, in absence of precise information, dead loads cannot be accurately predicted and may be in error by 15% to 20% or more.

To facilitate the estimate of the dead load for preliminary design purposes, as well as to develop some feelings for loads, some approximate material weights are given in Tables 2.1 and 2.2. Only typical values for rigid construction systems are shown; weights for flexible structures are discussed in Chapter 9.

For preliminary estimation purposes, typical weights for floor, roof, and wall systems have been selected in Table 2.2. They are averaged and given in terms of pounds per square foot (psf) of their own projected area. The weights of suspended ceilings together with the mechanical and electrical loadings usually range between 2 and 10 psf. In office buildings, where partition locations are subject to change, an equivalent distributed floor dead load of 20 psf should be used. Glass curtain walls weigh roughly 12 psf, in contrast to the much heavier precast or masonry facades of 40 to 80 psf and more.

The roof structure weights consisting of the primary bending elements (i.e., excluding roofing weight) increase with span although not in a linear manner. Since the various structural systems are economical for certain span ranges only, bending systems may be replaced by shell or tensile membrane systems where the profile pro-

TABLE 2.1
Typical Approximate Material Weights

Material	Weight (pcf)
Steel	490
Aluminum	171
Reinforced concrete: normal weight	150
lightweight	90–120
Marble	170
Granite	165
Earth, wet loose sand and gravel	125
Brick	120
Water	62.4
Snow: fresh	6
packed	12
wet	50
Lumber	35

vides axial force resistance. For example, the roof weight for a steel structure consists of the weight of the structure itself, ranging from 4 to 36 psf according to Table 2.2 (i.e., for short spans to long spans of 400 ft), and the weight of the roofing system, which ranges from 2 to 5 psf for exposed metal deck to 5 to 10 psf with built-up roofing, and up to 10 to 15 psf for cementitous decks unless lightweight fill is used.

Wood buildings are lighter than steel buildings, which, in turn, are lighter than concrete and masonry buildings. The overall average gross weight for ordinary wood buildings is roughly 40 to 50 psf (lb/ft^2); for steel buildings it is in the range of 50 to 80 psf or approximated as 5 to 8 pcf (lb/ft^3), while nonprestressed concrete buildings may have a density of twice as much. For example, a two-story light industrial building consisting of steel frame and open web steel joists, supporting a concrete slab over metal deck, with concrete block shear walls may have a density of about 5 pcf.

TABLE 2.2
Typical Approximate Design Dead Loads

Building Components	Dead Load (psf)
Roofs	
Primary steel members (beams and trusses)	
span: 40 ft	4–8
100 ft	9–12
150 ft	12–16
200 ft	15–20
300 ft	21–28
400 ft	27–36
Beams (joists) separate	2–4.5
Trusses (separate) 40 ft span	2–3.5
Roof bracing	0.5–1.5
Plywood sheathing (per inch of thickness)	3
2- to 4-in. solid timber decking	3–11
Nonwood decking per inch	2–7
Concrete floor per inch	8.5–12.5
Hollow-core concrete planks 6 in. thick	43–50
Metal deck	1–3

(continued)

TABLE 2.2 Continued

Building Components	Dead Load (psf)
Three-ply roofing	1
Three- to five-ply and gravel	5.5–6.5
Lightweight fill or insulation	0.2–2
Clay tile	10–20
Cement asbestos shingles ($\cong \frac{3}{4}$ in.)	4
Wood shingles (1 in.)	3
Asphalt shingles ($\cong \frac{1}{4}$ in.)	2
Hollow-core concrete planks with 2-in. topping	68–75
Example (design of purlins): composition roofing and insulation (6 psf) + metal decking (1.5 psf) + steel purlins (2 psf) + bracing (0.5 psf) + mech., etc. (2 psf)	12
Floors	
Reinforced concrete slab per inch of thickness	
Normal weight	12.5
Lightweight	9
Plywood per inch of thickness	3
$\frac{3}{4}$-in. tile on $\frac{1}{2}$-in. mortar bed	16
Linoleum or asphalt tile	1
Wood joist floor (2 × 8, 16 in. o.c.) and subfloor	6
3-in. concrete slab on steel deck and open web steel joists	60
8-In. precast planks plus $1\frac{1}{2}$-in. topping	94
Example: Metal deck (7 psf) + concrete floor (38 psf) + ceiling (7 psf) + mech. and elect. (6 psf) + floor beams (15 psf) + fire proofing (2 psf)	75
Ceilings	
Plaster on tile or concrete	5
Suspended metal lath and	
Gypsum plaster	10
Cement plaster	15
Acoustical fiber tile on rock lath and channel ceiling	5
Walls and partitions	
12-In. hollow concrete block	
Heavy	80
Light	55
Clay-brick (per inch of thickness)	10
Reinforced concrete (per inch of thickness)	12.5
Plywood (per inch of thickness)	3
4-in. brick plus 8-in. clay tile backing	75
Glass wall, large plate, heavy mullions	10–15
Window (glass, frame, and sash)	8
3- to 6-in. gypsum tile	10–18
3- to 6-in. clay tile	17–28
Plaster, 1-in. thick	10
Wood paneling, 1 in.	2.5
2-in. solid plaster on metal lath and studs	20
2 × 4 Studs, $\frac{1}{2}$-in. gypsum dry wall on two sides	8
Movable metal partitions	5–10

Should the live load be included, then the overall weight ranges roughly from 10 pcf for steel office buildings to about 14 to 18 pcf for concrete office buildings and 20 pcf for concrete apartment buildings. The use of high-strength materials results in less weight. This may be advantageous when strength, rather than stiffness, controls the structural design. The dead load of the structure can further be reduced by 10 to 20 psf when using lightweight concrete for the floors.

The weight of the structure itself only constitutes a relatively small portion of the total building dead load; it may be in the range of 20% to 50% for frame buildings, but varies with height. A 10-story steel frame structure, for example, may weigh as little as 6 psf, in contrast to a 100-story steel building with about 30 psf. The structure weight depends on the height, slenderness, loading conditions, and efficiency of the structure system (see Chapter 10); the structure weight of high-rise buildings is controlled by the vertical building planes, rather than the horizontal ones.

The fact that structure weight increases with span (e.g., roofs) and height (high-rise buildings) is due to change in scale (see also Section 1.4). This effect of scale is known from nature, where animal skeletons become much bulkier with an increase of size, since the weight increases with the cube, while the supporting area only increases with the square. The bones of a mouse make up only approximately 8% of the total mass, in contrast to about 18% for the human body.

Weight reduction is an important design criterion, since it will result in significant savings in materials, freight, foundations, and erection. On the other hand, weight may be beneficial and necessary when it must act as a stabilizing agent for slender buildings, that is, when the building layout takes advantage of gravity resistance in counteracting uplift forces. It should also not be forgotten that the building mass acts as a damping agent.

EXAMPLE 2.1

A 30-story, 450-ft-high office building has overall plan dimensions of 100×250 ft. The live load for the typical floors is 50 psf and 75 psf for the elevator lobbies. For this laterally braced rigid steel frame building, an equivalent average floor weight of roughly 92 psf may be assumed (i.e., 81 psf for the roof, 90 psf for office levels, and 240 psf for the mechanical level). For example, the equivalent average gravity loads for a typical floor consist of:

Lightweight concrete slab on deck	46 psf
Ceiling, finish, and mechanical	10 psf
Partitions	12 psf
Average weight of columns, beams, and walls around floor openings	22 psf
	90 psf

The average weight of the precast concrete cladding may be taken as 35 psf of wall area.

Based on the information given, the average approximate building density may be estimated as

$$[92(100 \times 250)30 + 35(100 + 250)2(450)]/(100 \times 250 \times 450) = 7.11 \text{ pcf.}$$

Live Loads

Gravity loads that are not part of the dead loads must be considered under live loads. Live loads are not permanent; they are variable and unpredictable. Live loads not only change over time but also depend on location and building type. *Floor live loads*

TABLE 2.3
Typical Minimum Uniform Live Loads

Examples of Occupancy or Use	Live Load (psf)
Residential (attics)	20
Residential (private dwellings, apartments, and hotel guest rooms), private rooms and wards in hospitals, classrooms in schools, dressing rooms in theaters	40
Office buildings (no computer office use), private passenger car garage, fixed seating areas in auditoriums	50
Laboratories in hospitals, reading rooms in libraries, balconies not exceeding 100 ft^2 for one- or two-family residences, orchestra floor in theater	60
Retail stores, light manufacturing, marquees	75
Corridors, court rooms	80
Assembly areas with movable seating, exterior exit balconies, terraces, public corridors, dance halls, fire escapes, restaurants, gymnasiums, lobbies, grandstands, stairs, repair garages, office buildings (office computer use), public dining rooms, public garages, skating rinks, wholesale stores, yards and terraces for pedestrians	100
Computer floor (load must be verified), stage areas, stackrooms in libraries, light storage, heavy manufacturing	125
Armories and drillrooms, stage floors in theaters, mechanical rooms (transformer rooms, elevator machine room, fan room, but weight or actual equipment should be used when greater)	150
Sidewalks and driveways with public access, heavy storage	250
Boiler room (but use weight of actual equipment when greater)	300

caused by the contents or objects are often called *occupancy loads,* while *roof live loads* may be due to construction loads, special use, water accumulation, ice, and snow.

Floor Live Loads. Floor live loads include the weights of people, furniture, books, filing cabinets, fixtures, and other semipermanent loads that were not considered under dead loads. Codes provide values for live loads (i.e., for the sustained portion based on regular use and the variable portion due to unusual events), mostly in terms of equivalent uniform loads distributed over the floor area, as given in Table 2.3.

The equivalent floor loads have evolved empirically from experience and not from systematic surveys of loading. For example, code provisions for office floor live loads vary widely from 50 to 100 psf. A survey taken on the actual occupancy load in various office buildings showed a maximum load of only 40 psf. Similarly, a load survey on apartments noted that the maximum load intensity measured in a 10-year period was about 26 psf, hence quite a bit less than the usual code value of 40 psf. When the live load acts on smaller areas, it may have to be considered as a concentrated load. Concentrated live loads given in codes indicate possible single-load action at critical locations such as stair treads, accessible ceilings, handrails, and due to mechanical and electrical equipment, elevators, cranes, as well as for certain building types (e.g., parking garages, manufacturing and storage buildings, libraries, hospitals). Concentrated live loads are generally assumed to act on an area of 2.5 ft^2 (unless otherwise specified in codes); they are located to produce maximum stress conditions. Stairway and balcony railings may have to be designed for 200 lb applied at any point and in any direction or for 50 plf (for other than dwelling units) applied in any direction along the top handrail member.

From the values in Table 2.3, it is apparent that public areas such as corridors must carry more live load than living or working areas, or that office buildings weigh more than apartment buildings, ignoring the difference in dead load. In other words, the live loads of approximately 80 psf for an office building are twice as high as for a residential building. Similarly, the live loads in public areas, including interior corridors, are at least twice as much as on living areas with 40 psf. Live loads for mechanical rooms are 150 psf and for plaza areas may be as high as 250 psf. The live load for office buildings may have to be further increased to say 100 psf, particularly adjacent to interior cores, to keep up with future developments in making buildings more intelligent. For spaces supporting computer equipment and backup batteries, the live load may be as high as 400 psf.

Although it may appear that some of the floor live loads are too conservative, there is always the unpredictable element to consider. For example, the live loads for exterior exit balconies of 100 psf are high, because the consequences of failure can be severe. The minimum regulated safety factors are warranted by such uncontrollable, extraordinary situations as people crowding because of ceremonies, parties, and fire drills or the overloading of parts of a building due to a change in occupancy that will exert more load on a specific area.

It is improbable that in a building the live load will fully cover a large tributary area to the same extent as a small area. For example, it is improbable that, in multistory structures, every floor simultaneously carries the full live load; in general, the larger the tributary area or the number of floors, the smaller the potential load intensity. Building codes take these conditions into account by allowing the use of floor live load reduction factors when the tributary area, supported by a structural member, is larger than 150 ft^2, for example, as based on the UBC, except for floor areas in places of public assembly. According to the UBC, floor live loads not exceeding 100 psf, supported by columns, piers, walls, foundations, trusses, beams, and two-way slabs, may be reduced as follows:

$$R = r(A - 150) \tag{2.1}$$

This reduction is not to exceed 40% for members receiving load from one level only (e.g., beams, single-story columns), or 60% for other members (e.g., multistory columns and walls, foundations) or R, as determined by the following expression:

$$R = 23.1(1 + D/L) \tag{2.2}$$

where R = reduction in percent
   r = rate of reduction = 0.08% for floors
   A = floor area supported by member
   D = dead load per square foot of area supported by the member
   L = unit live load per square foot of area supported by the member

A reduction of live load is not permitted for one-way slabs and when the live load exceeds 100 psf except that the design live load on columns may be reduced 20% according to the SBCCI.

According to the BOCA National Building Code, members having an influence area of 400 ft^2 or more may be designed for a reduced live load determined by the following equation:

$$L = L_o(0.25 + 15/\sqrt{A_i}\,) \geq \alpha L_o \tag{2.3}$$

where L = reduced design live load (psf)

L_o = unreduced design live load (psf)

A_i = influence area (ft^2) taken as four times the tributary area for a column, two times the tributary area for a beam, and equal to the panel area for a two-way slab $\geq$ 400 ft^2

α = 0.5 for members supporting one floor and 0.4 otherwise

The reduced design live load, however, cannot be less than 50% of the unreduced live load L_o for members supporting one floor and not less than 40% of L_o for members supporting more than one floor. Furthermore, live loads of 100 psf or less cannot be reduced for general-use parking structures, one-way slabs, roofs, or areas for public assembly. Live loads that exceed 100 psf and live loads in garages for passenger cars only, acting on members supporting more than one floor, may be reduced 20%; no reduction is allowed otherwise.

Live loads, in contrast to dead loads, became much more important after World War II. Whereas the buildings of the past were massive and heavy, the buildings of today are of light weight. Due to the intentions of the *Modern Movement,* a better understanding of material behavior, economical considerations, and the development of computers, among other criteria, buildings, as support and environmental control systems, have been optimized, thereby resulting in light, possibly minimal structures that are exposed to the outside or hidden behind only thin facade membranes.

The typical *live-to-dead (L/D) ratio* for steel buildings, for example, varies from about 4.0 for low-rise construction to 0.4 and less for high-rise structures. This is based on the weight of the steel structure itself, ranging from about 5 psf for low-rise construction to 30 psf and more for skyscrapers, or the corresponding gross dead load, ranging from about 20 to 100 psf. The service live loads for the primary floor areas vary approximately between 40 psf for apartment buildings to 80 psf for office buildings.

Although it seems efficient to reach a high L/D ratio according to strength and stiffness criteria, a floor system (e.g., open-web steel joists) may experience vibration problems that can be prevented by adding more weight or using other methods of damping.

EXAMPLE 2.2

The floor framing for a typical interior bay is shown in Fig. 3.4; it must support a dead and live load of 80 psf each. The live load reduction for the floor members is determined.

(*a*) *Beam loading*

The live load reduction, is determined according to the UBC.

The beam supports the following area

$$A = 8(25) = 200 \text{ ft}^2 > 150 \text{ ft}^2$$

Hence, live load reduction is permitted since the tributary area exceeds 150 ft^2 and the live load is not larger than 100 psf. The live load reduction is

$$R_1 = r(A - 150) = 0.08(200 - 150) = 4\%$$
$$R_2 = 40\%$$
$$R_3 = 23.1(1 + D/L) = 23.1(1 + 80/80) = 46.2\%$$

Therefore, the live load reduction is 4% or the reduced live load is

$$L_r = 80 - 0.04(80) = 0.96(80) \cong 77 \text{ psf}$$

(b) *Girder loading*
First, the live load reduction according to the UBC is determined.
The girder supports the area of

$$A = 2(16 \times 25/2) = 400 \text{ ft}^2 > 150 \text{ ft}^2$$

Since the tributary area is in excess of 150 ft^2 and since the live load is not larger than 100 psf, live load reduction is permitted.

$$R_1 = r(A - 150) = 0.08(400 - 150) = 20\%$$
$$R_2 = 40\%$$
$$R_3 = 46.2\% \quad \text{(see beam loading)}$$

The minimum reduction of 20% controls. Hence, the reduced live load is

$$L_r = 80 - 0.2(80) = 0.8(80) = 64 \text{ psf}$$

The live load reduction according to BOCA is as follows: The influence area of the girder is two times its tributary area.

$$A_i = 2(25 \times 16) = 800 \text{ ft}^2 \geq 400 \text{ ft}^2$$

Hence, the reduced live load according to Eq. (2.3) for a reduction of 22% is

$$L = L_o (0.25 + 15 /\sqrt{A_i})$$
$$= L_o(0.25 + 15 /\sqrt{800}) = 0.78\, L_o = 0.78(80) = 62.4 \text{ psf} \geq 0.5(80) = 40 \text{ psf}$$

Snow and Roof Live Loads. Under ordinary loading conditions, roof members are generally designed for snow or roof live loads. Occasionally, however, a roof may have to support heavy equipment and may have to function as public space (e.g., roof garden, driveway, parking) and therefore requires much greater loads. Domes over multipurpose arenas also have to support large scoreboards and other equipment loads possibly suspended from rigging grids as needed for different types of events.

Water loads may become important if a flat long-span roof is not properly drained, if it does not have sufficient slope, if the drains clog, or if it is too flexible. Rain, with a weight of 5.20 psf/in. of depth, will collect and form standing pools. That is, water from heavy rain storms, rain on snow, or snow meltwater may accumulate and concentrate as ponds and may cause a flexible roof structure to deflect, thereby attracting more water and causing a deeper pool. This process continues until either equilibrium is reached or collapse occurs. Ponded rain water on flat roofs is controlled by having sufficient slope and a proper drain, which should not become blocked, and by providing a sufficiently stiff roof structure to avoid ponding failure. Should the water freeze and be prevented from expanding, large lateral pressures are exerted on its boundaries. The situation is worsened when snow is added to the ice; the sequence of snowing, melting, and freezing without proper drainage causes heavy loading. One may also have to consider in the design heavy loads of icicles, which may form on protruding roof elements, and the formation of ice surfaces, which, in turn, attract wind forces.

For ordinary buildings, the roof live loads are much less than the floor live loads. A minimum roof live load of 20 psf on the horizontal projection is required by most codes for flat or low-pitched (less than 1:3) roofs and curved roofs with a rise less than one-eighth of span when the tributary loaded area of any structural member does not exceed 200 ft^2. This roof live load can be reduced to 16 psf when the tributary area is between 201 to 600 ft^2. For larger roof slopes or tributary areas over 600 ft^2, the live load can be reduced further; the absolute minimum live load is 12 psf.

The reader may want to refer to the respective codes for exact roof live load requirements. In this context, a minimum live load of 20 psf on the horizontal roof pro-

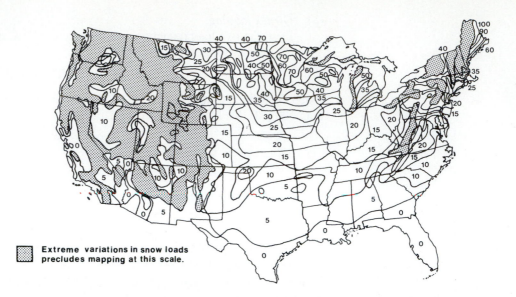

Figure 2.3 Ground snow loads, p_g (psf).

jection is assumed conservatively for preliminary design purposes to take into account unforeseen snow and construction loads. A roof live load of 10 psf is used for the design of greenhouses and 5 psf for fabric awnings and canopies.

The magnitude of *roof snow loads* depends on the geometry of the roof, its exposure, its insulation, and the direction of the wind. In most cases the roof snow loads are lighter than ground snow loads. The snow slides off pitched or curved roofs; wind blows the loose snow off flat roofs, and some of the snow melts and evaporates due to heat loss through the roof skin. However, there are occasions where the snow accumulates, as in the valleys of multiple folds, or due to drifting where low buildings are adjacent to higher ones, or at roof projections and parapet walls; for these cases heavier loads must be considered. Furthermore, asymmetrical loading conditions may arise as a result of drifting.

The distribution of maximum ground snow loads (p_g) in the United States, as recorded by the U.S. Weather Bureau, is shown in Fig. 2.3. The snow load map indicates values that typically range from 70 psf in the Northeast to 5 psf in the South. Special values must be used for the mountain regions where certain localities record snow loads of up to 300 psf. One inch of snow weighs approximately 0.2 to 1 psf depending on the moisture content. In other words, it takes about 6 in. of moist snow and as much as 30 in. of dry, fluffy snow to equal 1 in. of rain. Codes allow a reduction factor, C_e, to convert ground snow loads (p_g) to roof snow loads (p_f) for ordinary single buildings with low-slope roofs; this reduction factor is between 0.6 and 0.9 depending on the snow exposure conditions. In this context, it is assumed conservatively because of the unpredictable character of snow accumulations (e.g., excessive accumulation of snow in the vicinity of obstructions such as penthouses, signs, parapets, roof projections, and adjacent low roofs) that $C_e = 1$. Furthermore, the type of occupancy of the building is taken into account by the importance factor I (see Section 4.1); for standard occupancy structures, I is equal to 1. In other words, the flat-roof snow loads are taken conservatively as equal to the ground snow load ($p_f = p_g$).

The snow loads on roofs with slopes greater than 30° can be reduced (see BOCA and SBCCI) by multiplying the flat-roof snow load by the roof slope factor $C_s = 1 - (\theta - 30)/40$; other codes allow a reduction for roof slopes already over 20°. For curved roofs, the slope factor is determined by basing the slope on the vertical angle from the eave to the crown. In other words, the sloped-roof snow loads, acting on the horizontal projection of the roof surface, can be calculated according to the following formula:

$$p_s = C_s p_f = C_s C_e I p_g \cong (1 - (\theta - 30)/40)p_g \qquad (2.4)$$

where C_s = roof slope factor
          θ = roof slope (degrees)
          C_e = snow exposure factor
          I = importance factor
          p_g = ground snow load in psf (see Fig. 2.3)
          p_f = flat-roof snow load (psf)
          p_s = sloped-roof snow load (psf)

For roof slopes exceeding 70°, no snow loading has to be considered. Naturally, for continuously heated greenhouses, the slope factor will be lower. For multiple roofs (e.g., folded plate, sawtooth and barrel vault roofs) with parallel ridge lines, the roof slope factor may be assumed equal to 1 (C_s = 1.0) regardless of the roof slope.

According to the UBC, snow load, S (or p_g), in excess of 20 psf may be reduced by the *snow load reduction factor* R_s for each degree of pitch over 20°:

$$R_s = (p_g/40) - 1/2 \qquad (2.5)$$

Codes should be consulted with respect to unbalanced snow loads for sloped and curved roofs. For example, for roof slopes exceeding 20° (but less than 70°), the snow loads on the leeward side are greater than the ones on the windward side.

EXAMPLE 2.3

An inclined joist roof with a height-to-span ratio of 12:15 is located in a region where the snow load is 30 psf. Determine the snow load reduction if required.
    The roof slope is

$$\tan\theta = 12/15 \quad \text{or} \quad \theta = 38.66° > 30°$$

Since the roof pitch is larger than 30°, the snow load can be reduced.

$$p_s \cong (1 - (\theta - 30)/40)p_g$$
$$= (1 - (38.66 - 30)/40)\,30 = 24 \text{ psf} > 20 \text{ psf}$$

Hence, the reduced snow load controls the design since it exceeds the minimum roof live load of 20 psf.
    According to the UBC, since 30 psf > 20 psf, the snow load reduction factor is

$$R_s = (p_g/40) - 1/2 = (30/40) - 1/2 = 0.25$$

Hence, the reduced snow load is

$$p_s = p_g - (\theta - 20)R_s = 30 - (38.66 - 20)0.25 = 30 - 4.67 = 25.33 \text{ psf}$$

Construction Loads.   Although a structure is generally designed for the gravity and lateral load action on the finished building, some of its members and bays may be subject to larger loads during the erection process. Every contractor has developed a construction procedure proven economical to him. For instance, equipment and material may be stockpiled on a small area of the structure, which is especially true for city sites where little ground storage space is available and material is placed on the top floor. For this condition, shoring is required to distribute the weight, or the members have to be designed for these critical, concentrated loads. Furthermore, certain building portions, like cores or frame bays, may have to support the dead and live loads of derricks or cranes, which include the vertical and horizontal components of the derrick guy forces, and the gravity, uplift, and wind forces from a climbing tower crane.

Construction loads are also generated due to prevented volume change, as may be the case during winter construction, when the upper floor of a multistory building is cast under controlled temperature conditions, while the rest of the building may be exposed to freezing temperatures. Stresses are generated when the controlled temperature is stopped and a temperature drop occurs, resulting in differential movement. A major problem in concrete construction results when the contractor fails to allow sufficient curing time before the removal of shoring and formwork. Concrete increases in strength with time, particularly at the early stage of hardening; but since time is money to the contractor, he may remove the forms before the concrete has reached its minimum design strength. It must be remembered that the weight of the concrete, equipment, formwork, and workers on the upper floors must be supported by the lower floors and cannot exceed the live load for which they were designed. Usually, two stories of shores and one of reshores are required for high-rise structures for a rate of construction of one story per week. But for buildings designed for light live loads, such as apartment buildings, or for conditions of faster rates of construction, more floors have to be shored.

Construction loads must be considered for beams designed to act compositely with the concrete slab when no temporary shoring is used during the construction process. In this case, the beams have to be checked with respect to carrying construction loads in noncomposite action. Also, the lifting of prefab concrete components or stone slabs with cranes may generate much higher stresses during installation and handling than when the member is in place and may very well control the design of the member. For instance, a solid wall panel, when in place, may just have to carry axial loads, but when it is lifted horizontally, it will behave similarly to a flat plate, bending under its own weight; also critical is when the panel is being lifted from a flat to a vertical position. Naturally, its behavior depends on the specific rigging situation; the number of lifting points and their location must be given. Other types of stresses due to handling may be caused by accidental impact, vibration during transportation, and force fitting on site.

Wind and Earthquake Loads

Wind and earthquake loads are only briefly introduced in this section. For a more detailed discussion, refer to Section 4.1.

Wind and seismic loading cause horizontal force action on a building. They are dynamic loads, but can often be treated as quasistatic lateral forces. This approach is reasonable with respect to wind action as long as the building is not of unusual shape and as long as it is stiff enough so that it does not oscillate and give rise to accelerations, with the corresponding increase in force action. Naturally, the shape of the building as seen in plan and elevation considerably affects the lateral force resistance, remembering that the least resistance for a given wind direction is provided by the streamlined teardrop shape. Not only is the rigidity of a building improved by sloping the exterior columns, such as the truncated pyramid of the John Hancock Center in Chicago, but also the exterior lateral force resistance is reduced, thereby resulting in a large decrease of lateral drift.

A constant, uniform wind pressure may be assumed for the purpose of visualizing lateral force action on a building as a whole, realizing that the actual nonuniform pressure does generate torsion, which can, however, be treated as insignificant for symmetrical buildings, at least for preliminary design purposes. Typical average pressure values for the inland United States range from 15 to 50 psf for ordinary buildings. For inclined and curvilinear surfaces, the wind pressure may be taken as perpendicular to planes projected vertically from the building, as explained at the top of Fig. 4.8. It is also shown that the building shape has a substantial effect on design. For example, the efficient round building has to resist only 60% of the wind load on a comparable rectangular building.

While the wind exerts external lateral forces, the ground motion due to an earthquake causes internal lateral forces, besides vertical forces, which (however) are neglected. We can visualize the building as riding on an unstable earth. As the ground abruptly accelerates in a random fashion, the building portion above the ground will be left behind, thereby activating lateral *inertial forces*. In other words, the inertia of the mass tends to resist the movement, similar to the experience of a person in a car that suddenly increases in speed. The time it takes for the building to respond to the base-induced acceleration due to the fluctuating seismic ground loads becomes an important characteristic of the building; the fourth dimension, that of time, is introduced as a consideration of loading.

Assuming, for this introductory discussion, that the building is rigid by ignoring the effects of flexibility, structure type, mass distribution, location, and site geology, then the lateral inertial forces, according to Newton's second law [Eq. 1.1], are the product of the building mass M and the ground acceleration a. In this case, the mass is equal to the building weight W divided by the acceleration of gravity g, where the seismic base shear coefficient is $C_1 = a/g$

$$F = M(a) = W(a/g) = WC_1 \qquad (2.6)$$

This equation clearly expresses the fact that the magnitude of the lateral force F is directly related to the building weight. However, keep in mind that this magnitude may be less or more when the other, previously ignored factors are included; seismic codes take the reality of the building into account, for example, by replacing $C_1 = a/g = ZIC/R_w$ [see Eq.(4.5)]. It is common practice to express the magnitude of the seismic forces as a percentage of the building weight. Typical values for high-rise buildings in major seismic zones may range from about 5% for flexible rigid frame structures $(0.05W)$ to $0.20W$ for stiff bearing wall buildings. Typical values for low-rise buildings may range from $0.09W$ for flexible ductile structures to $0.14W$ for plywood wall structures and $0.18W$ for stiff structures using brittle material. Because of the heavier weight of concrete buildings, the lateral seismic forces are much higher than for steel buildings. In addition, since masonry and concrete walls have less reserve capacity than a steel frame, a higher safety factor results in an even larger lateral action. For preliminary design purposes, the building mass (weight) may be taken as uniform and proportional to the building volume.

For a typical rectangular building with a uniform mass distribution, the lateral forces due to seismic ground movement may be visualized as an equivalent static, triangular load, as indicated in Fig. 4.8. At the bottom of the same figure, the lateral force distribution for some other common building configurations is identified. For a uniform mass arrangement, this lateral force distribution is proportional to the shape of the building volume, thereby clearly demonstrating the pyramid as an efficient form.

It may be concluded that the building form and mass distribution, as reflected by the plan organization and vertical massing, determine the location of the resultant lateral seismic force. Furthermore, the form of the lateral force-resisting structure and its location within the building volume determine the type of action of the seismic force. When the centroid of mass does not coincide with the center of resistance, twisting is generated, as may be the case at floor levels with abrupt changes of stiffness. The effect of asymmetry, as seen in section and plan, is typical for the new breed of hybrid, compound building forms currently so much in fashion.

While earthquake forces constitute internal lateral loads generated by the mass and stiffness distribution, wind causes external forces on stiff buildings that depend on the exposed facade surface area. Seismic loading is usually critical with respect to the performance of stiff low- and mid-rise structures, while wind loading generally dominates the design of tall, slender buildings. The optimum design of high-rise buildings in areas of strong earthquakes conflicts with that for wind loading. Here, seismic action calls for ductility with much redundancy, while the wind resistance requires stiffness for occupant comfort.

Water and Earth Pressure Loads

Structures below ground must resist lateral and vertical loads due to earth and, possibly, hydrostatic pressure when also submerged in groundwater. For further discussion, refer to Section 4.3.

Loads Due to Restrained Volume Change: Hidden Loads

Entire buildings, parts of buildings, individual building components, and the materials, all move more or less in response to the direct force action of gravity and lateral loads or indirectly due to earth settlement or changes in material movement. Their response, that is, the degree of movement, depends on the stiffness of the building structure and its members, including the flexibility of the connections. Whatever the source is for the displacement and volume or shape change, if it should be held back from freely moving; additional forces, sometimes called *locked-in forces* or *reactive loading,* are induced in the structure element and the adjacent members, preventing the free deformation.

Each structural element in a building may bend and deform axially under force action. The horizontal floor framing deflects vertically under gravity loading, while the vertical structural building planes sway laterally under wind and seismic action. The recent development of increased strength of the major structural materials has resulted in reduced member sizes and a decrease of rigidity, hence an increase of member deflection. The dead loading causes a permanently deformed state to the structural members if they are not cambered, as in the case of floors. All materials more or less expand or contract as a result of changes in temperature. Some materials, such as brick and wood, swell and shrink with variations in moisture content. Other materials, like concrete and concrete masonry, go through the chemical process of drying shrinkage caused by air-drying during the early months of construction. The same materials also creep and shorten under sustained loading, which is especially true for prestressed concrete.

In general, a high degree of continuity should be avoided to prevent stress concentrations, and the exposure of elements should be reduced to a minimum by using sufficient insulation. For example, to control the cracking of low-tensile capacity material like masonry, *movement joints* should be provided at critical locations. These movement joints include *soft joints* (e.g., expansion joints, isolation joints), which may be used where independent parts of a structure are cast or placed directly against each other (see also Section 1.5) and where either of them may interfere with the freedom of movement of the other. They include *control joints* (e.g., contraction joints), which are made by weakening the section so that eventually controlled cracking along the joint results due to contraction, and *shrinkage strips,* which are temporary joints that are left open for a certain time to allow the early drying shrinkage for concrete. *Construction joints* may be required when, for instance, the process of concreting is interrupted so that the concrete pour is no longer continuous.

Interior members of buildings are faced with a relatively constant temperature of approximately 70°F, therefore not causing any change in length, while exterior members are exposed to weather variations ranging from the coldest temperatures in winter to the hottest in summer. In addition, it is not only the ambient air temperature that influences the temperature of the exposed surface, but also wind, solar radiation, and condensation. The average material temperature depends not only on the thermal resistance of the material covering a column, for example, but also on the duration of the exterior peak temperature. This change in temperature causes movement in the facade structure that is partially resisted by the inside structure.

The change in member length, ΔL, is proportional to temperature changes, ΔT, and is expressed by the coefficient of linear thermal expansion, α. It is equal to

$$\Delta L = \varepsilon_t L = \alpha L \Delta T \tag{2.7}$$

Here, L is equal to the original member length (in.), and the thermal strain $\varepsilon_t = \Delta L / L = \alpha \Delta T$. Notice that the displacement of the linear element increases directly with its length. The average coefficients of expansion for some materials are given later in Table 2.4 on page 126–127.

EXAMPLE 2.4

A low-rise building is enclosed along one side by a 100-ft-long clay masonry bearing wall. The structure was built at a temperature of 60°F and is located in the northern part of the United States where the temperature range is between –20° and +120°F.

The coefficient of expansion and the modulus of elasticity are taken from Table 2.4.

$$\alpha = 3.6 \times 10^{-6} \text{ in./in./°F}, \quad E = 2,400,000 \text{ psi}$$

First, the wall is assumed to move freely with no restraints from cross-walls and foundations.

The wall expansion in summer is.

$$\Delta L = \alpha(\Delta T)L = 3.6 \,(10)^{-6}(120 - 60)100(12) = 0.26 \text{ in.}$$

The wall contraction in winter is one-third larger.

$$\Delta L = \alpha(\Delta T)L = 3.6 \,(10)^{-6}[60 - (-20)]100(12) = 0.35 \text{ in.}$$

Now it is assumed conservatively that the free movement cannot happen ($\Delta L = 0$), because foundations (or substructure) and cross walls do not allow any movement. In other words, the force (or stress) in the wall is determined that is necessary to bring the member back to its original restrained position [see Eq. (2.54)]

$$\pm \Delta L = PL/AE = f_a\, L/E = \varepsilon_t L = \alpha L \Delta T$$

$$\pm f_a = \alpha E \Delta T \tag{2.8}$$

Therefore, the tensile stress due to shrinkage in wintertime is

$$-f_a = \alpha E \Delta T = 3.6 \,(10)^{-6}(2,400,000)80 = 691 \text{ psi.}$$

And the compressive stress due to expansion in summertime is

$$+f_a = 3.6 \,(10)^{-6}(2,400,000)60 = 518 \text{ psi.}$$

The tensile stresses are far beyond the material capacity of masonry! In reality, however, it is not as bad as it looks since the initial assumptions taken were unrealistic. In other words, the wall is not fully restrained from movement—the connection between wall and foundations is not rigid—and, in addition, the nearly constant temperature in the building has been ignored; that is, in this case the average wall temperature is not equal to the outside air temperature.

Often, it is assumed as a rule of thumb that stresses due to temperature and moisture changes are acceptable for ordinary brick masonry walls up to a length of 100 ft. For longer walls, expansion joints have to be used.

Dynamic Loads

In contrast to static loads, which are stationary or change slowly and cause a static *deflection,* dynamic loads vary more rapidly, or occur abruptly, and generate *vibrations,* thus introducing another dimension—that of time. The dynamic properties of the building that are activated are the mass, stiffness, and damping, in contrast to the static property of the building, which is only stiffness. Dynamic loads are not always cyclic; they can be sudden impact loads causing shock waves of short duration, such as blast loads due to explosions, sonic booms, or when moving loads like cars, trains, cranes, and elevators suddenly change their speed. They cause, in addition to more periodic vibrations, an impact on the supporting structure that results in longitudinal forces in the direction of movement, or a centrifugal effect on curved structures in a radial direction. Vibrational loads may come from within the building or from the outside. Internal sources are elevators, escalators, oscillating machinery, mechanical equipment, cranes, cars in a parking garage, helicopters landing on a roof, etc. Outside sources are the wind, earthquakes, noise, blasting, driving piles, traffic systems (e.g., streets, railways, subways, and bridges), and water waves, which are of hydrodynamic nature.

The building dead loads are stationary and fixed in magnitude, direction, and location and hence are static loads causing permanent deflections but when they are set in motion, they will generate dynamic loads due to inertia forces. Live loads, in contrast, are movable. They may be considered static, if they are applied slowly, as in the case of occupancy loads, although the natural period of the supporting structure may have to be considered, but they are dynamic when they are applied abruptly or change rapidly and cause vibrations in the building. Vibrations due to people walking or dancing may be generated in the floors in the vertical direction, while wind-induced vibrations are primarily in the horizontal direction, as are seismic oscillations, although earthquakes may also cause significant vertical motions.

But it is not just a question of the rate of application or of how fast a load fluctuates, but also one of how the building responds. The respective dynamic property of the structure is measured by the *natural period* of the building, that is, the time it takes for a building to freely swing back and forth to complete a full cycle of vibration without any external excitation. It is known from physics that, whenever a system is acted on by a periodic series of impulses having a vibrational period nearly equal to the natural period of the system itself, the oscillations in the system will gradually build up until it starts to *resonate,* which can lead to failure if undamped. For example, when soldiers cross a bridge, they must break step so that they do not march in rhythm with the bridge's natural frequency and cause the bridge to collapse. These vibrational forces are called *resonant loads;* they are quite different from the dynamic impact forces that produce large immediate effects.

The fundamental natural periods of typical buildings range from 0.1 sec for a single-story building, 0.6 sec for a 10-story bearing-wall brick building, 1 sec for a more flexible 10-story rigid frame building, 2 sec for a 20-story rigid frame building, 7 sec for the 59-story Citicorp building in New York, 7.6 sec for the 109-story Sears Tower in Chicago, to 10 sec for the 110-story World Trade Center in New York. The natural periods of long-span bridges are in the same range as for tall buildings.

While earthquakes apply sudden, violent, almost random forces with short periods, and the ground vibrates generally at a period of between 0.5 to 1.0 sec, the wind appears smooth in contrast: it grows to its strongest pressure and then decreases in a period of several seconds. Wind turbulence contains a wide range of periods, so the structure is only excited by a small portion of it; in other words, it is extremely variable in size and frequency. Comparing the natural periods of buildings with the those of the exciting sources, it becomes apparent that the state of partial resonance may be approached for stiff buildings under seismic action (although an amplification of the ground acceleration of more than fivefold at the top of tall buildings is not uncommon) and for flexible super-skyscrapers with their larger natural periods, possibly under smaller wind deflections.

It may be concluded that, when the period of the source is much longer (i.e, the frequency is much lower) than the period of the building, as is the case of stiff buildings under wind loading, the load can be treated as static. But when the period of the source is much shorter (i.e., the frequency is much higher) than the natural period of the building, as for seismic action or the waves caused by an explosion, and is experienced by people as vibrations, then the load must be considered dynamic with a corresponding increase in stresses. However, keep in mind that dynamic conditions may occur even when the loads vary slowly but with a period close, for example, to the one of a flexible, tall structure with long natural periods, thus causing resonant loading. Dynamic action causes larger loads than a comparative static one; the shorter the period of action (e.g., impact loads), the greater the load increase.

For the condition where the vibration of the structure is small, it is common practice to simply increase the static loads so as to take into account stress increase due to acceleration, while a dynamic analysis must be used for large vibrations. When members are subject to impact loads, it is usual practice (for typical conditions) to increase the live loads that induce the impact by *impact factors* to cover the dynamic effects; otherwise, dynamic analyses are performed to compute the maximum forces. In other words, because of the complexity of a dynamic analysis, codes allow the use of *equivalent static loads* for ordinary conditions. This approach is discussed for wind and seismic loading in Section 4.1.

Codes provide factors for common dynamic loading conditions as caused by impact and vibration. When moving loads such as cars, trains, cranes, and elevators change their speed (accelerate, decelerate), they cause an impact on the supporting structure that results in *longitudinal forces* in the direction of movement, or *centrifugal forces* on curved structures in a radial direction. For example, codes provide impact factors ranging from 1.1 to 2.0 to take into account the increase of load action due to machinery such as cranes and elevators. In other words, an elevator suddenly stopping is assumed to double its effective weight, or the elevator weight is increased by 100%.

Large dynamic forces must be controlled not just by determining their magnitude so that the building can be designed accordingly, but by changing the period of the source (e.g., driving motor), by isolating the source of the excitation, by damping, and by controlling the mass-to-stiffness ratio. The stiffening of skyscrapers to control oscillations is a most important design consideration. The amount of damping that will prevent all vibrations (i.e., critical damping) is proportional to the product of the stiffness and mass. We may conclude that, for a given structure, dynamic response can be reduced by the following:

- An increase in stiffness
- An increase in mass (for purposes of damping)
- Damping

However, since stiffening a structure results in an increase of material, as does the increase of mass, other methods of damping must be explored to possibly reduce costs.

Besides the *natural damping* of structural materials, *artificial damping* may be employed to control dynamic excitation, similar to shock absorbers on automoliles.

Abnormal Loads

While ordinary loading conditions have been discussed in the previous sections, *abnormal* or *accidental loads,* which are generally not considered in the design because of their low probability of occurrence and because they are assumed to be resisted by the reserve load capacity, are often the cause of building failure. Here are some examples:

- Explosions, external or internal blast loads (e.g., gas service system, bomb-ings)
- Collisions, impact loads (e.g., wind-blown debris, crane, vehicle, aircraft)
- Sonic boom
- Tornado
- Flooding
- Fire
- Vandalism and terrorism

Load Combinations

Many of the loads just discussed may act simultaneously and should be combined if they are superimposable. However, these loads are maximized where the probability of their combined action is less than their separate action, hence the separate design loads can be reduced when they are combined. Sometimes it is unreasonable to let loads act together; for instance, the probability of a wind with a 50-year recurrence interval occurring at the same time as a major earthquake, where most of the damage is done in a period of say 30 to 60 sec, is extremely small. Hence building codes do not require the design of structures for simultaneous action, but for that of either wind or seismic action. Similarly, for most roof structures the probability of maximum snow acting together with maximum wind is small; besides, 70 mph winds will blow at least part of the snow off the roof. To take conditions like that into account, codes allow an increase of allowable stresses or a reduction of loads for certain load combinations. For example, where dead load D acts together with live load L and wind load $W,$ the allow-able stresses can be increased by 33% or the loads reduced by 25%. This is equivalent to multiplying the loads by a *load combination probability factor* of 0.75.

For the various load combinations, the material design standards listed in Section 4.1 should be consulted. According to the codes, the following combinations of loads should be investigated to determine the most critical case:

1. Dead + floor live + roof live (or snow)
2. Dead + floor live + wind (or seismic)
3. Dead + floor live + wind + ½ snow
4. Dead + floor live + ½ wind + snow
5. Dead + floor live + snow + seismic
6. Dead + seismic (see BOCA)

Refer to the respective code for more specific requirements related to these load com-binations.

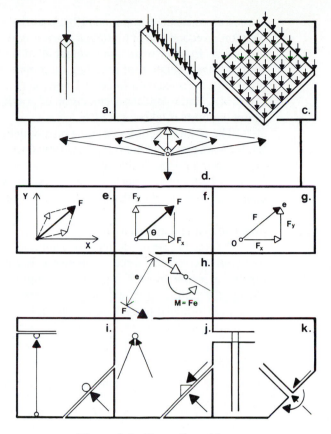

Figure 2.4 Properties of forces.

2.2 STATICS

Most of the loads discussed in the previous section exert *external forces* on a building structure. As they are transferred to the foundations and the soil, they generate *internal forces* or stresses along the way downward. Forces occur as *single* (concentrated) *forces, line loads,* and *surface loads* (Fig. 2.4a, b, c). A building as a *support structure* can be visualized as a complex system of force interaction where not just the building as a whole but also all of its parts are subject to many different types of forces that all must be in static equilibrium.

A force tends to change the state of rest of a body (structure); in other words, it causes translational and/or rotational movement, which, however, is arrested when the body is in static equilibrium and the forces acting on that body cancel each other. This body is considered rigid so that the magnitude, location, and orientation of the forces are not affected by the stiffness of the body.

Properties of Forces

The force is described by the following:

- Magnitude (e.g., pounds, kips, newtons)
- Direction as identified by the sense of the arrowhead and the angle of inclination
- Location or point of application

Forces are measured in pounds (lb), kips (k), newtons (N), and so on. According to Newton's second law of motion, *a force is equal to the product of mass, M, and acceleration, a*. In other words, the weight W of building components is the vertical force caused by acceleration of gravity g on a mass M [see Eq.(1.2)]. For example, 1-kilogram (kg) mass exerts a force of 1 newton (N) as caused by an acceleration of 1 m/s^2 [Eq.(1.4)]. For a standard acceleration of gravity $g = 9.81$ m/s$^2 = 32.2$ ft/s^2, the weight of one kg mass is taken as 9.81 N.

The point of action of a force on a structure is only important when the effect of the force on that structure (i.e., force flow through the structure) is investigated. In general, external forces (realizing that internal forces may become external forces in a free body, as is discussed later) can be moved anywhere along their line of action without changing their effect on that structure, this property of a force is called the *principle of transmissibility*. The assumption, however, is only true if the structure does not deform much and can be treated as a rigid body.

A force is a *vector* defined by magnitude and direction. It can be replaced by several equivalent scalar force components. For purposes of analysis, it is convenient to resolve a planar force F into two rectangular (orthogonal) components, F_x and F_y. Hence, the following relationships between the horizontal and vertical force components perpendicular to each other (Fig. 2.4f) can be established.

$$\sin\theta = F_y/F \quad \text{or} \quad F_y = F\sin\theta$$
$$\cos\theta = F_x/F \quad \text{or} \quad F_x = F\cos\theta \tag{2.9}$$
$$\tan\theta = F_y/F_x \quad \text{or} \quad F_y = F_x\tan\theta$$

$$F^2 = F_x^2 + F_y^2 \quad \text{or} \quad F = \sqrt{F_x^2 + F_y^2} \tag{2.10}$$

A force can also be resolved into component forces by the graphical equivalent to the analytical approach. The force components can be determined graphically with the *force parallelogram* (Fig. 2.4e) by letting the force to be resolved (and drawn to some scale) be the diagonal and the sides be the component forces of the parallelogram. The concept of the force parallelogram can be easily derived from a string supporting a single load as demonstrated in Fig. 2.4d. It is apparent from Figs. 2.4f and g that the parallelogram method can be simplified to the *triangle method,* or polygon method for more than two forces. The graphical addition of forces forming a polygon of forces is called a *force polygon*. Notice that only planar force systems have been introduced for the discussion of spatial force systems, refer to Eqs. (6.1) and (6.2).

Forces may cause translational movement along a straight line by pulling or pushing a structure, that is, by acting as tensile or compressive forces. Forces may also rotate a structure about some particular axis in the structure. This tendency of a force to rotate about a point (which is the center of the circular movement) is called the *moment* (or torque) of a force. The moment M is measured by the product of the force F and the perpendicular distance from the line of force action to the point of rotation (also called moment arm or lever arm) $e,$ as shown in Fig. 2.4h.

$$M = F(e) \tag{2.11}$$

The direction of a moment is either clockwise or counterclockwise. It has units according to the units of force and length, such as foot-kips (ft-k), inch-pounds (in.-lb), or newton-meters (Nm).

Force Systems

Building structures are three dimensional, and therefore the corresponding force systems must also be three dimensional. Ordinary buildings, however, can be usually considered as an assembly of independent horizontal and vertical planes (at least for preliminary design purposes) so that the force systems can be treated as *two dimensional* or *coplanar*. Hence, the investigation of various structure systems in this book is primarily concerned with two-dimensional structures; for discussion of ordinary three-dimensional or noncoplanar force systems, refer to Chapter 6 (Fig. 6.3).

Coplanar force systems can be classified as follows:

- *Collinear force systems:* all the forces act along the same line of action (Fig. 2.5)
- *Concurrent force systems:* the lines of action of all forces pass through the same point, the point of concurrence (Fig. 2.6)
- *Parallel force systems:* all the lines of action of the forces are parallel to each other (Fig. 2.7)
- General force systems or *nonconcurrent, nonparallel force systems* (Fig. 2.8)

These force systems will be briefly investigated with respect to the following properties:

- Determining the *resultant force* (i.e., the simplest force system) of a group of forces, which produces the same effect as the forces it replaces
- Establishing the *static equilibrium* of a force system

When the resultant of a force system is zero, the force system is in static equilibrium. In other words, a two-dimensional structure (e.g., xy plane) as well as all its parts (beams, columns, etc.) on which a group of forces acts is at rest when all the forces cancel each other. Or, according to Sir Isaac Newton's third law, *action and reaction are equal and opposite* (Section 1.2). This translational (vertical and horizontal) and rotational static equilibrium of a coplanar force system can be expressed as

$$\Sigma F_x = 0, \qquad \Sigma F_y = 0, \qquad \Sigma M_z = 0 \tag{2.12}$$

We may conclude that for a structure to be in static equilibrium the external forces acting on it must be balanced generally by at least three *reaction* or *support forces*. That is, the force system (consisting of external forces and reaction or internal forces) must be in static equilibrium. Rather than using the common equilibrium equations (2.12), others may be chosen [see Eq. (2.21)] For the discussion of equilibrium of spatial force systems, refer to Eq. (6.6).

Should the three equations of static equilibrium be sufficient for solving the unknown forces (which are usually the support forces), the structure is called *statically determinate*. When, however, the conditions of the structure are such that the three equations are not sufficient for the solution of the force flow, additional equations are required, and the structure is said to be *statically indeterminate* or *hyperstatic*. It is apparent that a structure must be in static equilibrium and stable, whether it is statically determinate or indeterminate; for further discussion of this topic, refer to Section 5.1.

To determine the magnitude of internal forces in a structure or part of a structure, the area of interest has to be isolated in a *free body* by cutting through or along that portion of the structure. Free-body diagrams may represent an entire building structure, a portion of a building structure, a structure component, a portion of a structure component, a joint, and so on. Each free body must be in static equilibrium, with the external forces balanced by the internal or reaction forces.

For establishing the nature of the reaction forces, the potential joint or support conditions must be known. There are three basic types in coplanar structures:

- *One-force potential joint forces* (e.g., rollers, pendulums, single-pinned columns, neoprene pads, and other sliding joints, as shown in Fig. 2.4i). For example, a roller can provide only one reaction force perpendicular to the surface upon which it is supported; hence, it allows freedom of translational movement parallel to the support, as well as rotational freedom of movement.
- *Two-force potential joint forces* (e.g., pins, hinges, as shown in Fig. 2.4j). For example, a pinned joint can provide two translational reaction forces, but allows freedom of rotational movement.
- *Three-force potential joint forces* (e.g., fixed, clamped, or continuous supports, as shown in Fig. 2.4k), also called *rigid joints.* They do not allow translational or rotational movements; they are capable of providing translational reaction forces (F_x, F_y) and rotational reaction forces (M_z).

Other special joint types will not be discussed in this brief introduction. Examples of connection types and forces are shown in Fig. 3.28. More complex support types and arrangements with the corresponding reaction forces are identified in Fig. 6.4 for three-dimensional structures.

In the following discussion the basic force systems are investigated with respect to determining their resultants and establishing static equilibrium. Because structures are generally designed with computers, graphical methods of analysis have been generally replaced by analytical ones. Therefore, the graphical approach is presented in this context, only if it helps us to understand structural concepts and thus is of educational value.

Collinear Force Systems.   Since all the forces are located along the same line of action (Fig. 2.5), the resultant force R can easily be obtained by simply adding the forces.

$$R = F_1 + F_2 + \cdots + F_n = \sum_{i=1}^{n} F_i \tag{2.13}$$

A positive sense is associated to the forces in one direction by labeling the magnitude of the forces positive, whereas the forces in the opposite direction are treated as negative. If the collinear force system is in static equilibrium, then there is no resultant force: $R = 0$.

$$R = \sum_{i=1}^{n} F = 0 \tag{2.12a}$$

Members carrying collinear forces, as is the case of floor loads on walls or columns, are either in tension or compression.

EXAMPLE 2.5

The four-story steel building in Fig. 2.5c is assumed to have an average weight of $D + L = 80 + 40 = 120$ psf. The total column load at the foundation level is to be found.

The typical column load at each floor level (including roof level) as based on 24- × 24-ft square bays is

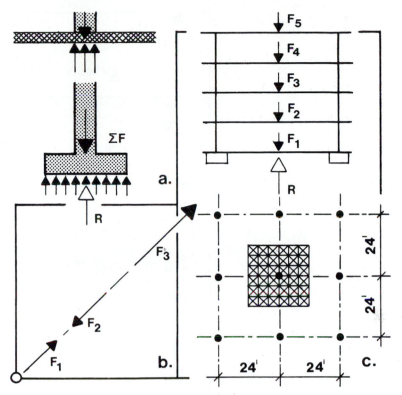

Figure 2.5 Collinear force systems.

$$F_1 = F_2 = \cdots = F_5 = 0.120(24 \times 24) \cong 69 \text{ k}$$

Therefore, the resultant force of all the collinear column loads acting on the foundation or the reaction force (in the opposite direction) at the base provided by the foundation, as based on vertical equilibrium of all the column loads, is

$$\Sigma F = 0 = 5(69) - R \quad \text{or} \quad R = 345 \text{ k}$$

Concurrent Force Systems. Whereas collinear forces can be added directly, concurrent forces (Fig. 2.6) can only be added vectorially where, in addition to magnitude, the direction of forces must also be considered. In the analytical approach, each force is resolved into two orthogonal components, F_x and F_y, according to the parallelogram method [Fig. 2.4f and Eq. (2.9)]. Then all the forces in the horizontal (ΣF_x) and vertical (ΣF_y) directions are added by assuming as positive, for instance, the upward direction of the vertical forces and the horizontal forces acting toward the right, so that the forces in the opposite directions are treated as negative.

The resultant force R of the forces $R_x = \Sigma F_x$ and $R_y = \Sigma F_y$ can now be determined according to the Pythagorean theorem [Eq.(2.10) and Fig. 2.6d] as

$$R = \sqrt{R_x^2 + R_y^2} = \sqrt{(\Sigma F_x)^2 + (\Sigma F_y)^2} \tag{2.14}$$

The angle of inclination of the resultant force (see Fig. 2.6d) is

$$\tan\theta = R_y/R_x = \Sigma F_y/\Sigma F_x \tag{2.15}$$

Since all the forces intersect at the point of concurrence, the resultant force must also pass through that point.

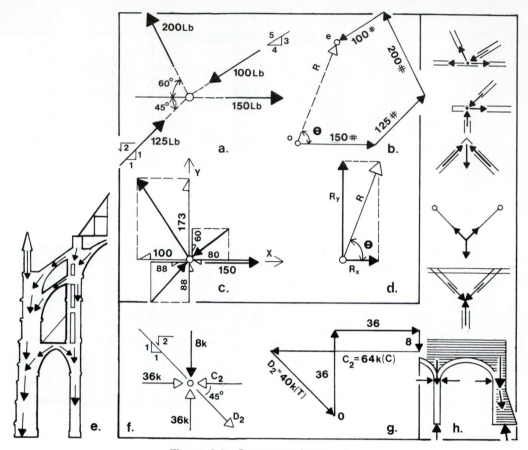

Figure 2.6 Concurrent force systems.

The forces can also be added graphically, which is equivalent to the analytical addition by using the *force polygon method* (Fig. 2.6b), an extension of the triangle method (Fig. 2.4g). The forces are drawn to some scale and added in a head to tail sequence parallel to the directions of the forces in the system. There is no specific order to the addition of forces; all sequences yield the same result.

The resultant force *R* closes the force polygon, which started at the origin *o* and ends at point *e*. In other words, rather than following the detour along the path of all the forces to point *e,* we could have taken the direct and shortest distance along the resultant force.

EXAMPLE 2.6

The magnitude and direction of the resultant force for the concurrent force system in Fig. 2.6a is determined analytically and graphically.

(a) *Analytical approach* (Fig. 2.6c and d)

The vertical resultant force is obtained by summing up the vertical force components, considering the upward direction as positive.

$$\Sigma F_y = R_y = 200 \sin 60° - (100/5)\, 3 + 125 \sin 45° = 173 - 60 + 88 = 201 \text{ lb}$$

Since the answer is positive, the vertical force acts upward.

The horizontal resultant force, assuming the forces acting to the right as positive, is

$$\Sigma F_x = R_x = -200 \cos 60° - (100/5)4 + 150 + (125/\sqrt{2})1$$

$$= -100 - 80 + 150 + 88 = 58 \text{ lb}$$

The horizontal resultant force acts to the right since the result is positive.

The resultant force of the concurrent force system is determined according to the Pythagorean theorem (Fig. 2.6d).

$$R = \sqrt{R_x^2 + R_y^2} = \sqrt{58^2 + 201^2} = 209 \text{ lb} \qquad (2.14)$$

The angle of inclination of the resultant is

$$\tan\theta = R_y/R_x = 201/58 \quad \text{or} \quad \theta = 73.9° \qquad (2.15)$$

Refer to Fig. 2.6d for identification of angle θ.

(b) *Graphical approach* (Fig. 2.6b)

The process of adding the forces vectorially according to the force polygon method is started at point *o* and ends with the arrowhead of the 100-lb force at point *e*. In this case, the forces have been added sequentially in a clockwise manner; any other sequence, however, yields the same result, although the force polygon figure will look different. Measuring the length of the resultant force and its angle of inclination with respect to the horizontal in Fig. 2.6b yields about the same results as the analytical approach.

A concurrent force system does not cause any rotation on a structure since all the forces pass through a common point of concurrence. Therefore, for a concurrent force system to be in static equilibrium requires only translational equilibrium of forces, or

$$\Sigma F_x = 0, \qquad \Sigma F_y = 0 \qquad (2.12b)$$

If the forces represent structural members and act toward the point of concurrence (e.g., joint), then they are in compression, but if they act away, they are in tension. Since, frequently, the sense of the unknown forces may not be known, it must be assumed at the start of the problem. The assumption is correct when the solution is positive. However, if the solution is negative, the sense of the force is in the opposite direction from the one assumed.

Using the *force polygon method* by adding the forces graphically rather than analytically, requires the polygon to be closed, since there cannot be any resultant force present.

A concurrent force system does not necessarily directly identify itself through the arrangement of members as indicated by the top three cases in Fig. 2.6h; it can be hidden, as, for example, within the exterior buttresses of the exterior arch (Fig. 2.6h, bottom) and the Gothic cathedral (Fig. 2.6e). In High Gothic cathedrals, the lower and upper flying buttresses transmit the outward thrust from the vaults and the wind pressure against the timber roof, respectively, to the heavy, stepped stone pier buttresses (where the pinnacles on top of the piers help to increase the weight), as shown for Amiens Cathedral in France.

A typical example of concurrent force systems in static equilibrium is the joints of trusses or the points where cables support loads. The analysis of planar, pin-connected trusses (see Fig. 5.51) can be approached by isolating each joint in a truss. Here, each joint free body represents a concurrent force system that must be in static equilibrium. Since only two equations of equilibrium are available for each concurrent force system, only two unknown member forces can be present at each joint so that the magnitude of the unknown forces can be determined directly at that joint. This method of analysis for trusses is called the *joint method.*

EXAMPLE 2.7

One of the joints of the truss in Example 5.16 (Fig. 5.53a) is investigated. The conditions for the isolated top chord joint, adjacent to the truss end, are identified in Fig. 2.6f. Since only two unknown member forces are present (diagonal member and top chord), the problem is statically determinate. The direction of the unknown forces is assumed as shown.

According to the analytical approach, vertical equilibrium of forces yields the magnitude of the diagonal member force. The vertical component of the diagonal force [Eq. (2.9)] is

$$D_{2y} = D_2 \sin \theta = D_2 \sin 45° = (D_2 / \sqrt{2})1$$

Hence, vertical equilibrium of forces according to Eq. (2.12) gives

$$\Sigma F_y = 0 = 8 - 36 + (D_2 / \sqrt{2})1 \quad \text{or} \quad D_2 = 39.60 \text{ k} \quad \text{(T)}$$

Since the solution results in a positive number, the assumed direction of the force as tensile was all right.

Horizontal equilibrium of forces according to Eq. (2.12) yields

$$\Sigma F_x = 0 = 36 - C_2 + 39.60 \cos 45° \quad \text{or} \quad C_2 = 64 \text{ k} \quad \text{(C)}$$

Hence, the assumed direction of the member in compression was right.

In the graphical solution, the force polygon must close if there is static equilibrium of forces (i.e., no resultant). It is common practice (but not necessary) to work in a clockwise direction around a joint when adding the forces vectorially.

First, a particular scale is selected in which the lengths of the force lines represent the magnitude of the forces (see Fig. 2.6g). In this case, the vertical 36-k force is drawn first and the horizontal 36-k force is added at the end (i.e., arrowhead) of the vertical force; then, in turn, the vertical 8-k force is added at the tip of the horizontal force so that all the forces flow in a continuous sequence. Now the unknown horizontal C_2 force is added at the arrowhead of the 8-k force and the diagonal force line is drawn (parallel to the diagonal force in the concurrent force system) through the starting point, *o,* of the vertical 36-k force. The intersection of the two unknown force lines represents the arrowhead of the horizontal C_2 force and the tail of the diagonal D_2 force. Transferring these forces back to the concurrent force system indicates that the horizontal force is in compression and the diagonal force in tension. Measuring the lengths of the force lines gives the magnitudes of the unknown forces, which check with the analytical solution.

In the general graphical solution of a truss, all member forces are determined. Here, the force polygons for each joint are combined to one force polygon for the entire truss, known as the *Maxwell diagram* of the truss. The graphical solution of trusses, generally using Bow's notation, is discussed in any book on statics. For further discussion of trusses in general, refer to Section 5.5.

Parallel Force Systems.   Parallel force systems (Fig. 2.7) can be composed of single loads arranged in some fashion or they can be line loads distributed along the length of a member and are measured, for instance, in plf (lb/ft), klf (k/ft), psf (lb/ft^2), pcf (lb/ft^3), and so on.

The magnitude of the resultant force of a parallel force system is easily obtained by adding the forces: the net force is equal to the resultant force.

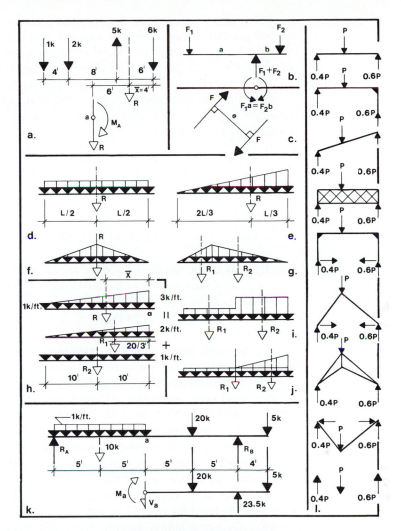

Figure 2.7 Parallel force systems.

$$R = F_1 + F_2 + \cdots + F_n = \sum_{i=1}^{n} F_i \qquad (2.16)$$

The resultant of a distributed load is equal to the area of the load diagram and acts through the centroid of the load diagram as is discussed for Figs. 2.11 and 2.15.

A *force couple* is a special case of parallel force systems. It consists of two equal but opposite forces (Fig. 2.7c). Here the resultant translational force is zero:

$$R = F - F = 0$$

But the couple exerts a moment

$$M = Fe \qquad (2.17)$$

where e = perpendicular distance between the forces. The moment of a force couple is constant regardless of the axis of rotation.

The location of the resultant force for a parallel force system is generally not known unless the forces are arranged in a symmetrical fashion so that the location can be determined by inspection. For example, the location of the resultant for equally spaced parallel forces of equal magnitude and sense, called a *uniformly distributed load* (as due to the weight of a member), is along the axis of symmetry or the center of gravity of the loads, that is, midway between the forces (Fig. 2.7d).

For the general condition of parallel forces, the location of the resultant force can be determined as based on *Varignon's theorem,* which states that *the moment of the resultant force about any point must be equal to the moment of all the forces (that are replaced by the resultant force) about the same point.*

$$R\bar{x} = \bar{x} \sum_{i=1}^{n} F_i = F_1 x_1 + F_2 x_2 + \cdots + F_n x_n = \sum_{i=1}^{n} F_i x_i$$

or in simplified form   (2.18)

$$\bar{x} = \Sigma F x / R = \Sigma F x / \Sigma F$$

The location of the resultant force for a triangularly distributed load can be shown to be at $\bar{x} = 2L/3$ measured from the zero load point (Figs. 2.7e and 2.16).

The location of resultants for other basic distributed line load systems is discussed later in this chapter in the section dealing with the geometrical properties of cross-sectional areas (Section 2.4).

Composite distribution load systems (Fig. 2.7h) can be broken up into simple shapes whose properties are known, as demonstrated in Example 2.9.

Using the graphical approach for finding the resultant force of a parallel force system is not very convenient. Although the force polygon yields the magnitude and direction of the resultant force, another polygon of forces is needed to determine the location of the resultant force. This second polygon is called a *string polygon* or *funicular polygon.* Here, the parallel (or nonparallel) forces are each replaced by two component forces in such a way that they form a continuous string of forces, where the intersection of the extension of the first and last strings represents the location that the resultant must pass through. This concept is similar to a single cable that adjusts its suspended form to the respective loading condition so that it can respond in tension. For example, under single loads the cable takes the shape of a string polygon or funicular polygon (see Fig. 9.5).

EXAMPLE 2.8

Replace the parallel force system in Fig. 2.7a by a single resultant force. Determine its magnitude and location, and find the force resultants at point *a.*

The net vertical force or resultant force is

$$R_y = \Sigma F_y = 1 + 2 - 5 + 6 = 4 \text{ k} \qquad (2.16)$$

The resultant force acts downward. The moment of the resultant force must be equal to the sum of the moments of the force components. Taking moments about the 6-k force yields the location of the resultant force.

$$4(\bar{x}) = -5(6) + 2(14) + 1(18) = 16 \quad \text{or} \quad \bar{x} = 4 \text{ ft} \qquad (2.18)$$

At any other location there will be a moment in addition to the resultant force. For example, at point *a* the moment is

$$M_a = 4(6) = 24 \text{ ft-k}$$

EXAMPLE 2.9

Determine the magnitude and location of the resultant force for the distributed load of trapezoidal configuration in Fig. 2.7h.
The total load or area of load diagram is.

$$R = 20 \, (1 + 3)/2 = 40 \text{ k}$$

To find the location of this force, the load diagram is split into a triangular and uniformly distributed load. The corresponding resultants of these load diagrams are

$$R_1 = 2(20/2) = 20 \text{ k}, \qquad R_2 = 1(20) = 20 \text{ k}, \qquad R = \Sigma R_n = 2(20) = 40 \text{ k}$$

The moment of the total resultant force about any point (in this case point *a)* must be equal to the moments caused by the partial resultants R_1 and R_2 about the same point. This relationship yields the location of the resultant force.

$$40(\bar{x}) = 20(10) + 20 \, (20/3) \quad \text{or} \quad \bar{x} = 8.33 \text{ ft} \qquad (2.18)$$

The most basic parallel system in static equilibrium is represented by the seesaw in Fig. 2.7b, where the lighter person balances the heavier one by selecting a larger lever arm so that the system is in rotational equilibrium or F_1 (a) $= F_2$ (b). In general, for a parallel force system to be in static equilibrium requires translational and rotational equilibrium of forces.

$$\Sigma F = 0, \qquad \Sigma M = 0 \qquad (2.19)$$

The equilibrium conditions can also be represented by two moment equations.

$$\Sigma M_a = 0, \qquad \Sigma M_b = 0 \qquad (2.20)$$

Since only two equations may be used, only two unknown forces can be present in the coplanar parallel force system unless it is statically indeterminate.
A typical example of parallel force systems is a beam with two unknown reaction forces. Notice that the vertical reaction forces for the various structure systems in Fig. 2.7l are identical in other words, they are independent of the force path.
The graphical solution of parallel, coplanar force systems necessitates that both the force polygon and the string polygon must be closed figures.

EXAMPLE 2.10

(a) Determine the magnitude of the two unknown reaction forces in the parallel force system in Fig. 2.7k.
First, the uniform load is replaced by its resultant force, $P = 1(10) = 10$ k. Then moments are taken, with all the forces about the unknown reaction force R_A assuming clockwise rotation as positive.

$$\Sigma M_A = 0 = 10(5) + 20(15) - R_B(20) + 5(24) \quad \text{or} \quad R_B = 23.5 \text{ k}$$

Establishing rotational equilibrium about the other reaction force yields

$$\Sigma M_B = 0 = R_A(20) - 10(15) - 20(5) + 5(4) \quad \text{or} \quad R_A = 11.5 \text{ k}$$

Check the vertical equilibrium.

$$\Sigma F_y = 0 = 10 + 20 + 5 - 23.5 - 11.5 \quad \text{OK}$$

(b) Determine the magnitude of the internal forces at point *a* (Fig. 2.7k) so that the free body is in static equilibrium.

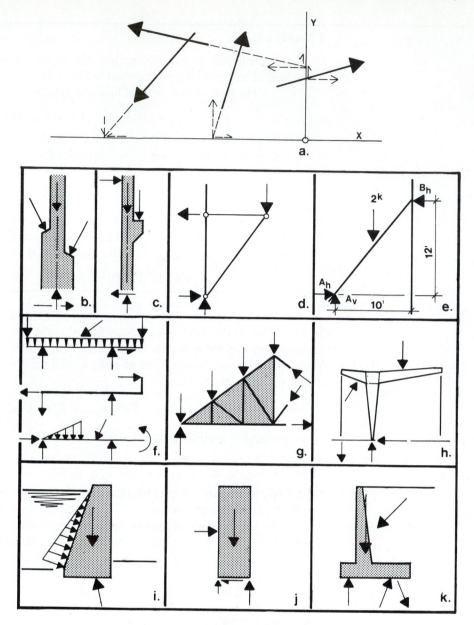

Figure 2.8 General force systems.

$$\Sigma F_y = 0 = V_a + 20 + 5 - 23.5 \quad \text{or} \quad V_a = -1.5 \text{ k} = 1.5 \text{ k} \quad \text{(upward)}$$
$$\Sigma M = 0 = 23.5(10) - 20(5) - 5(14) - M_a \quad \text{or} \quad M_a = 65 \text{ ft-k} \quad \text{(clockwise)}$$

General Force Systems. The coplanar force systems that have been studied up to now were special cases of the general, nonconcurrent, nonparallel force arrangements (Fig. 2.8a). The procedure for finding the resultant of a general force system is the same as used in the previous sections for the special conditions. In the mathematical approach, the magnitude, direction, and inclination of the resultant force are found exactly in the same manner as for a concurrent force system, whereas its location is determined as for a parallel force system using Varignon's theorem, where the moment of the resultant (or resultant components R_x and R_y) about any convenient point must be equal to the sum of the moments due to the individual forces.

In the graphical approach, the magnitude, direction, inclination, and location of the resultant force are determined by using the force polygon together with the string polygon as discussed in the previous section.

For general, coplanar force systems to be in static equilibrium, the forces acting on a structure must balance each other; there can be no translational (two degrees of freedom) and rotational (one degree of freedom) movement, as has been shown in the discussion of Eq. (2.12).

$$\Sigma F_x = 0, \quad \Sigma F_y = 0, \quad \Sigma M_z = 0 \tag{2.21a}$$

The three degrees of freedom of a planar structure, however, can be restrained by other combinations, that is, one force and two moment equations or three moment equations.

$$\Sigma F_x = 0 \quad (\text{or } \Sigma F_y = 0), \qquad \Sigma M_A = 0, \qquad \Sigma M_B = 0 \tag{2.21b}$$

$$\Sigma M_C = 0, \qquad\qquad \Sigma M_A = 0, \qquad \Sigma M_B = 0 \tag{2.21c}$$

Notice that the points of rotation cannot all lie on the same straight line.

In Fig. 2.8b to k, examples of structures in static equilibrium under general coplanar force systems are shown to give a first introduction to this type of loading condition. A typical example occurs for trusses when the *method of sections* is applied (Fig. 2.8g). Whereas the method of joints, as based on concurrent force systems, refers to finding every member force, which, however, is generally not required, the method of sections, in contrast, allows us to directly determine critical member forces. In this method a portion of a truss is isolated by cutting a section through the truss and the members to be investigated. The external forces in the isolated portion of the truss must be in equilibrium with the internal axial forces in the cut members. To find member forces directly, not more than three unknown members should be cut. For further discussion of trusses in general, refer to Section 5.5.

EXAMPLE 2.11

For the inclined roof beam in Fig. 2.8e, determine the unknown reaction forces due to a 2-k force at midspan.

Vertical equilibrium of forces indicates that the vertical reaction at support A must resist the roof load.

$$\Sigma F_y = 0 = A_v - 2 \quad \text{or} \quad A_v = 2 \text{ k}$$

Rotational equilibrium about support A, gives the horizontal reaction force at point B.

$$\Sigma M_A = 0 = 2(5) - B_h(12), \quad \text{or} \quad B_h = 0.83 \text{ k}$$

The horizontal equilibrium of forces clearly indicates that A_h balances B_h.

$$\Sigma F_x = 0 = A_h - 0.83, \quad \text{or} \quad A_h = 0.83 \text{ k}$$

The assumed directions of the reaction forces in Fig. 2.8e were correct.

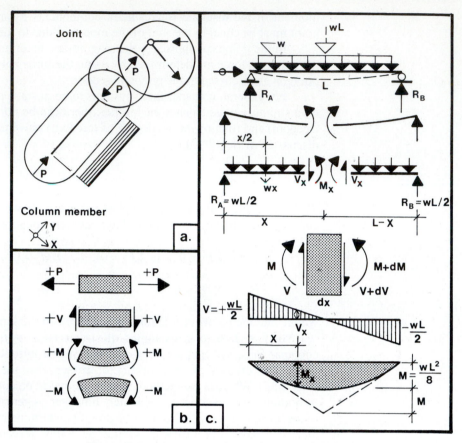

Figure 2.9 Internal forces in columns and beams.

2.3 INTERNAL FORCES IN BEAMS AND COLUMNS

A structure must be capable of resisting the external forces acting on it. As these forces flow along the structure to the ground, internal forces in the members are generated. The members, obviously, have to be sufficiently strong and stiff to support these internal forces. Therefore, they have to be known so that the various structure components can be designed. To determine the magnitude of these internal forces in the respective members to be investigated, the members have to be cut so that the internal forces can be exposed (Fig. 2.9).

The potential internal forces that occur in the basic linear elements of beams and columns are *normal forces, shear forces, bending moments and occasionally torsion,* as will be discussed in the subsequent sections. Once these forces are known, the minimum required member size can be found that should be capable of responding to the maximum internal stresses caused by the internal forces.

Columns or piers are generally vertical members acting mainly in compression. Nonvertical compression members are called *struts,* in contrast to *ties* that act in tension. The column in Fig. 2.9a carries a concentric force in purely axial action with no bending present. Cutting the column at any level indicates that the internal force must be equal to the external force ($\Sigma F_y = 0$) and that the axial force flow along the column is constant.

In contrast, a beam cannot transfer a force directly; it must bend in order to guide the external forces to the supports. The bending of the member in Fig. 2.9c, for example, indicates that bending moments must be present to cause the deflected member shape. The presence of the moment in beams can also be demonstrated from the free body in Fig. 2.9c using statics. In other words, the free body must be in static equilibrium, which necessitates the presence of a vertical internal force parallel to the section, called *shear force,* or simply *shear,* and a *bending moment,* or simply *moment.* In a beam–column, in addition, an internal normal force is present.

The direction of the internal forces shown in free bodies in Fig. 2.9b and c is considered positive, and the forces in the opposite direction are treated as negative. For example, cutting the beam in Fig. 2.9c in two halves, the internal forces at the section, according to the common sign convention, are treated as positive for a downward shear at the left free body and an upward shear on the right free body, as well as a counter-clockwise moment on the left free body and a clockwise moment on the right one. We can conclude that a positive bending moment produces an upward curvature of concave shape and a negative moment produces a downward curvature (Fig. 2.9b). In this context, the bending moments are shown along the tension side of a member.

In the following discussion of beams, it will be helpful to visualize the deflected shape of a beam, since the concave upward curvature corresponds to a positive moment and the downward curvature to a negative moment.

The Relationships among Load, Shear, and Moment

To demonstrate the relationship among load, w, shear force, V, and bending moment, M, the typical case of a simple beam carrying a uniformly distributed load (Fig. 2.9c) is investigated.

Each of the beam reactions must support one-half of the total load, $W = wL$, because of the symmetrical arrangement of the loads.

$$R_A = R_B = wL/2 = W/2$$

The beam is cut at a distance x from the left support so that the magnitude of the internal forces can be determined. Vertical equilibrium of forces in the left free body yields the shear force.

$$\Sigma F_y = 0 = \frac{wL}{2} - wx - V_x$$

or   (2.22)

$$V_x = \frac{wL}{2} - wx = w\left(\frac{L}{2} - x\right)$$

Notice that the mathematical expression represents a first-degree curve or inclined straight line with its maximum values at the supports $(x = 0, x = L)$ and zero value at midspan $(x = L/2)$. The graphical representation of Eq. (2.22) is the shear-force diagram, or simply *shear diagram,* showing the variation of the shear along the beam.

In more general terms, the relationship between load and shear can be expressed in terms of differential calculus by summing up the vertical forces on the beam section of length dx in Fig. 2.9c and assuming the shear forces in the opposite direction.

$$\Sigma F_y = 0 = V + wdx - (V + dV)$$

or   (2.23)

$$dV = wdx \quad \text{or} \quad w = dV/dx$$

In other words, the load at a point is equal to the slope of the shear diagram at that point. Rotational equilibrium of the forces in the left free body of Fig. 2.9c results in the magnitude of the internal bending moment.

$$\Sigma M = 0 = \frac{wL}{2}(x) - wx\left(\frac{x}{2}\right) - M_x$$

or
$$\tag{2.24}$$

$$M_x = \frac{w}{2}(xL - x^2)$$

Notice that the moment equation represents a second-degree curve or parabola that has its maximum value at midspan because of symmetry ($x = L/2$) and zero values at the supports ($x = 0$, $x = L$). The graphical representation of the moment equation is the bending-moment diagram, or simply *moment diagram,* showing the variation of the moment along the beam.

Because of symmetry, the moment is maximum at midspan ($x = L/2$) and is equal to

$$M_{max} = \frac{w}{2}\left[\frac{L}{2}L - \left(\frac{L}{2}\right)^2\right] = \frac{wL^2}{8} \tag{2.25}$$

Again, the relationship between moment and shear can be expressed in more general terms, based on differential calculus and rotational equilibrium of the forces on the beam section dx in Fig. 2.9c. Notice that dx is so small that the term $wdx^2/2$ approaches zero and can be dropped.

$$\Sigma M = 0 = M + Vdx - wdx(dx/2) - (M + dM)$$

or
$$\tag{2.26}$$

$$dM = Vdx \quad \text{or} \quad V = dM/dx$$

In other words, the shear at a point is equivalent to the slope of the bending-moment diagram at that point. For instance, for the given example, taking the derivative of Eq. (2.24) yields.

$$V_x = \frac{dM}{dx} = \frac{wL}{2} - wx, \qquad \text{OK}$$

Furthermore, where the slope of the bending-moment diagram is zero, the bending moment must be maximum. In other words, according to Eq. (2.26), *where the shear is zero the moment is maximum.*

$$V_x = \frac{dM}{dx} = 0 = \frac{wL}{2} - wx, \quad \text{or} \quad x = L/2$$

Naturally, in this case the location of the maximum moment was already known because of conditions of symmetry. Integrating Eq. (2.26) yields.

$$\int_{M_1}^{M_2} dM = \int_{x_1}^{x_2} Vdx \quad \text{or} \quad M_2 - M_1 = \int_{x_1}^{x_2} Vdx \tag{2.27}$$

The result shows that the difference between the moments between points 1 and 2, or ($x_2 - x_1$), is equal to the area of the shear diagram between those two points. For example, for the given case, the maximum moment can be found for $M_1 = 0$, $x_1 = 0$, and $x_2 = L/2$.

$$M_2 - 0 = M_{max} = \int_0^{L/2} Vdx = \frac{wL}{2}\left(\frac{L}{2}\right)\frac{1}{2} = \frac{wL^2}{8}$$

Beam Types: The Effect of Boundary Conditions

Beams may be organized according to their support types as follows:

- Simple beams
- Cantilever beams
- Overhanging beams
- Hinge-connected cantilever beams
- Fixed-end beams
- Continuous beams

Simple folded or curved beams are discussed in Fig. 2.27.

It is apparent that loads cause a beam to deflect. Since, according to the previous section, external loads initiate the internal forces, shear and moment (disregarding axial forces and torsion), deflection must be directly dependent on shear and moment. It is shown later in this chapter that typical beams are of the shallow type where deflection is generally controlled by moments. In contrast, the deflection of deep beams (e.g., wall beams) is governed by shear.

In the following discussion it is helpful to treat moment and beam deflection as directly related. Since the design of beams is primarily controlled by bending, emphasis is on the discussion of moments rather than shear.

The effect of the different boundary types (pin, hinge, overhang, fixity, continuity, and free end) on the behavior of beams is investigated using the typical uniform loading condition. It is shown in the previous section that a uniform load generates a parabolic moment diagram with a maximum moment of $wL^2/8$ at midspan. It is shown in the subsequent discussion how the moment diagram is affected by the various boundary conditions as demonstrated in Fig. 2.10.

Simple Beams.   A simple beam, for example in horizontal position, is supported at one end by a pin, providing a horizontal and vertical force resistance, and a roller at the other end, where it is free to move horizontally. The supports allow free rotation, resulting in a single deflected shape of upward curvature, hence causing only positive moments and thereby stressing the upper beam fibers in compression and the lower ones in tension. One must realize, however, that this ideal situation of a simple beam is rarely achieved since boundary restraints are always present.

A uniformly distributed load on a simple beam produces a moment diagram of parabolic shape with a maximum moment of $M = wL^2/8$ at midspan and zero values at the supports, as has been discussed previously (Fig. 2.10a). The maximum shear occurs at the reactions and is equal to one-half of the total load: $V_A = V_B = wL/2 = W/2$.

Cantilever Beams.   The most apparent cantilever structure is the tree, consisting of the vertical cantilevered trunks supporting the horizontal cantilevered branches; other examples include Alexander Calder's exciting mobiles. A pure cantilever beam has only a support at one end, while the other end is free. Hence, the support must be fixed so that it can provide the necessary rotational resistance. The cantilever deflection shows a single downward curvature (Fig. 2.10b), indicating that the moments must be negative throughout the length of the beam, thereby stressing the upper beam fibers in tension and the lower ones in compression. The reaction forces are obviously large since the external forces cannot be resisted by two supports as for simple beams.

$$-M_s = wL(L/2) = wL^2/2 = 4M, \qquad V_s = wL = W \qquad (2.28)$$

The shear is zero at the free end and maximum at the support. Notice that the support moment is four times larger in magnitude than the maximum moment M for a simple

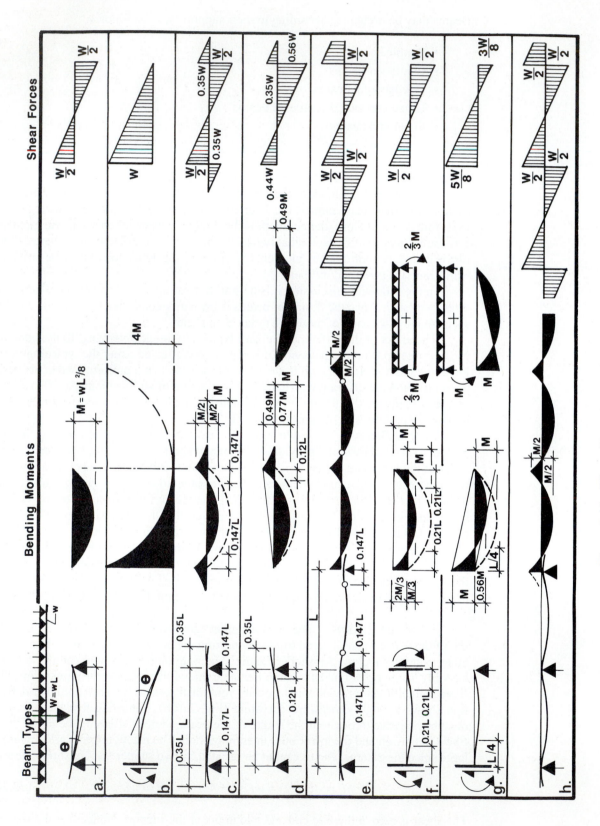

Figure 2.10 Effect of boundary conditions on beam behavior.

beam of equal span. It can also be shown that the maximum deflection is about 10 times larger.

Overhanging Beams. Usually, cantilever beams are natural extensions of beams; in other words, they are formed by adding to the simple beam a cantilever at one end or at both ends, which has a beneficial effect since the cantilever deflection counteracts the field deflection, or the cantilever loads tend to lift up the beam loads—the beam is said to be in double curvature (Fig. 2.10c); hence it has positive and negative moments. It is obvious that at the point of contraflexure or the *inflection point,* where the moment changes signs, the moment must be zero.

For demonstration purposes, a symmetrical overhanging beam with double cantilevers of $0.35L$ span has been chosen. The negative cantilever moments at each support are equal to

$$-M_s = w(0.35L)\,0.35L/2 \cong wL^2/16 = M/2 \tag{2.29}$$

These moments must decrease in a parabolic shape, in response to the uniform load, to a maximum value at midspan because of symmetry of beam geometry and load arrangement. We can visualize the moment diagram for the simple beam to be lifted up to the top of the support moments. Therefore, the maximum field moment, M_F must be equal to the simple beam moment, $M,$ reduced by the support moment $M_s.$

$$+M_F = M - M_S = \frac{wL^2}{8} - \frac{wL^2}{16} = \frac{wL^2}{16} = \frac{M}{2} \tag{2.30}$$

This value can be easily derived by simple statics. Notice that the lengths of the cantilevers were chosen so that the maximum field moment is equal to the support moments. The location where the moments are zero, as measured from the supports, is

$$wa^2/8 = wL^2/16 \quad \text{or} \quad a = 0.707L$$

or

$$x = (L - a)/2 = (L - 0.707L)/2 = 0.147L$$

$$\tag{2.31}$$

For further discussion of the effect of cantilevering, refer to Fig. 5.7.

Hinge-connected Cantilever Beams. Overhanging beams may be added to form beam chains, that is, continuous hinged cantilever beams, also called *Gerber beams* after the German bridge engineer, Heinrich Gerber, who pioneered the hinged cantilever beam system in 1866 in response to settlement problems. The placement of the hinges (i.e., zero moment points) in hinge-connected cantilever beams becomes most important. We should strive to have the negative support moments equal in magnitude to the positive field moments, as shown in Fig. 2.10e for typical interior span new conditions.

The beam types that have been presented so far are statically determinate, so the reaction forces can be obtained from simple statics. This, however, is not the case anymore for simple beams fixed at one or both ends and for continuous beams; these beams are statically indeterminate and therefore require additional equations for finding the reaction forces. The magnitude of reaction forces for statically indeterminate beams can be obtained from handbooks on building construction (e.g., *Manual of Steel Construction)*; some typical beam cases are identified in Table A.15.

Beams Fixed at One or Both Ends. A fixed-end beam behaves somewhat similarly to a simple beam with equal double cantilevers. Whereas the cantilevers provide some rotational restraint at the supports (which may be visualized as partial fixity), fixed ends do not allow any rotation by keeping the beam ends horizontal (i.e., the slope of the curvature is zero). The reaction moments necessary to resist any support rotation can be shown to be equal to two-thirds of that for the simply supported beam.

$$-M_s = wL^2/12 = 2M/3 \tag{2.32}$$

As for the overhanging beam, the moments decrease in a parabolic fashion toward midspan; in other words, the moment diagram for the simple beam is moved vertically upward to the support moments. We can also visualize the fixed beam to consist of a simple beam with the uniform load to which is added another simple beam carrying the support moments (Fig. 2.10f). Hence, the maximum field moment is equal to

$$+M_F = \frac{wL^2}{8} - \frac{wL^2}{12} = \frac{wL^2}{24} = \frac{M}{3} \tag{2.33}$$

It can be shown that the support moment for a beam fixed at one end only, often called a *propped cantilever beam* (Fig. 2.10g), is equal to

$$-M_s = wL^2/8 = M \tag{2.34}$$

Since it is difficult to fix a beam, often short beam spans are introduced at the ends (i.e., three-span continuous beam with large interior span) to provide partial fixity to the interior supports.

Continuous Beams   A continuous beam has more than two supports (two span, three span, etc.). Typical interior beams of continuous multispan beams of equal spans under uniform load action (Fig. 2.10h) can be visualized as an addition of fixed-end beams, hence having negative moments over supports and positive moments between supports. For further discussion, refer also to Fig. 2.27.

Load Types and Load Arrangements

Beam loads can be arranged symmetrically and asymmetrically. Remember, for symmetrical beams with symmetrical loading, the reactions can be determined directly—each reaction carries one-half of the total beam load. Beam loads can consist of concentrated loads, line loads, and any combination of the two. Line loads usually are uniformly or triangularly distributed; occasionally they are of curvilinear shape. The various types of loads, acting on a simple beam, are investigated in Fig. 2.11 for symmetrical conditions by keeping the total beam load W constant. We may conclude the following from Fig. 2.11 with respect to the shapes of the shear force and bending moment diagrams.

- The shear is constant between single loads and translates vertically at the loads.
  The shear due to a uniform load varies linearly (i.e, first-degree curve).
  The shear due to a triangular load varies parabolically (i.e., second-degree curve).
- The moment varies linearly between the single loads (i.e., first-degree curve).
  The moment due to a uniform load varies parabolically (i.e., second-degree curve).
  The moment due to a triangular load represents a cubic parabola (i.e., third-degree curve).

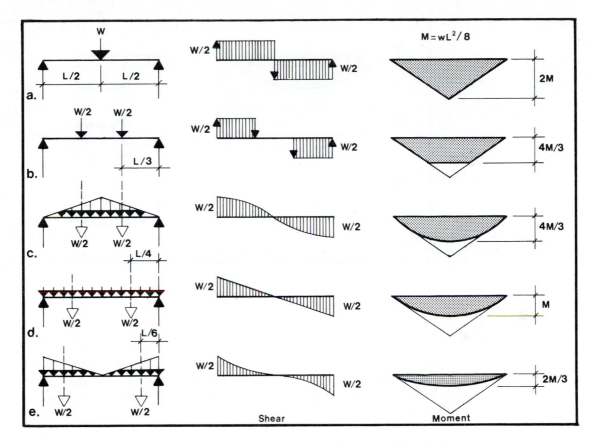

Figure 2.11 Load types and load distribution.

The relationship among load, shear force, and bending moment was discussed previously [Eqs.(2.23), (2.26), (2.27)]. One may conclude that the shear has a curvature one degree higher than the load, and the moment has a curvature one degree higher than the shear. For example, a uniform load has a zero-degree curve, the corresponding shear has a first-degree curve, and the moment a second-degree curve. The reader may also want to study in Fig. 2.11 the effect of load distribution. It is apparent that as the resultant load $W/2$ moves away from midspan the maximum moment decreases.

For general loading conditions, it is extremely helpful to derive the shape of the moment diagram by using the *funicular cable analogy* (Fig. 2.12). The single cable must adjust its suspended form to the respective transverse loads so that it can respond in tension. Under single loads, for example, it takes the shape of a string or funicular polygon, whereas under distributed loading, the polygon changes to a curve and, depending on the type of loading, takes familiar geometrical forms, such as a second- or third-degree parabola (see Fig. 9.5). For a simple cable, the cable sag at any point is directly proportional to the moment diagram of an equivalent beam on the horizontal projection carrying the same loads according to Eq. (9.2). In a rigid beam, the moments are resisted by bending stiffness, while a flexible cable uses its geometry to resist rotation in pure tension.

The various cases in Fig. 2.12 demonstrate how helpful it is to visualize the deflected shape of the cable (i.e., cable profile) as the shape of the moment diagram. Although the cases refer to simple beams, the effect of overhang, fixity, or continuity can easily be taken into account by lifting up the respective end of the moment diagram (see Fig. 2.10).

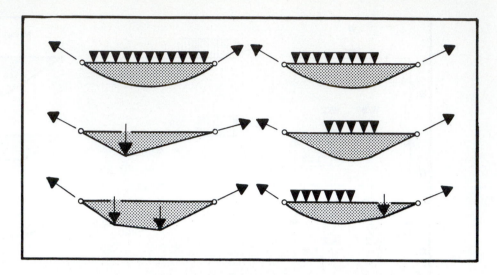

Figure 2.12 Funicular cable analogy.

A typical floor structure layout with a stair opening is investigated in Fig. 2.13 in order to study asymmetrical loading conditions in addition to setting up beam loading. The floor deck spans in the short direction perpendicular to the parallel beams that are 8 ft apart, as indicated by the arrows. Visualize the deck to act between the beams as parallel, 1-ft wide, simply supported beam panels or as joists spaced 1 ft apart that transfer one-half of the deck loads to the respective supporting beams. The contributing floor area each beam must support is shaded and identified in Fig. 2.13a; it is subdivided into parallel load strips that cause a uniform line load on the parallel beams. However, beam B7 is positioned on an angle and hence will have to carry a triangular tributary area. The loading diagrams with numerical values are given for the various beams as based on a hypothetical load of 100 psf including the beam weights; this load is also used for the stair area, but is assumed on the horizontal projection of the opening (see Fig. 5.26).

Beam B1 is supported by beam B2 framing the opening; its reaction causes single loads on B2 and G2. Beam B2, in turn, rests on beams B3 and B4; its reactions are equal to the single loads acting on these two beams. Since most of the beams are supported by the interior girders, their reactions cause single load action on the girders, as indicated for G1, where the beam reactions from the other side are assumed to be equal to the ones for B5; the girder weight is ignored.

For all practical purposes, one-way span action may be assumed for a rectangular slab having proportions of 1:2 or greater, even if it is also supported along the short sides. On the other hand, it is apparent that a square concrete slab with proportions of 1:1 and properly reinforced in both directions will span two ways and carry an equal load to each of the four supporting beams. In general, for a rectangular, two-way structure, the loads may be considered to be distributed in two perpendicular directions to the supporting beams, according to tributary areas formed by the intersection of 45° lines extending from the columns as shown in Fig. 2.13b.

In conclusion, the shear and moment diagrams for an asymmetrical beam with overhang are determined in Example 2.12 as based on the principles just introduced.

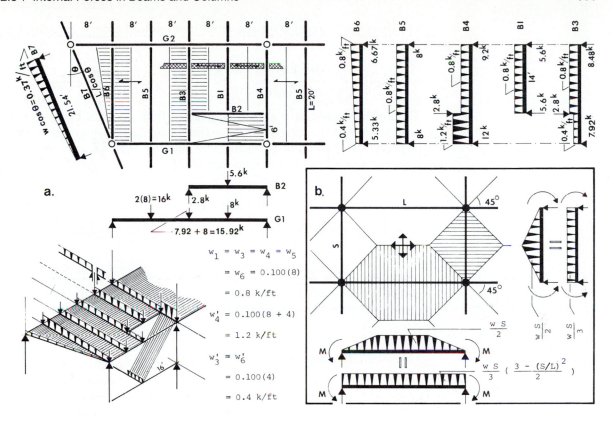

Figure 2.13 Typical floor framing.

EXAMPLE 2.12

The shear and moment diagrams are determined for the beam in Fig. 2.14. The reactions are found first. Taking moments about the unknown reaction R_B gives the magnitude of the other reaction R_A.

$$\Sigma M_B = 0 = R_A(20) - 1(12)14 - 6(4) + 4(5) \quad \text{or} \quad R_A = 8.6 \text{ k}.$$

Taking moments about reaction R_A yields

$$\Sigma M_A = 0 = 1(12)6 + 6(16) - R_B(20) + 4(25) \quad \text{or} \quad R_B = 13.4 \text{ k}$$

Checking if the vertical forces are in static equilibrium,

$$\Sigma F_y = 0 = 8.6 + 13.4 - 1(12) - 6 - 4 \quad \text{OK}$$

The horizontal reaction component at A is not activated since no horizontal force acts on the beam.

$$\Sigma F_x = 0 = R_{Ah}$$

The shear and moment values are determined as based on the free bodies in Fig. 2.14. The shear just to right of reaction R_A must be equal to the reaction; the moment at the pinned support obviously is zero.

$$V_A = R_A = 8.6 \text{ k}, \qquad M_A = 0.$$

The shear and moment at the end of the uniform load according to free body (b) is

$$\Sigma F_y = 0 = 8.6 - 1(12) - V_{12} \quad \text{or} \quad V_{12} = -3.4 \text{ k}$$

$$\Sigma M = 0 = 8.6(12) - 1(12)6 - M_{12} \quad \text{or} \quad M_{12} = 31.2 \text{ ft-k}$$

These values could have also been obtained from the right beam section [free body (c)]. Since the shear along the uniform load range changes from positive to negative, the location of the zero shear must be found because the maximum field moment occurs at that point [see also Eq. (2.26)]. From free body (d), the shear is zero at

$$\Sigma F_y = 0 = 8.6 - 1(x) \quad \text{or} \quad x = 8.6 \text{ ft}.$$

Now the maximum field moment can be obtained by taking moments about the reaction.

$$\Sigma M = 0 = 1(8.6)8.6/2 - M_{max} \quad \text{or} \quad M_{max} = 36.98 \text{ ft-k}.$$

Notice that the maximum moment is also equal to the triangular area of the shear diagram to the left of the moment [see Eq.(2.27)].

$$M_{max} = 8.6(8.6/2) = 36.98 \text{ ft-k}$$

The shear just to the right of reaction R_B [free body (e)], is

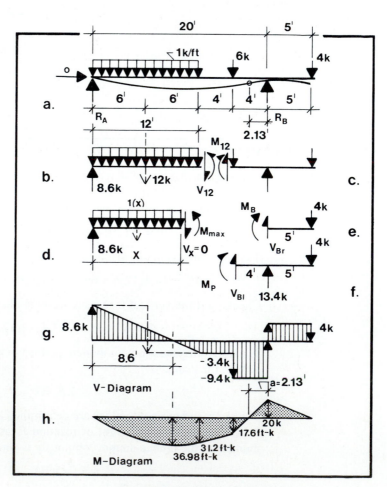

Figure 2.14 Example 2.12.

$$\Sigma F_y = 0 = 4 - V_{Br} \quad \text{or} \quad V_{Br} = 4 \text{ k}$$

The support moment M_B, according to the same free body, is

$$\Sigma M = 0 = 4(5) + M_B \quad \text{or} \quad M_B = -20 \text{ ft-k.}$$

The shear force just to the left of reaction R_B [free body (f)] is

$$\Sigma F_y = 0 = 4 - 13.4 - V_{Bl} \quad \text{or} \quad V_{Bl} = -9.4 \text{ k.}$$

The moment at the 6-k load, according to free body (f), is

$$\Sigma M = 0 = 4(9) - 13.4(4) + M_P \quad \text{or} \quad M_P = 17.6 \text{ ft-k}$$

The location where the moment is zero (inflection point of beam deflection) can be found from free body (f) by replacing the distance of 4 ft by the unknown distance a.

$$\Sigma M = 0 = 4(5 + a) - 13.4(a) \quad \text{or} \quad a = 2.13 \text{ ft}$$

The nature of beam behavior is discussed further later in this chapter (see Fig. 2.27).

2.4 PROPERTIES OF SECTIONS

Important geometrical properties of areas such as the location of the centroidal axes, the first and second moments of areas, as well as the section modulus and the radius of gyration, are introduced. They all represent significant characteristics necessary, for example, for the determination of stresses in a subsequent section. In this context, the area usually refers to the cross-sectional area of a structural member (of homogeneous material) cut perpendicular to its longitudinal axis.

Centroids

There are centroids of lines (straight, folded, curved), areas (simple, composite), and bodies. The centroids represent points of balance by treating the actually weightless lines and areas as thin sheets of constant thickness (Fig. 2.15).

Since a body (beam, slab, etc.) has mass, its centroid is equal to its *center of gravity* through which the weight of the body (i.e., resultant force) acts and is discussed with respect to Eq. (2.18): the location of the resultant force for a parallel force system. Although an area does not have mass, hence no center of gravity, it is still helpful for purposes of visualization to perceive an area A as equivalent to parallel forces ΣF and therefore its centroid as a center of gravity. When a body (e.g., plate) decreases in thickness until it becomes a plane with no mass so that area can be substituted for force ($F = A$ and $R = \bar{A} = \Sigma A$), then the location of the centroidal orthogonal axes can be obtained from Eq. (2.18) as

$$A\bar{x} = \bar{x}\Sigma A = \Sigma Ax \quad \text{or} \quad \bar{x} = \Sigma Ax / \Sigma A$$

$$A\bar{y} = \bar{y}\Sigma A = \Sigma Ay \quad \text{or} \quad \bar{y} = \Sigma Ay / \Sigma A \tag{2.35}$$

Hence, according to Varignon's theorem, *the moment of the whole area must be equal to the sum of the moments of its component areas about the same axis.* The intersection of the centroidal axes of an area represents the centroid of the area.

The *first moment of area* is an essential geometrical property of areas. It is not only directly related to the centroid of an area but also to shearing stresses in beams, as is discussed later. It is defined in more general terms, by using the process of integration, as

$$Q_x = \int y\,dA, \quad Q_y = \int x\,dA \qquad (2.36)$$

The *degree of symmetry* of areas is an important property of geometry (see also Fig. 2.15).

- For doubly symmetrical areas (|, I, H, □, O, etc.), the centroidal axes are the *axes of symmetry,* which can be determined by inspection. In other words, the location of the centroid is known.
- For areas with one axis of symmetry (T, U, A, E, etc.) the centroid lies on this axis; however, the centroidal axis in the perpendicular direction must be determined according to one of the expressions of Eq. (2.35).
- For unsymmetrical areas lacking symmetry, the position of the centroid is not known. Therefore, the location of the centroidal axes must be determined according to Eq. (2.35).

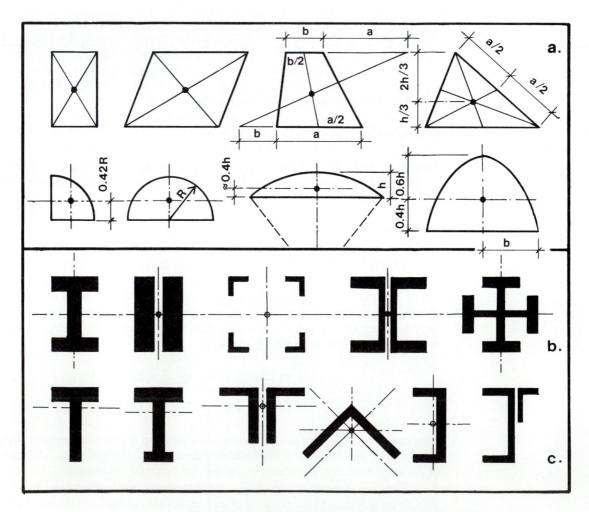

Figure 2.15 Centroids of simple and composite areas.

EXAMPLE 2.13

Determine the location of the centroidal axes for a triangle (Fig. 2.16).

$$A\bar{y} = \int_0^h y\,dA = \int_0^h y\,(x\,dy), \quad \text{where } x = y\,(b/h)$$

(2.37a)

$$\frac{bh}{2}\,(\bar{y}) = \frac{b}{h}\int_0^h y^2\,dy = \frac{b}{h}\left.\frac{y^3}{3}\right|_0^h = \frac{bh^2}{3} \quad \text{or } \bar{y} = 2h/3$$

We may conclude that the centroidal x_0 axis in Fig. 2.16 is located at

$$\bar{x} = b - (2b/3) = b/3 \tag{2.37b}$$

The location of centroids for basic areas can be obtained from handbooks on building construction (e.g., *Manual of Steel Construction*). The geometrical properties for some common, simple as well as more complex areas (such as curved ones) are identified in Fig. 2.15a and Table A.14.

Locating the centroid of a composite area, such as the ones shown in Fig. 2.15c, can be achieved by breaking the whole area up into simple ones whose general geometrical properties are known and then proceeding with the moment of area principle, as demonstrated in Example 2.14.

EXAMPLE 2.14

The centroid of the cross-sectional area for a wood T-beam, composed of two 3- × 6-in. members of the same material quality, is determined. As shown in Fig. 2.17a, the T-section can be divided either by treating web and flange as separate, by letting the web continue to the top face with two flange pieces attached to each side, or by assuming a rectangular block from which the nonexisting side webs must be subtracted.

First, the T-section is considered as consisting of the web and flange parts (Fig. 2.17b). The wood members each have a cross-sectional area according to Table A.9 of

$$A_1 = A_2 = 2.5 \times 5.5 = 13.75 \text{ in.}^2$$

The total cross-sectional area is

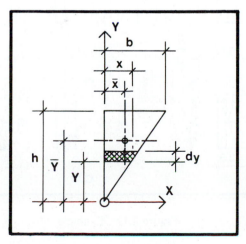

Figure 2.16 Determining the centroid for a triangle.

$$\Sigma A = A_1 + A_2 = 2\,(13.75) = 27.5 \text{ in.}^2$$

In this case, only the location of the horizontal centroidal axis has to be determined since the section is symmetrical about the vertical axis.

Any convenient x axis can be selected to take moments with the areas about; in this case, the axis of rotation is taken about the top face of the member. It is helpful to visualize the areas as horizontal forces acting at their respective centroids, thus representing a parallel force system (Fig. 2.17b). Hence, the location of the centroidal x_0 axis according to Eq. (2.35) is

$$\bar{y}\,\Sigma A = \Sigma Ay = A_1 y_1 + A_2 y_2$$

$$\bar{y}\,(27.5) = 13.75(1.25) + 13.75(5.25) \quad \text{or} \quad \bar{y} = 3.25 \text{ in.}$$

Rather than using an additive approach, we can also use a subtractive approach. For example, if the cross-sectional area of the member is treated as a rectangular section with a cross-sectional area of $A_1 = 5.5 \times 8 = 44 \text{ in.}^2$, then the nonexisting areas must be treated as negative; that is, $A_2 = -2(5.5 \times 1.5) = -16.5 \text{ in.}^2$. Hence, with respect to the top face, the location of the centroidal axis (Fig. 2.17d) is

$$\bar{y}\,\Sigma A = \Sigma Ay = A_1 y_1 + A_2 y_2$$

$$\bar{y}\,(27.5) = 44(4) - 16.5(5.25) \quad \text{or} \quad \bar{y} = 3.25 \text{ in.}$$

The Second Moment of Area or Moment of Inertia

The *second moment of area* is another important geometrical property of areas. It is called *moment of inertia* when it refers to the specific application of bending stresses and member deflections, as is discussed later in this chapter. The second moment, or the moment of inertia, of a finite area about a rectangular set of x and y axes, respectively, is defined as

$$I_x = \int y^2\, dA, \qquad I_y = \int x^2\, dA \tag{2.38}$$

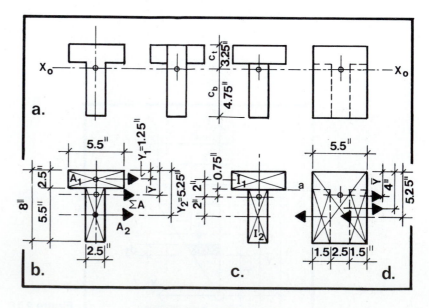

Figure 2.17   Example 2.14.

The *polar moment of inertia, $I_p = J_0$*, is defined as the second moment of area about the *z*-axis perpendicular to the *xy*-plane (Fig. 2.18d).

$$I_p = I_z = J_0 = \int r^2 dA = I_x + I_y \tag{2.39}$$

The *product of inertia* of a finite area with respect to a set of *xy* axes in the plane of the area is defined as

$$I_{xy} = \int xy \, dA \tag{2.40}$$

Notice that the product of inertia may be positive, negative, or zero.

Another property of the cross-sectional area A is the *radius of gyration, r,* which is directly related to its moment of inertia, *I*. It occurs in response to buckling of columns and is defined as

$$r = \sqrt{I/A} \quad \text{or} \quad I = Ar^2 \tag{2.41}$$

The radius of gyration can be visualized as the perpendicular distance between an imaginary long, thin strip of area and a parallel axis [see Eq. (2.50) and Fig. 2.18f].

Still another property of the cross-sectional area, related to the moment of inertia and bending stresses, is the *section modulus, S,* defined as

$$S = I/c \tag{2.42}$$

where c = distance from neutral axis of section to the extreme fiber as discussed later in this chapter.

EXAMPLE 2.15

Some geometrical properties of a rectangular cross-sectional area are investigated next (Fig. 2.18).

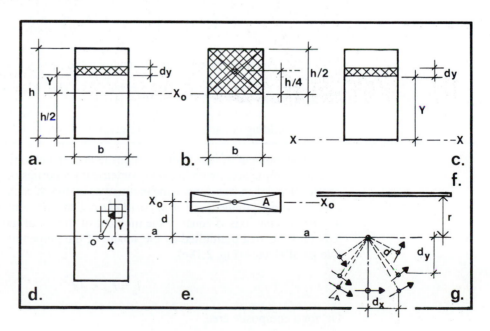

Figure 2.18   Second moment of area.

(a)   The first moment of area about the centroidal x_0 axis is determined, according to Eq. (2.36) (Fig. 2.18a and b).

$$Q_{x0} = \int y\,dA = \int_0^{h/2} y\,(b\,dy) = b\int_0^{h/2} y\,dy = b\left.\frac{y^2}{2}\right|_0^{h/2} = \frac{bh^2}{8}$$

or   (2.43)

$$Q_{x0} = \frac{bh}{2}\left(\frac{h}{4}\right) = \frac{bh^2}{8}$$

(b)   The moment of inertia about its centroidal x_0 axis is determined according to Eq. (2.38) in Fig. 2.18a.

$$I_{x0} = \int y^2\,dA = \int_{-h/2}^{h/2} y^2 b\,dy = b\int_{-h/2}^{h/2} y^2\,dy$$

(2.44)

$$= b\left.\frac{y^3}{3}\right|_{-h/2}^{h/2} = \frac{b}{3}\left[\frac{h^3}{8} - \left(-\frac{h^3}{8}\right)\right] = \frac{bh^3}{12}$$

Therefore, the moment of inertia about its centroidal y_0 axis must be

$$I_{y0} = hb^3/12 \qquad (2.45)$$

The corresponding section modulus about the x_0 axis [Eq.(2.42)] for $c = h/2$ is

$$S_x = I_x/c = (bh^3/12)/h/2 = bh^2/6 \qquad (2.46)$$

The radius of gyration with respect to the x_0 axis [Eq.(2.41)] is

$$r_x = \sqrt{\frac{I_x}{A}} = \sqrt{\frac{bh^3/12}{bh}} = 0.29h \qquad (2.47)$$

(c)   The moment of inertia of the rectangular section about its base (Fig. 2.18c) is

$$I_x = \int y^2\,dA = \int_0^h y^2\,(b\,dy) = b\int_0^h y^2\,dy = b\left.\frac{y^3}{3}\right|_0^h = \frac{bh^3}{3} \qquad (2.48)$$

(d)   The polar moment of inertia about the z_0 axis [Eq.(2.39)] is

$$I_{z0} = J_0 = I_{x0} + I_{y0} = (bh^3/12) + (hb^3/12) = A(h^2 + b^2)/12 \qquad (2.49)$$

The moments of inertia of common cross-sectional areas are provided in books on building construction and are given in Table A.14 for some simple cases.

The moment of inertia of a composite area is determined by dividing the whole area into simple areas with known geometrical properties (as was done for finding the centroid) and then transferring the moment of inertia of each area to a different parallel axis. According to the *theorem of parallel axes,* the moment of inertia of an area about any orthogonal axis is equal to the moment of inertia of the area about its centroidal axis plus the area multiplied by the square of the perpendicular distance between the two parallel axes (Fig. 2.18e).

$$I_a = I_{x0} + Ad^2 \qquad (2.50a)$$

Or, for a composite area,

$$I_a = \Sigma I_{x0} + \Sigma A d^2 \tag{2.50b}$$

The moment of inertia of fastener groups, pile groups, columns seen in building plan, and similar situations can be derived from Eq. (2.50) assuming that all the points have the same areas. For the condition where the areas of the points are small so that their moments of inertia are negligible ($I_{x0} = 0$), Eq. (2.50b) reduces to

$$I_x = A\Sigma(d_y)^2, \qquad I_y = A\Sigma(d_x)^2 \tag{2.50c}$$

where the terms are defined in Fig. 2.18g.

EXAMPLE 2.16

The moments of inertia about the centroidal axes of the T-section in Example 2.14 and Fig. 2.17 are determined.

(a)   First, the moment of inertia of the section about the x_0 axis is found by treating the web and flange cross-sectional areas as separate. The moments of inertia of the areas about their centroidal axes, according to Eq. (2.44), are

$$(I_{x0})_1 = bh^3/12 = 5.5(2.5)^3/12 = 7.16 \text{ in.}^3$$

$$(I_{x0})_2 = bh^3/12 = 2.5(5.5)^3/12 = 34.66 \text{ in.}^3$$

Hence, the moment of inertia of the composite area about its centroidal x_0 axis (Fig. 2.17c), according to the parallel axis theorem [Eq.(2.50b)], is

$$I_a = \Sigma I_{x0} + \Sigma A d^2$$

$$I_c = [7.16 + 13.75(2)^2] + [34.66 + 13.75(2)^2] = 151.82 \text{ in.}^4$$

Another approach may be used by first determining the moment of inertia of the composite area about axis a–a (interface of the two areas) using Eq. (2.48.)

$$I_a = \frac{5.5\,(2.5)^3}{3} + \frac{2.5\,(5.5)^3}{3} = 28.65 + 138.64 = 167.29 \text{ in.}^4$$

Now the moment of inertia about the a axis can be transferred to the centroidal axis of the composite section (Fig. 2.17c).

$$I_a = I_{x0} + A d^2$$

$$167.29 = I_c + 27.5(0.75)^2 \quad \text{or} \quad I_c = 151.82 \text{ in.}^4$$

The corresponding section modulus [Eq. (2.42)] with respect to the top face, for $c_t = 3.25$ in., is

$$S_{xt} = I_x/c_t = 151.82/3.25 = 46.71 \text{ in.}^3$$

And the section modulus with respect to the bottom face, for $c_b = 4.75$ in., is

$$S_{xb} = I_x/c_b = 151.82/4.75 = 31.96 \text{ in.}^3$$

(b) The parallel axis theorem is not needed for finding the moment of inertia about the centroidal y_0 axis since the centroidal y_0 axes of each area coincide with that of the composite area ($d_x = 0$). Hence, according to Eq. (2.44),

$$I_{y0} = \frac{2.5\,(5.5)^3}{12} + \frac{5.5\,(2.5)^3}{12} = 34.66 + 7.16 = 41.82 \text{ in.}^4$$

2.5 GENERAL MATERIAL PROPERTIES

The materials used in building construction are either of *inorganic* nature (e.g., metal, ceramics, glass, concrete, stone) or *organic* nature (e.g., wood, plastics with the exception of silicone).

To define all the properties of building material is extremely difficult. They include *mechanical, chemical* (composition, corrosion, reactivity, etc.), and *physical* (acoustical, electrical, magnetic, optical, thermal, etc.), characteristics. The mechanical properties are of great interest to the structural designer since they relate to the response of the material to loading, as well as to its endurance and workability. Of more interest in this context are the properties of *strength, stiffness, ductility,* and fracture *toughness* (resistance to impact).

Not all materials exhibit the same response in every direction of loading; the materials that do have that feature are called *isotropic* (metals, plastic film, coated fabrics, rubber membranes, etc.). Materials that do not have identical properties in all directions are called *anisotropic* (masonry, wood, some plastics, woven or gridded fabrics, etc.). Two-way grids may be *orthotropic* with different properties in each direction; in other words, orthotropic plates are a special case of anisotropic materials. Most materials on a microscopic scale show different properties from point to point and hence are nonhomogeneous; however, on a macroscopic scale they can be considered *homogeneous. Composite* materials, such as reinforced concrete, straw-reinforced clay, plaster board, plywood, coated fabrics, and fiber-reinforced plastics (FRP), obviously do not have this homogeneous quality.

Mechanical Material Properties

In statics, it is assumed that structural members can be treated as perfectly rigid bodies since the deformation of most materials is so small that it can be neglected. In reality, however, a member can only resist external forces by changing its shape and thereby activating internal forces in response to Newton's third law of *action equal to reaction* [Eq. (2.12)].

The most simple deformation is caused by axial force action. Visualize that a linear member is tested axially up to its failure in either tension or compression (assuming no prior failure due to instability). Its response to tension as expressed by the elongation ΔL is recorded with the corresponding load P. The resulting diagram is called the *stress–strain curve.* The *stress* is plotted along the ordinate and is equal to the axial force divided by the original cross-sectional area A of the member; in other words, stress is defined as load per unit area.

$$f = P/A \qquad\qquad (2.51)$$

Here, the ultimate *strength* of the member corresponds to the highest point on the curve (Fig. 2.19).

The *strain* is plotted along the abscissa. It is a measure of change in member length and is equal to the *axial deformation ΔL* divided by the original length L of the member and therefore has no units.

$$\varepsilon = \Delta L/L \qquad\qquad (2.52)$$

The maximum strains at the point of rupture vary extensively. The tensile strains of the metals in Fig. 2.19 range from a maximum for carbon steel to a minimum for cast iron. The elastic strains for typical construction materials vary from about 0.1% for stiff materials to about 800% for flexible materials, such as rubber.

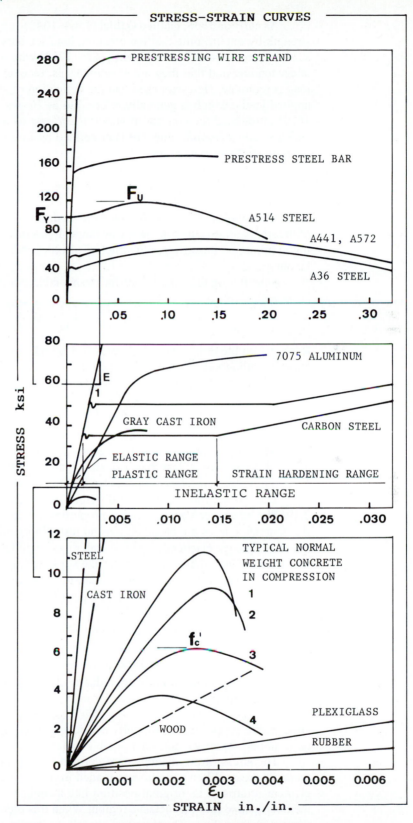

Figure 2.19 Stress–strain curves.

As early as 1638, Galileo Galilei (1564–1642) suggested the concept of breaking strength by determining that the breaking load for bars is proportional to their cross-sectional area. It was Robert Hooke (1635–1703) who found out that materials deform under loading and that they are *elastic,* that is, recover their original shape when the load is removed. He established that the deflection of structures is proportional to the applied loads, which is generally referred to as *Hooke's law.* Thomas Young (1773–1829) introduced the concept of *material stiffness* as a constant, which is sometimes called *Young's modulus* and was later defined by Louis Marie Navier (1785–1836) as the ratio of stress over strain.

$$E = f/\varepsilon \qquad (2.53)$$

Young's modulus, E, is usually called the *modulus of elasticity;* it represents the slope of the stress–strain curve in the elastic range. It is a measure of material stiffness, keeping in mind that the stiffness of structure also depends on its size, shape, and member layout.

Substituting the expressions for strain and stress into Eq. (2.53) yields

$$E = \frac{f}{\varepsilon} = \frac{f(L)}{\Delta L} = \frac{PL}{A \Delta L}$$

Rearranging the terms gives the expression for the axial deformation of a linear member of length L and cross-sectional area A:

$$\Delta L = \frac{PL}{AE} = \frac{fL}{E} = \varepsilon L \qquad (2.54)$$

The various stress–strain curves in Fig. 2.19 show distinctly different slopes in the elastic range. A material with a steeper slope or larger modulus of elasticity is stiff and will deform much less than a flexible material with a less steep slope and a small E, assuming the same force action, cross-sectional area, and member length.

The stress–strain curves for different materials vary widely, as shown in Fig. 2.19. The basic mechanical material properties are derived from these curves. All but the concrete curves are based on tension tests. Inorganic materials and many organic materials behave elastically, at least under moderate loads, but do not necessarily follow Hooke's law; that is, they do not show a linear stress–strain relationship where E is constant, such as the nonlinear viscoelastic properties of silk.

There seem to be three different types of curves:

- A continuous curve, as for cast iron and concrete (e.g., nonlinearly elastic materials)
- A straight line, as for Plexiglass
- A straight line abruptly or gradually changing into a curve, as for steel and aluminum

Other curve types, such as the fully elastic J-curves (e.g., for animal tissue), S-curves (e.g, rubber), or curves for other organic materials are not shown.

The stress–strain diagrams are defined by two distinct regions: the *elastic* and *inelastic* areas. When a member is loaded in the elastic range and then unloaded, the element returns to its original position like a rubber band. This is not the case for a member loaded into the inelastic region. When this member returns, it will have a permanent deformation. The range of the elastic strain is small in comparison to the inelastic strain. Materials that pass through large deformations before failing, such as steel and aluminum, are called *ductile,* while materials with relatively little deformation at failure, such as cast iron, concrete, and glass, are called *brittle.* For example, for a typical steel ($F_y = 50$ ksi), the plastic deformation is about 10 times that of the elastic, and the deformation at failure is nearly 100 times that of the elastic. In contrast, ordinary concrete will fail at about twice the elastic deformation of steel; in other words, its contraction at failure is about 2% that of steel. Brittle materials do not yield; they fail suddenly without warning. In brittle material at points of overstress, cracks form and propagate and then suddenly fail. In ductile materials, at points of stress concentration, the material yields locally and forces the surrounding material to take over, that is, the stresses are redistributed. Ductile material can absorb much more punishment and gives adequate warning of impending collapse by showing extensive plastic deformations. Some ductile materials become brittle at very low temperatures, such as steel at −30°F.

The majority of materials show stress–strain diagrams for compression and shear resembling those for tension, particularly for the elastic range. The moduli of elasticity in tension and compression are equal for many materials.

The material properties are derived from a simple test specimen, thereby neglecting that the properties vary with time, temperature, and possibly methods of testing, as well as the fact that real members are subject to two- and three-dimensional stresses. Only short-time loading is considered, but materials like concrete, masonry, and wood exhibit *time-dependent* features, such as shrinkage and creep.

For the measure of stiffness, E, of nonlinearly elastic brittle materials (e.g., concrete and cast iron), respective codes should be consulted. According to the ACI code, the modulus of elasticity for normal-weight concrete, E_c, and a concrete strength of f'_c (in psi) is

$$E_c = 57,000 \sqrt{f'_c} \quad \text{(psi)} \tag{3.1}$$

For example, for a typical 4000-psi concrete, the modulus of elasticity is 3605 ksi. While E for steels is constant and independent of strength, for concrete it is a variable, with the heavier and stronger concrete being stiffer. For lightweight concrete, the elastic modulus is generally 20% to 50% lower than for normal-weight concrete. Also, for wood, the modulus of elasticity varies according to grade and species.

The transition from the elastic to the inelastic range is not clearly defined by some materials (e.g., concrete, cast iron, stainless steel); in other words, the gradual yielding nature of some stress–strain curves is distinct from the ones of the sharp yielding type. For these materials the elastic limit, that is, the location where permanent deformations occur when the load is taken away, must be determined. Notice that some brittle materials, such as glass, have a nearly linear stress–strain curve up to fracture! For many materials the elastic limit is equal to the proportional limit and, for all practical purposes, may be considered equal to the location of the yield point. Beyond this point, stress and strain are no longer proportional to each other; the strain increases much more rapidly than the stress. Many steels demonstrate a perfect distinction between the different behavioral ranges of the stress–strain curve. The straight-line relationship of the elastic range clearly changes to the horizontal plateau of the *plastic range*, where only the strain increases without an increase of stress, before it changes to the *strain-hardening range*. Steel strands, prestressed wire, and aluminum do not exhibit a sharply defined yield point, as is characteristic of mild steels, and are also much less ductile.

Strains are induced not only in the direction of the force action (ε_L) but also in the two mutually perpendicular directions. This effect was evaluated by S. D. Poisson (1781–1840) and is known as *Poisson's ratio:*

$$v = -\varepsilon_t / \varepsilon_L \tag{2.55}$$

For most metals, it is in the range from 0.25 to 0.35. For structural steel, $v = 0.3$ may be assumed and for aluminum alloy, 0.33. For concrete it varies from about 0.11 for high-strength concrete to 0.21 for low-strength concrete; often an average value of 0.16 is used. For concrete shells, Poisson's ratio may be assumed equal to zero.

In contrast to axial forces, which change the length of a structural element, shear forces tend to deform the shape of a body as shown in Fig. 2.20 for, say, a typical long wall. Visualize the wall to consist of thin layers that under translational force action slide with respect to each other, resulting in a maximum shear deflection of Δ_s at the top. The average *shear strain,* γ, is defined by the angular change of the originally perpendicular horizontal and vertical faces as measured in radians:

$$\tan \gamma \cong \gamma = \Delta_s / L \tag{2.56}$$

According to Hooke's law, the *shear modulus G* measures the *elastic shear stiffness.* It represents the slope of the shear stress (f_v)–strain diagram in the elastic range:

$$G = \frac{f_v}{\gamma} = \frac{VL}{A\Delta_s} \tag{2.57}$$

The lateral shear deformation can now be given in terms of shear V acting along the shear area A, the shear modulus of the material, G, and the height of the element, L.

$$\Delta_S = \frac{VL}{AG} \tag{2.58}$$

The shear modulus can be expressed for isotropic materials in terms of the modulus of elasticity and Poisson's ratio:

$$G = \frac{E}{2(1+v)} \tag{2.59}$$

Often the shearing modulus for concrete and masonry is assumed as 40% of its modulus of elasticity: $G = 0.4E$.

Building Materials

Construction materials are classified according to their nature or according to their function. Their *function* is identified by their usage and their purpose (e.g., structural versus nonstructural). The *nature* of materials is organized as metallic (cast iron, steel, aluminum alloys, bronze, etc.), nonmetallic (stones, gravel, asbestos, glass, etc.), ceramic products (concrete, cement, brick, tile, etc.), chemical products (gypsum, paints, lubricants, etc.), polymers (thermoplastic and thermosetting materials, etc.),

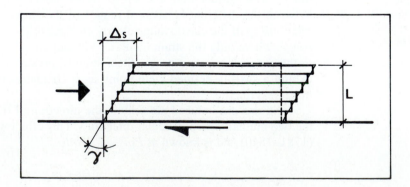

Figure 2.20   Shear deformation.

wood, natural fibers (cotton, flax, silk, etc.,), cellulosic products (paper, cardboard, felt, etc.), and other miscellaneous materials.

These materials, in turn, are selected according to their *suitability*. Some typical performance criteria are strength, flexibility, ductility, weight, durability, dimensional stability, workability, machinability, weldability, compatibility, availability, cost, toughness, hardness (i.e., resistance to penetration), resistance to fire or heat, sound resistance, weatherproofing, ease of jointing and handling, response to biological damage and other chemical attack, ease of maintenance, and many other criteria.

Besides the common *rigid* materials, textile fabrics have become most popular in building construction—they are *flexible,* collapsible, and easily transportable. Textile membranes not only occur as single-ply roofing, awnings, canopies, tents, and pneumatic structures, but are also used for storage facilities (containers, silos, etc.), transportation, and environmental protection. Most fabrics are produced from natural, synthetic, and mineral fibers such as cotton, aramid, polyamid (nylon), polyester, and fiber glass. They may be coated, for example, with polyvinyl chloride (PVC), polyvinyl fluoride (PVF, e.g., Tedlar), PTFE (e.g., Teflon), acrylic, silicone, or urethane rubber.

Various materials and their mechanical properties are listed in Table 2.4. The common materials (steel, wood, reinforced concrete, and coated fabrics) used for building structures are discussed in Sections 3.1 and 9.1; only aluminum alloys and plastics are briefly introduced here.

Aluminum as a structural material in building construction is used as an alloy, for example, for cladding systems, formwork for concrete, stud walls, frames, and space frames and geodesic domes. The most widely used alloy for structures is aluminum 6061-T6, and aluminum 6063-T5 is used for the extruded members of curtain walls. Care must be taken that aluminum is separated from dissimilar materials by painting or insulation. Electrostatic corrosion is caused by moisture when aluminum is in permanent contact with certain other metals, and chemical attack may occur when it is in contact with porous materials like wood and fiberboard that may absorb water or when it is in contact with wet alkaline materials such as mortar, plaster, and concrete.

Numerous different types of synthetic materials are marketed. They have been applied in the building construction field for nonstructural purposes as vinyl framed windows and siding, polybuthylene pipes, molded fiber glass, reinforced decorative polyester products (e.g., cornices) that are hardly distinguishable from plaster or wood, foam insulation, and so on. For structural purposes, they have been used as ropes (e.g., Kevlar 49-rope), reinforcing bars, fiber-glass-reinforced polyester frames using box sections, prestressed translucent Teflon-coated fiber-glass fabric tensile roofs, folded roofs with foam-core panels using polystyrene or urethane, and others.

Plastics derive their name from the fact that at some stage they are plastic and can be formed into their desired shape; most are derived from petroleum and various chemicals. They can be transparent and can be formed into complex shapes. Possible disadvantages that must be controlled when ordinary plastics are used as structural materials are their low values with respect to the modulus of elasticity, surface hardness, durability, and fire resistance and toxicity. The mechanical properties of plastics alter over time and with changes in temperature, resulting in loss of strength and stiffness. For example, the flexibility of a plastic roof panel can be overcome by shaping it into a curved or folded spatial structure where geometry provides most of the stiffness.

The two distinct classes of plastics are *thermoplastic* materials (acrylic, nylon, PVC, polyethylene, etc.) and *thermosetting* plastics (polyester, epoxy, polyurethane, silicone, phenolic, etc.). Whereas thermoplastics can be reheated and reused without significant change in properties, thermosetting plastics are in the plastic stage only once and then harden irreversibly. Thermosetting engineering plastics are generally reinforced with fibers mostly of inorganic material to improve the stiffness and strength and other mechanical properties. In reinforced composites, the plastic serves

TABLE 2.4
Approximate Average Mechanical Material Properties

Material	Unit Weight (pcf)	Ultimate Tensile Strength (ksi)	Approximate Allowable Stresses (ksi)		Modulus of Elasticity 10^3 (ksi)	Coefficient of Expansion 10^{-6} (in./in./°F)
			Tension	Compression		
Metals						
Carbon steel	490	58–80	22	22	29	6.5
Stainless steel 302	530	90	18	18	28	9.9
Cast iron	450	18–24	5–8	20–30	14–18.3	5.9
Bronze	510	60	12–30	12–30	15	10.1
Copper	555	32	8	8	16	9.3
Aluminum alloy						
6063-T5	171	27	9.5	9.5	10	12.8
6061-T6	171	45	15	14	10	12.8
Aircraft	171	70–80				
Lead	710	2.5				15.9
Titanium	282	80–145			15.5	5.4
Tungsten wire	1204	500			53	2.4
Reinforced concrete fibers						
Steel		50–250			29	
Stainless steel		300			23.2	
Glass (E)	156	500			10.4	
Synthetics						
Nylon	72	130			0.75	
Polypropylene	57	40–100			0.5–0.7	
Polyester	84	80–170			1.45–2.5	
Polyethylene		29–435			0.73–25	
Aramid (Kelvar 49)		525			17	
Acrylic		30–145			2.6	
Carbon (Type 2)	109	380			33.4	
Alumina whiskers		3000			62	
Cables						
Steel wire rope	490	200			20	
Steel structural strand	490	200			24	
Steel prestressing strand	490	270			28	
Kelvar 49-rope	90	400			19	
S-Glass rope	155	450			12.5	
Plastics						
Undirectional composites						
Glass/epoxy	112	160			5.7	
Graphite/epoxy	93	204			30.1	
Glass/polyester panel	100	15.5			1.25	17.5
Acrylic sheet	72	10			0.4	41
Glass-fiber reinforced cement panel	105	1.2–1.6			1.5	7
Ordinary concrete	145	0.3–0.5	0.1	2.0	3.2–5.5	5.5
Natural stone						
Granite	162	0.6–1.0		1.2	5.7–9.6	4.4
Limestone	144	0.3–0.7		0.8	3.0–5.4	4.2

TABLE 2.4 Continued

Material	Unit Weight (pcf)	Ultimate Tensile Strength (ksi)	Approximate Allowable Stresses (ksi)		Modulus of Elasticity 10^3 (ksi)	Coefficient of Expansion 10^{-6} (in./in./°F)
			Tension	Compression		
Masonry						
Brick	120		0.025	0.4	2.4	3.6
Granite				0.42		5.5
Limestone				0.35		4.3
Sandstone				0.28		6.5
Window glass (silicate)	156	4			7–13	5
Wood	35	±6	0.6	1.0	1.6	2.1
Soils						
Soft clay				0.02		
Loose gravel				0.06		
Sedimentary rock				0.2		
Bed rock				1.4		
Other materials						
Bone		20			4	
Cotton	93	42–125				
Diamond	220				170	0.67
Flax	93	100				
Gold lily		28			3.6	
Papyrus		29			1.9	
Leather	59	6				
Rubber	59				0.001	
Silk	84	45–83			14.5	
Spider's thread		35				
Wool	80	17–28			4.7	

as a matrix binder and continuous phase to protect the fibers (e.g., glass, asbestos, steel, Kevlar fibers) and to bond the much stronger fibers together. The most common of all reinforced plastics in the construction industry are the glass-fiber-reinforced polyester (GRP) composites; thermoplastics are often used without reinforcement (e.g., translucent acrylic domes, PVC films, pipes), For further discussion of other characteristics of plastics, refer to Sections 8.1 and 9.1.

Ordinary plastics are moderately strong but lack stiffness. To increase their structural properties, they may have to be reinforced with stronger and stiffer material, such as fibers to form a fiber-reinforced composite. The fibers, which are finer than human hair, come in various lengths, thicknesses, and geometries. They can be of *inorganic* nature (e.g., glass, steel, asbestos, carbon, boron) or *organic* nature (e.g., natural, such as cotton, jute and sisal, or synthetic, such as nylon, polyester, and polypropylene). The most common fibers in the building construction industry are glass fibers. *Fibers* are larger and weaker than whiskers; which are thin, needlelike crystals that can be grown from most substances. *Whiskers* may be up to 2 μm thick and possibly inches long; they are the strongest material (Table 2.4). In the late 1920s in England, A. A. Griffith showed that threadlike forms of material are by far stronger than the bulk in which the material is generally used; this is especially true for brittle materials. According to *Griffith's law,* for the fibers of brittle material the strength is inversely proportional to the diameter. Hence, the strength of the fibers increases as they get thinner. It makes sense that when the fiber material is used in bulk form, imperfections enter; *the strength of the whole is not equal to the strength of the sum of its parts.* Fibers cannot be structurally used by themselves; they may form cables or fabrics, or they may be embedded in a matrix material to form a composite. Glass-fiber reinforcement in plastics appears as chopped strand, chopped strandmat, roving, or cloths. In other words, the fibers may be placed in the plastics in a random manner or woven and possibly placed in several layers. Since the stiffness of fiber-reinforced plastics is still moderate, it is advantageous to use inherently stiff geometrical shapes like shells, folded and polyhedral structures, inflatable structures, and sandwich construction made up entirely of laminae or layers in which the stiffer material is placed along the outer faces.

Comparison of Materials

The primary mechanical and physical material characteristics, in this context, as related to the building structure are weight, strength, stiffness, and, to a certain degree, the effects due to temperature change. These properties are shown and compared for various materials in Table 2.4.

As the scale of a building enlarges, its weight increases faster than the live loads it is supporting. For instance, as a building gets higher, its weight grows with the volume (ft^3), while the live load only increases with the area (ft^2). It is thus obvious that material weight is an important consideration. Ideally, a minimum weight should support a maximum live load, ignoring economic factors and problems related to minimum-weight structures. There are occasions, however, when weight is a desired feature, as for reasons of stability.

As can be seen in Table 2.4 for typical structural building materials, the weights range from the very dense steel with 490 pcf to about 35 pcf for the weight of wood; steel is about 14 times heavier than wood, 5 times heavier than plastics, and roughly 3 times heavier than aluminum and concrete. Since wood is so light, large sections can be assembled and relatively easily handled. It is obvious that the comparison of the material weights must include the effect of strength and stiffness (as discussed later).

The elastic modulus determines how stiff or flexible a material is; it ranges from 170 ksi for diamond to 0.001 for rubber. Comparing the *stiffness* of the typical structural building materials, we find that steel is about 3 times stiffer than aluminum, nearly 10 times stiffer than ordinary concrete, and roughly 20 times stiffer than wood. It is apparent that organic materials are not as stiff as inorganic materials.

In long-span beam structures where the beam size is a function of the allowable deflection and where the superimposed loads are much larger than the dead load, steel may be a better choice, since aluminum, for instance, deforms about three times as much as a steel beam. Keep in mind, however, that long-span aluminum space frames may still be efficient since stiffness is less dependent on material but on the framework layout. In cases where the live load is small in comparison to the dead load, as for long-span roof structures, light wood and aluminum may offer certain advantages. The benefit of aluminum as structures for airplanes to carry a maximum of live loads is apparent. The comparison of member stiffness for different materials is based on the elastic modulus and short-term force action. However, certain materials like concrete, masonry, wood, and plastics deform in time, that is, *creep* under long-term loading.

The *tensile strength* for common building materials ranges from the high strength of steel and some composites to the low strength of wood. The allowable stress for steel is more than 30 times higher than for wood. The tensile strength of pure aluminum is relatively low, but the strength of some aluminum alloys is in the range of steels. It is apparent that materials like concrete, stone, and masonry are not tensile materials; they must be reinforced to resist tension. Notice that the tensile capacity of several synthetic fibers is quite compatible with that of conventional materials such as steel.

The allowable compressive stresses for common materials range roughly from a low of about 0.02 ksi for clay to 22 ksi for steel. In other words, as compared to clay, masonry is about 20 times stronger, wood is 50 times stronger, concrete about 100 times stronger, and steel about 1000 times stronger.

The *compressive strength* of columns is closely related to slenderness considerations and buckling criteria; that is, it is a function of the geometry and type of the structure. The relationship between tensile and compressive strength of materials is interesting. While steel is about equally strong in tension and compression, wood is 2 to 3 times stronger in tension, and concrete is 10 to 20 times stronger in compression. The compressive strength of brickwork is only about one-quarter to one-third that of brick; hence joints should be kept thin.

A comparison of the *coefficients of expansion* for the various materials shows that aluminum expands or contracts twice as much as steel and is 1.3 times that of bronze, while Plexiglass changes in length 20 times more than wood and about 6 times more than steel. This clearly indicates that care must be taken to allow for the large dimensional change of Plexiglass at its boundaries. For some reinforced plastics, the expansion approaches that of steel. Although the coefficient of thermal expansion of wood parallel to the grain is very low, as compared to other common materials, shrinkage and swelling must be taken into account.

To evaluate one aspect of structural efficiency, material weight must be related to its strength and stiffness. First, the length or rupture length L_u for the material is derived. The ultimate stress f_u of a suspended bar carrying its own maximum weight, $W_u = \gamma(V_u) = \gamma A L_u$, is

$$f_u = \frac{W_u}{A} = \frac{\gamma A L_u}{A} = \gamma L_u \text{ or } L_u = \frac{f_u}{\gamma} \quad \left(\frac{\text{lb ft}^3 \; 12^2 \text{ in.}^2}{\text{in.}^2 \; \text{lb ft}^2} \right)$$

(2.60)

$$L_u = f_u (12)^2/\gamma \text{ (ft)} \quad \text{or} \quad L_u = 0.0273 f_u/\gamma \quad \text{(miles)}$$

The rupture length is a constant. The *strength-to-weight ratio* f_u/γ is called the *specific strength*. Similarly, the *elastic modulus-to-weight ratio* E/γ is called the *specific elasticity*.

As the strength of a material increases and its weight decreases, its rupture length or specific strength increases. The specific strength for carbon steel is about 70% that of ordinary sawn lumber, whereas the ratio for aluminum alloy may be more than twice as much. Some structural plastics have outstanding strength-to-weight ratios. For example, certain synthetic cables have an ultimate length (while under short-term loading) about 7 to 10 times greater than steel strand. The rupture length for cotton fibers as used in tent construction is more than 4 times that of steel. Concrete has a low strength-to-weight ratio and is most effective when used in long-span roof structures in shell construction and prestressed concrete, resulting in a reduction of its high dead-to-live load ratio. High strength-to-weight structures are essential for mobile structures (e.g., car bodies, space vehicles, aircraft) where lightweight materials such as aluminum alloys, titanium, and reinforced plastics are essential. Keep in mind that the effect of high strength-to-weight ratios results in smaller member sizes for framed structures, which, in turn, may cause *instability* problems.

The elastic modulus-to-weight ratios of the typical steel and aluminum alloys are in the same range, while the one for wood is about one-quarter less. The specific elasticity of certain synthetic cables is about up to 4 times that of steel strand. Although reinforced plastic has a high strength-to-weight ratio, it also often has a relatively low specific elasticity and is therefore used in the building construction industry mostly for geometrically stiff spatial forms.

2.6 STRESS AND DEFORMATION

The external forces are resolved into internal forces as they flow along the building structure to the ground. This internal force flow stresses the members—the resultant forces of the stress distribution in the members represent the internal forces. The stress distribution, in turn, depends on the kind of force action, which is directly related to the member type. Stresses can be either *direct stresses* (normal stresses) or *shear stresses*. *Uniaxial stresses,* often called *simple stresses,* may be caused by translational forces (such as the axial stresses in columns), whereas the more complex *biaxial stresses* are caused by rotational forces, possibly resulting in bending stresses or torsional stresses. For example, the internal forces V and M in the transverse and longitudinal directions of a beam can be treated as the resultant forces of stresses. In other words, the longitudinal normal forces cause bending stresses and the transverse shear forces cause shear stresses.

Simple Stresses

The concepts of stress and strain or deformation, for simple translational force action that causes no rotation, have already been introduced in the previous section. A *simple stress, f,* is defined as a load uniformly distributed over the cross-sectional area A of a member or as load per unit area, similar to fluid pressure.

$$f = P/A \quad \text{(lb/in.}^2 = \text{psi, ksi, MPa} = \text{N/mm}^2) \qquad (2.51)$$

This stress can be a direct stress as for axial concentric forces acting parallel to the longitudinal axis of a member (e.g., column, Fig. 2.21a) causing *tensile* or *compressive stresses,* or as for internal forces acting perpendicular to a stressed plane causing *normal stresses* in compression (e.g., *bearing stresses*) or in tension. Simple stresses can also be *shear stresses* when the internal forces act parallel to a stressed plane or cross section of a member and tend to slide one section of a body past its adjacent part. Simple torsional shear stresses are discussed in Section 2.7, and simple thermal stresses were introduced in Eq. (2.8). Various examples of simple stress action are shown in Fig. 2.21.

EXAMPLE 2.17

The top chord of a wood truss frames into the bottom chord at the support using a notched heel joint as shown in Fig. 2.21e. The bearing and shear stresses due to the horizontal force component are determined.
The horizontal thrust force is equal to

$$P_h = P \cos \theta = 5 \cos 30° = 4.33 \text{ k}$$

Hence, the corresponding shear and bearing stresses are

$$f_s = P_h/A_s = 4330/6(8) = 90 \text{ psi}$$

$$f_n = P_h/A_n = 4330/6(1) = 722 \text{ psi}$$

Keep in mind, however, that simple stresses may be equivalent to a combination of other stresses. For example, the uniaxial compressive stress in a column is equivalent to maximum shear stresses at 45° and compressive stresses on all four sides as expressed in Fig. 2.21c may therefore result in shear failure of the column. Similarly, pure shear stresses are equivalent to a combination of tensile and compression stresses at 45° to the shear stresses, as shown in Fig. 2.23e.

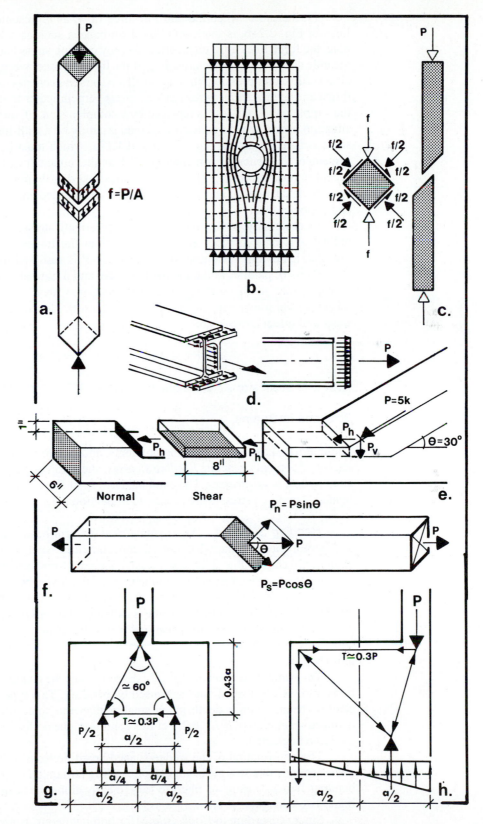

Figure 2.21 Introduction to simple stresses.

Furthermore, when the uniform force flow, as represented by the straight flow lines in Fig. 2.21b, is interrupted by disturbances such as a load concentration or a hole, the force lines must be diverted, thereby causing stress concentrations adjacent to the hole as reflected by the crowding of the lines. In other words, when the simple parallel vertical compression lines, also called *stress trajectories,* are pushed aside, lateral thrust and *tensile trajectories* (dashed lines) perpendicular to them are generated, and the simple uniaxial stress is replaced by a complex state of varying stresses. A similar situation occurs when a concentrated load is acting on a wall-like column, causing tension, as is explained in Figs. 2.21g and 1.17a, which must be resisted by steel reinforcement in concrete construction. When the concentrated load is moved to an eccentric position (Fig. 2.21h), the uniform stress distribution for symmetrical loading conditions changes to an asymmetrical one, causing tension forces along the top face and opposite face.

Occasionally, it may be required to investigate stresses at an angle to the longitudinal axis of a member at planes other than the transverse ones, as indicated in Fig. 2.21f. For this condition, the axial force P is resolved into a shear component, $P_s = P \cos \theta$, parallel to the inclined surface, and a normal component, $P_n = P \sin \theta$, perpendicular to the surface. The cross-sectional area of the inclined surface, as related to the transverse surface $A = bd$, is $A/\sin \theta$. Hence, the corresponding simple shear and direct stresses are

$$f_s = P \cos \theta \sin \theta / A \qquad f_n = P \sin^2 \theta / A \qquad (2.61)$$

Notice the shear stress f_s is maximum when $\theta = 45°$, $\qquad f_s = (P/A)/2$

Stresses in Beams

The internal shear forces, bending moments, and possibly axial forces in beams cause stresses. Moments generate *bending stresses,* shear forces generate *shear stresses,* and axial forces, *axial stresses.* The combined action of these stresses is also briefly investigated. Torsional stresses are discussed in Section 2.7.

Bending Stresses. As a symmetrical straight beam of constant cross section and homogeneous material is loaded concentrically, such as the rectangular simply supported beam in Fig. 2.22, compressive stresses (or shortening) of the upper beam portion and tensile stresses (or lengthening) of the lower beam portion are generated. At the transition of these two regions, there will be no change of length of the horizontal plane, hence no stresses; this plane is called the *neutral plane.* In other words, compressive bending stresses are generated above the neutral plane and tensile bending stresses below. The neutral plane occurs in the cross-section of a member as a neutral axis (N.A.).

The beam deflection in Fig. 2.22 demonstrates the change of the rectangular portions into wedge-shaped ones as caused by bending. This type of behavior is true for shallow beams with a depth much smaller than the span, often approximated as a depth-to-span ratio of $d/L < 0.2$. For this condition, the deflection due to shear can generally be neglected.

The deformation of the wedge-shaped beam portions indicates that parallel and plane sections in a straight beam before bending remain plane (but not parallel) during bending, hence causing the fibers to vary linearly along the beam depth, thereby resulting in two triangular *strain diagrams* (Fig. 2.22b).

Since, according to *Hooke's law* (for homogeneous, isotropic, and linearly elastic material), stress and strain are proportional to each other [Eq.(2.53)], and for the condition where the modulus of elasticity E in tension is the same as the one in compression, the bending stress diagram must have a triangular distribution similar to the strain diagram shown in Fig. 2.22b.

The external forces cause a bending moment that is equivalent to an internal force couple formed by compression and tension forces that are the resultants of the triangular compressive and tensile stress diagrams. These horizontal forces must be in equilibrium and therefore equal in magnitude. For example, the magnitude of the internal forces for the given rectangular beam section is

$$T = C = (d/2) f_b (1/2) \, b = f_b (db/4) \tag{a}$$

Resolving the bending moment into a force couple yields

$$M = Cz = Tz \tag{2.62}$$

The location of the internal forces, that is, the magnitude of the lever arm z, can be determined from the horizontal equilibrium condition of forces.

$$\Sigma Fx = 0 = C - T \quad \text{or} \quad T = C \tag{2.63}$$

For the given symmetrical condition of a rectangular beam, however, the locations of the stress resultants are known. They are located at the centroid of the triangular stress diagram at $(d/2)/3 = d/6$ from the top and bottom fibers, respectively. In other words, the internal lever arm is equal to $z = d - 2(d/6) = 2d/3$.

Substituting z and Eq. (a) into Eq. (2.62) gives the magnitude of the maximum bending stresses in tension and compression.

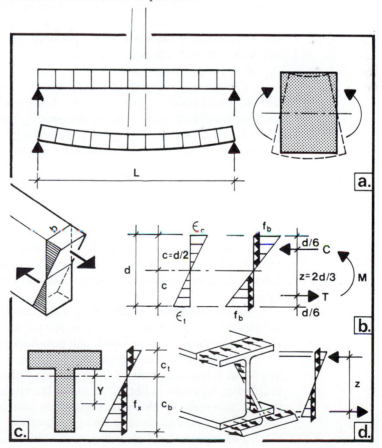

Figure 2.22 Bending stresses.

$$M = Cz = Tz = f_b(db/4)2d/3 = f_b(bd^2/6)$$

or

$$\pm f_b = M/(bd^2/6) = M/S \tag{2.64}$$

$$= M \frac{d/2}{bd^3/12} = \frac{Mc}{I}$$

Although this equation was derived for a particular beam of rectangular cross section, it holds true for all beam cross sections having a longitudinal plane of symmetry with forces acting along this plane.

Notice that the appearance of the moment of inertia with respect to the neutral axis I is a property of geometry in response to rotation of mass. Furthermore, notice that for simple bending the neutral axis is also a centroidal axis. The flexural stress formula, Eq. (2.64), can now be written in the more general form as

$$f_b = My/I \tag{2.65}$$

The flexural stress along the beam depth is proportional to the distance y from the neutral axis. Substituting $y = c_t$ or $y = c_b$ results in the maximum flexural stresses at the top and bottom fibers.

$$f_{bt} = \frac{Mc_t}{I} = \frac{M}{S_t}, \qquad\qquad f_{bb} = \frac{Mc_b}{I} = \frac{M}{S_b} \tag{2.66}$$

For symmetrical sections, the distance from the neutral axis to the extreme fibers is the same: $c_t = c_b = c$. Now Eq. (2.64) is applicable. The magnitude of the moment of inertia I or section modulus, $S = I/c$ [Eq. (2.42)], is a function of the shape of the beam cross-section as is discussed in the previous section; it is available for typical beam sections from steel and wood handbooks.

Combined Axial and Bending Stresses. Should the simple beam of the previous section, in addition to the transverse loading, also be subject to small axial loading, then the maximum stress may be obtained by simply superimposing the axial stress [Eq. (2.51)] on the bending stress [Eq.(2.64)] by assuming the beam to be relatively stiff so that the deflection has no effect on stress distribution.

$$f = P/A \pm M/S \tag{2.67}$$

Whereas beams usually carry comparatively small axial forces, columns support significant axial loads, possibly together with bending, where other design considerations must be taken into account. For further discussion of combined stress action, refer to the discussion of soil bearing pressure (Section 2.9), prestress forces (Sections 2.8 and 3.2), and the design of beam columns in Section 3.4.

Shear Stresses in Beams. The transverse shear forces in beams (often referred to as vertical shear forces as related to the typical horizontal beams) cause stresses, the distribution of which will now be determined.

In addition to the vertical shear forces, horizontal shear forces also exist in beams, as is apparent from the independent bending of planks loosely placed on top of each other, as shown in Fig. 2.23c. When the sliding of the planks with respect to each other is prevented by gluing them together, horizontal shear stresses between the layers in the longitudinal direction are generated.

The relationship between vertical and horizontal shear stresses can be determined by cutting an infinitely small square element out of the beam web (Fig. 2.23e);

the horizontal normal stresses balance each other and therefore are not shown. Notice that this element is in pure shear when isolated at the neutral axis where the bending stresses are zero.

The vertical shear forces acting on the element are in translational equilibrium. Rotational equilibrium, however, necessitates a horizontal shear force couple, causing an opposite moment to the vertical force couple. In other words, the vertical shear forces cause equal horizontal shear forces. We may conclude that the vertical and horizontal shear stresses are identical to each other anywhere along a beam.

$$f_{vv} = f_{vh} = f_v \qquad (2.68)$$

The shear force is the resultant force of the shear stresses distributed over the beam cross section. This shear stress distribution is derived from Fig. 2.23a using a simple rectangular cross section. Horizontal equilibrium of forces gives

$$C_2 - C_1 = V = f_v b dx \qquad (a)$$

The summation of the unit forces yields the resultant axial forces:

$$C_1 = \int_{y_1}^{c} f dA, \quad \text{where } f = M_1 y / I, \qquad \text{[see (Eq. 2.65)]}$$

$$= \frac{M_1}{I} \int_{y_1}^{c} y dA, \quad \text{where } Q = \int_{y_1}^{c} y dA, \qquad \text{[see (Eq. 2.36)]}$$

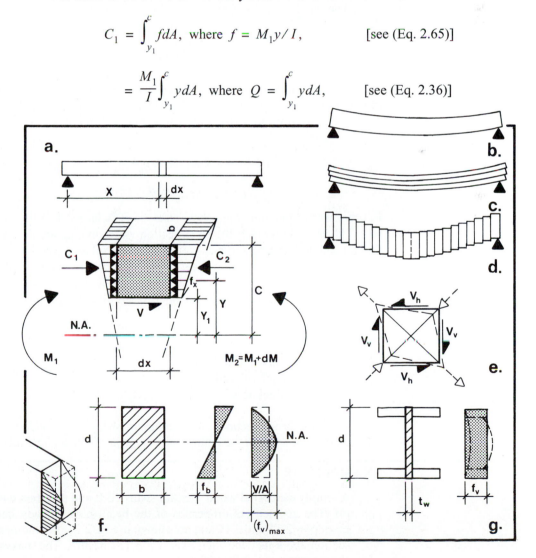

Figure 2.23   Shear stress in beams.

$$C_1 = M_1 Q/I, \qquad C_2 = M_2(Q/I) \tag{b}$$

Substituting Eq. (b) into Eq. (a) gives

$$C_2 - C_1 = (M_2 - M_1)\, Q/I = dM(Q)/I = f_v b\, dx$$

Hence, the general shear formula, letting $dM/dx = V$, according to Eq. (2.26), is

$$f_v = \frac{VQ}{Ib} \tag{2.69}$$

where f_v = horizontal or vertical shear stress
V = transverse shear force at section
Q = first moment of area about neutral axis [see Eq.(2.36)] using the area between top of beam and the level at which the shear stress is found
I = moment of inertia about neutral axis
b = width of beam at the level where f_v is determined

The shear flow or horizontal shear force per unit length of beam span, q, is defined as

$$q = f_v b = VQ/I \tag{2.70}$$

For a rectangular beam, the maximum shear stress occurs at mid-depth where the first moment of area is maximum. In Example 2.15, it was determined that $Q = bd^2/8$ and $I = bd^3/12$. Substituting these values together with the cross-sectional area of $A = bd$ into Eq. (2.69) yields

$$f_v = \frac{VQ}{Ib} = \frac{V(bd^2/8)}{(bd^3/12)\, b} = 1.5\, \frac{V}{A} \tag{2.71}$$

In other words, the maximum shear stress is 50% larger than the average shear stress V/A. The shear stress over the cross section varies in a parabolic manner from zero at the top and bottom fibers to maximum at the neutral axis (see Fig. 2.23f).

Similarly, the maximum shear stress for a beam with a circular cross section is 33% larger than the average shear stress. In an I-beam, the shear is resisted primarily by the web in an almost constant manner for all practical purposes (Fig. 2.23g).

$$f_v \cong V/A_w \cong V/dt_w \tag{2.72}$$

where V = transverse shear force
d = full depth of section
t_w = web thickness

For most beam sections, the maximum shear stress occurs at the neutral axis unless the beam width at the neutral axis is greater than at some other axis.

EXAMPLE 2.18

A simply supported wood T-beam spans 15 ft and carries a uniform load of 150 plf. The geometrical properties of the beam have already been determined in Examples 2.14 and 2.16 and are shown in Fig. 2.24; the moment of inertia of the section about its centroidal axis is $I_c = 151.82$ in.^{4}. The maximum bending and shear stresses are determined.

According to Fig. 2.11d, the maximum shear at the reactions and the maximum moment at midspan are equal to

$$V_{max} = W/2 = wL/2 = 150(15)/2 = 1125 \text{ lb}$$

$$M_{max} = WL/8 = wL^2/8 = 150(15)^2/8 = 4219 \text{ ft-lb}$$

Hence, the maximum bending stresses in compression at the top fibers and tension in the bottom fibers according to Eq. (2.66) are

$$+f_b = Mc_t/I = 4219(12)3.25/151.82 = 1084 \text{ psi}$$

$$-f_b = Mc_b/I = 4219(12)4.75/151.82 = 1584 \text{ psi}$$

The shear stresses at the reactions according to Eq. (2.69) are

$$\text{Bottom flange:} \quad f_v = \frac{VQ}{Ib} = \frac{1125\,(5.5 \times 2.5)\,2}{151.82\,(5.5)} = 37 \text{ psi}$$

$$\text{Top web:} \quad f_v = \frac{1125\,(5.5 \times 2.5)\,2}{151.82\,(2.5)} = 82 \text{ psi}$$

$$\text{Neutral axis:} \quad f_{v\,max} = \frac{1125\,(4.75 \times 2.5)\,2.375}{151.82\,(2.5)} = 84 \text{ psi}$$

The distribution of the stresses over the beam cross section are shown in Fig. 2.24.

Combined Shear and Bending Stresses.   Occasionally, the maximum stress due to the combined action of bending and shear stresses may have to be determined, as may be the case for low-tensile-strength materials (e.g., reinforced concrete beams in shear). This resulting stress, called the *principal stress,* is either in tension or compression.

The principal stresses vary from point to point along the beam; the paths they follow are called the *trajectories of principal stresses.* As an example, the stress trajecto-

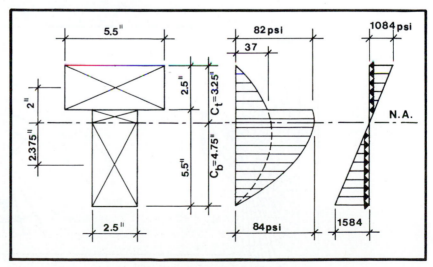

Figure 2.24   Example 2.18.

ries for a simply supported shallow beam under uniform loading are shown in Fig. 2.25a. These trajectories can be visualized as a series of funicular lines in tension (dashed lines) and compression (solid lines) or as sets of curves, each set consisting of a suspended cable intersecting with a funicular arch. For each set the trajectories intersect at right angles to each other at all points, demonstrating that the principal stresses at a given point are equal in magnitude but opposite in direction. Notice that the funicular orthogonal curves cross the neutral axis at 45°, thereby causing *pure shear* (Fig. 2.23e), which is maximum at the support; furthermore, the horizontal and vertical lines at top and bottom of the beam at midspan indicate *pure axial action.* The magnitude of the stresses or intensity of force flow (similar to the flow of water) is reflected by the spacing of the stress trajectories, indicating that the stresses are maximum at the outer beam fibers at midspan where the spacing of the lines is closest. The stress trajectories for a column with a branchlike, short beam projection are shown in Fig. 2.25b.

The stress trajectories for the simply supported beam remind one of A.G.M. Mitchell, who was among the first to study the optimization of structures or minimum-weight frameworks. He presented in 1904 the Mitchell framework (Fig. 2.25c), where he arranged tension and compression members in an optimal fashion to carry a single load from midspan to the supports, somewhat reflecting the patterns of stress trajectories.

Photoelastic modeling is among the most powerful methods of studying visually the overall structural behavior of a member (Fig. 1.4, middle right), that is, by using photoelasticity as an optical modeling technique for observing a transparent model (e.g., Plexiglas) in polarized light.

Elastic Deflections of Shallow Beams

Under external load action a symmetrical straight beam must deflect perpendicularly to its longitudinal axis. It is our concern, here, to identify some basic concepts related to beam deflections. The following relationships can be derived from Fig. 2.26b. It is assumed that the increase of member length is negligible; in other words, there is

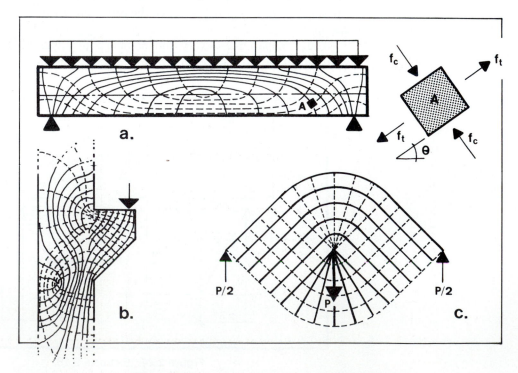

Figure 2.25   Principal stresses in rectangular beams.

hardly any difference between the original beam length and its deflected length, so $ds \cong dx$.

$$\tan d\theta \approx d\theta = \frac{dx}{R} = \frac{\varepsilon dx}{c} \qquad \text{or} \qquad d\theta/dx = 1/R \qquad \text{(a)}$$

Substituting Eq. (2.53), $\varepsilon = f_b/E$, and Eq. (2.64), $f_b = Mc/I$, yields the curvature of the beam deflection or the reciprocal of the radius of curvature.

$$\text{Curvature} = \frac{1}{R} = \frac{d\theta}{dx} = \frac{M_x}{EI} \qquad \text{(b)}$$

Letting y be the deflection of the beam at distance x (i.e., representing the beam deflection at any point), the slope of the curvature, $\tan \theta_x$, is easily found by taking the derivative of y with respect to x [see also the derivative of (Eq. 9.18)], or

$$\tan \theta_x \cong \theta_x = dy/dx \qquad \text{(c)}$$

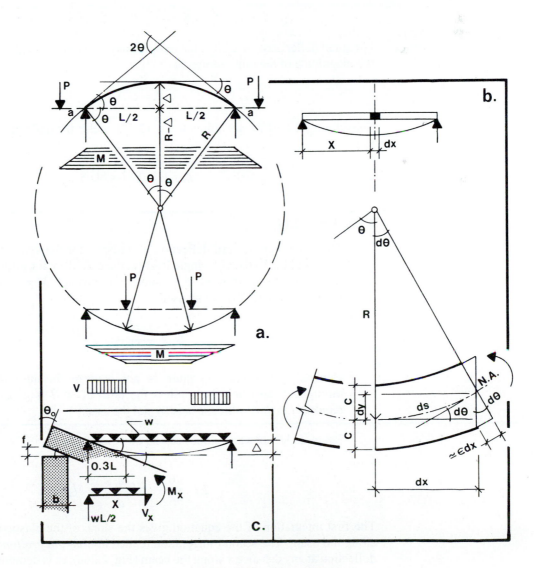

Figure 2.26 Beam deflection.

We may conclude from Eq. (c) that the rate of change of the slope $d\theta/dx = d^2y/dx^2$ must represent the curvature in Eq. (b). Hence, the equation for the elastic beam theory curvature is

$$\frac{1}{R} = \frac{d\theta}{dx} = \frac{d^2y}{dx^2} = \frac{M_x}{EI} \tag{2.73}$$

We notice that the curvature varies directly with the moment and inversely with *flexural stiffness EI*.

For the specific case of pure bending with no shear, where the moment is constant, $M_x = M$, the curvature must be constant, $d\theta/dx = \theta/L$; therefore, the deflected beam configuration must be circular, resulting in the following angular change according to Eq. (2.73):

$$\theta = ML/EI \text{ (rad. or deg} = 180 \text{ rad}/\pi) \tag{2.74}$$

The maximum deflection $y = \Delta$ for circular bending is derived from Fig. 2.26a as

$$R^2 = (L/2)^2 + (R - \Delta)^2 \quad \text{or} \quad \Delta^2 - 2R\Delta + L^2/4 = 0$$

For small deflections ($\theta \leq 20°$), the term Δ^2 may be assumed negligible as compared to the magnitude of the other terms, so

$$\Delta = L^2/8R \tag{2.75}$$

Substituting $R = EI/M$, according to Eq. (2.73), yields the deflection for circular bending:

$$\Delta = ML^2/8EI = \theta L/8 \tag{2.76}$$

EXAMPLE 2.19

The maximum upward deflection at midspan for the symmetrical overhanging beam in Fig. 2.26a with single loads at the end of each cantilever is determined. Since, for the main span, the moment is constant, $M = -Pa$, circular bending occurs, so Eq. (2.76) can be used.

$$\Delta = \frac{ML^2}{8EI} = -\frac{PaL^2}{8EI} \tag{2.77}$$

Generally, however, curvature is not constant because the moments vary throughout the beam span unless the ratio of M_x/I is constant. This, in turn, requires I to vary which is usually not the case.

Beam deflections and slopes can be obtained from Eq. (2.73) by integration.

$$EI\frac{d\theta}{dx} = EI\frac{d^2y}{dx^2} = M_x \tag{2.73a}$$

The first integration of the equation gives the slope of the elastic curve at any point along the beam. The second integration (called *double-integration method*) gives the deflection at any distance x along the beam (Fig. 2.26c), as is demonstrated in Example 2.20.

It must be emphasized that the purpose of this brief introduction to beam deflection is not the general discussion of the topic but solely to generate an awareness of basic concepts. Numerous other methods are available for determining beam deflections that are easier to handle than the double-integration method presented here.

EXAMPLE 2.20

Determine the maximum deflection of a simply supported beam carrying a uniform load w as shown in Fig. 2.26c. The general equation for the bending moment according to Eq. (2.24) is

$$M_x = \frac{wLx}{2} - \frac{wx^2}{2}, \quad 0 \le x \le L$$

But, according to the elastic beam theory,

$$M_x = EI\frac{d\theta}{dx} = EI\frac{d^2y}{dx^2} \tag{2.73a}$$

Equating the two expressions, gives

$$EI\frac{d\theta}{dx} = EI\frac{d^2y}{dx^2} = \frac{wLx}{2} - \frac{wx^2}{2}$$

This differential equation can be integrated to obtain the slope and deflection of the beam at any point. The first integration gives the slope of the elastic curve at any point.

$$EI\,\theta_x = EI\frac{dy}{dx} = \int_0^x \frac{wLx}{2}\,dx - \int_0^x \frac{wx^2}{2}dx = \frac{wLx^2}{4} - \frac{wx^3}{6} + C_1$$

The constant C_1 can be determined based on the known condition that the slope of the elastic curve must be zero at midspan because of symmetrical loading conditions.

$$EI\theta_x = 0 = \frac{wL\,(L/2)^2}{4} - \frac{w\,(L/2)^3}{6} + C_1 \quad \text{or} \quad C_1 = -wL^3/24$$

Hence, the general *slope equation* for the given case is

$$EI\theta_x = EI\frac{dy}{dx} = \frac{wLx^2}{4} - \frac{wx^3}{6} - \frac{wL^3}{24} \tag{2.78}$$

The maximum slopes at the reactions ($x = 0$, $x = L$) are

$$\pm\theta_0 = \frac{wL^3}{24EI} = \frac{M_{max}L}{3EI} \quad \text{(rad)} \tag{2.79}$$

The second integration of the differential equation or the integration of the slope equation yields the equation of the elastic curve.

$$EIy = \int_0^x \frac{wLx^2}{4}dx - \int_0^x \frac{wx^3}{6}dx - \int_0^x \frac{wL^3}{24}dx$$

$$= \frac{wLx^3}{12} - \frac{wx^4}{24} - \frac{wL^3x}{24} + C_2$$

The second constant C_2 can be determined as based on the known conditions of zero deflection at the supports ($x = 0$) as $C_2 = 0$.

Hence, the general beam *deflection formula* for the given case is

$$EIy = \frac{wLx^3}{12} - \frac{wx^4}{24} - \frac{wL^3 x}{24} \tag{2.80}$$

Because of symmetry of loading, the maximum deflection, $y_{max} = \Delta$, occurs at midspan, $x = L/2$.

$$\Delta = -\frac{5wL^4}{384EI} = -\frac{5M_{max}L^2}{48EI} \tag{2.81}$$

Usually, the downward deflection is treated as positive so that the negative sign in the equation may be disregarded.

Typical allowable deflection values for structural members are presented in Table 2.5 for ordinary situations where LL = live load, DL = dead load, and L = member span. For example, the maximum allowable deflection Δ_L of a 20-ft floor beam under superimposed design load is not to exceed

$$\Delta_L \leq L/360 = 20(12)/360 = 0.67 \text{ in.}$$

Or the maximum beam deflection Δ_{D+L} due to dead and live loads is not to exceed

$$\Delta_{D+L} \leq L/240 = 20(12)/240 = 1.00 \text{ in.}$$

The national material codes, however, should be consulted with respect to deflection limits for specific conditions.

Occasionally, the typical end rotation for an allowable deflection of $\Delta/L = 1/240$ is needed and can be determined as follows: The maximum bending moment according to Eq. (2.81) is

$$M_{max} = 48EI\Delta/5L^2$$

Substituting this expression into Eq. (2.79) gives

$$\theta_0 = \frac{M_{max}L}{3EI} = \frac{48EI\Delta}{5L^2} \frac{L}{3EI}$$

TABLE 2.5

Recommended Deflection Limits as a Function of Member Span L

	Type of Load	
Type of Construction	LL	DL+LL
Floor construction	L/360	L/240
Roof construction		
with plastered ceiling	L/360	L/240
with nonplastered ceiling	L/240	L/180
without ceiling	L/180	L/120
Vertical spans of walls		
interior walls	L/180	—
exterior walls	L/240	—
Farm buildings	—	L/180
Greenhouses	—	L/120

(2.82)

$$= \frac{16}{5}\left(\frac{\Delta}{L}\right) \cong \frac{\Delta}{0.3L}$$

Hence, the end rotation for the typical condition of $\Delta/L \le 1/240$ is

$$\theta_0 \le \frac{16}{5}\left(\frac{1}{240}\right) = 0.0133 \text{ rad} = 0.76° = 46 \text{ min}$$

(2.83a)

$$\cong \tan\theta_0 = 0.0133(12)/\text{ft} = 0.16 \text{ in./ft}$$

Hence, the end joint opening f (Fig. 2.26c) due to beam rotation according to $\tan\theta_0 = 0.0133 = f/b$ is

$$f \cong b/75 \tag{2.83b}$$

An important feature of a structural member is its *bending stiffness, k,* as a measure of resistance to rotation when it is subject to a bending moment M at one end.

$$k = M/\theta \propto EI/L \quad \text{(unit moment/radians)} \tag{2.84}$$

For example, for the situation where a moment is acting at one support producing a rotation θ, and the other support is hinged, according to Table A.15, the bending stiffness is

$$k = M/\theta = 3EI/L$$

Since the material does not generally vary, the bending stiffness of a member is proportional to the ratio of its moment of inertia-to-member length (I/L). For the condition where the moment acts at the pinned end but the other end is fixed, $k = 4EI/L$ (see Table A.15).

General Notes on Bending Members

Various beam types in an architectural context are described in Figs. 1.19, 1.20, and 3.1. Beams are distinguished in shape, that is, in cross section, and elevation, material, and support conditions. They may be part of a repetitive grid, such as joists, or they may constitute individual members; they may support a floor structure or span a stadium; they may form a stair, a bridge, or an entire building. From a structural point of view, typical beams may be organized as *shallow beams, deep beams, wall beams,* and *shell beams* (Fig. 2.28).

The direction, location, and nature of the loads as well as the member shape and curvature determine how the beam will respond to force action. In this context it is assumed that the beam material obeys Hooke's law and that a linear distribution of stresses across the member depth holds true. For deep beams, other design criteria must be developed. Only curved beams of shallow cross section that makes them only slightly curved (e.g., arches) can be treated as straight beams using linear bending stress distribution. Furthermore, it is assumed that the beam will act only in simple bending and not in torsion; hence there will be no unsymmetrical flexure. The condition of symmetrical bending occurs for doubly symmetrical shapes, such as rectangular and *W* shapes, when the static loads are applied through the centroid of their cross section, which is typical for most cases in building construction.

Since this special condition of simple bending due to symmetry may not be present because of asymmetry of the resisting structure and/or the eccentric action of the resultant force, some general concepts of structural behavior of bending members are briefly discussed.

In general, determining the stresses due to pure bending of an unsymmetrical section with no *axes of symmetry* requires complex calculations. First, the *principal axes* that are always mutually perpendicular and about which the moments of inertia are maximum and minimum, respectively, must be located; then the direction of the *neutral axis* has to be found. All these axes, together with the *centroidal axes,* pass through the centroid of the cross section. Naturally, for this general condition, the simple bending formula $f_b = Mc/I = M/S$, which applies only to symmetrical bending, cannot be used!

In addition, the loads must act through the shear center or center of twist, which is located at the intersection of the *shear axes,* in order not to generate torsion in addition to unsymmetrical bending (see Section 2.7). Therefore, this shear center must be located; it does not necessarily coincide with the centroid of the cross section. We may conclude that, when the load is applied at the centroid, the member may twist as it bends. Lack of symmetry results in eccentric loads, unsymmetrical bending, and torsion! Fortunately, cross sections usually have a certain degree of symmetry, which simplifies the understanding of the behavior and the stress calculations, remembering that an axis of symmetry is always a principal axis.

The cases in Fig. 2.27, center portion, describe some typical conditions with respect to the degree of symmetry of cross sections and load actions. They can be classified as follows:

- In doubly symmetrical shapes the centroidal axes, the principal axes, and the shear axes all coincide with the axes of symmetry. They are also neutral axes when the load acts through the centroid of the cross section, which is also the shear center; in other words, the bending axis is the centroidal axis of the beam. Should the force pass through the centroid of the section but not be parallel with one of the axes of symmetry, then the force is resolved into components in line with the principal axes and causes simple *biaxial bending,* where the maximum stresses may occur at any of the four extreme member corners.

$$f_b = \pm \, [(M/S)_x + (M/S)_y \,]_{1,2} \tag{2.85}$$

There will be torsion, together with simple bending, however, when the resultant force does not act through the centroid.

- In singly symmetrical shapes there is only one axis of symmetry; here the centroidal axes are also the principal axes. When the load acts parallel to one of the two principal axes, but through the shear center, simple bending occurs with no torsion: $f_b = M/S$. Should the load through the shear center not be parallel, it can be resolved into components parallel to the principal axes, causing simple biaxial bending with different maximum stresses at the extreme corners.

$$f_{b1} = + \, [(M/S)_x + (M/S)_y]_1$$

$$f_{b2} = - \, [(M/S)_x + (M/S)_y]_2 \tag{2.86}$$

When the load does not act through the shear center, but (for example) through the centroid, torsion is generated. The approximate response of a section to torsion is discussed further in Section 2.7. For example, for typical transverse load action through the centroids of common cross sections, as shown in Fig. 2.27, center portion, the following behavior can be identified:

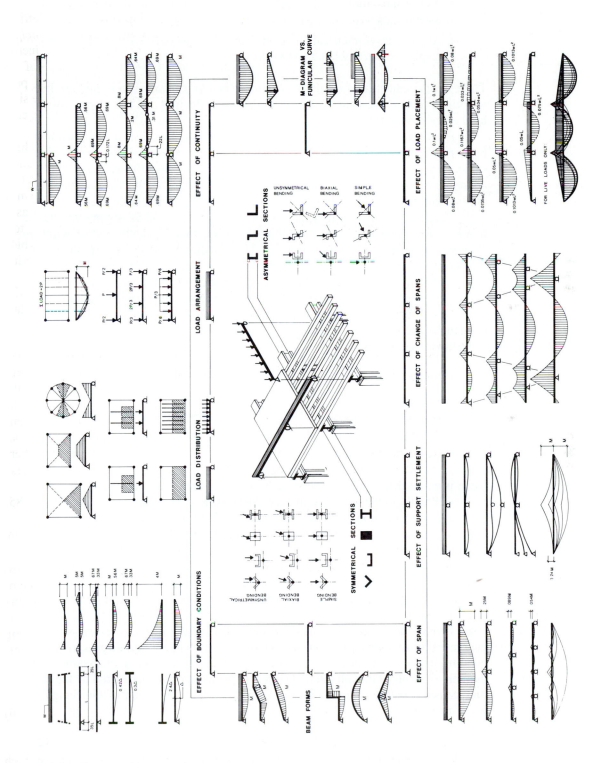

Figure 2.27 Nature of beam behavior.

145

- The doubly symmetrical I-section deflects vertically only.
- As the antisymmetrical Z-section deflects vertically, it also displaces laterally.
- As the singly symmetrical channel-section deflects vertically, it twists.
- As the nonsymmetrical angle with unequal legs deflects vertically, it not only displaces laterally but also twists.

These beams can still be designed by simple bending theory when they are laterally restrained from movement, that is, when lateral deflection and twisting are prevented by the beam boundaries.

For ordinary conditions it may be advantageous to select doubly symmetrical beam cross sections when the tensile and compressive capacities of the material are the same. However, when the compressive strength is larger than the tensile strength (e.g., concrete, masonry, cast iron), a singly symmetrical shape such as a T-section may be chosen so that the tensile strength is equal to the compression strength.

Beams do not necessarily have a constant cross section as is generally the case; it may vary along the beam (possibly abruptly), that is, depth and/or width may change in response to bending intensity. This condition is discussed in other parts of the book.

Furthermore, when ordinary beams are made of different materials to act *compositely,* the section must be transformed in order to use the simple bending theory, as is discussed further in Chapter 3. The nature of beam behavior, in general, which has been introduced partially in Section 2.3, is described in Fig. 2.27. It refers to the effects of following characteristics:

- Beam shapes
- Span
- Boundary conditions
- Continuity
- Beam cross section

- Support settlement
- Load distribution
- Load arrangement
- Load placement

Keep in mind that in contrast to straight beams, where loads are transferred in linear fashion directly to the supports, in curved beams (e.g., ring beams) the eccentricity of loads causes in addition torsional moments (Fig. 2.29u), as is discussed in the next section. For further discussion of beams, refer to Section 3.2.

Beams, in general, must be checked for the primary structural determinants of bending, shear, deflection, stability, and possibly bearing. The largest bending stresses appear along the top and bottom faces of ordinary beams, while the largest shear stress usually occurs at the neutral axis, where the flexural stress is zero. The geometry of the I-section is in direct response to this stress distribution, where the flanges primarily resist bending and the web resists the shear. In shallow beams, flexural stresses are generated by rotation of the entire section. In deep beams (e.g., girders, trusses), the bending capacity of the web may be neglected. Hence, we may assume that the moment is resolved into an internal couple of horizontal forces resisted by the flanges, thus causing mainly direct stresses (Fig. 2.28). Therefore, each of the flanges must resist an axial force H.

$$M = H(h) \quad \text{or} \quad \pm H = M/h \tag{2.87a}$$

This condition is even more apparent for the arch (Fig. 2.28) or cable, where the internal force couple consists of the compressive force carried by the arch at the crown and the tension force resisted by the tie at the base or the abutments at the supports. In other words, while the lever arm h is hidden or imprisoned within the member depth of the beam, for the arch and the other spatial surface structures, it breaks out of the restraint of the member and reflects geometry. Similarly, we may visualize the moment for a

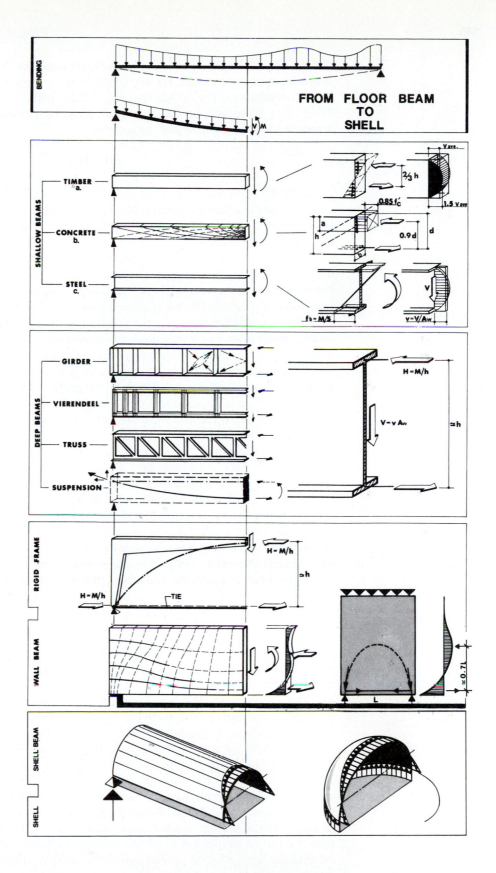

Figure 2.28 Bending member types.

147

shallow beam to be resolved into a couple represented by the resultants of the flexural stress diagram. For instance, for a rectangular beam section the distance z between the internal resultants is $0.67h$ (see also Fig. 2.22b), while for a round section it is only $0.59h$; for a W-section, on the other hand, it can be approximated as $z = 0.9h$, clearly demonstrating the efficiency of the section.

$$\pm H = M/z \tag{2.87b}$$

The behavior of a beam changes drastically as its height h increases for a given span L. It acts as a shallow beam when the beam height is much smaller than the span, approximately for a depth-to-span ratio of $h/L < 0.2$. Deep beam behavior with respect to flexure is introduced (i.e., plate theory) as the h/L ratio increases, and the stresses are not distributed linearly anymore. At approximately $h/L > 0.5$, loads are transferred to the supports primarily in direct arch action, with the corresponding thrust at the bottom. For this situation, the internal lever arm z may be approximated for preliminary design purposes as the lesser of the two values $0.7L$ or $0.7h$. For the concrete wall beam in Fig. 2.28, the compressive stresses in the imaginary arch are generally not critical; they are small in comparison to the magnitude of the tensile stresses concentrated at the bottom. The beam portion above the imaginary arch hardly assists in carrying loads.

When the mass of the bending member is transformed into the geometry of the shell, such as the *shell beam* in Fig. 2.28, the profile of the section is effectively used, since the rotation is resolved into direct longitudinal axial forces; rather than rotating the solid material, the entire section is rotated, thus allowing the thin-shell solution. Even more efficient from a force flow point of view is the *dome shell*. Here, the moment is transformed into direct forces that balance each other along the three-dimensional surface geometry. This dome is a true shell—not a beam anymore; its space geometry is most effectively used to transform rotation into membrane action.

The moment increases rapidly with the square of the span (L^2); thus the required member depth h (i.e., lever arm of resisting internal forces) must also correspondingly increase so that the stresses remain within the allowable range. In other words, the beam depth does not increase in proportion to the span, but to the square of the span when material strength controls! In addition to strength, the influence of flexibility, stability, vibration, geometry, and other criteria are discussed throughout this text.

2.7 TORSION

Torsion is generated when a member is twisted about its longitudinal axis. It is induced by the twisting moment, M_t, which may be due to a moment, a force couple, or eccentric force action Pe, where the load P is located at a distance e from the shear center. The lack of symmetry of a member section may initiate torsion when the force acts through the centroid of the section but not through the shear center unless the member is laterally restrained (e.g., by a floor slab) and not free to twist over its entire length.

For doubly symmetrical shapes, such as an I-section, the shear center coincides with the centroid of the section. For shapes with one axis of symmetry such as U- and T-shapes, the shear center lies on this axis. For example, when a vertical load is applied through the centroid of a channel parallel to its web (Fig. 2.29s), the member will twist as it bends downward since the shear center is not located at the centroid. In an asymmetrical section, the horizontal shear forces in the flanges, H, are not in balance and form a couple that twists the section. To balance the twisting moment $H(d)$, the external force must be located outside the section so that it can provide a counteracting moment $P(e_1)$. In other words, the location of the load at which the section does not twist, as based on $H(d) = P(e_1)$, is at

$$e_1 = (H/P)d \tag{2.88}$$

We may conclude that when the transverse loads are applied through the *shear center* of a section they produce only bending of a beam and no torsion.

The shear center of an angle lies at the juncture of its legs. When loaded through the centroid, parallel to one leg, the angle will not only twist, but will also deflect laterally in addition to the straight downward deflection (Fig. 2.29t).

The twisting moment may be resisted by torsional shear stress and bending torsion depending on the member type. The member shape is very important with respect to twisting resistance. Closed tubular shapes are most efficient; here, pure torsional shear strength provides nearly all the resistance. Open tubes or other open sections such as I- and U-shapes, on the other hand, offer low torsional shear stiffness; here, twisting resistance is resisted primarily by warping strength and to a lesser degree by pure torsional shear strength. Angles and thin plates are extremely weak in torsion and should be avoided.

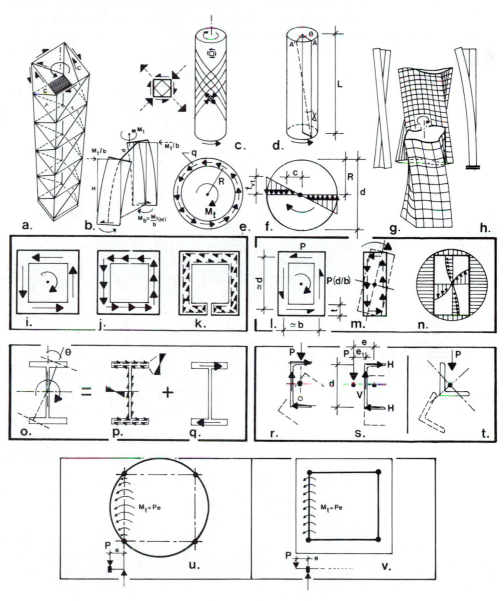

Figure 2.29 Torsion.

Typical examples of torsion in buildings are eccentric vertical cores cantilevering out of the ground, curved beams (Fig. 2.29u), and spandrel beams under one-sided eccentric loading (Fig. 2.29v), although the flexural stiffness of the floor structure may reduce or even prevent the twisting of the beams. Keep in mind that structural members in buildings are rarely designed solely for torsion but usually, in addition, for bending and possibly axial action.

Circular Sections

In a round shaft the shear flow at any point is proportional to its distance from the axis of rotation, and the resulting shear stress is one of pure torsional shear. In a thin-walled hollow shaft (Fig. 2.29e), the torsional shear stresses at a given section are uniform and thus represent a simple stress. When the hollow shaft is twisted, it does not displace laterally and change its cross section, its original straight vertical lines are forced into a helix (Fig. 2.29c). Hence, the principal stresses due to torsion form helices in compression and tension. These normal stresses, in turn, can be transformed into pure shear at 45° to the inclined planes resulting into uniform shear stresses in the transverse and vertical directions of a hollow shaft.

The constant torsional shear flow q in Fig. 2.29e balances the twisting moment M_t.

$$M_t = qR(2\pi R) \quad \text{or} \quad q = \frac{M_t/R}{2\pi R} = \frac{M_t}{2\pi R^2} \tag{2.89}$$

The corresponding uniform torsional shear stress, also called *St. Venant's torsion* after the French engineer Saint-Venant, who developed the theory of pure torsion in 1855, is

$$f_{vt} = \frac{q}{A} = \frac{M_t/R}{A} = \frac{M_t}{2\pi R^2 t} = \frac{M_t}{2tA_o} \tag{2.90}$$

Here, $A_o = \pi R^2$ is the area enclosed by the center line of the tube.

Letting the polar moment of inertia $J = I_x + I_y = 2I = 2\pi R^3 t$ and $R = c$ yields the general equation for torsional shearing stress at a distance c from the center of a solid or hollow circular shaft (Fig. 2.29f).

$$f_{vt} = M_t c/J \tag{2.91}$$

where M_t = twisting moment
c = radius at a distance c from the centroidal axis
J = polar moment of inertia = sum of rectangular moments of inertia
   = $I_x + I_y$ [see Eq.(2.39)]

Notice that the torsion formula is rather similar to the simple bending stress formula $f_b = Mc/I$ when the rectangular moment of inertia is replaced with the torsional moment of inertia, J.

In a *solid shaft,* the torsional shear stress varies linearly from zero at the center to a maximum at the outer surface (Fig. 2.29f). The maximum torsional shear stress at $c = R = d/2$, for $J = \pi d^4/32$, is

$$f_{vt} = \frac{M_t c}{J} = \frac{M_t}{\pi d^3/16} \tag{2.92}$$

The *angle of twist* or angular deflection θ of a round shaft of length L under constant torsion can be derived, according to Fig. 2.29d, as follows: From the geometry of the figure, letting $\tan \gamma \cong \gamma$ and $\tan \theta \cong \theta$, we may conclude that

$$AA' = L\gamma = R\theta \quad \text{or} \quad \gamma = R\theta/L$$

The shear stress according to Eq. (2.57) is

$$f_{vt} = \gamma G = R\theta G/L \tag{2.93}$$

Equating this expression with the torsion equation, gives the angle of twist.

$$\theta = \frac{M_t L}{GJ} \quad (\text{rad} \quad \text{or} \quad \text{deg} = (180/\pi)\,\text{rad}) \tag{2.94}$$

where G = shear modulus [Eq. (2.57)]
GJ = pure torsional stiffness

The fundamental equation for St. Venant's torsion often is also expressed for a rotation per unit length of member ($L = 1$) as $\theta/L = \theta = M_t/GJ$, or

$$M_t = GJ\theta \tag{2.95}$$

Noncircular Closed Sections

Only circular sections resist torsion in pure shear. The behavior of noncircular members is much more complex; when twisted they will not only rotate but also warp due to the deviation of the cross-section from the circle.

In *thin-walled, closed tubes* of any shape, however, the bending of the thin walls in slab action is so insignificant that the deformation of the cross section can be neglected. In other words, the torque may be assumed to be resisted primarily by torsional shearing stresses flowing in a continuous fashion like a liquid along the closed ring or tubular walls. This shear flow can be taken as nearly uniform for first-approximation purposes, and Eq. (2.90) can be used for determining the torsional shear stresses:

$$f_{vt} = \frac{M_t}{2tA_o} \tag{2.96}$$

Hence, for a *rectangular tube* (Fig. 2.29i) of constant wall thickness t, the approximate torsional shear stress using $A_o = bd$ is

$$f_{vt} = \frac{M_t}{2bdt} \tag{2.97}$$

In a similar fashion, the torsional shear stress for other thin-walled closed tubular members can be approximated by simply determining the cross-sectional area A_o of the full cross section.

It must, however, be emphasized that tubular sections may be weakened by openings as in the case of building cores, so St. Venant's torsion may be inadequate for analysis since axial stresses caused by bending torsion may have to be taken into account, as is discussed in the next section.

When a *solid rectangular beam section, bd,* is twisted, it warps, thereby causing a nonuniform torsional shear stress distribution as shown in Fig. 2.29n. The maximum shear stress does not occur at a point farthest removed from the axis of rotation, as at the corners of a square hollow shaft (which only rotates and where warping may be ignored), but surprisingly at the midpoint of the wide faces with no shear stress at the corners. These largest shear stresses are equal to

$$f_{vt} = \frac{M_t b}{J^*} = \frac{M_t b}{\alpha b^3 d} = \frac{M_t}{\alpha b^2 d}, \qquad b \leq d \tag{2.98}$$

Here, α is the shape factor, which depends on the depth-to-width (d/b) ratio of the member. It ranges from about one-fifth for a square member to one-third for a very thin plate. For approximation purposes, a shape factor of one-quarter may be assumed for typical beam sections, so the corresponding maximum torsional shear stresses are equal to

$$f_{vt} \cong \frac{M_t}{b^2 d / 4} \tag{2.99}$$

For concrete beams, the corresponding diagonal tension stresses (similar to the tensile stresses along the helix in a twisted circular beam, Fig. 2.29c) must be considered. Here, the space truss analogy may be used, where the steel cage covers the tension flow and the concrete the compression (Fig. 2.29a).

Open Sections

In open beam members, such as I-, U-, and L-sections, the continuous torsional shear flow cannot develop any more as in a ring. For example, the uniform shear flow along a closed tube is obviously interrupted by introducing a slot parallel to the longitudinal axis of the member (Fig. 2.29k). Now the shear flow cannot continue across the gap and must turn back, thereby causing a nonuniform torsional shear that results in warping of the section and a substantial increase in the maximum shear. When a member is free to warp (Fig. 2.29g), only *torsional warping shear stresses* are generated. But when the support restrains the member from warping, *bending torsion* occurs with the corresponding horizontal shearing stresses and bending stresses in the flanges for the case in Fig. 2.29h. We may conclude that the twisting moment M_t in open sections is no longer transmitted in pure St. Venant's torsion as for closed tubes, but also in *warping torsion.*

Open sections with at least two parallel flanges, such as I-, U-, and Z-sections, can effectively resist torsion in bending. In contrast, shapes without parallel flanges, such as T- and L-sections, have an extremely low torsional stiffness and should be avoided.

The response of an open section to torsion is explained in Fig. 2.29o to q, using a wide-flange section as an example. As shown, the section resists the twisting moment in torsional shear and bending torsion. For preliminary design purposes, it may be assumed that the twisting moment is fully resisted by bending torsion, particularly since the torsional bending stresses usually are by far larger than the torsional shear stresses, in contrast to closed tubes, where the twisting moment is assumed to be resisted entirely by simple torsional shear.

In other words, for open sections with parallel flanges, the twisting moment is resolved into a force couple that is resisted in bending by the flanges, which are assumed to act as independent beams. The design of I-beams with respect to bending torsion is discussed further in Fig. 3.6c and the corresponding text. The design for twisting of other section types with parallel flanges can be approached in a similar manner.

The flanges of simple span beams are free to warp under torsional loading at the support, while the webs are prevented from twisting and transmit the twisting moment. For fixed connections, the beams are prevented from warping and the flanges are assumed to transmit the twisting moment in bending and shear.

Open sections, especially asymmetrical ones, if at all possible, should be laterally braced to eliminate torsion, thus forcing horizontal members to deflect only vertically.

2.8 PRESTRESSING

The prestressing of structures is one of the great inventions of construction engineering. A structure is prestressed if it is resisting internal forces without any external load action, ignoring its own weight. The prestress forces induced under a controlled process cause stresses that oppose and reduce the critical stresses due to external loading, thereby resulting in a more economical internal force distribution and an increase of the structure's load-carrying capacity and stiffness.

The principle of prestressing is not new. An example in nature is the spider's web or the growth of a tree, which causes the outer fibrous layer of the trunk to be highly stressed in tension while most of the solid sapwood and the central heartwood are in compression. Other familiar examples are found in the iron bands over wooden wheels, which are heat shrunk, and the iron rings around wooden barrels, which are tensioned. The thin wire spokes of the bicycle wheel are prestressed with sufficient force so that they do not carry compression and buckle; the uniform radial tension produces compression in the outer circular rim (ring) of the wheel and tension in the inner ring.

Today, prestressing, willingly or unwillingly, is applied to all kinds of materials and structural elements. Undesired residual stresses are locked into steel members due to uneven cooling after hot rolling; they are also generated by uneven cooling due to welding or due to bending of cold-formed shapes from flat-rolled steel. On the other hand, glass may be prestressed deliberately by thermal or chemical tempering.

External posttensioning has been applied to the repair (e.g., sagging floor slabs, cracked beams) or retrofit of concrete and steel structures. High-strength bolts are pretensioned to such a level that the loads are transferred in friction along the contact surfaces. Prestressing is used to stabilize earth formations and sides of excavations to eliminate inside bracing. High-strength steel plates may be welded to the flanges of beams, which are bent with the help of jacks. When released, moments are induced in these beams and the plates are put into tension. A similar result is obtained by *cambering* the beam or placing a prestressed cable at the bottom face of a simply supported beam, for instance, which will cause the member to bend upward (Fig. 2.30g). Under its own weight the beam may then be in a horizontal position, resulting in a deflection only due to live loading. An additional strut support may be added to this beam at center span by using a pretensioned cable; this cable beam is called a *king-post truss* (Fig. 2.30g). Prestressing may be achieved through shortening cable members with a turn buckle or by forcing members that are too long into place or by tightening tie rods at the base of a frame. The characteristic of concrete to shrink can be used beneficially as a prestress agent by simply preventing the movement by adjoining structures.

Flexible structures must be prestressed so that they are stable. This can be done by air pressure, as for pneumatic structures, or by directly tensioning double-curvature anticlastic surfaces, as for tent structures.

Rigid structures strong in tension and compression, such as steel or wood systems, may be prestressed to reduce the deflection of long-span bending members and increase the buckling capacity of slender compression members, resulting in a reduction of weight and increase of stiffness. Rigid structures weak in tension but strong in

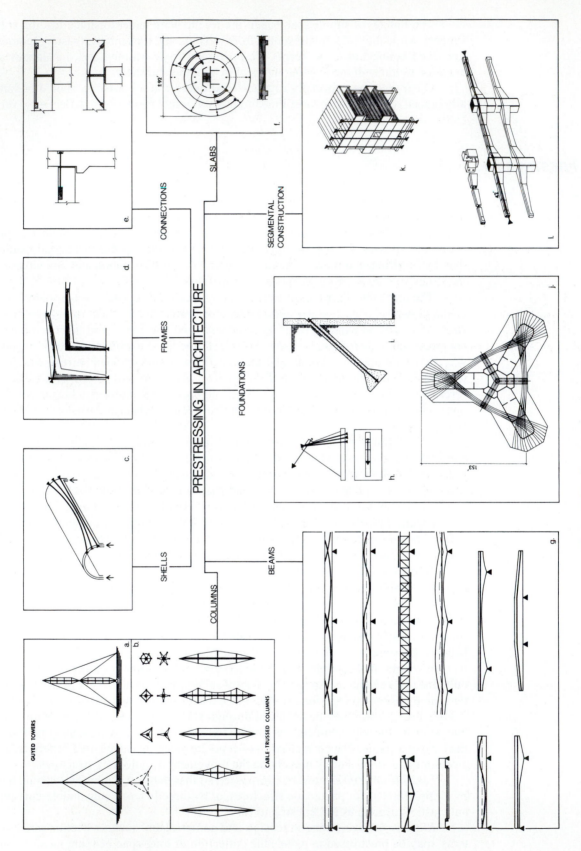

Figure 2.30 Prestressing in architecture.

compression, like masonry and concrete, are prestressed so that the compression induced cancels the tension due to external loading. In the past, this was achieved through increase of weight. The vaulting in Gothic cathedrals was kept in compression by an intricate system of buttresses and pinnacles. The 16-story Monadnock Building (1891) in Chicago required massive walls at its base to overcome the tensile stresses due to wind action. The weight of gravity dams and the weight of the earth backfill on cantilever retaining walls act as prestress agents in resisting overturning.

Today, the predominant application of prestressing is in concrete construction (Section 3.2), besides soft shell structures (Chapter 9). Because of the weakness of the material concrete to resist tension, cables are placed along the tension flow of a member. When prestressed, these cables induce compression in the surrounding concrete and thereby prevent the section from cracking if the member is *fully prestressed*. In ordinary reinforced concrete beams, only the cross-sectional area above the neutral axis is effective in resisting bending, while the rest below (i.e., cracked portion) is just a deadweight burden. When fully prestressed, the beam is uncracked and its entire cross section is available in resisting external loads, which in turn results in an increase in strength and stiffness or reduction of weight, and thus in material savings. Although prestressed concrete is more expensive than conventional reinforced concrete due to higher costs of the stronger materials, the necessary accessories and operations, the qualified labor, and the initial financial investments, the higher costs may still be offset by the speed of construction of precast units, the required larger-member span with low dead-to-live load ratio, reduction of building height because of less floor depth, and the prefabrication of complex structural shapes and assemblages under high-quality control.

The modern development of prestressed concrete is closely associated with the French engineer Eugene Freyssinet (1879–1962) in the first half of the twentieth century. He recognized that the prestressed member shortens due to shrinkage and creep, resulting in a loss of prestress, which he found can be reduced by applying much higher prestress forces; the large tension forces, in turn, require high-strength steel and concrete. Other engineers who made important early contributions to the practical application of prestressed concrete are Gustave Magnel of Belgium, Y. Guyon of France, and R. Morandi of Italy, and Franz Dischinger, Ulrich Finsterwalder, E. Hoyer, and Fritz Leonhardt of Germany. The rapid application of prestressed concrete started after World War II when steel was in short supply. T. Y. Lin of the University of California in Berkeley has distinguished himself since the early 1950s in making prestressed concrete a recognized and widely used method of construction in the United States. The first major prestressed concrete bridge in the United States was the Walnut Lane Bridge in Philadelphia built in 1949, while the world's first prestressed concrete high-rise building was the Diamond Head Apartments built in 1956 in Honolulu, Hawaii.

While the design of ordinary reinforced concrete is generally based on strength (ultimate) design, fully prestressed concrete members may be proportioned on the basis of linear elastic behavior, since prestressing is used to improve the member performance at service level and since the concrete is uncracked and steel and concrete are stressed at a relatively low level.

Two primary tensioning methods are used in concrete construction: *mechanical prestressing* using jacks and *chemical prestressing* using expansive cements. Another tensioning method is based on electrical prestressing, where tendons are lengthened by heating them.

The most common method is the mechanical one, which uses either pre- or post-tensioning. *Pretensioning* is usually associated with precast concrete produced in manufacturing plants. Tendons anchored to the forms or abutments outside the forms are tensioned before the concrete is cast. When the concrete has reached enough strength, the tendons are cut but prevented from returning to their original length by the concrete to which they are bonded, thus placing the surrounding concrete into compression.

Posttensioning of tendons is done by hydraulic jacks after the concrete has hardened by inserting the wire or rods in metal or plastic conduits, which were placed in the forms before the concrete was poured. The concrete element carries the compression usually in bearing through special anchoring devices at the end faces, rather than through bond as in pretensioning; however, end blocks must spread the concentrated prestress forces over the whole beam section. Since there is no bond between tendons and concrete, tension can be applied at different stages. Posttensioning techniques can be used on site with the member in place or for prefab concrete in a precast plant. It is usually applied to long-span structures and segmental construction. There are hundreds of patented posttensioning systems; they distinguish themselves by the method of anchoring, type of stressing tendon, and the grouting that is necessary to protect the steel from rust. For further discussion of prestress principles in concrete construction, refer to Section 3.2.

Chemically prestressed concrete was originated with the commercial development of expansive cements by the French engineer H. Lossier in the late 1930s. In the United States, Alexander Klein, at the University of California at Berkeley, reported in 1958 on a new compound, anhydrous calcium sulfoaluminate, which together with Portland cement in concrete causes an expansion of the concrete during the early stage of hydration. When the concrete attempts to expand, it is placed into compression because it is prevented from moving either by adjoining structures, subgrade friction, or the reinforcement. Expansive cement concretes are classified according to the level of prestress as follows:

- Shrinkage compensating (25 to 100 psi)
- Self-stressing (150 to 500 psi)

Chemical prestressing has been successfully applied to large-scale prefabricated units in box construction; it has been used in composite floor structures, where chemically prestressed precast corrugated plates serve as formwork for cast-in-place, shrinkage-compensating concrete; it has been applied to surface elements like walls, slabs, plates, and shells to prevent the development of shrinkage cracks.

There is no limit to the application of the prestress principle in building construction. It has been applied to individual members like piles, utility poles, railroad ties, and pavements, as well as to the large-scale building such as parking structures, residential construction, grandstands, vessels, storage tanks, water towers, ocean structures, ships, bridges, and long-span and high-rise buildings to name a few examples. Prestress tendons are not only arranged along linear members and surfaces, but also may be placed in a multidirectional fashion.

Application of the Prestress Principle to Various Structure Systems

Some typical applications of the prestress principle to architecturally designed elements or buildings are shown in Figs. 2.30 and 2.31. These examples are now briefly discussed.

Beams: Typical layouts of prestress cables are shown in Fig. 2.30g. As can be seen, tendons may be placed within the beam as for prestressed concrete or outside the beam as for steel or timber sections. They may consist of short pieces, overlapping tendons, or continuous cables. Prestressing can be applied to any beam types such as W steel sections, wood members, concrete beams of any shape, or trusses. The following methods of prestressing can be applied to steel beams:

- Stressing parts of beams, as for example by welding a pretensioned steel plate (flange) to a T-section or by welding a cover plate to a bottom flange of a cambered beam
- Casting a concrete slab in composite action to a deflected beam, called the *pre-flex* method of prestressing
- End anchoring pretensioned steel cables or bars

Cables can be arranged in a linear, draped, or curved manner. For the condition where the cable is straight and located at the centroidal axes of the beam section, the prestress force P only causes uniform axial stresses along the beam.

$$f_c = P/A_g \qquad (2.100)$$

When the tendon is placed with an eccentricity e away from the centroidal x axis but still located on the centroidal y axis, additional bending stresses due to $M = P(e)$ are generated on this symmetrical section (Fig. 3.17).

$$f = \frac{P}{A_g} \pm \frac{P(e)}{S_x} \qquad (2.101)$$

For the preliminary design of prestressed concrete beams, refer to Section 3.2.

While the tendon profile in posttensioned concrete beams of constant cross section may be curved and follow the tensile stress flow by responding as a suspended cable to gravity loading, the tendon profile in pretensioned beams may not be able to do so because of the process of fabrication. The straight tendons must be held down at certain points, and hence yield a draped profile. However, this shortcoming can be corrected by using arched, tapered, or haunched member forms with straight cables, yielding a result similar to curved tendons in straight members.

Unbonded prestressing tendons have also been applied outside concrete members. *External prestressing* may be a logical choice for rehabilitation and strengthening of existing structures.

Slabs: In slab construction, prestressing has been used for one- and two-way solid and joist slabs. One-way slabs can be visualized as consisting of parallel slab strips or 1-ft-wide beams. The rather flexible flat plates are especially suitable for posttensioning to obtain a wider column spacing, less floor depth, and a watertight roof deck. The two-way structural steel girder grid of the National Gallery in Berlin, Germany (Fig. 6.15g), was cambered at the center and the four cantilevered corners to counteract deflection. This cambering has a similar effect to prestressing. The thickness of a circular slab with the curved tendon layout shown in Fig. 2.30f was reduced by 33% through circular prestressing, resulting in overall economy because of reduction in weight. The Y-shaped cellular foundation slab of the CN Tower in Toronto (Fig. 2.30j) with the triangular tendon layout was prestressed to keep it crack free. For the preliminary design of a prestressed concrete slab, see Example 3.19.

Towers and Columns: In *cable-trussed columns* (Fig. 2.30b) often used for tent structures, the long slender compression member is laterally braced by intermediate struts or tensile elements, which, in turn, are held in place by prestressed cables. The critical buckling length of the slender column is reduced to the distance between the struts and/or tensile bracing. The areas of the guys are part of the column cross section and, since they are a substantial distance away from the centroidal axis, they yield a large moment of inertia, which lessens the significance of buckling. The guyed cables can be placed in several planes and are supported by open or closed ring strut systems. The composite action of the guyed column can be compared with the central spine and ribs of the human body resisting compression while the muscles carry the tension. Tall antenna structures (Fig. 2.30a) are guyed at intermediate levels with an initial tension high enough to ensure that the cable on the leeward side stays in tension under maximum wind pressure. The critical buckling length of the mast is the distance between the cable supports. Similarly, the masts of sailing ships are prestressed and laterally supported by rigging.

It is useful to prestress precast concrete columns that are subject to severe climatic conditions, as well as to cracking during handling, transportation, and erection, and that must resist bending so that the entire cross section can be used to resist rotation and reduce lateral sway. Since prestressing increases column stiffness, the buckling capacity of the slender column is also increased. The prestressing of tensile concrete columns offers the advantage of the steel tendons offsetting dead and live load elongations.

Prestressing has been applied to tall concrete towers used for water storage, rotating restaurants, and broadcast facilities. The world's tallest free-standing structure is the 1815-ft-high CN Tower in Toronto, which is fully prestressed (Fig. 1.9). The 144 tendons in the cantilevering tower shaft were posttensioned by a force of about 1000 tons so that no tensile stresses occur in the concrete under normal loading conditions.

Walls: Precast, prestressed wall panels are made from a variety of shapes, ranging from flat solid to ribbed panels of small and large scale; they range from plain interior concrete walls to perforated or framed facade panels. For instance, limestone facade panels have been posttensioned to allow them to overcome their low tensile capacity and to span the distance between the column supports.

Frames: The prestress tendons in frames (Fig. 2.30d) follow the tensile force flow as caused by the governing gravity loading case, as is further discussed for the load-balancing method of analysis in Example 3.19.

Segmental Construction: Prefab concrete sections are produced and possibly prestressed in temporary or permanent precasting plants. There is no limit set to the potential member shape. Typical standardized pretensioned sections (as mass produced by the precast concrete industry) are hollow-core slabs, single and double T's for floors and roofs, and inverted T, L, and I shapes and box sections as supporting beams. Precast elements may be used in combination with conventional construction or may be the building components of a fully prefabricated building system. The structural elements are assembled on site and tied together through posttensioning to form a rigid monolithic whole; the segments are supported by frictional forces between the contact surfaces as induced by the prestress forces. The continuity between the precast members can be established and moments transferred by posttensioned short tendons (Fig. 2.30e); long tendons provide better continuity. The interlocking of the prefab boxes for Habitat 67 in Montreal into a complex aggregate was made possible by posttensioning the units together with continuous cables.

The ICO Building System (Fig. 2.30k) developed by Sepp Firnkas of Boston consists of pretensioned floor panels and precast wall panels vertically posttensioned. The spandrel beams of the Gulf Life Building in Jacksonville, Florida (Fig. 2.30l), that cantilever 42 ft consist of precast concrete segments strung together with posttensioning cables. Another example are the 70-ft concrete I-beams of the Trade Group office building in Canberra, Australia (Fig. 2.31g), which are posttensioned and support the T-beam floor units that are also posttensioned, as the exposed ends of the tendons along the webs of the girders indicate.

Shells and Folded Plates: Although the stress level in shells is generally low, posttensioning is advantageous for stabilizing the shape, as well as to avoid shrinkage and tensile cracks, especially where water tightness is essential. To avoid cracking of the shell, it may be prestressed along the principal tensile stress trajectories. Shell beams (Fig. 2.30c) can be prestressed to increase their span capacity, or the cantilevering portion of a hyperbolic paraboloid (see Fig. 8.38d) may be tensioned along its suspended curvature to reduce deflection. Precast concrete panels may be laid on top of a network of posttensioning tendons, as for example, the 10- × 10-ft mesh for the 380-ft circular saddle roof of the Arizona State Fair Coliseum in Phoenix (1965), which are then prestressed to form a continuous shell surface. The tension rings of concrete domes that resist the lateral thrust due to gravity are, in general, prestressed. An example is the Toronto City Hall (Fig. 2.31e), where the tension ring along the edge of the dome and the two perimeter beams in the upper portion of the truncated cone are posttensioned. Ring beams are prestressed in a natural way when, instead of a dome, a suspended membrane or prestressed anticlastic net is used. The prestressing of the shells can also occur at the foundation level, where the ties may be pretensioned, thus, in addition, possibly eliminating abutments.

Trusses: Around the turn of this century, it was not unusual to use iron rods beneath wrought iron bridge trusses to apply an upward curvature. Steel rods along the upper chord of the 140-ft cantilever steel truss of the United Airlines Hangar at Chicago's O'Hare International Airport (Fig. 2.31c) are prestressed; the result was 20% less steel as compared to a conventional truss construction. In the Rock Island parking structure (Fig. 2.31b), 11-ft 10-in.-high precast Vierendeel trusses span 32 ft between the columns supporting the roof and second floor at the top and bottom chords, respectively. The trusses had to be post tensioned horizontally and vertically in the precaster's yard because of height limitations due to clearance requirements under bridges and weight considerations for reduction of transportation costs.

Suspended Concrete Hangers and Arches: The tension columns or hangers along the facade of a high-rise building supporting the floors can be made from prestressed concrete. The concrete is prestressed so that no tensile cracks form under full loading, and the member extension due to live loading is kept to a minimum. The loads of the truncated rectangular pyramidal dome of the Baltimore Convention Center (Fig. 2.31f) are carried by posttensioned suspended concrete arches located in the four inclined dome faces, where they are transferred to the corner columns. The perimeter tie beams along the base of the sloped sides are also posttensioned. A Lufthansa hangar at the Frankfurt Airport in Germany (Fig. 2.31d) employs twin suspended roofs, each spanning 440 ft. The roof consists of prestressed lightweight concrete strips 34.50 ft wide and 3.38 in. thick supported by a transverse prestressed concrete box girder at midlength and trussed abutments at the ends. Other types of prestressed tension members are tension rings for dome structures and tie beams for frames and shells.

Building Blocks: The potential application of the prestress principle is exemplified by the multidirectional layout of the posttensioning system of the Tax Court building in Washington, D.C. (Fig. 2.31a), where the curvilinear cable pattern follows the tensile force flow. The cantilevered building box located centrally above the entrance is anchored with posttensioning tendons to the four-story building block at its rear. The cables run in the vertical planes along the parallel shear walls. While some tendons continue in the walls to the base of the anchor block, others run into the top and third-floor slabs of the cantilevered and anchor blocks. Additional posttensioning cables are located in the roof and third-floor slabs.

Foundations: Retaining walls, or sheet pile walls as needed for excavation, may have to be anchored by prestressed tie backs directly into the rock or by posttensioned concrete piles deep in the soil (Fig. 2.30i). Foundation concrete piles of the bearing and friction type are often prestressed to increase their bending capacity. The slot-and-wedge type of foundations used for the anchorage of the guy cables for the Olympic

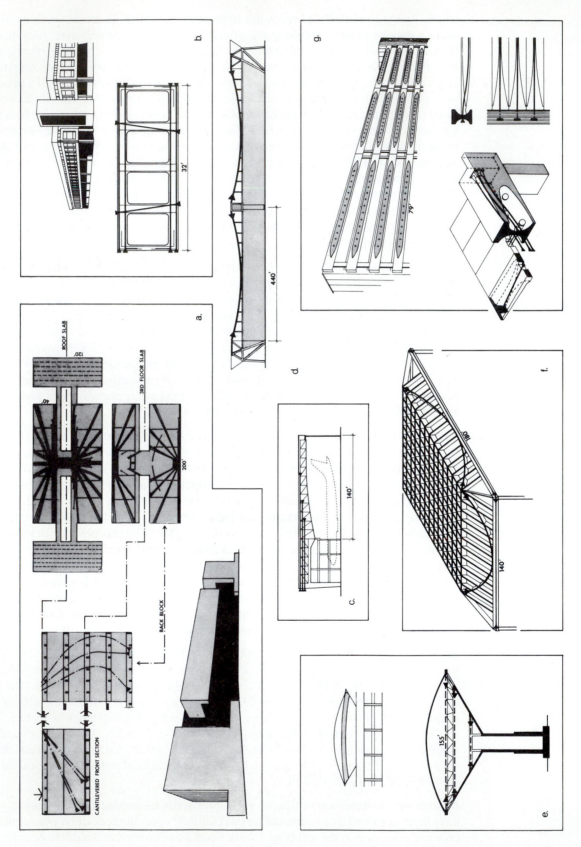

Figure 2.31 Prestressing: a case study.

160

membrane roof in Munich, Germany, are prestressed along the tensile force flow (Fig. 2.30h).

2.9   SOIL AND FOUNDATIONS

Foundations are necessary as transition structures from the building to the ground, since buildings are not usually founded directly on hard rock. The bearing capacity of the soil is generally much lower than that of structural materials; hence, *transfer structures* with *deep roots* or *flat roots* with a wide base, similar to base plates for steel columns, are required at the junction where columns and walls meet the soil. Foundations are either of the shallow type, which may consist of individual spread footings and continuous mats, or they are deep foundations of piles, piers, and caissons, as identified in Fig. 2.32. Building columns and walls are either supported by individual or combined footings directly bearing on the ground or they are supported by piles. Foundations on firm soils form a natural extension of the bearing elements of the superstructure, while foundations on soft soils may form a large mat to spread the loads over the entire footprint of a building, because the weak soil cannot support pressure concentrations.

Ordinary building foundations consist of either a collection of individual rectangular and strip footings or a large mat combining all the single footings. In seismic areas, the individual spread and pile foundations have to be linked and tied together by bracing struts so that the entire building foundation can act as a unit in sharing the load resistance. In crowded urban areas, the deep basements for tall buildings, particularly adjacent to other heavy buildings, cause excavation problems. In this case, special foundations, such as sheet piling, slurry walls, bracing of walls, or walls with tiebacks, along with underpinning of adjacent buildings and subways, are required. Pumps may be needed for conditions where the water table is high, which (in turn) may cause settlement problems with adjoining buildings.

The building foundations distribute the loads due to gravity and lateral force action to the ground. The resulting forces to be transferred are horizontal and vertical, depending on the structure systems (as indicated for various cases in Fig. 4.13). The vertical forces are generated by gravity (e.g., building weight) and overturning. When a building is loaded only concentrically, then the contact pressure at its base may be assumed as uniform for preliminary design purposes. But when overturning due to lateral force action, and less so due to asymmetrical gravity action, is generated in addition, then the overall base pressure is no longer constant. The type of force transition to the ground depends on the type of load-bearing element. While the foundations for a *gravity structure* of relatively symmetrical layout may cause a primarily uniform pressure distribution, the foundations for a *lateral force-resisting structure* may generate a nonuniform pressure due to rotation, thereby concentrating large forces on certain areas. The foundations of lateral force-resisting structure elements, such as walls, react in a nonuniform fashion (e.g., see Fig. 4.13a, c, g, k, l), while the frame columns may respond in a uniform manner (see other cases in Fig. 4.13). The eccentric loading of footings should be kept to a minimum; otherwise, the maximum contact pressure occurs only along the foundation edges, and the capacity of the soil is thus not efficiently used, besides causing nonuniform settlements. Considerations of nonuniform pressure distribution and stability are briefly discussed in the final section (see also Fig. 4.29). Occasionally, it may be necessary to employ under spread footings, which transfer compression forces, tension piles, or piers in order to resist uplift forces.

Horizontal forces (i.e., base shear), as from wind and unbalanced earth pressures, can be transmitted to the soil by one or more of the following methods:

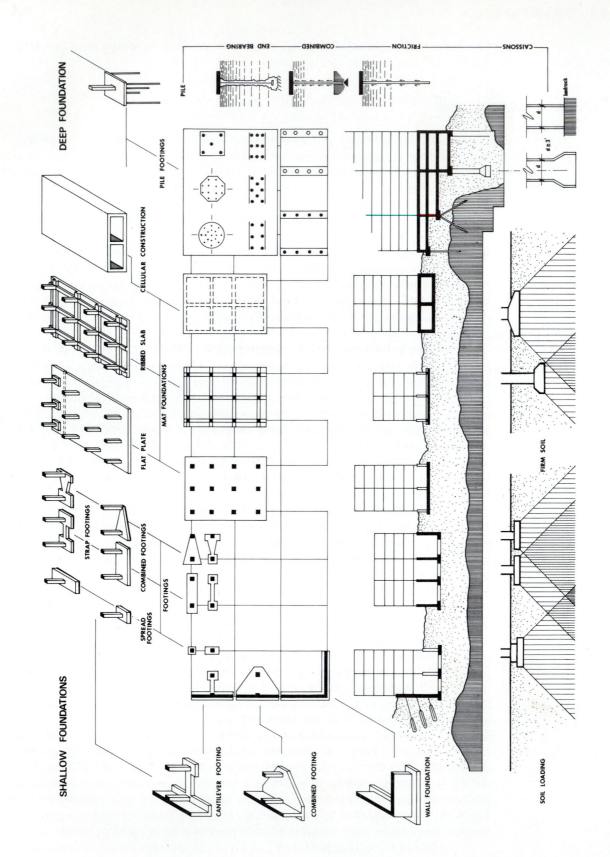

Figure 2.32 Foundation systems.

- Shear resistance at the base, such as provided by friction between the footings and soil, as well as by side friction
- Passive soil pressure, as provided by deep basements, piers, etc.
- Shear and bending resistance of piles or piers
- Axial resistance of battered piles
- Any combination of the preceding

For large lateral loads, isolated footings and pile foundations should be braced by struts so that the foundations are tied together and the force resistance is more evenly shared.

The horizontal thrust forces produced by the weight of long-span buildings, such as frames, arches, or domes, may be resisted by self-balancing structures such as ties and rings, or it may be absorbed by abutments (e.g., Fig. 9.2).

Soil Properties

This brief introduction to foundations calls attention to the necessity for investigation of the subsurface conditions of a building site so that appropriate foundations can be selected. This investigation includes (among other criteria) the nature of the soil, the soil stratification, and the thickness of the layers, as well as any existing underground structures. Soil engineers must determine the strength and settlement characteristics, as well as other criteria pertinent to the design of foundations. Often, only confirmatory soil explorations are required, especially in urban areas, where the local geology and groundwater conditions are known and subsurface maps are available and the performance of adjacent buildings can be evaluated. In more detailed investigations, *borings* are used to determine the soil profile, along with the location of the water table. Also, *test pits* of up to about 10-ft depth are often used to visually inspect the soil strata.

The spacing, number, and depth of borings depend on the complexity of the site, the type and size of building, code requirements, and other criteria. The testing of the soil samples obtained from the borings yields important material properties, such as shear strength, compressibility, and permeability, in addition to the necessary basic soil properties of density, porosity (void ratio), moisture content, grain size distribution, relative consistency (soft to hard), and so on.

The grain sizes range from boulders, gravel, and sand to clay and silt. The subsurface material is broadly divided into *rock* and *soil* (see Table 2.6), that is, cemented and loose materials. Soils, in turn, may be divided into *cohesive* and *noncohesive materials,* excluding organic deposits such as top soil.

Cohesive soils such as silts, clays, and clay mixtures contain a large proportion of fine particles. These expand and shrink with change of water content; they are compressible and may creep under constant load action and hence are prone to long-term settlements. They may have a low shear strength, which is derived primarily from cohesion (tension), and may lose part of it upon wetting and other disturbances. In contrast, granular soils, such as sand, gravel, and granular soil mixtures, are cohesionless, so that their shear strength must depend on the internal friction between grains; these are prone to immediate settlement. They do not exhibit elastic properties, and thus do not rebound when the load is taken away, as many other soils do. Critical, with respect to foundation design, are the following:

- The bearing capacity of the soil
- The control of excessive settlement
- The control of differential settlement between the various vertical support elements

The *ultimate bearing capacity,* that is, the shear failure of the soil, may only be controlling for soils such as rocks and certain clays. The tilting of the Transcona concrete grain elevator, Winnipeg, Canada, in 1913 at about 30°due to uneven loading, is a rare example of a complete ground rupture. Ground failure due to *slope instability* is treated as a special condition. In this case, a landslide may be initiated by a heavy structure, together with rainfall or seismic action, which makes the building slide downhill. An earthquake may cause saturated sands to liquefy so that soil flows out from under foundations and thereby may cause a building to rotate. For most soils (e.g., soft clays, sands, and mixtures), excessive *settlement* must be controlled, but may not necessarily be intolerable. For example, the Monadnock Building (1891) in Chicago has settled almost 2 ft with virtually no damage. Originally, the National Theater in Mexico City had sunk as much as 10 ft, due to the pumping of water from deep wells, but then was pushed up again by the weight of high-rise structures built nearby.

A primary reason for building failure is due to large *differential ground settlements;* hence, not the soil but the building fails. It is the stiffness of the superstructure and its tolerance for vertical movement which, in turn, influences the performance of the foundations, that is critical. Naturally, smaller building structures can be designed as a statically determinate system to allow large movements rather than to resist them with continuity and thereby to develop significant stresses. A spectacular example of differential settlement is the 179-ft Leaning Tower of Pisa (1350), which is currently about 18 ft out of plumb due to consolidation of the clayey soil under a layer of sand and which continues to tilt. Also, the heavy weight of a new skyscraper may cause settlement problems on existing adjacent buildings.

To control settlements and tilting, foundations on compressible soils should only be concentrically loaded; in other words, the column and wall bases should not be fixed to the footings, and lateral shear forces should not be transferred in bending. On the other hand, footings on highly compacted soils may be loaded eccentrically. Single footings should only be used on soils of low compressibility, because the independent displacements of the foundations may cause significant stresses in the superstructure. Columns may be joined by continuous footings to control vertical differential movements between them. Mat foundations are most effective in reducing differential movements on compressible soils.

TABLE 2.6
Typical Average Allowable Bearing Capacities of Various Soils

Soil Type	Bearing Capacity (ksf)
Silt (sandy or clayey silt)	1
Clay (soft), soft broken shale	3
Clay (stiff), wet sand, sand–clay mixture	4
Sand (fine, dry), sand–gravel mixture	6
Sand (compact coarse, dry), loose gravel, hard, dry clay	8
Gravel, gravel–sand mixture	12
Gravel–sand mixture well cemented	16
Soft rock (broken bedrock, compaction shale), hardpan	20
Sedimentary rock (hard shale, sandstone, limestone, siltstone)	50
Medium hard rock (slate, schist)	80
Hard rock (basalt, diorite, dolomite, gneiss, granite)	200

Climatic effects make it necessary to place shallow foundations at a depth below that of frost penetration and seasonal moisture change. In semiarid regions, *expansive soils* are found, which absorb rain water and swell in the rainy season and dry and shrink in the dry season. As the soil expands, it may create forces of up to 30 ksf and higher and movements of more than 6 in. with a corresponding uplift pressure; hence, perimeter footings should be founded on soil at a depth with constant moisture. Similar to the seasonal up and down movements, due to the *shrink–swell* cycles, are the *freeze–thaw* cycles in cold climates. It is apparent that footings should not be founded on frozen soil unless it is of a permanent nature or unless frostproof shallow foundations are used where insulation sheets trap the heat so that the soil beneath the foundations does not freeze. Interior footings that are not affected by frost may be placed higher than the frost line. However, some designers use as a minimum depth 2 ft below ground to be sure of encountering undisturbed soil.

Foundation Systems

Common foundation systems have already been briefly introduced at the beginning of this section. They may be broadly organized as systems that primarily transfer *vertical forces* to the ground and systems that resist large *horizontal forces*. They are classified as *shallow* and *deep foundations* as explained in Fig. 2.32. As the name suggests, shallow foundations transfer loads in bearing close to the ground surface; they may be further subdivided into *spread footings* and *mat foundations*.

Spread footings are divided into *isolated footings* (e.g., column footings), *strip footings* (e.g., for walls and rows of columns), *combined footings,* and *strap footings*. In conventional foundation design, bearing walls are supported on reinforced concrete spread footings of the strip type; footings for transverse partition walls are reinforced thickened areas of the grade slab. When spread footings and basements are below the groundwater table, the basement slab must be designed to resist hydrostatic *uplift pressure* (Fig. 4.33). Naturally, should the building stand on landfill, the footings would have to be supported on piles or piers.

Combined footings carry two or more columns or walls that are either so close to each other that their individual footings would overlap or where a column or wall is too close to the property line and would cause a large rotation on a single footing due to eccentric action. The designer must determine the shape of the combined footing, be it rectangular or trapezoidal, so that only uniform pressure is generated. In other words, to avoid rotation and unequal soil pressure, as well as unequal settlements, the centroid of the bearing area of the combined footing should coincide with the resultant of the loads acting on the footing. Should the distance between an eccentrically loaded perimeter column and an interior column be large and the bearing capacity of the soil be high, then the single footings could be linked by a strap beam rather than by a slab. For this *strap* or *cantilever footing,* the individual footings should be proportioned to generate only uniform soil pressure.

Individual footings are usually constructed in reinforced concrete, but for lighter loads they are also constructed as plain concrete footings, possibly of the pyramidal or stepped type. Occasionally, steel grillage foundations encased in concrete are necessary to spread heavy loads from steel columns to a wide base; in the past, timber grillages have been used to support masonry footings.

For poor soil conditions, an entire building may be placed on a *mat* or *raft foundation*. Here the upward acting soil pressure is in balance with the downward column and/or wall loads. The foundation can be visualized as an inverted floor structure using identical framing systems.

In the case of extremely poor soils and high groundwater level and to minimize settlements, the entire building substructure may be considered a cellular rigid foundation. The building is floating similarly to a ship, when the weight of the excavated earth is approximately equal to the weight of the building, thus keeping settlements to a minimum (see also Archimedes', principle, Fig. 4.33).

Deep foundations are used when adequate soil capacity is not available close to the surface and loads have to be transferred to firm layers substantially below the ground level. Deep foundations are basically columns. When settlement is a primary problem, then a pile length must be selected to minimize differential settlement.

The common deep foundation systems for buildings are *piles* and *piers* (caisson piles). While the small-diameter slender piles are normally driven into the ground, the large-diameter piers are placed by first excavating a hole; this distinction, however, may not always be that clear. Piers are made of treated timber, steel, and cast-in-place or precast concrete, or they may be composite systems. Piles can act in end bearing and/or skin friction. Whether a single building column is supported by only two piles or whether a group of columns (walls) is supported by a cluster of piles, a *concrete cap* is always necessary to distribute the loads from the superstructure to the piles. Where pile groups are subject to lateral forces and to avoid bending of piles, it may be advantageous to employ *batter piles;* typical batters vary from 1/12 to 5/12.

Other deep foundation systems occasionally used are *slurry walls* (i.e., a method of construction for earth-retaining walls) and *caisson foundations,* which are generally used for the construction of bridge piers and abutments. A caisson is a massive, cellular hollow box structure that is sunk into position and also provides the bracing for the excavation. The three major types are the *box caisson* or *floating caisson* (open at top and closed at bottom), the *open caisson* (open at top and bottom), and the *pneumatic caisson* (closed at top, open at bottom, and filled with compressed air to prevent water from entering the working chamber), as may be used for constructing an underground garage.

In seismic areas, the individual spread and pile foundations may have to be tied together by bracing struts so that the entire building foundation can act as a unit in sharing the load resistance. Special foundations are sometimes required, as for the building that cuts into the hill (Fig. 2.32). One solution could have been to let the building act as a cellular retaining wall in order to transfer the lateral earth pressure to the foundations. Another approach could have been a reinforced soil wall, where the earth is stabilized by some reinforcement technique. The solution shown employs tensile anchors or tiebacks to support the embankment.

It is apparent that for low-bearing soils the structural layout should be uniformly arranged with many supports and hence many small foundations so that the loads are distributed over a large area. Buildings with only few core supports and large foundations require firm soils.

Foundations for low-rise residential buildings of not more than two stories are usually selected and sized based on empirical methods and rules of thumb, since the loads in comparison to the bearing capacity of the soil are small. The typical foundation systems are as follows:

- Bearing walls with footings for buildings with basements
- Foundation walls or beam or pier (column, pilaster) foundations for buildings with crawl spaces
- Concrete slabs on ground and grade beams, which in turn may be supported on spread or pile foundations

Typical low-rise industrial buildings use piers to transfer the loads from the columns to the column footings; piers can be treated as short, stocky columns. *Grade beams* span between the column footings to support the exterior walls (i.e., wall loads are not transferred directly to the soil); the size of the grade beam depends on the weight of the wall. In warmer climates, where frost depth is not critical, shallow grade beams may be constructed as part of the floor slab to form a *thickened slab.*

Although self-balancing structures, such as ties and rings, may take care of the horizontal thrust produced by the weight of long-span buildings, horizontal forces due to wind and seismic action must still be transferred to the soil, as discussed in the introduction to this section. Large lateral forces are transmitted by *abutments,* such as the typical T-shaped ones (Fig. 9.2g), possibly with inclined footings and subgrade ties. The anchorage of the large tensile forces in tension structures is most critical. For the discussion of *tensile foundations,* refer to Section 9.10 (Figs. 9.32, 9.33).

In contrast to conventional houses, earth-sheltered residences (Fig. 4.31) must not only support an extensive earth weight on top of their roofs, yielding a high dead load of the structure as well as large foundations, but must also withstand lateral earth and possibly groundwater pressures. These lateral pressures may become critical when they do not balance each other and tend to move the building, as indicated for some cases in Fig. 4.31. Here, the lateral forces must be transmitted to the foundations and then into the ground. To reduce and control the magnitude of the lateral earth pressure, fully drained sand and gravel backfill should be used to prevent the possible swelling of clay and any loading due to frost. The heavy, relatively constant loads on underground buildings may make funicular (curvilinear) structures much more appropriate and effective.

Sizing Shallow Footings

As the name suggests, shallow foundations transfer loads in bearing close to the surface. They either form individual *spread footings* or *mat foundations,* which combine the individual footings to support an entire building or part of it. The two systems may also act in combination with each other, for example, where a service core is seated on a large mat while the columns are founded on pad footings.

The base size of column and wall footings or of mat foundations depends on the allowable bearing capacity of the soil, q_a. The allowable bearing capacity for shallow foundations supported on cohesive soils, such as stiff clays, depends on their shear strength, as determined by dividing the ultimate bearing capacity by a safety factor, for example, of 3. In this instance, the ultimate bearing capacity is derived from an average contact pressure that causes the soil mass to fail in shear. However, for cohesionless soils such as sands and gravel and cohesive soils of soft clay, the allowable bearing capacity is derived from the control of settlement, which is much lower than that obtained from the ultimate bearing capacity. The settlement for noncohesive soils, such as loose sands, is primarily of an immediate nature because they are permeable, and hence the water is squeezed out quickly; it consists of the contact settlement directly under the foundation displacing the soil laterally and changing the soil profile and of compression settlement. In contrast, compressible clays have a very low permeability, so a substantial part of the final settlement is due to long-term consolidation movements; in other words, settlement as a result of soil volume reduction due to the extrusion of water from the voids is a very slow process.

For the evaluation of the bearing capacity of the soil, the reader should refer to literature on soil mechanics. Building codes give allowable bearing capacities for various rocks and soils; typical values are shown in Table 2.6, which may be used for preliminary estimation purposes. They range from 3 ksf for medium-density soft clay, 8 ksf for loose gravel or compact coarse sand, and 20 ksf for compacted shale, to 200 ksf for hard rock.

The distribution of the contact pressure between the soil and bottom face of the foundation is an important consideration for the structural design of the foundation and for determining allowable bearing pressures. This distribution is rather complex, because a uniform force action does not generate a uniform response in the soil. The response of the soil to loading is not only dependent on the type of force action, but also on the stiffness of the foundation and that of the ground as well. For example, a rigid foundation concentrically loaded is supported at points that have deformed the least. These points of maximum pressure occur at the center for cohesionless sandy soils, but along the footing's outer edges for cohesive clayey soils. Nearly the opposite is true for flexible foundations deforming in a bowl shape. The flexible foundation on clay causes maximum pressure at the center, while on sand the pressure distribution is more uniform, with slightly larger values at the ends. Because of the nonuniform character of the soil and since foundations are generally neither rigid nor flexible, it is the common practice to treat the contact pressure as uniform, as indicated in Fig. 2.33.

A footing can be loaded by a translational force concentrically or eccentrically and possibly by rotation due to continuous boundary conditions. Typical moment-resisting footings are used for retaining walls, shear walls, frames, chimneys, and so on. Should the line of action of the force resultant P of all the loads acting on the foundation pass through the centroid of the base area A of the footing, then there is no rotation and the contact pressure is uniformly distributed. The average contact pressure q, which must not exceed the allowable bearing capacity q_a of the soil, is

$$q = P/A \leq q_a \tag{2.102}$$

Hence, the minimum required base area of the footing is

$$A = P/q_a \tag{2.103}$$

where P consists of all the vertical loads acting at the footing base, which includes the building loads, foundation weight, and the soil surcharge on top of the foundation.

Should the resultant of the superimposed loads not coincide with the centroid of the footing base area, the footing is loaded eccentrically and rotation is generated. This condition may be caused by placing the force resultant P out of line with the footing's center or by applying an external moment that is transferred through the continuous juncture of column or wall and foundation to the ground, as shown in Fig. 2.33g.

The distance between the resultant of the loads (the resultant of the pressure diagram) and the center of the base area is the eccentricity e. In the case where an external moment M is transferred to the ground, the moment can be replaced by a normal load acting eccentric with respect to the centroid of the contact area of the footing.

$$M = P(e) \quad \text{or} \quad e = M/P \tag{2.104}$$

The equivalent eccentricity e is equal to the ratio of moment to normal force. Should the eccentricity be zero, the moment is zero, and the footing is concentrically loaded.

For a footing eccentrically loaded about one axis, the soil contact pressure is no longer uniform. The maximum contact pressure q_{max} appears along the foundation edges where the translational pressure P/A and rotational pressure M/S are added, while the minimum pressure q_{min} appears at the opposite edge of the foundation where the tension due to rotation decreases the direct compression (Fig. 2.33b to d):

$$q = P/A \pm M/S \tag{2.67}$$

For a rectangular footing with a contact area of $A = Lb$, a section modulus of $S = bL^2/6 = AL/6$, and a moment of $M = P(e)$, Eq. (2.67) can be simplified to read

$$q = \frac{P}{A}\left(1 \pm \frac{6e}{L}\right) \tag{2.105}$$

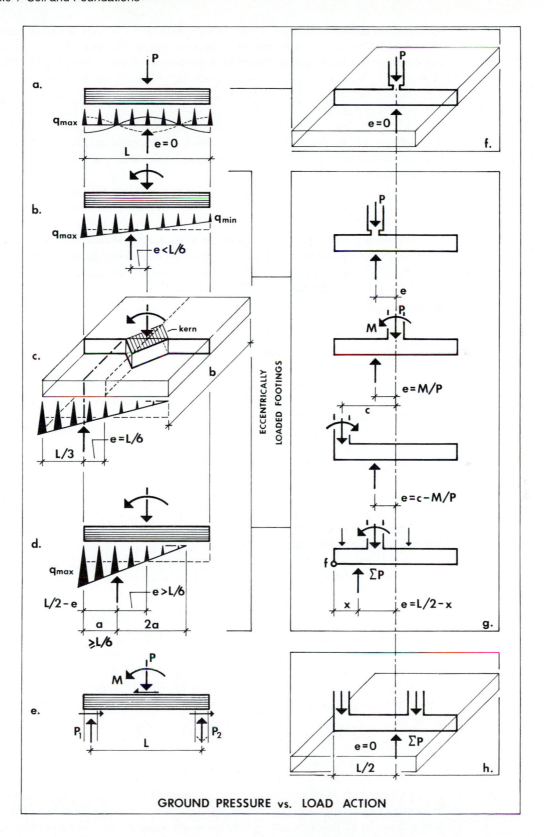

Figure 2.33 Load action on shallow foundations.

This equation can only be used if there is no tensile stress, which cannot be transferred from foundation to soil. The maximum eccentricity of the normal force not causing any tension can be easily determined for a rectangular footing from Eq. (2.105) as

$$q_{min} = 0 = \frac{P}{A}\left(1 - \frac{6e}{L}\right) \qquad \text{or} \qquad e = L/6 \qquad (2.106)$$

The same result is obtained from the location of the resultant force for a triangular pressure distribution (Fig. 2.33c), which acts at $L/3$ measured from q_{max} or $L/6$ measured from the centroidal axis.

We may conclude that the eccentricity e should be less than one-sixth of the footing width L if no tensile stresses are to appear. In general, the load should be applied in the middle third of the base area, which is called the *kern* (Fig. 2.33b). Should the resultant normal force be acting outside the kern, tensile stresses occur. The equation for the maximum compressive stresses under this condition is derived from Fig. 2.33d.

$$\Sigma F_y = 0 = P - q_{max}(3a/2)b$$

or
$$(2.107)$$

$$q_{max} = \frac{2P}{3ba} = \frac{2P}{3b(L/2 - e)} = \frac{P}{A}\frac{4L}{3L - 6e}$$

Note that when $e > L/2$ the footing is unstable. However, since a safety factor of 1.5 against overturning must be maintained, e must always be less than or equal to $L/3$, or $a > L/6$ (see also Fig. 4.29).

Should bending occur about both axes of the base area, an approach similar to that applied previously together with Eq. (2.85) is used to derive the following equations, which assume no tensile stresses.

$$q = \frac{P}{A} \pm \frac{M_L}{S_L} \pm \frac{M_b}{S_b} \qquad (2.108)$$

or for a rectangular footing $e_L \le L/6$, $e_b \le b/6$,

$$q = \frac{P}{A}\left(1 \pm \frac{6e_L}{L} \pm \frac{6e_b}{b}\right) \qquad (2.109)$$

For footings of nonrectangular shape, more laborious procedures (which are beyond the scope of this introductory discussion) are necessary.

In the case where the footing is to be supported by piles, the pile forces can be determined approximately by statics by assuming the piles to be elastic and to be hinged to a rigid foundation (Fig. 2.33e). Taking moments about the pile force P and summing up the forces yields

$$\Sigma M_{P1} = 0 = P(L/2) - M - P_2(L) \quad \text{or} \quad P_2 = P/2 - M/L$$
$$(2.110)$$
$$\Sigma F_y = 0 \qquad\qquad\qquad\qquad \text{or} \quad P_1 = P/2 + M/L$$

Occasionally, contact pressure may not control the design of a foundation. Even by causing the largest contact pressure, a soft strata below may still only be able to support a fraction of this amount. Similarly, when footings are very closely spaced or adjacent footings are located at different levels, overlapping soil stresses are created, which may govern the design (Fig. 2.32, bottom). Problems that may be generated by new foundations influencing nearby older ones should always be kept in mind.

Horizontal forces may be transferred to the soil either by friction between the footing and the soil, by passive soil pressure, or by using a key to anchor the foundation to the ground. Piles may resist horizontal forces in an axial manner (battered piles) and/or in shear and bending (vertical long piles).

EXAMPLE 2.21

A single-column footing carries an axial load of $P = 100$ k, which is assumed to include the footing weight and soil overburden. The allowable soil pressure is 4 ksf. The footing length L is determined for a constant footing width of $b = 5$ ft for the following different loading conditions:

(a)   No rotation, $e = 0$: The required footing size is

$$A = P/q_a$$

$$5L = 100/4 \quad \text{or} \quad L = 5 \text{ ft}$$

(2.103)

Try a 5- × 5-ft footing.

(b) $M = 100$ ft-k: The equivalent eccentric location of the normal force is

$$e = M/P = 100/100 = 1 \text{ ft.} \tag{2.104}$$

Assuming no tensile stresses and using Eq. 2.105 yields a minimum footing length of

$$q_{max} = \frac{P}{A}\left(1 \pm \frac{6e}{L}\right)$$

$$4 = \frac{100}{5L}\left[1 \pm \frac{6\,(1)}{L}\right]$$

or

(2.105)

$$L^2 - 5L - 30 = 0$$

$$L = \frac{5}{2} \pm \sqrt{\left(\frac{5}{2}\right)^2 + 30} = 8.52 \text{ ft}$$

Checking the assumption that there are no tensile stresses,

$$L/6 = 8.52/6 = 1.42 \text{ ft} > 1 \text{ ft.} \tag{2.106}$$

No tensile stresses are present. Try an 8-ft 7-in. × 5-ft footing.

(c) The tension stresses are exactly equal to zero:

$$q_{min} = 0 \qquad \text{or} \qquad e = L/6$$

$$q_{max} = \frac{P}{A}\left(1 + 6e/L\right) = \frac{P}{A}\left(1 + 6L/6L\right)$$

or

(2.111)

$$q_{max} = 2P/A$$

$$4 = 2\,(100)/5L \qquad \text{or} \qquad L = 10 \text{ ft}$$

Try a 10- × 5-ft footing. The equivalent moment acting upon the footing is

$$M = P(e) = 100\,(10/6) = 166.67 \text{ ft-k}$$

(d) $M = 200$ ft-k: The equivalent eccentric location of the normal force is

$$e = M/P = 200/100 = 2 \text{ ft}$$

Assuming that tensile stresses will be present, as based on the result of the investigation in part (c), since $M = 200$ ft-k > 166.67 ft-k.

$$q_{max} = 2P/3ba, \text{ where } a = L/2 - e = L/2 - 2$$

$$\tag{2.107}$$

$$4 = \frac{2(100)}{3(5)\,(L/2-2)} \qquad \text{or} \qquad L = 10.66 \text{ ft}$$

Checking the assumption that there is tensile action present independent of the investigation in part (c), $L/6 = 10.66/6 = 1.78$ ft < 2.00 ft; therefore, tension is present. Try a 10-ft 8-in. × 5-ft footing.

The multiple-column and asymmetrical column footings in Fig. 2.34 are investigated in Problems 2.30 and 2.31.

PROBLEMS

2.1   The typical floor framing of an office building consists of a 3.0-in. concrete slab on cellular steel decking that spans 8 ft between the beams as indicated in Fig. 3.4; it must support a partition dead load of 20 psf. Verify that the dead load a typical beam (BM) must support is approximately 60 psf, which includes 15 psf for ceiling, mechanical, and electrical systems, as well as fireproofing; the weight of the floor beams is considered insignificant as compared to the other loads.

2.2   A 20-story, 246-ft high, laterally braced flat slab concrete structure has plan dimensions of 100×200 ft. The live load for the office floor area is 50 psf. Estimate the average approximate building density, assuming an average equivalent floor weight of 164 psf (which includes the superimposed dead loads of 10 psf for ceiling, finish, and mechanical systems and 6 psf for partitions) and a curtain weight of 15 psf of wall area.

2.3   Determine the reduced live loads for the girder in Example 3.8 (Fig. 3.9) using a live load of 40 psf and a dead load of 15 psf.

2.4   Determine the live load reduction for the floor beam in Fig. 3.3f, considering a live and dead load of 60 psf each.

2.5   For a 15-story, flat slab office building, with a 20- × 20-ft column grid, determine the interior column loads at the first-floor level and then at the thirteenth-floor level. Assume a dead load of 140 psf and a live load of 80 psf. Consider the roof load as 75% of the floor load.

2.6   Determine the roof snow load on an ordinary inclined roof with a height-to-span ratio of 8/12 located in a region with a ground snow load of 40 psf.

2.7   Consider a steel beam (A36) with a span of $L = 20$ ft to be installed at 40°F and to be fixed between its boundaries. Determine the thermal stresses at 120°F for a mean beam temperature (i.e., axial action only).

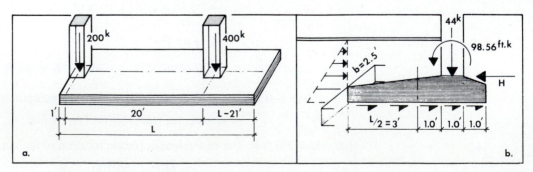

Figure 2.34 Foundation exercises.

2.8   Find the thermal stresses for the beam in Problem 2.7 at a temperature of –30°F and compare them with the allowable stresses of, say, 22 ksi.

2.9   Determine the maximum spacing of vertical expansion joints for a facade brick wall. Use 1-in. joints with a sealant that has a maximum allowable strain of 50%. The wall is built at a temperature of 35°F; consider a maximum future summer temperature of 115°F. The coefficient of expansion for the brick wall in the horizontal direction is 3.4×10^{-6} in./in./°F, and the factor for moisture expansion is roughly equal to 0.0002 in./in.

2.10 For the trusses in Fig. 5.51, determine the forces in the members that are identified by letters at their ends using the methods of joints or sections (see also Problem 5.32).

2.11 For the cases in Fig. 2.8, determine the reaction forces assuming a magnitude for the external loads as well as dimensions for the structures.

2.12 Determine the shear and moment diagrams with numerical values for the beams in Fig. 2.13 (see also Problem 3.9).

2.13 Determine the location of the centroidal axes and the moment of inertias about these axes for the sections in Fig. 2.15b and c; assume dimensions for the sections.

2.14 What would happen in reinforced concrete if the coefficients of the thermal expansion of steel and concrete are not approximately the same?

2.15 An aluminum frame of 6061 alloy for a window is 15 ft long and holds a piece of plate glass 14.99 ft long when the temperature is 60°F. In other words, the glass floats in the opening. At what temperature will the aluminum and glass be at the same length? Use

$$\alpha_{al} = 13 \times 10^{-6} \text{ in./in./°F} \quad \text{and} \quad \alpha_{gl} = 5 \times 10^{-6} \text{ in./in./°F}$$

2.16 What effect does material have on the deflection of beams? How does the deflection of beams made of steel, cast iron, aluminum, and timber compare if the beam is part of a lightweight structure where the superimposed loads are large in comparison to the dead load? Consider the beam moment of inertia as constant.

2.17 Repeat Problem 2.16, but consider the beam moment of inertia as a variable and leave the beam depth constant; in other words, evaluate the beam deflection as a function of the elastic flexibility EI.

2.18 To evaluate strength properties of different tensile materials, determine the ultimate length or rupture length of the following materials: carbon steel, wood, aluminum alloy 6061-T6, steel prestressing strand, steel wire rope, Kevlar 49-ropes, S-glass ropes, carbon and glass fibers, alumina whiskers, cotton, and silk. Assume the material to be of constant cross section and to hang vertically so that it is stressed in tension by its own weight and fails in rupture.

2.19 Investigate the materials in Problem 2.18, but now consider deflection as a criterion. Evaluate the deformation in terms of the specific elasticity E/γ. Derive this expression and compare the results.

2.20 Determine the structural efficiency of some materials, such as carbon steel, aluminum alloy, Plexiglass (acrylic sheet), concrete, limestone, window glass, and wood. Consider the specific strength and the specific elasticity; draw your conclusions.

2.21 For a given superimposed load, two beams, one of steel and the other of aluminum, will have the same moment of inertia if the two materials have the same strength. According to Problem 2.19, aluminum and steel have approximately the same specific elasticity. Does this mean that the two beams deform by the same amount?

2.22 Study the range of the following material characteristics: density, tensile strength, elastic flexibility, ductility, and expansion–contraction due to temperature change. Draw you own conclusions.

2.23 To visualize the thermal properties of materials, consider an element of 100 in. in length to pass through a temperature differential of 100°F. Compare the following materials: steel, aluminum alloy, tungsten, concrete, limestone, brick masonry, window glass, Plexiglass (acrylic sheet), and wood.

2.24 Compare the compatibility of the composite action of a reinforced concrete member 10 ft long for a temperature change of 100°F. Compare the concrete with steel and aluminum bars.

2.25 How much supporting area is required for a given compression load considering the following materials: carbon steel, wood, brick masonry, concrete, limestone, soft clay, loose gravel, sedimentary rock, and bedrock? Express the required area in terms of steel area.

2.26 How much more ductile is a mild steel as compared to a normal-strength concrete?

2.27 A steel pipe 8 Std column is filled with concrete and supports an axial load of 250 k. What are the stresses in the concrete and the steel? Neglect buckling.

2.28 Derive the equivalent uniform loads for the short beam as well as the long beam due to the uniform loads w on a two-way slab (Fig. 2.13b). Do not include the weight of the beam.

2.29 Determine the width b of a rectangular footing with a length $L = 5.40$ ft so that it can support a concentric axial load of $P = 50$ k and a moment of (a) $M = 0$, (b) $M = 22.5$ ft-k, (c) $M = 45$ ft-k, and (d) $M = 77.5$ ft-k. The allowable net soil pressure is 4 ksf for the given sandy soil conditions; the footing weight is considered already deducted from the soil capacity.

2.30 Determine the size of the multiple- column footing or, in this case, a combined footing supporting two columns as shown in Fig. 2.34a so that it does not rotate and uniform earth pressure is generated. Assume a net allowable soil pressure of 4 ksf.

2.31 Determine if the rectangular footing size shown in Fig. 2.34b is satisfactory for the given loading conditions. The lateral thrust H due to gravity loading is carried by a horizontal tie rod connecting the column bases of the frame, the lateral thrust due to the lateral wind forces is assumed to be taken by the passive soil resistance and frictional resistance along the base. Assume an allowable net soil pressure of 5500 psf. Also find the magnitude of the critical moment along the face of the column for which the footing should be designed.

Approximate Structural Design of Common Member Types

It is the intention to introduce or review the procedures needed for the preliminary design of the primary building structure components: beams, tension members, columns, beam columns, and column and wall footings. Considered are the common materials of steel, reinforced concrete, and wood, as well as prestressed concrete and composite construction. In addition, basic concepts of connection design are introduced.

The building design specifications of the American Institute of Steel Construction (AISC), the American Concrete Institute (ACI), and the American Forest and Paper Association are the basis for structural design; they have been incorporated into most building codes in the United States. For the preliminary design of steel and wood members, the traditional allowable stress approach has been used, while for the sizing of reinforced concrete elements the strength method has been employed.

The process of proportioning structural members according to the respective code requirements is often complex and time consuming. To be able to concentrate on the building structure as a whole, simple processes and formulas for the preliminary sizing of elements have been developed.

Although many handbooks are available for the sizing of beams and columns, it is the intention here to develop a feeling for structural behavior and self-confidence in being able to quickly proportion members and thus to be in full control of the design without having to depend completely on ready-made solutions as provided by handbooks and computer software (i.e., standardized computer methods). Simple equations have been developed that make the quick estimate of member proportions possible. This is especially important at various times during design and construction, for example, when a sense for the order of magnitude of forces and stresses must be developed at the early design stage when different structural layout systems are compared and evaluated, when member sizes must be known for esthetic purposes during the developmental stage of architectural design, when fast judgment for size is needed for checking drawings and computer outputs or for the designer to sense, on the construction site, when member sizes don't seem right, or finally, when first trial sections may be needed for the solution of indeterminate structures.

The disciplined approach required for the structural design of members in this chapter, however, should not be interpreted as if this consideration has a predominant position in the general design of buildings or that the general design of structures is fully controlled by structural designers. For example, the various beam–column cases in Fig. 1.20 convincingly express structure as architecture and articulate a clarity reflecting their own orders whether based on modularity, kits of elements, craftsmanship, machinelike high-precision technology, construction process, integrity of materials, logical disciplined simplicity, economy and function, biological or historical forms, minimal shapes, and so forth. Whatever the reason for the various forms, all are an integral part of the total design and are proof that the designer is in full control of the structure, which may include other dimensions of meaning than solely support.

3.1 MATERIALS

The mechanical properties of building materials, in general, have been introduced in Section 2.5. Here, only specific material characteristics are discussed so that structural members in steel, wood, and reinforced concrete can be designed according to the respective national material codes.

Structural Steel

In comparison to other structural materials in building construction steel is a very strong material with uniform properties; it offers quality and product consistency. Steel members can be easily produced and quickly erected. Steel buildings are lightweight, thus requiring only simple foundations.

Some common characteristics of steel have already been introduced in the general discussion of material properties (Section 2.5). The primary steel properties refer to strength, ductility, toughness against fracture (i.e., the ability to absorb energy before fracture occurs, which reflects a combination of strength and ductility), hardness (resistance to abrasion), weldability, fatigue resistance, and corrosion resistance. These mechanical properties are generally controlled by chemical composition, rolling process, and heat treatment. Certain properties, however, are common to all steels. They are elasticity ($E = 29,000$ ksi), density (490 lb/ft^3 or 3.4 lb/in.^{2}/ft of length), thermal expansion ($\varepsilon_t = 6.5 \times 10^{-6}$ in./in./$^\circ$F), and fire resistance.

Steel is a high-strength material but also heavy and expensive, therefore, the profile of steel shapes is optimized. Member cross sections consist of relatively thin parts, which, in turn, yield a high ratio of strength to weight (specific strength) and a relatively low ratio of cost to strength. The slender parts, however, also make members susceptible to buckling, so stability considerations become most important in steel design. When high-strength steels are employed, only small member sizes are needed, so stiffness must also be taken into account as a primary design determinant.

The main negative features of steel include *maintenance, fatigue,* and *fireproofing.* Exposed steel rusts and therefore must be painted or coated with nonrusting metal (e.g., zinc, aluminum). Fatigue must be considered for repeated loading conditions (i.e., repeated stress reversals), and the steel strength may have to be reduced. Most critical in steel construction is the fact that steel members must be protected against fire for a certain time period, since steel loses strength with an increase in temperature, and at approximately 1000°F the yield strength has almost reached the allowable stress levels. This is quite significant considering that ordinary building fires may reach temperatures on the order of 1300 to 1700°F! In addition, steel begins to creep at about 500° to 600°F; in other words, the modulus of elasticity does not remain constant but decreases with temperature increase! Also, the coefficient of expansion cannot be treated as a constant anymore. For further discussion refer to Sect. 1.1.

Steel consists of at least 95% iron combined with up to 1.7% carbon and other alloying materials depending on the type of steel desired. These other alloying elements may be chromium, copper, manganese, nickel, silicon, or others. The chemical composition, besides the manufacturing process, controls the main properties of strength, corrosion resistance, and weldability of steel. Carbon is the most important element. With an increase of carbon content, the strength and hardness of steel increase, but this increase affects adversely ductility, toughness, and weldability. It is well known that ductile carbon steels weld better than brittle steels. On the other hand, high-strength steels are more corrosion resistant than carbon steels. The difference between the ductile carbon steel and the brittle high-strength steels is clearly demonstrated by the stress–strain curves in Fig. 2.19. The yield stresses F_y and the tensile strengths F_u of the common structural steels vary from 32 to 58 ksi for mild carbon steels to 270 ksi for prestressing strand.

The specifications for the many types of steel are published by the American Society of Testing and Materials (ASTM). The ASTM standards have a common general format; they are used to control the quality of all steels and provide standard designations. The specifications of the American Institute of Steel Construction (AISC) for structural steel buildings conform to the ASTM grades of steel and are published in the AISC *Manual of Steel Construction*. AISC was founded in 1921 and published its first steel design specifications in 1926. These specifications have provided a basis for uniform practice in steel construction and are incorporated into most building codes in the United States. Other material specifications related to steel may also have to be consulted. They are published by the following agencies:

- American Association of State Highway and Transportation Officials (AASHTO)
- American Iron and Steel Institute (AISI)
- American Railway Engineering Association (AREA)
- American Welding Society (AWS)
- Concrete Reinforcing Steel Institute (CRSI)
- Post-tensioning Institute (PTI)
- Precast/Prestressed Concrete Institute (PCI)
- Steel Deck Institute (SDI)
- Steel Joist Institute (SJI)
- Welded Steel Tube Institute (WSTI)

The most common structural steels may be classified as follows and are described in the AISC Manual.

> **Carbon steels:** These steels range from low carbon steels (< 0.15% carbon) to high-carbon steels (0.60% to 1.70% carbon). ASTM A36 is a mild carbon steel (< 0.30% carbon) and the most common steel used in building construction. It has a minimum yield stress of $F_y = 36$ ksi and a minimum tensile strength $F_u = 58$ ksi.

High-strength, low-alloy steels: Among the many high-strength structural steels are ASTM A441 and A572. The minimum yield stress of these steels varies from 40 to 65 ksi and the minimum tensile strengths from 60 to 80 ksi depending on plate thickness and shape size. These steels are weldable and have an atmospheric corrosion resistance about twice that of carbon steel; they contain less than 5% alloying elements. ASTM A572 Grade 50 steel is generally the typical choice for building construction using high-strength steels.

Corrosion-resistant, high-strength, low-alloy steels (ASTM A242 and A588): These structural steels have minimum yield stresses ranging from 42 to 50 ksi and minimum tensile strengths from 63 to 70 ksi. They are self-weathering weldable steels, where the exposed unpainted surfaces oxidize (rust) and develops a protective coating (patina) that protects from further corrosion.

High-strength quenched and tempered alloy steels: These are very strong structural steels that obtain their high yield strengths through heat treatment by quenching (cooling) and tempering (reheating); they do not exhibit well-defined yield points as the carbon and high-strength, low-alloy steels do. ASTM A514 (high yield strength quenched and tempered alloy-steel plate) has a minimum yield stress of 90 to 100 ksi and a minimum tensile strength of 100 to 130 ksi depending on the thickness of the plate.

The various types of steel used in connections refer to, for instance, bolts, washers and nuts, rivets, anchor bolts, and threaded rods; filler metal and flux for welding; and stud shear connectors.

There are many other steels in building construction besides the ones just discribed and given by the AISC specifications, such as the following:

- Structural tubing
- Round pipes
- Sheet and strip steel
- Structural cold-formed members
- Stainless steels (e.g., curtain walls)
- Hot-rolled deformed bars (for concrete)
- Welded-wire fabric (for concrete)
- Steel wire (for prestressing)
- High-strength alloy bars (for prestressing)
- Strand and rope cable (for suspension structures)
- Steel castings and forgings (e.g., joints)

Most steel shapes are made of hot-rolled steel. However, keep in mind that other *hot forming* processes are being used for certain situations. For example, long elements can be formed by *drawing* (e.g., wires) or by *extrusion* (e.g., round and square bars), and complex shapes may be made by *castings* or *forgings*. Hot-rolled thin sheets or strips may later be cold formed into various types of structural shapes (i.e., light-gage steel products). The common rolled structural shapes range from simple plates and bars to complex cross-sectional sections. These shapes are listed and described in the Shape Tables of the AISC Manual. They are available in a number of groups, ranging from light to heavy shapes and from thin to thick plates or bars. The minimum yield stress varies with the size of shape and thickness of plate; as the member size or plate thickness increases, the yield stress tends to decrease.

The standard structural shapes and their designations are as follows:

- Wide-flange shapes (W shapes) are used as beams and columns; they are designated by their nominal depth and the weight per foot (e.g., W16 × 67) and are the most widely used structural members.
- American standard beams (*S* shapes)
- Miscellaneous shapes (*M* shapes)
- Bearing pile shapes (*HP* shapes)
- Channel shapes (*C, MC* shapes)
- Equal leg and unequal leg angles (*L* shapes)
- Structural tees split from *W, M* and *S* shapes (*WT, MT* and *ST* shapes)
- Round steel pipe
- Structural tubing (*TS* shapes)

For the preliminary structural design of the steel members, the allowable *stress design method* (working stress design or elastic design) is used. Here, the allowable stress is obtained by dividing the minimum specified yield stress by a safety factor as given by the AISC specifications and as is further discussed in the respective member design sections; the safety factor represents the ratio of strength to the maximum stress in the member. Usually, for brittle materials the safety factor is based on the ultimate strength, while for ductile materials, like steel, it is based on the yield stress (unless fracture occurs), which is treated as failure because of the large deformations that will occur in the plastic range. In other words, the allowable stress represents a certain percentage of the yield stress, such as for the typical conditions of $0.4F_y$ (shear), $0.60F_y$ (tension), $0.66F_y$ (bending), and $0.90F_y$ (bearing).

Structural Wood

Structural wood is used mostly in its natural state as sawn lumber, but also as manufactured products such as glued laminated timber, plywood, particleboard, and other wood-based fiber products. First, structural lumber will be briefly discussed.

Wood is an organic material and occurs in nature in many variations. It is comprised primarily of microscopic, vertical, long, tubular cells composed mainly of cellulose and cemented together by lignin. New layers of cells are formed each year in the growing tree along the outer surface in the cambium under the bark to form annual growth rings. In other words, the tree grows outward, thus increasing in diameter. The trunk cross section clearly shows a series of concentric rings that are organized in three zones: the bark, the light-colored sapwood, which is living tissue and functions as a food supply system, and the inactive dark-colored central core of heartwood, which is dead tissue and is used as support structure for the tree (Fig. 3.1). This structure of the wood clearly reflects the anisotropic nature of the material, which is very different from the crystalline structure of metals.

Trees are classified as hardwoods and softwoods. *Softwoods* are conifers, which have needles or scalelike leaves; they are usually evergreens. Common softwoods are cedar, fir, hemlock, larch, pine, redwood, and spruce. *Hardwoods* are deciduous trees or broadleafed trees that shed their leaves in the fall. Common hardwoods are beech, birch, locust, maple, oak, walnut, etc. The terms hardwood and softwood are misleading since they do not refer to the hardness or softness of the wood. The most widely used species of trees used as structural members in the U.S. building construction industry are Douglas fir and southern pine. It is not a simple task to identify the physical and mechanical properties of sawn lumber considering the many types of trees and location within the tree from where the wood has been cut. Among the important properties are specific gravity, moisture content, and defects that affect the strength of wood.

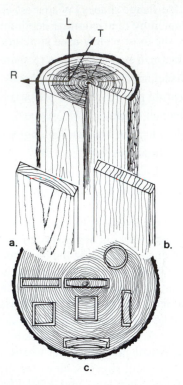

Figure 3.1 Wood characteristics.

The specific gravity G is a measure of the amount of cellulose present; it is defined as the ratio of oven-dry wood divided by a water weight of equal volume. In other words, specific gravity can be converted to oven-dry unit weight of wood, w_o, by multiplying it by the unit weight of water, γ_w. For example, for a specific gravity of $G = 0.49$, the corresponding oven-dry unit weight is

$$w_o = G\,\gamma_w = 0.49(62.4) = 30.58 \text{ lb/ft}^3$$

Correcting the unit weight by taking into account, say, a 15% moisture content yields the unit weight of the actual wood.

$$w = 1.15\,(w_o) = 1.15(30.58) \cong 35 \text{ pcf}$$

The unit weights of various wood species depend not only on density (i.e., percentage of summerwood present), but also on the moisture content, which can be quite large. Average unit *weights* at 15% moisture content for most species fall between 22 to 48 lb/ft^3, where Douglas fir–larch weighs about 34 lb/ft^3 and southern pine about 37 lb/ft^3. Often as an average value for lumber, in general, 35 lb/ft^3 is used, and the weight of treated wood is taken as 50 lb/ft^3.

The *moisture content* (MC) of wood is a most important consideration. It affects the strength, stiffness, durability, weight, withdrawal resistance of nails, and member dimensions. Furthermore, as wood responds to a change of humidity in the surrounding atmosphere, it shrinks and swells thereby causing hidden loads due to the accumulative effects that may result in differential movements causing distortion of openings, sagging floors, cracking of plaster, and so on. The moisture content is expressed as the weight of water in the wood as a percentage of the weight of the oven-dry wood. The moisture is contained as *absorbed* water within the cell walls (i.e., it is bound chemically) and as *free* water, which is held by capillary forces (surface tension) in the cell cavities. The point at which the free water has evaporated but the cell walls are still fully saturated is called the *fiber saturation moisture content,* which is on average about 30% for most species. The removal of moisture (seasoning) from the green wood through air or kiln drying below the fiber saturation point results in shrinkage that is proportional to the amount of water removed. The shrinkage occurs primarily perpendicularly to the grain (very little parallel to the grain) and mostly tangential to the growth rings (i.e., tangential shrinkage) and less so radially to the rings (i.e., radial shrinkage); tangential shrinkage may cause radial cracks (checks) in a log as it dries. *Dry* wood is defined as having not more than 19% moisture, otherwise the wood is considered as *green* or *unseasoned.* It is apparent that lumber should be dried under controlled conditions before it is installed, not only to increase its mechanical properties, that is, to make it stronger, stiffer, and more durable, but also to reduce shrinkage and thus to minimize dimensional changes (warping, distortion) of members. Further substantial shrinkage occurs as the wood is installed, say, at a moisture content of 19% and then dries to a 5% moisture content in winter. Naturally, it would be ideal if the lumber could be seasoned to the point where the moisture content is in equilibrium with that of the surrounding air so that it would neither shrink or swell. This, however, is not possible because of the daily and seasonal changes in humidity and temperature; there will always be moisture fluctuations in the wood. For example, in heated buildings wood loses moisture in winter and hence shrinks, but it swells as it gains the moisture back in summer. It is apparent that for situations where significant shrinkage may be expected (e.g., using green wood timbers) the connections between the various members should take shrinkage movement into account.

The longitudinal *thermal expansion* or *contraction* of wood parallel to the grain is only about 30% that of steel and is insignificant in comparison to the effect of shrinkage due to moisture content changes and, therefore, can generally be neglected. Also keep in mind that wood under long-term loads decreases in strength and, in contrast to steel, *creeps* (in a decreasing rate), which may be critical, especially for beams.

The main *deficiencies* of wood refer to defects in lumber, decay, insects, fire, and weathering. The defects in wood (holes, knots, shakes, splits, and checks) are taken care of by the grading rules for structural lumber. The other aspects refer to considerations affecting the durability of wood, which, in turn, can be controlled by using preservatives. The wood is protected by impregnating it with chemicals by pressure treatment; occasionally, surface treatment may be employed by painting or staining.

Decay of wood is caused by fungi, which depend on moisture, air, and favorable temperature. Fungal growth can be prevented by controlling one of these three necessary criteria, as, for instance, by using dry lumber and properly ventilating the respective spaces. Wood continuously submerged in water will not decay because of the absence of air. Instead of pressure-treated lumber, decay-resistant heartwoods (e.g., cedars, redwood) may be considered. Preventive measures must be taken against *insects* such as termites by keeping the wood away from the ground with a minimum clearance and by properly ventilating the space. Piles in salt waters should be pressure treated to resist the attack of marine borers. Weathering of exposed, untreated wood expresses in change of color (turning gray), warping (when boards warp they tend to pull out their fasteners), surface roughening, and checking or cracking.

Because wood is a combustible material, codes limit the floor area of wood buildings and the height to a maximum of four stories, depending on the type of occupancy and the locality. Although lumber burns, heavy members will retain strength under fire longer than some unprotected metals. To achieve the necessary fire rating, wood members should be oversized, known as *heavy timber,* thus allowing the outer layer of char to reduce further burning. An extra fire protection may be provided by using fire-retardant-treated wood.

Wood is utilized by converting the tree logs into sawn lumber and then grading it in an orderly fashion. In sawmills the logs are either cut tangent to the annual rings called *plainsawing* (Fig. 3.1a), or radially to the rings, called *quartersawing* (Fig. 3.1b). Large logs may be cut both ways. Quartersawed lumber produces an edge grain or vertical grain appearance, while plainsawed lumber produces a flat grain or cross grain (where the fibers do not run parallel with the edges of the lumber; the cross grain is measured by the slope of the grain). Here, the term grain refers to the appearance of the annual rings on the end face of a board or the longitudinal fibers on the surface. Plain sawed lumber is most vulnerable to distortions due to shrinkage. The effect of tangential and radial shrinkage on the deformation of lumber pieces as a function of their location within the tree is shown in Fig. 3.1c.

There is an immense variety in the quality of sawn lumber pieces. They are organized with respect to their use to guarantee minimum standards. The grading of structural lumber (i.e., stress-graded lumber) specifies minimum quality standards with respect to strength and use, taking physical and mechanical material properties into account, including strength-reducing characteristics such as natural defects (knots, splits, and checks, etc.), slope of grain, size of member, and so on. Generally, *visual grading* is the basis for structural design values as controlled by various grading rules agencies in accordance with standard ASTM procedures; machine stress-rated lumber will become more important in the near future. Each grade has a commercial designation, such as Dense Select Structural, Select Structural, Dense No. 1, No. 1, Dense No. 2, No. 2, No. 3, and so forth. Each grade has specific allowable unit stresses assigned to it depending on the species, the direction of load action and the member size classification. Keep in mind that the same stress grade has different allowable stresses in the various size categories.

Structural lumber is classified according to its size and use as dimension lumber, timber, and decking. Dimension lumber is used for light frame construction and timber for post and beam construction.

> *Dimension:* Lumber 2 in. to 4 in. thick and 2 in. to 4 in. wide, light framing and studs (e.g., 2×4 joist); 2 in. to 4 in. thick, 4 in. wide as based on appearance grades rather than structural grades; 2 in. to 4 in. thick, 5 in. and wider, joist and plank grades (e.g., 2×6 joist). Here the designations *wide* and *thick* refer to a horizontal member orientation.
>
> *Beams* and *stringers:* Members that are at least 5 in. in width and are 2 in. deeper than wide; they are graded with respect to loading on the narrow side (e.g., 6×10).
>
> *Posts* and *timbers:* Members of nearly square cross-sections that are at least 5×5 in. in size with a depth not more than 2 in. greater than the width; they are graded primarily for column action where bending is secondary (e.g., 6×6, 6×8).
>
> *Decking:* Solid or laminated lumber from 2 in. to 4 in. thick, and 6 in. and wider (e.g., 3×6 in.), tongued and grooved on the narrow face; it is used as a roof, floor, or wall membrane for span ranges from 6 to 20 ft for roofs.

Lumber sizes are identified by their *nominal* dimensions (e.g., 2×4); for stress analysis, actual dimensions must be used (e.g., dressed size $1\frac{1}{2} \times 3\frac{1}{2}$. For the sectional properties of standard dressed lumber, refer to Table A.9.

The allowable unit stress (design) values for visually graded structural lumber are published by the American Forest & Paper Association (the *National Forest Products Association* before 1993) as a supplement to the *National Design Specification for Wood Construction* (NDS). They are based on seasoned wood species, material density, size classification, and the direction of loading. Some typical approximate design value ranges for Douglas fir–larch and southern pine are given in Table A.8 for preliminary design purposes.

The allowable stress values take into account the complex anisotropic nature of wood; wood is nonhomogeneous and not an isotropic material like steel. Wood is often described as *orthotropic,* that is, as a material that has distinctly different mechanical properties in each of three principal directions: the *longitudinal axis* parallel to the grain, the *tangential axis* perpendicular to the grain and tangential to the annual growth rings, and the *radial axis* perpendicular to the grain and normal to the rings. Hence, the strength of wood is not the same along the main axes; for example, wood has a much higher compressive capacity parallel to the grain (F_c) than perpendicular to the grain ($F_{c\perp}$). Some other wood characteristics are that the compressive strength parallel to the fibers is larger than tensile strength parallel to the fibers (F_t); wood is extremely weak in tension perpendicular to the fibers, thus requiring large joints; the horizontal shear strength (F_v) is only a small percentage of the bending strength (F_b), quite in contrast to steel; the modulus of elasticity (E) along the grain only is significant (it is for short-term loading only); it is low in comparison to steel, indicating the flexible nature of wood and the fact that deflection is often a critical design determinant for beams. The wide variation in strength between the various wood species is expressed by the range of allowable compressive stresses parallel to the fibers varying from extremes of about 400 psi and less to 2000 psi and more. Since the strength of wood depends on so many variables, the safety factors for the working stresses show a wide range from approximately 1.25 to 5, often using an average factor of 2.5. Typical wood has a high strength-to-weight ratio (about equal to that of carbon steel) and a low cost-to-strength ratio.

The tabulated design values in Table A.8 may have to be modified for certain conditions by multiplying them by *adjustment factors* to obtain the allowable design values. The various modifiers refer to load duration, member size, member form, member orientation, repetitive member use, orientation of load action, lateral stability, member curvature, moisture content, fire-retardant treatment, interaction stress, temperature, and other special situations. Some adjustment factors are briefly introduced now, while others are discussed in the context of member design later in this chapter.

Since wood exhibits the special characteristics of carrying substantially greater loads for short durations than long ones, the tabulated design values (with the exception of modulus of elasticity and compression perpendicular to grain) can be corrected, since they are based on normal load duration of approximately 10 years. The typical *load duration factors* C_D given in Table 3.1 should be used for the shortest duration in the combination of loads being considered; that is, for a load combination, the load type with the least duration determines the selection of the factor. For example, for $D + L + W$, the factor is 1.6, while it is 1 for $D + L$, but 0.9 for D alone.

The design values specified in NDS are based on a maximum moisture content of 19%. When the moisture content of the wood exceeds this value for an extended period of time, the design values must be multiplied by the *wet service factors,* C_M.

Furthermore, the NDS design values refer to ordinary ranges in temperature not exceeding 100°F and occasionally heated to temperatures up to 150°F. Wood decreases in strength under sustained temperatures above 100°F; up to a temperature of 150°F, it will regain its strength when cooled. Wood members exposed to sustained temperatures above 100°F must be reduced by the *temperature factors,* C_t.

TABLE 3.1

Typical Load Duration Factors, C_D

Load Duration for Typical Design Loads	C_D
Permanent (e.g., dead load)	0.90
Ten years (e.g., occupancy live load)	1.00
Two months (e.g., snow load)	1.15
Seven days (e.g., construction load)	1.25
Ten minutes (e.g., wind or earthquake loads)	1.60
Impact (e.g., impact load)	2.00

For the preliminary design of structural lumber, the following allowable stress values may be used:

$$F_b = 1200 \text{ psi}, \qquad F_b = 1400 \text{ psi (for repetitive member uses)}, \qquad F_v = 85 \text{ psi},$$

$$F_c = 1000 \text{ psi}, \qquad F_{c\perp} = 500 \text{ psi}, \qquad F_t = 600 \text{ psi}, \qquad E = 1,600,000 \text{ psi}$$

Briefly introduced in the following sections are glued laminated timber and plywood; these wood products represent the most important structural components in the building construction industry. Other structural composite lumber includes laminated veneer lumber (LVL), parallel strand lumber (PSL), and laminated composite lumber.

Glued Laminated Timber.   Sawn lumber members are limited in size, length, and strength by the diameter of the log and the wood species. Glued laminated timbers (glulam timbers), on the other hand, can be fabricated in any size, length, and shape to respond to large loads and long spans (currently up to about 500 ft). They give much more freedom to form giving and allow a variety of shapes, such as tapered, as well as singly and doubly curved arched members, or members of other profiles. Glulam timbers do not depend on the size of a tree; they use the lumber from smaller trees. Furthermore, as heavy timber they offer the advantage of fire resistance. The weight of glulam sections may be taken as 36 pcf.

Glulam timbers consist of relatively thin wood laminations bonded together with adhesives. The grain of all the laminations runs approximately parallel to the length of the member. Usually the laminations are of 2-in. nominal thickness (1½ in. actual thickness), although for certain situations, as for arches and sharply curved members, 1-in. laminations (¾ in. actual thickness) will be used. Therefore, the overall depth of glulam members is a multiple of the number of laminations. The standard nominal widths (where the actual widths are given in parentheses for western species), are as follows:

3 (2½), 4 (3⅛), 6 (5⅛), 8 (6¾), 10 (8¾), 12 (10¾), 14 (12¼), and 16 in. (14¼ in.)

Refer to Table A.10 for section properties of a selected group of structural glulam timber.

Glulam construction permits an efficient use of several wood species (mostly Douglas fir and southern pine) and lumber grades. The desired quality of the wood can be selected and the laminations can be seasoned to the desired moisture content, to increase the strength and to control warping and dimensional stability. In addition, a high-quality control is provided by the production process in the plant. High-grade laminations are usually located at the top and bottom of glulam timber, where the bending stresses are maximum as controlled by the outer laminations in the tension zone; lower-grade wood with the necessary shear strength can be placed near the neutral axis. This layout of the laminations is of full advantage when the beam is loaded perpendicular to (the wide faces of) the laminations, which, naturally, should be the primary direction of loading. When the beam is loaded in the other direction (i.e., bends about the weak axis), much of the advantage of glulam timber is lost. For glulam columns, same–grade laminations are distributed uniformly across the member cross section when the force action is primarily axial.

The allowable stresses for glulam member design are given in the *Standard Specifications for Structural Glued Laminated Timber of Softwood Species,* published by the *American Institute of Timber Construction* (AITC) in the *Timber Construction Manual.* The design values are based on, among other criteria, the use of visually graded and machine-stress–rated laminations, members not more than 12 in. deep, normal duration of load, dry conditions of use, that is, a moisture content less than 16% as for most covered structures, relatively straight prismatic members, and adequate lateral support. Should these conditions not be met, the tabulated design values may have to be modified for depth, member slenderness, moisture content (wet conditions of use), load duration, pressure treatment, tapered and curved members, beam columns, lack of lateral support, and when members are loaded parallel to the wide faces of laminations (i.e., flat use factor, C_{fu}).

The standard allowable bending stresses F_b for glulam timber range from 1600 to 2400 psi, indicating that the variation in strength is less by far than for sawn lumber, besides demonstrating that its average allowable stress is substantially higher than an ordinary allowable bending stress, say of 1400 psi for sawn lumber. Notice also the difference in allowable shear stress F_v of, say, 165 psi for glulam beams to 85 psi for sawn lumber.

For the preliminary design of glulam members, the following allowable stress values may be used conservatively:

$$F_b = 1800 \text{ psi}, \qquad F_v = 165 \text{ psi}, \qquad F_c = 1500 \text{ psi}, \qquad F_{c\perp} = 500 \text{ psi}$$

The recently developed fiber-reinforced glulam where the laminated layers are replaced with a high-strength fiber reinforced plastic, increases the overall design strength of beams by about 50 to 80%.

Plywood.   The most popular use of plywood is as roof, floor, and wall sheathing in residential construction, although it is also found in commercial and industrial buildings. Among the numerous other applications in the building construction field are complex roof forms such as folded plate and shell structures. It is used as concrete formwork and in the fabrication of structural members such as flat or curved stressed skin panels, sandwich panels, and plywood beams. These beams may have an I-shaped cross section or they may be box beams consisting of plywood webs and lumber flanges (i.e., composite plywood–lumber beams), or they may be plywood-glued laminated beams. Plywood panels are usually made up of an odd number of thin wood veneers or plies of at least three layers (in general, 3-ply, 5-ply, and 7-ply panels), ranging in thickness from 1/16 to 5/16 in. and bonded together with adhesives. The plies are placed with their grain at right angles to each other at alternate layers so that they form crossbands with respect to the face plies, thereby improving the strength of the plywood panels and controlling shrinkage. An odd number of wood veneers is generally used so that the grain of the face plies runs in the same direction, which provides a symmetrical layout and thus minimizes warping. The standard panel sizes are 4 ft × 8 ft and range in nominal thickness from 1/4 to $1\frac{1}{8}$ in.

Plywood panels, particularly the thicker ones, can be treated as having nearly isotropic properties. Their thickness usually depends on the loads acting normal to the surface as due to gravity and wind. The panels transfer these loads in bending as one-way slabs using the *1-ft strip beam method* for purposes of analysis. The orientation of thin plywood panels should be such that they span with their grain perpendicular to the supports (i.e., parallel to the span) to be more effective. In diaphragm action, plywood panels act as thin webs (plates) where the shear parallel to the surface must be resisted by the plywood panels, but realizing that nailing requirements related to the transfer of this shear become most critical.

Plywood can be produced from many wood species, which have been organized in five strength groups. The strongest and stiffest group with the highest structural grade level includes Douglas fir and southern pine, which are commonly used in building construction. Plywood panels are manufactured according to their resistance to moisture as *exterior* (exposed to weather or high-moisture conditions), using glue that is 100% waterproof together with the appropriate ply grades, and as *interior* types with highly moisture resistant glue. The strength of plywood is naturally a function of the quality of the veneers, which, in turn, are classified in five grades. For a more detailed discussion of plywood classification and mechanical properties, as well as the effective section properties of plywood, refer to the *Plywood Design Specification* (PDS) as published by the *American Plywood Association* (APA).

The design values for plywood are similar to those for lumber; plywood has a high ratio of strength to weight. For preliminary design purposes the following allowable stresses for plywood, as based on normal duration of loads, may be used:

$$F_b = 1650 \text{ psi}, \qquad F_c = 1540 \text{ psi}, \qquad F_t = 1200 \text{ psi}$$

$$F_s = 75 \text{ psi rolling shear in the plane of plies}, \qquad E = 1,800,000 \text{ psi}$$

These values are for grade-stress-level S-2 using group 1 species and dry conditions as applied to roof decking, subflooring, and wall sheathing.

For typical situations, plywood floor sheathing spans up to 4 ft and roof sheathing up to 6 ft. In contrast, *particleboard* panels are of much less strength than plywood, but are sufficient for ordinary floor, roof, and wall sheathing for spans generally up to 2 ft.

Reinforced Concrete

Reinforced concrete is a nonhomogeneous, composite material that consists of concrete and steel reinforcement. The reinforcement is needed because the concrete capacity in tension is very low. In contrast to ductile carbon steel, concrete is brittle and fails suddenly without warning—it is not a truly elastic material; in other words, its behavior in the elastic range is difficult to predict. Concrete shrinks as it dries and may crack; it creeps under sustained loading, and it forms microcracks under higher stresses. It is apparent that the structural design of reinforced concrete is more complex than that of ordinary steel or wood.

The various branches of concrete, besides common reinforced concrete, include plain concrete, fiber-reinforced concrete, ferrocemento, prestressed concrete (see Section 3.2), precast concrete, and new cement-based materials. Here, only the properties of cast-in-place normal-weight reinforced concrete are briefly discussed.

The Material Concrete.   Concrete consists of a mixture of aggregates bound together by cement, water, and admixtures. Critical with respect to its strength is the proportioning of the concrete, that is, the gradation of aggregates, the amount of cement paste needed to fill the voids, and the amount of water used.

Aggregates should be closely packed by using the right combination of sizes to reduce the amount of cement paste needed to fill the voids. The close packing due to gradation, the strength, surface texture (related to bond), and durability of the aggregates are all related to the strength of concrete. Aggregates (natural or manufactured) are graded in size from fine, which pass through a No. 4 sieve (e.g., sand), to coarse (e.g., gravel); usually, the maximum gravel size for ordinary buildings is ¾ in. The maximum aggregate size is controlled by the member size and should not be larger than one-fifth of the narrowest dimension of the forms, one-third of the slab depth, or three-quarters of the minimum clear spacing between the reinforcing bars.

Whereas the aggregates occupy about 70% to 75% of the concrete volume, the cement paste uses approximately 7% to 15%. The hydraulic cement bonds the aggregates chemically together, requiring water for this chemical reaction, called *hydration*. Portland cement is the most common hydraulic cement. In 1824, Joseph Aspdin obtained in England a patent for the first artificial cement, which he called *Portland cement*.

The primary raw materials, from which cement is produced, consist of calcareous components (e.g., limestone) and siliceous components (e.g., shale, clay). These materials are blended and fired (i.e., changed into cement clinker) before they are ground into Portland cement. The various cement types, besides the normal one, may have either of the following characteristics: high early strength, low heat of hydration, and sulfate resistance.

Water is not only required for hydration but also for workability so that the concrete can easily be placed into forms. About 25% of the cement weight in clean water is needed for complete hydration and an additional 10% to 15% for mobility. In other words, the typical minimum water–cement ratio by weight is about 0.35 to 0.40, or 4 to 4.5 gallons of water per 94-lb sack of cement. As the *water–cement ratio* increases, the strength of concrete decreases because the water above about 25% is not required for chemical reaction and produces pores. Whereas additional water improves the workability of concrete, this free water also increases the porosity of the concrete; more voids, in turn, reduce the concrete strength. The *slump test* is used to measure the concrete consistency.

The proportions of concrete mix can be defined either by volume or weight. For example, 1:2:4 stands for 1 part cement, 2 parts sand, and 4 parts gravel.

Chemical admixtures are used to improve workability, to minimize the water–cement ratio, to reduce shrinkage cracking, to improve durability, to speed hydration or retard hardening, to add color, and so on. The common admixtures are *air-entraining agents* that introduce tiny air bubbles to improve workability and durability, as well as provide resistance to freezing and thawing cycles (i.e., air voids act as reservoirs for excess water pressure); *accelerators* that speed up hydration (e.g., for cold weather construction); and *plasticizers* to improve workability, thereby reducing excess water and allowing a minimum water–cement ratio.

After the concrete has been proportioned and mixed, it is conveyed to and placed in the formwork. During this process the segregation of the aggregates from the sand and water must be avoided, as well as the bleeding of the concrete (i.e., the movement of the water to the surface). After the concrete has been placed, it is compacted with vibrators to prevent air pockets (honeycombing) from forming. Then it must be cured for at least 7 days.

The specified strength and structural design are based on 28 days after the concrete has been placed. The concrete has reached about 70% of its strength in 7 days and gained approximately 85% to 90% in 14 days. The final concrete strength depends largely on maintenance of proper moisture and temperature conditions during the initial week, called *curing*. In other words, since concrete hardens through chemical reaction between cement and water, concrete must be prevented from loss of water or from drying too fast, which may require covering with plastic or wet sheets or the use of sprinklers. The concrete must also be protected from freezing in the initial period. In general, concrete gains strength as long as drying of concrete is prevented.

Normal-weight concretes, that is, concretes made from stone or gravel, are assumed to weigh 145 pcf, or 150 pcf when steel reinforcement is included. Structural lightweight concretes, for instance, made from expanded shales and slags, weigh from 90 to 120 pcf. Heavyweight concretes used for dams or nuclear reactors, on the other hand, may weigh more than 200 pcf depending on the nature of the aggregate (e.g., iron ores).

Mechanical Properties of Concrete.   Briefly discussed next are the properties of concrete related to compression, tension, and the time dependent volume changes: shrinkage, creep, and thermal movement.

Typical stress–strain curves for common concrete strengths are shown in Fig. 2.19, bottom. Notice that there are different curves with different shapes for the various concrete strengths, and also notice that low-strength concretes exhibit more ductility. These curves are based on testing the compressive strength of standard cylinder specimens (6-in. diameter by 12-in–high concrete cylinders) according to ASTM specifications after they have hardened for 28 days. The highest compressive stress at the top of the curve is denoted by f_c'; in other words, f_c' represents the 28-day cylinder compressive strength. All ordinary concretes up to about 5000 psi reach their maximum strength at a strain of about 0.002. Although the ultimate strains at the end of the descending branches (at failure) for the concretes vary, the ACI Code recommends a maximum usable ultimate strain of $\varepsilon_u = 0.003$ for all strengths of concretes.

The most common compressive strengths in reinforced concrete construction are between 3000 to 4000 psi and between 4000 to 6000 psi in the prestressed concrete field. It is apparent, however, that as a building gets taller the strength of the columns must increase toward the base. For example, in a 70-story reinforced concrete building, concrete with strengths up to 12,000 psi may be used in the low- and mid-range levels and from 10,000 to 4000 psi for the high levels. The highest strength ever used in a building currently is 19,000 psi. Researchers, however, are able to make concrete in the 25,000- and 35,000-psi range in laboratory context; a superhigh-strength 100,000-psi concrete was tested by J. F. Young at the University of Illinois in 1991.

To determine the modulus of elasticity E_c for concrete is not a simple task. As the various stress–strain curves indicate, it differs for each concrete strength (i.e., each curve is different). Furthermore, the stress–strain relationship in the elastic range is not a straight line, or possibly only at a low stress. At normal stress levels it is a curve; hence, stress and strain are not proportional as for other materials. Concrete is usually assumed to behave elastically up to about $0.5f_c'$ so that a line drawn from the origin to that point, the slope of which is E_c, is often referred to as a *secant modulus*. The ACI Code specifies the modulus of elasticity for concretes weighing between $w_c = 90$ and 155 pcf as based on short-time tests. The empirical equation, Eq. (3.1), reflects that it is not a constant as for steel but depends on its strength and its weight. The expression can further be simplified for normal-weight concrete as

$$E_c = w_c^{1.5}\, 33\sqrt{f_c'} \;= 57,000\sqrt{f_c'} \quad \text{(psi)} \tag{3.1}$$

For example, E_c is 3605 ksi for 4000-psi concrete, or its range for ordinary concretes is from about 3000 to 4000 ksi.

It must be emphasized, however, that not only the curved stress–strain relationship in the elastic range, reflecting the nonlinear behavior of concrete, makes the prediction of the modulus of elasticity difficult, but, in addition, the type of concrete, type of loading, shape of member, age of concrete, and other long-term effects have a large influence on its magnitude.

In addition to the initial instantaneous elastic deformations that occur when the loads are applied, concrete also undergoes time-dependent deformations due to creep, shrinkage, and temperature changes. Concrete creeps under sustained loads over a period of time (approximately 2 to 3 years). The magnitude of the *creep* strain in the elastic range is approximately proportional to the elastic strain (which is on the order of one to three times the elastic strains), but it increases much faster at higher stresses. It is apparent that long-term deflection must be considered for the design of flexural members.

The drying *shrinkage* that occurs during the early months of construction (i.e., nearly 90% during the first year) is, like creep, a time-dependent deformation. As the excess water that has not been used for hydration evaporates, the concrete dries, causing the exterior surfaces to shorten or shrink and thereby initiating tensile stresses, since movement is generally prevented by the boundaries. The shrinkage strain is affected by the water–cement ratio, the relative humidity of the environment, and the composition and strength of the concrete.

In shrinkage-compensating concrete as often used for slab-on-grade construction, an expansive agent (e.g., anhydrous calcium sulfoaluminate) creates expansion or compressive forces that offset the tensile stresses due to shrinkage. Similarly, synthetic fibers (e.g., nylon, polyester, polypropylene) are often added to concrete as secondary reinforcement to reduce shrinkage cracking while the concrete is still plastic.

The tensile strength of concrete is about 10% to 15% of the compressive strength: 0.1 to $0.15f_c'$. It is approximately proportional to the square root of the compressive strength and is determined indirectly by the *split-cylinder test* or directly by the uniaxial tensile test as approximately equal to $6.5\sqrt{f_c'}$. The tensile strength of concrete is neglected for the design of reinforced concrete members. For the design of plain concrete members, such as footings, the allowable flexural stress in tension with respect to factored loads (See Eq. (3.3)) is taken as

$$F_{tu} \leq 5\,\phi\sqrt{f_c'}\,, \quad \text{where } \phi = 0.65 \tag{3.2}$$

Reinforcement. Concrete must be reinforced for most situations because of its low tensile strength. Besides the common steel reinforcement, fiber reinforcement consisting of fibers of steel, glass, carbon, polypropylene, or other synthetic material have recently become popular. The common types of steel reinforcement for ordinary cast-in-place concrete are hot-rolled deformed bars and welded-wire fabric; whereas for prestressed concrete, they are steel wires, strands, and bars. The mechanical properties of structural steel, in general, have already been discussed at the beginning of this chapter and are not further considered here.

Steel reinforcement for concrete consists largely of bars. They are generally deformed to provide mechanical anchorage with the concrete, rather than relying only on bonding by chemical adhesion or surface roughness as for smooth bars. Full bonding between concrete and steel reinforcement is essential for complete composite action—the bars are not allowed to slip.

The steel bars are available in 11 sizes and are manufactured in diameters from 3/8 to 2¼ in.; they are designated by numbers from 3 to 18 (see Table A.1). For bars #3 through #8, the bar number divided by 8, represents the bar diameter. For example, a #6 bar is 6/8 = 3/4 in. in diameter. The actual bar diameter across the deformations is approximately 1/8 in. larger.

The types of reinforcing bars available are billet steel, rail steel, axle steel, and low-alloy steel; they are listed in Table A.6. Grade 60 steel (i.e., $F_y = 60$ ksi) is most commonly used in reinforced concrete construction. In contrast, strands for prestressing have a minimum ultimate strength of 250 and 270 ksi.

Rolled onto the surface of the reinforcing bars are identification marks that show the steel type, grade of steel, bar size, and producing mill. The pattern of deformation varies with the manufacturer.

Welded wire fabric consists of smooth or deformed cold-drawn wires welded together in rectangular patterns; they are supplied in sheets or rolls depending on the wire size. The wires range from 0.134 to 0.628 in. in diameter. The minimum yield strengths range from 65 ksi (for smooth wire fabric) to 80 ksi (for deformed wire fabric). Welded wire fabric (WWF) is designated first by grid size (the first number gives the spacing of the longitudinal wires), followed by a letter (W) for smooth wires or (D) for deformed wires, and a number that denotes the cross-sectional area of the wire in hundredths of a square inch. For example, WWF6 × 6 - W2.0 × W2.0 represents a 6- × 6-in. grid of smooth wires where each of the wires has a cross-sectional area of 0.02 in.².

The minimum clear cover to the reinforcement for cast-in-place concrete as protection against corrosion by weather or loss of strength from fire exposure is as follows:

- Concrete cast against and exposed to earth   3 in.
- Concrete exposed to weather
    #6 through #18 bars   2 in.
    #5 bar and smaller   1½ in.

- Concrete not exposed to weather

Slabs, walls, joists:	#11 bar and smaller	¾ in.
	#14 bar and larger	1½ in.
Beams and columns		1½ in.
Shells and folded plates:	#5 bar and smaller	½ in.

The spacing of reinforcement is discussed in the context of member design.

Structural Design.   The structural design of reinforced concrete members is based on the *Building Code Requirements for Reinforced Concrete* as published by the American Concrete Institute (ACI), which, in turn, has been adopted by the major model building codes. Some other agencies that provide material specifications and information relevant to concrete construction are the following:

- American Society for Concrete Construction (ASCC)
- Concrete Reinforcing Steel Institute (CRSI)
- National Precast Concrete Association (NPCA)
- Portland Cement Association (PCA)
- Post-tensioning Institute (PTI)
- Precast/Prestressed Concrete Institute (PCI)

Reinforced concrete can be designed by either using the working stress method (i.e., elastic design) or the strength design method. In *working stress theory,* members are designed as based on stresses derived from elastic analysis, with the corresponding material behavior maintained within the elastic range. The service loads cause stresses that must be kept below the allowable ones. The permissible stresses are specified by codes as a fraction of the yield stress, for example, for steel beams as $F_b = 0.66F_y$, or as a fraction of the crushing strength of the material, such as $F_b = 0.45f_c'$ for reinforced concrete beams, or as a fraction of the buckling load for slender columns.

In contrast, the *strength design method* recognizes the uncertainty in loading between the load types and the uncertainty of resistance. Here, the combined effect of the loads is not to exceed the structural resistance to particular failure modes. In other words, the design strength must be at least equal to the required strength. In the strength design method the service loads are increased by *load factors* and the member strengths are reduced by *strength reduction factors*. The factors reflect the degree of uncertainty, that is, inaccuracies in the theory, variations in the material properties and member dimensions, possibly the importance of a member in relation to the overall structure, and uncertainties in the determination of the loads.

Elastic design, in general, is based on a clear understanding and prediction of stresses under service loads in the elastic range, realizing that it does not take into account the uncertainties of various load types. Safety factors are related to the linear elastic range (i.e., up to the proportional limit) and not the ultimate load capacity. Concrete, however, is not a homogeneous material; stress and strain are not proportional to each other under normal stress level. The safety factors for the allowable stresses are based on a straight-line theory and do not reflect the curvilinear (nearly parabolic) shape of the stress–strain diagram; that is, the true behavior of concrete cannot be constant. The behavior of reinforced concrete as a composite material is complicated further by the effects of shrinkage, creep, and cracks, among other criteria. We may conclude that the elastic stresses determined according to the working stress method cannot represent the true stresses but only an estimate.

It is apparent that it is much more reliable to base the design of reinforced concrete on failure or the ultimate load capacity rather than attempting to predict elastic behavior. This is the reason why working stress design for concrete is rarely used anymore. Keep in mind, however, that for the calculation of deflections and cracks at service loads it is still required. For the design of thin shells and folded plates, however, elastic design is an accepted approach by considering reinforced concrete ideally elastic, homogeneous, and isotropic with identical properties in all directions.

According to the ACI Code, in the strength method, the design strength must be at least equal to the required strength. The design strength provided by a member is equal to the nominal strength multiplied by a strength reduction factor ϕ. The required strengths are the service load effects (w, P, V, M, and T) increased by the respective load factors (see Eq. 4.19). For example, the required strength U to resist the dead load D and the live load L shall be at least equal to

$$U = 1.4D + 1.7L$$

Or, for instance, the required moment strength shall be at least equal to

$$M_u = 1.4\,M_D + 1.7\,M_L$$

The magnitude of the load factors depends on the type and combination of loads reflecting the accuracy with which the loads can be predicted. The ACI load factors for other loads and load combinations than dead and live loads (such as wind, earthquake, earth and fluid pressures, differential settlement, creep, shrinkage, and temperature change) are given in Eq. (4.19).

In general, the basic requirements for strength design are

$$\text{Design strength} \geq \text{required strength } U$$

or   (3.3)

$$\phi\,(\text{nominal strength}) \ \geq \ U$$

For example,

Bending strength:	$\phi M_n \geq M_u\ (= 1.4\,M_D + 1.7\,M_L$, for example)
Axial strength:	$\phi P_n \geq P_u$
Shear strength:	$\phi V_n \geq V_u$
Torsional moment strength:	$\phi T_n \geq T_u.$

The strength reduction factors ϕ for the given situations are as follows:

Axial tension, and bending with or without axial tension	0.90
Shear and torsion	0.85
Axial compression with or without bending:	
With spiral reinforcement	0.75
With ties	0.70
Bearing on concrete	0.70
Flexure in plain concrete	0.65

3.2 DESIGN OF BEAMS

Beams are distinguished in *shape* (e.g., straight, tapered, curved), *cross section* (e.g., rectangular, round, T-, or I-sections), *material.* (e.g., homogeneous, mixed, composite), and *support conditions.* (simple, continuous, fixed, etc.). They may be *shallow beams, deep beams,* or *wall beams* depending on their depth-to-span ratio. They may be not only the common planar beams, but spatial beams such as folded plate beams, shell beams, or space trusses. Beams may be part of a repetitive grid (e.g., parallel or two-way joist system) or may represent individual members; they may support ordinary floor and roof structures or span a stadium; they may form a stair, a bridge, or an entire building. In other words, there is no limit to the application of the beam principle.

The effect of load action (eccentric versus concentric) on beam behavior in response to member shape and profile may be in *simple bending, biaxial bending,* or *unsymmetrical bending, a*s discussed in Chapter 2 (Fig. 2.27).

Beams, in general, must be checked for the primary structural determinants of bending, shear, deflection, possible load effects of bearing, and lateral stability. Usually, short beams are governed by *shear,* medium-span beams by *flexure,* and long-span beams by *deflection.* The moment increases rapidly with the square of the span (L^2), thus the required member depth (i.e., lever arm of resisting internal forces, or moment of inertia I) must also correspondingly increase so that the stresses remain within the allowable range. The deflection, however, increases with the span to the fourth power (L^4), clearly indicating that with increase of span deflection becomes critical. On the other hand, with decrease of span or increase of beam depth (i.e., increase of the depth-to-span ratio), the effect of shear must be taken into account, which is a function of the span (L) and primarily dependent on the cross-sectional area of the beam (A). Deflections in the elastic range are independent of material strength and are only a function of the stiffness EI, while shear and bending are dependent on the material strength.

The various beam types and the many possible loading conditions together with the type of bending were introduced in Chapter 2. Here, only typical beams in steel, wood, reinforced concrete, and composite construction as part of floor or roof structures are designed approximately according to the national material specifications. Common framing arrangements and layouts are discussed in Chapter 1 and Fig. 3.2 together with load flow, whereas typical gravity roof and floor loading conditions are presented in Chapter 2. The effects of various structural design criteria on floor and roof framing are studied in Fig. 3.3 and numerically evaluated in the problems (see also Fig. 2.13). They include the following effects:

- Span direction: (a), (b), (k), (m), (n), (s), (z)
- Bay proportion: (b), (c)
- Cantilever beam construction: (t)
- Beam spacing: (d), (e), (f)
- Framing floor openings for stairs, elevators, or other vertical shafts: (g), (h), (y)
- Column layout: (k), (n), (u)
- Cantilevering: (l), (j)
- Slanted corner: (o)
- Scale: (p)
- Corner or intersecting building units: (q)
- Partition wall or other heavy load layout: (r)

For all practical purposes, one-way span action may be assumed for a rectangular concrete slab having the proportions of 1:2 or greater, even if it is also supported along the short sides. On the other hand, it is apparent that a square concrete slab with the proportions of 1:1 and properly reinforced in both directions will span two ways and carry an equal load to each of the four supporting beams.

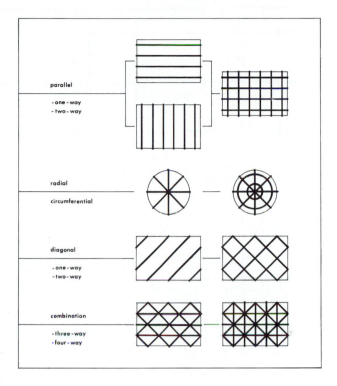

Figure 3.2 Typical floor–roof structure layouts.

In general, for a rectangular, *two-way structure* the loads may be considered to be distributed in two perpendicular directions to the supporting beams according to the tributary areas formed by the intersection of 45° lines extending from the columns, as shown in Fig. 2.13b.

The bending moments for indeterminate continuous beams may be approximated by an equivalent uniform load per lineal foot, as given in Fig. 2.13b and derived in Problem 2.28. The approximate behavior of the two-way slab is investigated in Fig. 6.16 and the respective discussion.

Steel Beams

Typical steel beams for floor or roof framing in building construction are the common rolled sections, cover-plated W-sections, open web steel joists, trusses, castellated beams, stub girders, plate girders, and tapered and haunched-taper beams.

For the design of open web steel joists, the reader should refer to the loading tables of the respective manufacturers or the specifications of the Steel Joist Institute (SJI). The institute adopted the first standard specifications in 1928, followed in 1929 by the first load table. Steel joists have the appearance of shallow trusses, mostly of the Warren-type configuration, and are designed as simply supported uniformly loaded beams assumed to be continuously laterally supported, at the top chord. The standard open-web joist designation consists of the depth, the series designation, and the particular chord type (e.g., 22K6). Three series are available for floor and roof construction:

- K-series standard steel joists range from 8 to 30 in. in depth; they have a span range of 8 to 60 ft.
- LH-series longspan steel joists range in depth from 18 to 48 in.; they have a span range from 25 to 96 ft.

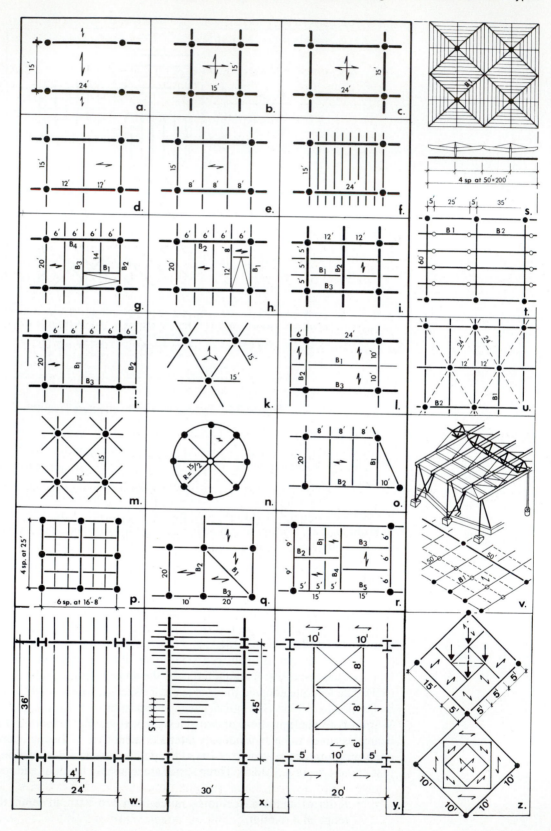

Figure 3.3 Floor–roof framing systems.

- DLH-series deep longspan steel joists range in depth from 52 to 72 in.; they have a span range from 89 to 120 ft for floors and 144 ft for roofs.

The typical joist spacing varies from 2 to 4 ft to provide efficient use of the corrugated steel deck. The joists are stabilized by either diagonal or horizontal bridging. For the preliminary estimate of the joist depth, a depth-to-span ratio of 1/24 may be used, which is about one-half of the typical value for the wider spaced long-span trusses. Therefore, the nominal depth d (in.) of an open web steel joist is approximately equal to one-half of its span L (ft).

$$d = L/2$$

In this section the structural design of the common rolled W-, M-, or S-shapes is briefly discussed. Rolled beam sections are produced in a very wide range of sizes; currently, the largest beam is a W36 × 848 with a section modulus of $S_x = 3170$ in.^{3}. For long spans, say beyond 100 ft, deeper and lighter beam sections should be considered, such as plate girders (single web or double web) or trusses, that is, *built-up members*. The design of ordinary floor and roof beams is generally controlled by bending and then, for certain situations, checked for shear, deflection, and web crippling; for higher grade steels serviceability should be considered. Usually, the axial forces in frame beams are rather small and can be neglected for preliminary design purposes.

Flexure.   The familiar flexural stress for symmetrical rolled beam sections for simple bending according to Eq. (2.64) is

$$f_b = M/S \leq F_b \quad \text{or} \quad S \geq M/F_b \tag{3.4}$$

The section modulus necessary to resist the given moment can easily be looked up in the S_x tables of the AISC Manual and Table A.11 in this book. Here, the various beam sections are arranged in groups, where the bold-printed top section in each group has the least weight and largest S_x of that group.

The allowable stress F_b depends on the lateral support of the compression flange and whether the beam shape can develop its plastic moment capacity under overload conditions before failure due to local buckling occurs. According to the AISC specifications, local buckling of the compression flange and web, as well as the lateral support of the compression flange, must be considered.

A *compact* beam section is one that is capable of reaching its plastic moment capacity before any local buckling occurs. According to AISC, the limiting width-to-thickness ratio for an unstiffened compression flange of an I-shaped rolled beam is

$$\frac{b_f/2}{t_f} \leq \frac{65}{\sqrt{F_y}} \quad \text{or} \quad F_y = F_y' = \left(\frac{65}{b_f/2t_f}\right)^2$$

Similarly, limiting d/t_w ratios are given for webs under combined flexural and axial compression. Here F_y''' identifies the theoretical yield stress above which the shape turns noncompact.

In other words, the theoretical yield stresses F_y' and F_y''' are hypothetical values above which the flange and web are noncompact. They are listed in the W-Shapes Table of the AISC Manual, and F_y' is also listed in our Table A.11. Notice in Table A.11 that all A36 sections, with the exception of one (W6 × 15), have $F_y \leq F_y'$ and thus are compact; but seven sections are noncompact for $F_y = 50$ ksi. A dash in the F_y' column indicates that F_y' is greater than 65 ksi; currently, no shapes are produced above this yield stress.

The compression flanges of beams bending about their strong axis (i.e., deflect vertically) tend to buckle laterally (somewhat similarly to columns, although the tension flanges of beams tend to keep the compression flange straight) about their weak axis, depending on the spacing and nature of the lateral support system. For typical floor framing conditions, beams may be considered fully laterally supported by the concrete slab. In other words, the *unbraced length* of the compression flange is equal to $L_b = 0$. When lateral support is provided only at intervals of L_b such as by cross beams, then they should be properly connected to the compression flange. In addition, the lateral support system should be braced and adequately stiff. There are, however, borderline cases that are difficult to evaluate because only partial lateral support is provided. For example, metal decking for roof framing and wood flooring are rather flexible and may have to be disregarded as lateral support for preliminary design purposes. The allowable bending stress for compact sections is a function of the unbraced length L_b of the compression flange as compared to the maximum theoretical lengths L_c and L_u that are derived from stability criteria.

The design parameters of F_y', F_y''', L_c, and L_u are listed in the Allowable Stress Design Selection Table and the W-Shapes Properties Table of the AISC Manual of Steel Construction, as well as in Table A.11 of this book except F_y'''.

The allowable bending stress for simple bending about the strong axis of compact rolled sections ($F_y \leq F_y'$) when $L_b \leq L_c$ is

$$F_b = 0.66\,F_y, \qquad \text{where } F_y \leq 65 \text{ ksi} \tag{3.5a}$$

For example, for A36 steel, $F_b = 0.66(36) = 23.76$ ksi, which is often taken as 24 ksi. For *weak-axis bending* of doubly symmetrical compact sections, including solid rectangular and round bars, the major axis stiffness resists lateral movement. Therefore, the allowable bending stress is larger.

$$F_b = 0.75\,F_y \tag{3.5b}$$

When members bend about their minor axis, they seldom need to be braced because of their superior lateral stiffness; the same reasoning applies for the bending of compact box sections (e.g., tubing).

At times the compression flange may not be braced as is typical for building columns that may have an unbraced length L_b larger than the theoretical value of L_c, but which should not be larger than L_u for preliminary design purposes, $L_c < L_b \leq L_u$. Furthermore, should the beam–column web be noncompact ($F_y > F_y'''$) or the flange be noncompact $F_y' < F_y \leq 2.14F_y'$, then the allowable bending stresses for those conditions may be taken as

$$F_b = 0.60\,F_y \tag{3.5c}$$

For example, for A36 steel, $F_b = 0.6(36) \cong 22$ ksi. To determine the allowable bending stress for the situation where the unbraced length of the compression flange is $L_b > L_u$ is very complex. Now it is not constant anymore, as assumed in the previous discussion, but varies with the unbraced length; in other words, as L_b increases F_b decreases. Since the beam design for this condition is not common, there will be no further discussion of the topic, and the reader is referred to the Allowable Moments in Beams Charts in the AISC Manual (as is illustrated in Example 3.3).

Shear.   The shear stress in the web usually does not control the design of ordinary rolled beam sections, as do the bending stresses in the flanges. This is due to the high-shear capacity of the webs of rolled sections (i.e., sufficient web thickness is available) and due to the high-shear strength of steel, which is about 61% of its bending strength, which is large in contrast to other materials like wood, concrete, and masonry. Shear may become a primary design factor and should be checked for the following typical conditions:

- Relatively short span beams with heavy loads (approaching deep beam action)
- Beams supporting heavy concentrated loads near the reactions (thus causing hardly any moment), a condition that may be caused by the setback of a building profile, where columns are terminated and are picked up by transfer beams
- Beams with webs weakened by notching, coping, or holes for pipes and ducts, particularly at locations where shear forces are high, as is the case near reactions

Shear stresses are not evenly distributed across W-, M-, or S-beam sections under simple bending; they are zero along the flange faces and are maximum at the neutral axis where the bending stresses are zero [see discussion of Eqs. (2.69) and (2.72)]. Since the flanges hardly resist any shear and since nearly all shear is carried by the web in a fairly uniform fashion, specifications often use an *average web shear* approach; in actuality the maximum shear stress is about 12% larger than the average shear stress. This unconservative approximation is corrected in the allowable stresses.

The allowable shear stress F_v should be equal to or larger than the average shear stress f_v in the webs of rolled sections which have a typically low depth-to-web thickness ratio, thus ensuring shear yielding rather than buckling as a mode of failure, as is the case for the slender webs of plate girders:

$$f_v = V/dt_w \leq F_v = 0.4\,F_y \tag{3.6}$$

where f_v = shear stress (ksi)
V = shear force (k)
d = depth of beam (in.)
t_w = web thickness of beam (in.)

Holes in Beams.   The size and location of holes in rolled beams and girders that may be necessary to allow, for instance, conduits, pipes, and ducts to pass through must be controlled. Large holes in the web influence the shear capacity and large holes in the flanges the flexural strength of the beam. It is apparent that holes in the web should occur where the shear is small and holes in the flanges where the moment is small. The AISC specifications do not require deduction of holes in compression and tension flanges unless the reduction of the area of either flange exceeds 15% of the gross flange area, and then only the area in excess of 15% shall be deducted.

Deflection.   For long spans and shallow beam sections, enough member stiffness must be provided; that is, deflection must be considered. It is apparent that when beams are too flexible people may feel uncomfortable and unsafe. Furthermore, nonstructural components attached to flexible beams such as ceilings, partitions and curtains may be stressed and damaged; in addition, differential movement between structural members must be controlled so as not to influence the load flow. When a dead load deflection Δ_D is large, the designer may have to camber the beam by the amount of Δ_D, thereby also eliminating the appearance of sagging. Ponded rainwater on flat roofs should be controlled by having sufficient roof slope (e.g., 1/4 in./ft), sufficient roof stiffness, and proper drainage.

Deflection limitations are generally given in terms of some fraction of the span length (Δ/L) and in terms of span-to-member depth ratio (L/d). The AISC specifications limit the live load deflection of beams and girders (that support a plaster ceiling, which should not be fractured) to 1/360 of the span length; for the same conditions, the total deflection due to dead and live load is limited to $L/240$ according to some codes (see Table 2.5). The maximum live load deflection of roof structures (without plaster ceiling) is in the range of $L/180$ to $L/240$ depending on the sensitivity of the structure. The Commentary on the AISC Specs proposes as deflection limits for fully stressed floor beams and girders $L(F_y/800)$, or $L/22$ for A36 steel as a minimum beam depth. Thus, a 30-ft beam should have a depth of at least $30(12)/22 = 16.36$ in. for the deflection not to control. Furthermore, the depth of fully stressed roof purlins (except for flat roofs because of ponding) should not be less than $F_y/1000$ times the span, or $L/28$ for A36 steel. Where the floor framing is subject to vibrations with no sources of damping available, a beam depth of a least $L/20$ is suggested.

In contrast to deep beams for which the shear deformation controls, for shallow beams it can be neglected and only the flexural deflection has to be considered. It may be expressed in general terms for maximum deflection conditions as

$$\Delta = \text{coef.} \ (WL^3/EI) \tag{3.7}$$

where $\quad W$ = total beam load (k)

$\qquad\qquad L$ = beam span (in.)

$\qquad\qquad E$ = modulus of elasticity = 29,000 ksi for steel

$\qquad\qquad I$ = moment of inertia (in.^{4})

$\qquad$ coef. = depends on loading and boundary conditions of beam

For the special condition of a simple span beam carrying a uniform load the coefficient is equal to 5/384 [see Eq. (2.81)] or as can be obtained, for example, from the Beam Diagrams and Formulas Tables in the AISC Manual as well as Table A.15. This deflection for a beam with a uniform moment of inertia can be conveniently expressed, by using $M = wL^2/8$ and $f_b = (Md/2)/I$, as

$$\Delta_s \ = \ \frac{5wL^4}{384EI} \ = \ \frac{5ML^2}{48EI} \ = \ \frac{10f_bL^2}{48Ed} \tag{3.8}$$

where $\quad M$ = maximum field moment due to a uniform load (k-in.) = $wL^2/8$

$\qquad\quad f_b$ = bending stress (ksi)

Letting $E = 29,000$ ksi for steel yields the following expression:

$$\Delta_s \ = \ \frac{ML^2}{161I} \tag{3.9}$$

where units are M (k-ft), L (ft), I (in.^{4}).

The required moment of inertia can easily be obtained from Eq. (3.9) when the limiting deflection is known. For example, for the typical maximum allowable total load deflection of $L/240$ and the maximum allowable live load deflection of $L/360$ (Table 2.5), the following simple expressions can be used.

$$I = 0.186 \ M_L L = 0.0233 \ W_L L^2, \qquad I = 0.124 \ M_T L = 0.0155 \ W_T L^2 \tag{3.10}$$

where $\quad M_L$ = live load moment = $W_L L/8$ (ft-k)

$\qquad\quad M_T$ = dead and live load moment = $W_T L/8$ (ft-k)

$\qquad\quad I$ $\ \ $ = moment of inertia of beam (in.^{4})

$\qquad\quad W_L$ = total live load = $w_L(L)$ (k)

W_T = total dead and live load = $w_T (L)$ (k)
L = beam span (ft)

EXAMPLE 3.1.

A 30-ft simple span beam (A36) supports a uniform dead load of 1 k/ft and a uniform live load of 0.8 k/ft. Determine the beam size as based on typical deflection criteria.

$$I = 0.0233 \ W_L L^2 = 0.0233(0.8 \times 30)30^2 = 503 \text{ in.}^4$$
$$I = 0.0155 \ W_T L^2 = 0.0155(1.8 \times 30)30^2 = 753 \text{ in.}^3$$

The required section is obtained from the Moment of Inertia Selection Table in the Manual of Steel Construction. Try W21 $\times$ 44, $I_x = 843$ in.^{3}.

Equation (3.8) can be obtained in the more popular form often used in practice by letting $E = 29{,}000$ ksi for steel and L in (ft).

$$\Delta_s = \frac{f_b}{967}\left(\frac{L^2}{d}\right) \cong \frac{f_b}{1000}\left(\frac{L^2}{d}\right) \tag{3.11}$$

where units are f_b (ksi), L (ft) and beam depth d (in.)

This expression can be used for preliminary design purposes of simply supported beams not only for the uniform loading case but also for other common loading conditions, as is found in typical floor and roof structures, as long as the maximum stress occurs near midspan. The equation is tabulated in the AISC Manual (Camber and Deflection Coefficient Table) for various maximum stress and span/depth conditions.

Equation (3.11) can be further simplified for A36 steel by assuming conservatively that $f_b = F_b = 23.8$ ksi.

$$\Delta_s = 0.0246L^2/d \tag{3.12}$$

where: L = beam span (ft)
d = beam depth (in.)
Δ_s = beam deflection due to full loading (in.)

Rearranging the terms in Eq. (3.12) clearly demonstrates that the ratio of deflection-to-span is proportional to the depth-to-span ratio of the beam.

$$\Delta/L = (0.0246)(L/d) \tag{3.13}$$

EXAMPLE 3.2

Find the maximum possible deflection of a uniformly loaded W21 $\times$ 44 beam spanning 30 ft.

$$\Delta_s = 0.0246L^2/d = 0.0246 (30)^2/20.66 = 1.07 \text{ in.} < L/240 = 30(12)/240 = 1.5 \text{ in.}$$

For preliminary design purposes, Eq. (3.12) can also be used for floor or roof framing in A36 steels where the beams carry the concentrated loads of the supporting filler beams. Notice that the deflection is primarily dependent on the square of the span and the beam depth for a maximum bending stress F_b; otherwise, the deflection increases with an increase in bending stress. Hence, beams of the same span and depth have the same deflection when they are stressed to their allowable limit under full loading conditions.

The corresponding deflection for a fixed beam is $\Delta = 0.2\Delta_s$, and the maximum deflection for the interior bay of a continuous beam may be assumed to be about 25% of that of the simple beam:

$$\Delta = 0.25\Delta_s \qquad\qquad (3.14)$$

EXAMPLE 3.3

The floor framing for a typical interior bay of a multistory braced steel skeleton structure is shown in Fig. 3.4. The preliminary beam and girder sizes are determined for a dead and live load of 80 psf each using A36 steel and flexible connections.

a) *Beam design.* The compression flange of a typical filler beam is fully laterally supported by the floor slab ($L_b = 0$). In Example 2.2, the reduced live load has been determined as $L_r \cong 77$ psf. Therefore, the beam must support the following uniform load where the beam weight is included in the floor dead load:

$$w = w_D + w_L = 8(0.080) + 8(0.077) = 0.64 + 0.62 = 1.26 \text{ k/ft}$$

The maximum moment is

$$M_{max} = wL^2/8 = 1.26\ (25)^2/8 = 98.44 \text{ ft-k} \qquad (2.25)$$

The required section modulus is

$$S_x = M_x/F_b = M_x/0.66\ F_y = 98.44\ (12)/0.66(36) = 49.72 \text{ in.}^3 \qquad (3.4)$$

Try W18×35, $S_x = 57.6$ in.^{3}, $I_x = 510$ in.^{4}, $F_y' \geq F_y$.

The maximum live load deflection is within the allowable limits as shown by the following calculations:

$$\Delta_L = \frac{5wL^4}{384EI} = \frac{5(0.62/12)(25 \times 12)^4}{384(29,000)510} = 0.37 \text{ in.} \qquad (3.8)$$

$$\leq L/360 = 25(12)/360 = 0.83 \text{ in.}$$

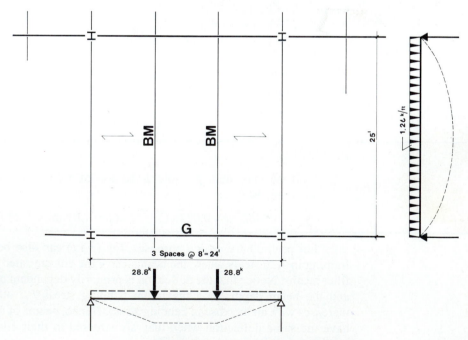

Figure 3.4   Typical floor framing bay.

Shear stresses and web crippling are not critical for this ordinary floor beam.

Now, for reasons of comparison, the *load and resistance factor method* is used for the design of the beam.

$$w_u = 1.2D + 1.6\,L = 1.2(0.64) + 1.6(0.62) = 1.76 \text{ k/ft}$$

$$M_u = w_u L^2/8 = 1.76(25)^2/8 = 137.5 \text{ ft-k}$$

The required plastic section modulus is

$$Z_x = M_u/\phi_b F_y = 137.5(12)/0.9(36) = 50.93 \text{ in.}^3$$

Select as a trial section according to the *Plastic Design Selection Table* in the AISC Manual

$$W16 \times 31, \ Z_x = 54.0 \text{ in.}^3$$

b) For the sake of exercise, it is assumed that the compression flange of the beam is only laterally supported at the ends by the girders.

From the Allowable Moments in Beam Charts in the AISC Manual,

$$M = 98.44 \text{ ft-k}, \qquad L_b = 25 \text{ ft}$$

Try $W12 \times 53$, $M_R = 112$ ft-k, $S_x = 70.6$ in.^{3}.

Therefore the allowable bending stress is

$$f_b = F_b = M/S = 112(12)/70.6 = 19.04 \text{ ksi} = 0.53 F_y$$

c) *Girder design.* In Example 2.2, the reduced live load has been determined as $L_r = 64$ psf.

In this first estimate of the girder size the girder weight is ignored. The magnitude of the beam loads acting on the girder (see Fig. 3.4) is

$$P = (0.080 + 0.064)(25 \times 8) = 28.8 \text{ k}$$

The maximum moment is

$$M_{max} = P(L/3) = 28.8(8) = 230.4 \text{ ft-k}$$

The required section modulus is

$$S_x = M_x /0.66 F_y = 230.4(12)/0.66(36) = 116.36 \text{ in.}^3$$

Try $W24 \times 62$, $S_x = 131$ in.^{3}, $M_R = 259$ ft-k, $L_c = 7.4$ ft, $L_u = 8.1$ ft.

The additional moment due to the girder weight is

$$M = wL^2/8 = 0.062(24)^2/8 = 4.46 \text{ ft-k}$$

The total moment is

$$M_{total} = 230.4 + 4.46 = 234.86 < 259 \text{ ft-k}, \quad \text{OK}$$

Therefore, the trial section is all right.

It is assumed, for the sake of exercise, that the compression flange of the girder is only laterally supported by the beams. Since the unbraced length ($L_b = 8$ ft) is larger than $L_c = 7.4$ ft but less than $L_u = 8.1$ ft, the allowable bending stresses are reduced and equal to $F_b = 0.6 F_y$. Therefore, the capacity of the selected beam is only

$$M_R = (259 /0.66)0.6 = 235.46 \text{ ft-k} \le 234.86 \text{ ft-k}, \quad \text{OK}$$

The section is still all right. Notice that the $W21 \times 62$ could have been also selected; it has an allowable bending stress of $0.66 F_y$ since $L_c = 8.7$ ft. ≥ 8 ft.

Shear stresses, deflection, and web crippling are not critical for the preliminary design of ordinary floor beams (for a check, see Problem 3.10).

Web Crippling.   Occasionally, a heavy concentrated load is applied to the top flange or bottom flange of a beam, as may be the case for a column sitting on the top flange or a beam resting on a base plate (or seat) transferring a high load (Fig. 3.5). For this condition a high compressive stress concentration may result in the web close to the junction with the flange unless the web is thick or stiffened. That is, localized failure or web crippling may occur at the critical web toe of the fillets (at the distance k from the beam face). According to the AISC Specs, the load is assumed to spread out from the bearing or contact length N over a distance of $N + 5k$ or $N + 2.5k$, as shown in Fig. 3.5. In other words, the concentrated load R, when it is applied at a distance from the member end greater than the member depth, is resisted by the critical web area $t_w(N + 5k)$. The resulting bearing stress must be less than the allowable web crippling stress unless bearing stiffeners are used.

$$\frac{R}{t_w\,(N + 5k)} \le 0.66F_y \tag{31.15a}$$

But, when the concentrated load R is applied at or near the end of the member,

$$\frac{R}{t_w\,(N + 2.5k)} \le 0.66F_y \tag{3.15b}$$

EXAMPLE 3.4

Determine the preliminary beam size using $F_y = 50$ ksi for the condition in Fig. 3.5, where the beam must transfer a column load of 300 k; the beam is laterally supported at the load.

The reaction R_A is equal to

$$\Sigma\,M_A = 0 = 300(7) - \text{R}_A(10) \quad \text{or} \quad R_A = 210 \text{ k}$$

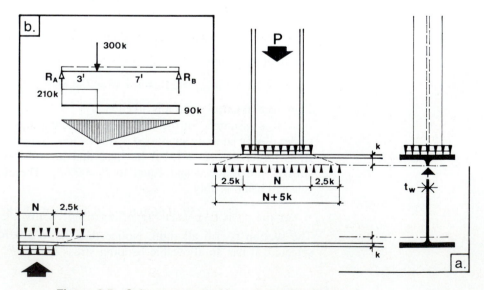

Figure 3.5   Column supported by a short-span beam.

The maximum moment is

$$M_{max} = 210(3) = 630 \text{ ft-k}$$

The required section modulus is

$$S_x = M_x/0.66F_y = 630(12)/0.66(50) = 229.09 \text{ in.}^3$$

Try W30 × 90, $S_x = 245$ in.^{3}, $M_R = 674$ ft-k, $L_c = 7.2$ ft > 7 ft, $d = 29.53$ in., $t_w = 0.470$ in., and $k = 1\frac{5}{16}$ in.

The beam weight may be ignored for this preliminary investigation since it is negligible in comparison to the column load.

Checking the shear stress

$$f_v = V/dt_w = 210/29.53(0.470) = 15.13 \text{ ksi}$$

$$\leq 0.4F_y = 0.4(50) = 20 \text{ ksi}, \quad \text{OK}$$

The length of the column base plate is determined, so web crippling is not critical. Notice that the concentrated load is located 3 ft from the end of the beam, which is more than the required d distance of $30/12 = 2.5$ ft.

$$\frac{R}{t_w(N+5k)} = \frac{300}{0.47[N+5(1.313)]} = 0.66(50) \quad \text{or} \quad N = 12.78 \text{ in.}$$

Try a base plate length of $N \geq 13$ in.

The length of bearing at the critical beam support must be

$$\frac{R}{t_w(N+2.5k)} = \frac{210}{0.47[N+2.5(1.313)]} = 0.66(50) \quad \text{or} \quad N = 10.26 \text{ in.}$$

Try $N \geq 11$ in. This bearing length can be reduced if bearing stiffeners are used.

Fast Approximations. To quickly estimate the size of typical simply supported braced floor or roof beams (A36), the following rules of thumb (with mixed units) are often found in practice.

- The section modulus S_x of a W-section is roughly equal to the product of beam weight w (lb/ft) and its nominal depth d (in.) divided by 10.

$$S_x \cong wd/10 \quad \text{or} \quad I_x \cong S_x(d/2) \cong wd^2/20 \qquad (3.16)$$

Similarly, the plastic section modulus Z_x of a W-section can be approximated as

$$Z_x \cong wd/9$$

For example, for a W16 × 36

$$S \cong 36(16)/10 = 57.6 \text{ in.}^3, \quad \text{actual } S \text{ is } 56.5 \text{ in.}^3$$
$$Z \cong 36(16)/9 = 64 \text{ in.}^3, \quad \text{equal to the actual } Z_x$$

- The nominal depth d (in.) of a W-section is approximately equal to one-half of its span L (ft).

$$d \cong L/2 \qquad (3.17)$$

- For the primary beams supporting filler beams, the nominal depth is often assumed as

$$d \cong L/1.5$$

- The beam section weight w (lb/ft) is roughly 1.25 times the total load W (k) that the beam must support.

$$w \cong 1.25W \tag{3.18}$$

- The absolute maximum deflection Δ (in.) is one-tenth of the nominal beam depth d (in.).

$$\Delta_{max} \cong d/10 \tag{3.19}$$

EXAMPLE 3.5

A simply supported floor beam spans 20 ft and carries a total load of 20 k. Estimate the beam size using A36 steel.

Nominal beam depth: $d \cong L/2 = 20/2 = 10$ in.

Required beam weight: $w \cong 1.25\,(W) = 1.25\,(20) = 25$ lb/ft

Maximum deflection: $\Delta \cong d/10 = 10/10 = 1$ in.

The hypothetical section is a $W10 \times 25$, which does not exist. Try the section closest to it, a $W10 \times 26$.

Unsymmetrical Bending of Doubly Symmetrical Rolled Shapes. When compact W-, M-, and S-shapes are subject to simple *biaxial bending*, as for the familiar roof purlins on inclined roofs (see discussion Fig. 3.8), that is, when the load P passes through the centroid of the cross section as demonstrated in Fig. 3.6a, the following interaction equation may be used:

$$\frac{f_{bx}}{F_{bx}} + \frac{f_{by}}{F_{by}} = \frac{(M/S)_x}{0.66F_y} + \frac{(M/S)_y}{0.75F_y} \leq 1$$

or $\tag{3.20}$

$$S_x \geq \frac{M_x}{0.66F_y} + \frac{M_y}{0.75F_y}\left(\frac{S_x}{S_y}\right) = \frac{M_x}{0.66F_y}\left[1 + 0.88\left(\frac{M_y}{M_x}\right)\left(\frac{S_x}{S_y}\right)\right]$$

Since the beam section cannot be determined directly from this equation because of the two unknowns S_x and S_y, the ratio of S_x/S_y may be assumed. The ratio ranges from about 2 for heavy W-column sections with wide flanges to roughly 12 for light beam members. For ordinary beam sections, an estimate of $S_x/S_y = 5$ is reasonable, especially when the lateral loads constitute only a small portion of the vertical loads (i.e., M_y/M_x is small). Therefore, the following equation may be used as a first estimate

$$S_x \geq \frac{M_x}{0.66F_y} + \frac{5M_y}{0.75F_y} = \frac{M_x}{0.66F_y}\left(1 + 4.4\,\frac{M_y}{M_x}\right) \tag{3.21}$$

When the load P does not pass through the centroid of the wide-flange section, then in addition to biaxial bending *twisting* (i.e., torsion) is generated. Since the evaluation of *unsymmetrical bending* is complex, the following simplified approaches are often used in practice for preliminary design purposes of relatively stiff W-beams with wide flanges when torsion is not excessive. In other words, this approximate approach can be safely applied to short beams, but will be overly conservative for long-span members. Should the load P be applied directly to the flange (Fig. 3.6b), then its horizontal component, P_y, is assumed to be resisted entirely by that flange in bending; here the section modulus of the flange may be approximated as $S_y/2$ of the beam section. This loading situation occurs, for instance, for crane girders, where a moving crane causes lateral thrust forces. For the condition where P acts parallel to the web (Fig. 3.6c), the twisting moment $M_T = P(e)$ is assumed to be resisted (entirely in bending) by a couple formed by the two flange forces $P_y = M_T/d$. It should also not be forgotten that torsion is much more efficiently resisted in nearly pure shear by a closed box shape. We may conclude that the following interaction equation holds true for cases (b) and (c) in Fig. 3.6, assuming $L_c < L_b \le L_u$.

$$\frac{f_{bx}}{F_{bx}} + \frac{f_{by}}{F_{by}} = \frac{(M/S)_x}{0.6f_y} + \frac{2(M/S)_y}{0.75F_y} \le 1 \qquad (3.22)$$

This expression can be simplified for first estimation purposes as before assuming $S_x/S_y = 5$.

$$S_x \ge \frac{M_x}{0.6f_y}\left[1 + 8\left(\frac{M_y}{M_x}\right)\right] \qquad (3.23)$$

Typical examples for this loading condition occur on spandrel beams, for instance due to only a one-sided slab support or a one-sided connection of a filler beam or due to the eccentric action of the curtain wall.

Naturally, whenever possible the detailing should provide proper bracing to take the torsion out of the beam rather than expecting the beam to resist it in twisting.

EXAMPLE 3.6

A 16-ft simple span W14 beam carries at midspan a 20-k concentrated eccentric load that acts at 2 in. parallel to the web (see Fig. 3.6c). The flanges are laterally simply supported only at their ends (i.e., restrained from warping) so that the

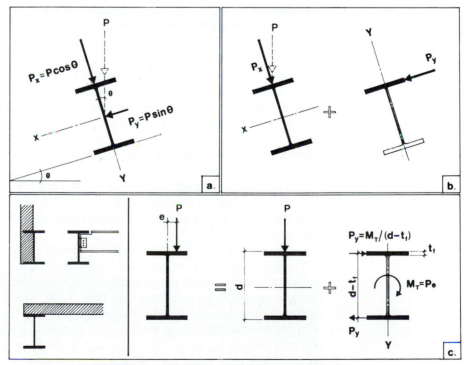

Figure 3.6 Unsymmetrical bending of W-sections.

flanges are assumed to resist the torsion entirely in bending. Estimate the beam section, ignoring the beam weight. Use A36 steel. The beam moment is

$$M_x = PL/4 = 20(16)/4 = 80 \text{ ft-k}$$

The twisting moment is $M_T = P(e) = 20(2) = 40$ in.-k.

The horizontal flange forces P_y at midspan, replacing the twisting moment, as based on the nominal 14-in. depth for this first approximation, are

$$P_y = M_T/d = 40/14 = 2.86 \text{ k}$$

Hence, the maximum flange bending moments at midspan are

$$M_f = M_y = P_y L/4 = 2.86(16)/4 = 11.44 \text{ ft-k}$$

For first estimation purposes, Eq. (3.23) is used.

$$S_x \geq \frac{M_x}{0.6F_y}\left[1 + 8\left(\frac{M_y}{M_x}\right)\right] = \frac{80(12)}{0.6(36)}\left[1 + 8\left(\frac{11.44}{80}\right)\right] = 95.29 \text{ in.}^3$$

Try W14×61, $S_x = 92.2$ in.^{3}, $S_y = 21.5$ in.^{3}, $d = 13.89$ in.

$$L_c = 10.6 \text{ ft} < L_b = 16 \text{ ft} \leq L_u = 21.5 \text{ ft}, \quad F_b = 0.6 F_y$$

Check the interaction equation.

$$\frac{(M/S)_x}{0.6F_y} + \frac{2(M/S)_y}{0.75F_y} = \frac{80(12)/92.2}{0.6(36)} + \frac{2(11.44)12/21.5}{0.75(36)} =$$

$$= 0.48 + 0.47 = 0.95 < 1, \text{ OK}$$

Steel Plate Girders

Plate girders are built-up, single-web, I-shaped beams or double or multiweb box girders. They are composed of heavy flanges and relatively thin web plates, which are stiffened by vertical and possibly horizontal plates; special bearing stiffeners are required under concentrated loads. Today, the plate assembly is generally welded together as shown in Fig. 3.7d.

For heavy loading conditions the plate girders may provide a larger moment of inertia than is available for rolled beams. For ordinary loading they may be more economical than rolled beams; common simple span ranges are from about 50 to 130 ft. But even at spans as short as about 35 ft they may become competitive with rolled beams or built-up rolled beams since the designer has the freedom to proportion the cross section of the girder. Average depth/span ratios are in the range of 1/10 to 1/12 but generally less than 1/20.

The thin web and the corresponding buckling considerations make the design of plate girders complex; their behavior is somewhat between rolled beams and trusses. After the web of a plate girder has buckled under shear action and has become ineffective, which actually represents buckling along the compression diagonal of the web plate, the tension diagonal in the opposite direction must resist the entire shear (Fig. 3.7a). This tension field in each web panel must be stabilized by the flanges and the vertical stiffeners, which act as compression struts. Hence, the behavior of the plate girder is similar to that of a Pratt truss (Fig. 3.7b), where the girder flanges constitute the chords and the webs may be designed as diagonal tension-field members together with the vertical stiffeners as compression members. In this context, the truss analogy will only be used for the preliminary design of the flanges, which are assumed to resist the entire moment, while the web carries all the shear. It is assumed that instability in the flange and web together with the stiffeners does not develop prior to yield (Fig. 3.7c); otherwise, instability criteria such as vertical and lateral buckling of the compression flange and lateral buckling of the web will require a reduction of the allowable stresses.

The following approach may be used to proportion an I-shaped welded plate girder (A36) for first trial purposes.

- Assume a typical plate girder depth d of one-tenth to one-twelfth of its span unless other considerations like headroom or esthetic reasons are given.
- Determine the web thickness as based on having no reduction in flange stress ($F_b = 0.6F_y$). Thus, the corresponding web thickness according to the permissible depth–thickness ratio $h/t_w \leq 760/\sqrt{F_b} \leq 760/\sqrt{22} \leq 162$, is

$$t_w \geq h/162 \tag{3.24}$$

But the web thickness should never be less than about one-half of this value (for A36), which is based on the assumption that transverse stiffeners are spaced at a distance a equal to more than one and a half times the distance between the flanges ($a > 1.5h$).

- Determine the flange area A_f by assuming the entire moment to be carried by the flanges using the web depth of the section as an internal lever arm (flange-area method, Fig. 3.7c).

$$f_c = P/A_f = (M/h)/A_f \leq F_b = 0.6F_y$$

or

$$A_f \geq bt_f = (M/h)/F_b \tag{3.25}$$

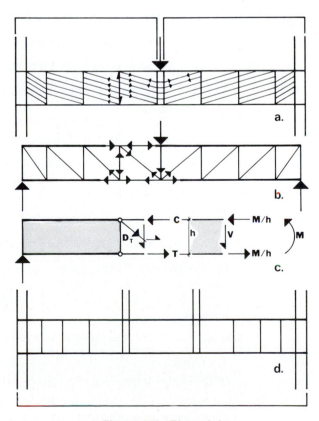

Figure 3.7 Plate girder.

where $F_b = 0.6(36) \cong 22$ ksi, assuming proper vertical and lateral support of the compression flange.

Check the compression flange thickness so that it is adequate against local buckling according to the permissible width–thickness ratio.

$$(b/2)/t_f \leq 95/\sqrt{F_y/k_c} = 15.83\sqrt{k_c} \quad \text{or} \quad b/t_w \leq 32\sqrt{k_c} \qquad (3.26)$$

where $\qquad k_c = \dfrac{4.05}{(h/t_w)^{0.46}} \qquad$ if $h/t_w > 70$

- Check the shear stress in the web (disregarding the tension field action and also the limiting h/t_w ratio) with respect to the use of the simple allowable shear stress, $F_v = 0.40F_y$.

$$f_v = V/A_w \leq F_v = 0.4F_y \qquad (3.27)$$

where: $A_w = t_w h$ and $h = d - 2t_f$. Now the trial girder section can be checked by using the familiar moment of inertia method; the size as well as the location of the stiffeners must also be determined.

EXAMPLE 3.7

The 80-ft span welded plate girder of A36 steel in Fig. 3.7 carries a concentrated column load of 100 k at midspan and a uniform load of 4 k/ft, which includes the assumed girder weight of about 0.35 k/ft. Do a preliminary girder design.

The maximum girder shear and moment are

$$V_{max} = wL/2 + P/2 = 4(80)/2 + 100/2 = 210 \text{ k}$$

$$M_{max} = wL^2/8 + PL/4 = 4(80)^2/8 + 100(80)/4 = 5200 \text{ ft-k}$$

(a) *Preliminary web design.* The overall girder depth is in the range of about

$$d = L/10 \text{ to } L/12 = 80/10 \text{ to } 80/12 = 8 \text{ to } 6.67 \text{ ft} = 96 \text{ to } 80 \text{ in.}$$

The corresponding web depth is in the range of 92 to 76 in. assuming a flange thickness of 2 in.

As a first trial, a web depth of 78 in. will be investigated. The required web thickness for no reduction of flange stress, according to Eq. (3.24), is

$$t_w = h/162 = 78/162 = 0.48 \text{ in.}$$

The minimum thickness is about one-half of this value. Try a web plate $3/8 \times 78$ in., $A_w = 29.25$ in.^{2}, $h/t_w = 208$. Keep in mind, however, that the allowable bending stress in the compression flange must be be reduced. Checking the shear stress,

$$f_v = V/A_w = 210/29.25 = 7.19 \text{ ksi} \leq 0.4 F_y = 0.4(36) = 14.4 \text{ ksi}$$

(b) *Preliminary flange design.* Assuming the moment to be resisted by the flanges and disregarding the reduction of the allowable compression stress due to buckling for this preliminary design approach, yields the following approximate flange area:

$$A_f = (M/h)/F_b = (5200(12)/78)/22 = 36.36 \text{ in}^2$$

Try flange plates $1\frac{3}{4} \times 21$ in., $A_f = 36.75$ in.^{2} Check if the flange is adequate with respect to local buckling [see Eq. (3.26)].

$$b/t_f = 21/1.75 = 12 \leq 32\sqrt{\frac{4.05}{208^{0.46}}} = 18.87, \text{ OK}$$

Check the assumed girder weight (steel weighs 490 lb/ft^3).

$$w = [(29.25 + 2(36.75))/12^2](1)\ 490 = 349.64\ \text{plf} < 350, \quad \text{OK}$$

Keep in mind that the weight of the vertical stiffeners must still be added. Other girder cross sections should be investigated, and then the final trial girder section must be checked accurately.

Wood Beams

Briefly introduced in this section is the design of solid sawn lumber beams, glued laminated timber beams, prefabricated wood beams, and composite beams (e.g., flitch beams). For beams made from laminated veneer lumber, parallel strand lumber, and composite lumber, the reader may want to refer to the respective literature of the manufacturers. Sheet material such as decking (plywood panels, particleboard panels, etc.) also acts as beams.

The structural design of ordinary rectangular solid floor and roof wood beams (e.g., joists, rafters, lintels, purlins, stringers, girders) is relatively simple since they usually can be treated as statically determinate structures. Similar to the design of steel beams (or beams of other materials), the laterally supported beam section is usually selected based on bending and then checked for shear stress, deflection, and possibly bearing.

Sawn Lumber Beams. Most common are the rectangular sawn lumber beams. Although the procedure for the approximate design of wood beams is presented in this section, a nearly same process is applicable to joist and glulam beam design.

A. Flexure: The familiar expression for the basic flexural stress is

$$f_b = M/S \le F'_b \tag{3.28}$$

where

$$F'_{bx} = (F_b)\ C_D\ C_M\ C_t\ C_L\ C_F\ C_r \cong F_b\ C_D, \qquad F'_{by} = (F_b)\ C_D\ C_M\ C_t C_F\ C_{fu} \cong F_b\ C_D$$

The tabulated bending stresses F_b, however, may need to be adjusted to take into account the effect of load duration, wet service, temperature, stability, size, flat use, and repetitive member use, as represented by the adjustment factors. Load duration, wet service, and temperature factors have been discussed in the introduction of this chapter. In this context, for ordinary conditions and full lateral support of beam, only load duration is considered for bending about the strong x axis, for preliminary design purposes.

As for steel beam design, lateral stability must also be considered for the design of wood beams. In other words, the tendency of the compression edge of the beam to buckle and the beam to deflect laterally has to be taken into account. The *beam stability factor* C_L reduces the tabulated bending stresses due to stability considerations. In typical wood construction, beams are generally continuously, laterally supported along the compression edge throughout their length by sheathing, and at their ends they are prevented from rotation by bridging so that $C_L = 1.0$. The following approximate rules may be applied in providing lateral restraint as based on the ratio of depth to thickness (or width) using nominal dimensions so that flexural stress reduction is not necessary because $C_L = 1.0$.

- If the ratio does not exceed 2 to 1, no lateral support is needed.
- If the ratio is 3 to 1 or 4 to 1, only the ends should be held in position and prevented from rotation by blocking, bridging, or proper connections to adjacent members.

- If the ratio is 5 to 1, one edge should be held in place for the entire length.
- If the ratio is 6 to 1, the beam should be laterally braced by bridging or full-depth solid blocking at intervals not exceeding 8 ft, unless both edges are held in position, or the compression edge should be held in place for the entire length, as by adequate sheathing or subflooring, and the ends at the support should be prevented from rotation.
- If the ratio is 7 to 1, both edges should be held in line for the entire length.
- For beam column action, the depth-to-width ratio of the member is not to exceed 5 to 1 if one edge is held in place.

In the previous edition of the NDS, *size factor*, C_F, was required for rectangular wood beams 5 in. or thicker exceeding a depth d of 12 in., applied to the allowable bending stress. This practice is continued here for preliminory design purposes, although the 1991 NDS has extended the range of the factor and changed its base stresses.

$$C_F = (12/d)^{1/9} \le 1.0 \tag{3.29}$$

The *flat use factor*, C_{fu}, for bending about the weak axis may be conservatively taken equal to 1. In contrast to single members responding individually in bending to load action, repetitive members joined together, say by bridging or sheathing, share the load resistance. Therefore, for repetitive members, such as joists, truss chords, rafters, studs, planks, decking, and similar members that are at least three in number, spaced not more than 24 in. apart, and are joined by floor, roof, or wall load-distributing systems, tabulated bending stresses for dimension lumber 2 to 4 in. thick for single-member use are increased by the *repetitive member factor*, $C_r = 1.15$.

The *form factor*, C_f, ensures that the moment capacity of beams with nonrectangular cross section is the same as for bending members with square cross-section and the same cross-sectional area. For example, for a beam with a circular cross-section $C_f = 1.18$, and for a square section beam loaded diagonally, $C_f = 1.414$.

Most of the adjustment factors just discussed, with exception of the load duration factor C_D, however, may be considered as equal to 1 for ordinary conditions and where lateral buckling of the rectangular beam is prevented.

Besides the common simple bending cases, occasionally *biaxial bending* occurs as for the pitched roof in Fig. 3.8. Here the roof skin is assumed to prevent the twisting $P(e)$ of the purlins ($e = 0$). The maximum stresses are found by superimposition of the bending stresses about the x and y axes and appear at the extreme top (1) and bottom (2) corners of a purlin at its midspan.

$$\pm f_{b\max} = f_{b\max x} + f_{b\max y} = M_x/S_x + M_y/S_y \le F_b' \tag{3.30}$$

As an example for this loading condition, refer to Problem 5.20. For the case where the allowable bending stresses about the two principal axes are different, as for glulam beams for which the allowable bending stress about the y-axis is much less than for bending about the major axis, the following relationship holds true:

$$f_{bx}/F_{bx}' + f_{by}/F_{by}' \le 1 \tag{3.31}$$

For further discussion of the topic, the reader may want to refer to the section titled **Unsymmetrical Bending of Doubly Symmetrical Rolled Shapes** (Steel Beams).

B. Shear: For wood beams, the shear along the grain (horizontal shear) is a most critical design consideration. The horizontal shear strength is only a small percentage of

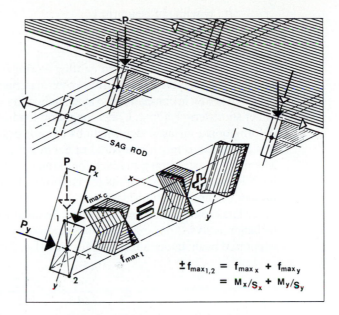

Figure 3.8 Example of biaxial bending.

the bending strength, in contrast to steel; therefore, the shear stress must always be checked. The shear stress for sawn lumber beams and glulam beams, according to Eq. (2.69), is

$$f_v = VQ/Ib \leq F'_v \tag{3.32}$$

where

$$F'_v = (F_v)C_D C_M C_t C_H \cong (F_v)\,C_D \quad \text{(for preliminary design purposes)}$$

The adjustment factors for load duration, wet service, and temperature have already been discussed. For the particular situations where the *shear stress factor* C_H can be applied, the reader should refer to NDS.

For wood beams with rectangular cross sections, the maximum shear stress, according to Eq. (2.71), is

$$f_v = 1.5V/A \leq F'_v \tag{3.33}$$

where V = vertical shear; the maximum shear in bending members is located at distance d from the support face

$A = bd$ = cross-sectional area (in.^{2})

Notching beams at their supports severely reduces their shear capacity. The shearing strength of a rectangular wood beam notched on the lower tension face at the ends is reduced depending on the depth of the notch in relation to the depth of the member (Fig. 3.9c). The actual shear stress parallel to the grain can be computed as follows:

$$f_v = \frac{1.5V}{bd'} \frac{d}{d'} \leq F'_v \tag{3.34}$$

where V = vertical shear force (lb)
F'_v = allowable shear stress (psi)

b = width of beam (in.)
d' = depth of beam above notch (in.)
d = depth of beam (in.)

C. Deflection: The deflection is often a critical design determinant for wood beam design since wood is not as stiff as other materials. The modulus of elasticity E along the grain is low in comparison to steel, indicating the flexible character of wood beams under short-term loading. Under long-term loading, however, in addition to the elastic deformations, inelastic deformations (creep) occur, which may result in deflections that are twice as much compared to the initial deflection when the load was applied. Therefore, it is common practice for permanent loading conditions to use one-half of the allowable deflection limits for unseasoned lumber and two-thirds when seasoned lumber or laminated timber is used.

The elastic deflections can easily be obtained following the standard procedures of beam analysis. For example, the maximum *flexural deflection* of a typical simply supported beam under uniform loads [see (Eq. 2.81)] is

$$\Delta_s = \frac{5wL^4}{384EI} = \frac{5ML^2}{48EI} = \frac{10f_bL^2}{48Ed} \tag{3.8}$$

Here, the modulus of elasticity E has to be adjusted by the wet service factor C_M, the temperature factor C_t, and the buckling stiffness factor C_T: $E' = (E)\,C_M\,C_t\,C_T \cong E$. For preliminary design purposes, let $E = 1500$ ksi and $f_b = 1.50$ ksi so that the following simple expression is obtained [see also Eqs. (3.11) and (3.12)]:

$$\Delta_s = \frac{f_b}{50}\left(\frac{L^2}{d}\right) = 0.03L^2/d \tag{3.35}$$

where: the units are f_b (ksi), L in (ft), and beam depth d (in.) For ordinary conditions, *shear deflections* can be ignored unless the beam has a small span-to-depth ratio of $L/d \le 15$, as is often assumed for the preliminary design of rectangular beams.

The usual recommended deflection limits for floor beams of ordinary usage are $L/360$ for live load deflection and $L/240$ for total load deflection. For roof beams of residential, commercial, and institutional buildings with plaster ceilings, the allowable deflection for live loading is $L/360$ and for total loading $L/240$, but for roof beams without plaster ceiling the corresponding deflection limits are $L/240$ and $L/180$. Similarly, the deflection limits for roof beams of industrial buildings are $L/180$ and $L/120$ (see Table 2.5).

D. Bearing: At points where loads are transferred to the beam, the wood fibers perpendicular to the grain are compressed. These critical locations may occur at the support where the beam rests on other members or where a concentrated load, such as a column, sits on a beam. It is apparent that the area of beam bearing A_b must be large enough to transfer these loads without overstressing the material. Usually, the minimum bearing length required is 1½ in. on wood or metal and 3 in. on masonry. The tabulated compression stresses perpendicular to grain apply to bearings of any length at the end of a member and to all bearings 6 in. or more in length to any other locations. For shorter bearing lengths and not nearer than 3 in. to the end of a member, the tabulated design values can be increased by the *bearing area factor*, C_b. Refer to NDS for the definition of the factor. The simple bearing stress perpendicular to the grain, for example, due to the reaction R at the support is

$$f_{c\perp} = R/A_b \le F'_{c\perp} \tag{3.36}$$

where $F'_{c\perp} = (F_{c\perp})C_MC_tC_b \cong F_{c\perp}$ (for preliminary design purposes).

NDS should be consulted for allowable bearing values at an angle to grain, as is the case of an inclined roof rafter sitting on a stud wall (or Fig. 2.21e).

EXAMPLE 3.8

Design the laterally supported rectangular sawn lumber girder for the residential floor framing in Fig. 3.9, which supports beams at the third points of the span, for a normal temperature range and dry-service conditions. Assume a live load of 40 psf and a dead load of 15 psf. The live load deflection is limited to $L/360$ and the total load deflection to $L/240$. Use *Douglas fir–larch* with the following design values for beams and stringers:

$$F_b = 1600 \text{ psi}, \qquad F_v = 85 \text{ psi}, \qquad F_{c\perp} = 625 \text{ psi}, \qquad E = 1,600,000 \text{ psi}$$

The influence area of the girder is two times the tributary area [see (Eq. 2.3)].

$$A_i = 2(10 \times 16) = 320 \text{ ft}^2 < 400 \text{ ft}^2$$

Hence, there will be no live load reduction for the girder. In this first estimate of beam size the girder weight is ignored. The magnitude of the beam loads acting on the girder (Fig. 3.9) is

$$P = P_D + P_L = (15 + 40)(10 \times 8) = 1200 + 3200 = 4400 \text{ lb}$$

The girder span of 24 ft is assumed from center support (center of bearing area) to center support. Because of the symmetrical load arrangement, it is apparent that the reactions $R = P$. The maximum moment is

$$M_{\max} = R(L/3) = PL/3 = 4400 \,(24/3) = 35,200 \text{ ft-lb}$$

The required section modulus, assuming $F'_b = F_b$, is

$$S \geq M/F_b = 35,200 \,(12)/1600 = 264 \text{ in.}^3$$

According to Table A.9, try 6- × 18-in. beam with $S_x = 280.729$ in., with the minimum cross-sectional area of $A = 96.250$ in.^{2} (i.e., minimum weight). Since the section exceeds 12 in. the bending stresses must be reduced by the size factor C_F which, however, may be ignored for preliminary design purposes.

$$C_F = (12/d)^{1/9} = (12/17.5)^{1/9} = 0.959 \tag{3.29}$$

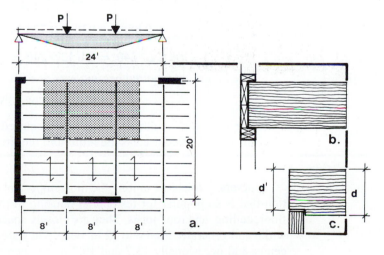

Figure 3.9 Floor framing (Example 3.8).

The approximate weight of the girder, using $\gamma = 34$ pcf, is

$$w = \gamma V = \gamma A(1) = 34 \, (96.25/12^2) \cong 23 \text{ plf}$$

This uniform load causes an additional maximum moment of

$$M = wL^2/8 = 23(24)^2/8 = 1656 \text{ ft-lb}$$

The new required section modulus, as based on $F'_b = F_b \, C_F$, is

$$S \geq M/C_F F_b = (35{,}200 + 1656)12/0.959(1600) = 288.24 \text{ in.}^2$$

Try 6- $\times$ 20-in. beam section with $S_x = 348.563$ in.^{3}, $A = 107.25$in.^{2}, and $I_x = 3398.484$ in.^{4}. This section should be all right, taking into account the larger beam weight and size factor.

The shear stress is checked by assuming conservatively that the beam weight is concentrated at the single loads.

$$R = V_{max} = 4400 + 23(107.25/96.25)24/2 = 4708 \text{ lb}$$

$$f_v = 1.5 \, V/A = 1.5 \, (4708)/107.25 = 65.85 \text{ psi} \leq F'_v = F_v = 85 \text{ psi}, \quad \text{OK}$$

Check the shear stress for the condition where the beam is notched 2 in. on its lower sides at the supports.

$$f_v = (1.5V/bd')d/d' = (1.5(4708)/5.5(17.5))19.5/17.5$$

$$= 81.76 \text{ psi} \leq F'_v = F_v = 85 \text{ psi}, \quad \text{OK}$$

The maximum deflection due to the concentrated live loading is

$$\Delta_L = \frac{23P_L L^3}{648EI} = \frac{23(3200) \, (24 \times 12)^3}{648(1{,}600{,}000) \, 3398.484} = 0.5 \text{ in.}$$

$$< L/360 = 24(12)/360 = 0.8 \text{ in.}, \quad \text{OK}$$

The maximum deflection at midspan due to full loading, using a beam weight of $23(107.25/96.25) = 26$ plf, is

$$\Delta = \frac{23PL^3}{648EI} + \frac{5wL^4}{384EI}$$

$$= 0.5(4400/3200) + \frac{5 \, (26/12) \, (24 \times 12)^4}{384 \, (1{,}600{,}000) \, 3398.484}$$

$$= 0.688 + 0.036 = 0.724 \text{ in.}$$

$$\leq L/240 = 24(12)/240 = 1.2 \text{ in.}$$

The minimum required bearing area at each support, as based on $F'_{c\perp} = F_{c\perp}$, is

$$A_b \geq R/F_{c\perp} = 4708/625 = 7.53 \text{ in.}^2$$

For the beam width of 5.5 in., the required bearing length is $7.53/5.5 = 1.37$ in. < 3 in., often used as the minimum required length.

Joists.   Joists are among the most important components of the wood construction that is so popular for residential housing in the United States. They are used as floor, ceiling, and roof joists. They are closely spaced small members of nominal 2-in. width and generally from 6 to 12 in. deep. They are usually spaced 12, 16, or 24 in. on center and occasionally 13.7 and 19.2 in. to accommodate lengths of sheet material. The spacing of the joists depends on the deck capacity (e.g., thickness and strength of plywood sheathing). The joists may just be simply supported members bearing on the exterior wall and the center ridge beam, or they may be continuous beams being supported at intermediate points on walls or girders.

Typical wood floor framing consists of the supporting joists and bridging, subflooring (e.g., 4- × 8-ft standard plywood sheathing) and finish flooring. The slender joists are laterally stabilized by cross-bridging or solid blocking between the joists at intervals $S \leq 8$ ft and by solid blocking at the supports. The bridging allows continuity between the closely spaced joists and redistribution of loads. For this condition of repetitive member use (where not less than three members are spaced not more than 2 ft on center and are joined by floor or roof structure), codes permit the allowable bending stresses for single members to be increased by 15%. In other words, closely spaced joists (or rafters, studs, trusses, etc.), spaced not more than 24 in. apart and joined adequately, distribute loads in bending to adjacent members.

From a structural point of view, wood-joist floor systems represent a complex composite system that is a multilayer planar structure with interlayer slip, which is difficult to analyze. Because of this complex interaction of the components, the rather crude *piece-by-piece method* of design is generally used where the interaction between the various floor components is ignored, and joists are designed as single beams.

The typical sloped roof joist behaves like a simple inclined beam with only vertical reaction forces under uniform gravity loading (Fig. 5.29d) causing no lateral thrust at the base. Since the joist does not resist any axial forces at the location of maximum moment, we may conclude that it is basically equivalent in behavior to a simple horizontal beam, which can be designed rather easily (for further discussion of inclined roof joists, refer to Section 5.3).

Typical live loads in residential housing for floor joists are 40 psf for living areas and 30 psf for sleeping rooms and attic floors used for storage. For ceiling joists (used as attic floor joists where the attic is not used as storage space), 10 psf are generally used. The typical dead load for wood-joist floor systems is about 10 psf.

The typical dead load for roof joist construction is between 10 to 15 psf, depending on the type of roof covering and if a ceiling is used. The snow or roof live loads are generally in the range of 20 to 40 psf. The stiffness of floor joists supporting a plaster ceiling is controlled by an allowable live load deflection of $L/360$. Where stiffness is of lesser importance, as for ceiling joists and low-sloped roof joists with drywall ceilings, the live load deflection is limited to $L/240$. For high-sloped roof joists with no ceiling and where the live loads are generally of short duration, $L/180$ is often set as a limit.

EXAMPLE 3.9

In a residential building, floor joists span 16 ft and, for this first investigation, are assumed to be spaced 16 in. apart. They must support a live load of 40 psf and a dead load of 10 psf (estimated as 2.5 psf for wood flooring, 2.1 psf for floor plywood, 2.9 psf for floor joists and bridging, and 2.5 psf for drywall ceiling). Use Douglas fir–larch with the following design stress values: $F_b = 1450$ psi, $F_v = 95$ psi, and $E = 1,700,000$ psi. Assume normal temperature range, dry service conditions, and full lateral support of joists.

The floor plywood acts as a continuous slab and distributes the following loads to the joists, realizing that there is no live load reduction because the tributary area is so small.

$$w = (16/12)(40 + 10) = 53.33 + 13.33 \cong 67 \text{ plf}$$

The reaction, or maximum shear force (ignoring conservatively the critical shear at distance d from the support face) for the simply supported joist is

$$R \cong V_{max} = wL/2 = 67(16/2) = 536 \text{ lb}$$

The simple span joists generate the familiar maximum moment of

$$M_{max} = wL^2/8 = 67(16)^2/8 = 2144 \text{ ft-lb}$$

The required section modulus as based on an allowable bending stress of $F'_b = F_b = 1450$ psi for repetitive member use, is

$$S \geq M/F_b = 2144(12)/1450 = 17.74 \text{ in.}^3$$

Try 2×10 joists, 16 in. o.c., $A = 13.875$ in.^{2}, $S = 21.391$ in.^{3}, and $I = 98.932$ in.^{4} according to Table A.9. Checking the shear stress, although generally not critical for ordinary residential floor joist construction, yields

$$f_v = 1.5V/A = 1.5(536)/13.875 = 57.95 \text{ psi} \leq F'_v = F_v = 95 \text{ psi} \qquad \text{OK}$$

Checking the maximum deflection, as usually based on the allowable live load deflection, yields

$$\Delta_L = \frac{5w_L L^4}{384EI} = \frac{5(53.33/12)(16 \times 12)^4}{384(1,700,000)98.932} = 0.468 \text{ in.}$$

$$\leq L/360 = 16(12)/360 = 0.533 \text{ in.,} \quad \text{OK}$$

Glulam Beams.   The characteristics of glulam timber were introduced in Section 3.1. Ordinary laminated beams are designed in a similar fashion as laterally supported solid, sawn lumber beams with the exception of a new coefficient C_V (volume factor), which is introduced for strong axis bending but is not to be used simultaneously with the stability factor C_L; the lesser of the two shall be used. The allowable bending stress about the strong axis is

$$F'_{bx} = (F_{bx}) \, C_D \, C_M \, C_t \, (C_L \text{ or } C_V) \cong (F_{bx}) \, C_D C_F$$

For a beam with full lateral support, $C_L = 1.0$. In this context the size factor C_F of the previous NDS is used for preliminary design purposes rather than factor C_V. For members over 12 in. deep, the *size factor* C_F must be applied to the allowable bending stresses; this factor, however, is already integrated in the modified section modulus values in Table A.10, but as based on the previous design standards. Furthermore, the design values must be modified for duration of loads other than normal, when the moisture content will be 16% or more, when loads are applied parallel to wide faces of laminations (i.e., flat use factor, C_{fu}), and for beams with variable depth.

Typical glulam beams have a constant rectangular cross section. Although straight beams are most common, tapered and curved glulam beams are often desired to provide the necessary roof slope or interior clearance or for esthetic reasons. Typical beams with variable depth are shown in Fig. 3.10. They can be of the following configuration: double-tapered straight (a), single-tapered straight (b), double-tapered pitched (c), double-tapered pitched with constant cross section (d), double-tapered curved (e), and curved (f).

A *curvature factor,* C_c, is required for the curved portions of glued laminated timber bending members to modify the allowable flexural stresses.

$$C_c = 1 - 2000 \, (t/R)^2 \qquad (3.37)$$

where $t =$ thickness of lamination (in.)
$\qquad\quad R =$ radius of curvature of inside face of lamination (in.)
$\qquad\quad t/R \leq 1/100$ for hardwoods and southern pine
$\qquad\qquad\;\; \leq 1/125$ for other softwoods

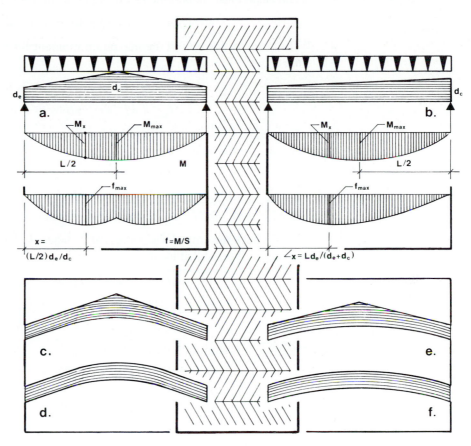

Figure 3.10   Glulam beams.

The design of beams with variable depth is not as simple as for ordinary straight beams. Here a complex stress interaction takes place where the combined effect of shear, bending, compression, and tension parallel to the grain and compression and tension perpendicular to the grain must be taken into account. Furthermore, the maximum bending stress, for example, for uniformly loaded tapered beams, occurs not at the maximum moment location but at a distance x, as indicated in Figs. 3.10a and b.

Long-span glulam beams are often cambered to control deflection. A minimum camber of 1.5 times the dead load deflection is recommended. In addition, flat roofs should provide an adequate drainage to prevent ponding, therefore, requiring a minimum pitch or camber of not less than 1/4 in. per foot of slope.

EXAMPLE 3.10

Glulam beams for an industrial building roof are arranged in a parallel fashion and spaced at 16 ft. The beam span from center support (center of bearing area) to center support is $L = 60$ ft; the clear span may be assumed as $L_c = 59.5$ ft for this first investigation. The beam may be considered laterally braced against rotation of ends, and the top is held in place by the roof deck. Roof slope is a minimum of $\frac{1}{4}$ in. per foot to prevent ponding. Assume a snow load of 30 psf and a dead load of 20 psf, which includes the weight of the glulam beam. Use the following design values for normal temperature range and dry conditions of use:

$$F_b = 2400 \text{ psi}, \quad F_v = 165 \text{ psi}, \quad F_{c\perp} \text{ (bottom)} = 650 \text{ psi}, \quad E = 1{,}700{,}000 \text{ psi}$$

The beam must support the following uniform dead and live load:

$$w = w_D + w_L = 16(20 + 30) = 320 + 480 = 800 \text{ plf}$$

The reaction forces and the maximum moment for the simply supported beam are

$$R = wL/2 = 0.8(60/2) = 24 \text{ k}$$

$$M_{max} = wL^2/8 = 0.8(60)^2/8 = 360 \text{ ft-k}$$

The required section modulus, taking into account load duration $C_D = 1.15$ as based on snow, and size $F_b' = F_b C_D C_F$, is

$$S \geq M/F_b C_D = 360(12)/2.4(1.15) = 1565 \text{ in.}^3$$

Try $6\frac{3}{4} \times 40\frac{1}{2}$ in. section with a modified section modulus $S_x = 1612$ in.^{3}, $A = 273.4$ in.^{2} and $I = 37367$ in.^{4} (Table A.10).

Check the critical shear stress at distance d adjacent to the face of the support. The shear at that location is

$$V_{max} = wL_c/2 - wd = w(L_c - 2d)/2 = 800 [59.5 - 2 (40.5/12)]/2 = 21,100 \text{ lb}$$

$$f_v = 1.5 \; V/A = 1.5(21,100)/273.4 = 116 \text{ psi} \leq F_v' = F_v \, C_D = 165(1.15), \quad \text{OK}$$

Check the bearing stresses perpendicular to the grain at the support (i.e., at bottom tension face), assuming a bearing length of 6 in. at each end.

$$f_{c\perp} = R/A_b = 24,000/6.75(6) = 593 \text{ psi} \leq F_{c\perp}' = F_{c\perp} = 650 \text{ psi}, \quad \text{OK}$$

Check the beam deflection as based on recommended deflection limitations for industrial roofs of $L/120$ for total load action and $L/180$ for live load action.

$$\Delta_{TL} = \frac{5wL^4}{384EI} = \frac{5 \, (800/12) \, (60 \times 12)^4}{384(1,700,000) \, 37367} = 3.67 \text{ in.}$$

$$\leq L/120 = 60(12)/120 = 6 \text{ in., OK}$$

$$\Delta_L = (w_L/w) \, \Delta_{TL} = (480/800) \, 3.67 = 2.20 \text{ in.}$$

$$\leq L/180 = 60(12)/180 = 4 \text{ in.}$$

Determine the camber of the beam using 1.5 times the dead load deflection.

$$\Delta_D = (w_D/w) \, \Delta_{TL} = (320/800)3.67 = 1.47 \text{ in.}$$

$$\text{Camber} = 1.5 \, \Delta_D = 1.5(1.47) = 2.21 \text{ in.}$$

EXAMPLE 3.11

Replace the beam in Example 3.10 by a double-tapered straight beam using a roof slope of 1/12. This is just a first layout of the beam profile, assuming a beam width of $6\frac{3}{4}$ in.

The end depth of the beam is determined as derived from shear stress requirements. Assuming for this first trial a depth of 25 in. yields the following maximum shear at distance d from the support face.

$$V_{max} = 800 [59.5 - 2(25/12)]/2 = 22,133 \text{ lb}$$

$$f_v = 1.5V/A = 1.5(22,133)/6.75d_o = 165(1.15), \quad \text{or} \quad d_o = 25.92 \text{ in.}$$

The assumption was close enough; try an end depth of 26 in. Check the following center-line depth using a roof slope of 1/12:

$$d_c = 26 + [30(12)]/12 = 56 \text{ in.}$$

Built-up Wood Beams

Besides the common rectangular sawn lumber and glulam beams representing solid pieces of one material, beams are also produced from a combination of various components and materials, such as sawn lumber, glulam timber, plywood, structural composite lumber, steel, and so on. In other words, these composite beams may consist of a combination of lumber–plywood, glulam–plywood, composite lumber–plywood, wood–steel, and so forth.

The most common types of build-up wood beams are the following:

- I-joists and I-beams
- Plywood beams and plywood panels
- Steel-reinforced beams

Wood I-beams are engineered lightweight beams with a web panel glued to top and bottom flanges (Fig. 3.11c). For the design of I-beams, the literature of the I-joist industry should be consulted.

Plywood Beams and Panels.   Plywood panels and lumber (or glulam timber), joined together with adhesives, nails, or both, can be used in composite action in many ways. The most common systems are as follows:

Lumber–plywood beams

Plywood stressed-skin panels

Plywood sandwich panels

Plywood beams consist of single or multiple plywood webs attached to single or multiple lumber flanges, as well as of vertical stiffeners along the beam (between the flanges) to prevent web buckling and to redistribute concentrated loads. The common plywood–lumber beam shapes are shown in Figs. 3.11d to h; they may form I- or box-beams with extra webs if needed for shear reinforcement. Bearing stiffeners are required at concentrated load action, which includes the beam ends (supports).

Plywood box girders may span 100 ft for lightly loaded structures, although the more common span range is between 20 to 60 ft.

Plywood beams can be proportioned quickly for first trial purposes (similar to steel plate girders) by ignoring the composite action and assuming the following:

- A beam depth of one-eighth to one-twelfth of the span may be selected.
- The cross-sectional area of the lumber flanges A_f may be determined by assuming that the moment is resisted entirely by the flanges.

$$A_f = \frac{M/d_1}{F_t'}$$

where M = bending moment

F_t' = controlling allowable axial stress parallel to the grain of the flange lumber, which is usually the allowable tension stress [Eq. (3.89)]

$$d_1 = \text{center-to-center distance between the flanges}$$

- The plywood web thickness is determined by assuming all the shear to be carried by the web.

$$f_v = 1.5V/d \, \Sigma t \le F'_v \tag{3.39}$$

For example, for a plywood box beam the required thickness of each web should approximately be equal to

$$t \ge 3V/4dF'_v \tag{3.40}$$

- The deflection of plywood beams may have to include the effect of shear. For rough, first-approximation purposes, the bending deflection may be increased by a certain percentage to include shear deflection which is dependent on the span-to-depth ratio of the beam. Shear deflections can be ignored for $L/d = 20$ but should be increased by 50% for $L/d = 10$; for the range between, the respective percentage can be interpolated linearly.

EXAMPLE 3.12

Determine the preliminary size of a 28-ft-span roof plywood box beam that must support a uniform load of 400 plf, which includes snow loading. Use Douglas fir–larch; select structural flanges and structural interior plywood panels with $F_t = 1200$ psi and $F_v = 190$ psi.

Estimate a beam depth of

$$d \cong L/12 = 28(12)/12 = 28 \text{ in.}$$

The maximum moment is equal to

$$M_{\max} = wL^2/8 = 400(28)^2/8 = 39{,}200 \text{ ft-lb}$$

Assuming that 2×6 flanges are needed, $h_1 = 28 - 5.5 = 22.5$ in.

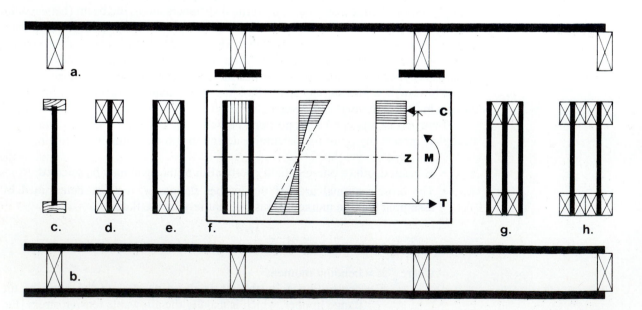

Figure 3.11   Built-up wood beams.

The required flange area, taking load duration (snow) into account, according to Eq. (3.38), is

$$A_f = \frac{T}{F_t'} = \frac{M/d_1}{F_t C_D} = \frac{39{,}200\,(12)\,/22.5}{1.15\,(1200)} = 15.15 \text{ in.}^2$$

Try two $2 \times 6s$ for each flange, $A_f = 2(8.25) = 16.5 \text{ in.}^2$.

The maximum shear force is conservatively assumed equal to

$$V_{max} = wL/2 = 400(28)/2 = 5600 \text{ lb}$$

The required plywood thickness according to Eq. (3.40) is

$$t = 3V/[4dF_v C_D] = 3(5600)/[4(28)1.15(190)] = 0.687 \text{ in.}$$

Try a 3/4-in. structural I plywood web along each face.

Plywood stressed-skin panels are used as floor, roof, and shear wall panels. They are usually 4-ft-wide panels that consist of longitudinal lumber stringers, headers at the ends, and possibly blocking (for lateral support of stringers) and are covered on top, or top and bottom, with plywood sheathing of at least 5/16 in. thickness. The skin(s) must be rigidly connected to the longitudinal ribs so that they can act together as a composite unit allowing no slippage between them. We can visualize a panel to behave under perpendicular loading as a series of parallel built-up I-beams, where the skins (flanges) take most of the moment (besides acting as slab in the other direction), and the ribs perform as webs carrying most of the shear. The common panel types are as follows:

- Two-sided stressed-skin panels (with skins on top and bottom, Fig. 3.11b).
- One-sided stressed-skin panels (with skin on top, Fig. 3.11a).
- T-flange stressed-skin panels (with skin on top and lumber strips on the bottom of the stringers, Fig. 3.11a, center).

Since the modulus of elasticity of the stringers is less than that of the plywood, the *transformed-section* approach may be used for design, as is typical for composite construction (for discussion of that method, see the next section).

When in the two-sided stressed-skin panel the ribs are replaced by a continuous lightweight core (e.g., paper honeycombs, plastic foams such as expanded polystyrene or urethane, formed metal), then this structure is called a *sandwich panel*. Again, proper bonding between facing and core is most critical for true composite action with no slippage. The behavior of the panel may be compared to that of a shallow, wide I-beam, where the axial forces due to bending are resisted by the thin facing (stabilized by the core against buckling and wrinkling) that acts as flanges and provides the proper bending stiffness; the shear is carried by the core.

Steel-reinforced Beams. To increase the strength and stiffness of wood beams, they may have to be reinforced with steel flitch plates or rolled steel sections such as channels. The components must be bolted together so that the assembly can act as a unit. *Flitch beams* are occasionally used in wood construction. Side plates may be used to reinforce existing beams while single, thin steel plates may be placed between wood sections so that long-term deflections are reduced and the beam strength is increased without increase of beam depth (Fig. 3.12).

The size of the flitch beam is determined by assuming that the various material components deflect by the same amount, in turn requiring that they are properly connected. The sizing and placing of the bolts depend on the position of the plates. For example, lateral bracing for the thin steel side plates must be considered so that they

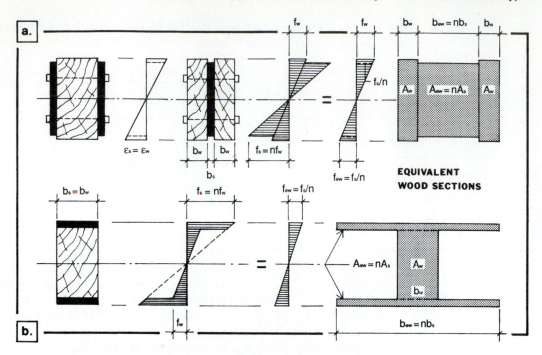

Figure 3.12 Composite wood–steel beams.

do not buckle, or bolts may be needed for vertical load transfer should the loads not be applied simultaneously to all sections, and similarly bolts may be required for horizontal shear transfer.

As the composite (flitch) beam deflects, the strain varies linearly across the section so that the strain in both materials must be the same, assuming no slippage between the components.

$$\varepsilon_w = \varepsilon_s \tag{a}$$

Then the elastic range stress and strain are proportional to each other ($E = f/\varepsilon$).

$$(f/E)_w = (f/E)_s$$
or $$\tag{3.41}$$
$$f_s = (E_s/E_w)f_w = nf_w \quad \text{or} \quad f_{ew} = f_s/n$$

where $n = E_s/E_w$ = modular ratio $\cong$ 16 to 29
f_{ew} = the stress in the equivalent wood section

This expression clearly demonstrates that the stiffer material takes a much higher load under the same strain.

In the next step, the composite section is visualized as a transformed homogeneous section. Therefore, in this equivalent wood section, the axial bending forces P_s at a distance from the neutral axis in the steel fibers must be equal to the forces P_w in the wood fibers at the same distance from the neutral axis.

$$P_s = P_w \quad \text{or} \quad (fA)_s = (fA)_w \tag{b}$$

Substituting Eq. (3.41) into (b) yields

$$nf_w A_s = f_w A_w \quad \text{or} \quad A_{ew} = nA_s \tag{3.42}$$

In other words, the steel plates can be transformed into equivalent wood beams simply by multiplying the cross-sectional area of the steel section, $A_s = d_s b_s$, by the modular ratio n.

Since the depth for both materials in the flitch beam is the same ($d_w = d_s$), Eq. (3.42) can be simplified further.

$$b_{ew} = nb_s \tag{3.43}$$

Therefore, the steel plates can easily be converted to wood members by multiplying the steel plate thickness, $b_s = t_s$, by n (Fig. 3.12a). Once the cross-sectional areas of the various materials of a composite member have been transformed into an equivalent material, the stresses can be determined by the familiar design approach for homogeneous members.

EXAMPLE 3.13.

A simply supported floor beam spans 12 ft and carries a uniform load of 800 plf. Use southern pine with $F_b = 1450$ psi and $E = 1,700,000$ psi, and use A36 steel with $F_b = 0.6F_y$. First, estimate the preliminary beam size assuming only joist sizes to be available.

The maximum moment is equal to

$$M_{max} = wL^2/8 = 800(12)^2/8 = 14,400 \text{ ft-lb}$$

The required section modulus for $F'_b = F_b$ is

$$S \geq M/F_b = 14,400 \ (12)/1450 = 119.17 \text{ in.}^3$$

Four 2×12s are needed: $S_x = 4(31.641) = 126.56$ in.^{3}. This solution is considered not acceptable, so two 2×12s with a thin steel plate (A36) between, will be investigated.

The section modulus for the two 2×12s is

$$S_w = 2(31.641) = 63.28 \text{ in.}^3$$

The section modulus for the steel plate is

$$S_s = td^2/6 = t(11.25)^2/6 = 21.09t$$

The modular ratio is

$$n = E_s/E_w = 29,000/1700 = 17.06$$

Now it has to be determined which of the allowable stresses in the transformed section controls the design. In other words, when the steel is transformed into an equivalent wood section, its allowable bending stress according to Eq. (3.41) is

$$f_{ew} = f_s/n = 22,000/17.06 = 1290 \text{ psi} < 1450 \text{ psi}$$

where $f_s = F_s = 0.6F_Y = 0.6(36) \cong 22$ ksi. Hence the allowable steel stress controls the design of the transformed section.

The acting moment is resisted by the material components as follows:

$$M = M_w + M_s = (fS)_w + (fS)_s$$
$$14,400(12) = (1290 \times 63.28) + (22000 \times 21.09t) \quad \text{or} \quad t = 0.197 \text{ in.}$$

Try a $\frac{1}{4}$- in. by $11\frac{1}{4}$- in. steel plate.

Reinforced Concrete Beams

In cast-in-place concrete construction the beams form an integral part of the floor framing systems. With respect to gravity loading they constitute T-sections (or L-sections for the spandrel beams) with respect to the positive bending along the midspan region, but only rectangular sections for negative bending close to the supports (Fig. 3.13). Simple rectangular sections or inverted T-sections are also typical for precast concrete construction, where the slab may rest on the beams without any continuous interaction. Since the axial forces in frame girders are relatively small, they are ignored for preliminary design purposes.

The preliminary sizing of flexural members depends on stiffness and strength considerations. Usually, stiffness controls the design of flexible elements such as slabs, joists, shallow beams, and long-span beams. For this condition the minimum member depth can be determined from flexibility considerations. The ratio of deflection to span Δ/L of a beam can be expressed in terms of its span-to-depth ratio L/t multiplied by a constant C [e.g. (Eq. 3.13)].

$$\Delta/L = C(L/t)$$

To avoid the complex deflection calculations in reinforced concrete, limiting L/t ratios are given. Hence, the following approximate minimum member thicknesses with t (in.) and L (ft) for various cases, as derived in Problem 3.21, are

$$
\begin{aligned}
&\text{Cantilevers:} && L(12)/t = 8 \quad \text{or} \quad t_{min} = 1.5L \\
&\text{Simple-span beams:} && t = L/16 = L(12)/16 = 3L/4 \\
&\text{Continuous-span beams:} && \text{End: } t = L/18.5 = L(12)/18.5 \cong 2L/3. \\
& && \text{Interior: } t = L/21 = L(12)/21 = 4L/7.
\end{aligned}
\tag{3.44}
$$

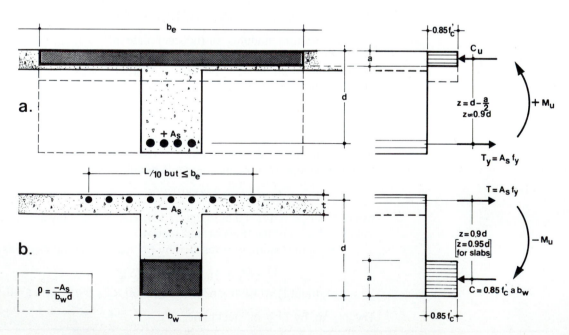

Figure 3.13 Equivalent stress distribution for typical singly reinforced concrete floor beams at ultimate load. (Reproduced with permission from *The Vertical Building Structure*, Wolfgang Schueller, copyright © 1990 by Van Nostrand Reinhold.)

These values are based on normal-weight concrete and Grade 60 steel. For Grade 40 reinforcement, the beam depth may be reduced by 20%, but it must be increased by 20% or 10% for 90 pcf or 120 pcf lightweight concretes, respectively.

The size of ordinary beams (i.e., not short- or long-span beams or beams with large concentrated loads near the support) that do not need compression reinforcement is controlled by flexural compression. Shear is rarely critical, with the exception of members with unreinforced webs (e.g., joists, foundations) and of punching shear around columns in flat plates and slabs. The following expressions for sizing of members and for finding the flexural reinforcement have been derived from the bending strength of the beam, while the shear reinforcement has been obtained from its shear strength (see Problem 3.21). The size of concrete beams is dependent on the amount of reinforcement used in the section as measured by the steel ratio $\rho = A_s/b_w d$ at the critical support location where the continuous beam acts only as a rectangular member. In other words, the size of ordinary floor beams is controlled by the flexural compressive strength at the maximum moment location, where the resisting compressive cross-sectional area is the smallest. In this context, the equation for selecting the concrete beam depth is based on the typical condition of Grade 60 steel together with an average reinforcement ratio of $\rho = \rho_{max}/2 \cong 1\%$ so that the bars can be easily placed; this also provides a reasonable deflection control. For this condition the coefficient of resistance is $R_u = \phi R_n = M_u/bd^2 \cong 0.52$ ksi [see also (Eq. 3.51)]. This expression is now changed in form as given next but using mixed units; it is applicable to beams with common concrete strengths of 3000 to 6000 psi. The equation can also be represented in terms of the service moment M by assuming an average load factor of 1.5 as is derived in Eq. (3.55b).

$$bd^2 = 23M_u \cong 35M \quad \text{or} \quad d = \sqrt{\frac{M_u}{R_u b}} = \sqrt{\frac{M_u}{23b}} \tag{3.45}$$

where M_u = ultimate bending moment (ft-k) $\cong 1.5M$
 b = beam web width (in.)
 d = effective depth of beam (in.)

Notice that the expression is in mixed units. From this equation, the beam depth can easily be found by assuming a typical beam width. Similarly, rather than using $\rho \cong 1\%$, the corresponding expression for $\rho \cong 2\%$ (as often used for column designs that act primarily as beams) is

$$bd^2 = 14M_u \tag{3.46}$$

Equation (3.45) also reminds us of the approximate maximum concrete stress f_c in the working stress approach for the balanced condition of a rectangular section as if it were derived from an uncracked section. Using the same material strength as before results in

$$f_c \cong M/S = M/(bd^2/6) \leq 0.45f_c' \quad \text{or} \quad bd^2 = 40M \tag{3.47}$$

If Eq. (3.45), however, is also used for Grade 40 steel, a minimum of about 1.6% of reinforcement is required, and 1.3% for 50-ksi steel, rather than the assumed 1% for 60-ksi steel. Assuming a continuous beam with the critical support moment due to gravity loading to control its design, Eq. (3.45) can further be simplified (see Problem 3.21) to

$$d = 1.45l_n\sqrt{w_u/b_w} \tag{3.48}$$

where w_u = uniform ultimate load (k/ft) l_n = clear beam span (ft)
         d = effective beam depth (in.), b_w = beam stem width (in.)

For preliminary design purposes, the following simplified expression is derived from Eq. (3.48) for typical beam widths of $b = 10$ to 16 in. with at least 1% of Grade 60 steel, but using the span length L from center to center of supports.

$$d = 0.4L \sqrt{w_u} \tag{3.49}$$

Keep in mind when selecting the beam proportions that wide, shallow beams may be more economical from an overall construction point of view than the narrow, deep beams one is more used to. For this condition, as for slabs under normal loads, the minimum member thickness is mostly determined by deflection limitation rather than strength, besides having to consider fire resistance requirements.

When the cross section is known, the moment reinforcement can be found as follows. Horizontal equilibrium of the forces in Fig. 3.13b yields the depth a of the stress block.

$$\Sigma F_x = 0 = C_u - T_y = 0.85 f_c' ab - A_s f_y$$

or

$$a = \frac{A_s f_y}{0.85 f_c' b} = \frac{p f_y d}{0.85 f_c'} \tag{3.50}$$

Rotational equilibrium in Fig. 3.13 necessitates the balance of the acting moment M_u and the steel strength $A_s f_y$ which is then reduced by the capacity reduction factor $\phi = 0.9$.

$$M_u \le \phi M_n = \phi T_y z = \phi A_s f_y (d - a/2) = \phi A_s f_y j_u d$$

$$\tag{3.51}$$

$$= bd^2 \left[\phi p f_y \left(1 - \frac{p f_y}{1.7 f_c'} \right) \right] = R_u bd^2$$

In the last expression, reflecting the moment resisting capacity of the concrete, the coefficient of resistance R_u is tabulated in most books on reinforced concrete as a function of ρ. From Eq. (3.51), the approximate moment reinforcement for rectangular beams, joists, and T-beams in ordinary buildings can be found by using an average internal lever arm length of $z = j_u d = 0.9d$.

$$A_s = M_u/(0.8 f_y d) \ge \rho_{min} b_w d = 0.2(b_w d)/f_y \tag{3.52}$$

where A_s = moment reinforcement (in.^{2})
         M_u = ultimate moment (in. k)
         f_y = yield stress of steel (ksi)
         d = effective beam depth (in.)
         ρ = $A_s/b_w d$ = steel ratio

This equation gives reasonable results for typical beam steel ratios of 1% to about 1.6%. For this range it is quite insensitive to the various common concrete strengths and the variation of steel ratio, as is apparent from the closely bundled and nearly straight lines of the strength curves for a given steel (see Problem 3.21). For fast approximation purposes, however, this equation may even be used for its entire permitted range as long as the steel ratio is less than ρ_{max}, which is given in Table A.4 for various steel–concrete combinations. Attention must be given to 3000-psi concrete together with 60-ksi steel, where the equation becomes less precise beyond a steel ratio of about 1.1%.

Equation (3.51) is also used for the T-beam behavior of the composite beam-slab at midspan (Fig. 3.13), where the flexural capacity of the concrete is so large that the stress block depth a usually lies within the flange (slab), so the T-section can be treated as a wide, shallow rectangular beam section of width b_e for preliminary design purposes; here only the minimum steel ratio ρ_{min} must be checked.

For ordinary solid concrete slabs, which have a much lower steel ratio than beams, often as a first approximation $z = 0.95d$ is taken, which results in the following required moment reinforcing:

$$A_s = M_u/(0.85f_yd) \geq A_{s\,min} \tag{3.53}$$

Generally, Grade 60 bars are used; Grade 40 requires 50% more steel. Substituting Grade 60 steel in Eq. (3.51) yields the following simplified expression for beam design, which, however, is often also used for slab design when fast estimates are needed, but taking into account the different $A_{s\,min}$; the equation is in mixed units.

$$A_s = M_u/4d \geq b_wd/300 \tag{3.54}$$

where M_u = ultimate moment (ft-k)
$\quad\quad A_s$ = moment reinforcement (in.^{2})
$\quad\quad d$ = effective beam depth (in.)

Notice that the minimum reinforcement for this condition is $\rho = 0.33\%$. The maximum reinforcement ratio ρ_{max}, which ensures that the beam is underreinforced and fails in tension, does not have to be checked if the moment reinforcement equation is only applied to beams with reinforcement ratios up to about 1.6%; but we must watch out for the combination of 3000-psi concrete and Grade 60 steel with $\rho_{max} = 1.61\%$.

According to the working stress method, Eq. (3.52) can be expressed by using $z = jd \cong 7d/8$ for first-approximation purposes, as

$$A_s = \frac{M}{f_s z} = \frac{M}{f_s jd} \cong \frac{M}{0.875df_s} \tag{3.55a}$$

Here, the tensile stresses in the reinforcement are limited to $f_s = 20$ ksi for Grade 40 and 50 steel, and $f_s = 24$ ksi for Grade 60 and higher-strength steel. Another equation often used for the approximate design of bending reinforcement as based on the service loads and ordinary loading conditions of $D/L = 2/1$ can be derived as follows:

$$M_u \cong 1.4(2M/3) + 1.7(M/3) = 1.5M$$

$$A_s = M_u/(0.8f_yd) = 1.5M/[0.8(60)d] = M/32d \tag{3.55b}$$

Notice that this equation does not have mixed units; the moment is in in. k and d in inches.

The effective depth d from the compression face to the centroid of the steel reinforcement can be approximated and related to the member thickness t as follows:

Beam   (interior; for exterior exposure subtract 0.5 in.)
$\quad\quad\quad$ single layer (always for top steel of T-beam) $d = t - 2.50$
$\quad\quad\quad$ double layer $d = t - 3.50$

Joists: $d = t - 1.25$ (3.56)
Slabs: one-way slabs: $d = t - 1.00$
          two-way slabs (center of upper layer): $d = t - 1.50$

A simplified approach to shear design is discussed next. Although the size of ordinary beams is usually not controlled by shear, shear may become critical for short-span beams that carry heavy loads and beams with unreinforced webs, such as joists.

The shear capacity of reinforced concrete consists of the strength of the concrete ϕV_c and of the shear reinforcement ϕV_s. It obviously must be at least as large as the ultimate shear force V_u as for instance $V_u = 1.4 V_D + 1.7 V_L$ at the point of investigation (Fig. 3.14).

$$V_u \le \phi V_n = \phi V_c + \phi V_s \tag{3.57}$$

A conservative definition of the shear strength of concrete in beams and one-way slabs is

$$\phi V_c = \phi (2 \sqrt{f_c'}) b_w d \tag{3.58}$$

Letting the capacity reduction factor for shear be $\phi = 0.85$ and the typical concrete type be $f_c' = 4000$ psi yields $\phi V_c = 0.85(2\sqrt{4000}) b_w d / 1000$, or

$$\phi V_c = 0.11 b_w d \tag{3.59}$$

where b_w = web width of beam (in.)
        d = effective beam depth (in.)
        ϕV_c = shear strength of concrete (k)

The allowable working shear stress for beams, walls, and one-way slabs is $1.1\sqrt{f_c'}$. Keep in mind that for beam column design the shear capacity of the concrete is very much increased by the axial compressive stress! The shear reinforcement is located in the beam web and is provided by the inclined (bent-up)

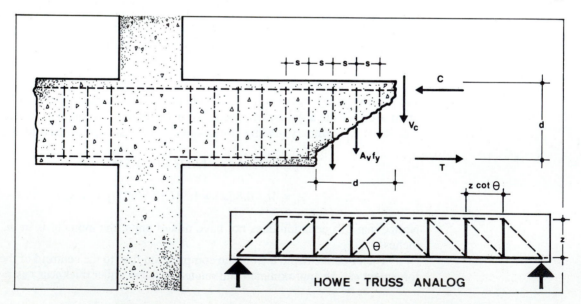

Figure 3.14 Shear force resistance of vertical stirrups (Reproduced with permission from *The Vertical Building Structure*, Wolfgang Schueller, copyright © 1990 by Van Nostrand Reinhold.)

portion of the longitudinal steel, which is approximately in line with the diagonal tension close to the support and/or is provided by the U-shaped vertical stirrups. Here only the latter are considered to resist the shear; the Howe-truss analog in Fig. 3.14 reflects convincingly the idealized tension and compression action of the concrete beam. The strength of the shear reinforcement for relatively shallow beams with transverse failure through the entire section is independent of the concrete member dimensions and the concrete strength. The capacity of the stirrups along the section in Fig. 3.14 is

$$\phi V_s = \phi (A_v f_y) n = \phi (A_v f_y) d/s, \quad \text{or} \quad s = A_v f_y d/V_s \tag{3.60}$$

where $n(A_v f_y)$ is the entire tensile capacity of n number of stirrups along the diagonal crack. The number of stirrups n can be expressed in terms of the stirrup spacing s as $n = d/s$ by assuming conservatively the horizontal projection of the diagonal crack equal to the effective depth d. The stirrup spacing is thus conveniently represented as a function of the effective beam depth d. Using the usual maximum stirrup spacing of $s = d/2$, we can continue with smaller spacing intervals for standard conditions of $d/3$ and for most conditions the closest spacing of $d/4$, considering that the stirrups should not be closer than about 3 inches. The stirrup spacing should be made in not less than 1/2-in. increments

$$s = d/n, \, d/2, \, d/3, \text{ and } d/4.$$

For typical conditions of $f_y = 60$ ksi, $\phi = 0.85$, and No. 3 stirrups with two legs of $A_v = 2(0.11) = 0.22$ in.^{2}, the following stirrup capacity is achieved:

$$\phi V_s = \phi A_v f_y n = 0.85(0.22)60n = 11n \tag{3.61}$$

where $n = 2, 3,$ and 4.

For wide beams, use multiple stirrups, as, for example, four legs when $24 < b_w \leq 47$ in. The following general design process may be used:

• No shear reinforcement is required for beams if

$$V_u \leq \phi \, V_c /2 = \phi \sqrt{f_c'} \, b_w d \tag{3.62}$$

Or, expressed in terms of the nominal ultimate shear stress for 4000-psi concrete:

$$v_u = V_u/b_w d \leq \phi \sqrt{f_c'} = 54 \text{ psi}$$

• A minimum shear reinforcement is required for beams (but not for shallow beams, slabs, footings and joist construction) if

$$\phi V_c/2 < V_u \leq \phi V_c \tag{3.63}$$

or for 4000-psi concrete when $v_u = V_u/b_w d \leq 107$ psi

• Shear reinforcement is required if

$$V_u > \phi V_c \tag{3.64}$$

When the ultimate shear force V_{ux} at any point x along the beam exceeds the shear capacity of the concrete, shear reinforcement must resist the excess shear $V_u - \phi V_c$.

According to Eqs. (3.57) and (3.60), the required shear capacity of the stirrups ϕV_s and the corresponding stirrup spacing are

$$V_u - \phi V_c \leq \phi V_s \quad \text{or} \quad s \leq A_v f_y d/(V_u/\phi - V_c) = A_v f_y d/V_s \tag{3.65}$$

For instance, for the special condition of $\phi V_s = 11n$,

$$n \geq (V_u - \phi V_c)/11 = d/s \geq 2, \quad \text{or} \quad s \leq 11d/(V_u - \phi V_c) \leq d/2 \qquad (3.66)$$

Not only should the maximum spacing or the minimum stirrup reinforcement be $s \leq d/2$, but also the web reinforcement should be able to transfer a shear stress of $\phi 50 = 42$ psi.

$$s \leq A_v f_y / 50 b_w \qquad (3.67)$$

This expression does not control ordinary beam sizes of 4000-psi concrete. The following restriction for $s_{max} = d/2$ is required:

$$V_u - \phi V_c \leq \phi 4 \sqrt{f_c'}\, b_w d, \quad \text{where} \quad \phi V_c = \phi 2 \sqrt{f_c'}\, b_w d,$$

or
$$V_u \leq \phi 6 \sqrt{f_c'}\, b_w d \leq 3\,(\phi V_c) \qquad (3.68a)$$

Let $f_c' = 4000$ psi and $\phi = 0.85$; then $V_u \leq 0.32 b_w d$ or $v_u = V_u/b_w d \leq 322$ psi. Hence, the minimum beam depth d required for typical conditions is approximately

$$d_{min} \cong 3V_u/b_w \qquad (3.68b)$$

Keep in mind, however, that the code does allow smaller beam sections for a maximum stirrup spacing of $d/4$ and the average ultimate shear stress not to exceed $10\sqrt{f_c'}$

Additional shear is generated by torsion, which may have to be covered with closed stirrups; the corresponding additional bending is resisted by longitudinal bars around the beam perimeter. Torsion may be critical for spandrel beams in monolithic floor structures, where torsion reinforcement is required. The twisting of interior beams due to asymmetrical arrangement of live loads is generally not critical. Torsion is not further treated here, but the designer should always keep in mind the importance of attempting to reduce the member twisting through special detailing so as to minimize torsional effects on structural components or by sizing the member so that torsion can be neglected.

With respect to the various approximate beam sizes, we may conclude the following:

- Any beam proportion is possible as long as the depth is at least equal to the one required by the deflection control [Eq. (3.44)].
- The shear strength is directly related to the cross-sectional area of beams [Eq. (3.58)] and thus does not influence the beam proportion, only its size; it only becomes critical for short-span beams under heavy loads or beams with unreinforced webs. The moment capacity, however, is primarily affected by the square of the depth [Eq. (3.45)], thus making narrow, deep beams more efficient from a local material point of view as reflected by the lower steel ratios. The usual depth-to-width ratio for shorter spans is 1.5 to 2, while for larger spans the ratio may be 2.5 to 3 and larger. But keep in mind that, from an overall point of view, it may still be more economical to use wide, shallow beams rather than narrow, deep beams. For instance, in pan joist construction the supporting beams often have the same depth as the joists in order to reduce formwork costs rather than material costs and to reduce the overall building height. Furthermore, by changing the width of beams only the bottom forms are affected but not the side forms and shores. Shallow, wide beams are first checked with respect to depth according to deflection control, and then the width is found from flexural requirements. The beam widths are usually multiples of 2 or 3 in. Often, constant beam sizes are used for one building story by only changing the reinforcement according to the span and load variations.

- The beam proportions are also influenced by the placement of the flexural reinforcement. In narrow beams, several layers of longitudinal steel may be required. Refer to Table A.3 for the minimum beam widths for various bar combinations. Keep in mind that for situations where large quantities of steel bars in beams (or columns) are difficult to fit up to four bars can be bundled together under certain conditions. Also, the shear reinforcement has an effect on the beam sizes; small beams, for instance, may need very close stirrup spacing. Additionally, it may be advantageous to make beams wider than narrow columns by at least 2 in. so that the bars in the beam corners can pass unobstructed.

EXAMPLE 3.14

Determine the moment capacity of the 12- $\times$ 24-in. reinforced concrete beam with 4 #8 bars at midspan (see Fig. 3.16, section A–A) using $f_c' = 4000$ psi and $f_y = 60,000$ psi.
  Check the steel ratio ρ for $A_s = 3.16$ in.^{2}.

$$\rho = A_s/bd = 3.16/12(24 - 2.5) = 0.0122 = 1.22\%$$

$$< \rho_{max} = 2.14\% \quad \text{(see Table A.4)}$$

Hence, all the steel can be used in figuring the moment capacity from Eq. (3.51).

$$M_u = \phi p f_y b d^2 \left(1 - \frac{pf_y}{1.7f_c'} \right)$$

$$= 0.9(0.0122)60(12)\,21.5^2 \left(1 - \frac{0.0122\,(60)}{1.7\,(4)} \right)$$

$$= 3261 \text{ in.-k} = 271.75 \text{ ft-k}$$

EXAMPLE 3.15

Determine the cross section for a reinforced concrete beam, assuming a beam width of 12 in. as based on a steel ratio of approximately 1%. The beam must resist an ultimate moment of $M_u = 380$ ft-k. Use $f_c' = 4000$ psi and $f_y = 60,000$ psi.
  The beam depth according to Eq. (3.45) is

$$bd^2 = 23M_u$$

$$12d^2 = 23(380) \quad \text{or} \quad d = 26.99 \text{ in.}$$

Hence, the total beam depth, assuming a single layer of steel bars, is approximately

$$t \cong 26.99 + 2.5 = 29.49 \text{ in.}; \quad \text{try} \quad t = 30 \text{ in.}$$

EXAMPLE 3.16

The continuous 16- $\times$ 24-in. reinforced concrete girder in Fig. 3.16 (section B-B) must resist an ultimate moment of $M_u = 341$ ft-k at midspan Determine the approximate steel reinforcing required to resist the moment, using $f_c' = 4000$ psi and $f_y = 60,000$ psi.
  The approximate required steel reinforcing according to Eq (3.52) is

$$A_s = \frac{M_u}{0.8f_y d} = \frac{341\,(12)}{0.8\,(60)\,(24 - 2.5)} = 3.97 \text{ in.}^2$$

Check the steel ratio.

$$\rho = A_s/bd = 3.97/16(21.5) = 1.15\% < 1.6\%, \quad \text{OK}$$

Try 4 #9 in one layer, $A_s = 4.00$ in.^{2} (Tables A.1 and A.3).

One-way, Solid Reinforced Concrete Slabs. The primary action of the slab is generally in response to gravity loading and thus, for preliminary design purposes, can be treated locally without having to consider the entire building structure.

Most floor slabs are of the one-way type. They span from cross-beam (filler beam) to cross-beam, which in turn are supported by the frame girders. In cast-in-place concrete construction the slabs form a monolithic whole with the floor framing. They are narrow panels with the long sides at least twice the length of the short sides. They bend in a nearly cylindrical surface under uniform gravity loading, with the single curvature reflecting the one-way action of the slab. The dual function of the slab should be remembered: it not only spans from beam to beam, but also forms the flanges of the beams in the perpendicular direction (T-beam action).

In load-bearing masonry buildings, the slabs may span from wall to wall in a continuous fashion if cast in place, or they may be precast units with simple spans. Wall footings and retaining walls are other examples of primary one-way slab action, as are the curtain panels with respect to lateral force pressure.

The one-way slab is treated as a series of independent 1-ft-wide shallow beams. The primary flexural reinforcement is positioned along the beam action, but the secondary steel must be placed perpendicular to control cracks due to shrinkage and temperature drop, as well as to redistribute possible concentrated loads. This temperature reinforcement is also the minimum flexural steel required for the slab. It is equal to

$$A_{s\min} = 0.002bt \qquad \text{for } f_y = 40 \text{ and } 50 \text{ ksi}$$

$$A_{s\min} = 0.0018bt, \qquad \text{for } f_y = 60 \text{ ksi} \qquad (3.69)$$

This temperature reinforcement is located on top of the bottom bars but under the top bars, and shall not be spaced farther apart than **5*t* or 18 in.** The primary flexural reinforcement in one-way slabs and walls other than concrete joists shall not be spaced farther apart than **3*t* or 18 in.**

In two-way slabs the spacing of the bars shall not exceed **2*t*** at critical sections, but otherwise **3*t*.** The normal span range for one-way concrete slabs is 10 to 20 ft with a typical slab thickness of 4 to 7 in. The slab thickness for residential and office construction is usually controlled by deflection limitations and by fire-resistance requirements; flexural and shear stresses rarely control except for heavy loading conditions. For a preliminary estimate of the slab thickness t (in.) as a function of the center-to-center span L (ft), but ≥ 4 in. for fireproofing, use the following:

- Simply supported $t = (L/20)/1.25 = L(12)/20(1.25) \cong L/2 \geq 4$ in.
- Cantilever $t = (L/10)/1.25 = L(12)/10(1.25) \cong L \geq 4$ in.
- Continuous both ends $t = (L/28)/1.25 = L(12)/28(1.25) \cong L/3 \geq 4$ in.
- Continuous one end $t = (L/24)/1.25 = L(12)/24(1.25) = 0.4L \geq 4$ in.

The continuous slab for exterior bays may have to be larger! The above values are based on Grade 40 steel and normal-weight concrete ($w_c = 145$ pcf). For Grade 60 steel, as for 90-pcf lightweight concrete, the slab thickness should be increased by 25%; for 110- and 120-pcf lightweight concrete the slab thickness should be increased by 10%. But keep in mind that a thinner slab may be more economical although deflection may have to be checked.

The monolithic character of concrete structures makes the design of slabs highly indeterminate and requires a frame analysis. To simplify the design of continuous beams and one-way slabs under gravity action, the ACI Code permits the use of approximate moment and shear coefficients as well as bar cutoffs (Fig. 3.15), taking into account the critical live load arrangement for the following conditions:

- There are two or more spans.
- Spans are approximately equal, with the larger of the two adjacent spans not greater than the shorter by 20%,
- Loads are uniformly distributed.
- Unit live load does not exceed three times unit dead load.

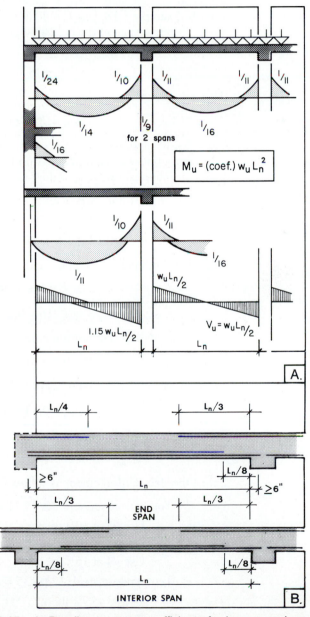

Figure 3.15 A. Bending moment coefficients for beams and one-way slabs. B. Recommended bar cutoffs for ordinary beams and slabs.

The coefficients are given in terms of clear span l_n at critical locations for various boundary conditions in Fig. 3.15A. A typical factored moment is

$$M_u = (\text{coef.})w_u l_n^2 \tag{3.70}$$

But the negative moment at the face of all supports for slabs with spans not exceeding 10 ft may be taken as $w_u l_n^2/12$

EXAMPLE 3.17

A six-story concrete frame office building consists of 30- × 34-ft bays with a floor framing as shown in Fig. 3.16. The concrete slab supports 5 psf for ceiling and floor finish and a 20-psf partition load, as well as a live load of 80 psf. Here a typical interior slab is investigated; for the design of the beams and girders, refer to Example 3.18. Use $f_c' = 4000$ psi and $f_y = 60,000$ psi.

The clear span of the typical interior bay using a supporting beam width of 12 in. is

$$l_n = 15 - 12/12 = 14 \text{ ft}$$

The slab thickness is estimated as

$$t = (L/3)1.25 = (15/3)1.25 = 6.15 \text{ in.}$$

Use a 6¼-in. slab, where $d \cong t - 1.0 = 6.25 - 1.00 = 5.25$ in.

For a one-way slab, there is no live load reduction. The uniform ultimate load the slab must carry, according to Eq. (4.19), is

$$
\begin{aligned}
w_u &= 1.4w_D + 1.7w_L \\
&= [1.4(6.25(150/12) + 5 + 20) + 1.7(80)]/1000 \\
&= 1.4(0.103) + 1.7(0.080) = 0.28 \text{ ksf or klf/ft of slab}
\end{aligned}
$$

The critical moments at support and midspan, according to Fig. 3.15A, are

$$-M_u = w_u l_n^2/11 = 0.28(14)^2/11 = 4.99 \text{ ft-k/ft}$$

$$+M_u = w_u l_n^2/16 = 0.28(14)^2/16 = 3.43 \text{ ft-k/ft}$$

The corresponding slab reinforcement, according to Eq. (3.53), is

At support: $-A_s = M_u/(0.85f_y d) = 4.99(12)/[0.85(60)5.25] = 0.224$ in.²/ft.

At midspan: $+A_s = 3.43(12)/[0.85(60)5.25] = 0.154$ in.²/ft

$$\geq A_{smin} = 0.135 \text{ in.}^2/\text{ft}$$

Temperature reinforcement is

$$A_{smin} = 0.0018bt = 0.0018(12)6.25 = 0.135 \text{ in.}^2/\text{ft}$$

Select the slab reinforcement as follows (or see Table A.2):

Top bars at support: $12/0.224 = s/0.2$, $s = 10.71$ in.
  Use #4 at 10½ in. o.c., $A_s = 0.229$ in.²
Bottom bars in the field: $12/0.154 = s/0.2$, $s = 15.58$ in.
  Use #4 at 15½ in. o.c., $A_s = 0.155$ in.²
Temperature steel: $12/0.135 = s/0.11$, $s = 9.78$ in.
  Use #3 at 9½ in. o.c., $A_s = 0.139$ in.²

The critical steel ratio is $\rho = A_s/bd = 0.229/12(5.25) = 0.36\% < 1.6\%$

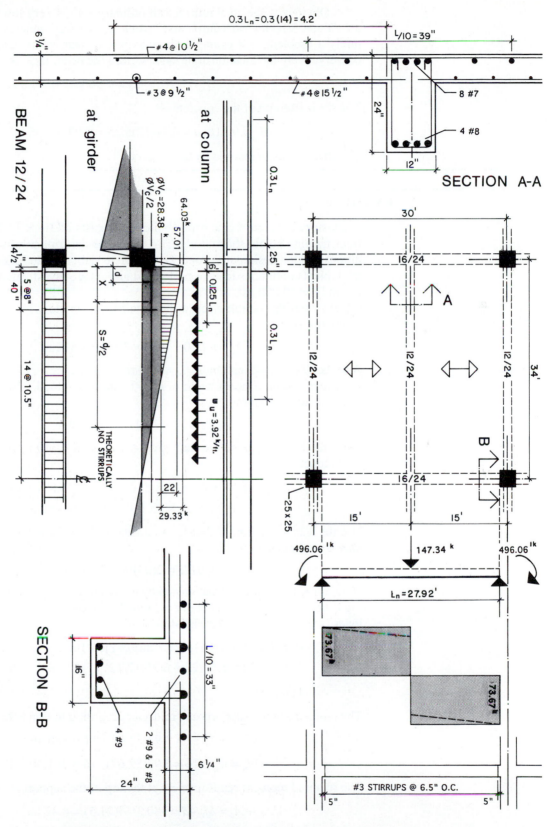

Figure 3.16 Design of concrete floor structure (see Examples 3.17 and 3.18)

Obviously, the steel ratio for an ordinary slab is very low! Refer to Figs. 3.15B and 3.16 for bar cutoffs and layout. In one-way concrete slabs, shear is rarely a problem. For example, the maximum shear in this exercise, ignoring conservatively the location at distance d from the beam face, is

$$V_u = 0.28(14/2) = 1.96 \text{ k/ft}$$

The shear strength of the concrete is

$$\phi V_c = 0.11 b_w d = 0.11(12)5.25 = 6.93 > 1.96 \text{ k/ft} \tag{3.59}$$

The shear is easily resisted by the concrete.

EXAMPLE 3.18

A six-story concrete frame office building consists of 30- × 34-ft bays with the floor framing layout shown in Fig. 3.16. For the description of the loading conditions and the design of the one-way concrete slab, refer to Example 3.17. Here the typical continuous interior beams and girders are investigated. Use $f'_c = 4000$ psi and $f_y = 60,000$ psi.

(a) The interior column sizes at the base of the six-story building are roughly estimated as

$$A_g = nA/10 = 6(34 \times 30)/10 = 612 \text{ in.}^2 \tag{3.117}$$

Assume 25- × 25-in. columns, $A_g = 625 \text{ in}^2$

(b) *Beam design.* A beam width of $b_w = 12$ in. is assumed. The clear span for a girder width of 16 in. is.

$$l_n = 34 - 16/12 = 32.67 \text{ ft}$$

Since the beam supports the floor area $A = 15 \times 34 = 510 \text{ ft}^2 > 150 \text{ ft}^2$, the live load reduction is

$$R_1 = r(A - 150) = 0.08(510 - 150) = 28.8\% < 40\% \tag{2.1}$$
$$R_2 = 23.1(1 + D/L) = 23.1(1 + 103/80) = 52.84\% \tag{2.2}$$

Use the least reduction of 28.8%. As a first trial for the beam weight, its depth may be estimated as

$$t = 2L/3 = 2(34)/3 = 22.67 \text{ in.} \tag{3.44}$$

Here, for reasons of construction and by using increments of 2 in. the floor framing is kept to a constant depth of $t = 24$ in. The beam must support the following loads (see Example 3.17 for slab dead load):

Slab loads: $[1.4(103) + 1.7(80)0.71]15/1000$	$= 3.61 \text{ k/ft}$
Stem weight: $1.4 [(24 - 6.25)12/12^2](1)0.150$	$\underline{= 0.31 \text{ k/ft}}$
Total load:	$= 3.92 \text{ k/ft}$

The moments at the supports and midspan according to Fig. 3.15 are

$$-M_u = w_u l_n^2/11 = 3.92(32.67)^2/11 = 380.36 \text{ ft-k}$$
$$+M_u = w_u l_n^2/16 = 3.92(32.67)^2/16 = 261.50 \text{ ft-k}$$

The required moment reinforcement at the top of the support is

$$-A_s = M_u/4d = 380.36/4(21.5) = 4.42 \text{ in.}^2 \tag{3.54}$$

Try 8 #7, $A_s = 4.80 \text{ in.}^2$

Where the flanges of T-beam construction are in tension, part of the flexural reinforcement is to be distributed over the effective flange width assumed as

$L/10 = 32.67(12)/10 = 39.20$ in. Place 4 #7 inside the steel cage and 2 #7 on each side in the slab (Fig. 3.16). The corresponding steel ratio indicates that the solution can only be treated as an approximation since it is larger than 1.6%, unless this beam section is enlarged and thus the steel ratio reduced.

$$\rho = A_s/b_w d = 4.8/12(21.5) = 1.86\%$$

The required bottom reinforcement at midspan is

$$+A_s = M_u/4d = 261.50/4(21.5) = 3.04 \text{ in.}^2 \tag{3.54}$$

Try 4 #8, $A_s = 3.16$ in.^{2} (see Table A.3).

Because of the high compression block resistance of the flanges in the field, only the minimum amount of reinforcing or ρ_{min} is checked; according to the ACI Code, just the web width is to be used.

$$A_{smin} = b_w d/300 = 12(21.5)/300 = 0.86 \text{ in.}^2 \leq 3.16 \text{ in.}^2 \tag{3.54}$$

Notice that for a simply supported T-section ρ_{max} almost never presents a problem since the large resisting flange width lowers the compressive stresses.

Next the shear will be investigated; #3 stirrups are used with $f_y = 60$ ksi. The maximum beam shears at the girder faces are

$$R = 3.92(32.67/2) = 64.03 \text{ k}$$

The maximum shear acts at distance d adjacent to the face of the girder.

$$V_{umax} = 64.03 - 3.92(21.5/12) = 57.01 \text{ k}$$

Check if the beam depth is satisfactory.

$$d_{min} = 3V_u/b_w = 3(57.01)/12 = 14.25 \leq 21.5 \text{ in.,} \quad \text{OK} \tag{3.68}$$

The concrete strength is

$$\phi V_c = 0.11 b_w d = 0.11(12)21.5 = 28.38 \text{ k} \tag{3.59}$$

The stirrup strength is

$$\phi V_s = 11n \tag{3.61}$$

Establish a pattern of stirrup spacing of say $d/2$, $d/2.67$, and $d/3$ with their corresponding capacities.

$$s = d/2, \quad \text{or} \quad \phi V_s = 11(2) = 22 \text{ k}$$
$$s = 3d/8, \quad \text{or} \quad \phi V_s = 11(2.67) = 29.33 \text{ k}$$
$$s = d/3, \quad \text{or} \quad \phi V_s = 11(3) = 33 \text{ k}$$

The closest stirrup spacing occurs at the support, where the maximum excess shear is

$$V_u - \phi V_c = 57.01 - 28.38 = 28.63 \text{ k}$$

The stirrup spacing of $3d/8$ with a capacity of 29.33 k covers the excess shear of 28.63 k.

The location x must be found where the excess shear is equal to 22 k, that is, where the spacing 3d/8 can be changed to $s_{max} = d/2$. From similarity of triangles in Fig. 3.16, we obtain

$$R/(l_n/2) = (\phi V_s + \phi V_c)/[(l_n/2) - x]$$

or

$$x = [1 - (\phi V_s + \phi V_c)/R]l_n/2$$

$$64.03/(32.67/2) = [22 + 28.38]/[(32.67/2) - x] \quad \text{or} \quad x = 3.48 \text{ ft}$$

Use the following stirrup spacing:

$$s = 3d/8 = 3(21.5)/8 = 8.06 \text{ in.,} \quad \text{say #3 at 8 in. o.c.}$$
$$s = d/2 = 21.5/2 = 10.75 \text{ in.,} \quad \text{say #3 at } 10\tfrac{1}{2} \text{ in. o.c.}$$

The stirrup layout is shown in Fig. 3.16, but keep in mind that many other groupings are possible. The intention here was solely to present a fast approximation.

Theoretically, no stirrups are required beyond $\phi V_c/2$ close to midspan. But under a partial live load on one-half of the beam span, worse shear conditions will be generated along the center beam portion, probably requiring a stirrup spacing of $s_{max} = d/2$; in addition, practical considerations of construction will require stirrups for the steel cage.

c) *Girder design.*

Assume the girder proportions as $b_w/t = 16/24$. Since the girder span is less than the beam span but the depth is the same, deflection will not be a consideration. The live load reduction is assumed not to change from the one for the beam. The clear span of the girder is

$$l_n = 30 - 25/12 = 27.92 \text{ ft}$$

The concentrated load reactions from the beams are

$$P_u = 3.92(32.67) = 128.07 \text{ k}$$

Girder weight:	$1.4(24(16)/12^2)(1)0.150$	$= 0.56$ k/ft
Additional live load:	$1.7(0.080.\times.0.71)16/12$	$= 0.13$ k/ft.
Total uniform girder load:		$= 0.69$ k/ft

The single load is transformed into an equivalent uniform load by equating the support moments for the two loading cases.

$$M_s = Pl_n/8 = wl_n^2/11$$
$$w_{eq} = 11P/8l_n = 11(128.07)/8(27.92) = 6.31 \text{ k/ft}$$

The total uniform load is

$$w = 0.69 + 6.31 = 7.00 \text{ k/ft}$$

The critical moments and the corresponding flexural steel are at the following points:

- Support:

$$-M_u = w_u l_n^2/11 = 7.00(27.92)^2/11 = 496.06 \text{ ft-k}$$
$$-A_s = M_u/4d = 496.06/4(21.5) = 5.77 \text{ in.}^2$$

Try 2 #9 and 5 #8, $A_s = 5.95$ in.^{2}, distributed across $27.92(12)/10 = 33.5$ in. The corresponding steel ratio is close to 1.6% so the girder proportions can be considered satisfactory from a flexural point of view for preliminary design purposes.

$$\rho = A_s/b_w d = 5.95/16(21.5) = 1.73\%$$

- Midspan:

$$+M_u = w_u l_n^2/16 = 7.00(27.92)^2/16 = 341.04 \text{ ft-k}$$
$$+A_s = M_u/4d = 341.04/4(21.5) = 3.97 \text{ in.}^2$$

Try 4 #9, $A_s = 4.00$ in.^{2} (see Table A.3).

- Check:

$$A_{smin} = b_w d/300 = 16(21.5)/300 = 1.15 \text{ in.}^2 < 4.00 \text{ in.}^2$$

Next the shear is investigated using No. 3 stirrups with $f_y = 60$ ksi. Because the uniform load is much smaller than the concentrated load and also to reduce the calculations, it is conservatively assumed that it is replaced by its resultant at midspan (Fig. 3.16).

The maximum beam shears at the column faces or at distance d away are

$$V_{umax} = (128.07 + 0.69(27.92))/2 = 147.34/2 = 73.67 \text{ k}$$

First check if the given beam depth is satisfactory with respect to using $s_{max} = d/2$

$$d_{min} = 3V_u/b_w = 3(73.67)/16 = 13.81 \text{ in.} < 21.5 \text{ in.,} \quad \text{OK}$$

Since the shear is nearly constant, the stirrup spacing will be constant!

$$V_u \le \phi V_c + \phi V_s \quad \text{or} \quad V_u - \phi V_c \le \phi V_s$$

$$V_u - 0.11\, b_w d \le \phi A_v f_y n$$

$$73.67 - 0.11 \,(16 \times 21.5) \le 11n, \qquad 35.83 \le 11n = 11d/s$$

or

$$s \le 11d/(V_u - \phi V_c) = 11(21.5)/35.83 = 6.60 \text{ in.}$$

(3.66)

Use #3 stirrups at 6½ in. o.c.

Prestressed Concrete Beams

The general concept of prestressing was introduced in Section 2.8. Here, only the principle of prestressing concrete beams is briefly discussed. Prestressing has been used for some time in segmental construction where precast elements are assembled on site and tied together through posttensioning to form a rigid, monolithic whole. It has also been widely employed in the precast concrete industry for mass producing standardized pretensioned floor elements such as hollow-core slabs and double-tee sections. Only recently, unbonded posttensioning of cast-in-place floor framing systems for high-rise buildings including garages has developed on a larger scale, although the posttensioning of slab-on-grade construction (i.e., floating slabs) has been in use for some time. The span ranges of various concrete floor framing systems are limited by economic considerations. With increase of span, floor structures become too heavy, thereby leaving only a small portion of their strength available for added service loads; in addition, long-term deflections due to creep become critical. By prestressing conventional concrete floor framing systems, not only can their depth be substantially reduced but also their span range may be increased by roughly 30% to 40%.

The two mechanical prestressing methods are pretensioning and posttensioning; *pretensioning* is usually associated with precast concrete. Here, tendons are anchored to molds or abutments outside the forms and are tensioned before the concrete is cast. When the concrete has reached enough strength, the tendons are released or cut in the long-line production process, but prevented from returning to their original length by the concrete to which they are bonded, thus placing the surrounding concrete in compression.

Post-tensioning of tendons is done by hydraulic jacks after the concrete has hardened and reached about 75% of the design strength by inserting the wires, strands, or bars in metal or plastic conduits, which were placed in the forms before the concrete was poured. The concrete element carries the compression generally in bearing through special anchoring devices at the end faces, rather than through bond as in pretensioning, although the tendons may also be bonded to the concrete by grouting within the conduit.

The purpose of prestressing concrete is to improve the behavior of the composite material and to use less of it because of the high strength of the individual components. Here an external prestress force is applied, which bends the member opposite to the bending resulting from loading. Whereas in nonprestressed reinforced concrete, the section has to substantially crack before the steel reinforcement can fully act, in the prestressed design approach the composite interaction of concrete and steel is very much improved by prestretching the high-strength steel so that the section is uncracked under service loads and the full cross-sectional concrete area is available for resistance. The locked-in, constant compressive stresses efficiently resist the tensile stresses due to external loading. The result yields shallower and stiffer sections with a corresponding lower story height and a better control of deflection, which is especially important for long spans. However, keep in mind the higher cost of prestress steels and anchorages, the high installation costs, and the necessary high quality control, together with a more sophisticated construction process, that makes posttensioning primarily economical only for larger spans. Since a prestressed concrete member shortens due to shrinkage and creep, part of the prestress force is lost; this loss may be in the range of, say, 40 ksi. As it is common practice to limit prestress losses to approximately 20% of the initial prestress force, a 200-ksi steel is required for a prestress loss of 40 ksi. Hence, it is apparent that high-strength steels must be used for posttensioning; they may be wires, strands, or alloy bars. There are two grades of the popular posttensioning strands with strengths of 250 and 270 ksi and a modulus of elasticity of 27,500 ksi (see Table A.5). For the geometrical properties of alloy bars of Grades 145 and 160, refer to the corresponding bars for nonprestressed reinforcement (Table A.1).

Typical span-to-depth (L/t) ratios for prestressed floor framing systems are roughly 45 for solid slabs, 32 for wide beam bands, 30 for joists and joist slabs, and 24 for beams. The depth of prestressed concrete members varies between approximately 60% and 85% of that of equivalent nonprestressed members. For the design of prestressed, continuous floor beams in typical high-rise buildings, the following rules of thumb may be used, where the span L is in feet and the beam thickness t in inches.

- Solid slabs $t = L/3.5$ to $L/4$
- Wide band beams $t = L/2.5$ to $L/3$
- Joist slabs $t = L/2$ to $L/2.5$
- Beams $t = L/1.7$ to $L/2$
- Girders $t = L/1.3$ to $L/2.25$

For preliminary design purposes the lower-range values may also be applied to simple span conditions.

The area of prestress steel required to resist an effective prestress force P in a concrete member under an external moment M can be roughly approximated according to Eq. (3.55a), but only for quick first trial purposes, as

$$A_s = P/f_s = (M/z)/f_s, \quad \text{where } z \cong 0.8t, \quad f_s = 0.56f_{pu}$$

The required cross-sectional area of the concrete member can be estimated from

$$A_c = P/f_c = (M/z)/f_c, \quad \text{where } z \cong 0.8t$$

Here the precompression stress f_c may be taken as approximately 350 psi for slabs and twice as high for beams.

In this context it is assumed that the concrete is *fully prestressed* so that the entire cross section is effective in resisting the external forces. It sometimes may be more economical, however, to allow some cracking so that the beam is only *partially pre-*

stressed, where the tendons are stressed below the usual level or where the member contains a significant amount of nonprestressed steel. Since the section is not cracked in the fully prestressed design, it can be treated as any other homogeneous composite member, where the allowable stress approach can be used as based on elastic behavior:

$$f = P/A \pm M/S \tag{2.67}$$

Two critical loading stages must be investigated.

- The initial stage, where the beam bends upward as caused by the prestress force and possibly only counteracted by its own weight
- The final stage, where the beam deflects downward as governed by the full gravity condition

Generally, the final loading stage controls the magnitude of the prestress force; however, the critical maximum tensile and compressive stresses in the concrete section may very well occur during the initial loading stage. Since emphasis here is placed on the preliminary design of common posttensioned floor framing members, only the typical uniform loading is considered to act on beams and slabs, although a loading case due to beam action on a girder with straight-line tendon profile is shown in Fig. 3.17d. The uniform load causes a parabolic moment diagram, remembering that a flat parabolic draped tendon or a compressed concrete arch represents the funicular shape of uniform loading and thus responds in pure axial action. When a parabolic tendon profile is selected (Fig. 3.17a, b) and the cable is tensioned, a uniform upward load is generated that partly balances the downward gravity load. In the *load-balancing concept* of design, as developed by T. Y. Lin in the early 1960s, *the cable profile corresponds to the arrangement of the applied loads and mirrors the moment diagram of the gravity loads;* the tensioned tendons generate equivalent forces acting opposite to the externally applied ones and produce a moment diagram exactly opposite to the one caused by part of the external loads. The concept of two-dimensional load balancing by posttensioning a flat slab (Fig. 3.17a) is rather similar to the one-directional beam approach.

The designer must determine the portion of the applied load to be balanced by the prestress force, which is not a simple decision. Often, it is convenient as a first cycle to balance the dead load for the condition where the live loads arc in the range of the dead loads; others assume an average compression stress for the initial estimate of the balanced load. These precompression stresses are in the range of 175 to 400 psi for slabs and 600 to 800 psi for beams so as to avoid excessive creep.

Here it is assumed that the balanced load w_p is equal to the dead load w_D so that the dead load deflection is equal to the camber due to prestressing, thus theoretically resulting in no deflection. Since the loads balance each other and the section is under constant compression (Fig. 3.17b, c), the moment caused by the dead load at midspan must be equal to the prestress moment $P(e)$. Treating the tendons as suspended cables similar to case (c), we obtain from statics that the resisting moment $P(e)$, as provided by the cable force, must balance the rotation due to the external loads $w_p = w_D$.

$$P(e) = w_p L^2/8 = w_D L^2/8$$

Hence, for the general condition the load w_p to be balanced by the prestress force P is

$$w_p = 8Pe/L^2 \tag{3.71}$$

Or the required prestress force P to balance the dead load w_D is

$$P = w_p L^2/8e = w_D L^2/8e \tag{3.72}$$

At this stage the parabolic tendon is tensioned with such a magnitude compressing the concrete beam as to cause a uniform upward force exactly equal to the uniform dead load. The result is a net zero load and a constant compression along the beam as

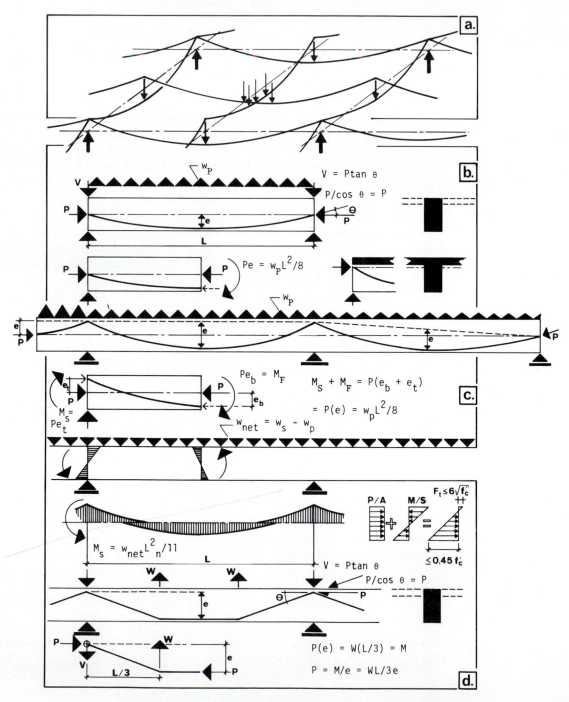

Figure 3.17   The Load-balancing method is prestressed concrete (Reproduced with permission from *The Vertical Building Structure*, Wolfgang Schueller, copyright © 1990 by Van Nostrand Reinhold.)

long as the prestress tendon is anchored at the centroidal axis of a simply supported beam (b), or above the centroidal axis for the continuous beam portion (c) in Fig. 3.17. For the preliminary design of floor beams, the flange action of the slabs may be ignored and the T-beams may simply be treated as rectangular sections where the pressure line coincides with the centroidal axis, so there are only axial stresses and no bending due to the prestress force P.

$$f_a = P/A \tag{3.73}$$

For a continuous beam the cables are not just hanging over the supports with sharp breaks in curvature as has been assumed, but form a reverse transitional smooth curvature with points of inflection close to the support. However, it has been shown that the load-balancing method is sufficiently accurate without taking into account the effect of the reverse tendon curvature. Although the final loading stage controls the magnitude of the prestress force, the initial stress state must also be considered for the tendon design since there will be a loss of the initial prestress force P_i due to creep, drying shrinkage, elastic shortening, slippage of steel, etc. It is beyond this discussion to deal with the complex issue of prestress loss. As a first approximation, often 15% to 25% is taken; here an average of 20% reduction of the initial prestress force is assumed.

$$P = 0.8P_i$$

The allowable stresses f_{pi} in the posttensioning tendons immediately after tendon anchorage or prestress transfer is 70% of the tensile strength f_{pu}.

$$f_{pi} = 0.70f_{pu}$$

Thus, the required prestress tendon area is

$$A_s = P_i/f_{pi} = (P/0.8)/0.7f_{pu} = P/0.56f_{pu} \tag{3.74}$$

The analysis of prestressed beams under the loads that are not balanced can now be made as for nonprestressed beams. The unbalanced portion of the loads, that is, the live loads for this case, will cause the following flexural stresses:

$$f_b = +M_L/S$$

The maximum combined compressive stresses due to prestressing and live load bending after all losses should be less than the allowable stresses.

$$f_c = f_b + f_a = M_L/S + P/A \le 0.45f'_c \tag{3.75}$$

The maximum tensile stress in the concrete is

$$f_t = f_b - f_a = M_L/S - P/A \le 6\sqrt{f'_c} \tag{3.76}$$

The initial stress state in the concrete after prestress transfer and before prestress losses is not checked in this preliminary design; the reader may refer to the ACI Code to obtain the respective allowable compressive and tensile stresses for the concrete.

EXAMPLE 3.19

A continuous one-way slab of an office building spans 25 ft from center to center beams and has clear spans of 23 ft. The slab must support a live load of 80 psf. Estimate the required posttensioning strands for a typical interior bay, realizing that the end spans require more prestressing. Because live and dead loads are nearly equal to each other, base your design on the assumption of zero deflection under full dead load. Use $f_c' = 4000$ psi and Grade 270 strands.

Estimate the slab thickness as

$$t \cong L/4 = 25/4 = 6.25 \text{ in.; try a } 6\frac{1}{2}\text{-in. slab}$$

The slab weighs $(150/12)$ $6.5 = 81.25$ psf, and adding floor finish yields a final dead load of 85 psf.

For a parabolic tendon layout, the maximum cable drape is

$$e = t - 2d' = 6.5 - 2(1) = 4.50 \text{ in.}$$

The required prestress force to balance the dead load is

$$P = w_D L^2/8e = 0.085(25)^2/[8(4.5)/12] = 17.71 \text{ k/ft of slab} \qquad (3.72)$$

This causes an average precompression stress of

$$f_a = P/A = 17,710/6.5(12) = 227 \text{ psi} \qquad (3.73)$$

This value is within the typical range of 175 to 400 psi.

Now the flexural stresses due to the live loads must be checked. The critical support moment, according to Fig. 3.15A, is

$$-M_s = w_L l_n^2/11 = 0.080(23)^2/11 = 3.85 \text{ ft-k/ft}$$

The corresponding bending stresses are

$$f_b = +M/S = 3850(12)/[12(6.5)^2/6] = 546 \text{ psi}$$

The combined compressive stresses according to Eq. (3.75) are

$$f_c = f_b + f_a = 546 + 227 = 773 \text{ psi} \le 0.45 f_c' = 0.45(4000) = 1800 \text{ psi}$$

The compressive stresses after all prestress losses are low as compared to the allowable ones. The tensile stresses according to Eq. (3.76) are

$$f_t = f_b - f_a = 546 - 227 = 319 \text{ psi} \le 6\sqrt{f_c'} = 6\sqrt{4000} = 380 \text{ psi}$$

Since the tensile stresses are OK, the prestress tendons can be designed. The required strand area according to Eq. (3.74) is

$$A_s = P/0.56 f_{pu} = 17.71/0.56(270) = 0.117 \text{ in.}^2/\text{ft}$$

This yields a steel weight per square foot of slab of

$$w_s = A_s \gamma = A_s(490/12^2) = A_s(3.4 \text{ lb/in.}^2/\text{ft}) = 0.117(3.4) = 0.398 \text{ lb/ft}^2$$

This value is typical for one-way slabs. For a strand with 0.5-in. diameter, the spacing is

$$12/0.117 = s/0.153 \quad \text{or} \quad s = 15.69 \text{ in.}$$

Use 0.5-in.-diameter Grade 270 strands spaced at 15.5 in. on center, but some minimum bonded, nonprestressed reinforcement must be added according to the ACI Code. In the final precise design, the prestress member must be checked not only as an uncracked section for the elastic conditions just investigated, but also for the initial design stage; also the complex stress distribution at the anchorage points must be considered. Additionally, the member must be checked for deflection due to the live loads, remembering that the dead loads do not cause any deflection, and for deflection due to creep, which is critical in prestressed members, and it must be checked as a cracked section for the condition of ultimate strength for flexure and shear. The shear reinforcement is determined as for ordinary reinforced concrete beams, but also taking into account the axial stresses due to posttensioning.

In Example 3.19, the live load and dead load were nearly equal to each other, and the balancing of the dead load gave a reasonable result. For many loading conditions, however, the live load is much less than one-half the total load. Here balancing all the dead load may be uneconomical and require too much prestress force; designers often assume as a first trial 80% of the dead load to be balanced by the equivalent prestress load w_p.

$$w_p = 0.8 w_D \qquad (3.77)$$

Should the live load be high in comparison to the dead load, then part of the live should also be balanced. For this condition, we may assume as a first trial

$$w_p = w_D + w_L/2 \qquad (3.78)$$

But we must keep in mind that the upward camber may be excessive when the live load and superimposed dead load are not acting; hence, only the portion of the live load that occurs frequently should be balanced. As a rule of thumb, we may assume that dead and live loads are approximately equal for the design of slabs in office buildings; for slabs in apartment buildings and for beam design, however, the dead load will usually exceed the live load.

For the general condition where dead and live loads are not of approximately the same magnitude, a somewhat time-consuming trial and error procedure is required to determine the magnitude of the balancing load that represents an economical solution. For this condition, the following iterative process may be used as based on the tensile concrete capacity by starting with an initial trial prestress load w_p.

The flexural stresses are only caused by the loads that are not balanced, that is, the service loads minus the equivalent prestress loads.

$$w_{net} = w_s - w_p$$

The critical moment for a typical continuous interior span may be taken as

$$M = w_{net} (l_n)^2/11$$

The magnitude of the prestress force P is estimated from the critical tensile stresses due to the full service loading.

$$f_t = f_b - f_a \le 6\sqrt{f_c'}$$

$$f_a \ge f_b - 6\sqrt{f_c'} \quad \text{or} \quad P/A \ge M/S - 6\sqrt{f_c'} \qquad (3.79)$$

or

$$P \ge (A/S)M - 6A\sqrt{f_c'}$$

After the prestress force has been found, the equivalent uniform prestress load according to Eq. (3.71) is

$$w_p = 8Pe/L^2 \qquad (3.71)$$

This equivalent load should be equal to the load assumed at the start of the design. If this is not the case, a new value must be assumed and the process continued until the values have converged; this convergence is usually rapid.

Composite Steel Concrete Beams

Composite beam action may be achieved by either encasing the steel beam in concrete, where the natural bond between steel and concrete is sufficient to provide resistance to horizontal shear and which is usually done for purposes of fireproofing, or by connecting the steel beam to the concrete slab with mechanical shear connectors. The composite action of steel beams and concrete slabs with the aid of shear connectors is a common construction practice in high-rise buildings today for spans larger than 25 to 30 ft. In general, it is efficient with heavy loading, long spans, and widely spaced beams; it usually increases the ultimate strength substantially and the deflection will be about 33% to 50% less than the deflection of noncomposite beams. The bonding between the interface of the two materials is generally achieved by stud or channel connectors, which are welded to the flanges and resist the horizontal shear, thus prohibiting slippage; limited slippage is allowed in partial composite action, which may be more economical. In noncomposite floor systems, the slab and beam act independently in opposite directions, while in composite construction the steel beam together with a portion of the slab forms a T-beam as in monolithic concrete construction (Fig. 3.13). When corrugated steel deck spans perpendicularly to the supporting beam, only the concrete above the deck is considered as structurally effective (in other words, the concrete below the top of the steel deck is ignored), while for the parallel condition the full depth of the slab is effective. For continuous beams there will be composite action along the positive moment region, but noncomposite action along the negative moment region if no special reinforcement is provided.

Composite beams can be designed according to the *transformed area method* [Eq.(3.43)] by converting the concrete area into an equivalent steel area. In other words, the effective concrete slab width b is divided by the modular ratio $n = E_s/E_c$, which is generally taken as the closest integer value, where E_s is the modulus of elasticity of steel (29,000 ksi) and E_c is the modulus of elasticity of the concrete (Eq. 3.1). The effective width of the concrete slab flange for a typical interior beam is not to exceed one-fourth of the beam span nor the spacing s of the adjacent beams (Fig. 3.18). For preliminary design of composite beams with shear connectors, a rough rule of thumb is that the capacity of the steel beam in composite action is increased by one-third; that is, the steel beam alone can be designed for 75% of the moment. Often the steel beam depth is, estimated as 80% of the noncomposite section. The designer must keep in mind, however, that during the construction stage the beam alone, if it is not shored, must be able to resist the floor dead load and construction loads in noncomposite action.

The number of connectors needed to resist the shear at the interface of the slab and steel beam can be estimated as follows. It is assumed for preliminary design purposes that the neutral axis (N.A.) falls within the concrete slab of the composite section so that the concrete slab acts in compression and the steel beam in tension, which is the case for ordinary conditions in buildings. Hence, for full composite action, the shear connectors must be able to transfer the tensile capacity of the steel beam $T_y = A_sF_y$ (see Fig. 3.18); in other words, T_y is balanced by the shear load capacity of the connectors $V_{hu} = 2V_h$ by using a load factor of 2.

$$T_y = A_sF_y = 2V_h \quad \text{or} \quad V_h = A_s F_y/2 \qquad (3.80)$$

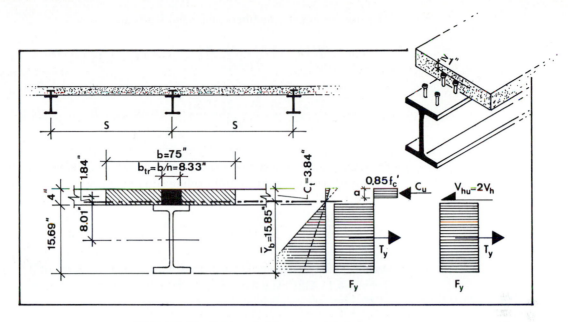

Figure 3.18 Composite steel concrete beams.

where V_h is the total horizontal shear to be resisted between the points of maximum positive moment and zero moment and A_s = area of the steel beam cross-section.

EXAMPLE 3.20

The beams for the floor framing in Example 3.3 are investigated for composite action. As a first trial, it is assumed that the steel beam alone can be designed for 75% of the bending moment; $0.75(98.44) \cong 74$ ft-k. Hence, a $W16 \times 26$ with the following properties is checked: $d = 15.69$ in., $A = 7.68$ in.^{2}, $I = 301$ in.^{4}; 3000-psi normal-weight concrete is used for the 4-in. concrete slab with an allowable compressive stress of $f_c = 0.45f'_c = 0.45(3000) = 1350$ psi. The loading conditions are considered not to change for this preliminary investigation. Shored construction is assumed.

The modulus of elasticity of concrete is

$$E_c = 57,000\sqrt{f'_c} = 57,000\sqrt{3000} = 3.12 \times 10^6 \text{ psi} \qquad (3.1)$$

The modular ratio is $n = E_s/E_c = 29 \times 10^6/3.12 \times 10^6 = 9$. The effective width of the concrete flange for typical interior conditions is

$$b \leq L/4 = 25\,(12)/4 = 75 \text{ in.} < s = 8(12) = 96 \text{ in.}$$

The transformed effective width is $b_{tr} = b/n = 75/9 = 8.33$ in. The transformed concrete area is $A_{ctr} = tb_{tr} = 4\,(8.33) = 33.33$ in.^{2}. The location of the centroidal axis [Eq.(2.35)] is

$$\bar{y}_b = \frac{\Sigma Ay}{\Sigma A} = \frac{7.68\,(15.69/2) + 33.33\,(15.69 + 4/2)}{7.68 + 33.33} = 15.85 \text{ in.}$$

The transformed moment of inertia [Eq. (2.50b)] is

$$I_a = \Sigma I_{xo} + \Sigma Ad^2$$

$$I_{tr} = 301 + 7.68(8.01)^2 + (8.33(4)^3/12) + 33.33(1.84)^2 = 951 \text{ in.}^4$$

The stresses at extreme fibers under full loads are

$$f_c = \frac{Mc_t}{nI_{tr}} = \frac{98.44\,(12)\,3.84}{9\,(951)} = 0.53 \text{ ksi} < 1.35 \text{ ksi}$$

(3.41)

$$f_s = \frac{M\bar{y}_b}{I_{tr}} = \frac{98.44\,(12)\,15.85}{951} = 19.69 \text{ ksi} < 0.66F_y \cong 24 \text{ ksi}$$

Should a shallower beam be required, then a W12 × 26 should be checked. The maximum deflection due to full loading [Eq.(3.9)] is

$$\Delta_s = \frac{ML^2}{161I_{tr}} = \frac{98.44\,(25)^2}{161\,(951)} = 0.4 \text{ in.} < \frac{L}{240} = \frac{25\,(12)}{240} = 1.25 \text{ in.}$$

Assume 3/4-in-diameter headed studs with a shear capacity of $q = 11.5$ k (from Table I 4.1, AISC Manual). Hence, the total number of shear connectors [(Eq. (3.80)] is

$$V_h/q = A_sF_y/2q = 7.68(36)/2(11.5) = 12 \qquad (3.81)$$

Use twelve 3/4-in. studs on each side of the beam center line, evenly spaced since no concentrated loads are present. Notice that the *Composite Beam Selection Tables* in the AISC Manual can be conveniently used for the design of composite beams.

3.3 DESIGN OF TENSION MEMBERS

Tension members constitute an important part of modern building construction in its search for lighter and more efficient structures. They are the primary elements in cable and membrane roofs as well as tensegrity structures (see Chapter 9). In high- rise and low-rise construction, tension members are found as primary structural elements in trusses and for suspension buildings where they are used as vertical hangers and catenaries. They also occur as diagonal lateral wind bracing, vertical supports for wall girt systems (sag rods), cable stays for girders (cable beams), guy wires for hoists and derricks, cable supports for elevators, and support for bridges. Most tension members are steel, such as cables with socket ends, rods with threaded ends, eyebars with forged ends for pin connections, pin-connected plates, and rolled shapes, including built-up members, usually with bolted or welded connections. Typical wood tension members are found in trusses. Prestressed concrete tensile members may be used, for example, as tension rings for domes and as hanging columns.

The design of tension members is relatively simple ($f = P/A$) in comparison to column and beam design, although the member shape is very much dependent on the detailing of the connections. The primary concern, in this section, is with respect to the preliminary design of typical steel members, such as simple structural shapes, rods and flat bars, and cables. Wood tension members are briefly discussed at the end of this section. For the design of tensile membranes, refer to Chapter 9.

Single Structural Steel Shapes and Built-up Steel Members

Rolled sections are selected when some rigidity is required to resist bending and reversal of loading, as for some truss members since cables and rods are flexible. It is clear that structural shapes must possess a minimum slenderness to be considered rigid. The typical rolled member shapes are angles, channels, tubes, pipes, and wide-flange sections. Built-up shapes are assembled from single members and tied together with lacing, solid or perforated plates, or tie plates to form, for example, closed box shapes or open shapes like double angles or channels and star forms. The connection details for tension members are most important as they often are the weakest part. Typical connections are of the welded or bolted angle or channel type.

Usually, the shape of the cross section has little effect on the tensile capacity of a member, that is, not considering the connection influence. Tension tends to straighten the member in contrast to compression, which causes the member to laterally displace and buckle.

The allowable stresses for concentrically loaded tensile members, with the exception of pin-connected members (e.g., eyebars), take into account two possible modes of failure.

$$F_t = 0.6\,F_y \qquad \text{yielding of the gross section}$$

$$F_t = 0.5\,F_u \qquad \text{fracture at highly stressed locations as at} \\ \text{the net section } (F_u = \text{ultimate tensile stress})$$

In other words, the following critical design conditions of excessive deformation causing yielding of the entire gross area of the section and of localized fracture at its weakest effective net area at connection points must be considered.

$$f_t = P/A_g \;\leq\; F_t = 0.6\,F_y \tag{3.82a}$$

$$f_t = P/A_e \;\leq\; F_t = 0.5\,F_u \tag{3.82b}$$

The effective cross-sectional area A_e is defined as the net cross-sectional area A_n corrected for *shear lag,* that is, shear stress concentrations near fasteners. These stress buildups usually occur at connection points, especially where not all parts of a section are connected, or they may occur at splices where eccentric load action is generated because the force resultants do not line up. Often the center of gravity of a member does not coincide with the resultant of the gage lines along which the forces are transmitted, in turn causing rotation. For example, when only one leg of an angle is bolted to a gusset plate to transfer the entire load, eccentric force action together with stress concentration in the connected leg is generated, while leaving part of the other leg unstressed. These stress concentrations are taken into account by the U factor. According to the AISC Specs, the effective net area for bolted and riveted connections when the load is transmitted through some but not all of the cross-sectional elements of a member is

$$A_e = UA_n \tag{3.83}$$

For welded connections, A_n is replaced by the gross area A_g. When the load is transmitted directly through each of the cross-sectional elements by connectors, then $U = 1$ or $A_e = A_n$. The following U values may be used for bolted and riveted connections.

a. W, M, or S shapes with flange widths not less than two-thirds the depth, and structural tees cut from these shapes, provided the connection is to the flanges. Bolted or riveted connections shall have no fewer than three fasteners per line in the direction of stress: U = 0.90.

b. W, M, or S shapes not meeting the conditions of subparagraph (a), structural tees cut from these shapes and all other shapes, including built-up cross sections. Bolted or riveted connections shall have no fewer than three fasteners per line in the direction of stress: $U = 0.85$

c.   All members with bolted or riveted connections having only two fasteners per line in the direction of stress: $U = 0.75$.

With respect to welded connections, refer to the AISC Specs for U values.

     The net cross-sectional area (or simply net area) A_n refers to the gross area A_g of the section minus the hole area that is the product of the thickness and the net width of that member. The holes are either lined up or staggered. When the holes extend across a member as a straight line, the net width is simply the gross width minus the hole diameters. The net width of a staggered hole arrangement, on the other hand, is much more complex. It is obtained, according to the AISC Specs, by deducting from the gross width the sum of the diameters of all the holes in the chain, but adding for each gage space in the chain the quantity $s^2/4g$. This empirical expression represents an approximation for the complex stress state for staggered hole arrangements. In other words, the net area is equal to

$$A_n = A_g - \Sigma t d_h + \Sigma s^2 t / 4g \tag{3.84a}$$

Or, for members of uniform thickness,

$$A_n = t w_n = t(w_g - \Sigma d_h + \Sigma s^2 / 4g) \tag{3.84b}$$

where A_n = net area of section (in.^{2})
           A_g = gross area of section (in.^{2})
           w_g = gross width (in.)
           w_n = net width (in.)
           d_h = width of hole (in.)
           s = longitudinal center-to-center spacing (or pitch) of any two consecutive holes (in.)
           g = transverse center-to-center spacing (or gage) of the same two holes (in.)

     The critical net area is obtained from the chain that gives the least net width, as is illustrated in Example 3.21. Standard-size bolt holes are taken as the fastener diameter plus 1/16 in., whereas oversized holes are 1/8 in. larger than the fastener diameter. Furthermore, short-slotted or long-slotted holes may be desired to improve the speed of erection, as well as for situations where field adjustments are required.

     For preliminary design purposes, the minimum longitudinal spacing between the centers of holes, called the *pitch, s,* may be assumed as $3d$ and the minimum edge distance as $1.5d$. For the transverse spacing of the gage lines in members, called the *gage, g,* refer to the AISC Manual.

     Besides the primary failure types of yielding of the gross section and fracture at the net section, also *tearing failure* at the ends of members along fasteners due to shear or a combination of shear along a plane through fasteners plus tension along a plane perpendicular to it must be checked. This *block shear* is assumed not to be critical for preliminary design purposes and to be taken care of by the appropriate minimum edge distances and minimum spacings of the fasteners.

     To provide some minimum stiffness for tension members (other than rods) for erection purposes and to avoid undesirable lateral movement and vibrations, the AISC Specs recommend a maximum slenderness ratio of $L/r = 300$. Hence, the minimum radius of gyration of a section should be

$$r_{min} = \sqrt{I / A_g} = L/300 \tag{3.85}$$

where r = least radius of gyration of section (in.)
          L = length of member (in.)

The elongation of tension members in the elastic range according to Eq. (2.54), is

$$\Delta L = f_t L/E = PL/A_g E \tag{2.54}$$

When tension members are also subject to bending, instability is less critical, so yielding may be assumed to control the design. They can be proportioned for preliminary design purposes according to the following expression (in this case about one axis only):

$$A_s = (P + BM)/F_t = (P + BM)/0.6F_y \tag{3.86}$$

Refer to the discussion of steel column design, Eq. (3.100), for the derivation of the equation and the definition of the terms.

The choice of connections clearly affects the design of tension members. Therefore, it makes sense that the preliminary design of tension members (with the exception of pin-connected members) may be based conveniently on yielding of the entire gross area of the section, rather than on fracture at its weakest effective net area at connection points, which obviously must also be taken into account in the final design when the connection detail is known.

$$f_t = P/A_g \leq F_t = 0.6\,F_y \quad \text{or} \quad A_g = P/0.6F_y \tag{3.82a}$$

EXAMPLE 3.21

Determine the tensile capacity of a 15-ft single-angle member ($L6 \times 4 \times 5/8$) made of A36 steel. The location of the gage lines in the short and long legs as well as the staggered arrangement of the 7/8-in. bolts in oversized holes are shown in Fig. 3.19a.

The angle has the following properties: $A_g = 5.86$ in.^{2}, $r_{min} = r_z = 0.864$ in. The gross area of an angle is equal to the product of member thickness and gross width. This gross width, in turn, is the sum of leg widths minus the angle thickness. Therefore, for determining the net width for angles, according to the AISC Specs, the gage for holes in opposite adjacent legs is the sum of the gages from the back of the angle less the thickness (see Fig. 3.19a). The critical net width is determined by the following various paths (chains) along one, two, and three bolt holes [see Eq. (3.84)].

One hole: path ABC not critical (by inspection)
Two holes: path $DEFG$: $w_n = 9.375 - 2\,(\frac{7}{8} + \frac{1}{8}) = 7.375$ in.
Three holes: path $DEHFG$:

$$w_n = 9.375 - 3\,(1) + \frac{3^2}{4\,(4.125)} + \frac{3^2}{4\,(2.5)} = 7.82 \text{ in.} > 7.375 \text{ in.}$$

Therefore, the critical net area is equal to

$$A_n = w_n t = 7.375\,(5/8) = 4.61 \text{ in.}^2$$

The effective net area according to Eq. (3.83) using $U = 1.00$ (since each leg is connected to transmit the tension force) is

$$A_e = UA_n = 1.00\,(4.61) = 4.61 \text{ in.}^2$$

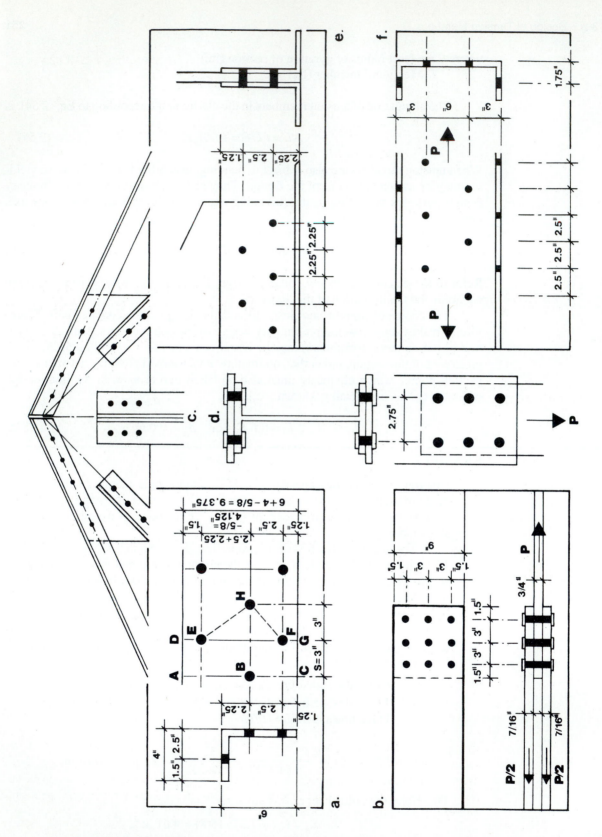

Figure 3.19 Connections for tension members.

The tensile capacity as based on yielding of the gross section [Eq. (3.82a)] is

$$P \leq 0.6\, F_y A_g = 0.6(36)5.86 = 126.58 \text{ k}$$

The tensile capacity as based on fracture at the least net area [Eq. (3.82b)] is

$$P \leq 0.5 F_u A_e = 0.5(58)4.61 = 133.69 \text{ k} > 126.58 \text{ k}$$

Check slenderness of member

$$L/r_{min} = 15(12)/0.864 = 208 < 300, \quad \text{OK}$$

Hence, failure due to yielding of the gross section controls the capacity, $P_{all} = 126.58$ k.

EXAMPLE 3.22

Design a 20-ft-long hanging W10 column of A36 steel to carry 150 k. Assume the member to be connected by two lines of 3/4-in. bolts (in oversized holes) in each flange (see Fig. 3.19d) with at least three bolts in a line.

The required gross area of the section as based on yielding [Eq. (3.82a)], is

$$A_g \geq P/0.6 F_y = 150/0.6(36) = 6.94 \text{ in.}^2$$

Try W10 × 26, $A_g = 7.61$ in.^{2}, $t_f = 0.44$ in., $r_{min} = 1.36$ in., $d = 10.33$ in., $b_f = 5.77$ in. $< 2d/3 = 2(10.33)/3 = 6.89$ in.; therefore, $U = 0.85$.

The required gross area as based on fracture at the net section [Eq. (3.82b)], etc., is

$$A_e = UA_n = U(A_g - \Sigma t d_h) \geq P/0.5\, F_u$$

or

$$A_g \geq P/(0.5\, F_u U) + \Sigma t d_h = 150/(0.5(58)0.85) + 4(0.44)\,7/8$$

$$= 7.63 \text{ in.}^2 > 7.61 \text{ in.}^2$$

Hence, the next larger section must be selected. Checking the slenderness ratio for the trial section yields

$$L/r = 20(12)/1.36 = 177 < 300, \quad \text{OK}$$

Use W10 × 30.

Round and Square Steel Rods and Flat Steel Bars

Rods and bars are used as wind bracing, hangers, sag rods for roofs, and support for girts. To limit the slenderness of rods, often a minimum diameter of 5/8 in. or $L/500$ is taken. They may be pretensioned to prevent the sagging of diagonal rods under their own weight due to lack of stiffness. Rods are connected by employing welding or threading the ends and using bolts, turnbuckles, or clevises, while flat bars may be welded, bolted, or pin connected (e.g., eyebars). The AISC Specs allow for the reduced area through the threaded portion of a rod by conveniently incorporating the gross area A_D. The allowable tension stress for threaded rods is equal to $0.33F_u$, where the tensile strength F_u for A36 steel may be taken as 58 ksi.

$$f_t = P/A_D \leq 0.33\, F_u \quad \text{or} \quad A_D = 3P/F_u \tag{3.87}$$

Here the tension force P should be at least 6 k.

EXAMPLE 3.23

(a) A tie rod at the base of a frame must resist an outward thrust of $H = 110$ k. Determine the diameter of the threaded rod using A36 steel.

$$A_D = 3\ P/F_u = 3(110)/58 = 5.69\ \text{in.}^2$$

$$A_D = \pi d^2/4 \quad \text{or} \quad d = \sqrt{4A_D/\pi} = \sqrt{4\,(5.69)/\pi} = 2.69\ \text{in.}$$

Use a 2 3/4-in-diameter rod.

(b) Determine the magnitude of the tensile stress for a steel rod that is prestressed 1/16 in. for each 20 ft. As based on Eq. (2.53), the following result is obtained.

$$f_t = E\varepsilon = E(\Delta L/L) = 29{,}000(1/16)/20(12) = 7.55\ \text{ksi}$$

Steel Cables

Cables (wire ropes and bridge strands) are used for suspension roofs, bridges, hoists and derricks, possibly for suspension buildings, and occasionally as diagonal wind bracing where they are pretensioned so that they can accept compression and will not vibrate. Special fittings are required for the connection of cables.

   Since the capacity of steel cables is roughly six times higher than that of mild steels (with an ultimate tensile strength F_u in the range of 200 to 220 ksi depending on the coating class and 270 ksi for prestressing strand), the resulting material will be a minimum. This yields small cross-sectional cable areas and, together with a lower effective modulus of elasticity of 20,000 ksi for ropes to 24,000 ksi for strands, causes cables to be flexible and form-active and makes them vulnerable to excessive elongations, $\Delta L = PL/AE$. The cable capacity can be obtained from the manufacturer's catalogues or Table A.13, but for rough preliminary design purposes an allowable tension stress F_t of about $F_u/3$ together with a resisting metallic cable area A_n of roughly two-thirds of its nominal gross area A_s may be assumed. Hence the required nominal cross-sectional cable area is in the range of

$$A_s \cong 1.5A_n = 1.5(P/F_t) \cong 4.5P/F_u \tag{3.88a}$$

The corresponding cable diameter can be easily obtained from $A_s = \pi d^2/4$ as

$$d = \sqrt{4A_s/\pi} = 1.13\sqrt{A_s} \cong 2.4\sqrt{P/F_u} \tag{3.88b}$$

For a more detailed discussion of cable structures and their approximate structural design, refer to Section 9.1.

Wood Tension Members

Wood members in tension are found in trusses, in bottom chords of simply supported lumber–plywood beams, and as chords in wall and floor diaphragms. The weakest point along a tension member is usually at the connection where fasteners (with the exception of nails and screws) have reduced and weakened the cross-sectional area of the member.

   The tensile stress parallel to the grain of wood is based on the net section area A_n obtained by subtracting from the gross section area, the projected area of all materials removed by boring, grooving, dapping, notching, or other means, ΣA_h; it cannot exceed the tabulated tensile stress parallel to the grain as specified for sawn lumber or glulam timber modified by the following adjustment factors:

$$f_t = P/A_n = P/(A_g - \Sigma A_h) \leq F_t'$$

where $\quad F_t' = (F_t) \, C_D \, C_M \, C_t \, C_F \cong (F_t)C_D \quad$ (for preliminary design purposes)

The load duration (C_D), wet service (C_M), and temperature factors (C_t) have been defined in the introduction to this chapter, whereas the size factor (C_F) has been discussed in the section on **wood beams**

The allowable tension stress for many lumber grades is substantially lower than the allowable stresses for bending F_b. Although wood has a relatively high tensile strength, keep in mind that the member size may be a function of the connection type. When a tension member is also subject to bending, instability is less critical, and the following two linear interaction equations should be checked with respect to tensile and compressive stresses.

$$f_t/F_t' + f_b/F_b^* \leq 1.0 \quad \text{and} \quad (f_b - f_t)/F_b^{**} \leq 1.0 \qquad (3.90)$$

where $\quad F_b^* \;\; = F_b'$, except beam stability factor, C_L

$\qquad\quad F_b^{**} = F_b'$ except volume factor, C_V

EXAMPLE 3.24

Do a preliminary design of the bottom chord of a roof truss that must carry a tension load of 5 k due to dead and snow loads. Assume bolted joints are used with a single row of 3/4-in.-diameter bolts at each end. Use southern pine with $F_t = 725$ psi.

The required net area is

$$A_n \geq P/F_t' = P/F_t C_D = 5000/(725)1.15 = 6.00 \text{ in.}^2$$

For bolted joints, the bolt holes may be assumed 1/16-in. larger than the bolt diameter. Assuming a 1.5-in. member thickness gives the following gross area:

$$A_g = A_n + A_h = 6.00 + 1.5(3/4 + 1/16) = 7.22 \text{ in.}^2$$

Try 2×6 in. chord section with $A = 8.25$ in.^{2}

3.4 DESIGN OF COLUMNS AND BEAM COLUMNS

Columns are the primary components of skeleton structures; they are also found as masts, posts, pillars, pilasters, caissons, and piers, possibly in mixed construction. But columns, or compression members, do not necessarily have to have a vertical position; they may be the horizontal or diagonal chords of trusses, the compression flanges of beams, diagonal bracing members, or the struts of cable beams.

In contrast to tension members where the forces tend to straighten the members, the concentric compression forces in columns cause members to laterally displace and bend, which is called *buckling,* a form of instability.

Compression columns are not always necessarily the standard rectangular, solid sections as in concrete and wood construction or the W-sections as used in steel; many variations in cross section and elevation are possible, as exemplified in Fig. 3.20. In cast-in-place concrete, the formwork allows the building of any desired shape, as for instance, expressed by the branching principle, as do prefab systems, where members can be combined in infinitely many ways. However, keep in mind that the column as a whole and its thin parts locally must resist buckling, that is, the bending which is caused by purely concentric force action.

The cross section of a column may be varied and may decrease toward zero-moment locations at midheight or top and/or bottom in x or y directions. On the other hand, the column cross section may be enlarged at midheight to strengthen its buckling resistance at this critical location.

Some of the many possibilities for various column forms in steel are investigated in Fig. 3.20; these forms are not necessarily based on structural considerations of efficiency but may be determined by other requirements. They range from single members, such as ordinary pipe or rolled W-sections and cover-plated columns, to the multiple-member built-up systems composed of angles, tees, channels, or W-sections.

These multiple members are generally arranged in an open cruciform or closed tubular fashion so as to yield the same buckling resistance about both major axes. However, the context of the column may require other arrangements of the elements. The vertical multiple-column components may be connected into one unit by employing continuous or perforated cover plates or by batten plates or diagonal lacing.

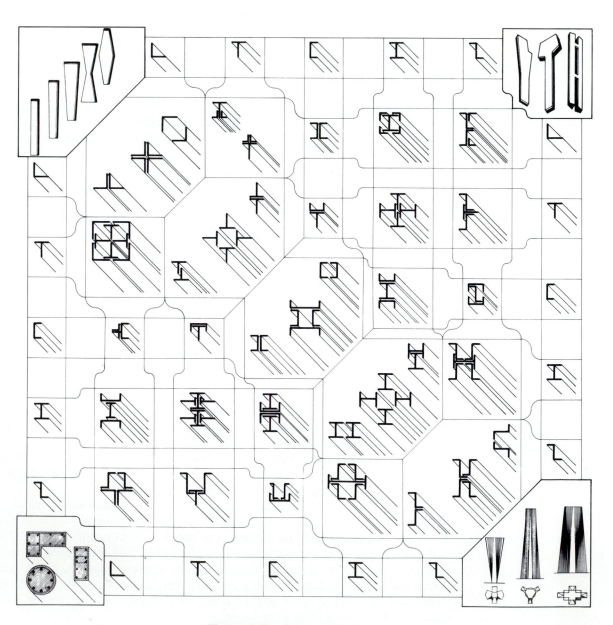

Figure 3.20 Column shapes.

A column concentrically loaded by an axial force P causes a uniform compressive stress that should not exceed the allowable axial compressive stress F_a.

$$f_a = P/A \le F_a \tag{3.91}$$

The allowable compressive stress is not a constant; however, it is dependent on the column slenderness and other parameters, which are discussed later.

Only for short *stocky columns* or for columns that are prevented from buckling is the allowable compressive stress constant and based on material strength. Long, *slender columns* fail in elastic buckling, as the Swiss mathematician Leonhard Euler presented in 1757. He was the first to recognize that instability and not crushing was the mode of failure. He derived the critical load P_{cr}, or the *buckling load* (using differential and integral calculus), which caused the column to suddenly bend without a moment present to initiate bending and independent of the column strength, as

$$P_{cr} = \pi^2 EI/L^2 \tag{a}$$

Equation (a) is usually presented as critical buckling stress in introducing the term *slenderness ratio* L/r by substituting $I = Ar^2$ and $f_{cr} = P_{cr}/A$.

$$f_{cr} = \frac{\pi^2 E}{(L/r)^2} \tag{b}$$

Euler's formula, however, represents ideal conditions of a truly centrally loaded, perfectly straight slender column with hinged end conditions for materials with a linear stress–strain response. In reality, columns are restrained at their boundaries from rotation. Therefore, the slenderness ratio L/r is replaced by l_e/r, where the effective column length, l_e, represents the distance between the inflection points of the buckled curve. For example, a column fixed at both ends forms inflection points (i.e., zero moment points) at $L/4$ measured from top and bottom, thus having an effective length of $L/2$, that is, behaving like a pin-ended column of $L/2$ length. In material specifications, l_e is generally referred to as Kl; in other words, the column length is multiplied by the *effective length factor* K to obtain the effective length. For example, for the preceding fixed column the K-factor is 0.5. Now, Euler's formula can be rewritten and put into the more general form of

$$f_{cr} = \frac{\pi^2 E}{(l_e/r)^2} = \frac{\pi^2 E}{(Kl/r)^2} = \frac{\pi^2 EI}{(Kl)^2 A} \tag{3.92}$$

where E = material modulus of elasticity (ksi)

   f_{cr} = critical uniform compressive stress at which the column buckles (ksi)

   r = $\sqrt{I/A}$ = least radius of gyration (in.)

   l = length of column (in.)

   l_e = effective column length (in.) = Kl

   A = gross cross-sectional area (in.^{2})

   I = least moment of inertia of cross-section (in.^{4}.)

   K = effective length factor

The allowable axial compressive stress F_a, as based on elastic buckling, is equal to the buckling stress f_{cr} divided by a safety factor.

The effective length factor K measures the amount of restraint at the column boundaries, that is, the degree of fixity at top and bottom, which is not always easy to determine. The ranges of K factors for various typical building conditions are given in Fig. 3.21. There are two basic types of columns we must distinguish:

- Columns in *braced buildings* are assumed not to experience sidesway, so they are usually in the low-slenderness range. The K values for braced columns cannot exceed 1 ($K \leq 1$). For this condition, single curvature is the worst possible configuration, and the actual column length can be safely taken as equal to the effective length, $l_e = L,$ or ($K = 1$).
- In *nonbraced buildings* such as rigid frame and flat slab construction, sidesway must be resisted by the beams/slabs and columns. Here the columns will sway laterally, resulting in an effective column length equal to or larger than the actual column length ($K \geq 1$).

The problem lies in defining the degree of column end restraint and the effect of sidesway. For instance, for the case where a simple two-hinge portal frame is laterally braced, the K factor is 0.7 if the girder is rigid, but it is 1.0 if the girder is flexible and does not provide any restraint to the column (Fig. 3.21, bottom). The real condition, however, will fall between these extreme cases. Should the frame not be braced, then it may be very unstable, depending on the flexibility of the girder ($2 \leq K < \infty$). In multistory construction, the relative sizes of the columns and girders at each floor level control the amount of rotation and thus govern the effective column length. As the girder sizes increase (or composite floor action is introduced) or the column sizes decrease because of higher strength, less rotation occurs and the K factor is reduced. The reader may also refer to the section on Frame Stability in the AISC Manual. Here, K values are given for six idealized conditions where the end conditions can be fixed or pinned (i.e., free to rotate) in combination with or without sidesway (i.e., lateral translation). Since true joint fixity is rarely fully realized, the recommended K-values are slightly higher than the theoretical equivalent ones. The selection of K-factors is discussed further in the context of the numerical examples.

Unfortunately, Euler's simple formula does not predict the strength of ordinary building columns. These columns generally are not long and slender and do not buckle in the elastic range, nor are they short and stocky and fail in crushing or yielding with no buckling. These columns are in the intermediate range and their capacity depends on the strength of the material, among many other criteria, and much less so on the modulus of elasticity. The formulas for the design of intermediate columns are discussed in the context of the various materials in the following sections.

Rarely do columns solely support axial forces. In most instances, they must also resist bending caused by transverse loading (wind, earthquake), continuous boundary conditions, or minor bending effects due to eccentric axial force application (beam connections, etc.). This combined action causes a direct stress f_a and a flexural stress f_b for bending about one axis only or for biaxial bending. Refer to Fig. 3.22 for common beam column loading conditions and the corresponding moment diagrams.

Before discussing steel, wood, and reinforced concrete columns, the following points common to all columns may be reemphasized. It has been shown that the degree of column slenderness is measured by Kl/r and that with increasing slenderness the strength of axially loaded columns decreases. According to their mode of failure, columns can be roughly grouped as follows:

- Short, *stocky columns* that fail in crushing and/or yielding (i.e., *material failure)* when the slenderness is approximately $Kl/r \leq 40$.

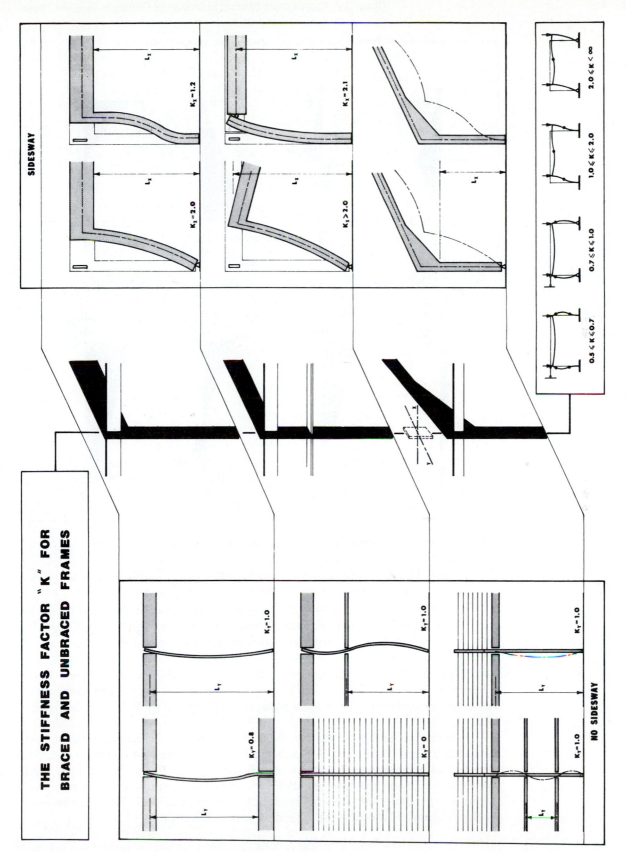

Figure 3.21 Effective length factor.

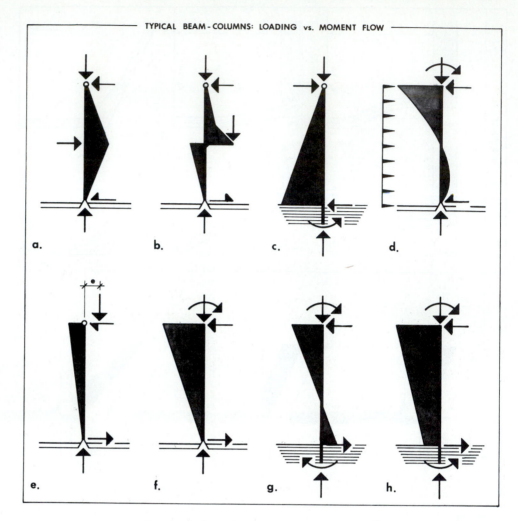

Figure 3.22 Typical beam–column loading.

- Long, *slender columns* fail in elastic buckling (i.e., *stability failure)* at approximately $Kl/r \geq 120$.
- *Intermediate long columns,* covering the range between short and long columns, fail by a combination of crushing or yielding and buckling (i.e., material and stability failure), such as inelastic buckling for steel columns.

Typical building columns can generally be considered short and in the low range of intermediate long columns. Cross bracing for steel frames and the web members of trusses are examples of long column behavior as described by Euler's buckling formula.

From a structural point of view, the size and shape of a column depend on both the magnitude and type of force action, as well as its slenderness. A *column that* carries primarily axial loads and that is laterally braced by the floors should have equal moments of inertia in both principal directions. A *beam column,* on the other hand, will require a larger dimension about its bending axis and may possibly be braced between the floors against buckling in the other direction of less stiffness.

Steel Columns

Typical column shapes in ordinary multistory frame construction are W10, W12, and W14 sections; occasionally pipes and structural tubing are found. The W14 series is widely used; it provides an extensive range of sections, with some having wider flanges to balance the radii of gyration about both major axes. This series furnishes areas ranging from 6.49 in.^{2}(W14 × 22) to the jumbo sizes of 215 in.^{2} (W14 × 730). Keep in mind, however, that the deeper W-beam sections rather than the standard W14 ones may result in savings for peripheral rigid frame systems where shear wracking (i.e., bending) is a controlling design criterion.

Built-up columns may be needed for large loading conditions such as occur at the base of tall buildings and/or for long unbraced columns to increase their buckling capacity (i.e., moment of inertia), as may be the case for the first two floors of a structure. Built-up columns may also be required to provide stiffness for drift control. They consist basically of multiple members such as plates, angles, channels, and W-sections. They may form typical shapes such as the following:

- Coverplated W-sections
- Cruciform columns built, for instance, from two or four W-sections
- Box columns (open or closed, with or without interior webs) forming single-, double-, or triple-shaft columns. The multiple-shaft columns may be connected by lacing, batten, or perforated or solid coverplates.

The arrangement of elements and shape of these built-up columns is not just a function of construction (connection) and esthetic considerations, but must also provide the largest possible radii of gyration in response to local and overall buckling conditions.

Column Action. It was already discussed that the uniform compressive stress in a column cannot exceed the allowable axial compressive stress F_a.

$$f_a = P/A \le F_a \tag{3.91}$$

The allowable axial compressive stress F_a is a function of the slenderness ratio Kl/r as shown in Fig. 3.23. The shape of the column curve reflects the obvious fact that with increase of the effective column length the column capacity decreases. There are three distinct regions:

1. The initial relatively flat portion of the curve clearly indicates that a *short column* fails in yielding with no buckling present.
2. The end portion of the curve is derived from elastic buckling of *long columns.*
3. The transition portion between the two end conditions represents the behavior of *intermediate columns,* which fail in inelastic buckling, that is, yielding together with buckling.

The AISC specs have integrated short column behavior into their inelastic column formula; therefore, they distinguish only between two column types, as discussed next.

The allowable axial compressive stress for long columns is based on Euler's formula [Eq. (3.92)] using a safety factor of 23/12 = 1.92, where the slenderness ratio is not to exceed 200.

$$F_a = \frac{f_{cr}}{23/12} = \frac{12\pi^2 E}{23\,(Kl/r)^2} \cong \frac{149{,}000}{(Kl/r)^2}, \qquad \text{when } \frac{Kl}{r} \ge C_c \tag{3.93}$$

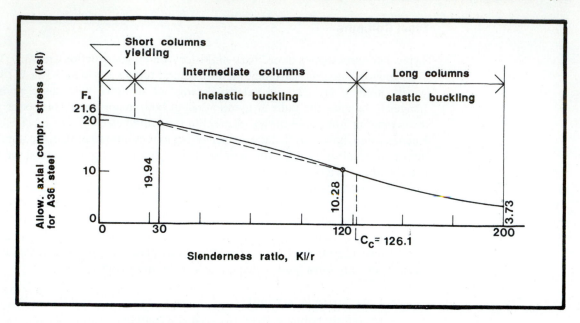

Figure 3.23 AISC column curves for A36 steel.

The elastic buckling formula is applicable up to a stress level where Euler's buckling stress is equal to $F_y/2$. The slenderness ratio at which this occurs is called C_c.

$$\frac{F_y}{2} = \frac{\pi^2 E}{(Kl/r)^2} \quad \text{or} \quad C_c = \frac{Kl}{r} = \sqrt{\frac{2\pi^2 E}{F_y}} \tag{3.94}$$

In other words, a *slender column* in A36 steel can be designed as based on Euler's formula if its slenderness ratio is equal to or larger than 126.1 and preferably not exceeding 200.

When the largest effective slenderness ratio is less than C_c, the column fails in inelastic buckling. In other words, intermediate steel columns will fail by both yielding and buckling. The strength of columns in this range is a function of several parameters, in addition to their length, such as grade of steel, degree of end restraint, magnitude of the initial crookedness, type of load action (axis of bending), shape and size of cross section, manufacturing method (residual stresses), and other imperfections of material and dimensions. The AISC column formula (see AISC Manual, Eq. E2-1) attempts to take all these factors into account, as reflected by the rather complex mathematical expression. Since F_a depends on the slenderness ratio, the allowable stress values are conveniently tabulated in the AISC Manual for Kl/r values from 1 to 200 and are reproduced in Table A.12 for $F_y = 36$ ksi.

The AISC allowable stress formula for *intermediate and short columns* as based on inelastic buckling can be simplified for the intermediate range, for preliminary design purposes using A36 steel, by replacing the curve with a straight line between $Kl/r = 30$ and 120 (see Fig. 3.23).

$$F_a \cong 23 - 0.1 \, Kl/r, \quad \text{for } 30 \leq Kl/r \leq 120 \tag{3.95}$$

where F_a = allowable axial compressive stress (ksi)
         K = effective length factor
         l = actual unbraced length of member (in.)
         r = governing radius of gyration (in.)

For example, for $Kl/r = 44$, $F_a \cong 23 - 0.1(44) = 18.60$ ksi, which is close to the true value of 18.86 ksi (see Table A.12).

For *short columns,* below a slenderness of about 30, $0.6F_y$ is only decreased by approximately 10%. Therefore, for preliminary design purposes, short columns can be designed for $F_a = 0.6F_y$.

The typical column slenderness for laterally braced high-rise buildings is rather low. For instance, for an unbraced column height of $1 = 11$ ft, the slenderness ratio and the corresponding allowable axial stresses (Table A.12) for typical sections are as follows:

$$W12 \text{ section with } r = 3 \text{ in.}: Kl/r = 1(11)12/3 = 44 \quad \text{or} \quad F_a = 18.86 \text{ ksi}$$
$$W14 \text{ section with } r = 4 \text{ in.}: Kl/r = 1(11)12/4 = 33 \quad \text{or} \quad F_a = 19.73 \text{ ksi}$$

Notice there is only about 5% difference in the allowable stresses for the two cases. Due to the slenderness, $0.6F_y$ is decreased by only about 14% indicating the short-column character of the typical building column in a braced structure.

The design of various column types is demonstrated in the following examples. Keep in mind that the larger Kl/r must be determined for conditions where the effective column lengths about the major and minor axes are not the same. For a column, however, where the conditions about both axes are identical, obviously the least radius of gyration controls, which is usually r_y, with the exception of r_z for single angles.

EXAMPLE 3.25

Estimate the size of a steel column at the first-floor level of a 15-story braced frame building. The interior column has an unbraced length of 13 ft and supports a 24-× 24-ft bay. Assume a 80-psf dead load and a reduced live load of 50 psf. Investigate a W14-section with $r \cong 4$ in.; use A36 steel.

The total axial load is

$$P = 15(24 \times 24)(0.080 + 0.050) = 1123.20 \text{ k}$$

Since the building is braced, the K value is conservatively assumed equal to 1.

$$Kl/r = 1(13 \times 12)/4 = 39, \text{ at which } F_a = 19.27 \text{ ksi (Table A.12)}$$

The required cross-sectional area is

$$A \geq P/F_a = 1123.20/19.27 = 58.29 \text{ in.}^2$$

Try W14 × 211, $A = 62$ in.^{2}, $r_{min} = r_y = 4.07$ in. $\geq r_{ass} = 4$ in.,   OK.

EXAMPLE 3.26

Assume the column in Example 3.25 is braced about its weak axis at mid-height. First assume the weak axis controls the design.

$$(Kl/r)_y = 1(6.5 \times 12)/4 \cong 20, \quad \text{at which } F_a = 20.60 \text{ ksi} \quad \text{(Table A.12)}$$

The required cross-sectional area is

$$A \geq P/F_a = 1123.20/20.60 = 54.52 \text{ in.}^2$$

Try W14 × 193, $A = 56.8$ in.^{2}, $r_x = 6.50$ in., $r_y = 4.05$ in.

Checking the allowable stress with respect to buckling about the x axis yields

$$(Kl/r)_x = 1(13 \times 12)/6.5 = 24 > 20, \quad F_a = 20.35 \text{ ksi}$$

$$A \geq P/F_a = 1123.20/20.35 = 55.19 \text{ in.}^2 < 56.8 \text{ in.}^2, \quad \text{OK}$$

Use $W14 \times 193$.

EXAMPLE 3.27

Assume the building of Example 3.25 is unbraced. What W14 section is required if the K value is estimated as 1.8 for first trial purposes.

$$Kl/r = 1.8(13 \times 12)/4 \cong 70, \quad \text{or} \quad F_a = 16.43 \text{ ksi. (Table A.12)}$$

Hence, the required cross-sectional area is

$$A \geq P/F_a = 1123.20/16.43 = 68.36 \text{ in.}^2$$

Try $W14 \times 233$, $A = 68.5 \text{ in.}^2$, $r_y = 4.10 \text{ in.} \geq r_{ass} = 4 \text{ in.}$, OK

EXAMPLE 3.28

What is the axial capacity of a 20-ft-high $W12 \times 65$ column (A36) that acts like a flagpole (i.e., fixed at its base and free to sway and rotate at the top)?
    A $W12 \times 65$ has the following properties: $A = 19.1 \text{ in.}^2$, $r_y = 3.02 \text{ in.}$

$$Kl/r = 2.1(20 \times 12)/3.02 = 166.89 \geq C_e = 126.1$$

Therefore, Euler's formula [Eq. (3.93)] applies:

$$F_a \cong \frac{149,000}{(Kl/r)^2} = \frac{149,000}{166.89^2} = 5.35 \text{ ksi} \quad \text{(see also Table A-12)}$$

Hence, the column can carry the following force:

$$P = AF_a = 19.1 (5.35) = 102.19 \text{ k}$$

The previous examples could have been solved more easily by using the column load tables in the AISC Manual. The tables are set up for W- and S-sections, pipes, structural tubing, double and single angles, and structural tees. The column loads are arranged according to the effective column length Kl usually with respect to the minor axis. Refer to Fig. 3.24 for the design of other column types.

Beam–Column Action. Most columns must resist bending in addition to axial action. It is nearly impossible in building construction to truly center loads on columns, besides, columns may not be perfectly straight. In addition, columns may be bent by wind or other transverse forces. Examples of beam-column action include the chords of trusses that act as axial members with respect to the overall truss behavior, but also may have to act as continuous beams to transfer loads to the truss joints.

In hinged braced frames, bending is caused by the eccentric beam reactions along the column faces, inducing relatively small local moments. In rigid frames, moments are directly transferred to the columns because the beams are continuously connected. Here bending is caused not only by the direct effect of the end moments induced by lateral loads and gravity, but possibly also by the secondary effect of the axial loads due to eccentric action along a laterally deflected slender column (i.e., P–Δ effect).

Ignoring the P–Δ effect for the time being by assuming the column as relatively stiff and laterally braced (i.e., low slenderness ratio) or assuming that the axial action is small in comparison to the bending action, then the combined compressive stresses

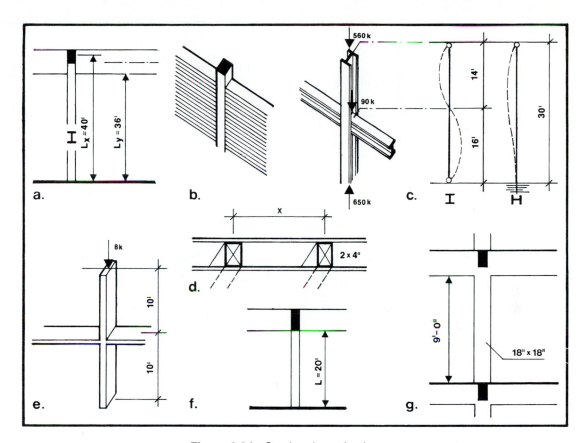

Figure 3.24 Steel and wood columns.

are easily obtained by simply adding the stresses due to axial compression and biaxial bending. The resulting linear interaction equation is

$$f_{max} = f_a + f_{bx} + f_{by} = P/A + (M/S)_x + (M/S)_y < F_{all} \qquad \text{(a)}$$

or

$$\frac{f_a}{F_{all}} + \frac{f_{bx}}{F_{all}} + \frac{f_{by}}{F_{all}} \le 1.0$$

Since the allowable stresses for column action are different from the allowable bending stresses, Eq. (a) is corrected and takes the following form:

$$\frac{f_a}{F_a} + \frac{f_{bx}}{F_{bx}} + \frac{f_{by}}{F_{by}} \le 1.0 \qquad (3.96)$$

The AISC Specs recommend the use of this linear interaction equation for predominant bending members with small axial forces when $f_a/F_a \le 0.15$. Many of the low-rise building frames investigated in this book are primarily bending systems where this simple interaction equation is applicable or can be used for preliminary design purposes.

For stocky members with a low slenderness ratio, the allowable axial compressive stress will be controlled by yielding rather than buckling: $F_a = 0.6\,F_y$. For this condition, Eq. (3.96) takes the following form (see Fig. 3.25),

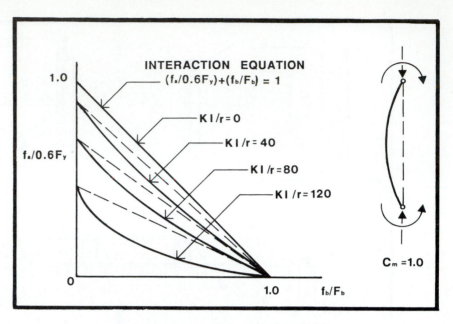

Figure 3.25 Beam–column interaction.

$$\frac{f_a}{0.6F_y} + \frac{f_{bx}}{F_{bx}} + \frac{f_{by}}{F_{by}} \le 1.0 \tag{3.97}$$

The interaction between the axial load P and the bending moment M cannot always be considered as a simple addition or percentage process as was just done. This approach is reasonable for predominate bending members where axial loads are small or for short columns of low slenderness (say $Kl/r \le 40$), where the interaction between P and M can be treated linearly and secondary effects can be ignored. For columns with higher slenderness, however, the interaction between P and M causes a new situation: the interaction line becomes a curve due to the secondary P–Δ bending effects (Fig. 3.25). The straight-line equation must therefore be modified to take into account these secondary effects due to the lateral displacement of the column. In practice, this is done by magnifying the moment consisting of the nonsway moment due to gravity and the sway moment due to the lateral loads. These secondary effects are especially critical for tall flexible buildings with large lateral loads, as well as for bridge-type frames with only few columns and relatively flexible long-span girders.

These second-order bending moments due to the P–Δ effect are taken into account by the AISC Specs with the magnification factor $1 - f_a/F_e'$, modified by the C_m factor. Hence, the corrected version of Eq. (3.96) for this new situation is

$$\frac{f_a}{F_a} + \left[\frac{C_{mx}}{1 - f_a/F_{ex}'}\right]\frac{f_{bx}}{F_{bx}} + \left[\frac{C_{my}}{1 - f_a/F_{ey}'}\right]\frac{f_{by}}{F_{by}} \le 1.0 \tag{3.98}$$

where $f_a = P/A$ = actual axial compressive stress (ksi)
$f_b = M/S$ = actual maximum compressive bending stress (ksi)
F_a = allowable compressive stress for column action alone, regardless of the plane of bending (ksi)
F_b = allowable compressive bending stress for beam action alone (ksi)
F_e' = Euler's formula divided by a safety factor [Eq. (3.93)] with respect to buckling about the bending axis

C_m = modification factor, which varies with loading and end conditions

= 0.2 to 1.0

- For unbraced frames: $C_m = 0.85$
- For braced frames: reverse curvature: $C_m = 0.2$ to 0.6

  single curvature: $C_m = 0.6$ to 1.0

- For braced columns with transverse loads assume:

  pinned ends $C_m = 1.0$

  restrained ends $C_m = 0.85$,

Refer to AISC Specs for a more precise definition of the factor!

According to the AISC Specs, beam–columns must be proportioned to satisfy Eqs. (3.97) and (3.98) when $f_a/F_a > 0.15$. For the condition where $f_a/F_a \leq 0.15$, Eq. (3.96) must be satisfied.

Since the proportioning of beam–columns is rather time consuming, the following approximate method of design may be used. It is assumed for the preliminary design of beam–columns that the lateral building drift is kept within reasonable limits so that the columns will be within a low slenderness range where the P–Δ effect can be ignored and the moments do not have to be significantly amplified. We may conclude that the column capacity can be represented by a straight-line interaction of P_a and M_a (or their respective stresses) for this low slenderness range. The familiar simple interaction equation for bending about one axis only is

$$f_a/F_a + f_b/F_b \leq 1.0$$

It is assumed conservatively that $F_b = F_a = P_a/A$ for preliminary design purposes.

$$P/A + M/S \leq P_a/A$$

Let $B = A/S = bending\ factor$, which transforms the moment into an equivalent axial force P', that is $P'/A = M/S$, or $P' = (A/S)M = BM$

$$P + P' = P + BM \leq P_a$$

Since the assumptions made were conservative, the equation tends to overestimate the column size. Therefore, rather than selecting the section required by this calculation, the next smaller one is chosen as a first trial section. In general, considering biaxial bending, the equivalent axial force P_{eq} is

$$P_{eq} = P + P_x' + P_y' = P + B_x M_x + B_y M_y \tag{3.99}$$

Now the column tables in the AISC Manual can be conveniently used to select a section as based on $(Kl)_y$ or $(Kl)_x/(r_x/r_y)$, whichever controls. As a first trial use $B_x = 0.18$ (for W14), $B_x = 0.21$ (for W12), $B_x = 0.26$ (for W10), $B_x = 0.33$ (for W8), $B_y = 3\ B_x$, and $r_x/r_y = 1.7$. When more than one-half of the column capacity is used in bending, however, the column tables may not provide an economical solution since the bending is more efficiently resisted by deeper W-sections rather than the typical column members in the tables.

EXAMPLE 3.29

Determine the preliminary size of a W14 column (A36) that is 12 ft long and is not braced about its strong axis, but is braced about its weak axis and hence does not sway in this direction. The column carries an axial load of 500 k and a moment of $M_x = 200$ ft-k (Fig. 3.26a). As a first trial, assume $K_y = 1.0$ and $K_x = 1.8$.

For preliminary estimation purposes, it may be assumed that the braced weak axis controls the design using $B_x = 0.18$.

$$P_{eq} = P + P' = P + B_x M_x$$
$$= 500 + 0.18(200)12 = 500 + 432 = 932 \text{ k}$$

From the Column Load Tables of the AISC Manual for $Kl = 1(12)$, select not the section required but the next one lower. Try $W14 \times 159$, $P_a = 911$k, $B_x = 0.184 \cong 0.18$, and $r_x/r_y = 1.60$.

Check the slenderness ratio about the x axis: $(Kl)_x/(r_x/r_y) = 1.8(12)/1.6 \cong 13.5 > 12$. Hence, the effective length for the x axis controls. For $Kl = 13.5$ ft, try $W14 \times 159$, $P_a = 894.5$ k.

EXAMPLE 3.30

Find the approximate size of the W14 column (A36) in Example 3.29, but now for an unbraced building. As a first trial assume $K_y = 1.8$, which may be quite conservative, and $Kl = 1.8(12) = 21.6$, say 22, together with $P_{eq} = 932$ k. From the AISC Column Tables, try $W14 \times 176$, $P_a = 874$ k.

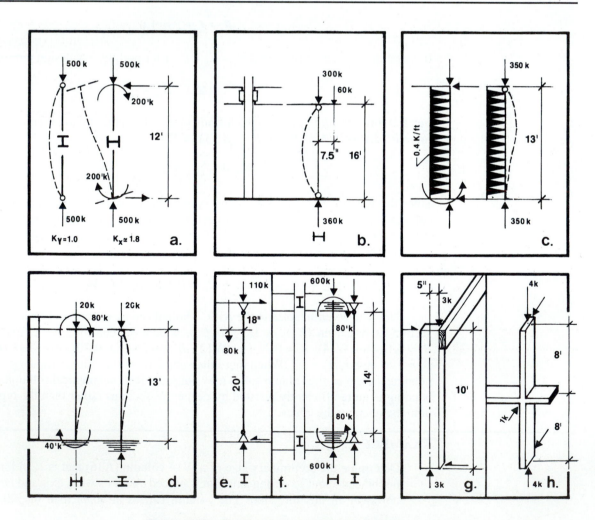

Figure 3.26 Beam–columns in steel and wood.

Should the AISC Steel Manual not be available, approximate the bending factor for W-sections as $B_x = 2.5/t$ for heavy sections as for taller buildings, $B_x = 2.6/t$ for mid-rise structures, and $B_x = 2.7/t$ for lower buildings (Problem 3.41) using the nominal depth t; also assume that $B_y = 3B_x$. For the special condition where the moment is very large in comparison to the axial load, a deep beam section may be more economical than the typical W14 and W12 column sections; for this situation, estimate $B_x = 3/t$.

Substituting this information into the beam column equation yields for heavy loads

$$P + P' = P + B_x M_x = P + 2.5M_x/t = A_s F_a$$

or   (3.100)

$$A = (P + B_x M_x)/F_a = (P + 2.5M_x/t)/F_a$$

The column area A_s can easily be estimated by assuming for a W14 ($t = 14$ in., $r_y = 4$ in.), W12 ($t = 12$ in., $r_y = 3$ in.), W10 ($t = 10$ in., $r_y = 2.6$ in.), W8 ($t = 8$ in., $r_y = 2.0$ in.), W6 ($t = 6$ in., $r_y = 1.5$ in.), and $r_x = 1.7r_y$. In addition, with respect to this equation when bending action is substantial, do not select as a first trial the section required but rather the next lower one because of the conservative assumptions made.

EXAMPLE 3.31

Determine the preliminary column size for the case in Example 3.29 by applying Eq. (3.100) assuming $K_x = 1.8$ and $K_y = 1.0$. Check which axis controls.

$$(Kl/r)_y = 1(12)12/4 = 36$$

$$(Kl/r)_x = 1.8(12)12/1.7(4) = 38.12 \geq 36$$

Hence the strong axis barely controls the design. Using the straight-line column formula approximation [Eq. (3.95)] yields

$$F_a \cong 23 - 0.1(Kl/r) = 23 - 0.1\,(38.12) = 19.19 \text{ ksi}$$

Thus, the required minimum cross-sectional area for the column is

$$A = (P + 2.5\,M_x/t)/F_a$$

$$= [500 + 2.5(200(12)/14)]/19.19 = 48.39 \text{ in.}^2$$

Try W14 × 159, $A = 46.7$ in.^{2}, $r_x = 6.38$ in., $r_y = 4$ in.

Notice that it would have made hardly any difference for the column design if the known effective length of the y axis rather than the estimated length of the x axis had been used for this preliminary investigation, as was pointed out in Example 3.29.

EXAMPLE 3.32

Check whether the W14 × 159 in Example 3.29 is satisfactory. The section has the following properties: $A = 46.7$ in.^{2}, $S_x = 254$ in.^{3}., $r_x = 6.38$ in., $r_y = 4.0$ in., and $L_c = 16.4$ ft.

The axial compressive stress is

$$f_a = P/A = 500/46.7 = 10.71 \text{ ksi}$$

The slenderness ratios for the x and y axes are

$$(Kl/r)_x = 1.8(12 \times 12)/6.38 \cong 41, \qquad (Kl/r)_y = 1.0\,(12 \times 12)/4 = 36$$

The slenderness ratio of the strong axis controls. The respective axial compressive stress (Table A.12) is $F_a = 19.11$ ksi.

$$f_a/F_a = 10.71/19.11 = 0.56 > 0.15$$

Therefore, Eqs. (3.97) and (3.98) must be checked.

The actual compressive bending stress is

$$f_b = M_x/S_x = 200(12)/254 = 9.45 \text{ ksi}$$

Since the unbraced column length $L_b \leq L_c = 16.4$ ft, and since $F_y \leq F_y'$ and F_y''',

$$F_b = 0.66 F_y = 0.66(36) \cong 24 \text{ ksi}$$

The elastic buckling formula [Eq. (3.93)] is a function of $(Kl/r)_b = (Kl/r)_x \cong 41$.

$$F_e' = \frac{12\pi^2 E}{23(Kl/r)_b^2} = \frac{12\pi^2 (29000)}{23(41)^2} = 88.83 \text{ ksi}$$

Because of joint translation (sidesway),

$$C_m = 0.85$$

Now Eqs. (3.97) and (3.98) can be checked.

$$\frac{f_a}{0.6F_y} + \frac{f_b}{F_b} = \frac{10.17}{0.6(36)} + \frac{9.45}{24} = 0.5 + 0.39 = 0.89 \leq 1.0, \quad \text{OK}$$

$$\frac{f_a}{F_a} + \left(\frac{C_m}{1 - f_a/F_e'}\right)\frac{f_b}{F_b} = 0.56 + \left(\frac{0.85}{1 - 10.71/88.83}\right)0.39$$

$$= 0.56 + 0.38 = 0.94 \leq 1.0, \quad \text{OK}$$

The W14 × 159 section is satisfactory.

Wood Columns

Typical wood compression members are found in post-beam construction. They are used in stud walls and as truss members; they can be straight or tapered. Wood columns generally have a rectangular or round cross section. They form either *solid members* (e.g., sawn lumber, poles, or glued laminated timber), *spaced columns* (where individual members are separated by spacer blocks at their ends and midpoints), *box columns,* or *built-up columns* (solid pieces side by side connected, for instance, by nails or bolts).

Column Action. As for steel columns, wood columns may be classified according to their slenderness ratio as short columns, intermediate columns, and long columns. A *short column* fails by crushing of its fibers, a *long column* fails by elastic buckling, and an *intermediate column* fails by a combination of crushing and buckling. Whereas in the past NDS clearly separated column behavior according to slenderness ratio into the familiar three column types, the new column formula, in contrast, covers the full slenderness range in a continuous manner.

The concept of slenderness ratio $l_e/r = Kl/r$ has already been discussed under steel column design (see Fig. 3.21). The effective length factors, called K_e in wood design, for various end conditions are published by NDS. Their theoretical values are naturally equal to the ones for steel design; however, for the top-right case in Fig. 3.21, the recommended K_e value for wood construction is 2.4, whereas in steel construction it is 2.0.

Wood columns used in ordinary residential construction do not sidesway because buildings are generally laterally braced. Since it is difficult to evaluate the degree of end fixity of the column, it is usually conservative to assume pinned ends and an effective length equal to the actual unsupported column length, $l_e = l$.

The uniform compressive stress f_c in a column cannot exceed the allowable axial compressive stress parallel to the grain F_c', which, in turn, is equal to the allowable compression stress based on crushing, F_c^*, adjusted by the column stability factor C_P, as introduced later.

$$f_c = P/A \leq F_c' = F_c^* \, C_P \cong (F_c) \, C_D \, C_P \qquad (3.101)$$

where $F_c^* = (F_c) \, C_D \, C_M \, C_t \, C_F \cong (F_c) \, C_D$ (for preliminary design purposes).

Here, the tabulated compression design value F_c is modified by the applicable adjustment factors. The load duration (C_D), wet service (C_M), and temperature (C_t) factors have already been defined in the introduction to this chapter, whereas the size factor (C_F) is only applied to dimension lumber sizes but is disregarded here, since the 1986 NDS base stresses are used for preliminary design purposes. The allowable compressive stress parallel to the grain F_c' is a function of the slenderness ratio l_e/r similar to the design of steel columns (see Fig. 3.23). For long columns with large l_e/r, it is based on Euler's formula [Eq. (3.92)] using a safety factor of 2.74.

$$F_{cE} = \frac{\pi^2 E'}{2.74 \, (l_e/r)^2} = \frac{3.60 E'}{(l_e/r)^2}, \quad \text{where} \quad l_e/r \leq 173 \qquad (3.102a)$$

For typical solid rectangular columns, it is convenient to replace l_e/r by l_e/d by substituting $r = \sqrt{I/A} = \sqrt{bd^3 / [12\,(bd)]} = d/\sqrt{12} = 0.29d$, so that the critical buckling stress is

$$F_{cE} = \frac{0.30 E'}{(l_e/d)^2}, \quad \text{where} \quad l_e/d = K_e l/d \leq 50 \qquad (3.102b)$$

Notice that the allowable modulus of elasticity E' is obtained by adjusting the tabulated values: $E' = (E)\,C_M\,C_t \cong E$ for ordinary conditions.

Euler's column formula may be used for the preliminary estimating of sizes for *long columns* of approximately $l_e/d \geq 35$. In other words, the first trial cross-sectional area, as based on $C_P = F_{cE}/F_c^*$, is

$$A \geq P/F_{cE} \qquad (3.103)$$

The allowable compressive stress parallel to the grain, F_c', for *short columns* with a very small l_e/d ratio is equal to the tabulated compressive stress parallel to the grain multiplied by all applicable adjustment factors except the column stability factor, which is nearly 1, $C_P \cong 1.0$. According to the previous NDS, this condition applied when $l_e/d \leq 11$. In other words, the first trial cross-sectional area for a typical short column is

$$A \geq P/F_c C_D \qquad (3.104)$$

Similar to steel column design (see Fig. 3.23), the NDS column formula does not distinguish between short, intermediate, and long columns by using separate formulas as in the past. The column stability factor C_P in Eq. (3.101) covers the full slenderness range in a continuous manner and is defined for rectangular solid sawn lumber columns as

$$C_p = \frac{1 + (F_{cE}/F_c^*)}{1.6} - \sqrt{\left[\frac{1 + (F_{cE}/F_c^*)}{1.6}\right]^2 - \frac{F_{cE}/F_c^*}{0.8}}$$
(3.105)

The terms in the equation have already been defined previously. For glulam columns replace 1.6 by 1.8 and 0.8 by 0.9. Round columns or poles of diameter D can be treated as equivalent square columns of the same cross-sectional area.

$$\pi D^2/4 = d^2 \quad \text{or} \quad d = 0.886D$$
(3.106)

EXAMPLE 3.33

Determine the axial load capacity of a 15-ft-long, 6- × 8-in., Douglas fir–larch column that supports a roof structure with snow loads. The column is assumed pin ended and not to sway; it is laterally braced at midheight about its y axis (similar to Fig. 3.24e).

The design values are assumed as $F_c = 1000$ psi, $E = 1,600,000$ psi for Post and Timber size classification. The cross-sectional area of the 6 × 8 column with an actual size of $5\frac{1}{2} \times 7\frac{1}{2}$ is 41.250 in.^{2} (see Table A.9).

The slenderness ratios about both axes for $l_e = K_e l = (1) l = l$ are

$$(l_e/d)_x = 15(12)/7.5 = 24 \le 50, \qquad (l_e/d)_y = 7.5(12)/5.5 = 16.36$$

The larger slenderness ratio about the x-axis controls the design. Hence, the critical buckling stress, according to Eq. (3.102b) for $E' = E$, is

$$F_{cE} = \frac{0.3E'}{(l_e/d)^2} = \frac{0.3 (1,600,000)}{24^2} = 833 \text{ psi}$$

The tabulated compression stress multiplied by the applicable adjustment factors is

$$F_c^* = F_c C_D = 1000 (1.15) = 1150 \text{ psi.}$$

Now the column stability factor, according to Eq. (3.105), can be determined as follows:

$$C_p = \frac{1 + (833/1150)}{1.6} - \sqrt{\left[\frac{1 + (833/1150)}{1.6}\right]^2 - \frac{833/1150}{0.8}} = 0.57$$

Therefore, the axial load capacity of the given column, according to Eq. (3.101), is.

$$P = F_c'A = F_c C_D C_P A = 1.000(1.15)0.57(41.25) = 27.04 \text{ k}$$

EXAMPLE 3.34

Design a 9-ft-long laterally supported square Douglas fir–larch column to carry an axial load of 17 kips, which includes snow loading.

The design values are assumed as $F_c = 1200$ psi, $E = 1,700,000$ psi. for the Posts and Timbers category. The design of columns cannot be done directly since the allowable compressive stress is a function of the slenderness ratio (i.e., size of column), which is not known. A trial-and-error approach must be used. For first-approximation purposes, the column is considered as short. Therefore, the required cross-sectional area, according to Eq. (3.104), but ignoring the load duration factor, is

$$A \geq P/F_c = 17/1.200 = 14.17 \text{ in.}^2$$

Try a 5×5 member with $A = 20.25$ in.^{2}
The slenderness ratio for $l_e = K_e l = (1)l = l$ is

$$l_e/d = 9(12)/4.5 = 24 \leq 50$$

Hence, the critical buckling stress, according to Eq. (3.102b) for $E' = E$, is

$$F_{cE} = \frac{0.3E'}{(l_e/d)^2} = \frac{0.3 \, (1,700,000)}{24^2} = 885 \text{ psi}$$

The required cross-sectional area as based on Euler's formula [Eq. 3.103], is

$$A \geq P/F_{cE} = 17/0.885 = 19.2 \text{ in.}^2 < 20.25 \text{ in.}^2$$

The required cross-sectional area is very close to that of the trial section. But, since Euler's formula is not applicable for this intermediate slenderness range and gives allowable stresses that are too large, a 6×6 member with $A = 30.25$ in.^{2} is investigated.
The slenderness ratio for this member is

$$l_e/d = 9(12)/5.5 = 19.64$$

The critical buckling stress is

$$F_{cE} = \frac{0.3E'}{(l_e/d)^2} = \frac{0.3 \, (1,700,000)}{19.64^2} = 1322 \text{ psi}$$

The tabulated compression stress multiplied by the applicable adjustment factors is

$$F_c^* = F_c \, C_D = 1200 \, (1.15) = 1380 \text{ psi}$$

Now the column stability factor can be determined according to Eq. (3.105) as follows:

$$C_p = \frac{1 + (1322/1380)}{1.6} - \sqrt{\left[\frac{1 + (1322/1380)}{1.6}\right]^2 - \frac{1322/1380}{0.8}} = 0.68$$

Hence, the axial load capacity for this trial column, according to Eq. (3.101), is

$$P = F_c C_D C_P A = 1.20(1.15)0.68(30.25) = 28.39 \text{ k} \geq 17 \text{ k}$$

This trial section is satisfactory. Check if the 5×5 member works.

Beam–Column Action.   Similar to steel beam–column design, second-order moments due to the P–Δ effect may have to be taken into account with a magnification factor. For the condition where the member is subject to axial action and bending about only one principal axis due to transverse loading, NDS proposes that the following interaction equation must be satisfied.

$$\left[\frac{f_c}{F_c'}\right]^2 + \frac{f_b}{F_b'}\frac{1}{1-(f_c/F_{cE})} \leq 1.0 \tag{3.107}$$

where $f_c = P/A$ = actual axial compressive stress parallel to grain (psi) $< F_{cE}$

$\quad\;\; f_b = M/S$ = actual flexural stress (psi)

$\quad\;\; F_c'$ = allowable compressive stress parallel to grain (psi) that would be permitted if an axial force alone existed

$\quad\;\; F_b'$ = allowable bending stress (psi) that would be permitted if the bending moment alone existed

$\quad\;\; F_{cE}$ = buckling stress about the bending axis (psi)

When bending is caused by eccentric axial action rather than only by side loads and/or applied moments, flexural stresses can be computed for short pin-end columns of rectangular cross section as follows:

$$f_b = \frac{M}{S} = \frac{Pe}{bd^2/6} = \frac{6Pe}{Ad} = f_c\,(6e/d) \tag{3.108}$$

NDS amplifies this bending stress to consider the P–Δ effect and adds it to the bending stress due to transverse loading in Eq. (3.107).

$$f_b = f_c\,(6e/d)\,[1 + 0.234\,(f_c/F_{cE})] \tag{3.109}$$

where f_c = compression stress parallel to grain $< F_{cE}$

$\quad\;\; F_{cE}$ = buckling stress about bending axis

$\quad\;\; e$ = eccentricity of axial load P to center line of column (in.)

$\quad\;\; d$ = face dimension of rectangular column in the plane of bending (in.)

EXAMPLE 3.35

Determine whether a 21-ft-long, 4- × 8-in., Douglas fir–larch column, braced about the weak axis at 7-ft intervals, is satisfactory in supporting an axial load of 4 k (of normal load duration) that has an eccentricity of 4.88 in. about the strong axis. The pin-end column does not sidesway since the building is laterally braced.

The design values for the given grade (category 2 in. to 4 in. thick, 5 in. and wider), are $F_b = 1500$ psi, $F_c = 1250$ psi, $E = 1,800,000$ psi. The cross-sectional area is $A = 25.375$ in.^{2} (Table A.9). The critical slenderness ratio is

$$(l_e/d)_{\max} = (l_e/d)_x = 21(12)/7.25 = 34.76 \leq 50, \qquad (l_e/d)_y = 7(12)/3.5 = 24$$

The critical buckling design value, according to Eq. (3.102b) for $E' = E$, is

$$F_{cE} = \frac{0.3E'}{(l_e/d)^2} = \frac{0.3\,(1,800,000)}{34.76^2} = 447 \text{ psi}$$

The actual compressive stress is

$$f_c = P/A = 4000/25.375 = 158 \text{ psi} < 447 \text{ psi}.$$

The allowable axial compressive stress, according to Eq. (3.101), as based on $F_c^* = F_c C_D = 1250 \, (1.0) = 1250$ psi, is

$$F_c' = F_c^* C_P = F_c C_P = 1250 C_p$$

The column stability factor [Eq. (3.105)] is

$$C_p = \frac{1 + (F_{cE}/F_c^*)}{1.6} - \sqrt{\left[\frac{1 + (F_{cE}/F_c^*)}{1.6}\right]^2 - \frac{F_{cE}/F_c^*}{0.8}}$$

$$= \frac{1 + (447/1250)}{1.6} - \sqrt{\left[\frac{1 + (447/1250)}{1.6}\right]^2 - \frac{447/1250}{0.8}} = 0.326$$

Therefore, the allowable axial compressive stress is

$$F_c' = F_c^* C_P = F_c C_P = 1250(0.326) = 408 \text{ psi} > 158 \text{ psi}$$

Notice, because of the high slenderness of the column (i.e., long column), the allowable axial stress is close to the critical buckling stress F_{cE}, which therefore could have been used for fast approximation purposes. The bending stress due to eccentric force action, according to Eq. (3.109), is

$$f_b = f_c(6e/d)[1 + 0.234(f_c/F_{cE})] = 158(6 \times 4.88/7.25)[1 + 0.234(158/447)]$$
$$= 158(6 \times 4.88/7.25)1.08 = 690 \text{ psi}$$

Substituting the various stress values into the interaction equation [Eq. (3.107)], and letting $F_b' = (F_b)C_D C_L \cong F_b$, by assuming for this quick check that the bracing about the y axis provides enough lateral support so that the beam stability factor $C_L = 1.0$, yields

$$\left(\frac{f_c}{F_c'}\right)^2 + \frac{f_b}{F_b'} \frac{1}{1 - (f_c/F_{cE})} = \left(\frac{158}{408}\right)^2 + \frac{690}{1500} \frac{1}{1 - (158/447)}$$

$$= 0.150 + 0.46 \, (1.55) \; = \; 0.150 + 0.713$$

$$= 0.863 \le 1.0, \; \text{OK}$$

The column section seems to be satisfactory for the given loading conditions.

Reinforced Concrete Columns

Most concrete columns in ordinary buildings are rectangular or round. Occasionally, they are of polygonal, T- or L-shape as for corner columns, or they form flat wall columns.

The typical building column contains at least 1% but not more than 8% vertical reinforcement ($0.01 A_g \le A_{st} \le 0.08 A_g$), which is held in position by the lateral ties or spirals to form a stiff steel cage. The practical limit for fitting the bars, if they are not bundled together, is about 5% to 6% of the column area A_g; when more than 4% reinforcement is used, however, the beam–column bar clearances at the floor level should be checked. In general, the steel areas are kept less than 3% of the gross concrete area. Concrete columns may not just be tied or spiral columns, but interact with other materials in composite action. The composite column may be either a concrete-filled steel pipe or tube (e.g., jumbo column scheme), or it may be a reinforced concrete column with an encased steel shape as is typical for high-rise composite frame construction.

Because of the inherent continuity of cast-in-place reinforced concrete construction, concrete columns must always be treated as beam–columns. They will be capable of achieving *material failure* in ordinary braced frames, while *stability failure* may occur in unbraced frames or for very slender braced columns. Hence, we may distinguish between two groups of columns from a behavioral point of view:

- *Short columns,* where slenderness is ignored.
- Slender or *long columns,* where the effects of slenderness must be considered. Here, the axial load capacity of a column may be significantly reduced due to the moments resulting from lateral deflection of the column (P–Δ effect).

Columns of laterally braced buildings can generally be treated as short; here an effective clear height of $Kl_u = 1.0l_u$ can safely be assumed. The slenderness effects for the design of braced frames can be ignored in ordinary monolithic concrete construction if the column thickness t is larger than one-fourteenth of the clear column length l_u measured between the floor slabs, or top of floor slab and bottom of beam or column capital (see solution to Problem 3.60). In general for columns of rectangular cross section

$$t \geq l_u/14 \quad \text{for typical columns above first-floor level}$$

$$\text{(3.110)}$$

$$t \geq l_u/10 \quad \text{for first-floor columns with zero end restraint}$$

Thus, for typical net column heights of $l_u = 8$ to 12 ft, the corresponding minimum column dimensions are in the range of 8 to 10 in., which is usually also the minimum as based on fire resistance, although most designers observe a minimum dimension of 12 in. Surveys have shown that for more than 90% of columns in braced frames, buckling can be ignored and the columns can be treated as short!

Selecting column proportions in frames not braced against sidesway such that slenderness criteria do not control may be unreasonable because of the large column sizes thus required. According to the ACI Code, unbraced columns can be considered short if $Kl_u/r \leq 22$. Using a minimum effective length factor of $K = 1.2$, which is based on a nearly equal column to beam stiffness or nearly rigid beam action, results in the following expression for columns of rectangular cross section for which slenderness considerations could be ignored under certain conditions:

$$t = l_u/6 \quad \text{(3.111)}$$

This limiting condition represents roughly the changeover from short to long columns and may be particularly applicable to ordinary low-rise buildings and the upper floors of high-rise buildings. Notice that for slenderness effects to be neglected the column for the unbraced frame would have to be about twice as thick as the one for the braced frame, which obviously does not make sense, although it is clear that a laterally braced column may be much more slender than the equivalent unbraced column since it has to resist much less lateral force. Surveys have shown that 40% of columns in unbraced frames are actually short columns! Keep in mind, however, that with the development of high-strength concretes much smaller cross sections are possible, which will in turn result in more slender members that are vulnerable to stability problems and secondary loading effects.

For purposes of the preliminary design of columns, it is assumed that in ordinary braced buildings slenderness can be ignored. The same is assumed for unbraced buildings, but here the column sizes are selected as based on a low percentage of steel so that more reinforcement can be added later in the final design stage when slenderness will have to be taken into account.

The columns are treated as short in the following discussion so that material failure may be assumed. In cast-in-place reinforced concrete construction, a minimum moment for every column must be considered because of the inherent continuity and rigidity of the monolithic connections. In other words, a column must always be treated as a beam–column. Its behavior and mode of failure depend on the relative magnitude of the ultimate forces P_u and M_u or the distance e, that is the statically equivalent representation of the force P_u acting eccentric with respect to the centroidal axes of a cross section where the eccentricity e is measured by $e = M_u/P_u$.

Most concrete columns of ordinary buildings fail in compression as surveys have shown; that is, the eccentricity e of the axial force is less than the eccentricity e_b of the balanced loading case at which compression and tension failure occur simultaneously. When $e > e_b$; in other words, when the moment is very large in comparison to the axial force, P_u lies well outside the cross section and the column fails in tension.

In the compression range it holds true that the larger the axial load, the smaller the moment the column can sustain. In the tension zone the opposite is true: with an increase of axial action, larger moments can be carried because of the axial load prestressing the column and suppressing the moment-induced tension, as shown in Fig. 3.27.

In conclusion, the following three short column types can be distinguished as based on loading:

1. Columns that carry primarily axial forces and only small moments; they are predominantly axial members that fail in compression. This condition is typical for interior columns of laterally braced buildings with regular bay layouts.

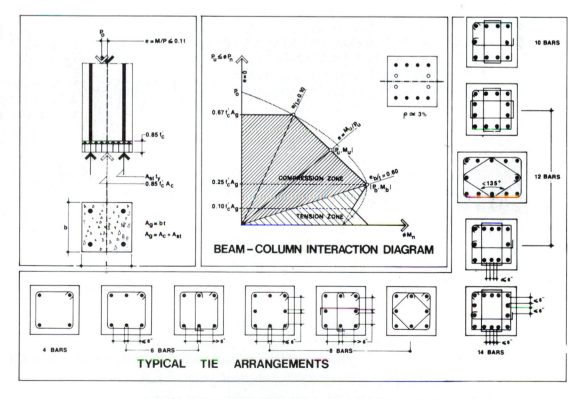

Figure 3.27 Reinforced concrete columns (Reproduced with permission from *The Vertical Building Structure*, Wolfgang Schueller, copyright © 1990 by Van Nostrand Reinhold.)

2. Columns that must resist axial forces and bending, with each possibly using nearly equal portions of the column capacity; these are truly beam–columns, which fail mostly in compression. This condition is typical for flexible unbraced buildings.

3. Columns that carry large bending moments but small axial forces. They are predominantly bending members (beams) where failure is initiated in tension.

For the preliminary design of concrete columns, primary emphasis here is on the first two types, which are more common in regular buildings; that is, solely compression failure is considered. Only normal-weight concrete will be used.

For the condition where the column moments are small, the ACI Code (318-71) used to require, for tied columns, a minimum eccentricity of 10% of the overall depth of the section.

$$e \geq 0.1t \tag{3.112}$$

This case is typical for the interior columns of ordinary laterally braced buildings with regular bays (i.e., where beam gravity moments balance each other). Here the columns do not resist much lateral force with the exception of the upper floors, where, however, the combined action of gravity and wind or earthquake is not critical. Nearly the same column sizes as derived for the interior may also be used for the exterior columns, which carry less axial load but larger moments due to the girder rotation as caused by the unbalanced gravity loads.

This case can also be used as a first approximation for the upper floors of nonbraced buildings or for massive building blocks where, as has been discussed, lateral force action does not control the design, but slenderness must still be considered. For this condition the column section may, for instance, be selected as based on the minimum vertical reinforcement of 1% of the gross area of the column ($\rho_{min} = A_{st}/A_g = 0.01$). The effects of slenderness will then be covered by additional steel in the final design stage.

The compressive capacity $\phi P_n = P_o$ of a short concrete column under concentric loads only consists of the sum of the concrete strength $\phi 0.85 f'_c A_c$ and the steel strength $\phi f_y A_{st}$, as shown in Fig. 3.27. This expression is further reduced for tied columns by a factor of 0.8 to take into account the nature of continuity in cast-in-place concrete structures; here the moment action of $M_{umin} = P_u e_{min}$ is roughly equal to the minimum eccentricity concept of $e_{min} = 0.1t$ of previous codes. Hence, the compressive strength of a short, tied rectangular column is

$$P_u \leq \phi P_n$$

$$\leq \phi\, 0.80[0.85 f'_c (A_g - A_{st}) + f_y A_{st}] \tag{3.113a}$$

$$\leq \phi\, 0.80 A_g [0.85 f'_c (1 - \rho_g) + \rho_g f_y] \tag{3.113b}$$

$$\leq \phi\, 0.80 A_g [0.85 f'_c + \rho_g (f_y - 0.85 f'_c)] \tag{3.113c}$$

where P_u = factored axial load (k)

A_g = area of column cross section (in.^{2}) = cross-sectional area of concrete A_c plus area of longitudinal reinforcement A_{st} (i.e., $A_g = A_c + A_{st}$)

f'_c = compressive strength of concrete (ksi)

ρ_g = A_{st}/A_g = column reinforcement ratio

f_y = yield strength of longitudinal reinforcement (ksi)

From this equation the required cross-sectional area of the column can easily be derived. For the typical conditions of Grade 60 steel, realizing that the effect of the concrete strength in decreasing the steel strength ($f_y - 0.85 f_c'$) is minimal [see Eq. (3.113c)] and using a capacity reduction factor for tied columns of $\phi = 0.7$, the following simple approximation can be used for the first sizing of short concrete columns:

$$A_g \geq P_u/0.5(f_c' + \rho_g f_y) \geq P_u/(0.5 f_c' + 0.3 \rho_g), \qquad \text{for } e = M_u/P_u \leq 0.1t \qquad (3.114)$$

where $\rho_g = A_{st}/A_g$ = column reinforcement ratio; only $0.3\rho_g$ in %.

For instance, for a minimum steel ratio of $\rho_g = 1\%$ and 4000-psi concrete, $A_g = 0.44P_u$, but for a higher steel ratio of $\rho_g = 4\%$, $A_g = 0.31P_u$. This clearly shows that by increasing the amount of reinforcement by four times the column cross section can be reduced by about 30%. For spiral columns the capacity is 14% larger than for tied columns or, alternatively, the cross-sectional area of a column can be reduced by 12% if spiral steel is used instead of ties. However, the cost of spiral steel is about twice that of tie steel. Another even simpler expression, often used in practice, can be derived from Eq. (3.113c) for $\rho_g \cong 1\%$ by dividing the axial load by an average stress.

$$A_g = P_u/0.55 f_c' \qquad (3.115)$$

The axial capacity of a round tied column can be approximated as roughly 80% of the capacity of a square tied column with the sides equal to the diameter and with the same number of vertical bars. In other words, the diameter of the round column should be about 13% larger than the sides of the square column to achieve the same strength. On the other hand, the axial capacity of a round spiral column is about 90% the capacity of a square tied column using the same main reinforcement. Regular polygonal columns can approximately be designed as based on a circular section enclosed within the column boundaries. Other column shapes, such as L-shaped columns, can be treated as intersecting rectangular columns.

As based on the *working stress approach*, an expression similar to Eq. (3.115) can be derived for 1% steel and an allowable compressive stress of $0.22f_c'$.

$$A_g = P/0.25 f_c' = 4P/f_c' \qquad (3.116)$$

Equation (3.115) can be further simplified for the following typical loading conditions of 100-psf dead load and a reduced live load of 50 psf:

$$P_u = nA[1.4(0.100) + 1.7(0.050)] = 0.225nA$$

Substituting this expression into Eq. (3.115) by letting $f_c' = 4$ ksi yields approximately

$$A_g = nA/10 \qquad (3.117)$$

The column area A_g (in.^{2}) is equal to the total floor area nA(ft^2) that the column supports, divided by 10, where the floor area is equal to the supported area of one floor A multiplied by n stories. For apartment buildings with smaller loads, the following expression is often used for estimation of concrete column sizes:

$$A_g = nA/12 \qquad (3.118)$$

For the preliminary estimation of laterally braced columns in office buildings as based on 1% percent steel, for 8000-psi concrete use a required cross-sectional area of $A_g = nA/20$, and for 6000-psi concrete, $A_g = nA/15$. Keep in mind that these equations are only applicable to interior columns of laterally braced buildings, with regular column layouts satisfying the preceding loading conditions.

Similar to Eq. (3.113a), the compressive strength of a *composite column* is shared by the steel section, the longitudinal reinforcement, and the concrete. Hence, the compression strength of a stocky composite column, where the design yield strength of the steel core is limited to 52 ksi, can be expressed for the following conditions as

Concrete filled pipes or tube: $$P_u \leq A_s F_y + A_{st} F_{yr} + 0.85 f_c' A_c \tag{3.119}$$

Concrete-encased structural steel: $$P_u \leq A_s F_y + 0.7(A_{st} F_{yr} + 0.85 f_c A_c) \tag{3.120}$$

where shear transfer between the concrete and steel must be considered.

EXAMPLE 3-36

A 15-story laterally braced concrete frame building is organized on 20- × 20-ft bays, approximately satisfying the discussed loading conditions. Estimate the column sizes assuming the same column to continue for three stories. An interior column supports a typical floor area of

$$A = 20(20) = 400 \text{ ft}^2$$

or the floor area of three stories is 1200 ft².

13th to 15th floor:	$A_g = 1200/10 = 120 \text{ in.}^2$	Try 12 × 12 in.
10th to 12th floor:	$A_g = 2(120) = 240 \text{ in.}^2$	Try 16 × 16 in.
7th to 9th floor:	$A_g = 3(120) = 360 \text{ in.}^2$	Try 20 × 20 in.
4th to 6th floor:	$A_g = 4(120) = 480 \text{ in.}^2$	Try 22 × 22 in.
1st to 3rd floor:	$A_g = 5(120) = 600 \text{ in.}^2$	Try 25 × 25 in.

At the building base, roughly $A_g/A = [(25/12)^2/400]100 \cong 1.1\%$ of the floor area is taken by the columns! Often engineers guess first column sizes that are subject primarily to axial action by starting with a minimum 12-in. square column and adding 1 in. per story. The gross area of rectangular columns is taken as roughly equal to the one of square columns.

In general, as based on formwork costs, it is more economical not to change column sizes from floor to floor but rather to alter only the material strength and the percentage of reinforcing steel. In multistory buildings, the largest column size is determined at the base of a building for a high-strength concrete of, say, 10,000 psi and a reasonable maximum percentage of Grade 60 steel. As the column progresses upward, the percentage of reinforcing bars decreases to a minimum of 1%. Then the 10,000-psi column is replaced by a lower-strength concrete of, say, 8000 psi. This process is continued until, after several floors, the column size can be substantially reduced, but then consideration should be given to changing only one dimension of the column. It is often economical to change column sizes in increments of at least 30% to 50% or more.

Keep in mind that slab openings for pipe chases at the columns may require larger column sizes to provide an adequate slab capacity for flat plates.

EXAMPLE 3.37

Investigate an interior column of Example 3.36 at the first floor level. Use a steel ratio of about 3% rather than the minimum amount of 1% as in Example 3.36, together with $f_c' = 4000$ psi and $f_y = 60$ ksi. The clear height of the column is 10 ft. The ultimate floor load, according to Eq. (4.19), is

$$w_u = 1.4D + 1.7L = 1.4(0.100) + 1.7(0.050) = 0.225 \text{ ksf}$$

The column load is equal to the floor load multiplied by the floor area that the column must support.

$$P_u = 0.225(20 \times 20)15 = 1350 \text{ k}$$

The required column area, according to Eq. (3.114), is

$$A_g = P_u/(0.5f'_c + 0.3\,\rho_g) = 1350/[0.5(4) + 0.3(3)] = 465.52 \text{ in.}^2$$

Try a 22- × 22-in. column, $A_g = 484$ in.^{2} The column can be treated as short and the effect of slenderness ignored, since

$$l_u/t = 10(12)/22 = 5.46 < 10 \qquad\qquad (3.110)$$

Determine the true ρ_g ratio from Eq. (3.114).

$$484 = 1350/[0.5(4) + 0.3\rho_g], \text{ which yields } \rho_g = 2.63\%$$

The required vertical steel is

$$A_{st} = \rho_g A_g = 0.0263(484) = 12.73 \text{ in.}^2$$

Use 14 #9, $A_{st} = 14$ in.^{2}, by placing five bars on two faces and the rest along the two other faces.

According to the ACI Code, all nonprestressed bars shall be enclosed by lateral ties at least #3 in size for longitudinal bars #10 and smaller, and at least #4 in size for larger bar sizes and bundled bars. The vertical spacing of the ties is the least of the following values:

16 longitudinal bar diameters:	$16(1.128) = 18.05$ in.
48 tie bar diameters:	$48(0.375) = 18$ in.
least column dimension:	22 in.

Use #3 ties at 18 in.

Furthermore, according to the ACI Code, ties shall be arranged such that every corner and alternate longitudinal bar shall have lateral support provided by the corner of a tie with an included angle of not more than 135°, and no bar shall be farther than 6 in. clear on each side along the tie from such a laterally supported bar. The clear distance between the longitudinal bars must not be less than 1.5 times the vertical bar diameter nor $1\frac{1}{2}$ in. For typical tie arrangements, refer to Fig. 3.27. Checking the critical column faces with the four bars and the wider spacing as shown in Fig. 3.27 yields.

$$[22 - 2(1.5 + 0.375 + 2(1.128))]/3 = 4.58 \text{ in.} < 6 \text{ in.,} \quad \text{OK}$$

Remember that when the ties also have to act as stirrups to carry excess shear in beam–columns, then $s_{max} = d/2$ must be considered! Usually, inadequate shear reinforcement in the columns is the primary cause of most failures in earthquake regions (i.e., shear failure in the columns).

In tall rigid frame buildings the columns must resist the lateral forces in shear wracking, that is in bending, which is especially critical in the bottom portion of the building. In addition, slenderness considerations become important since they magnify the moment action. The predominant column action of laterally braced buildings is replaced by beam–column behavior in nonbraced buildings!

Beam columns can be roughly designed for a typical reinforcement ratio of about 2.5% to 3% by treating the moment action M_u as an equivalent eccentricity of the axial force P_u: $M_u = P_u$ (e) or $e = M_u/P_u$. The required cross-sectional column area can be derived for Grade 60 steel as follows:

- The required cross-sectional area for a column that carries an axial load and only a small moment, and with a steel ratio of nearly 2.5% can be derived from Eq. (3.114) as approximately

$$A_g = 1.5P_u/f_c', \quad \text{for } e \cong 0.1t \tag{a}$$

- For the special balanced condition at which the column fails simultaneously in tension and compression due to axial action and extensive bending, it is derived in the solution to Problem 3.61 that the axial load P_b at which this type of failure occurs is approximately equal to

$$P_b = 0.3bdf_c'$$

By letting $P_u = P_b$ and $d \cong 0.83t$, the balanced column area for this boundary stage can be roughly expressed as

$$A_g = P_u/0.25f_c' = 4P_u/f_c' \tag{b}$$

For a typical steel ratio of 3%, the tension failure occurs at about $e_b = 0.6t$ (see the solution to Problem 3.61).

- Now the cross-sectional column area for a low but typical reinforcement ratio of about 2.5% to 3% can be approximated from Fig. 3.27 by a straight-line interpolation between the upper and lower boundaries of the compression failure range of $1.5P_u/f_c'$ and $4P_u/f_c'$, respectively, for Grade 60 steel (see Problem 3.62). This simplified version of the beam–column equation refers only to the special condition where the eccentricity does not exceed about six-tenths of the depth of the column as based on 3% steel, keeping in mind that e_b varies with the steel ratio (e.g., $\rho \cong 2.5\%$, $e \leq t/2$).

$$A_g = P_u[1 + 5(e/t)]/f_c', \quad \text{for } e \leq 0.6t \tag{3.121}$$

Notice the similarity of this equation to Eq. (2.105). It has been assumed that the bars are concentrated on the two end faces perpendicular to bending, which is reasonable when large bending moments are present. Should the bars be placed along all four faces, then the ones on the side faces parallel to the moment action (Fig. 3.27 top center) should be ignored for this preliminary design approach. Generally, when $e/t > 0.2$, the reinforcement may be placed most efficiently on the end faces of a rectangular column. It is apparent that a round column is inefficient for moment resistance.

For the condition where the moments are very large and the axial forces are small so that tension failure controls the design of the column ($P_u < P_b$, $e > e_b$), the column may be treated as a beam for preliminary design purposes, assuming $\phi = (0.7 + 0.9)/2 = 0.8$, since the axial forces cause a reduction of the tension force or steel reinforcement.

$$M_u = \phi A_s f_y z \cong 0.8 A_s f_y (0.9d), \quad \text{or} \quad A_s \cong \frac{M_u}{0.7f_y d} \tag{3.122}$$

For primary beam action, the required moment reinforcing may be concentrated along the tension face only [see Eq.(3.51)], while on the compression face compression reinforcement will be needed when $\rho > \rho_{max}$; refer to Example 5.3 for further discussion.

In response to biaxial bending, the design of symmetrical square corner columns with bars along all four faces can be approximated for preliminary design purposes by simply adding the eccentricities $e_x + e_y$ or the moments M_{ux} and M_{uy}, rather than adding them vectorially.

As a rough first estimate of beam–column sizes, designers often use Eq. (3.114) but ignore the capacity of the reinforcement. The effects of bending and slenderness will then later be accommodated with the longitudinal bars in the final design stage.

$$A_g = P_u/0.5f_c' = 2P_u/f_c' \qquad (3.123)$$

Notice that this expression also applies to the special case of $e = t/5$ in Eq. (3.121). Should this eccentricity be exceeded but the steel ratio of 3% be kept, then the cross-sectional column area may be estimated as $(2.5 \text{ to } 3) P_u/f_c'$, depending on how large the eccentric force action is. When the working stress method is used, Eq. (3.123) can be replaced by the following expression for rough estimation purposes:

$$A_g = P/0.22f_c' \cong 5P/f_c' \qquad (3.124)$$

This equation represents a load capacity of 40% of that computed in accordance with Eq. (3.123), as required by the ACI Code.

EXAMPLE 3.38

An unbraced concrete column has an unsupported length of $l_u = 10$ ft; it carries $P_u = 640\ k$ and $M_u = 320$ ft-k. Determine the column size using a typical steel ratio of 3% assuming that secondary load effects can be covered with additional reinforcement. Use $f_c' = 4000$ psi and $f_y = 60$ ksi.

The equivalent eccentricity of the axial force is

$$e = M_u/P_u = 320(12)/640 = 6 \text{ in.}$$

As a first estimate from Eq. (3.123),

$$A_g = P_u/0.5f_c' = 640/0.5(4) = 320 \text{ in.}^2$$

which yields about 18×18 in.; use Eq. (3.121) and try $t = 18$ in.

$$A_g = P_u [1 + 5(e/t)]/f_c' = 640 [1 + 5(6/18)]/4 = 427 \text{ in.}^2$$

This yields about 18×22 in. Try $t = 22$ in.

$$A_g = 640 [1 + 5(6/22)]/4 = 378 \text{ in.}^2$$

Use an 18- $\times$ 22-in. column, $A_g = 396$ in.^{2}, $e/t = 6/22 = 0.27 < 0.6$

The approximate longitudinal steel, using conservatively $\rho_g = 3\%$, is

$$A_{st} = \rho_g A_g = 0.03(396) = 11.88 \text{ in.}^2$$

Try 10 #10, $A_s = 12.70$ in.^{2} placed on the two end faces, and #3 ties at 18 in. (see Example. 3.37).

Check the slenderness:

$$l_u/t = 10(12)/18 = 6.67 \text{ in.} \qquad \text{(see Eq. 3.111)}$$

Hence, depending on the lateral stiffness of the building, the slenderness may not be critical for this case since it is only slightly above the limiting value of 6, which, however, was based on a nearly equal column-to-beam stiffness, or rigid beam action.

3.5 DESIGN OF CONNECTIONS

In this context only one characteristic of joint performance, stress resistance, is briefly reviewed. Individual structural members must be connected so that forces can flow from one component to the other. There are many different connection types resulting from joining a multitude of member shapes and sizes. They not only depend on the material, member shapes, and degree of continuity (strength) required, but also on detail considerations (e.g., number and inclination of members), fireproofing, geometric fit, handling and erection, service conditions, cost of installation, fabrication precision, construction expertise, appearance, and so on. A joint should allow for correction in dimensional variations and should be assembled easily. The connection type influences the behavior of the joined members; its capacity should be at least equal to those of the members.

Typical structural building details for steel, wood, and precast concrete are standardized and can be obtained from publications by the respective national associations (e.g., AISC, NDS, PCI). Some common connection systems are shown in Fig. 3.28.

There are many ways of organizing connection systems; here a brief review is approached as based on the following criteria.

A. *Location of joint* (Fig. 3.28): Joints for typical skeleton or bearing wall construction appear in the horizontal floor planes and the vertical supporting planes.
   1. Collinear connections: column to column, beam to beam, wall to wall, slab to slab, etc.
   2. Planar 90° connections: beam to beam, beam to column, slab to wall, slab to beam, wall to beam, wall to foundation, etc.
   3. Other angular connections: truss connections, etc.

   In steel frame construction we may distinguish basically between beam connections, base plates, and connections for axial systems.

B. *Connection methods*
   1. Single-plane joints without connector plates: butt, scarf, finger, key, and other interlocking joints, as well as monolithic joints, such as for cast-in-place concrete and masonry (wet joints).
   2. Single-plane joints with connector plates: gusset plates, seating angles and web angles in steel construction, and joist metal hangers, framing anchors, post caps, metal straps, and steel or wood plates in wood construction.
   3. Lap joints: single and multiple laps.

   Most connection methods of larger scale in component construction employ some type of fastener system.

C. *Fastener types*
   1. Steel: rivets, bolts (unfinished and high strength), pins, fusion welding, forging, etc.
   2. Wood: nails and spikes, staples, wooden pegs, screws and lag screws, bolts, adhesives and glues, connectors (e.g., split rings, shear plates, spike grids, sheet-metal nail plates).
   3. Precast concrete: prestress cables, lapped reinforcing bars, metal studs, inserts, anchored steel shapes, etc.
   4. Masonry: mortar (mixtures of lime and/or cement with sand and water), wire ties, steel reinforcing, etc.

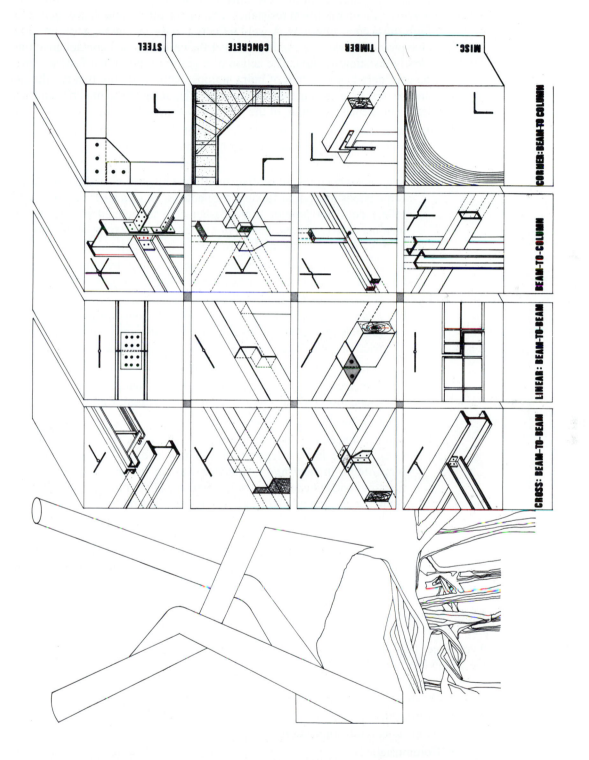

Figure 3.28 Connections.

5. Fabrics: welding, gluing, sewing, clamping, bolting, using ropes, and anchoring (e.g., sleeve expansion and adhesive chemical anchors).

From a structural point of view, only a single pin provides a true *point connection* with no rotational resistance, but, on the other hand, it also generates high stress concentrations. A weld reflects a *line connection,* while an adhesive provides a *surface connection* over the entire bonded contact area, thus clearly resisting rotation. The action of a group of individual fasteners may be considered as a system of point actions, while realizing that the density and pattern of arrangement as well as the character of the material and fasteners influence the nature of the force flow across the joint.

D. *Structural behavior:* According to the type of forces that must be transferred across the joint, we may distinguish between the following connection systems.

1. *Simple connections* allow free rotation. They are classified as sliding connections and pin or flexible connections. A *sliding connection* (e.g., roller) does not provide any resistance to movement parallel to the member, whereas a *pin connection* does not allow translational movement at all. Typical examples are the beam-to-beam cases in Fig. 3.28.

   a. *Shear connections* (or bearing type): the load transmission causes the fasteners to be in shear and bearing; it may only cause direct shear and, in addition, torsion.

   b. *Friction-type connections:* high-strength bolts are prestressed and thus clamp joining plates together so that forces are transmitted in friction, rather than by the connectors in shear and bearing.

   c. *Tension connections:* the loads cause the fasteners to be in tension.

   d. *Compression-type connections:* typical examples are column to foundation (bearing plates), column to column, wall to wall, etc.

   e. *Combined shear and tension connections.*

2. *Fixed connections,* also called rigid connections or moment connections. As the name suggests, in addition to direct forces, rotation is transmitted. In *semirigid connections,* as for low-rise frame construction, the moment capacity is only a fraction of the full restraint. Moment connections are typical for rigid frame construction (see also continuous portal knee connections in Fig. 3.28).

In general, it is reasonable to assume that the shear is transferred along the web faces of bending members, while rotation is carried along the flange faces.

Steel Connections

Typical connection systems for steel frames are identified in Fig. 3.29. They include the following:

- Beam-to-beam connections
- Beam splices
- Beam-to-column connections
- Column splices
- Beam-to-concrete wall connections in mixed construction

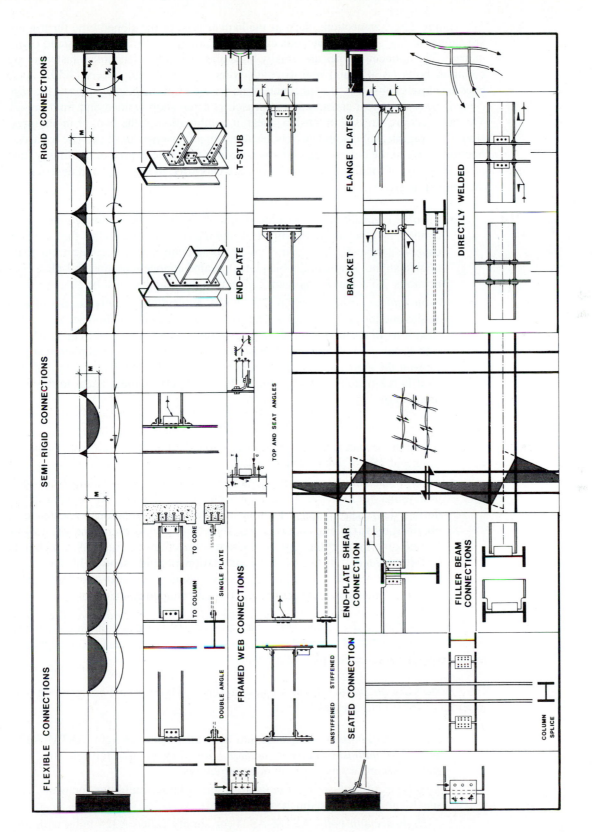

Figure 3.29 Steel connections for frames (Reproduced with permission from *The Vertical Building Structure*, Wolfgang Schueller, copyright © 1990 by Van Nostrand Reinhold.)

Some other connections, such as the hanger type or the sloped type for trusses, are shown in Fig. 3.19. Connections can be one sided (eccentric) or two sided (symmetric). The connector types are usually bolts and/or welds. Mechanical fasteners are common bolts (A307) and high-strength bolts (A325, A490); rivets (A502) have become close to obsolete. Bolted connections are either of the *bearing type* or the *slip-critical type.* A typical construction approach is often shop welded and field bolted. The fasteners are stressed in tension, shear, a combination of the two, or in compression. For example, splices and gussets usually subject fasteners to shear. The basic distinction between the connection systems is according to the degree of continuity, that is, their degree of rotational restraint. They are classified in Fig. 3.29 and described in the following list.

- *Flexible connections,* also called simple, hinged, or shear connections. These connection types are so flexible in acceptance of end rotation that they are only capable of transferring shear. Typical connections are web connections (e.g., double angles, single or double plates), seated connections, and end-plate connections. To achieve a moment connection, the flanges must be connected.

- *Semirigid or partial restraint (PR) beam connections* usually consist of web angles and top and seat angles. They do allow some end rotation because of the flexibility of the angles and thus develop partial-end moments. In this case, the web connection is assumed to resist the shear, while the top and bottom angles transfer the moment. Semirigid joints allow the redistribution of the moments, so it is reasonable to design a joint for a support moment equal to the field moment of

$$\pm M_s = (wL^2/8)/2 = wL^2/16$$

so that the section modulus for the beam is one-half of that for a simply supported beam and 75% of that for a fixed beam.

- *Rigid (FR) connections* are provided by the nondeformability of the beam–column joints; for example, a beam connected to a column at 90° remains at 90° during deformation under loading. The necessary rotational resistance is obtained by stiff flange connections; therefore, a rigid frame analysis can be carried out. Some of the typical moment connections in Fig. 3.29, which may be fully welded, fully bolted, or a combination, possibly shop welded and field bolted, are end-plate connection, T-stub connection, two-sided beam-stub bracket, flange plate connection, and directly welded connection.

In the following discussion, briefly introduced are the design of simple connections using bolts and welds, as well as the design of steel base plates for concentrically loaded columns and bearing plates for beams.

Bolted Connections.   There are several types of bolts. The most common ones in building construction are as follows:

- *Carbon steel bolts,* 60-ksi minimum ultimate tensile strength, ASTM A307, also called *unfinished bolts* or *common bolts;* they are used for light structures and for secondary members in other structures.
- Two types of *high-strength bolts* are used in larger buildings for which first specifications were established in 1951.

    High-strength bolts, ASTM A325, or simply called A325 bolts with ultimate tensile strength of 120 ksi.

    Heat-treated steel structural bolts ASTM A490, or simply called A490 bolts with ultimate tensile strength of 150 ksi.

These bolts range in diameter from 1/2 to $1\frac{1}{2}$ in. in 1/8-in. increments.

High-strength bolts, in contrast to common bolts and rivets, are tightened (pre-tensioned) to such a degree that the forces in shear connections are transmitted across the connected members by friction rather than by bolts in shear and bearing. For the case where high-strength bolts are fully tightened to their required tensions (which is equal to 70% of the minimum tensile strength of bolts), the connection is called a *friction-type* or *slip-critical connection.* When bolts are not fully tensioned, then the connections are called *bearing-type connections;* here movement (i.e., slippage) is expected. It is apparent that friction-type connections have a higher load-carrying capacity than bearing-type connections. High-strength bolts in direct tension connections must be fully tensioned so that the contact pressure between the connected material does not decrease so much and the parts separate.

In bearing-type connections, when bolt threads are included in the shear plane, the bolt capacity is less as compared to the case where the bolt threads are excluded. For ordinary conditions, however (i.e., normal bolt and member sizes), the threads are excluded from the shear plane.

The following classification of high-strength bolts, as based on the type of application, is used:

A325-SC	A490-SC	Slip-critical connection
A325-N	A490-N	Bearing-type connection with threads included in shear plane
A325-X	A490-X	Bearing-type connection with threads excluded from shear plane

In bearing-type connections, the frictional resistance due to the tensioning of the bolts is conservatively neglected. In other words, it is assumed that the members slip slightly, thereby putting the bolts in shear (usually single or double shear), and the connected material adjacent to the bolts in bearing as shown in Fig. 3.30a.

Although the shear forces in slip-critical connections are resisted by friction along the contact surface(s) (that is, since there is no slippage, bolts do not bear against the members and do not have to resist shear), the AISC Code still assumes the bolts to be stressed in shear but not in bearing. Slip-critical connections are required, for example, where loads are fluctuating, possibly causing fatigue problems, where slotted holes are used in the direction of loading or for beam–column connections of tall, rigid frame structures where high-strength bolts and welds must share the loads.

Possible failure modes of bolted (or riveted) connections are as follows:

- Shear failure of bolts, controlled by sufficient number of bolts
- Shear failure of plate, controlled by sufficient edge distance
- Bearing failure of plate, controlled by sufficient number of bolts and/or plate thickness
- Bearing failure of bolts, rarely the case, because bolts are of higher strength than members
- Tensile failure of bolts, controlled by bolt thickness or number of bolts
- Bending failure of bolts, only critical for long bolts and controlled by bolt thickness
- Tensile failure of member, controlled by sufficient cross-sectional area

Considered are generally the *shear failure of bolts* and the *bearing failure of plates,* which must be checked analytically. Minimum spacing of bolts is given by the AISC Code to prevent other types of failure. For example, for preliminary design purposes, the minimum spacing of bolts in shear connections in the line of force is taken as three diameters (3d) between bolts and 1.5d for the end spacing.

The *shear capacity* of a bolt in single shear is equal to the cross-sectional area of the shank times the allowable shear stress (Fig. 3.30a).

$$P_s = A_b F_v = (\pi d^2/4) F_v \qquad (3.125)$$

The allowable shear stress F_v on fasteners can be obtained from the AISC Code. For example, for the following common cases using standard round holes, it is equal to

$$
\begin{aligned}
&A307: && F_v = 10 \text{ ksi} \\
&A325\text{-}SC: && F_v = 17 \text{ ksi} && A490 - SC: && F_v = 21 \text{ ksi} \\
&A325\text{-}N: && F_v = 21 \text{ ksi} && A490\text{-}N: && F_v = 28 \text{ ksi} \\
&A325\text{-}X: && F_v = 30 \text{ ksi} && A490\text{-}X: && F_v = 40 \text{ ksi}
\end{aligned}
\qquad (3.126)
$$

The *bearing capacity* of a bolt is equal to the bearing area [which is taken as the projected area of the bolt, or the bolt diameter d times the thickness of the connected plate(s) t] multiplied by the allowable bearing stress of the connected parts, which is equal to $F_p = 1.2 F_u$ for standard holes.

$$P_b = A_b F_p = dt(1.2 F_u) \qquad (3.127)$$

It is assumed for bolted and riveted connections under direct force action (no eccentricity) that the load is shared equally by all the bolts. This assumption may be justified by visualizing plastic behavior rather than elastic; in other words, as bolts and plates deform plastically, forces are redistributed. Only simple shear joints, where the loads pass through the center of resistance of the bolt arrangement, are briefly investigated in the following examples. Besides the simple but asymmetrical lap joints in sin-

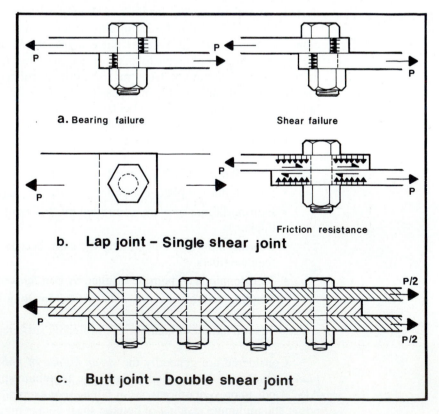

Figure 3.30   Simple bolt shear connections.

gle shear (Fig. 3.30b) and symmetrical butt joints in double shear (Fig. 3.30c) for tension members, standard beam–web connections are designed.

EXAMPLE 3.39

The A36 steel plates in Fig. 3.19b are connected with high-strength 3/4-in. diameter A325-X bolts (i.e., with threads excluded from shear plane) in standard holes. The capacity of the bolts is determined as based on a bearing-type connection and a slip-critical connection. The member capacity has already been checked in Problem 3.35 as $P = 138.62$ k.

(a) The capacity of the connectors is determined for a bearing-type connection. The allowable shear stress on a bolt is $F_v = 30$ ksi [see Eq. (3.126)]. Hence, the capacity of a 3/4-in. diameter bolt in double shear according to Eq. (3.125) is

$$P_s = 2(A_b F_v) = 2[(\pi(3/4)^2/4)30] = 2(0.442)30 = 26.52 \text{ k}$$

The bearing capacity of a bolt on the 3/4-in.-thick plate (which is thinner than the two 7/16-in. plates together), according to Eq. (3.127), is

$$P_b = A_b F_p = dt(1.2F_u) = 0.75(0.75)1.2(58) = 39.15 \text{ k}$$

Since the shear capacity of the bolt is less than the bearing capacity, the total capacity of the nine bolts is

$$P_{tot} = 9(26.52) = 238.68 \text{ k} > 138.62 \text{ k}$$

The minimum edge distance of bolts is $1.5d = 1.5(3/4) = 1.125$ in., and the spacing of the bolts should be at least $3d = 3(3/4) = 2.25$ in. for preliminary design purposes. Notice that the capacity of the bolt in double shear and bearing can also be obtained directly from the Steel Manual.

(b) The capacity of the connectors is determined for a slip-critical connection. The allowable shear stress on a bolt for a standard hole is $F_v = 17$ ksi [see Eq. (3.126)]. Hence, the capacity of one bolt in double shear according to Eq. (3.125), is

$$P_s = 2(A_b F_v) = 2[(\pi(3/4)^2/4)17] = 2(0.442)17 = 15.03 \text{ k}$$

The same result can be obtained directly from the Steel Manual. Since in a slip-critical connection the bolt is assumed not to bear against the plates, the total capacity of the nine bolts is solely based on shear and is equal to

$$P_{tot} = 9(15.03) = 135.27 \text{ k} < 138.62 \text{ k}$$

For the remote possibility that the connection could slip into bearing, the bearing capacity of 39.15 k as determined in part (a) is larger than the shear capacity of 15.03 k.

(c) For purpose of comparison, the shear capacity of a common unfinished A307 bolt is determined. The allowable shear stress on the fastener is $F_v = 10$ ksi [Eq. (3.126)].

$$P_s = 2(A_b F_v) = 2(0.442)10 = 8.84 \text{ k} < 39.15 \text{ k}$$

The allowable bearing capacity does not change. Hence, the total capacity of the bolts is

$$P_{tot} = 9(8.84) = 79.56 \text{ k}$$

EXAMPLE 3.40

The number of 3/4-in.-diameter A325 bolts for the beam-to-column connection in Example 3.3 (Fig. 3.4) using a simple double-angle web connection (see Fig.

3.29) is determined. The W18 × 35 beam (A36) with a web thickness of $t_w = 0.30$ in. must transfer a reaction force of $1.26(25/2) = 15.75$ k. A 5/16-in. thickness for the connection angles (A36) is assumed.

Beam web connections are designed for shear only. The slight rotational restraint due to the vertical spacing of fasteners is neglected. For the design of standard beam–column connections with web angles, the following criteria must be considered:

- Bearing on column flange or angle
- Single shear on column
- Bearing on beam web or angles
- Double shear on web
- Shear on the net area of the connecting angles and web tear out (block shear), which, however, are not checked in this preliminary investigation

(a) The number of A325-N bolts (i.e., with threads included in a shear plane) for a bearing-type connection is determined. First the beam–web connection is investigated. The bolts are in double shear. For an allowable shear stress of $F_v = 21$ ksi [Eq. (3.126)], the capacity of a 3/4-in-diameter bolt is

$$P_s = 2(A_b F_v) = 2(\pi(3/4)^2/4)21 = 2(0.442)21 = 18.56 \text{ k}$$

Since the web thickness of 0.3 in. is less than the total angle thickness of $2(5/16) = 0.625$ in., bearing on web controls. Hence, the maximum load for a 3/4-in. diameter bolt bearing on a 0.3-in. web is

$$P_b = A_b F_p = dt(1.2F_u) = (3/4)0.3(1.2)58 = 15.66 \text{ k} < 18.56 \text{ k}$$

The bolts connected to the columns are in single shear. Since the angle thickness is less than the column flange thickness, bearing on angles controls, which, however, is less critical than bearing on beam web. We may conclude that bearing on web controls. Therefore, the number of A325 bolts required is

$$n = P/P_b = 15.75/15.66 = 1.01 \tag{3.128a}$$

Use two 3/4-in. diameter A325-N bolts.

The capacity of the connection as based on shear on the net area of the connecting angles, however, must still be checked. The vertical spacing of bolts is estimated as

$$1.5d = 1.5(3/4) = 1.125, \quad \text{use 1.25 in. end spacing}$$
$$3d = 3(3/4) = 2.25, \text{ use 3-in. spacing between bolts}$$

Therefore, the length of the angle may be estimated as $2(1.25) + 3 = 5.5$ inches. The preceding result can also be obtained directly from the Steel Manual.

(b) The number of A325-SC bolts for a slip-critical connection is determined. For an allowable shear stress of $F_v = 17$ ksi, the shear capacity of one bolt in double shear is

$$P_s = 2(A_b F_v) = 2(0.442)17 = 15.03 \text{ k}$$

For the remote possibility that the connection could slip into bearing, the bearing capacity of 15.66 k determined previously is larger than the shear capacity of 15.03 k. Therefore, the load capacity is controlled by shear.

The number of bolts required is

$$n = P/P_s = 15.75/15.03 = 1.05 \tag{3.128b}$$

Use two 3/4-in. diameter A325-SC bolts.

Welded Connections.   In the welding process, metallic parts are joined by heating the material at the interface so that it melts and fusion can take place. Although several welding processes are available in building construction, generally arc welding is used which was developed by Sir Humphry Davy in 1801.

In *arc welding,* heating is achieved by means of an electric arc between an electrode (held manually or by an automatic machine) and the materials being joined. In other words, one terminal of an electric circuit is attached to the parts to be connected and the other terminal to an electrode. As the two terminals are brought close together, an electric arc and intense heat are formed, initiating the melting of some of the base material and the end of the electrode so that fusion can take place.

Of the major arc welding processes, the manual *shielded metal arc process* (SMAW) is most common. The *submerged arc process* (SAW) is primarily a shop-welding process performed by an automatic or semiautomatic method. It provides the deepest penetration (i.e., the depth of base material at which fusion stops), resulting in a larger load capacity of the welds.

The most common types of electrodes used with A36 steel are E60XX and E70XX, where the electrode numbering system identifies the minimum tensile strength and the following letters give some other information, such as the welding position. The minimum yield strength of the E70 electrode is 60 ksi, which is substantially higher than the A36 base material.

The most common *weld types* in building construction are the *fillet weld* and various types of *groove welds* (butt welds) that are either full penetration or partial penetration welds. Occasionally, *slot* or *plug welds* are used. Groove welds join two butting pieces directly in the same plane, as shown in Fig. 3.31b, c, and e. They are organized according to their particular shape as square, V, bevel, U, J, flare V, and flare bevel welds (see Steel Manual for further explanations). The nearly perfect fit required for groove welds can usually not be achieved on the construction site; therefore, for most connections, fillet welds are used. Fillet welds generally join two surfaces perpendicular to each other. They are controlled by shear stresses, in contrast to groove welds, which usually act in tension or compression. Welds are not only classified according to *weld types,* but also according to their position during welding (as flat, horizontal, vertical, and overhead welds), and according to the *joint types* (e.g., lap, T-, butt, corner joints), as indicated in Fig. 3.31.

The strength of a full penetration groove weld is proportional to its cross-sectional area and the strength of the filler metal, which, in turn, is stronger than the typical A36 base material. We may conclude that complete penetration groove welds are stronger than the base material when subjected to shear, tension, and compression. Therefore, the allowable stresses for groove welds can conservatively be assumed the same as for the base material.

The strength of a fillet weld is proportional to the leg size, the length, and the strength of the filler metal. When tested to failure, *equal-leg fillet welds* fail by shear at an angle of about 45° through the throat regardless of the direction of the applied load (Fig. 3.31a), although transverse fillet welds have been found substantially stronger than longitudinal welds. Based on the conservative assumption of a shear weld, the capacity of a 1-in. fillet weld for a throat length of $0.707a = 0.707(1)$, is

$$P_w = A_{th}F_v = 0.707F_v, \qquad \text{where } F_v = 0.3F_u \qquad (3.129)$$

The allowable shear stress (for an A36 base material) is a function of the tensile strength of the weld. In other words, for

E60XX electrodes: $F_v = 0.3(60) = 18$ ksi

E70XX electrodes: $F_v = 0.3(70) = 21$ ksi

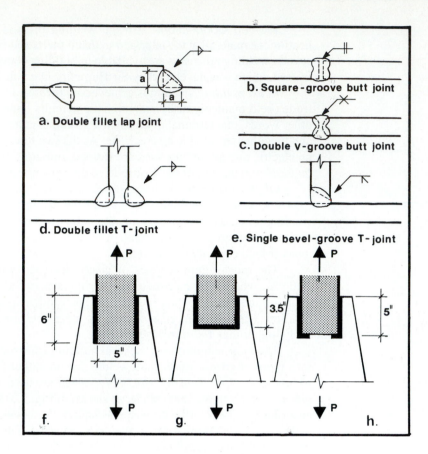

Figure 3.31   Welds.

The results indicate that the weld metal is stronger than the material it connects.

We may conclude that the capacity of a 1/16-in. fillet weld using E70XX electrodes for force action in any direction is

$$P_w = 0.707(1/16)21 = 0.928 \text{ k/in.} \tag{3.130}$$

The load-carrying capacity of any other fillet weld size for E70XX electrodes can now be easily obtained. For example, a 1/4-in. (or 4/16-in.) fillet weld has a capacity of

$$P = 0.928(4) = 3.71 \text{ k/in.}$$

Fillet welds made by the submerged arc process (SAW) have a larger load capacity than the welds made by other arc welding processes. According to the AISC specifications, the effective throat thickness is taken equal to the leg size for 3/8-in. and smaller fillet welds (see Steel Manual for other requirements).

Other code provisions related to the control of weld strength address the length and size of fillet welds.

- The *maximum size of fillet welds* for material 1/4 in. or more in thickness cannot be greater than the thickness of the material minus 1/16 in.; otherwise, it can be equal to the material thickness.
- The *minimum size of fillet welds* depends on the thicker of two parts joined but cannot exceed the thickness of the thinner part. The minimum size of fillet welds is 1/8 in. for a material thickness of 1/4 in. or less, 3/16 in. for material

thickness over 1/4 to 1/2 in. of the thicker part joined, 1/4 in. for material thickness over 1/2 to 3/4 in., and 5/16 in. for material thickness over 3/4 in.

- The *minimum effective length* l_w of fillet welds cannot be less than four times the nominal size t_w unless the weld size does not exceed one-quarter of its effective length.

$$l_w \geq 4t_w, \quad \text{or} \quad t_w \leq l_w/4$$

- The length of longitudinal fillet welds used alone in end connections of flat bar tension members should be at least equal to the perpendicular distance between them.
- The minimum length of intermittent fillet welds should be 1.5 in.
- The length of end returns around corners should be not less than twice the nominal weld size.

For other fillet weld limitations, refer to the AISC specifications or the Structural Welding Code of the American Welding Society (AWS). The reader may want to refer to the Steel Manual for the welding symbols that have been developed by AWS so that information can be efficiently presented on drawings.

EXAMPLE 3.41

A 5- × 1/2-in. flat tension bar is welded to a 3/4-in.-thick gusset plate as shown in Fig. 3.31f, g, and h. A36 steel is used and E70XX electrodes by employing the SMAW process.

For preliminary design purposes, the allowable tensile capacity of the plate according to Eq. (3.82a) is determined as

$$P = A_g F_t = A_g(0.6F_y) = 5(1/2)0.6(36) = 54 \text{ k}$$

The maximum weld size is $1/2 - 1/16 = 7/16$ in. The minimum weld size is 1/4 in. A fillet weld size of 5/16 in. is selected. The capacity of this weld according to Eq. (3.129) is

$$P_w = 0.928(5) = 4.64 \text{ k/in.}$$

The total length of the 5/16-in. fillet weld required is

$$l_w = P/P_w = 54/4.64 = 11.64 \text{ in.}$$

Use 12 in. of 5/16-in weld.

This length can be distributed as follows:

(a) Using fillet welds along each of the longitudinal plate edges (Fig. 3.31f): In other words, use $12/2 = 6$-in. welds along each edge. This length satisfies the code requirement for flat bar tension members to have a length of each fillet weld not less than the perpendicular distance between them.

(b) Using fillet welds along the longitudinal plate edges with end returns (Fig. 3.31h): The minimum end returns are equal to $2t = 2(5/16) = 0.625$ in., say 1 in. Therefore, the lengths of the longitudinal welds are $l_w = [12 - 2(1)]/2 = 5$ in.

(c) Using fillet welds along the longitudinal and transverse plate edges (Fig. 3.31g): The lengths of the longitudinal welds, as based on a 5-in. transverse weld, are $l_w = (12 - 5)/2 = 3.5$ in.

Bearing Plates.   Bearing plates are required at the junction where the strong material, steel, transmits loads to a weaker material, such as stone, concrete, or masonry. In other words, base plates for steel columns and bearing plates for steel beams are needed as transfer structures to materials of lower bearing capacity, such as concrete piers, pedestals and footings, or masonry walls. Bearing plates can be compared to shallow foundations that distribute column or wall loads to the soil, as is discussed in the next section.

Column base plates may have to transfer bending moments and axial forces (e.g., Fig. 5.8a and l) or solely axial forces (e.g., Fig. 5.8c to j) for concentrically loaded columns. Base plates may be shop welded directly to columns of smaller size (Fig. 3.32a), or angles may be used to connect the two (refer to Steel Manual for typical details). Once in place, anchor bolts tie the plates, for example, to the concrete base. To ensure proper contact and full bearing, the base plate is placed on cement grout unless the plates are straightened by pressing or milling of bearing surfaces.

In the following discussion, only base plates for columns concentrically loaded are briefly investigated. Furthermore, only the design of base plates for columns under average size loads is introduced. For this condition, the base plate acts as a two-way cantilever slab under the assumed uniform upward pressure (Fig. 3.32a). With respect to the design of base plates under light loads, where the plate is just large enough to accommodate the column profile, the reader may want to consult the Steel Manual.

The plate area A required, depends on the allowable bearing strength F_p of the material that gives support:

$$f_p = P/A \leq F_p \quad \text{or} \quad A = BN \geq P/F_p \tag{3.131}$$

Typical allowable bearing stresses are

Sandstone and limestone:	$F_p = 0.40$ ksi	(3.132)
Brick in cement mortar:	$F_p = 0.25$ ksi	
On full area of concrete support:	$F_p = 0.35 f_c'$	

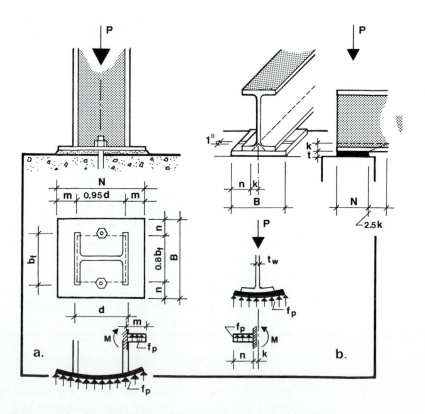

Figure 3.32   Bearing plate behavior under direct force action.

On less than full area of concrete support: $\quad F_p = 0.35 f_c' \sqrt{A_2/A_1} \leq 0.7 f_c'$

where f_c' = compressive strength of concrete (ksi)

         A_1 = area of base plate (in.^{2})

         A_2 = area of portion of supporting surface that is geometrically similar and concentric with the loaded area (in.^{2})

Notice that for the case where the full supporting concrete area is not covered by the plate the bearing strength of the concrete is larger. For this condition, the base plate area in Eq. (3.131) becomes.

$$A_1 \geq \frac{1}{A_2} \left(\frac{P}{0.35 f_c'} \right)^2 \geq \frac{P}{0.7 f_c'} \qquad (3.133)$$

Once the base plate size has been determined, its thickness is derived from bending by visualizing the cantilever portions of the plate to bend upward under the uniform pressure as shown in Fig. 3.32a. Because the behavior of the two-way action of the plate is complex, it is recommended in the Steel Manual as an approximation to design the plate as parallel, 1-in.-wide cantilever beams in perpendicular directions using the following cantilever projections (as defined in Fig. 3.32):

$$n = (B - 0.80\, b_f)/2, \qquad m = (N - 0.95d)/2 \qquad (3.134)$$

The maximum cantilever moment, for example, for the m projection using a 1-in.-wide strip is.

$$M = f_p(m)m/2 = f_p m^2/2$$

This moment causes a maximum bending stress f_b that must be less than the allowable bending stress, $F_b = 0.75 F_y$.

$$f_b = M/S \leq F_b = 0.75 F_y, \quad \text{where} \quad S = 1\,(t_p)^2/6$$

$$= \frac{f_p m^2/2}{1\,(t_p)^2/6} \leq 0.75 F_y$$

or

$$t_p \geq 2m \sqrt{f_p/F_y} \qquad \text{or} \qquad t_p \geq 2n \sqrt{f_p/F_y} \qquad (3.135)$$

where f_p = actual bearing pressure (ksi)

        F_y = yield stress of steel (ksi)

        m, n = cantilever projections [Eq.(3.134)]

        t_p = thickness of plate (in.)

The expression for the minimum plate thickness as a function of the n-projection can be derived in a similar fashion.

EXAMPLE 3.42

A W14 × 68 steel column (A36) must transmit a load of 360 k to a 22- × 22-in. concrete pier ($f_c' = 3$ ksi). Estimate the A36 steel base plate size.

     The column dimensions are $d = 14.04$ in., and $b_f = 10.035$ in. The required base plate area according to Eq. (3.133) is

$$A_1 \geq P/0.7f'_c = 360/0.7(3) = 171.43 \text{ in.}^2$$

$$\geq \frac{1}{A_2}\left(\frac{P}{0.35f'_c}\right) = \frac{1}{22^2}\left(\frac{360}{0.35(3)}\right)^2 = 242.87 \text{ in.}^2 > 171.43 \text{ in.}^2$$

As based on the column proportions, try a 14- × 18-in. plate, $A_1 = 14(18) = 252 \text{ in.}^2$

The actual bearing pressure is

$$f_p = P/A_1 = 360/252 = 1.43 \text{ ksi}$$

The m and n cantilever projections (see Fig. 3.32a) are

$$n = (B - 0.80b_f)/2 = (14 - 0.80(10.035))/2 = 2.99 \text{ in.}$$

$$m = (N - 0.95d)/2 = (18 - 0.95(14.04))/2 = 2.33 < 2.99 \text{ in.}$$

To minimize the plate thickness, m should be approximately equal to n; therefore, other plate proportions should be investigated.

Hence, the required plate thickness according to Eq. (3.135) is

$$t_p = 2n\sqrt{f_p/F_y} = 2(2.99)\sqrt{1.43/36} = 1.19 \text{ in.}$$

Try a base plate: PL $1\frac{1}{4} \times 14 \times 18$ in.

Beam bearing plates for simply supported beams are designed similarly to the base plates for concentrically loaded columns. The plates are assumed to distribute the beam reactions as uniform pressure, although this is not true in reality. Based on this assumption of an average pressure, the area of a bearing plate is found from Eq. (3.131) or (3.133). For the selection of the N dimension of the plate (Fig. 3.32b), not only the thickness of the wall must be considered but also web crippling (Fig. 3.5).

The plate can be visualized to act as a one-way cantilever slab under the upward uniform pressure (Fig. 3.32b). The location of the critical cantilever moments depends primarily on the stiffness of the beam flanges. It is surely conservative to neglect the effect of the beam flanges and to assume that the maximum moment occurs at a distance k from the beam center line, as suggested in the Steel Manual (see Fig. 3.32b). Therefore, the plate thickness can be determined from Eq. (3.135), letting the cantilever projection be equal to

$$n = (B/2) - k \tag{3.16}$$

EXAMPLE 3.43

A W18 × 35 beam is supported on a 8-in. brick wall and transmits a reaction force of 16 k. Design the beam bearing plate. The beam dimensions required for the calculations are $k = 1\frac{1}{8} = 1.125$ in. and $t_w = 0.3$ in.

The required bearing plate area according to Eq. (3.131) as based on an allowable bearing pressure of 0.25 ksi for brick [see Eq. (3.132)], is

$$A \geq P/F_p = 16/0.25 = 64 \text{ in.}^2$$

Taking into account the thickness of the wall, try a $7 \times 9\frac{3}{16}$-in. plate, $A = 64.31 \text{ in.}^2$

The actual bearing pressure is

$$f_p = P/A = 16/64.31 = 0.25 \text{ ksi}$$

Checking the plate proportions (i.e., dimension N) for web crippling, according to Eq. (3.15b), yields

$$\frac{R}{t_w(N+2.5k)} \leq 0.66 F_y$$

$$\frac{16}{0.3\,[\,7+2.5\,(1.125)\,]} = 5.44 \text{ ksi} \leq 0.66\,(36) = 23.76 \text{ ksi, OK}$$

The cantilever projection is $n = (B/2) - k = (9.19/2) - 1.125 = 3.47$ in. Hence, the required plate thickness according to Eq. (3.135) is

$$t_p = 2n\sqrt{f_p/F_y} = 2\,(3.47)\sqrt{0.25/36} = 0.58 \text{ in.}$$

Try a steel bearing plate: $5/8 \times 7 \times 9\frac{3}{16}$ in. PL.

Wood Connections

Wood connection methods include metal fasteners, timber joinery, and adhesives. Adhesives are generally applied for glulam, plywood, and I-joist manufacture in the plant. Most common are metal fasteners. For light connections, they are nails, wire staples, wood screws, and framing connectors made from sheet metal for fast assembly purposes. In heavy connections, bolts, lag screws (lag bolts), and other connectors such as split rings and shear plates are used. Typical mechanical connection hardware, besides the common tension straps and gusset plates, for example, is beam-to-column U-bracket, beam-to-girder saddle connection, beam face hanger connection, beam face clip connection, hinge connector for cantilever beam, etc.

Although metal fasteners have generally replaced traditional wood joints, timber joinery still offers a pleasing appearance with its strong sense of craftmanship. Here, adjoining members are shaped to interlock and mate by using, for example, the familiar mortise-and-tenon joint or dovetail joint, possibly inserting wooden pegs to increase the strength and rigidity.

Usually, fasteners resist loads in shear. Occasionally, screws and nails are used in tension that is, in withdrawal direction from *side grain;* withdrawal from *end grain* of wood is not permitted (Fig. 3.33b). In this context, only simple concentric force action on joints is briefly investigated.

The load-carrying capacity of a fastener depends not only on the fastener type and the species (e.g., density) and condition of wood, but also on the size and orientation of members, location of fastener related to the wood grain and to the next fastener, edge and end distances, and type of load (e.g., duration of load, eccentric load).

Generally, loads are resisted in single or double shear (Fig. 3.33a) similarly to simple steel connections. Since bolts and lag screws bear against the members that they connect, member thickness becomes a critical consideration. It is apparent that fasteners bearing parallel to the grain have a higher capacity than the ones acting perpendicular to the grain, such as the horizontal member in Fig. 3.33d. Should bolts have to transfer loads at an angle to the grain (Fig. 3.33e) by acting somewhere in the range between parallel and perpendicular to the grain, then the *Hankinson formula* (see NDS) or respective graphs must be used to determine the fastener capacity. The load-carrying capacity of most wood fasteners has been determined experimentally and is available in tables or graphs (e.g., NDS, Timber Construction Manual).

The number of fasteners n required to resist a given load P depends on the load-carrying capacity of one fastener, P_F, obtained from NDS, Timber Construction Manual, codes, etc., multiplied by the applicable adjustment factors.

$$n = P/P_F \qquad (3.137)$$

It is apparent that timber connectors have a larger capacity than nails; that is, more nails are required to resist a given load. Keep in mind, however, that special conditions of load distribution are applicable to *group action* of fasteners.

The minimum spacing of various fastener types is given in the respective references. As an example, the typical spacing of bolts is shown in Fig. 3.33d for preliminary design purposes. The *loaded edge distance* for loading perpendicular to the grain and the *end distance* for loading parallel to the grain are critical.

EXAMPLE 3.44

Two 2- × 6-in. tensile members are connected with 3/4-in.-diameter bolts. For this initial investigation, determine the number of bolts required to transfer a load of 5 k due to dead and snow loads. The wood member has already been designed in Example 3.24.

The allowable bolt loads, according to the respective NDS 1991 table for southern pine and a bolt length of 1.5 in. in single shear parallel to the grain, is $Z_{\parallel} = 800$ lb, as based on normal load duration; perpendicular to the grain, the load is $Z_{\perp} = 460$ lb. The *group action factor, C_g,* is a reduction factor that

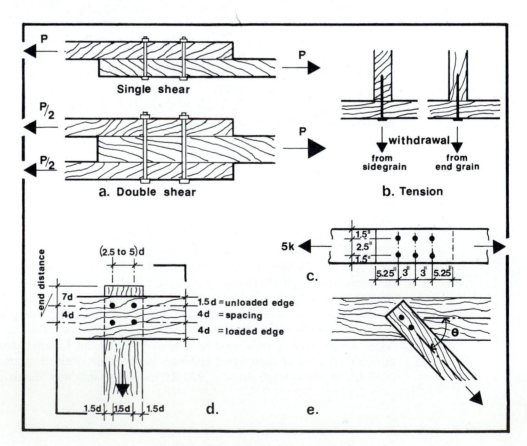

Figure 3.33   Simple wood connections.

accounts for the nonuniform loading of fasteners; it decreases as the number of fasteners in a row increases. Since usually more fasteners are selected than exactly required, the reduction due to group action (which may be in the range of 5% for this problem) should be taken care of for preliminary design purposes. The *geometry factor,* C_Δ, is equal to 1 when the spacing requirements for bolts are satisfied. The number of bolts required, taking into account two months duration of snow as well as dry service and normal temperature range, is

$$n = P/P_F = P/Z(C_D C_M C_t C_g C_\Delta) \cong P/Z C_D = 5000/1.15(800) = 5.43, \text{ use 6 bolts}$$

The bolt spacing for this parallel-to-grain loading condition is shown in Fig. 3.33c. The minimum end spacing for softwood species should be $7d = 7(\frac{3}{4}) = 5.25$ in. The minimum spacing between bolts in a row parallel-to-grain loading is $4d = 4(\frac{3}{4}) = 3$ in. The spacing across the grain should be at least $1.5d = 1.5(\frac{3}{4}) = 1.125$ in.

Special considerations must be given to *nailed joints,* which are used primarily in light frame construction. The nail load-bearing capacity is greatly dependent on wood density: the denser the wood, the more load a nail can carry. Commercial wood species are rated for hardness and divided into four groups, where group I is hardwoods and group II is the strongest softwoods (e.g., Douglas fir and southern pine). Nail load-bearing capacity is also dependent on type and size (diameter and length) of nail, penetration depth, and nail arrangement. The typically used *common steel wire nails (common nails)* are stronger, for example, than *box nails.* Typical sizes of common nails range from eight penny (8d) with 2.5-in. length and a 0.131-in. diameter to 20d nails 4 in. long and 0.192 in. in diameter. As a rule-of-thumb, nails are spaced at penetration length to achieve their maximum capacity.

Reinforced Concrete Connections

The joining of members in cast-in-place concrete structures is, by their very nature, continuous; monolithic beam-to-column joints are briefly discussed later (see Fig. 5.16). The joints for precast concrete frames are usually hinged; moment connections in the field are expensive. Continuous connections can be achieved by posttensioning the precast elements together or by using precast shell units with cast-in-place cores, where the shells are the formwork for the wet concrete. Typical soft or pin beam-to-column connections for multistory precast concrete construction are the following:

- Haunches projecting from the columns, possibly using elastomeric bearing pads
- Embedded steel shapes (e.g., vertical plates, W-sections), projecting from the column
- Seated connections using horizontal plates or angles

Precast concrete joints may be organized according to the forces that they transfer as compression, tension, and shear joints as discussed previously. They are either of the dry or wet type. The dry or cold joints, reinforced with mild steel, usually form hinged joints for member splices. Wet splices, possibly with interlocking hooked bars may, however, provide moment resistance.

3.6 DESIGN OF SIMPLE REINFORCED CONCRETE COLUMN AND WALL FOOTINGS

Foundations are usually made of concrete and are necessary to transfer forces from the building structure to the ground if the building is not directly founded on solid rock. The capacity of the soil is generally much less than that of the structural materials; thus, at the junction where columns and walls meet the soil, transitional structures with a wider base are generally required. The resulting spread footings have a similar purpose as base plates for columns, although they must, in addition, limit settlement, particularly differential settlements of the various vertical support elements.

Foundation systems and properties of soil were introduced in Section 2.9 and are not further treated here. In this context only the design of ordinary, single, reinforced concrete column and wall footings is presented.

Typical shallow foundations for multispan frames are shown in Fig. 3.34. While in the top case only one large footing stabilizes the entire vertical building plane laterally, in the second case the footings of the central rigid portal unit do so. These footings should be completely restrained from rotation so that the moments can be transmitted to the ground in a controlled manner, which in turn requires reliable soil conditions. Often the soil is poor, and the more elaborate foundations required are not economically feasible; therefore, it may be advantageous to have hinged connections between column base and footing. Furthermore, it is apparent that eccentrically loaded

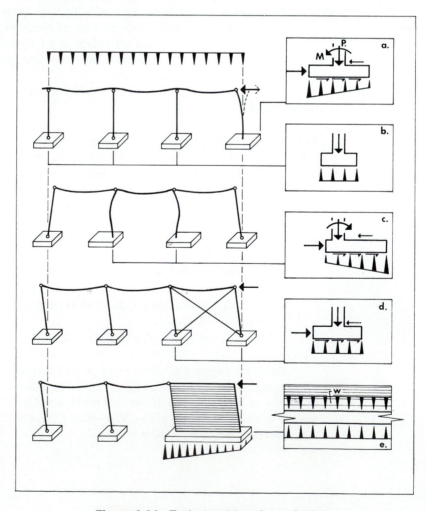

Figure 3.34 Typical multibay frame footings.

footings are of much larger size than the centrally loaded ones. Pin-connected conditions can be obtained, for instance, by bracing one bay, as indicated in Fig. 3.34 by the last two cases, or by stabilizing adjacent bays. In this way the footings carry primarily axial forces, if the relatively small rotational effect due to the lateral base forces is neglected.

In the following preliminary design of typical spread footings, only centrally loaded wall footings and single column footings are investigated. It is the intention here to quickly proportion plain and reinforced wall and column concrete footings under concentric loads for moderate height buildings. As has been discussed in the section on foundations (Fig. 2.33), the contact soil pressure distribution is assumed uniform, as being based on rigid foundation and truly elastic soil behavior.

The base size of a column or wall footing depends on the allowable bearing capacity of the soil q_a (Table 2.6). When factored loads are used, a load factor of 1.6 may be conservatively assumed to increase the allowable soil pressure. The allowable bearing pressure q_a should obviously be equal to or greater than the actual contact pressure q.

$$q = P/A_f + w_t \le q_a \qquad \text{(a)}$$

Here P represents the column or wall load and w_t the foundation weight as well as the surcharge, possibly consisting of the soil weight on top of the foundation and surface loads. Shear and bending in the footing, however, are only caused by P, which must be redistributed by the footing to a uniform load; the uniform downward pressure due to w_t does not cause internal forces in the footing. Therefore, it is convenient to work with the *net soil pressure* q_{an} concept rather than the *gross soil pressure q*.

$$q_{an} = q_a - w_t \ge P/A_f \qquad \text{(b)}$$

Hence, the required base area A_f of a footing, taking into account typical loading conditions, is

$$A_f \ge P_{D+L}/q_{an} \quad \text{or} \quad A_f \ge 0.75\, P_{D+L+W}/q_{an} \qquad \text{(3.138a)}$$

Should factored loads be used, then the solution can be approximated as

$$A_f \ge P_u/1.6 q_{an} \qquad \text{(3.138b)}$$

where typical factored column or wall loads, according to Eq. (4.19), are

$$P_u = 1.4D + 1.7L, \quad \text{or} \quad P_u = 0.75(1.4D + 1.7L + 1.7W) \qquad \text{(3.139)}$$

The thickness of reinforced footings is usually controlled by shear, since shear reinforcement is seldom used. The footings of ordinary, laterally braced, moderate-height buildings, however, contain a low percent of steel reinforcement (i.e., are only lightly reinforced). Therefore, it is reasonable to select the minimum reinforcement, in which case flexural considerations generally determine the footing depth rather than shear.

The footings are treated as inverted slabs where the wall or column loads balance the uniform soil pressure. This soil pressure causes upward bending of the cantilever slabs and tension at the bottom, therefore requiring reinforcement at the bottom of the footings (Fig. 3.35). The necessary reinforcement to resist the critical cantilever moments is determined from Eq. (3.51).

The following expressions for quickly finding the footing depth are based on the typical foundation material of $f'_c = 3000$-psi concrete and Grade 60 steel. The minimum reinforcement ratio, letting $d/t = 0.9$, is

$$\rho_{min} = 0.0018 = A_s/bt, \quad \text{or} \quad A_s = 0.002bd \qquad (3.69)$$

Substituting the given conditions into Eq. (3.53) yields

$$M_u = \phi A_s f_y z \cong 0.9 A_s f_y (0.95d)$$
$$= 0.9(0.002bd)60(0.95d) = 0.103bd^2, \quad \text{let } b = 12 \text{ in.}$$
$$M_u = 1.23d^2 \text{ (k-in.)}$$
$$M_u = 0.103d^2 \text{ (k-ft)} \quad \text{or} \quad d^2 = 9.709 M_u \qquad (c)$$

The uniform soil pressure $q_s = P_u/A_f$ causes a maximum moment at the following critical location measured by the distance c from the edge of the footing (Fig. 3.35):

- At the face of columns, pedestals, or walls for footings supporting concrete columns, pedestals, or concrete walls
- Halfway between the middle and the edge of a masonry wall for footings supporting masonry walls
- Halfway between the face of a column and the edge of a steel base plate for footings supporting columns with steel base plates

Although shear generally does not have to be considered as based on the previous assumptions, the reader should still be reminded of some basic considerations. The location of the critical section for *one-way beam shear* is at the distance d from the critical location of the bending moment (see Fig. 3.35c). In contrast to the one-way slab action of the wall footing, the flat plate action of the single-column foundation is usually controlled by *two-way punching shear* rather than beam shear. Here the footing's shear capacity is provided along a perimeter concrete area measured at a distance of $d/2$ from the concrete column faces. The punching shear stresses along the given critical perimeter area can only be considered nominal, as actual shear failure would not occur along that surface, but along the surface of a truncated pyramid, as shown in Fig. 3.35b for the special condition of a square column.

Now the process is continued to determine the footing depth as based on flexural considerations by using the minimum amount of reinforcement. The maximum factored cantilever moment, according to Fig. 3.35c, is

$$M_u = q_s(c)c/2 = q_s c^2/2 \qquad (d)$$

Substituting Eq. (d) into (c) yields

$$d^2 = 9.709\, q_s c^2/2 \quad \text{or} \quad d = 2.2c\sqrt{q_s} = 2.2c\sqrt{P_u/A_f} \qquad (e)$$

The footing thickness t is obtained by adding to the depth d from the compression face to the centroid of the steel the minimum concrete cover of 3 in. and one-half bar diameter for a single-layer wall footing or one bar diameter for a double-layer single column footing.

$$t \geq 2.2c\sqrt{q_s} + d' \geq 10 \text{ in.} \qquad (3.140)$$

where t = overall thickness of footing (in.)

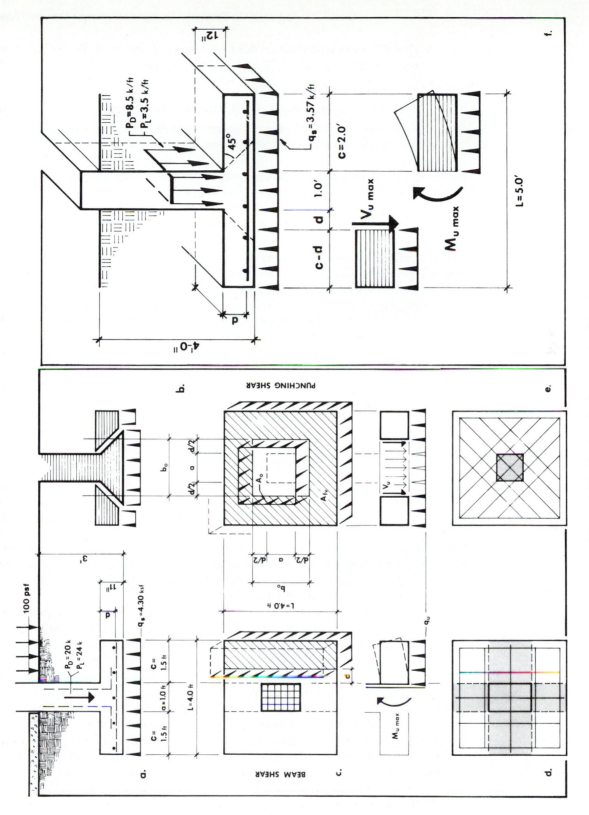

Figure 3.35 Design of column and wall footings.

$q_s = P_u/A_f =$ the soil reaction due to the factored loading (ksf)

P_u = factored concentric column or wall load (k)

A_f = base area of footing (ft^2)

$d' \cong 3.5$ in. for wall footings or 4.0 in. for column footings

c = distance from the edge of the footing to the critical location of the maximum moment as defined previously (ft)

Notice that the expression is in mixed units. Once the footing depth has been determined, the steel reinforcement can easily be found from Eq. (3.69).

$$A_s = \rho_{min} \, bt = 0.0018bt \tag{3.141}$$

Note that bar sizes must be small enough to be fully developed by anchorage between maximum moment location (e.g., face of concrete column) and outer end of bar; therefore, this development length of the steel must be checked.

It may be more economical to use plain concrete footings for light loading conditions rather than providing moment reinforcement. The depth of plain concrete footings is controlled by the flexural strength of the concrete. For a permissible flexural tension stress of $F_t = 5\sqrt{f'_c}$ and $\phi = 0.65$, the required footing depth is

$$M_u \le \phi M_o = \phi \, SF_t = \phi \, S(5)\sqrt{f'_c}$$

or

$$q_s \, c^2/2 \le (0.65)\,(1(d)^2/6)5\sqrt{3000} \quad \text{or} \quad d \ge 4.1c\sqrt{q_s} \tag{f}$$

Since footings are cast against soil, 2 in. of concrete should be added to the thickness d required for strength to allow for unevenness of excavation and contamination of the concrete in contact with the soil.

$$t \ge 4.1c\sqrt{q_s} + 2 \ge 8 \text{ in.} \tag{3.142}$$

For the definition of the terms, refer to Eq. (3.140). Furthermore, notice that the thickness of plain footings is not quite twice that of reinforced footings with minimum reinforcement! It may still be necessary to provide minimum reinforcing parallel to the wall to control differential ground movement.

At the intersection of concrete column and footing, the axial compression is transmitted in bearing across the joint. Here the allowable bearing on the footing as well as the allowable bearing on the base of the column must be considered. For ordinary conditions, that is, when the column strength does not exceed twice the footing strength, the column bearing strength controls the design of ordinary square footings. When the concrete is adequate to transmit the forces, a minimum area of dowels equal to at least 0.005 times the column area A_g is required using at least four bars (of the same size or smaller than the column bars) for rectangular columns. For the required compression development length of the dowels into column and footing, refer to the ACI Code. Also keep in mind that dowels may also have to transfer horizontal forces and tension; however, these loading conditions are not investigated here.

EXAMPLE 3.45

Design a concrete footing for a 12-in. exterior concrete wall of a one-story building. The footing is placed at a depth of 4 ft below the ground in order to avoid movement of the soil due either to heaving (uneven rise of the ground surface) or thawing of the frozen soil. The wall loads are shown in Fig. 3.35f. The allowable soil-bearing capacity at the footing base is $q_a = 3$ ksf. Consider the weight of concrete as 150 pcf and the weight of the earth on top of the footing (backfill material) to be 100 pcf.

(a) Design a reinforced concrete footing using Grade 60 steel and 3000-psi concrete.

Rather than estimating the thickness of the footing, which is in the range of one to one and a half times the wall thickness, an average weight of soil and concrete above the footing base of 130 pcf is assumed. Therefore, the net permissible soil pressure is

$$q_{an} = 3 - 4(0.130) = 2.48 \text{ ksf}$$

Or the required base area of the footing is

$$A_f = \frac{P_{D+L}}{q_{an}} = \frac{8.5 + 3.5}{2.48} = 4.84 \text{ ft}^2/\text{ft of length} \qquad (3.138a)$$

Try a 5-ft footing width. The soil reaction to the factored loading is

$$q_s = \frac{P_u}{A_f} = \frac{1.4\,(8.5) + 1.7\,(3.5)}{5.0\,(1)} = 3.57 \text{ ksf}$$

The distance of the cantilever portion from the edge of the footing to the face of the concrete wall is

$$c = (5 - 1)/2 = 2 \text{ ft}$$

The required footing thickness according to Eq. (3.140) is

$$t \geq 2.2c\sqrt{q_s} + 3.5 = 2.2\,(2.00)\sqrt{3.57} + 3.5 = 11.81 \text{ in} \geq 10 \text{ in.}$$

Try a 12-in.-thick footing. Checking the assumed average weight of soil and concrete above the footing base yields

$$w_t = [0.150(12/12) + 0.100(36/12)]/4 = 0.113 \text{ pcf} < 0.130 \text{ pcf}, \quad \text{OK}$$

The required footing reinforcement in the main transverse direction as well as the temperature reinforcement on top, in the longitudinal direction, is

$$A_s = \rho_{min}bt = 0.0018(12)12 = 0.259 \text{ in.}^2/\text{ft} \qquad (3.141)$$
$$s/0.2 = 12/0.259 \quad \text{or} \quad s = 9.27 \text{ in.}$$

Try #4 at 9 in. o.c., $A_s = 0.267$ in.^{2} (Table A.1).

The development length of the transverse bars must be checked. The longitudinal temperature steel also should assist in controlling any differential movement (i.e., possible cracking) in continuous footings.

(b) Design a plain concrete footing.

It is assumed that the base area of the plain footing is the same as for the reinforced footing. In this case, the required footing thickness according to Eq. (3.142) is

$$t \geq 4.1c\sqrt{q_s} + 2 = 4.1\,(2.00)\sqrt{3.57} + 2 = 17.49 \text{ in} \geq 8 \text{ in.}$$

Try an 18-in.-thick footing.

Checking the assumed average weight of soil and concrete above the footing base yields

$$w_t = [0.150(18/12) + 0.100(30/12)]/4 = 0.119 \text{ pcf} < 0.130 \text{ pcf}, \quad \text{OK}$$

Some minimum amount of steel reinforcing may still be needed in the longitudinal direction parallel to the wall to control possible differential settlement along the continuous footing.

EXAMPLE 3.46

The reinforced concrete footing for the steel column on a 12- × 18-in. concrete pedestal in Example 5.2 is to be designed. The column is hinged to the foundation and thus transfers only the axial forces $P_D = 20$ k and $P_L = 24$ k (Fig. 3.35a). For the given loading condition, the lateral thrust is balanced by the tie rod connecting the column bases of the frame and does not influence this preliminary foundation design. The exterior curtain walls and interior walls are assumed to be carried by independent foundations. The footing base is located 3 ft below the floor slab. The soil weight is 120 pcf, and the surcharge loading is 100 psf which is assumed to include the weight of the concrete slab. Use an allowable soil-bearing pressure of 3500 psf.

(a) Design a reinforced-concrete square spread footing using Grade 60 steel and 3000-psi concrete.

Rather than estimating the footing thickness, which is in the range of one to two times the column width, an average weight of soil and concrete above the footing base of 130 pcf is assumed. Hence, the net permissible soil pressure is

$$q_{an} = 3.50 - 3(0.130) - 0.100 = 3.01 \text{ ksf}$$

Or the required base area of the footing is

$$A_f = \frac{P_{D+L}}{q_{an}} = \frac{20 + 24}{3.01} = 14.62 \text{ ft}^2 = 3.82^2 \text{ ft}^2 \qquad (3.138a)$$

Try a 4- × 4-ft square footing.

To simplify the complex behavior of the flat slab type of foundation, visualize the slab to be cut into radial segments (Fig. 3.35e) so as to form cantilever beams projecting from the central located column. Based on this assumption, the main reinforcement could theoretically be placed in a similar radial manner. However, the design of single footings has been simplified by the ACI Code, which treats such foundations as two intersecting wall footings placed perpendicular to each other (Fig. 3.35d). In other words, the behavior of a single-column footing may be visualized as an inverted, isolated flat plate uniformly loaded and supported by a central column, where the entire load is carried in one-way action in both directions, thus requiring two identical layers of reinforcement (for a square footing) placed perpendicular to each other at the bottom.

The soil reaction to the factored loading is

$$q_s = \frac{P_u}{A_f} = \frac{1.4\,(20) + 1.7\,(24)}{4\,(4)} = \frac{68.8}{16} = 4.30 \text{ ksf}$$

The maximum length of the cantilever portion of the footing from the face of the pedestal (measured in this case along the critical short side) is

$$c = (4 - 12/12)/2 = 1.5 \text{ ft}$$

The required thickness of the footing, according to Eq. (3.140), is

$$t \geq 2.2c\sqrt{q_s} + 4 = 2.2\,(1.5)\,\sqrt{4.30} + 4 = 10.84 \text{ in} \geq 10 \text{ in.}$$

Select an 11-in.-thick footing. The required footing reinforcement, according to Eq. (3.141), is

$$A_s = \rho_{\min}\, bt = 0.0018(4 \times 12)11 = 0.95 \text{ in.}^2$$

Try 5 #4 bars $\times$ 3 ft-6 in., $A_s = 1.0$ in.^{2}, each way, or #4 at 10.5 in. o.c., each way. The bar spacing is less than the maximum of 18 in. [and $< 3t = 3(11) = 33$ in.]. The tension development length of the bars must be checked. The minimum area of column dowels required is

$$A_{sd} = 0.005\, A_g = 0.005\,(12 \times 18) = 1.08 \text{ in.}^2$$

In other words, 4 #5 dowels, $A_{sd} = 1.24$ in.^{2} (Table A.1) are required. Determine whether additional dowels must be provided to carry excess compression.

Check the required compression development length of the dowels into column and footing.

(b) Because of the light loading, design a plain concrete square spread footing using 3000-psi concrete.

The footing size remains the same as in part (a). The footing thickness according to Eq. (3.142) is

$$t \geq 4.1\,c\sqrt{q_s} + 2 = 4.1(1.5)\sqrt{4.30} + 2 = 14.75 \text{ in.} \geq 8 \text{ in.}$$

Select a 15-in.-thick footing and 4 #5 dowels; see calculations in part (a).

PROBLEMS

3.1   For the typical floor or roof framing systems of Fig. 3.3, with the exception of cases (g) to (j), (l), and (r) to (v), determine the maximum moment of the primary beams (not filler beams); consider the beams to be simply supported. Use a uniform live and dead load of 100 psf, which includes the weight of the beams.

3.2   For the framing plans of cases (g) to (j), (l), and (r) to (v) of Fig. 3.3, determine the size of the simply supported beams as indicated. Assume a floor and stair loading of 100 psf, which includes the beam weight and live load reduction; consider the stair span equal to the length of the stairwell opening. Use A36 steel. For this preliminary design, bending is assumed to control and the compression flange is considered to be laterally supported by the floor structure.

3.3   Determine the placement of the filler beams for a typical structure bay (Fig. 3.3d, e, f, etc.) to yield the most economical section for the supporting beam; that is, find the smallest possible moment in the beam by comparing the following filler beam spacings: 2, 3, 4, 5 equal spacings and 1 ft on center.

3.4   Determine the most economical open-web steel joist to span the typical interior bay (Fig. 3.3x) using the standard K series. Assume a dead load of 50 psf, a live load of 80 psf, and an allowable live load deflection $L/360$. Neglect the change in dead load as the joist spacing changes.

3.5   Compare the floor structure in Fig. 3.3w, where the joists are spaced at 4 ft and span the long direction, with the case where the joists span the short direction and are also 4 ft apart. Use the same gravity loading as for Problem 3.4. Which of the two framing systems is the lighter one? Select the open-web steel joists and W-beam sizes, ignoring the effect of live load reduction.

3.6   Determine the preliminary size of a continuous beam with two equal spans of 25 ft carrying 1.5 k/ft. Use A36 steel and assume the beam is laterally supported. The beam deflection should not exceed 1/240 of the span.

3.7   Design the 20-ft-span roof purlins (A36) that are spaced at 10 ft for a single-story industrial building as based on bending. Assume a dead load (including purlin weight) of 12 psf and a live load from snow of 20 psf. The purlins are fully laterally supported by the roof skin.

3.8   Design the beams and girders for the typical interior floor framing bay in Fig. 3.3e using A36 steel. Assume that bending controls this preliminary investigation. The beams support a 4-in. reinforced-concrete slab (reinforced concrete weighs 150 pcf). Use a super-imposed dead load of 10 psf and a live load of 60 psf. The compression flanges of the beams are laterally supported by the concrete slab!

3.9   Select the most economical sections for the beams B3, B4, B5, and B7 as well as girder G1 of the floor framing in Fig. 2.13 assuming full lateral support of the compression flanges. Use A 36 steel. The dead load (including beam weights) is 60 psf and the live load is 40 psf; ignore live load reduction. Assume that bending controls this preliminary design of the beams.

3.10 Check the girder in Example 3.3 for a maximum deflection of $L/240$, shear stresses, and web crippling at the support.

3.11 Determine the allowable bending stresses for a W30 × 132 for simple spans of 10, 15, and 30 ft if lateral support is provided at the beam ends only. Use A36 steel.

3.12 A 15-ft cantilever beam supports a single load of 10 k at its end; the beam is laterally supported only at the ends. Design the beam using A36 steel.

3.13 A 6-ft span, simply supported W24 × 146 section (A36) carries a 300-k load at 2 ft from the left support. Check the beam for bending, shear, and web crippling if the length of bearing of the load is 12 in. and at the support is 8 in.

3.14 Design a 26-ft-span, simply supported steel beam (A36) that supports a uniform load of 4 k/ft. The beam deflection is limited to 1/1200 of the span.

3.15 The floor joists in Example 3.9 must carry in addition an 8-ft high partition wall that weighs 20 psf and runs perpendicularly to the joists at midspan. Determine if the joists are still OK.

3.16 Are the joists in Example 3.9 still satisfactory if they have to cantilever 5 ft to provide a balcony that must support a live load of 60 psf and a snow load of 30 psf. The dead load may be reduced to 5 psf.

3.17 Roof joists are spaced 16 in. apart and span 22 ft. They must support a dead load of 15 psf and a snow load of 30 psf. The allowable deflection is $L/240$ for the live load and $L/180$ for the total load. Determine the joist size using southern pine with $F_b = 1400$ psi, $F_v = 90$ psi, and $E = 1,600,000$ psi.

3.18 Change the conditions in Example 3.10 to 80-ft- span beams spaced 14 ft, carrying a live load of 50 psf. Assume the roof is for a commercial building without a plaster ceiling, with an allowable live load deflection of $L/240$ and an allowable total load deflection of $L/180$. All other conditions in Example 3.10 remain the same.

3.19 Determine the allowable uniformly distributed load (that includes snow loading) for a 19-ft span southern pine 6- × 12-in. roof beam. The allowable deflection for total loading is $L/240$. Find also the required bearing area at each support of the beam. The allowable design values are $F_b = 1100$ psi, $F_v = 95$ psi, $E = 1,400,000$ psi, and $F_{c\perp} = 375$ psi.

3.20 A flitch beam of 12-ft span consists of a 6- × 12-in. wood beam of Douglas fir–larch reinforced with a 1/4-in. × 11-½ in. steel (A36) side plates. The design values for the wood are $F_b = 1300$ psi and $E = 1,600,000$ psi. Determine the allowable uniformly distributed load for the beam.

3.21 Derive the equations that approximate the concrete beam size and the moment and shear reinforcement.

3.22 Determine the moment capacity of a 12- × 20-in. rectangular reinforced-concrete beam with 4 #9 bars in one row, and $f_c' = 4000$ psi and: $f_y = 60,000$ psi. First use the accurate method of analysis and then the approximation presented in section 3.2.

3.23 A 14- × 22-in. reinforced concrete beam with $f_c' = 4000$ psi and $f_y = 40,000$ psi must resist a design moment $M_u = 134$ ft-k. First determine the approximate amount of steel required to resist the moment, and then find the precise amount.

3.24 A simply supported reinforced concrete beam spans 20 ft and must support uniform service loads of 1.5-k/ft dead load and 3.5-k/ft live load. Use a concrete strength of $f_c' = 4000$ psi and Grade 60 steel. Determine the beam cross section using a beam width of 14 in. so that deflection is not critical, and find the required reinforcing.

3.25 Design the beam in Problem 3.24 approximately by using the working stress method.

3.26   A 20-ft-span, simply supported rectangular reinforced concrete beam must support a uniform service dead load of 0.8 k/ft and a concentrated service live load of 15 k at midspan. Use $f_c' = 4000$ psi and $f_y = 60,000$ psi. Design the beam using $d \cong 2b$ and $\rho \cong 1\%$.

3.27   Determine the required dimensions and reinforcing steel for a continuous reinforced-concrete beam that must resist a critical support moment of $M_u = 145$ ft-k and a maximum field moment of 105 ft-k. Use $\rho \cong 1\%$, $d \cong 1.75b$, $f_y = 60$ ksi, and $f_c' = 4$ ksi.

3.28   Determine the moment capacity of a simply supported reinforced-concrete beam with 3 #11 bars that forms a T-beam with an effective flange width of 30 in. and that has a web width of 14 in., the slab thickness is 4 in. and the effective depth to the steel is 20 in. The concrete strength is $f_c' = 4000$ psi and $f_y = 60,000$ psi. Compare the approximate solution with the accurate one.

3.29   A concrete slab is supported by a rectangular beam that must resist a maximum shear force of $V_u = 30$ k. Use $f_c' = 4000$ psi and $f_y = 40,000$ psi. Assume the ratio of beam width to beam depth as about 1/2. Find the beam size b/t and the stirrup spacing, if required, for the following cases: (a) no shear reinforcing, (b) minimum shear reinforcing, (c) absolute smallest section as based on $d/2$ maximum spacing.

3.30   A continuous concrete girder ($b/d = 14/24$) with a clear span $l_n = 18$ ft supports floor beams at the third points. These beams each cause concentrated loads of $P_u = 80$ k. For preliminary design purposes, the weight of the girder stem is already included in the single loads. Use $f_c' = 4000$ psi and $f_y = 60,000$ psi. Determine the spacing of the #3 stirrups.

3.31   A continuous beam ($b/d = 12/24$) with a clear span of 21 ft supports a uniform ultimate floor load of $w_u = 8$ k/ft. Determine the #3 stirrup spacing for $f_c' = 3000$ psi and $f_y = 60,000$ psi.

3.32   Design approximately the typical interior 12-ft span of a continuous one-way concrete slab supported by 10-in.-wide floor beams. It carries a live load of 80 psf besides its own weight. Use $f_c' = 4000$ psi and $f_y = 40$ ksi. Select the slab thickness so that no deflection check is required.

3.33   Determine the preliminary size of a continuous 20-ft concrete slab. Because of the very large span, use lightweight concrete (110 pcf) of $f_c' = 4000$ psi and $f_y = 40,000$ psi. For this concrete the required depth, as based on no deflection check for normal-weight concrete, must be increased by 10%. What will be the maximum steel ratio if the slab must support a live load of 80 psf and a ceiling load of 5 psf? Assume the beam stems to be 1 ft wide.

3.34   Investigate the exterior span of a continuous one-way concrete warehouse slab that rests on 12-in. wide-beams that are 12 ft apart; the slab is supported along the facade by spandrel beams. Assume a dead load of 25 psf in addition to the slab weight and a live load of 200 psf. Allow an extra 1/2 in. of slab thickness to serve as a wearing surface. Use $f_c' = 4000$ psi and $f_y = 40,000$ psi.

3.35   Determine the tensile capacity of the lapped plate joint (A36) in Fig. 3.19b with three lines of 3/4-in. diameter bolts. Assume that the bolt capacity is larger than the member capacity.

3.36   Determine the tensile capacity of the channel (C 12×25, A36 steel) in Fig. 3.19f for the 7/8-in.-diameter bolt arrangement shown.

3.37   A pair of A36 bottom chord truss angles ($L6 \times 3\frac{1}{2} \times t$) in Fig. 3.19e are connected to a gusset plate with 3/4-in.-diameter bolts and must resist a tensile force of 140 k. Determine the angle thickness for a member length of 12 ft.

3.38   Design the simple 9-ft-long diagonal truss angle (A36) in Fig. 3.19c. The angle is connected to a gusset plate with three 7/8-in. bolts and must carry a force of 35 k.

3.39   Select a threaded round tie rod (A36) at the base of an arch to resist a thrust force of 60 k.

3.40   Approximate the allowable stress formula for A36 intermediate-long steel columns by assuming a straight-line relationship between allowable stress and slenderness. Use the range from $Kl/r = 30$ to 120.

3.41   Show that the bending factor B_x for a W-section is roughly equal to $2.5/t$ and explain the meaning of the bending factor.

3.42   Estimate the size of an interior, 11-ft-long W12 column (A36) that carries a concentric axial load of 600 k. The column is braced about its weak axis but is free to sway about its strong axis.

3.43   Estimate the size of a 12-ft W14 column (A36) that supports 1000 k and a wind moment of 275 ft-k. Consider the column to be braced about both axes.

3.44   Estimate the size of a corner column in a skeleton building that is braced in both directions. The column has an unbraced length of 12 ft with respect to both axes and is subject to the following loads: $P = 500$ k, $M_x = 60$ ft-k, and $M_y = 10$ ft-k. Use A36 steel.

3.45   A W8 × 67 (A36) is used as a warehouse column. The bottom clip angle connection is equivalent to pin connection in each direction. Deep plate girders frame into the web, thereby fixing the weak axis of the column at the top; small bracing beams are clipped to the flanges and cause pinned condition about the strong axis (Fig. 3.24a). Determine the capacity of the column; assume that the column does not sway.

3.46   A tubular steel column is 20-ft long and has to support an axial load of 50 k (Fig. 3.24f). The column can be considered hinged at the lower end and fixed by the deep trusses at the top. The building does not sway. Select a standard steel pipe column (A36).

3.47   Design a rectangular structural tubular column to carry an axial load of 40 k (Fig. 3.24b). The column is 20-ft long; it is pin connected at its ends and fully braced by a masonry curtain wall about its weak axis. Assume that the column does not sway; $F_y = 46$ ksi.

3.48   Design the A572 (Grade 50) column in a braced building (Fig. 3.24c). The 30-ft-long column is braced about its weak axis by a spandrel beam located 14 ft down from the top. Treat the laterally braced column as pinned about its weak axis and about its strong axis fixed at bottom and pinned at top. The column must support a total load of 650 k.

3.49   Determine the thickness of a 16-ft-long, braced, 8 × 6 double-angle member (long legs 3/8 in. back to back of angles) to carry a load of 300 k. Use A36 steel and assume $K_x = K_y = 1.0$.

3.50   Design an A36, W10 column that is 16 ft long and has pinned ends. The column must carry the loads shown in Fig. 3.26b. Assume no sidesway about both column axes. Check if the solution is adequate.

3.51   Is the laterally braced W12 × 79 (A36) adequate for the given conditions in Fig. 3.26c if the boundaries about both axes are identical? The 13-ft column is pinned at the top and fixed at the base; it bends about the strong axis.

3.52   Check whether a W12 × 40 (A36) rigid frame column (Fig. 3.26d) is satisfactory for the given conditions. The 13-ft column about its strong axis is treated as fixed at top and bottom and sways laterally; about its weak axis, the column does not sway and is considered hinged at top and fixed at bottom.

3.53   Determine the approximate size for a W14 (A36) column that is 20 ft long and has pinned ends about both axes and has to resist the loads given in Fig. 3.26e. Notice that the eccentric load causes bending about the weak axis of the laterally braced column.

3.54   A 14-ft-long column has girders framing into it at top and bottom with moment resisting connections (Fig. 3.26f). It must carry an axial load of 600 k, including the girder and beam reactions at the top. Live load imbalance causes a potential maximum moment at the top and bottom of 80 ft-k as shown in the load diagram. The k value for the weak axis is 1.0 (i.e., hinged ends and no lateral sway), and the k value for the strong axis is estimated as 1.5 (i.e., some end fixity and lateral sway). Determine the approximate size of a W12 section. Use A572 (Grade 50) steel.

3.55   Determine the column size for an axial load of 8000 lb (normal duration). The rectangular column is 20 ft long and braced about its weak axis at midheight (Fig. 3.24e). Use Douglas fir–larch with $F_c = 1250$ psi and $E = 1,800,000$ psi.

3.56   An 8-ft high stud wall carries a load of 2 k/ft (normal duration). What is the spacing of 2 × 4s (i.e., 12, 16, or 24 in.)? The wall is finished on both sides (Fig. 3.24d). Use southern pine with $F_c = 975$ psi, $F_{c\perp} = 565$ psi, and $E = 1,600,000$ psi.

3.57   Determine the size of a rectangular, 9-ft-high wood column at a building corner that is continuously supported by a wall about its weak axis and must support a load of 14 k (normal duration). Use Douglas fir–larch with $F_c = 600$ psi and $E = 1,300,000$ psi.

3.58 Is a 10-ft-long, 4- × 8-in. column adequate in supporting a load of 3000 lb (normal duration) that is applied eccentric by 5 in. (see Fig. 3.26g). Use Douglas fir–larch with the following properties: $F_c = 925$ psi, $F_b = 1300$ psi, and $E = 1,600,000$ psi.

3.59 Check if a 4 × 10, 16-ft-high column that is braced about its weak axis at midheight is satisfactory. It must support a wind load of 1 k at midheight in bending about the strong axis and an axial force of 4 k of normal duration (Fig. 3.26h). Use Douglas fir–larch with the following allowable stresses: $F_c = 925$ psi, $F_b = 1300$ psi, and $E = 1,600,000$ psi.

3.60 Determine the minimum dimensions for rectangular concrete columns in laterally braced and nonbraced buildings so that slenderness does not have to be considered in the structural design. Derive also the column equations (3.114) to (3.118).

3.61 Determine the approximate magnitude of the balanced axial force P_b at which the column fails in tension and compression. Assume the placement of the reinforcement parallel to the bending axis and symmetrically arranged. Use $f_y = 60$ ksi.

3.62 Derive the beam–column equation (3.121) that represents the compression failure range by using a straight-line approach for reinforcement ratios between about $\rho_g = 2.2\%$ and 3.0% as based on $f_c' = 4000$ psi and $f_y = 60,000$ psi.

3.63 An interior column of a braced flat plate building with a clear height of 9'–6" carries an ultimate load of $P_u = 400$ k. The flat plate design requires a 12- × 12-in. column. Determine the approximate amount of steel using $f_c' = 4000$ psi and $f_y = 60,000$ psi.

3.64 Select the size of a square, tied concrete column that must carry $P_D = 200$ k, $P_L = 125$ k, $M_D = 50$ ft-k, and $M_L = 25$ ft-k. Use $f_c' = 4000$ psi and $f_y = 60,000$ psi. Also estimate the steel, considering sidesway not critical.

3.65 Size the column in Problem 3.64 so that it acts primarily in axial action and find the reinforcement.

3.66 What column sizes do you select for the case in Problem 3.64 if the eccentricity of the axial load increases to $e = 8$ in. and then to $e = 12$ in.?

3.67 Determine the preliminary size of an interior concrete column with about 2% steel for a laterally braced building. The column has an unsupported height of 8 ft 6in. and carries axial loads of $P_D = 300$ k and $P_L = 200$ k. Use $f_c' = 4000$ psi and $f_y = 60,000$ psi.

3.68 Estimate the size of an exterior column at the first-floor level of a 50-story laterally braced building. The column supports a typical floor area of 10×20 ft and an average dead load of 165 psf and a live load of 80 psf. Assume a live load reduction of 60%. The average floor height is 12 ft6 in. Use $f_c' = 5000$ psi and $f_y = 60,000$ psi. Keep in mind that the unbalanced gravity moments must be taken care of in the final design.

3.69 How much does the column of Problem 3.68 shorten due to elastic deformations under the full loading and due to creep of say 0.1% in/in.? Do you foresee any structural problems?

3.70 Find the size for an interior column of a laterally braced rigid frame concrete building at the top floor. Assume the column has a clear height of $l_u = 10$ ft and carries $P_D = 100$ k and $P_L = 40$k. The live load moments due to gravity balance each other. Use $f_c' = 4000$ psi and $f_y = 60,000$ psi, and use the minimum amount of reinforcement.

3.71 Repeat Problem 3.70, but consider the column not laterally braced and part of a rigid frame.

3.72 The typical interior bays of a 12-story flat plate apartment building are 18×18 ft with a typical story height of 9 ft. The lateral forces are resisted by shear walls. Assume a live load of 40 psf and a dead load of 100 psf, which includes the weight of the lightweight $7\frac{1}{2}$-in. concrete slab, the partitions, and columns. Estimate the size of a typical interior column at the first floor level. Use $f_c'/f_y = 5/60$ and $\rho_g = 2\%$ of reinforcement.

3.73 Determine the typical interior column size at the first-floor level for an unbraced five-story flat plate building using a story height of 12 ft. The three-bay frames, with each bay 20 ft wide, are spaced 20 ft apart; thus, each interior column supports a 20-ft square bay. The loading for each floor including roof is 8-in. concrete slab (100 psf), column weight assumed as 10% of slab weight, partition (20 psf), 50-psf live load, and 30 psf for the exterior wall. Assume a constant wind pressure of 20 psf. Use $f_c'/f_y = 4/60$.

3.74 A 12-story frame-shear wall building is subdivided into three bays of 22 ft in the cross direction and seven bays of 26 ft in the long direction. Two shear walls are provided in the transverse direction. The typical story height is 12 ft. Estimate the interior column size for the rigid frame at the second-floor level. Use an average dead load of 160 psf, which includes the weight of the partitions (20 psf), ceiling and mechanical (10 psf), columns, beams and slabs. The live load is 50 psf which may be reduced by 60%. Keep in mind that the axial gravity action may be assumed to control the preliminary design of the column. The gravity moments can be ignored because of the interior position of the columns. The lateral force action is probably less critical since the building is not very high, and in addition the lateral shear may be assumed to be resisted primarily by the shear walls near the base of the building; we may hence conclude that the column moments are relatively small. Use $f_c'/f_y = 4/60$ and about 2% of reinforcement.

3.75   Do a preliminary design of an interior square column with a clear height of 12 ft at the second floor of an 11-story rigid frame office building in concrete. The loading is as follows: $P_D = 350$ k, $P_L = 130$ k, $M_D = 45$ ft-k, and $M_L = 35$ ft-k. Use $f_c'/f_y = 4/60$.

3.76   A square, rigid-frame concrete building with a central structural core is subdivided into nine structural bays 36×36 ft each. The building is 10 stories high; typical floor height is 13 ft and the bottom floor is equal to 15 ft. Determine the size of a typical facade column at the first-floor level using Grade 60 reinforcing and 5-ksi concrete. Use an average dead load of 170 psf, of floor area and an exterior wall load of 30 psf of wall area. The roof live load is 20 psf while the typical office live load is 50 psf and the partition live load is 20 psf. Reduce the live load by 60% for the column design. For this preliminary design approach, assume that for the given low building height the wind loading case is not critical, especially since the concrete structure provides such a high gravity load. Also, ignore the gravity moment due to the unbalanced beam moment that the exterior column must carry; additional reinforcing will resist this moment.

3.77   Composite simply supported floor beams are spaced 8 ft apart and span 30 ft. Estimate their size, assuming a total floor load of 175 psf, A36 steel, and 4000-psi concrete.

3.78   A shallow 14- $\times$ 60-in. posttensioned continuous beam supports a 30-ft, $7\frac{1}{2}$-in.-thick, one-way concrete slab on each side. The beam span from center to center of the 16- $\times$ 16-in. columns is 30 ft. The superimposed dead loads consist of 10 psf for ceiling and 20 psf for partitions and a reduced live load of 50 psf. Estimate the size of the prestress tendons and check the concrete stresses. Use $f_c' = 4000$ psi and Grade 160 prestress bars. For this preliminary investigation, ignore the effective flange width of the slab and treat the beam as an isolated rectangular section.

3.79   A 20-in.-wide $\times$ 30-in. deep, 24-ft-span continuous concrete beam is to be posttensioned with parabolic cables and a sag of 6 in. For this case about 40% of the total 3-k/ft load is assumed to be balanced. Determine the required prestress force and the necessary tendons and check the stresses. Use Grade 270 strands and 4000-psi concrete. Ignore the effect of the slab flanges or T-beam action.

3.80   Determine the upward loading for the cantilever and end span conditions of the beam in Fig. 3.17c due to the prestress force found in Problem 3.78, which was derived from the typical interior span of a continuous beam. Assume the interior beam span conditions do not change; use a cantilever length of 10 ft and the exterior span equal to the interior one. Draw your conclusions.

3.81   Determine the size of a reinforced-concrete footing to support a 12-in. masonry wall that transmits to its footing a dead load of 6 k/ft and a live load of 4 k/ft. Find also the approximate amount of steel and show the layout of the reinforcing. Use 3000-psi concrete and Grade 60 steel. The allowable soil pressure is 2500 psf.

3.82   Design the wall footing of Problem 3.81 as a plain concrete footing.

3.83   A single reinforced-concrete footing must carry a 15-in. square concrete column transferring a dead load of 200 k and a live load of 150 k. The column is located at the center of the square footing. Determine the size of the foundation and the approximate amount of reinforcement for an allowable soil pressure of 5 ksf, using 3000-psi concrete and 60-ksi steel. The height from foundation base to ground level is 4 ft.

3.84   A square footing supports a 16- $\times$ 18-in. column of 4000-psi concrete that carries a service dead load of 190 k and a service live load of 160 k. The footing base is located 4 ft below the ground level. The soil weight is 120 pcf and that of the concrete 150 pcf. Use a permissible soil pressure of 4500 psf.

    a.   Design a reinforced-concrete square footing using 3000-psi concrete and Grade 60 steel. Show how you place the reinforcement and the column dowels.

    b.   Design a plain concrete square footing using 3000-psi concrete.

3.85 A wall footing supports a 12-in. concrete block masonry wall that carries a service dead load of 20 k/ft and a service live load of 10 k/ft. The footing base is located 3 ft 10 in. below the finished ground line. The weight of the soil on top of the footing is 120 pcf. Use an allowable soil pressure of 5000 psf.

    a.   Design a reinforced-concrete footing using 3000-psi concrete and Grade 60 steel.

    b.   Design a plain concrete footing using 3000-psi concrete.

3.86 A W18 × 50 steel beam (A36) is supported on a 12-in. brick wall. Determine the size of the bearing plate if a load of 49 k must be transmitted.

3.87 A W8 × 24 steel column (A36) is supported on a brick pier. Determine the base plate size if a load of 35 k must be transmitted.

3.88 Determine the number of 3/4-in. diameter A325-N bolts required to connect the hanging W10 column in Example 3.22 using a bearing-type connection. Assume 6- × 5/8-in. connection plates (A36). How many bolts are needed if A325-X bolts should be used?

3.89 Design the simple double-angle web connection for a W36 × 230 (A36) beam that must transfer a reaction force of 340 k. Use 7/8-in. diameter 490-X bolts in 15/16-in.-diameter holes.

3.90 Investigate the situation where the hanging column in Example 3.22 (and Problem 3.88) is connected directly to its boundaries with fillet welds using E70XX electrodes.

chapter 4

The Lateral Stability of Buildings

Buildings generally consist of parallel horizontal planes or floor structures and the supporting vertical planes, such as walls and/or frames. Gravity and lateral forces are dispersed through the floor and roof structures to the vertical planes and from there to the ground. The magnitude, direction, and type of action of the static force flow depend on the geometry and stiffness of the vertical planes and on their arrangement within the building volume. The distribution of the lateral forces along the floor and roof diaphragms to the vertical resisting structures is studied here for typical conditions, together with the primary lateral forces of wind, earthquake, and earth and water pressure.

The response of a building to lateral force action is most clearly expressed by exposed diagonal bracing, as is so boldly demonstrated by the 100-story truncated pyramid of the John Hancock Center in Chicago (1968) and the 420-ft-span Oakland–Almeda County Coliseum (1966, Fig. 9.15c, bottom), both designed by SOM. The cross-bracing of lightweight hinged steel assemblies is a common method of stiffening, as has already been used for the Crystal Palace in London (1851) and been celebrated in the Pompidou Center in Paris (1977).

The cross-bracing with thin rods may articulate the minimal character of the structure, whereas heavy diagonal members may express some other esthetics. The bracing may occur in every structural bay or only in selected ones, it may cross the entire bay from corner to corner or only part of it (i.e., be concentric or eccentric), it may form X-bracing or single diagonal bracing, changing direction from bay to bay. The cross-bracing may be replaced by V-columns or trussed columns serving the same purpose of lateral bracing. The reader may want to study further the richness of architectural expression in Fig. 4.1, especially the importance of the connection details.

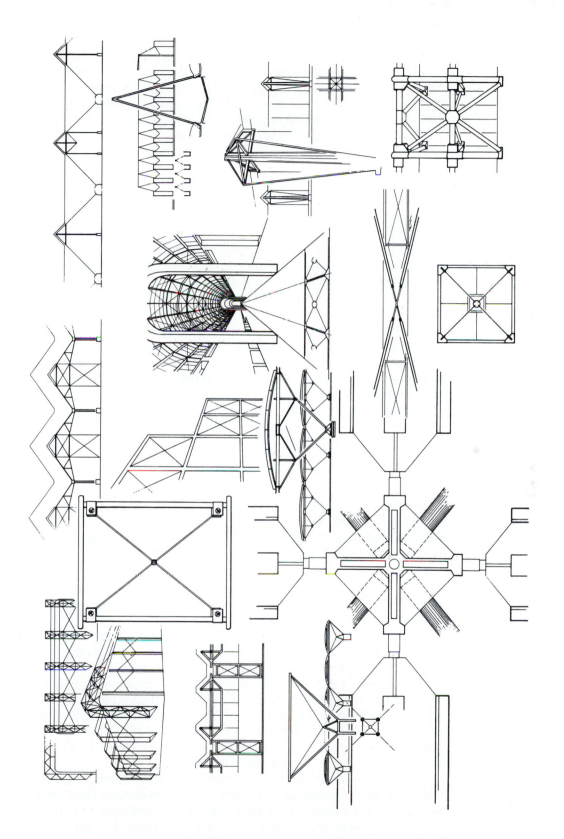

Figure 4.1 Lateral stability of buildings.

4.1 LATERAL LOAD ACTION

A structure must be strong enough to resist the many types of physical forces imposed on it. The magnitude and direction of these forces vary with the material, type of structural system, purpose of the building, and the locality. The most obvious loads are due to gravity action, as caused by the self-weight of the building, snow, and occupancy. Lateral forces are exerted on the structure by wind and earthquakes, as well as by earth and hydrostatic pressure. The lateral forces tend to slide and rotate the building block, and the wind attempts to lift up the roof; gravity, in contrast, will counteract and stabilize the structure.

Wind and seismic loading cause horizontal force action on a building. They are dynamic loads, but can often be treated as quasistatic lateral forces. This approach is reasonable with respect to wind action as long as the building is not of unusual shape and is stiff enough so that it does not oscillate and give rise to accelerations, with the corresponding increase in force action. Naturally, the shape of the building, as seen in plan and elevation, considerably affects the lateral force resistance, remembering that the least resistance for a given wind direction is provided by the streamlined teardrop shape. Not only is the rigidity of a building improved by sloping the exterior columns, such as the truncated pyramid of the John Hancock Building in Chicago, but also the lateral force resistance is reduced, thereby resulting in a large decrease of lateral drift.

Whereas wind exerts external lateral forces, the ground motion due to an earthquake causes internal lateral forces, besides vertical forces, which (however) are neglected. We can visualize the building as riding on an unstable earth. As the ground abruptly accelerates in a random fashion, the building portion above the ground will be left behind, thereby activating lateral inertial forces. In other words, the inertia of the mass tends to resist the movement, similar to the experience of a person in a car that suddenly increases in speed. The time it takes for a building to respond to the base-induced acceleration due to fluctuating seismic ground loads becomes an important characteristic of the building; the fourth dimension, that of time, is introduced as a consideration of loading.

While earthquake forces constitute internal lateral loads generated by the mass and stiffness distribution in response to motion, wind causes externally applied forces on stiff buildings that depend on the exposed facade surface area. Seismic loading is usually critical with respect to the performance of stiff low- and mid-rise structures, while wind loading generally dominates the design of tall, slender buildings. The optimum design of high-rise buildings in areas of strong earthquakes conflicts with that for wind loading. Here, seismic action calls for ductility with much redundancy, while the wind resistance requires stiffness for occupant comfort.

Not only wind and earthquake, besides the centrifugal outward effect of cars upon curved bridges and other loading conditions, but also gravity together with the respective geometry may cause lateral force action on a building. When columns are inclined, gravity will cause lateral thrust, which increases as the column moves away from the vertical supporting condition. Horizontal floor beams at the top act as ties in tension when the columns lean outward, but as struts in compression at the bottom. For a symmetrical structure, the thrust due to dead load will self-balance, but the horizontal forces due to asymmetrical live loads must still be resisted, as for an asymmetrical building where also the weight causes thrust.

Among the many examples of lateral-force action are the large lateral pressures and impact forces generated by a crane. A traveling elevator causes pumping action, particularly in a single-elevator shaft, together with pressures from the wind entering through the vent shafts; the shaft walls must be designed to resist these forces. Similarly, the walls of the stair shaft for a tall building must resist the lateral air pressure due to pressurization in case of fire, especially close to the fan location. A fire in an interior bay of a multistory building will cause an expansion of the concrete floor, with corresponding horizontal forces. This thermal thrust must be resisted by the cooler adjacent frames and/or shear walls; it acts similar to an eccentric external prestressing force that increases the moment capacity. Other examples of lateral-force action include, possibly, excessive stresses in columns resulting from the internal hydrostatic pressure of liquid used for fireproofing due to the height of the columns. Should a large-panel structure be designed to avoid a progressive collapse, then the building must be able to withstand an internal blast pressure of 5 psi.

In this section only the primary lateral loads due to wind and earthquake action are briefly discussed. Water and earth pressure loads are introduced in the last section of this chapter.

Wind Loads

Wind forces are among the most violent sources of destruction in nature. It is extremely difficult to predict their complex behavior, since they are not constant and static like dead load, but are dynamic and fluctuate in an unpredictable manner, not only in magnitude but also in direction. Wind behavior is influenced by the *topography* (open, wooded, rolling, hilly, urban, vegetation, roughness, etc.), *building type* (shape, size, height, texture, flexibility, degree of tightness, openness, etc.), and the *nature of the airflow* (air density, direction, velocity, degree of steadiness, etc.).

Tornadoes are the most devastating winds. In the United States, they appear mostly between the Rocky and Appalachian Mountains, with a tangential wind velocity as high as 500 mph within the tornado tunnel. The probability of a particular building being hit by a tornado is extremely small. Therefore, a structure is usually not designed for tornadoes. Buildings designed for the 200-mph range should generally be safe 95% of the time and this should also take care of severe hurricanes. While tornadoes only act over a relatively small area of about 3 square miles, hurricanes affect many thousands of square miles and thus are much more destructive. Hurricanes occur in the United States primarily along the Gulf and South Atlantic coastal zones. They cause damage not only because of the high wind velocity but also because of the waves (erosion) and rainfall (flooding). The maximum winds appear close to the eye of the hurricane. Typical design values for strong hurricanes are 150 mph and 90 mph for average ones. The radius of curvature of the rotating winds of hurricanes is so large that their path may be considered straight. Similar in behavior to the hurricanes in the Atlantic and South Pacific are the typhoons in the Western Pacific and the tropical cyclones in the Indian Ocean, Arabian Sea, and offshore Australia. Keep in mind that 70- to 90-mph windstorms can lift off roofs and uproot trees.

As air moves along the earth, it is retarded close to its surface due to frictional drag. The wind near the ground behaves in an erratic manner; although the wind velocity will generally be less in cities, the gustiness may be much greater. In the boundary layer where the buildings are located, the mean wind velocity increases with height, with the rate of increase being a function of the ground roughness. The greater the interference by surrounding objects (i.e. trees, land forms, buildings), the higher the altitude at which the maximum and steady velocity occurs. When the airstream passes the building, its behavior is drastically altered. It is deflected and then rejoins the original flow pattern, exerting pressure on the front side and suction along all other sides. The building must resist the total wind force that is the sum of the pressures. The degree of disturbance of the original flow pattern depends, among other criteria, on the building shape and the nature of the surface; a sawtooth face may cause significant wind drag. It is apparent that, for a given wind direction, the streamlined teardrop shape (as exemplified by aircraft wings and the body profile of the dolphin) provides the least resistance to the airflow.

Since the behavior of an airstream is harder to visualize than the flow of water, imagine a simple solid object submerged in flowing water, rather than in an airstream in a wind tunnel. This simple object placed perpendicular to the current may have various forms, such as a flat plate, a round cylinder, or an airfoil. Each shape generates a different complex flow pattern. It is obvious that the streamlined teardrop shape, ignoring its surface friction, provides the least resistance to the original flow pattern; examples are aircraft wings, the bodies of some birds, and fishes. Nature provides in the body profile of the dolphin a convincing statement of the efficiency of form relative to the least resistance in water. Consider now ordinary building shapes like those given in Fig. 4.2A to be submerged. When the stream passes the obstruction, its behavior is drastically altered; it is deflected and then rejoins the original flow pattern behind the object. The degree of disturbance of the original flow pattern that is the crowding of the streamlines indicates the increased speed and the corresponding intensity of load action. The flowing water exerts pressure on the front side and suction on all other sides, that is, *drag forces* in the direction of flow and, like airplane wings, *lift forces* perpendicular to it. Turbulence is generated at abrupt changes of geometry, as is the case in the vicinity of building corners and ridges. In other words, as the wind strikes a structure, its shape, size, texture, openness, and flexibility have an influence on the magnitude of the pressure distribution. A building is subjected to *normal* and *transverse forces*. The transverse action is caused by side and lift forces, such as the familiar lift action on aerofoil shapes. The normal forces are also called *drag forces* because the body is dragged in the direction of the flow; they are caused on buildings primarily through direct normal pressure, that is, positive pressure on the windward face and simultaneously negative pressure (suction) on the leeward face, but also through surface friction tangential to the building's surface.

When a *steady* streamline airflow of velocity V is completely stopped by a rigid body, the stagnation pressure (or velocity pressure) q on this body is related to the square of its velocity according to Daniel Bernoulli (1700–1782)

$$q = \frac{1}{2} p V^2 \qquad \text{(a)}$$

where the air mass density p is the air weight divided by the acceleration of gravity $g = 32.2$ ft/sec^2. For the special condition of a standard atmosphere with a temperature of 15°C (59°F) at sea level, the air weighs 0.0765 lb/ft^3, which yields the following value, corresponding to a wind speed in mph by using 5280 ft/mile and 3600 sec/hr. Hence, Eq. (a) can be expressed for V (mph) as

$$q = \frac{1}{2}\left(\frac{0.0765}{32.2}\right)\left(\frac{5280}{3600}V\right)^2 = 0.00256 V^2 \text{ (psf)} \qquad \text{(b)}$$

Thus, the stagnation pressure is proportional to the square of the magnitude of the instantaneous wind velocity. Also, the air mass density varies with locality, altitude, time, and weather; a different numerical coefficient may be derived if respective climatic data are known.

Up to this point, only the local pressure on a rigid, nonmovable large body has been considered. In reality, the airflow does not come to a complete halt, but only changes as it passes the building object and deviates from its original path in response to the structure. Until now, the nature of the building had no effect on the wind pressure distribution; this obviously cannot be true. As the wind strikes a structure, its shape, size, texture, openness, and flexibility have an influence on the magnitude of the pressure distribution. Furthermore, the wind velocity is not steady; it consists of the constant mean velocity (steady component) and the varying gust velocity (dynamic component). Therefore, a building deflects along the direction of the wind due to the mean wind pressure and vibrates from this position due to gust buffeting, which can be larger than the static sway! It must be emphasized that wind velocities are nonuniform

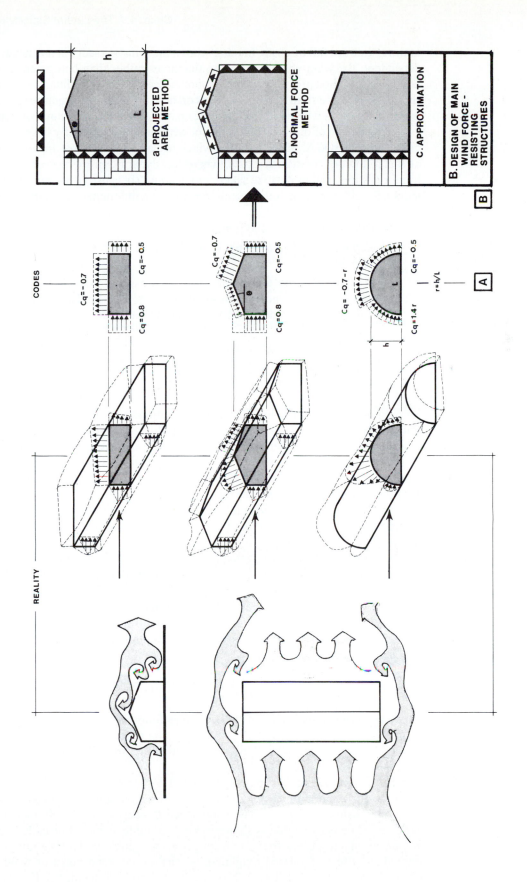

Figure 4.2 Wind pressure distribution as related to ordinary building shapes.

in time; they are full of gusts. In other words, the wind pressure is not steady, it is time dependent; the wind velocity pressure fluctuates rather than being constant. In addition, the stagnation pressure acts on a more or less flexible building, which may not be simply treated as a rigid body. It depends on how the building structure responds to the wind fluctuations whether the wind forces have to be treated as static or dynamic.

The design of ordinary buildings for wind action is generally based on one of the model codes (BOCA, SBCCI, UBC), which use a *static pressure approach* unless one of the following conditions exists, where the dynamic properties of the main wind force-resisting system must be taken into account (because of its sensitivity to wind-induced oscillations) and other conditions where wind tunnel testing may be required:

- *Tall buildings* over 400 ft high
- *Slender tower buildings* with a height that exceeds five times the least horizontal dimension
- *Flexible buildings* that are prone to wind-excited oscillations due to other reasons. Some designers claim that structures with fundamental periods greater than 1 should be designed for dynamic loads. Since ordinary buildings, not higher than about 10 stories, have generally fundamental periods of less than 1 sec, they can be designed statically.
- *Hybrid buildings* and *buildings of unusual shape,* which include open structures

In contrast to static loads, which are stationary or change slowly and cause a static deflection, dynamic loads vary more rapidly (e.g., cyclic loads) or occur abruptly (e.g., impact loads) and generate vibrations (thus introducing another dimension, that of time) with the corresponding increase in stresses. These vibrational forces may result in *resonant loads* when the period of the wind is nearly equal to the natural period of the structure, as may be the case for flexible buildings (e.g., towers, suspension bridges, long-span roofs) and for building components. When a dynamic analysis is necessary, national standards must be consulted and wind tunnel testing may be required.

Codes convert the basic velocity wind pressure [Eq. (b)] to a static design wind pressure by using coefficients that take into account *shape* (e.g., height, width, length), *form,* and *type* of building, as well as *exposure* conditions. The approximate wind pressure values used in this book can be derived from any of the codes. For reasons of convenience, the relatively simple version of the Uniform Building Code (UBC), 1991 edition, will be the basis of discussion. According to the UBC, the *static design wind pressure* is

$$p = C_e C_q q_s I \quad \text{(psf)} \tag{4.1}$$

where the terms in the equation are defined as follows:

$$q_s = 0.00256 V^2 = \text{stagnation pressure, Eq. (b)}$$

The basic wind speed, V(mph), is measured by weather stations across the United States as the maximum *average wind speed* (by averaging the gusts), which is also called the *fastest-mile speed.* The wind speed increases with height (according to an exponential velocity profile as shown in Fig. 4.3) up to an elevation called the *gradient height,* at which ground friction no longer effects airflow. To take into account the immediate affect of surface roughness on airflow, all recordings of wind speed are taken at a height of 33 ft (10 m). The extreme fastest-mile wind, $V = V_h = V_{33}$ (mph), is based on a 1-, 25-, 50-, and 100-year mean recurrence interval. Ordinarily, wind veloc-

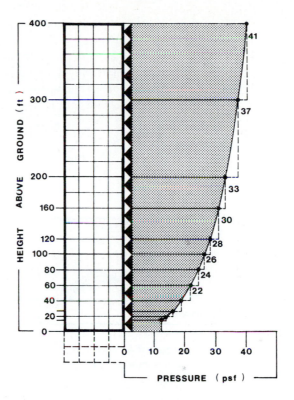

Figure 4.3 Typical wind pressure distribution for a basic wind speed of 80 mph (Exposure B).

ities with a mean recurrence period of 50 years should be used (that is the fastest-mile wind speed has a 2% probability of occurring in any year or once every 50 years), if there is no high degree of risk to life and property in case of failure, thus allowing some damage for stronger wind occurring once in 100 years. Minimum basic wind speeds (mph) for 50-year mean recurrence interval and Exposure C (as is defined later) are presented in Fig. 4.4; the basic wind speed for the Virgin Islands and Puerto Rico is 110 mph. It can be seen that the wind speeds range from 70 to 90 mph for inland areas and up to 110 mph along the Gulf and South Atlantic coasts, reflecting the intensity of hurricanes. It should be noted, however, that for some special wind regions with particular topography or other conditions local wind records must be consulted.

I = importance factor; it depends on the occupancy category

   = 1.15 for *essential facilities* (e.g., hospitals, fire and police stations, emergency shelters) and *hazardous facilities* (e.g., structures containing toxic or explosive substances dangerous to the general public)

   = 1.00 for all other special and standard occupancy structures

C_e = combined height, exposure, and gust factor. In other words, the UBC combines conveniently the *velocity pressure coefficient* and the *gust response factor* of other codes into a single factor. This factor takes into account that the wind velocity generally increases with height, reflecting an exponential velocity profile (Fig. 4.3), the shape of which is dependent on the exposure conditions, that is, the roughness of the surrounding terrain. The factor also takes into account the dynamic character of the airflow; in other words, wind pressure is not steady, it is time dependent—the wind velocity pressure fluctuates rather than being constant and its behavior depends on how the structure responds to these fluctuations. The C_e factor depends on the height above the average level of adjoining ground (ft)

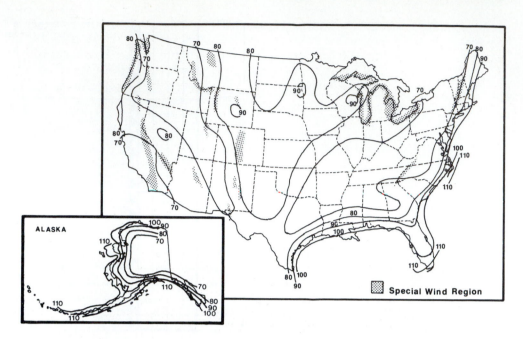

Figure 4.4 Minimum basic wind speeds (mph) for 50-year mean recurrence interval (Exposure C).

and exposure conditions. Typical values from 15 to 400 ft for various exposure conditions are:

C_e = 1.39 to 2.34 for Exposure D (open, flat terrain facing large bodies of water)

= 1.06 to 2.19 for Exposure C (flat, open terrain)

= 0.62 to 1.80 for Exposure B (terrain with buildings, forest, or surface irregularities 20 ft or more in height covering at least 20% of the area extending 1 mile or more from the site), as is typical for most built-up urban and suburban areas.

C_q = *pressure coefficient* for the entire lateral force-resisting structure or for portions of the structure. The pressure coefficients define the magnitude of the wind forces as affected by the building shape; they represent average loads for the primary lateral force-resisting structure but local loads on elements and components of the structure (e.g., wall cladding, roof purlins). It is apparent that the local pressures may exceed the average pressure on the structure considerably. The pressure coefficients take into account, for instance, the following situations:

- The direction of pressure (e.g., direct pressure on wall as positive and negative pressure on leeward wall or uplift on a flat roof)
- The average pressure on the primary wind-force-resisting structure (e.g., walls, roofs)
- The local pressure on individual elements and components of structure not located in areas of discontinuity (e.g., cladding, parapet walls, roof elements)
- The local pressure on elements and components of structures in areas of discontinuities (e.g., wall corners, roof eaves, rakes or ridges with or without overhangs, canopies)
- Pressures on other structures (e.g., chimneys, tanks, open frame towers, signs, flagpoles, ladders, cable structures, open cantilever structures, open structures)

Pressure coefficients for various structure types are given in the major building codes. Some typical values for low-rise buildings are shown in Fig. 4.2A. For example, the pressure coefficient for the windward wall of a flat-roofed rectangular building is 0.8 regardless of the building proportions, while the suction value on the leeward wall, depending on the plan dimensions, is between –0.2 and –0.5 (and –0.7 on side walls). Most codes give a factor of $0.8 + 0.5 = 1.3$ for typical rectangular prismatic buildings, which reflects the resultant effect of pressure and suction as required for the overall building design; in other words, about 60% of the entire wind force on the building is assumed to act on the windward wall. It must be emphasized that ordinary structures of rectangular plan with flat, gabled, and arched roofs were assumed in Fig. 4.2A, which can be considered long enough so that the typical wind distribution along the long faces is not disturbed by the airflow around the corners and edges where large local suctions with updrafts occur.

The UBC allows the horizontal pressure coefficient of 1.3 for primary frames and systems of 40 ft or less in height, but requires $C_q = 1.4$ for structures over 40 ft in the *projected area method*. According to the UBC, for the design of the primary lateral load-resisting structures, less than 200 ft high, the projected area method may be used except for gabled rigid frames, where the *normal force method* should be applied (Fig. 4.2B). In the projected area method, the wind pressures are assumed to act on the full vertical projected areas (i.e., the horizontal pressure) and on the full horizontal projected areas (i.e., the vertical pressure) simultaneously, as indicated in Fig. 4.2B.

Some codes do not distinguish between *external* and *internal wind pressures*. We must realize, however, that buildings are not airtight. There is always leakage through joints, cracks, and vents. Doors and windows may be open or portions of a wall may have large openings. When the openings are in the windward side, the internal pressure will be positive and will approach the external pressure value as the wall openings increase. Openings in the leeward face will cause internal suction.

Internal wind pressure may be visualized as having a balloonlike effect with a uniform pressure acting outwardly on all internal surfaces when the wind enters primarily on the windward side. But openings in the leeward or side walls allow air to be pulled out of the interior, thereby causing inward acting pressure.

Codes that do not distinguish between external and internal pressure coefficients have combined these factors to appropriate pressure coefficients to generate the critical load for *enclosed buildings*. The UBC treats a partially enclosed structure as an open one when it has a greater area of exterior wall openings (e.g., doors, windows) on any one wall than the sum of the areas of the openings on all the other walls and has more than 15% of the wall area open. Special pressure coefficients are provided by codes for *open structures*.

Although the wind pressure in Eq. (4.1) is assumed to increase with height, in urban areas the grouping of buildings may cause funneling of the air mass through narrow spaces between the buildings and will influence the wind pressure distribution, possibly resulting in a maximum pressure at the base or center of the windward face, rather than at the top as assumed by codes.

The wind pressure distribution for large-scale special structures must be obtained from model testing in a wind tunnel. The magnitude of the pressure coefficients reflects the efficiency of the building shape with respect to wind action; streamlined buildings have the smaller coefficients.

Approximate Wind Pressure on Primary Lateral Force-resisting Structures. For the preliminary design of ordinary, relatively stiff buildings, not of unusual shape or exposure, where the effects of turbulence in urban areas may be neglected, a pressure coefficient of $C_q = 1.3$ may be used. Keep in mind, however, that this value applies to rectangular prismatic buildings and may be reduced for other prismatic and round forms. Therefore, Eq. (4.1) can be simplified to

$$p = C_e C_q q_s I = C_e(1.3)0.00256V^2(1) = 0.00333C_eV^2 \qquad (4.2)$$

Various wind pressure values are given in Table 4.1 for basic wind speeds of 70 and 80 mph and Exposures B and C taking into account $C_q = 1.4$ when $H > 40$ ft ≤ 400 ft Equation (4.2) is plotted in Fig. 4.3 for a basic wind speed of 80 mph and Exposure B. The wind pressure distribution demonstrates the exponential profile.

For the preliminary design of ordinary low-rise buildings, $C_e = 1.0$ may be assumed by using a constant uniform wind pressure rather than the stepped pressure distribution in Fig. 4.3. This approach should be reasonable for a single-story building (Exposure C) and buildings not higher than about 70 ft (Exposure B), although this approach may also be used for higher buildings for fast estimation purposes. Hence, Eq. (4.2) can be simplified further to

$$p = 0.00333V^2 \qquad (4.3)$$

For example, for a wind speed of $V = 80$ mph, the static, horizontal wind pressure is

$$p = 0.00333(80)^2 = 21 \quad \text{psf}$$

We may conclude that for typical wind speeds between 70 and 80 mph the minimum lateral wind pressures for ordinary low-rise buildings are in the range of 16 to 21 psf.

For buildings with sloped or arched roofs, the *projected area* method is used in which the wind pressures are assumed to act simultaneously on the projected horizontal and vertical areas of the building (Figs. 4.2B and 5.26B). In this context, for the preliminary design of the main wind-force-resisting structure, however, the vertical pressure on the horizontal projected roof area is neglected.

For the design of gabled rigid frames, UBC requires the usage of the *normal force method* (Fig. 4.2b). The lateral wind pressure on the walls of a long gabled rigid

TABLE 4.1
Approximate Design Wind Pressures p (psf) for Ordinary Wind Force-Resisting Building Structures[a]

Height above Grade (ft)	Exposure B		Exposure C	
	Basic Wind Speed (mph)			
	70	80	70	80
0–15	10	13	17	23
20	11	14	18	24
25	12	15	19	25
30	12	16	20	26
40	14	18	21	28
60	17	22	25	33
80	18	24	27	35
100	20	26	28	37
120	21	28	29	38
160	23	30	31	41
200	25	33	33	43
300	29	37	36	47
400	32	41	38	50

[a]Values for intermediate heights above 15 ft may be inderpolated.

frame building, due to wind action perpendicular to the ridge, is the same as for flat or very slightly pitched roofs. Whereas the leeward roof pressure is always outward, the wind pressure distribution on the windward slope depends on the angle of inclination of the roof and on the configuration and relative dimensions of the structure. For a very slightly pitched roof ($\theta \leq 10°$), the wind pressure is outward for the windward slope, and for a high pitched roof of about $\theta \geq 37°$ it will have an inward pressure, as long as the height-to-width ratio of the building does not exceed approximately 1.5. On the other hand, a low pitched roof (say, $\theta = 20°$) will have an outward pressure when the width (or length) of the structure is small relative to the height (say, $L/h \leq 2$), but will experience inward pressure if the width (or length) of the structure is large.

We may conclude that as the roof slope increases the suction on the windward slope changes into pressure; this change is often assumed to occur at roughly $\theta = 30°$ for ordinary long, low-rise buildings. Roof pressure coefficients may be used as follows for preliminary design purposes:

Leeward slope $C_q = -0.7$

Windward slope $C_q = -0.7$ ($\theta < 10°$ or slope < 2:12)

$C_q \cong -0.9$ to $+0.3$, ($10° \leq \theta \leq 37°$ or slope 2:12 to less than 9:12)

$C_q \cong 0.010\,0$ ($\theta > 37°$)

In the context of this book, for the preliminary design of gabled rigid frames only a uniform wind pressure is considered on the vertical projection of the roof; in other words, the uniform wall pressure is continued up to the top of the building for the sizing of A-frames. For roof purlins and roof deck, however, suction and pressure normal to the roof must be taken into account.

The wind pressure distribution along arched roofs is very complex and depends, among other criteria, on the height-to-width ratio of the structure and whether the structure is elevated or springs from the ground level. The pressure distribution is simplified as shown in Fig. 4.2A, where radial pressure appears on the windward side while maximum suction perpendicular to the surface occurs at the crown. As for gable frames, a uniform wind pressure on the vertical projection of the structure is also assumed for the preliminary design of arched roof structures.

The wind pressure distribution for flexible buildings, such as for pneumatic and tent structures, is much more complex than for rigid ones since they deform under load and change their shape, thereby influencing the wind action. A static analysis may not be appropriate since they may flutter and vibrate so that dynamic loading must be included. The wind pressure is assumed to act perpendicular to the membrane surface. For open tent structures, the wind forces below the membrane must be added to the external ones. For pneumatic structures, the internal air pressure, in addition to the criteria for rigid structures, influences greatly the magnitude of the pressure coefficients. For further discussion of this topic, refer to Chapter 9.

Approximate Wind Pressure on Elements and Components of Structure.   It is apparent that the local wind pressure on individual areas can be by far larger than the average wind pressure acting on the building as a whole. Therefore, local pressure coefficients must be larger than the average pressure coefficients used for the design of the main lateral force-resisting structure. Furthermore, the local pressure coefficients at areas of discontinuity (e.g., corners, ridges, and eaves) must be larger than the ones not located in these areas.

For the design of typical *wall elements* not in areas of discontinuity (and not for open structures), a pressure coefficient of $C_q = \pm 1.2$, and $C_q = \pm 1.3$ for parapet walls, may be used. Hence, the typical wind pressure for ordinary low-rise building wall elements using $C_e \cong 1.08$, according to Eq. (4.1), is

$$p = C_e C_q q_s I = \pm (1.08)(1.2)0.00256 V^2 (1) = \pm 00333 V^2 \qquad (4.3a)$$

We may conclude that for the preliminary design of low-rise buildings, the local wind pressure for the design of the wall elements may be assumed equal to the average wind pressure on the entire building. Similarly, for the design of *roof elements* not in areas of discontinuity (and not for open structures), the pressure coefficient is $C_q = -1.3$ ($\theta <$ 30° or slope < 7:12) and $C_q = \pm 1.3$ ($30° \leq \theta < 45°$). We may conclude that for the preliminary design of roof elements (e.g., purlins, decking) of ordinary low-rise pitched buildings the wind pressure values perpendicular to the roof may be assumed equal to the wall pressure values. Hence, for the typical 70- and 80-mph winds, the wind pressure (or suction) values are in the range of 16 to 21 psf. It should be kept in mind that light roofs are most vulnerable to wind suction; the possible uplift of the roof skin must be prevented by adequate anchorage.

Conclusion.   We should not overlook that wind does not necessarily act uniform, as is generally assumed by codes. In reality, its nonuniform pressure distribution, together with the fact that wind can come from any direction, will cause asymmetrical action with respect to the center of rigidity of the structure and thus will generate torsion and twist the building. Wind forces from different directions must be considered in the design of buildings.

Wind action on ordinary low, rigid buildings rarely controls the design of the primary structure, as is exemplified later in this book. It will influence, however, the sizing of the secondary structural elements.

The information on extreme winds, as provided by building codes, must be used with caution since local meteorological and topographic features may require higher values. Furthermore, wind maps do not show tornado winds, since building regulations do not require taking them into account in the design. If a building is located in a tornado-prone area and not designed to withstand it because of economic reasons, a protective shelter should be incorporated into the design.

Seismic Loads

Earthquakes are among the most awesome natural forces; the Xian quake of 1556 in China killed 830,000 people, by far the most destructive quake ever. Earthquakes occur suddenly, without warning, and within 10 to 20 seconds can turn cities into wastelands, make islands disappear, alter the flow of rivers, and give birth to new land and lakes. Building designers in the United States often automatically relate earthquakes to California and Alaska; they do not realize that the earthquakes of Charleston, South Carolina (1886), and of New Madrid, Missouri, near Memphis (1811, 1812) were of nearly the same magnitude as the most severe one in San Francisco (1906) and that more than one-third of the U.S. population lives in areas of high to moderate seismic risk. Not only the large western cities of Los Angeles and San Francisco, but also metropolitan areas of Buffalo, Providence, Boston, Charleston, Memphis, St. Louis, Salt Lake City, Seattle, and Anchorage are regions prone to major earthquakes. Severe earthquakes have occurred in many other parts of the world, such as around the rim of the Pacific Ocean (e.g., Chile, Peru, Nicaragua, Guatemala, Mexico, California, Japan, New Zealand), along the Mediterranean Sea (e.g., Morocco, Greece, Italy, Yugoslavia, Romania, Turkey), around the Himalayan Mountains (e.g., China, India), and in Iran. Building designers must be acquainted with the effects of quakes so they can make buildings earthquake resistant and safeguard life.

According to the theory of *plate tectonics*, the earth's crust consists of separate plates that float on the earth's molten interior. Each of the plates moves (creeps) a few inches every year. At points of convergence, the plates want to slide past each other, as (for instance) along the San Andreas fault, where the Pacific plate and the North American plate move in the same northwestward direction but at a different rate. If the plates are prevented from doing so at certain locations by friction or by being locked into each other, elastic strain is stored and accumulates until the forces can no longer be resisted by the material. When the capacity is exceeded, sudden rupture and slippage occur, which may cause the upper crust of the earth to fracture and form a fault. This abrupt release of strain energy results in complex vibrations propagating at high speeds from the source (*focus* or *hypocenter*) in all directions through the earth and along its surface, reaching a given building at different time periods with different velocities from different directions (Fig. 4.5), causing buildings to be pushed and pulled from side to side and up and down, as well as to be tilted. The more significant seismic wave types are the faster longitudinal *P-waves* (primary waves, compressional waves, push–pull waves), which compress the earth in front and move building foundations back and forth in the direction of travel. The later and slower transversal *S-waves* (shear or secondary waves) oscillate in a plane perpendicular to the direction of propagation and tend to move the building foundations up and down and side to side at right angles to the P-waves. The P- and S-waves are body waves traveling through the interior of the earth, keeping in mind that only compressional waves are possible through gases and liquids. In contrast, the *Q-waves* (Love waves), with no vertical movement, and the *R-waves* (Rayleigh waves), with both vertical and horizontal movement, are relatively slow moving surface waves that propagate along the earth's crust.

Since the high-frequency components of the seismic waves, in contrast to the low-frequency, larger-period waves, tend to weaken rapidly as they propagate away from the source, short-period, low-rise and massive buildings tend to be excited more near the epicenter, while long-period tall buildings are activated more by the low frequency waves that are transmitted over larger distances. Naturally, the nature of the regional geology that is the transmission path to the site influences the behavior of the seismic waves. Usually, the most destructive part of an earthquake is over in 10 to 30 sec.

Earthquakes are broadly classified as *shallow-focus earthquakes,* where the focus is located near the epicenter, thus causing more intense local effects, and *deep-*

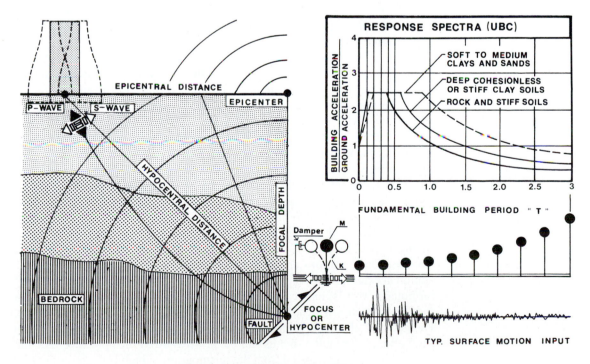

Figure 4.5   Seismic force action.

focus earthquakes affecting much larger areas. Most western earthquakes are generated along the plate boundaries near the surface (usually 4 to 10 miles in California). However, the potential earthquakes east of the Rocky Mountains are of the midplate type, originating deep in the earth. They do not occur along recent active fault lines; in this case, the weight of the overlying rock, sand, and clay squeezes the understructure of the earth. For example, the area in southeastern Missouri that represents a critical seismic zone around New Madrid is located over an ancient subterranean fracture called the New Madrid fault zone. Eastern earthquakes are less frequent, but because of the more ancient geology with fewer faults to slow the travel of the waves relatively quickly, shock waves travel much farther than in the West and hence can damage a much wider area, thus making them potentially by far more destructive. Recently, it has been discovered that in California not only the familiar *surface faults* exist, but also a network of *hidden faults* 4 to 10 miles beneath the surface. These hidden faults do not break the earth's crust but form folds and eventually hills, causing earthquakes during this process.

A major percentage of earthquakes are of tectonic origin, but a second cause may be due to volcanic eruptions or subterranean movements of magma. Besides the natural tectonic and volcanic earthquakes, there may be artificially induced ones, such as due to underground nuclear explosions and large water reservoirs. The primary effects of earthquakes causing possible building damage are the following:

- *Ground rupture* in the fault zone.
- *Ground failure* due to landslides, where ground displaces horizontally and/or vertically without ground rupture; mudslides, avalanches, and ground settlement; shaking of ground resulting in loss of bearing capacity as due to liquefaction, where saturated granular soil is transformed from a solid state into a liquefied state due to porewater pressure, causing the soil to behave like quicksand
- *Tsunamis,* which are large sea waves generated by the sudden displacement of land at the ocean bottom.
- *Ground shaking:* its effect is influenced by the magnitude of energy released, the location of the focus, its duration, the geology of the site, and building characteristics.

Among the secondary and often more critical effects of earthquakes are fire, disease, looting, explosion, flooding, and disruption of economic and social life.

The remaining discussion of earthquake-resistant building design will be concerned only with *ground shaking* as the source of damage, since the chances for ground rupture under a building are extremely small. The structural design for earthquake forces, because of their random character, cannot be considered as an exact engineering science; some engineers even claim it to be more of an art. The design is based on assumptions rather than on precise data; the prediction of ground motion, together with the response of the building riding on the earth along three directions, may be impossible to determine accurately.

A network of seismic stations is distributed across the United States in regions having a past history of earthquakes to record and attempt to predict earthquake motions. These stations are equipped with instruments such as seismographs, strong-motion accelerographs, and tiltmeters. This monitoring grid can locate the *focus* of an earthquake, that is, its location deep in the earth's crust where the rupture originated, and the *epicenter,* which lies on the earth's surface directly above the focus (Fig. 4.5). The network also records the point of highest wave intensity, which is generally centered around the fault and thus can be far away from the epicenter. The magnitude of the earthquake is calculated from the seismogram, which is the response of the seismograph to the motion of the ground recording a zigzag line reflecting the varying amplitude of the vibrations. The Chinese not only use seismological instruments for the prediction of earthquakes, but also pay close attention to the oddities of animal behavior.

Earthquakes are classified either according to the magnitude of energy they release or to their intensity, that is, destructiveness. The *Richter scale,* invented by Charles F. Richter of the California Institute of Technology in 1935, is a measure of the energy released at the focus. It ranges from 3 to 9, based on a logarithmic scale, where each unit increase reflects an increase of about 32 times more energy. Comparing the energy release of the Hiroshima atom bomb (about 6.4 on the Richter scale) with that of the 1964 Alaska earthquake of 8.4 shows that about 1000 (i.e., 32×32) times more energy was freed by the earthquake. The largest known earthquake is 8.9 (Ecuador, 1906); the 1906 San Francisco earthquake measured 8.25, and the 1812 New Madrid one has been estimated as 8.2. In one of the most devastating quakes in recent history, nearly a quarter of a million people died in the 1976 Tangshan earthquake (China) of Richter magnitude 7.8.

Earthquakes are classified as *moderate* from 6 to 7, as *major* from 7 to 7.75, and above that as *great.* The Richter scale does not concern itself with the effect of the earthquake. An earthquake of Richter magnitude 6 may be far more destructive when it hits a densely populated region directly than one of magnitude 8 with its focus far away from inhabited areas. Nearly five times more people were killed in the Armenian earthquake of 1988, which registered 6.9 on the Richter scale, as compared with Mexico City's 8.1 quake of 1985, where 10,000 people died.

The m*odified Mercalli intensity scale (MMI),* as initially developed by Guiseppe Mercalli (1850–1914) for use in Italy, and then modified by H. O. Wood and F. Neumann in 1931 for the conditions in the United States, is of a subjective nature. It describes the degree of damage based on 12 intensity divisions. The *seismic risk map* (see Fig. 4.6) is used by the Uniform Building Code and correlated with the MMI Scale. **Buildings should be able to resist minor earthquakes without damage, moderate earthquakes without structural damage but possibly with some nonstructural damage, and major earthquakes without collapse but possibly with some structural as well as nonstructural damage.** In addition, building components

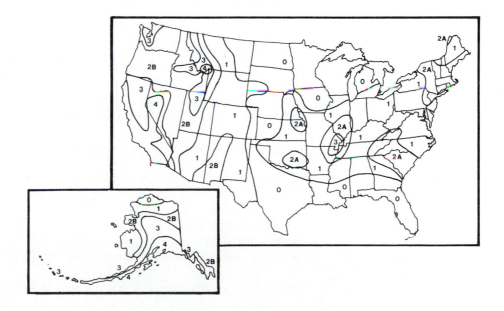

Figure 4.6 Seismic zone map.

should not cause any personal injury or damage to the structure resulting in its collapse.

The Building Response to Seismic Loads. For the sake of simplicity, the complex, random ground vibrations can be visualized as known horizontal movements traveling back and forth. Ignored are the vertical vibrational displacements, since the building is already designed for gravity in this direction. As the earth abruptly accelerates horizontally in one direction, taking the building foundation along, but leaving the portion of the building above the ground behind, it causes lateral inertia forces to act. This phenomenon is similar to that experienced by a person traveling in a car that suddenly increases in speed or the vertical inertia force experienced when an elevator abruptly rises. Assuming initially that the building and its foundation were rigid by ignoring the properties of the structure, as well as the particular character of the exciting motion, which includes the effect of the site geology, the acceleration of the building would be equal to the ground acceleration. The lateral *inertial forces* would then, according to Newton's second law (Eq. 1.1), be the product of the building mass M and the ground acceleration a, where the mass, in turn, is equal to the building weight, W divided by the acceleration of gravity, $g = 32\ ft/s^2$; in other words, $W = Mg$ (Eq. 1.2)

$$F = Ma = W(a/g) = WC_1 \qquad (2.6)$$

Letting $a/g = C_1$, a *seismic base shear coefficient,* results in ground acceleration expressed as a percentage of g: $a' = C_1\ (g)$. For example, a ground acceleration of $0.2g$ is equal to $0.2(32) = 6.4$ ft/sec^2, or the lateral inertia force is 20% of the weight of the rigid building. Ground shaking accelerations vary from $0.2g$ to $0.3g$ for a moderate earthquake up to $0.4g$ and greater for a severe earthquake.

To obtain the magnitude of the inertial forces, the building has been assumed to be rigid and to rock and heave similarly to a ship at sea, which obviously is not true. The inertia of the building mass resists ground movement and causes the building to deform and hence activates the building stiffness; the random shaking of the base results in a series of complex oscillations of the building. However, it has been found that the elastic response of multistory buildings is largely determined by the first or fundamental mode of vibration, so the building can be treated for preliminary design purposes as a *single-degree-of-freedom* (SDF) *system* or cantilever pendulum. One-story buildings actually are single-degree-of-freedom systems where the mass of the building m can be visualized as lumped together at a single point, the roof level (Fig. 4.5, center bottom). If the seismic motion is assumed harmonic with a circular period of vibration $\omega = 2\pi/T$ (i.e., the distance of one full wavelength of harmonic motion $\omega t = 2\pi$) and if damping is neglected, it can be shown that the time it takes for this ideal, elastic building to go through one full cycle of one back and forth free motion, called the *natural period of vibration, T,* is given by

$$T = 1/f = 2\pi\sqrt{m/k} = 2\pi\sqrt{W/kg} \qquad (\text{sec}) \qquad (4.4)$$

where f = natural cyclic frequency of vibration = $1/T$ = number of cycles of vibration
        per second (cps), often also called *hertz* (Hz)
    m = mass = W/g, see Eq. (1.2)
    W = weight of building
    g = acceleration due to gravity = 32.2 ft/sec^2
    k = $1/\Delta$ = lateral stiffness of structure [see Eq. (4.31)]

The equation clearly expresses that the natural period of the building is dependent on its mass and stiffness. A building has a shorter natural period or higher frequency if it is either lighter or stiffer. The natural period of the building should be different from the ones of the existing sources (e.g., earthquake or wind) to prevent resonance, as is discussed in Section 2.1. For the application of Eq. (4.4), refer to Example 5.4.

Naturally, the seismic motion is not harmonic and damping must be taken into account, in addition to the many other considerations that are now discussed; in other words, Eq. (4.4) is only helpful for introducing basic concepts. To evaluate the effect of the building stiffness on the magnitude of the lateral inertia force, or building acceleration, several simple regular buildings of decreasing stiffness are considered as inverted pendulums (see Fig. 4.5, right), visualizing an effective mass for each case. The increase of the pendulums in height indicates the increase of flexibility and also the increase in the natural period T of the building, where the natural period represents the time it takes for the structure to swing back and forth, going through one full cycle of free motion. The pendulums selected show the typical fundamental natural periods of 0.1 to 0.4 sec for low-rise buildings and roughly 1 to 3 sec for ordinary high-rise buildings. An average damping value, as provided by natural damping, is included. The structure is assumed to behave linearly, although in reality some members may be stressed beyond their elastic limit, resulting in a nonlinear response with permanent deformations.

The pendulums are attached to a movable base and are shaken back and forth under a simulated ground acceleration, which is based on the maximum one experienced as obtained from an actual or simulated earthquake accelerogram. If the maximum response (e.g., building acceleration divided by ground acceleration) of each of the buildings is plotted, a *response spectrum* is formed for the given seismic movement, keeping in mind that every site actually has its own response spectrum, although the shapes of the various response spectra are very similar to each other. The UBC provides three response spectra for three different soil conditions in Fig. 4.5. The following conclusions may be drawn from the selected response spectrum:

- A rigid (infinitely stiff) building will move together with the ground without deflection (i.e., its fundamental period of vibration is zero, $T = 0$), so that its acceleration is the one of the ground, or the ratio of the two is equal to 1.
- A relatively stiff building with a natural period of about 0.3 sec reaches a lateral acceleration much larger than the one of the ground. This is the state where the soil period coincides with the one of the building, thus theoretically resulting in resonance (infinitely large forces) if there no damping is present. In other words, the ground motion is amplified as it moves through the building, reaching an acceleration on the upper floors much higher than that at the ground. According to the UBC, this peak acceleration of the building is reached at about 0.5 sec (depending on the soil conditions) and is 2.5 times larger than the maximum ground acceleration. In addition, structures located on soft soils will experience a larger acceleration than the ones on stiff soils.
- As a building becomes more flexible, its acceleration decreases and eventually, at about 1.5 sec for stiff clay soils, for example, shows an acceleration of less than the ground. However, it must be kept in mind that these most intense waves occur during the first few seconds of the earthquake and that the later long-period waves may come close to the natural period of the building and thus be much more critical.

It may be concluded from our discussion that flexible buildings have longer natural periods of vibration with a lower maximum acceleration and therefore attract less lateral force than stiff buildings with short periods and a larger acceleration. Also, keep in mind that buildings on soft soils may experience a larger acceleration than ones on stiff soils. The principle of the response spectrum is defined mathematically by modifying the coefficient, C_1 in Eq. (2.6), and obtaining a numerical coefficient, $C = 1.25S/T^{2/3}$ [see Eq.(4.6)] that takes into account soil characteristics (S) and the natural period of the building (T).

The assumption in the response spectrum that the behavior of a multistory building can be simply represented by its fundamental period as a single-degree-of-freedom system is oversimplistic, especially for flexible structures. Visualize a high-rise structure to be modeled as lumped masses at each floor level along a vertical column (e.g., Fig. 4.7), where a typical lumped mass consists of the floor and the one-story high columns, walls, and partitions. When this lumped-mass, multiple-degree system is excited by ground vibrations, many different types of motion, with their corresponding deflected shapes, are possible. There will be many degrees of freedom (i.e., patterns of deformation) where a natural period is associated with each mode. However, for stiff multistory buildings, it has been found that the first or fundamental period is the dominant one and that even for more flexible buildings it contributes the largest influence.

The UBC modifies Eq. (2.6) by considering, in addition to the building weight, the dynamic characteristics of the building (i.e., the fundamental period of vibration) and the geology of the site, the seismic risk zone, the type of structure, the importance of the building, and the building configuration (size and shape of building, size and arrangement of structure and mass). According to the UBC, the minimum design seismic forces in critical earthquake regions may be based on the *static force procedure* (equivalent static lateral force method) only for regular structures with certain structure types limited in height and for certain occupancies only. The configuration of *regular buildings* refers to relatively symmetrical building shapes (in plan and elevation) and the uniform distribution of the lateral force-resisting structure with a continuous path of force flow, as well as to a relatively symmetrical arrangement of massing. In other words, regular building shapes should have no significant vertical and horizontal irregularities in geometry, stiffness, strength, and mass so that torsion and stress concentrations are kept to a minimum. But, also, the size of regular buildings must be considered; large horizontal building slabs may not be able to act as a unit and fracture, or tall slender building towers may tend to turn over under seismic force action.

The seismic forces for irregular buildings in critical seismic zones must generally be determined by using a *dynamic lateral force procedure,* where the building's response T to given ground vibrations is found together with the corresponding seismic forces.

Irregular buildings refer to a significant asymmetrical arrangement and discontinuities of geometry, mass, or the lateral force-resisting structure. Physical discontinuities (e.g., setbacks, offsets, cantilevers, reentrant corners, cutouts, large openings, and other abrupt changes) cause interruption of force flow and stress concentrations. Asymmetrical arrangement of mass (i.e., where the center of mass does not coincide with the center of rigidity) may result in large torsional forces, which may be especially critical when changes of arrangement occur within the building. *Vertical irregularities* may occur, for instance, through change of structure system or structure location. This change may result in discontinuity of capacity (e.g., *weak story* with less strength than the story above) or in discontinuity of stiffness (e.g., *soft story* with less lateral stiffness than the story above). *Horizontal irregularities refer,* for example, to asymmetrical plan shapes (e.g., F-, L-, T-, and U-shaped plans) or discontinuities within the floor diaphragms such as cutouts, large openings, reentrant corners, out-of-plane offsets, and other abrupt changes.

The static lateral force procedure can be used for the design of regular buildings with certain height limits (Table 4.2) and irregular buildings not more than five stories or 65 ft high. According to the 1991 edition of the UBC static force procedure, the total design base shear V in a given horizontal direction may be obtained as follows:

$$V = \frac{ZIC}{R_w} W \tag{4.5}$$

Here, the *seismic zone factor*, *Z*, represents one of the six seismic zones (0, 1, 2A, 2B, 3, and 4) into which the United States is divided and which are identified in the seismic zone map (Fig. 4.6). The seismic zone factor represents an approximate value of the effective horizontal peak ground acceleration, expressed as a fraction of the acceleration of gravity *g* and varies from zero in zone 0 to 0.4*g* in zone 4 with the highest seismic risk. (Notice that in 1994 the new edition of the UBC was published when this book was already in production.)

The terms in the basic seismic formula [Eq.(4.5)], may be defined approximately, for preliminary design purposes, as follows:

V = total lateral seismic force, or shear at the base
W = total dead load and applicable portions of live load such as:
    for storage and warehouses, 25% of the live load
    where a partition load is used for floor design, 10 psf must be included
    where the snow load is greater than 30 psf, the entire snow load may have to be
        included
    the total weight of permanent equipment must be included
Z = seismic zone factor (for zone, refer to seismic zone map, Fig. 4.6)
    = 0 in zone 0 (no seismic risk)
    = 0.075 in zone 1
    = 0.15 in zone 2A
    = 0.2 in zone 2B
    = 0.3 in zone 3
    = 0.4 in zone 4 (greatest seismic risk)
I = importance factor (depends on the occupancy category)
    = 1.25 for *essential facilities* (e.g., hospitals, fire and police stations, emergency shelters) and *hazardous facilities* (e.g., structures containing toxic or explosive substances dangerous to the general public)
    = 1.00 for all other special and standard occupancy structures
C = numerical coefficient, often called the design response spectrum value [Eq.(4.6)]
S = site coefficient for soil characteristics
    = 1.0 for rocklike formations or stiff soil condition, where the soil depth is less than 200 ft
    = 1.2 for deep stiff soil condition, where the soil depth exceeds 200 ft
    = 1.5 for a soil profile at least 70 ft in depth that contains more than 20 ft of soft to medium stiff clay, but not more than 40 ft of soft clay; this factor is usually assumed if the soil properties are unknown
    = 2.0 for a soil profile containing more than 40 ft of soft clay
T = fundamental period of vibration (seconds) of the structure in the direction under consideration [Eq.(4.7)]
R_w = numerical coefficient (Table 4.2)

The *C* coefficient is defined as

$$C = \frac{1.25S}{T^{2/3}} \leq 2.75 \tag{4.6}$$

where $C/R_w \geq 0.075$, to assure that a long-period building is designed for adequate seismic load action.

The fundamental period of vibration T of the building in the direction under consideration may be approximated, according to the UBC, as

$$T = C_t (h_n)^{3/4} \tag{4.7}$$

where h_n = building height (ft) above base

C_t = 0.035 for steel moment-resisting frames

= 0.030 for reinforced concrete moment-resisting frames and eccentrically braced steel frames

= 0.020 for all other buildings

In the case of moment-resisting frames that resist 100% of the lateral forces, for fast estimation purposes, a natural period of 0.1 sec per floor may be assumed.

$$T = 0.10N \tag{4.8}$$

where N = the total number of stories above grade.

The influence of the building structure on seismic action and its ability to resist lateral forces is taken into account by the R_w *factor*. A partial list of R_w coefficients together with building height limits for various structure systems is given in Table 4.2; the UBC should be consulted for a complete list of values. The coefficients range from 4 to 12, as based on past performance of the structure type reflecting the differences in ductility or energy-dissipation capacity, as well as the degree of structural redundancy.

A high factor ($R_w = 12$) is assigned to the ductile *special moment-resisting space frame* (SMRF). In this continuous, three-dimensional, rigid-frame structure, the members and their joineries have the required strength and deformation capacity to resist gravity and lateral forces without the help of other structure types (e.g., bearing walls, diagonal bracing). *Ductility* is the ability of members, together with connections within the structure, to go through a number of inelastic cycles of deformation without significant loss of strength, thus clearly establishing the high redundancy of the structure and its capacity for carrying overloads. In contrast, in nonductile *ordinary moment-resisting frames* (OMRF), the connections cannot provide necessary reserve strength, so R_w may be as low as 5.

Low factors (e.g., $R_w = 6$) are given to *bearing wall systems,* which carry both gravity and lateral forces (e.g., bearing shear walls); for example, the diagonals in trussed frames not only support lateral forces but also gravity loads. Bearing concrete and masonry shear walls, being brittle materials, form cracks and lose their load-carrying capacity rather than redistribute stresses, should they not be properly reinforced; they are incapable of deforming much beyond the elastic range and lack reserve strength. Therefore, structures that lack ductility must have more strength, and they must remain within the elastic range of the materials, allowing no inelastic deformations. The height limits for bearing wall structures in seismic zones 3 and 4 are 65 or 160 ft, respectively.

R_w factors of intermediate magnitude (e.g., $R_w = 8$) are given to laterally *braced frames,* that is, frames that are braced by trussing or by shear walls. In these structure systems, the frames carry the gravity loads, whereas the bracing resists only lateral loads. These systems can be visualized as stiff boxes (similar to bearing wall structures) responding in shorter periods and thus causing larger lateral forces. They lack ductile behavior because of their inability to deform much (e.g., compression steel diagonals buckle, in other words fail through instability, or brittle walls crack), although, in case of failure of the bracing elements, the frame will enter as a backup system and will provide some lateral resistance. The height limits in seismic zones 3 and 4 vary from 65 to 240 ft.

In buildings that employ the *braced moment-resisting frame system* (which is a typical example of a *dual system)*, the frame carries the gravity loads while the lateral forces are resisted by the frame together with the shear walls or diagonal bracing. The moment-resisting ductile frame must be capable of resisting at least 25% of the total lateral load, according to the UBC, in order to act as a backup system (as based on its ductile character) in case of failure of the bracing elements. Because of the varying degree of ductility of the dual systems, that is, depending on the combination of the structure types, the R_w coefficients range from high values of 12 for eccentrically braced ductile steel frames and ductile rigid frames braced by concrete shear walls to low values for other systems.

TABLE 4.2
Partial List of R_w Coefficients for Various Structure Systems[a]

Structural Systems	R_W	H (ft)
Bearing wall system		
Light-framed walls with shear panels		
Plywood walls for structures three stories or less	8	65
All other light-framed walls	6	65
Shear walls		
Concrete	6	160
Masonry	6	160
Braced frame system (using trussing or shear walls)		
Steel eccentrically braced ductile frame	10	240
Light-framed walls with shear panels		
Plywood walls for structures three stories or less	9	65
All other light-framed walls	7	65
Shear walls		
Concrete	8	240
Masonry	8	160
Concentrically braced frames		
Steel	8	160
Concrete (only for zones 1 and 2)	8	—
Heavy timber	8	65
Moment-resisting frame system		
Special moment-resisting frames (SMRF)		
Steel	12	N.L.
Concrete	12	N.L.
Concrete intermediate moment-resisting frames (IMRF) (only for zones 1 and 2)	8	—
Ordinary moment-resisting frames (OMRF)		
Steel	6	160
Concrete (only for zone 1)	5	—
Dual systems (selected cases are for ductile rigid frames only)		
Shear walls	12	N.L.
Concrete	8	160
Masonry		
Steel eccentrically braced ductile frame		
Steel	12	N.L.
Concentrically braced frame		
Steel	10	N.L.
Concrete (only for zones 1 and 2)	9	—

[a]*Partial reproduction from Uniform Building Code*, 1991 Edition, courtesy of International Conference of Building Officials.

The following approximations may be derived from Eq. (4.5) for the preliminary design of ordinary low-rise buildings (e.g., about five stories for rigid-frame steel buildings and seven stories for masonry buildings) in regions of greatest seismic risk using $Z = 0.4$, $I = 1$, and $C = 2.75$.

$$V = \frac{ZIC}{R_w} W = \frac{0.4 \, (1) \, 2.75}{R_w} W = \frac{1.1 \, W}{R_w} \tag{4.9}$$

For example,

$$\text{Ductile moment-resisting frames } (R_w = 12)\text{: } V = 0.09W \tag{4.9a}$$

$$\text{Masonry or concrete bearing walls } (R_w = 6)\text{: } V = 0.18W \tag{4.9b}$$

It may be concluded that the maximum lateral inertia forces for ordinary regular low-rise buildings in highly seismic regions range from approximately 9% of the building weight for a more flexible ductile structure to about 18% for a stiff structure using brittle materials. In other words, the lateral seismic forces for bearing wall structures are roughly double as much as for ductile rigid frame structures. Notice that in this approximate approach the height of the building has no direct effect on the magnitude of the lateral force, in contrast to wind action.

For high-rise buildings, however, height must be taken into account in determining the natural period of vibration [see Eq. (4.7)]. For example, the inertial force for a flexible, 30-story, ductile rigid frame structure is only about 4% of the building weight.

It is common practice to express the magnitude of seismic forces as a percentage of the building weight. Typical values range from about 5% for flexible, ductile, high-rise frame structures to approximately 20% for stiff bearing wall buildings. In addition, because of the heavier weight of concrete buildings, the lateral seismic forces are much higher than for steel buildings.

Lateral Distribution of Base Shear.   The formula $V = ZICW/R_w$ does not indicate how the shear force is distributed throughout the height of the structure. The shear force, at any level, depends on how the structure deforms, that is, on the mass at that level and the amplitude of oscillation, which may be assumed to vary linearly with the height of the building. Earthquake forces deflect a structure into certain shapes, known as the natural modes of vibration. Only the most important first three modes are shown in Fig. 4.7, but it must be realized that a high-rise building actually is a multiple-degree-of-freedom system with many possible patterns of deformation. For example, a 30-story, rigid- frame building has a first period of vibration of roughly $T_1 = 0.1N = 0.1(30) = 3$ sec and, according to the linear approximation of the first three modes of vibration in Fig. 4.7, has the following second and third periods of approximately $T_2 = T_1/3 = 3/3 = 1$ sec, and $T_3 = T_1/5 = 3/5 = 0.6$ sec. In any case, the first or fundamental mode, as exemplified by the pendulum, has still been found to contribute the largest influence, especially for stiff or short-period buildings responding more abruptly. Flexible long-period buildings respond in slower, longer, and more complex movements; in this case, the higher modes of vibration indicate the whiplash effect, which is taken into account by the concentrated load F_t at the top of the building (Fig. 4.7). Related to the shape of each mode is a certain distribution of lateral forces. As long as there is no large inelastic deformation, instantaneous lateral forces are found by superposition of the forces resulting from each vibration mode; sometimes the forces add and sometimes they cancel one another. The resulting maximum shear envelope (Fig. 4.7) can be visualized as being generated by an approximately triangular load. The Uniform Building Code uses a triangular lateral load configuration for a building with uniform mass distribution along its height, that is, for a building structure of regular rectangular shape with nearly equal floor weights and heights and no

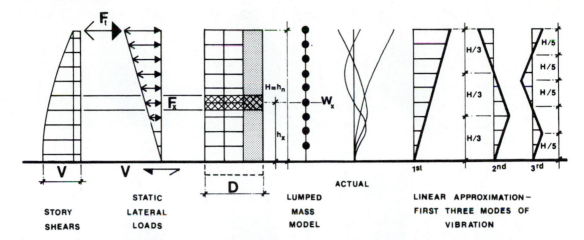

Figure 4.7 Equivalent lateral seismic load distribution (Reproduced with permission from *The Vertical Building Structure,* Wolfgang Schueller, copyright © 1990 by Van Nostrand Reinhold.)

vertical and horizontal irregularities in geometry, stiffness, strength, and mass. For the effect of the mass arrangement of other building shapes on the lateral force distribution, refer to Fig. 4.8.

Recognizing the whiplash effect of flexible buildings, the code places part of the total lateral base shear V as a concentrated load F_t at the top of the structure, while the balance $(V - F_t)$ is to be distributed in a triangular fashion over the entire building height. The top load F_t is only present if the fundamental period of vibration $T > 0.7$ sec, but does not have to be larger than $T = 3.57$ sec. It is equal to

$$F_t = 0.07TV \le 0.25V \tag{4.10}$$

According to the UBC, the balance of $(V - F_t)$ is distributed over the entire building height (generally as concentrated loads at the floor levels) in a triangular fashion, assuming the floor weight constant for every floor level, as shown in Fig. 4.7.

$$F_x = \frac{(V - F_t)\,h_x}{h_1 + h_2 + \cdots + h_n} = \frac{(V - F_t)\,h_x}{\sum\limits_{i=1}^{n} h_i} \tag{4.11a}$$

Taking into account that the weight at the floor levels may not be constant, the magnitude of the distributed forces F_x is given by

$$F_x = \frac{(V - F_t)\,w_x h_x}{\sum\limits_{i=1}^{n} w_i h_i} \tag{4.11b}$$

where w_i, w_x = that portion of W that is located at or is assigned to level i or x, respectively

h_i, h_x = height in feet above the base to level i, n, or x, respectively; the level n is the uppermost level in the main portion of the structure

To take into account the unpredictable character of earthquakes, the UBC requires, in addition to the calculated seismic loads, *accidental torsion.* It requires that

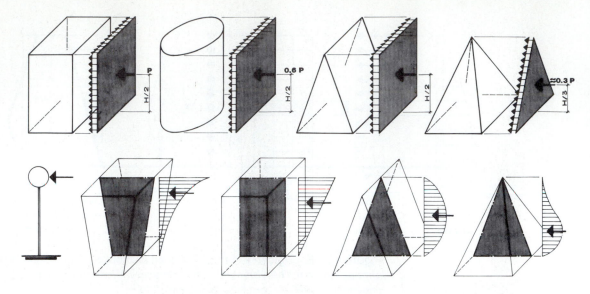

Figure 4.8 Effect of building form on wind and seismic load distribution.

the center of mass at each floor level be displaced laterally (in each direction) from its original position by 5% of the building dimension (at that floor level) perpendicular to the force direction.

It may be concluded that for a regular rectangular building with a uniform mass distribution the lateral forces due to seismic ground movement may be visualized as an equivalent static, triangular load, as indicated in Figure 4.8. At the bottom of the same figure, the lateral force distribution for some other common building configurations is identified. For a uniform mass arrangement, this lateral force distribution is proportional to the shape of the building volume, thereby clearly demonstrating the pyramid as an efficient form.

It may also be concluded that the building form and mass distribution, as reflected by the plan organization and vertical massing, determine the location of the resultant lateral seismic force. Furthermore, the form of the lateral force-resisting structure and its location within the building volume determine the type of action of the seismic force. When the centroid of mass does not coincide with the center of resistance, twisting is generated, as may be the case at floor levels with abrupt changes of stiffness. The effect of asymmetry, as seen in section and plan, is typical for the new breed of hybrid, compound building forms currently so much in fashion.

EXAMPLE 4.1

A one-story industrial building located in a suburban area with a basic wind speed of 80 mph and in seismic zone 3 is investigated for lateral force action; its dimensions are shown in Fig. 4.9. The building is a bearing wall structure with 8-in. exterior masonry walls that act as shear walls to resist the lateral forces. The roof structure consists of wood framing with joists supported on girders, which, in turn, are resting on the long walls.

The layout of the vertical structure is simplified by arranging it symmetrically about both major axes and ignoring the effect of wall openings for doors and windows, which may, however, weaken and reduce the rigidity of the walls and, when asymmetrically placed, cause torsion. In other words, there is assumed to be no eccentricity between the center of mass and the center of rigidity that would cause torsion—accidental torsion is ignored. The soil properties are not known.

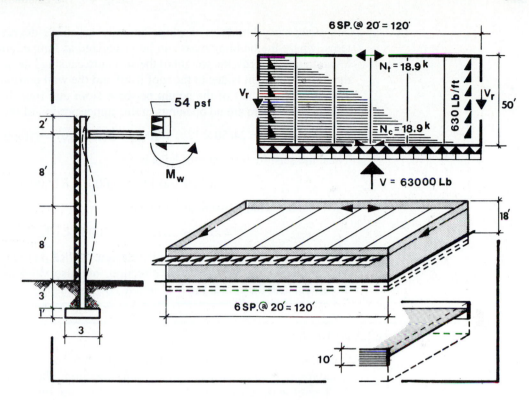

Figure 4.9 Example 4.1.

The loads are as follows:

Felt, three-ply with gravel	5.5 psf
1/2-in. plywood sheathing	1.5 psf
Joists and bridging estimated as:	3.0 psf
Insulation, lights, and misc.	5.0 psf
Suspended ceiling	5.0 psf
Roof framing	20.0 psf
Girders estimated as	4.0 psf
Roof weight	24.0 psf
Live load	20.0 psf
Average load for 8-in. masonry walls and pilasters	90.0 psf

A uniform wind pressure of 21 psf is assumed according to Eq. (4.3).

Here, only the lateral force action perpendicular to the long walls is checked; the force action in the other direction is easily resisted by the long walls.

According to the equivalent static lateral force method, the seismic load is

$$V = \frac{ZIC}{R_w} W = \frac{0.3(1)2.75}{6} W = 0.14W$$

where $Z = 0.3$, $I = 1$, $R_w = 6$, $S = 1.5$

$T = C_t(h_n)^{3/4} = 0.020(16)^{3/4} = 0.16 \text{ sec} \leq 0.7 \text{ sec}$

$C = 1.25 \, S/T^{2/3} = 1.25(1.5)/0.16^{2/3} = 6.36 > 2.75$

One-story buildings displace laterally as single-degree-of-freedom systems, where the building mass can be visualized as lumped together at the roof level. In other words, the weight of the structure causing lateral seismic action on the roof diaphragm is due to the roof itself and the wall portion consisting of the parapet and one-half of the height between floor and roof. It is conservatively assumed that there are no openings in this upper portion of the wall.

$$W = 24(50 \times 120) + 90(120 + 50)2(10) = 450,000 \text{ lb}$$

Thus, the seismic force at roof level is

$$V = 0.14W = 0.14 \, (450,000) = 63,000 \text{ lb}$$

The resultant wind pressure at roof level is

$$P_w = 21(8 \; + \; 2)120 = 25,200 \text{ lb} < 63,000 \text{ lb}$$

The seismic force clearly controls the design. Notice, that the 63-k force is not only applicable in the transverse direction, but also valid for the longitudinal direction since the values for T, C, and R_w do not change.

EXAMPLE 4.2

Determine the approximate critical lateral loading for a 25-story, ductile, rigid-space frame concrete structure in the short direction. The rigid frames are spaced 25 ft apart in the cross direction and 20 ft in the longitudinal direction. The plan dimension of the building is 175×100 ft, and the structure is $25(12) = 300$ ft high (see Fig. 4.10a). This office building is located in an urban environment with a wind velocity of 70 mph. and in seismic zone 4. For this first investigation, an average building dead load of 16 psf is used. Accidental torsion is neglected. The soil conditions are unknown.

The total building weight is

$$W = 0.016(100 \times 175 \times 300) = 84,000 \text{ k}$$

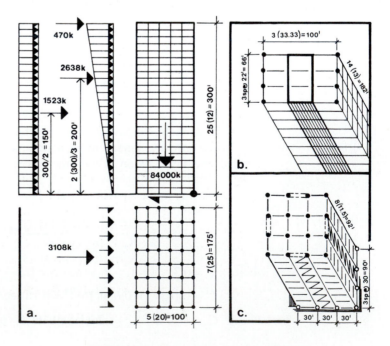

Figure 4.10 Example 4.2 and exercises.

Design resistance $\geq$ effect of design loads, or required resistance.

$$\phi\, R_n \geq \Sigma\, y_i\, Q_i \tag{4.17}$$

The term *resistance* includes both strength limit states and serviceability limit states. The load factors γ and the resistance factors ϕ reflect the degree of uncertainty, that is, inaccuracies in the theory, variations in the material properties and member dimensions, and uncertainties in the determination of the loads. Probabilistic methods have been used for the selection of the factors by the AISC LRFD Specification. The resistance factor ϕ is equal to or less than 1.0, because there is always the chance for the actual resistance to be less than the nominal value R_n. The load factors γ are equal to or larger than 1.0, reflecting the fact that the actual load effects may deviate from the nominal mean loads. The factored resistance R_n depends, among other criteria, on the material and limit state, the mode and consequence of failure, and possible errors in the model used for the analysis.

The AISC LRFD Specification for Structural Steel Buildings permits the use of both elastic and plastic structural analyses. It lists the following load combinations with the corresponding load factors, as based on the ASCE 7-88 Specs for the design of regular buildings:

(1) $1.4D$

(2) $1.2D + 1.6L + 0.5(L_r \text{ or } S \text{ or } R)$ $\tag{4.18}$

(3) $1.2D + 1.6(L_r \text{ or } S \text{ or } R) \ + \ (0.5L \text{ or } 0.8W)$

(4) $1.2D + 1.3W + 0.5L + 0.5(L_r \text{ or } S \text{ or } R)$

(5) $1.2D + 1.5E + (0.5L \text{ or } 0.2S)$

(6) $0.9D - (1.3W \text{ or } 1.5E)$

Exception: The load factor on L in combinations (3), (4), and (5) shall equal 1.0 for garages, areas occupied as places of public assembly, and all areas where the live load is greater than 100 psf.

When the structural effects of F, H, P, or T are significant, they shall be considered in designs as the following factored loads: $1.3F$, $1.6H$, $1.2P$, and $1.2T$.

The more common resistance factors range from $\phi = 0.9$ for tension yielding and for beams in bending and shear, $\phi = 0.85$ for columns, $\phi = 0.75$ for tension fracture, to $\phi = 0.6$ for bearing on A307 bolts.

The ACI Building Code for Reinforced Concrete lists the following load combinations with the corresponding load factors γ for the design of regular buildings:

$$1.4D + 1.7L$$
$$1.4(D + T)$$
$$0.75[1.4D + 1.7L + 1.7(W \text{ or } 1.1E)] \tag{4.19}$$
$$0.75(1.4D + 1.4T + 1.7L)$$
$$0.9D + 1.3(W \text{ or } 1.1E)$$

Should earth pressure H and/or fluid pressure F have to be considered in the design, use $1.7H$ and/or $1.4F$. The resistance or strength reduction factors ϕ for several situations are defined in Section 3.1, Eq. (3.3).

4.2   THE RESPONSE OF BUILDINGS TO LATERAL FORCE ACTION

A building structure can be visualized as consisting of horizontal planes or floor and roof framing, as well as the supporting vertical planes of walls and/or frames. The horizontal planes tie the vertical planes together to achieve a box effect and a certain

STABILITY OF BASIC VERTICAL STRUCTURAL BUILDING UNITS

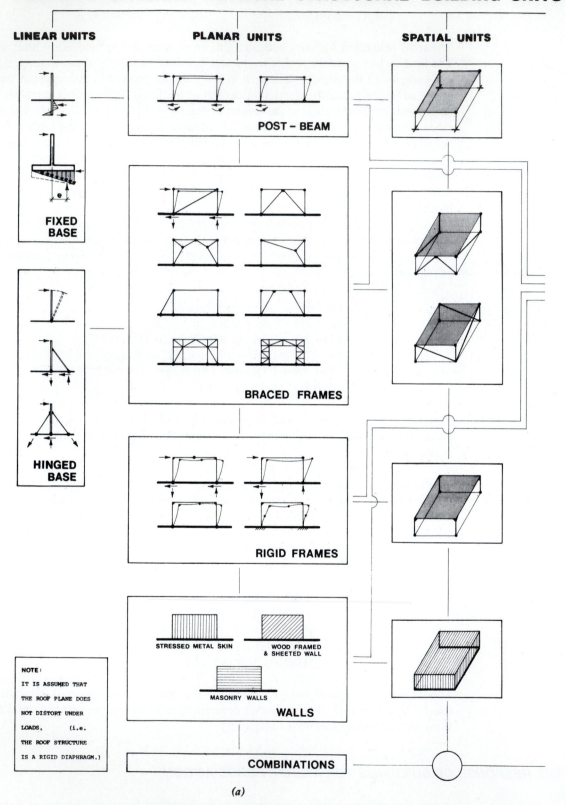

(a)

Figure 4.11 Stability of basic lateral force-resisting structure systems.

POSSIBLE LOCATION OF UNITS IN BUILDING

SYMMETRICAL ARRANGEMENT

ASYMMETRICAL ARRANGEMENT

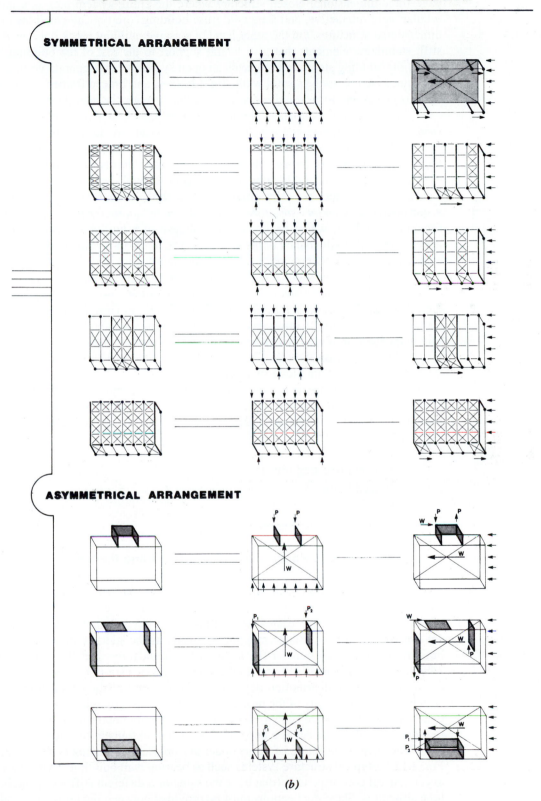

(b)

Figure 4.11 Continued

degree of compactness. It is obvious that a slender, tall tower building must be a compact, three-dimensional closed structure where the entire system acts as a unit. The tubular, core interactive, and staggered truss buildings are typical examples of three-dimensional structures. On the other hand, a massive building block only needs some stiff, stabilizing elements that give lateral support to the rest of the building. In this sense, the building structure represents an open system where separate vertical planar structure systems, such as solid walls, rigid frames, and braced frames, are located at various places and form *stand-alone systems* that provide the lateral stability.

Every building consists of the *load-bearing structure* and the *nonload-bearing structure.* The main load-bearing structure, in turn, is subdivided into the *gravity structure,* which carries only the gravity loads, and the *lateral force-resisting structure,* which supports gravity forces, but also must provide stability to the building. For the condition where the lateral bracing only resists horizontal forces (or a partition acts as a shear wall), but does not carry gravity loads with the exception of its own weight, it is considered a *secondary structure.* Failure of secondary members is not as critical as for main members, where an immediate collapse of a building portion may occur, depending on the redundancy of the structure. The nonload-bearing structural building elements include wind bracing, as well as the membranes and skins, that is, the curtains, ceilings, and partitions that cover the structure and subdivide the space.

The lateral force-resisting structure in a building tower may be concentrated entirely in the central core, for instance, when on optimal view and thus a light perimeter structure is desired. Conversely, rather than hiding the lateral force-resisting structure in the interior, it may be exposed and form the perimeter structure, as for tubes. In low-rise buildings, on the other hand, the lateral force-resisting structure may form interior or exterior stand-alone systems.

Structure systems, in general, are discussed in Chapters 1 and 10, as well as throughout this book. With respect to lateral force-resisting structures, as discussed previously, we may distinguish the following three basic types, as identified in Fig. 4.11:

- Moment-resisting frames
- Braced frames
- Shear walls
- Combinations of these (e.g., dual systems)

Diaphragm Action of Horizontal or Sloped Building Planes

The horizontal forces are transmitted along the floor and roof planes, which act as deep, flat beams spanning between the vertical lateral force-resisting structures. For example, for the basic building shapes in Fig. 4.11, the curtain panels are assumed to act similarly to one-way slabs spanning vertically as the lateral wind forces strike the building facade. They transfer the loads to the foundations and the roof level from where, in turn, they are distributed to the lateral force-resisting structural systems. The force flow or force distribution depends on the location of the lateral force-carrying structures and the rigidity of the roof and floor structures. Possible locations of the vertical stable structures are shown in Fig. 4.11b. The horizontal roof framing is trussed to indicate its ability to act as a horizontal beam and to transfer the lateral forces.

The deep beam action of floors under uniform lateral loading is demonstrated in Fig. 4.12 for typical structure systems such as bearing wall buildings, symmetrical and asymmetrical core structures (that have the same or a different stiffness), and bundled tube structures. Shear connections must be provided between the horizontal and vertical planes to transmit the lateral forces; this is especially important for precast, large-panel, and skeletal construction.

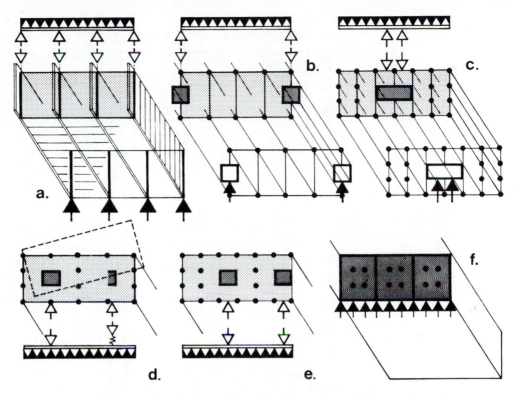

Figure 4.12 Horizontal force flow (Reproduced with permission from *The Vertical Building Structure,* Wolfgang Schueller, copyright © 1990 by Van Nostrand Reinhold.)

Should the resultant force action not pass through the center of rigidity, twisting is generated. This torsion is efficiently resisted by a perimeter structure as in Fig. 4.12f, rather than by a concentric core (as in Fig. 4.12c). Because of the large lever arms, it is beneficial to let solid walls at least wrap around the exterior corners of a building.

Once the lateral forces are distributed to the resisting vertical structure planes, these systems must act as vertical cantilevers to carry the forces down to the ground. The various two- and three-dimensional structure systems in Fig. 4.13 demonstrate how the overturning moment due to wind action is resisted at the base by different cantilevers, consisting of solid walls or skeleton construction. In this case, the skeletons may act either as *flexural systems* when they are rigid frames (Fig. 4.13f), where the members react in bending, or they may be predominantly *axial systems,* as for braced frames where the forces are effectively resisted directly in tension and compression (Fig. 4.13b, d, e).

The behavior of roof and floor diaphragms under lateral loads depends on their shape, depth-to-span ratio, structure, and support conditions (e.g., lateral stiffness). In contrast to the thin-skin construction for many roofs of low-rise buildings, the typical floor structures for high-rise buildings are treated as rigid; that is, concrete slabs, metal decking with concrete fill, cast-in-place slab–beam systems, and precast floor systems with cast-in-place topping may be considered rigid, whereas wood and metal (without concrete fill) decking may be treated as flexible. To control the flexibility of plywood diaphragms, the UBC limits the span-to-depth ratio to 4:1.

There are situations, however, where the floor or roof framing must be strengthened and stiffened to perform properly as a diaphragm. These situations occur when the floor is weakened by numerous openings or abrupt changes of form where lateral

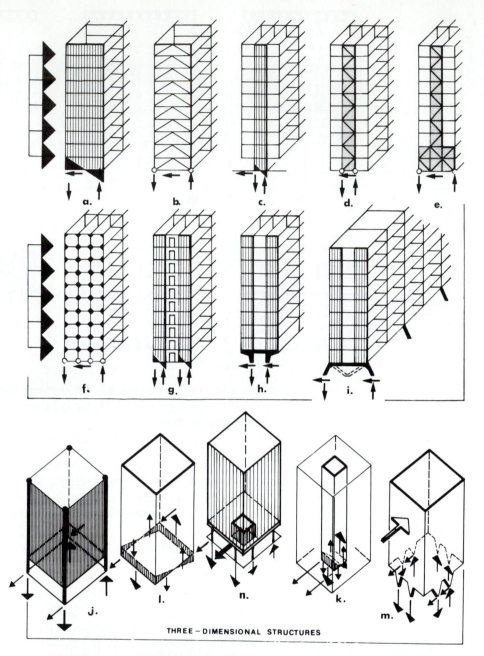

Figure 4.13 Resistance of various structure systems to overturning. (Reproduced with permission from *The Vertical Building Structure*, Wolfgang Schueller, copyright © 1990 by Van Nostrand Reinhold.)

forces have to be redistributed or when stress concentrations occur due to sudden changes of mass and structure systems.

When floors are constructed from precast concrete decks, the panels must be connected so that the entire floor can act as a unit, without allowing any slippage between the panels and to achieve diaphragm action. The floor structure is tied together by continuous bond beams around the perimeter and across the building, possibly together with a cast-in-place topping to form cell-like compartments.

When facade columns are not laterally supported by floor beams and when it is necessary to increase the torsional resistance of the spandrel beams, floor bracing may be employed along the periphery of the building. When structure systems within a building change, horizontal transition systems may be required, for example, when the lateral forces are transferred from the interior core to the exterior perimeter structure or when a building sits on top of another building.

When roof decking for low-rise buildings does not provide sufficient lateral support, the purlins may have to be braced. Then truss action may even be necessary for the distribution of the lateral loads to the respective vertical lateral force-resisting systems (e.g., Fig. 5.28). Bracing effectively resists the racking of the roof, especially under lateral loads from a direction other than parallel or perpendicular to the building, thereby providing torsional stiffness for the roof plane. Bracing may also be needed during the construction stage. It consists, generally, of single steel angles or rods capable of acting in tension only, although they will carry compression if they are prestressed. They are usually arranged in crossing pairs so that if racking puts one of the elements into compression (which it is incapable of resisting) its complement will resist the racking by being placed in tension. In the case of deep roof beams, the bracing may be located along the upper or lower chord planes, or the planes may be linked, forming a kind of space truss. Bracing may also be necessary to prevent the lateral buckling of the entire roof plane and to keep the filler beams in place so that they give the proper lateral support to the compression flange when the major beam bends (Fig. 4.14).

The specific arrangement of the bracing depends on the position of the lateral force resisting elements, the required stiffness of the roof structure, and the magnitude of the force flow. Various layouts of trusses as a function of their supporting bents are shown in Fig. 4.11b. The reader can easily follow the flow of forces from roof level to foundations as caused by lateral loading along the major building axes.

A floor diaphragm acts as a deep beam at each story with respect to lateral force action and can be visualized as the thin web of a huge, flat floor girder primarily resist-

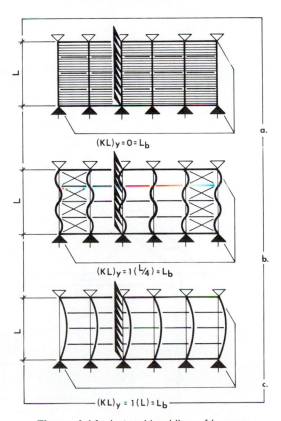

Figure 4.14 Lateral buckling of beams.

ing shear, while the boundary members (e.g., spandrel beams) perpendicular to the load action carry the rotation in an axial manner similar to girder flanges, as demonstrated in Fig. 4.15.

EXAMPLE 4.4

The wood roof diaphragm of Example 4.1 is investigated. In this flexible thin-skin-framed structure, the roof sheathing acts as a diaphragm when properly fastened with shear connectors to the purlins and perimeter members so that the skin does not buckle and shear can be transferred.

The wood diaphragm carries the following uniform seismic load in simple beam action (see Fig. 4.9):

$$w = W/L = 63,000/120 = 525 \text{ plf}$$

The maximum shear the diaphragm must resist (which is equal to the wall reactions) is

$$V_r = W/2 = wL/2 = 63,000/2 = 31,500 \text{ lb}$$

Considering this shear to be distributed along the end wall evenly yields the following unit shear:

$$v_{max} = V/d = 31,500/50 = 630 \text{ lb/ft}$$

For the selection of the plywood diaphragm, refer to the respective tables of the lumber industry. The allowable shear values are a function of the plywood thickness and layout, the type (size) and spacing of nails, and whether blocking for connecting the edges of the panels is used. Blocking may be provided along the diaphragm boundaries in regions of large shear, rather than over the entire

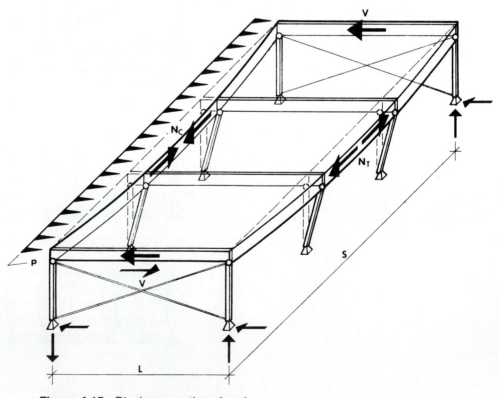

Figure 4.15 Diaphragm action of roof.

roof. Anchorage must be provided so that the shear can be transferred from the roof plane to the walls.

The uniform seismic load causes the following maximum moment for simple beam action:

$$M_{\max} = wL^2/8 = WL/8 = 63.00(120)/8 = 945 \text{ ft-k}$$

This rotation is resisted along the diaphragm edges perpendicular to the load action in compression and tension (see Fig. 4.15), that is, by the flanges or chords of the huge, horizontal roof plate girder. Hence, the maximum axial chord forces P at midspan are

$$\pm P = M/d = 945/50 = 18.9 \text{ k}$$

The wood chord size is estimated by assuming an allowable tensile stress of $F_t = 655$ psi, as

$$A_n = P/C_D F_t = 18,900/1.60(655) = 18.03 \text{ in.}^2 \tag{3.89}$$

Try two 2×8s, $A = 21.75$ in.^{2} that must be properly attached to the diaphragm and have the following net area assuming 3/4-in.-diameter bolts for splicing:

$$A_n = 1.5[7.25 - (3/4 + 1/16)]2 = 19.32 \text{ in.}^2 > 18.03 \text{ in.}^2$$

Keep in mind that more chord members may be required depending on the type of splicing.

Lateral Building Deflection

The lateral sway of buildings is primarily caused by wind and earthquake. Keep in mind, however, that under dynamic loading low buildings may vibrate primarily in their first mode, whereas taller ones will also deflect in the higher modes similar to the movement of a snake (Fig. 4.7). The sidesway produced by unsymmetrical vertical loading usually balances itself over several stories and can be neglected for relatively stiff buildings.

The tall buildings in Fig. 4.16 respond to lateral forces primarily as *flexural cantilevers* if the resisting structure consists of shear walls or braced frames. The behavior of these systems is controlled by rotation rather than shear; they have a high shear stiffness as provided by the solid wall material or axial capacity of the diagonals, so the shear deformations can be neglected. But tall buildings act as *shear cantilevers* when the resisting elements are rigid frames, since the shear can only be resisted by the girders and columns in bending. In this context, the effect of rotation (i.e., axial shortening and lengthening of columns) is secondary and may be ignored for preliminary design purposes. The combined action of different structure systems, such as rigid frames, together with a braced core (depending on the relative stiffness of each system), may have the appearance of a flat S-curve, with a shear-type frame building sitting on top of a flexural cantilever-type structure.

In the preceding discussion it has been assumed that the structure systems were for tall buildings and of the same height. It is apparent that when the shear wall or braced frame is no longer shallow and slender, as for the extreme case of a horizontal panel in a low-rise building, they do behave like shear cantilevers and not flexural cantilevers.

The design of slender buildings is controlled by stiffness rather than strength, which includes not only translational deflections, but also torsional deflections at building corners; hence the concern for torsional stiffness as well. The wind causes a dynamic response in flexible buildings, especially in the across-wind direction, thereby generating twisting and oscillations. Furthermore, secondary moments due to

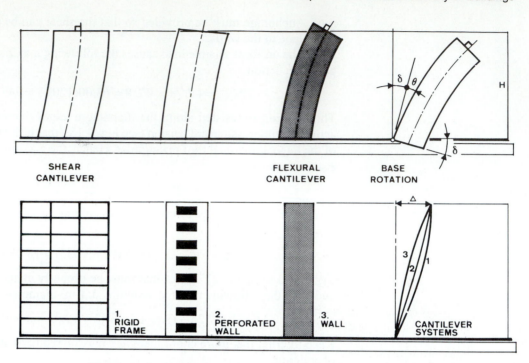

Figure 4.16 Effect of structure type on cantilever action. (Reproduced with permission from *The Vertical Building Structure,* Wolfgang Schueller, copyright © 1990 by Van Nostrand Reinhold.)

large drift (*P*–Δ effect) must be considered. In general, the stiffness of buildings or their lateral drift must be controlled for the following reasons:

- **Architectural integrity**: A flexible building may distort so much that it stresses the architectural subsystems or the curtains and partitions, as well as the mechanical systems.

- **Occupant comfort**: Excessive lateral sway, together with oscillations (i.e., acceleration) due to gusty winds, may not be acceptable for human comfort. People may be motion sensitive and not be able to work under high wind storms, and some may even experience motion sickness. Naturally, the following simple drift ratios do not control the performance of the structure with respect to the building occupants' sensitivity to motion.

- **Structural stability:** Under lateral loads, the mass of a flexible building will shift and move the centroid of each floor, and thus the weight of the structure increases the tendency of overturning. The resulting large lateral deflections cause the *P*–Δ effect, where the eccentric action of gravity loads results in secondary moments due to first-order displacements. The consideration of the *P*–Δ phenomenon is essential for the design of flexible tall buildings.

Unless it can be demonstrated that greater drift can be tolerated, codes limit the lateral deflections of buildings or *story drift* (that is the lateral displacement of one level relative to the level above or below) to $\Delta_{max} = h/400 = \mathbf{0.0025h}$ for wind. For earthquake displacement, the limits according to the UBC are

$$T < 0.7\text{sec} \quad \Delta \le h/200 = 0.005h \quad \text{or} \quad 0.04h/R_w$$
$$T \ge 0.7\text{sec} \quad \Delta \le h/250 = 0.004h \quad \text{or} \quad 0.03h/R_w$$

In other words, for relatively stiff buildings (with a fundamental period of vibration T of less then 0.7 sec), the allowable lateral deflection for seismic action may be double that for wind action since wind occurs more often and lasts longer; it should not exceed 0.005 times the story height h. For example, the allowable interstory displacements for the 25-story building in Example 4.2, with respect to seismic action, can be determined as follows:

$$0.03/R_w = 0.03/12 = 0.0025 < 0.004$$
$$\Delta \leq 0.0025h = 0.0025(12)12 = 0.36 \text{ in.}$$

Or the allowable deflection for the building is $\Delta \leq 0.36(25) = 9$ in. In case portions of the building structure are not constructed to act as an integral unit, they should be adequately structurally separated to avoid contact under the deflections of each portion.

In the following discussion, the lateral deflections for the basic structure systems in Fig. 4.17 are introduced.

(a) Frames: The lateral deflection for some typical regular, single-bay and multibay frames are discussed in Chapter 5 (see Figs. 5.17 and 5.20). For example,

- For single-story post-beam structures with identical cantilever columns (Fig. 4.17b) and for frames with rigid girders and hinged bases (Fig. 4.17a), the lateral deflection is

$$\Delta = \frac{Ph^3}{3E\Sigma I_c} \tag{5.16}$$

- For single-story rigid frames in general, refer to Eqs. (5.20) and (5.39).

(b) Braced Frames: The bracing of frames can take many configurations, as shown in Fig. 4.11a. Here, only the lateral deflection of a diagonally braced, single-story bay

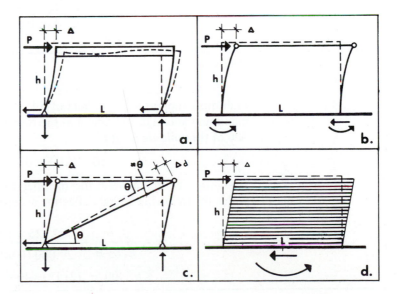

Figure 4.17   Basic lateral force-resisting structures.

in Fig. 4.17c is briefly investigated. For this situation the lateral distortion is generally controlled by *web drift* due to the tensile diagonal only.

The axial tension in the diagonal is found from geometry to be $P_d = P/\cos\theta$. This force causes an increase of the diagonal member length $L_d = L/\cos\theta$, according to Eq. (2.54), and is equal to

$$\Delta_d = \left(\frac{PL}{AE}\right)_d = \frac{f_d L}{E\cos\theta} \tag{a}$$

The horizontal component of the elastic elongation of the diagonal is approximately equal to

$$\Delta_2 = \frac{\Delta_d}{\cos\theta} = \frac{f_d L}{E\cos^2\theta} \tag{b}$$

where f_d = average stress in diagonal member due to horizontal force action
$= P_d/A_d = P/A_d \cos\theta$

Equation (b) can now be expressed in terms of the cross-sectional area A_d.

$$\Delta = \frac{f_d L}{E\cos^2\theta} = \frac{PL}{EA_d \cos^3\theta} = \frac{Ph}{EA_d \cos^2\theta \sin\theta} \tag{4.20}$$

However, keep in mind that for determining the lateral deflection of braced frames in high-rise buildings the deformations of the horizontal web members (with respect to web drift) and chord drift must be taken into account (see Schueller, 1990).

For the condition where the lateral sway should not exceed 1/400 of the building height and for $E = 29,000$ ksi, the required cross-sectional area of the steel diagonal bracing to resist a lateral force P (kips) should be

$$\Delta = \frac{Ph}{EA_d \cos^2\theta \sin\theta} \le \frac{h}{400}$$

or

$$A_d \ge \frac{0.014P}{\cos^2\theta \sin\theta} \quad (\text{in.}^2) \tag{4.21}$$

EXAMPLE 4.5

A diagonally braced, hinged steel frame (A36), spanning 25 ft and being 12 ft high must support a lateral load of 5 k at the roof level. Estimate the absolute maximum lateral displacement of the frame; the lateral sway is not to exceed 1/400 of the building height.

$$\tan\theta = 12/25 \quad \text{or} \quad \cos\theta = 0.902.$$

Assuming, conservatively, the member to be fully stressed, $f_d = F_d = 22$ ksi, yields the following lateral deflection according to Eq. (4.20):

$$\Delta = \frac{f_d L}{E\cos^2\theta} = \frac{22\,(25 \times 12)}{29,000\,(0.902)^2} = 0.28 \text{ in.}$$

$$\le h/400 = 12(12)/400 = 0.36\text{in.} \quad \text{OK}$$

(c) Concrete and Masonry Walls: The lateral deflection of a solid cantilever wall (in this case due to single load action at the top, as shown in Fig. 4.17d) consists of the flexural deformation Δ_f and the shear deformation Δ_s. The deflection formulas can be obtained from structural engineering handbooks (by letting the shearing modulus G for concrete and masonry be 40% of its modulus of elasticity E, $G \cong 0.4E$) as

$$\Delta = \Delta_f + \Delta_s = Ph^3/3EI + 1.2\,Ph/GA \qquad (4.22a)$$
$$\cong Ph^3/3EI + 3Ph/AE$$

Letting $A = tL$ and $I = tL^3/12 = AL^2/12$,

$$\Delta = \frac{P}{Et}\left[4\left(\frac{h}{L}\right)^3 + 3\left(\frac{h}{L}\right)\right] \qquad (4.22b)$$

A similar lateral deflection expression can be derived for uniform load action, as is typical for high-rise building cantilevers (see Schueller, 1990), as

$$\Delta = \frac{1.5V}{Et}\left[\left(\frac{h}{L}\right)^3 + \left(\frac{h}{L}\right)\right], \quad \text{where } V = wh \qquad (4.23)$$

where t = wall thickness (in.)
   L = effective length of wall (in.)
   P = lateral force on top of wall (lb)
   w = uniform cantilever load (plf)
   h = wall height (ft)
   E = modulus of elasticity (psi)

For simple wall cantilever action, we may draw the following conclusions from Eqs. (4.22b) and (4.23) for preliminary design purposes as demonstrated in Fig. 4.18:

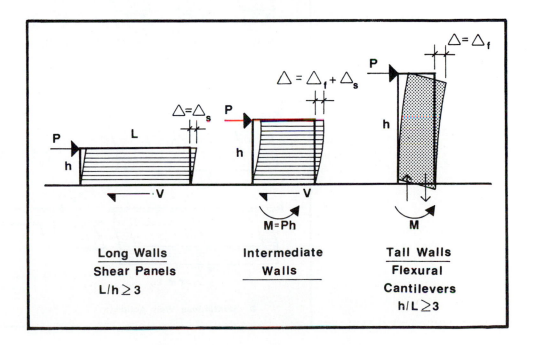

Figure 4.18 Lateral deflection of solid walls.

- *Long solid walls* that are at least three times longer than high ($L/h \geq 3$) behave like stiff shear panels where shear deflections control, which are related directly to the wall proportions (h/L); in other words, shear panels wrack in a linear fashion by distorting as a parallelogram.
- *Tall, slender, solid walls* that are at least three times higher than wide ($h/L \geq 3$) may be treated as deep cantilever beams where flexural deflection clearly controls.
- *Intermediate solid walls* represent the transition stage from the long to the high slender wall. For this condition, shear together with flexure causes the lateral deflection.

The lateral deflection of a solid wall that is fixed at top and bottom due to single load action at the top (Fig. 4.19b), as is assumed for story walls in multistory structures, can be shown to be equal to

$$\Delta = \frac{P}{Et}\left[\left(\frac{h}{L}\right)^3 + 3\left(\frac{h}{L}\right)\right] \tag{4.24}$$

Notice that the shear deflection Δ_s is identical in both Eqs. 4.22b and 4.24 having the same loading condition and hence is not affected by the boundary conditions of the wall!

From Eq. (4.24), the following rough preliminary approximations may be derived:

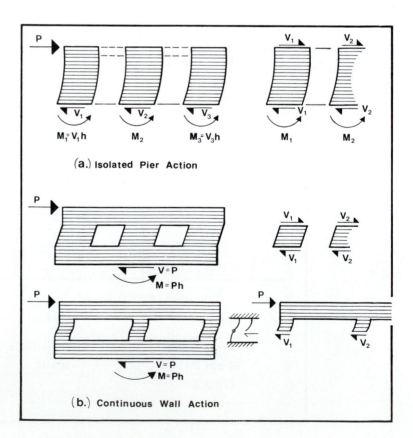

Figure 4.19   Lateral deflection of walls with openings.

- *Long, solid walls* that are approximately one- and-a-half times longer than high ($L/h \geq 1.5$) may be assumed to be controlled by shear deflections.
- *Tall, slender, solid walls* that are approximately six times higher than wide ($h/L \geq 6$) behave primarily as flexural cantilevers.
- For *intermediate solid walls* ($0.67 < h/L < 6$), shear together with flexure must be considered.

The lateral deflection of a shear wall with openings may be assumed (for preliminary design purposes) as equal to the deflections of the individual piers (Fig. 4.19). We may want to distinguish between the following two extreme situations:

- A single-story wall with door openings or large window openings that allow only shallow spandrel beams can be treated as *isolated piers,* in other words, as individual cantilevers (Fig. 4.19a).
- A wall with relatively small window openings that allow deep spandrel beam action may be broken down into three horizontal strips. Here, the lateral deflection of the wall panel may be based (for rough preliminary design purposes) solely on the stiffness of the piers between the openings by treating the piers as fixed along the horizontal top and bottom beam strips (Fig. 4.19b). It is apparent that this approach, in general, is not conservative since it neglects the nature of the boundaries (e.g., rotation of piers), in addition to the axial shortening of the piers as well as the flexibility of the spandrels and foundation.

Should the *n* wall piers be all identical (width *b),* each provides the same shear resistance: $V_1 = V_2 = V_n = P/n$, and the unit shear is constant, $v = V_i/b$. Hence, the approximate lateral deflection can be easily obtained from Eqs. (4.22b) or (4.24) as based on the preceding assumptions. Should, however, the various piers be of different length (i.e., lateral stiffness), the unit shear is no longer constant. Its magnitude is dependent on the stiffness of the piers, keeping in mind that the stiffer elements attract more force (as is discussed later in this chapter). Once the shear forces for piers are known, the lateral deflection for the wall panel can be determined as previously discussed.

EXAMPLE 4.6

The shear stresses and the approximate lateral deformations for the 8-in. brick wall (7.625 in. actual thickness) in Fig. 4.20 are determined. Assume $F_v = 17$ psi and $E = 2,400,000$ psi.
Each pier carries the following shear:

$$V_1 = V_2 = V_3 = V_i = V/n = 30/3 = 10 \text{ k}$$

Hence, the critical shear stress in the wall is

$$f_v = V_i/bt = 10,000/5(12)7.625 = 21.86 \text{ psi}$$
$$\leq 1.33 \, F_v = 1.33(17) = 22.61 \text{ psi, OK}$$

First, the lateral deflection of the piers (i.e., middle strip) is estimated. Since $h/L = 5/5 = 1 > 0.67$, shear and flexure are considered.

$$\Delta = \frac{P}{Et}\left[\left(\frac{h}{L}\right)^3 + 3\left(\frac{h}{L}\right)\right] = \frac{10}{2400\,(7.625)}\left[(1)^3 + 3\,(1)\right]$$

$$= 0.00055 + 0.00164 = 0.00219 \text{ in.} = 2.19 \times 10^{-3}\text{in.}$$

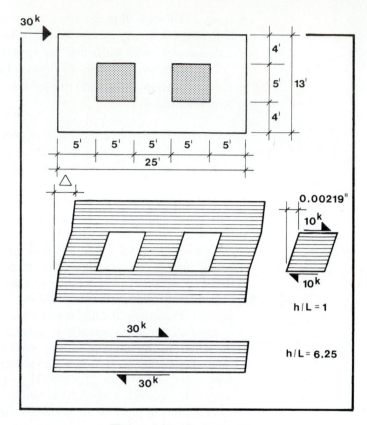

Figure 4.20   Example 4.6.

The lateral deflection of the solid top and bottom strips is clearly controlled by shear ($L/h = 25/4 = 6.25 > 1.5$).

$$\Delta = \frac{3P}{Et}\left(\frac{h}{L}\right) = \frac{3\,(30)}{2400\,(7.625)}\left(\frac{4}{25}\right) = 0.00079\,\text{in.}$$

Hence, the total lateral wall deflection is roughly

$$\Delta = 0.00219 \,+\, 2(0.00079) = 0.0038 \text{ in.}$$
$$\leq h/400 = 13(12)/400 = 0.39 \text{ in.,}\quad \text{OK}$$

(d) Wood-frame Shear Walls: The lateral deflection of wood-frame shear walls as commonly used for light-frame residential construction, in contrast to solid walls, is very difficult to determine. It depends not only on the type, size, and spacing of studs and the type, thickness, and placing of surfacing material (e.g., plywood, particleboard, wood boards, drywall, plaster), which may be provided on one side or both sides, as well as the presence of blocking and whether the panels are staggered, but also on the response of nailing (or stapling) and the height-to-width ratio (h/L) of the shear panel.

The plywood sheathing, which is widely used, is either applied horizontally (perpendicular to the studs) or diagonally. The *transversely sheathed wall* is obviously weaker and more flexible than the *diagonally sheathed wall,* which allows truss action. Stud walls may be diagonally braced by using thin metal strips or boards flush with the frame face, called *let-in bracing,* especially when panels are installed vertically.

The lateral deflection of wood-frame walls is generally controlled by shear, that is, they wrack like shear panels in the configuration of a parallelogram (Fig. 4.21a).

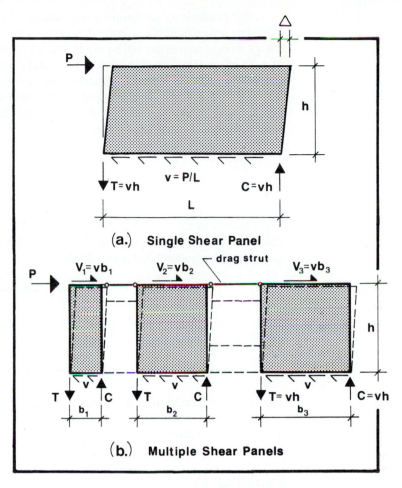

Figure 4.21   Wood shear walls.

Usually, the shear is assumed uniformly distributed along the wall so that the unit shear is equal to

$$v = P/L \quad (plf) \tag{4.25}$$

The sheathing provides the necessary shear resistance to lateral force action, besides preventing the studs from buckling about their weak axis.

The studs at the shear wall edges are treated generally as chords that resist the bending in tension and compression (Fig. 4.21a).

$$M = Ph = (vL)h = T(L) = C(L)$$

or

$$T = C = M/L = vh \tag{4.26}$$

To control lateral deflections and to ensure deep beam action, usually an upper limit is set to the shear wall proportions at $h/L = 3.5$ for plywood and particle panels with *blocking at the horizontal sheathing joints,* and $h/L = 2$ for unblocked conditions. For other sheathing materials, the respective codes should be consulted (e.g., UBC); for preliminary design purposes, however, an allowable height-to-length ratio of $h/L \leq 1.5$ may be assumed. Should the wall proportions become slenderer than permitted by the maximum h/L ratios given, then the shear walls can no longer be treated solely as stiff deep beams.

For preliminary design purposes, a wood shear wall with openings may be perceived as linked multiple-shear panels rather than a continuous wall (Fig. 4.21b); multistory wood-frame shear walls may be treated as single-story shear walls sitting on top of each other. Because of the nature of lateral distortion of wood walls, the shear distribution across the individual panels may be assumed constant, in contrast to solid walls. Hence, the unit shear v in every wall panel can be obtained by simply dividing the lateral force P at the roof level (or the total shear at the floor level above) by the sum of the panel lengths, Σb, (Fig. 4.21b).

$$P = v(b_1 + b_2 + b_3) = v\Sigma b$$

or   (4.27)

$$v = P/\Sigma b$$

The rotation in the various panels is assumed to be resisted by the respective outer studs in tension and compression (Fig. 4.21b). It can be shown that these axial forces are of constant magnitude for all panels.

$$T_1 = C_1 = M_1/b_1 = V_1 h/b_1 = vb_1 h/b_1 = vh$$

Since the unit shear v and the wall heights h are constant, $T_1 = C_1 = T_2 = C_2 = \cdots$; therefore,

$$T = C = vh \qquad (4.28)$$

The lateral deformation of wood-frame panels is generally not controlled by the stiffness of studs and sheathing but primarily by *nail slip;* therefore, its magnitude cannot be that easily determined as for some solid walls—the respective literature for wood structures (e.g., UBC) should be consulted.

EXAMPLE 4.7

The single-story wood building shown in Fig. 4.22 with shear walls along its perimeter must resist a wind pressure of 30 psf against its long face. The unit shear and chord forces in the transverse studwalls are determined so that the walls can be designed.

The roof diaphragm carries the following uniform wind load:

$$w = 30 (12/2) = 180 \text{ plf.}$$

Hence, each of the transverse end walls must resist the load P which is equal to the maximum shear $V_{\max}$ in the roof diaphragm.

$$P = V_{\max} = 180(65)/2 = 5850 \text{ lb}$$

The transverse walls consist of multiple panels, where the critical panel proportion is $h/L = 12/6 = 2 < 3.5$. Therefore, the unit shear can be determined, according to Eq. (4.27) as

$$v = P/\Sigma b = 5850/(8 + 10 + 6) = 244 \text{ plf}$$

Refer to the respective literature on wood construction or to codes (e.g., UBC) for allowable shear values of plywood nailed walls to select the appropriate plywood sheathing (with blocking), as well as the type and spacing of nailing.

The rotation in the transverse walls is resisted in the outer studs of the individual wall panels. The chord forces, according to Eq. (4.26), are

$$T = C = vh = 244(12) = 2928 \text{ lb}$$

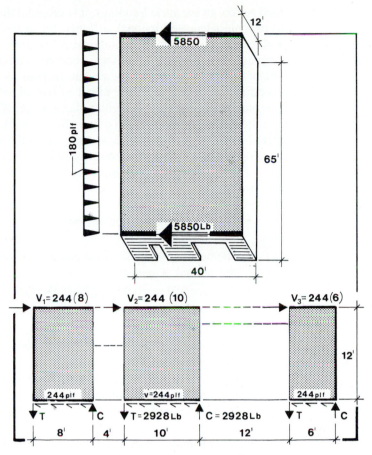

Figure 4.22 Example 4.7.

For the design of the chords (end studs in wall panels), in addition, gravity loads must naturally be taken into account; compression and uplift are to be considered separately.

The Distribution of Lateral Forces to the Vertical Lateral Force-resisting Structures

The layout of the resisting vertical systems can take many different forms, varying from symmetrical to asymmetrical arrangements (Figs. 4.11b and 4.23) or range from a minimum of three planes to a maximum of a cellular wall subdivision. The resisting system may be located within the building as a single spatial core unit or as separate individual planes, or it may be located along the periphery.

In a symmetrical building with a regular arrangement of vertical structures, where the line of action of the resultant of the applied lateral loads passes through the center of rigidity (center of resistance), the supporting systems deflect equally in a purely translational manner. In a symmetrical building, the geometric center coincides with the center of mass and the center of rigidity; it is located at the intersection of the axes of symmetry. However, even if a symmetrical ordinary structure should be loaded eccentrically, the torsional effects are relatively small and may be ignored for preliminary design purposes.

Asymmetry in buildings is caused by geometry, stiffness, or mass distribution; the applied resultant load does not act through the center of rigidity. In this case, the rigid floor diaphragms not only translate, but also rotate in the direction of the lateral load action. In eccentric buildings, it is essential that the vertical elements be arranged to provide adequate torsional stiffness, for instance, along the periphery for curtain wall construction to control rotational deflections. In asymmetrical building shapes of stepped or sloped configurations, torsional response is generated that varies with height and corresponds to the change of plan forms. It is beyond the scope of this investigation to deal with the effect of hybrid, compound building shapes, and the complexity of composite, irregular plan and elevation forms, as well as their response and their effect on lateral load action. In this work, only basic concepts of action and reaction of simple, but typical, rectangular building forms are briefly investigated to develop some feeling for the behavior of buildings.

The following simplifying analysis assumptions are used for preliminary design purposes:

- Rigid diaphragm action of floors; they must be stiff enough to go through rigid body displacement.
- The bending stiffness of the planar, vertical structure about its weak axis (i.e., slab action) is neglected.
- The torsional resistance of the individual planar vertical elements is neglected.
- Intersecting planar units are treated as separate; there is no shear flow around the corners.
- The shear flow along shear walls is considered constant, as based on average shear stresses.
- Hinged connections between the floor diaphragms and the lateral force-resisting vertical planes are assumed (i.e., interaction of floor systems and shear walls is ignored).

Although floor structures for buildings can generally be treated as rigid diaphragms, there are situations for which this may not be the case, such as the following:

- Closely spaced shear walls in relatively narrow buildings are stiffer in comparison to the floor diaphragms.
- For low-rise buildings, the floor or roof diaphragms are often more flexible than the supporting shear walls (as in light wood-framed construction); therefore, for deep diaphragms (where shear deformations control) the diaphragms may be treated as flexible and supported by rigid walls.
- Floor diaphragms in long, narrow buildings with deep beam proportions of greater than, say, 3:1 that span large distances across the building.
- Floor diaphragms that are weakened by cutouts and openings for atriums, escalators, stairs, etc., unless they are braced.
- Wood and metal deck (without concrete fill) roofs as well as prefabricated floor systems without cast-in-place topping are to be treated as flexible, unless the floor or roof framing is braced to allow the planes to act as horizontal or inclined trusses (see Fig. 5.28).

The following presentation of typical cases is approached from the point of view of increase in the degree of complexity by, first, dealing with statically determinate conditions, proceeding to symmetrical buildings, then introducing torsion, and finally discussing irregular structures with mixed construction.

Statically Determinate Conditions. When a skeleton building is laterally braced by not more than three planar, lateral force-resisting systems, then the lateral force distribution to these elements can be treated as statically determinate, based purely on equilibrium conditions and independent of the stiffness of the resisting systems.

$$\Sigma F_x = 0, \qquad \Sigma F_y = 0, \qquad \Sigma M_z = 0 \qquad\qquad (2.12)$$

Notice that there is no need to distinguish between direct and torsional force action, and no need to determine the location of the center of resistance. The three vertical planes can be arranged in the building in any fashion as long as they are neither *parallel,* since they could not resist perpendicular loads, nor is their line of action *concurrent,* since they could not provide rotational stability.

    The typical placement of the vertical elements is along the perimeter, or as a core in the interior, or any other combination (Fig. 4.23). Should the three structural planes form a spatial unit, such as a channel or wide flange section, or should the arrangement consist of an L- or T-section together with a separate plane, the assembly can be

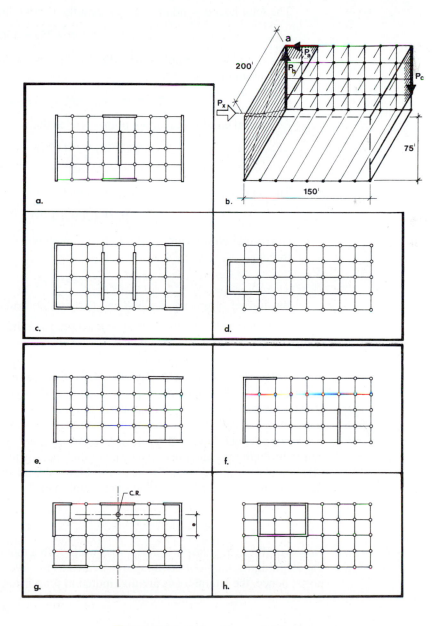

Figure 4.23 Lateral force distribution.

treated (for preliminary design purposes) as consisting of separate planes, as shown in Fig. 4.11b and as is further discussed in the subsection here on torsion. Treating a spatial section as consisting of separate elements is usually conservative, because the additional strength due to continuity is ignored. However, keep in mind that an earthquake may cause much larger forces in a connected system than in one where the planes are separated by joints that allow freedom of independent movement! Furthermore, the flanges of T-, U-, and I-sections improve only the flexural rigidity, but hardly offer any shear rigidity, as is needed for lower buildings.

EXAMPLE 4.8

For the 20-story building in Fig. 4.23b, with an average floor height of 10 ft (or a building height of 200 ft), the force distribution to the walls is found as caused by an average wind pressure of 30 psf; the hinged skeleton does not provide any lateral force resistance.

The total lateral wind pressure against the narrow face is

$$P_x = 0.030(200 \times 75) = 450 \text{ k}$$

Horizontal equilibrium in the x direction gives the force on wall A at the base of the building.

$$\Sigma F_x = 0 = 450 - P_A, \qquad P_A = 450 \text{ k}$$

Rotational equilibrium about point a yields the wall forces that resist the torsion.

$$\Sigma M_a = 0 = 450(75/2) - P_c(150), \quad \text{or} \quad P_c = 112.5 \text{ k}$$
$$\Sigma F_y = 0 = 112.5 - P_B \quad \text{or} \quad P_B = 112.5 \text{ k}$$

The total wind pressure against the broad face is

$$P_y = 0.030(200 \times 150) = 900 \text{ k}$$

Horizontal equilibrium in the y direction shows that wall A does not resist any forces.

$$\Sigma F_y = 0 = P_A$$

Because of symmetry, walls B and C share the load equally.

$$P_B = P_c = 900/2 = 450\text{k}$$

Statically Determinate Conditions Due to Symmetry. Even when there are more than three braced vertical planes, the lateral force flow may still be treated as statically determinate for certain conditions of symmetry of plan layout, structure systems, material, and force action. Some examples are shown in Fig. 4.23.

For a regular layout of rigid frame bents or *widely spaced* cross walls with force action through the center of resistance (so that there is no torsion) and where the floor diaphragms are stiffer than the resisting elements, all the identical n walls or rigid frame bents carry an equal portion of the total load P.

$$P_i = P/n \tag{4.29}$$

However, for many *closely spaced* cross walls with a much larger stiffness than that of the diaphragms, the floor diaphragm is treated as *flexible* with unyielding vertical supports; hence, the lateral loads are distributed in proportion to the tributary facade area.

EXAMPLE 4.9

The single-story building in Fig. 4.24 has a symmetrical layout of four parallel shear walls of the same material and thickness and must resist a total lateral load of 40 k.

(a) For the condition of four equally stiff shear walls and rigid diaphragm action of the floors, each wall carries the same force.

$$P_w = P/n = 40/4 = 10 \text{ k}$$

(b) For the condition where the interior shear walls are less stiff than the exterior ones (e.g., same wall thickness but less length or less thickness but the same length), it is apparent that the stiffer walls attract more force than the flexible ones, assuming rigid diaphragm action of the floors (see Example 4.10).

(c) In contrast, when the floor diaphragm is treated as flexible (e.g., wood construction), it is assumed to distribute the lateral loads as two independent, simply supported, deep beams to the rigid wall structures. In other words, the short interior walls carry double as much force as the long exterior walls because their tributary areas are twice as large as those for the exterior walls. Notice that the effect of stiffness does not enter the analysis since the walls are considered infinitely rigid. Hence, each interior wall carries

$$P_{wi} = P/(n-1) = 40/(4-1) = 13.3 \text{ k}$$

Here the term $n-1$ is equal to the number of bays. Each exterior wall resists

$$P_{we} = P_{wi}/2 = 13.3/2 = 6.7 \text{ k}$$

(d) The behavior of *semirigid* diaphragms is highly indeterminate; they are assumed to distribute the loads like a continuous beam to the vertical, flexible supports.

Indeterminate Force Distribution for Conditions of Symmetry. The lateral force distribution becomes indeterminate when more than three vertical planes are stabilizing the building and when special conditions of symmetry, as in the preceding subsection, are not available.

In this section, the special case of a *symmetrical layout* of different structure systems (e.g., Fig. 5.24b) that are not identical, as in Fig. 5.24a, is investigated. In this

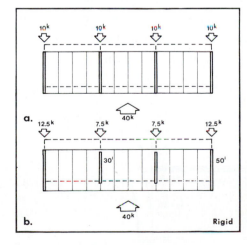

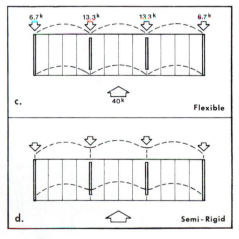

Figure 4.24 Horizontal diaphragm action.

instance, the loads cause only translational movement; in other words, the center of rigidity coincides with the resultant load action and the center of mass. The designer must distinguish, however, the difference between the construction that employs only one structure system of the same material, such as the bearing wall building of Fig. 4.24, and mixed construction, such as a rigid frame braced by concrete walls.

The horizontal loads cause the rigid floor diaphragms to laterally displace, thereby forcing all the vertical structural planes to deflect by the same amount under direct load action. Therefore, the lateral forces are distributed in proportion to the relative stiffness of the supporting structural systems. It is apparent that the stiffer structure attracts a larger load than the more flexible one, since it takes more force to deflect the more rigid element by the same amount as the less rigid one.

For example, both concrete piers in Fig. 4.25 can be treated as flexural cantilevers, where shear deformations can be ignored (i.e., $h/L \geq 3$); each one deflects by the same amount (see Table A.15).

$$\Delta_1 = \Delta_2 = \Delta = P_1 h^3/3EI_1 = P_2 h^3/3EI_2 \quad \text{or} \quad P_2 = P_1(I_2/I_1) \tag{a}$$

The total load P is shared by both piers.

$$P = P_1 + P_2 \tag{b}$$

Substituting Eq. (a) into (b) yields

$$P_1 = P\frac{I_1}{I_1 + I_2}, \quad P_2 = P\frac{I_2}{I_1 + I_2}$$

These expressions can be generalized for a structural system i that will resist the following portion of the total load P:

$$P_i = P(I_i/\Sigma I) \tag{c}$$

Here, the ratio, $I_i/\Sigma I$, is called the *relative stiffness* of the structural system i, which is expressed for any situation as $k_i/\Sigma k$. In other words, the lateral force P is distributed to the structural system i in direct proportion to the relative stiffness of that system.

$$P_i = P(k_i/\Sigma k) \tag{4.30}$$

where P = entire lateral load at given floor level
          P_i = the lateral load to be resisted by the structure system i
          k_i = absolute stiffness of system i
          $k_i/\Sigma k$ = relative stiffness of system i

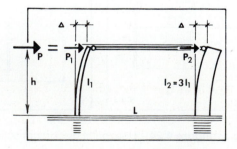

Figure 4.25 Effect of stiffness on lateral force distribution.

The preceding approach was assumed in the y direction. A similar approach is used for the force action in the x direction.

The *stiffness* is a measurement of the resistance of the structure to lateral deformation. It is defined by a spring constant k as a function of the lateral deflection Δ caused by a unit force $P = 1$; in this case, the stiffness is the reciprocal of flexibility.

$$k = P/\Delta = 1/\Delta \tag{4.31}$$

For the condition of n connected parallel or in a series arrangement lateral force-resisting structure systems, where each has its own absolute lateral stiffness k_i, the lateral sway can be expressed as

$$\Delta = P/\Sigma k_i \tag{4.32}$$

Since the distribution of the lateral force is not dependent on the absolute, but rather the relative stiffness of the connected elements, Eq. (4.30) can be further simplified for certain situations. However, first some of the approximate absolute stiffnesses for rigid frames, braced frames, and solid walls are briefly defined. The lateral deflection for some common simple structure systems has already been introduced previously in this chapter.

- The total stiffness for a single-story multibay frame with rigid girders and hinged bases or for post-beam structures with cantilever columns is equal to the sum of the individual column stiffnesses [see Eqs.(5.16) and (5.39)].

$$\Sigma k = 3E\Sigma I_c/h^3 \tag{4.33}$$

- The stiffness of an x-braced frame panel, as based on web drift due to the tensile diagonal only according to Eq. (4.20), is

$$k = (EA_d \cos^3 \theta)/L = (EA_d \sin\theta \cos^2\theta)/h \tag{4.34}$$

- The stiffness of a solid cantilever wall in response to flexure and shear according to Eq. (4.22b) is.

$$k = Et/[4(h/L)^3 + 3(h/L)] \tag{4.35}$$

As has already been stated, it is not the absolute stiffness given here, but the relative stiffness that is needed, because the various systems are only compared to each other. Simpler expressions are now developed for the relative stiffness of typical cases.

The deflection of a wall structure can be that of a flexural cantilever similar to a shallow beam that is proportional to the bending moment, or it can be that of a shear cantilever, similar to a deep beam where the slope of the curvature is proportional to the shear force, or it can be a combination of these.

For the distribution of the lateral forces to the long cross walls of a low-rise building often the flexural deflections can be neglected and analysis can be solely based on the shear deformations. Therefore, the relative rigidity of the wall is then directly proportional to its cross-sectional web area A_i, assuming the same wall material ($E = $ constant); or for equally thick walls, the lateral forces can be distributed simply in proportion to the wall lengths L.

$$P_i = P(k_i/\Sigma k) = P(A_i/\Sigma A) = P(L_i/\Sigma L) \tag{4.36}$$

For high-rise wall structures where the solid walls are at least three times higher than wide, the shear deflections may be neglected, so the wall stiffness is proportional to the moment of inertia when $E = $ constant; or, for equally thick walls, the wall stiffness is proportional to the cube of the wall length. Therefore, the lateral forces can be distributed as follows:

$$P_i = P(k_i/\Sigma k) = P(I_i/\Sigma I) = P(L_i^3/\Sigma L^3) \qquad (4.37)$$

Should different materials be used, such as concrete block together with clay brick walls, then Eq. (4.37) becomes

$$P_i = P[(EI)_i/\Sigma(EI)] \qquad (4.38)$$

The distribution of lateral forces to coupled shear walls is very complex and beyond the scope of this discussion.

EXAMPLE 4.10

Case (b) in Fig. 4.24 (Example 4.9) is investigated.
It is reasonable to assume that the shorter interior walls also act as shear panels in this single-story building so that the relative wall stiffness can be conveniently expressed in terms of wall length. Hence, the total lateral relative wall stiffness is

$$\Sigma k = \Sigma L = 2(30 + 50) = 160 \text{ ft}$$

Each exterior wall carries

$$P_e = P(L_e/\Sigma L) = 40(50/160) = 12.5 \text{ k}$$

Each interior wall resists.

$$P_i = P(L_i/\Sigma L) = 40(30/160) = 7.5 \text{ k}$$

Check: $\Sigma P = 0 = 2(12.5 + 7.5) - 40$,   OK

EXAMPLE 4.11

The single-story industrial building in Fig. 4.26, with a symmetrical layout of its perimeter shear walls about the x axes, is assumed to resist a uniform wind pressure of 20 psf against the long facade for preliminary investigation purposes. The effect of foundation deformation and soil settlement are considered insignificant. The lateral force distribution to the 8-in.-thick solid concrete walls (case A) is investigated.
The rigid roof diaphragm must support the following total wind pressure:

$$P_t = 0.020(160 \times 24/2) = 38.4 \text{ k}$$

Hence, each of the cross-wall systems must resist

$$P_w = 38.4/2 = 19.2 \text{ k}$$

This load, in turn, is shared by the cantilever pier ($h/L = 24/8 = 3$) and the shear panel ($h/L = 24/24 = 1$). For initial estimation purposes, the effect of the pier, because of its lack of lateral stiffness as compared to the stiff shear wall, could clearly be neglected.
The total relative wall stiffness according to Eq. (4.35) is

$$\Sigma k = k_1 + k_2 = 1/[4(3)^3] + 1/[4(1)^3 + 3(1)] = 0.009 + 0.143 = 0.152$$

Hence, the stiff wall carries

$$P_2 = P(k_2/\Sigma k) = 19.2(0.143/0.152) = 0.94P = 18.06 \text{ k}$$

whereas the pier only resists

$$P_1 = P(k_1/\Sigma k) = 19.2(0.009/0.152) = 0.06P = 1.14 \text{ k}$$

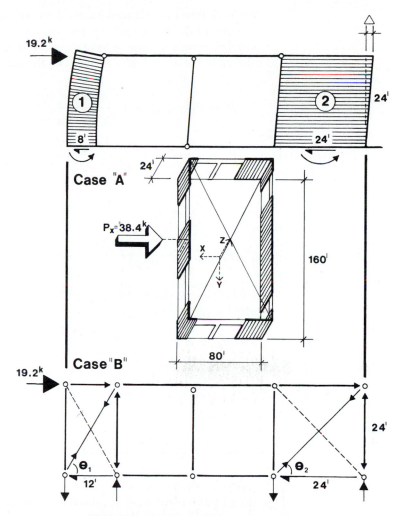

Figure 4.26 Examples 4.11 and 4.12.

Check: $\Sigma F_x = 0 = 18.06 + 1.14 - 19.2$, OK. It is apparent that the stiff, long wall resists nearly the entire lateral force.

The average shear stress in the concrete wall is

$$f_v = P_2/A_2 = P_2/L_2 t = 18{,}060/(8 \times 24 \times 12) = 7.84 \text{ psi}$$

This stress is easily resisted by the concrete!

EXAMPLE 4.12

Rather than the concrete walls in Example 4.11, braced framing, as shown in Fig. 4.26 (case B), is investigated using A36 steel.

The geometrical properties of the diagonal tension braces are

$$\tan\theta_1 = 24/12 = 2, \qquad \cos\theta_1 = 0.447, \qquad \sin\theta_1 = 0.894$$
$$\tan\theta_2 = 24/24 = 1, \qquad \cos\theta_2 = \sin\theta_2 = 0.707$$

The total lateral stiffness according to Eq. (4.34) is

$$\Sigma k = \Sigma (EA_d \sin\theta \cos^2\theta)/h$$

The force flow to the braced frames, however, cannot be determined easily, since the sizes of the braces are not known. For this preliminary investigation, it is assumed that all the braces have the same size. Therefore, the forces can be distributed to the braced bays as based on the inclination of the diagonals only.

$$P_i = P(k_i/\Sigma k) = P\left[\frac{\sin\theta_i \cos^2\theta_i}{\Sigma \sin\theta\cos^2\theta}\right] \tag{4.39}$$

$$P_1 = P\left[\frac{0.894\,(0.447)^2}{0.894\,(0.447)^2 + 0.707\,(0.707)^2}\right]$$

$$= P\left[\frac{0.1786}{0.532}\right] = 0.336P = 0.336\,(19.2) = 6.45\text{k}$$

$$P_2 = P\left[\frac{0.707\,(0.707)^2}{0.532}\right] = 0.664P = 0.664\,(19.2) = 12.75\text{k}$$

In this case, the size of the steel rod is controlled by the larger bay. Its size is estimated according to Eq. 3.87.

$$f_t = P_d/A_d \le 0.33F_u, \qquad \text{where } P_d = P_i/\cos\theta_1$$
$$= P_i/(A_d \cos\theta_i) \le 0.33\,F_u$$

$$A_d = P_i/[1.33(0.33\,F_u)\cos\theta_i]$$

$$= 12.75/[1.33(0.33)58(0.707)] = 0.708 \text{ in.}^2$$

$$A_d = \pi d^2/4 = 0.708 \quad \text{or} \quad d = 0.95 \text{ in.}$$

Try 1-in.-diameter cross-bracing for the bays (see also Example 3.23).

Torsion.   In the following case studies, torsion will be included in the analysis of the lateral force distribution as caused by asymmetry in geometry, stiffness, or mass distribution in the building plan or as caused by eccentric force action. The concept of torsion has already been introduced in Section 2.7.

Torsion is generated by asymmetrical buildings, eccentric and asymmetrical layout of the bracing systems, when the center of gravity of the floors does not coincide with the center of rigidity, or by the direction of the load when the resultant force action at the center of wind pressure does not pass through the center of rigidity, that is, by eccentric force action. In addition, torsional effects are caused by nonuniform wind pressure, accidental seismic eccentricities, and torsional ground motion. Torsion is most effectively resisted at points furthest away from the center of twist, such as at the corners and perimeter of the building. Closed tubular shapes are clearly efficient because of their inherent torsional stiffness. Also keep in mind that asymmetry of irregular building plans can be avoided by separating individual wings with joints so that the building does not act as a unit, but rather as an assembly of separate symmetrical structures.

In the first two parts of this section, simple core structures are investigated *where the center of rigidity is known;* in this instance, emphasis is placed on the behavior of the building core under torsion. In the last part, the force distribution to the various resisting structure systems of walls, frames, and cores, due to torsion, is studied as based on floor systems behaving as rigid diaphragms.

Cores can form closed or open shafts and single- or multicell systems. Typical core cross sections are round, square, or rectangular, or they are open tubes composed of wide flange and channel shapes. The rigid floor diaphragms stiffen the cores and are assumed to prevent the distortion of the cross section at each floor level. In this simplified investigation, the effect of wall perforations and coupling between the various structure systems is ignored. For the approximate lateral force distribution to the core components, we will distinguish between closed and open shafts.

A. Closed Shafts: Typical closed shafts are round, rectangular, or any other polygonal shape; they can form single or partitioned cells. The closed circular tube, because of its complete symmetry, is the ideal shape for resisting torsion. For a solid round shaft, the torsional shear stress at any point is proportional to its distance from the axis of rotation; hence, the resulting stress state for a thin-walled hollow tube is one of pure uniform shear. When a hollow shaft is twisted, it does not laterally displace and change its cross section. The most convincing way for understanding the stress flow is achieved by twisting the shaft and noticing that the original straight vertical lines are forced into a helix; hence, the principal stresses due to torsion form helices in compression and tension. These normal stresses can be transformed into pure shear at 45° to the inclined planes, yielding uniform shear stresses in the transverse and vertical directions of a thin-walled shaft (Fig. 2.29c).

Also, in noncircular, closed, thin-walled tubes of any shape, the torque is resisted primarily by the shearing stresses along the walls of the tube, although a noncircular section (when twisted) will not only rotate, but also warp, due to the deviation of the cross section from the circle. However, in a thin-walled, closed tube the bending of the thin walls in slab action is so insignificant that the deformation of the cross section can be neglected. Because of the low thickness-to-width and width-to-height ratios of the high-rise building core walls, the tube can be treated as a thin-walled beam. It can be assumed that the twisting is primarily resisted by the torsional shearing stresses flowing in a continuous fashion like a liquid along a closed ring. This uniform torsional shear stress is also often called *St. Venant's torsion* [see Eq.(2.90)].

Since the torsional shear stress along a thin-walled rectangular tube can be treated as constant, the wall shear forces (Fig. 4.27c) are proportional to their wall lengths.

$$f_{vt} = P_L/bt = P_s/at \quad \text{or} \quad P_L/b = P_s/a \quad \text{or} \quad P_s = P_L(a/b) \tag{a}$$

The torsion M_t is resisted by the two-wall couples (Fig. 4.27c):

$$M_t = P_s(b) + P_L(a) \tag{b}$$

Substituting Eq. (a) into (b) yields

$$P_s = (M_t/a)/2, \quad P_L = (M_t/b)/2 \tag{c}$$

The *direct shear* (or translational forces) are resisted entirely by the web walls parallel to the force action.

$$P_L = P/2 \tag{d}$$

Therefore, the maximum shear force due to direct and rotational actions occurs along the inside wall.

$$P_{max} = (P + M_t/a)/2 \tag{4.40}$$

For a square tube $(a = b)$, the torsional wall forces are constant: $P_L = P_s = P_t$. The maximum shear due to translation and rotation is given by Eq. (4.40).

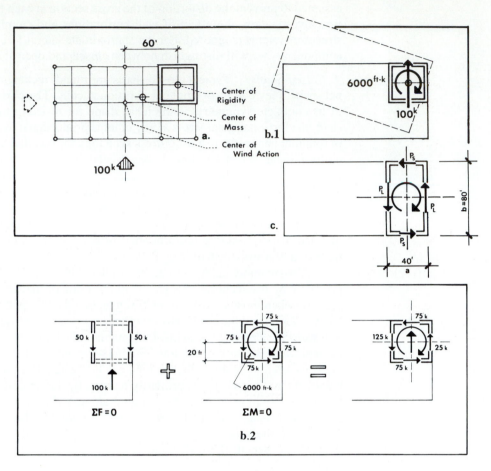

Figure 4.27 Example 4.13.

EXAMPLE 4.13

The average wind pressure of 15 psf against the long facade of the 41 ft 8 in.-high four-story building in Fig. 4.27a, with plan dimensions of 80 ft × 160 ft, is resisted by an eccentric 40-ft closed, square core with 8-in.-thick concrete walls. The total wind pressure the core must resist is approximately

$$P = 0.015(160 \times 41.67) = 100 \text{ k}$$

This pressure causes twisting of the core of (see Fig. 4.27b.1).

$$M_t = 100(60) = 6000 \text{ ft-k}$$

The torsion is resisted by the two parallel wall systems.

$$6000 = 2(P_t)40 \quad \text{or} \quad P_t = 75 \text{ k}$$

The direct shear is resisted by the web walls.

$$P_d = 100/2 = 50 \text{ k}$$

Hence, the maximum shear due to translation and rotation (Fig. 4.27b.2,) is

$$P_{max} = P_d + P_t = 50 + 75 = 125 \text{ k}$$

The same result could have been obtained from Eq. (4.40).

$$P_{max} = (P + M_t/a)/2 = (100 + 6000/40)/2 = 125 \text{ k}$$

The maximum shear stress is

$$f_v = P_{max}/at = 125,000/(8(40)12) = 32.55 \text{ psi}$$

This stress can be resisted by the concrete wall. The same result could have been obtained immediately by incorporating Eq. (2.96).

$$f_v = \frac{P}{at} + \frac{M_t}{2a^2t} = \frac{50}{40(0.67)} + \frac{6000}{2(40)^2 0.67} = 4.66 \text{ ksf} = 32.39 \text{ psi}$$

B. Open Shafts: For open building shafts such as cantilevering I-, E-, H-, and U-sections, the twisting moment M_t is no longer transmitted primarily in simple uniform torsional shear as for closed tubes, but in *bending torsion* together with *nonuniform torsional shear* (see also Section 2.7).

Open shapes cause significant warping under torsion, thereby generating large torsional bending and shear stresses. Visualize a wide flange section that is not restrained at its boundaries to be twisted in such a manner that the section is free to warp (i.e., rotated flanges form straight lines and are not bent); then the member is subject to uniform warping and only torsional shear stresses (Fig. 2.29g). However, when this beam is fixed at one end, similar to a vertical building cantilever shaft restrained by the foundations, the warping is prevented at the fixed end and the torsion is entirely resisted by the bending resistance of the flanges (Fig. 2.29h), which is called *warping torque;* in this case, the St. Venant's shear stresses are zero. The effect of the end restraint reduces as one moves toward the free end or top of the core. In other words, the rate of twist varies along the member and increases from zero at the base to a maximum at the top, indicating that warping is nonuniform and that the torsional resistance of the section changes from pure warping torque to almost pure torsional shear at the free end.

Although pure torsional shear may predominate near the top of a building, it is also relatively small and less critical for preliminary investigation purposes. The twisting is primarily resisted by the bending of the flanges in the lower building portion due to the restraint of the foundations and the coupling to the adjacent building components. The torque is assumed to be replaced by a force couple along the flanges, ignoring the bending of the web slab due to its lack of stiffness. It may be concluded that the twisting is resisted by the lateral bending of the flanges with the corresponding shear and normal stresses. It is beyond the scope of this investigation to further discuss the complexity of the interaction of St. Venant's torsion and warping torsion along the building between the extremes of fixed and free ends, particularly the effect of access openings.

Symmetrically Arranged Multiple Lateral Force-resisting Systems. Buildings with a symmetrical layout of structure under torsion will not be investigated in this context. The approximate distribution of lateral forces to the various resisting structure systems of walls, braced frames, rigid frames, and cores is based on rigid diaphragm action of the floors (see Schueller, 1990).

Asymmetrically Arranged Multiple Lateral Force-resisting Systems. For the general condition where the lateral force-resisting systems are arranged in an asymmetrical fashion, as for the simple shear wall arrangement in Fig. 4.28, the center of rigidity must be found first. Its location depends not only on the layout, but also on the

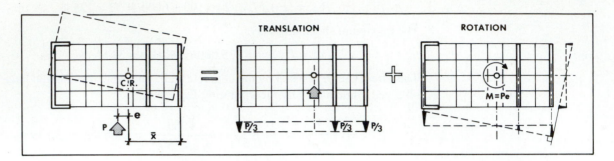

Figure 4.28 Asymmetrically arranged lateral force-resisting system.

stiffness distribution of the resisting systems. For further discussion of the topic, refer to Schueller, 1990.

Overturning

When a building is loaded concentrically only with gravity, the contact pressure at its base is uniform (see Fig. 2.33a). However, when rotation is generated, in addition, by lateral force action, the base contact pressure is no longer constant. For a trapezoidal contact pressure diagram, no tensile stresses are present (see Fig. 2.33b). In this instance, the compression prestresses the building, so the tension due to lateral force rotations is suppressed, and the resultant force due to the vertical gravity and the horizontal force (see Fig. 4.29) falls within the middle third or the *kern* of the supporting base (see Fig. 2.33c). This resultant force is transferred through the foundations to the ground similarly to the roots of a tree. When the resultant falls outside the kern, as is the case for a triangular contact pressure distribution, tensile stresses or partial uplift occur. Naturally, when the resultant force falls outside the base, as for narrow buildings with a minimum of weight, the building will topple if it is not anchored to the ground. For this condition, the building base may also be widened. It is apparent that gravity loads should be collected to the lateral force-resisting elements in a building, which should be located, in turn, where the base is broadest.

   Using weight to stabilize or prestress a building or building component is not an invention of the modern age. The medieval master builders sensed the kern concept when they suppressed the lateral arch thrust from the roof and vaulting on the flying

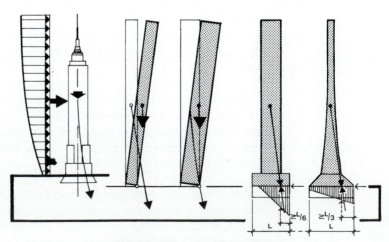

Figure 4.29 Lateral stability.

buttresses of the Gothic cathedrals by increasing the weight of the supporting piers through addition of small spires (Fig. 2.6e).

The overturning moment must be resisted by the weight of the lateral force-resisting structure above the level of investigation and the capacity of the material or anchorage establishing the continuity with the structure below. Considering, conservatively, that only the weight or minimum dead load counteracts, then according to the codes the stabilizing, resisting moment M_{react} must be at least 50% larger than the acting moment M_{act} caused by the lateral loads (keeping in mind that some codes require higher safety factors), unless the structure is anchored to resist the excess moment.

$$SF = \frac{M_{react}}{M_{act}} \geq 1.5 \tag{4.41}$$

This minimum requirement yields the location of the resultant force at a minimum distance of $L/6$ away from the base edges (see Fig. 4.29). Critical, with respect to the stability of a skyscraper, is the relation of its height to the width at the base parallel to the wind direction, assuming that the lateral force-resisting structure is located along the building perimeter. It is apparent that a wider base spreads out the loads. The maximum aspect ratios for skyscrapers are in the range of 6 to 8, although the *sliver* apartment buildings in New York, which are squeezed (for example) into a 20-ft-wide space between town houses, are extremely narrow slab-type structures possibly 20 stories high or more. These slivers have reached a slenderness of above 10! The most slender structure among the world's tallest buildings is believed to be the 810-ft-high, 72-story City Spire (1988) in New York. This mixed-use slim concrete tower, with a footprint width of only 80 ft, has a slenderness ratio of 10:1.

Keep in mind that the lateral forces also tend to slide a building horizontally, although most codes do not require a particular safety factor against sliding. For retaining walls, a minimum factor of safety of 2.0 should exist against sliding; for wind and earthquake, it may be reduced to 1.5. Should the friction force be insufficient to resist the translational movement, anchorage must be provided to absorb the excess sliding force.

EXAMPLE 4.14

Determine whether the entire building in Example 4.2 (Fig. 4.10a) is safe with respect to base overturning. The seismic moment about the base is

$$M_{act} = 2638(2/3)300 + 470(300) = 668,600 \text{ ft-k}$$

The wind moment about the base (for the sake of exercise, since it obviously is smaller than the seismic moment) is

$$M_w = 1523(300/2) = 228,450 \text{ ft-k}$$

The dead load resisting moment is

$$M_{res} = 84,000(100/2) = 4,200,000 \text{ ft-k}$$

Hence, the safety factor against overturning is

$$SF = M_{res}/M_{act} = 4,200,000/668,600 = 6.28 > 1.5, \quad \text{OK}$$

4.3 WATER AND EARTH PRESSURE LOADS: BASEMENT WALLS AND RETAINING WALLS

Structures below ground, including buildings with floating foundations resting on compressible soil, are subject to loads that differ from those encountered above grade. The substructure of a building must support lateral pressures caused by earth, earthquakes, ground water, expansive thrust from frozen earth, and surcharge loads on the ground surface. The magnitude and distribution of lateral earth pressures perpendicular to the substructure walls are highly indeterminate. They not only depend on the nature of the soil, but also on the flexibility of the resisting structure. The lateral pressure ranges from an *active stage,* where the earth is allowed to move laterally (as is true for the more flexible cantilever retaining walls), to a *state of rest* for absolutely rigid structures (as is usually applicable to the basement wall condition, assuming that the opposite sides balance each other), to the *passive stage,* where the wall moves toward the earth. In this context, the minimum pressure is caused by the active state and the maximum one by the passive state, with the earth at rest causing a pressure between these two extremes.

For preliminary design purposes, the lateral earth pressure may be treated as an equivalent liquid pressure due to some percentage of soil weight. For *cohesionless soils,* sands, and gravels, the minimum active pressure values range from 30 pcf for dry granular soils and a horizontal ground surface to twice as much for wet soils approaching the weight of water. In contrast, *cohesive soils* (such as clays and silts) swell when they absorb water, thereby causing large lateral forces; in this case, the equivalent liquid pressure may be as high as their unit weight of approximately 120 pcf, generating a pressure four times larger than for dry granular soils. It is apparent that the wall backfill should consist of cohesionless material and not saturated soils so that proper drainage can be provided. It may be concluded that cohesionless soil may be treated as a kind of fluid weighing twice as much as water and exerting about half the pressure.

The equivalent, minimum lateral liquid pressure on the wall in Fig. 4.30 for a dry granular soil is roughly equal to the equivalent fluid unit weight of 30 pcf (or, according to the Rankine theory, equal to the *coefficient of active earth pressure, $k_a \cong 0.3$,* multiplied by the unit weight of the soil, $w \cong 100$ pcf) multiplied by the distance from the top of the soil to the depth in question, thus causing a simple linear pressure variation as a function of height according to the familiar facts of hydrostatics.

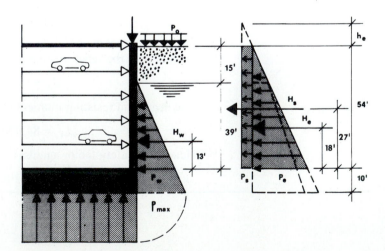

Figure 4.30 Lateral earth and water pressure on building substructure (Reproduced with permission from *The Vertical Building Structure,* Wolfgang Schueller, copyright © 1990 by Van Nostrand Reinhold.)

$$p_e = k_a w h_i = 0.3(100)h_i = 30h_i \qquad (4.42)$$

Therefore, the maximum lateral pressure at the intersection of wall and foundation in Fig. 4.30, is

$$p_e = 30(54) = 1620 \text{ psf/ft of wall}$$

Alternatively, the total resultant earth pressure for the given conditions is

$$H_e = 30h(h/2) = 15h^2$$
$$= 1.62(54/2) = 43.74 \text{ k/ft of wall}$$

In cases where groundwater is present, a lateral pressure is caused by the unit weight of water of 62.4 pcf.

$$p_w = wh_i = 62.4h_i \qquad (4.43)$$

Due to the hydrostatic pressure, the weight of the soil below the water level is reduced, as the *buoyant weight* of the soil is equal to its normal dry weight reduced by the weight of the water it has displaced. It is conservative for approximation purposes to neglect the buoyant weight of the soil by considering the dry weight. Also, as a first approach, an equivalent fluid pressure of 80 pcf is often used for the combined action of the noncohesive soil and water, or a 20% reduction of the water pressure may be taken for the combined action.

$$p_w = 0.8(62.4)h_i \cong 50h_i \qquad (4.44)$$

Therefore, the maximum water pressure, as based on a groundwater table 15 ft below grade, is approximately

$$p_w \cong 50(39) = 1950 \text{ psf/ft of wall}$$

Alternatively, the total lateral water pressure is

$$H_w = 1.95(39/2) = 38.03 \text{ k/ft of wall}$$

The maximum lateral water pressure at the base of the foundation is equal to the buoyancy pressure attempting to lift the building. In the early stages of construction, the upward lift is of major concern (see Example 4.16). The basement floor slab has to be designed for the uplift force.

$$p_{max} = wh_i = 62.4(49) = 3057.6 \text{ psf}$$

Additional pressure that must be resisted by the wall may be caused by loads acting on top of the backfill due to people, cars, slabs, and buildings. These loads are called *surcharge* and may be approximated by uniform strip, line, or point loads. The lateral pressure p_s due to a uniform surcharge p_o may be approximated by transforming it into an equivalent height h_e of earth backfill. The equivalent height of the soil is assumed to be equal to the surcharge p_o divided by the weight of the soil w: $h_e = p_o/w$.
The following relationship can be derived from Fig. 4.30:

$$h_e/p_s = h/p_e$$

or
$$p_s = p_e(h_e/h) = k_a wh(p_o/w)/h = k_a p_o$$

The equivalent lateral pressure due to the surcharge from fixed or moving loads depends on the type of soil. It may be treated as constant along the wall and may be approximated for noncohesive dry soil conditions, $k_a \cong 0.3$ as

$$p_s = 0.30p_o \qquad (4.45)$$

For a vehicular surcharge load of $p_o = 150$ psf, the uniform lateral pressure is

$$p_s = 0.30(150) = 45 \text{ psf/ft of wall}$$

Alternatively, the total lateral pressure due to the surcharge load is

$$H_s = 0.045(54) = 2.43 \text{ k/ft of wall}$$

As the soil expands, causing an active lateral pressure, it may encounter the resistance of backfill on the opposite side of the wall. This resisting soil contracts and provides a *passive earth pressure, p_p*. The passive earth pressure is always larger than the active. For the given soil conditions, it can be shown that the *coefficient of passive earth pressure, k_p,* is the inverse of the coefficient of active earth pressure k_a. Hence, the maximum passive earth pressure at the base of the wall (Fig. 4.34) can be shown to equal

$$p_p = k_p wh \cong wh/k_a = 100h/0.3 = 333h \qquad (4.46)$$

Hence, the lateral passive earth pressure may be approximated by using an equivalent fluid pressure of 333 pcf.

Basement Walls.   Basement walls, exterior walls of underground structures (tunnels and other earth-sheltered buildings), or retaining walls must resist lateral earth pressure as well as additional pressure due to other types of loading. Basement walls carry lateral earth pressure generally as vertical slabs supported by floor framing at the basement level and upper floor level(s). The axial forces in the floor structure are, in turn, either resisted by shear walls or balanced by the lateral earth pressure coming from the opposite side of the building.

Basement walls of low-rise buildings can be treated as primarily flexural systems, where the horizontal forces act perpendicularly to the plane of the wall and where the relatively small axial forces (which absorb some of the tension due to bending) may be ignored for preliminary design purposes.

For the condition in Fig. 4.32c, only part of the basement is below ground, thus causing smaller bending stresses. Notice that the wall in Fig. 4.32b is not supported by the first floor slab; hence it cannot span vertically, assuming that it is not designed as a cantilever wall; this wall must act as a horizontal slab supported by the cross walls!

Although basement walls act as vertical slabs supported by the horizontal floor framing, keep in mind that during the early construction stage when the upper floor has not yet been built the wall may have to be designed as a cantilever.

In contrast to conventional houses, earth-sheltered residences (Fig. 4.31) must not only support an extensive earth weight on top of their roofs, yielding a high dead load of the structure as well as large foundations, but must also withstand lateral earth and possibly groundwater pressures. The heavy, relatively constant loads on underground buildings may make funicular (curvilinear) structures much more appropriate and effective.

Lateral pressures may become critical when they do not balance each other and tend to move the building, as indicated for some cases in Fig. 4.31. Here the lateral forces must be transmitted to the foundations and then into the ground. To reduce and

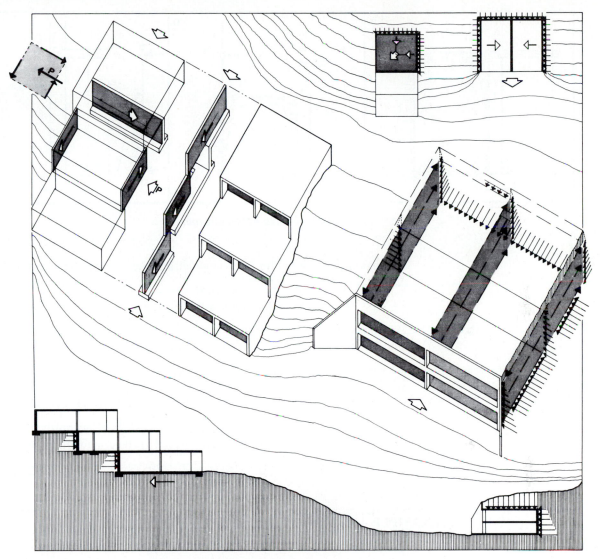

Figure 4.31 Earth-sheltered buildings.

control the magnitude of the lateral earth pressure, fully drained sand and gravel back-fill should be used to prevent the possible swelling of clay and any loading due to frost.

EXAMPLE 4.15

The masonry basement wall in Fig. 4.32a must support a granular backfill with an equivalent fluid weight of 30 pcf. Determine the maximum bending stresses the wall must resist.

The wall carries the lateral earth pressure as a simple one-way slab supported by the basement and first-story floors, respectively. The maximum pressure at the bottom of the wall, according to Eq. (4.42), is.

$$p_e = 30h = 30(9) = 270 \text{ psf/ft of wall}$$

The resultant force of the triangular load is

$$H_e = 0.270(9)/2 = 1.22 \text{ k/ft}$$

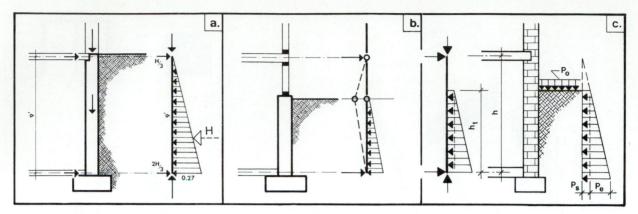

Figure 4.32 Simple basement walls.

The maximum moment for this triangular loading case can be found from Table A.15 or any other beam table as

$$M_{\max} = p_e h^2/15.6 = H_e h/7.8 = 1.22(9)/7.8 = 1.41 \text{ ft-k/ft of wall}$$

The corresponding critical tensile stress due to flexure in a 12-in-thick solid wall, ignoring conservatively the effect of the wall weight (P_D/A), is

$$f_b = \frac{M}{S} = \frac{1410\,(12)}{12\,(12)^2/6} = 59 \text{ psi}$$

In this example, the surcharge loads were considered insignificant. Should a street, however, be close to the wall, special surface loads must be taken into account (Fig. 4.32c).

EXAMPLE 4.16

The phenomenon of buoyancy can be exemplified by evaluating the effect of sudden flooding during the construction of a watertight residential basement. The water table rises up to 4 ft above the basement floor level, or 1 ft below ground surface. The building basement is 40×60 ft in plan; for other information about the building layout, refer to Fig. 4.33. Assume a weight of 100 psf for exterior walls, 60 psf for the interior wall, and 80 psf for the basement slab. The approximate weight of the basement at the stage of flooding is

$$W = 2[8(60 + 40)]0.100 + 8(60)0.060 + 40(60)0.080 = 380.80 \text{ k}$$

The maximum lateral hydrostatic unit pressure at the base of the wall is equal to the uniform uplift pressure of

$$p_w = wh = 62.4(4)/1000 = 0.25 \text{ ksf} \tag{4.43}$$

The total uplift force caused by the unit pressure across the whole basement floor area is

$$P_w = 0.25(60)40 = 600 \text{ k}$$

The same buoyant force is obtained by using *Archimedes' principle,* which states that a body wholly or partly immersed in a fluid is buoyed up with a force equal to the weight of the fluid displaced by the body. For the given condition, the volume of the displaced water is

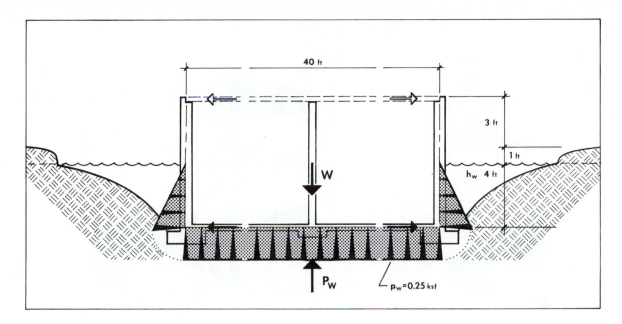

Figure 4.33   Uplift pressure.

$$V = 60(40)4 = 9600 \text{ ft}^3$$

The weight of the displaced water is

$$P_w = wV = 9600(62.4)/1000 = 599.04 \text{ k}$$

The uplift force is larger than the basement weight. The building will be lifted to a level at which it will float like a ship; at that level, the buoyant force will be equal to the basement weight (see Problem 4.18).

This problem could have been eliminated by using a drain and pump operation to keep the water away from the building. However, this method only works if the soil has a low permeability; for high-permeability soil, the pumps will not be able to keep up with the inflow of water.

Retaining Walls.   Retaining walls, in contrast to basement walls, are freestanding; they are needed at sudden changes of ground level. Their stability depends on their own weight and/or the weight of the soil on top of the footing. Usually, masonry is used for low retaining walls and reinforced concrete for tall walls.

Typical retaining walls are shown in Fig. 4.34. The most common types are the following:

- *Gravity wall,* made of plain concrete, stability depends on its weight, no tension forces are in the wall.
- *Cantilever wall* (T and L shaped), consists of three projecting elements (wall or stem, heel, and toe) each of which acts as a cantilever; in other words, the wall cantilevers off the base, and the heel and toe cantilever off the wall. Stability depends on the weight of the wall and backfill and on the strength of the projections in bending. The maximum concrete wall thickness at the base may be estimated as $h/10$; the maximum concrete base thickness may be assumed equal to the maximum wall thickness. The length of the base is approximately $\pm h/2$ and should be located below the frost line.

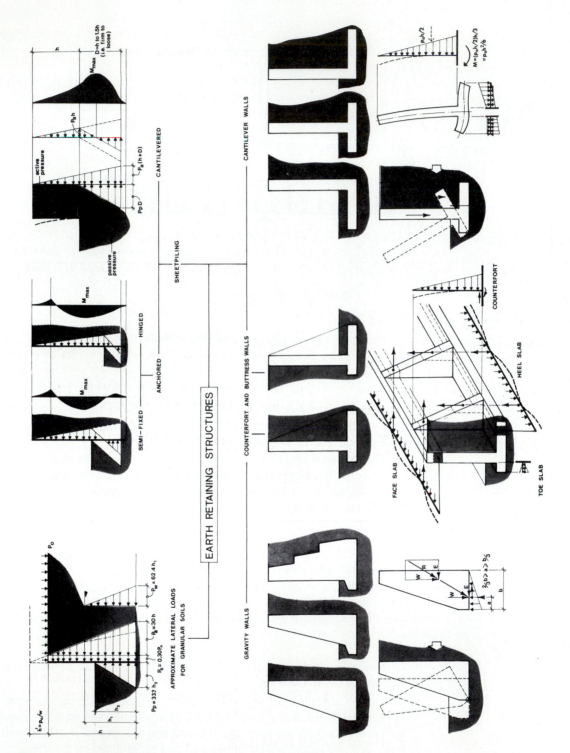

Figure 4.34 Retaining walls.

- *Counterfort* and *buttress walls* (i.e., cross walls): counterforts (in tension) and buttresses (in compression), spaced at regular intervals, support the wall (stem) and footing, which act as continuous horizontal slabs.

Retaining walls must be checked for *overturning, sliding,* and *maximum soil pressure* as demonstrated in the following example; it is assumed that the walls have been properly designed so that they will not fail.

Retaining walls are designed for lateral earth pressure and possibly surcharge. Hydrostatic pressure, however, must be taken into account unless adequate *drainage* is provided, which may consist of the following:

- Gravel (i.e., cohesionless soil) as backfill behind the wall
- Perforated drainage pipes along the base of the wall
- Weep holes through the wall
- Combination of these

The lateral earth pressure causes nonuniform soil pressure under the base; the maximum soil pressure at the toe (Fig. 4.35) cannot be larger than the allowable value. In general, it is desirable to keep the resultant force within the middle third of the base area (kern) so that no uplift occurs and the entire footing is utilized.

EXAMPLE 4.17

Investigate the earth pressure against the cantilever wall in Fig. 4.35. Check the stability of the retaining wall and the foundation, which are resisting a granular backfill 20 ft high and a surcharge of 200 psf. The weight of the backfill is 100 pcf and of the concrete, 145 pcf. The allowable soil pressure is given as 5 ksf.

The maximum wall and base thicknesses are estimated as $h/10 = 20/10 = 2$ ft. The vertical loads per foot of wall due to the weight of the stem (P_3), the weight of the base (P_2), and the weight of the soil on top of the heel including the surcharge (P_1), ignoring the soil above the toe, are

$$
\begin{aligned}
P_1 &= 0.100(18)8(1) + 0.200(8)1 & &= 16.00 \text{ k} \\
P_2 &= 0.145[2(12)1] & &= 3.48 \text{ k} \\
P_3 &= 0.145[18(1)1 + 18(1)/2] = 2.61 + 1.31 = & &3.92 \text{ k} \\
\hline
P_T & & &= 23.40 \text{ k}
\end{aligned}
$$

The maximum equivalent active liquid pressure p_e at the base foundation is

$$p_e = 30h = 30(20) = 600 \text{ psf} \tag{4.42}$$

The uniform lateral horizontal pressure p_s due to the surcharge $p_o = 200$ psf is equal to

$$p_s = 0.3p_o = 0.3(200) = 60 \text{ psf} \tag{4.45}$$

The passive earth pressure is conservatively neglected. With respect to the stability of the wall, the surcharge live load must be considered, since its lateral action constitutes a critical condition although its vertical action is beneficial as a stabilizing element.

The horizontal load resultants due to backfill (H_2) and surcharge (H_1) are

$$
\begin{aligned}
H_2 &= 600(20/2)/1000 = 6.00 \text{ k} \\
H_1 &= 60(20)/1000 = 1.20 \text{ k} \\
\hline
H_T & = 7.20 \text{ k}
\end{aligned}
$$

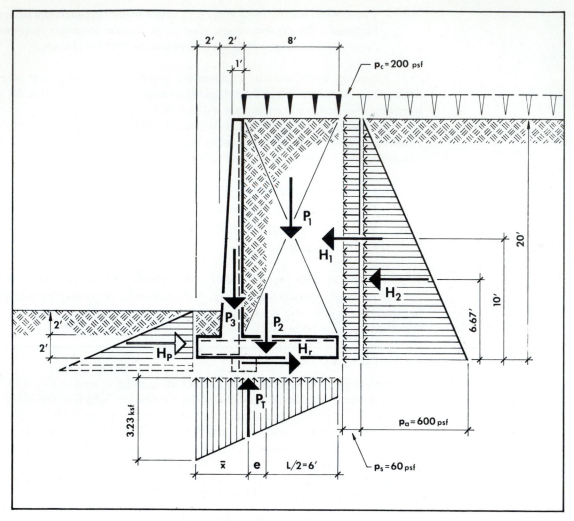

Figure 4.35 Example 4.17.

The lateral loads tend to rotate the wall about its toe. This overturning moment is

$$M_0 = 1.20(10) + 6.00(6.67) = 52.02 \text{ ft-k}$$

The vertical loads are attempting to resist the overturning. The resisting moment is equal to

$$M_r = 16.00(8) + 3.48(6) + 2.61(3.5) + 1.31(2.67) = 161.51 \text{ ft-k}$$

To have adequate safety, the stabilizing moment should be at least 50% larger than the acting moment for granular soils. For cohesive backfill, the resisting moment should be at least double the overturning moment.

$$SF = \frac{M_r}{M_o} = \frac{161.51}{52.02} = 3.11 > 1.5 \qquad (4.41)$$

The proportions of the retaining wall are adequate with respect to tilting.

The lateral pressure does not only cause rotation, but also translational movement. The resistance to *sliding* is provided by the friction and/or adhesion between the base and the soil, if we conservatively neglect the passive earth pressure. The magnitude of the horizontal resisting force H_r can be approximated by using the *coefficient of friction* μ. It can be shown that the maximum frictional force that can be developed is proportional to all the normal forces P_T between the two surfaces. The ratio of H_r/P_T is defined by the static coefficient of friction. Hence, the resisting frictional force can be expressed as

$$H_r = \mu P_T \tag{4.47}$$

For preliminary approximation purposes, the following coefficients of friction may be used:

Silt	0.35
Coarse-grained soil with silt	0.45
Coarse-grained soil without silt	0.55
Sound rock with rough surface	0.60

For this exercise, a coefficient of friction $\mu = 0.50$ is assumed, which is considered conservative. The total resisting friction is

$$H_r = \mu P_T = 0.50(23.40) = 11.70 \text{ k}$$

The safety factor against sliding is usually taken as at least 2.0; that is, the horizontal resistance H_r should be at least double the horizontal force action

$$SF = \frac{H_r}{H_T} = \frac{11.70}{7.2} = 1.63 < 2.0, \quad \text{NG}$$

The retaining wall proportions are not satisfactory with respect to sliding although the passive earth resistance was neglected. Therefore, either the wall base should be widened, or a key should be formed under the base of the footing (called a shear block) to increase the passive earth resistance.

Finally, the proportions of the foundation have to be checked so that the vertical contact pressure between soil and bottom face of the base slab is less than the allowable bearing capacity of the soil. The rotational equilibrium of all the forces about the toe in Fig. 4.35 yields the location of $\bar{x}$ of the resultant of the bearing pressure, which is equal in magnitude to the sum of all the vertical forces $\Sigma P_i P_T$ (see also Fig. 2.33g).

$$\sum M_F = 0 = \sum_{i=1}^{n} P_i x_i - \sum_{i=1}^{n} H_i y_i - \sum_{i=1}^{n} P_i (\bar{x}) = M_r - M_0 - P_T(\bar{x})$$

Notice that the magnitudes of the resisting moment M_r and acting moment M_0 have already been determined. Hence, the location of the resultant soil pressure force is.

$$\bar{x} = \frac{M_r - M_o}{P_T} = \frac{161.51 - 52.02}{23.40} = 4.68 \text{ ft} \tag{4.48}$$

The eccentricity e as measured from the center of the base area is

$$e = \frac{L}{2} - \bar{x} = \frac{12}{2} - 4.68 = 1.32 \text{ ft} < \frac{L}{6} = \frac{12}{6} = 2 \text{ ft}$$

The resultant force is acting within the kern or middle third of the base area, which shows that there will be no tensile stresses. The maximum soil pressure at the toe for this case can be determined from Eq. (2.105).

$$q_{max} = \frac{P}{A}\left(1 + \frac{6e}{L}\right) = \frac{23.40}{12(1)}\left(1 + \frac{6(1.32)}{12}\right) = 3.24 \text{ ksf} < 5 \text{ ksf}, \quad \text{OK}$$

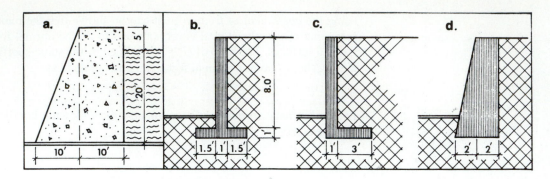

Figure 4.36   Retaining wall exercises.

The allowable soil pressure is larger than the actual maximum contact pressure.

Now the assumed thickness of the three projecting cantilever elements, the stem, and base (heel and toe) must be checked for shear forces and moments that they will have to withstand.

For example, with respect to the cantilever stem, the maximum shear at the top of the footing is

$$V_{max} = 0.060(18) + 0.030(18)(18/2) = 5.94 \text{ k/ft}$$

The corresponding ultimate shear force, according to Eq. (4.19), is

$$V_u = 1.7V = 1.7(5.94) = 10.10 \text{ k/ft}$$

The shear strength of concrete, according to Eq. (3.58), using 3000-psi concrete and letting the effective cantilever depth $d = 24 - 3.0 - 0.5 = 20.5$ in., is

$$\phi V_c = \phi 2 \sqrt{f'_c} \, bd = 0.85(2) \sqrt{3000} \, (12)20.5 = 22{,}906 \text{ lb} > 10{,}100 \text{ lb}$$

Hence, the stem thickness is satisfactory. The maximum cantilever stem moment is

$$M_{max} = 0.060(18)18/2 + 0.030(18)(18/2)18/3 = 9.72 + 29.16 = 38.88 \text{ ft-k/ft}$$

The corresponding ultimate moment is

$$M_u = 1.7M = 1.7(38.88) = 66.10 \text{ ft-k/ft}$$

This critical moment is covered by Grade 60 steel using Eq. (3.53).

$$A_s = M_u/(0.85f_y d) = 66.10(12)/(0.85(60)20.5) = 0.759 \text{ in.}^2/\text{ft}$$

$$A_{s\,min} = \rho_{min} \, bd = (0.2/f_y)bd = (0.2/60)12(20.5) = 0.82 \text{ in.}^2/\text{ft} > 0.759 \text{ in.}^2/\text{ft}$$

$$s/0.79 = 12/0.82, \qquad s = 11.56 \text{ in}$$

Try #8 bars at 11½ in. o.c., $A_s = 0.82$ in.^{2}.

Notice that the steel reinforcing must be placed inside, as indicated in Fig. 4.35. With respect to some simple exercises, the reader may refer to Fig. 4.36 and the corresponding exercises.

PROBLEMS

4.1   A seven-story office building is located in an urban environment with a wind speed of 70 mph and in seismic zone 2B. The building, 75 ft × 90 ft in plan view, is divided into three bays in each direction; it is 7(11.5) = 80.5 ft high. The lateral forces are resisted by ductile rigid steel frames along the perimeter of the building. The soil conditions are unknown. Use the following loading conditions for this preliminary investigation:

Floor weight, including girders, columns, and spray-on fireproofing: 85 psf

Curtain wall, including column and spandrel covers: 15 psf

Determine the total critical base shear for wind and seismic action. Check the building stability against overturning. Ignore accidental torsion.

4.2   Estimate the magnitude of the seismic base shear as a percentage of the dead load for a three-story plywood bearing wall apartment building in seismic zone 3.

4.3   Investigate a 14-story commercial building, located in San Francisco, across its critical short direction (Fig. 4.10b). A central braced steel core system carries 75% of the lateral load, while the rest of the loads are equally shared by the ductile, moment-resisting facade steel frames. Assume a typical average equivalent floor dead load of 130 psf and an average wind pressure of 24 psf for this preliminary investigation. The average floor height is 13 ft. Ignore e_{min}, that is, the accidental torsion.

    a.   Determine the seismic shear each lateral force-resisting system must carry.

    b.   Determine the total wind shear and rotation. Compare the results with the seismic action.

    c.   Determine the safety factor against overturning for the controlling case by treating the building as a single block for first-approximation purposes, rather than checking each structure system.

4.4   The eight-story office building structure shown in Fig. 4.10c is located in an urban area (80-mph wind) in a seismic probability zone 2B. Assume the weight of the floor structure and the vertical framing to be equivalent to 100 psf of floor area, and the weight of the facade structure to be 15 psf of facade area. Consider the roof weight to be equal to a typical floor weight; ignore the accidental torsion.

    a.   The braced frames at the central bay of each facade are designed to resist the lateral forces. Determine the total base shear as well as the total wind force.

    b.   How is the seismic force distributed to the braced frames? Give the numerical value for the force each wall carries.

    c.   According to the code requirements, is the building safe against overturning? If it should turn out not to be safe, what corrections do you propose?

4.5   A 10-story ductile rigid, space-frame concrete building has an average dead weight of 16 pcf. The frames are spaced 30 ft apart along the short direction and 25 ft along the long direction; the building is 10(12) = 120 ft high. Determine the total critical lateral shear on the 200- × 90-ft structure, which is located in an urban area of high seismicity, with a critical wind velocity of 80 mph, and is not an essential facility. Check the building stability.

4.6   Estimate the wind pressure distribution for a box-type, single-story building caused by a tornado. Assume a wind velocity of 250 mph, with all doors and windows closed. When the tornado passes over the building, it causes a lower outside pressure than inside the house for a short time, resulting in a uniform internal pressure attempting to burst outward. Consider the pressure drop as 0.1 atm = 212 psf.

4.7   The exterior 8-in., 12-ft high reinforced-concrete walls of a single-story industrial building, located in seismic zone 4, extend 3 ft above the roof line. First determine the cantilever wall moment, and then determine the maximum field moment by disregarding the effect of the wall cantilever. Draw your conclusions.

4.8   A one-story shopping center, located in seismic zone 4, is constructed of exterior masonry walls with an average weight of 75 psf and a plywood panelized roof with glued-laminated beams and purlins weighing about 16 psf (which includes allowance for the composition roof). The 200- × 60-ft building in plan is stiffened in the cross direction by three 35-ft-long shear walls along each exterior wall and one at the center of the building.

The lateral seismic shear for this low-rise masonry building, according to Eq. (4.9b), may be approximated as 18% of the building weight, which, in turn, may be estimated as 35 psf for this type of building. Investigate the 60-ft deep roof diaphragm spanning 100 ft between the shear walls. Determine the diaphragm, drag strut, and chord forces.

4.9   For the building in Example 4.8 (but with a different shear wall layout) as defined in Fig. 4.23c, where the interior walls are 50 ft long, whereas the exterior ones are 75 ft, the lateral force flow to the walls is determined as caused by the wind pressure against the long facade. The resistance of the flange walls perpendicular to the wind is conservatively ignored.

4.10 Determine how much load each pier in Fig. 4.25 carries.

4.11 Determine the approximate magnitude of the forces the shear walls in Fig. 4.23 cases (a) to (f) must resist. Assume a constant wind pressure of 20 psf to act against the 25-ft-high facade walls. First, evaluate the wall reactions to the wind action parallel to the short direction of the building and then the reactions to the wind action parallel to the long side. The square grid in Fig. 4.23 represents a dimensional network of 25×25 ft. Assume rigid diaphragm action and neglect the columns in resisting any lateral forces. The wall thicknesses are constant. Consider the curtain to carry one-half of the pressure to the roof plane.

4.12 Determine the approximate magnitude of the forces in the shear walls of cases (a) and (c) in Fig. 4.23 assuming flexible roof diaphragm action. The loading is the same as for Problem 4.11.

4.13 For the building in Example 4.1, determine the magnitude of the additional shear forces in the cross walls at the roof level due to torsion using the 5% minimum eccentricity (UBC).

4.14 Determine the approximate magnitude of the forces the walls in case (d), Fig. 4.23, would have to resist if the core is closed and forms a square tube. For further information, refer to Problem 4.11. Also find the maximum shear stresses for the 8-in.-thick masonry walls.

4.15 Determine the approximate magnitude of the forces along the walls of case (h), Fig. 4.23. For the loading conditions, refer to Problem 4.11.

4.16 Investigate case (g) of Fig. 4.23 for the same loading conditions as in Problem 4.11. Consider only the wall resistance parallel to the force action.

4.17 Investigate case (g) of Fig. 4.23, but without considering the front facade walls.

4.18 Determine the height of the water level at which the building of Example 4.16 will float. What would you do to prevent the lifting of the building?

4.19 A concrete dam of the dimensions shown in Fig. 4.36a must support a water depth of 20 ft. Determine if the resultant force lies within the central third of the base (kern) so that seepage under the dam may not be a problem.

4.20 For the retaining walls shown in Fig. 4.36b, c, and d, consider the lateral earth action as an equivalent liquid pressure of 30 pcf. The soil is assumed to weigh 100 pcf and the concrete 145 pcf. The soil-bearing capacity is 5 ksf. The coefficient of friction between concrete and soil is assumed as 0.4. Passive earth pressure may be neglected. Investigate overturning, sliding, and maximum soil pressure for each retaining wall. Show the primary reinforcing for the retaining walls given and also for a counterfort wall.

4.21 Determine the maximum stem moment for the cantilever retaining walls of Problem 4.20. Show where to place the steel reinforcing in the wall.

4.22 Determine the maximum moment the wall in Fig. 4.32a must resist, if in addition to the backfill, it must also support a surcharge load on the ground of 100 psf.

4.23 Determine the maximum moment for the basement wall in Example 4.15 if the groundwater level rises to 6 ft from the ground surface. Neglect conservatively the buoyant weight of the earth below the water table.

4.24 Determine the maximum moment for the basement wall in Example 4.15 if only part of the wall is in the ground. Consider the height of the backfill to be 5 ft.

Frames, Arches, and Trusses

Two-dimensional skeleton structures composed of linear members are briefly investigated in this chapter. This most common group of planar structure systems includes post–beam structures, bent beams, rectangular portal frames, cantilever frames, braced frames, pitched frames, arches, and trusses. Here, the term portal frame suggests a large entrance in the abstract sense; as a frame, it implies the continuity between beam and column.

These structure types may form short-span or long-span, single-story or multi-story, as well as single-span, multispan, or cantilever systems. They range from low-rise to single, open, large-volume buildings to skyscrapers. They may support flat, tilted or arched roofs as single units or repetitive bents; they may be open or closed rings; they may represent beam, frame, arch, or truss bridges. The linear elements may also form megamembers, that is, building portions or entire buildings, such as beam buildings and megaframes.

The functional organization of a building determines the column layout. In Fig. 1.15 various column arrangements are shown for a typical rectangular building; a similar approach could be used for buildings of any other plan configuration. The systems range from multibay structures, such as for industrial buildings, to single-unit long-span structures where the columns are concentrated at or close to the perimeter to allow for unobstructed interior space.

Primary emphasis in this chapter is on the preliminary design of simple, but common single-story enclosure systems to develop a feeling for the behavior of structures. Multistory structures are introduced in Chapter 10.

5.1 INTRODUCTION TO FRAMES

The spirit of the skeleton is reflected in the Gothic cathedrals, where the master builders overcame the massive construction of the previous periods with the light skeleton derived from the interplay of rib thrusts, weight, columns, flying buttresses, and abutments, that is, the balance of forces (Fig. 1.8, bottom).

It was not until the turn of the nineteenth century that the metal skeleton began to slowly replace masonry and wood construction. The development of the frame is closely related to the nineteenth-century engineering and the invention of first cast iron, followed by wrought iron and then steel. The *iron frame* is often treated as a symbol of the Industrial Revolution, which is accompanied by urbanization and rapid population growth. During this period, new building types were born, ranging from the long-span structures for the great exhibition halls and train sheds to the multibay framing for mills, factories, and warehouses. But also for traditional building types, the new technology of structural ironwork was occasionally expressed as architecture. Henri Labrouste exposed the slender iron columns and iron arches in the Bibliothèque Ste. Geneviève, Paris (1843), to achieve a powerful interior space.

Multibay skeleton construction reached its high point in 1885 with the 10-story Home Life Insurance Company Building in Chicago by William Le Baron Jenney, which is considered the birth of the skyscraper. There were no bearing walls; the entire building weight was carried solely by the metal skeleton consisting of cast iron columns, wrought iron beams for the first six floors, and steel beams for the remaining floors. The masonry facade was hung from the frame like a curtain for the first time.

A landmark for the expansion of structure concepts and the inventive mind of engineers with respect to the large scale of immense enclosures is Joseph Paxton's Crystal Palace of 1851 in London built entirely of cast iron, wood, and glass. This glass–iron structure, the world's largest building at that time, is not only the first example of a large-scale lightweight structure, but also employed new design concepts related to modular coordination of the standardized parts the building consisted of, which included mass production and rapid assembly. These design considerations became an important part of the Modern Movement when, for example, Walter Gropius 60 years later talked about the industrialization of housing through machine-produced standardized building parts.

An important milestone in the development of long-span frame structures is William Henry Barlow's 240-ft-span St. Pancras station of 1868 in London; it was the longest span structure in the world at that time. The vaulted train shed is supported by 6-ft-deep pointed wrought iron lattice arches fixed at their base and spaced at 29 ft.

About two decades later the unprecedented 375-ft span of the legendary Galerie des Machines at the 1889 Paris Exhibition marks a new era for long-span structures. The glazed roof is supported by 10-ft-deep three-hinged pointed lattice arches. Probably for the first time, the concept of the hinged joint or point support was clearly exposed, thereby helping to articulate the structural behavior of the frame and giving expression to the potential of steel strength.

The application of the skeleton concept can only be suggested by the cases in Fig. 5.1. They have been selected to demonstrate the endless range of structures and to indicate that they cannot simply be organized as types, as is often done in structural engineering with line diagrams.

Pier Luigi Nervi treats the internal staircases for the large Sports Palace in Rome, Italy (1960, Fig. 5.1, center), as powerful structural organisms. In contrast, Adrien Fainsilber's simple post and beam structure for the Cité des Sciences at La Villette, Paris (1986, Fig. 5.1, top right), not only intends to express structural action but is also used in a more decorative sense to articulate the almost Neoclassical appearance of the huge exposed steel trusses sitting on the granite-clad concrete towers. The primary

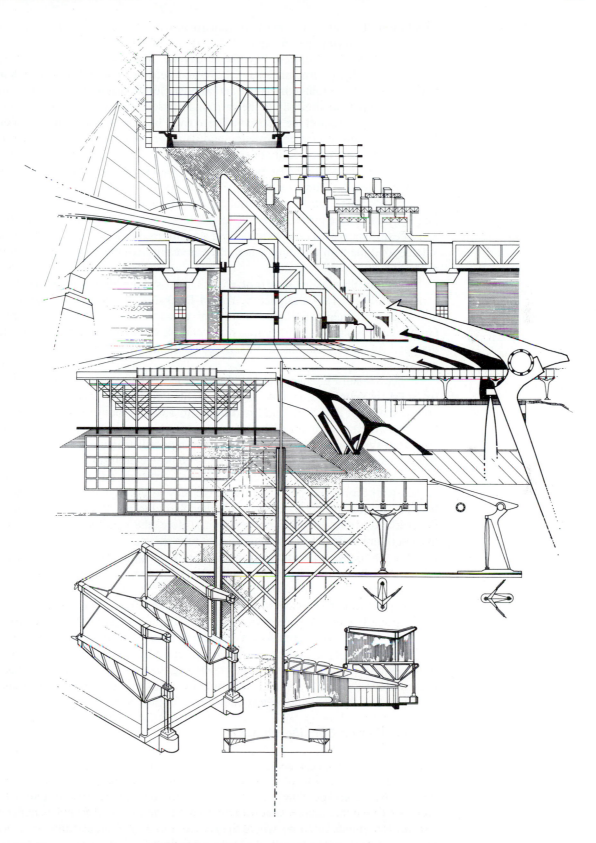

Figure 5.1 Frames, arches, trusses.

structure in this building consists of four rows with five 132-ft-high hollow concrete towers, each carrying the 16 exposed, 214-ft-span steel trusses.

In Fay Jones's Pinecote Pavillon, Picayune, Mississippi (1987, Fig. 5.1, center left), in contrast, the structure establishes a geometric theme with respect to the whole building, which is organic in nature and almost Gothic inspired. The slender, vertical columns branch out similarly to the surrounding trees and become interwoven as they grow upward to support the stepped pitched roof. It is the purity of the all-wood structure, that is, the revealed construction and its response to light, that expresses simply and honestly the poetic quality of space.

Nick Grimshaw clearly expresses the structure of the Sainsbury supermarket, Camden Town, London, U.K. (1988, Fig. 5.1, bottom). The main parallel 132-ft-span frames consist of slightly arched roof trusses suspended from tapered cantilever steel plate girders. These girders form the long interior arms of asymmetrical double-cantilever beams supported on concrete-filled stanchions, while the short arms project outside beyond the wall cladding, where they are each tied down by four vertical tension rods.

The saillike tensile fabric roof of San Diego's Convention Center (1988, Fig. 5.1, center top, Fig. 9.18a), designed by Arthur Erickson, spans the enormous distance of 300 ft and is anchored to giant triangular concrete fins recalling the buttresses of Gothic cathedrals. Each fin represents a complex frame structure composed of the diagonal fin beam, columns, beams, diagonal members, arches, and slabs, all of various sizes.

The 10-story Exchange House, London, U.K. (1989, Fig. 5.1, top left), designed by SOM, is supported on four parallel, segmented, seven-story high, tied parabolic arches spanning 256 ft over railroad tracks to concrete buttresses. The gravity loads flow from the floor trusses to the hangers or columns to the arches, as the order of the structural systems clearly demonstrates. The exposed structure projects more than 6 ft from the exterior fire-rated wall cladding to be protected from fire.

Santiago Calatrava interprets the skeleton structure quite differently. He is concerned with formal precedents and with the play of gravity, as suggested in the cantilevered glazed platform canopy of the Stadelhofen Railroad Station, Zürich, Switzerland (1990, Fig. 5.1, bottom right). Here, the cantilever beams carrying the glass roof are welded to a continuous steel tube that acts as a beam and torsion ring to transfer the loads to the inclined, branching Y-columns of triangular cross section, which, in turn, are stabilized by vertical hinged columns. The two-legged composite columns are spaced approximately 20 ft.

Calatrava is fascinated with how structure works and how the loads are carried to the ground, which he demonstrates by articulating its tactile quality and the organic nature of the skeleton comprised of sculptural, bony-shaped elements asymmetrically arranged. He emphasizes the dynamics of structure by making the potential movement of forces visible. He achieves that by expressing the unbalance of forces. It seems as if the cantilevers are on the verge of rotating by articulating the hingelike tubular beam and by letting the slanted columns be just caught in time by the vertical struts.

General Design Considerations

Typical frame shapes can be derived in direct response to primary loading as indicated in Figs. 5.2 and 9.5. Visualize a suspended cable to respond to a given load by adjusting its shape correspondingly and then to be frozen and inverted so that it forms an arch that is in pure compression with no shear and moment. This arch is called a *funicular arch* or *funicular frame* where the *pressure line* (or funicular line) coincides with the central axis of the frame. However, keep in mind that usually the loads do not remain constant, and, as they change, the rigid frame cannot adjust its shape and hence is subject to flexure.

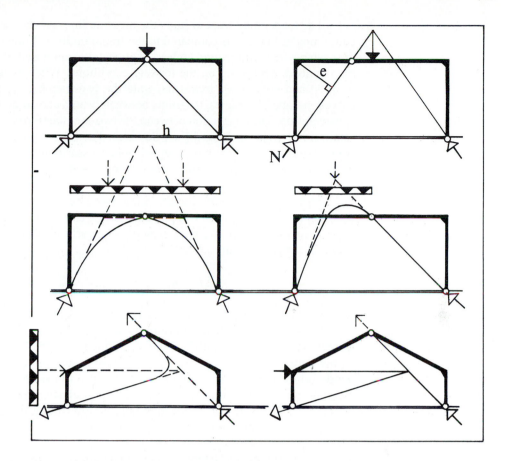

Figure 5.2 Pressure line responses to various load actions: funicular shapes.

Comparison of the ideal funicular shape (i.e., pressure line) with the real frame profile, as for the cases in Fig. 5.2, helps to identify the formal efficiency from a force flow point of view, since the difference between the shapes is equivalent to rotation, that is, eccentricity e of the axial force N along the pressure line. In other words, the bending moment at any point is proportional to the distance e perpendicular to the pressure line and the frame.

$$M = N(e)$$

Since the bending moments are zero at the hinges, the pressure line must touch or pass through these zero-moment points ($e = 0$). Hence, the location of the hinge determines the rise of the pressure line. Notice that with increase of height of the funicular arch the reactions become steeper and the thrust force components decrease.

When the eccentricity e is minimal, as may be the case for an arch, and the pressure line falls within the *kern* of the section (i.e., central third of a rectangular section), there may be no tension since the moment is so small. This feature becomes an important consideration for materials of weak tensile capacity such as masonry; typical examples are the vault arches of the Gothic cathedrals.

When, however, the eccentricity is large, such as for the rectangular portal frame in Fig. 5.2, which differs substantially from the parabolic pressure line in response to the typical uniform loading condition, the moment is clearly the dominant design determinant. The largest moment occurs at the knee of the frame where e is maximum.

When the pressure line lies within the frame, the moments are negative, but when the pressure line is located on the outside, the bending moments are positive.

The reader may want to study the pressure lines for the various common loading conditions in Fig. 5.2 as compared to the frame outlines. Funicular frames do not distinguish between frames and arches. They may be composed of linear and/or curvilinear members depending on the loading conditions. Whereas frames are made up of beam(s) and column(s), arches integrate the two-member types into a continuously curved member. In frames, the joints between beams and columns are a critical design consideration. The connection between the two-member types may be continuous (i.e., semirigid or rigid) or hinged (i.e., flexible). The hinges are simple field connections not transferring any moments. They are not necessarily placed at the columns, but may be located more efficiently away from them, as is discussed in this section.

Behavioral Considerations.   Some basic behavioral considerations for simple frames are studied in Fig. 5.3. A planar structure can be at rest only if there is no translational or rotational movement present, as clearly reflected by the following three equations of equilibrium:

$$\Sigma F_x = 0, \qquad \Sigma F_y = 0, \qquad \Sigma M_z = 0 \qquad\qquad (2.12)$$

It should be remembered that alternative equation sets are possible (see Eq. 2.21).

These conditions of equilibrium do not apply only to the external conditions of a member or member assemblage (beam, column, frame, truss, entire building, etc.), that is, the equilibrium between the external loading and the reactions of the respective nondeformable, perfectly rigid members, but also to the internal force flow along the member, now taking into account its behavioral elastic properties. The nature of the internal force flow at a given location is found by cutting the member; at any cut not more than three force components occur (Fig. 5.4).

Redundancy and Stability.   The behavior of the overall structure (of adequate strength) depends not only on how it is supported, but also on how its members are arranged and connected to each other. For instance, the various types of rectangular frames in the top portion of Fig. 5.3 demonstrate clearly unstable conditions.

The first frame is externally unstable because only two roller supports are provided; the two vertical reaction forces are not sufficient and cannot prevent horizontal movement. Equilibrium of planar, nonparallel, nonconcurrent force systems requires at least three reactive conditions. But the reactions should be placed so that they are neither concurrent nor parallel, since otherwise the structure would be geometrically unstable. The other frames in the top row of Fig. 5.3 are internally rather than externally unstable, due either to missing members or because of too many hinged connections. A structure that is unstable does not necessarily collapse, since it may reach a state of equilibrium after it has moved.

When the external and/or internal conditions of the structure are such that the three equations of equilibrium are not sufficient for the solution of the force flow, additional equations are necessary, and the structure is called *hyperstatic* or *statically indeterminate*. Hence, planar structures are only statically determinate if the three equations of equilibrium can solve for the magnitude of the internal and external forces due to any type of loading. It should be obvious that a structure must be stable whether the method of analysis is determinate or indeterminate.

The degree of indeterminacy may be found by making the structure statically determinate and stable by removing supports and/or cutting members. The number of force-resisting conditions taken away is equal to the degree of indeterminacy or *redundancy,* so named after the number of forces that are not needed to ensure static equilibrium. For instance, the continuous rigid frame (Fig. 5.3, bottom) can be made determinate by cutting the beam, thus forming two independent, cantilevering, deter-

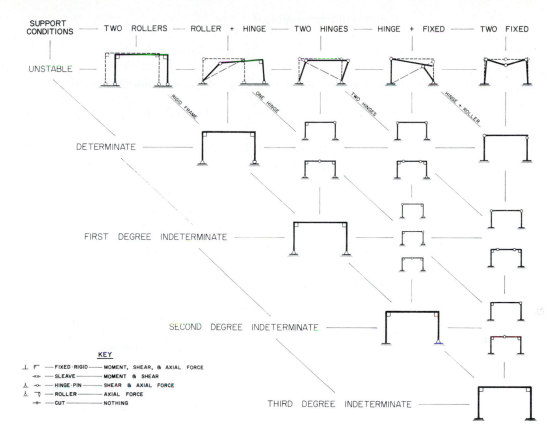

Figure 5.3 Stability and redundancy of simple frames.

minate tree units. Three restraints have been removed: the normal, shear, and moment resistance. Hence, the structure is three times indeterminate or has three redundants.

Redundancy is an important phenomenon because it allows the force flow to take an alternative path if the structure should be failing at a certain location, thus not necessarily resulting in a progressive total collapse of the building.

The process of determining the possible redundancy and stability of a structure can be formalized as follows. The unknown forces in a planar frame of (m) members are the shear (V), the axial force (N), and the moment (M) for each member, as well as the support reaction components (r):

$$\text{Unknown conditions:}\quad 3m + r \qquad\qquad\text{(a)}$$

Each rigid joint (j) is capable of transferring translational and rotational forces and thus furnishes three equations of equilibrium ($3j$). However, frame members are not necessarily all connected with moment resisting joints; some may be pinned to each other, or internal hinges may appear along members. These special conditions (c) release some of the unknown internal forces.

$$\text{Known conditions:}\quad 3j + c \qquad\qquad\text{(b)}$$

We may conclude that, if the number of unknown conditions is equal to the number of known conditions, only then is the planar frame *statically determinate* and *stable*.

$$3m + r = 3j + c \qquad\qquad\text{(5.1a)}$$

The structure is unstable if

$$3m + r < 3j + c \qquad (5.1b)$$

The structure is *hyperstatic,* i.e., *statically indeterminate,* if

$$3m + r > 3j + c \qquad (5.1c)$$

The condition of the three-hinge portal frame (Fig. 5.3) can be evaluated as follows, realizing that a pin support prevents horizontal and vertical movement ($r = 2$) and that the hinge in the beam releases one unknown, the moment ($c = 1$):

$$
\begin{aligned}
3m + r &= 3(3) + (2 + 2) = 13 \\
3j + c &= 3(4) + 1 \qquad = 13 \\
\hline
13 &= 13
\end{aligned}
$$

The frame is determinate and stable.

The degree of indeterminacy for the continuous frame with a hinged joint can be determined as follows. A fixed-end support provides resistance to horizontal, vertical, and rotational movement ($r = 3$). In general, a pinned joint with n members framing into it can be visualized as consisting of $n - 1$ members with hinged ends.

$$
\begin{aligned}
3m + r &= 3(3) + (3 + 3) = 15 \\
3j + c &= 3(4) + 1 \qquad = 13 \\
\hline
15 &> 13
\end{aligned}
$$

The frame is twice indeterminate. Although this comparison of known and unknown conditions is necessary, it is not sufficient with respect to the stability of framed structures. For further discussion of this topic, the reader may refer to the section on trusses.

The force flow in indeterminate structures, in contrast to determinate ones, depends on the stiffness of the members, as is discussed in the next section. In other words, the designer has the opportunity to control the distribution of internal forces not only through change of geometry, but also through member stiffness.

A structure with redundant members or support conditions is stronger and stiffer; it is less vulnerable to collapse if some of the members should be failing. Redundancy is a particularly important consideration for long-span structures, which cannot rely on adjacent structures for support as short-span structures do; they are extremely vulnerable not to localized but total building collapse, as so clearly exemplified by the failure of the Hartford Civic Center Coliseum in 1978. Statically determinate structures, on the other hand, can absorb material changes and movements (shrinkage, creep, temperature and moisture changes, settlement, lack of fit, etc.) without causing additional stresses.

Internal Force Flow.   To design a member, the internal force flow along that member must be known so that the minimum required size can be found that will be capable of responding to the maximum internal stresses with an appropriate safety margin. The internal force flow is determined by cutting the structural element that is under investigation. The member portions to the left or to the right of the section are called *free bodies.* Typical free bodies are shown in Fig. 5.4.

The internal forces that occur at the point of section represent the equivalent internal actions and identify the minimum required material resistance to shear, axial action, and rotation; they are determined from equilibrium with external loads and reactions of the respective free body.

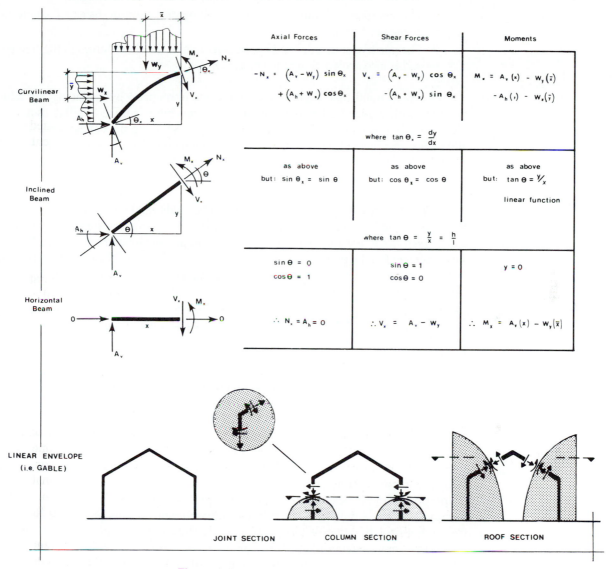

Figure 5.4 Internal force flow for line elements.

It is helpful to derive general expressions for the magnitude of the internal forces as based on the typical loading conditions given in Fig. 5.4. For example, the following equation for the shear V_x can be derived from the arch free body by summing all the forces parallel to V_x.

$$\Sigma F_y = 0 = A_v \cos\theta_x - A_h \sin\theta_x - W_z \sin\theta_x - W_y \cos\theta_x - V_x$$

$$V_x = (A_v' - W_y) \cos\theta_x - (A_h + W_z) \sin\theta_x$$

The other equations for N_x and M_x in Fig. 5.4 can be derived in a similar manner. The same equations can also be obtained from the free bodies of the right structure portion rather than the left; for this condition only the shear equation will be negative.

The following conventions are used in plotting the force flow along the frame members:

- The moments are shown along the tension side of the members.
- The compression forces are shown along the exterior faces of single-unit enclosure systems.
- The positive shear (defined by the direction of the shear forces in the free bodies) is also shown along the exterior faces.

Keep in mind that slight inaccuracies are introduced by replacing the actual structure with an idealized model, the *line diagrams*. For example, a wall does not provide just a single load reaction to a beam sitting on it; along the contact area of beam and wall a bearing pressure is generated, the resultant of which can be visualized as the reaction. Rarely does a simply supported beam rest on a roller and pin, as is assumed; for example, a joist may sit on opposite masonry walls of exactly the same nature.

Preliminary Design Considerations.   The loads can be selected by referring to Sections 2.1 and 4.1; reasonable preliminary dead loads can be estimated from Table 2.2. *Typical load distributions* for a simple building envelope are shown in Fig. 5.5. The placement of the gravity loading depends on the spacing of the roof joists (purlins, beams, etc.) that span the distance between the frames. It is shown in the discussion of floor framing systems in Section 3.2 that the beam spacing has hardly any influence on the magnitude of the moment that acts on the supporting frame beam, but only on the shear distribution, which for preliminary investigations may be considered as not critical. Hence, for preliminary design purposes, gravity loading can be considered uniformly distributed. Similar reasoning can be applied to the lateral force action that the building curtain transfers to the frames, keeping in mind that direct uniform pressure on the column will cause larger bending moments in the column (Fig. 5.10c).

Only the primary critical load arrangements and combinations are given in Fig. 5.5 and are studied further in the context of the design problems. Usually, for single-story frames of ordinary height (but not for post–beam systems with cantilever columns), lateral forces do not have to be considered, at least for preliminary design purposes, since they, in combination with other loads, do not control the design of the primary structural frame units. In other words, the primary design determinant for the proportioning of the beams and columns of typical single-story frames is bending due to gravity action, similarly to the behavior of continuous beams, that is, quite in contrast to high-rise buildings where axial action is obviously an important consideration. The axial compression in addition to bending, especially in the columns, is rather small and often not significant enough to be considered in the preliminary design; however, if axial action has to be included, the simple interaction equation [Eq. (3.96)] needs only to be considered for preliminary design purposes.

The spacing of parallel frames is in the range from 16 to 20 ft and up to about 40 ft, depending on the building type. The purlins span the distance between the frames and support the deck or sheathing. For pitched roofs it is assumed that the force component parallel to the roof plane is transferred by the roof skin in diaphragm action to the frames, so the purlins neither have to twist nor to bend about their weak axis (Figs. 3.6 and 3.8).

To prevent sidesway buckling of the columns about their weak axes, as well as to resist lateral force action perpendicular to the frames, shear walls or diagonal bracing in certain bays along the facade usually is provided (Fig. 4.1). Furthermore, depending on the stiffness of the roof diaphragm, bracing may have to be added along the roof plane to prevent lateral buckling of the frame beams (Fig. 4.14), which should be determined by the spacing of the purlins. The lateral stability of buildings is discussed in Section 4.2; refer to Figs. 4.11 and 5.28 for quick clarifications.

Most frames are pin connected to their foundation, realizing that for normal soil conditions (no rock) the footings probably will rotate anyway and thus form equivalent hinges independent of the degree of fixity between the columns and their foundations.

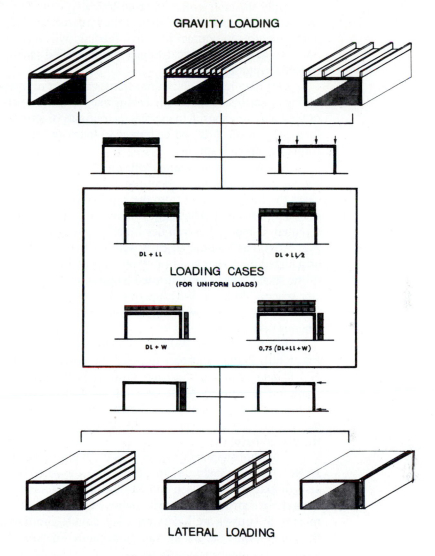

Figure 5.5 Typical loading conditions.

The lateral thrust due to gravity action, especially for frames with small height-to-span ratios or spans larger than about 60 to 80 ft, is so large that it must be resisted by tie rods between the column bases, rather than directly by the soil.

5.2 RECTANGULAR FRAMES

Rectangular frames may form single-bay or multibay and one-story or multistory structures. They may also be beams and then are called *Vierendeel trusses.* As one-story units, they may provide enclosure as individual units or repetitive bents, or they may support an entire building on top, as for Le Corbusier's l'Unité d'Habitation (apartment block) at Marseille, France (1952). It is clear that there is no limit set to the application of the frame concept.

One-story, Single-bay Frames

Typical single-span structures are portal frames or cantilever frames (e.g., tree frames). Naturally, there are numerous other possibilities as can only be suggested by the cases in Fig. 5.11.

Only the more common basic frame types in Fig. 5.8, used as roof enclosures, will be investigated in detail in the following section. While the geometry of these support structures is symmetrical, live loading will introduce conditions of asymmetry. Some typical asymmetrical frames are identified in Fig. 5.11 and are studied in the end-of-chapter problems.

Ordinary rectangular portal frames are part of a repetitive structure system providing an enclosure where the beams may be hidden between the roofing and ceiling, or they may be exposed to express the purpose of structure. The beams may just sit on their column supports and be hinged to them, or together with the columns they may form continuous frames.

Some historically important, more recent single-bay post–beam structures are shown in Fig. 5.6, all of them requiring column-free interior space. In some of the cases the roofs are hung from the exposed beams; in other cases, the beams are integrated into the roof. The frames are arranged in a parallel, radial, two-way or circumferential fashion. Mies van der Rohe's famous Crown Hall at IIT in Chicago (1955, Fig. 5.6a), has become an early symbol for the celebration of the portal frame. He articulated the power and beauty of the post–beam structure by exposing the lightness of the steel skeleton as contrasted by glass surface and making possible the huge interior space (220 ft × 120 ft) free of columns. The roof platform is suspended from the welded plate girders, which are 60 ft apart and, together with the columns, form continuous portal frames.

For the Munsun–Williams–Proctor Institute Museum, Utica, New York (1957, Fig. 5.6d), Philip Johnson expressed the power of span and enclosure with the four 123-ft-long (and 10.75.ft-deep girders) posttensioned intersecting concrete portal frames. The two-way frame structure not only supports the roof, but also the cantilevering exterior concrete wall beams and the interior balcony hung from it. The enclosed building box supported by the frames seems to float, which is emphasized by the glazed base.

The roof supporting system for the Philips Exeter Academy Athletic Facility in Exeter, New Hampshire (1969, Fig. 5.6c), designed by Kallmann & McKinnell, consists of exterior 17-ft-deep and 10-ft-wide triangular-shaped steel pipe trusses, which are simply supported on laced, double-steel columns that form a frame to resist wind; most of the trusses are 135 ft long. The double-chord truss members are on top where the compression occurs for simply supported beams. Also, here the roof framing is suspended from the space trusses, which are spaced 40 ft on center.

Because of the enormous span of 324 ft, the rigid bents of the Kansas City Kemper Memorial Arena (1975, Fig. 5.6f), designed by Helmut Jahn, were made of tubular triangular space trusses (27 ft deep with a single upper chord, double lower chord, spaced 153 ft on center), giving wind resistance in the transverse as well as longitudinal directions. The separation of the bents from the suspended roof substructure allows for the independent movement of the exposed trusses. The roof steel structure weighs 23.5 psf of floor area.

From the six stainless steel trussed frames for the Planetarium in Stuttgart, Germany (1977, Fig. 5.6e) by W. Beck-Erlang, which are arranged in a radial manner, is hung a hexagonal step-pyramidlike dome clad in glass. The four primary roof steel girders of the Reading Assembly Centre, U.K. (1976, Fig. 5.6b), by M. Johnson Marshall, are supported by 10 inclined Y-shaped concrete columns, which, in turn, form a ring system of interlocking portal frames following the plan of the elongated hexagon.

The structure of Norman Foster's Sainsbury Centre for Visual Arts, Norwich, U.K. (1978, Fig. 5.6h), consists of parallel, closely spaced, more than 8 ft deep and nearly 6 ft wide triangular, welded, tubular steel trusses. The space trusses, with a double upper chord and a single lower chord, span 122 ft to the lattice towers of similar shape. The trussed columns act as vertical cantilevers in the transverse direction. Cross-bracing provides lateral stability in the longitudinal direction.

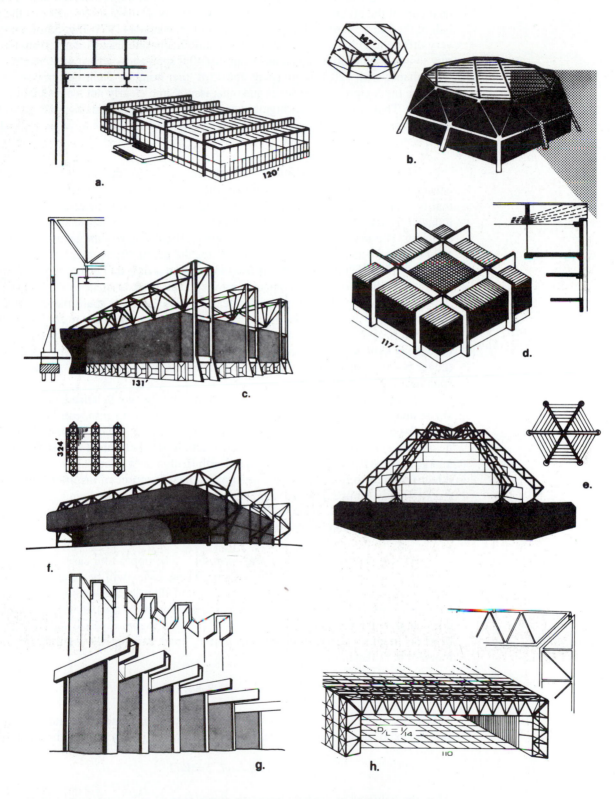

Figure 5.6 Single-bay portal frames.

In contrast to the previous cases, where structural expressionism is a predominant part of the architecture, Arthur Erickson's frames covering the space of the great hall of the Museum of Anthropology, Vancouver, Canada (1977, Fig. 5.6g), express a very different architectural position. The parallel concrete frames range from tall narrow shapes with cantilever channel beams of 40 ft span and piers of 45 ft height to low wide shapes with 120-ft beam spans and 15 ft pier heights. Their design was derived from the Indian longhouse frames and other issues, and clearly not from supporting the building. There is a visual ambiguity because beams and columns not only seem to be overdesigned but are also all of the same size, expressing an enormous redundancy from a structural point of view.

The Effect of Cantilevering. With increase of span, the simply supported beam concept becomes less efficient because of the rapid increase in moment and deflection, that is, increase in dead weight. The magnitude of the bending stresses is very much reduced by the cantilever type of construction, as the graphical analysis in Fig. 5.7 explains. The maximum moment in the symmetrical double cantilever beam is only 17% of that for the simply supported case for the given arrangement of supports and loading! Often this arrangement is used to achieve a minimal beam depth for conditions where the live load, in comparison to the dead load, is small so that the effect of live load arrangement becomes less critical. As the cantilever spans increase, the cantilever moments increase, and the field moment between the support decreases. When the beam is cantilevered by one-half of the span, the field moment at midspan is zero because of symmetry and the beam can be visualized as consisting of two double-cantilever beams. For this condition the maximum moment is equal to that of a simple span beam.

A powerful design concept is demonstrated by the two balanced, double-cantilever structures carrying a simply supported beam, which, in turn, can be treated as a beam fixed at both ends for the given situation in Fig. 5.7. This *balanced cantilever beam concept* is often used in bridge construction. It was applied for the first time on a large scale to the 1708-ft-span Firth of Forth Bridge in Scotland (1890). Its form is in direct response to force flow intensity and has been a source of inspiration for numerous designers. Notice, that the shape of the trusses for the example in Fig. 5.7, bottom, conforms to that of the bending moment diagram. The truss can be treated as a beam fixed at both ends and hinged at the points of zero moment (i.e., inflection points).

Simple, Statically Determinate Frames. The upper four cases in Fig. 5.8 represent typical statically determinate frames. These cases will be briefly investigated first with respect to uniform gravity loading and then lateral loading. In the simple *post–beam* structure (case a) the beam is pin connected to the columns and therefore does not transfer any moments to the columns. It represents a typical simply supported bending member carrying a uniform gravity load causing the familiar maximum moment at midspan of

$$M = wL^2/8 = WL/8 \tag{2.25}$$

The beam only transfers shears (i.e., vertical reactions) since it rotates freely, thus causing axial forces in the columns; the columns act as pure axial members supporting

$$N = wL/2 = W/2$$

Also, the columns of the *bent beam* structure (case c) only carry axial forces since the roller support allows lateral movement and hence does not induce bending to the columns.

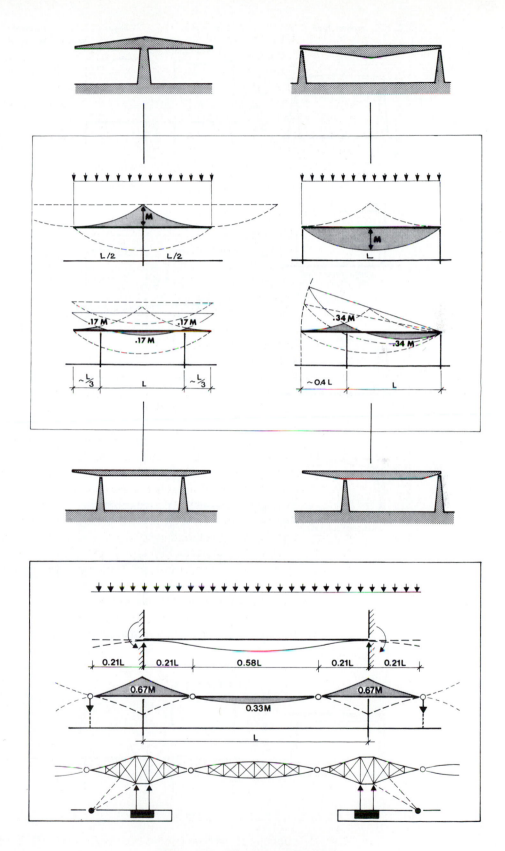

Figure 5.7 Effect of cantilevering.

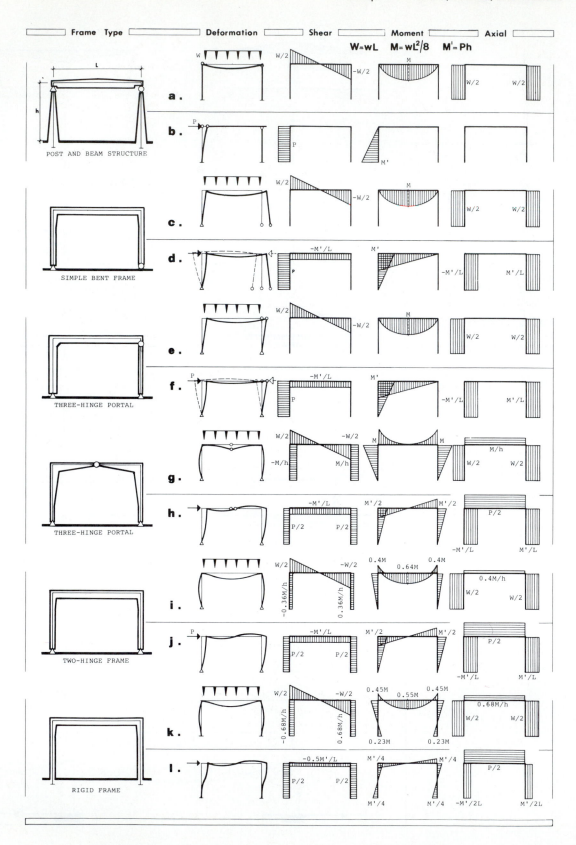

Figure 5.8 Internal force flow for basic portal frame units.

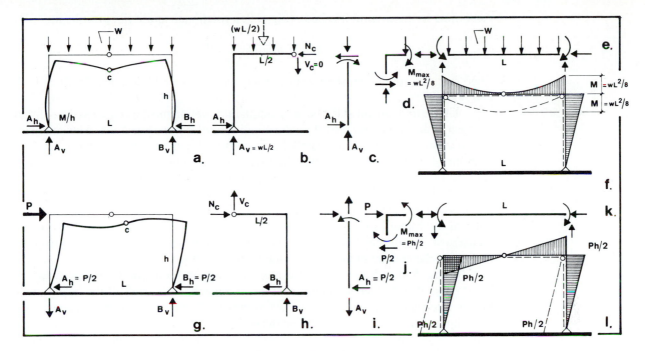

Figure 5.9 Three-hinge frame.

$$N = W/2$$

As there is no bending transferred from the columns to the beam, the beam behaves like a simply supported member carrying a uniform load, although columns and beam are rigidly connected.

$$M = wL^2/8 = WL/8$$

Similar conditions occur for the *L-frame* supported by the *hinged column* (case *e*) Also, here the columns carry only axial forces because of the pinned-column situation. Therefore, the beam behaves like a simply supported member, as for the preceding cases. Again, the rigid connection between frame column and beam has no effect on the force flow. The cases that have just been discussed do not represent true frames with respect to gravity loading since they are not capable of providing resistance to lateral movement at the base, as is clearly identified by the respective displacement patterns of the frames in Fig. 5.8; they are basically all post–beam structures not taking advantage of their profile. In contrast, the *three-hinge portal frame* (case *g*) is a true frame. Here, horizontal thrust or kickout forces are generated by the gravity loads at the base (since the boundary conditions do not allow outward displacement), which, in turn, cause bending of the columns and beam (i.e., bending is generated through shear wracking).

Due to symmetry of loading and frame geometry, each column (as for the other cases) carries the following axial load:

$$N = W/2$$

The thrust forces H at the base are determined from the free body in Fig. 5.9b.

$$\Sigma M_A = 0 = (wL/2)(L/4) - N_c(h) \quad \text{or} \quad N_c = A_h = H = wL^2/8h$$

According to the column free body (Fig. 5.9c), rotational equilibrium results in

$$M = N_c h = wL^2/8 = WL/8$$

Therefore, the column must be treated as a *beam–column* and designed for $M = WL/8$ and $N = W/2$.

The beam moment at the beam–column intersection must be equal to the column moment at the corner because of rotational joint equilibrium (Fig. 5.9d). This moment, $M = WL/8$ must decrease in a parabolic shape from the maximum to zero at the midspan hinge. We may conclude that the beam must also be treated as a beam–column (Fig. 5.9e) and designed for $M = WL/8$ and the axial thrust force $N_c = H = M/h$.

The behavior of the three-hinge frame can also be derived from Fig. 5.27, where the pressure line in response to the uniform load is a parabolic funicular line. The deviation of the frame geometry from the parabolic pressure line is largest at the corners and zero at the midspan hinge. Therefore, the maximum moment of $M = WL/8$ must be resisted at the corners of the beam–column intersections.

In Fig. 5.9f, it is shown that the moment diagram for the post–beam structure is lifted up so that the maximum moment $wL^2/8$ appears at the ends of the beam rather than at midspan. From the point of view of gravity loading, we may conclude that the three-hinge frame is less efficient than the post–beam structure since the columns must carry beam moments as well as axial forces; it is assumed that strength, not stiffness, is the basis of the design. Keep in mind that the costs of a continuous beam–column connection are much greater than for a simple hinged connection.

Now the four cases will be investigated under the lateral loading P. It is assumed that in the *post–beam structure* (Fig. 5.8b) only one column resists the load in vertical cantilever action by allowing a roller support for the beam. The column acts as a pure beam that must resist a maximum base moment of

$$M = Ph$$

Naturally, the load can be shared by both columns (assumed identical) by pin connecting the beam so that

$$M = Ph/2$$

Notice that the restraint due to pin connecting the beam, in turn, will cause bending of the columns under gravity loading. However, for approximation purposes, these relatively small column moments may be neglected.

Only the pinned column (not the column on rollers) of the *bent-beam structure* (Fig. 5.8d) is capable of resisting the lateral force. Therefore, the maximum moment at the beam–column intersection due to shear wracking is

$$M = Ph$$

The rotation caused by the lateral force is resisted by the axial column forces

$$Ph = NL \quad \text{or} \quad N = Ph/L = M/L$$

The force flow due to lateral force action in the *L-frame* supported *on a hinged column* is analogous to the previous case. Also, here only the frame column can transfer the shear in bending to the base since the other column is hinged at the top.

The *three-hinge portal frame* resists the rotation due to the lateral force action in axial action of the columns similar to some of the other cases.

$$Ph = NL \quad \text{or} \quad N = Ph/L = M/L$$

The symmetrical structure resists the lateral shear equally at each of the reactions (Fig. 5.9h) since

$$\Sigma \, M_c = 0 = (Ph/L) \, L/2 - B_h(h) \quad \text{or} \quad B_h = P/2$$

Therefore, the maximum moments M_s at the beam column intersections due to shear wracking are

$$M_s = (P/2)h = Ph/2 = M/2$$

Again, comparing the three-hinge portal with the post–beam structure (Fig. 5.9L), we notice that in one case the maximum moment is carried at the beam–column intersection, while in the other case it is resisted at the column–foundation intersection. From a lateral loading point of view, the post–beam structure is less efficient because of the large foundation necessary to transmit the rotation to the ground, realizing that settlement effects must be taken into account.

EXAMPLE 5.1

Various other important frame characteristics are studied in Fig. 5.10. The response of the frame to change in geometry and loading is explained with moment diagrams. The themes are presented in three groups, and the cases within each of them are discussed from the left to the right.

    A. We investigate the effect of *hinge location* along the beam portion of a rectangular bent (pinned to the base) carrying a uniform load w (refer also to Fig. 5.8). When the hinge is located at the beam–column intersection, the beam responds like a simply supported beam where the maximum moment is located at midspan. On the other hand, when the hinge moves to the midspan, the moment must be zero at this point, but occurs now at the beam–column intersections. We may conclude that a certain hinge location, somewhere between beam midspan and support, causes a field moment equal to the support moment.

$$M/2 = (wL^2/8)/2 = wL^2/16$$

This moment is generated by an equivalent simply supported beam of span x.

$$wL^2/16 = wx^2/8 \quad \text{or} \quad x = 0.707L$$

In other words, the hinge location is at $(L - 0.707L)/2 = 0.147L$ from the support.

    For the condition where there is no hinge, the portal frame is indeterminate, so the moment distribution depends on the stiffness of beam and columns, as is discussed in the following sections (see Fig. 5.12 and Fig. 5.27, right).

    B. The behavior of a three-hinge portal frame under uniform gravity loading is studied as one column is shortened, as the location of the hinge is moved, and as a cantilever is added. When the frame supports are not at the same level, the horizontal thrust forces at the base will influence the magnitude of the vertical reaction forces. Rather than applying the usual procedure of taking moments about one of the reactions and about the beam hinge (which yields two equations with two unknowns), the following simplified approach is used.

    It is apparent that the horizontal reaction forces must balance each other since no external lateral force action is present.

$$\Sigma H = 0 = A_h - B_h \quad \text{or} \quad A_h = B_h = H$$

The horizontal thrust forces H are resolved into H' forces so that they pass through both supports. They must resist at center span $M = wL^2/8$ as generated by the uniform loading.

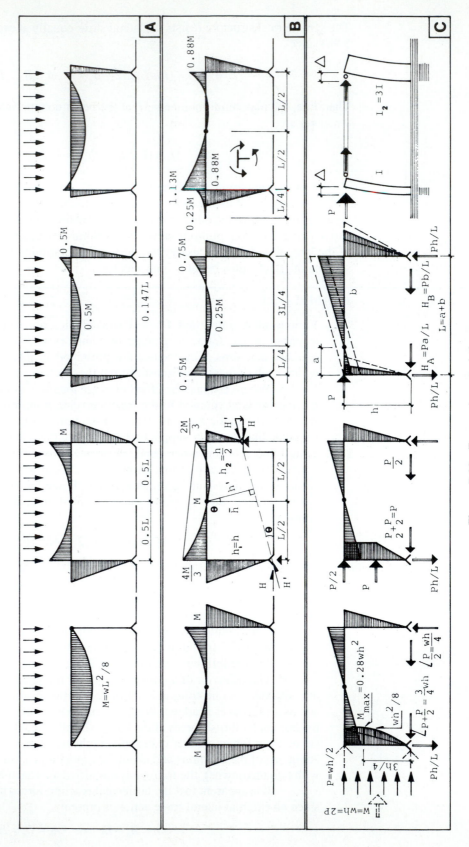

Figure 5.10 Frame characteristics.

$$H'(h') = wL^2/8 \quad \text{or} \quad \frac{H}{\cos\theta}\,(\overline{h}\cos\theta) = wL^2/8$$

or

$$H = wL^2/8\overline{h}, \qquad \text{where } \overline{h} = (h_1 + h_2)/2$$

For this specific case, where $h_1 = h$, $h_2 = h/2$, and $\overline{h} = 3h/4$, the magnitude of the thrust action is

$$H = wL^2/6h$$

Now the vertical reaction forces can be found by taking moments about the supports. The moments at the top of the columns are

$$\text{Long column:} \quad M_c = H(h) = wL^2/6 = 1.33M$$

$$\text{Short column:} \quad M_c = H(h/2) = wL^2/12 = 0.67M$$

We may conclude that the shortening of one column reduces the moment in that column but increases the moment in the long column.

In the next portal frame case, the hinge location is at $L/4$ from the support. Therefore, the support moment must be equal to

$$M_s = wL^2/8 - w(L/2)^2/8 = 3wL^2/32 = 0.75M$$

And the field moment is

$$M_f = M - M_s = M - 0.75M = 0.25M$$

Notice that the hinge location could be altered to generate a more even distribution of the moment M.

Adding a cantilever of $L/4$ length to the three-hinge portal frame causes asymmetry. Therefore, first the vertical reactions must be found to resist the total load of $1.25wL$.

$$\Sigma M_b = 0 = A_v(L) - 1.25wL(0.625L) \quad \text{or} \quad A_v = 0.78wL$$

$$\Sigma V = 0 = B_v + 0.78wL - 1.25wL \quad \text{or} \quad B_v = 0.47wL$$

Taking moments about the hinge, using the right free body, gives the magnitude of the thrust force.

$$\Sigma M_c = 0 = H(h) - 0.47wL(L/2) + (wL/2)L/4 \quad \text{or} \quad H = 0.11wL^2/h$$

This thrust force causes a maximum column moment of

$$M_c = H(h) = 0.11wL^2 = 0.88M$$

The maximum cantilever moment is

$$M_{ca} = w(L/4)(L/8) = wL^2/32 = 0.25M$$

The rotational equilibrium at the beam–column joint yields the maximum beam moment of

$$M_b = 0.25M + 0.88M = 1.13M$$

We may conclude that the addition of the cantilever reduces the column moment but increases the beam moment at the cantilever side.

C. The effect of lateral force action on frame behavior is investigated for some specific conditions. In the first two cases the effect of lateral force distribution on a three-hinge portal frame is studied. Usually, the curtain wall is assumed to span vertically, thereby transferring one-half of the total lateral load, $W = wh = 2P$, to the spandrel beam at roof level and the other half directly to the foundations. The spandrel beam, in turn, causes a single load, $P = wh/2$, at the beam–column intersection (see Fig. 5.9g and L); for this condition the maximum moment in the frame is $M = Ph/2 = 0.25wh^2$.

In the first case it is assumed that one column must resist the lateral pressure directly as a uniform load (e.g., the curtain spans horizontally from column to column). This case can be visualized as consisting of a vertical beam carrying a uniform load $W = wh = 2P$, which causes a maximum moment at mid-height of $wh^2/8$. The reaction of this beam $W/2 = P$ at the top, in turn, must be carried in the usual frame action. The superposition of the two cases results in a new maximum column moment at the windward side of $M_c = 0.28wh^2 = 1.12M$ at $3h/4$ from the base. It may be concluded that only the moment in the windward column is increased by 12% (i.e., local effect), while the beam–column moments remain the same as for single–load action at roof level.

In the next case, the curtain panels are supported in addition by a horizontal beam at mid-height so that the column must resist the beam reaction at this location. Again, it can be shown, using a similar reasoning as before, that the magnitude of the frame moments is not changed. The reader may want to study the two cases in more detail to verify the statements presented.

In the third case the effect of hinge location along the beam portion in response to lateral force action (similar to the investigation in part **A** for gravity loads) is studied. It is shown that the magnitude of the horizontal reactions is proportional to the hinge location. Only for the specific condition where the hinge is located at midspan ($a = b = L/2$) are the horizontal reactions equal to each other, $H_A = H_B = P/2$. But should the hinge move to one of the supports (say, $a = 0$), then the entire lateral load must be resisted by the frame column ($H_B = P$, $H_A = 0$), as has already been investigated. The magnitude of the frame moments is directly related to the horizontal reaction forces (e.g., $M = Hh$).

Finally, a simple post–beam structure is investigated. Since the cantilever columns of this structure are not identical, each column must carry a certain portion of the total lateral load P. In case of simple flexural deflections (i.e., shear deformations can be ignored), equal column heights, and same material (E = constant), Eq. (4.30) can be presented in the simple version of

$$P_i = P(I_i/\Sigma I)$$

For this specific example, where $\Sigma I = I_1 + I_2 = I + 3I = 4I$,

$$P_1 = P(I/4I) = P/4, \qquad P_2 = P(3I/4I) = 3P/4$$

Therefore, the larger column resists 75% of the total load while the smaller column resists only 25%.

Besides the more common frame enclosure systems under typical loading conditions, which have been discussed up to now, some other frame types including *cantilever frames* are shown in Fig. 5.11. These frames are loaded in various ways and suggest many different possibilities of frame action; they are investigated in the problems.

An Approximate Solution of Simple, Statically Indeterminate Portal Frames.
Only typical rectangular bents are investigated. When a portal frame is pinned to its base, the frame is once indeterminate (Fig. 5.3). However, if it is fixed so that rotation is transferred to the foundation, it is three times indeterminate. Besides the three equations of static equilibrium, other methods of analysis must be developed to determine the magnitude of the force flow along these frames. Here, the deflected shape of the frame under controlling loading conditions yields the approximate location of the inflection points of the curves, which, in turn, give additional information so that the

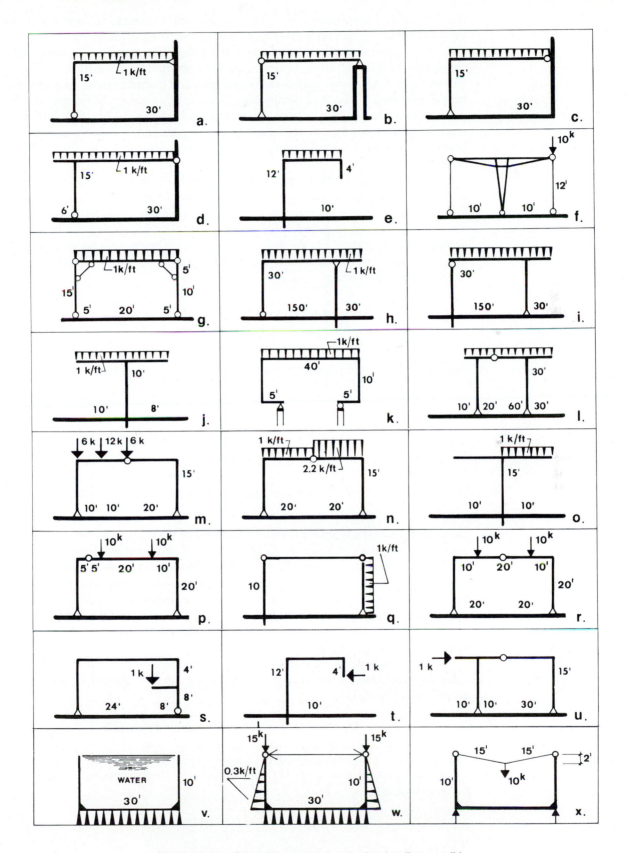

Figure 5.11   Other frame types and other loading conditions.

frame analysis is made determinate. The response of the frame to uniform gravity loading and lateral force action is investigated in the following discussions.

A. In general, uniform force action along the beam portion of a frame may be
considered as the controlling loading case for a preliminary investigation of *gravity
action*. For this condition, frames can be unfolded and treated as three-span continuous
beams, as shown in Fig. 5.12 (center top). Because of symmetry of loading and frame
geometry, the small effect of beam–column joint displacement is ignored. It can also
easily be shown that the moment diagram for the continuous, determinate three-span
beam with a hinge at midpoint of the center span is the same as for the three-hinge portal frame.

The two-hinge frame is once indeterminate; hence one more equation (in addition to the three static equilibrium equations) is needed for finding the force flow. In
this case the *three-moment theorem* (derived by the French engineer B. P. E. Clapey-
ron in 1849) is used to determine the support moment M_S so that the beam can be
solved by simple statics. For the three-span continuous beam, with uniform loading on
the central span and constant beam and column moments of inertia, the three-moment
equation takes the form of

$$M_A \frac{h}{I_c} + 2M_S \left(\frac{h}{I_c} + \frac{L}{I_g} \right) + M_s \frac{L}{I_g} = -\frac{wL^2}{4} \frac{L}{I_g}$$

Because of the hinged boundary conditions the support moments are zero ($M_A = 0$), so
the equation further simplifies to

$$M_S \left(\frac{2h}{I_c} + \frac{2L}{I_g} + \frac{L}{I_g} \right) = -\frac{wL^2}{4} \frac{L}{I_g}, \qquad \text{let } k = \frac{(I/L)_g}{(I/h)_c}$$

$$M_S = -\frac{wL^2}{12} \left(\frac{1}{1 + (2/3)(I_g/L)\, h/I_c} \right) = -\frac{wL^2}{12} \left(\frac{3}{3 + 2k} \right) \qquad (5.3)$$

This expression clearly shows that the magnitude of the support moment M_S is dependent on the stiffness of beam (I_g/L) and column (I_c/h); for the definition of member
bending stiffness, refer to Eq. (2.84). As the column stiffness decreases, the support
moment M_S approaches zero or the beam approaches simply supported conditions. On
the other hand, as the column stiffness increases the beam approaches fixed beam conditions with a boundary moment of $M_S = -wL^2/12$. The effect of the stiffness of beam
and column on moment flow or the transition from a simply supported beam to a fixed
beam is described in Fig. 5.12.

It is reasonable to assume for first-approximation purposes that the beam and
columns are of equal stiffness, or $k = 1$. Hence the support moment for this condition
according to Eq. (5.3) is

$$M_S = -wL^2/20 = -0.4M, \qquad \text{where } M = wL^2/8 \qquad (5.4)$$

The maximum field girder moment at midspan can easily be found by subtracting from
the absolute maximum moment M (i.e., moment of a simply supported beam) the support moment M_S.

$$M_g = M - 0.4M = 0.6M \cong wL^2/13$$

We may conclude that 60% of the total moment M is carried by the beam, while
the remaining 40% is carried across the continuous corner junction into the columns.

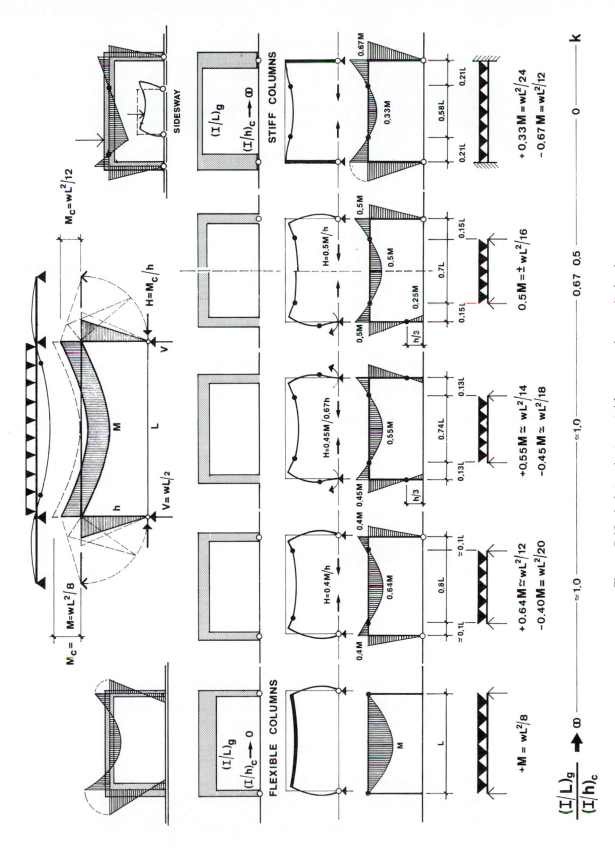

Figure 5.12 Indeterminate portal frames under gravity loads.

417

For this situation the inflection points are located at $0.11L$ from the support. Usually, however, typical member stiffness conditions result in inflection points between **0.10L** **and 0.12L.** Therefore, for the preliminary design of the frame girder, inflection points are conservatively assumed at $L/10$ to yield larger beam sizes. Considering the portion of the beam between the assumed inflection points (i.e. imaginary hinges) as a simply supported beam of span $0.8L$ results in the following familiar moment expression:

$$M_g = w(0.8L)^2/8 = 0.64\,M \cong wL^2/12 \tag{5.5}$$

The lateral thrust forces at the reactions (as well as the axial girder force) of the two-hinge portal frame can be obtained from the column free body.

$$H(h) = M_S \quad \text{or} \quad A_h = B_h = H = M_S/h \tag{5.6}$$

The vertical reactions are each equal to one-half of the total, symmetrically arranged load.

$$A_v = B_v = wL/2 \tag{5.7}$$

It is not only the stiffness of the members that influences the location of the inflection points, but also the type of loading. For instance, for the case of a single, vertical load acting at midspan of the two-hinge frame, it can also be shown that inflection points form at $0.15L$ for equal beam and column stiffnesses.

If the hinged support conditions of the rectangular bent are replaced by fixed ones, then not only are additional inflection points introduced in the columns, but also the locations of the zero-moment points in the beam are influenced. It can be shown that the equation for the moment M_S at beam–column intersection is equal to Eq. (5.3) if the factor of $2/3$ in the denominator is replaced by a factor of $1/2$. It can be further shown that the support moments at the fixed base are equal to $-M_S/2$, causing inflection points in the columns at $h/3$ measured from the base.

The continuous rigid frame is three times indeterminate. In this specific case, due to symmetry, only two more conditions have to be known besides the three equations of static equilibrium. As the frame deflects under the gravity loading, inflection points form within the beam and column ranges. The location of these points of zero moment will be assumed so that the approximate force flow along the frame can be determined.

By fixing the two-hinge frame to the base, more moment is attracted to the stiffer columns, thereby decreasing the field moment in the beam. The beam inflection points for the given uniform loading will move farther away from the columns as compared to the location of $0.1L$. Their location may be assumed at $0.13L$ measured from each column (Fig. 5.12) as based on equal column and beam stiffnesses.

The maximum moment at midspan for an equivalent simply supported beam of a length equal to $l = L - 2(0.13L) = 0.74L$, is

$$M_g = w(0.74L)^2/8 = 0.55M \cong wL^2/14 \tag{5.8}$$

The beam support moment at the column intersection is the difference between the moment for a simply supported beam and the actual beam moment at midspan.

$$M_S = M - 0.55M = 0.45M \cong wL^2/18 \tag{5.9}$$

The inflection points in the columns form at one-third of the column height measured from the base as caused by equal column and beam stiffnesses. At this location the moment is zero and increases linearly to its maximum value M_S at the beam–column intersection and decreases to one-half of that maximum moment at the base.

$$M_A = M_B = 0.45M/2 = M_S/2 \qquad (5.10)$$

The continuous frame can be visualized as a two-hinge frame sitting on column stubs $h/3$ high. Hence, a similar approach may be used for its solution, as already described for the two-hinge portal frame. We may conclude that the thrust forces at the reactions must therefore be equal to

$$H(2h/3) = M_S \quad \text{or} \quad A_h = B_h = H = 1.5M_S/h \qquad (5.11)$$

The vertical reactions are each equal to one-half of the total, symmetrically arranged load.

$$A_v = B_v = wL/2 \qquad (5.12)$$

B. The response of the statically indeterminate portal frames to *lateral loading* is investigated in Fig. 5.13.

Under lateral force action the two-hinge portal deflects by forming an inflection point at beam midspan, assuming the beam not to be rigid and equal column sizes. Since the location of the inflection point is assumed to be at $L/2$, this case is identical with the lateral loading of the three-hinge portal. Therefore, refer to the lateral force analysis for the three-hinge portal frame for the approximate behavior of the two-hinge frame.

If the pinned ends of the bent are fixed, additional inflection points will form in the columns. The location of these points of contraflexure depends on the relative stiffnesses of columns and beam as indicated in the bottom portion of Fig. 5.13. For deep girders, which do not allow any rotation, inflection points will be at $0.5h$. For flexible beams, which act more like ties and are unable to resist rotation, the zero-moment point moves to the beam–column intersection; that is, the column behaves as a vertical cantilever. For equal stiffness of beam and columns, the point of contraflexure is at $0.57h$. For typical frames the inflection point is in the range of **$0.5h$ to $0.6h$** measured from the base. In addition, the possible rotation of the foundations due to soil settlement should be taken into account, as well as the fact that column base connections are rarely truly fixed, which will cause the inflection points to move down, say to $0.5h$. We may conclude that it is reasonable, for preliminary design purposes, to assume inflection points at $h/2$. Therefore, the wracking of the frame by the lateral force P causes inflection points to form at mid-height of the columns and at midspan of the beam as based on the simplifying assumptions just discussed. We may visualize a two-hinge frame of height $h/2$ sitting on column cantilever stubs. Hence, the vertical reaction forces can be found as for the three-hinge portal frame.

$$A_v(L) = P(h/2) \quad \text{or} \quad A_v = -B_v = Ph/2L \qquad (5.13)$$

The horizontal reaction forces must be equal because of conditions of symmetry since the inflection point forms at beam midspan, as was discussed previously.

$$A_h = B_h = H = P/2$$

These horizontal shear forces at column mid-height wrack the frame and cause equal moments at the top (and ends of beam) and the bottom of the columns.

$$M_S = (P/2)h/2 = Ph/4 = 0.5M \qquad (5.14)$$

where $M_S = Ph/2$ as based on the two-hinge frame.

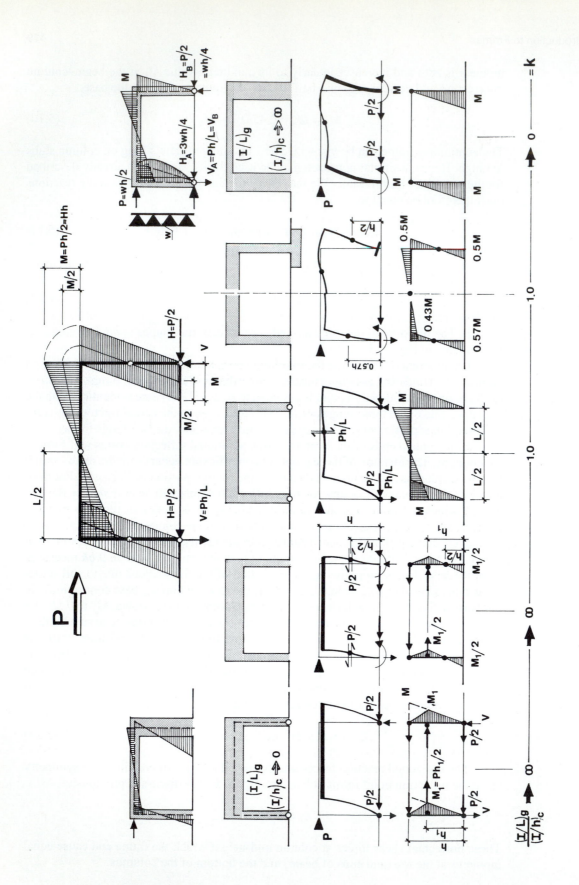

Figure 5.13 Indeterminate portal frames under lateral load action.

In comparison with the two-hinge frame, the rigid frame carries moments, axial column forces, and maximum beam shear of one-half the magnitude. While the two-hinge frame attracts the total moment to the beam–column intersection and the post–beam structure to the column–foundation–earth intersection, the continuous frame, being between the two systems, distributes the moment equally to top and bottom (Fig. 5.13, center top). Should the columns not have the same stiffness, the column moments will not be equal because of nonsymmetry of the frame. The stiffer column will attract a greater share of the load and carry a larger moment, as is shown in Figs. 5.10C, 5.12, and 5.13 (top left).

C. The traditional knee-braced timber structures (Fig. 5.14) behave similarly to rigid frames. The analysis of these bents (with identical columns) for vertical loading is indeterminate. The knee braces do not provide a rigid support to the beam, as the deflected configuration of the pin-based bents in Fig. 5.14c and g indicates; the degree of support settlement depends on the stiffness of the columns. For preliminary design purposes, the beams may be treated as simply supported, as indicated in Fig. 5.14. The columns may be visualized as vertical beams loaded by the brace force P, where bending is caused by the horizontal component.

$$P_x = P\cos\theta = P_Y \cot\theta \tag{5.15}$$

For the analysis of a knee-braced portal frame with respect to lateral loading, an inflection point at beam midspan may be assumed (Figs. 5.14a and b) and we then proceed as for the two-hinge portal frame. The reader may want to check the solution of the case in Figs. 5.14g and h.

The approximate methods just presented ignore the additional moments due to the first-order displacements (P–Δ effect). Under unsymmetrical conditions such as due to different boundaries, member sizes (i.e. stiffnesses), hinge location, vertical loading (see Fig. 5.12, top right) and obviously lateral load action, the frames displace laterally in an effect referred to as *sidesway*. Obviously, these displacements cause additional bending, which cannot be neglected in an exact engineering analysis. This also clearly indicates that the superposition of critical values due to lateral forces and symmetrical vertical loading as presented here can only be approximate, since the combined force action of the cases generates a new condition; that is, the sum of the individual actions is not equal to the combined action.

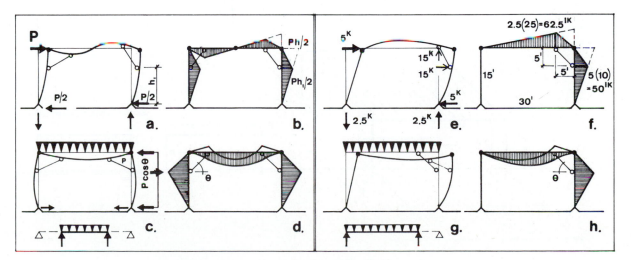

Figure 5.14   Knee-braced portal frames.

EXAMPLE 5.2

Different framing structures are investigated for enclosing the building volume shown in Fig. 5.15. The following roof loads must be supported: 25-psf dead load, 30-psf live load, and 20-psf wind load. The columns do not sway about their weak axes, because the building is laterally braced in the long direction. The preliminary beam and column sizes are determined for cases b and d using A36 steel; the other cases are investigated in Problem 5.7.

A typical interior frame must support a roof area of one-half of a bay on each side; that is, the joists on each side of the frame transmit to the frame one-half of the total roof gravity load that they are carrying.

$$w = 40(0.025 + 0.030) = 1.0 + 1.2 = 2.2 \text{ k/ft}$$

Similarly, the curtain panels transmit the lateral wind load to the spandrel beams and roof diaphragm, which in turn apply a single load P to the frame at the beam–column intersection.

$$P = 40(15/2)0.020 = 6 \text{ k}$$

(a) *Investigation of the Two-hinge Frame (Case d).* The preliminary design of ordinary, single-story, single-bay, low-rise frames, where column moments are generated by the gravity beam loads, can be based on the uniform gravity loading case (i.e., Fig. 5.15, cases c, d, and e). The frame beam can be treated as a predominant bending member since the column action is negligible (i.e., axial force action, due to thrust, is insignificant and can be ignored for preliminary design purposes).

Assuming the beam stiffness to be roughly equal to the column stiffness yields the following maximum moment (see Fig. 5.12).

$$M_g \cong wL^2/12 = 2.2(40)^2/12 = 293.33 \text{ ft-k}$$

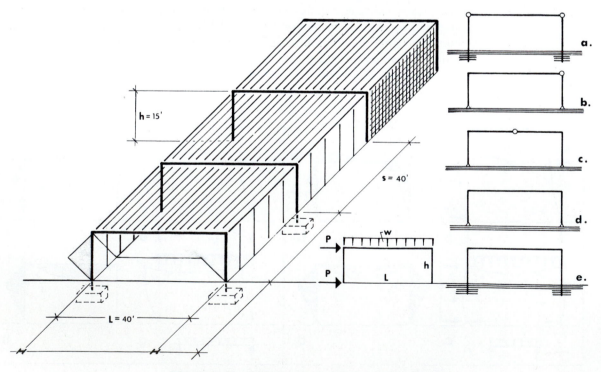

Figure 5.15 Different rectangular enclosure frames.

Therefore, the required section modulus for a laterally supported compact beam can be estimated for $F_b = 0.66F_y = 0.66(36) \cong 24$ ksi [Eq. (3.5a)] as

$$S = M/F_b = 293.33(12)/24 = 146.67 \text{ in.}^3 \qquad (3.4)$$

Select as a trial section (Table A.11)

$$W24 \times 68, \qquad S_x = 154 \text{ in.}^3, \qquad L_c = 9.5 \text{ ft}, \qquad F_y' \geq F_y$$

The beam is assumed to be laterally supported by the joists at a spacing of not more than 9.5 ft.

The gravity case controls the design of the column as it did for the beam, although we must keep in mind that when the height increases the full gravity and wind cases will govern the column design; see part (b).

For preliminary design purposes, the column can be treated as a beam since the effect of the axial forces is relatively small. Therefore, a trial section is estimated based on bending only by selecting not the section required, but the next larger one, to take axial action into account. The maximum column moment according to Fig. 5.12 is

$$M_c \cong wL^2/20 = 2.2(40)^2/20 = 176 \text{ ft-k}$$

Since the compression flange of the column is laterally supported only at its ends, the allowable bending stress is assumed as $F_b = 0.6F_y$, according to Eq. (3.5c). Therefore, the required section modulus can be estimated as

$$S = M/F_b = M/0.6F_y \quad \text{and} \quad L_u \geq 15 \text{ ft} \qquad (3.4)$$

$$= 176(12)/0.6(36) = 97.78 \text{ in.}^3$$

Try W14 $\times$ 68, $S_x = 103$ in.^{3}, $L_u = 23.9$ ft ≥ 15 ft, $L_c = 10.6$ ft < 15 ft, and $F_y' \geq F_y$.

Checking the ratio of girder stiffness to that of the columns (see AISC Steel Manual for I-values),

$$k = \frac{(I/L)_g}{(I/h)_c} = \frac{1830/(40 \times 12)}{723/(15 \times 12)} = \frac{3.81}{4.02} = 0.95$$

The solution is close to 1; therefore, the estimated member sizes should be reasonable, keeping in mind that as k decreases below 1, the beam field moment decreases while the support (column) moments increase.

(b) *Investigation of the Three-hinge L-frame (Case b).* As has been previously discussed for Fig. 5.8e, the frame beam is solely a bending member under vertical loading and its design is obviously controlled by gravity action. According to Fig. 5.8e, the maximum moment is

$$M_g = wL^2/8 = 2.2(40)^2/8 = 440 \text{ ft-k}$$

The required section modulus for a laterally supported compact beam is

$$S = M/F_b = 440(12)/24 = 220 \text{ in.}^3 \qquad (3.4)$$

Select as a trial section (Table A.11)

$$W30 \times 90, \qquad S_x = 245 \text{ in.}^3, \qquad L_c = 10 \text{ ft} \geq L_b$$

The preliminary design of the fixed-base cantilever columns and the *L*-frame column (see also Example 5.6) of cases a and b in Fig. 5.15 must be based on the combined loading case of gravity and lateral force action, since gravity loading by itself does not generate direct column moments. For ordinary conditions (i.e., not for long-span systems), the design is controlled by dead load together with wind (or earthquake). For preliminary design purposes, the frame column can be treated as a beam, since it must resist in bending the entire wind, by selecting not the section required, but the next larger one, to take the relatively small axial action into account. The column moment according to Fig. 5.8f is

$$M_c = Ph = 6(15) = 90 \text{ ft-k}$$

$$S \cong M/F_b = M/0.6F_y = 90(12)/0.6(36) = 50.0 \text{ in.}^3 \quad \text{and} \quad L_u \geq 15 \text{ ft}$$

Try W12 ×45, $S_x = 58.1$ in.^{3}, $L_u = 17.7$ ft ≥ 15 ft, and $L_c = 8.5$ ft < 15 ft.

The simple hinged pendulum column supporting the L-frame is designed for axial gravity action only.

$$N = 2.2 \, (40/2) = 44 \text{ k}$$

Because of the high slenderness of the column, elastic buckling may be assumed. Furthermore, the known buckling condition about the laterally braced weak axis is only considered for this preliminary investigation. For a W6 section try $r_Y = 1.4$ in.

$$F_a = 149,000/(Kl/r)^2 \tag{3.93}$$

$$= 149,000/[1(15 \times 12)/1.4]^2 = 149,000/(129)^2 = 9.01 \text{ ksi}$$

Therefore, the required cross-sectional area of the column is

$$A_c = N/F_a = 44/9.01 = 4.88 \text{ in.}^2 \tag{3.91}$$

Try W6 × 20, $A_c = 5.87$ in.^{2}, $r_Y = 1.50$ in. (see AISC Steel Manual).

The same result is obtained by using the Column Tables in the AISC Steel Manual for $KL = 15$ and $N = 44$ k.

EXAMPLE 5.3

The two-hinge, reinforced-concrete portal frame in Figs. 5.15d and 5.16 is approximately designed. The following roof loads are assumed: 100-psf dead load (where the long-span joist slabs are assumed to be 12 in. deep), 30-psf live load, and 20-psf wind load. A typical interior frame is investigated using $f_c' = 4000$ psi and $f_y = 60,000$ psi. The wind loading case can be ignored for this preliminary study, as has been discussed previously.

To estimate the beam size, first only the slab loads are considered. The ultimate load, using the load factors according to Eq. (4.19), is

$$w_{us} = [1.4(0.100) + 1.7(0.030)]40 = 7.64 \text{ k/ft}$$

The beam size is controlled by the compressive strength at the column support since at midspan the slab (acting as beam flanges) provides a high compression resistance. The support moment may be estimated according to Eq. 5.4 (for $k \cong 1$) as

$$M_s = wL^2/20$$

$$M_{us} = w_u L^2/20 = 7.64 \, (40)^2/20 = 611.2 \text{ ft-k}$$

The beam size may be determined as based on an average reinforcement ratio of $\rho \cong 1\%$ using a beam-to-width ratio of about 2 (see Example 3.15).

$$bd^2 = 23M_u$$

$$b(2b)^2 = 23(611.2) \quad \text{or} \quad b = 15.20 \text{ in., say 16 in.} \tag{3.45}$$

$$16(d)^2 = 23(611.2) \quad \text{or} \quad d = 29.64 \text{ in.}$$

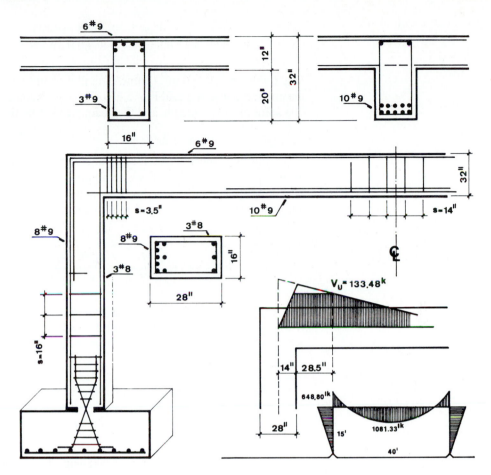

Figure 5.16 Approximate design of a two-hinge, reinforced-concrete portal frame.

Assuming a double layer of steel [see (Eq. 3.56)] yields a beam depth of

$$t \cong d + 3.5 = 29.64 + 3.5 = 33.14 \text{ in.}$$

Try a 16-ft × 32-in. beam section. This beam also satisfies the minimum depth requirement, so no deflection check is needed.

$$t = L/21 = 40(12)/21 = 22.86 \text{ in.} < 32 \text{ in.} \tag{3.44}$$

The beam must support the following loads by adding the stem weight of the beam to the known slab loads.

$$w_u = 7.64 + 1.4[(32 - 12)16/12^2](1)0.150 = 8.11 \text{ k/ft}$$

The approximate field and support moments according to Fig. 5.12 (for $k = 1$) are

$$M_{ug} = w_u L^2/12 = 8.11(40)^2/12 = 1081.33 \text{ ft-k}$$

$$M_{us} = w_u L^2/20 = 8.11(40)^2/20 = 648.80 \text{ ft-k}$$

The axial beam force is equal to the horizontal reaction forces of the frame.

$$N_{ug} = H = M_{us}/h = 648.80/15 = 43.25 \text{ k}$$

In comparison to the beam moment, the axial force is negligible for preliminary design purposes, as has been discussed previously. The required moment reinforcement at the column support, assuming the reinforcement in two rows, is approximately

$$A_s = \frac{M_u}{0.8f_y d} = \frac{648.80(12)}{0.8\,(60)(32-3.5)} = 5.69\,\text{in.}^2 \qquad (3.52)$$

Try 6 #9 bars, $A_s = 6$ in.^{2} (Table A.1) placed in two layers at the top face of the beam (see Table A.3 and Fig. 5.16). The corresponding steel ratio indicates that the solution is reasonable and the beam size is OK from a flexural point of view.

$$\rho = A_s/b_w d = 6/(16)28.5 = 1.32\% < 1.6\%$$

For the T-beam behavior due to composite beam–slab action at midspan, it is assumed that the stress block depth lies within the flange, so that the T-section can be treated as a wide, shallow rectangular beam section for preliminary design purposes. Hence, the required reinforcement at midspan is approximately equal to

$$A_s = \frac{M_u}{0.8f_y d} = \frac{1088.33(12)}{0.8\,(60)\,28.5} = 9.49\ \text{in.}^2 \qquad (3.52)$$

Try 10 #9 bars, $A_s = 10$ in.^{2}, placed in two layers at the bottom of the beam (see Table A.3 and Fig. 5.16). The maximum shear acts at distance d from the face of the column (see Fig. 5.16). Assuming a column depth of 28 in., the shear is

$$V_u = 8.11[40 - 2(14 + 28.5)/12]/2 = 133.48 \text{ k}$$

The shear strength of the concrete may be approximated conservatively as

$$\phi V_c = \phi(2\sqrt{f_c'})\,b_w d = 0.85\,(2\sqrt{4000})16(28.5)/1000 = 49.03\text{k} \quad (3.58)$$

Since the ultimate shear force V_u exceeds the shear capacity of the concrete, shear reinforcement must resist the excess shear $V_u - \phi V_c$ (Eq. 3.57).

$$\phi V_s \geq V_u - \phi V_c = 133.48 - 49.03 = 84.45 \text{ k} \quad \text{or} \quad \phi V_s/\phi = 84.45/0.85 = 99.35 \text{ k}$$

The required stirrup spacing s, using No. 3 stirrups with two legs, is

$$s = \frac{A_v f_y d}{V_s} = \frac{2\,(0.11)60(28.5)}{99.35} = 3.79 \text{ in.,} \quad \text{say 3.5 in.} \qquad (3.60)$$

For the condition where the maximum shear is less than three times the concrete strength, the minimum stirrup spacing should be $s \leq d/2 = 28.5/2 = 14.25$ in., say 14 in.

$$V_u = 133.48 \text{ k} \leq 3(\phi V_c) = 3(49.03) = 147.09 \text{ k} \qquad (3.68a)$$

We may conclude that the stirrup spacing changes from 3.5 in. at the support to 14 in. at midspan; various grouping patterns of the stirrups are possible.

The column design is based on $M_{us} = 648.80$ ft-k and $N_u = 8.11(40/2) = 162.2$ k. This column can be treated as a beam for preliminary design purposes, as has been discussed previously; it is assumed to have about 2% of reinforcement, so its depth can be determined using the frame width of 16 in.

$$bd^2 = 14M_u$$
$$\qquad\qquad\qquad\qquad\qquad\qquad\qquad (3.46)$$
$$16d^2 = 14(648.80) \quad \text{or} \quad d = 23.83 \text{ in.}$$

In other words, assuming two bar layers, $t \cong d + 3.5 = 23.83 + 3.5 = 27.33$ in.; try $t = 28$ in. The equivalent eccentricity of the axial force is

$$e = M_u/P_u = (648.8/162.2)12 = 48 \text{ in.} = 1.71t$$

Naturally, tension clearly controls since the axial force falls far outside the column cross section. Furthermore, the ACI Code specifies for beam–columns, which are actually beams because they are only carrying small axial forces, that the strength reduction factor ϕ may be increased linearly from 0.7 to 0.9 in the range from $\phi P_n = 0.1 f_c' A_g$ to zero for tied columns with symmetric reinforcement.

$$\phi P_n = 0.1 f_c' A_g = 0.1(4)(16 \times 28) = 179.2 \text{ k} > 162.2 \text{ k}$$

The beam character of the column is apparent. The required moment reinforcing is estimated by using $\phi = 0.8$.

$$A_s \cong \frac{M_u}{0.7 f_y d} = \frac{648.80(12)}{0.7(60)24.5} = 7.57 \text{ in.}^2 \qquad (3.122)$$

Try 8 #9 bars, $A_s = 8$ in.^{2} (Tables A.1 and A.3), in two rows along the outside face as indicated in Fig. 5.16. Some of the bars may be discontinued since they only have to be used as corner reinforcement. The steel ratio is equal to

$$\rho = A_s/bd = 8/16(24.5) = 2.04\% < \rho_{max} = 2.14\% \quad \text{(Table A.4)}$$

Therefore, theoretically no compression reinforcement is needed, although usually at least a quarter of the critical tension steel is provided along the compression face; in this case try 3 #8 bars. It is apparent that a reduction of column depth t should be considered since compression steel is present. The vertical spacing of the ties is the least of the following values (see Example 3.37).

16 Vertical bar diameters	16(1.128)	$\cong$ 18 in.
48 Tie bar diameters	48(3/8)	= 18 in.
Least column dimension		16 in.

Try #3 ties at 16 in. The design shear strength of the column at the joint is well above the applied shear of 43.25 k, so the shear requirement is met.

Lateral Deflection. The lateral deflections of some typical structural systems due to direct (external) loading are shown in Fig. 5.17A. The sway of the identical columns with fixed bases in case c can easily be determined by assuming each cantilever column to resist an equal portion of the load P. In other words, for the general condition of a multibay post–beam structure, the lateral deflection Δ (see Table A.15) is

$$\Delta = \frac{Ph^3}{3E\Sigma I_c} \qquad (5.16)$$

where P = the total lateral force located at roof level
  h = story height
  E = modulus of elasticity of columns
  ΣI_c = moments of inertia of all the columns

For the special case of a single-bay structure with $\Sigma I_c = 2I_c$, the lateral deflection is equal to

$$\Delta = \frac{Ph^3}{6EI_c} \qquad (5.17)$$

The lateral deflection of a multibay two-hinge frame with deep (i.e., rigid) girders or trusses (case a) is equal to the cantilever column construction just discussed by ignor-

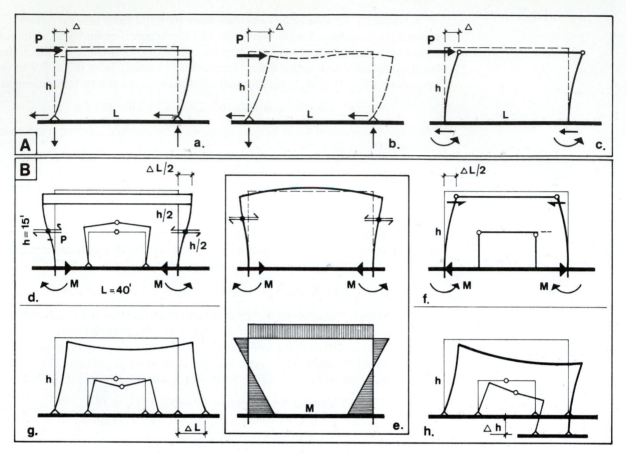

Figure 5.17 Lateral deflection of simple frames.

ing the negligible effect of chord drift (i.e., axial column deformations). Should the columns of a multibay frame with deep girders have fixed supports, the lateral deflection is equal to

$$\frac{\Delta}{2} = \frac{P(h/2)^3}{3E\Sigma I_c} \quad \text{or} \quad \Delta = \frac{Ph^3}{12E\Sigma I_c} \tag{5.18}$$

For the lateral deflection of frames in general, however, beam flexibility must be considered (case b.) Therefore, the effect of beam rotation must be added to the effect of column bending discussed previously. It can be shown (see Problem 5.8) that the lateral sway due to beam rotation for a multibay frame hinged at its base is approximately equal to

$$\Delta_b = \frac{Ph^2 L}{12E\Sigma I_b} \tag{5.19}$$

Adding this expression to Eq. (5.17) and letting $\Sigma I_b = (1)I_b$ yields the approximate lateral sway of a single-bay, two-hinge portal frame due to shear wracking (ignoring chord drift).

$$\Delta = \Delta_c + \Delta_b = \frac{Ph^2}{6E}\left(\frac{h}{I_c} + \frac{L}{2I_b}\right) \tag{5.20}$$

where L = width of bay
I_b = moment of inertia of the portal beam

The other terms have already been defined.

EXAMPLE 5.4

Determine the approximate natural period of the one-story, two-hinge steel frame (case d) in Example 5.2 using $I_b = 1830$ in.[4], $I_c = 723$ in.[4], $E = 29,000$ ksi, and $g = 32.2$ ft/sec[2].

The entire mass is considered to be lumped together at the roof level by treating the walls as massless. The weight is equal to

$$W = 1.0(40) = 40 \text{ k}$$

The lateral stiffness according to Eqs. (5.20) and (4.31) is

$$k = 1/\Delta = 1/[(h^2/6E)(h/I_c + L/2I_b)]$$
$$= 1/[((15 \times 12)^2/6(29,000))(15(12)/723 + 40(12)/2(1830))] = 14.13 \text{ k/in.}$$

The approximate natural period of vibration, according to Eq. (4.4), is

$$T = 2\pi\sqrt{W/kg} = 2\pi\sqrt{40/(14.13(32.2)12)} = 0.538 \text{ sec} \quad \text{or} \quad f = 1/T = 1.86 \text{ Hz}$$

Deformations are not only generated by direct force action due to gravity and wind or earthquakes but also indirectly by hidden loads as caused by material movement and earth settlement (Fig. 5.17B). Should these deformations be prevented (or partially restrained) by adjoining members, stresses are generated. These conditions result from such factors as elastic deformations, environmental changes in temperature and humidity, shrinkage, constant loading (creep), loss of prestress forces, and support movement due to spreading and settling of the foundation. In statically determinate structures, externally induced deformations cause no additional stresses, because the members are free to rotate and translate in response. In indeterminate structures, however, the continuity between members does not allow for movement without causing additional stresses. To develop an awareness for the effects of some typical hidden forces, several frame examples with their response to different types of deformations are shown in Fig. 5.17B to demonstrate how horizontal and vertical support displacement causes bending in a two-hinge frame, while no bending stresses are imposed on the statically determinate three-hinge frame. For preliminary designs of frames and walls, the effect of material volume changes can usually be ignored, and small foundation movements can be tolerated.

EXAMPLE 5.5

For the rigid frame of case d in Fig. 5.17, the maximum column stress as caused by expansion of the deep roof steel beam is determined. The rigid beam had a length of 40 ft when the building was built at a temperature of around 40°F. When the temperature rises in and outside the building to 120°F, the temperature change is $\Delta T = 120 - 40 = 80$°F. The W16 × 67 steel columns with $S_x = 117$ in.[3] and $I_x = 954$ in.[4] resist this movement.

The coefficient of expansion for steel is $\alpha = 0.0000065$ in./in./°F. Therefore, the beam expansion is

$$\Delta L = \alpha L(\Delta T) = 6.5 \times 10^{-6}(40 \times 12)80 = 0.25 \text{ in.} \tag{2.7}$$

Each column resists the force caused by one-half of the total expansion.

$$\frac{\Delta L/2}{2} = \frac{P(h/2)^3}{3EI} \quad \text{or} \quad P = 6\,(\Delta L)\,EI/h^3 \tag{5.21}$$

Letting $M = P(h/2)$ yields

$$M = 3(\Delta L)EI/h^2 \tag{5.22}$$

Hence, the induced moment in the column is equal to

$$M = \frac{3(0.25)29{,}000(954)}{(15 \times 12)^2} = 640.42 \text{ k-in.} = 53.37 \text{ k-ft}$$

The moment causes a bending stress equal to

$$f = M/S = 640.42/117 = 5.47 \text{ ksi}$$

In cases where the temperature stress is added to the stresses caused by gravity action, the allowable stresses can be increased by 33%. For concrete frames we must also consider a temperature drop, together with shrinkage, which is often treated as an equivalent temperature change of about 33°F for estimation purposes.

A similar approach can be used for the post–beam structure in Fig. 5.17f or for a two-hinge frame with a rigid girder. Here the beam shortens due to temperature drop and/or shrinkage and to creep for a prestressed concrete beam.

$$\frac{\Delta L}{2} = \frac{Ph^3}{3EI} \quad \text{or} \quad P = 1.5(\Delta L)EI/h^3 \tag{5.23}$$

Letting $M = Ph$ yields

$$M = 1.5(\Delta L)EI/h^2 \tag{5.24}$$

Note that the maximum moment is twice as much for the rigid frame because of its higher degree of indeterminacy.

One-story, Multibay Frames

For industrial buildings and certain types of commercial buildings, often multibay column grid systems are employed, as demonstrated by the cases in Fig. 5.18. Prefabrication construction methods are used for these modular structures that should be adaptable to changes and future expansions. The column spacing should be wide enough so as to cause the least interference with the production or service process, that is, the placement of equipment and work stations, transportation requirements, and assembly areas; it should be adjustable toward future relocation and adaptable to new production methods about either building axis. The roof shape of these horizontally spread buildings depends a great deal on the treatment of ventilation and lighting (i.e., natural versus artificial); possible roof shape cross sections are shown in Fig. 5.19p and q.

The typical multispan frames, as exemplified by the cases in Fig. 5.18, are organized in a parallel manner on square or rectangular column grids. Among the pioneering efforts in industrial architecture is Albert Kahn's Chrysler Half-Ton Truck Assembly Plant near Detroit, Michigan (1937), which was chosen in 1944 by the Museum of Modern Art as one of the outstanding examples of American architecture. Important industrial buildings of the 1960s include the following cases, which also clearly demonstrate the principles of support structure. The Reliance Controls Electronics Plant at Swindon, U.K. (1965, Fig. 5.18a), designed by Norman and Wendy

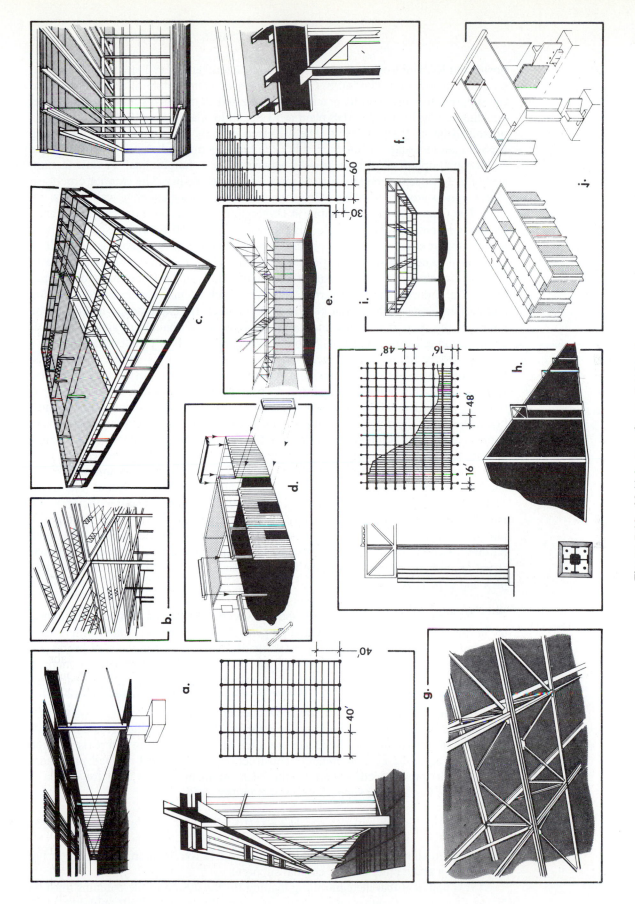

Figure 5.18 Multiple-span frame structures.

Foster and Richard Rogers, together with the structural engineer Tony Hunt, is laid out on a square column grid of 40 ft. Clearly expressed is the nature of the assembly and the support structure with its bracing, as contrasted to the enclosing corrugated steel cladding. Craig Ellwood's SDS plant at El Segundo, California (1966, Fig. 5.18h) is organized on 48-ft square bays. The 4-ft-deep primary trusses consist of 64-ft-long trusses cantilevering 8 ft on each side (i.e., hinge-connected cantilever beams); the cantilevers support 32-ft-span trusses between. These longitudinal trusses, in turn, carry the 2-ft-deep secondary trusses spaced at 8 ft on center. The exterior cruciform steel columns are free standing outside the enclosing precast concrete wall panels.

The Cummins Engine Factory at Darlington, U.K. (1966, Fig. 5.18f), designed by Kevin Roche and John Dinkeloo, is planned on 30- × 60-ft bay sizes. The purpose of the support structure, with its precise detailing and the fully glazed curtain wall, is exposed. The celebration of the steel skeleton and glass, that is, the purity and clarity of structure, surely is in the classical tradition of Mies van der Rohe. Angelo Mangiarotti in Italy, during the same period, was very much occupied with the construction and prefabrication of industrial buildings in concrete. He expressed wonderfully the nature of the support structure and its assembly as a well-balanced, almost classical composition integrating the articulation of the detail (e.g., Fig. 1.20, bottom right).

Multibay frames form either post–beam systems such as simple span beams, hinge-connected cantilever beams, and continuous beams (Fig. 5.19b to e) or frames, besides any combination of the above. In post–beam construction the roof plane is hinged to the columns. Here the lateral forces are resisted by either cantilever columns, braced bents, or shear walls. In contrast, in frame construction, beams and columns are connected by rigid joints or partial restraint connections to form continuous vertical structural planes capable of lateral force resistance. These multibay frames may consist of any of the basic simple portal bents in Fig. 5.8. Naturally, the beams and frames can be combined in many different ways, as exemplified in Fig. 5.19i to o. The frame systems in Fig. 5.19 should be arranged and members connected so that they provide lateral stability and are at least statically determinate (see Fig. 5.3), unless the roof diaphragm is capable of transferring the lateral forces to stable elements elsewhere in the building. Keep in mind that, in contrast to steel construction, in precast concrete construction moment connections in the field are usually avoided because they are complex and expensive.

Post-Beam Structures. First, the post–beam structure of symmetrical geometry in Fig. 5.20B is briefly investigated. In general, the analysis of beams with respect to gravity loading may be statically determinate if simple or cantilevered beams are employed (e.g., Fig. 5.19b, c, and d), or it may be statically indeterminate for continuous beams. For the preliminary design of a continuous beam in Fig. 5.20e, the uniform gravity loading controls the design. It would be questionable to consider a critical live load arrangement for flat roofs where snow does not follow such patterns, assuming a constant building height and no effect of parapets, that is, assuming areas do not collect snow. Furthermore, roof live loads are often relatively small in comparison to the dead load, as is the case in concrete construction, so the effect of load placement becomes less pronounced. Therefore, the beam moment usually used for design is based on the first interior support, where it is largest. It is

$$M = wL^2/10 \qquad (5.25)$$

This moment should also cover the effect of possible live load arrangements during construction at the other interior column supports.

It is apparent that the controlling loading for the design of the columns must be lateral force action together with gravity, since gravity action by itself only introduces axial forces. The lateral loads are distributed by the beams (that act as struts) to the columns in proportion to their stiffness. The columns, in turn, act as vertical cantilevers to

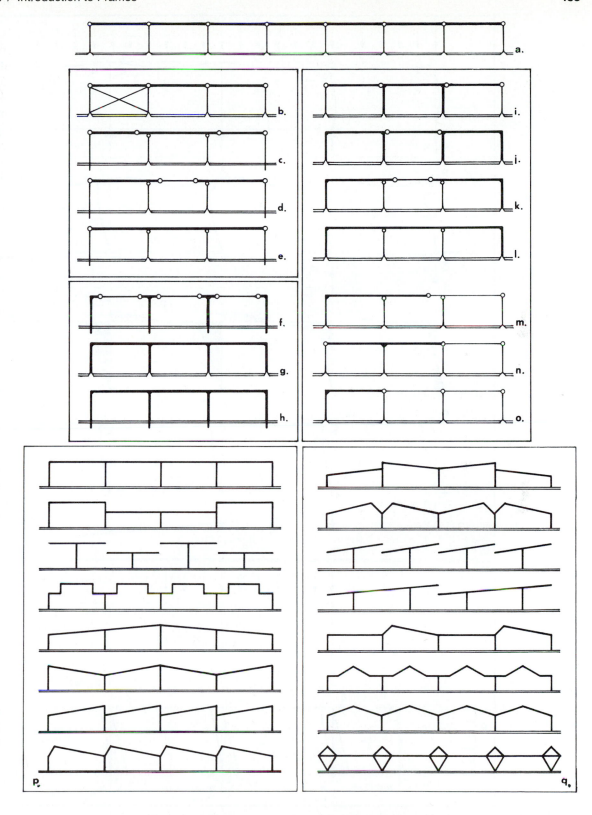

Figure 5.19 Single-story, multibay frame systems.

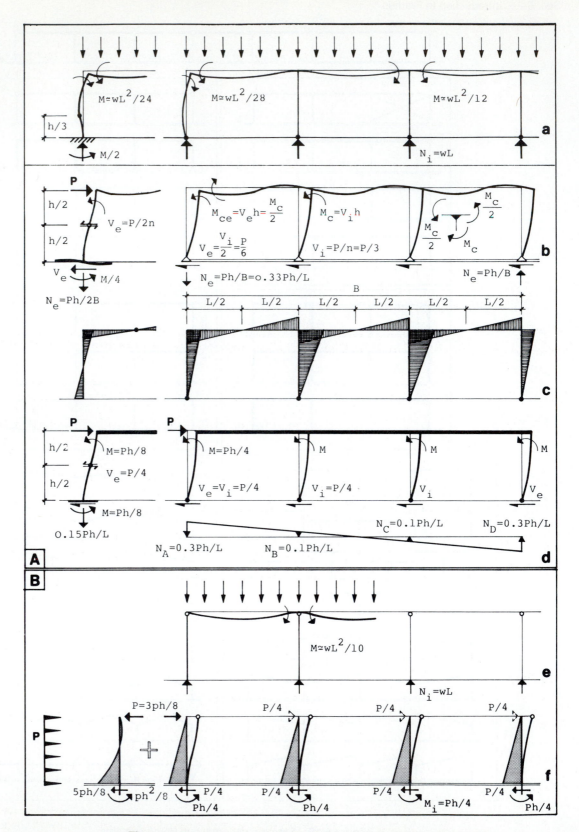

Figure 5.20 Behavior of typical multibay frames under vertical and horizontal loading.

carry the lateral loads in bending to the ground. When the cantilever columns are of equal size, length, and material (i.e., equal stiffness), the lateral loads are shared equally among them. For example, the maximum moment due to a lateral load P at the roof level in any of n columns at their base is

$$M = (P/n)h \qquad (5.26)$$

Should the wind pressure be uniformly distributed as in Fig. 5.20f, the exterior column can be visualized as a vertical beam fixed at the base, with an elastic support at the top. The reactions for the free body according to Table A.15 are indicated in Fig. 5.20f (left). The horizontal reaction at the top, in turn, is equal to the single load P, which is shared by all four cantilever columns equally (i.e., the columns, including the exterior ones, are assumed to be of the same stiffness for preliminary design purposes). Therefore, each column, with the exception of the exterior one on the windward side, must resist a maximum base moment of

$$M_i = (P/4)h = 3ph^2/32, \qquad \text{where } P = 3ph/8 \qquad (5.27)$$

In addition to the cantilever moment as caused by P, the exterior column on the windward side must also resist the vertical beam moment.

$$M_e = 3ph^2/32 + ph^2/8 = 7ph^2/32 \qquad (5.28)$$

Notice that for the given condition and for lateral force action from either direction the exterior columns must carry more than twice as much moment as the interior ones.

Portal Frames. In contrast to post–beam construction, where the roof plane is hinged to the vertical supports (i.e., the primary beams are pin jointed to the columns), in frame construction beams and columns are continuous and attached by rigid joints. Any bending induced in one member causes bending in all other members continuously connected to it. There are many types of frames such as the more flexible hinged column tree systems on the one hand (Fig. 5.19f) and truly stiff rigid frames on the other (Fig. 5.19h).

Briefly introduced here are the continuous symmetrical frames with hinged and fixed column bases in Fig. 5.20A. The analysis of multibay rigid frames is highly indeterminate. To promote an understanding of the force distribution along the frame, gravity and lateral loading are investigated separately and then the critical moment values are combined, ignoring the magnification of the moment due to sidesway (i.e., P–Δ effect).

In the following discussion, only the special case of rectangular bents of equal span and height having at least four bays is considered. The building is assumed to be of uniform strength where all columns are of nearly the same size. It was shown in the discussion of single indeterminate portal frames that the force flow distribution depends on the stiffnesses of the beams and columns. Although multibay rigid frames are highly indeterminate, certain simplified assumptions can be made so that an approximate analysis for gravity loading can be performed. To consider various live load arrangements for flat roofs is questionable, as was already discussed previously for the approximate design of post–beam structures. Therefore, it is reasonable to treat the gravity loads as uniform. We may conclude that for symmetrical frame geometry the interior columns will hardly resist any rotation and hence allow a continuous beam action equivalent to the behavior of fixed beams (Fig. 5.20a). In other words, the maximum moments for the design of typical interior frame beams (at the interior column supports) can be assumed as equal to

$$M = wL^2/12 \qquad (5.29)$$

Because of the symmetrical arrangement of the loading and the symmetrical geometry of the frame, the interior columns (with hinged or fixed base) may be assumed not to carry any gravity moments and thus are rather slender, since also the wind moments, in general, are rather small. This condition, obviously, is not true for unsymmetrical frames, where the columns must balance the difference in beam moments at the beam–column intersection, nor is it true for the exterior columns of the symmetrical frames, which must resist the beam rotation of the end bays. The maximum moments at the top of the exterior columns can be approximated by still assuming the first interior beam support fixed, while the exterior rigid beam–column junction only provides partial fixity to the beam (Fig. 5.20a, left). For this condition it can be shown that, for equal beam and column stiffness, the maximum column moments are

$$\text{For hinged base:} \quad M = wL^2/28 \tag{5.30a}$$

$$\text{For fixed base:} \quad M = wL^2/24 \tag{5.30b}$$

The behavior of the multibay frame under lateral loading (Fig. 5.20b and d) depends on the stiffness of the girders and columns, as has been discussed for Fig. 5.13 for simple single-bay portal frames. Rigid deep girders do not allow any rotation; thus the columns (hinged to the base) behave like vertical cantilevers fixed to the girders. For the given condition in Fig. 5.20d, each identical column resists an equal portion of the lateral load P; therefore, the horizontal reaction forces (i.e., shear forces) that each column must resist are

$$V_i = V_e = P/(n + 1) \tag{5.31}$$

where V_i, $V_e =$ interior and exterior horizontal column shear forces at the base
         $n + 1 =$ number of bays plus one = number of identical columns

These shear forces wrack the frame and cause maximum column moments at the top of

$$M = [P/(n + 1)]h = Ph/(n + 1) \tag{5.32}$$

Or, for this specific example, $M = Ph/4$.

The vertical reactions can be determined approximately by assuming the multibay frame to act like a vertical flexural cantilever, where the axial stresses in each column are assumed to be proportional to the distance from the neutral axis of the bent. Therefore, for this example,

$$f_A/1.5L = f_B/0.5L = -f_C/0.5L = -f_D/1.5L \tag{a}$$

where $f_A = (N/A)_A$, etc.

For equal column cross-sectional areas, $A_A = A_B = A_C = A_D = A$, Eq. (a) can be simplified to

$$N_A = -N_D = 3N_B = -3N_C \tag{b}$$

Rotational equilibrium about the neutral (centroidal) axis of the multibay bent using the relationships in Eq. (b) yields the axial column forces.

$$\Sigma M = 0 = Ph - 2[3N_B(1.5L) + N_B(0.5L)]$$

$$N_B = 0.1Ph/L = -N_C, \qquad N_A = 3N_B = 0.3Ph/L = -N_D \tag{5.33}$$

For the condition where the columns are fixed at the base, inflection points will form at mid-height, so we can visualize a hinged frame of height $h/2$ to sit on cantilever column stubs. Hence, the vertical reactions (axial column forces) can be found using a similar process as before. The maximum column moments at top and bottom are equal to

$$M = (P/4)h/2 = Ph/8 \qquad (5.34)$$

For the condition where the girders are not rigid, the regular multibay frame, hinged to the base, may be assumed to deflect under lateral loading by forming inflection points at the midspan of beams, as was discussed for two-hinge portal frames. Here the *portal method* of analysis may be used for preliminary design purposes, which treats the structure as a shear cantilever dominated by shear wracking. The method provides reasonable results for single-story frames with flexible girders and relatively stiff columns. The approximate behavior of a multibay system is derived from the sum of the independent actions of the individual portal units. For a structure of n equal bays, each frame unit carries an equal portion P/n of the lateral force P. Based on this assumption and on the discussion of the individual portal frame units under lateral loading, we may draw the following conclusions:

(i) The total lateral shear P is equally shared by the interior columns, while an exterior column only carries one-half the amount resisted by an interior one.

$$V_i = P/n, \qquad V_e = V_i/2 = P/2n \qquad (5.35)$$

For the specific case in Fig. 5.20b, where the number of bays is $n = 3$, the corresponding column shears are

$$V_i = P/3, \qquad V_e = P/6$$

(ii) The rotation due to the lateral force P is only carried by the exterior columns (because of the girder flexibility); that is, the force P is resisted by a force couple composed of the axial forces in the exterior columns. For a hinged frame, these axial forces are

$$P(h) = N_e(B) \quad \text{or} \quad N_e = Ph/B \qquad (5.36)$$

For the specific case in Fig. 5.20b, where $B = 3L$, the axial forces in the exterior columns are

$$N_e = Ph/3L$$

Should the pinned ends of the frame be fixed, additional inflection points form in the columns. The location of these points of contraflexure depends on the relative stiffnesses of columns and beams. For deep (i.e., rigid) girders, inflection points form at $h/2$ (Fig. 5.20d, left). But, as the girders become more flexible, these inflection points move upward. On the other hand, taking into account that base connections are rarely truly fixed (e.g., rotation due to soil settlement), which will cause the inflection points to move down, then it still may be reasonable for preliminary design purposes to assume inflection points at $h/2$ even when the girders are not rigid. We may conclude that a continuous frame with fixed bases may be visualized as a hinged frame of height $h/2$ that sits on column stubs of the same height $h/2$. Hence, the axial forces in the exterior columns of the rigid frame are only one-half as large as those of the hinged frame.

The column moments can easily be derived by visualizing the shear forces to wrack the frame or by visualizing the shears at the inflection points to cause cantilever action of the members, as has already been discussed for the response of individual portal frame units (Fig. 5.13). For the hinged frame, the maximum interior column moments at the top are

$$M_c = V_i(h) = Ph/n \qquad (5.37)$$

Note that the interior columns carry twice as much moment as the exterior ones. For the continuous frame the maximum column moments at the top and bottom are only one-half as large as those for the equivalent hinged frame.

$$M_c = V_i(h/2) = Ph/2n \qquad (5.38)$$

The maximum beam moments at the supports are caused by the wracking of the vertical shear. This shear is equal to the axial forces in the exterior columns and thus is constant across the building width B. The constant shear at the assumed hinges at midspan causes the following maximum beam support moments for hinged frames:

$$M_{ce} = (Ph/B)L/2 = V_e h$$

Note that the beam moments can also be derived from the maximum exterior column moments, to which they are equal, as the beam and column moments must balance each other across the corner joint (see Fig. 5.20b).

For the preliminary design of single-story, multibay frames of normal height, we may assume that the full gravity loading case controls the design of beams and exterior columns, while the design of the interior columns is governed by the lateral force action together with gravity loading (full gravity loading for long-span beams or dead load only for ordinary spans). The interior columns usually can be treated as beam columns using the bending factor approach. For the column design of frames with flexible beams, an effective length factor $K_Y = 1.0$ may be used by assuming the weak column axis to control and by assuming that there is no sidesway about the weak axis (i.e., the building is laterally braced in this direction). Naturally, for the condition where deep girders (or trusses) are used, the buckling condition about both axes is known (i.e., $K_X = 2.1$ and $K_Y = 1.0$). The preliminary design of exterior columns may be based on bending only (i.e., the column is treated as a beam) by selecting not the section required, but the next larger one, to take axial action into account, keeping in mind that, when $L_c < L_b \leq L_u$, $F_b = 0.6F_Y$ (see also Problem 5.18).

EXAMPLE 5.6

The approximate member sizes are determined for a typical interior, four-bay steel frame (A36) consisting of a continuous roof girder supported by fixed-base cantilever columns (Fig. 5.21, case c). Long-span steel joists are spaced at 6 ft on center span between the girders. The roof skin consists of roofing, metal deck, and insulation, which together weigh 12 psf. An equivalent joist and frame girder weight of 3 psf each is assumed. The snow loading is 30 psf and the wind load is 15 psf. The size of the open-web joists can be easily selected from manufacturers' catalogs.

A typical interior girder supports the following uniform load (ignoring the actual load distribution due to the joists at 6 ft on center):

$$\begin{aligned} w_D &= 30(12 + 3 + 3)/1000 = 0.54 \text{ k/ft} \\ w_L &= 30(30)/1000 \qquad\quad = 0.90 \text{ k/ft} \\ \hline w &\qquad\qquad\qquad\quad\; = 1.44 \text{ k/ft} \end{aligned}$$

The curtain panels transmit the lateral wind load to the spandrel beams and roof diaphragm, which in turn apply a single load P to the frame at the beam–column intersection.

$$P = 0.015(30 \times 20/2) = 4.5 \text{ k}$$

The girder design is clearly controlled by the gravity-loading case. The critical moment occurs at the interior support of the outer span and may be approximated as

$$M_g \cong wL^2/10 = 1.44(60)^2/10 = 518.4 \text{ ft-k} \qquad (5.25)$$

Therefore, the required section modulus for $F_b = 0.6F_y = 0.6(36) \cong 24$ ksi is

$$S = M/F_b = 518.4(12)/24 = 259.2 \text{ in.}^3$$

Select as a trial section (Table A.11)

$$\text{W30} \times 99, \qquad S_x = 269 \text{ in.}^3, \qquad L_c = 10.9 \text{ ft} > 6 \text{ ft}, \qquad F_y' \geq F_y$$

The design of the cantilever columns is controlled by full gravity loading (because of the relatively heavy load from the large girder spans) and wind; remember that gravity does not generate any direct bending into the columns. The axial column force is equal to

$$N_{D+L} = wL = 1.44(60) = 86.4 \text{ k}, \qquad P_D = 0.54(60) = 32.4 \text{ k}$$

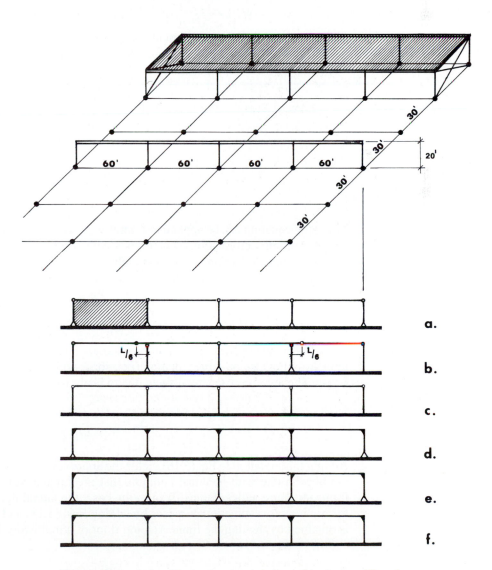

Figure 5.21 Approximate structural design of typical multibay frame structures.

The critical wind moment at the base of the columns is

$$M_c = (4.5/5)20 = 18 \text{ ft-k} \tag{5.26}$$

For this case, the columns cannot be treated simply like beams, as usually is the case for typical one-story enclosures. Here they behave like beam columns since the effect of bending is reduced as five columns resist the wind. A W10 section is investigated with $B_X = 0.26$. The equivalent axial force approach is used for this preliminary design approach, taking into account 25% reduction of loading because of the combined loading case.

$$P_{eq} = P + P' = P + B_x M_x \tag{3.99}$$

$$= 0.75[86.4 + 0.26(18 \times 12)] = 0.75[86.4 + 56.16] = 106.92 \text{ k}$$

$$\geq 32.4 + 56.16 = 88.56 \text{ k}$$

It is assumed first that the braced weak column axis controls the design: $(KL)_Y = 1(20) = 20$. Therefore, from the Column Tables of the AISC Manual the following section is obtained by selecting not the section required, but the next lower one.

$$\text{W10} \times 33, \qquad P_a = 95 \text{ k}, \qquad B_x = 0.277, \qquad r_x/r_Y = 2.16$$

Checking the slenderness of the strong axis: $(KL)_x/(r_x/r_y) = 2.1(20)/2.16 = 19 < 20$. Hence, the strong axis does not control and the section should be satisfactory for first-estimation purposes.

Lateral Sway.   The lateral deflection for some of the typical regular, multibay frames in Fig. 5.20 has already been discussed in Fig. 5.17. For example, for post–beam structures with identical cantilever columns, the lateral deflection is

$$\Delta_c = \frac{Ph^3}{3E\Sigma I_c} \tag{5.16}$$

The same equation can be applied to frames with rigid girders (Fig. 5.20, case d) ignoring the column deformations. For frames with rigid girders and fixed bases (case d, left) the lateral deflection is approximately

$$\Delta_c = \frac{Ph^3}{12E\Sigma I_c} \tag{5.18}$$

The lateral deflection of multibay frames, in general, depends not only on column sway but also on girder rotation (i.e., shear wracking, ignoring column deformations). The approximate lateral sway due to beam rotation (see Problem 5.8) is

$$\Delta_b = \frac{Ph^2 L}{12E\Sigma I_b} \tag{5.19}$$

Adding this equation to Eq. (5.16) yields the approximate lateral deflection of a regular multibay frame with identical columns that are hinged to the base. For a multibay frame structure with columns fixed to the base, the lateral deflection can be obtained for preliminary estimation purposes by adding Eqs. (5.18) and (5.19). The same result is obtained by treating the frame as a top floor of a multistory building (see Schueller, 1990).

$$\Delta = \Delta_c + \Delta_b = \frac{Ph^2}{12E}\left(\frac{h}{\Sigma I_c} + \frac{L}{\Sigma I_b}\right) \tag{5.39}$$

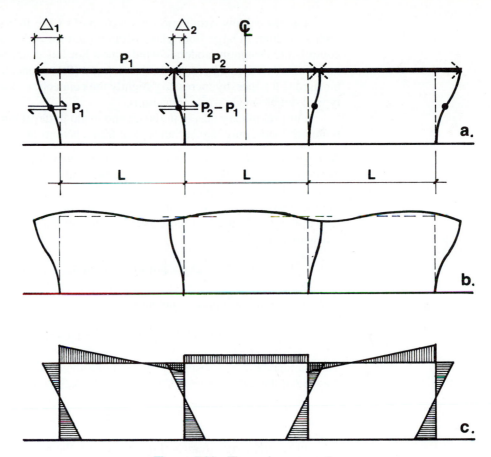

Figure 5.22 Thermal movement.

To determine the thermal stresses in the three-bay rigid frame structure with deep girders in Fig. 5.22a, an approach can be used similar to the single-bay case in Example 5.5. The lateral deflection of the exterior column Δ_1 and the deflection of the interior column Δ_2 can be found as follows, ignoring the negligible effect of the axial girder deformation.

$$\Delta_1 = \frac{P_1 h^3}{12\,(EI)_c} = \alpha\Delta T(L + L/2), \qquad \text{hence } P_1 \text{ is known} \tag{5.40}$$

$$\Delta_2 = \frac{(P_2 - P_1)\,h^3}{12\,(EI)_c} = \alpha\Delta T(L/2), \qquad \text{hence } P_2 = 1.33P_1 \tag{5.41}$$

Refer to Example 5.5 for further investigation of the problem.

Multistory Rigid Frames

Because of the high degree of indeterminacy of rigid-frame buildings, approximate methods of analysis have been developed. Since the rigid frame will respond to gravity loading in a distinctly different manner than to lateral force action, as is expressed in the deformed configurations of the frames, each loading case will be investigated separately, as has been done for one-story single-bay and multibay frames. These deflected frame shapes will be the basis for the following approximate analysis by making assumptions as to the locations of the inflection points where the moments are zero, thereby transforming the indeterminate structure into a determinate one.

For typical low-rise frame construction of two to four stories and regular layout with a medium to large floor plan, as often used for office buildings, gravity loading controls the design in moderate wind zones. For this situation, all frame columns can be efficiently utilized by using partial restraint (PR) connections, rather than the regular rigid (FR) ones, by increasing slightly the connection stiffness, for example, of the typical double-angle web connections.

For preliminary design purposes and to aid in developing a feeling for the distribution of loads, only regular but typical frame layouts are considered (see Fig. 5.23).

A. *Gravity Load Action:*   In *single-bay* multistory frames, inflection points may be assumed to form at $0.1L$ from the beam ends and at mid-height of columns (although they may be taken at $0.3h$ from the base of the column at the top floor and at $0.6h$ from the fixed base of the column at the first floor) for rough preliminary design purposes (see Fig. 5.23b). Based on these assumptions, the maximum beam moment occurs at midspan and is equal to

$$+M = w(0.8L)^2/8 = 0.08wL^2 \cong wL^2/12 \tag{5.42}$$

The corresponding support moment is

$$-M = (0.125 - 0.08)wL^2 = 0.045wL^2 \cong wL^2/22 \tag{5.43}$$

This moment will be resisted by the columns above and below equally, assuming that they are of the same stiffness. Therefore, the column moments can be approximated as

$$\pm M_c = (wL^2/22)/2 = wL^2/44 \tag{5.44}$$

In *multibay* rigid frames (Fig. 5.23a left), the largest moment in the beams may occur in the field or at the supports; it may be roughly approximated (see Schueller, 1990) as

$$\pm M = wL^2/12 \tag{5.45}$$

For multibay low-rise frames with semirigid (PR) connections, as previously discussed, the support and midspan beam moments may be assumed as equal to

$$\pm M = (wL^2/8)/2 = wL^2/16 \tag{5.46}$$

For the preliminary design of the columns, uniform gravity loading on the entire building may be assumed. Thus, interior columns will carry only axial loads for relatively symmetrical layout conditions, since the balanced loading causes the beam moments at the columns to cancel each other. However, the situation is different for the exterior columns, which are rotated by the beams; for their preliminary design, Eq. (5.44) may be used (i.e., they are treated as the columns of single-bay frames). For low-rise frames with PR connections, the column sizes may be estimated as based on the gravity-loading case using a preliminary K factor of 2.

B. *Lateral Load Action:*   For the preliminary design of typical, regular rigid frames, equal beam and column stiffnesses may be assumed with points of contraflexure at mid-length of all members. According to the *portal method* of analysis, the total lateral shear on the frame is distributed at every floor level to the columns in proportion to the width that each column supports. In other words, it is assumed that a frame bent acts as a series of independent portals, where the total frame shear is taken by each portal in proportion to its span. For equal bays this results in dividing the shear equally between the number of bays so that the interior columns carry twice as much load as

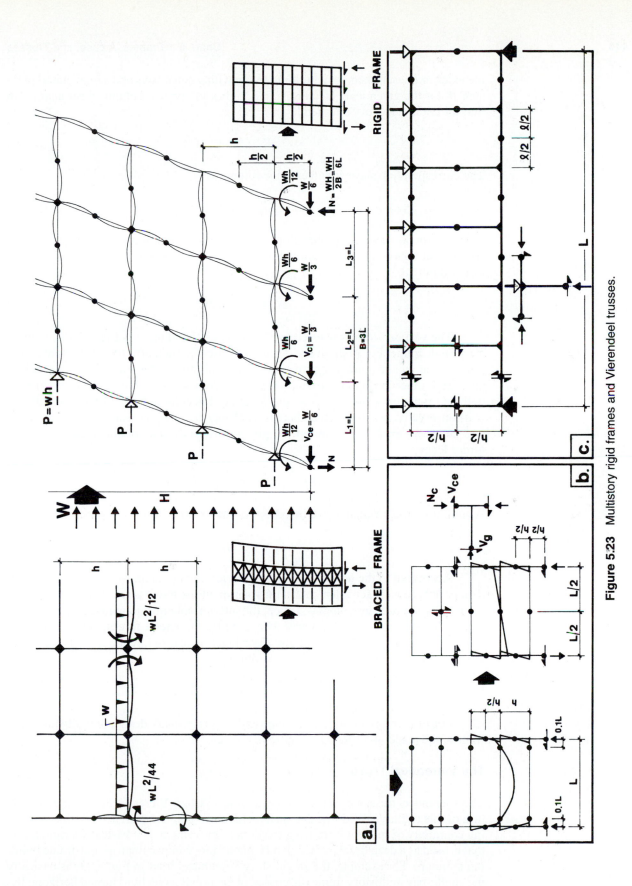

Figure 5.23 Multistory rigid frames and Vierendeel trusses.

443

the exterior ones. Hence, for a building frame with n equal bays and a total lateral pressure W for the given level (see Fig. 5.23a, right), the typical column shear against an interior column at story mid-height is

$$V_{ci} = W/n \qquad (5.47a)$$

The corresponding shear for an exterior column is

$$V_{ce} = V_{ci}/2 \qquad (5.47b)$$

This horizontal shear at mid-height $(h/2)$ wracks the column and causes moments at the top and bottom of each column. For example, the maximum moments for an interior column are

$$M_c = \pm V_{ci}(h/2) \qquad (5.48)$$

The vertical shear is resisted by the girders in bending; it is found by isolating a story-high frame. For preliminary design purposes, particularly for tall buildings, the girder shear can be quickly approximated by letting the girder and column shears balance each other in rotation around a typical interior story-high cruciform module, as defined by the inflection points at mid-length of the columns and girders; in addition, it is conservatively assumed that there is no change in column shear for the given story, and hence no axial force flow in the girders, by using the column shear below the girder level.

$$V_{ci}(h) \cong V_g(L) \quad \text{or} \quad V_g \cong V_{ci}(h/L) \qquad (5.49)$$

A similar moment equation can be set up for the exterior beam–column joint:

$$V_g \cong V_{ce}(2h/L) \qquad (5.50)$$

According to the portal method the interior axial column forces of the independent portals cancel each other so that the rotation of the frame is resisted in axial action entirely by the exterior columns; hence, the interior columns do not carry any vertical forces. In other words, the moment M caused by the lateral forces, acting about the inflection points at mid-height of any floor level, is resisted by the axial forces N_c in the exterior columns across the frame width B.

$$N_c(B) = M \quad \text{or} \quad N_c = M/B \qquad (5.51)$$

For further discussion of the theme and the preliminary design of multistory rigid frames, the reader may want to refer to Schueller, 1990.

The Vierendeel Truss

The Vierendeel truss, which the Belgian engineer Arthur Vierendeel popularized at the turn of this century, avoids the diagonal members of conventional trusses, thus providing continuity of space and allowing freedom of movement. It is basically a rigid frame that is used as a deep beam, although it is much more compact than a rigid-frame building structure. For example, the single-story Vierendeel truss in Fig. 5.23c is basically the single-bay multistory frame (discussed in the previous section) turned horizontally.

Since the horizontal frame beam resists uniform gravity loads similar to the way the vertical cantilever frame resists lateral loads, the *portal method* of analysis can be used for preliminary design purposes, assuming uniform member distribution. The

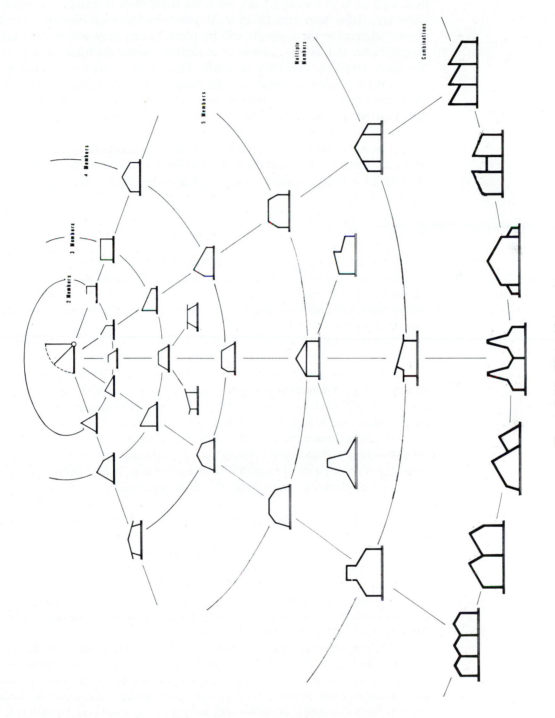

Figure 5.24 Pitched frame shapes.

force flow can thus be approximated by estimating inflection points at the mid-length of all members for each bay of the Vierendeel truss.

One should realize, however, that the frame-beam is considerably less efficient as compared to an ordinary truss, since the shear must be resisted in bending by the members, rather than directly in axial action by the truss diagonal members. The chords adjacent to the supports will be much larger; they will behave primarily as *beams,* rather than *beam–columns* as at midspan where the truss moment is large but the shear wracking generally is small. They must resist large bending moments because of the high shear adjacent to the support, but only a little axial force due to the small truss moment. The chord size next to the support can be quickly estimated by treating it simply as a beam and by selecting not the section required, but the next larger one, to take axial action into account.

The truss columns act primarily as *beams,* especially adjacent to the support where the shear is maximum; the horizontal shear wracks the columns and causes large moments. For the approximate design of Vierendeel trusses, refer to Schueller, 1990.

5.3 PITCHED FRAMES

The application of the pitched frame concept ranges from typical usage as the primary structure for roofs to the special case of a single A-shaped steel tower from which a bridge roadway is hung, the free-standing transmission tower with its slender cantilevering arms, and finally the unique application as a support for a structure that cantilevers daringly from a cliff (Fig. 5.25f).

The potential occurrence of inclined members for various shapes of roof enclosures is studied in Fig. 5.24. Shown are frames with single or multiple inclined members and with inclined columns and/or inclined beams. They may consist of just two members (e.g., A-frame) or of many. They may be symmetrical or asymmetrical, continuous or split, of single pitch, double pitch, or repeated shapes, and so on. The reader may want to refer to Fig. 5.29 (top left) for the description of common roof shapes.

In vernacular architecture, flat roofs, including quasi flat roofs only slightly tilted for drainage, are often found in warm climates, whereas pitched roofs seem to occur in areas experiencing a lot of rainfall or with much snow. In architecture, the popular flat tops of the modern movement have often been replaced by the crowns of the postmodern era.

Pitched roofs can be constructed with frames, including braced and trussed frames, trusses, and any combination, as indicated in Fig. 5.25. It is interesting to note that the Chinese in their traditional triangulated roof construction used progressively shorter, propped-up, horizontal beams on top of each other so that the roof supports rise in steps; in other words, each of the stacked beams rests on a pair of short posts, in turn carrying a shorter beam. This type of roof construction, such as for the familiar pagodas, had a special meaning to the far eastern world, in contrast to the efficient truss or braced-frame approach of the West.

The possible application of the pitched frame concept can only be suggested by the cases in Fig. 5.25. Examples range from the typical roof structures, or the 18-ft span glulam cantilever structure (case L), to the grandiose entry in case (d) and the suspended railway station in case (h), as well as the unique application of the frame as a support for a structure that cantilevers daringly from a cliff (f). In case (o), the entire building on the hillside is a pitched roof consisting of parallel, asymmetrical A-frames. Radially arranged welded steel gable frames were used for the polyhedral half-octagon dome in case (p).

As a support structure for the exterior envelope, the folded frame concept occurs on the large scale of the International Congress Center, Berlin, Germany (1979, case n), designed by R. and U. Schüler, where the 230-ft-span transverse frame trusses are

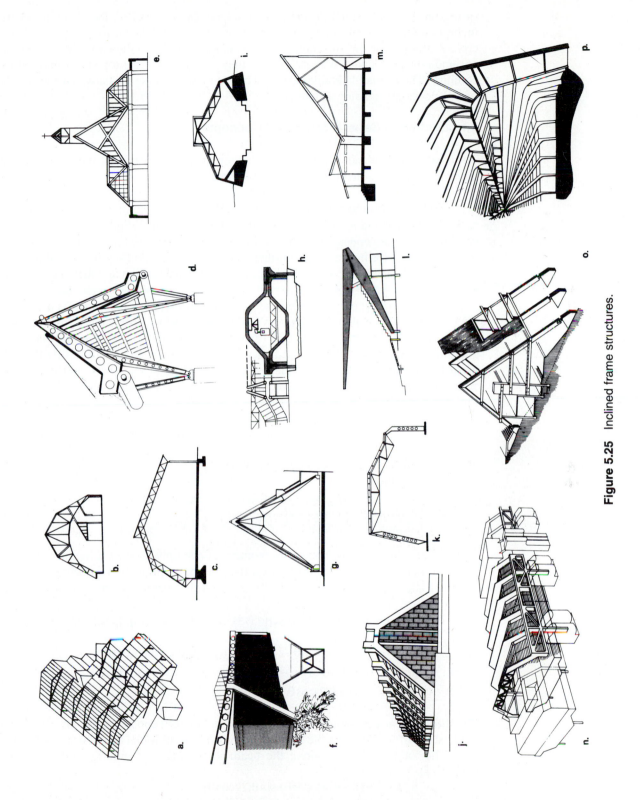

Figure 5.25 Inclined frame structures.

sitting on the longitudinal facade trusses, which, in turn, rest on the concrete pillars incorporated in the stair towers. On a smaller scale, we find the 88-ft-wide precast concrete A-frames for the Rosenthal Glass Factory, Amberg, Germany (1970, case j), designed by TAC under Walter Gropius; they are hinged at the top and supported by columns. The ribs are laterally stabilized by the inclined concrete slabs, which provide for the necessary shading as louvers.

Introduction to General Structural Concepts

To develop understanding about the behavior of sloped members, the nature of the inclined beam is studied first and then the frame.

Simply Supported Inclined Beams. The beam to be investigated is pin supported at the reaction A and roller supported at the reaction B (Fig. 5.26). The reaction conditions at A are considered to be constant, while three different orientations of the roller support B are investigated. Furthermore, the effects of different types of loading on the beam behavior are studied. First, uniform gravity loading and then uniform lateral loading are investigated.

The uniform gravity loading case (such as for a typical sloped roof) consists of the weight w_D acting along the structure and the live load w_L (e.g., snow), which is usually given on the horizontal projection of the beam (Fig. 5.26a). To determine the shear and moment flow along the inclined beam, it is convenient to transform first the live load w_L into a uniform load along the beam, $w_L(L)/(L/\cos\theta) = w_L\cos\theta$, so that it can be added to w_D (Fig. 5.26b). Then the component $w_{y'}$ acting perpendicular to the beam is found, which causes beam action.

$$w_y' = (w_D + w_L\cos\theta)\cos\theta = w_D\cos\theta + w_L\cos^2\theta \tag{5.52}$$

The maximum moment caused by this load is

$$M_{max} = (w_D\cos\theta + w_L\cos^2\theta)(L/\cos\theta)^2/8$$
$$= (w_L + w_D/\cos\theta)L^2/8 = wL^2/8 \tag{5.53}$$

We may conclude that the maximum moment of a simply supported inclined beam is equal to that for an equivalent beam on the horizontal projection carrying the same loading as the inclined beam (Figs. 5.26c, e, h). Notice that the moment is independent of beam slope and support conditions.

The maximum shear at the reactions is caused by $w_{Y'}$ (Eq. 5.52) and is equal to

$$V_{max} = w_Y'(L/\cos\theta)/2 = (w_D + w_L\cos\theta)\cos\theta(L/\cos\theta)/2$$
$$= (w_D + w_L\cos\theta)L/2 = (w_L + w_D/\cos\theta)\cos\theta(L/2) \tag{5.54}$$
$$= (wL/2)\cos\theta = A_v\cos\theta$$

The shear depends on the beam inclination but not on the support conditions. It is equal to the shear of the beam on the horizontal projection multiplied by $\cos\theta$ (Fig. 5.26g).

The total axial force along the beam is obtained by summing up w_x' (Figs. 5.26b,d, f).

$$\Sigma w_x' = (w_D + w_L\cos\theta)\sin\theta(L/\cos\theta)$$
$$= L(w_L + w_D/\cos\theta)\sin\theta = wL\sin\theta \tag{5.55}$$

The distribution of this total axial force depends on the reaction condition at B.

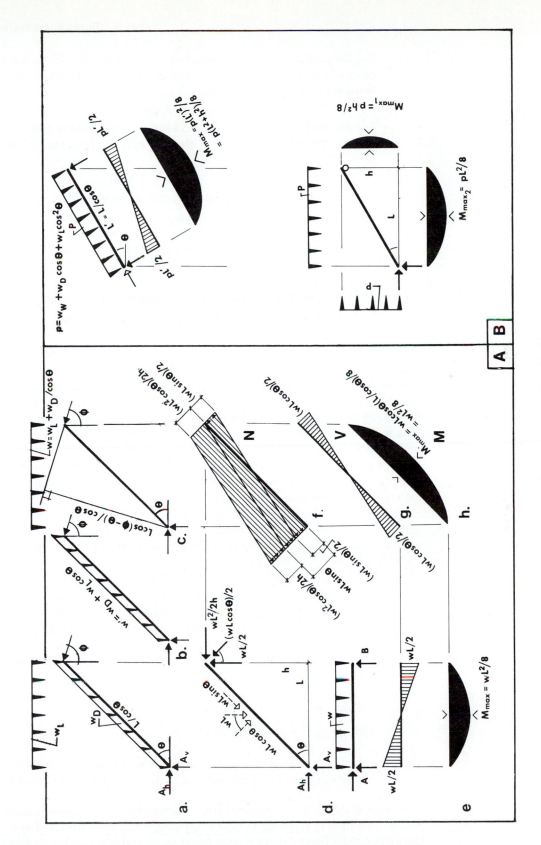

Figure 5.26 Simply supported inclined beams.

449

- For the condition where the reaction at B is perpendicular to the beam, there cannot be any axial force component at that point. Therefore, the entire axial force must be resisted at reaction A.

$$N_A = wL \sin\theta, \qquad N_B = 0 \tag{5.56}$$

- For the condition where the reaction at B is vertical (as is typical for roof joist construction), the reactions will share equally the resistance of the total axial force.

$$N_A = -N_B = (wL \sin\theta)/2 = A_v \sin\theta \tag{5.57}$$

Notice there is no lateral thrust force at A under gravity action!

- For the condition where the reaction at B is horizontal, not only the entire $\Sigma w_x'$ must be resisted at A, but in addition a uniform axial load $H_B \cos\theta$ as generated by the reaction force H_B. The horizontal reaction H_B is equal to

$$H_B(h) = wL(L/2) \quad \text{or} \quad H_B = wL^2/2h \tag{5.58}$$

Therefore, the maximum axial force at reaction A is

$$N_A = wL \sin\theta + (wL^2/2h) \cos\theta = A_v \sin\theta + A_h \cos\theta \tag{5.59}$$

The most efficient support condition in this context is the vertical one, since the total external axial force component $wL \sin\theta$ is carried equally at A in compression and at B in tension. For the perpendicular support case, the whole force component $wL \sin\theta$ must be resisted at the reaction A. In the horizontal reaction case the axial forces are very much increased. Not only does the axial load component due to the external loads have to be resisted, but also an additional force component due to the orientation of the reaction B.

In general, as the reaction B rotates away from the vertical position, the axial force flow in the beam increases, especially when it has rotated through an angle larger than the angle of inclination θ of the beam.

The uniform lateral loading case is analogous to the uniform gravity loading case just discussed; visualize span L and height h to be exchanged. For example, the maximum moment due to q on the vertical projection of the inclined beam is

$$M_{\text{max}} = qh^2/8 \tag{5.60}$$

The case where uniform loads p act perpendicular to an inclined roof due to wind or due to the respective components of dead and live loading (Fig. 5.26B) can easily be handled in the familiar fashion. Sometimes, however, it may be more convenient to deal with the wind load components on the horizontal and vertical beam projections. The maximum beam moment is

$$M_{\text{max}} = p(L')^2/8, \qquad \text{where } (L')^2 = L^2 + h^2$$
$$M_{\text{max}} = p(L^2 + h^2)/8 = pL^2/8 + ph^2/8 \tag{5.61}$$

Thus, the maximum beam moment may be obtained from equivalent beams on the horizontal and vertical projections carrying the same loading as the inclined beam.

Frame versus the Pressure Line.   Various types of pressure lines (thrust lines) or funicular lines in response to typical building loads are introduced in Fig. 5.2. It is shown that the difference between the frame shape and pressure line is directly proportional to the magnitude of the bending moment along the frame, which therefore immediately identifies the formal efficiency of the frame from a force flow point of view. The effect of various three-hinged pitched frame profiles, as related to the controlling uniform gravity loading case, is studied further in Fig. 5. 27. Shown is the transition from

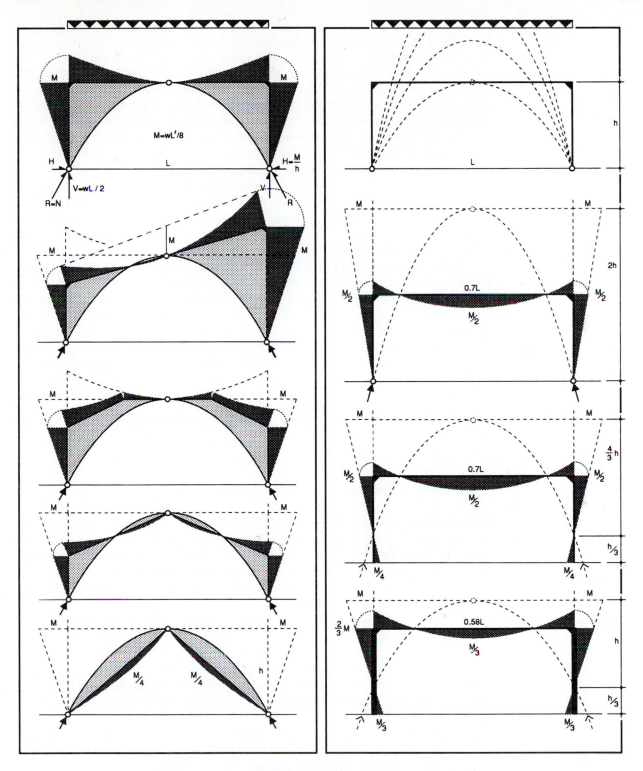

Figure 5.27 Effect of frame profile (left); effect of beam and column stiffness for continuous frames (right).

the rectangular portal frame to the gable frame and A-frame, all resisting the same thrust force H. It is apparent that the eccentricity from the parabolic pressure line is large for the rectangular frame, with a maximum at the frame knee, and small for the gable frame, which approaches the shape of the arch. The largest moment in the gable frame at the knee depends on the height of the column, as indicated, $M_s = Hh$. The maximum positive moment in the inclined beams occurs near the top hinge and is much less than the negative support moment and almost never has any effect on the preliminary design of the frame. It is apparent from this discussion that the gable frame is far more efficient than the rectangular portal frame from a force distribution point of view.

As the columns become shorter, the negative moment at the beam–column junction, $M_s = Hh$, reduces correspondingly, while the positive field moment increases. It becomes zero for the A-frame where there is only positive bending with maximum moments at midspan of the inclined beams. The reader may want to compare the cases in the left column of Fig. 5.27 in more detail to see how the magnitude of the moment M for the rectangular portal frame is affected by the frame profile.

In the right column of Fig. 5.27, the effect of member stiffness for indeterminate frames is shown. The influence of the beam-to-column stiffness ratio on the moment flow is discussed in more detail in Fig. 5.12. Whereas for determinate three-hinged frames, the parabolic pressure line is a constant (assuming the location of the hinges is not changed), for indeterminate frames it must be a variable in response to the member stiffness conditions. Here, for reasons of convenience, the rectangular frame has been chosen and kept constant to demonstrate the deviation of the various pressure lines. A similar approach can be used for pitched frames. Notice that as the height of the pressure arch increases the column moment decreases, but the beam field moment increases, which is equivalent to decrease of column stiffness and/or increase of beam stiffness. Furthermore, with an increase of height of the funicular arch, the reactions become steeper and the thrust force components smaller, thereby causing smaller column moments.

Pitched Roof Structures in Residential Construction

The traditional inclined roof forms in residential construction are the shed roof, gable roof, and gambrel roof; from these three basic types are derived the hip and Mansard roofs (Fig. 5.29). Other shapes, in turn, can be developed from these five major forms either by transformation or by allowing them to intersect with each other as a result of irregular plan layout. Basic principles of roof construction have already been briefly reviewed and are not further dealt with in this context.

The sloped roof may be chosen for esthetic reasons, as an environmental protection against heavy rains, for the purpose of integrating a solar collector, or for functional reasons to utilize the interior space more efficiently. Whatever the reason for its choice, structure must be provided to support the form.

The typical structural skeletal systems including some detailing are defined in Fig. 5.28. The most common residential roof structures are generally in wood. They are as follows:

- Joist roofs, including post–beam roofs (e.g., plank–beam, purlin–beam)
- Rafter roofs (e.g., A-frame construction)
- Truss roofs (see Section 5.5)
- Any combination of these (e.g., truss–rafter)

Some less frequent construction systems used in the housing field are the following:

- Rigid-frame (see gable frame structures), pole-type construction
- Panelized roofs (see Chapter 7)

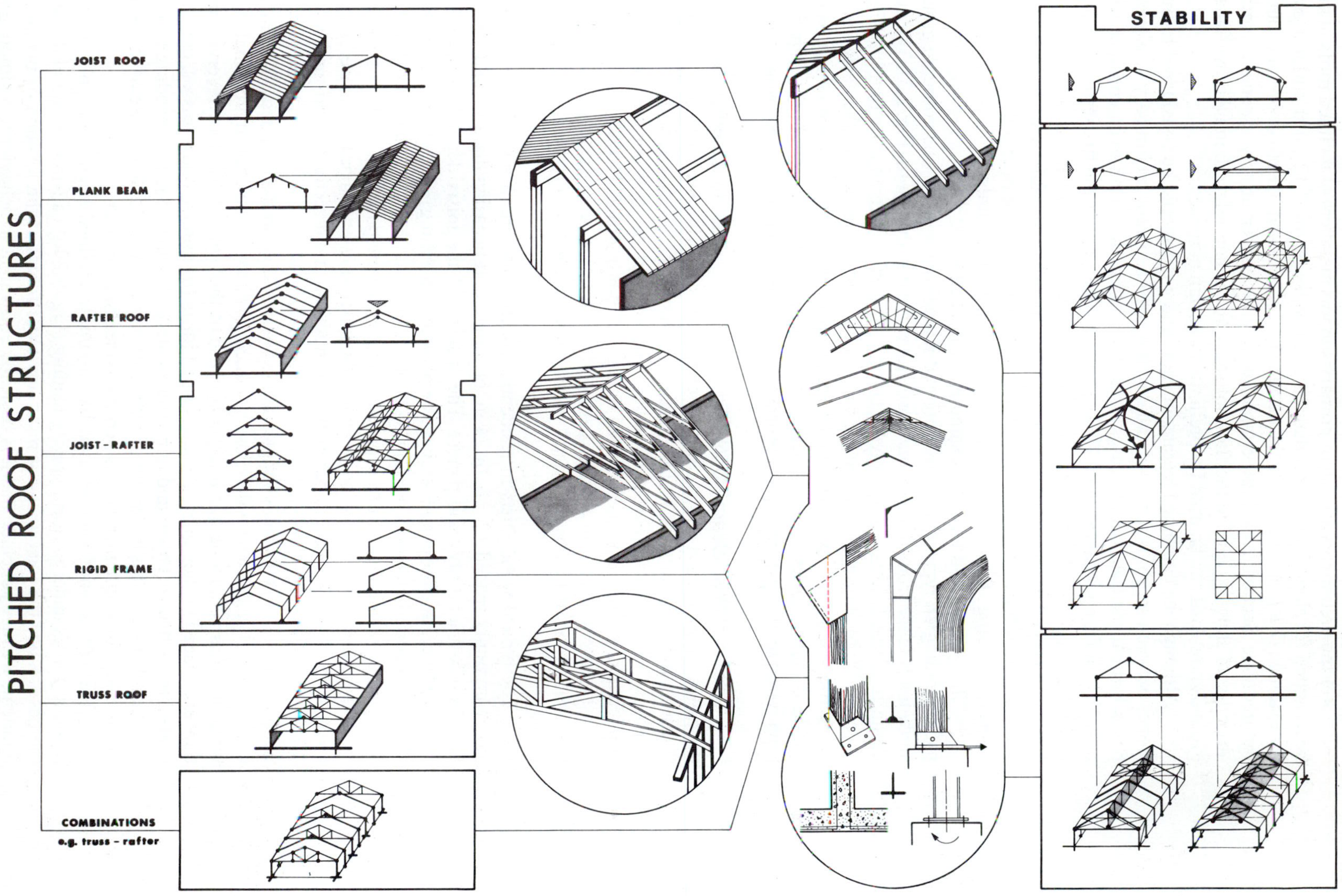

STABILITY

PITCHED ROOF STRUCTURES

JOIST ROOF

PLANK BEAM

RAFTER ROOF

JOIST – RAFTER

RIGID FRAME

TRUSS ROOF

COMBINATIONS
e.g. truss – rafter

Figure 5.28 Pitched roof structures.

453

For the preliminary design of primary support members of inclined roofs, it may be assumed for slopes not greater than 30° (or not exceeding 7:12), that uniform gravity loading ($D + L$ or S) controls the design (Fig. 5.26a) and the effect of wind can be ignored (see Section 2.1). Furthermore, it may be conservatively assumed that the sloped-roof snow is equal to the flat-roof snow and is not reduced; in other words, the ground snow load can be used for preliminary design purposes.

Although the combined loading of snow and wind should be considered for roof slopes between approximately 30° and 45°, it is reasonable for preliminary design purposes to treat the two loading cases separately. Some designers even consider wind only for roof slopes larger than 37° (i.e., larger than 9 : 12) for first-estimation purposes (see Section 4.1). For roofs steeper than 45°, snow and wind do not have to be considered together under ordinary conditions. Therefore, the following loading cases should be investigated for first-estimation purposes: $D + L$ or S, and $D + W$. For roof slopes exceeding 70°, snow loading generally can be disregarded.

The typical dead load, for example, for pitched wood roofs is between 10 to 15 psf depending on type of roof covering and if a ceiling is used. The common ground snow loads range from 70 psf in the Northeast to 5 psf in the South (Fig. 2.3). The minimum roof live load is conservatively taken as 20 psf to take construction loads into account. Wind pressure values on the vertical building projection for the design of the gable frames, as well as normal to the roof slopes (both in tension and compression) for the design of joists, beams, purlins, and decking are assumed roughly as 16 psf (70-mph wind) and 21 psf (80 mph) for ordinary low-rise buildings (see Section 4.1).

Joist and Post–Beam Roofs (Fig. 5.29).   In contrast to the rafter roof, where the rafters support each other at the crown to form a continuous frame unit, the joists function independently as inclined simple bending members, supported by beams spanning in the perpendicular direction. The concept of joist construction can be applied to any roof shape, as indicated in Fig. 5.29 for traditional as well as free forms. It must be mentioned, however, that in wood construction often the term *rafter* is used for sloped roof structures in general, while the term *joist* is only associated with flat roofs.

The typical wood joists in residential construction are of nominal 2-in. dimension commonly spaced 12, 16, or 24 in. on center and occasionally 13.7 and 19.2 in. based on 8-ft-long plywood panels. The spacing of the joists depends on the deck capacity (e.g., thickness and strength of plywood sheathing). The joists may just be simply supported members bearing on the exterior wall and the center ridge beam, or they may be continuous beams supported at intermediate points on walls or beams.

In post–beam construction, visualize the joists, now called beams, spaced further apart, such as 6 to 8 ft for timber plank construction, and directly supporting the deck or, if too far apart, the purlins. From a structural point of view the two frameworks are quite similar; in both systems simple inclined beams support the roof deck. The difference is one of scale, where the closely spaced joists are of comparatively small size, while the beams must be much larger because of the higher loads they receive, since they are spaced farther apart. Closely spaced joists (or rafters, studs, trusses, etc.), not spaced more than 2 ft apart and joined adequately, distribute loads in bending to adjacent members. For this condition of repetitive member use, 15% increase of the bending stresses for single members is permitted (see Section 3.2).

The typical roof joist behaves like a simple inclined beam with only vertical reaction forces under uniform gravity loading (Fig. 5.29d) causing no lateral thrust at the base. Since the joist does not resist any axial forces at the location of maximum moment, we may conclude that it is basically equivalent in behavior to a simple horizontal beam, which can be designed rather easily.

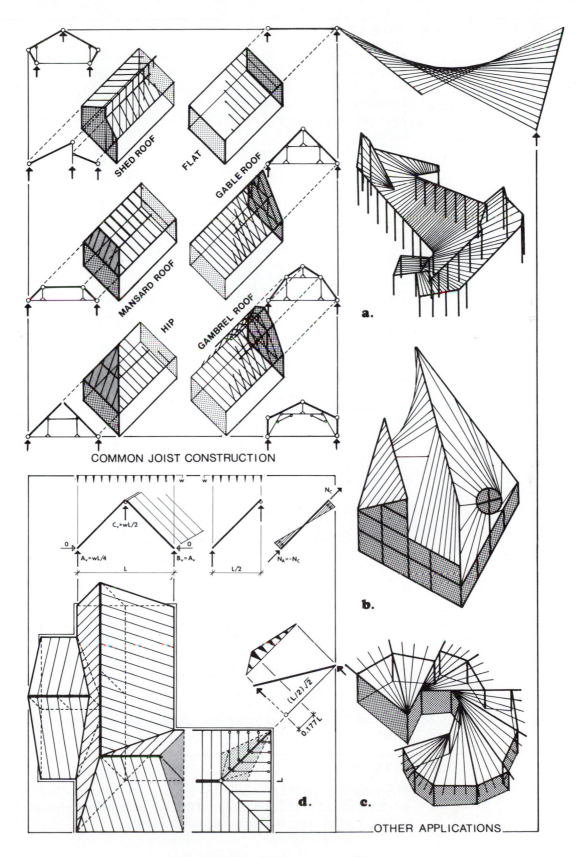

SHED ROOF

FLAT

GABLE ROOF

MANSARD ROOF

GAMBREL ROOF

HIP

COMMON JOIST CONSTRUCTION

a.

b.

$C_v = wL/2$

$A_v = wL/4$

$B_v = A_v$

L

$L/2$

0

0

w w

N_C

$N_A = -N_C$

$(L/2)\sqrt{2}$

$0.177L$

d.

c.

OTHER APPLICATIONS

Figure 5.29 Joist roof construction.

EXAMPLE 5.7

An inclined joist roof with a slope of 4:12 has a span of 15 ft measured along the horizontal projection. The joists are assumed to be spaced 24 in. on center for this first investigation. They must support a dead load of 15 psf, which includes a plaster ceiling, and a snow load of 30 psf. For this low-slope roof, the effect of wind can be ignored with respect to the design of the joists. Use Douglas fir–larch with $F_b = 1450$ psi, $F_v = 95$ psi, and $E = 1,700,000$ psi (see Table A.8).

The roof slope is equal to

$$\tan\theta = 4/12 \quad \text{or} \quad \theta = 18.44° < 30°$$

The load along the horizontal projection of the joist (Fig. 5.26c) is

$$w = (w_L + w_D/\cos\theta)s = (30 + 15/\cos\theta)2 = 60 + 32 = 92 \text{ plf}$$

The maximum moment according to Eq. (5.53) is

$$M_{max} = wL^2/8 = 92(15)^2/8 = 2588 \text{ ft-lb}$$

The maximum shear according to Eq. (5.54) is

$$V_{max} = (wL/2)\cos\theta = (92(15)/2)\cos\theta = 655 \text{ lb}$$

The required section modulus as based on the combined loading case (since the live load is about twice as large as the dead load), taking into account load duration (Table 3.1), is

$$S = M/F_b C_D = 2588(12)/1450(1.15) = 18.62 \text{ in.}^3 \qquad (3.28)$$

Try 2×10 joists, $S_x = 21.391$ in.^{3}, $A = 13.875$ in.^{2}, and $I_x = 98.932$ in.^{4} (Table A.9).

The shear stress, according to Eq. (3.33), is

$$f_v = 1.5V/A = 1.5(655)/13.875 = 70.81 \text{ psi} < 95(1.15) = 109.25 \text{ psi}, \quad \text{OK}$$

It is apparent that the live load deflection, because of the relatively high magnitude of the live loads, will control stiffness considerations. The live load perpendicular to the beam (Fig. 5.26) is equal to

$$p = w_L \cos^2\theta = 60 \cos^2\theta = 54 \text{ plf}$$

The live load deflection, according to Eq. (3.8), is

$$\Delta_L = \frac{5wL^4}{384EI} = \frac{5(54/12)\left[15(12)/\cos\phi\right]^4}{384(1,700,000)98.932} = 0.45 \text{ in.}$$

$$\leq \frac{L'}{360} = \frac{15(12)/\cos\phi}{360} = 0.53 \text{ in., OK}$$

Although the 2×10 joists are satisfactory, a different spacing should be investigated.

EXAMPLE 5.8

Investigate the design of the roof joist if the height of the roof in Example 5.7 is increased from 5 to 12 ft. It is assumed that the loading conditions do not change, and the wind pressure normal to the roof is 16 psf.

The roof slope is

$$\tan\theta = 12/15 \quad \text{or} \quad \theta = 38.66° > 30°$$

This roof must be considered steep where the effect of wind should be considered. Therefore, a joist must support a normal wind pressure of

$$w_n = 16(2) = 32 \text{ plf}$$

According to Fig. 5.26B, this normal pressure is equal to the pressure on the horizontal and vertical projection of the sloped surface.

The dead load along the horizontal projection of the roof is

$$w = w_D/\cos\theta = 2(15/\cos\theta) = 38 \text{ plf}$$

The reduced snow load on the horizontal projection of the beam, according to Eq. (2.4), is

$$w_s = [(1 - (\theta - 30)/40)p_g]s = [(1 - (38.66 - 30)/40)\ 30]2 = 47 \text{ plf}$$

The maximum moment due to the $D + S$ case, taking into account load duration, is

$$M_{D+L} = [(38 + 47)/1.15]\ 15^2/8 = 2079 \text{ ft-lb}$$

The maximum moment due to the $D + W$ case, taking into account load duration (Table 3.1), is

$$M_{D+W} = [(38 + 32)/1.60]15^2/8 + (32/1.60)12^2/8 = 1590 \text{ ft-lb}$$

The uniform gravity case clearly controls! The required section modulus is equal to

$$S = M/F_b = 2079(12)/1450 = 17.21 \text{ in.}^3$$

Try 2×10 joists 24 in. o.c. (Table A.9); check shear and deflection.

Rafter Roof (Fig. 5.30). In the simplest form of residential application, a continuous three-hinge A-frame is formed by the inclined beams, called rafters, which act together with the floor joists. Here the rafters support each other and are often joined by a ridge board, unlike the independent roof joists, which are carried by the ridge beam. The lateral thrust that is absent from roof joists, but typical for frames, is resisted conveniently by the floor joists, which behave like tie rods across the width of the building. For steeper roofs, more intricate rafter framing systems may be necessary, such as the three statically indeterminate *collar frames* in Fig. 5.30g to i.

By placing a *collar strut* into the A-frame (Fig. 5.30b) to reduce the bending in the rafters, the lateral thrust at the base is increased by the amount of compression induced into the collar tie; the frame is now statically indeterminate to the first degree. For the strut to be truly effective, floor decking or horizontal bracing should connect the various collar ties together to form a deep, relatively stiff, horizontal beam, which must be supported by some shear walls (c). For this method of construction, the collar strut provides a rigid support to the rafters; otherwise, a flexible support (d) would only be available under unsymmetrical load action, requiring much larger rafter sizes.

As for joist construction, the rafters may be spaced closely at 12, 16, or 24 in. on center (Fig. 5.30b) supporting the deck directly, or they may be spaced farther apart so that purlins are needed to span between them (Fig. 5.30a). The four sloped surfaces of the simple hip roof support each other laterally and thus provide natural stability (Fig. 5.29). Hip roofs can be of joist or rafter construction. In joist construction the diagonal hip members along the junction of the surfaces (similar to the valley members along the junction of two pitched roofs) perform as inclined beams supported by a column, which for instance may be located at the intersection of the hip and ridge; the loading diagram for a typical hip beam is shown in Fig. 5.29d. Note that without the column supports the hip and ridge members must act as a spatial rafter framework. In rafter construction, the shortened rafters (called jack rafters) that meet along the hip rafters support each other structurally, thereby placing the hip rafters in tension. Similar loading conditions occur at the intersection of the wings of L-, T-, or U-shaped houses.

The behavior of a typical three-hinge A-frame, which includes the simple rafter roof, is investigated now for the most common loading conditions as based on parallel frame arrangement. Investigating the symmetrical and asymmetrical uniform gravity loading cases in Fig. 5.31, it is shown that for preliminary design purposes only uniform gravity loading acting across the full width of the frame can be considered, since this case yields the maximum shear, moment, and axial forces.

The maximum moment according to Eq. (5.53) is

$$M_{max} = w(L/2)^2/8 = wL^2/32 \qquad (5.62)$$

The maximum shear according to Eq. (5.54) is

$$V_{max} = (w(L/2)/2) \cos\theta = (wL/4) \cos\theta \qquad (5.63)$$

The maximum axial force at the beam base according to Eq. (5.59) is

$$N_A = w(L/2) \sin\theta + (w(L/2)^2/2h) \cos\theta$$
$$= (wL/2) \sin\theta + (wL^2/8h) \cos\theta = A_v \sin\theta + A_h \cos\theta \qquad (5.64)$$

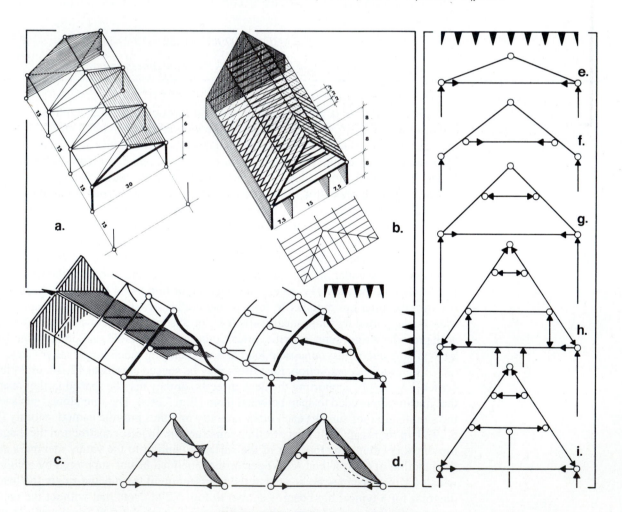

Figure 5.30 Rafter roof construction.

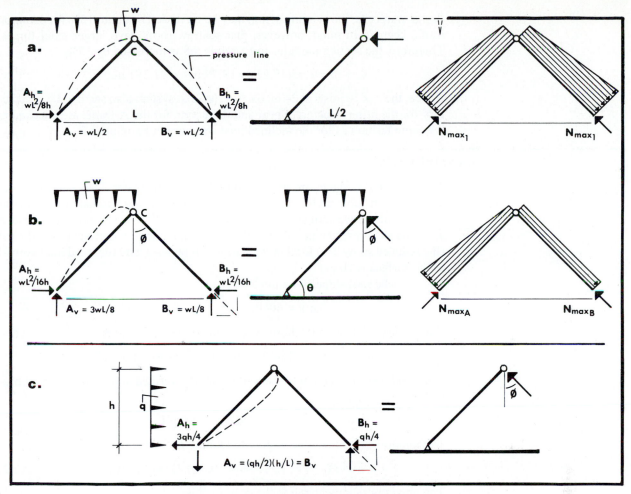

Figure 5.31   A-frame response to typical loading.

The axial force at quarter-span, where the moment is maximum, is equal to

$$N_L/4 = [(wL/2)\sin\theta]/2 + (wL^2/8h)\cos\theta$$
$$= (wL/4)\sin\theta + (wL^2/8h)\cos\theta \tag{5.65}$$

With respect to wind loading, refer to Eq. (5.61).

It can be concluded that the uniform gravity loading case may be assumed to generally control the preliminary design of roof joists as well as rafters. While the shear and moment flow is the same in both construction systems, the axial force distribution is different. Since the joist does not carry any axial force at the point of maximum moment, we may conclude that the joist is basically *a simple beam*, while the rafter is a *beam–column*.

EXAMPLE 5.9

Investigate the roof system in Example 5.8 by using rafter construction. It was shown that the gravity case controlled the preliminary design. Although the maximum moment and shear are the same for both construction systems, the axial force action at the rafter midpoint must be taken into account. In other words, the rafter must be designed as a beam–column. Since the effect of axial action on member design is secondary in comparison to bending, often, for preliminary design purposes, the bending action may be increased by about 15% for common roofs, keeping in mind, however, that shallow roofs have larger axial forces. Therefore, the section modulus in Example 5.8 is increased by 15%.

$$S = 1.15(17.21) = 19.79 \text{ in.}^3 < 21.391 \text{ in.}^3$$

Hence, the 2×10 joists at 24 in. o.c. for the joist system also seem to be satisfactory for the rafter structure. Shear and deflection for the A-frame are checked in the same fashion as for the inclined joist structure in Example 5.7.

EXAMPLE 5.10

Three-hinged A-frames are spaced parallel at 15 ft on center. They span 80 ft and are 60 ft high. Do a preliminary design of the frame using laminated timber construction of the western species type with $F_b = 2400$ psi. The roof dead load is assumed as 15 psf, the minimum live load on the horizontal projection is 12 psf (the reduced snow load for this steep roof will be less), and the wind load normal to the surface is 20 psf.

The roof slope (for a pitch of $18:12$) is

$$\tan\theta = 60/40 = 1.5 \quad \text{or} \quad \theta = 56.31°$$

The gravity load along the horizontal projection of the roof is

$$w = w_L + w_D = 15(12 + 15/\cos\theta) = 180 + 406 = 586 \text{ plf}$$

The wind pressure normal to the roof (or on the vertical and horizontal roof projection) is

$$w_n = 15(20) = 300 \text{ plf}$$

The maximum moment due to gravity loading is

$$M_{D+L} = wL^2/32 = 0.59(80)^2/32 = 118 \text{ ft-k} \tag{5.62}$$

The maximum moment due to the dead load is

$$M_D = M_{D+L}(w_D/w) = 118(406/586) = 81.75 \text{ ft-k}$$

The maximum moment due to the $D + W$ case is

$$M_{D+W} = w_n h^2/8 + w_n L^2/32 + M_D$$
$$= 0.3(60)^2/8 + 0.3(80)^2/32 + 81.75 = 135 + 60 + 81.75 = 276.75 \text{ ft-k}$$

The required section modulus for the gravity-loading case taking beam–column action into account (see Example 5.9) is approximately

$$S \cong 1.15 (M_{D+L}/F_b C_D) = 1.15(118)12/2.4(1.15) = 590 \text{ in.}^3$$

Since the wind action is normal to the surface and causes beam action, the preliminary design of the rafter is based on bending only, ignoring the effect of the axial action due to the relatively small dead load.

$$S = M_{D+W}/F_b C_D = 276.75(12)/2.4(1.60) = 865 \text{ in.}^3$$

Wind action clearly controls the design of this steep A-frame. Try a $6\frac{3}{4}- \times 30$-in. section (Table A.10) with $S_x = 914.5$ in.^{3} and $d/b = 4.44 < 5$.

Gable Frame Structures

There are infinitely many types of frames incorporating inclined members, including the A-frame and post–beam systems just discussed. Some typical cases are shown in Fig. 5.32; they can only give an indication of the potential occurrence. These frames

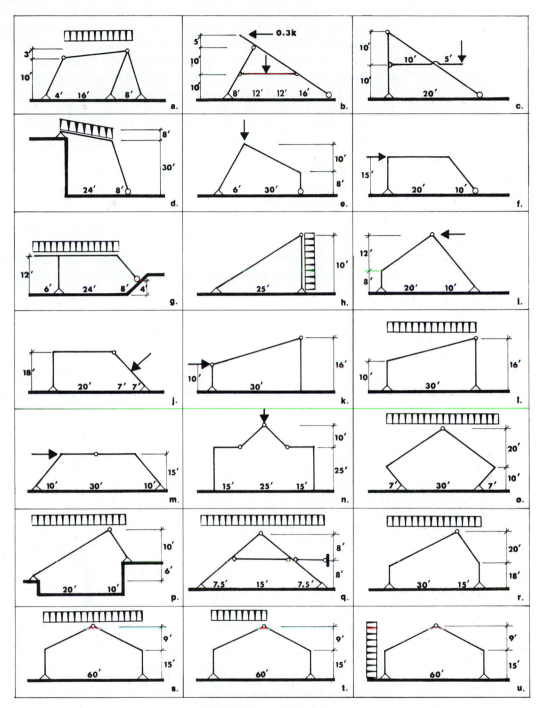

Figure 5.32 Problem 5.19.

are investigated in Problem 5.19 for the various loading conditions shown. Here, only the most common symmetrical gable frames (Fig. 5.33) are studied further.

Gable frames are usually either of the two- or three-hinge type. While the two-hinge frame is once indeterminate, the three-hinge frame is statically determinate and can be analyzed by using the three equations of equilibrium. The typical span ranges for timber frames are from 50 to 250 ft, with a spacing from 16 to 30 ft.

The design of the frame members depends on the critical loading conditions of uniform gravity loading, dead load together with live load on one-half the span, and dead load plus wind load, as has been discussed previously. For the preliminary design of typical frames with slopes not greater than about 30°, however, it is reasonable to only consider the uniform gravity-loading case for ordinary wind conditions. But keep in mind that as the frame gets taller the wind will become a primary design determinant; here the wind load may be assumed as uniform on the entire vertical projection of the frame (Fig. 4.2c). Furthermore, the asymmetrical gravity-loading case yields the larger field moment and larger center-span shear for the gable beam; this obviously has to be considered in the final structural design.

The reader may refer to Problem 5.19 and Fig. 5.32s, t, and u, and compare the force flow for the typical gable frame acted on by different types of loading.

Three-hinge Gable Frame (Fig. 5.33).   It was shown that the reactions for the three-hinge frames are independent of the frame shape. The reactions for the uniform gravity-loading case are

$$A_v = B_v = wL/2, \qquad A_h = B_h = H = M/h = (wL^2/8)/h \qquad (5.66)$$

The internal axial and shear force flow along the range of the vertical column is equivalent to the reactions in magnitude. The column moment varies with height and is maximum at the knee.

$$M_e = M_d = H(h_c) \qquad (5.67)$$

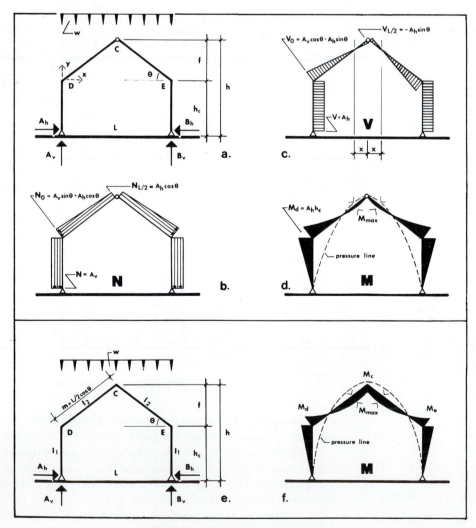

Figure 5.33   Three-hinge gable frame.

For example, letting the knee height be equal to $0.8h$ yields

$$M_e = M_d = [(wL^2/8)/h]0.8h = wL^2/10$$

The force flow is shown in Fig. 5.33. The internal force flow along the gable beam is found by using the equations derived in Fig. 5.4. In general, the magnitude of the axial forces is

$$-N_x = (A_v - W_y)\sin\theta + (A_h + W_x)\cos\theta$$

$$x = 0: \quad -N_0 = A_v \sin\theta + A_h \cos\theta \tag{5.68}$$

$$x = L/2: \quad -N_{L/2} = (A_v - wL/2)\sin\theta + A_h \cos\theta = A_h \cos\theta \tag{5.69}$$

In general, the magnitude of the shear force, according to Fig. 5.4, is

$$V_x = (A_v - W_y)\cos\theta - (A_h + W_x)\sin\theta$$

$$x = 0: \quad V_0 = A_v \cos\theta - A_h \sin\theta \tag{5.70}$$

$$x = L/2: \quad V_{L/2} = (A_v - wL/2)\cos\theta - A_h \sin\theta = -A_h \sin\theta \tag{5.71}$$

The maximum field moment in the beam occurs at the distance x (but this time measured from the crown and not from the support as previously) where the shear is zero.

$$x = ? \quad V_x = 0 = A_h \sin\theta - wx \cos\theta = (wL^2/8h)\sin\theta - wx \cos\theta$$

$$x = (L^2/8h)\tan\theta, \quad \tan\theta = f/(L/2)$$

$$x = Lf/4h \quad \text{and} \quad y = (2f/L)x$$

The maximum moment at that location is

$$M_{max} = Hy - wx(x/2) = \frac{w}{32}\left(\frac{Lf}{h}\right)^2 \tag{5.72}$$

The field moment M_{max}, in comparison to the knee moment M_d (i.e., M_{min}), is rather small and does not have to be considered for preliminary design purposes. Furthermore, the asymmetrical live load arrangement causes a larger field moment.

Two-hinge Gable Frame (Fig. 5.33e). The frame is once indeterminate; one additional known condition is needed. The relative sizes of the frame members influence the force flow along the frame. For instance, replace the pin at support B with a roller. Now the frame is statically determinate and the column will roll outward under the load action. The amount of lateral displacement depends on the stiffness of the frame members. It can be shown that the horizontal base force necessary to push the column back to its original position (i.e., to have a two-hinge frame again) is approximately

$$A_h = B_h = \frac{wL^2}{8h_c N}(8 + 5Q) \tag{5.73}$$

where $\quad Q = \dfrac{f}{h_c}, \quad k = \dfrac{I_2/m}{I_1/h_c} = \dfrac{I_2 h_c}{I_1 m}$

$$N = 4(k + 3 + 3Q + Q^2), \quad m = (L/2)\cos\phi$$

Note that for rectangular portal frames $Q = 0$, [see Eq. (5.3)].

The relative member sizes can be estimated as the ratio of rafter inertia to column inertia, I_2/I_1. Once the reactions are known, the force flow along the frame can be obtained by statics. The moments at the knee and ridge are

$$M_d = M_e = A_h(h_c), \quad M_c = A_v\frac{L}{2} - \frac{wL}{2}\frac{L}{4} - A_h(h) = wL^2/8 - A_h(h)$$

The maximum field moment can be found in the same manner as for the three-hinge frame.

EXAMPLE 5.11

The laminated wood gable frame shown in Fig. 5.34 spans 80 ft and is 30 ft high at the crown; the eave height is taken as $2h/3 = 20$ ft. The frames are spaced 16 ft on center. Dead and snow load are 15 and 30 psf, respectively. The effect of wind does not have to be considered for the preliminary design of this shallow roof. Try western species combination with the following allowable stresses under normal duration of loads:

$$F_b = 2200 \text{ psi}, \qquad F_c = 1650 \text{ psi}, \qquad F_v = 165 \text{ psi}, \qquad E = 1,800,000 \text{ psi}$$

The roof slope is

$$\tan\theta = 10/40 = 0.25 \quad \text{or} \quad \theta = 14.04° < 30°$$

The uniform loads acting on the horizontal projection of the rafter are

$$w = w_L + w_D = 16(15 + 30/\cos\theta) = 735 \text{ plf} = 0.74 \text{ klf}$$

The reactions are

$$A_v = B_v = wL/2 = 0.74(80)/2 = 29.6 \text{ k}$$

$$A_h = B_h = H = wL^2/8h = 0.74(80)^2/8(30) = 19.73 \text{ k}$$

The size of the column at the base is controlled by shear; the axial stresses are rarely critical (see Schueller, 1983). Try a member width of 8.75 in.

$$f_v = 1.5V/A \le F_vC_D$$

$$= 1.5(19.73)/8.75d_b \le 1.15 (0.165) \quad \text{or} \quad d_b \ge 17.83 \text{ in.}$$

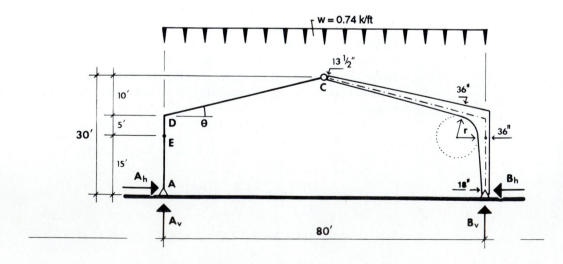

Figure 5.34   Example 5.11.

Try $8\frac{3}{4}$- $\times$ 18-in. section, $A = 157.5$ in.^{2} (Table A.10). A constant arch width of $8\frac{3}{4}$ in. is found to be reasonable.

The critical section in the tapered column occurs at point E where the curve transforms into a straight line; the moment at the frame knee will not be needed for the preliminary design. The curved portion at the knee requires special design considerations; here the allowable flexural stresses must be reduced because of the curvature. In other words, as a bending member becomes curved, the neutral axis no longer coincides with the center of gravity but shifts toward the inner face as, for example, for the given conditions in Fig. 5.46b, thereby increasing the bending stress along the inner face and decreasing the stress in the outer face. As the radius of curvature decreases, the stress increase at the inner face becomes more severe. In addition to the flexural stresses, radial tensile or compressive stresses perpendicular to the grain [see Eq. (5.88)] have to be checked.

The location of the tangent points can be estimated at this first stage of design. Normally, haunch radii of 9 ft 4 in. are used for western species laminated timber and 7 ft 0 in. for southern pine lumber. The depth at the critical section E can also be quickly estimated as

$$d \cong 1.33 \text{ to } 2d_b \cong 2(18) = 36 \text{ in.}$$

The moment at point E, assuming a distance of $20 - 5 = 15$ ft, is

$$M_E = Hy = 19.73(15) = 295.95 \text{ ft-k}$$

At that location not only bending occurs but also axial forces, which are, however, relatively small. Therefore, for preliminary design purposes the beam–column can be treated like a beam, but increasing the bending action by about 15% to take axial action into account. Hence, the approximate section modulus is

$$S \cong 1.15(M/F_b C_D) = 1.15(295.95)12/2.2(1.15) = 1614 \text{ in.}^3$$

Assuming a constant arch width of $8\frac{3}{4}$ in. yields a member depth at point E of 36 in. according to Table A.10. Try $8\frac{3}{4}$- $\times$ 36-in. section with $S_x = 1672.8$ in.^{3} and $b/d = 1/4.11 > 1/5$.

For fast trial purposes, the depth of the section at the crown is generally taken as 50% larger than the width. The member size at that location is seldom controlled by force action (i.e., shear and axial stresses); the arch depth must be proportioned to meet the purlin dimensions framing into it.

$$d \cong 1.5b = 1.5(8.75) = 13.125 \text{ in.}$$

Try $8\frac{3}{4}$- $\times$ $13\frac{1}{2}$-in. section, $A = 118.1$ in.^{2}. The layout of the gable frame as determined by this preliminary design is shown in Fig. 5.34. The depth of the three-hinged glulam gable frames at the crown is usually in the range of $L/50$ to $L/70$. For the design of tapered members, refer to the discussion of Fig. 3.10.

Conclusion. As conclusion to the investigation of the various pitched roof enclosures, several structure systems are compared in Fig. 5.35. The maximum moment of $M = wL^2/8$ for the bent beam is taken as the basis for this comparison. It appears that the A-frame (d) with a maximum rafter moment of only 25% of M and pure axial action in the columns is the most efficient solution if the tie rod (or buttress) is acceptable. This tie, however, is not required for frames (c) and (f), which are continuous across the rafter–column intersection and resist the thrust as maximum bending at the junction, thereby throwing a large moment into the columns. Obviously, the

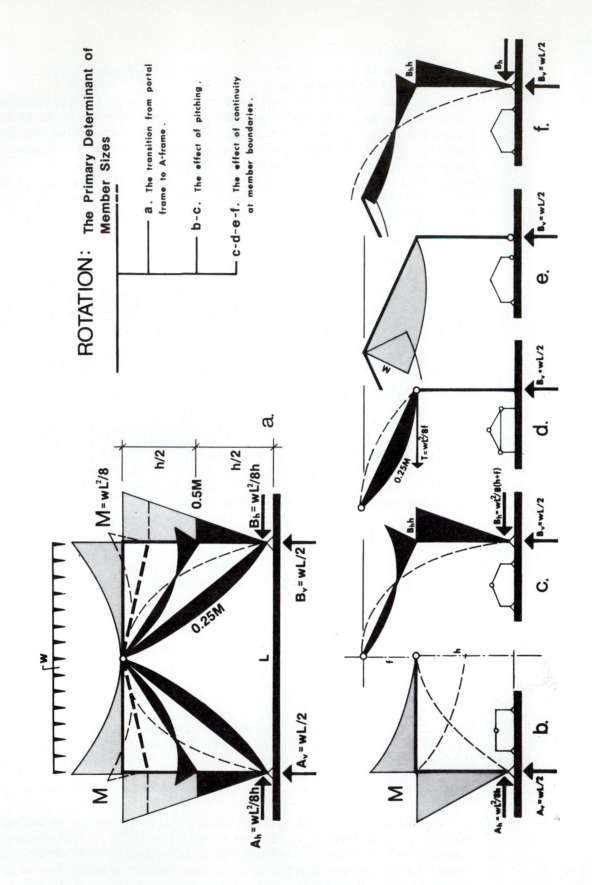

ROTATION: The Primary Determinant of Member Sizes

a. The transition from portal frame to A-frame.

b-c. The effect of pitching.

c-d-e-f. The effect of continuity at member boundaries.

Figure 5.35 Response of typical gable frame roof enclosures to gravity loading.

ideal solution from the force flow point of view is the parabolic arch in pure compression, reflecting the funicular response to uniform gravity loading, realizing, however, that asymmetrical loading now becomes the governing design determinant.

5.4 ARCHES

The arch is part of the frame family, but it distinguishes itself from the other members by providing a continuous one-member enclosure without having any abrupt kink points along its geometry. The internal forces flow smoothly along the arch and are not concentrated at points of sudden change of form, assuming the external loads evenly distributed and not concentrated either unless they are located at kink points.

Arches may be composed of many different types of curves, as demonstrated in Fig. 5.36. The early Greek mathematicians were familiar with common basic curves, which Apollonius, who lived during the third century B.C., described in his book, *The Conics*. He showed that the basic curves are the circle, ellipse, parabola, and hyperbola (Fig. 9.5, center), which can all be derived from *conic sections*. It is the cycloid the Greeks were not familiar with. Galileo, in the seventeenth century, suggested it as a good shape for bridge arches. A cycloid is generated by a point on the perimeter of a circle rolling along a straight path.

Curves can be used for arch construction as single-, double-, or multiple-curvature systems. With respect to circular curves, arches can be one-centered (e.g., semicircular, segmental), two-centered (e.g., drop, equilateral, or acute pointed arches), and multicentered [e.g., three-centered depressed arch, four-centered Tudor arch, and ogee (onion) arch]. Other arch shapes may not be derived from geometry but from structural behavior, such as funicular arches. The reader may want to study in Fig. 5.36 the innumerable ways basic curves may be combined to yield various arch profiles.

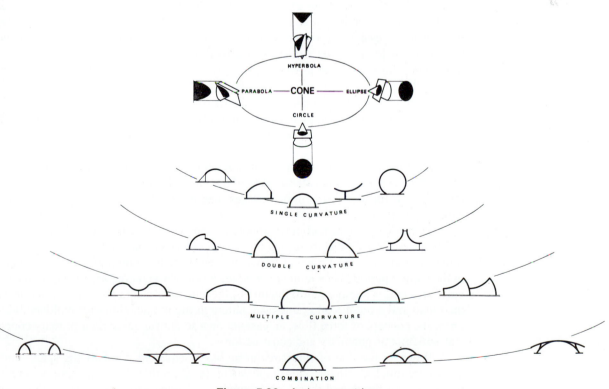

Figure 5.36 Arch geometries.

The more abstract investigation of arches in Fig. 5.37 attempts to stimulate architectural design. The many examples suggest how arches may define building spaces and that they are not just used for bridges or arranged parallel to define industrial or public spaces. Arches may be used as single or multiple interactive systems; they may make a wall opening possible, they may form directly the roof enclosure, they may support another structure, or they may form chords of trusses by being rigid or prestressed cables (e.g., cable-supported strutted arch). Toward the end of last century, the use of hollow tile arches had become quite common for floor construction of high-rise buildings in the United States. These floor arches were either of the flat or segmental type (i.e., shallow vaults). Arches do not have to stand vertically but can be inclined or may be horizontal and possibly warped to form open or closed rings as used for domical structures. The reader may also refer to Fig. 1.5, where the richness of curvilinear grids is investigated, and Fig. 1.8, where it is demonstrated how the outward movement due to lateral arch thrust is prevented by abutments.

Historical Development

In the past, the arch together with the barrel arch and the archlike vault were among the few structural systems that made it possible to span larger distances by using masonry with its low tensile capacity. Probably the first arches built were based on the corbelling principle, where horizontal masonry courses are projected slightly beyond the previous course (e.g., Fig. 8.2a). These *corbeled arches* are false arches that do not develop lateral thrust, which is a basic characteristic of true arches.

As early as several centuries B.C., the Assyrians and Babylonians included arches in their architecture (e.g., Ishtar Gate in Babylon, circa 575 B.C.). Although the Egyptians and Greeks were familiar with the arch concept, they did not make it part of their building design language. Only the Romans developed the arch as an important element of their architecture, as expressed by triumphal arches and amphitheaters or as demonstrated by the remarkable aqueducts in Nîmes and Segovia; they not only used its engineering potential but gave it a special architectural meaning. Their arches and barrel vaults generally were semicircular in shape. They selected this arch type because it was easier to construct and possibly because they thought there were no horizontal thrust forces at the vertical supports. Since a cable takes the shape of a circle only under uniform radial loads (Fig. 9.9), we may conclude that the circular arch is not an efficient shape from a force flow point of view since external load action almost never is radial, with the exception of culverts deep in the earth.

In contrast to the Roman semicircular arches, the imposing barrel vault of the palace of Ctesiphon, now Tag-e-Kisra, near Bagdad, Iraq, attributed to the Sassanian King Chosroes I (A.D. 531–579), is of a near catenary or almost parabolic configuration. The funicular shape of this 120-ft-high archlike vault, spanning 84 ft, was necessary because of the low tensile capacity of the mudbrick material, keeping in mind that the Romans, for their semicircular arches, had available much stronger masonry, as well as concrete.

In Europe, the semicircular arches of the Romans were adopted again as an essential part of architecture during the Romanesque period. Transformation from the round arch to the slender pointed arch happened at the beginning of the twelfth century in the Gothic period. The Gothic *pointed arch* is a two-centered arch composed of circular arcs and is a good approximation of the funicular shape for a uniformly distributed load and a point load at midspan, although the medieval master builders did not know the concept of force flow, at least not on a scientific basis; their primary concern was structure as geometry and construction.

Whereas the rise of semicircular arches is equal to one-half the span, Gothic arches are much slenderer. This emphasis on height rather than span results in reduction of lateral thrust and therefore, in smaller columns and buttresses. Gothic cathe-

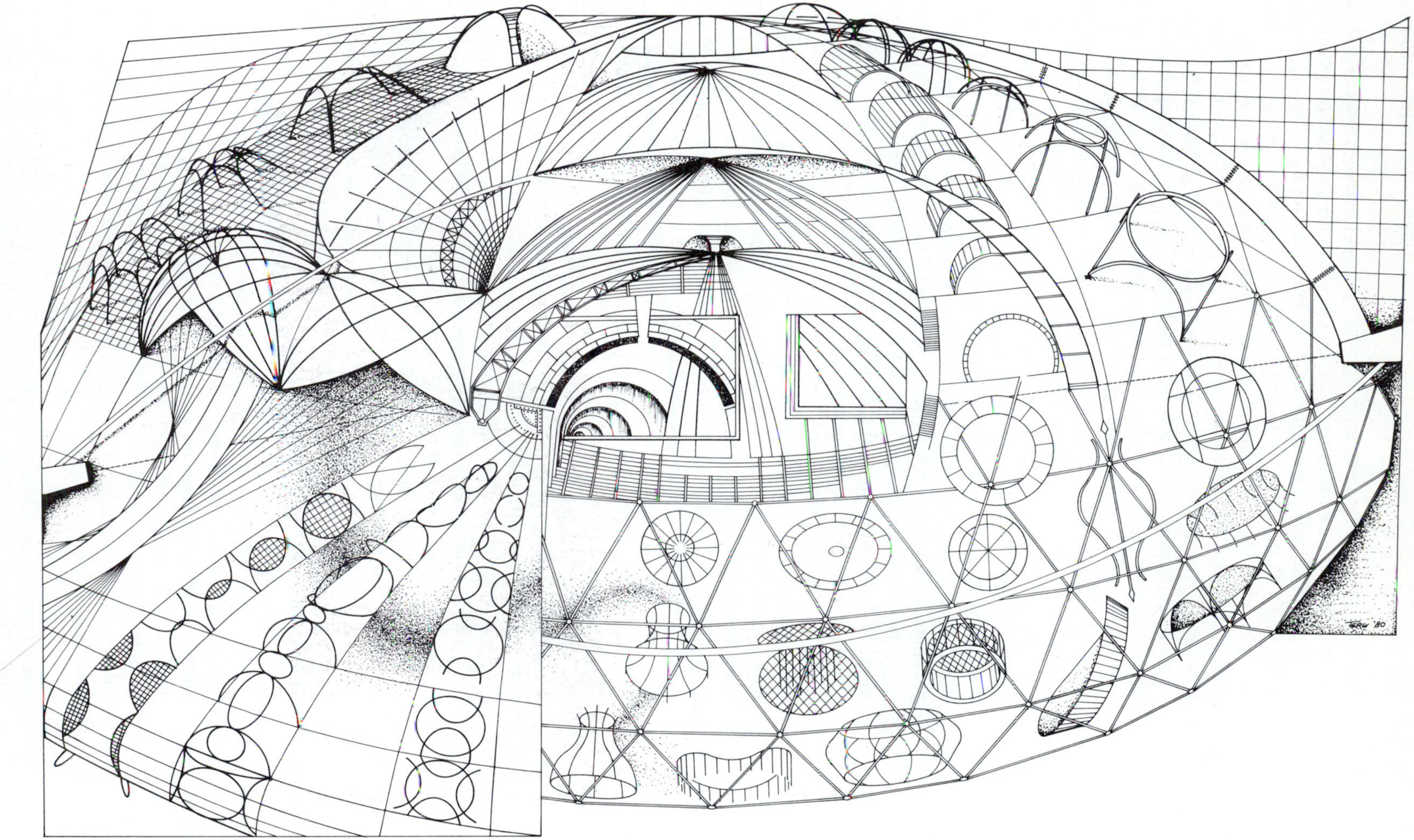

Figure 5.37 Arches as enclosures.

drals are admired for their seemingly weightless interior spaces, which are achieved not only through record horizontal spans, but by span-to-height proportions. For example, the central aisle of the cathedral at Reims (1290) is 123 ft high to the vaults and is 48 ft wide, that is, a span-to-height ratio of 1:2.56, with Amiens (ca. 1220) and Cologne (ca. 1332) cathedrals reaching 1:3, and finally Beauvais (ca. 1272) the daring proportion of 1:3.33, quite a development from the typical ratio of 1:2 for Romanesque churches. These proportions together with the effects of light and articulation of the slender skeleton structure convey a feeling of antigravity and dematerialization of space, in turn resulting in a never-ending upward surge as is so powerfully and daringly expressed in the French cathedrals of the High Gothic period (Fig. 8.2d). As opposed to the Gothic arches, the pointed arches of Islamic architecture often take the form of *ogee arches* which are composed of S-shaped curves.

In contrast to high-pointed arches are the flat arches of the Renaissance bridges, surely influenced by the shallow medieval Ponte Vecchio in Florence (1367) with its three segmental arches and an unbelievable rise-to-span ratio for the central 86-ft span of only 1:7.5. The ingenious Renaissance engineers replaced in their bridge design the previous semicircular arch forms by segmental or multicentered depressed curves, as so clearly expressed by the beautiful Rialto Bridge in Venice (1591), consisting of a segmental arc spanning 89 ft and rising 21 ft, and the Ponte Santa Trinita in Florence (1569) with its three arches composed of two parabolic arcs with a shallow profile and a rise-to-span ratio of 1:7.

It may have been Robert Hooke, in 1670, who was the first to realize from a scientific point of view that the catenary is the funicular response of the arch weight. Christopher Wren introduced the concept of the catenary dome shape with the conical brick dome supporting the cupola of St. Paul's Cathedral in London (1710). But Giovanni Poleni was the first to actually use a model of string and lead weights to obtain the thrust line of St. Peter in Rome in 1743 and thus was able to make his recommendations for the number of tension rings required to prevent further cracking of the cupola.

Antonio Gaudi (1852–1926), around the turn of this century, revived the idea of the funicular curves of the loads in his search for the true nature of form. He derived arch shapes from suspended scale models so as to achieve purity of form or, from an economical point of view, maximum efficiency of materials. Gaudi also used parabolic arches as an approximation for catenary curves, for example, in the Colegio Teresiano, Barcelona, Spain (Fig. 5.38, center).

The nearly 100-ft-span cast iron arch bridge at Coalbrookdale, U.K., across the river Severn, completed in 1779, is often considered as the change to a new era or as the turning point from stone and brick, as the dominant materials for arches, to iron. The new material of iron and later steel made long spans and new building types possible.

While the correct arch shape is of utmost importance for materials weak in tension or where weight is an important consideration, the strong new materials, iron and then steel, which were developed throughout the nineteenth century, did not have to place that much emphasis on this aspect, as they can easily resist bending. The continuous (fixed) trussed, wrought iron arches for St. Pancras Station in London (1868) span the enormous distance of 240 ft. Twenty years later, a new era in scale was established with the immense 375-ft span of the three-hinged, trussed steel arches for the Galerie des Machines of the 1889 Paris Exhibition.

Gustave Eiffel's great 530-ft-span, crescent-shaped, parabolic two-hinge, trussed arch for the Garabit viaduct was completed in 1884. A symbol of its time is the Eiffel Tower in Paris (1889) with its unbelievable height of 984 ft. This vertical cantilever consists of four latticed box corner columns forming exponential arched shapes, which lean against each other at the top and are tied together by four square platforms at different levels.

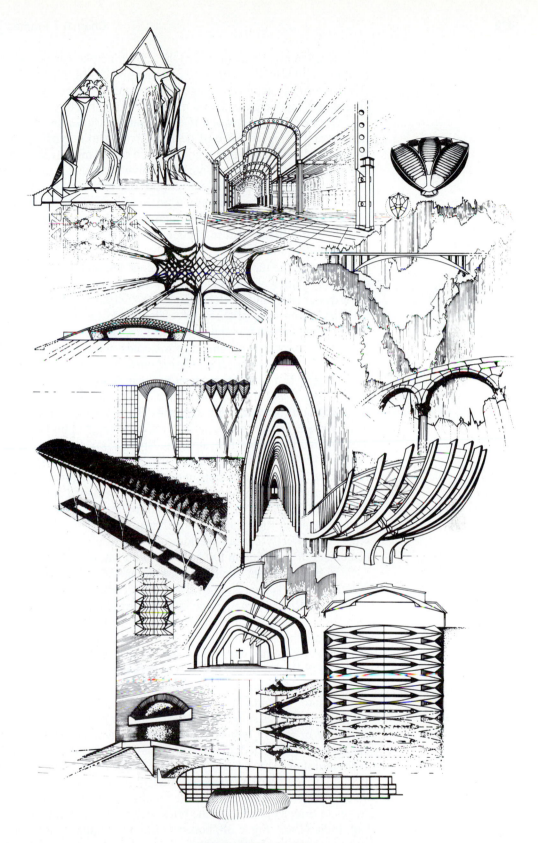

Figure 5.38 Arches.

Keep in mind, however, that the arch principle was not just applied to long-span buildings and bridges. In the reading room of the Bibliothèque Nationale, Paris (1868), Henri Labrouste achieved a high-quality interior space by allowing the latticed iron arches carrying the glass domes to be part of architecture.

Thonet's first bentwood rocking chair of 1860 must also be mentioned, a most famous and complex piece of design. Alvar Alto's experiments with the sculptural potential of bent laminated wood resulted in his much celebrated armchair of 1930, where a continuous seat-and-back bent plywood sheet is supported by bent laminated wood frames on each side, which reminds one of the arches in his church in Riola, Italy (1978, Fig. 5.38 center, bottom). Modern structural glued laminated timber was used in the U.S. building construction industry for the first time in 1935, although it was already in use before that in Europe. Today, wood arches can be bent and shaped to almost any required form.

Around the turn of this century, concrete arches were rediscovered on a larger scale and exciting new forms were developed, especially in bridge construction by the great designers François Hennebique (1842–1921) and Robert Maillart (1872–1940). Maillart became one of the most celebrated designers of reinforced concrete bridges and has been very influential in architecture. He pioneered the optimization of form by carrying loads with a minimum of materials, thereby not only achieving economical solutions but also esthetically pleasing ones. Among his masterpieces is the Salginatobel Bridge over a deep mountain gorge near Schiers, Switzerland (1930, Fig. 5.38, top right). Here, a 292-ft-span, three-hinged arch supports thin vertical walls upon which the continuous road platform slab rests. The varying arch depth increases toward the quarter-spans and decreases again toward the hinge at midspan. In other words, where arch and roadway deck fuse along midspan, a box section is formed, and where deck and arch separate at approximately the quarter span, a U-section is generated, which changes to a shallow rectangle at the abutments. The arch shape reflects the funicular response to maximum moments caused by single-load action at the quarter-spans due to live loads (Fig. 5.40c). Here, live loads represent a significant proportion of the total load because of the relatively small span of bridge. Maillart's funicular arch form seems to perfectly control the force flow with a minimum of effort through correct placement of mass and in full harmony with nature.

In bridge design, rapid progress in the sciences, material strength, and construction techniques is expressed by the world's currently largest steel arch span of 1700 ft of the New River Gorge Bridge in West Virginia (1978) and the world's longest concrete arch of 1280 ft of the Krk Bridge in Croatia (1979).

In architecture, the early work of the pioneers Robert Maillart and then Pier Luigi Nervi, together with the creative experimentations of Modernism and the formability of the major materials (steel, reinforced concrete, and wood), has made a limitless formal expression possible as the various arches in Figs. 5.37 and 5.38 clearly demonstrate. The arch principle can be employed in many different ways within the building context, as suggested in Fig. 5.38. For instance, arches may form the building enclosure or roof: a typical parallel arrangement yields a cylindrical shape, while a radial one results in a dome. The arch may provide support for another structure, such as for a roof or roadway as is typical for bridges; or it may be the compression ring for a stadium roof. Eero Saarinen's Gateway Arch in St. Louis (1963), derived from an inverted catenary curve, soars like an abstract sculpture to 630 ft. Louis Kahn uses arches to bridge openings in the brick wall for the Institute of Management, Ahmedabad, India (1963, Fig. 5.38, bottom left). He articulates how the thrust of the shallow arches is balanced by the concrete ties.

In contrast to the single Roman arch or Maillart's Salginatobel Bridge is the late Gothic mesh vault of St. Lorenz in Nürnberg, Germany (Fig. 5.38, center left), where the arched ribs grow naturally out of the piers to form an intricate grid of ribs or a ribbed masonry vault. The 10-ft-deep trussed steel arches for the Convention Center in Niagara Falls, New York (1974, Fig. 5.38, center left), designed by Johnson/Burgee, rise only 34 ft from the abutments and span 355 ft; they are spaced 63 ft on center. The very shallow arch configuration was selected to give the impression of a light, almost floating roof structure.

A rather complex arrangement of arches is used for the spherical roof segment of the marvelous 564-ft-span velodrome in Montreal (1976, Fig. 5.38, top right), which rests on four abutments; it was designed by Roger Taillibert. The primary roof framing consists of three pairs of prestressed concrete arches and two edge arches. The arch pairs, in turn, are tied together transversely by Y-shaped prestressed beams that give support to curved translucent plastic panels. The precast concrete arch elements were glued together before they were posttensioned so that the roof can act as a true monolithic surface structure.

The wooden shell for the auditorium at the University of Reims, France (1976, Fig. 5.38, center right), designed by A. and D. Dubard de Gaillarbois, consists of 16 glulam ribs bent upward, which rest on reinforced concrete frames at the base. These cantilevering timber arches (that have a maximum span of 43 ft) support the slightly inclined roof at the top and the seating at the bottom.

Impressive are the long, glass, vaulted corridors of the United Airlines Terminal at O'Hare International Airport, Chicago (1987, Fig. 5.38, center top), designed by Helmut Jahn. The corridor roofs are supported by parallel, exposed, welded steel, arched bents spaced 30 ft on center and spanning a maximum distance of 50 ft and are resting on multipipe, battened, column assemblies, that is, steel pipe column clusters of sculptural appearance. The 24-in.-deep arched girders are perforated with 12-in. holes 24 in. on center to make them light and airy. The huge bulbous roof shape for Renzo Piano's shopping center in Bercy-Charenton, near Paris, France (1990, Fig. 5.38, center bottom), is defined by radially arranged glulam arches, similar to the ribs in a body, covering the concrete frame building structure.

In Santiago Calatrava's anatomical structures, the arch concept is always an essential part of the formal language. In his project for the transepts of St. John the Divine Cathedral in New York (1991, Fig. 5.38, top left), he proposes bony arched frames of organic Gothic expression. The glass vault of Toronto's 90-ft-high Galleria (1992, Fig. 5.38, center left) receives its support from slender tree columns leaning inward, each carrying four lower parabolic arches at 12-ft intervals and upper circular arches between. The columns split twice into two branches to carry the alternating parabolic and circular arches.

For the Concert Hall at Sur, near Zürich, Switzerland (1988, bottom right), for roof support structure, Calatrava uses parallel, leaflike arches resting on slender columns at the ends; each arch is tied together with four stretched cables. The composite, 82-ft-span, three-hinged arches consist of steel box girders of V-shaped cross section stiffened by vertical plates that extend to support the roof structure above; the trussed parabolic top chords touch each other at midspan. The depth of the V-shaped webs increases from a minimum at midspan to a maximum and closed shell, where the cables penetrate through the sculpted opening. Here, Calatrava articulates the play between compression and tension by letting the tension cables disappear. The overall image of the sculptured arch seems to be that of a huge stringed instrument.

Response of the Arch to Loading

The most common funicular cable shapes are shown in Fig. 9.5; they are the catenary and parabola for typical gravity-loading conditions. When these shapes are "frozen" and inverted, they become *funicular arches* responding in pure compression. However, the assumed loads do not remain constant. When they change, the rigid arch cannot readjust its shape and hence is subject to flexure. We may conclude that, although the arch is primarily a compressive structural system, it also must resist bending, which for roof structures generally controls its design.

In addition, boundary conditions of the arch are to be considered, since large thrust forces may have to be resisted, with the necessary capacity being provided by the soil, tie rods, or buttresses. Light, steep arches cause less thrust than corresponding heavy, flat arches! An arch must be connected to its base so that any lateral displacement is prevented. Should one of the supports be equivalent to a roller, then the arch loses its identity and degenerates to a curved beam with its corresponding large bending moments. This fact clearly shows that much care must be taken with the design of the anchorage system, which must resist the lateral thrust.

From a material point of view, the arch should have a funicular shape, but arches are rigid and cannot adjust their geometry under variable load action as flexible cables do in order to stay in tension. True funicular arches in building construction generally do not exist; they are only funicular for one loading condition. There are rare situations, however, where the nonchanging dead weight comprises the largest portion of the total load, such as for underground structures and possibly the medieval Gothic vaults (Fig. 2.6). For these cases the arch geometry can be selected so that the pressure line can be kept within the middle third of the member cross section, resulting in no tension along the entire arch length. In general, an optimal geometrical arch form with a minimum of moment action can be derived by superimposing the influence of various loading conditions and by developing a critical loading envelope. Rather than going through this complex optimization process, the pressure line, due to the constant dead load, is often used as the arch form so that moments are only generated by live load action. For building types where the superimposed dead load acts along the arches in a uniform manner, the funicular response is the *catenary*. Because of the complex mathematical nature of this form, it is often approximated by a second-degree parabola, as discussed in the section on cables. This parabola represents the funicular response to uniform load action on the horizontal roof projection. Hence, dead load action may be approximated as a horizontal uniform load and considered similar to snow loading.

Axial forces may be beneficial for materials weak in tension such as masonry and concrete, since they act as natural prestress agents trying to overcome the cracking (tension) of the section as caused by bending. Although the arch is primarily an axial system, assuming that its shape is funicular to the predominant loading in contrast to the beam, which is primarily a bending system, the relative small arch moments still determine the size of the arch, as shown later.

The selection of the height-to-span ratio of the arch is an important consideration. As the height increases for a given span, the lateral thrust decreases but the arch length increases; in other words, with increase of height the support costs decrease, while the arch costs increase. It is apparent that there is some optimal height-to-span ratio from a material efficiency point of view.

Most arches are either three hinged, two hinged, or fixed. The three-hinged arch is investigated first because it is statically determinate. **For preliminary design purposes, two-hinged parabolic and shallow circular arches can be treated as three-hinged ones. Even for fixed arches, the lateral thrust is not much larger and may be roughly approximated as equal to the three-hinged ones.** The three-hinged arch finds wide application, especially in glued-laminated timber construction, but is also used in precast concrete and steel construction. Some of the advantages of its application lie not only in the simpler structural analysis and in its adaptability to any movement without being stressed, but also in easier transportation, erection, and assemblage procedures.

The analysis of the three-hinged arch is approached in the same manner as for the frame. The reactions do not change under constant loading on the horizontal roof projection; they are independent of the frame shape that bridges the given space. However, the intensity of force flow obviously varies with the shape of the structure. In the introduction to frames it is shown that, for a given loading condition, the difference between the ideal funicular form and the actual frame shape constitutes rotation (bending), which is the primary determinant of member size.

For the case where an arch is loaded only by a single external force P, such as a concentrated vertical or horizontal load or a moment (Fig. 5.39a to c), the reactions can be found directly by balancing the three forces R_A, R_B, and P about the point of concurrency d, where all the three forces meet when translated along their line of action. The reaction R_B must pass through the hinge at the crown since there is no external force interference along this side of the arch and the moment is zero at the hinge; the reaction force is collinear and in equilibrium with the resultant force at the crown (Fig. 5.39a). Hence, where R_B intersects with the force P is the location of the point of concurrency; this point connected to the other support establishes the direction of the reaction R_A. The lines of force flow for R_A and R_B directly balancing the external load are analogous to an imaginary cable supported at the reactions that has adjusted its shape to the force action so that it can resist in axial manner. This cable is "frozen" and inverted to form the funicular shape, also called the *pressure line* as discussed previously.

The axial forces along the funicular lines must be relocated to the real structure. In general, the displacement of a translational force to any other place in space must in addition cause rotation about this new force location. Furthermore, the translational force must be resolved into components perpendicular and parallel to the member curvature, resulting in shear and axial action, respectively. From this discussion we may conclude that the moment at any given arch section can be determined by the product of the axial funicular force and the corresponding perpendicular distance to the location of the section. For instance, the maximum moment at load P (Fig. 5.39a), causing tension along the inside face of the member, is

$$M_{max} = R_A(e_1) = R_B(e_2)$$

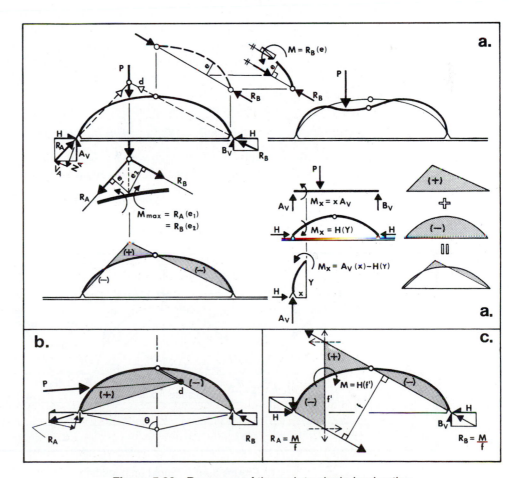

Figure 5.39 Response of the arch to single-load action.

It is apparent that the largest moment appears at an arch section that is farthest away from the pressure line (e_{max}) if the small effect in change of magnitude of the funicular force is ignored; notice that for the investigated case the funicular forces are constant.

Although the moments at any location along the arch can be found as just discussed, it is considered simpler to determine them in the familiar fashion, by cutting the member at the point to be investigated and then establishing equilibrium of all the forces acting on the free body. For example, for the special case in Fig. 5.39a, the arch moment at a distance x is equal to

$$M_x = A_v(x) - H(y) = M_{beam} - M_{geometry} = M_{arch}$$

This expression shows that the moment of a simply supported beam, $M_{beam} = A_v(x)$, with the same span and the same loading as the arch is reduced by the effect of geometry or thrust, $M_{geometry} = H(y)$. Since the bending moment diagram of the simply supported beam has the same shape as the pressure line, and the moment diagram of the thrust has the shape of the arch, we may conclude that deviation of the arch geometry from funicular form is defining and equivalent to the arch moment diagram.

EXAMPLE 5.12

The simple loading case of a single load at the crown of a parabolic and a circular arch in Fig. 5.40a, b is studied. Derivation of the geometry for a parabola and circle is presented in Section 9.2.

The funicular shape for single-load action is the A-frame, where the load flows directly to the supports without any detour; in other words, at the crown hinge, the zero moment location, the forces are in direct equilibrium. The reaction forces balance the load P at the point of concurrency.

It is apparent that the largest moment appears at an arch section that is farthest away from the pressure line (e_{max}) if the small effect in change of magnitude of the funicular force due to weight is ignored; notice that for the investigated case the funicular forces are constant.

The reactions for the parabolic and circular arches in Fig. 5.40a, b are

$$A_v = B_v = P/2, \qquad A_h = B_h = H = M/h = PL/4h$$

For the particular parabolic arch with $h = L/2$ and the semicircular arch with $h = R = L/2$, however, the thrust force H is equal to the vertical reaction forces.

$$A_v = B_v = H = P/2$$

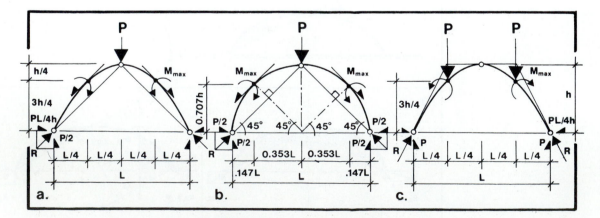

Figure 5.40 Simple arch loading.

The resultant forces at the reactions or the axial funicular forces in general are

$$R_A = N_A = \sqrt{A_v^2 + H^2} = (P/2)\sqrt{1 + (L/2h)^2} \qquad (9.5)$$

But for the particular arches with $h = R = L/2$, the resultant forces are

$$R = N = (P/2)\sqrt{2} \qquad (9.5a)$$

The maximum shear and the axial arch forces at the supports can be determined by using the general equations of Fig. 5.4.

$$V_A = A_V \cos\theta - A_h \sin\theta, \qquad -N_A = A_v \sin\theta + A_h \cos\theta$$

Notice that not only the moment of an equivalent simply supported beam but, even more, the beam shear is reduced. On the other hand, an axial force is introduced due to arch action; now buckling criteria must be considered.

For the semicircular arch with a vertical slope at the reactions and $\tan\theta_o = \tan 90° = 0$, the shear and the axial force are each equal to

$$V_A = -N_A = P/2$$

For the parabolic arch with $h = L/2$, on the other hand, the slope at the supports according to Eq. (9.15) is

$$\tan\theta_A = 4h/L = 2 \quad \text{or} \quad \theta_A = 63.44°$$

Hence, the shear and axial force at the support are clearly different from the radial arch because of the difference in curvature.

The maximum moment for the parabolic arch occurs at $x = L/4$, where the perpendicular distance e between the pressure line and the arch section is maximum. At that location the shear is zero and the axial force is maximum as well as tangential to the curvature. The y distance, where $x = L/4$, can be determined from the equation of the parabola [Eq. (9.19)] as

$$y = 4h(Lx - x^2)/L^2 = 4h[L(L/4) - (L/4)^2]/L^2 = 3h/4$$

The maximum moment is obtained by taking moments with the reactions about this point.

$$M_{max} = H(3h/4) - A_v(L/4) = (PL/4h)(3h/4) - (P/2)L/4 = PL/16 \qquad (5.74)$$

This moment causes tension along the outside of the member so that it is treated negative according to sign convention. The maximum moment for the semicircular arch is found in a similar manner. It is apparent that it occurs at an angle of 45°, as indicated in Fig. 5.40b. The location is found as follows:

$$\sin 45° = y/R, \qquad y = 0.707R = 0.707h$$

$$\cos 45° = (R - x)/R, \qquad x = 0.293R = 0.147L$$

Rotational equilibrium about this point yields the maximum moment.

$$M_{max} = (P/2)0.707R - (P/2)0.293R = 0.207PR \cong PL/10 \qquad (5.75)$$

Again, this moment causes tension along the outside of the member and is treated as negative according to sign convention. Notice that the maximum moment generated by the parabola is only about 60% of that for the semicircular arch.

For the loading case where two single loads act at $L/4$ on a parabolic arch (Fig. 5.40c), the maximum shear and maximum axial forces occur at the supports while the maximum moments occur at the loads. It can easily be verified that the maximum moment causes tension along the inside of the member and is equal to

$$M_{max} = PL/16 \qquad (5.76)$$

In the following discussion, the arches are considered to carry uniformly distributed loads as is the case for cylindrical roof forms, where parallel arches provide the support of the roof skin or barrel arches form the enclosure structure. The same loading cases investigated for frame design will also be analyzed for the arches.

With respect to the assumed uniform dead load, we ignore the fact that the load actually varies because of the arch curvature and that the arch may not be of constant size. The wind pressure distribution along curved roofs is complex and depends, among other criteria, on the height-to-width ratio of the structure (Fig. 4.2). For preliminary design purposes of ordinary roofs, an equivalent slope may be used that is equal to the slope of a line from the eave to the crown. Then it may be proceeded with the calculation of the loads, as discussed in the previous section for pitched roofs. In this context, wind loads of ordinary magnitude are only considered for arches with a rise-to-span ratio of larger than 1:3; they are treated as uniform lateral loading against the vertical arch projection.

We may conclude that for preliminary design purposes as based on the assumptions above, wind may be neglected if $h/L \leq 1/3$, or as the arch height increases to more than about one-third of its span, wind should be considered, keeping in mind that the assumption of wind distribution may be rather conservative. Roof snow loads may be reduced for arched structures steeper than 1:3 for first investigations.

In this context only parabolic and circular arches are considered, the geometry of which is discussed in Section 9.2. Other geometries can be treated as parabolas if their height-to-span ratio is less than about 1:5. The reader may want to refer to Fig. 5.41 and Problem 5.24 for other types of arches under different kinds of loading. Under uniform gravity action, notice that the horizontal thrust forces of the symmetrical arch assembly in case m (Fig. 5.41) balance each other for the interior arches but are unbalanced for the end arches and must be resisted by massive corner piers that act as abutments, unless all arches are held together by horizontal ties.

Parabolic Arch (Fig. 5.42). Under uniform, vertical load action (Fig. 5.42a), the pressure line coincides with the centroidal axis of the arch, or the parabolic arch is the funicular shape for the given loading. There is no bending and no shear along the arch; the forces are resisted in purely axial manner. This momentless arch is often called a *funicular arch*. For a more detailed discussion on funicular behavior, the reader may refer to Section 9.2, the parabolic cable.

The reactions are obtained in the same manner as for the three-hinged portal and the A-frame and are equal to these cases:

$$A_v = B_v = wL/2, \qquad A_h = B_h = H = M/h = wL^2/8h \qquad (5.77a)$$

For this special case, the resultant reactions and the axial forces at the supports are equal to each other because of the absence of shear.

$$N_{max} = N_B = N_A = R_A = \sqrt{A_v^2 + A_h^2} = \frac{A_h}{\cos\theta} = \frac{A_v}{\sin\theta} \qquad (5.77b)$$

where the slope of the arch at the reactions is

$$\tan\theta_o = A_v/A_h = 4h/L \qquad (9.15)$$

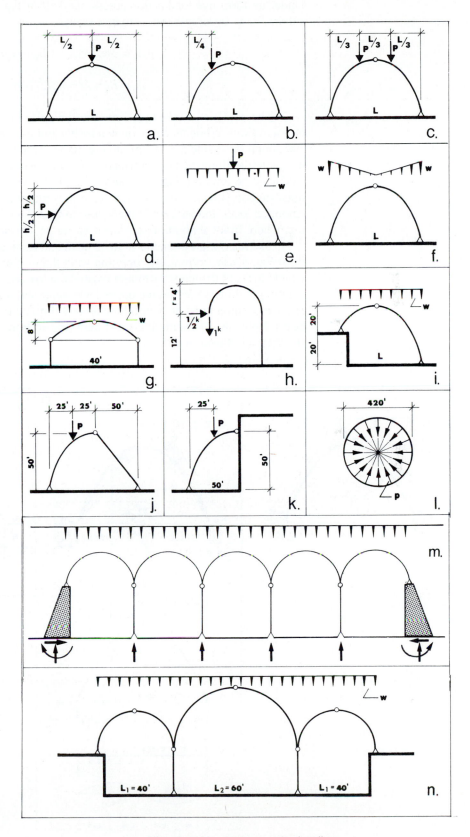

Figure 5.41 Arches under loading.

Under uniform live load action across one-half of the arch (Fig. 5.42b), the reactions can be found by simple statics to be equal to

$$A_v = 3w_L L/8, \qquad B_v = w_L L/8 \qquad (5.78)$$

$$A_h = B_h = H = w_L L^2/16h \qquad (5.79)$$

The deviation of the arch from the funicular line indicates the magnitude of the moment flow. While the moment is positive along the loaded side where it causes tension along the inside face of the arch section (as the member deformation also shows), it is negative along the nonloaded portion. The position of the critical moments on each side must be found so that M_{max} and M_{min} can be determined. For cases where an arch must support single loads, the location of the critical moment may be at one of the concentrated loads and hence relatively easy to find. This is not the case for uniform loading, where first the point of zero shear has to be determined.

The maximum moments at the loaded and nonloaded sides can easily be derived from Fig. 5.42b, bottom, by superimposing the uniform loading case $w/2$ (Fig. 5.42a, which does not develop any moments) and the loading condition of $w/2$ acting in alternating directions to obtain the desired asymmetrical loading case. The loading case in alternating directions does not develop any thrust and can therefore be treated like a beam with the corresponding maximum and minimum moments located at quarter spans ($x = L/4$ from reactions).

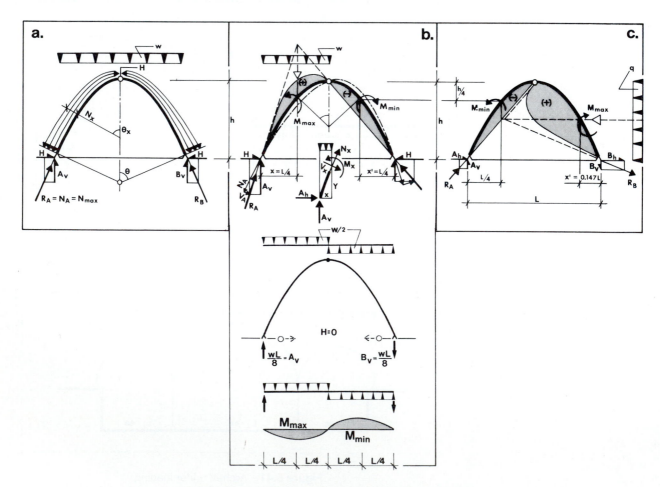

Figure 5.42 Parabolic arch under uniform load action.

$$M_{max} = (w_L/2)(L/2)^2/8 = w_L L^2/64, \qquad M_{min} = -w_L L^2/64 \qquad (5.80)$$

The maximum moment at the nonloaded side is opposite to the one on the loaded side and of the same magnitude but reverse direction (see Fig. 5.42b). Note that the axial forces at the points of maximum moment are equal to each other in magnitude. It can be seen that the axial compressive force at the nonloaded side must be equal to the resultant reaction R_B, which acts along the funicular line and moves parallel so that it is tangential to the arch at the maximum moment location where the shear is zero.

$$N_{L/4} = N_{3L/4} = R_B = \sqrt{B_v^2 + B_h^2} \qquad (5.81)$$

The maximum shear forces appear at either the crown hinge or at the support, as discussed for inclined beams. The parabolic arch responds to uniform lateral loading, as indicated in Fig. 5.42c by the difference between the funicular line and the arch, assuming conservatively the load to act on the vertical projection of the roof. But, since the wind pressure distribution for an arch is considered less critical than for an A-frame of the same height-to-width ratio, the curvilinear beam may be replaced for preliminary design purposes by an imaginary straight inclined beam so that the maximum moment due to the lateral load is approximately equal to

$$M_{max} = qh^2/8 \qquad (5.60)$$

Ribbed domes consist of radially arranged arches that are laterally supported by a tension ring at the base and a compression ring at the crown. For preliminary design purposes, these arches can be treated as three-hinge ribs carrying a triangular gravity load as shown in Fig. 5.43a. The maximum moment for parabolic arches (see Problem 5.24 and Fig. 5.41f) is equal to

$$M_{max} = wL^2/162 \qquad (5.82)$$

It is clear that asymmetrical live loads must also be considered during the preliminary design stage. When a ribbed dome is supported by two main diagonal three-hinged arches (crossing at right angle at the center), which support secondary two-hinged arches, the main arches must not only support the gravity loads transferred by the secondary arches but also the thrust generated by them (see Fig. 5.43b). The reader may want to investigate the preliminary sizes of glulam arches assuming a roof dead load of 30 psf for the design of the main arches and 25 psf for the design of the secondary arches. For further discussion of domes, refer to Section 8.3.

Circular Arch (Fig. 5.44).   Under uniform, vertical loading across the full width of the arch, the parabolic pressure line does not coincide with the geometry of the circular arch. As indicated for the right arch portion in Fig. 5.44a, the funicular line is located inside, thus causing negative moments along the entire arch.

The reader may want to refer first to the section on the circular cable (Section 9.2), introducing the geometry of the circle.

The reactions for the full gravity-loading case have already been determined for other frame cases, but are expressed here in terms of the angle θ (see Fig. 5.44a).

$$A_v = B_v = wL/2 = wR \sin\theta \qquad (5.83)$$

where $\quad \sin\theta = L/2R \qquad$ or $\quad L/2 = R \sin\theta$
$$\cos\theta = (R - h)/R \quad \text{or} \quad h = R(1 - \cos\theta)$$

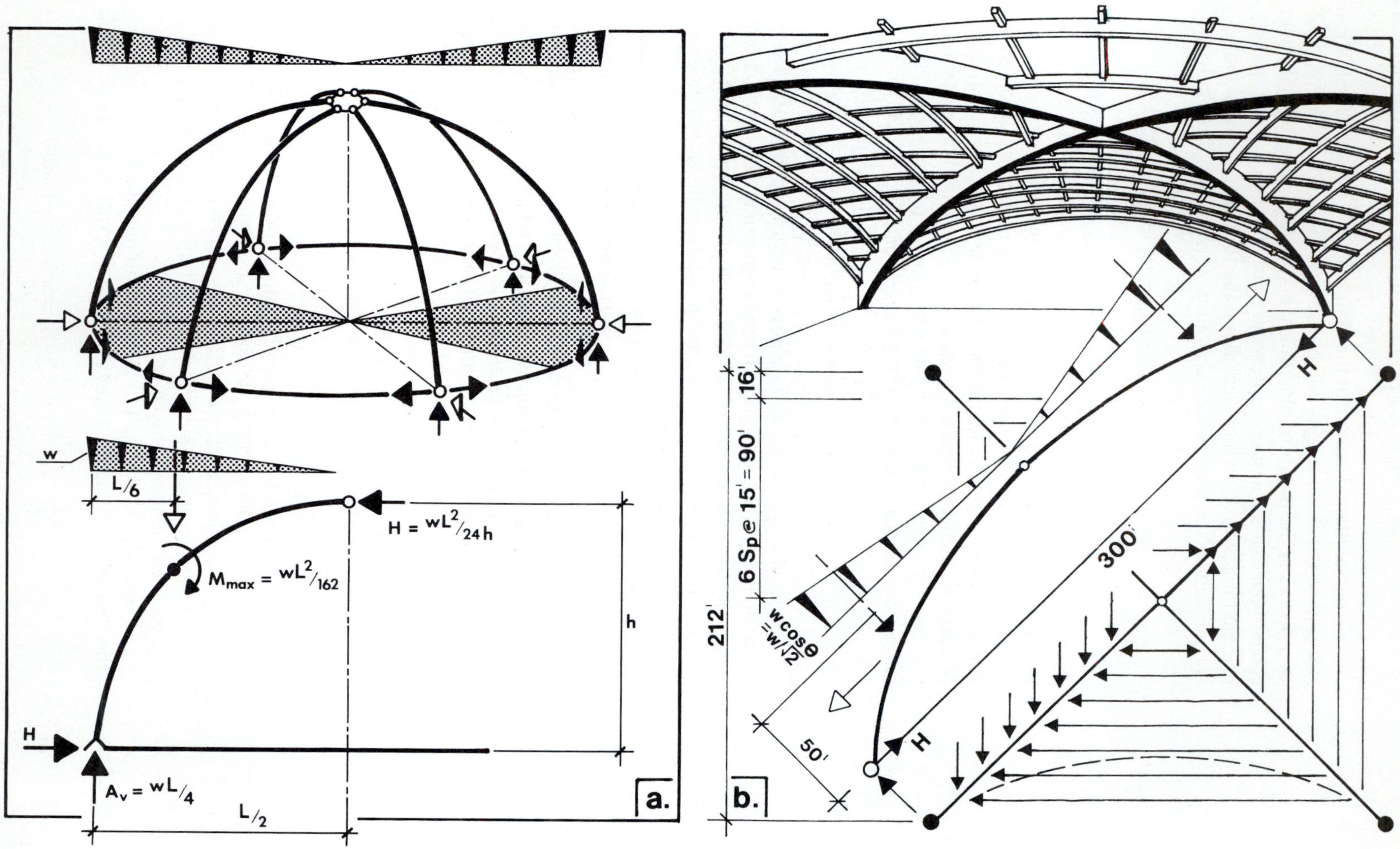

a.
w
L/6
H = wL²/24h
M_max = wL²/162
h
H
A_v = wL/4
L/2
b.
16'
6 Sp@ 15' = 90'
212'
w cosθ = w/√2
300'
H
50'

Figure 5.43 Ribbed domes.

$$A_h = B_h = \frac{wL^2}{8h} = \frac{w}{2} \frac{(R\sin\theta)^2}{R(1-\cos\theta)}$$

but $\quad \sin^2\theta = 1 - \cos^2\theta = (1 - \cos\theta)(1 + \cos\theta)$

$$A_h = B_h = H = \frac{wL^2}{8h} = \frac{wR}{2}(1 + \cos\theta) \qquad (5.84)$$

It can be shown that the maximum moment is equal to (see Schueller, 1983)

$$M_{max} = -wh^2/8 \qquad (5.85)$$

Note that the maximum moment is equal to the one for a simply supported beam of a span equal to the arch height, which in turn means that it must appear at mid-height.

For a semicircle, the maximum moment at $y = h/2 = R/2$ and at a horizontal distance of $x = 0.067L$ from the support is

$$M_{max} = -wR^2/8 = -wL^2/32 \qquad (5.86)$$

The maximum moment for the semicircular arch is twice as large as for the parabolic arch by considering only live loading. Should the dead load be one-third of the total loading, the M_{max} for the semicircular arch is three times larger than for the parabolic arch! As the circular arch becomes flatter, the effect of the height h as the moment determinant decreases and the effect of the span L increases, and the arch approaches the geometry of the parabola, indicating that the design will be controlled by the asymmetrical, uniform loading case.

We may conclude that *steep circular roof arches,* often roughly approximated as $h/L > 1/3$, are designed for full uniform gravity loading, keeping in mind, however, that the effect of wind must be checked when $h/L > 1/3$, although it does not have to be considered together with snow for preliminary design purposes. Wind loading is treated in the same fashion as for parabolic arches.

Shallow circular roof arches with a rise-to-span ratio of $h/L \leq 1/8$ can be treated as parabolic arches where moments are only generated by one-half live loading.

For the rise-to-span ratios in the range between steep and shallow roof arches, circular arches may be considered as parabolic arches for first-approximation purposes, although the effect of dead load causing bending must be considered. This change in geometry greatly simplifies the calculations, since the live load moment is a constant and equal to $-w_L L^2/64$, while the dead load moment is conservatively

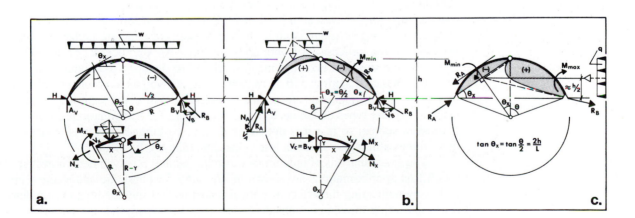

Figure 5.44 Circular arch under uniform load action

PARABOLIC   ARCHES   OR   SHALLOW   CIRCULAR   ARCHES (h/L $\leq$ 1/5)

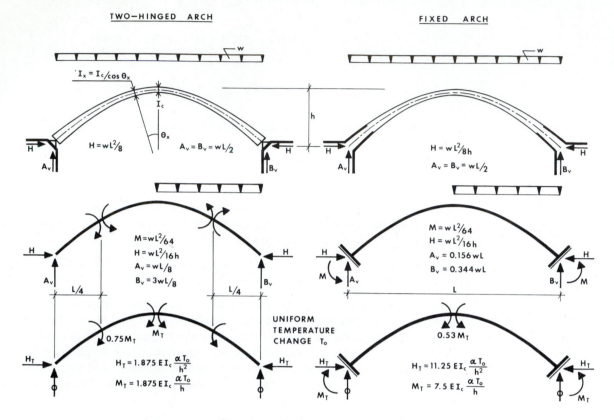

Figure 5.45   Indeterminate arches.

assumed as $-w_D h^2/8$, ignoring safely the difference in location between the two moments.

The axial force at the maximum moment location for the asymmetrical loading case (Fig. 5.44b) is equal to the resultant reaction R_B.

$$N_{\text{at} M_{\min}} = R_B = \sqrt{B_v^2 + H^2} \tag{5.87}$$

Indeterminate Arches.   When the hinge at the crown of the three-hinged arch is eliminated, a once indeterminate, two-hinged arch, redundant to the first degree, is generated. When, in addition, the support hinges are made rigid, a three times indeterminate, hingeless or fixed arch is created. Many steel and some timber and prefab concrete arches are two hinged, while most cast-in-place concrete arches are hingeless because of the process of continuous casting. As the degree of indeterminacy increases, the structure becomes increasingly more sensitive to boundary movements such as foundation settling and tie rod spreading, as well as to volume changes due to variations in temperature, creep, and shrinkage. While the determinate arch can respond to these types of actions by moving freely (i.e., ideal situations are assumed as defined in statics) so that it is not stressed, the two-hinged arch must resist some of these forces although it is much less vulnerable than the fixed arch.

The exact analysis of indeterminate arches cannot be given here; however, some rough first approximations are shown in Fig. 5.45 for parabolic and shallow circular arches by neglecting the effect of critical placement of live loading, elastic deformations, and other movement considerations. It is surprising that the continuity at the crown has hardly any effect on the lateral thrust of the two-hinged arch. For the asymmetrical loading case, the point of contraflexure appears approximately at the crown so that the moment at this location is zero, indicating that the thrust force has hardly changed. We may conclude, for preliminary design purposes, that the two-hinged parabolic and flat circular arches can be treated as three-hinged ones. Even for the fixed arch the lateral thrust is not much larger and may be considered to be equal to the one in the three-hinged arch.

Preliminary Design of Common Arches

The rise-to-span ratio has a great effect on the structural design of roof arches. As the ratio decreases, the arch behavior changes from a predominant *bending system* for steep arches to a predominant *beam–column system* (with much larger axial force action) for shallow arches. For parabolic arches, the critical moments are only a function of span (e.g., $M = w_L L^2/64$). For circular arches, the effect of height on the magnitude of the moments decreases as they get flatter and is replaced by the span as the critical moment determinant (i.e., approaching parabolic arches). It is apparent from $H = M/h$ that the lateral thrust and hence the axial force action increase with decrease of height, thereby intensifying the beam–column action. Generally, bending (and not lateral thrust) causes the largest stresses in roof arches and controls the design at critical sections; no moment magnification is required for preliminary arch design. The size of light steel arches may, however, be controlled by buckling about the main axis.

For the preliminary design of steep roof arches under uniform load action (say $h/L \geq 1/3$), the critical moment may be increased by about 10% to take axial action into account. But for fast estimation purposes for shallow roof arches ($h/L \leq 1/8$), the moment may be increased by about 30% to take the much larger thrust forces into account. For arches in the intermediate rise-to-span ratio, the moment may be increased by 20% to obtain a first estimate of member size at the critical section.

For three-hinged arches, the member sizes at the hinges are smaller since they do not have to resist bending (see Example 5.11). For material with a low tensile capacity, such as masonry, the dead load (or arch depth) must be increased to decrease the live load-to-dead load ratio, thereby making it possible to keep the pressure line within the middle third or *kern* of the section and thus preventing tension from developing (Fig. 5.47). *Funicular arches*, naturally, are designed as columns for pure axial action.

The allowable bending stresses for glued laminated curved members must be slightly reduced because of the effect of curvature, as discussed previously [Eq. (3.37)]. In addition to flexural stresses, radial stresses are generated in curved members. It is shown in Fig. 5.46a that the abrupt change in direction of the members causes diagonal forces in a square or tapered haunch (which are resisted by a diagonal stiffener, for example) and uniform radial forces in a curved haunch (Fig. 5.46c). In curved glulam members, radial stresses perpendicular to the grain are generated. They are in tension when the bending moments tend to straighten the member, as in Fig. 5.46d (i.e., tend to increase the radius of curvature), but otherwise are in compression. The *radial stress* caused by a moment in a curved member of constant cross section can be derived as follows:

The resultant tension and compression forces due to bending in a W-section and rectangular timber section, as shown in Fig. 5.46a and d, respectively, are

$$T = C = M/d, \qquad T = C = \frac{M}{2d/3} = \frac{3M}{2d} \tag{a}$$

The resultant radial force for a tapered haunch according to Eq. (9.70), letting $\alpha = \theta/2$, is

$$F_r = 2T \sin \alpha \quad \text{(lb)} \tag{b}$$

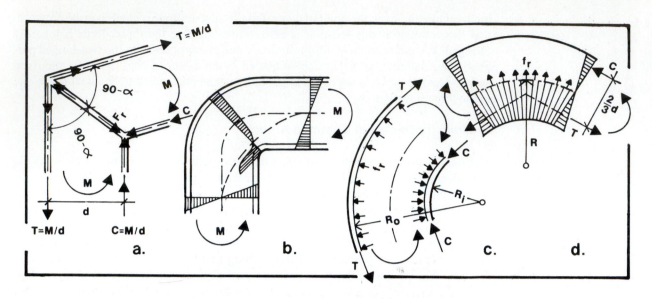

Figure 5.46 Radial forces and stresses.

The resultant radial unit force for a curved member according to Eq. (9.57) is

$$F_r = T/R \quad \text{(lb/in.)} \tag{c}$$

Substituting Eq. (a) into Eq. (c) and dividing the radial unit force by the resisting cross-sectional area at the center line of the section yields the radial tension or compression stress f_r (at that location), which is caused by a bending moment in a curved member of constant rectangular cross section (Fig. 5.46d).

$$f_r = \frac{F_r}{1(b)} = \frac{T}{Rb} = \frac{3M}{2Rbd} = \frac{3M}{2RA} \tag{5.88}$$

where R = radius of curvature at center line of member (in.)
   $A = bd$ = cross-sectional area of member (in.^{2})
   M = bending moment (in. lb)

When M tends to increase the curvature (decrease R), $f_r \leq F'_{c\perp}$. When M tends to decrease the curvature (increase R), $f_r \leq F'_{rt}$. The critical radial tensile stress for southern pine, for instance, is limited to one-third the usual design value in horizontal shear for all loading conditions: $F'_{rt} = F'_v/3$.

The thickness-to-span ratio of arches varies with the arch profile and the live load-to-dead load ratio. Typical wood arches seem to be in the range of $d/L \cong 1/40$ to $1/50$, whereas steel arches may be estimated as roughly $d/L \leq 1/70$. The proportions of ordinary reinforced-concrete arches may be assumed between the two materials. Typical height-to-span ratios for roof arches seem to be in the range of $1/3$ to $1/5$. Keep in mind that the costs involved to resist large thrust forces for shallow roofs, requiring buttressing and proper foundations, may be substantially higher than savings in the roof area.

EXAMPLE 5.13

A cylindrical roof is supported by parallel, three-hinged laminated wood arches of constant cross section. The arches are 33 ft high, span 90 ft, and are spaced 16 ft apart. They are laterally braced by purlins spaced at 8-ft intervals measured along the arch. They must support a dead weight of 15 psf, which includes the arch weight, a snow load of 30 psf, and a wind load of 20 psf assumed to act on the vertical roof projection.

For preliminary estimation purposes, the approximate size of a typical interior arch is determined using laminated timber construction of the western species type with $F_b = 2400$ psi and $F_r = 15$ psi (radial tension stress for loading conditions other than wind and earthquake). The arch must support the following uniform loads:

$$w_D = 15(16) = 240 \text{ lb/ft} = 0.24 \text{ k/ft}$$
$$w_L = 30(16) = 480 \text{ lb/ft} = 0.48 \text{ k/ft}$$
$$w_W = 20(16) = 320 \text{ lb/ft} = 0.32 \text{ k/ft}$$

First, a circular arch will be investigated and then a parabolic one.
(*a*) *Circular arch.* The rise-to-span ratio of the arch is

$$h/L = 33/90 = 1/2.73 > 1/3$$

Hence, the arch can be treated as steep, and its preliminary design can be assumed to be controlled by full gravity loading (where the snow load reduction is conservatively neglected), although the effect of wind must be checked.

First the needed geometrical relationships for the circular arch are determined. The radius of curvature according to Eq. (9.64) is

$$R = (L^2 + 4h^2)/8h = (90^2 + 4(33)^2)/8(33) = 47.18 \text{ ft}$$

The slope at the reactions according to Eq. (9.63) for $x = 0$ is

$$\tan\theta_o = (L/2)/(R - h) = 45/(47.18 - 33) = 3.17 \text{ or } \theta_o = 72.51°$$

The arch length according to Eq. (9.59) is

$$l = \pi R\theta°/90° = \pi(47.18)72.51/90 = 119.42 \text{ ft}$$

The dead load is acting along the arch and not normal to the member axis as for horizontal beams. It is convenient to project this load onto the horizontal roof plane. As has been shown for the inclined beam, the transformed load is equal to $w = w'/\cos\theta_x$. Since $\cos\theta_x$ is a variable and dependent on the arch geometry, the dead load on the horizontal roof projection must be curvilinear. To simplify the calculations for the preliminary design, first the total dead weight is found by multiplying the load per foot by the arch length l and then resolving the total load again into a uniform load on the horizontal roof projection by dividing it through the span L.

$$w_D = w_D'(l/L) = 0.24(119.42/90) = 0.32 \text{ k/ft} \tag{5.89}$$

The maximum moment due to full gravity loading, Eq. (5.85), taking into account load duration, is

$$-M_{D+L} = wh^2/8 = [(0.32 + 0.48)/1.15]33^2/8 = 94.7 \text{ ft-k}$$

The approximate maximum moment due to wind pressure, ignoring conservatively the minus moment due to dead load $-w_D h^2/8$ (and considering the effect of load duration), and treating the windward face as critical, is

$$M_W = qh^2/8 = (0.32/1.60)33^2/8 = 27.23 \text{ ft-k}$$

The full gravity-loading case clearly controls the preliminary design. The required section modulus, taking axial action into account by adding about 10%, is

$$S \cong 1.1(M/F_b) = 1.1(94.7)12/2.4 = 520.85 \text{ in.}^3$$

Try $6\frac{3}{4}$- $\times$ $22\frac{1}{2}$-in. section (Table A.10) with $S_x = 531.1$ in.^{3}, $A = 151.9$ in.^{2}, and $b/d = 6.75/22.5 = 1/3.33 > 1/5$.

(b) Parabolic arch. The slope at the reactions according to Eq. (9.15) is

$$\tan\theta_o = 4h/L = 4(33)/90 = 1.46 \quad \text{or} \quad \theta_o = 55.71°$$

The arch length according to Eq. (9.21) is

$$l = L[1 + (8/3)(h/L)^2] = 90[1 + (8/3)(33/90)^2] = 122.27 \text{ ft}$$

There is only about 2% difference in arch length between the circular and parabolic arch, which can be ignored for this first design stage. In addition, the dead load does not enter the preliminary design of the arch since it is assumed not to generate any moments. The design is controlled by bending due to one-half live loading. The maximum moment for this condition according to Eq. (5.80) is

$$\pm M_{max} = w_L L^2/64 = (0.48/1.15)90^2/64 = 52.83 \text{ ft-k}$$

The approximate maximum moment generated by the lateral wind pressure is

$$M_w \cong qh^2/8 = (0.32/1.60)33^2/8 = 27.23 \text{ ft-k}$$

The gravity case clearly controls the design. The required section modulus is

$$S_x \cong 1.1(M/F_b) = 1.1(52.83)12/2.4 = 290.57 \text{ in.}^3$$

Try $5\frac{1}{8}$- $\times$ $19\frac{1}{2}$-in. section (Table A.10) with $S_x = 307.7$ in.^{3} and $b/d = 5.125/19.5 = 1/3.81 > 1/5$.

EXAMPLE 5.14

Reduce the height of the arch in Example 5.13 from 33 ft to 18 ft. Compare parabolic and circular arches. The rise-to-span ratio of the arches is

$$h/L = 18/90 = 1/5$$

This ratio indicates that the arch is in the typical range between steep and shallow arches. For this condition, ordinary wind action can be ignored and the snow loads are not reduced. For preliminary design purposes, the circular arch can be treated as a parabolic one although the effect of dead load must be considered.

The arch length for a parabolic configuration, according to Eq. (9.21), is

$$l = L[1 + (8/3)(h/L)^2] = 90[1 + (8/3)(1/5)^2] = 99.6 \text{ ft}$$

The uniform gravity loads are

$$w = w_L + w_D = 16[30 + 15(99.6/90)]/1000 = 0.48 + 0.27 = 0.75 \text{ k/ft}$$

The critical moment for the parabolic arch is only a function of span. The decrease of arch height increases, however, the magnitude of the axial forces! The maximum moment due to live loads, according to Example 5.13, is

$$\pm M_{max} = w_L L^2/64 = 52.83 \text{ ft-k}$$

With respect to the preliminary design of the parabolic arch, the moment should be increased not by 10% as for steep arches, but by 20% to take the larger axial forces into account.

The maximum moment for the circular arch, as based on $D + L/2$ on the nonloaded side, is roughly approximated as

$$-M_{D + 1/2L} \cong w_D h^2/8 + w_L L^2/64 = (0.27/1.15)18^2/8 + 52.83 = 62.34 \text{ ft-k}$$

Notice that the critical moment for the circular arch is about 18% larger than for the parabolic arch. The required section modulus, using a 20% increase of the moment for the given rise-to-span ratio to take into account axial action, is

$$S \cong 1.2(M/F_b) = 1.2(62.34)12/2.4 = 374.04 \text{ in.}^3$$

Try $6\frac{3}{4}$- $\times 19\frac{1}{2}$-in. section (Table A.10) with $S_x = 405.3$ in.^{3}, $A = 131.6$ in.^{2}, and $b/d = 1/2.89 > 1/5$.

The maximum moment due to one-half snow load on the loaded side (ignoring conservatively the negative moment due to dead load) is positive and therefore causes a radial tensile stress. First the radius of curvature at the center line of the arch must be found so that the critical radial tensile stress can be determined according to Eq. (5.88).

$$R = (L^2 + 4h^2)/8h = (90^2 + 4(18)^2)/8(18) = 65.25 \text{ ft} \qquad (9.64)$$

$$f_r = \frac{3M}{2RA} = \frac{3(52,830)12}{2(65.25)12(131.6)} = 9.23 \text{ psi} < 15(1.15) = 17.25 \text{ psi}$$

Notice, that radial compressive stresses are generated by $M = -62.34$ ft-k.

Minor Masonry Arches

Masonry arches are often a continuous part of a wall rather than freestanding and therefore can be treated as fixed and three times statically indeterminate; they are difficult to analyze and beyond the scope of this introduction. Simplifying approximations, however, may be used for *minor arches*, that is, the typical short-span, shallow segmental and flat jack arches. These arches are considered to have spans of up to 6 ft and a rise-to-span ratio of $r/s \leq 0.15 = 1/6.67$.

Minor wall arches (see Fig. 5.47) are similar to lintels; they carry a small portion of wall weight as a triangular load, possibly a uniform floor or roof load, and occasionally light concentrated forces. Because these loads are of relatively small magnitude and because of the low tensile capacity of masonry, it may be assumed that the bending stresses in these shallow arches are small in comparison to the axial stresses and do not generate any tensile stresses. Furthermore, since the loads are arranged usually in a symmetrical manner and because of the high redundancy of the wall arch, we may conclude that it is reasonable to assume a pressure line for minor arches that is contained within the middle third or kern of the section, thus preventing tension from developing, as shown in Fig. 5.47.

The location of the pressure line at the upper edge of the kern at the crown identifies the maximum positive moment of the fixed arch, whereas the pressure line at the lower edge of the kern at the reactions identifies the maximum negative arch moments. At these critical locations the tensile stresses are assumed zero. In addition, because of the shallow nature of the minor arch, the maximum axial force N at the reactions is assumed equal to the thrust force H for preliminary design purposes.

$$f_t = f_a - f_b = H/A - M/S = 0 \qquad (a)$$

From the resulting triangular stress block represented by Eq. (a) as shown in Fig. 5.47, the maximum compressive stress can be easily derived.

$$H = [f_m(d)1/2]b \quad \text{or} \quad f_m = 2H/A \qquad (5.90)$$

where f_m = maximum compressive stress (psi)

A $= bd$ = cross-sectional area of the arch (in.^{2})

The magnitude of the thrust force can be obtained from the familiar expression $H = M/h$ by visualizing the pressure line as the resultant arch, where generally the

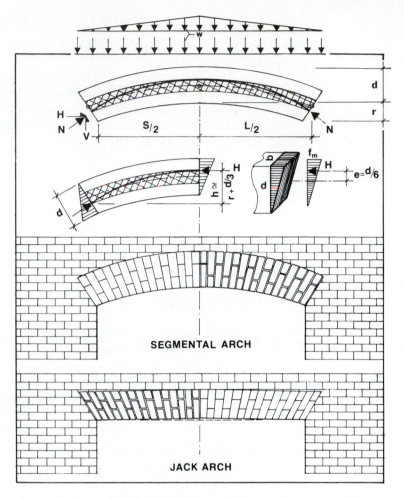

Figure 5.47 Minor masonry arches.

clear span of the opening S rather than the true span L is used as arch span. Therefore, for uniform loading the familiar thrust force is equal to

$$H = wS^2/8h = WS/8h \qquad (5.91)$$

where $h \cong r + d/3$ for shallow segmental arches
$h = d/3$ for flat jack arches

In the case of triangular loading w (due to the wall weight), the horizontal thrust force for uniform loading w must be increased by one-third.

Also, the maximum shear stresses along the skewbacks must be checked. Because of the shallow nature of the segmental arch, it may be treated as a parabola so that the slope of curvature at the reactions may be approximated according to Eq. (9.15) or (5.91) for uniform loading.

$$\tan\theta_o = 4h/S \cong W/2H \qquad (5.92)$$

Now, the shear V_o at the reactions can be determined as based on the expression in Fig. 5.4. The corresponding average shear stress is

$$f_v = V_o/A = V_o/bd$$

With respect to jack arches, literature on masonry should be consulted to determine the angle that the skewback makes with the vertical.

EXAMPLE 5.15

Do a preliminary design of a 12-in.-deep segmental arch in an 8-in. masonry wall that bridges an opening of 5 ft and has a rise of 6 in.; the arch must support a uniform load of 1140 plf (see Schueller, 1990, Example 6.6). Use a compressive strength of brick masonry of $f_m' = 2500$ psi, that is, an allowable compressive stress of $f_m = 0.2 f_m' = 0.2(2500) = 500$ psi.

The thrust force, according to Eq. (5.91), is equal to

$$H \cong \frac{WS}{8(r + d/3)} = \frac{1140\,(5)\,5}{8(6 + 12/3)/12} = 4275\,\text{lb}$$

The maximum compressive stress is approximately equal

$$f_m = 2H/A = 2(4275)/(12 \times 8) = 89 \text{ psi} \leq 500 \text{ psi}, \quad \text{OK}$$

The shear stresses along the skewback must be checked, as well as the shear stresses due to the lateral thrust in the adjacent supporting walls.

5.5 TRUSSES

Trusslike support structures for timber bridges and pitched roofs have existed for a long time. According to Vitruvius in the first century B.C., the Romans already used inclined rafters for some of their roofs, tying them together at the base to eliminate the lateral thrust on the wall supports. In the Middle Ages the steeply pitched timber roofs consisted of rafters with horizontal collar beams, vertical posts, and diagonal bracing that give the appearance of trusslike structures with incomplete triangulation. An outstanding example is the 69-ft-span *hammer beam truss* at Westminster Palace (1402). This complex, highly indeterminate roof structure can be visualized as the composite interaction of the trusslike A-frame and arched ribs. Leonardo da Vinci seems to be the first to actually propose a truly triangulated truss, but Andrea Palladio about 100 years later is usually credited as the inventor of modern trusses. He described triangulated timber trusses for bridges in *The Four Books of Architecture* (1570), used for spans up to 100 ft.

The domestic triangulated pitched roof trusses were developed from the simple *king-post truss* (i.e., triangular arch), with a central post (Fig. 5.49a) acting as a tie to prevent the sagging of the horizontal bottom chord, as well as from the *queen-post truss* with two posts. An example of the development of the king post concept is Christopher Wren's unique truss solution for the Sheldonian Theatre in Oxford (begun 1664). The 72-ft-span trusses, spaced 18 ft apart, have three posts, but not perfect triangles as suggested by Palladio; they resemble a modern Howe truss, similar to Fig. 5.49d but with shallower top chords for the center portion. The truss profile matches closely the bending moment diagram due to uniform gravity loading, the benefit of which, however, Wren could not have realized.

Parallel chord beamlike trusses were developed during the first half of the nineteenth century for bridges. During the early part of the century, wood and iron were often combined by letting the timber carry compression and iron the tension before timber was replaced in bridges by wrought iron and later steel.

The true evolution of trusses began in the first half of the nineteenth century, initiated primarily by railway construction in the United States and Russia. Some pioneering design engineers were W. Howe (1841), T. Pratt (1844), A. Fink (1852), and

S. Whipple in North America, as well as J. Warren (1848) in England, just to name a few. The development of the truss theory follows closely and is credited to S. Whipple (1847) and D. Jourawski (1850) for the *method of joints,* A. Ritter (1862) for the *method of sections,* K. Culmann (1864), C. Maxwell (1864), L. Cremona (1872), and R. Bow (1873) for the *graphical solution of trusses,* and A. Möbius (1837) and O. Mohr (1874) for the *determination of the statical determinacy of trusses.*

In the second half of the nineteenth century, engineers employed iron and steel truss beams and truss arches extensively. They were not only used on bridges, but for other long-span structures such as market halls, exhibition buildings, and railway stations, as is discussed elsewhere in this chapter. The imaginative application of the truss principle probably reached its high point with Gustave Eiffel's tower for the Paris Exhibition in 1889. But, the truss concept was not only applied to engineering structures. For example, the latticed iron arches carrying glass domes over the reading room of the Bibliothèque Nationale in Paris (1868), designed by Henri Labrouste, are an integral part of the architectural space.

Basic Truss Characteristics

Some important truss characteristics are identified in Fig. 5.48. They are organized according to geometry, application, and statical considerations. There is no limit to the geometrical layout of trusses. Standard trusses are named after the engineers who introduced them. Among the important first truss types are the *Howe truss* (1841) using compression diagonals and vertical tension members as derived from Palladio's truss configuration, the *Pratt truss* (1844) with the diagonals in tension and the verticals in compression, and the *Warren truss* (1848) in which the diagonals are alternately in tension and compression under uniform loading along the top chord (see also Fig. 5.49i to l). The original *Fink truss* (1851) had no bottom chord, and the tension had to be carried by diagnol members similar to overlapping king-post trusses (e.g. multi-strut cable-support beams).

Application of the truss ranges from the small scale of a joist to the large scale of a deep truss supporting a stadium roof. Trusses may replace any solid elements such as beams, columns, arches, or frames. From a shape point of view, they may be classified as *truss cantilevers, truss beams* of various profiles (flat, tapered, pitched, curved, crescent, etc.), *truss arches,* and *truss frames* forming single or multibay structures. Trusses may be organized according to the arrangement of members and according to behavioral considerations:

- *Simple trusses* are formed by the addition of triangular member units and can be further subdivided as shown in Fig. 5.48.
- *Compound trusses* are formed by the addition of simple trusses. Again, different variations are identified according to the way in which the trusses are connected. Keep in mind that these compound trusses are not necessarily composed of only triangles, indicating that stable trusses can be generated by figures other than triangles.
- *Complex trusses* are neither simple nor compound. Special methods of analysis must be applied to these trusses even if they are statically determinate, since more than three members are attached to each joint, and sections cut through at least four members result in more unknowns than available equations at the location to be investigated.

The selection of truss geometry is not a simple undertaking, even from a purely functional point of view. It is closely related to economical considerations, where least weight does not necessarily yield an optimum solution, since with increase of span, fabrication, transportation, and erection become controlling cost determinants. From

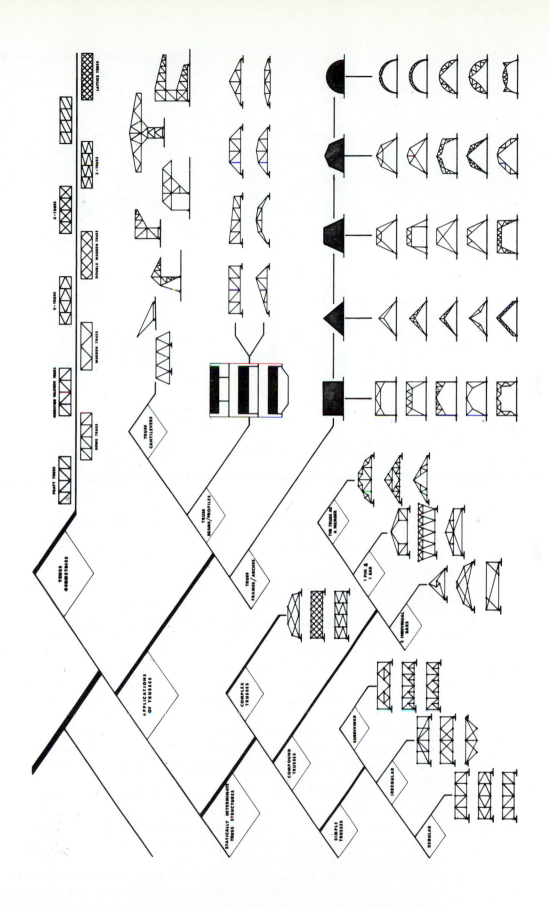

Figure 5.48 Introduction to trusses.

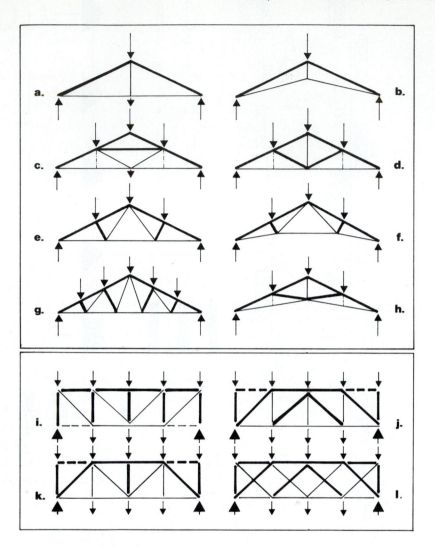

Figure 5.49 Basic triangular and rectangular trusses.

the point of view of optimum weight, the shape of the truss should be *funicular* with respect to the critical external loading case so that the chords carry all the loads and the web members are zero and thus are primarily used for lateral bracing of the compression chord; this condition is obviously only true for one loading case. For example, under typical uniform gravity-loading the curved bowstring truss and the crescent truss are examples of funicular trusses (Fig. 1.19). Keep in mind, however, that the shape of the roof, in turn, influences the external load action; a lower pitch results in a larger snow load but less wind pressure as compared to a steep roof with less snow but a higher wind pressure.

The truss system widely used for *flat roofs* and floors under ordinary loading conditions are the Pratt truss and Warren truss. Flat roofs are pitched at least 1/4 in./ft to allow proper drainage of water. This may be achieved by building up the roof surfacing material, by slightly sloping the top chord of the trusses, or by tilting the whole truss slightly when no ceiling is required.

Pitched roofs are introduced in Section 5.3. There it is shown that the three-hinge A-frame or the simple rafter roof is the most basic support structure for inclined roofs of short spans. It is also shown that for larger spans and steeper roofs more intricate rafter framing systems are required, as indicated in Fig. 5.30. Rather than bracing the A-frame internally with collar beams and vertical posts, full internal triangulation, that is, truss action, can be introduced (Fig. 5.49).

The simplest triangulated truss is naturally the A-frame (when loaded at the joints), which has no internal web members. When a vertical tie is added (Fig. 5.49a) to control the sagging of the light bottom chord member, the truss is called *king-post truss*. When the bottom chord is raised (Fig. 5.49b) to provide extra central clearance (e.g., vaulted ceiling), the truss is called a *Swiss truss*. As the span of the top chords increases, support may be provided by diagonal struts at midspan (Fig. 5.49c and d). Complete truss action is achieved for these cases by adding a horizontal compression strut or a vertical tension tie. A common support structure for pitched roofs is the *Fink truss* (Fig. 5.49e). The bottom chord of the Fink truss may be cambered as indicated in Fig. 5.49f. Another configuration of raised bottom chord construction is the *scissor truss* (Fig. 5.49h). It should be kept in mind that with the increase of truss span the number of web members must obviously also increase. For long-span pitched roofs, the *Belgian* (or double Fink, Fig. 5.49g), Pratt, and Howe trusses (Fig. 5.49i and j) are often used.

Besides, the typical parallel chord and symmetrical pitch and arch-type truss profiles, almost any other symmetrical or asymmetrical truss configuration is possible, as can only be suggested by the cases in Figs. 5.48, 5.50, and 5.51. Special truss types may have to be developed because of esthetical and unusual functional requirements or due to particular loading conditions.

In contrast to custom-designed trusses, allowing virtually any shape, are the ones produced by truss manufacturers. Most of the trusses used in the United States for standard construction are prefabricated and sold as patented industrial products. Typical are the open-web steel joists and the light-frame wood trusses used for residential and commercial applications with spans up to about 60 ft, made from dimension lumber where the members are connected by toothed connector plates. Common for flat roofs and floor construction are trusses of the Warren-type configuration, whereas Fink trusses are often used in wood roof construction.

The arrangement of the truss members determines which are in tension and compression. For instance, for the simply supported truss beams (Fig. 5.49, bottom) under uniform gravity loading along the top chord, the diagonals of the Pratt truss (i) are in tension and the verticals in compression, while for the Howe truss (j) this situation is reversed, and for the modified Warren truss (k), the diagonals are alternately in tension and compression, whereas the lattice truss (l) can be visualized as two superimposed single Warren trusses (i.e., double Warren truss).

The proximity of the web members depends on the unsupported length of the compression bars, which should not be too slender because of buckling criteria. We may conclude that the shorter truss bars should be in compression, while the longer ones should be in tension. The bending stresses in the compression chord should be kept small in comparison to the axial stresses by reducing the member length, or they should be eliminated by placing the purlins directly at the panel points. A deeper truss section will result in smaller chord forces, and smaller bays will yield steeper diagonals carrying less force. Their slope is kept in the range of 40° to 50° for flat trusses and 30° to 60° for pitched trusses.

Should the compression chord not be laterally braced by the roof structure, the chord must be laterally supported by transverse roof bracing, or space trusses (Fig. 6.13) may be used. The spacing of the trusses depends on the span capacity of the roof skin and determines the magnitude of the loads the trusses must support. Under normal conditions, lightweight materials should be taken to reduce the costs.

When continuous chord members are used, their sizes can easily be determined as based on the critical load action, as shown in the following example. Should cables be used for tension members in composite wood–steel trusses, care must be taken that under no loading condition will the members be in compression, since the structure may collapse. In other words, for truss members that only experience tension under all loading conditions, steel cables can be used such as for cable-supported beams (see Fig. 9.12). Also refer to Fig. 9.11 for the discussion of prestressed cable trusses.

Figure 5.50 Trusses and buildings.

Case Study. The application of trusses to architecture is suggested by the cases in Fig. 5.50. The trusses range from simple beams with flat or pitched profiles to complex framing systems: (b), (c), and (j). They range from the relatively small scale building (j) to the large-scale stadium with a span of 440 ft (k); they range from standard geometries to individual designs (h). Case (g) demonstrates cantilever trusses; folded truss configuration is used in case (b); the building (case j) is raised on steel cantilever trusses supported on concrete piers. These steel trusses carry the building including the roof wood trusses.

The roof for the Port Authority Bus Terminal in New York (1962, Fig. 5.50i), designed by Pier Luigi Nervi, consists of the main diagonally placed, triangular, reinforced-concrete trusses supported by latticed side walls and central columns. The Air Force Academy Chapel at Colorado Springs, Colorado (1961, Fig. 5.50a), by Walter Netsch of SOM, is constructed of 75.ft-long trussed steel pipe tetrahedrons, which form in the assembled stage two inclined, triangular folded, spatial surface structures of A-frame cross section 151 ft high. The typical tetrahedron bay is 14 ft × 84 ft.

The exposed skeletal steel structure for the four-level parking facility on top of the Veterans Memorial Coliseum in New Haven, Connecticut (1972, Fig. 5.50e), designed by Kevin Roche John Dinkeloo & Associates, consists of nine 340-ft-long, 34-ft-deep cantilever trusses spanning the distance of 184 ft between the massive reinforced concrete piers, which are spaced 62 ft on center except over the arena.

The skylight roof structure of Gund Hall, Harvard's Graduate School of Design (1972, Fig. 5.50f), designed by John Andrews, is supported by nine 11-ft-deep, 134-ft-span inclined steel pipe trusses spaced 25 ft apart. These primary trusses are connected by smaller ones that carry the roof panels and glazing.

The asymmetrical stepped profile of the Rainbow Center Winter Garden at Niagara Falls, New York (1977, Fig. 5.50c), designed by Cesar Pelli, is formed by five transverse, continuous steel roof trusses cantilevering at each end. They are supported on four post-tensioned, reinforced-concrete columns, which also act as vertical cantilevers to resist lateral loading. Twelve secondary trusses span 36 ft between the transverse trusses in the longitudinal direction.

The roof of the Joe Louis Sports Arena in Detroit, Michigan (1980, Fig. 5.50k), is supported by two huge 40-ft-deep, 440-ft-span steel trusses that form a stiff 58-ft-wide spine carried by end towers. The main trusses, in turn, support the side trusses with a maximum span of 134 ft.

An example of the new generation of megatrusses is the huge 60-ft-deep, approximately lens-shaped ones for the 493- × 378-ft roof of Chicago's United Center (1994), which is supported on 36 concrete columns along the perimeter of the building. The primary roof structure consists of a two-way multipost tied arch system where four queen-post, tied arch trusses span in the short transverse direction and two multipost ones in the longitudinal direction. The top chords consist of typically 13.5.ft-deep arched steel trusses and the bottom chords of downward sloping wide flange tension ties strutted at eight intersection points of the trusses by 60-ft-long compression web members. Secondary 13.5-ft-deep trusses span between the tied arch trusses and the perimeter columns, in turn supporting infill beams. The weight of the roof structure is 25 psf. Structural engineers for the project were Thornton–Tomasetti Engineers.

Most of the examples in Fig. 5.50 make the truss an important part of architecture, although usually building trusses are treated from a purely technological and functional point of view as a means of bridging space, as is typical of floor joists and roof trusses for housing, industrial buildings, and gymnasiums.

Assumptions for Truss Analysis. Planar trusses are rigid structures composed of straight bars that are connected to one another primarily to transmit external loads in axial force action; for the discussion of space trusses, refer to Chapter 6. The interpretation of the behavior of trusses, as expressed in their analysis, is based on the following simplified assumptions:

- Hinged connections with frictionless single pins: the partial rotational capacity of the connections and the fact, for example, that the bars are bolted or welded to thin gusset plates of steel trusses is ignored (e.g., Fig. 3.19c).
- Weightless members: the weight of the members causes bending, which is small in comparison to the axial load; therefore, it can be neglected.
- Load application at joints: loads, such as purlins, are often applied to the members and thus cause bending. The bending stresses for typical conditions are in the range of 10% to 20% of the axial stresses.
- Forces act in the same plane (e.g., single lap joint causes eccentricity).
- Members are assumed straight and of constant cross section; their centroidal axes coincide with the centroid of the joint.
- Truss members are slender with a negligible bending capacity.
- Small deflection theory: the truss displacement is small and hardly influences the magnitude of the force flow.

In an *ideal truss*, members only resist forces in an axial fashion and not in bending. The small flexural stresses in the *real truss* are called secondary stresses.

A truss is generated by first connecting three members to form a base triangle. To this triangle other triangles are added, that is, two bars for each joint. Hence, the total number of truss members m is equal to the initial three of the base triangle with three joints j, plus two members for each of the remaining $(j-3)$ joints.

$$m = 3 + 2(j-3) = 2j - 3 \qquad \text{(a)}$$

The number of bars given in the equation must be provided for the truss to be internally stable and determinate. Should there not be enough bars, the truss is unstable, but if there are more bars then needed, the truss analysis is indeterminate. Although this condition of the number of bars is necessary, it is not sufficient since the members must be properly arranged and, in most cases, form triangles. To prevent the collapse of the structure, either members and sometimes reactions may be added or frame action (i.e., rigid joints) must be introduced.

For a planar structure to be externally stable and determinate, three reaction forces ($r = 3$), which can be visualized as three equivalent axial members that should be neither parallel nor concurrent, must be provided. Should $r < 3$, the structure is externally unstable; if $r > 3$, the structure is externally indeterminate. The term 3 in Eq. (a) can be replaced by r so that the equation takes into account both internal and external conditions.

$$\begin{array}{lll} m + r = 2j, & \text{statically determinate; check stability} & \\ m + r < 2j, & \text{unstable truss} & \text{(5.93)} \\ m + r > 2j, & \text{statically indeterminate truss; check stability} & \end{array}$$

Checking the conditions for the Pratt truss in Fig. 5.53,

$$41 + 3 = 2(22) \quad \text{or} \quad 44 = 44$$

Since the number of bars and reactions is equal to twice the number of joints and since the members as well as reactions are properly arranged, the truss is stable and statically determinate internally and externally. Note that a truss that is internally unstable ($m < 2j - 3$) but externally redundant ($r > 3$) may still be stable as a total system ($m + r = 2j$), since the additional reactive conditions may, but not necessarily do, prevent the collapse of the structure.

Trusses, in general, are made of steel or wood and sometimes of reinforced concrete. Members of steel trusses are standard rolled sections such as angles, channels,

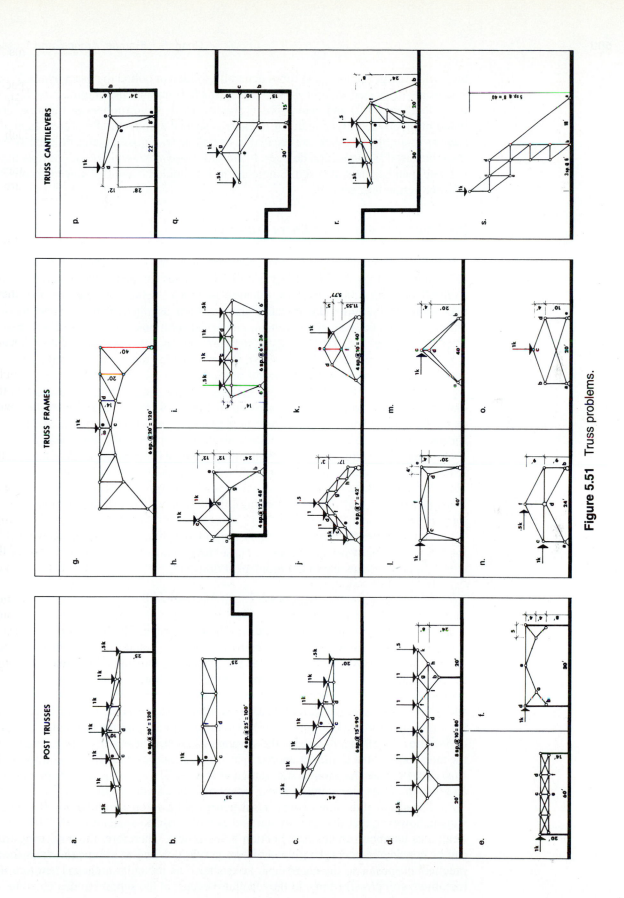

Figure 5.51 Truss problems.

499

tees, wide flange sections, and tubes generally welded or bolted to gusset plates. Wood trusses are organized as *light-frame trusses* (made from dimension lumber with spans of up to about 60 ft) and *heavy-timber trusses* (made from timbers or manufactured wood products) for larger spans or to express special architectural features. In *mono-chord trusses,* the members are located in one plane with gusset plates on each side of the joints used as connection devices. In *double-chord trusses*, single or multiple internal web members may be used. Single lap joints should be avoided since eccentricity of force action is generated.

Preliminary Design of Ordinary Trusses

It is not the intention here to introduce the analysis of trusses in general, which is done in Section 2.2 (Figs. 2.6f and 2.8g). Simple flat trusses can generally be solved by the *methods of joints* and/or *sections*, whereas graphical methods in the past may have been quicker for more complicated shapes, such as pitched and arched trusses, although today, because of computers, numerical solutions only are used. The reader interested in a review of truss analysis may refer to the truss enclosure systems in Fig. 5.51 and Problem 5.32. In this context, the primary interest is in the quick sizing of critical members of ordinary trusses for preliminary design purposes.

From a structural point of view, the truss configuration should reflect the funicular shape due to loading as is discussed in Fig. 1.19, Fig. 5.2, and elsewhere. For example, a truss should be curvilinear in response to uniform load action and pitched for point loading (e.g., triangular with respect to a single load). For these *funicular trusses*, the chord profile matches the bending moment (or pressure line) of the applied loading, thereby causing all interior members, independent of their arrangement, to be zero. Naturally, these trusses are only funicular under one set of loading. As the live loads change (i.e., asymmetrical loading), the web members are activated and are no longer zero.

Next we investigate typical simply supported ordinary trusses that carry uniform loads for preliminary design purposes. These trusses carry the familiar maximum shear at the support and the maximum moment at midspan.

$$V_{max} = wL/2, \qquad M_{max} = wL^2/8 \qquad (2.25)$$

The distribution of force flow depends on the truss shape. The basic truss profiles of flat, pitched, and curved trusses are shown in Fig. 5.52, and the efficiencies of their profiles are compared. The curvilinear profile can be considered optimal since it is nearly funicular for the given uniform loading. The chords of the flat truss are only used efficiently in the midspan range, as is indicated by the deviation of the truss profile from the pressure line (Fig. 5.52a). This, in turn, suggests the benefit of the shape for larger spans where bending controls. On the other hand, for the triangular truss, the chords are only efficiently used at the supports where the shear is maximum, indicating the advantage of the shape for shorter spans (i.e., shear beam action). The trapezoidal truss falls between the parallel and triangular trusses. The approximate design of chord and web members at maximum stress locations is briefly discussed next.

The shape of the *bowstring truss* is generally based on a circular arc. It has been shown, however, that shallow circular arches with a rise-to-span ratio $h/L \leq 1/8$ can be treated as parabolic arches (see Section 5.4, Arches). Therefore, the bowstring truss can be considered under the given uniform gravity loading as a funicular truss for all practical purposes, so the chords can be designed as funicular arches. Therefore, the maximum compressive force in the top chord occurs at the support and is equal to

$$N_{max} = R = \sqrt{V^2 + H^2} = H\sqrt{1 + 16\,(d/L)^2} \qquad (9.16)$$

And the constant tension chord force is

$$H = M/d = wL^2/8d \tag{5.94}$$

The web (or internal) members must resist the shear that is caused by asymmetrical live loading, as is discussed further later.

The maximum chord forces for *triangular trusses* occur at the supports (see Fig. 5.52c). Balancing the shear at that location (or using the *method of joints*, Example

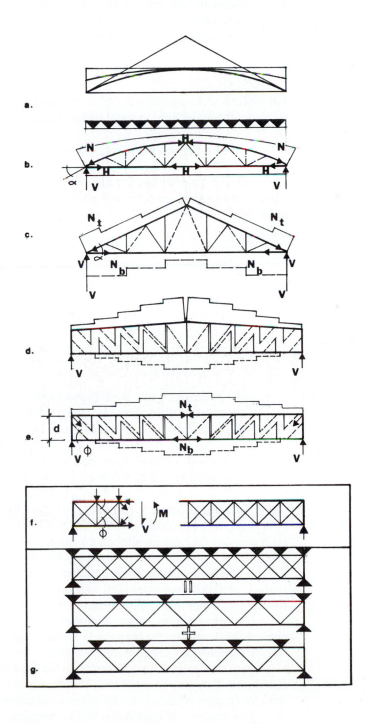

Figure 5.52   Response of ordinary trusses to uniform gravity loading.

2.7) yields the compressive force in the top chord N_t and the tensile force in the bottom chord N_b.

$$\Sigma F_y = 0 = V - N_t \sin\alpha \quad \text{or} \quad N_t = V/\sin\alpha = (wL/2)/\sin\alpha \qquad (5.95)$$

$$\Sigma F_x = 0 = V - N_b \tan\alpha \quad \text{or} \quad N_b = V/\tan\alpha = VL/2d = wL^2/4d \qquad (5.96)$$

For the preliminary design of *parallel chord* (or almost parallel chord) *trusses*, it is convenient to treat trusses as beams, where the chords are assumed to resist the moments and the web members the shear. Therefore, the critical chord forces N occur at midspan where the moments are maximum.

$$N_t \cong -N_b \cong M/d = wL^2/8d \qquad (5.97)$$

The critical web members occur at the reactions where the shear is maximum. For example, the maximum diagonal tensile force N_d is

$$\Sigma F_y = 0 = N_d \sin\phi - V \quad \text{or} \quad N_d = V/\sin\phi = (wL/2)/\sin\phi \qquad (5.98)$$

For the approximate design of indeterminate, *multiple-web, parallel-chord trusses*, such as cross-braced and lattice trusses, the following approach may be used, as based on the method of shears and moments:

Cross-braced trusses (Fig. 5.52f)
a. When the diagonals are long and slender, they may be assumed to carry only tension. Therefore, the diagonal in compression buckles and may be treated as a zero-force member.
b. When both diagonal members are stiff enough to resist compression then it may be assumed that the panel shear is resisted equally by the vertical components of the compression and tension diagonals, respectively (Fig. 5.52f).

$$\Sigma F_y = 0 = V - 2N_d \sin\phi \quad \text{or} \quad \pm N_d = (V/2)/\sin\phi \qquad (5.99)$$

Lattice trusses: Again, it can be assumed that the shear is resisted by the number of diagonal web members cut along the respective vertical section, in the same fashion as for the cross-braced truss. The chords resist the moments. The chord forces can also be obtained by visualizing the lattice truss decomposed into individual determinate trusses (such as into two single Warren trusses in Fig. 5.52g) and then adding the chord forces of the individual trusses.

The average height of ordinary steel and wood trusses can be estimated from the typical depth-to-span ratios for trusses, which are in the range of about 1:5 for triangular trusses to 1:12 for parallel chord trusses. As an approximation, an average truss depth of $L/10$ is often used for ordinary loading conditions, resulting in a ratio of $L/5$ for the maximum height of a symmetrical, triangular truss. The depth for closely spaced, prefabricated, open-web steel roof joists spanning up to about 150 ft and spaced not farther than approximately 10 ft apart is about $L/20$.

It may be necessary to compute the truss deflection to check the truss stiffness and possibly to determine the magnitude of the camber. Usually, trusses spanning further than 40 ft should be cambered (often camber is taken as about 1/1000 of the span), or for long spans the camber should be equal to the dead load deflection, although the bottom chord alone may only be cambered. The allowable live load deflection of ordinary long-span roof trusses is taken as $L/180$ unless elements sensitive to deflection are attached.

The analysis of truss deflection is complex, but it can be estimated for simple span trusses with parallel chords as follows. The deflection of a truss is caused by bending and shear, in other words, it is due to chord deflection and web deflection. The *web deflection* is caused by the axial deformations of the vertical and diagonal members and, according to Eq. (2.54), is equal to

$$\Delta_{web} = \Delta_v + \Delta_d = \Sigma f_v d/E + \Sigma f_d l_d/E \cos\theta \tag{a}$$

For example, for the Pratt truss in Fig. 5.53, $l_d = d/\cos\theta$. The sums of the web deflections in Eq. (a) represent the deflections of the individual truss panels between the support and maximum deflection at midspan (i.e., of $n/2$ panels). Hence, the maximum web deflection for a multipanel truss is approximately [see Schueller, 1990, Eq. (7.18), for derivation of the equation]

$$\Delta_{web} = (nd/2E) (f_v + f_d/\cos^2\theta) \tag{b}$$

The *chord deflection* can be estimated by treating the truss like a beam where the moment of inertia is provided by the horizontal chords (which are assumed of constant cross section for the entire length). According to Eq. (3.8), the beam (or chord) deflection is equal to

$$\Delta_{chord} = \frac{10 f_b L^2}{48 E d} \tag{c}$$

Hence, the maximum deflection of the flat, uniformly loaded, multipanel Pratt truss (with at least $n = 5$ panels) can be estimated as

$$\Delta = \Delta_{web} + \Delta_{chord} \tag{5.100}$$

$$\cong \frac{nd}{2E} (f_v + f_d/\cos^2\theta) + \frac{10 f_b L^2}{48 E d}$$

where f_v = axial compressive stress in vertical member
  f_d = axial tensile stress in diagonal member
  f_b = average axial stress in chord members = (tensile bottom chord stress + compressive top chord stress)/2
  E = modulus of elasticity

A quick deflection check can be performed by using conservatively the allowable design stresses for the members, rather than the actual member stresses.

EXAMPLE 5.16

A gymnasium roof is supported by steel trusses spanning 80 ft and arranged parallel (Fig. 5.53). The trusses must carry a snow load of 30 psf; the dead load of 20 psf includes the truss and bracing weight, which can be approximated roughly as 10% of the load it is supporting and is considered as 4 psf for this case. The wind pressure on the vertical surfaces is 20 psf. The preliminary, critical truss member sizes and column sections will be estimated in this example using A36 steel. For typical conditions, the spacing of the trusses is often considered roughly to be between one-quarter and one-fifth of the truss span, but not more than 20 ft. In this case, the spacing selected is $L/4 = 80/4 = 20$ ft.
(a) *Truss Design.* The depth of the trusses is estimated as $L/10 = 80/10 = 8$ ft. The small pitch of the trusses is neglected for this preliminary design. The design of the critical truss members is based on the uniform gravity-loading case for this flat roof structure. Remember, for steep roofs the combined wind and gravity-

loading case may be controlling, as was discussed for A-frames. The uniform load that the truss supports is equal to

$$w = (0.030 + 0.020)20 = 1 \text{ k/ft}$$

This load causes single forces at the interior joints.

$$P = 1(8) = 8 \text{ k}$$

The truss, as a whole, is subject to beam action, while the individual truss members only respond as axial elements. Here, for fast approximation purposes, the truss is treated as a beam, where the chords, similar to the flanges of a girder, are assumed to resist the moments, while the web members carry the shear as the stirrups and bent up bars do in reinforced concrete beams.

The stiffness (depth) of the trussed beam is so high in comparison to the supporting columns that its behavior is not restrained by the continuous columns and is allowed to rotate freely under gravity loading; hence, it can be assumed to act as a simply supported beam with a maximum moment at midspan of

$$M_{\max} = wL^2/8 = 1(80)^2/8 = 800 \text{ ft-k}$$

It is assumed that the moment is fully resisted by the chords and resolved into a force couple according to Eq. (5.97), as shown in Fig. 5.53b.

$$N_t \cong -N_b \cong M/d = 800/8 = 100 \text{ k}$$

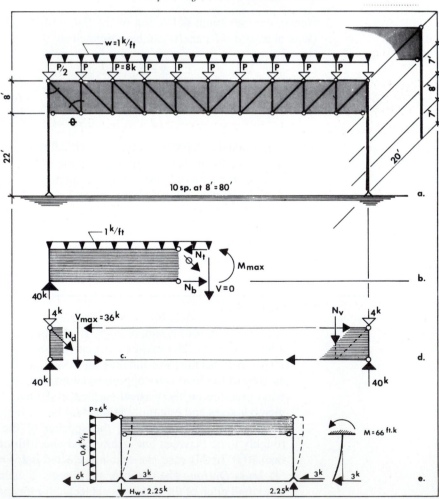

Figure 5.53 Example 5.16.

Considering the top and bottom chord forces to be equal to one another ignores the effect of the web members or the fact that the loads are acting at the nodal points (see Problem 5.31).

Note that in this approach the layout of the web members does not have an effect on the magnitude of the chord forces at midspan where the shear is zero. Hence, even indeterminate multiple-web trusses can be analyzed by this approximate method.

The top chord, in addition to the axial force, must also resist local bending as caused by the roof skin. It acts as a continuous beam supported at the joints and carrying the uniform roof skin load to the joints. The maximum beam support moment is approximated, according to Eq. (5.25), as

$$M = \frac{wl^2}{10} = \frac{(0.030 + 0.016)\, 20\, (8)^2}{10} = 5.89 \text{ ft-k}$$

The bending can be eliminated, if the purlins are placed directly at the truss joints, although this may be uneconomical because of the wide spacing.

Since the magnitude of the moment is relatively small as compared to the axial force, it is transformed by the bending factor into a fictitious axial load. The top chord is laterally supported about its weak axis by the roof skin, and the small effect of buckling about its strong axis is neglected. Hence, the section can be assumed to fail in yielding rather than buckling. According to Eqs. (3.99) and (3.100), the cross-sectional area is approximately equal to assuming a W8 section.

$$A = (P + P')/F_a = (P + B_x M_x)/F_a$$
$$= (P + B_x M_x)/0.6F_y = (100 + 0.33(5.89 \times 12))/0.6(36) = 5.71 \text{ in.}^2$$

Since the bending stresses for typical conditions are in the range of 10% to 20% of the axial stresses, for fast approximation purposes, the effect of bending may be taken into account by increasing the axial stresses conservatively by 20%.

$$A \cong 1.2P/0.6F_y = 1.2(100)/0.6(36) = 5.56 \text{ in.}^2 \qquad (5.101)$$

Try W8 $\times$ 21, $A = 6.16$ in.^{2}, $B_x = A/S_x = 6.16/18.2 = 0.338$.

In this case, the bottom chord does not have to be designed for bending, since it does not carry loads between panel points. As based on tension and yielding of the gross section (for preliminary design purposes), the following gross area according to Eq. (3.82a) is required:

$$A = P/0.6F_y = 100/0.6(36) = 4.63 \text{ in.}^2$$

Try W8 $\times$ 18, $A = 5.26$ in.^{2}.

The maximum shear adjacent to the supports is resisted fully by the web members. Its magnitude is equal to

$$V_{max} = 40 - 4 = 36 \text{ k}$$

This shear is carried by the diagonal and vertical web members. The vertical component of the diagonal force must be equal to the shear when using the beam approach (Fig. 5.53c). Hence, the diagonal tension force according to Eq. (5.98), simplified for the square truss bay, is

$$N_d = V/\sin\theta = V/(1/\sqrt{2}) = V\sqrt{2} = 36\sqrt{2} = 50.91 \text{ k} \quad (T)$$

The approximate cross-sectional area according to Eq. (3.82a) as based on yielding of the gross section is

$$A = P/0.6F_y = 50.91/0.6(36) = 2.36 \text{ in.}^2$$

Try $L3 \times 3 \times 7/16$, $A = 2.43$ in.^{2}.

As based on the method of sections, the vertical web member carries directly the shear in compression (Fig. 5.53d).

$$\Sigma F_y = 0 = N_v - 40 + 4, \qquad N_v = 36 \text{ k } (C)$$

For a slenderness of $Kl = 1(8) = 8$, according to the AISC column tables try $2L3 \times 3 \times 1/4$, $P_{all} = 36$ k, and $A = 2.88$ in.^{2}. At this stage some detailing should be done to see how the members fit together.

The approximate truss deflection can be estimated from Eq. (5.100) for $\cos\theta = 1/\sqrt{2}$ or $\cos^2\theta = 0.5$. The design stresses for the members are $f_d = 50.91/2.43 = 20.95$ ksi, $f_v = 36/2.88 = 12.5$ ksi, and $(f_b)_{av} = 100[(6.16 + 5.26)/2] = 17.51$ ksi.

$$\Delta = \frac{nd}{2E}(f_v + f_d/\cos^2\phi) + \frac{10f_bL^2}{48Ed}$$

$$= \frac{10(8)12}{2(29,000)}(12.5 + 20.95/0.5) + \frac{10(17.51)(80 \times 12)^2}{48(29,000)8(12)}$$

$$= 0.207 + 0.694 + 1.208 = 2.109 \text{in.} \leq L/240 = 80(12)/240 = 4\text{in.}$$

The truss deflection is OK considering the use of $L/240$ as allowable deflection is conservative.

(b) *Column Design.* For the case where the trussed portal acts as a transverse bent resisting lateral forces, the preliminary design of the columns is based on the dead load case together with wind loading under ordinary loading and span conditions, since under gravity loading only the deep truss is assumed to behave as a simply supported beam. The typical structural unit (Fig. 5.53e) must resist a lateral wind force of

$$P = 0.020(20 \times 15) = 6 \text{ k}$$

The response of the column to this force depends on how it is connected to the trussed beam. Typical possibilities are illustrated in Fig. 5.54. The column is either a truss, a single solid, or a built-up member. In case (a) the column and beam form a continuous truss unit. The magnitude of the axial force flow can be

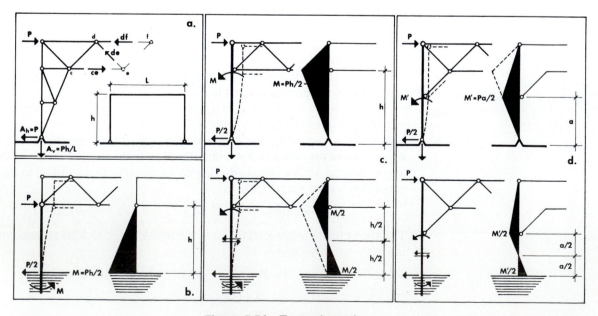

Figure 5.54 Trussed portals.

determined by any of the known analytical or graphical methods of solution for trusses. In all the other cases of Fig. 5.54 the trussed beam is attached to single-column members. To obtain an approximate solution for these cases, it is assumed that the trussed, deep beam is rigid and does not deform under lateral force action. It carries one-half of the wind or seismic load to the leeward column; in other words, the columns are considered to be of the same size and to share the loads equally. The lateral load P wracks the columns (Fig. 5.53e and Fig. 5.54c) and generates a moment of

$$M = (6/2)22 = 66 \text{ ft-k}$$

This moment may be resisted at the base by the cantilever columns of the post–beam structure (Fig. 5.54b) or at the location where the bottom chord of the truss is attached to the continuous column of the two-hinge frame unit (case c, top). Should the column be fixed to the base and continuous up to the top, similar to a fixed portal unit (case c, bottom), the moment can be assumed to be shared approximately equally at the base and the point of intersection with the bottom chord, since rigid beam behavior causes inflection points at midway between the column base and bottom of truss.

$$M = 66/2 = 33 \text{ ft-k}$$

The magnitude of the column moment can be further reduced by introducing knee braces (case d), which are diagonals extended from the truss to the column in order to decrease the column span or reduce the wracking distance of the column (i.e., its cantilever length). The points of contraflexure are approximated at midway between the column base and the bracing support.

The preliminary column size will be determined using A36 steel for the two-hinge portal unit (Fig. 5.54c, top). The column must carry an axial dead load of

$$N_D = 20(0.020)80/2 = 16 \text{ k}$$

The dead load moment in the columns caused by the deep beam is small compared to the wind load moment and is neglected. The axial column forces in tension and compression due to wind action are found by taking moments about one of the reactions.

$$6(30) = N_w(80) \qquad N_w = 2.25 \text{ k}$$

The critical column is on the leeward side, where the axial forces due to wind and gravity add. Hence, the column has to support the following axial load and moment:

$$N = 16 + 2.25 = 18.25 \text{ k}, \qquad M = 66 \text{ ft-k}$$

It has been shown in Example 5.2 that a column of a single-story frame can be treated as a beam for preliminary design purposes by selecting not the section required, but the next larger one, to take axial action into account. Since the compression flange of the column is laterally supported only at $L_b = 8$ ft, it is assumed that $F_b = 0.6F_y$. A W14 section is desired as column.

$$S = M/F_b = 66(12)/0.6(36) = 36.67 \text{ in.}^3$$

Try W14 × 30, $S_x = 42$ in.^{3}, $L_c = 7.1$ ft, and $L_u = 8.7$ ft $> L_b = 8$ ft (see Schueller, 1983, for further discussion of approximate column design).

5.6 LONG-SPAN SKELETON STRUCTURES

Any of the two-dimensional skeleton systems discussed in this chapter may be used for long spans of at least 100 ft; the spans, naturally, depend on the material and the structure system. Several special building types that employ the principle of the cantilever or the beam (bridge) are now investigated briefly.

The large scale of long-span structures may require unique building configurations quite different from traditional forms, as well as other materials and systems with more redundancy and nonconventional detailing techniques as compared to small-scale buildings. It requires a more precise evaluation of loading conditions than just provided by codes. This includes the placement of expansion joints as well as the consideration of secondary stresses due to the deformation of members and their interaction, which cannot be ignored any more, as for small-scale buildings or structures of high redundancy. Furthermore, a much more comprehensive field inspection is required to control the quality during the erection phase; postconstruction building maintenance and periodic inspection are necessary to monitor the effects of loading and weather on member behavior in addition to the potential deterioration of the materials.

We must also consider that the potential consequence of failure to a large number of people makes it mandatory that special care be taken in the design of long-span structures. It must be realized that the linear increase of dimensions does not result merely in the linear increase of building size and in the corresponding methods of construction, but that the larger scale generates new complex design determinants, as is discussed in Section 1.4.

Cantilever Structures

Cantilever structures range from the small scale of brackets, overhanging eaves, canopies, cantilevered balconies, corbeled arches, branches of trees, and other types of projections to the large scale of stadium and hangar roofs. Various cantilever beam types are introduced in Section 2.3 as cantilever beams, overhanging beams, hinge-connected cantilever beams (Gerber beams), and propped cantilevers. The effect of cantilevering is discussed briefly in Fig. 5.7 and the corresponding text.

The effort of cantilever action seems almost overcome by the daring, minimal bent tubular steel cantilever chairs of the late 1920s, such as Mies van der Rohe's *Mr Chair* (1927), and Marcel Breuer's *Cesca Chair* (1928). Similarly, Mies van der Rohe's famous *Barcelona Chair* of 1929, made of stainless steel bands in single and reverse curvature to form cross-frames, has become a lasting statement of the visual power of the cantilever. Pioneers of the *Modern Movement* elegantly expressed with the cantilever principle the floating character of support structure.

It was the spectacular Firth of Forth Bridge near Edingburgh, U.K., designed by John Fowler and Benjamin Baker (completed in 1890), that demonstrated for the first time the large-scale potential of the cantilever principle (similar to Fig. 5.7b). With its 1710-ft main span, it was the world's longest-span bridge at the time and symbolized a major achievement of Victorian engineering; its 680-ft-span cantilever arms are still among the largest ever built. This arched truss steel bridge expresses powerfully the cantilever beam principle; its form has been a source of inspiration for numerous designers.

Some typical examples of buildings using the cantilever principle are shown in Fig. 5.55. Frank Lloyd Wright considered the building as an integral part of nature, as an organism. He perceived the cantilever projecting out from the central building core as being similar to the branches of a tree. Continuity of the horizontal planes, so convincingly expressed in his Prairie architecture, such as the Robie House, Chicago (1909, Fig. 5.55a), made it possible to balance opposite loads in a natural manner. These overhanging horizontal surfaces generate an openness and lightness; they some-

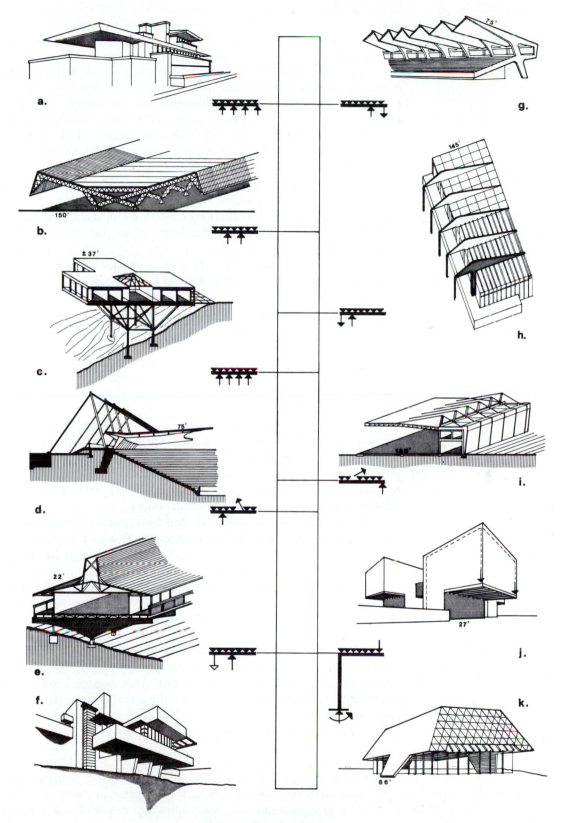

Figure 5.55 Cantilevers.

how allow the inner space to move outward. The integration of structure and space, or architecture and engineering, is most perfectly expressed in Wright's famous house *Fallingwater* (1936, Fig. 5.55f). Here the cantilevered concrete slabs projecting over the waterfall seem to be part of nature; they seem to grow out of the earth.

In general, house construction on steep hills is not economical; it may require extensive site corrections (excavations, grading, retaining walls, etc.) and possibly impair the stability of the site. Often, however, the cantilever principle provides an efficient solution. In case (c) the cross-shaped cantilevering building rests on four central columns with inclined arms, while in case (e), each of the parallel double cantilever plywood box beams sits on the main central concrete foundation, with a pipe column at the uphill side resisting unequal loading.

In case j, each of the cantilevering masses is supported by its side walls, which in turn act as vertical cantilevers with respect to gravity action. The continuous wall–roof space truss envelope of case (k) is only edge supported at three locations; it allows a column-free interior space and flexibility in the placement of openings and large overhangs.

The single- or double-cantilevered roofs for the airline hangars eliminate interior supports, thus resulting in an unobstructed floor area necessary for service of the airplanes. Similar are the roofs of grandstand structures, where the column-free cantilever system is an apparent solution. In case (g) the thin-shell sawtooth, hyperbolic paraboloid roof spans between the cantilevering frames, while in case (d) the cantilevering roof girders are supported from above by steel pipes in tension.

Hangars. The cantilever type of construction for aircraft hangars was developed during the 1950s; particularly, the double-sided cantilever or wing theme became popular (Fig. 5.56). At the beginning of the 1970s, large-scale hangars were built to shelter the giant Boeing 747s.

The cantilever concept permits maximum flexibility of operation since the interior space is free of any obstructions and since most of the columns along the perimeter are eliminated to provide the necessary freedom for movement. The facade doors are ground supported on base tracks, while the runners at the roof spandrel give the lateral support; the free deflection of the cantilever roof must, obviously, be taken into account in the design of the sliding hangar doors.

Hangars are built as single- or double-cantilever systems. In the symmetrical double-cantilever design, the dead load balances, which is not the case for a single-cantilever structure, where extensive anchorage must be provided to take care of the overturning moment. The central spine for double cantilevers contains the shops, stores, office facilities, and so on; it supports the roof and acts as counterweight for any asymmetrical loading. Often, the floors are suspended from the continuous roof structure, thereby relieving some of the stresses caused by the cantilever action.

The roof structure must not only resist the gravity and lateral loading, as well as effects due to temperature change, but also uplift forces. Special care must be taken in suspended construction when the roof weight is less than the uplift forces, with the cables not being able to carry compression. This consideration is not a problem for pure cantilever beam structures; however, these systems require deep members, resulting in much dead space.

There are basically two main groups for the typical cantilever systems shown in Fig. 5.56.

- Pure cantilever structures, possibly prestressed, using as members (or roof skin) tapering trusses, girders, folded plates, or shells
- Cable-supported cantilevers supporting any of the members just mentioned (see also Section 9.4)

In case g, cables support the corrugated shell roof near the outer edges and are anchored to deep concrete wall beams at the top of the central building portion. From

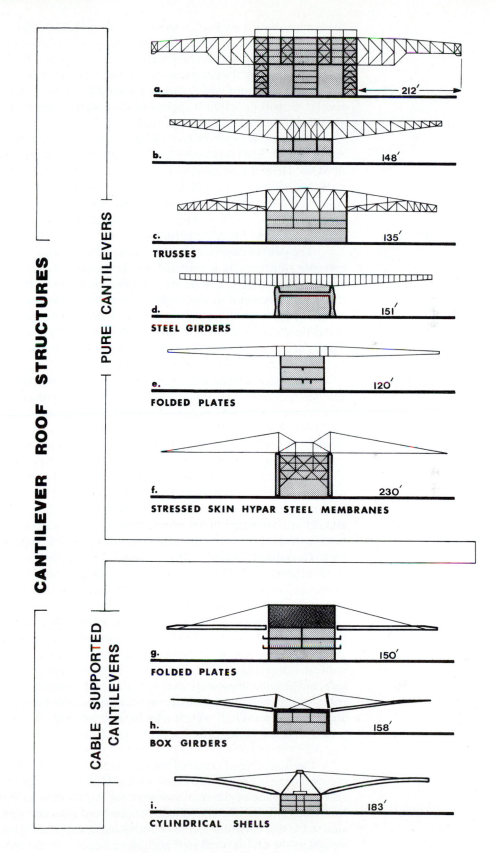

Figure 5.56 Air hangar cantilever structures.

these anchor walls are suspended the two floors below, thus causing the stresses due to the roof system to counteract. Corrugations give the thin-shell beam its necessary flexural rigidity. The cantilevering steel girders spaced at 32.5 ft of case (h) are pin connected to the central steel frame core and supported by stayed cables at 50 ft from the outer edge. The roof was designed for 40-psf live load and 30-psf uplift pressure. A similar support principle is applied to the concrete structure of case (i), where curved cylindrical shell beams are employed.

A powerful expression of cantilever architecture is the United Airlines maintenance hangar, San Francisco, California (1958, case d), designed by Myron Goldsmith of SOM. Here, the continuous welded plate girders are spaced at 51.5 ft and pin connected to the columns, which in turn are tied together and fixed to the ground. The form of the tapering columns corresponds to the moment variations under combined vertical and horizontal loading, similar to the tapering steel girders, which change in depth from 14 to 5 ft at the perimeter.

The double-cantilever hangar of case (e) employs a nonprestressed, folded plate concrete structure that is continuous over the central core portion. The roof portion of case (a) forms nearly a semicircle in plan view. Its 52-ft-deep radial cantilever steel trusses are anchored to the 11-story central circular core, which serves as counterweight to the huge projecting roof wings. Under hurricane conditions the hangar doors act as tie-down for the uplift forces. Other examples of continuous double-cantilever trusses are given in cases (b) and (c). The trusses increase in depth from the perimeter to inner column supports, following the magnitude of the moment flow.

Enormous spans were achieved by the cantilevered roofs of the maintenance hangars for American Airlines in Los Angeles and San Francisco, designed by Lev Zetlin (1971, case f). The folded plate roof cantilevers a distance of 230 ft on each side from the 100-ft-wide central steel trussed core. It consists of flat corrugated steel decking twisted into hyperbolic paraboloid shapes and welded to the restraining, rigid edge members along the valleys and ridges. The warped light-gauge metal acts as a shear membrane similar to the web of an I-beam, keeping in mind that shell behavior is very different from beam behavior; the unique geometry of the hyperbolic paraboloid carries the loads primarily in pure shear action within its surface, resulting in a much stronger and stiffer system. Furthermore, about 20% of the loads is resisted by prestressed cables running from the ridge at the core to various points of support along the valley members. The typical folded hypar module is of triangular cross section that tapers from a depth of about 40 ft. This stressed-skin membrane type of construction yielded a weight of about 17 psf, resulting in a saving of 40% over conventional truss systems.

Grandstands.   Grandstand structures must provide the necessary support for the roof and seating, as well as the access and exit passages. The required unobstructed view makes the cantilever system an apparent solution. As open structures, they are not just vulnerable to the static loading of downward gravity and upward wind pressure, but also to temperature changes and to the dynamic loading of wind gusting. For cable-stayed roofs, the tensile forces must be anchored within the supporting structure, and the downward roof weight may have to be larger than the potential upward wind pressure. Some historically important grandstand cantilever structures are shown in Fig. 5.57.

Pier Luigi Nervi attracted international attention for the first time with the asymmetrical concrete support structure for the Municipal Stadium of Florence, Italy (1932, case e). He planned the roof geometry for the stadium so that the resultant of all the roof loads falls between the primary support columns and thus does not cause any tension in the rear columns, thereby eliminating costly foundation anchorages. Should the weight of the cantilevered roof portion be larger than the one of the back portion, the rear support is in tension, causing the inner support to resist the entire weight and the tie load, thus requiring extensive massive piers. Also, for the Flaminio Stadium in

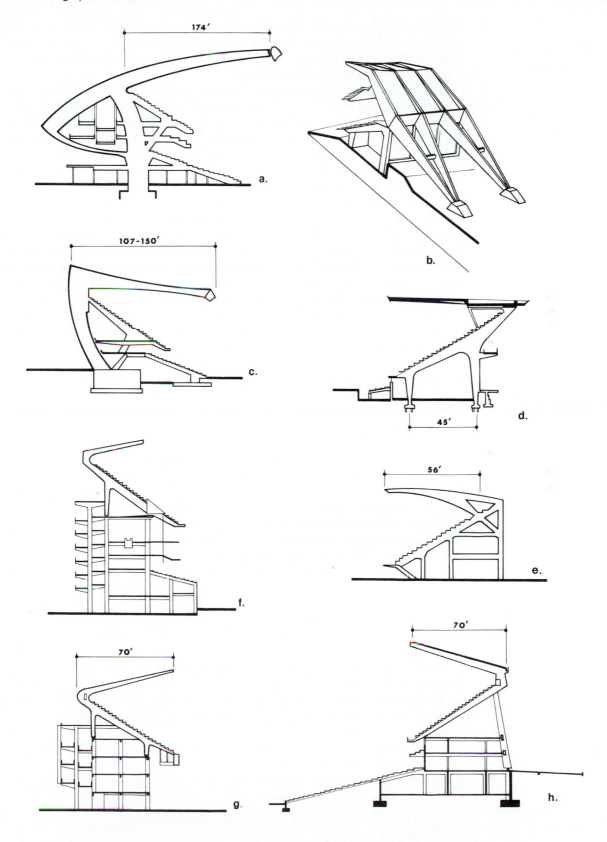

Figure 5.57 Grandstand cantilever structures.

Rome, Italy (1959, case d), Nervi selected the dimensions of the cantilevering folded plate roof, tapered front to back, as well as the location and inclination of the concrete-filled steel pipe columns, so that roof loads do not increase the stresses in the overhanging portion of the grandstand frame.

The cantilevering radial concrete ribs of the inverted L-shape for the Parc des Princes Stadium in Paris, France (1972, case c), designed by Roger Taillibert, are fixed to the ground and span the striking maximum distance of 150 ft; stands and other facilities underneath are structurally independent. The columns of these bents are of variable parabolic profile arched outward and are of a varying triangular cross section. The entire bent rib is composed of precast segments epoxy glued and posttensioned together as based on the Freyssinet system.

The Olympic Stadium in Montreal, Canada (1976, case a), designed by Roger Taillibert, is derived from similar structural and construction principles. The difference lies in the larger scale, the heavier loads due to the wider spacing of the radial bents, and the rear cantilever, with its hanging ramps that provide the counterballast to the roof rotation. The precast ribs support the spectator seating as well as the roof steel beams with the metal deck. As in Paris, the two-story compression ring at the top, which contains the lighting, sound, heating, and ventilation equipment, is not used as a lateral support for the cantilevering ribs; the large temperature differences in Montreal would have caused excessive stresses due to the continuity between the ring and cantilever arches. The ring, which supports a retractable fabric roof, causes a point load as high as 150 t at the ends of the cantilevers. Giant bent steel frames are formed by the cantilevering roof and the upper seating deck of cases (g) and (h); they are supported by the building below.

Beam Buildings

Considering an entire building to span between vertical cores, thus behaving as a gigantic beam, immediately brings to mind the urban megastructures proposed by the futurists of the 1960s. Architectural groups such as the Metabolists in Japan and Archigram in England or designers such as Yona Friedman, E. Schulze-Fielitz, and Paolo Soleri, just to name a few, employed the principle of bridge building for the design of their multilayer cities. Some of their design concepts can be found, on a smaller scale, in the offshore oil-drilling towns and the interstitial systems in hospital design.

The idea of using the bridge not only for traffic but also for buildings is not new. The Old London Bridge (1209, demolished in the nineteenth century) and the Ponte Vecchio (1367) in Florence are famous examples, where shops, housing, etc., were and are an integral part of the bridge. Present technology makes it possible that a bridge need not support the building, but, instead, that one beam building constitutes the bridge.

Bridge-type buildings are not necessarily part of a macro scale; they may only house a restaurant or provide a corridor for pedestrian movement. Again, there is no set limit to application of the concept. There may be many reasons for designing a building or a portion of a building as a beam structure, such as the following:

- To preserve special site conditions, such as existing buildings, or to leave the terrain below undisturbed
- To allow space for other activities, such as plazas, exhibition spaces, roads, and railways
- To link other buildings or to bridge valleys and rivers
- Marine structures

The necessity for the bridge-type building may be due to intensive use of land because of high real estate costs, or because of the specific location, or the desire for a view and open spaces. Beam buildings may be of various forms, ranging from platforms to linear tubular forms. The major structural elements may be wall beams, trusses, girders, frames (Vierendeel trusses), monocoque shells, or any other structural system. The beam buildings may be one or multistory systems; they may be supported by a single core or many cores that are arranged in many different ways; they may be continuous beams, cantilever beams, tree units, or simply supported individual beams. The supporting cores contain the vertical transportation and energy supply systems.

Some typical examples of beam buildings are shown in Fig. 5.58 and are discussed briefly. The building in case (d) consists of two parallel facade frames, each being made up of two columns and 6-ft-deep cantilevering floor and roof plate girders, which are tied together by the window mullions that serve as struts to equalize deflections between floor and roof level. The steel girders, in turn, carry the 90-ft trusses that are spaced 9 ft on center and support the concrete slabs. In case (a) the two huge, cantilevering, posttensioned concrete box beams contain the corridors and support the roof and floor framing between them. The fenestration pattern in the box beam reflects the layout of the prestress cables. Two-story Vierendeel facade trusses that are laterally hinged to the columns support the floor and roof structures in case (c).

The bridge wing of the two-story Art Center College of Design, Pasadena, California (1976, case e), designed by Craig Ellwood, is carried by four parallel 16-ft-deep steel trusses that span across the roadway and tie the adjacent buildings together. The building box (i) sits on only four trussed pylons. Its primary structure consists of four giant interior cantilevering trusses that support the outer facade trusses of the same 57-ft height. Three of the supporting towers contain the stairwells and the fourth one houses the elevator shaft. Paolo Soleri's organic, bonelike cantilevered bridge (g) has sculptural qualities; its depth and member density follow the intensity of the moment flow.

The circular, treelike house of case (h) sits on a central concrete pedestal and revolves. While its cantilever action is achieved by hanging the floor framing from the radially arranged roof trusses, the building box in (k) rests on cantilevering frames. The two-story building in case (j) is hung from two huge cantilevering steel plate girder frames.

PROBLEMS

5.1   Investigate the frames in Fig. 5.19f to L with respect to stability and redundancy.

5.2   For the statically determinate frames (with the exception of cases g and x) in Fig. 5.11, draw the moment diagrams and give numerical values at critical locations. Also show how the frames deform under the external loading; indicate clearly how you obtain the solutions.

5.3   Show the loading cases you will investigate for the preliminary design of the simple shelter in Fig. 5.11j. Show moment diagrams for each loading case.

5.4   If the form of the frame members of simple rectangular bents is to respond to the magnitude of the force flow (tectonics), how would you shape the frames? Discuss at least four different bent types.

5.5   Compare the force flow of a three-hinge frame with constant beam and column moments of inertia under uniform loading with the one under asymmetrical live loading (Fig. 5.11n). Verify that the asymmetrical loading case does not influence the preliminary design of the frame. Do you know of any condition where the asymmetrical loading case may have an influence on the design?

5.6   The frames, cases a and c of Fig. 5.11, are considered as support structures for an addition to an existing building. Which of the two proposals do you select if the deciding factor is economy of materials, and gravity loading is assumed to control this preliminary design? Determine the preliminary member sizes (A36) for the case you select. The roof joists are spaced 5 ft on center. The columns do not sway laterally.

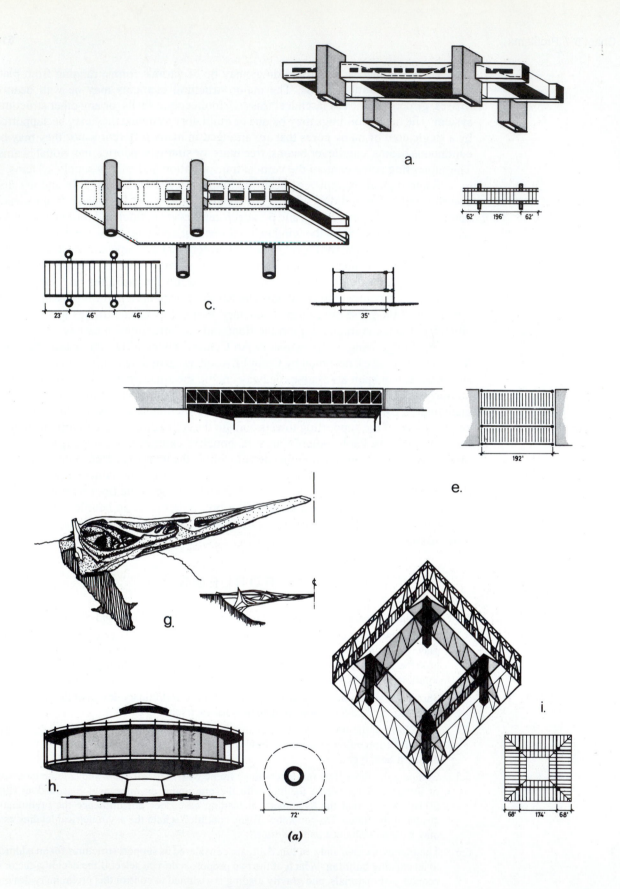

a.

c.

62' 196' 62'

23' 46' 46'

35'

192'

e.

g.

h.

72'

(a)

i.

68' 174' 68'

Figure 5.58 Beam buildings.

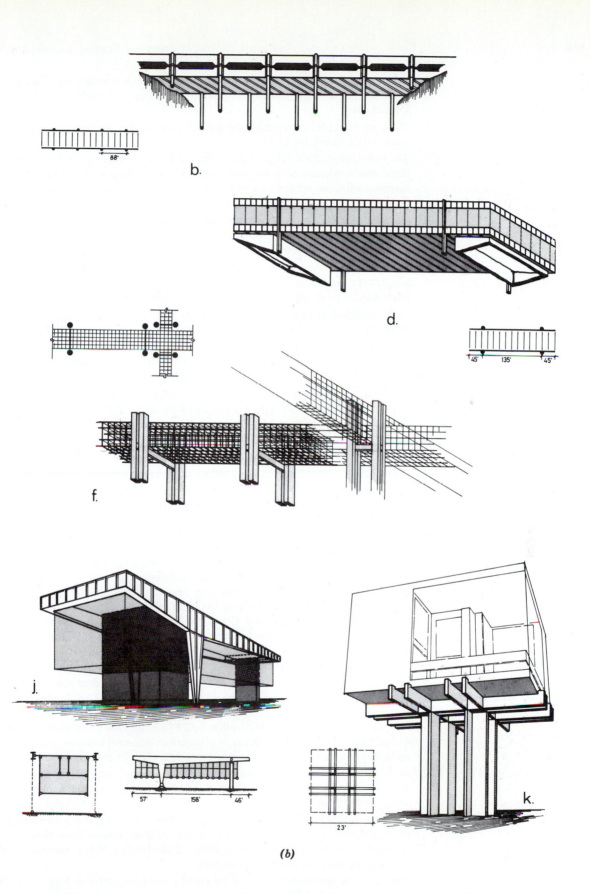

b.

d.

f.

j.

k.

(b)

5.7 Investigate the cases in Fig. 5.15 that have not been approximately designed in Example 5.2. Find the preliminary beam and column sizes using A36 steel as based on the critical loading cases.

5.8 Show the effect of beam rotation on the lateral deflection of two-hinge portal frames; in other words, derive Eq. (5.19).

5.9 The foundations of a continuous portal frame fixed at the base rotate slightly under lateral force action because of weak soil conditions, generating a partially fixed joint with assumed inflection points in the column at $h/3$ measured from the base. Show the moment diagram for a unit load action and the deformation of the frame.

5.10 Assume the footing of the frame in Example 5.2(a) with W14 × 68 columns to spread 1/2 in. horizontally due to tie elongation. Determine the bending stresses in the columns assuming rigid beam action.

5.11 Determine the moments in 16- × 16-in. concrete columns (5000-psi concrete, normal weight) of a post–beam structure 12 ft high as caused by shortening of a prestressed, pre-fab beam due to shrinkage and creep; assume the beam shortening as 0.24 in.

5.12 One foundation of a rigid frame (single bent), fixed at the base, settles vertically. It does not form any inflection points in the columns, only at midspan of the beam. Show the deformation of the frame and the moment diagram.

5.13 If the frame of Problem 5.12 is pinned to its base, how does the frame deform as one support settles vertically? Show also the moment diagram.

5.14 Show the moment and axial force diagrams with numerical values for the framing systems in Fig. 5.21 as caused by a lateral unit load acting at the roof level. Also draw the deflected frame. Compare the solutions and draw your conclusions.

5.15 Repeat Problem 5.14, but now consider uniform gravity loading, that is, dead and live load rather than lateral force action.

5.16 Determine the approximate sizes of the major structural members for the framing system shown in Fig. 5.21b. Assume the roof skin to consist of 2- × 6-in. tongue-and-groove wood planks 12 ft long supported by timber beams spaced 6 ft apart. The beams are hinged to the girders. The roof skin consists of decking, insulation, and built-up roofing, altogether weighing 12 psf. Assume an equivalent weight for the beams of 4 psf and girders of 3 psf. Only gravity loading is needed, since the wind forces are carried by the roof diaphragm to shear walls along the facade. Use glued laminated southern pine for the design of the beams and girders, which has the following properties: $F_b = 2.4$ ksi, $F_v = 0.2$ ksi, and $E = 1800$ ksi. Design the girders as based on bending alone, while for the design of the beams also consider deflection. The allowable axial stresses for the interior column are $F_c = 1.5$ ksi and $E = 1800$ ksi. Check the 2-in. deck, which has the following properties: $F_b = 1.31$ ksi for repetitive members, $F_v = 0.08$ ksi, and $E = 1500$ ksi. For the estimate of member sizes, ignore the effect of load duration and member size.

5.17 Investigate the structural framing system in Fig. 5.21c, but assuming the interior columns hinged (i.e., not resisting any lateral forces). Use 8-in. hollow core slabs to span between the post–beam concrete structure. The roofing material weighs 10 psf and the precast slabs 57 psf; the snow load is 30 psf. Check if the prefab slab is satisfactory. Estimate the size of the girder by assuming the beam depth to be 2.5 to 3 times the beam width. Also check the required amount of steel reinforcement at the critical moment location. Determine the size of the central column. Use 4000-psi normal-weight concrete and Grade 60 steel.

5.18 Determine the approximate member sizes for the typical, interior, four-bay steel frame structure hinged to the foundations in Fig. 5.21d. Refer to Example 5.6 for the description of the loading conditions.

5.19 For the statically determinate frames shown in Fig. 5.32, draw shear, moment, and axial force diagrams and show numerical values; indicate clearly how these values are obtained. Also, show how the frames deform under loading. Where numerical values for the loads are not given, assume a unit load.

5.20 Determine the approximate size of the purlins and roof rafters in Fig. 5.30a. The roof framing is spaced at 15 ft and supports the purlins running perpendicularly to it. The building is laterally stabilized by a horizontal, rigid plane, such as a trussed floor level, and by vertical, rigid diaphragms in the end bays, which act as shear walls. Assume 10 psf for sheathing, roofing, and purlins, 2 psf for rafters, and a snow load of 20 psf on the horizontal roof projection. Use structural lumber with the following properties: $F_b = 1550$ psi, $F_v = 85$ psi, $F_c = 1000$ psi, and $E = 1,200,000$ psi.

(a) Design the simply supported timber purlins, spanning between the rafters, by arranging them in eight equal spaces across the full width of the roof. First, assume that the thrust component parallel to the roof plane is carried by the skin, such as planking to the rafters. Then design the purlins by considering that the skin, such as plywood sheathing, is only able to resist the twisting and not the lateral movement parallel to the roof slope (Fig. 3.8).

(b) Select the preliminary rafter size as based on gravity bending only. Check shear stresses. Check deflection of rafter for an allowable live load deflection of $L/240$ and a total load deflection of $L/180$.

5.21 Design the building in Fig. 5.30a and Problem 5.20 as a joist roof. Assume the joists to be spaced at 16 in. on center.

5.22 For a preliminary estimate, determine the size of a typical rafter (Fig. 5.30b) spaced at 2 ft on center. Notice that this case has already been analyzed in Problem 5.19, case q. Use the following loads: 13 psf for roof tiles and laths, 2 psf for rafter weight, 20 psf for snow load on the horizontal projection, and 14 psf wind load normal to the roof surface. Use the lumber given in Problem 5.20, the only difference being that the rafters are assumed to have a plaster ceiling, which results in an allowable live load deflection of $L/360$ and an allowable total load deflection of $L/240$.

5.23 In an industrial building, three-hinge gable frames of A36 steel are spaced 20 ft apart. The frames must support the following loads:

Dead load:	roofing	4 psf	Live load:	snow 20 psf
	metal deck	2 psf		
	insulation	3 psf		
	purlins	4 psf	Wind load:	neglected for this
	other	2 psf		preliminary design
		15 psf		
	frame	5 psf		
	Total	20 psf		

The symmetrical gable frames span 100 ft, they are 35 ft high at the center, and the height of the columns is 20 ft.

Determine the preliminary sizes of column and rafter by considering only bending. The critical moment locations at the haunch are identified in the column at 5 ft down from the beam–column intersection and in the beam at a horizontal distance of 8 ft. The haunch itself will not be designed at this initial stage. The rafter is laterally stabilized by the purlins, which are spaced at 5.8 ft on center, and the column is laterally supported about its weak axis by the girts.

5.24 Determine M_{max}, V_{max}, N_{max}, and N at the location of M_{max} for the arches in Fig. 5.41. All arches are parabolic with the exception of cases g, h, l, and m which are circular. Show pressure lines, moment diagrams, and deformed shapes of arches. For the cases where dimensions and loads are not given, assume the following values: $P = 10$ k, $w = 1$ k/ft, $L = 100$ ft, and $h = 20$ ft.

5.25 Do a preliminary design of the three-hinged, circular, laminated wood arch of constant cross section that was analyzed in Problem 5.24(g). The arches are spaced 16 ft apart. The arch thrust is carried by steel tie rods (A36 steel). Consider the roof loads to be supported by closely spaced roof joists causing a uniform load. Assume the following loading: 30 psf for snow, 20 psf for wind on the vertical roof projection, and a dead load of

Estimated weight of arch	3.0 psf
Roof joists	2.0 psf
1-in. Sheathing	2.5 psf
Acoustical ceiling	2.5 psf
Three-ply built-up roofing	4.0 psf
	14.0 psf

Try glued laminated southern pine combination with the following allowable stresses for dry conditions: $F_b = 2400$ psi, $F_c = 1750$ psi, $F_v = 200$ psi, and $E = 1,600,000$ psi.

5.26   Select a steel wide-flange section (A36) for the arch in Problem 5.25, assuming the same loading conditions and the arches to be laterally supported by purlins spaced 4 ft apart.

5.27   Three-hinged parabolic steel arches (A36) arranged in parallel are used as the primary structural elements for a factory building, similar in appearance to case g in Fig. 5.41. The arches are 40 ft high, span 200 ft, and are 40 ft apart. They are laterally braced by trusses spaced at 10 ft. Assume the buckling about the weak axis to control the allowable axial stresses. Use the following loading:

Dead load:	beams	7 psf
	cladding	3 psf
		10 psf
Snow load on horizontal roof projection		15 psf
Wind load on vertical roof projection		15 psf

Determine the preliminary arch size by considering it to be of constant cross section.

5.28   A ribbed parabolic dome has a diameter of 240 ft and is 40 ft high. It consists of 36 equally spaced radial steel ribs (A36) meeting at the compression ring at the crown, which has a small torsional rigidity and thus can be considered equivalent to a spherical hinge. The ribs are tied together by 19 horizontal, equally spaced ring purlins. The spatial arch interaction as caused by ring purlins and roof skin is ignored; they are assumed to be individual members acting as independent, three-hinged arches. Use a dead load of 18 psf, which is to change linearly from zero at the crown to a maximum at the base. The small constant dead load component due to the rib weight does not cause any moment and is conservatively included in the triangular load distribution. The live load (snow) is considered as 20 psf to act on the horizontal roof projection. Determine the approximate arch size.

5.29   The dome in Problem 5.28 must carry at the apex an additional concentrated load of 20 k due to lighting and ventilating equipment. Determine the approximate member size by superimposing conservatively the maximum moments and axial forces of the two loading cases, although they are not acting at the same location.

5.30   Determine the preliminary size of the circular concrete compression ring for the suspended cable roof of Problem 9.17. Approximate the individual radial cable force action by changing it to a uniform radial pressure p. Assume the vertical cable force component to be carried directly by the facade walls. Use 3000-psi concrete.

5.31   Determine the exact magnitude of the forces in the chord members at midspan for the uniform gravity loading case of Example 5.16.

5.32   For the truss enclosure structures in Fig. 5.51, determine the forces in the members that are identified by letters at their ends (i.e., at joints).

5.33   Estimate the truss member sizes if the two-way roof structure of Example 6.3a (Fig. 6.16a) is replaced by a parallel, one-way truss system, using the same truss geometry and spacing and the same loading as for the two-way structure.

5.34   Consider the columns of the trussed portal in Example 5.16 to be fixed at their base. Determine the axial column forces and critical moments, and find the preliminary column section.

5.35   A roof is bridged by 80-ft-span glulam bowstring trusses that have a radius of curvature of 80 ft and are spaced 16 ft apart. Assume a dead load of 15 psf and a snow load of 30 psf; there are no ceiling loads. The allowable stresses are $F_c = 2000$ psi and $F_t = 2400$ psi. Estimate as fast as possible the depth of the $3\frac{1}{8}$-in.-wide glulam top and bottom truss chords by assuming the uniform loads to act on the horizontal roof projection.

chapter 6

Space Frames

Although buildings are three dimensional, their support structures can often be treated as an assembly of two-dimensional structures (e.g., horizontal and vertical planes), unless they are truly of three-dimensional nature. Space frames represent a special group of three-dimensional structures that includes shells, folded plates, stressed skin systems, polyhedral structures, tubular structures, and others.

Space frames consist of linear members that do not all lie in the same plane. The members are placed in such a manner that forces are transferred in a spatial fashion, rather than the usual planar one. Since the members are arranged in three-dimensional space, the joints become a most critical consideration, particularly because many more members may have to be connected. The type of joint also determines whether it is rigid or hinged. Space trusses have hinged joints, whereas space frames have rigid joints, although the term *space frame* generally refers also to three-dimensional trusses.

Space frames can form single- or multilayer grids. Single-layer space frames are of the folded or bent envelope type, in other words, three-dimensional surface structures. Like thin shells, they are axial systems that have hardly any bending capacity normal to their latticed surface, and only the individual grid members spanning from joint to joint can behave as beams. *Single-layer space frames* obtain their strength through spatial geometry (i.e., their profile) by being either folded or curved. They may be organized according to the shape of their surfaces, based on the pattern formed by the member network, as is discussed in Chapter 8. The bar grids may be acting independently or in composite action together with the skin. In this chapter, only the special envelope-type structures of *polyhedral*, domelike roof systems shown in Fig. 6.1 are briefly discussed.

Multilayer space frames are generated by adding spatial, polyhedral units to form new three-dimensional building blocks. In contrast to single-layer systems, the multilayer structure has bending stiffness and does not need curvature; therefore, it can behave as a flat plate. A familiar example is the horizontal double-layer space frame roof where depth is a determinant of strength. It should be kept in mind that stability considerations may require a single-layer dome to become a double-layer system, as Buckminster Fuller did for some of his geodesic domes, where profile together with surface depth provide the strength and stiffness.

The transformation of two-dimensional thinking in structural design to the spatial treatment of force flow is now possible due to computer technology; computers are programmed to deal with the complex interaction of physical phenomena. Similar to planar trusses, space frames for preliminary design purposes are considered to be hinged so that their members resist only axial forces, and no moments are transferred across the joints. Since the state of stress is primarily axial, the loads are assumed to act directly at the hinges, and the effect of member weight and truss deformation may be treated as negligible for preliminary design purposes.

6.1 THE DEVELOPMENT OF SPACE FRAMES

The source of space frame structures can be found on both the micro and macro scale in the organic and inorganic world around us, where the physical order is reflected in visible geometry. Nature is full of examples showing the principles of surface grids and spatial packing characteristics. The hexagonal prisms of the honeycomb, the cell structures of plant organisms, the porous structure of a sponge skeleton, and the uniform repetitive order of crystals, just to name a few, all exemplify the packing of spatial units and hence three-dimensional geometry. The architectures of crystals such as the beautiful structures of snowflakes or the internal arrangement of the atoms, which determine the external visible crystalline shapes, are examples of that spatial geometrical order typical of space frame construction. Nature forms its structures over time, thus allowing them to respond with a minimum of effort; hence, through principles of economy, nature can achieve an optimal strength in a lightweight skeletal structure. It is apparent that the designer should study geometry in nature and learn, for instance, from the structural patterns in a leaf, where the network of veins branch out from the stem and interconnect the entire surface, or from the microscopically small grid surfaces of the radiolaria with their extremely delicate skeletal cell walls (see Fig. 8.1).

D'Arcy Thompson's book *On Growth and Form* (first edition in 1917), in which biological structures and processes are studied, had a marked influence on the experimental designers of the 1960s by relating architecture and biology. In his Institute of Lightweight Structures at the University of Stuttgart, Germany, Frei Otto united biologists, architects, and structural engineers to study natural structures from an esthetical, functional, morphological and biostatics point of view. Johann-Gerhard Helmcke of the institute gained fame in the late 1960s with his investigations of diatom shells and radiolaria skeletons as ideal lightweight structures.

The origin of space frames from a geometrical or spiritual point of view may be traced back to classical Greece. Pythagorean and Platonic thinking interpreted the universe in terms of the number and form of geometrical figures. Plato related the world of ideas to the material world by proposing the five basic solids: the cube, the tetrahedron, the octahedron, the icosahedron, and the dodecahedron to correspond to earth, air, fire, water, and the cosmos. The Islamic geometric ornament evolved out of the interplay of Platonic figures and proportions and later influenced the geometric architecture of the Gothic period. The revival of Platonic thought in the Middle Ages resulted in a daring challenge to Euclidean geometry. The intricate ribbed vaulting of the late Gothic churches clearly predicts Nervi's hangars, first built in 1940 in Italy using prefabricated lattice ribs as lamella vaults. They represent some of the few reinforced concrete space frame structures ever built.

The concept of spatial construction has evolved out of the various framing systems for huts or houses of the nomadic tribes and agricultural societies; it has evolved out of planar truss, medieval vaulting, and dome construction. From the structure of the radially ribbed masonry domes of the Renaissance was developed a more elaborate expression of geometry and structure in the Baroque period. As in the Gothic epoch, Guarini expressed the skeleton construction in the intricate, intersecting arches that form the dome and lantern support of S. Lorenzo at Turin, Italy (1687). The development of concrete and steel in the nineteenth century allowed framed dome structures of much less weight and much larger spans to replace the traditional timber trusses and solid masonry vaulting. To the early forms of only radial ribs were added horizontal rings to tie the arches together; with advances in material and fabricating technology and methods of analysis, the dome slowly developed into a latticed shell structure. In 1863, J. W. Schwedler built in Berlin the first braced dome, and the structural system became known as the Schwedler dome, possibly representing the first lightweight surface structure. The geometrical layout of the building clearly reflects the repetitive use of standardized elements identifying the new concepts of prefabrication and mass production. These considerations of lightness, prefabrication, and repetitive member use are all basic design determinants of space frame construction.

The concept of three-dimensional structures also evolved out of the medieval timber framing for walls and especially for steeply pitched church roofs, which was developed to a high degree of sophistication in the trusslike structural systems. In 1570, Palladio was probably the first designer to actually build truly triangulated beam trusses for a bridge. The high point of truss design, especially from a geometrical point of view, was reached in the nineteenth century. The truss was not only imaginatively applied in bridge design, but also in architecture as support of roofs for functional spaces such as railway stations, exhibition halls, warehouses, winter gardens, and market halls. The trussed gable frames for the Galeries des Machines for the 1889 Paris Exhibition reached an unbelievable span of 375 ft, thereby setting up a new dimension for column-free, long-span enclosure systems. Engineers with bold new design concepts integrated the new materials of iron, glass, steel, and concrete with fabrication and construction considerations. Paxton's Crystal Palace in London (1851) is often considered as a model. It established the independence of the frame by expressing the extreme lightness of the glass–frame structure and represents the new design concepts of mass production, mechanization, standardization, and modularity, that is, building systems, features that are all related to space frames with their many repetitive elements.

The multilayer space frames evolved directly out of the planer trusses of the nineteenth century. The Eiffel Tower in Paris (1889) is a famous example of a three-dimensional structure, although it actually consists of an assembly of two-dimensional trusses. Also, the innovative Firth of Forth cantilever railway bridge near Edingburgh, Scotland (1890), indicates the trend toward spatial truss construction; however, its riveted tubular steel sections are still quite massive.

In 1881, August Föppl published his treatise on space frames, which was the basis for Gustave Eiffel's analysis of the tower. For his work around the turn of the century, the universal genius Alexander Bell is credited as the inventor of the space frame. He was preoccupied with tetrahedral forms to obtain strength with a minimum of material weight. His space frames were composed of tetrahedra and octahedra and were used in his man-carrying kites, a multilayered aeroplane wing, and in an observation tower. To fly the first aeroplane in 1905, the Wright brothers used a construction based on spatial trusses. Buckminster Fuller entered the limelight in the 1920s with his inventions of the first tensegrity structure (1927) and the Dymaxion House (1928).

The French engineer Robert Le Ricolais proposed double-layer grids in the mid-1930s as derived from studies of three-dimensional structures in nature, thereby being one of the first to provide evidence of the economy of space frames. In 1940, he introduced his famous three-way double-layer grids. He continued his studies of spatial experimental structures at the University of Pennsylvania School of Fine Arts in Philadelphia, when he became a Senior Fellow of Architecture in 1955. He considered topology, which he defined as "the geometry of rubber," as the backbone of structure.

Max Mengeringhausen built his first space frame structure in Berlin, Germany, in 1940, using his famous MERO joint. In North America the idea of mass production of space frame structures was initiated in the early 1950s by Á. E. Fentiman in Canada with his Triodetic system and by C. W. Attwood in the United States with the Unistrut system. The French designer Stéphane du Château introduced his first space frame systems in the late 1950s. During the same period, Konrad Wachsmann designed for the U.S. Air Force the famous space frame cantilever hangars with extraordinary esthetical qualities that excited many architects, in a way anticipating the spirit of Philip Johnson's Crystal Cathedral of 1980. His book of 1961, *The Turning Point of Building,* has been most inspiring to many architects of the 1960s, in which he described the new architecture of industrialization and mass production.

Another pioneer of space frame structures is Zygmunt Makowski, whose analytical and experimental studies at the University of Surrey, U.K., together with his numerous publications, have contributed greatly to a better understanding and further development of the field. He was responsible for one of the first long-span space frame structures, the 453-ft clear-span Boeing 747 jet hangar at London's Heathrow Airport completed in 1970.

Space frames, however, were not only celebrated during the late 1950s and 1960s as roofs, that is, as efficient horizontal-span systems from an engineer's point of view, but also as geometrical organizers of urban spaces, as well as architectural spaces to live in as derived from the packing of polyhedra as part of the organic architecture of the 1960s. For example, in 1957 Louis Kahn and Anne Griswold Tyng proposed the megaspace frame concept for a 616-ft-high building where the primary floors are part of the space frame and are supported by 66-ft-high tetrahedrons. Yona Friedman and Eckhard Schulze-Fielitz developed their urban megastructures based on multilayer space frames. Alternate environments in the utopian tradition, for example, were represented by Warren Chalk's *Underwater City* and Peter Cook's *Plug-in-City,* integrating polyhedra and spatial frames in the spirit of the Archigram group. In 1969, Alfred Swenson proposed a 150-story megatube using double-layer perimeter space frames as wall structures.

Quite popular was the polyhedric architecture of Alfred Neumann, Zvi Hecker, David Emmerich, Moshe Safdie, and the theoretical investigations of Keith Critchlow, among many other designers, as typically applied to high-density communal architecture and emergency shelters. We should also not forget the counterculture architecture of the hippie communes using polyhedral and geodesic domes, as inspired by Buckminster Fuller.

The spirit of three-dimensional structures as celebrated during the 1960s is reflected by the daring space frame structures at Expo '67 in Montreal and at Expo '70 in Osaka. The experiments of the pioneers, possibly reaching a high point with Buckminster Fuller's U.S. Pavilion geodesic dome at Montreal's Expo '67, are being continued by a new generation of designers in the 1990s, for instance, as represented by Renzo Piano's composite tensegrity structures and Wendel R. Wendel with his projects of free-flying spatial structures in outer space, such as space stations, work platforms, reflector telescopes, and large antennas.

6.2 SIMPLE SINGLE-LAYER SPACE FRAMES

Typical polyhedral buildings or roof units are shown in Fig. 6.1. The various forms are generated by transformation as well as trial and error. From the four basic pyramids at the center of the drawing with different regular polygonal bases, a wealth of shapes is derived using rational processes such as translation, rotation, truncation, and member placement, but at the same time being intuitive about the selection of the shapes.

Some typical examples of architectural buildings are presented in Fig. 6.2. These cases clearly express the richness of the various spatial framing systems. They range from the small-scale shelter (e.g., case a) to the large-scale stadium (case i). They

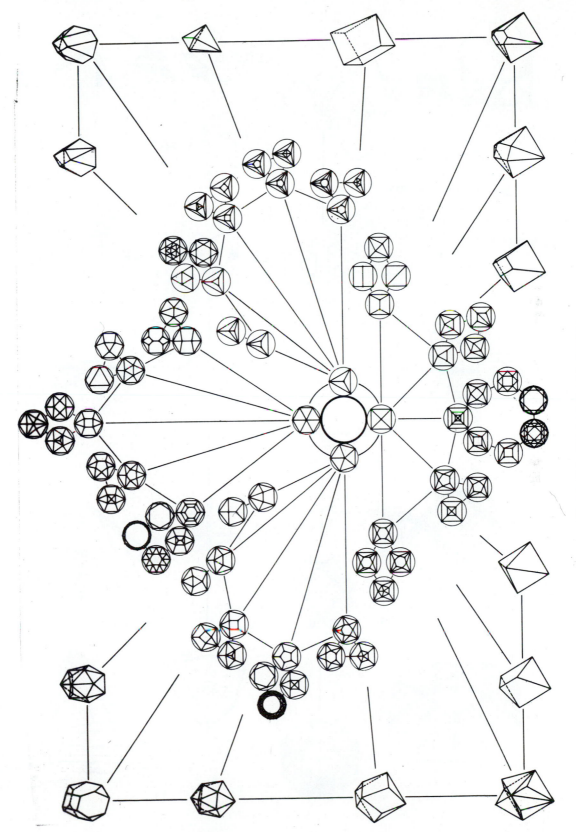

Figure 6.1 Polyhedral roof structures.

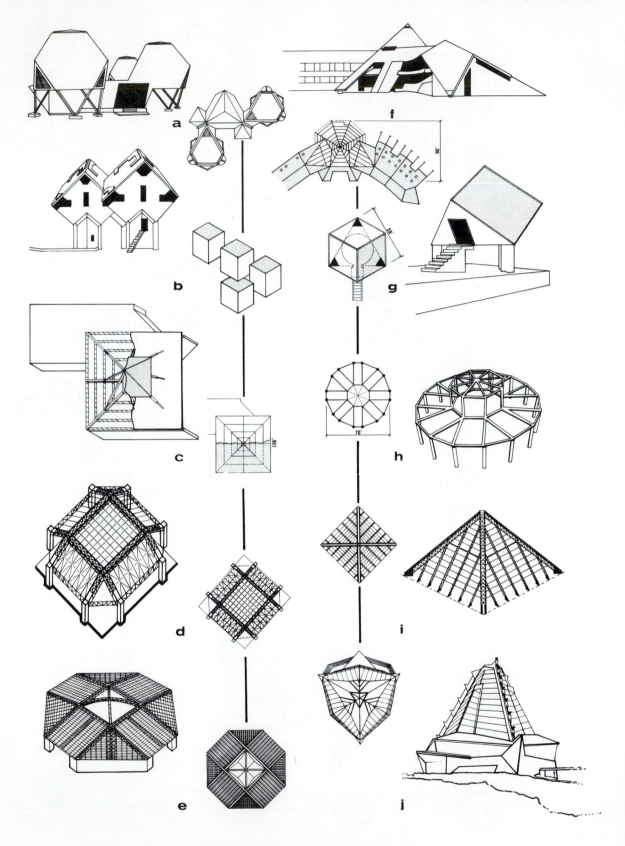

Figure 6.2 Single-layer space frame roofs.

range from simple three-dimensional shapes (cases *c, d, e, h,* and *i*), unusual polyhedral roofs (cases *a, b,* and *g*), and complex folded roof shapes (case *f*) to Frank Lloyd Wright's Beth Sholem Synagogue organic glass tower on a distorted hexagon plan in Philadelphia (1954, case *j*).

Next, some of the cases in Fig. 6.2 are briefly discussed in historical sequence. Eduado Torroja's operating ampitheater of the University City Hospital, Madrid, Spain (1934, Fig. 6.2h), represents an early modern example of a three-dimensional structure. Here, radially arranged horizontal beam ribs and the central 30-ft polygonal, vertical skylight framing form a continuous structure, where the top polygonal ring is in compression and the bottom one in tension. The 70-ft-span concrete roof structure is hinged to the perimeter columns.

The pyramidal roof structure for the one-story main library section of the Chappaqua Library, Chappaqua, New York (1977, Fig. 6.2c), designed by Kilham Beder & Chu, is supported by a single central column in maypole style. The column carries eight horizontal radial girders cantilevered from sloping hangers similar to a modified king-post system. Piet Blom's "tree houses" for community housing in Helmond, Holland (1977, Fig. 6.2b), consist of rotated, three-level, timber-framed cubes supported on concrete trunks that provide vertical access.

The double-pitched domical roof of the Brendan Byrne Arena at New Jersey's Meadowlands is a 372-ft × 414-ft clear-span structure designed by the engineers Skilling, Helle, Christiansen, Robertson (1981, Fig. 6.2d). The structural framing system consists of the primary two-way, 16.5-ft-deep steel arched box trusses (providing a catwalk system) supported by eight steel box towers containing the stairs and elevators. The towers are connected by 50-ft-deep cantilever wall trusses that give support to the secondary roof trusses and exterior walls. Posttensioned cables act as a tension ring to balance the outward thrust of the arches and to control lateral deflection.

The University of North Carolina's basketball arena at Chapel Hill, North Carolina (1986, Fig. 6.2e), is covered by a nearly 400 ft span arched steel roof with a fabric center; the structure was designed by Geiger Associates. The roof structure consists of four diagonal trussed box arches (which are used as catwalks), subdividing the roof into four rectangular and four triangular bays along the perimeter and a slightly skewed rectangular bay at the center. The lateral thrust of the arches is resisted by the perimeter tension steel beam at the base; refer to Fig. 9.19 for further discussion of funicular tension or compression rings. The steel weight for the roof structure is 15 psf. The fabric lantern is approximately 122 ft on each side and consists of Teflon-coated fiber glass carried on crisscross triangular pipe arches and is tied down to the compression ring by cables.

The 290-ft-tall Memphis Pyramid, home of the Memphis State University basketball team, has a 450-ft square base (1991, Fig. 6.2i). It is a 60% scale version of the Great Pyramid of Cheops in Egypt. Its structure was designed by the engineers Walter P. Moore and Associates and consists of 14- × 18-ft box trusses along the four edges, each about 400 ft long; two of them contain stairs to provide access to the skylight observation deck at 254 ft. The primary corner trusses support six one-way secondary trusses on each face (spaced at 56 ft), which in turn carry the brace beams (spaced 30 ft up the slope) and the 32-in.-deep long-span joists. The joists are held in position by diagonal bridging. A compression ring at about three-quarter height gives additional stiffness to the structure. The structural weight is 18.3 psf of projected roof area.

Introduction to Space Statics

A typical force F can be resolved into F_x, F_y, and F_z components as shown in Fig. 6.3a. These force components form a box in which the diagonal is equal to the force resultant. Visualize these force components to be carried by simple axial members of length *x, y,* and *z,* as indicated by the smaller box contained in the larger one. Since the members act only axially, the magnitude of the forces must be proportional to their respective member length.

$$F_y/y = F_x/x = F_z/z = F/L \quad \text{or} \quad F_y = F(y/L) = F_x(y/x) = F_z(y/z), \quad \text{etc.} \quad (6.1)$$

A similar relationship can be set up for the other force components. The resultant force F_{xy} of the forces F_x and F_y, according to the Pythagorean theorem, is equal to

$$F_{xy}^2 = F_y^2 + F_x^2 \tag{a}$$

The final resultant, in turn, is obtained by substituting Eq. (a):

$$F^2 = F_{xy}^2 + F_z^2 = F_x^2 + F_y^2 + F_z^2 \quad \text{or} \quad F = \sqrt{F_x^2 + F_y^2 + F_z^2} \tag{6.2}$$

A similar relationship can be established between the member lengths.

$$L^2 = x^2 + y^2 + z^2 \quad \text{or} \quad L = \sqrt{x^2 + y^2 + z^2} \tag{6.3}$$

For the special case where the spatial force system is concurrent (Fig. 6.3b), the forces at the point of concurrency must be in translational equilibrium. The sum of the $x, y,$ and z components of all the forces must be zero.

$$\Sigma F_x = 0, \qquad \Sigma F_y = 0, \qquad \Sigma F_z = 0 \tag{6.4}$$

For this condition, only three independent equations may be written, although the force equations can be replaced by moment equations. Only three unknowns should be present at the concurrent joint, since otherwise the solution is indeterminate.

For the special case where the spatial force system consists only of parallel forces, such as the gravity loads acting on the roof supported by three vertical columns (Fig. 6.3c), the forces must be in translational and rotational equilibrium. The moments are taken about axes, rather than about points as in planar systems. If, however, the spatial system is considered as consisting of side, front, and plan views, then the axes of rotation become points of rotation again. For the case shown, the forces can be solved by using the following equations of equilibrium:

$$\Sigma F_z = 0, \qquad \Sigma M_x = 0, \qquad \Sigma M_y = 0 \tag{6.5}$$

Obviously, these basic equations can be replaced by other equations of equilibrium, since rotation must be zero about any point in space, such as

$$\Sigma F_z = 0, \qquad \Sigma M_y = 0, \qquad \Sigma M_{y'} = 0$$

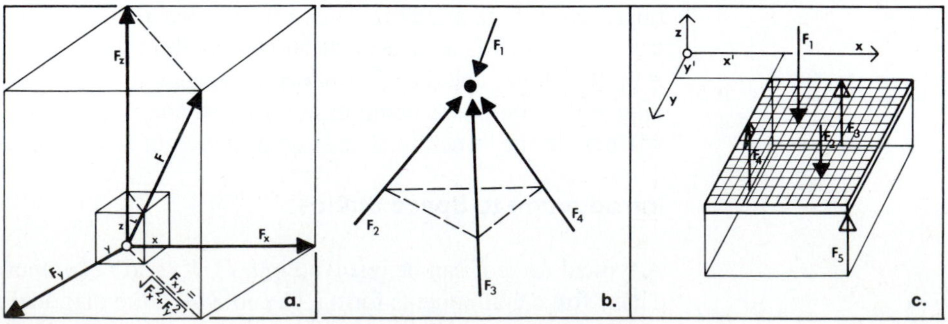

Figure 6.3 Equilibrium in space.

or

$$\Sigma F_z = 0, \qquad \Sigma M_x = 0, \qquad \Sigma M_{x'} = 0$$

or

$$\Sigma M_x = 0, \qquad \Sigma M_{x'} = 0, \qquad \Sigma M_y = 0, \text{ etc.}$$

Only three independent equations of equilibrium are available and necessary for a solution, as is apparent from Eq. (6.5); hence only three unknowns should be present for the structure to be determinate. Should, for instance, the roof be supported by four columns and the gravity loads not be arranged in a symmetrical manner, the structure is indeterminate.

In general, when spatial force systems are neither concurrent nor parallel, the translational and rotational equilibrium along and about each axis, respectively, yield the following six independent equations necessary for solution:

$$\begin{array}{lll} \Sigma F_x = 0, & \Sigma F_y = 0, & \Sigma F_z = 0 \\ \Sigma M_x = 0, & \Sigma M_y = 0, & \Sigma M_z = 0 \end{array} \qquad (6.6)$$

Note not more than six unknown forces should be present to have a determinate structure. Any of the force equations can be replaced by additional moment equations. The proper selection of the axes of rotation may save a great deal of calculation.

Examples of nonparallel, nonconcurrent force systems are shown in Fig. 6.4d to i, where a building or roof (treated as rigid bodies) is supported by a group of six columns ($r = 6$), each carrying only axial forces. Different support arrangements are given to establish stability under any type of load action. Rather than drawing a three-dimensional view of the structure to be investigated, it may sometimes be more convenient to take planar views from front, side, and top by treating the spatial structure as a series of interacting planar systems. Forces are projected onto the planes, and the ones normal to the planes that cannot produce reactions perpendicular to their line of action are now shown. Force components in each plane can now be treated in the familiar manner, remembering that the proper number of force components should be in each plane and that they should be arranged so that they are neither parallel nor concurrent. In Problem 6.2, the cases in Fig. 6.4d to i are further investigated, and the loca-

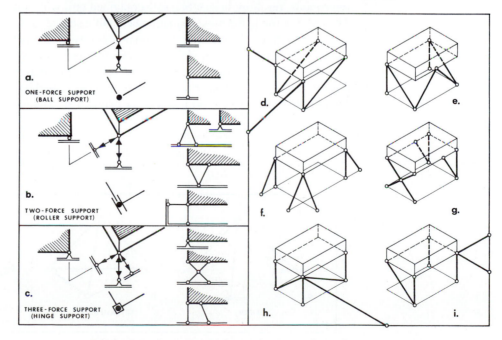

Figure 6.4 Support conditions for three-dimensional structures.

tion of any members is corrected if their arrangement should not provide the necessary resistance. Should the structure not behave as a rigid body, however, additional supports are required!

In the discussion of concurrent force systems, it is shown that at least three equations or three forces, that is, a minimum of three members m, not lying in the same plane, are necessary to prevent a point P from moving in any direction under load action. This basic stable, three-dimensional structure is a *tripod* (Fig. 6.5a); space frames can be formed by simply adding tripods to one another. We may conclude that for each joint j there must be at least three members m, not lying in the same plane, for a simple truss to be internally stable; more members result in redundancy, while less than three members yields instability.

$$\begin{aligned} m &= 3j, & \text{statically determinate} \\ m &< 3j, & \text{unstable} \\ m &> 3j, & \text{statically indeterminate} \end{aligned} \qquad \text{(b)}$$

Furthermore, it is shown that for nonconcurrent, nonparallel spatial force systems, six independent equations of equilibrium are available. We may conclude that for a space frame to be externally stable at least six independent reaction components, properly arranged, must exist. Hence, for a rigid structure to be externally stable and determinate, the number of reactive conditions must be

$$r = 6 \qquad \text{(c)}$$

The reaction components can be visualized as equivalent to members. Then the total number of members, considering the stiff space truss externally and internally, can be derived from Eqs. (b) and (c) as

$$m + r = 3j \quad \text{or} \quad m = 3j - r = 3j - 6 \qquad \text{(6.7)}$$

The same conclusion can also be found by studying the growth of a space truss as derived from its base unit, the *tetrahedron* (Fig. 6.5b). It is obtained by simply adding tripods, resulting in a stable, self-contained unit with four joints and six members. The tetrahedron is the most fundamental, stable spatial unit, similar to the triangle for planar structures. Its multidimensional stability and independence can easily be checked by squeezing a model of the structure between one's hands. The three base members in the tetrahedron can be visualized as equivalent to the external thrust reactions of the tripod. A space truss of j joints can be formed by adding three members for each of the remaining $(j-4)$ joints to the original six members of the tetrahedron with four joints.

$$m = 6 + 3(j-4) = 3j - 6$$

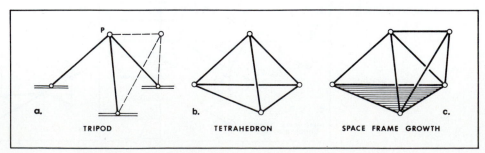

Figure 6.5 The Basic, stable, three-dimensional structure.

It was shown before that the term $3j$ in the equation represents the internal stability, while the term 6 is equal to the necessary reaction conditions for external stability. The equation can be rewritten by letting $r = 6$, as was already derived.

$$m + r = 3j, \quad \text{statically determinate; check stability,}$$
$$m + r < 3j, \quad \text{unstable space truss} \tag{6.8}$$
$$m + r > 3j, \quad \text{statically indeterminate; check stability}$$

In general, it is necessary that the total number of members and reaction components be equal to three times the number of joints so that the structure is externally and internally determinate and the forces can be found by statics. Although this condition is necessary, it is not sufficient, as already discussed in relation to planar trusses [Eq. (5.93)]; the members must be properly arranged. A more precise treatment of the spatial stability of compound and complex space trusses is beyond the scope of this work.

Some building or roof forms are shown in Fig. 6.6. It is apparent that the hinged cube can only be stable if each of its rectangular faces is stabilized by a diagonal, while polyhedras with triangular faces, such as the tetrahedron, octahedron, and icosahedron, are self-contained stable units. On the other hand, the half-cuboctahedron and the half-octahedron are unstable. Keep in mind that instability of the building unit does not necessarily mean that the whole structure is unstable, as additional external reaction components may correct this situation.

In the discussion of external supports, only columns have been considered. A more general treatment of support organization is according to the number of translational forces that can be transmitted from one structural system to another (Fig. 6.4a to c):

- *One-force support,* such as a single column or a spherical ball, which can only transfer forces perpendicular to the plane upon which it is sitting.

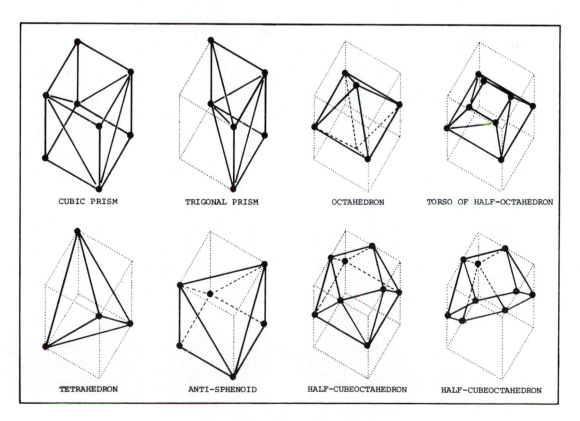

Figure 6.6 Common polyhedra derived from cube.

• *Two-force support,* such as a two-member column support or a roller, which cannot transfer forces in the direction it is allowed to roll.

• *Three-force support,* such as a pin or ball socket or a three-member support with the members not in the same plane.

Next, the simple polyhedral roof forms in Fig. 6.7 are investigated mathematically. Framed domes (see Fig. 8.25) are not studied further here because of the complexity of analysis necessary to solve for the member forces. Some of the braced domes may be treated as membranes, as is discussed in the section on dome shells.

EXAMPLE 6.1

For the hinged, four-sided pyramid (Fig. 6.7a), reactions and bar forces are found as caused by a vertical unit load.

There are two reactive force components at supports *a, b,* and *d* and one at *c.* Hence, the total number of unknown external forces is seven, but only six equations of equilibrium are available.

$$r = 3(2) + 1 = 7 > 6$$

Hence, the structure is once indeterminate externally! Checking the truss as a whole for five joints with eight members and seven equivalent reaction members yields.

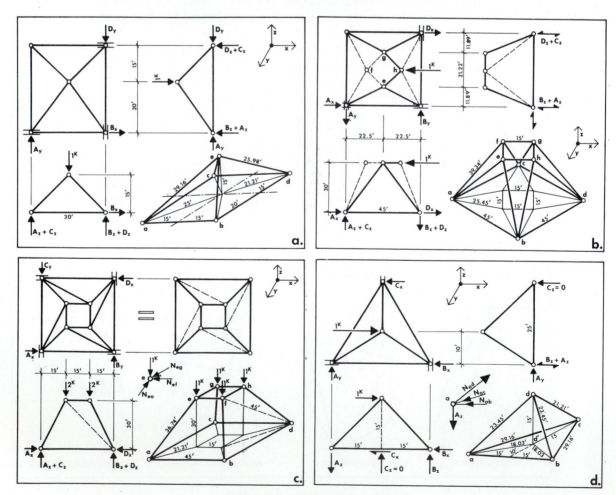

Figure 6.7 Examples 6.1 and 6.2.

$$m + r = 3j, \qquad 8 + 7 = 3(5), \quad \text{or} \quad 15 = 15 \qquad (6.8)$$

The structure as a whole is determinate. Although the pyramid as a self-contained unit is unstable, the additional member needed is provided by the extra external reaction. Because the structure is externally indeterminate, the reaction components cannot all be solved directly, but only in conjunction with the bar forces. In this specific case, due to symmetry of geometry about the y-axis and due to the type of loading, the support forces can be obtained directly. The following vertical reactions must be equal to each other:

$$A_z = B_z, \qquad C_z = D_z$$

From the side elevation the vertical force components can now be computed.

$$\Sigma M_{dc} = 0 = 1(15) - (B_z + A_z)35$$
$$A_z + B_z = 0.43 \text{ k}, \qquad A_z = B_z, \qquad A_z = B_z = 0.43/2$$
$$\Sigma F_z = 0 = (D_z + C_z) + 0.43 - 1$$
$$D_z + C_z = 0.57 \text{ k}, \qquad C_z = D_z, \qquad D_z = C_z = 0.57/2$$

From the plan view, the y and x force components can be found.

$$\Sigma F_x = 0 = B_x, \qquad B_x = 0$$
$$\Sigma M_a = 0 = D_y(30), \qquad D_y = 0$$
$$\Sigma F_y = 0 = A_y, \qquad A_y = 0$$

First, the lengths of the diagonal members are found so that the bar forces can then be computed.

$$L_{ae} = L_{be} = \sqrt{15^2 + 20^2 + 15^2} = 29.16 \text{ ft}$$

$$L_{ce} = L_{de} = \sqrt{15^2 + 15^2 + 15^2} = 15\sqrt{3} = 25.98 \text{ ft}$$

For this case, the bar forces at any joint can be determined. Starting at support a and using the method of joints yields

$$\Sigma F_z = 0 = N_{ae}(15/29.16) - 0.43/2 \qquad N_{ae} = 0.42 \text{ k} \quad (C)$$
Because of symmetry: $\qquad\qquad\qquad\qquad N_{ae} = N_{be}$
$$\Sigma F_x = 0 = 0.42(15/29.16) - N_{ab} \qquad N_{ab} = 0.22 \text{ k} \quad (T)$$
$$\Sigma F_y = 0 = 0.42(20/29.16) - N_{ac} \qquad N_{ac} = 0.29 \text{ k} \quad (T)$$
Because of symmetry: $\qquad\qquad\qquad\qquad N_{ac} = N_{bd}$

At support c, bar forces are computed as follows:

$$\Sigma F_z = 0 = N_{ce}(15/25.98) - 0.57/2 \qquad N_{ce} = 0.50 \text{ k} \quad (C)$$
Because of symmetry: $\qquad\qquad\qquad\qquad N_{ce} = N_{de}$
$$\Sigma F_x = 0 = 0.50(15/25.98) - N_{cd} \qquad N_{cd} = 0.29 \text{ k} \quad (T)$$
$$\Sigma F_y = 0 = 0.50(15/25.98) - 0.29 \qquad \text{OK}$$

The results are checked at joint e.

$$\Sigma F_z = 0 = 1 - 2\left(0.42\,\frac{15}{29.16} + 0.50\,\frac{15}{25.98}\right) \qquad\qquad \text{OK}$$

In the following example, a complex space truss is investigated briefly. Here more than three members are attached to a joint and the bar forces cannot be solved directly anymore with the three equations of equilibrium available for a concurrent force system. A determinate complex space frame can still be analyzed by setting up the three equations of statics for every joint and then by solving these equations simultaneously by computer. However, some approaches may make it still possible to solve complex space trusses directly under certain conditions:.

- Conditions of symmetry of building geometry and loading yield additional information about reactions and member forces.
- Before starting any analysis, zero members should be identified. The elimination of bars for a given loading case reduces the number of members at a joint.
- Sometimes it may be helpful to sum the forces at a joint normal to or within the inclined trussed planes by rotating the coordinate system so that the $x-y$ plane is now the inclined trussed plane. For instance, for the case where one of the three bars at a joint does not act in the plane defined by the two others, this bar must be zero, if no external force is applied to the point of concurrency.

The special considerations just identified are applicable to any spatial truss system; they may reduce the computations and make solutions easier to obtain.

EXAMPLE 6.2

The reactions and bar forces are found for the square, truncated pyramid (Fig. 6.7c) under symmetrical gravity load action at the crown joints. At each of the supports two reaction components are present: $r = 4(2) = 8 > 6$.

As can be seen, the structure is externally twice indeterminate. However, the space truss, as a whole, with eight joints, sixteen members, and eight equivalent reaction bars, is determinate.

$$m + r = 3j, \qquad 16 + 8 = 3(8), \quad \text{or} \quad 24 = 24 \qquad (6.8)$$

In general, the bars should be determined first, since there are too many support forces but not enough equations available. In this specific case, due to the axisymmetric conditions of the roof geometry and the symmetrical arrangement of the loads, it is apparent that each vertical reaction carries an equal share of the total loading.

$$A_z = B_z = D_z = C_z = 1 \text{ k}$$

Because of the absence of any rotation in the $y - x$ plane, we can conclude that all reaction forces in this plane must be zero. The lateral thrust due to gravity is self-contained and is carried by the tension ring at the base of the roof.

$$A_x = D_x = 0, \qquad C_y = B_y = 0$$

For general loading conditions the solution of this structure is very complicated, but for the given case causing no twisting, the funicular shape is defined by the edge members only (Fig. 6.7c), so the diagonal members are not needed and are equal to zero.

$$N_{ag} = N_{be} = N_{df} = N_{ch} = 0$$

The lengths of the members are as follows:

$$L_{ae} = L_{bf} = L_{dh} = L_{cg} = \sqrt{15^2 + 15^2 + 30^2} = 36.74 \text{ ft}$$

$$L_{ag} = L_{be} = L_{df} = L_{ch} = \sqrt{15^2 + 30^2 + 30^2} = 45.00 \text{ ft}$$

Applying the method of joints yields the following member forces:

Joint e: $\Sigma F_z = 0 = N_{ea}(30/36.74) - 1 \qquad N_{ea} = 1.23 \text{ k} \quad (C)$

Because of symmetry: $N_{ea} = N_{fb} = N_{hd} = N_{gc} = 1.23$ k   (C)

$\Sigma F_x = 0 = 1.23(15/36.74) - N_{ef}$ $N_{ef} = 0.5$ k   (C)

Because of symmetry: $N_{ef} = N_{fh} = N_{hg} = N_{ge} = 0.5$ k   (C)

Joint a: $\Sigma F_x = 0 = 1.23(15/36.74) - N_{ab}$ $N_{ab} = 0.5$ k   (T)

Because of symmetry: $N_{ab} = N_{bd} = N_{dc} = N_{ca} = 0.5$ k   (T)

It is apparent that the forces in the tension ring at the base must be equal to the forces in the compression ring at the crown.

6.3 MULTILAYER SPACE FRAMES

Space frame roofs are used for glazed atriums, pavilions, and entry canopies or to cover large spans for buildings such as assembly halls, exhibition spaces, gymnasiums, sports facilities, stadiums, hangars, factories, warehouses, multipurpose halls, shopping centers, and airport terminals.

Design Considerations

Space frame structures are usually not hidden behind cladding but are visible. Therefore, a high-quality control of construction is required, and the esthetical quality of the many members and the joints holding them together is celebrated.

To develop some understanding of multilayer space frames, the nature of spatial geometry should be investigated first. The order of the three-dimensional member network must be recognized and understood from a geometrical as well as behavioral point of view. The path of the force flow not only depends on the load action, shape of the space frame, arrangement of the bars, cross-sectional shapes of the components, type of connections, and stiffness of the entire plate structure, but also on the type and location of the support system. Aspects of mass production expressed in modularity, such as repetition of members, details, and assemblages, together with the easiness of fabrication, transportation, and erection procedures, are essential criteria that must be considered in the design of space frames.

Geometry.   The geometry of space frames is often interpreted either from the close packing of polyhedra or from the interplay of surface tesselations. For a double-layer space frame, the top and bottom surface grids are connected at their nodes by web members to form an aggregate of polyhedra or a three-dimensional network. The web members are generally diagonals; however, for cases where the nodes of the two layers are exactly above each other, vertical struts may be employed. There are infinite variations of overlapping planar tesselations and many ways of joining the two horizontal planes. The relationship of typical surface meshes is studied in Fig. 6.8, where, as an example, the square and the equilateral triangle are used as base units and processed through different transformations to indicate the unlimited possibilities of spatial network formation. Familiar are the antiprisms shown in the corners, which have identical top and base figures like prisms, except that they are instead rotated against each other, thereby generating triangular faces for the vertical planes.

The nature of two-dimensional tesselations is introduced in Fig. 1.4, where the three regular grids (triangle, hexagon, square) and the eight semiregular or Archimedean grids are shown. Common single-layer grids are the following:

- Two-way grids: rectangular, diagonal
- Three-way grids: triangular, hexagonal, triangular/hexagonal
- Four-way grids: rectangular/diagonal

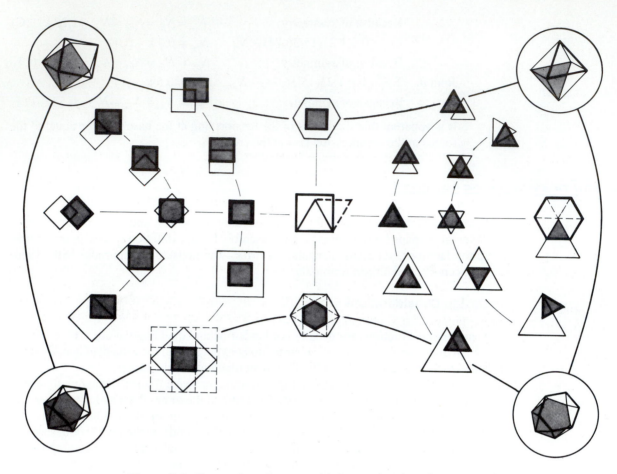

Figure 6.8 Generation of space grids by overlapping planar networks.

Most double-layer space frames are formed by rectangular and/or diagonal grids so that top and bottom chords can either be parallel or skewed to each other or to the edges of the support structure. Typical solutions are shown in Figs. 6.9 and 6.10, keeping in mind that there are infinitely many other possibilities.

Direct Grids or Truss (Lattice) Grids (Fig. 6.9): Top and bottom grids are identical and directly above each other, such as the following:

- Two-directional square grids either parallel (a) or diagonal (skew) to the boundaries
- Three-directional triangular grids (b)
- Four-directional quadruple grids generated by superimposing square and skew grids

Space Grids (Figs. 6.9 and 6.10): Two-directional square grids, not necessarily identical, are offset with respect to each other.

- *Offset grids:* top and bottom grids are identical, but offset with respect to each other, such as square-on-square (e) generated by oblique translation and modifications such as openings in top and bottom grids formed by the process of elimination, resulting in a square-on-larger-square grid (f)
- *Differential grids:* top and bottom square grids differ, such as diagonal square grid set on smaller square grid (g) and diagonal square grid set on larger square grid (h), or square-on-diagonal, or rectangle-on-diamond, etc.

Three-directional, triangular network, or modifications generate offset or differential grids, such as the following:

- Triangular grid on triangular grid offset generated by oblique translation (c)
- Triangle on hexagon
- Modifications such as internal openings in the top or bottom grid, resulting in a combination of hexagonal and triangular grid on triangular grid (d)

The horizontal grid layers are connected by web systems consisting of continuous or staggered, vertical, or inclined trusses, such as those of the Pratt or Warren type.

The space frame geometry can be derived from the all-space-filling polyhedra, where edges and vertices form a spatial network. The polyhedra are packed face-to-face, leaving no free space. Double-layer space trusses can either be produced by the additive process of packing or by the subtractive process of cutting the given spatial aggregate with two horizontal, parallel planes. Before continuing the discussion of close packing systems, the nature of polyhedra is briefly reviewed in order to develop an understanding for the degree of regularity of these elements.

A polyhedron is defined as a three-dimensional figure composed of at least four faces (F) intersecting along the edges (E), which in turn intersect at the vertices (V). The angle formed by two faces along the common edge is called the *dihedral angle*. Leonhard Euler (1707–1783) proved that the number of vertices, edges, and faces bear a constant relationship to each other.

$$F + V = E + 2 \qquad (6.9)$$

For example, for a tetrahedron with four triangular faces, four vertices, and six edges (Fig. 6.11a),

$$4 + 4 = 6 + 2$$

Polyhedra may be organized according to their degree of regularity as expressed by the number and type of polygonal surfaces, how the polygons meet at their common edges, the dihedral angles at the edges and the number of polygons meeting at a vertex. Of special interest are the *uniform polyhedra* with identical vertices, which all touch a circumscribed sphere. They are composed of regular polygons meeting in pairs at a common edge; the dihedral angle at the edge is always less than 180°. It can be shown that not more than five regular polyhedra exist composed of only one type of face each; they are known as the five *Platonic solids* in honor of Plato, who was the first to mention them in his book *Timaeus* (c.a. 300 B.C.) They are the tetrahedron, octahedron, and icosahedron, all composed of equilateral triangles; the hexahedron (cube) composed of squares; and the dodecahedron composed of pentagons. The concept of the regular polyhedron is important since it consists of regular polygons for all its faces with vertices and dihedral angles identical. This feature of repetition is a fundamental consideration for the design of space frames.

Although all the Platonic polyhedra can be arranged inside each other, they are shown in Fig. 6.11 only inside the cube for purposes of visualization. Their important properties, including dihedral angles, are as follows:

Tetrahedron	4F (triangles), 4V, 6E, 70°32', stable
Hexahedron (cube)	6F (squares), 8V, 12E, 90°, unstable
Octahedron	8F (triangles), 6V, 12E, 109°28', stable
Dodecahedron	12F (pentagons), 20V, 30E, 116°34', unstable
Icosahedron	20F (triangles), 12V, 30E, 138°11', stable

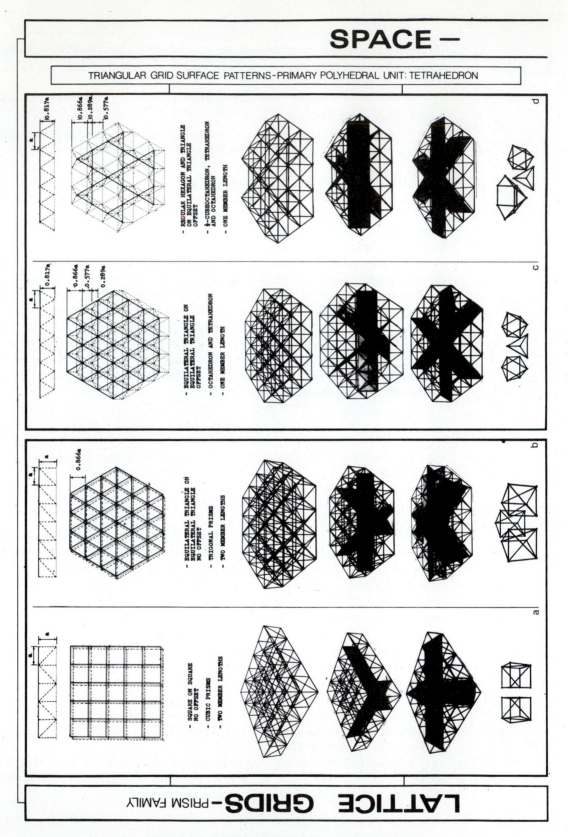

TRIANGULAR GRID SURFACE PATTERNS-PRIMARY POLYHEDRAL UNIT: TETRAHEDRON

LATTICE GRIDS—PRISM FAMILY

d
- REGULAR HEXAGON AND TRIANGLE ON EQUILATERAL TRIANGLE OFFSET
- ½-CUBOCTAHEDRON, TETRAHEDRON AND OCTAHEDRON
- ONE MEMBER LENGTH

c
- EQUILATERAL TRIANGLE ON EQUILATERAL TRIANGLE OFFSET
- OCTAHEDRON AND TETRAHEDRON
- ONE MEMBER LENGTH

b
- EQUILATERAL TRIANGLE ON EQUILATERAL TRIANGLE NO OFFSET
- TRIGONAL PRISMS
- TWO MEMBER LENGTHS

a
- SQUARE ON SQUARE NO OFFSET
- CUBIC PRISMS
- TWO MEMBER LENGTHS

Figure 6.9 Double-layer space frame systems 1.

SQUARE GRID SURFACE PATTERNS-PRIMARY POLYHEDRAL UNIT: HALF-OCTAHEDRON

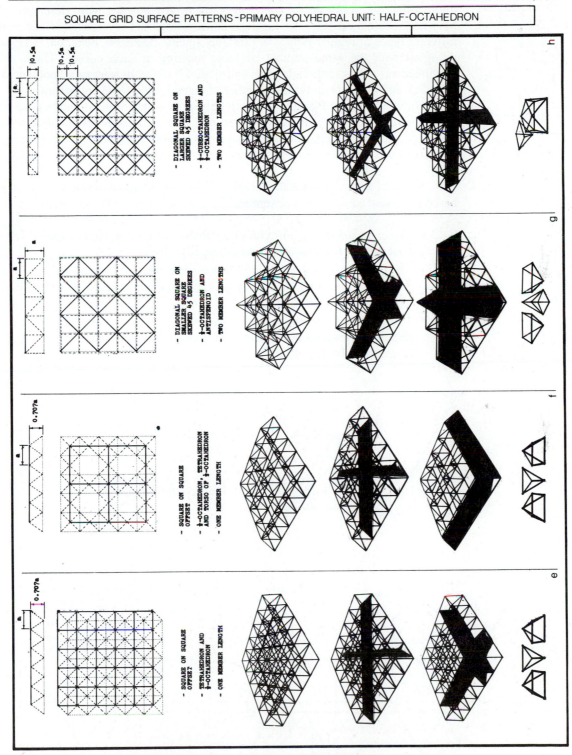

Figure 6.10 Double-layer space frame systems 2.

There are two *quasiregular polyhedra* not having identical regular faces: the cuboctahedron (dymaxion) and the icosidodecahedron. The *semiregular polyhedra*, often called the *Archimedean solids* in honor of Archimedes, who described them, have neither identical faces nor identical dihedral angles. They consist of two or three different types of regular polygons (triangle, square, pentagon, hexagon, octagon, or decagon) with the same number of faces meeting at each corner. There are 11 Archimedean solids, although often taken as 13 if the quasiregular polyhedra are included, and there are an infinite number of regular prisms and antiprisms.

Only three of the Platonic solids (tetrahedron, octahedron, and cube) and six of the Archimedean solids (truncated tetrahedron, truncated octahedron, cuboctahedron, truncated cuboctahedron, rhombicuboctahedron, and truncated dodecahedron) fill space (or form three-dimensional lattices) either by themselves or in combination with each other.

Some of the more typical examples of interest to space frame construction as based on spatial equipartition or semiregular equipartition are the following:

- The eight *single self-closepacking polyhedra:*(a) cube, (b) hexagonal and trigonal prisms, (c) truncated octahedron, (d) others
- The ten *dual close packing polyhedra:* (a) octahedron + tetrahedron, (b) octahedron + cuboctahedron, (c) octahedron + truncated cube, (d) tetrahedron + truncated tetrahedron, (e) others
- The eight *triple close packing polyhedra:* (a) truncated tetrahedron + cuboctahedron + truncated octahedron, (b) truncated tetrahedron + truncated cube + truncated cuboctahedron, (c) others

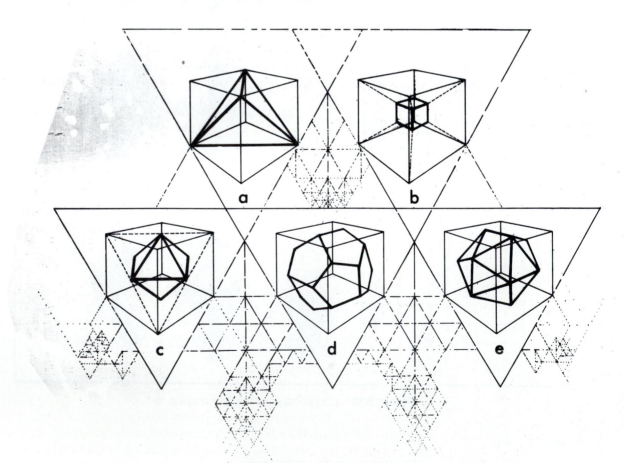

Figure 6.11 Platonic solids.

Any of these space-filling polyhedra systems can be cut by parallel horizontal planes to form double-layer space frames such as the ones shown in Figs. 6.9 and 6.10. Notice that cases c to f only employ one member length, *a,* with the truss depths adjusted accordingly, while the other cases use different member lengths. The space frames shown in the two figures can also be generated by packing the polyhedra identified in Fig. 6.6. All the solids in this figure are derived from the cube; hence all dimensions such as height and member length are a multiple of $\sqrt{2}$.

Double-layer space frames can also be visualized as consisting of a collection of prefabricated, skeletal pyramidal units with a square (one-half octahedron), triangular (tetrahedron), pentagonal, or hexagonal base. These pyramids can be connected in many different ways to produce the desired spatial network. For instance, the space frames based on square space grids (Fig. 6.10) consist of *square pyramids* joined along their top and bottom faces, while the one based on triangular space grids (Fig. 6.9c) is composed of *triangular pyramids.*

Even though standardization is a basic requirement of mass production, the all-space-filling systems just discussed are not necessarily the most useful; less regular ones may be more desirable, depending on the situation, because of function, availability of materials, or construction considerations, for example. Furthermore, the force flow along the space frame varies from a maximum to a minimum. Thus, the usage of identical member lengths and possibly sizes, together with the same connectors, may not be economical from a material point of view.

Structural Behavior.   The stability of single-layer, envelope-type space frames, such as those of the basic polyhedra units in Fig. 6.6, is checked by Eq. (6.8). While the stable, single-layer systems have triangulated surfaces for statically determinate boundary conditions, the multilayer space frames usually do not. Hence, to investigate the stability of multilayer aggregates, new mathematical expressions will have to be developed, a subject that is beyond the purpose of this presentation. However, certain conclusions can be drawn from the building blocks forming the space frames. Of the closepacking polyhedra in Fig. 6.6, only the trigonal prism, octahedron, and tetrahedron are stable, while the other units are unstable (see Problem 6.1).

It is apparent that if a double-layer space frame is composed of stable polyhedral units, then the entire assemblage must also be internally stable, such as the triangular lattice grid (Fig. 6.9b) and the triangular grid on triangular grid offset (Fig. 6.9c) composed of tetrahedra and octahedra. For this situation only, the support system must be stable.

Should the hinged space frames be composed of stable and unstable units, the total structure may still be internally stable if the stable units are connected to each other in a manner such that new stable assemblages are generated. This, however, is not the case for the space frame square on square offset (Fig. 6.10e), where the four tetrahedra do not join to form a new stable unit. But it is stable when supported along all four sides as discussed later.

For the condition where space frames are composed of unstable polyhedral units, such as the square lattice grid (Fig. 6.9a) or the diagonal-on-square cases (Fig. 6.10g, h), the entire structure must also be internally unstable. It is shown in the discussion of polyhedral domes that even for internally unstable structures, when sufficient external supports are provided to compensate, the overall structure can be stable. The same reasoning, obviously, is applicable to space frames or any other structure. For larger-scale structures, the spatial network is internally stabilized by adding members, usually in the compression plane (Fig. 6.12), rather than externally at the supports. A square grid may be stabilized by additional face diagonals causing torsional stability by either enclosing the structure along its periphery (a), by subdividing the structure plane into large-scale triangular units (b, c), or by trussing each bay (d). In addition, the face diagonals create a more direct load path and improve the stiffness. If the roof decking is properly fastened to the frame, it also can provide the necessary torsional rigidity of a diaphragm in a stressed skin fashion.

From a behavioral point of view, space frames can be organized according to their response to loading, using as example flat double-layer plates (Fig. 6.13), as follows:

- *Truss grids* (Figs. 6.9a, b; 6.13a to c): Vertical trusses arranged in a two- or three-way fashion form truss grids. The two-way rectangular grids can be placed parallel/perpendicular or diagonal (skew) to the boundaries; the skew grids, often called *diagrid* structures, provide more rigidity, especially at the corners, thereby generating additional support in these areas and a reduction of span. While the load in case *a* of Fig. 6.13 is carried along one direction only, it is distributed by the truss grids of cases *b* and *c* in a two- and three-way manner. When the grid members are placed closer together and trusses are replaced by girders or beams, the system is called a beam grid and forms a *grillage-type* structure. For closely spaced concrete ribs acting compositely with the slab, the plate structure forms a slab grid known as the *waffle slab*.

- *Folded plate trusses* (Fig. 7.5d): This condition of one-way spatial surface action is not shown in Fig. 6.13, but is discussed in Chapter 7. Visualize the parallel trusses of case a to lean against each other to form a corrugated surface.

- *Space trusses* (Fig. 6.13d): Two folded plate trusses are tied together at their base to form a new linear spatial beam, which resists forces in all directions, including torsion.

- *Space grid trusses* (Fig. 6.13e-h): By letting the stiff space trusses intersect in a two-way grid fashion (e, f, g), a space frame of the square-on-square offset type is formed by connecting square pyramids, providing considerably more torsional stiffness than the two-way truss grids. This structure can also be visualized as a two-way folded plate structure. The interaction of the linear space trusses is shown in (g); the compression member is assumed to consist of two members so as to improve the buckling capacity of the top chord.

For the condition where one of the planar grids is diagonal, such as the diagonal square on smaller square (h), the force flow can be visualized more along three directions, with one of the trusses being staggered. This surface structure is stiffer than case (g) and should come closer to the behavior of the two-way solid slab. However, when tetrahedra and octahedra are connected, the much stronger and stiffer, true three-way space grid results. In other words, the stiffest double-layer space frame is the one formed by triangular grids offset with respect to each other (Fig. 6.9c). Here, three intersecting, continuous, inclined trusses distribute the loads truly in three directions.

Planning Considerations. Space frames can be built from all kinds of materials, although the ones discussed here and used mostly in practice are made of steel or

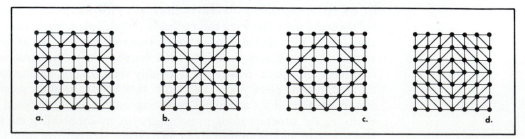

Figure 6.12 Location of face diagonals.

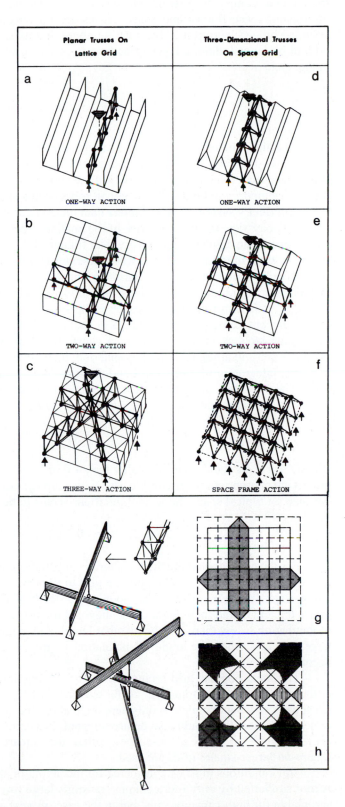

Figure 6.13 Structural behavior of double-layer space frames.

aluminum. They can behave as hinged skeletons (i.e., space trusses), as rigid space frames, or as composite systems, where skin and skeleton act as a unit. In the following discussion, space frames and space trusses are no longer differentiated; joints are assumed hinged, permitting free rotation about any axis. The typical depth-to-span ratio of horizontal, double-layer space frames used as roofs for ordinary conditions is in the range of

$$\frac{1}{12} \leq \frac{d}{L} \leq \frac{1}{25} \quad \text{or} \quad \frac{1}{9} \text{ for cantilever spans}$$

When the flat roof plate is supported around the edges, use $d \cong L/20$ for three-way space grids, $d \cong L/15$ for two-way spans, and $d \cong L/10$ to $L/12$ for one-way spans. When the space frame is supported only at the corners, use $d \cong L/15$. Furthermore, for the same support conditions, the diagonal grid systems and the dense layout systems (e.g., proprietary systems) are shallower than rectangular grid systems.

The bay size, a, or modularity of the spatial structure can be derived directly, should one of the regular or partially regular frames of Figs. 6.9 and 6.10 be used, since for these structures truss depth and bay size are related and often approximated as equal to each other for preliminary design purposes. However, other spatial geometries may be more economical for certain span and loading conditions. In this case, the bay size is dependent on considerations such as the location and type of supports, the building dimensions, the capacity of the deck or roof joists, the relationship of axial stresses to bending stresses in the compression chord, the density of members (number of members per unit area), repetition, orientation, inclination, and type of members, type of joints, integration of mechanical and electrical equipment, and other depth limitations. The maximum *bay size* is often assumed not larger than one-fifth of the span or 30 ft, or a typical module size in the range of $L/6$ to $L/4$.

$$L/5 \geq a \leq 30 \text{ ft.}$$

But the bay sizes should not be smaller than 4 ft or larger than 12 ft for ordinary conditions when proprietary systems are used with a typical range of 5 to 7 ft. An optimum bay size for typical space frame configurations seems to be in the range of.

$$a \leq d \text{ to } 2d, \quad \text{or} \quad 0.5 \leq d/a \leq 1.0$$

For the condition where the top chord is in compression, it may be advantageous to use a top grid smaller than the bottom grid to improve the buckling capacity of the compression chord members. The module and depth of the space frame also determines the length and inclination of the diagonal members. Generally, when the angle of inclination is less than 30°, the member length and magnitude of force become too large, whereas when the angle of inclination is greater than 60°, the member grid becomes too dense.

The most popular grid arrangements for rectangular plan shapes use diagonal grids either as top or bottom layer as, for example, the typical square-on-larger diagonal square, or when the plate is only supported at the four corners, a square-on-square offset set diagonally should also be investigated. For composite plan shapes made up of rectangular areas (e.g., L-shape), often the square-on-square offset or the square-on-larger-square grids are used.

Soil conditions must be considered for the layout of space frame roof structures. A roof supported by only four columns transmits large forces to the ground, which is acceptable for high-capacity soils, but in the case of weak soils, the cost of installing the large foundations that are required may make it economically infeasible.

Joints. Clearly, the joints in space frames are much more sophisticated than in planar trusses since they are in three-dimensional space and may have to accept a large number of members. The economy of the space frame is closely related to the method of joining. The number of joints per square foot should be kept to a minimum. Larger modules with fewer members and nodes are generally more economical. The selection of the design of the connection depends on the scale and shape of the structure, the magnitude of the loads, and the joint capacity needed. It depends on whether it is assembled in the field or in the shop; it is a function of the shape and size of the bars as well as the geometry of the spatial member network, that is, the angle of intersection. In addition, it depends on requirements of accessibility and workability, easiness of construction, adaptability to attachments of other materials, ease of dismantling (if so desired), and esthetic criteria. Too many members joining at a node cause congestion, requiring the member ends to be tapered. A typical joint has approximately six to ten members attached to it. Connections should allow dimensional tolerances so that the members will fit together properly.

We distinguish generally between proprietary and nonproprietary systems. Most proprietary systems are based on having a patent on their joints. Nonproprietary systems are custom-designed space frames that vary with each building and are used for special situations or where large spans are required of more than approximately 200 ft, where commercial systems may not be strong enough or economically competitive. They usually consist of standard rolled steel shapes with welded multiplanar gusset plates as joints, fewer members and larger modules, and simple geometry so that the number of members framing into the joint is kept to a minimum.

Proprietary systems, that is space frame systems marketed by commercial manufacturers, are based on repetitive use of members, particularly joints as well as structure shapes. They are typically used in the range of 50- to 100-ft spans or more. In this context, only planar double-layer space frame systems are briefly introduced. Most of these systems are adaptable to changes in depth, grid size, and magnitude of loading, as well as various support conditions.

Some of the more common proprietary space frame systems currently available are the British *Nodus* and *Space Deck* systems, the French *Unibat* system, the German *MERO* and *Oktaplatte* systems, the Canadian *Triodetic* system, the Dutch *Octatube* system, the Japanese *Tamoe Unitruss*, *NS Truss*, and *TM Truss* systems, and the American *Unistrut, Power-Strut, IBG, PG, OMNI* HUB*, and *Multihinge* systems.

The history of prefabricated space frame structures started in 1940 with Max Mengeringhausen's invention of the ingenious MERO connector, now well known all over the world. In 1944, Charles W. Attwood erected the first building using the Unistrut concept, which was marketed in 1955. The Triodetic system was developed by A. F. Fentiman in the early 1950s, and Stéphane du Château introduced his first space frame systems, which developed into the Unibat system around 1970. The Space Deck system was introduced in the early 1950s, the Nodus system in 1972, and Mike Eekhout developed the Octatube system in 1973. Fujio Matsushita is often considered the originator of prefabricated steel space structures in Japan.

It is not a simple task to organize space frames based on the connection of their members, since infinitely many solutions are possible. The members can be directly attached to each other by welds, bolts, threads, keys, adhesives, grooves, pins, press, and so on. The connectors may be flat, bent, or built-up gusset plates; they may be clamps as used for scaffolding, or they may have the shape of hubs, stars, balls, or polyhedra internally threaded or with threaded projecting shanks. They may be a single piece or consist of several parts; they may have to be assembled or they may be a finished product. Any member types can be used, such as solid, tubular, or open sections (angles, channels, wideflanges, tees), where the members may include special end pieces. The end pieces may be a tapped pipe, flattened pipe, solid rod, tapped solid rod, flat sheet, forged cone, etc. The group of stressed skin type of space structures using thin metal sheets of steel or aluminum or skins of plywood, plastics, paper, ferrocement, and so on, are not further discussed here.

Members can be connected by the familiar methods of bolting, welding, gluing, keying, and by other special techniques. Some typical connection nodes that can be made from hot rolled steel sections, hot forging, cast steel, or aluminum alloys, etc., are the following:

- Connector plates (e.g., flat, dished): *Power-Strut, Unistrut* systems
- Hollow sphere: *PG, IBG, Nodus, Oktaplatte, NS Space Truss, Tomoe Unitruss* systems
- Solid balls: *MERO, Omni* Hub, TM Truss* systems
- Extruded hubs: *Triodetic* system
- Welded multiplanar gusset plates
- Nodeless joints using struts with ears or tags: *Multihinge* system
- A double-layer space frame system may also consist of prefabricated pyramids with triangular, square, or hexagonal bases bolted together, possibly leaving empty intermediate spaces: *Space Deck* and *Unibat* systems

Some of the more typical connection types are shown in Fig. 6.14. Information with respect to the strength (capacity) of the proprietary systems is provided by the respective manufacturers.

- *IBG System, Roper IBG International, case g*. The system consists of hollow ball nodes with openings for inserting fasteners into the round aluminum tubes.
- *MERO System, MERO Structures, Inc., case k*. The system consists of solid forged steel spheres into which threaded holes are bored, round tubular members with cone-shaped ends, and protruding threaded bolts fitting the tapped holes in the nodes.
- *Unistrut System 1 or previous Moduspan System, Unistrut Corp., case e*. The system consists of bent connector plates with contact faces in the direction of the struts and cold-formed, U-shaped members bolted to them. Reinforcing parts are added if a higher load-carrying capacity is required.
- *Nodus System, British Steel Corp., case c*. The system is comprised of tubular steel sections, where the chords are clamped into the joint (consisting of two half-castings) by a single, high-strength center bolt. The web members have forked end connectors, which are secured to lugs on the joint by stainless steel pins.
- *Power-Strut System, Van Huffel Tube Corp.,case b*. The system consists of flat steel connector plates to which the rectangular tubular members can directly be bolted, while the web members are bent at their ends so that they can also be attached.
- *Triodetic System, Triodetic Space Frames, Inc., case f*. The system consists of slotted hub connectors (aluminum extrusions) and usually aluminum tubing pressed at the ends so that it can be force fitted into the slots.
- *Multihinge System, Pearce Structures*. The system has no plates, hubs, or nodes; strut member ends are simply bolted to each other. Each strut has small flanges (called hinge elements) welded to both ends of the round steel tubing. Paired hinge elements join strut ends together with a special bolted connection.
- *OMNI*HUB System, Starnet Structures*. The system consists of solid aluminum spheres with threaded holes and round tubing with tapered solid plugs

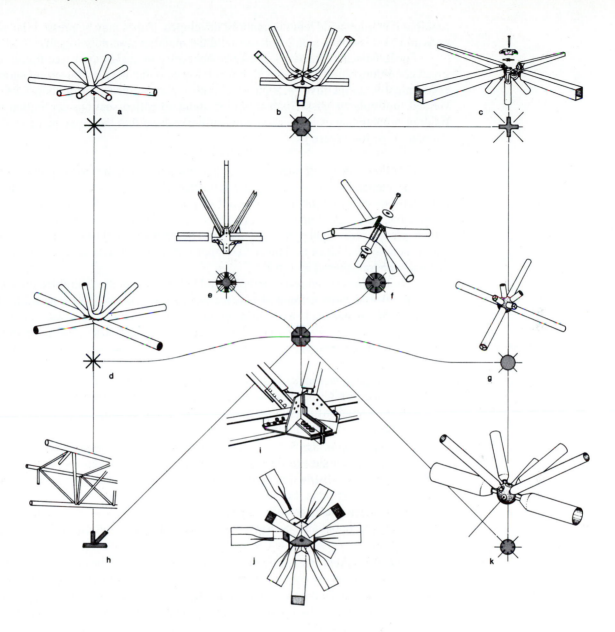

Figure 6.14 Common space frame joints.

welded to the ends of the struts. Recessed within the strut plug is a steel bolt that is screwed into the hub.

- *PG System,* PG Structures. The system consists of hollow spherical nodes with a removable portion and nuts inside, round or square tubes, and high-strength steel rods. The connection is achieved by rods that run from node to node within the tubes and are prestressed, thereby placing all the struts into compression.

No connector piece is necessary if the members are welded directly to each other, such as the bent diagonal rods or the tubular webs to the top and bottom chord members (cases d and h), keeping in mind that welding does not allow for any member tolerances. For cases where pipe members intersect, additional joint pieces may be required if enough weld length cannot be developed. Tubes may be pressed flat at their ends and bent so that a single bolt can hold the member assembly together (case a).

For large-scale structures, it may be necessary to design the space frame for the given condition, rather than use a commercial system that may not be strong enough or economically competitive. A typical solution (case i) consists of a shop-fabricated, welded, multiplanar gusset unit to which standard rolled sections are bolted in the field; or tubular shapes are necked down at the ends so that they may be connected to the plate-type joint (case j).

Erection.   Another important design consideration relates to the process of construction. Depending on the scale and sequence of erection, temporary interior supports may be required, or the entire roof or portions of it may be assembled on the ground and then jacked up or lifted into position, rather than constructing it in the air (i.e., in place). Individual pyramidal units can be stacked and thus easily shipped to the site. Frame assemblages no longer than approximately 60 ft or wider than 15 ft can generally be transported by truck.

Space frame components can be assembled in place or preassembled on site or in the shop. Special construction methods include deployable (foldable/extensible) structures that are transported in a collapsed compact state and then deployed into their service shape, as is most important for space technology. The typical methods of construction are as follows:

> *Scaffold methods:* Individual space frame components are assembled in place at high elevations requiring scaffolding and temporary supports, but only small lifting equipment. Moving scaffolds may be used to reduce the amount of scaffolding.

> *Block methods:* Sets of components are preassembled on ground (in the shop or on site) to form subassemblies or blocks, therefore allowing more efficient construction control and less work at high elevations and less temporary supports, but requiring heavier lifting equipment, such as cranes or winches attached to the building framework or temporary column supports. Blocks could also be assembled on temporary scaffolding and slid into position.

> *Lift-up methods:* Large portions of the structure or the complete structure is lifted by heavy equipment (e.g., hydraulic jacks) into place; the structure can also be pushed up by jacks. Since most of the member assembly is done on or near the ground, the work quality is easier to control and the work flow is more efficient.

Flat Space Frame Roofs

Konrad Wachsman designed his ambitious long-span space frame hangars during the 1950s (e.g., Fig. 5.55b), and at the same time Mies van der Rohe investigated structures allowing vast interior space completely column free with supports concentrated on a few points along the perimeter. Mies van der Rohe demonstrated his ideas powerfully in 1958 with the proposal for the administration building for the Bacardi Company, Santiago, Cuba. He proposed a 177-ft square coffered concrete plate supported by only eight cross-shaped, tapered concrete columns on the perimeter, which were to occur 50 ft in from each of the four corners so that the corners would be cantilevered.

Typical examples of horizontal space frame roof structures are shown in Fig. 6.15. They are either supported along their edges or composed of a multibay system. The space frame geometries range from two-way Vierendeel trusses, two-way trusses diagonally braced (F) or not braced (C), and the girder grillage (G), to the hexagonal Vierendeel frame (I); they range from four-sided pyramids interconnected in single or

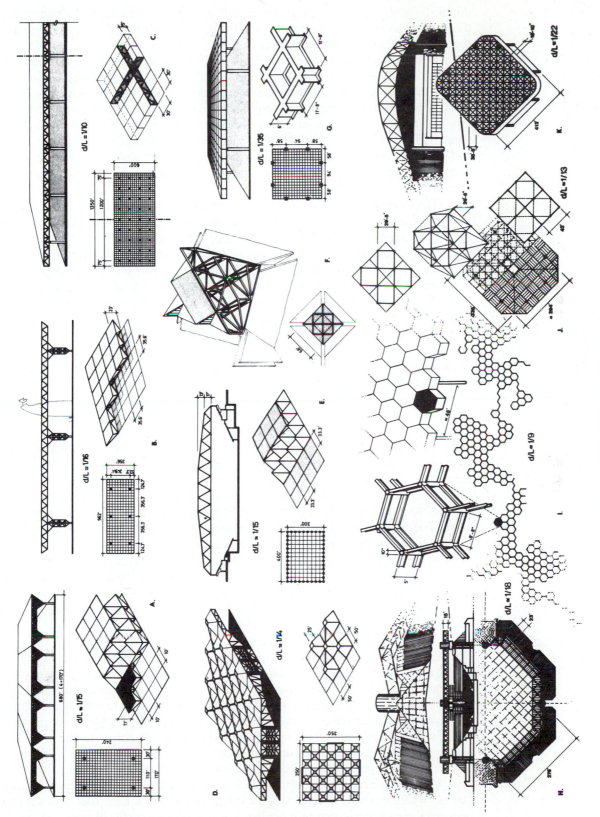

Figure 6.15 Flat space frame roofs.

double layers (B, E, A) to truncated pyramids arranged in a checkerboard pattern with diagonal rods connecting the pyramids in the upper chords (D). The cases range from the small-scale to long-span structures.

In the spirit of the never built Barcardi building in Cuba, Mies van der Rohe designed the famous New National Gallery in Berlin, Germany (1968, Fig. 6.15G). Here, a 210-ft square, flat, two-way, 6-ft-deep steel girder system that forms a nearly 12-ft square grid is pin connected to eight flanged cruciform columns on the perimeter.

Quite different is the roof structure of Stanley Tigerman's St. Benedict's Abbey Church in Benet Lake, Wisconsin (1972, Fig. 6.15F). Its triangular sloping roof planes and triangular profiles suggest a complex support structure, which, however, is not the case. Two-way laminated wood trusses placed diagonally cover the 68-ft square plan area.

The following cases are all of the long-span type and are briefly presented in historical sequence. The Pauley Pavilion at the University of California in Los Angeles is one of the largest horizontal-span space frame structures in the United States (1965, Fig. 6.15E). The 300- × 400-ft clear span plate is supported along all four edges and is of a hip-roof configuration with a 30-ft depth at the center and a 17-ft depth at the perimeter to efficiently respond to bending moments and to provide drainage. The space frame consists of a series of four-sided pyramids of varying height with their tops (vertices) connected by a rectangular member grid forming a square-on-square offset space grid of 33.33-ft square bay size. The steel framing, weighing 15 psf, consists of W10 and W14 members with three-dimensional gusset plates as connections. Architects for the project were Welton Becket and Associates and the structural engineer was Richard R. Bradshaw.

Kenzo Tange's magnificent Pavilion of Expo '70 in Osaka is a giant 23-ft-deep cantilever space frame structure supported on six columns that covers an area of 962 × 356 ft (Fig. 6.15B). The space frame grid is based on the square-on-square offset using 35.63-ft bay sizes. The framing consists of steel tubes ranging from about 24 to 16 in. in diameter, necked down at their ends to be connected to steel balls of 30.5-in. diameter as based on the MERO system; the structure weight is 19 psf.

Denver's Currigan Hall space frame roof (1970, Fig. 6.15A) consists of four 170- × 240- ft sections, where each is supported at its corners by inverted pyramid columns. The 14.14-ft-deep space frame is composed of three horizontal grid layers of the double square-on-square offset type, which can also be viewed as a two-way system of inclined double Warren trusses. The framing, which weighs 13.2 psf, consists of equal length, 10-ft, double-angle steel members as based on a 10- × 10 ft-module. Chicago's enormous McCormac Place, destroyed by fire in 1967, was reconstructed in 1971 using a two-way, 15-ft-deep steel truss system supported on columns spaced 150 ft on center, with a 75-ft cantilever on the perimeter (Fig. 6.15C).

Atlanta's Omni Coliseum space frame roof (1974, Fig. 6.15D) covers a clear-span, 350-ft square area and is supported by giant cantilever wall trusses along the perimeter. The roof, requiring 16 psf of structural steel, consists of 25-ft-high truncated pyramid po ds made of heavy steel plates, 50-ft square at their bases and 15-ft square at their tops, that alternate with flat steel framed areas in a checkerboard pattern. The pyramids touch each other at the corners so that the bottom edges form the space frame's lower chord, while the upper chord is formed by diagonal members that connect the pyramid's corners. This unique space frame system was patented by the structural engineers Prybylowski & Gravino and called the *Ortho-quad Truss*.

One of the largest horizontal space frames is the roof for the Reunion Arena in Dallas, which opened in 1980 (Fig. 6.15K). The 420-ft square roof spans 412 ft between the eight pin-supporting slender, 6-ft-diameter concrete columns and cantilevers 70 ft at all four corners. The structure is of the square-on-square offset type with a module of 36 ft 6 in. and a depth of 20 ft ($d/L = 1/22$), weighing 22 psf, turned diagonally at 45° to the edges. Skewing the truss grid in plan stiffens the corners and causes reversals across them, effectively reducing the midspan moments and deflections. The roof structure was designed by Paul Gugliotta.

A 420- × 378-ft rectangular horizontal space frame with clipped corners covers the Carver–Hawkeye Sport Arena at the University of Iowa, Iowa City, Iowa (1984, Fig. 6.15J). It is resting on just eight concrete columns with clear spans of about 336 ft × 294 ft and is cantilevered beyond the columns ($d/L \cong 1/14$); the plate is floating on the columns to relieve stress buildup. The 25-ft deep space frame, weighing 17.5 psf, is based on the patented Takenaka truss principle, that is, a space truss consisting of a bottom square grid with 42-ft square bays and a top chord frame turned 45° or skewed in relation to the bottom chord, in other words, a diagonal square-on-square space grid. The upper chord and diagonal members are 12-in.-diameter steel pipes or tubes, while the bottom chord members are wide flange sections. In winter, stresses due to temperature difference between the exposed upper portion of the space frame and the inside bottom chord carrying the roof joists are most critical design considerations. The roof structure was designed by Geiger Associates.

Also shown is the magnificent Palais Omnisport at Paris–Bercy (1984, Fig. 6.15H), designed by the architects M. Andrault, P. Parat, and A. Guvan, together with the engineer Jean Prouvé, which is covered by a 15-ft deep two-way truss grid, as based on 23-ft square modules, spanning the arena on the diagonal. The space frame uses the MERO system. It covers an area of about 329 × 329 ft and is supported at each of its four splayed corners by cylindrical concrete shafts of 20-ft diameter, which are set 263 ft apart.

Approximate Design of Flat Double-layer Space Frames

Double-layer space frames, in general, are highly indeterminate; however, should they be determinate, the structure can be solved by statics, using for each joint the three equations of equilibrium [Eq.(6.4)]. Since usually more than three members are attached to a node, the bar forces cannot be found directly from the joint to which they are connected. For instance, the space frame with triangle-on-triangle offset (Fig. 6.9c) has 74 joints with nine members attached to each interior node. To determine the bar forces, $3(74) = 222$ equations are needed, which obviously is not feasibly solved by hand; computers must be used. The situation is by far more complex for indeterminate space frames, where initial member sizes must first be assumed before a structural analysis can be performed. In the mid-1960s, even computers had trouble handling extremely large structures, as was the case for Buckminster Fuller's double-layer geodesic dome for Expo '67 in Montreal (see Fig. 8.34, top), which had about 6000 nodes, 24,000 members, and 27 miles of tubes.

However, to develop an intuitive feeling for behavior and also to be able to estimate the preliminary member sizes, the *plate analogy* is applied to double-layer space frames. In other words, the flat double-layer space frame is assumed to behave like a solid plate, keeping in mind that the force flow must follow the grid prescribed by the bar layout and that it travels along the stiffest members to the respective supports. This, however, is not the case for a solid homogeneous slab, where the direction of the force flow changes with loading. As the density of the spatial member network increases (i.e., for a high number of nodes) and also with the use of certain inherently strong geometries like the triangular grid on the offset triangular grid, the space frame approaches the behavior of the solid plate. The axial bar forces are obtained from the shear and moment values of the analogous slab multiplied by the respective member spacing. But we must keep in mind that depending on the type of spatial grid the maximum tension and compression in the chords do not necessarily appear at the same location as the maximum moment.

The behavior of flat space frame roofs (i.e., slabs) depends on the type and location of support system. Typical slab structures are shown in Fig. 6.16a to i. The horizontal surface structures may form only a single bay (a to f) or they may be multibay systems (g, i). Single plates may be square, rectangular, triangular, etc., in other words, of any configuration. The roofs may be supported by walls or columns, which in turn may be arranged in various ways, but requiring a minimum of three columns to be stable. If walls or closely spaced columns are placed parallel to the long sides (b), the space frame can be considered to transfer loads directly from wall to wall like a rectangular one-way slab. However, if the supports are positioned along the periphery of a roof, the roof structure will respond similarly to a two-way slab if the ratio of the longer side to the shorter one is not larger than 2. To obtain an effective two-way action of the slab, its proportions of long span to short span (L_L / L_s) should be between 1 and 1.5.

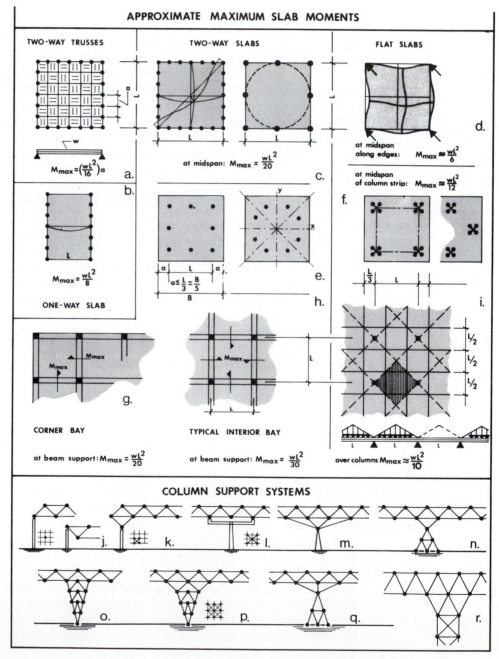

Figure 6.16 Slab analogy and slab support.

In the following presentation a square plate under uniform gravity loading, as indicated in Fig. 6.16, with different support locations is investigated to develop an understanding of stress flow, keeping in mind, however, that the slab analogy can only be considered a rough, preliminary design approach.

Plate Types.   Double-layer space frames are usually supported along the edges or at the corners with or without overhangs.

A. Supports Uniformly Distributed around Edges: First, two-way grid structures of rectangular and diagonal layout, such as single-layer concrete waffle slabs or double-layer trusses, are investigated. After the direct grid structures, true space grid structures are presented.

The behavior of a square beam grid, continuously supported along its circumference (Fig. 6.16, case a), depends on the spacing of the ribs. Should the two-way beams consist of relatively deep members of equal size, such as trusses, and should these members be spaced relatively far apart, this highly indeterminate, intersecting grid structure may be replaced for approximation purposes by statically determinate vertical trusses, where each carries one-half of the respective bay loading, assuming the joist layout as shown in Fig. 6.16a. The uniform load for a typical truss is $w(a/2)$ with a maximum moment at midspan of approximately

$$M_{max} \cong \frac{(wa/2)L^2}{8} \cong \frac{wL^2}{16}a \tag{6.10}$$

Similarly, it can be shown that the maximum bending moment for a 45° diagrid (Problem 6.15) is approximately

$$M_{max} \cong (wL^2/16)a\sqrt{2} \tag{6.11}$$

However, should the beams be closer together, the behavior of the roof structure is that of a grillage and may be roughly approximated by considering it as a solid slab, ignoring its difference in torsional rigidity. In true space frames, this torsional strength is provided by the spatial diagonal web members; hence, a space frame roof continuously supported and hinged along its edges to the walls or the closely spaced columns (case c) acts very much like a simply supported two-way slab, which has its maximum moment at *center span* with a magnitude, according to Table A.7, of approximately

$$M_{max} = wL^2/20 \tag{6.12}$$

Notice that this moment is 25% less than the one used for the truss grid due to the continuity of the solid slab providing twisting moments, thereby relieving the bending moments. There are diagonal, negative moments of smaller magnitude in the corner region that generate a clamping effect.

The shear is not uniformly distributed along the periphery as is the case only for a circular slab; it is maximum at the middle of the plate sides; in the corners the plate has the tendency to lift up. Generally, the load to the edge beams is based on the subdivision of a rectangular plate into triangular tributary areas for the short beams (with spans L_s)and trapezoidal areas for the long beams (Fig. 2.13b). For this condition, the maximum shear on the slab occurs at the ends of the center strips and may be conservatively approximated as

$$V_{max} = wL_s/2 \tag{6.13}$$

For approximation purposes, it can be assumed that as long as there is at least one support at midlength of the edges (case c), the space frame can still be considered as a two-way slab simply supported along its perimeter. However, if these midspan columns along the edges are taken away, the behavior of the slab drastically changes, approaching that of a *flat slab* (case d). For further discussion of plates simply supported along all edges, refer to Example 6.3 where direct grid and true space grid structures are investigated.

B. Supports at Four Corners Only: The response of a slab, simply supported at its corners, to uniform gravity loading only is extremely complex. The slab must carry the loads directly to the corner columns, which is quite different than the condition where the two-way slab transmits the loads to the stiff edge beams (or columns along the edges), which in turn bring them to the corner columns. In a flat slab there are no beams and the slab itself must provide this additional task; hence the magnitude of the moments is in the same range as for one-way slabs and much larger than for equivalent two-way slabs continuously supported. Here, the maximum moments occur at midspan along the edges and may be roughly approximated, according to S. Timoshenko and S. Woinowsky-Krieger (*Theory of Plates and Shells,* 2nd ed., McGraw-Hill, 1959), as for a square plate

$$M_{\max} = 0.153 \, wL^2 \cong wL^2/6 \tag{6.14}$$

Notice that the moment is more than three times larger than that of a two-way slab supported by beams, but the increase in deflection is by far higher. In addition, the effect of shear in the flat slab is much more pronounced because of the punching effect at the columns, where a quarter of the total roof load must be resisted by the slab in shear.

$$V_{\max} = w(L/2)^2 = wL^2/4. \tag{6.15}$$

It is apparent that, as the member sizes along the edges of the flat slab space frame increase and approach beam action, the stiffness as well as strength of the structure increases, eventually behaving as a two-way slab system on beams.

C. Cantilever Plate on Four-corner Supports: When the four-corner supports of the previous case are moved inward to form a two-way cantilever slab (case f), then the maximum field moments and deflection are very much reduced. For example, assuming for preliminary design purposes a cantilever span of $L/3$ as based on equal maximum positive and negative moments for a symmetrical double cantilever beam (i.e., column strip) under uniform loads (Fig. 5.7), the approximate maximum moment can be determined as follows:

The sum of the positive and average negative moments is

$$M_o = wL^2/8 = 0.125wL^2$$

The average negative cantilever moment is.

$$M_s = wa^2/2 = w(L/3)^2/2 = 0.056wL^2 = 0.45M_o$$

The average positive midspan moment is

$$M_F = M_o - M_s = 0.125wL^2 - 0.056wL^2 = 0.069wL^2 = 0.55 \, M_o$$

It is unrealistic, however, to treat the moments as average across the entire slab width since the stiffer column strips attract more load than the more flexible middle strip. Therefore, the column strips of $2(L/3 + L/4) = 7L/6$ width are assumed to resist

60% of the positive moment for the entire slab width and the middle strip of $L/2$ width, the remaining 40% (see Problem 6.10). In other words, for preliminary design purposes the column strips may be assumed to carry $60 - 40 = 20\%$ more than the average positive moment and the middle strip 20% less. Hence, the average maximum field moment in the column strips is approximately equal to

$$M_F = 1.2(0.069 \, wL^2) \cong wL^2/12 \tag{6.16}$$

Similarly, the stiffer column strips are assumed to resist 70% of the entire cantilever moments, or $70 - 30 = 40\%$ more than the average support moment.

$$M_s = 1.4(0.056 \, wL^2) = wL^2/12.76 \tag{6.17}$$

Notice that the average negative moments over the columns are nearly equal to the maximum field moments in the column strips, indicating the benefit of the $L/3$ cantilever span as based on uniform loading. Although the effect of cantilevering results in smaller chord sizes, it hardly influences the magnitude of the shear, that is, the web member sizes.

D. Continuous Multibay Plate with Column Supports at Bay Corners: As the cantilever lengths of the preceding case increase, the two-way slab approaches the interior panel of a multibay flat slab system (case i). The support moment of $0.45M_o$ for the cantilever slab changes to $0.65M_o$ for the interior slab. In other words, the maximum moments now occur over the columns. The behavior of a multibay flat slab system is very complex. The maximum moment for a typical interior square plate occurs over the columns and is roughly approximated (see Problem 6.10) as

$$M_{\max} = wL^2/10 \tag{6.18}$$

The magnitude of this moment is not necessarily conservative. The effective width of the imaginary beam (column strip) is dependent on the size of the column head. The support width and the rigidity of the slab determine how the moments are distributed.

Slab Analogies. For the preliminary design of double-layer grids, it is assumed that the chords resist the bending moments and the web members, the shear. Hence, the moment is resisted axially by the top, N_t, and bottom, N_b, chord members (see Fig. 6.17a).

$$M = N_t(d) = N_b(d) \quad \text{or} \quad N_t = N_b = M/d \tag{6.19}$$

The critical web members occur at the reactions where the shear is maximum. The forces in the diagonal members can easily be obtained by equating the vertical force components to the shear [e.g., Eq. (5.98)]. To determine the deflection of space frames, the flexural stiffness EI must be known. For preliminary design purposes, the moment of inertia can be roughly approximated by ignoring conservatively the effect of the web members. For the condition where the cross-sectional areas of the bottom chord members, A_b are equal to the ones of the top chord members, A_t, the moment of inertia is equal to

$$I \cong 2[A(d/2)^2] = Ad^2/2, \quad \text{where } A_t = A_b = A \tag{6.20}$$

But for the condition where $A_t \neq A_b$,

$$I \cong \frac{A_t A_b}{A_t + A_b} d^2 \tag{6.21}$$

Types of Supports. Space frames may be supported by space frame walls or other wall types and/or columns; they may also be cable supported. A space frame column must support gravity loads and possibly lateral forces. Typical column types are shown in Fig. 6.16j to r. They may be either simple one-point columns, two-point columns with one-way column arms, four-point columns with crosscolumn arms, or pyramidal supports; or they may be multipoint support systems. For point-supported, flat, slab-type space frames, punching shear around the supports (as for the familiar column footings) causes large forces in the web members. These forces may be reduced by providing shear heads, crossarms, or treelike supports similar to the drop panels or column capitals in concrete construction. Thus, a point support may be changed to a four-point support (k, l) or to a nine-point support (p) to increase the perimeter support area for resistance of punching shear and reduction of the positive field moments, especially deflections. Columns are either separated from the roof plane by a pin, allowing free rotation of the much stiffer roof structure and only resisting axial and shear forces, or they may be an integral part of the space frame so as to form a two-hinge frame in cross section, with the roof plane being rigid and not deforming under lateral loading. The pin supports are either at the ground level (o, p) or at some location along the column (k, m, n, q). The roof can be supported at the top or bottom chord level.

Lateral Force Resistance. The lateral loads must be carried by the columns or walls in either cantilever action or together with the roof in frame action, as is investigated for trusses in Fig. 5.54. However, should there be an independent shear wall structure to resist the lateral forces, the exterior columns can be constructed as pendulums supporting axial loads and allowing free expansion or contraction of the roof plane as caused by elastic behavior and temperature change. The upward acting wind can be ignored for preliminary design purposes since, in general, the roof weight is larger than the suction force. For the design of the cladding and roof membrane, the wind must be considered.

Preliminary Design. The space frame geometry is selected to minimize the number of joints (labor costs) and the weight of the structure by reducing the range between maximum and minimum member forces. For proprietary systems, it may be more economical to select a denser grid than to achieve less weight, whereas for long-span structures a sparser grid with fewer members and nodes is generally more economical. It is more efficient from a force flow point of view to support the structure along the boundaries either at the nodes of the lower or upper layers.

The weight of ordinary steel space frames of approximately up to 100- × 100-ft span is about 2 to 4 psf, whereas an equivalent aluminum space frame is about one-half of it. For the preliminary design of space frames, only uniform gravity loads are considered. Wind loading including uplift, differential live loads, ponding loads, and forces due to thermal movement are ignored. The top chord members are assumed to be only in axial action by neglecting local bending. In other words, concentric joints are assumed with the loads acting at the joints, keeping in mind, however, that the top chords may have to support the roofing structure directly (e.g., purlins, which are alternated in checkerboard pattern), thereby causing local bending in addition to the primary axial action (see Example 5.16). Sometimes the top chord members are made shorter than the bottom chord members to reduce buckling and bending. A separate roofing substructure may also be used to transfer the loads directly to the joints, thus eliminating bending.

EXAMPLE 6.3

Different space frame roof systems with $F_y = 36$ ksi are investigated for a gymnasium. The two-way roof covers a column-free space of 100- × 100-ft and is simply supported along its four edges. The structure must support a live load of 20 psf; the weight of the roof skin is assumed as 15 psf, the self-weight of the space frame as 4 psf, and other superimposed dead load as 6 psf. The change of weight for the different structural systems to be investigated is ignored in this preliminary study

(a) Two-way Truss Structure (Fig. 6.16a). The roof plane is subdivided in each direction into six equal spaces, yielding a square grid of $100/6 = 16.67$ ft (Fig. 6.17a, b). For a depth-to-span ratio of $d/L = 1/15$, the typical truss depth is $d = L/15 = 100/15 = 6.67$ ft.

Because of the relatively wide spacing of the trusses [see Eq.(6.10)], each may be considered to individually carry the load of $15 + 4 + 6 + 20 = 45$ psf or, for one half-bay,

$$w = 0.045(16.67/2) = 0.38 \text{ k/ft}$$

The maximum moment at midspan is

$$M_{max} = wL^2/8 = 0.38(100)^2/8 = 475 \text{ ft-k}$$

For the preliminary design of trusses, refer to Example 5.16 and Problem 6.12.

(b) Two-way Beam Grid. A square grid of 5 ft is selected. The beams are spaced relatively close so that this grillage structure may be considered analogous to a two-way slab simply supported along its periphery, ignoring the fact that its torsional rigidity is very small. The slab moment due to a load of $w = 10 + 15 + 20 = 45$ psf is approximately

$$M_{max} = wL^2/20 = (0.045)100^2/20 = 22.50 \text{ ft-k} \tag{6.12}$$

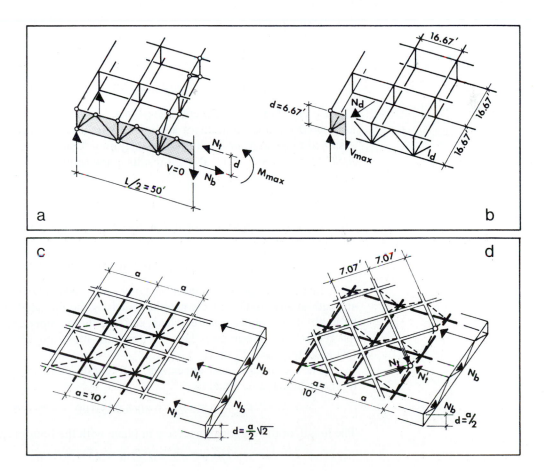

Figure 6.17   Example 6.3.

The critical beams along the center of the plate must resist the following moments as based on their spacing:

$$M_{BM} = 5(22.5) = 112.50 \text{ ft-k}$$

The required section modulus, by considering the compression flange laterally supported, is

$$S = M/0.66F_y = 112.5(12)/0.66(36) = 56.82 \text{ in.}^3$$

Try W18 × 35, $S_x = 57.6$ in.^{3}. (Table A.11).

The shear does not have to be checked since it will not be critical, but deflection must still be considered for this preliminary approach possibly cambering the structure.

(c) Diagonal Square-on-Square Space Frame (Fig. 6.10h). A truss depth of $d = L/20 = 100/20 = 5$ ft is selected. Depth and bay size are related by $a = 2d$, thus yielding a system with only two member lengths. Hence, the bottom square module is $a = 2(5) = 10$ ft, subdividing the roof into 10 equal bays in each direction. The top skewed square grid, as identified in Fig. 6.17, is

$$(a/2)\sqrt{2} = a/\sqrt{2} = d\sqrt{2} = 5\sqrt{2} = 7.07 \text{ ft}$$

The maximum forces in the chord members occur at the center of the grid. The maximum tensile force in the bottom chord at midspan is directly resisted and equal to

$$N_b = (M_{max}/d)a = (22.5/5)10 = 45 \text{ k} \tag{6.19}$$

The maximum compression force in the top chord at midspan is found by resolving the moment component N'_t into the diagonal forces N_t (Fig. 6.17d).

$$N'_t = 2N_t/\sqrt{2} = (M_{max}/d)a \tag{6.22}$$

$$N_t = (M_{max}/\sqrt{2})a/d = N_b/\sqrt{2} = 45/\sqrt{2} = 31.82 \text{ k}$$

The top chord members are considered to act only axially; that is, local bending is ignored for this preliminary approach, which means that the roof loads are assumed to be transferred directly to the joints. Using the column tables of the *AISC Manual* for $Kl = 1(7.07) \cong 7$ results in the following preliminary steel pipe section:

$$\text{TS 3 OD} \times 0.216, \quad P = 36 \text{ k}$$

The approximate cross-sectional area for the bottom tension chord as based on yielding of the gross section is

$$A = P/0.6F_y = 45/0.6(36) = 2.08 \text{ in.}^2 \tag{3.82a}$$

In case of special connection types, it is assumed here that the tensile strength at any portion of the member, such as the threaded end piece, and the strength of the connectors are at least equal to the member capacity as based on its typical cross section. This assumption must be checked. Try TS 3 OD × 0.216, A = 2.23 in.^{2}.

The maximum shear appears at mid length of the sides and is conservatively assumed to be equal to

$$V_{max} = 0.50wL = 0.50(0.045)100 = 2.25 \text{ k/ft} \tag{6.13}$$

The length of the diagonal members in plane with the bottom chords is equal to

$$l_d = a/2\sqrt{2} = d\sqrt{2} = 5\sqrt{2} = 7.07 \text{ ft}$$

The maximum compression in the diagonals is found by letting the vertical components of the two diagonals be equal to the bay shear.

$$2(N_d/l_d)d = V_{max}(a), \qquad \text{where} \quad d = a/2 \tag{6.23}$$

$$N_d = V_{max}/\sqrt{2} \; a = V_{max}(l_d) = 2.25(7.07) = 15.91 \text{ k}$$

The section cannot be directly taken from the *AISC Manual*; hence, try 2½ OD $\times$ 0.203, $A = 1.70$ in.^{2}, $r = 0.947$ in.
For the slenderness ratio, the allowable axial stresses are

$$Kl/r = 1(7.07)12/0.947 \cong 90, \quad \text{or} \quad F_a = 14.20 \text{ ksi}, \quad \text{(Table A-12)}$$

$$f = N_d/A = 15.91/1.70 = 9.36 \text{ ksi} \leq 14.20 \text{ ksi}, \quad \text{OK}$$

(d) Square-on-Square Offset Space Frame (Fig. 6.17c). A square grid pattern of 10 ft is investigated. For economical reasons, it is decided to use only one member length, a; therefore, the truss depth has to be adjusted accordingly.

$$d = \sqrt{a^2 - (a/\sqrt{2})^2} = (a/2)\sqrt{2} = a/\sqrt{2} = 10/\sqrt{2} = 7.07 \text{ ft}$$

The maximum tensile and compressive chord forces occur at the center of the grid at midspan for $M = 22.5$ ft-k (see case b)

$$N_t = N_b = (M/d)a = M\sqrt{2} = 22.5\sqrt{2} = 31.82 \text{ k} \tag{6.19}$$

The required steel pipe section for the top compression chord as based on $Kl = 1(10) = 10$, according to the *AISC Manual* column tables, is

$$\text{TS } 3\tfrac{1}{2} \text{ OD} \times 0.226, \qquad P = 38 \text{ k}$$

The approximate cross-sectional area of the tensile bottom chord is

$$A = P/0.6F_y = 31.82/0.6(36) = 1.47 \text{ in.}^2.$$

Try TS 2½ OD $\times$ 0.203, $A = 1.70$ in.^{2}. Two diagonal members of length $l_d = a$, must resist the maximum shear in compression.

$$V_{max}\, a = 2\,(N_d/l_d)d = 2(N_d/a)a/\sqrt{2} = N_d\sqrt{2}$$

or   (6.24)

$$N_d = (V_{max}/\sqrt{2})a = (2.25/\sqrt{2})10 = 15.91 \text{ k}$$

Try TS 2½ OD $\times$ 0.203, $A = 1.70$ in^2 and, $r = 0.947$ in.
    Check: $Kl/r = 1(10)12/0.947 \cong 127$, $\quad F_a = 9.26$ ksi   (see Table A-12).

$$f = N_d/A = 15.91/1.70 = 9.36 \text{ ksi} \approx 9.26 \text{ ksi},$$

which is close enough since shear is conservative.

Notice that the square-on-square offset space grid can also be treated as intersecting triangulated trusses (i.e., two-way space trusses, Fig. 6.13g) so that a much simpler analysis can be used!

For the conditions where the roof structure is only supported at the four corners (with high forces along the edge members) and where it represents a two-way cantilever plate, refer to Problems 6.13 and 6.14. For this situation of only four-corner supports at least one chord layer should be located on the diagonal to obtain a more uniform stress distribution.

Other Double-layer Space Frames

Double-layer space frames are not only used as horizontal plates(Fig. 6.15), but can take any shape. Besides the common regular geometry of polyhedra, folded plates, and shells, as discussed in the following two chapters, space frames are ideally suited as irregular enclosure systems often required for lobbies, courtyards, and atriums in urban contexts. Typical shapes are sloped plates, stepped plates, and folded plates, that is, multiplate systems as described in Fig. 6.18. Space frames can adjust easily to multifunctional uses. The rich architectural possibilities are expressed by the community center at Maranello near Modena, Italy (1978, bottom left).

Examples of space frames as an important part of architecture include the triangular central courtyard of I. M. Pei's East Building, National Gallery of Art, Washington, D.C. (1978, center left bottom). The large space frame roof structure from which the huge Calder Mobile is hung measures 225 ft on two sides along the diagonals and 150 ft on the other side. It is comprised of 25 tetrahedronlike steel modules forming a triangular lower chord grid of $45 \times 45 \times 30$ ft, with the diagonal members rising 12 ft to the upper chord (i.e., asymmetrical triangular pyramids). Structural engineers were Weiskopf & Pickworth.

The terraced Law Courts building in Vancouver, Canada (1980, center right bottom), designed by Arthur Erickson, is covered by an enormous glazed, folded, space frame canopy supported by parallel, stepped concrete frames. Often glazed space frames are used in urban context for plaza atriums not only to extend the outdoor public space into the building, but also to connect buildings and provide a landscape plaza and communal space. They open the interior space to natural light and exterior views, but shelter against the external environment, and modulate the interior climate through solar gain and venting by creating a microclimate. Philip Johnson and John Burgee were among the first architects to introduce the concept. The Crystal Court at the IDS Center in Minneapolis (1975) consists of an irregular pentagonal plan covered by a stepped glazed space frame, and the eight story atrium at the Pennzoil Place in Houston (1976) is enclosed by glass-covered sloping tubular steel space frames. A typical example is the unique irregular-shaped space frame for an atrium in Shinagawa-ku, Tokyo, Japan (1991, bottom right). It covers a triangular plan surrounded by three high-rise and two low-rise buildings. The folded complex space frame consists of welded 6.22-ft-deep two-way trusses based on 10.53-ft square direct grids braced in the chord planes. The trusses span as stepped arches between two high-rise buildings in one direction and as one-half arches about 170 ft between the third high-rise and the low-rise buildings in the perpendicular direction.

The following three buildings are of enormous scale, reminding one of London's Crystal Palace of 1851. The monumental Crystal Cathedral at Garden Grove, California, is surely one of Philip Johnson's masterpieces (1980, top right). The building's plan is shaped like an elongated, four-pointed star measuring 207×415 feet along the main axes. The roof slopes in three planes and rises to 128 ft at its apex. The building is covered by a continuous shell-like, column-free, folded wall-roof space truss and enclosed with a reflective glass curtain wall. The structure consists of parallel, triangular, folded space trusses spanning across the short direction. To provide lateral stability to the building, the top chords had to be braced in the longitudinal direction whereas the bottom chords act independently. The exterior structural planes of walls and roofs must act as diaphragms to withstand a major earthquake which is not an easy task for this complex column-free folded plate structure. For typical conditions, the pipe truss members are connected by welding slotted ends to gusset plates. The chord member sizes range from 3 to 6 inches diameter and the diagonal pipes from 2 to 3 inches in diameter. Structural engineers for the space frame structure were Severud–Perrone–Sturm–Bandel.

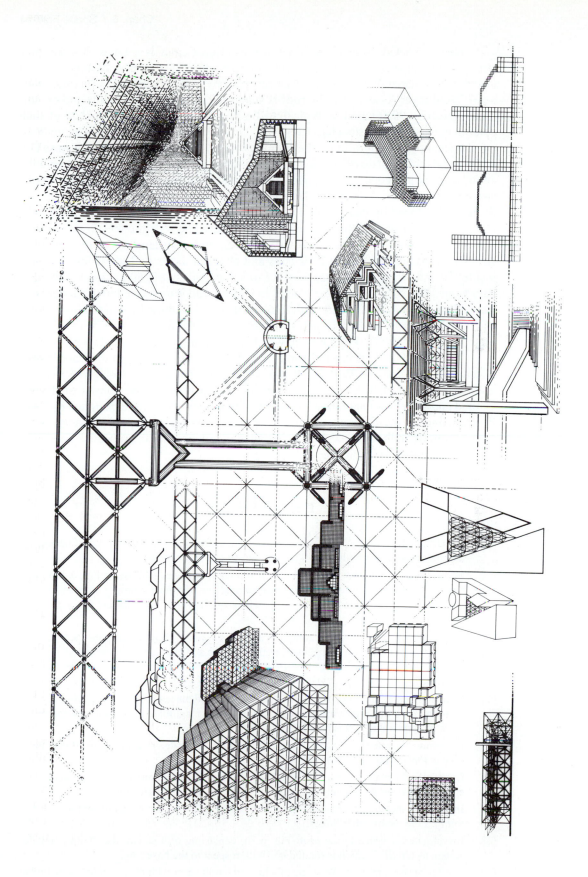

Figure 6.18 Other double-layer space frame types.

New York City's Jacob K. Javits Convention Center is a huge, 20-acre, pre-stressed steel pipe roof and wall space frame structure 15 stories at its highest level (1986, top center and Fig. 1.2d); it was designed by I. M. Pei and the structural engineers Weidlinger Associates. The roof is supported on space frame walls and on columns spaced at 90 ft based on typical 90-ft square bays. The bottom portion of each column consists of four steel legs filled with concrete. Welded to the top is an inverted tubular steel pyramid forming a 10-ft square capital. The space frame grid is based on a 10-ft square-on-square offset, that is, a tetrahedral geometry of 5-ft depth. Along the column strips a second space frame layer is added to form a stiff diamond-shaped space truss, and a third layer is added above the columns to act as drop panels. The PG System, developed by Paul Gugliotta, was used; it consists of nodes, tubes, and rods. Here, the tubes carry all the compression forces and the rods the tensile forces. The rods run from node to node and provide prestress to the system according to the rod size. The rods are threaded at their ends and attached to the hollow nodes with nuts inside. The rods vary in diameter from 0.5 to 3.3 in. and are of 150-ksi steel. The tubes vary from 3 to 8.5 in. in diameter. The glass curtain wall is connected to the nodes. The basic glass panel is 10 ft square with four squares of 5-ft lights framed in aluminum.

Another enormous large-span structure is the space frame pyramid at California State University (1994), Long Beach, which is 186 ft high and 345 ft at its base.

The glazed space frames of Biosphere II, a miniature model of the Earth's eco-systems, located north of Tuscon, Arizona, enclose an area of more than 3 acres (1991, middle left). The glass houses reach a height of up to 85 ft and consist of domes, vaulted and folded surface forms, and stepped pyramids. They were designed by Peter Pearce using his Multihinge system. The structures are sealed with glass panels without the help of mullions since they were installed directly into the space frame's outer chords and sealed with silicone.

PROBLEMS

6.1   Determine the stability of the self-contained space trusses shown in Fig. 6.6. If necessary, add members at proper locations so that the structures are stable.

6.2   Show conceptually how you obtain the reactions for the cases shown in Fig. 6.4d to i by first considering a resultant lateral load parallel to the short building side and then a resultant gravity load to be acting somewhere on the rigid roof. Correct the location of support members if they are not properly placed.

6.3   Change the orientation of the rollers at supports a and c of the hinged tetrahedron (Fig. 6.7d) by placing the one at a parallel to the one at b and changing the one at c so that it is parallel to the original position of the roller at a. What can you say about this support structure?

6.4   If the horizontal unit load in Fig. 6.7d is relocated and placed parallel to the projection of member $\overline{dc}$ so that it acts toward joint c, what is the magnitude of the reaction forces for the tetrahedron due to this new loading condition?

6.5   Determine the reactions and bar forces for the hinged tetrahedron (Fig. 6.7d) if the lateral load is replaced by a unit gravity load.

6.6   Determine the reactions and bar forces for the four-sided pyramid (Fig. 6.7a) under a lateral unit load acting at joint e toward cd, parallel to the long base sides.

6.7   Investigate the square, truncated pyramid in Fig. 6.7c by placing a unit gravity load at joint f. Find the member forces.

6.8   Investigate the space frame roof (Fig. 6.7b) consisting of a square base with a smaller square as top ring, which is rotated 45° with respect to the base ring.

6.9   For a roof structure with the shape of a hinged tetrahedron with its geometry identified in the isometric view of Fig. 6.7d, find the reactions and bar forces as caused by a lateral unit load.

6.10 Determine the approximate bending moment for a typical interior square plate of a multi-bay flat slab structure.

6.11 Determine the critical maximum bending moment for a square cantilever space frame plate first with $L/4$ overhangs and then with $L/2$ overhangs. Assume the spacing of the four-corner columns as 20 ft. Use 30-psf dead load and 20-psf live load.

6.12 Do a preliminary design of the two-way truss structure in Example 6.3a.

6.13 Determine the critical chord and web member sizes if the roof of Example 6.3c is supported at its four corners only, but increase the space frame depth to about $L/14$.

6.14 Determine the critical chord and web members if the corner columns in Problem 6.13 are moved inward to form 20-ft cantilevers around all edges. Use the square-on-square offset grid (Example 6.3d) and a depth of $d \cong L/14$.

6.15 Determine the approximate maximum bending moment for a rectangular two-way dia-grid plate structure simply supported around the edges.

6.16 Investigate the space frame roof supported at its four corners in Problem 6.13 using an equilateral triangle on equilateral triangle offset grid, where the space frame is formed by a series of regular tetrahedra. In this system there is only one member length of $a = 10$ ft. Show how you determine the truss depth and other geometrical requirements. For this first approximation, only estimate the critical compression chord member size by selecting standard steel pipes ($F_y = 36$ ksi); disregard the inclination of the bars in the chord planes for this first estimate.

chapter 7

Folded Plate Structures

The capacity of a flat, thin surface structure is rather limited in the scale range for which it can be used; however, its strength and stiffness are very much improved if it is folded or bent, which increases its depth and thus its moment of inertia. The effectiveness of a folded plate structure in terms of material requirements approaches that of shells; although not quite as efficient, it has the advantage of straight-line construction.

Besides roof structures, familiar examples of folded plate structures are concrete stairs and, on a small-scale, corrugated light-gauge metal panels, sheet piling with their various cross-sections, folded lampshades and lanterns, facade panels, etc. The richness of fold geometries has preoccupied the Chinese and Japanese for many centuries; they consider paper folding, called *origami* by the Japanese, as a form of art.

Complex ribless folded vaults were built by the late Gothic designers, especially along the Baltic coast, Bohemia, and Saxony. The Muslim architects of the fifteenth century also employed the folded vault principle. Among the first modern folded plate structures are Freyssinet's concrete airship hangars at Orly near Paris, France (1916), and the German coal bunkers of the early 1920s.

Wassili Luckhardt's artistic designs in Germany, around 1920, suggest folded structures. The visions of the Expressionist architects of Bruno Taut's circle and others, at this period, were realized much later by Gottfried Böhm in the free-form folded structures of the pilgrimage church at Neviges (1968, Fig. 7.2a) and the town hall tower at Bensberg, Germany (1969). He modeled the buildinglike sculptures and used concrete as a plastic rather than structural material. In contrast to this world of fantasy and mystics, away from function and technology, is the position of the structuralists of the 1950s and 1960s. They are concerned with tectonics, construction, experimentation with new structural forms, and geometries.

Pier Luigi Nervi surely influenced the development of the folded surface concept with the famous 328-ft-span undulating corrugated vault for Salone B of the Turin Exhibition Hall in Italy completed in 1948. He applied the concept of ferrocemento forms, which were developed during the mid-1940s; that is, he used ferrocemento elements together with cast-in-place concrete ribs to shape the folded vault.

Long-span folded plate concrete structures occur for the first time in air hangar construction during the late 1950s (see Fig. 5.56); they were developed by Ammann &

Whitney. The TWA hangar in Kansas City (1956) features cable-supported folded plate concrete cantilevers extending 150 ft on each side of the central spine. In contrast, the National Airlines hangar in Miami (1958) uses pure 111-ft span, nonprestressed, folded plate concrete cantilevers.

Through the use of *folded hypar shells* made of light gage steel decking in stressed-skin construction, cantilever structures reached their high point in 1971 with the Los Angeles and San Francisco American Airlines hangars (Fig. 5.56f), designed by Lev Zetlin, with their enormous 230-ft spans on each side of the central core.

Another early, impressive example of the long-span folded plate concept is the corrugated concrete dome of the University of Illinois Assembly Hall at Urbana (1962, Fig. 8.27h), designed by Harrison & Abramovitz, architects, and Ammann & Whitney, structural engineers. The branching triangular folds convincingly express the required buckling strength for this 400-ft-span shell.

Among the many designers of the 1960s who have dealt with the folding principle in more general terms are David Emmerich in France with his experiments in constructive geometry, especially polyhedra, and Alfred Neumann and Zvi Hecker with their polyhedric architecture in Israel. Folded, prestressed cable roof structures with curvilinear edges and bent surfaces (Fig. 9.1) and folding tensegrity systems (Fig. 9.21) became popular in the late 1980s. The application of the folding principle to architecture with respect to ordinary rigid roof structures is discussed in the next section together with the cases in Fig. 7.2.

7.1 FOLDED PLATE STRUCTURE TYPES

In this context only rigid corrugated surfaces with linear edges are briefly investigated, in other words, situations where flat surfaces are replaced by ordinary *folded surfaces*. For the discussion of *polyhedral surface structures* and folds with curvilinear edges, refer to Fig. 6.1 and 9.1 and the respective text.

Whereas the *flat surface* acts as a slab to resist forces perpendicular to it, for the *folded surface* in contrast, these forces are resolved into components parallel and perpendicular to the fold surfaces, thereby activating slab and deep beam action, as is discussed in the next section (Fig. 7.5, top).

We distinguish between prismatic and nonprismatic folded surfaces of regular or irregular geometry. Prismatic folded surfaces consist of parallel linear fold lines and rectangular surfaces; that is, their cross section remains constant throughout the length. When the cross section is variable, the folded surface is nonprismatic. A special case for this condition is the antiprismatic folded surface. Antiprisms have identical top and bottom polygonal figures like prisms, except that they are rotated against each other, thereby generating triangular faces for the connecting planes (see Fig. 6.8).

The following geometrical features are important characteristics for folded plate surfaces as partly identified in Fig. 7.1.

- *Types of folds:* plane or curved; rectangular, trapezoidal, rhombic, triangular, pentagonal, saddlelike, etc. with straight or curvilinear edges (e.g., parabolic chords, Fig. 9.1c); regular and irregular prismatic, antiprismatic, or nonprismatic folded surfaces. When the folded surfaces can be formed by a folding operation, they are called *developable*.

- *Fold arrangement:* parallel, two-way, three-way, radial, circumferential, or any combination.

- *Fold cross-section:* V, W, M, Z, and U types, northlight forms (saw tooth roof), cellular, shell simulation (polyhedral), in other words, continuous or segmental folds of open or closed cross sections. Numerous other modifications are possible.

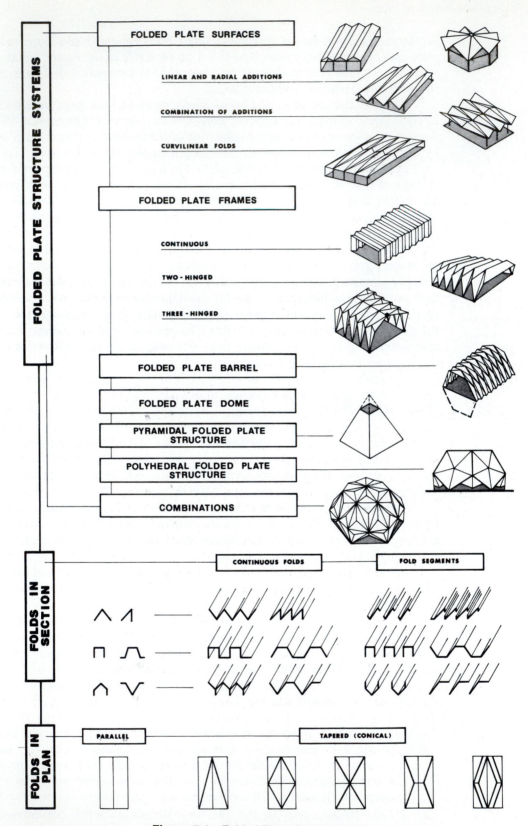

Figure 7.1 Folded Plate Structure systems.

The structure of folded surfaces can be solid, framed, trussed, corrugated, etc., using reinforced concrete, steel, wood, plastics, paperboard, composites, and so on (see Fig. 8.8). They can be constructed as cast-in-place, assembled as prefab components, or made possibly through a folding operation.

The behavior or type of folded plate roof structure depends on the plan configuration and the support conditions, as well as on the geometry of the folds, as shown in Fig. 7.1. For example, when the folded plate structure represents a rectangular surface structure, it acts as a one-way system when it consists of parallel folds (i.e., prismatic folded surface), but as a two-way system when it is made up of triangular or trapezoidal folds (i.e., nonprismatic folded surface). When the folded surface is folded or bent, it forms spatial structures such as adjacent frames, vaults, pyramids, domes, saddles, and so on. The fold arrangement determines where hinges form and whether the structural action is predominately linear, surfacelike, or spatial. The folds of cylindrical surfaces are usually derived from antiprisms as, for instance, the folded plate barrel in Fig. 7.1, which is hexagon based.

It is beyond the scope of this discussion to investigate the wealth and complexity of fold geometries. However, an appreciation for formal and technical possibilities is developed from the study of typical architectural buildings, which includes the wider range of architectural concerns. The few examples of folded plate structures in Fig. 7.2 express the limitless formal potential of the principle. Various fold shapes constitute either beams, plates, or entire enclosures.

The use of folded plates as part of architectural language was developed during the late 1950s. An important example of this early period is the UNESCO Headquarters Conference Building in Paris, France, designed by Marcel Breuer, Bernard Zehrfuss, and Pier Luigi Nervi, which represents a folded plate concrete structure on trapezoidal plan (1957, Fig. 7.2j). The continuous roof corrugations fold into the exterior walls in front and back with interior column supports; they represent a two-bay frame with hinged supports at the base. The magnitude of bending moments required a variation in the height of the undulations, but was not sufficient in resisting compressive stresses, so a concrete slab had to be added to provide the necessary strength. The maximum positive moment occurs at the center of the longer, 130-ft span, where the top of the corrugated structure is in compression, and the maximum negative moments at the supports (especially the interior one), where the bottom of the roof structure is in compression. Therefore, the solid concrete slab is curved to follow the pressure line from the lower edge to the upper edge of the undulations, thus effectively resisting the compressive bending stresses, besides laterally stabilizing the folds.

Minoru Yamasaki used the folded plate principle for the ACI Headquarters Building in Detroit, Michigan (1957, Fig. 7.2d) as a showpiece of concrete construction. The thin, double-cantilever, tapered, folded, slab roof is based on a 4-ft module and is only supported on the walls along the central corridor with 19-ft overhangs on each side.

Other examples of this early period include the diamond-shaped concrete roof planes and folded walls of the Christ Church at Bochum, Germany (1959, Fig. 7.2c), by Dieter Oesterlen. Folded plate concrete bents enclose St. John's Abbey Church, Collegeville, Minnesota, by Marcel Breuer (1961, Fig. 7.2g). The auditorium roof structure of the Technical University at Delft, Holland (1966, Fig. 7.2k), by Van den Broek & Bakema, consists of folded concrete plates resting on the dish-shaped, sculptured concrete auditorium, which, in turn, is supported on two massive columns and the central four-story building portion. The Air Force Academy Chapel at Colorado Springs, Colorado (1961, Fig. 5.50a), by Walter Netsch of SOM, is another striking early example of the folding principle. Here, two inclined triangular folded spatial steel truss structures form an A-frame in cross section. Foldable low-cost housing enclosures were successfully developed by Hirshen and Van de Ryn in 1966 for migrant workers in California using 3/8-in-thick, prefolded paperboard vaults (see Fig.

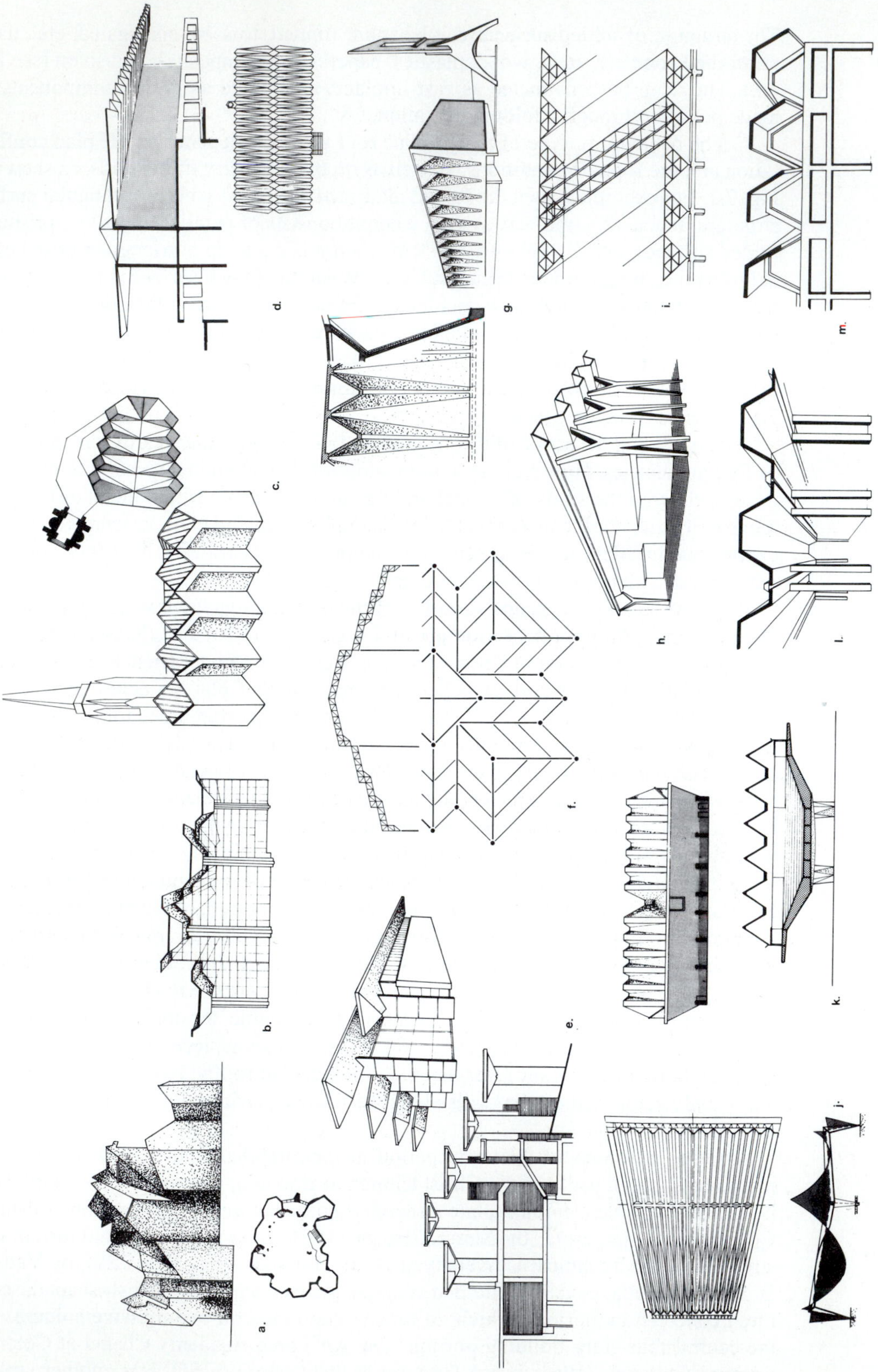

Figure 7.2 Folded plate structures.

8.10c), which consisted of sandwich elements of coated kraftboard with a polyure-thane foam core.

Among the long-span structures should be mentioned the folded, 21-ft-deep Warren Steel trusses of the Roe Bartle Exhibition Hall in Kansas City, Missouri (1976, Fig. 7.2i), by Helmut Jahn, spanning nearly 310 ft. The roof structure of the city hall in Memmingen, Germany (1984, Fig. 1.11b), designed by Peter Faller, consists of parallel, folded spatial trusses with a maximum span of 112 ft. The top chords of the trusses are glulam members, and the curved bottom chords are three steel ties of parabolic shape.

In contrast to the continuous surface folds of the examples in Fig. 7.2, cases (b, e, and m) employ individual folded beams. Very different fold patterns occur for the irregular roof structure of the Berlin Free University Philosophical Institute by H. and I. Baller (1983, Fig. 1.11c).

The addition to the National Air and Space Museum, Washington, D.C. (Fig. 7.2f), by Hellmuth, Obata & Kassabaum, besides the other cases presented, indicates that there are no limits set to the application of the folding principle!

7.2   STRUCTURAL DESIGN OF ORDINARY FOLDED PLATE ROOF STRUCTURES IN CONCRETE, STEEL, AND WOOD

First, the more common prismatic folded surfaces with triangular and trapezoidal cross sections under vertical load action are investigated, since this is the typical situation needed for preliminary design purposes. The common characteristics of folded plates are identified in Fig. 7.3 and should be referred to in the following discussion.

A folded plate roof structure may be formed by just one single folded plate unit, or it may consist of a multiple-bay system. Visualize a folded plate structure as a system of inclined beams leaning against each other. These tilted beams are carried by transverse supports such as frames and walls at locations close to their ends. These supports transmit the vertical load components to the ground and also act as stiffeners, preventing the structure from unfolding or flattening out. The lateral thrust that causes this tendency to unfold is often carried by tie rods attached directly to the folded plates, in which case the support structure must resist only the vertical forces. Intermediate transverse diaphragms (i.e., panels above and/or below the folds or ties) are sometimes used in order to reduce deformation, as well as to ensure that the folded surface is stiff enough to behave as a total unit. To increase the stiffness in beam action, longitudinal supports may be placed along the valley fold lines; that is, beams and trusses can be used to this effect. While the interior folded plates of a multiple-bay system can support each other laterally, the free longitudinal edges of the exterior plates are vulnerable to large displacements and, in general, need to be stiffened by edge beams.

To develop some understanding about the structural behavior of folded plates, the vertical load action is resolved into components perpendicular and parallel to the plates. The normal force components cause the plate to respond as a one-way slab with supports along the fold lines. These supports are critical since they are, to some degree, flexible. For the purpose of approximate study, we may assume that the multibay roof is uniformly loaded, thus, for a typical interior unit no differential displacements between the supports along the valley and ridge edges are caused; the supports along the fold lines can be considered rigid. To have true one-way slab action, using standard concrete construction, the plate span L must be at least twice the plate width b ($L \geq 2b$). The load components parallel to the plates, that is, the reactions of the transverse slab, cause the plates to act as deep beam structures resting on transverse supports; it is assumed that the inclination of the plates is steep enough to allow for this type of behavior.

The analysis can be assumed statically determinate for folded plates hinged along the fold lines, such as steel and timber trusses or stressed-skin plates that lean

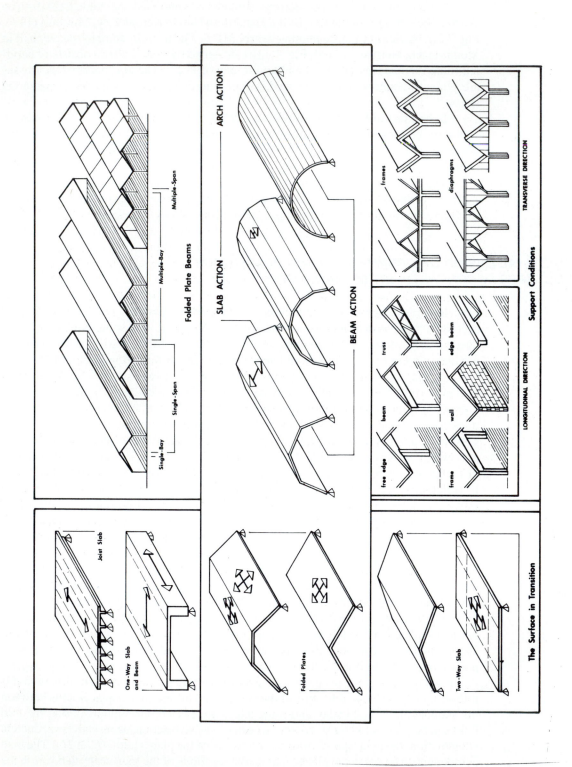

Figure 7.3 Slab action versus beam action.

against each other and are covered, for example, with light-gage metal deck or plywood. In this case, the deck behaves similarly to the one-way slab, and the inclined *trusses* or *girders* act as independent deep beams. However, the analysis is highly indeterminate for monolithic, continuous, folded plate structures such as cast-in-place concrete construction. Here, the plates act as both deep beams and slabs.

The following discussion attempts to develop an understanding of the behavior of monolithic folded plate structures. Previously, it was stated that the assumption of one-way slab action is justifiable only for plates having a proportion of $L \geq 2b$. As the plate span decreases and approaches the same dimension as its width, the plate begins to act more as a two-way slab. On the other hand, as the plate span increases, the slab action becomes less important. These conditions are encountered in such various examples as corrugated metal panels used as decking, monolithic concrete joist slabs, and wide–flange metal beams, where the flange width, in comparison to the beam span, is so small that only the beam action is considered.

The interplay of slab and beam action is investigated from a geometrical point of view in Fig. 7.3 (top left). The ribs in the concrete joist slab are so closely spaced that its primary behavior is its beam action; the slab behavior perpendicular to the joists is so small that usually only temperature reinforcing is required. As the ribs move farther apart, slab action becomes more and more pronounced, resulting in a one-way slab system sitting on beams. The independent action of beam and slab is apparent not just from the location of the beams, but also from the relative thickness of the slab as compared to the beam size. However, due to the continuity between slab and beam, part of the slab acts together with the beam, establishing the familiar concept of the *effective flange width*.

Next, the widths of the edge beams are reduced to the same thickness as the slab and then inclined (i.e., W-shaped plates), thus decreasing the span of the slab. Now a new structural system is generated: a continuous folded surface. Here the horizontal and inclined plates both act as slabs and beams. The plate thickness depends on its width, that is, the magnitude of the transverse bending stresses depends on the span of the slab. The total height of the folded beam is a function of the longitudinal span; the usual height-to-span ratio for folded plate structures is in the range h/L 1/10 to 1/15. Further inclination of the plates eliminates the horizontal portions and results in V-shaped folding (Fig. 7.3, center). Now the direct slab action is decreased, since the force components normal to the surface, which only cause slab action, are reduced. On the other hand, the overall effect may not cause a reduction in slab action because the inclined slab span has been increased. As the fold height decreases and the structure becomes flatter, the force components normal to the plate due to gravity action become larger; hence, the slab action increases, approaching a two-way slab (Fig. 7.3, bottom left). Although the force components parallel to the plate decrease, the moment of inertia of the folded plate beam decreases much faster, resulting in an increase in beam action and hence in longitudinal stresses. It is obvious from this discussion that the slope of the plates must be reasonable so that the assumed behavior is possible and the material is efficiently used. For V- or W-shaped folding, the slope of the plates is typically in the range from 25° to 45°.

It is of interest to note the changes in structural behavior in a given folded plate system as the number of folds is increased to the point where the structure becomes the equivalent of a single curvature shell beam (Fig. 7.3, center right). Although the effect of the number of folds on the beam behavior in the longitudinal direction is less pronounced, the slab behavior in the transverse direction is very much influenced by the change. As the number of folds increases, the slab width decreases, hence the slab moments decrease, keeping in mind that the slabs are considered to support each other along the fold lines. When the number of folds has become infinite, fold discontinuities and slab moments disappear. Now the forces normal to the curvilinear surface are carried in the transverse direction in arch action, that is, by axial forces and bending (e.g., similar to Fig. 5.44). It is the lack of curvature that distinguishes folded plates from shells. Folded plates are much more vulnerable to bending stresses than shells and thus less efficient from a force resistance point of view.

Approximate Design of Prismatic Folded Plates.   A typical V-folded roof structure is generally constructed either of wood, steel, or reinforced concrete. Wood folded plates consist of lumber or glulam framing and plywood skin similar to box girders (Fig. 3.11) leaning against each other. The frame is made up of the longitudinal chord members along the ridges and valleys and the intermediate rafters as indicated in Fig. 8.8. The thickness of the plywood plate is generally derived from membrane shear (i.e., web action) and then checked for sheathing action in bending. The plywood web must be fastened continuously to its framing so that it can resist the shear caused by deep beam action. The spacing of the rafters depends on the sheathing capacity. The rafters behave as simply supported beams transferring the loads to the top and bottom chords. In stressed-skin construction, however, rafters and skin act as a composite unit rather than separately. At points of support, bearing rafters are required to carry the plywood shear to the supports. The ridge and valley flange members resist the bending in the longitudinal direction; their sizes are determined from the inclined girder action. Refer to Problems 7.4 and 7.5 for further discussion of plywood folded plates.

In steel construction, membrane-type or stressed skin type folded plate roof structures are generally used for shorter spans, possibly up to 100 ft, whereas the longer-span, truss-type folded plates are used as investigated in the following example. In steel folded plates, light-gage steel deck panels span transversely between the fold-line flange members at ridges and valleys. In other words, they not only act as simply supported slabs in the transverse direction, but also as members or webs of inclined girders to resist shear in the longitudinal direction.

In the following examples a typical V-folded roof structure (Fig. 7.4a) will be examined by studying the approximate behavior of an interior bay unit. Two different construction systems are analyzed: first, hinged, tilted steel trusses and then continuous reinforced concrete plates. This folded plate roof is a single-span, multibay structure with the following dimensions: longitudinal span; $L = 100$ ft; fold height; $h = 10$ ft; bay width; $2a = 30$ ft.

For this ordinary structure, it is reasonable to only consider gravity loading for preliminary design purposes. The roof must support a snow load of 30 psf on the horizontal projection (keeping in mind that the true load distribution is very different). The height-to-span ratio of the roof is typically in the range of

$$h/L = 10/100 = 1/10$$

The fold angle measured from the horizontal is

$$\theta = \arctan(10/15) = 33.69°, \qquad \cos\theta = 0.832, \qquad \sin\theta = 0.555$$

The width of a typical plate is

$$b = a/\cos\theta = 15/0.832 = 18.03 \text{ ft}$$

The span-to-width ratio of the plate is

$$L/b = 100/18.03 = 5.55 > 2$$

Since the span-to-width ratio is greater than 2, the folded plate acts as a one-way slab in the transverse direction.

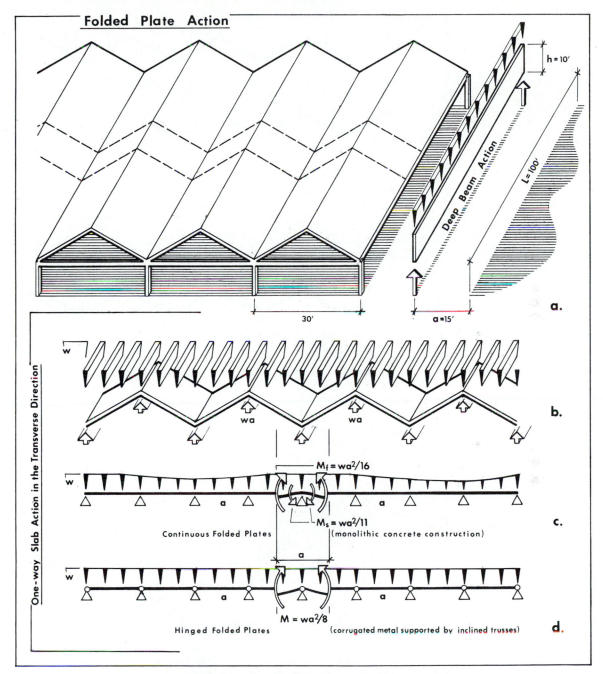

Figure 7.4 Examples 7.1 and 7.2; slab action.

EXAMPLE 7.1

Assume the roof to be constructed of tilted steel Pratt trusses subdivided into ten equal bays of 10-ft width each (Fig. 7.5d). The dead load is of the following magnitude:

Built-up roofing and insulation	5 psf
Roof decking material	5 psf
	10 psf
Truss weight	10 psf
Total weight	20 psf

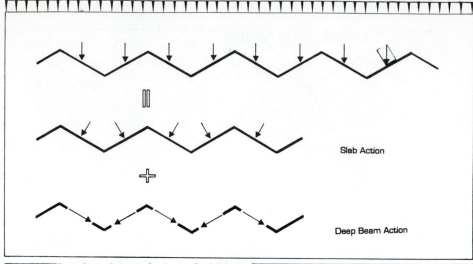

Slab Action

+

Deep Beam Action

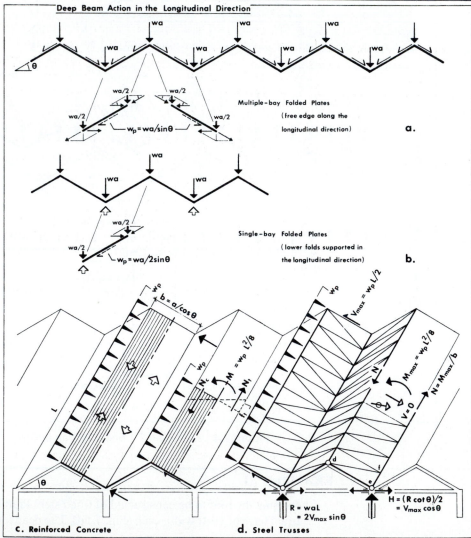

Figure 7.5 Examples 7.1 and 7.2; beam action.

(a) Design of Roof Deck. The deck supports 30 psf of snow load on its horizontal projection and 10 psf of dead load along its own plane. By transferring the dead load onto the horizontal projection, the two loads may be added (see also Fig. 5.26).

$$w = 30 + 10/\cos\theta = 42 \text{ psf}$$

The deck acts as a simply supported one-way slab spanning from top chord to bottom chord with a maximum moment of

$$M_{max} = wa^2/8 = 0.042(15)^2/8 = 1.18 \text{ ft-k/foot width}$$

Using $F_y = 33$ ksi with an allowable bending stress $F_b = 0.6F_y$ yields the required section modulus of

$$S = M/F_b = 1.18(12)/0.6(33) = 0.72 \text{ in.}^3/\text{foot width}$$

The steel decking may now be selected from manufacturers' catalogs and then checked for resistance to an allowable live load deflection of $\Delta_L \leq L/240$ (see Example 5.7).

(b) Design of Inclined Trusses (A36 Steel). The trusses carry the reactions of the simply supported steel decking to the transverse supports located at both ends of the structure. Two typical interior trusses are isolated and separated from the remainder of the roof as shown in Fig. 7.5a. The deck reactions are resolved into two force components: horizontal components, which balance and cancel each other under symmetrical loading conditions, and forces parallel to the truss plane, which must be carried in bending to the end supports through axial member action. It is assumed that the trusses carry forces only parallel to their surface.

Resolving the reactions of the steel decking into linear loads parallel to the truss plane (Fig. 7.5a) yields

$$w_p = wa/\sin\theta \tag{7.1}$$

Should the trusses be supported longitudinally along the valley chords by walls or beams, as in Fig. 7.5b, only one-half of the deck loads is carried by the trusses, while the other half is transferred by the deck directly to the longitudinal supports.

The uniform load on the horizontal roof projection, including the truss weight, is

$$w = 30 + 20/\cos\theta = 54 \text{ psf}$$

The linear loads parallel to the truss plane are equal to

$$w_p = wa/\sin\theta = 0.054(15)/0.555 = 1.46 \text{ k/ft}$$

The maximum moment for the simply supported truss is

$$M_{max} = w_p L^2/8 = 1.46(100)^2/8 = 1825 \text{ ft-k}$$

This moment occurs at midspan and is considered to be resisted by the axial action of the top and bottom chords, according to Eq. (5.97), as shown in Fig. 7.5d; the effect of the steel decking in resisting bending in the longitudinal direction is conservatively neglected. The forces in the chords are

$$N = M_{max}/b = 1825/18.03 = 101.22 \text{ k}$$

In this approach it is assumed that a truss chord acts independently of the one adjacent; should two adjacent trusses share a chord, the axial force is twice as large.

In contrast to trusses, where local bending of the chord elements due to continuous load action must be taken into account, the chords of trusses for folded plate roofs are braced by steel decking; hence, only axial action is considered in the preliminary design. The allowable compressive stress for the top chord is equal to $0.6F_y$ or about 22 ksi. This allowable stress is not reduced for buckling because adjacent trusses, as well as the steel decking, are assumed to provide enough lateral support. The required steel area for the top chord member for both adjacent trusses, similar to Eq. (5.101), is

$$A = N/F_c = N/0.6F_y = 101.22(2)/22 = 9.20 \text{ in.}^2$$

For preliminary design purposes, select an angle section $L8 \times 8 \times 5/8$, $A = 9.61$ in.^{2}. The detailing of the truss connections must be carefully considered, especially when inclination of the plates is some angle other than 45°.

The approximate size of the bottom chord member (again acting as flange for both adjacent trusses) is based on yielding of the gross section and, according to Eq. (3.82a), is equal to the top chord size. Therefore, the member sizes for top and bottom chords may be assumed equal to each other for preliminary design purposes.

The vertical reaction at the ends of the truss is carried by inclined columns, such as column $\overline{de}$ (Fig. 7.5d). The axial compressive force in this member is equal to

$$N_d = 1.46(50) = 73.00 \text{ k} \quad (C)$$

For a slenderness about the strong axis (since $l_y = 0$) of $(Kl)_x/(r_x/r_y) = 1(18.03)/1.77 \cong 10$, according to the column tables in the AISC Manual, try W6 × 20.

The largest shear force acts adjacent to the reactions and is carried by the diagonal $\overline{df}$. Balancing the forces at joint d yields the diagonal tension force.

$$N_{df} = (73/18.03)20.62 = 83.49 \text{ k}$$

where the length of the diagonal $\overline{df}$ is equal to $\sqrt{18.03^2 + 10^2} = 20.62$ ft.

The approximate cross-sectional area as based on yielding of the gross section, according to Eq. (3.82a), is

$$A = N/0.6F_y = 83.49/0.6(36) = 3.87 \text{ in.}^2$$

Try W6 × 15, $A = 4.43$ in.^{2}, and $r_y = 1.46$ in. Check the slenderness of member,

$$(L/r)_{min} = 20.62(12)/1.46 = 170 < 300 \quad \text{OK} \tag{3.85}$$

The gravity loads acting on the roof cause a column load in the transverse support structure (Fig. 7.5d) of

$$R = w(2a)L/2 = wa\,L = 0.054(15)100 = 81 \text{k} \tag{7.2}$$

The transverse support structure must also resist the horizontal thrust of the end bays (and thrust due to asymmetrical loading) by means of tie rods or frame action.

$$H = (R/2)\cot\theta = (R/2)a/h = (81/2)\,15/10 = 60.75 \text{ k} \tag{7.3}$$

The approximate cross-sectional area as based on yielding of the gross section is

$$A = N/0.6F_y = 60.75/0.6(36) = 2.81 \text{ in.}^2$$

Try $L4 \times 3 \times 7/16$, $A = 2.87$ in.^{2}.

EXAMPLE 7.2

The folded plate truss structure was considered to be a determinate system where the decking behaved independently as a slab in the transverse direction, while the truss acted as a plate or deep beam in the longitudinal direction. Reinforced concrete folded plates, however, are monolithic; the plates act not only as deep beams but also simultaneously as slabs. The approximate analysis and design of concrete folded plates may be carried out by the same methods as those for independently acting truss systems. The plates are first designed as continuous one-way slabs supported by the adjacent plates along the fold lines and then as inclined deep beams carried by the transverse supports. These two behavioral systems are superimposed on each other, resulting in a structure that responds as a unit to both force actions.

(a) Design of the Slab Action. Since the plates are continuous across the fold lines, the one-way slab action is also continuous (Fig. 7.4b). The ACI moment coefficients (Fig. 3.15A) are used to approximate the magnitude of the moments at the supports (fold lines) and at midspan for a typical interior plate. The net span for such a plate is conservatively assumed to be the distance between the edges of the fold lines.

The thickness of the plate is primarily dependent on the slab action. For continuous one-way slabs, the thickness (in inches) is assumed to be one-third of the span (in feet) for this first rough approximation (i.e., $L/t = 36$), but at least 3 in. Using this rule of thumb, and considering conservatively the inclined slabs as equivalent slabs on the horizontal projection, but ignoring the effect of steel grade (i.e., increase of slab thickness), result in a preliminary average slab thickness of

$$a/3 = 15/3 = 5 \text{ in.}$$

The dead load of the 5-in. slab for normal-weight concrete together with roofing and insulation materials is

$$w_D = 5 + 150(5)/12 = 68 \text{ psf}$$

The dead load is transferred to the horizontal roof projection (see also Fig. 5.26) and added to the live load.

$$w = w_D/\cos \theta + w_L = 68/0.832 + 30 = 112 \text{ psf}$$

The slab moments according to Fig. 3.15 are as follows: At the fold lines

$$M_s = wa^2/11 = 0.112(15)^2/11 = 2.29 \text{ ft-k/ft}$$

At slab midspan

$$M_f = wa^2/16 = 0.112(15)^2/16 = 1.58 \text{ ft-k/ft}$$

Checking the compressive stresses by treating reinforced concrete as ideally elastic and homogeneous [see also (Eq. 3.47)] yields

$$f_c \cong M/S = M/(bd^2/6) = 2.29(12)/(12(4)^2/6) = 0.86 \text{ ksi}$$
$$\leq 0.45 f_c = 0.45(4) = 1.8 \text{ ksi}$$

Using 60-ksi steel and 4-ksi concrete yields the following approximate reinforcing at the fold lines for $d \cong h - 1 = 5 - 1 = 4$ in:

$$A_s \cong M/(0.875 df_s) = 2.29(12)/(0.875(4)24) = 0.327 \text{ in.}^2 \text{ /ft of slab} \qquad (3.55a)$$

Or the required spacing of No. 4 bars is $12/0.327 = s/0.2$, or $s = 7.34$ in. Try No. 4 bars at 7 in. on center, $A_s = 0.343$ in.^{2}/ft (Table A.2). The required steel at midspan is

$$A_s = 0.327(1.58/2.29) = 0.226 \text{ in.}^2/\text{ft}$$

Or the required spacing of No. 4 bars is $12/0.226 = s/0.2$, or $s = 10.62$ in. $\leq 3t = 3(5) = 15$ in. ≤ 18 in. Try No. 4 bars at $10\frac{1}{2}$ in. on center, $A_s = 0.229$ in.^{2}/ft (Table A.2). The required steel for temperature reinforcing is:

$$A_s = 0.0018bt = 0.0018(12)5 = 0.108 \text{ in.}^2/\text{ft} \qquad (3.69)$$

The maximum bar spacing is $5t = 5(5) = 25$ in. ≤ 18 in. Try No. 3 bars at 12 in. on center, $A_s = 0.11$ in.^{2}/ft (Table A.2). Refer to Fig. 7.6 for the placement of the reinforcing bars in the slab.

(b) Plate Design. The plate action in the longitudinal direction is highly indeterminate. To develop understanding about the behavior, the plates are assumed to be hinged along the fold lines, and then the same process of analysis as for the inclined trusses of Example 7.1 is followed. The treatment of the panels as individual units causes incompatibility along the plate boundaries, since the stresses and deformations along the fold lines do not coincide as they should for continuous boundary conditions. The process of correcting these errors by the method of successive balance, for instance, is typical for the solution of statically indeterminate structures. However, no correction is necessary for the special condition of an interior plate of a multibay V-shaped folded plate structure where all the bays are identical and subject to the same uniform loading. For this condition the stresses of the individual plates along the fold lines are equal to each other, and the hinged boundary conditions are identical to the continuous ones.

Rather than using the approach employed for the truss example, the *beam method* is taken by treating the folded plate structure as a series of parallel beams or as a corrugated slab. This statically determinate approach is simpler for preliminary design purposes; it is not restricted to just V-shaped folded structures, but can be applied to other systems, such as the common folded plate roof with a trapezoidal cross section.

Considering the cross section of a folded plate structure as being equivalent to the cross section of a beam can only be taken as a preliminary estimate and as an approximate structural analog because the effect of any distortion of the section is neglected. In reality, the thin plate elements are much more susceptible to deformations than ordinary beam sections; the resulting stress pattern is not distributed linearly, as assumed in beam theory. However, if the folds are stiffened, the beam model becomes more realistic, especially as the span of the folded plate structure increases, as has already been discussed (Fig. 7.3). The beam method is

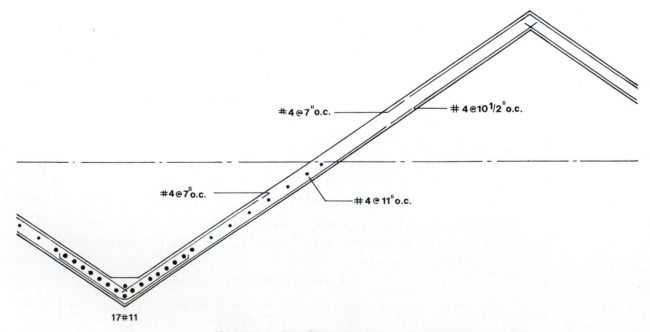

Figure 7.6 Main slab and plate steel reinforcing at midspan.

sometimes used by structural engineers as a possible final design solution for interior plates of a multiple folded plate structure if the span-to-height ratio is greater than 10 ($L/h \geq 10$), which is the case for this example.

In Fig. 7.7 a typical interior cross section of the beam is shown. Because of symmetry of geometry and loading, only one inclined beam is considered. It is demonstrated in Problem 7.1 that the plate analysis using the equivalent vertical beam approach yields the same result as the inclined beam approach in Example 7.1. The equivalent line load on the horizontal roof projection is

$$wa = 0.112(15) = 1.68 \text{ k/ft}$$

This yields a maximum moment at midspan of

$$M = (wa)L^2/8 = 1.68(100)^2/8 = 2100 \text{ ft-k}$$

It is assumed for folded plates and thin shells that the concrete is ideally elastic, homogeneous, and isotropic (see Section 8.1); hence, the entire section modulus for the concrete cross section can be used. The equivalent section modulus of a vertical beam placed parallel to the gravity loads (Fig. 7.7 and Problem 7.1) is

$$S_{CA} = (t/\sin\theta)h^2/6 = (5/0.555)(10 \times 12)^2/6 = 21,622 \text{ in.}^3 \qquad (7.4)$$

As based on the assumption of ideally homogeneous, isotropic behavior, the maximum compressive stress at the top fold lines at midspan is

$$f_c = \frac{M}{S} = \frac{2100(12)}{21,622} = 1.17 \text{ ksi} < 0.45 f'_c = 0.45\,(4) = 1.80 \text{ ksi}$$

The maximum compressive stress at the top is easily resisted by the strength of the concrete. The steel reinforcement will resist all tension; any tensile capacity of the concrete is neglected. The approximate required steel reinforcement, using an internal lever arm of the tension–compression couple of $z \cong 2h/3$ as based on an ideally elastic homogeneous rectangular concrete section, is

$$A_s = \frac{M}{zf_s} = \frac{M}{f_s\,(2h/3)} = \frac{2100\,(12)}{24(2)10(12)/3} = 13.13 \text{ in.}^2 \qquad (3.55)$$

According to the ACI Code, where tensile stresses greatly vary, it may be desirable to concentrate tension reinforcement in regions of maximum tensile stress, rather than distributing it over the entire zone of the varying tensile stress. Try 17 #11 bars, $A_s = 17(1.56) = 26.52 \text{ in.}^2 > 2(13.13) = 26.26 \text{ in.}^2$ (Table A.1). These

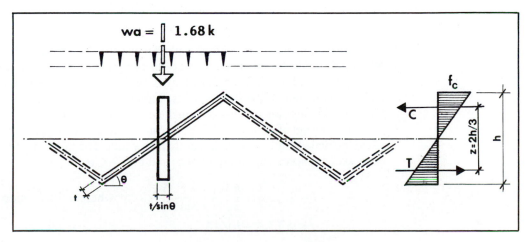

Figure 7.7 Equivalent beam.

bars are concentrated at mid-thickness of the plate in the valley fold region as indicated in Fig. 7.6. Since the major longitudinal reinforcing is concentrated in the valleys, the remaining portions of the tension area should, according to the ACI Code, have a minimum reinforcing of

$$A_s = 0.0035bt = 0.0035(12 \times 5) = 0.21 \text{ in.}^2/\text{ft} \tag{8.7}$$

Try #4 bars at 11 in. on center, $A_s = 0.22$ in.^{2}/ft, $s_{max} = 3t = 3(5) = 15$ in. ≥ 11 in.

The maximum shear stresses appear close to the supports as diagonal tension at 45° to the longitudinal axis. Shear reinforcing at 45° must be provided if the diagonal tension cannot be carried by the concrete itself. The approximate maximum shear at the supports due to the direct gravity loading only is

$$V_{max} = w_p L/2 = 1.68(100/2) = 84 \text{ k}$$

The maximum nominal average shear stress is

$$f_v = \frac{V}{h(t/\sin\phi)} = \frac{84,000}{(10 \times 12)5/0.555} = 77.7 \text{ psi}$$

The allowable shear stress for the concrete in beams and one-way slabs is

$$v_c = 1.1\sqrt{f_c'} = 1.1\sqrt{4000} = 69.57 \text{ psi} < 77.7 \text{ psi} \tag{7.5}$$

The shear stresses are larger than what the concrete is expected to be able to support. Therefore, diagonal reinforcing at the supporting ends must be provided. The diagonal tensile force is equal to

$$N_v = f_v A = f_v(12 \times t) = 77.7(12)5/1000 = 4.66 \text{ k/ft}$$

The diagonal reinforcement, disregarding conservatively the shear capacity of the concrete, is approximately equal to

$$A_s \geq N_v/f_s = 4.66/24 = 0.194 \text{ in.}^2/\text{ft}$$

Try #4 bars at 12 in. on center at the supports, $A_s = 0.20$ in.^{2}/ft (Table A.2). The spacing of the diagonal bars increases as the shear decreases away from the support. For a typical layout of reinforcing, refer to the section on cylindrical shell beams (see Fig. 8.20).

The thickness of the plates may be controlled by steel placement and concrete cover. For this condition, a cover of 3/4 in. is used since a waterproof membrane protects the concrete roof. Hence, the minimum slab thickness is

$$t_{min} = 0.75 + 0.50 + 1.41 + 0.50 + 0.75 = 3.91 \text{ in.} < 5 \text{ in.}$$

Although the V-shaped folded plate is the simplest form, it has the disadvantage of not providing much flange resistance in compression on top, and at the bottom limits the placement of reinforcing steel, in this case, for simple beam action. It is apparent that by adding horizontal flange plates at top and bottom, in other words, by introducing W-folds, this situation is very much corrected.

The design of trapezoidal folded beam sections can be approached as for V-folds when the plate length is much larger than its width. The moment of inertia of the beam cross section, according to Eq. (2.50b) and Fig. 7.8, is

$$I = 2\left(\frac{a_1}{2}\right)t\left(\frac{h}{2}\right)^2 + \left(\frac{t}{\sin\phi}\right)\frac{h^3}{12}, \quad \text{where} \quad b = h/\sin\phi$$

$$I = th^2(3a_1 + b)/12$$

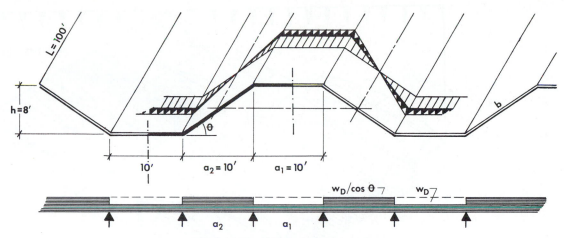

Figure 7.8 W-shaped folded plate structure.

Or the section modulus is equal to

$$S = I/(h/2) = th(3a_1 + b)/6 \qquad (7.6)$$

The approximate required steel reinforcement can be determined according to Eq. (3.55a) using an internal lever arm of the compression–tension couple of $z \cong h$, similar to a plate girder section.

$$A_s = M/zf_s \cong M/hf_s \qquad (7.7)$$

Triangular Folded Plate Surface Structures.   The response of nonprismatic folded plate surface structures to loading is very complex. For the special case shown in Fig. 7.9, the behavior of the roof under uniform gravity loading can be visualized as previously as the combined action of transverse slabs and longitudinal plates.

- First, the vertical loads are carried in slab action in the transverse direction to the fold lines. Notice that, because of the triangular shape of the plates, the slab spans are not constant anymore as for prismatic folds. We can visualize that the fold line members form *triangular (spatial) three-hinge arches* that cause lateral thrust in the transverse direction which must be carried by the trusses to the building ends.

- The transverse thrust is carried by the two *inclined trusses* spanning in the longitudinal direction in deep beam action, with the ridge member serving as compression chord and the tie rods along the exterior faces as bottom chords, while the inclined valley intersections form diagonal members; the thrust caused by the two trusses is resisted by ties across the building ends.

A more general approach would be to subdivide the folded surface into a grid of beams. After the variable inertia of the typical beam has been determined, critical stresses are found. This approach overestimates the magnitude of the stresses, since the real surface action in the transverse direction was ignored.

PROBLEMS

7.1   Show that the plate analysis using the equivalent vertical beam approach [Example 7.2b, Eq. (7.4)] does yield the same solution as the plate analysis in Example 7.1.

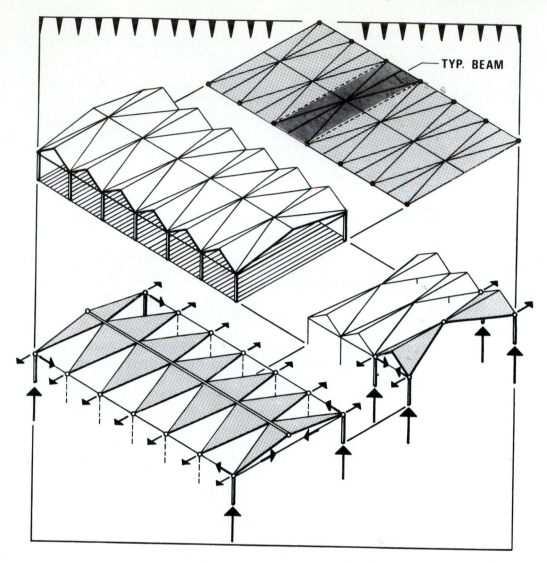

TYP. BEAM

Figure 7.9 Triangular folded plates.

7.2   Determine the longitudinal tensile stresses for a typical interior plate by considering the folded plates in Example 7.2 as hinged along the fold lines. Use the same approach as for the solution of the inclined trusses in Example 7.1.

7.3   Investigate a typical interior fold of a multibay, 120-ft-span folded plate, welded truss structure similar to the one shown in Fig. 7.5d. The Pratt truss is subdivided into eight spaces, each at 15 ft; a typical bay width is $2a = 20$ ft, and the roof height is 10 ft. Assume a dead load of 11 psf for the roof deck, insulation, and the built-up roof, as well as 10 psf for the truss. The snow load on the horizontal roof projection is 20 psf. Use A36 steel for the truss and an allowable bending stress of $F_b = 20$ ksi for the steel deck. Determine the approximate sizes for the critical truss members, the type of roof deck, and the size of the tie along the end supports.

7.4   Do a preliminary design of a folded plate plywood roof of a construction similar to the V-shaped, single-skin plate structure shown in Fig. 8.8. The 64-ft simple span structure has a typical bay width of $2a = 40$ ft and a roof height of 10 ft. Assume a dead load of 10 psf and a snow load of 15 psf, both to act on the horizontal roof projection. The allowable stresses for the standard plywood sheathing are $F_v = 190$ psi and $F_b = 1200$ psi. The allowable stresses for the other members are $F_b = 1250$ psi, $F_c = 1100$ psi, and $F_t = 1050$ psi. Determine the approximate sizes of the sheathing, the rafters, which are spaced at 32 in. on center, and the valley and ridge chords.

7.5   Consider the folded plate plywood structure in Problem 7.4 to be a single-fold unit (or the outer plates of a multibay system) supported on longitudinal wall enclosures. Determine the preliminary member sizes for the sheathing, rafters, and chords.

7.6   Determine the approximate amount of the main reinforcing in the longitudinal and transverse directions for a typical $3\frac{1}{2}$-in.–thick interior concrete plate of a V-type folded plate roof of 4 ft height and a bay width of 10 ft (i.e., $a = 5$ ft) spanning 60 ft. Assume a dead load of 50 psf and a snow load of 30 psf. Use 60-ksi steel and 4-ksi concrete.

7.7   For the W-shaped, 4-in.-thick folded concrete plate in Fig. 7.8, determine the main reinforcing at midspan in the transverse and longitudinal directions. The folded plate structure acts as a simply supported beam spanning 100 ft; the effect of the small cantilevers at the ends is neglected. Use 60-ksi steel and 4-ksi concrete. Assume a snow load of 30 psf and 5 psf for built-up roofing and insulation.

chapter 8

Shell Structures

The domes and vaults of the past are the forerunners of the surface structures of today. The low tensile strength of materials such as mud, brick, stone, and concrete influenced designers to use thick surfaces, possibly reinforced with ribs, resulting in heavyweight structures. In contrast to the traditional surface forms of domes, vaults, cloister vaults, and groin vaults, the present surface structures can take almost any shape. The surface shape is the primary characteristic of the shell; its capacity is by far higher than a structure made up of individual linear elements of similar dimensions.

Surface structures are classified as rigid or soft shells, besides the traditional vaults. Soft shells are prestressed tensile membrane structures that are flexible and with very little weight, as discussed in Chapter 9. Rigid shells can be thick or thin. Whereas thick shells provide bending stiffness, thin shells resist loads in purely axial and shear action. They are thicker than tensile membranes since they carry loads in compression but thinner than vaults since they don't bend. For low-tensile-strength materials, tension is resisted by reinforcement, as in the case of reinforced-concrete shells.

In contrast to flexible membranes, thin shells want to resist loads in compression as based on funicular arch action, unless they are tensile shells. Rigid *tensile shells* are identified by their hanging curves and are often found in wood construction using, for instance, pre-bent tension joists. Reinforced-concrete tensile shells are usually prestressed so that the concrete remains in compression and uncracked (see Sections 9.5 and 9.6). This, however, was not the case for the suspended roof of the terminal building at Dulles Airport in Washington, D.C., which forms a rigid tensile vault. *Compressive shells*, on the other hand, possibly generated by inverting hanging membranes so that they are in compression under their own weight and prestressed to take care of asymmetrical loading causing tension, have the advantage that they remain uncracked and are very tight, so they require hardly any coating and waterproofing.

Shells may consist of curved solid surfaces, as is typical for ordinary ribless concrete shells, or they may form skeletal shells, as is usually the case in metal construction. A skeletal surface structure may be treated as a *reticulated shell* when the member arrangement is dense and uniformly distributed (i.e., in a regular networklike fashion).

For example, a radially ribbed and ringed dome represents a transitional stage to a reticulated shell, such as a three-way grid dome. On the other hand, a *ribbed dome*, where the material is concentrated in the ribs, has to be treated as consisting of individual radial arches; therefore, the dome has no relationship to a shell. The geometrical configuration of the member arrangement in skeletal shells is discussed under the various shell types.

Shells are not only used as roofs in architecture, but also for many other purposes, such as ship hulls, car bodies, car fenders, pipes, piles, foundations, dams, tunnels, offshore structures, towers, bridges, storage tanks, and pressure vessels.

8.1 INTRODUCTION TO THIN-SHELL AND SKELETAL-SHELL STRUCTURES

To arrive at an initial understanding of rigid bent surface structures, an awareness of their origins in an architectural historical context is developed. Further appreciation is generated by pointing to examples in nature as analogs for curvilinear surface forms. The organization of various common surface geometries is presented. The membrane forces for common surfaces under typical loading are derived; they establish an essential basis for understanding rigid shells.

Bent Surface Structures in Nature

Living organisms are constantly changing and adjusting to new external pressures; they are transforming in time and space. Their formal response has always been intriguing to designers and a constant source for new discoveries, although we can never fully understand the forces and principles that shape the organism. Some mathematicians, architects, scientists, poets, and artists have fully concentrated on only the geometry in nature in order to recognize the conformity and order of proportions defining space as expressed by lines, surfaces, and shapes; they may have tried to unveil the *Divine Order*. For instance, the formal features of symmetry and regularity are always reflected by the beautiful snowflake crystals, although they are never the same. Their patterns remind us of the minute skeletal shell structures of the diatoms (marine algae) and radiolaria (unicellular organisms). These skeletons reveal an extraordinary complexity and delicacy of geometry, as well as a nearly endless variety of shapes and surface structures. They seem to correspond to architectural constructions and to illustrate basic building concepts, such as least weight by employing a minimum solid ribbing, that is, accepting the structural engineering analog. Some radiolaria even encompass properties of triangulated, stressed skin shells.

There is an abundance of surface structures in nature that are not just to be found on the microscopic scale; some typical examples are shown in Fig. 8.1. Common rigid shell forms are the shells of eggs, snails, turtles, mussels, skulls, hollow horns (e.g., goat, sheep), clay nests of ovenbirds, nests of the weaverbirds, etc.; all these shells express an unbelievable richness of surface forms as well as the strength of the bent surface structure. The various spiral geometries of seashells, especially the beauty and perfection of the nautilus, have always been inspiring. The domelike shape of the skull uses minimum material to achieve maximum strength so that vital organs can be protected. The bone itself, on a microscopic scale, consists of an intricate three-dimensional rigid lattice network that contains the soft tissue elements. Depending on the type of skull capsule, the number of layers ranges from one to multilayer systems similar to sandwich construction.

There are many types of tensile membranes in nature. The lightweight wings of insects and bats respond with the necessary flexibility and mobility; they can be described as folding soft shell structures. They may be reinforced with a delicate network of ribbing, as for the much described dragonfly, reminding us of the branching grid structures of leaves.

Pneumatic forms are found in sea foam, soap bubbles, and organic flexible cells stabilized by fluids. The hydrostatic skeleton and the lack of stiff components, typical for some worms, clearly respond to flexibility.

There are infinitely many types of spider webs, ranging from two- to three-dimensional net structures; they may be the familiar vertical, sheetlike radial webs or suspended tentlike membranes.

The radially stiffened membrane of Victoria, the giant water lily of the Amazon, has fascinated many architects, probably starting with Joseph Paxton, who was intrigued by the beauty and strength of the branching rib pattern that supports the larger than 5-ft pad at the underside. Its structure influenced his design of the barrel vaulted

Figure 8.1 Surfaces in nature.

iron–glass structure of London's Crystal Palace of 1851, which, in turn, had an extensive impact on subsequent architecture.

The Development of Bent Surface Structures in Architecture

The great dome structures of the past (Fig. 8.2), together with cylindrical barrel vaults and the intersection of vaults (cloister and groin vaults), are the forerunners of present-day shells, keeping in mind that the modern thin membranes, with their formal flexibility, may directly express structure, in contrast to the sculptural forms, for instance, of the Renaissance and Baroque periods, where the dome is part of the total architecture. Still, the study of various construction methods for dome structures throughout history will strengthen our understanding of current technology.

The structural design of the great domes of the past did not have a scientific origin. They were sized as based on empirical knowledge derived over time from material characteristics, construction processes, correction of failures, geometric symbolism, and so on, and were contained in design rules as described, for instance, by the Roman Vitruvius in the first century B.C. and later in the Renaissance by Leone Battista Alberti (1485). Gothic master masons equated structure with geometry, possibly reflecting an intuitive feeling for force flow, although it was never formulated. Renaissance architects, on the other hand, were absorbed by the geometry of form in general; they tried to reflect the harmony of nature in architectural proportions through mathematics. Structural concerns, as reflected by the revived interest in Roman engineering, hardly had any influence on the form-giving process of architecture. However, during this period of humanism, scholars began to search for an order in the universe separate from religion, thereby developing the basis for the modern sciences. Leonardo da Vinci (1452–1519) clearly represented the ideals of this period; he recognized and defined several effects of theoretical mechanics. The evolution of structural mechanics started about 100 years later with Galileo Galilei and then progressed with the development of scientific thought through the Age of Reason so that by the nineteenth century the basics of mechanics and elastic material behavior were clearly formulated. The necessary theory for the structural design of thin shells was then further developed at the beginning of this century.

The earliest dome forms probably are derived from the *corbeling* principle, where rings of horizontal masonry layers project slightly beyond the ring of the previous layer, usually yielding conical outlines. The most famous example in the western hemisphere is the Treasury of Atreus in Mycenae (circa 1325 B.C., Fig. 8.2a).

Corbeled domes were not used anymore in Europe for large-scale enclosures; however, in India the Islamic architecture continued to employ the principle. A sensational example is the great hemispherical dome of Gol Gumbaz at Bijapur in southern India (A.D. 1625–1656), which is 125 ft in diameter and thus larger than the Hagia Sofia in Constantinople. Astonishing is the fact that the dome uses much less thickness than the Pantheon to resist the tensile hoop forces. India's famous Taj Mahal, a 58-ft-span mausoleum, was finished in 1647. A double dome system was used where the outer bulbous dome is nearly 200 ft high. Also mentioned in this context should be *cribbed*, nonthrusting *domes*, which consist of a series of horizontal regular polygons of beams (successively rotated by some angle) by superimposing progressively shorter cribbed beams and thereby generating the inner dome shape.

Whereas the Islamic dome structures in India have their origin in Persia as well as in the Hindu corbeling techniques, the western dome structures were very much influenced by Roman construction. The Romans had achieved immense spans of 90 ft and more with their vaults, and as so powerfully demonstrated by the 143-ft span of Hadrian's Pantheon in Rome (circa A.D. 123., Fig. 8.2b), which was unequaled in

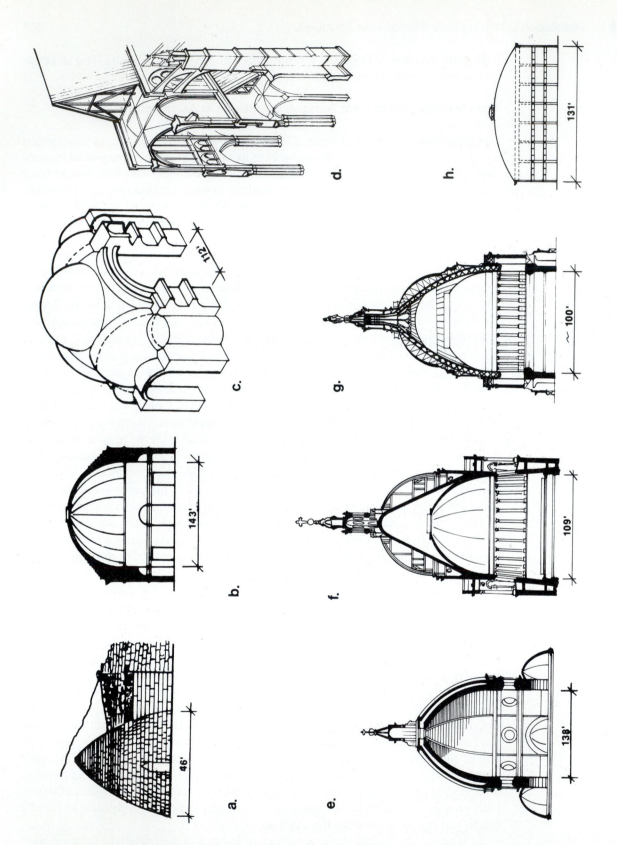

Figure 8.2 Development of long-span roof structures.

a.

b. 143'

c. 112'

d.

e. 138'

f. 109'

g. ~ 100'

h. 131'

Europe until the second half of the nineteenth century. The reactions of the hemispherical concrete dome due to gravity are tangential to the support and hence do not cause a horizontal thrust component, as do shallow domes. However, the familiar thrust due to the geometry of semicircular arches and vaults must be balanced in dome vaults internally by tensile hoop forces along the lower dome portion, as will be discussed later in more detail. These tensile circumferential forces are resisted by the massive dome thickness, which increases toward the base, as well as by stronger material. To keep these forces to a minimum, the weight of the concrete is successively reduced toward the crown by employing lighter-weight aggregates and reducing the vault's thickness (besides using "coffers"), thereby also reflecting the magnitude of the force flow along the arches in the radial direction. The dome vault is further stiffened by eight relieving arches embedded in the concrete.

The series of domes of Justinian's Hagia Sofia in Constantinople, used by the master builders Anthemius of Tralles and Isodore of Miletus (A.D., 537, Fig. 8.2c) for the first time, cause a rather dynamic flow of solid building elements together with an interior spaciousness that is very different from the more static Pantheon. The shallow main brick dome of 112-ft span is reinforced with ribs and is almost entirely in compression; thus, it evades the tensile stresses in semicircular domes and the necessary increase in shell thickness, which might not have been feasible because of the low tensile capacity of the brick and the lost art of Roman concrete technology. The dome sits on four gigantic *pendentives* that convert the round base to the square space below. The pendentives, in turn, are vertically supported by four huge circular arches. The lateral thrust, which is large for shallow domes (but not present for semicircular domes where it is transformed into circumferential tensile stress bands along the bottom part of the dome), is resisted by two semidomes in one direction and by massive corner buttresses in the other direction. The action of these buttresses was not fully understood by the architects and thus could not prevent several collapses of the roof. Only in the cathedrals of the Middle Ages was the art of *buttressing* developed to a high-level of sophistication.

About 1000 years later the legendary Turkish architect, Sinan Abdur-Mennan (1489–1588), during the Golden Age of Ottoman architecture, inspired by the Byzantine Hagia Sophia, transformed its architecture into the floating antigravity structures so typical for Ottoman mosques. Here, the arches supporting the central dome are of the same stiffness, and the dome thrust is resisted by half-domes in both major directions, thus evading the support problem of the Hagia Sophia. The Selimiye Mosque (1575) in Edirne is Sinan's largest dome structure with a 103-ft diameter.

Gothic cathedrals (Fig. 8.2d) are admired for their seemingly weightless interior spaces, which are not achieved through record horizontal spans, but by span-to-height proportions (e.g., cathedral at Reims: 123-ft height to vault, 48-ft width for central aisle), with Amiens and Cologne cathedrals reaching about 1:3 and finally Beauvais the daring proportion of 1:3.33, quite a development from the typical ratio of 1:2 for the Romanesque churches! These proportions, together with the effects of light and articulation of the slender skeleton structure, convey a feeling of antigravity and dematerialization of space, in turn, resulting in a never-ending upward surge, as is so powerfully and daringly expressed in the French cathedrals of the High Gothic period.

Gothic master builders used the *pointed arch* (composed of circular arcs) as primary support structure, which approximates efficiently the catenary arch. Furthermore, in contrast to the semicircular arches of the Romanesque period, the smaller span-to-height ratio of the Gothic arches causes smaller lateral thrust forces.

The disadvantage of the continuous thrust of the cylindrical vault of earlier periods was overcome with the *cross vault,* where the loads are concentrated along the diagonal ribs and guided to point supports so that the walls are free for large window openings. The typical Gothic cross vault is obtained by intersecting two pointed cylinders and placing transverse ribs and diagonal ribs along the intersection lines (see Fig. 8.22). The ribs act together with the stone webbing as a composite vault. In the late Gothic period the solid pointed vault is replaced by an intricate network of ribs. The vaults are supported vertically by interior piers, while their thrust and the thrust caused by the steep timber roof are usually transferred by two separate *flying buttresses* to the exterior massive vertical piers. These huge pier buttresses are topped by pinnacles, which can be visualized as a prestress agent adding weight, so that the resultant force due to thrust and weight is kept within the middle third of the horizontal pier cross section (kern), thereby developing no tension (Fig. 2.6e). The high point and limit of the Gothic construction method was reached by Beauvais Cathedral (1347) with an incredible vault height of 158 ft.

The 138-ft-span dome of S. Maria del Fiore in Florence (1434, Fig. 8.2e) can be considered the first modern dome structure and an important guideline for future domes. Be aware that the dome is the first long-span structure of large scale since the Hagia Sophia was built nearly 900 years ago. The geometry of the *octagonal pointed dome* is generated from the intersection of four circular cylinders. The primary supporting sandstone arches are placed along the four ridges, but in addition, two intermediate secondary ribs are located in each of the cylindrical sectors. These radial ribs, together with horizontal circumferential stone rings, tie the inner and outer masonry shells of this *double-layer dome* together. In contrast to the previous domes, which show predominantly vault (surface) action, here the radial steep arches can be assumed to be the primary structural components; they are not just the ribbing of a double vault. The arches cause lateral thrust even if their curvature is tangential to the supports; this thrust is further increased due to the weight of the heavy lantern, but on the other hand, reduced through the large dome height. Here, the hoop tension is not resisted by the stronger and thicker vault portion along the lower part of the dome as for the Pantheon, nor by buttressing as for the Hagia Sophia, but by hidden tension rings that consist of several layers of stone chains (sandstone blocks joined by iron clamps) and a timber chain. *Filippo Brunelleschi*, the inventive designer of this polygonal dome, developed the method of composite action between ring and arches as well as the idea of double-vault construction, which can be traced back to the seventh century Islamic "Dome of the Rock" in Jerusulem and in Europe to the thirteenth century Byzantine domes of St. Mark's in Venice; these domes employ outer timber framing. In addition, Brunelleschi's real invention lies in the erection of the dome. He did not use any central temporary shoring to support the high dome, but employed the horizontal sandstone rings to prevent the arches from tilting inward during construction.

Michelangelo's dome for St. Peter in Rome (1590), with nearly the same span, is based on similar construction principles, although its double brick vault is thinner, no horizontal circumferential stone rings are used, and less radial ribs are employed. The entire dome thrust was resisted first by only three iron chains along the base, which proved to be insufficient, so five more tension rings had to be installed in 1743 as derived by Giovanni Poleni. He used a model of string and lead weights to obtain the thrust line and thus was able to make his recommendations.

A revival of the idea of high Gothic ribbed vaulting is seen in Guarini's S. Lorenzo in Turin (1666–1687). Here, the solid domical surface is ingeniously resolved into eight intersecting arches, forming a starlike pattern in plan view. During the Baroque period, frequently the vaults and domes only appear as support structures, but are plastered ceilings suspended from load-bearing timber framing.

Christopher Wren introduced the concept of the catenary dome shape with the conical brick dome of 109-ft span and only 18-in. thickness (that is tied together at its base by a double iron chain) to support the cupola of St. Paul's Cathedral in London

(1710, Fig. 8.2f); the dome is a surface, not a ribbed structure. This loadbearing brick cone is not visible, it is located between the inner, self-supporting brick dome and the hemispherical, outer, trusslike timber structure, which is partly carried by the conical shell. The shape of the cone, only slightly curved along its inclined portions, comes very close to the funicular form responding to the single load due to the heavy 700-ton masonry lantern and the uniform roof loading, thus causing mainly compression and allowing this extremely light structure, the first of its kind, to be built.

The 91-ft-span dome of the Church of the Invalides in Paris (1680–1691), built by Jules Hardouin-Mansart at about the same time, is constructed quite differently. It consists of an outer load-bearing, complex, trussed, timber framework and a self-supporting inner, elliptical brick dome, which splits at about midheight into two vaults. Jacques Germain Soufflot followed Wren's concept for the Pantheon in Paris (1755–1792), but replaced the trussed outer skin with a third masonry vault. His building marks the high point and the end of masonry vaulting. Now iron skeleton structures slowly start to gain in importance.

While the dome architecture of the early part of the nineteenth century was controlled by neoclassicism and in the later part by eclecticism where different styles were combined, it was at the beginning of this century that the developments of the Modern Movement made the expression of the support structure possible.

Skeletal Domes.   The transition to modern dome structures occurred during the 1800s, especially the second half, when iron and glass domes together with cylindrical surfaces were extensively used for galleries, arcades, greenhouses, exhibition halls, railway stations, market halls, and so on. It may have been the Crystal Palace in London (1851) that had a major influence on the development of the shell-like glass–iron surface structures all over Europe during the nineteenth century.

Cast iron dome structures started with the hemispherical ribbed dome of the Halle-au-Blé in Paris (1811), designed by F. Bélanger and the engineer F. Brunet. Thomas U. Walter's 99-ft-span dome for the U.S. Capitol Building in Washington, D.C. (1865, Fig. 8.2g), can be seen as a typical example of the new material iron. The main structural elements for the ribbed dome are the inner trussed elliptical arches of nearly uniform depth. They support the crescent-shaped trusses on top, with the upper chords defining the elliptical profile of the cupola, and also carry the inner hemispherical plaster dome.

Johann W. A. Schwedler is credited as one of the first designers to have introduced true shell grid structures in 1863. He replaced the traditional ribbed dome with the *braced dome* concept. Schwedler's 207-ft-span dome in Vienna (1874) is often considered a milestone for dome development. It was among the first true domes based on surface behavior rather than on the concentrated arch action, as for conventional ribbed domes.

The cylindrical building of the Galeries des Machines, for the 1889 Paris Exhibition, reached an unprecedented and an unheard of 375-ft span by employing parallel, three-hinged, trussed gable frames made of steel.

Schwedler influenced the development of other bracing types for skeletal domes during this time. In 1922, Walter Bauersfeld introduced the three-way steel grid dome, which functioned both as formwork and reinforcement for the thin concrete shell (Zeiss–Dywidag system). Important also was the introduction of the lamella roof concept to the United States in 1925 by G. R. Kiewitt, which had been developed by the German Zollinger in 1908 and led to the immense 642-ft-span Houston Astrodome in 1964 and the 680-ft Louisiana Superdome in New Orleans in 1975.

Braced dome construction became very popular during the 1950s through Buckminster Fuller's geodesic domes and his writings. An important historical turning point is the 384-ft, stressed-skin, space grid geodesic dome for the Union Tank Car Company in Baton Rouge, Louisiana, designed by Fuller and completed in 1958.

When the Civic Auditorium in Pittsburgh opened in 1961, the 415-ft-span stainless steel dome, supported by a huge 200-ft-span trussed steel cantilever frame, which is subject to large vertical and lateral loads, was the first modern retractable roof of this scale; it was engineered by Ammann & Whitney. The dome is divided radially into eight leaves; six of them are movable and rotate around a pin at the top and are supported mainly by the cantilever frame. Currently, the Toronto Sky Dome (1989) with a clear span of 674 ft is the largest operable dome structure. It consists of four independent steel roof segments, with three of them traveling. The two central, sliding, parabolic barrel sections consist of trusses spanning across the field. Of the hemispherical shells at the ends, one is fixed and the other rotates. The roof opens in only 20 minutes.

Thin-shell Structures.   The Spanish architect Antonio Gaudi, in the late nineteenth century, introduced important new concepts of design. He was fully absorbed in the relationship of form and structure. He searched for efficient structural shapes that exerted a minimum of bending and lateral thrust; he studied the intricacy of force and form from an experimental and empirical point of view. For instance, for the Colonia Guell Chapel (1898–1914) near Barcelona, he derived in 1908 the funicular shape of the vaults and their inclined piers from a suspended string–sheet scale model with the proper weights hanging on it, to simulate the inverted rigid structure fully loaded. Thus, he introduced the catenary surface shape, a concept that has its origin in Wren's brick cone dome for St. Paul's Cathedral in London. Gaudi may be considered the unacknowledged forerunner of modern shell construction; he employed the traditional Catalan thin-tile vaulting, also called *timbrel vaults*, for the unusually complex curved surfaces that he developed. Although not widely recognized, Gaudi may be the first ever to have used hyperbolic paraboloids in building construction.

Among the first important reinforced concrete structures is the Church of Jean de Montmartre in Paris (1897), designed by Anatole de Baudot, who used reinforced concrete ribs and brick shells for the dome construction.

The tradition of ribbed masonry domes was revived at the beginning of the twentieth century with reinforced concrete, obviously stimulated by steel skeleton construction. It was much easier for the reinforced cast-in-place concrete, than for the masonry, to resist the critical hoop tension along the bottom portion of a steep dome. The most impressive concrete dome of this period is the Centennial Hall in Wroclaw, Poland (formerly Breslau, Germany) designed by Max Berg (1912) and engineered by Dyckerhoff and Widmann (Fig. 8.3f). This ribbed dome is supported on four skew arches and consists of 32 radial ribs, which are tied together horizontally by five rings. This concrete dome has a span of 213 ft and thus was the first concrete dome to surpass the span of the Pantheon in Rome. Since the weight of concrete domes, which are based on traditional rib construction, increases rapidly with span, it was just a question of time for thin concrete shells to be developed.

In 1916, Eugene Freyssinet began to build the two famous concrete airship hangars of parabolic cylindrical shape at Orly near Paris. These arched, 18-ft-deep, undulating thin shell vaults had a span of 262 ft and were 184 ft high with a maximum shell thickness of only $3\frac{1}{2}$ in., thus yielding a very light structure. These arched vaults, however, do not have a true shell form; the arch moments due to wind are merely efficiently resisted by the depth of the folds, thus allowing the thin concrete corrugations.

While engineers were in the process of developing the thin shell concept, architects were also experimenting with the new-found freedom of form giving using the new material, reinforced concrete. For example, the Duldeck Residence (Fig. 8.3c) designed by the Austrian Philosopher, artist, and architect Rudolph Steiner as part of the Anthroposophical colony in Dornach, Switzerland, in 1915, reflects this trend. The

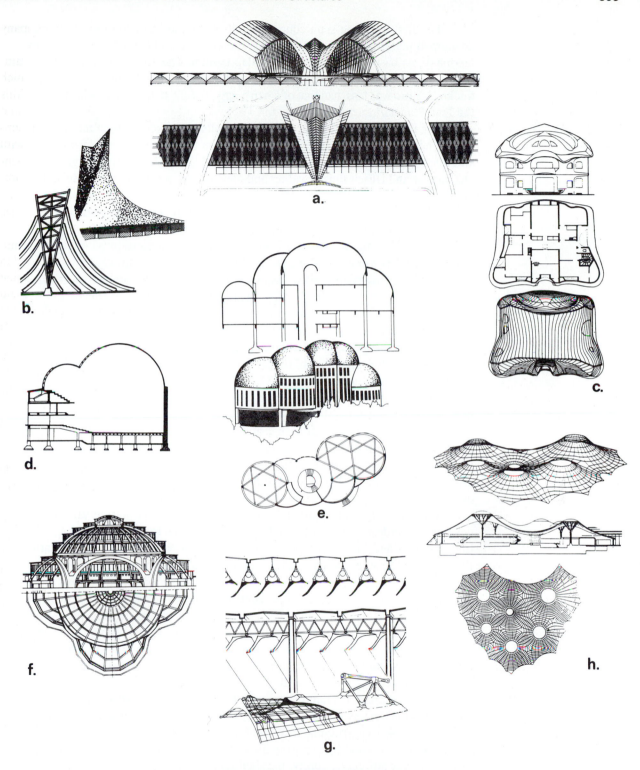

Figure 8.3 Bent surface structures.

form of the building is surely in anticipation of Goetheanum II, the famed school of spiritual science. It represents early German expressionism and reminds one of Eric Mendelsohn's Einstein Tower of 1920. The plasticity of the roof allows a rich interplay of form. The experimentation with hanging concave shapes in reinforced concrete was rather novel for its time and predicted the later development of prestressed suspended concrete shells.

The first true long-span concrete shells of cylindrical and domical shapes were developed under Franz Dischinger and Walter Bauersfeld and built in 1924 in Jena, Germany, as based on the Zeiss–Dywidag system. The 131-ft-span, shallow, ribless dome for the Schott Company in Jena (Fig. 8.2h) was only $2\frac{3}{8}$ in. thick. Its radius-to-thickness ratio was less than that of a hen's egg. The 82-ft-span hemispherical dome for the Jena Planetarium was completed in 1925.

The first large-scale, long, cylindrical shells were used for the Frankfurt am Main market halls (Germany) in 1927. The shells were 46 ft wide and had large edge beams and a span of 121 ft, with a thickness of only $2\frac{3}{4}$ in. They were designed by Franz Dischinger and Ulrich Finsterwalder and constructed according to the Zeiss-Dywidag system.

The first truly large span concrete shell structure was the enormous 217-ft-span Leipzig market hall, Germany, completed in 1929 and engineered by F. Dischinger. This octagonal dome was derived from the cloister vault concept by intersecting four elliptic cylinders using ribs along the fold lines.

The first long-span shell structures in the United States are the short cylindrical shells for the Sports Arena in Hershey, Pennsylvania, covering an area of 233×342 ft, and designed by Anton Tedesco of Roberts and Schaefer in 1936 as based on the Zeiss–Dywidag system.

To the basic geometrical shapes of the cylinder and the dome, Giorgio Baroni added the hyperbolic paraboloid (hypar) with the roof shells for the Alfa-Romeo automobile factory in Milan, Italy, in 1934.

Whereas the shell pioneers Franz Dischinger, Ulrich Finsterwalder, and Dyckerhoff and Widmann A.G. of Germany were very much involved in the development of shell construction with respect to span and least weight (i.e., minimum shell thickness), it was the engineer–designers Eduardo Torroja of Spain and Pier Luigi Nervi of Italy who introduced a new formal language that stimulated both engineers and architects, especially after World War II, to become involved in the exciting formal potential and the challenge of the bent surface.

With the Algeciras Market Hall, Spain, in 1934, the daring and imaginative designer Eduardo Torroja modified the traditional dome form. He built one of the shallowest domes and for the first time a spherical dome with polygonal boundaries. The flat dome is 156 ft in diameter, has a radius of curvature of 145 ft, and rests on eight equally spaced columns along the periphery. In the Frontón Recoletos in Madrid, Spain, in 1935 (Fig. 8.3d), Torroja dared to place two circular cylindrical segment shells of different size adjacent to each other, thereby giving the illusion of instability in section since the pair of arches seem to be missing a central support. In reality, the shells behave as beams spanning in the longitudinal direction. For the Zarzuela Hippodrome Grandstand (1935), Torroja uses shallow hyperboloidal sections with varying rise to cantilever an effortless 41 ft.

Robert Maillart, the famous concrete bridge designer, gave a special architectural meaning to the Cement Industries Hall at the Swiss Exposition in Zürich (1939, demolished, Fig. 8.14g). He articulated the lightness of the parabolic cylindrical shell through its spatial qualities and by exposing along the edges the thinness of the shell without using any edge stiffeners. Surely, the shell represented a true advertisement of the formal potential of concrete.

Pier Luigi Nervi attracted first international attention with the Municipal stadium of Florence in 1932, where he addressed new space concepts and suggested new construction methods using primarily concrete frame construction. But it was the magnificent long-span airplane hangar at Orvieto, Italy, in 1936 that made him famous. He resolved the 328- $\times$ 131-ft span cylindrical shell vault into a delicate skeletal surface structure of the lamella type to control buckling, thereby also achieving a decorative experience. This poured-in-place concrete structure was destroyed during World War II. A second modified hangar version was built in 1940 (Fig. 8.14e), which was lighter and more elegant than the earlier one. This time the lattice roof was prefabricated and only supported by six buttresses arranged symmetrically to equalize the force flow due to the vault's weight. Nervi built a series of lamella vaults, for example, the Exhibition Hall, Salone C, in Turin, Italy (1950, Fig. 8.14c). This 165- × 215-ft lamella vault, which is 45 ft high, consists of precast ferrocemento units and rests on four inclined arches so as to efficiently resist the thrust of the roof vault. Nervi possibly reached the high point of his lamella roof construction in 1957 with the magnificent Small Sports Palace in Rome (Fig. 8.27j), exposing the lateral thrust with the external buttressing. Nervi, similar to Torroja, was able to design structures of human properties and spatial qualities that will always be appreciated as masterpieces of architecture. He derived the power of expression based on the right engineering forms and well-balanced proportions directly reflecting the efficiency of force flow.

After World War II, an explosion of concrete shell architecture occurred, as demonstrated by Felix Candela in Mexico, who unveiled the true potential of the hyperbolic paraboloid in his numerous buildings, such as the magnificent church of La Virgen Milagrosa in Mexico City of 1955 (Fig. 8.38 h).

Engineers continued in their search for larger spans and new construction methods. In 1953, the warped, oval, 234-ft-span saddle shell roof for the Schwarzwaldhalle in Karlsruhe, Germany, designed by U. Finsterwalder, was the first large-scale prestressed concrete saddle shell structure.

The search for the limits of concrete shell construction may have been reached in 1958 with the CNIT Exhibition Hall in Paris, France (Fig. 8.43f), designed by Nicholas Esquillan. This double-layer shell structure, consisting of an intersection of three parabolic cylinders, spans the enormous distance of 720 ft and is still so far the largest-span reinforced concrete structure. The largest concrete dome is currently the 661-ft-span Kingdome, Seattle, Washington (Fig 8.27g), designed by J. Christiansen and completed in 1976.

The tradition of the great engineering designers Torroja and Nervi was continued by the structural engineer Heinz Isler of Switzerland during the 1960s and his beautiful minimal, free-form, thin-shell structures (e.g., Fig. 8.44u).

The work of Torroja and Nervi has also challenged architects to integrate shell structures in their building design. During the 1950s, architects like Eero Saarinen, Kenzo Tange, Marcel Breuer, Le Corbusier, and many others experimented with new geometrical forms that were not solely controlled by engineering thinking. Among some of the famous early examples of shell structures are Yamasaki's Air Terminal Building, St. Louis (1954), Saarinen's Kresge Auditorium at MIT in Cambridge, Massachusetts, U.S. (1955), Le Corbusier's magnificent chapel at Ronchamp (1955), Jørn Utzon won the competition for the Sydney Opera House in 1957, Saarinen's TWA Terminal Building at New York's JFK Airport (1962), and Kenzo Tange's St. Mary's Cathedral in Tokyo (1964).

Domes became very popular in the second half of the 1960s and early 1970s through the counterculture architecture of the hippie movement as inspired by Buckmister Fuller. This alternative architecture is exemplified by the commune Drop City, Colorado, 1965, which was designed to a large extent by Steve Baer using modified geodesic dome concepts (see *Domebook* 2. 1971).

But not only architects and engineers were involved in experimentation with surface structures and using the new-found freedom of conquering space made possible by science or technology; artists also searched for new forms and their meaning. This is demonstrated by the constructivists' sculptures such as Naum Gabo's *Torsion* of

1936, Antoine Pevsner's *Developable Column* of 1942, and Max Bill's *Endless Loop* of 1949.

Early examples of experimentation with shells are found in chair designs, such as Alvar Aalto's famous armchair of 1933 using plywood bent in one direction, Charles Eames's bent plywood chair molded in two directions of 1946, and Saarinen's "Womb" chair, a fiber-glass molded shell with metallic legs, of 1948.

The New Generation of Shell Structures.   The new shell architecture does not necessarily celebrate structural rationality as so powerfully expressed by Nervi, or based on the principles of Viollet-Le-Duc's engineering esthetics, or constructivism. In other words, surface forms do not necessarily articulate economy of construction and efficiency of force flow and strive for the perfection of support structure. In contrary, structure may be hidden and treated as subordinate to form. It seems that in many of the new bent surface forms structure is not only seen as support but also as an ornament, giving a more complex meaning to space, as exemplified by some of the cases in Fig. 8.3.

Fumihiko Maki's Fujisawa Gymnasium (1984, Fig. 1.2b) transmits a strong symbolic meaning reminding one of a Japanese warrior helmet. Although its form is obviously not derived from structural minimalism, engineering is wonderfully integrated into the design of the supporting structure. The roof of the main arena building is resting on two giant, parallel, 11.5-ft-deep trussed steel arches of triangular cross section spanning 265 ft. The space perpendicular to them is bridged by latticed steel arches forming vaults.

The complex sculptural church roof at the ancient marketplace in Rouen, France (1979, Fig. 8.3b), is composed of hyperbolic paraboloid surfaces. The steep roof structure is supported by a triangulated trussed steel arch forming a central spine, which, in turn, supports the wood ribs that act as suspended glulam arches.

The intersection of two unequal spherical domes for the first Gotheanum in Dornach, Switzerland (1920), designed by Rudolph Steiner and Carl Schmid-Curtius, must have influenced Yasufumi Kigima's Forest Museum in Kuma-gun, Japan (1984, Fig. 8.3e). Here, an irregular ensemble of seven interlocking concrete domes tops the building, which is anchored to the steep mountain slope.

The contours of the undulating roof shell for the Solemar in Bad Dürrheim, Germany (1987, Fig. 8.3h), reminds one of nature, where the upper, truncated, cone-shaped portions form the hills and the lower saddle shapes the valleys. Five tree columns of various heights, with light domes on top, support the ribbed wood roof membrane. The shape of the roof was optimized based on the flexible tensile net structure so that the hard shell acts primarily as a thin membrane. The contour lines represent the layout of the primary radial, suspended, meridional arches, which are tied together by the secondary horizontal ring beams and are anchored to the edge arches.

The unique delicate roof–ceiling leaf shapes for the Menil Museum in Houston, Texas (1987, Fig. 8.3g) give a dramatic expression to the simple interior spaces. Sophisticated studies were required by Renzo Piano, the architect, and Peter Rice, the engineer, involving material research, structural behavior, and optimization of lighting angle to develop the shape of the sinuously curved thin leaves and their production. As sunlight-control systems, they act as reflectors to diffuse the light and exclude direct sunlight from entering. As structures, the ferrocement leaves seem to be suspended from triangulated ductile iron trusses, but, in reality, act as a composite system placing the 40-ft-span trusses in compression and the ferrocement leaves in tension.

The Spaniard Santiago Calatrava presents this new generation of structures as architecture, in the tradition of Torroja and Nervi, by giving it a much richer meaning. His central hall for the railway station in Lyons, France (1993, Fig. 8.3a), seems to express the flight of a bird or butterfly with its powerful dihedraled wings cantilevering from a thoraxlike trussed arch.

Development of the Long-span Structure.   The change from the traditional dome structures of the past through the transition period of the second half of the nineteenth and first half of the twentieth centuries, and then to the present-day bent surface structures, with their limitless formal potential, is enormous, as becomes apparent from Table 8.1 and the discussion in this chapter and the next.

The record in span was held by the Pantheon for nearly 1700 years before steel made possible, during the second half of the nineteenth century, the Galeries des Machines in Paris, which reached the record span of 375 ft in 1889, more than double the Pantheon span. The first concrete structure to break the Pantheon's record was the Centennial Hall in Breslau in 1912 with 213 ft. Record spans were achieved after 1950, first in concrete shell construction with the 720-ft-span CNIT Exhibition Hall in Paris (1958), in steel construction with the 680-ft-span Louisiana Superdome in New Orleans (1973), in pneumatic construction with the 722-ft-span Pontiac Stadium (1975), and finally, in cable construction with the 770-ft-span Georgia Dome in Atlanta (1992). We may conclude that, from a technical point of view, *span* does not impose limits anymore on architecture, at least not in the traditional sense; 350-ft spans are common today. Buckminster Fuller even proposed to cover Manhattan with a geodesic dome 2 miles in diameter to produce a controlled climate.

As the span of a structure increases, the weight does also, while the live load remains constant. It is apparent that to achieve large spans weight must be reduced to a minimum by employing efficient structural systems. The development of the long-span structure is therefore directly related to the development of new lightweight structure systems and may be divided into the following three stages:

- *Heavyweight domes:* This period ends with the Centennial Hall in Breslau (1912).
- *Thin concrete shells* and *skeletal steel domes:* This period ends with the Louisiana Superdome (1973) and the King County Stadium (1975).
- *Fabric domes:* Air domes were popular up to the late 1980s for long spans and then were replaced by cable domes. Currently, the Georgia Dome is the world's largest-span structure at 770 ft (Figs. 9.21e and 8.26).

Differences between the heavy traditional domes and the modern thin shells of stronger materials and more efficient stress distribution, as expressed in span and weight, become apparent from Table 8.1. The weight of the flat concrete shell for the Schott Company (Fig. 8.2) has nearly the same span as St. Peter's, but only about 5% of its weight. Some of the domes built later are much lighter in weight proportionally, although they have much larger spans. A cable dome may be considered almost weightless. Although air domes can be treated as weightless, the rigidity of the soft shell must be provided by mechanical support, in other words, they are not passive roof structures but active ones.

The weight of the structure is directly related to its thickness. The traditional heavy-weight domes have a thickness-to-span ratio of up to about 1:40, whereas egg shells are in the range of 1:130. In contrast, concrete shells can go up to about 1:1800, clearly identifying the superiority of man-made shells! A typical t/L ratio for concrete shells may reach about 1:600. Notice that the t/L ratio for beams is about 1:24, or about 25 times larger than for concrete shells, indicating the material efficiency of membrane systems as compared to bending systems, ignoring other considerations of structural design such as the effect of scale.

TABLE 8.1
Historically Important Long-span Dome Structures

Date	Name	Span (ft)	Approximate Average Thickness (ft)	Span/Thickness	Approximate Weight (psf)	Percent
123	Pantheon, Rome (concrete dome)	144	13	11	1475	277
1434	S. Maria del Fiore, Florence (double-shell masonry dome)	138	7	20		
1585	St. Peter's, Rome (double shell brick dome)	137			533	100
1710	St. Paul's, London (brick cone supporting outer and inner shells)	109	3	36		
1912	Centennial Hall, Breslau (ribbed reinforced-concrete dome)	213			393	74
1924	Schott Co., Jena, Germany (reinforced-concrete shell)	131	0.2	655	27	5
1929	Leipzig Market Hall (ribbed reinforced concrete shell)	217	0.3	723	78	15
1953	Schwarzwaldhalle, Karlsruhe (prestressed concrete saddle shell)	234	0.19	1232		
1958	CNIT Exhibition Hall, Paris (reinforced concrete double shell)	720	0.39	1846	58	11
1964	Astrodome, Houston, Texas (steel lamella dome)	642			16	3
1969	Convention Center, Ohio University, Athens, Ohio (steel Schwedler dome)	328			9	1.7
1973	Louisiana Superdome, New Orleans (steel lamella dome)	680			26	5
1975	King County Stadium, Seattle, Washington (reinforced concrete, 5-in. shell between 6-ft-deep radial arches)	661	0.41	1612	85	16
1977	Stadium Northern Arizona University, Flagstaff (triangular grid timber dome)	502			17	3
1975	Pontiac Stadium, Pontiac, Michigan (pneumatic dome)	722			1	0.2
1989	Florida Suncoast Dome, St. Petersburg, Florida (cable dome)	688			5	0.9
1992	Georgia Dome, Atlanta (hypar tensegrity dome)	770				
	Hen's egg	0.13	0.001	130		

The preceding comparisons are based purely on technical and not esthetic considerations. The values of scale and engineering achievement as represented by the Houston Astrodome, for example, may not be superior at all to the spatial qualities of the *dome architecture* of the past! Furthermore, bent surface structures should not be associated only with engineering technology and efficient long spans, but should evolve as an integral part of the entire architecture, as is discussed elsewhere.

Surface Classification

Thin shells are form-resistant structures with sufficient curvature that carry loads primarily in *membrane action,* that is, in pure axial and shear action along the middle plane of the shell. In contrast to prestressed tensile membranes (Section 9.6), thin shells like to resist loads in pure compression (ideal situation), but generally shear and tension do occur; bending stresses are usually restricted to the boundaries.

In the traditional design approach, the shell geometry is given, and the magnitude of the membrane forces is determined. It is apparent that the selected geometry may be less than ideal for the given loading condition. On the other hand, the structural shape may be derived analytically or experimentally as based on loading, membrane stress state, and certain geometrical parameters (e.g., span, height, boundary conditions) to yield an optimal shell surface. For example, a catenary shell carries its weight in pure compression, or a spherical membrane resists air pressure in uniform tension.

Before discussing various shell shapes in more detail, basic features of geometry are identified first. The *curve* is the most fundamental property of the surface. Important characteristics of planar curves are discussed in more detail in Section 9.2, such as the shape and the length of the curve, as well as the normals and tangents to the curve. Furthermore, the area, location of the centroid, and moment of inertia of the area below the curve are familiar characteristics.

Surfaces can be defined by many different curves; therefore, some special curvatures must be identified: the *principal curvatures*, the *Gaussian curvature,* and the *mean curvature* as explained in Fig. 8.4. These curvatures characterize the surface as a single- or double-curvature system, where the double-curvature surface is further subdivided into synclastic and anticlastic surfaces. The curvatures for *synclastic surfaces* are all downward (or upward) as for domical shapes, whereas for *anticlastic surfaces*, oppose each other, as for saddle shapes.

The geometry of surfaces can often be identified as *surfaces of translation* and *surfaces of rotation* (revolution), that is, when they are generated by translating a curve over another curve or rotating a curve about an axis. A shell is a *ruled surface* if it can be generated by a series of straight lines; a synclastic surface cannot be a ruled surface. Only single-curvature surfaces are developable; that is, their surfaces can be flattened without stretching or tearing. Refer to Fig. 8.4 for a more precise definition of the terms just introduced.

Shell shapes may be classified (as shown in Figs. 8.4 and 8.5) as follows:

A. Geometrical, Mathematical Shapes

   A.1 *Conventional* or *basic shapes* as defined in books on analytical geometry

   • Single-curvature (developable and ruled) surfaces

   Surfaces of translation: cylinder

   Surfaces of rotation: cone

   • Double-curvature (nondevelopable) surfaces

   Synchlastic surfaces

   Surfaces of translation: elliptic paraboloid (elpar)

   Surfaces of rotation: domes

   Anticlastic (ruled) surfaces

   Surfaces of translation: hyperbolic paraboloid (hypar, h.p.), conoid

   Surfaces of rotation: hyperboloid of revolution

   A.2 *Segments of basic shapes:* additions of segments, etc.

   A.3 Translation and/or rotation of lines or surfaces

   A.4 Construction process

   A.5 *Corrugated surfaces*

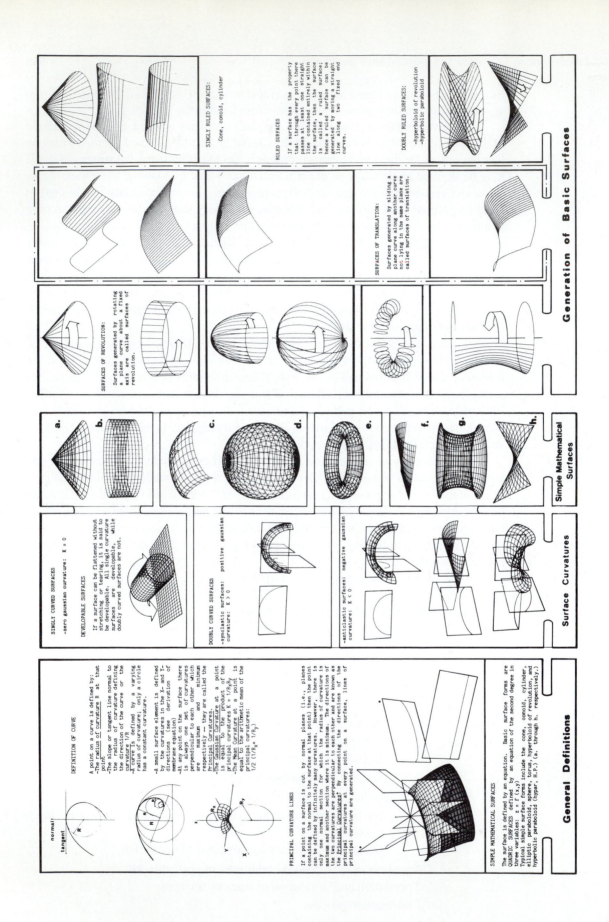

Generation of Basic Surfaces

SINGLY RULED SURFACES:

Cone, conoid, cylinder

RULED SURFACES

If a surface has the property that through every point there passes at least one straight line contained entirely within the surface, then the surface is called a ruled surface; hence a ruled surface can be generated by moving a straight line along two fixed end curves.

DOUBLY RULED SURFACES:

- hyperboloid of revolution
- hyperbolic paraboloid

SURFACES OF TRANSLATION:

Surfaces generated by sliding a plane curve along another curve not lying in the same plane are called surfaces of translation.

SURFACES OF REVOLUTION:

Surfaces generated by rotating a plane curve about a fixed axis are called surfaces of revolution.

Simple Mathematical Surfaces

a.
b.
c.
d.
e.
f.
g.
h.

Surface Curvatures

SINGLY CURVED SURFACES

- zero gaussian curvature: K = 0

DEVELOPABLE SURFACES

If a surface can be flattened without stretching or tearing, it is said to be developable. All single curvature surfaces are developable, while doubly curved surfaces are not.

DOUBLY CURVED SURFACES

- synclastic surfaces: positive gaussian curvature: K > 0

- anticlastic surfaces: negative gaussian curvature: K < 0

General Definitions

DEFINITION OF CURVE

A point on a curve is defined by:
- The radius of curvature R at that point.
- The slope or tangent line normal to the radius of curvature defining the direction of the curve or the curvature 1/R

- A curve is defined by a varying radius of curvature; only a circle has a constant curvature.

- A small surface element is defined by the curvatures in the X- and Y-directions (see derivation of membrane equation.)
- At any point on the surface there is always one set of curvatures perpendicular to each other which are maximum and minimum respectively — they are called the Principal Curvatures.
- The Gaussian Curvature at a point is equal to the product of the principal curvatures $K = 1/R_x R_y$
- The Mean Curvature at a point is equal to the arithmetic mean of the principal curvatures: $1/2 \, (1/R_x + 1/R_y)$

normal
tangent
R
R_x
R_y
z
x
y

PRINCIPAL CURVATURE LINES

If a point on a surface is cut by normal planes (i.e., planes containing the normal to the surface at that point) then the point can be defined by infinitely many curvatures. However, there is only one normal section for which the radius of curvature is maximum and another section where it is minimum. The directions of the two curvatures are perpendicular to each other and are known as the Principal Curvatures. By connecting the directions of the principal curvatures at every point on a surface, lines of principal curvature are generated.

SIMPLE MATHEMATICAL SURFACES

The surface is defined by an equation. Basic surface forms are QUADRIC SURFACES defined by an equation of the second degree in three variables: z = f (x,y).
Typical simple surface forms include the cone, conoid, cylinder, elliptic paraboloid, sphere, torus, hyperboloid of revolution, and hyperbolic paraboloid (hyper. H.P.) (a. through h. respectively.)

Figure 8.4 Surface classification 1.

600

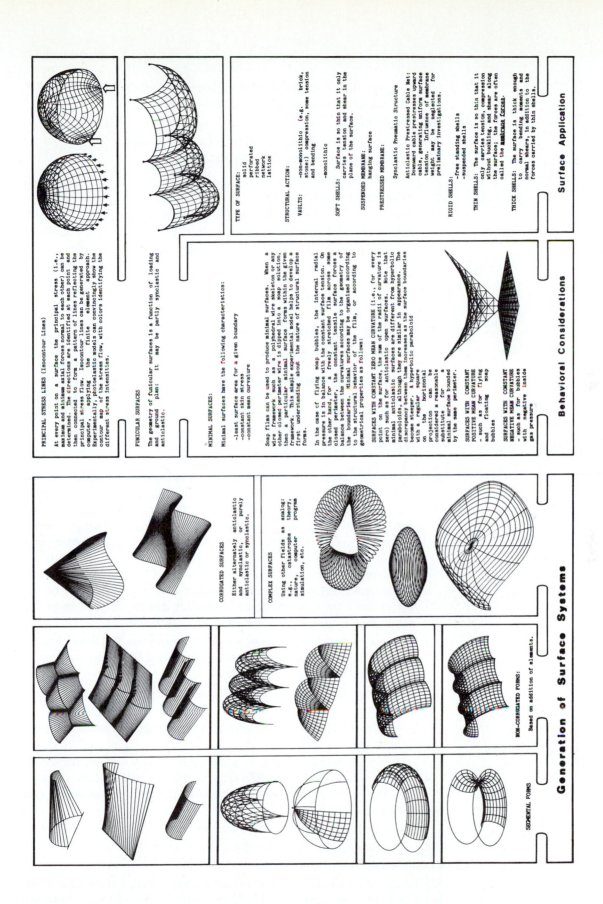

Surface Application

TYPE OF SURFACE:
- solid
- perforated
- ribbed
- network
- lattice

STRUCTURAL ACTION:

VAULTS: −non-monolithic (e.g., brick, stone:) compression, some tension and bending
−monolithic

SOFT SHELLS: Surface is so thin that it only carries tension and shear in the plane of the surface.

SUSPENDED MEMBRANE: hanging surface

PRESTRESSED MEMBRANE: Synclastic Pneumatic Structure

Anticlastic Prestressed Cable Net: Downward cable prestresses upward cable, generating uniform surface tension. Influence of membrane weight may be neglected for preliminary investigations.

RIGID SHELLS: −free standing shells
−suspended shells

THIN SHELLS: The surface is so thin that it only carries tension, compression without buckling, and shear along the surface; the forces are often called the membrane forces.

THICK SHELLS: The surface is thick enough to carry bending moments and normal shears, in addition to the forces carried by thin shells.

Behavioral Considerations

PRINCIPAL STRESS LINES (Isocontour Lines)

At every point on a surface the principal stress (i.e., maximum and minimum axial forces normal to each other) can be determined. The directions are identified at each point and then connected to form a pattern of lines reflecting the principal stress flow. Isocontour lines can be generated by computer, applying the finite element approach. Experimentally, photoelastic models can convincingly show the contour map of the stress flow, with colors identifying the different stress intensities.

FUNICULAR SURFACES

The geometry of funicular surfaces is a function of loading and ground plan; it may be partly synclastic and anticlastic.

MINIMAL SURFACES:

Minimal surfaces have the following characteristics:
- least surface area for a given boundary
- constant skin stress
- constant mean curvature

Soap films can be used to produce minimal surfaces. When a wire framework—such as a polyhedral wire skeleton or any other closed perimeter wire is dipped into a soap solution, then a particular minimal surface forms within the given framework. This simple experimental model helps to develop a first understanding about the nature of structural surface forms.

In the case of flying soap bubbles, the internal radial pressure is in balance with the constant surface tension. On the other hand, for a freely stretched film across some closed perimeter, the constant tensile surface forces a balance between the curvatures according to the geometry of the boundaries. Minimal surfaces may be organized according to the structural behavior of the film, or according to geometrical properties as follows:

SURFACES WITH CONSTANT ZERO MEAN CURVATURE (i.e., for every point on the surface, the sum of the radii of curvature is zero) such as for anticlastic open surfaces. Note that minimal anticlastic surfaces are different from hyperbolic paraboloids, although they are similar in appearance. The distances between them increases as the surface boundaries become steeper. A hyperbolic paraboloid with a regular square on the horizontal projection can be considered a reasonable substitute for a minimal surface bounded by the same perimeter.

SURFACES WITH CONSTANT POSITIVE MEAN CURVATURE - such as for flying and floating soap bubbles

SURFACES WITH CONSTANT NEGATIVE MEAN CURVATURE - such as for bubbles with negative inside gas pressure.

Generation of Surface Systems

CORRUGATED SURFACES

Either alternately anticlastic and synclastic, or purely anticlastic or synclastic.

COMPLEX SURFACES

Using other fields as analog: e.g., catastrophe theory, nature, computer simulation, etc.

NON-CORRUGATED FORMS:

Based on addition of elements.

SEPENTIAL FORMS

Figure 8.5 Surface classification 2.

A.6 *Complex surfaces*: for example, catastrophe surfaces representing discontinuous phenomena as part of chaos theory

B. Structural Shapes

B.1 *Minimal surfaces* have the following characteristics:

Least surface area for a given boundary

Constant skin stress

Constant mean curvature

The membrane equation (8.1) in the next section expresses the magnitude of the axial membrane forces, N as proportional to their curvatures $(1/R)$ in response to pressure p perpendicular to the membrane. For constant skin stress, $N_y = N_x = N$, the membrane equation simplifies to

$$\frac{P}{N} = \frac{1}{R_y} + \frac{1}{R_x} \tag{a}$$

Letting $R_y = R_x = R$, for a sphere, yields

$$p/N = 2/R \tag{9.101}$$

This is the equation for the spherical membrane in response to air pressure, which represents a minimal surface. When $p = 0$ in the membrane equation, the particular minimal surface for an anticlastic liquid soap film surface with constant surface tension is established:

$$-N_y/R_y = N_x/R_x \tag{9.85}$$

B.2 *Funicular surfaces:* The funicular form is determined under the predominant load. Usually, dead load is taken as the form-defining loading using mathematical or experimental methods (e.g., catenary surface). Antonio Gaudi experimented with the inverted hanging membrane concept as the source of generating form rendering pure compression. Funicular shells are especially applicable to underground structures and earth-sheltered homes with a low live-to-dead load ratio. For example, conventional roof loads are about 50 psf but may be 300 psf or more for earth-sheltered residences due to the permanent soil cover, thereby prestressing the shell in compression.

B.3 *Optimal surfaces,* resulting in weight minimization, are developed for various load and boundary conditions by taking into account not only surface geometry but shell thickness.

B.4 *Free-form shells* may be derived from experimentation, such as the following:

Wire mesh as self-supporting formwork using shotcrete and/or urethane foams

Pneumatic formwork; e.g., by spraying rigid polyurethane foam insulation on the inside to stiffen the airform skin, which then can serve as the base for steel reinforcing bars and layers of shotcrete (see Example 9.12)

Inverted hanging fabric membrane or steel reinforcing mesh

Flowing viscous fluid (developed by H. Isler)

C. Composed or Sculptural Forms

Shell shapes do not have to articulate engineering esthetics or constructivism but may have other meanings. The shapes may be derived from historical context, art, analogs in nature or other fields, or different concerns so as to be of more decorative nature.

Membrane Forces

Membrane structures are continuously curved and so thin that they carry uniform loads primarily in axial action; bending can be treated as insignificant, disregarding the effect of boundary members. In *thin shells,* tensile and compressive membrane forces occur, whereas *tensile membranes* only resist tensile forces. In tensile membranes, compressive forces are resisted by prestressing the flexible membrane so that it always remains in tension (see Chapter 9).

It is convenient in this introduction to first visualize tensile membrane forces in a flexible surface structure as caused by normal pressure. Flexible membranes respond to external force action in pure tension by adjusting their geometry in a manner similar to single-cable systems, or inverted rigid membranes (funicular shells) respond in pure compression. Naturally, thin shells must be thick enough so that they do not buckle. The membrane's force reaction to **normal** pressure applied to its surface is investigated here. The free body of a surface element is shown in Fig. 8.6. The element is cut along the principal curvatures so that only axial membrane forces (N_x and N_y) resist the external load p. There are no tangential shear forces along the membrane edges because of the absence of skewed curvatures. The membrane curvatures, along which the forces act, are equal to $1/R_x$ and $1/R_y$. It is assumed that the free body is sufficiently small so that the curvatures are constant along the edges; they are considered to be circular arcs. The arc lengths of the edges are

$$a = \beta R_x \quad b = \alpha R_y$$

where α and β are expressed in radians. The total resultant pressure P acting on the surface is equal to

$$P = pab = p(\alpha R_y)(\beta R_x) \tag{a}$$

The components of the membrane forces parallel to the resultant pressure P (Fig. 8.6b) are

$$N_{yv} = N_y \sin\frac{\alpha}{2}, \qquad N_{xv} = N_x \sin\frac{\beta}{2} \tag{b}$$

The resultant pressure P must be resisted by the sum of the vertical force components acting along the perimeter of the free body.

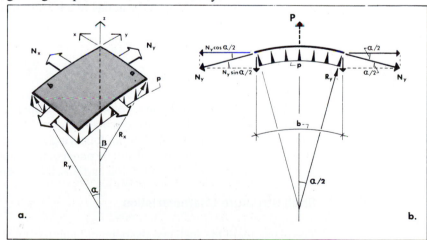

Figure 8.6   Membrane free body.

$$\Sigma V = 0 = P - 2[N_{yv}(a) + N_{xv}(b)] \tag{c}$$

Substituting Eqs. (a) and (b) into (c) and considering that the angles α and β to be very small [$\sin(\alpha/2) \cong \alpha/2$] yield

$$pR_xR_y = R_xN_y + R_yN_x \quad \text{or} \quad p = \frac{N_y}{R_y} + \frac{N_x}{R_x} \tag{8.1}$$

This formula is known as the *membrane equation*. It shows that under load pressure normal to the membrane's surface the axial membrane forces are proportional to their curvatures.

The membrane equation can be simplified for the special case of an axisymmetrical form or *surface of revolution*. Axisymmetrical surfaces are formed by rotating a line, called the meridian, with a varying radius of curvature R_1, about a fixed axis. As the meridian rotates, each point along its length describes a circle or hoop having a radius R_2 about the fixed axis. The radii R_1, and R_2 are the *principal radii of curvature* (Fig. 8.7a, b, and c). Due to the symmetry of form of all axisymmetrical surfaces, the uniform pressure loading normal to the surface is resisted by axial forces only, and the principal force flow coincides with the principal curvatures.

The membrane equation can be rewritten by using the expressions for the hoop or circumferential forces N_θ and meridional or arch forces N_ϕ as

$$p = \frac{N_\phi}{R_1} + \frac{N_\theta}{R_2} \tag{8.2}$$

Independent equations for the membrane forces can be derived from Fig. 8.7c and d. Visualize the membrane to be cut by a horizontal section perpendicular to the axis of rotation. The vertical components of the meridional membrane force along the circumference must resist the vertical components of the pressure normal to the surface. The vertical pressure components are equal to the pressure acting on an imaginary circular plate at the level to be investigated: $P = p\pi R_0^2 = 2\pi R_0 (N_\phi \sin\phi)$.

$$N_\phi = \frac{P}{2\pi R_0 \sin\phi} = \frac{pR_0}{2\sin\phi} = \frac{pR_2}{2} \tag{8.3}$$

The meridional membrane forces at a given level are constant and are proportional to the radius defining the hoop curve at that level. Substituting N_ϕ into Eq. (8.2) and solving for the hoop force yields

$$N_\theta = pR_2\left(1 - \frac{R_2}{2R_1}\right) = R_2\left(p - \frac{N_\phi}{R_1}\right) \tag{8.4}$$

The membrane forces for some common dome shells, such as spherical, parabolic, elliptical (i.e., ellipsoid), conical or conoidal domes, under various symmetrical loading conditions are investigated in the following sections. It is apparent that these surface systems in all likelihood will not be pure compression shells (i.e., funicular shells). They must resist loads also in tension and possibly in shear.

Shell Structure Characteristics

The plasticity of the shell and the potential uninterrupted flow of the bent surface have opened a new dimension in architectural design. Shells, as the name suggests, are closely related to nature and thus express the dynamism of change that characterizes

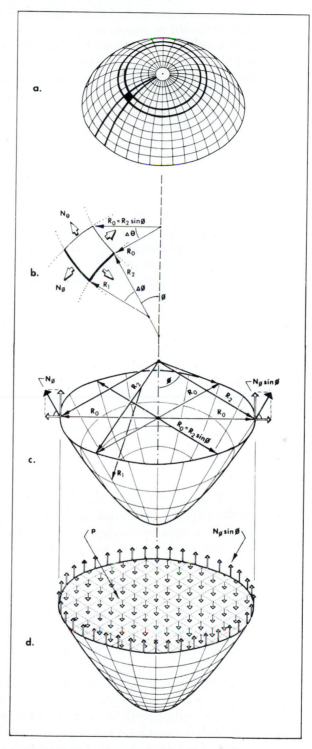

Figure 8.7 Membrane forces due to load p normal to surface.

living organisms. In the past, emphasis in shell design has often been only on scale and geometry and the techniques of structural design, rather than the integration of other design determinants and response to a much broader scope of cultural considerations.

The behavior and some of the characteristics of the shell can best be visualized by first observing a suspended flexible membrane, as discussed in the previous section. This membrane is under pure tension in response to the loads. When it is frozen and inverted, it becomes a shell in pure compression. However, as the live loads change, the flexible membrane adjusts its shape so as to remain in tension. The shell, on the other hand, cannot do so; it is rigid. It not only has to react in compression but also in tension and in tangential shear; however, the shell is a compressive membrane when it is prestressed. The shell must be very thin, as exemplified by an eggshell, and thus does not resist any bending, normal shear, or twisting, but still must be thick enough so that it does not buckle. Hence, it is apparent that the thin shell, because of its double curvature, can resist uniform loading in direct force action within its surface plane. Since many *skeletal shells* also resist loads primarily by in-plane forces, and not in bending, they are treated in this chapter as shells for approximate design purposes.

The membrane theory introduced in the previous section is taken as the basis for the design of thin shells, but it does not necessarily reflect the true force distribution, because it only assumes statically determinate conditions. However, simple mathematical expressions make it possible to develop a feeling for the overall magnitude of the force flow. The approach here can only be considered a crude approximation, but it is a valid introduction to a field of structures otherwise not accessible to the general designer. The actual analysis of shells is extremely complex and can only be dealt with by a selected group of structural engineers specializing in this field.

In the derivation of membrane theory, it is assumed that symmetrical uniform loads act on a continuously curved surface, since concentrated loads can only be reacted in bending, assuming that the shell curvature does not have a kink point at the location of the load. It is further assumed that the forces along the membrane edges are compatible with the support conditions, which is rarely the case. This incompatibility along the boundaries between the thin shell and the rigid support causes bending in the shell, which fortunately is only significant near the edges in most cases; in this area the shell thickness is increased.

Concrete shells of large span may have to be *prestressed* to reduce tensile cracks and control deflection. *Posttensioning* may also be necessary, if the shell is constructed from precast components, so that they can be tied together.

The selection of the shell shape evolves out of a complex synthesis of many architectural design determinants. The shell form reflects the spirit that the designer wants to express, besides having to serve the function of the building. Its shape must provide the load capacity for the given scale with a minimum of bending and must respond to construction considerations. Its geometry may be derived as was described previously. The shell is not necessarily a continuous solid surface; it may consist of precast ribbed shell units with very thin slabs or be formed by a network of members. The shell material ranges from reinforced-concrete, steel, aluminum, wood (e.g., plywood), and plastics to reinforced ceramic shells; common materials are discussed in more detail in the next section. The cost of reinforced concrete shells is primarily in the formwork, since in many countries the cost of labor is more critical than the cost of materials. Hence, the shape of large-scale shells must lend itself to a mechanized construction process, such as mobile scaffolding and possibly the utilization of prefab surface units, particularly where intricate forms are required. It is obviously advantageous if the shell is constructed from a series of self-supporting parts so that the formwork is repetitive and can be reused. For example, compared with monolithic concrete dome construction, *segmental shell construction* may offer a greater forming economy because identical triangular shell segments are cast individually so that forms can be reused by supporting the forms on a movable shoring system. When prestressed cable nets are used as a construction platform (e.g., saddledome, Fig. 9.20d, bottom), no shoring is required. Later, under final service conditions, these cables become the main reinforcing of rigid shells.

Most of the great engineering pioneers of concrete shell design were less theoreticians than builders. The designers Candela, Christiansen, Dischinger, Esquillan, Finsterwalder, Freyssinnet, Isler, Nervi, Tedesko, Torroja, and Tsuboi, just to name a few, have all integrated structural design with the actual building of shells. In the context of shell development, Dyckerhoff and Widmann in Germany must also be mentioned, as well as Roberts and Schaefer and Ammann & Whitney in the United States.

For preliminary design purposes, shells may be organized into three groups according to their static behavior:

1. *Shell beams,* or long shells, are singly or doubly curved (or folded) beams similar to corrugated panels with the beam span by far larger than the beam width. Here, bending is the primary feature, which is resisted for simple beam action by a tension zone along the bottom edges and a compression zone at the crown. The approximate design of shell beams is further discussed under cylindrical shells and folded plates.

2. *Shell arches* are singly or doubly curved shells; they are short shells and span primarily in the cross direction rather than the longitudinal direction, thus acting predominantly as arches and not as beams.

3. *True shells* are doubly curved surface structures, each with particular geometrical characteristics. Common forms are discussed in the previous sections, and typical applications to architecture are further investigated in the following discussion. Double-curvature shells are obviously stronger than single-curvature ones, assuming proper support is provided so that full advantage can be taken of arch action in two directions. For example, the hypar shell transfers uniform symmetrical loading by tangential shear to the supporting edge ribs, which are assumed to be placed parallel to the form generators. Hence, the sum of the vertical components of the shear along the shell periphery is equal to the total load.

Shell Material

Various construction and design methods using reinforced concrete, precast concrete, ferrocement, wood, steel, and plastics are now briefly discussed. Other materials are used in traditional rigid surface construction. In many parts of the world, masonry vaults are built from stone, sun-dried mud brick, burnt brick, or tiles laminated in layers. The Catalan thin-tile masonry vaulting technique, or *timbrel vaults,* consist of 50% or more of mortar, and the thin clay tiles that function as aggregate thereby cause the vault to act more like a concrete shell, rather than a conventional masonry vault.

Eskimos build their igloos from compacted snow blocks similar to corbel dome construction. Among other materials that have been proposed are fiber-glass-reinforced ice domes that may be able to span up to 500 ft. They could be constructed by spraying water onto spherical inflated membranes.

The currently popular glass-covered roofs may be treated as skeletal shells. Attempts are even made to build all-glass surfaces without framing by employing the glass as load bearing.

Reinforced-concrete Shells. Unless precast technology is applied, wood is generally used as formwork for cast-in-place construction. Simple curves in formwork sheathing can be made with dry plywood down to a 24-in.-radius of curvature. Occasionally, plastic foam forming and inflated forms (see Example 9.12) are used, besides corrugated steel culverts or other patented curved steel sheets with ribbed corrugations that function both as formwork (possibly without shoring) and reinforcement for the concrete. Especially for tubular earth-sheltered and cut-and-cover construction (e.g., earth-sheltered residences, arched bridges, drainage structures, silos, tunnels, bunkers), inflated forms, ribbed metal lath, and corrugated metal culverts are used, where the form is covered, for example, with low-slump concrete or shotcrete. The corrugations may be filled with foaming plastic (polyurethane) in liquid form, which expands and hardens instantly.

For the design of thin concrete shells, two types of internal force actions must be considered: the *membrane* and *bending actions.* The two axial membrane forces and membrane shears are assumed to act tangentially along the centroidal surface of the shell; in other words, they are assumed to be uniformly distributed over the shell thickness. The internal stresses are determined according to elastic behavior, which is an accepted design approach for shells because the stresses are generally very low and of direct nature, by considering reinforced concrete ideally elastic, homogeneous, uncracked (disregarding the effect of reinforcement), and isotropic with identical properties in all directions. Therefore, the bending stresses are assumed to be linearly distributed. We may conclude that the *compressive stresses* in the shell and auxiliary members (e.g., ribs, edge beams), according to elastic theory, are and must be limited to

$$f_c = \frac{N_c}{A} + \frac{M}{S} = \frac{N_c}{bt} + \frac{M}{bt^2/6} \leq 0.45 f_c' \qquad (8.5)$$

The *tensile stresses* are resisted entirely by reinforcement; the concrete is assumed to carry no tension. The allowable tensile stress for Grades 40 or 50 steel is 20,000 psi, and for Grade 60 reinforcement or welded wire fabric it is 24,000 psi. Thus, the required steel area to resist the tensile stress for Grade 60 steel (in ksi), for example, is

$$f_t = N_t/A_s \leq F_t = 24 \quad \text{or} \quad A_s \geq N_t/F_t = N_t/24 \qquad (8.6)$$

For the case where tensile stresses vary greatly in magnitude, reinforcement resisting total tension may be concentrated in regions of maximum tensile stress. But minimum reinforcement throughout the tensile zone should still be not less than,

$$A_{s\min} = 0.0035 A_g = 0.0035 bt \qquad (8.7)$$

Minimum shell reinforcement must be provided in two orthogonal directions, which should not be less than shrinkage or temperature reinforcement for slabs. For example, for Grade 60 deformed bars or welded wire fabric, the minimum shell reinforcement shall be

$$A_{s\,min} = 0.0018 A_g = 0.0018 bt \qquad (3.69)$$

The reinforcement shall not be spaced farther apart than five times the shell thickness, nor shall the spacing exceed 18 in. It is assumed to act at the middle of the shell surface and is placed either parallel to the line of principal stress (a deviation of up to 15° is still permitted) or in two or three component directions by providing additional reinforcement. The membrane reinforcement is typically placed in two orthogonal directions with a third diagonal direction added in areas where tension is high. Usually, the directions of the reinforcing bars do not coincide with the direction of the principal membrane tension.

The shell thickness is rarely dictated by strength requirements but rather by stability considerations for large-scale surfaces, by the required compatibility with stiff edge members (i.e., bending), and most often by the minimum concrete cover for the reinforcement. Where the concrete is neither exposed to weather nor in contact with the ground, the minimum concrete cover for shells and folded plates for #5 bars and smaller should be 1/2 in. for nonprestressed cast-in-place construction (3/4 in. for #6 bars or larger), and 3/8 in. for precast concrete as well as prestressed concrete. The minimum shell thickness for the typical condition of three layers of reinforcement (e.g., transverse, longitudinal, and diagonal) is

Concrete cover: $2 \times 1/2$	$= 1.00$ in.
Three layers of #4 bars: $2 \times 4/8$	$= 1.00$ in.
Minimum spacing between top and bottom layer	$= \underline{1.00 \text{ in.}}$
	$t_{min} = 3.00$ in.

The typical shell thickness in the United States is in the range of 3 to 4 in. Much thinner shells, however, have been built. The 52-ft hemispherical dome of the Zeiss Planetarium at Jena, Germany (1922), is only $1\frac{3}{16}$ in. thick. It consists of a self-supporting triangular mesh of reinforcement (similar to a high-frequency geodesic dome) as based on the Zeiss–Dywidag system; here the concrete surface was formed by shotcreting the self-supporting mesh. Similar methods of construction have been used later by other builders to eliminate the cost of formwork. This first Zeiss Planetarium dome may be considered the world's first ferrocement large-scale structure, although not exactly corresponding to the definition of ferrocement. This construction method yields extremely thin shells; here cement mortar is pressed (instead of sprayed under pressure as for the planetarium) into several layers of wire mesh (welded or woven as chicken wire or expanded metal lath). Ferrocement consists of a much higher proportion of reinforcement and only fine aggregates (sand), in contrast to the conventional reinforced concrete with coarse aggregates and a low steel ratio.

Ferrocement shells are quite strong due to the relatively monolithic character of the composite material. In the early 1940s, Pier Luigi Nervi pioneered the application of ferrocemento from which the word ferrocement is derived. He used the principle for the first time on a large scale for the corrugated roof of the 1948 Exposition Hall in Turin, Italy, where $1\frac{1}{2}$-in. thick prefab ferrocement panels are tied together by cast-in-place concrete ribs.

The advantage of ferrocement construction lies in the saving of material, equipment, formwork (only occasional shoring may be required), and skilled labor; it also lies in its design flexibility (formability), strength, and watertightness. On a large scale, a certain advantage may lie in its light weight, which lends itself to prefabrication, as was demonstrated by Nervi. Thin, semirigid ferrocement ribbons are also used as suspended roof membranes, especially for long-span structures. In general, however, the use of ferrocement in the United States is still in the developmental stage. This is due to the excessive construction time and because it is not easily adaptable to industrialized processes of mechanization. Presently, in this country, ferrocement seems to be mainly employed for boat building. For instance, 25- to 45-ft-long boats have a shell thickness that varies from 1/2 to 3/4 in. depending on the strength and number of mesh layers.

More recent developments in concrete shell design are glass-fiber and steel-fiber reinforced concrets. Since no conventional reinforcing steel is needed, the shell can be reduced in both weight and thickness, in turn requiring a less strong formwork. These lightweight shells clearly lend themselves to prefabrication. Concrete has also been reinforced with polymer grids rather than steel mesh. Fabric-reinforced concrete shells are often found when inflatable forms are used.

Wood Shells.   In Fig. 8.8, typical shell surfaces in wood are identified. Three basic structural systems are derived from the nature of the surface structure that resists the membrane forces.

A. Skeletal Shells: Here the membrane forces are resisted by a member grid that makes up the skin (i.e., *skin frame structure*), rather than by a solid surface. The sheathing and purlins are not considered to act compositely with the member network comprising the shell, although they may be needed as diaphragms depending on the layout of the shell framing, especially if it is not triangulated. Various framework patterns are discussed later in this chapter. While for most framed surfaces the placement of the sheathing is apparent, for some grid shells it is placed according to the geometry of its member layout.

B. Solid Laminated Shells: Molded plywood is the simplest form of a solid shell on a small scale. Its technology was perfected during World War II in the aircraft industry (aircraft fuselages, wings, etc.) and the shipbuilding industry (laminated ship and boat parts such as boat hulls). Just after World War II, Charles Eames introduced his now classic molded plywood chairs.

Plywood is a composite material made up of several veneers with opposing grains, where the durability of the plywood is very much dependent on the quality of the adhesive. In conventional roof construction, plywood sheathing (standard size 4 ft × 8 ft × 1/4 to $1\frac{1}{8}$ in.) may span up to about 6 ft in bending.

Thin, solid laminated wood shells are constructed from layers of lumber sheathing or plywood. They are glue-nailed or stapled together to prevent buckling of the individual layers and are supported along their perimeters by edge beams. Since the membrane stresses are relatively low, high-grade material is not required. Boards or plywood are laid in various layers in different directions depending on the shape of the shell. For shallow surfaces, the tongue and groove boards can easily be bent or twisted, as the radii of curvature are generally much larger than the thickness of the membrane; usually, they are placed along the principal curvatures and/or at 45° to them. For instance, in a four-layer glue-stapled hypar shell, the boards in the bottom layer follow the first principal curvature, and the second and third layers are placed parallel to the straight-line generators, but perpendicular to each other, while the fourth layer is positioned along the other principal direction.

For the condition where three layers of plywood are glue stapled, the panels are arranged in a staggered fashion at a 45° angle to the layer beneath so that no joints are in line and on top of each other. Although plywood has a much higher shear capacity than lumber and is assembled faster, lumber sheathing has the advantage of narrow board width, which adjusts more easily to the doubly curved hypar shape and thus has only a slight twist when placed along the straight-line generators.

In a shallow cylindrical shell, the boards in the bottom layer may be placed along the principal curvature, while the second and third layers are laid 45° and perpendicular to each other. For steeper slopes, curved members have to be prefabricated by applying some type of softening or plasticizing treatment. Curved plywood can also be produced by bending and gluing the plies simultaneously, similar to the production of curved laminated members.

Laminated wood shells are lightweight. Usually, two to three layers of sheathing of 1- to $1\frac{1}{2}$-in. thickness are adequate for the coverage of a square area of roughly 60 × 60 ft with a shell weight of about 5 psf.

We may approximately design solid wood shells for the condition where the sheathing is placed along the principal curvatures, as for a two-layer hypar shell, for example, by assuming the unit tensile (or compressive) stress to be equal to the principal tension (or compression) force per inch divided by only the thickness of the sheathing along which the respective force acts. This approach of considering two

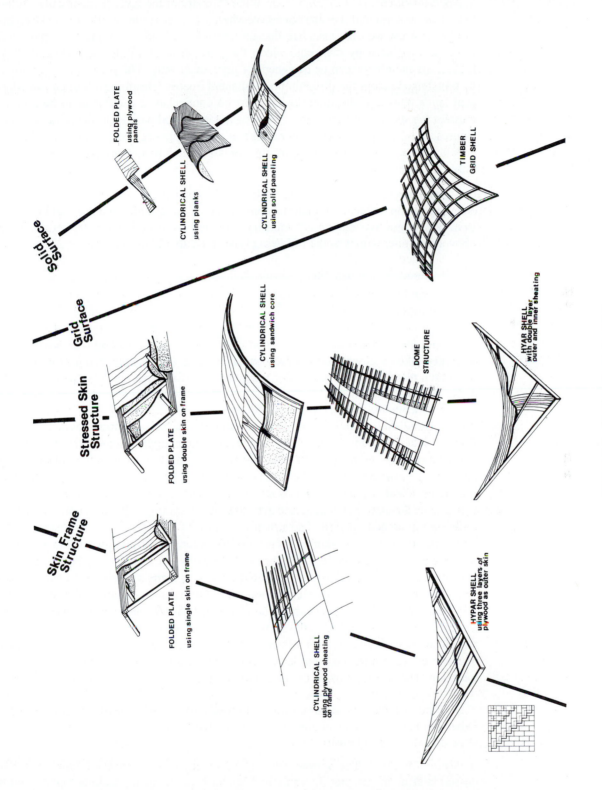

Figure 8.8 Shell surfaces in wood.

FOLDED PLATE
using plywood
panels

CYLINDRICAL SHELL
using planks

CYLINDRICAL SHELL
using solid paneling

TIMBER
GRID SHELL

Solid
Surface

Grid
Surface

Stressed Skin
Structure

Skin Frame
Structure

FOLDED PLATE
using double skin on frame

CYLINDRICAL SHELL
using sandwich core

DOME
STRUCTURE

HYAR SHELL
with double layer
outer and inner sheating

FOLDED PLATE
using single skin on frame

CYLINDRICAL SHELL
using plywood sheating
on frame

HYPAR SHELL
using three layers of
plywood as outer skin

independent layers is conservative since the capacity of the layer perpendicular to the force action is ignored. For the condition where the boards run parallel to the straight-line generators, we may visualize that under uniform load action the principal axial forces are replaced by their equivalent, the pure shear at 45° to them (Fig. 8.39b). Hence, the membrane can be designed for pure shear only. The shear, however, must be transferred across the discontinuous adjacent boards (sheathing) through the adjacent layer; thus, the sheathing layers must be connected to each other to be able to transfer the shear. But should the sheathing not be placed parallel to the principal curvatures, the principal forces do not act parallel to the grain, but at some angle. Now, new elastic properties, such as allowable stresses, must be determined for the antisotropic material lumber.

C. Composite Shell Systems.   The concept of composite action of various structural elements is applicable to all materials on a small or large scale. The principle is taken from the automotive and aircraft industry. In wood construction it is applied to glued plywood-lumber panels of the following types (see also Fig. 3.11):

- Stressed-skin panel (double or single skin)
- Sandwich panel
- Solid core panel

In general, these panels are prefabricated similarly to glued plywood I- or box beams. They can take any form, such as flat panels (rectangular, triangular, trapezoidal, etc.) for folded plate structures, curved or warped panels for hypar shells, or arched panels as for cylindrical surfaces.

Conceptually, *stressed-skin* surfaces are obtained by bonding the sheathing to the frame or grid shell, thereby generating a composite surface structure that has a significantly higher strength and stiffness. The prefab panel approach as derived from aircraft fabrication techniques in the 1940s came only into commercial building application in the late 1950s. The panel component is produced by integrally connecting the plywood panels to the stringers so that it acts as a composite system where no slippage occurs between panels, as can be caused, for example, by horizontal shear due to flexure in arched systems. The structural action of a flat panel can be visualized to be similar to a series of adjoining built-up I-beams for double-skin systems or T-beams for single-skin panels. Here, the flanges or the plywood facings may be assumed to carry the axial forces and bending, while the webs or stringers resist the shear.

In *sandwich* or *solid-core* construction the lumber stringers are replaced by some other core material, such as foamed plastic or paper honeycomb for sandwich panels, as will be discussed further in the section on plastic shells. The structural behavior of the surfaces is similar to that of stressed-skin panels. The facings can be thinner because the core material provides continuous support to the skin and thus prevents buckling. The shear is carried by the core material.

Steel Shells.   Most large-scale steel shells, like domes, are of the skeleton type. Other steel shells may be composed of corrugated steel panels or they may be of the stressed-skin type of construction.

A. Skeletal Shells.   The various framing systems for different shell forms are discussed later in this chapter. In a *trussed shell* surface, the metal decking together with the purlins is not needed as diaphragms. It is a separate load-carrying system only acting as the roofing surface.

B. Steel Sheet Shells.   Water tanks, ship hulls, car bodies, pipelines, boilers, and containers clearly show the application of continuous metal sheets developed into closed shell units. In building construction, corrugated steel panels are frequently used because of their higher stiffness, especially against buckling, and because of their one-way bending capacity.

In barrel vaults the corrugations are generally placed along the arch action similarly to folded plates, where they provide the bending capacity due to transverse slab action. Should the corrugations extend longitudinally, transverse stiffeners and ribs must be provided to maintain the shape.

In hypar shells the formed steel decking is conveniently placed parallel to the straight-line generators by only having to slightly warp them. Most long-span hyperbolic paraboloids are constructed from two layers of mutually perpendicular plates plug welded at the common intersections of the flat surfaces, as the typical example in Fig. 8.9 indicates. Here, the four hypar surfaces, each one 33.5-ft square, are supported by four-corner and one-center columns, as well as by edge beams along the column lines.

For approximate design considerations, refer to the discussion of *solid laminated wood shells.*

C. Stressed-skin Shells.   The strength and stiffness of the skeleton-type shell is very much improved if the roof skin becomes an integral part of the shell membrane. The strength may further be increased by developing the flat or slightly bent skin into folded polyhedral shapes or into bent hyar panels. For this condition, the formed spatial panels may replace some of the members of the framed shell. Going one step further yields a shell surface composed of three-dimensional panels attached to each other along their edges, which now form the member framework.

Plastic Shells.   Plastic shells immediately bring to mind the richness and unprecedented expression of the continuously floating curved surfaces in furniture design of the 1960s, or they may remind us of the intricate, spatial shell shapes of car bodies or sculptured curtain walls. When, in the early 1940s, Charles Eames and Eero Saarinen (influenced by the military and aerospace engineers who developed fiber-

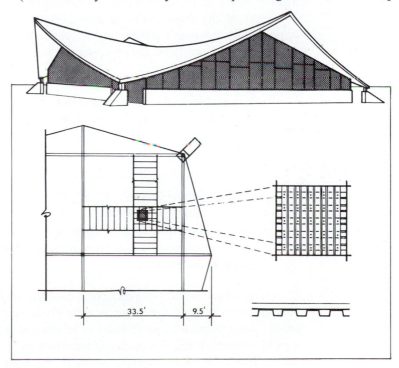

Figure 8.9 Steel sheet hypar shell.

reinforced composites for use as curved surface composites) invented the molded fiber-glass shell technique and then mass produced their chairs, nobody could have predicted the explosion of the new era of material plastics about 15 years later and the corresponding development of a wealth of innovative forms. In Italy, some architects converted to industrial design and created a new design esthetics and art form beautifully expressed in furniture, which exposed the true nature and plasticity of the continuous shell. In 1955, I. Schein, Y. Magnant, and R. A. Coulon in France built the first all-plastics house in the shape of a snail's shell. Two years later, the famous Monsanto House (Fig. 8.10n) was developed by Hamilton and Goody and exhibited for many years in Disneyland. The cantilevering wings of the cross-shaped module clearly express the strength of the three-dimensional shell and the logic of the plastic form. Today, industrialized, cellular, containerlike, intricate shell shapes for shelters, mobile units, building cores (kitchen and bathroom modules), etc., are very common, as are the shell-shaped cladding panels for framed or trussed structures. Some early pioneers who developed an architecture of plastic shells are Ionel Schein, Rudolf Doernach, and Wolfgang Döring in Germany, Renzo Piano and Angelo Mangiarotti in Italy, and Arthur Quarmby in England.

But plastic shells do not necessarily have to be factory produced; they can be constructed directly on site using urethane foams. Already in the 1950s, architects Felix Drury and John M. Johansen in the United States had convincingly expressed how the process of construction may evolve into free-form shell structures that are in close contact with nature and its organic forms.

The major groups of modular panel construction, sandwich shells and foam shells, as identified in Fig. 8.10, are briefly discussed now.

A. Modular Plastic Panels. These types of panels are factory produced and used primarily as secondary or contributing structural elements, such as fill-ins for frames and trusses (e.g., corrugated roof panels, curtains, skylights). For larger-scale surfaces, they may be fabricated in segments and assembled on site to form a complete folded plate or shell, as in the twelve 3/8-in.-thick glass-fiber reinforced polyester segments making up the 32-ft-diameter onion dome in case f.

Almost any shape can be generated by placing plastics, generally liquid epoxies or polyesters, or layered resin-saturated reinforced material, usually fiber-glass fabric, over a mold. Among the many processing techniques, the more common for generation of shells are the following:

- Lay-up (fiber-glass boats, wing skins, large unconventional components, etc.)
- Pultrusion (rods, I-beams, channels, etc.)
- Continuous process with conveyor belt (sheeting material, corrugated panels, etc.)
- Injection molding (e.g., shell-type furniture)
- Blow molding (e.g., skylight domes)
- Cold or hot press molding
- Matched-die molding (e.g., automotive applications)
- Filament winding (case i) (aerospace components, room modules, tanks, pipes, poles, etc.)

The actual method of production depends on the scale, shape, and number of components (custom tailored versus mass production), cost, and other required properties. Some positive characteristics of reinforced plastics are their formability, light weight, strength, toughness, and light transmission. Their disadvantages lie in their relatively low stiffness and surface hardness, the high coefficient of expansion (Table 2.4), creep behavior, unknown durability, flammability, and cost. The stiffness of plas-

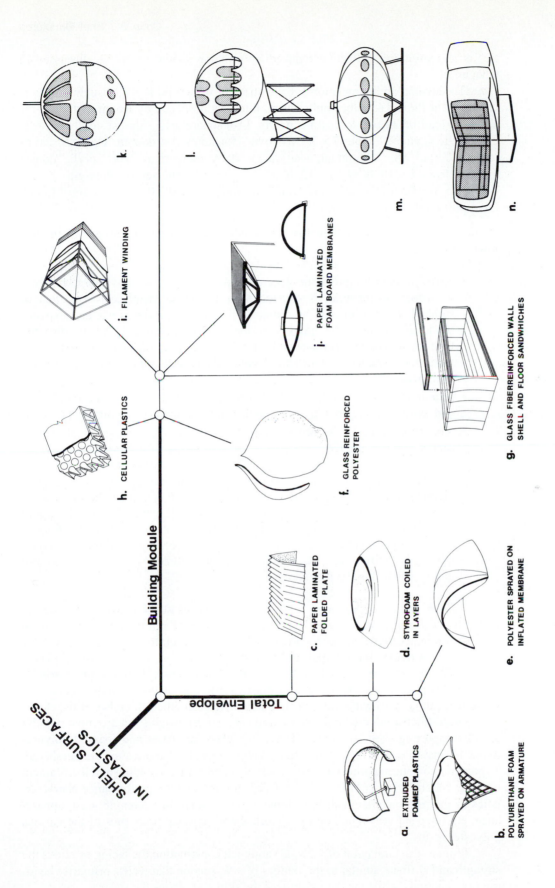

SHELL SURFACES IN PLASTICS

Building Module

Total Envelope

a. EXTRUDED FOAMED PLASTICS

b. POLYURETHANE FOAM SPRAYED ON ARMATURE

c. PAPER LAMINATED FOLDED PLATE

d. STYROFOAM COILED IN LAYERS

e. POLYESTER SPRAYED ON INFLATED MEMBRANE

f. GLASS REINFORCED POLYESTER

g. GLASS FIBER-REINFORCED WALL SHELL AND FLOOR SANDWICHES

h. CELLULAR PLASTICS

i. FILAMENT WINDING

j. PAPER LAMINATED FOAM BOARD MEMBRANES

k.

l.

m.

n.

Figure 8.10 Shell surfaces in plastics.

615

tics is increased and deflection reduced by utilizing spatial forms, as discussed in Chapter 2.

To develop an appreciation for the strength of acrylic shells, visualize spherical segment Plexiglas domes only 1/4 in. thick, as produced by Rohm and Haas for a design load of 20 psf, to cover a nearly 14-ft square area. Almost 4600 transparent acrylic skylights (about 7×3.3 ft) cover Houston's Astrodome. The U.S. Pavilion at "Expo '68, " Montreal, was enclosed by more than 2000 acrylic domes varying from 8×10 ft to 10×12 ft in size. Translucent acrylic Plexiglas panels (9.5×9.5 ft $\times$ 0.16 in.) cover the enormous roof area (808,000 ft^2) of the Munich Olympics tentlike net structures.

B. Foam Shells.   Foam shell enclosures may be generated on site by one of the following methods:

- *Spray process:* Liquid foam, generally polyurethane, which is better known as an insulation material, is sprayed onto a flexible membrane [e.g., tent or inflated skin (e)] or onto an armature (b); it then produces gas and expands while, at the same time, quickly hardening. The shell may be weatherproofed by a rubber coating on the outside. This process of construction, in addition to using the foam shell as structure, generally is only applied to smaller-scale buildings.
- *Extrusion process:* In this method, a truck-mounted boom with a mold at its end extrudes layer after layer of foam, which solidifies immediately (a).
- *Spirogeneration:* In this process, developed by Dow Chemical Co., thin-shell domes are constructed from long blocks or strips of expanded polystyrene (Styrofoam) similar to the building of Eskimo igloos (d). Successive layers are fused by employing a rotation boom anchored at the center of the dome, which carries the heat welding equipment. The foam shell is used as thermal insulation as well as formwork for the thin concrete shell, which has a thickness of about 3/4 in. for up to a 130-ft span and about $1\frac{3}{8}$ in. for up to a 200-ft span, with a foam shell thickness of 8 in. The structural concrete shell is a high bonding latex modified concrete, which is sprayed in three layers onto a wire mesh that is attached to the foam dome.

C. Sandwich Shells.   Laminated sandwiches consist of several layers of different materials. The rigid, *thin facing sheets* may be made of reinforced plastics, metal (steel, aluminum), plywood, hardboard, cement asbestos, reinforced concrete, etc. The *core*, which provides the thermal and acoustical insulation, may consist of plastic foams (PVC, polystyrene, polyurethane, phenolic plastic), honeycomb made of impregnated paper [or made of metal or plastics (h)], particleboard, lumber core, plywood (e.g., three plies with grains at right angles to each other), member grids, etc.

Glass-fiber-reinforced polyester composites in sandwich form are most popular as self-supporting shell enclosures. Here, two glass laminates and foamed polyurethane, for instance, with localized stiffeners at points of load application are employed. Typical manufacturing processes are the simple open contact molding and the filament winding for shells under large stresses. Keep in mind that the performance characteristics of the material (flammability, smoke release, durability, strength, cost, appearance, etc.) depend very much on the chemical and physical properties of the various polyester resins.

From a structural point of view, in sandwich construction the facing provides the strength and stiffness similar to the flanges of a W-section that resists primarily bending (axial action), while the core stabilizes the skin and carries the shear.

Sandwich shell construction is applied to panel systems (facade panels, roof cladding), to mass-produced, single-unit containers (cabins, bus shelters, retail sales kiosks, telephone booths, kitchen and bathroom units, etc.), and to modular aggregates for housing [room modules, sectional units, wall and roof panels for mobile homes, individual components (g), etc.].

The cantilevering shell buildings in Fig. 8.10, cases (k) to (n), clearly express their monocoque character, that is, the fact that the total structure participates in resisting the loading as already practiced in the automotive and space industries. The shell of the Futuro II house (m), which has the shape of an ablate spheroid that is roughly 26 ft in diameter and 12 ft high, consists of only a 2-in.-thick stressed fiber-glass sandwich with polyurethane foam insulation. The folded plate barrel-type structure (c) is a paper board construction consisting of 3/8-in.-thick sandwich elements of polyethylene impregnated Kraftboard with polyurethane foam core. The sandwich dome for the information center at the Hannover Fair, Hannover, Germany (1970), spans nearly 130 ft and weighs only 6.75 psf. For the sake of comparison, a Dow dome for the same span with an 8-in. foam shell and a 3/4-in. reinforced concrete shell weighs about 10.40 psf.

8.2 CYLINDRICAL SHELLS

The development of cylindrical shell forms originated in simple masonry barrel vaults and was perfected by the Romans as expressed so powerfully in the Maxentius Basilica in Rome (313 A.D.), where the vaults reached a span of 85 ft. Similarly, more than 200 years later, the parabolic brick vault of Ctesiphon achieved a span of 83 ft (see Section 5.4, Arches). The art of vaulting reappeared in the church architecture of the early Romanesque period. The evolution from heavy, tunnellike masonry vaulting to groined vaulting, and finally to the elegant, floating, intricate-ribbed vaulting of the Gothic cathedral took more than three centuries; some beautiful rib patterns are shown in Fig. 8.22. The transformation from the vault to the shell, however, is an achievement of the early twentieth century.

In 1916, Eugene Freyssinet began to build the two famous concrete airship hangars of parabolic cylindrical shape at Orly near Paris (Fig. 8.14a). These arched, 18-ft-deep, undulating, thin-shell vaults had a span of 262 ft and were 184 ft high, with a maximum shell thickness of only $3\frac{1}{2}$ in., thus yielding a very light structure. These arched vaults, however, do not have a true shell form; the arch moments due to wind are merely efficiently resisted by the depth of the folds, thus allowing the thin concrete corrugations.

Franz Dischinger and Walter Bauersfeld developed the necessary theory for the design of cylindrical concrete shells and built, in 1924, the first reinforced cylindrical concrete shell for a factory building of the Zeiss Company in Jena, Germany, as based on the Zeiss–Dywidag system.

The first large-scale, long cylindrical shells were used in 1927 for the Frankfurt am Main market halls in Germany. The shells were 46-ft wide and had large edge beams; their span was 121 ft and the shell thickness was only $2\frac{3}{4}$ in. They were designed by F. Dischinger and U. Finsterwalder and constructed according to the Zeiss–Dywidag system.

The further development and modification of the early cylindrical shell shapes can only be suggested by the Frontón Recoletos in Madrid, Spain (1935, later demolished, Fig. 8.3d), where the great designer Eduardo Torroja dared to place two circular cylindrical segment shells of different size (40-ft radius for the larger one) side by side, with portions reticulated to admit light. In section, they give the illusion of instability since the pair of arches seem to be missing a central support. Naturally, in reality, they span as shell beams the 175 ft in the longitudinal direction. For the Zarzuela Hippodrome Grandstand (1935), Torroja uses 16-ft-wide shallow hyperboloidal sections with varying rise and thickness to cantilever effortlessly 41 ft.

Robert Maillart, the famous concrete bridge designer, gave a special architectural meaning to the Cement Industries Hall at the Swiss Exhibition in Züirich (1939, demolished, Fig. 8.14g), which was a short, parabolic, cylindrical shell of $2\frac{3}{8}$ in.-thickness that was 38 ft high, 53 ft wide, and 70 ft long.

Pier Luigi Nervi attracted international attention with the magnificent long-span airplane hangar with a span of 328×131 ft at Orvieto, Italy, in 1936. The roof was a poured-in-place skeletal, cylindrical, shell vault in the lamella tradition; it was destroyed during World War II. A second modified hangar version was built in 1940 (Fig. 8.14e), which was lighter and more elegant than the earlier one. This time, the lattice roof was prefabricated and only supported by six buttresses arranged symmetrically to equalize the force flow due to the vault's weight. Nervi built a series of lattice vaults, for example, the Exhibition Hall, Salone C, in Turin, Italy (1950, Fig. 8.14c). This 165- $\times$ 215-ft lamella vault, which is 45 ft high, consists of precast ferrocemento units and is resting on four inclined arches so as to efficiently resist the thrust of the roof vault.

The first long-span shell structures in the United States are the short $3\frac{1}{2}$-in.-thick cylindrical shells for the Sports Arena in Hershey, Pennsylvania, covering an area of 233×342 ft. The shells were designed by Anton Tedesco in 1936 as based on the Zeiss–Dywidag system.

The potential of the single-curvature bent form was also suggested by Alvar Aalto's famous armchair of 1933 by its continuous seat and back in bent plywood.

Cylindrical Shell Types

Cylindrical shells are not just characterized by the form of their cross section, but also by the types of support in the longitudinal and transverse directions, by the kind of diaphragms and edge beam conditions, and by the continuity of the shells across several bays and spans. Because of the similarity of support conditions for the folded plate beam structure and the cylindrical shell beam, the reader may refer to Fig. 7.3 for further identification of these characteristics.

Single-curvature cylindrical shells may be composed of many different types of curves, as shown in Fig. 8.11. The basic curves range from the defined geometries of the circular segment, the parabola, ellipse, hyperbola, and cycloid, to the responsive geometry of the funicular line. These basic shapes may be joined in innumerable ways to yield cross sections of various forms, which may be identified as follows:

- Single shell constructed from single segment or multiple segments
- Single shell versus multiple shells (corrugated forms)
- Repeating versus nonrepeating units
- Convex versus concave versus undulating forms
- Continuous versus discontinuous (Y-shapes, tilted S-shapes, etc.) forms
- Symmetrical versus asymmetrical shells (e.g., north light shell), etc.

The cylindrical shell units may be arranged in a parallel, radial, or intersecting manner; they may be straight, folded, or bent. Intersecting shells are discussed later in this chapter; here, only linear, parallel cylindrical shells are briefly investigated.

Shell profiles may be shallow or steep. Steep interior shells of a multiple-shell system may have free edges, whereas the end shells may have horizontal or upturned edge beams; shallow shells usually have edge beams (Fig. 8.21).

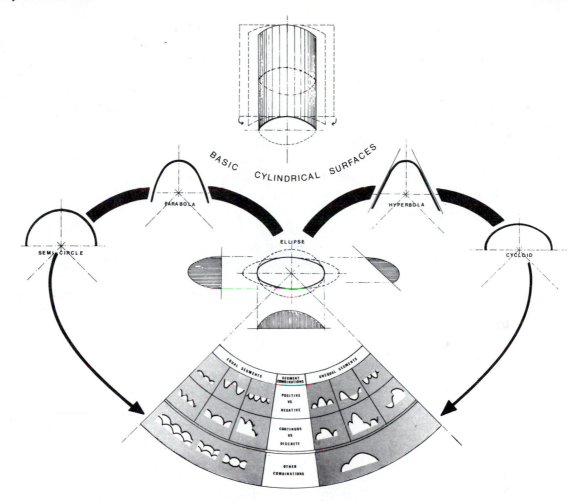

Figure 8.11 Basic concepts related barrel shells.

The behavior of a simple, linear, cylindrical unit depends on its geometry, material, loading conditions and the type and location of its supports. In the following discussion, uniform symmetrical gravity load action is assumed, as wind stresses are generally small and can be neglected for preliminary design purposes.

The influence of support location is apparent from Fig. 8.12. Should the shell be continuously supported along its longitudinal edges by deep beams, frames, walls, or foundations, the forces are carried directly in the transverse direction to the supports. Its behavior may be visualized as the response of parallel arches, each 1 ft wide. These arches must be relatively thick, since they respond to forces by bending as well as axial force action. Since bending is a primary design consideration, these single-curvature surface structures are not considered shells, because their primary structural response is not membrane-type action. They are called *vaults* and may be approximately designed as if they were arches.

On the other hand, if there are no supports in the longitudinal direction, but only in the transverse direction, the shell must behave as a beam spanning in the longitudinal direction; the forces can no longer be carried in arch action directly to longitudinal supports. For a cylindrical shell with a small chord width in comparison to its span, the primary response will be longitudinal beam action. These types of shells are called *long barrel shells* or *shell beams;* they can be visualized as shallow beams with curvilinear cross section and may form simply supported or continuous beams. However, when the

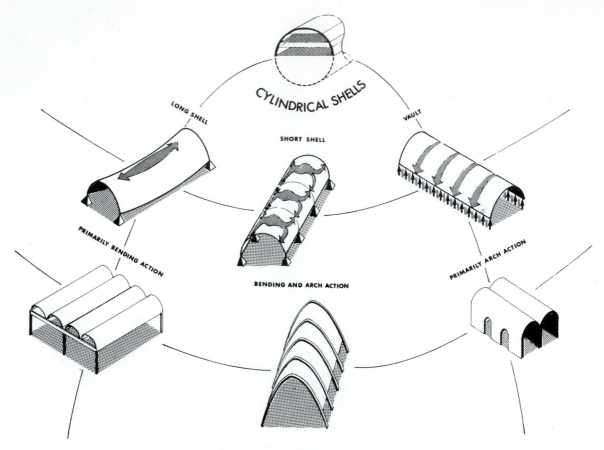

Figure 8.12 Barrels.

chord width is large in comparison to the shell span, transverse arch action is the dominating behavior; these shells are called *short barrel shells*.

Case Study. Typical barrel shells are shown in Figs. 8.13 and 8.14. Whereas short barrels are often used for large spaces, such as auditoriums with transverse spans possibly over 300 ft and longitudinal spans of up to approximately 50 ft, shell beams, with spans of up to about 160 ft and chord widths of up to 50 ft, are often used for warehouses and similar spaces.

Some typical architectural cases exhibiting the shell beam principle are shown in Fig. 8.13. The usual span of cylindrical concrete shell beams is in the range from 50 to 160 ft with a chord width of 20 to 50 ft and a span-to-depth ratio of between 1:10 and 1:15. The shell thickness is usually 3 in. and may be increased near ribs and edge beams.

The cross section for the various cases range from the undulating concrete shells (f) to the north-light shells (d), to the channellike precast concrete units with bowed webs (b). The shells act not only as simple or continuous beams, but also as cantilevers, as reflected by the 50-ft cantilever shells (m) tapered from front to back, following the intensity of the moment flow.

The parallel concrete cycloid shells for Louis Kahn's Kimbell Art Museum in Fort Worth, Texas (h), are separated by 7-ft-wide channel-type beams. The 23-ft-wide, 4-in.-thick posttensioned shells span 104 ft and are supported at each end by arches and columns. The 3-ft skylight openings at the shell crowns are framed similarly to a Vierendeel truss to transfer the shell forces so that the shell can behave as a continuous unit.

In case (e), individual inverted plywood shell beams, 1¼ in. thick, span 30 ft and support arched glass-fiber skylights. The roof in case (j) consists of parallel 3-ft 3-in.-

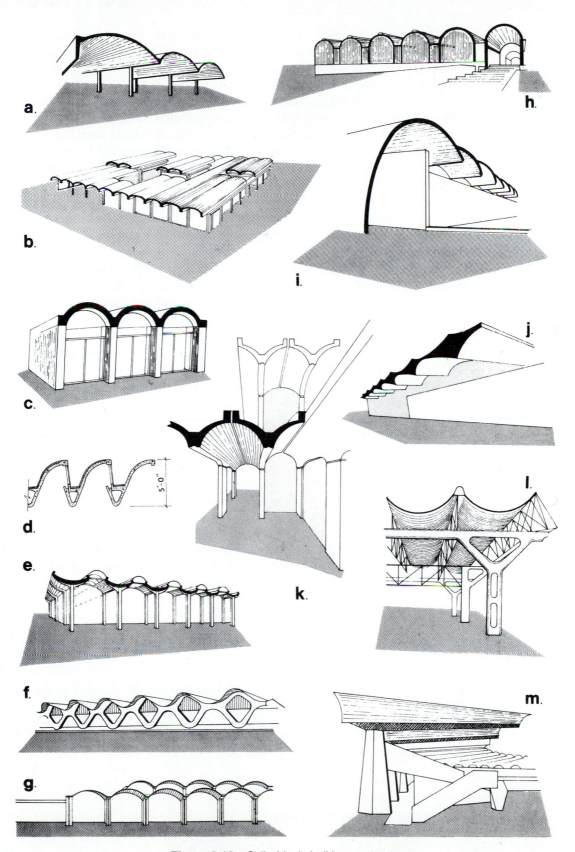

Figure 8.13 Cylindrical shell beam structures.

deep plywood box girders, 11 ft 8 in. apart, to which are attached, on top and bottom, respectively, arched and suspended double layers of 1/4-in. plywood sheets. The roof in case (l) has the appearance of an inverted cylindrical shell, where, in reality, corrugated steel sheeting is suspended from parallel trusses spaced at about 13 ft apart.

One of the largest shells covers the New Orleans' Rivergate Exhibition Center. The roof consists of six main, two-span, slightly arched shell beams that have spans of 139 of 253 ft and 30-ft cantilevers at each end. A typical barrel is 18 ft deep ($d/L = 1/14$), 60 ft wide, and posttensioned along its full length of 452 ft.

Some examples that express the concept of short cylindrical shells and vaults are shown in Fig. 8.14. They range from the small scale of arched corrugated sheet metal (f) as found in industrial construction and single-layer geodesic aluminum frame barrels with tension membranes as enclosure (d), to the large scale of parabolic barrel vaults consisting of parallel folded concrete arches (a) or prefabricated welded pipe space frames (i).

In case (b), the $2\frac{3}{4}$-in.-thick cylindrical concrete shells are inclined for light purposes. The lateral thrust of the hipped lamella vaults (c, e), of the arches supporting the short shells (h, k), and of the vault (j) is resisted by inclined buttresses. This is powerfully expressed by the fan-shaped buttresses at the crossing of the supporting arch structure in (c). Nervi used his precast concrete system for the 219-ft clear span parabolic vault of the field house at Dartmouth College (1962). The double-layer, trussed steel barrel of the Waldstadion Sports Complex, Frankfurt, Germany (1982), spans 230 ft. The arched roof consisting of triangular and hexagonal cells transfers the loads directly to the ground through abutment frames.

Membrane Forces in Circular Cylindrical Shells

A cylindrical surface can be generated by translating a straight line the meridian (Fig. 8.7) with $R_1 = \infty$, along a fixed curve having a variable radius of curvature R_2. For a circular cylindrical membrane, the radius of curvature is constant, that is, $R_2 = R$ (Fig. 8.15a). There is no force action in the longitudinal direction of the open-ended cylinder; the meridional forces are zero, since the loads acting perpendicular to the single curvature membrane must be resisted along the circumferential direction (Fig. 8.15b). The circumferential or hoop forces due to the radial force action normal to the surface are derived from the membrane equation [Eq.(8.2)] by substituting $R_1 = \infty$ and $R_2 = R$.

$$p = \frac{N_\phi}{R_1} + \frac{N_\theta}{R_2} = \frac{N_\theta}{R}$$

$$N_\theta = pR \tag{8.8}$$

Note that the compressive hoop forces are constant along the surface and are twice as large as for the spherical membrane, which uses two equal curvatures to resist the same loading [see also Eq. (9.57)].

The hoop forces as caused by the weight of the membrane, when $0° \le \theta \le 90°$, are found by expressing w in terms of p (Fig. 8.28b).

$$p = w \cos\theta$$

Substituting p into Eq. (8.8) yields

$$N_\theta = wR\cos\theta \tag{8.9}$$

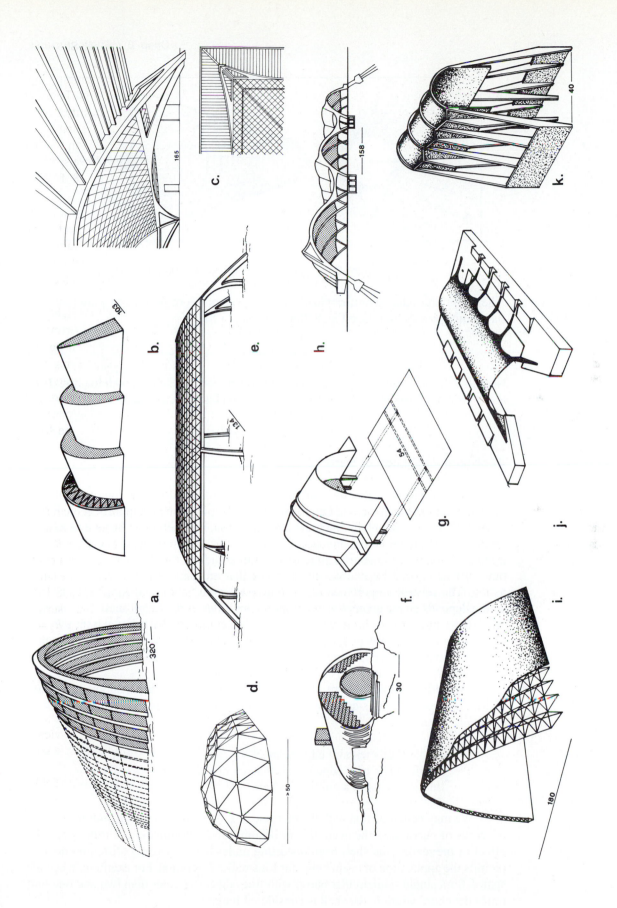

Figure 8.14 Vaults and short cylindrical shells.

623

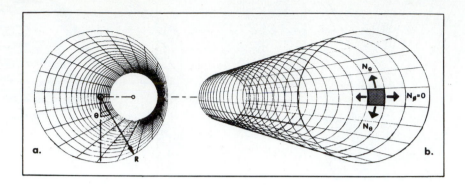

Figure 8.15 Circular Cylindrical Membrane.

The hoop forces due to uniform loads on the horizontal roof projection q are found by substituting $p = q\cos^2\theta$ (see Fig. 8.28b) into Eq. (8.8)

$$N_\theta = qR\cos^2\theta \tag{8.10}$$

Note that the maximum compression due to arch action appears for both loading cases at the crown of a standing membrane where $\theta = 0$ (Fig. 8.19); hence,

$$N_{\theta max} = (w + q)R \tag{8.11}$$

Approximate Design of Barrel Shells

It is shown in Fig. 8.12 that the vault transfers forces directly to the longitudinal supports in transverse arch action. On the other hand, if there are no supports in the longitudinal direction, but only in the transverse direction, the shell must behave as a beam spanning in the longitudinal direction. In other words, the arch action in the transverse direction cannot be transferred directly to the longitudinal rigid supports as for vaults, but must be carried by internal shear forces that are part of the longitudinal beam action. The relative importance of the transverse arch action and longitudinal beam action depends on the geometry and support conditions of the barrel shell. Barrels with long spans and narrow chord widths ($b \ll L$) are controlled by beam action; they are called *long barrel shells* or *shell beams*. On the other hand, when the chord width b (i.e., arch span) is large in comparison to the span L, transverse arch action becomes predominant; these barrels are called *short barrel shells*.

Cylindrical shells that are of symmetrical cross section and support uniformly distributed loads are considered long when they meet the following *span-to-radius of curvature* conditions:

Single shells without edge beams	$L/R > 5$
Single shells with edge beams that are not too deep	$L/R > 3$
Interior shells of a multiple system	$L/R > 2$

We may conclude that with the increase of L/R (increase of span and/or decrease of radius of curvature), the beam theory becomes more dominating. Furthermore, the effect of preventing the shell from distorting under loads by using stiffening devices permits the application of beam behavior for smaller L/R ratios. For example, when the span L for a single semicircular barrel with free edges is greater than two and one-half times the chord width b, the shell is considered long.

In conclusion, long barrel shells are shell beams that can be visualized as beams with curvilinear cross section. They are assumed not to distort under load action so that linear stress distribution can be used.

The potential warping of the shell (i.e., deflections of the crown and particularly of the edges) is reduced by introducing longitudinal edge beams, transverse ribs, ties, diaphragms, or adjacent shells. The usual *height-to-span ratio* for long shells varies from 1:10 to 1:15 and is 1:20 for prestressed shells.

As L/R decreases and the transverse supports move closer to each other, or as the dimension of the longitudinal span becomes smaller than the dimension of the shell width, the shell behavior changes from a primary action in the longitudinal direction (beam action) to a primary action in the transverse direction (arch action). For example, when the chord width b of an interior semicircular cylindrical shell is greater than the span L, it is called a *short barrel shell*. In other words, the type of cylindrical shells that are neither beams nor vaults are called *short shells*; their structural behavior is complex and very much dependent on their geometrical proportions. The forces in the longitudinal direction are not distributed anymore in a linear manner, but in a curvilinear one. It is known from elementary mechanics that the behavior of shallow beams ($h/L < 0.2$) with linear stress distribution changes drastically as their height increases for a given span by introducing deep beam behavior with nonlinear stress distribution.

The transition from a linear stress distribution (long shell) to a curvilinear one (short shell) is shown in Fig. 8.16. Notice that in the short shell more than one tensile zone exists, as the tension along the crown portion indicates. The maximum compression does not occur at the shell crown anymore. Critical transverse bending moments occur at the stiff edge beam supports. The stress flow in the transverse direction may be approximately evaluated for preliminary design purposes by using membrane arch action at the crown supported by inclined beams near the valley, that is, by plate action in the steeper shell portion (Fig. 8.17b), which, in turn, may be replaced by imaginary arches and cables (Fig. 8.17a).

The three single-curvature linear surface structures—vault and short and long shells—can be further studied by investigating the transition of a one-way, beam-supported slab to a shell beam (Fig. 8.18). For the case where the horizontal slab is supported by beams, the loads are first carried in slab action transversely and then in beam action longitudinally (see also Fig. 7.3). If the slab is bent, the slab action is replaced by arch action, assuming that the longitudinal edge beams are very stiff in comparison to the shell, so the conditions are similar to a vault sitting on longitudinal supports.

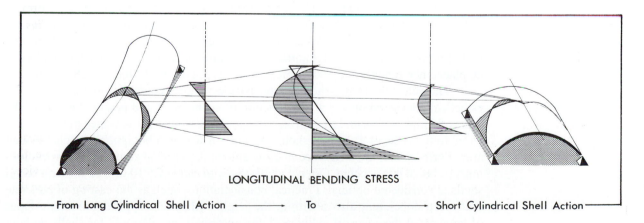

LONGITUDINAL BENDING STRESS

From Long Cylindrical Shell Action ← → To ← → Short Cylindrical Shell Action

Figure 8.16 Long versus short barrel shell.

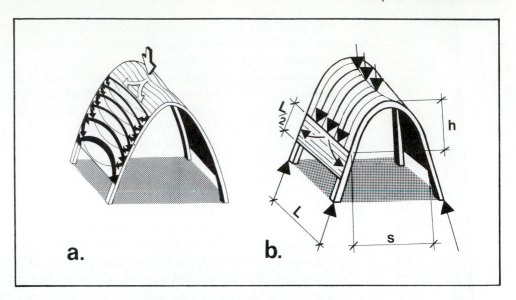

Figure 8.17   Behavior of short barrel shells.

However, if the edge is relatively flexible, edge beams and shell can act together as a total unit: the primary forces in this shell with edge beams are of axial nature, caused by beam action in the longitudinal direction rather than bending due to arch action in the transverse direction.

The size of the edge beam depends on the ratio of shell height to shell span, among other considerations. In the longitudinal direction, the shell acts as a beam; its cross-sectional height is the primary determinant of the moment of inertia, that is, the magnitude of the axial stresses. For a shallow long shell, large edge beams may be necessary to obtain the required moment of inertia. As the shell height increases, the moment of inertia also increases, thereby decreasing the size of the edge beam, which eventually will no longer be needed (Fig. 8.18, bottom).

Similar reasoning is true for the behavior of the shell in the transverse direction. The flat shell causes large lateral thrust forces under vertical loading due to transverse arch action; this thrust must be carried by edge beams. As the height of the shell increases, the amount of transverse thrust decreases; transverse bending is treated as a function of chord width only. Remember that the loop forces, due to membrane action for a semicircular cylindrical membrane, are zero along the longitudinal edges (Fig. 8.19). We may conclude that as shell height increases the size of the edge beams decreases, keeping in mind that the conditions for a single shell beam are very different from those of a multiple-shell grouping, where the shells support each other laterally.

For long interior cylindrical shells, the edge beams are vertical since their primary purpose is to resist longitudinal forces, while for short shells the edge beams can be placed horizontally, since they must resist the lateral thrust due to arch action.

The selection of the shell height, besides being dependent on functional and esthetical considerations, is also a function of economical and practical aspects. The costs of a high shell with large surface area and low weight per unit area must be compared to the costs of a shallow shell with less surface area but higher weight per unit area. Furthermore, the difference in costs among the varying construction procedures must be taken into account. Often a *width-to-depth ratio* of 1:10 to 1:12 for long barrel shells is considered optimal. Practical considerations, such as the casting of concrete onto steep slopes, may also determine the height of the shell. Typical maximum slopes of tangents at the edges of cylindrical concrete shells are $\theta_0 = 45°$ for shells without edge beams and $\theta_0 = 30°$ for shells with edge beams.

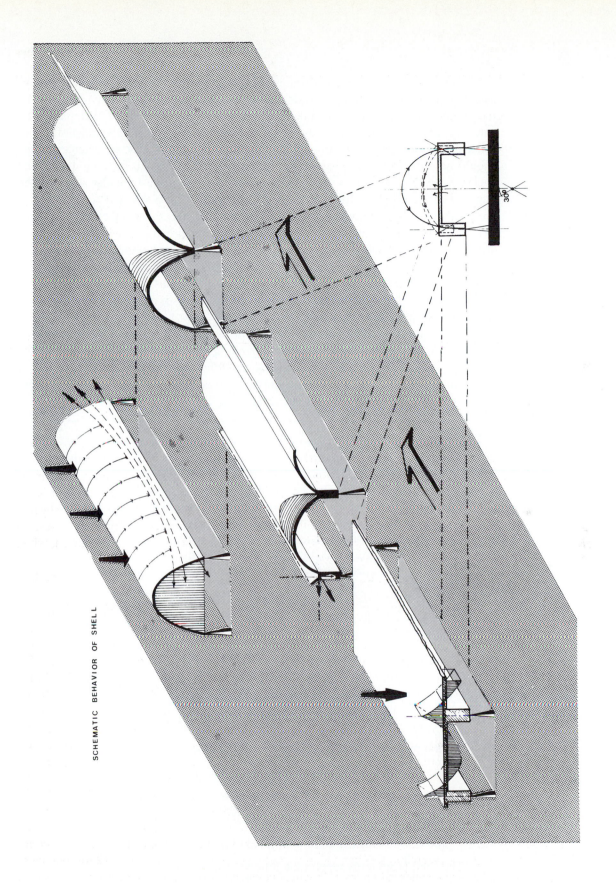

SCHEMATIC BEHAVIOR OF SHELL

Figure 8.18 From the joist slab to shell beam.

The selection of the shell curvature should depend on the primary shell behavior, assuming that other criteria do not determine the shell form. Should arch behavior be the primary action, such as for vaults and very short cylindrical shells, then parabolic, catenary, or other funicular shapes derived from load action should be most efficient. Since dead load is, often the primary load for short concrete shells, the catenary shape should be selected. These shapes are not, however, appropriate for long cylindrical shells, since the arch forces are directly transferred along the funicular shape to the edges, where the lateral thrust causes large deformations, especially if no edge beams are provided. These deformations make the use of beam theory questionable, as it assumes that the beam cross section does not warp under loading. Typical cylindrical shapes used, in general, are segments of circles, cycloids, and ellipses. Notice that the ellipse and cycloid provide a greater stiffness than the circle because of the larger curvature close to the edges. Furthermore, the tangent to the curvature along the edges of the cycloid and ellipse is steeper and so reduces the effect of transverse bending along the edges as caused by arch action.

At the junction of the relatively rigid transverse support structure or the transverse ribs with the thin-shell membrane, bending disturbances are generated in the shell. These disturbances are pronounced in the short shell, since they are propagated across the width of the shell, but have only a localized effect on the long shell, where the overall beam behavior is not influenced.

EXAMPLE 8.1

A typical interior, 3-in.-thick cylindrical semicircular barrel of a multiple shell system without edge beams is 20 ft wide and spans 100 ft from support to support (similar to Fig. 8.21b). The shell loading consists of 5 psf for roofing and insulation and 30-psf snow load. A concrete strength of 4000 psi and Grade 60 steel are used. Refer to Fig. 8.19 for the general behavior of long barrel shells.

The span-to-depth ratio is $L/h = 100/10 = 10$, which is within the typical range of long barrel shells. In this case the span-to-radius ratio $L/R = L/h = 10 > 2$ clearly indicates that the barrel shell can be treated as a beam in the longitudinal direction.

(a) Beam Action in the Longitudinal Direction. The various geometrical properties for the semicircular cross section of a shell beam (see Table A.14 and Schueller, 1983) are as follows:

$$\text{Arc length:} \quad L = \pi R = \pi(10) = 31.42 \text{ ft} \tag{9.59}$$

$$\text{Area:} \quad A = Lt = \pi Rt = \pi(10)(3/12) = 7.85 \text{ ft}^2$$

$$\text{Location of centroidal axis:} \quad c_t = 0.363R = 0.363(10) = 3.63 \text{ ft}$$

$$c_b = h - c_t = 10 - 3.63 = 6.37 \text{ ft}$$

First moment of the area about the centroidal axis:

$$Q_{\max} = 0.42tR^2 = 0.42(3/12)10^2 = 10.5 \text{ ft}^3$$

Moment of inertia about the centroidal axis:

$$I = 0.3tR^3 = 0.3(3/12)10^3 = 75 \text{ ft}^4$$

The concrete shell weighs $w_D = 3(150/12) + 5 = 43$ psf

The shell beam must support the following total load:

$$w = 0.043(31.42) + 0.030(20) = 1.35 + 0.6 = 1.95 \text{ k/ft}$$

The maximum simple-span moment at center is

$$M_{\max} = wL^2/8 = 1.95(100)^2/8 = 2437.5 \text{ ft-k}$$

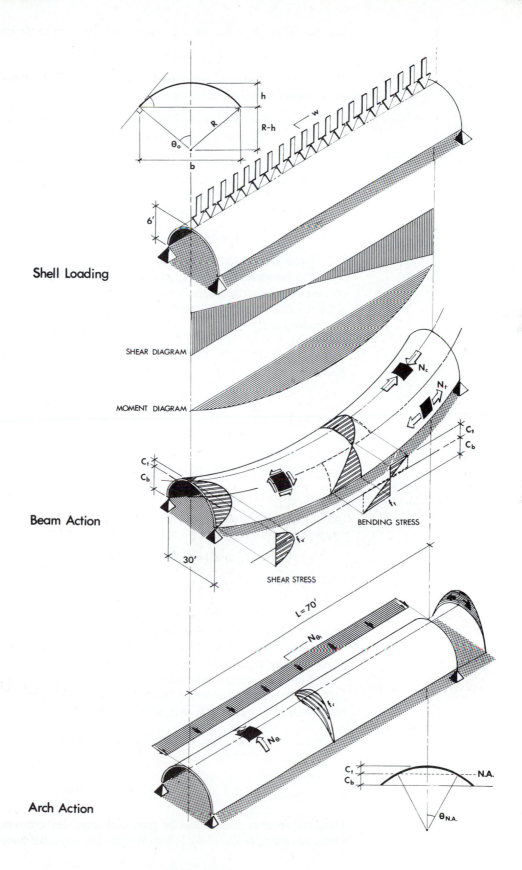

Figure 8.19 Behavior of long barrel shell.

The maximum shear is assumed conservatively equal to the reaction forces.

$$V_{max} = wL/2 = 1.95(100/2) = 97.5 \text{ k}$$

The maximum longitudinal compressive stresses in the crown portion at midspan, according to Eq. (2.66), are

$$f_c = Mc_t/I = 2437.5(3.63)/75 = 117.98 \text{ ksf} = 0.82 \text{ ksi}$$
$$\leq 0.45 f_c' = 0.45 (4) = 1.80 \text{ ksi}$$

The actual compressive stress in the concrete is by far less than the allowable compressive stress. Hence, the selected slab thickness is more than adequate and clearly not dependent on strength considerations. The maximum tensile stress in the bottom edge at midspan, according to Eq. (2.66), is

$$f_t = Mc_b/I = 2437.5(6.37)/75 = 207.03 \text{ ksf}$$

According to the ACI Code, the reinforcement to carry the total tension may be concentrated in the region of maximum tensile stress. Here, however, for preliminary design purposes, a rectangular stress distribution over a height of 1 ft is used, which is then assumed to decrease toward the neutral axis. Hence, the maximum tensile force is

$$N_t = f_t A = 207.03(3/12)1 = 51.76 \text{ k/ ft}$$

The corresponding steel area to resist this tensile force is

$$A_s \geq N_t/F_t = 51.76/24 = 2.16 \text{ in.}^2/ \text{ ft} \qquad (8.6)$$

Try 5 #6 bars, $A_s = 2.20 \text{ in.}^2/ \text{ ft}$ (Table A.1). These longitudinal bars are to be placed along the bottom shell portion over about a 1-ft range. Then the amount of reinforcement may be decreased toward the neutral axis, keeping in mind that the minimum reinforcement in the tensile zone should be

$$A_{smin} = 0.0035A_g = 0.0035(12)3 = 0.126 \text{ in.}^2/ \text{ ft} \qquad (8.7)$$

In the compression zone, only the minimum amount of steel needed to control temperature and shrinkage must be provided.

$$A_{smin} = 0.0018A_g = 0.0018(12)3 = 0.065 \text{ in.}^2/ \text{ ft} \qquad (3.69)$$

The maximum bar spacing is $5t = 5(3) = 15 \leq 18$ in. Try #3 bars at 15 in., $A_s = 0.088 \text{ in.}^2/ \text{ ft}$.

Should longitudinal edge beams and shallower shells be used, the neutral axis will be close to the junction of the shell and the beam, and the steel reinforcement carrying longitudinal tension forces may be concentrated in the beams, since primarily the beams will be carrying the tension.

Maximum diagonal tensile stresses, as caused by shear, will appear close to the transverse shell supports. The shear causes a maximum tangential diagonal tension stress at the neutral axis, according to Eq. (2.69), of

$$f_v = \frac{VQ}{I(2t)} = \frac{97.5\,(10.5)}{75\,(2)\,3/12} = 27.3 \text{ ksf} = 190 \text{ psi}$$

$$\geq 1.1\sqrt{f_c'} = 1.1\sqrt{4000} = 70 \text{psi}$$

Diagonal reinforcement must be provided since the shear capacity of the concrete, according to Eq. (7.5), is insufficient. The diagonal tensile force is equal to

$$N_v = f_v A = 0.190(12)3 = 6.84 \text{ k/ ft}$$

The diagonal reinforcement at the neutral (centroidal) axis, disregarding conservatively the shear capacity of the concrete, is

$$A_s \geq N_v / F_t = 6.84/24 = 0.29 \text{ in.}^2/\text{ft}$$

Try #4 bars at 8 in. on center at the support (Table A.2). The spacing of the diagonal bars can be increased as the shear decreases away from the support.

(b) Arch Action in the Transverse Direction. Arches must carry loads in the transverse direction. But, since they are not supported independently, the beam action in the longitudinal direction must do so. The stiffness along the shell edges influences very much the magnitude of the transverse bending moments, which are very difficult to determine. The behavior of an interior long barrel of a multiple shell group can be conservatively approximated for a 1-ft-wide shell strip as based on arch analysis (Fig. 5.45). However, for this condition, the transverse moments are relatively small and may be ignored for preliminary design purposes.

The magnitude of the axial compression forces in the transverse direction can be found from the equation for the maximum hoop forces at the crown [Eq.(8.11)] as derived for membrane action.

$$N_{\theta max} = (w + q)R \cong (1.95/20)10 = 0.975 \text{ k/ft}$$

$$f_c = N_\theta/A = 975/12(3) = 27 \text{ psi}$$

This compressive stress is by far less than the allowable stress of $0.45f'_c = 0.45(4000) = 1800$ psi, keeping in mind, however, that the compressive stresses due to transverse bending $[\cong M/(t^2/6)]$ must be added.

(c) Conclusion and Investigation of Support Conditions. The typical conceptual layout of the steel reinforcing is shown in Fig. 8.20. The steel bars follow the tension trajectories of the principal stresses, with large bars at midspan, close to the bottom edges, to carry the longitudinal tension and smaller bars, placed at 45° diagonal to the transverse supports, which carry the shear (diagonal tension). Small bars are placed in the transverse direction to carry the tension caused by bending in arch action. At other places in the longitudinal and transverse directions, the minimum required temperature and shrinkage steel is provided.

Additional reinforcing steel, as well as the thickening of the shell, may be necessary at the continuous junction of the thin shell and the rigid end supports, due to local bending disturbances in the longitudinal direction. These types of secondary moments are always generated at the intersection of the shell membrane and stiffer structural elements such as ribs and edge beams, due to the incompatibility of edge conditions (displacements) along the shared boundaries of elements providing different types of structural action (see Fig. 8.31).

The transverse supports at the ends of the shell beams must carry these uniformly distributed loads.

$$w = 97.5/20 = 4.88 \text{ k/ft}$$

where a typical interior column carries 97.5 k.

The horizontal thrust at the shell support can be roughly evaluated by treating the arch as a three-hinged one [see Eq.(5.84)]. Hence, the approximate lateral thrust force is equal to

$$H = wb^2/8h = 4.88(20)^2/8(10) = 24.4 \text{ k}$$

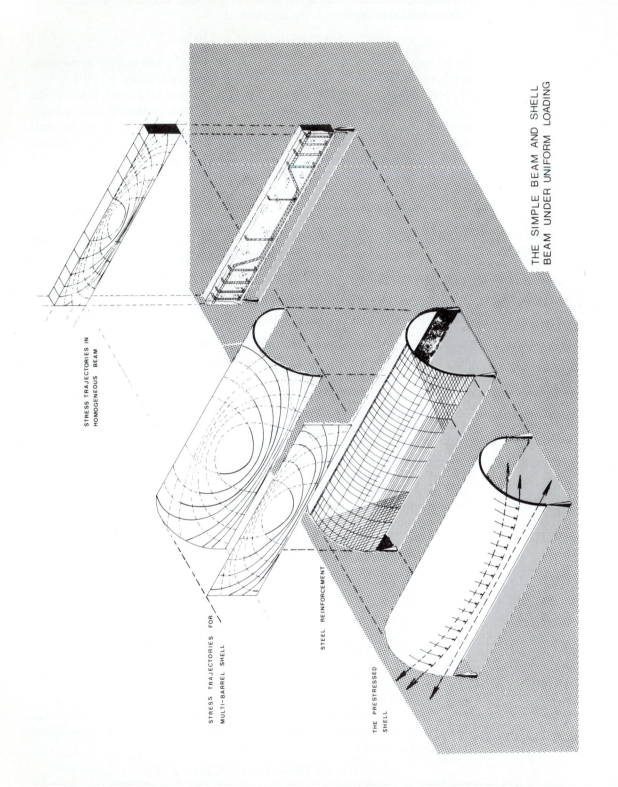

STRESS TRAJECTORIES IN HOMOGENEOUS BEAM

STRESS TRAJECTORIES FOR MULTI-BARREL SHELL

STEEL REINFORCEMENT

THE PRESTRESSED SHELL

THE SIMPLE BEAM AND SHELL BEAM UNDER UNIFORM LOADING

Figure 8.20 Rectangular beam versus shell beam.

EXAMPLE 8.2

The barrel shell in Example 8.1 is investigated quickly as a prestressed beam using Grade 270 strands. Prestressing is usually applied to long-span shells to reduce deflection and cracking. It generates an almost pure membrane with uniform longitudinal compression.

Since the live load is much less than the dead load, 80% of the dead load is balanced for this first approach [see Eq.(3.77)]. The maximum cable drape for this simply supported beam is estimated as

$$e \leq c_b - d' \cong 6 \text{ ft}$$

The prestress force required to balance $0.8w_D$ according to Eq. (3.72) is

$$P = M/e = 0.8w_D L^2/8e = w_D L^2/10e = 1.35(100)^2/10(6) = 225 \text{ k}$$

The required prestress tendon area according to Eq. (3.74) is

$$A_s = P/0.56f_{pu} = 225/0.56(270) = 1.49 \text{ in.}^2$$

Along each longitudinal edge, use five 0.5-in. diameter, Grade 270 strands with $A_s = 2[5(0.153)] = 1.53$ in.^{2} (see Table A.5). In prestressing of cylindrical shells, the cables follow the major tensile stresses; we can visualize the shell as sitting on the cables (Fig. 8.20).

The average precompression stress [Eq.(3.73)] is

$$f_a = P/A = 225,000/7.85(12)^2 = 199 \text{ psi}$$

The concrete section is sufficient since this average stress is low, compared to a typical range from about 200 to 500 psi or higher.

EXAMPLE 8.3

A typical interior, long, cylindrical, circular shell (Fig. 8.19) of a multiple-shell group is investigated. It spans 70 ft and is considered to be simply supported and not to have any edge beams. The chord width of the shells is 30 ft and the height is 6 ft. The shell must support the same loads as in Example 8.1. The wind load, in general, may be neglected for first-design purposes. A preliminary shell thickness of 3 in. has been assumed. The thickness rarely depends on strength considerations, but is generally selected to ensure adequate coverage of the reinforcement. Only for very large shells, buckling may have to be considered as a determinant of shell thickness, although buckling of cylindrical shells is not as critical as it is for domes because a relatively smaller portion of the shell surface is in compression. A concrete strength of 4000 psi and Grade 60 steel are used here.

The height-to-span ratio of the shell beam should be satisfactory since it is in the typical range from 1:10 to 1:15.

$$h/L = 6/70 = 1/11.67$$

The radius of curvature according to Eq. (9.64) is

$$R = (s^2 + 4h^2)/8h = (30^2 + 4(6)^2)/8(6) = 21.75 \text{ ft}$$

The semicentral angle or the maximum slope along either edge is derived from Eq. (9.58) as

$$\sin\theta_o = \frac{s/2}{R} = 15/21.75 = 0.69 \quad \text{or} \quad \theta_o = 43.60°$$

The arc length according to Eq. (9.59) is

$$l = \pi R(\theta_o/90°) = \pi(21.75)(43.60/90) = 33.10 \text{ ft}$$

The span-to-radius ratio $L/R = 70/21.75 = 3.22 > 2$ indicates that the barrel can be treated as a beam.

The shell beam must support the following uniform load:

$$w = 0.043(33.10) + 0.030(30) = 2.32 \text{ k/ft}$$

The maximum moment at midspan is

$$M_{max} = wL^2/8 = 2.32(70)^2/8 = 1421 \text{ ft-k}$$

Since the geometrical properties of the shell beam are difficult to determine (and too lengthy for a fast estimate), in contrast to the stress approach for the semicircular shell in Example 8.1, the *internal couple method* will be applied as has been used elsewhere in this book for other beam types. The location of the center of the longitudinal steel reinforcement is assumed at 1 ft from the base (valley), or as $d = 6 - 1 = 5$ ft. The lever arm for the compression–tension couple is conservatively estimated as $z = 0.9d$ [see discussion at Eq. (3.52)]. The moment at midspan is resisted by the internal force couple

$$T(z) = C(z) = M$$

Or the maximum tensile force is

$$T = M/z = wL^2/8z = 1421/0.9(5) = 315.78 \text{ k}$$

Hence, the area of longitudinal reinforcement for one half-shell at midspan is

$$A_s \geq T/F_t = (315.78/2)/24 = 6.58 \text{ in.}^2$$

Try 11 #7 bars, $A_s = 6.61$ in.^{2}

Should an edge beam be used, the steel reinforcement from each shell [i.e., 2(11 #7) bars] can be concentrated in the beams, although fewer bars of larger size should be considered.

The compressive stresses do not have to be checked since they are generally not critical for ordinary conditions.

To determine the maximum shear stresses in the same fashion as in Example 8.1 would be very complex because the shell is not semicircular. Here, for preliminary design purposes, the following approach is used. From the free body in Fig. 8.21a, it is apparent that at the neutral axis the total horizontal shear V_h must be equal to the axial tension force T (or compression force C), where V_h is the resultant of the triangular shear diagram for the shell beam under uniform load action. Denoting the maximum unit shear force at the support by N_v, we obtain

$$V_h = T = N_v(L/2)/2, \quad \text{or} \quad \text{for one-half-arc} (T/2)$$
$$N_v = 4T/L = 4(315.78/2)/70 = 9.02 \text{ k/ft} \tag{8.12}$$

Or the diagonal reinforcement at the neutral axis at the support is

$$A_s \geq N_v/F_t = 9.02/24 = 0.38 \text{ in.}^2/\text{ft}$$

Try #5 bars at 9.5 in. on center at the support, $A_s = 0.39$ in.^{2}/ft.

For the transverse steel reinforcement, refer to the discussion in Example 8.1.

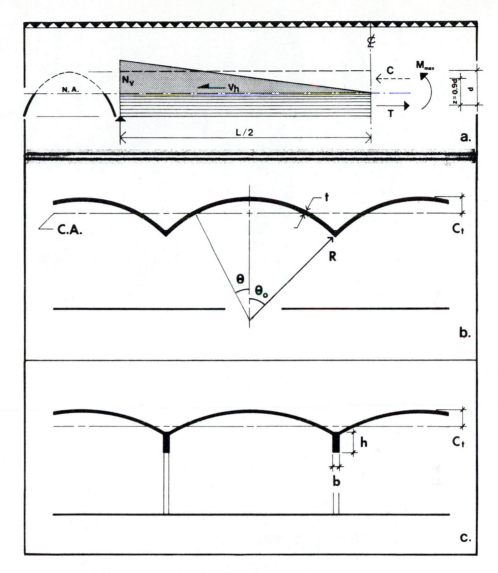

Figure 8.21 Barrel shells with or without edge beams.

Cylindrical Grid Structures

The solid shell surface can be replaced by a grid structure; this aspect is discussed in more detail in Chapter 6 and in the next section. Some of the various tessellations possible for the cylindrical shell are simply trussed, rectangular, diagonal, three way, or geodesic, as indicated in Fig. 8.22. From a formal point of view, the grid shells have evolved from the *ribbed vaulting* of the late Gothic cathedrals; these ribbed vaults, in their simplest form, consist of diagonal ribs along groins and the masonry webbing. The richness of various Gothic rib patterns is expressed in Fig. 8.22. The *iron–glass vaults* of the first half of the nineteenth century in England, used as winter gardens, may be considered the most recent forerunners of ribbed cylindrical shells.

The popular diamond-patterned lamella roof was developed in 1908 by Zollinger in Dessau, Germany, and was further advanced by G. Kiewitt in the United States. The roof structure may be visualized as a system of intersecting diagonal arches made up of short members of uniform length, called *lamellas*. The diagonal member grid is indirectly triangulated by the roof decking or purlins, thus making the roof surface stable, even though the members may be hinged to each other. The predominant construction

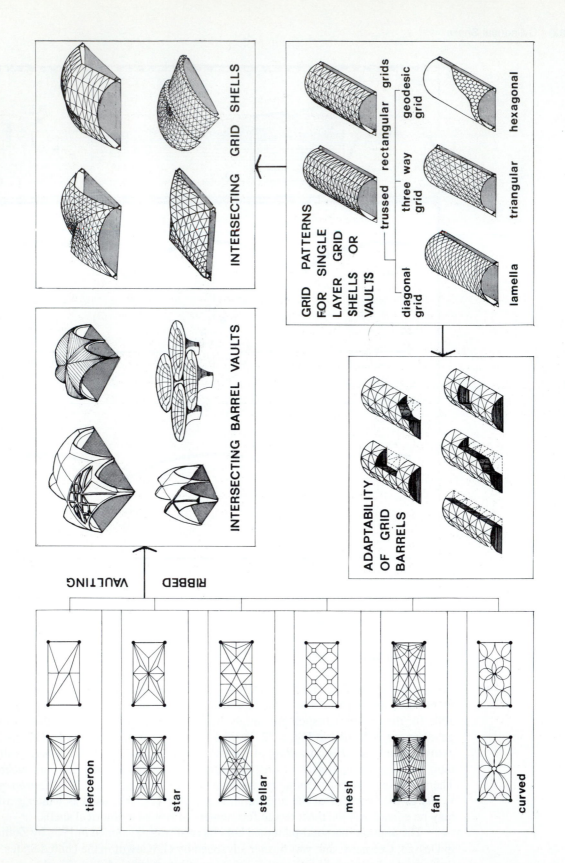

Figure 8.22 Cylindrical grid structures.

material is wood, but for larger-span structures, steel and concrete are used. Luigi Nervi's first hangars in Italy (1936, 1940) are famous examples of the principle applied for the first time to large-scale concrete construction (Fig. 8.14e).

While in steel construction lamellas may be moment connected, in wood structures only thrust and shear connections are used. Here, usually straight members of twice the length of the diamond sides are arranged in a staggered manner.

For preliminary design purposes, the lamella vault may be treated as a monolithic arched surface, which, in turn, can be considered as parallel three-hinge arches, rather than the actual two-hinge ones, spanning in the short direction of the building. The imaginary arches are first assumed as 1 ft wide and then corrected for the true lamella spacing (see Schueller, 1983).

Asymmetrical Shell Beams

Asymmetrical shell beams, such as the north-light cylindrical shell (Fig. 8.23), are much more intricate in behavior and complex to analyze as compared to symmetrical shell beams. The beam method, however, can still be used for rough, first-approximation purposes by taking into account unsymmetrical bending.

First, imagine a series of cylindrical shells to be separated at their crown (Figs. 8.23a and b) and to support each other with struts. When these shell units act independently of each other, each unit is a Y-shape with twin cantilevers, often called a *butterfly shell*; these shell types are stiffened on top. Next, visualize the butterfly shells missing one of their balancing cantilevers, resulting in nonsymmetrical half-shells (c). At this stage the shell units can be seen as separate inclined cylindrical shells supported in the transverse direction by a small top beam and, at the bottom, a much larger gutter beam. These two beams are tied together by mullion struts to form a slightly slanted frame (d). The frame, however, is considered not to generate continuous transverse action; any potential continuity in this direction is conservatively neglected.

Since a typical shell unit is inclined, the gravity load P must be resolved into components P_x and P_y parallel to the principal axes. The bending stresses caused by the P_y force about the x axis are found in similar fashion as for a normal symmetrical beam (e, a). Notice the relatively low moment of inertia I_x of a half-shell, as compared to that of the symmetrical shell composed of the two half-units (a). In other words, the effective depth of the north-light shell is very flat and much less than the height of the full cylindrical shell unit, thus yielding much larger bending stresses.

The moment of inertia I_y is by far larger than I_x, thus resulting in fewer critical stresses. The moment of inertia about the y axis can be approximated by ignoring the effect of the shell and by assuming that the top and bottom beams (i.e., flanges) carry the entire moment as caused by the P_x force.

The usual span range of north-light cylindrical concrete shell beams is in the range from 40 to 100 ft and up to 150 ft if prestressed. Typical bay widths are 25 to 40 ft with a shell thickness of 3 to $3\frac{1}{2}$ in.

8.3 THIN-SHELL DOMES AND SKELETAL DOMES

In historical context, domes represent an ideology of their own with a specific symbolic meaning. They were associated in the Roman, Christian, and Islamic worlds with temples, churches, mosques, memorials, baptistries, tombs, baths, and other meeting places; they have been seen not just as functional solutions for covering large spaces. This monumental architecture has derived from the domical shelters of the indigenous architecture that used primitive methods of construction, as is still found today in many

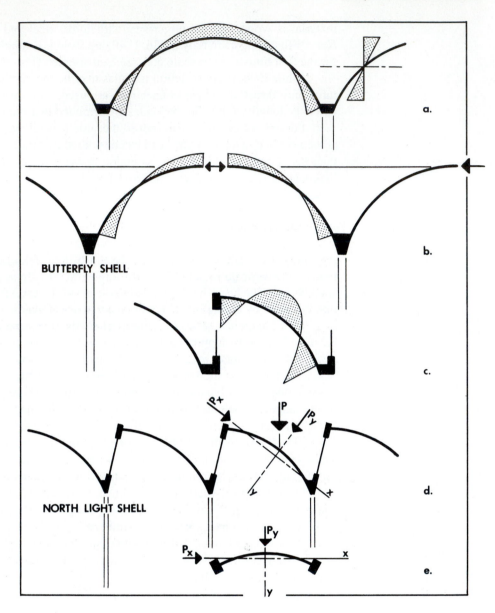

Figure 8.23 Various cylindrical shell types.

regions of the world. They range from the stone trulli of Apulia in İtaly, mud brick domes in Egypt, the yurts of the Mongol nomads of the central Asian steppe, the igloos of the Eskimos, and the wealth of dome shapes of the African tribal architecture, to the geodesic domes of the hippie communes during the late 1960s in the United States. The historical development of domes is discussed in the introduction to this chapter.

Dome Types

Most domes are surfaces of revolution, that is, synclastic forms such as of spherical, parabolic, elliptical, and conical configuration, often covering circular or nearly circular areas. Other dome shapes, like the onion, melon, and bulbous domes, are found in Byzantine and Islamic architecture. Domes on polygonal bases may be segmental surfaces (see Fig. 8.30) with straight or curvilinear edges. To cover a rectangular floor plan, it may be more convenient to use a section of a torus or a translational surface, such as an elliptical paraboloid (Fig. 8.35); in the past, *pendentives* (i.e., spherical triangles) were often employed to make the transition from the circular dome to the rectangular base. Dome surfaces may be continuous or pointed, as for conoidal and conical domes. They may consist of a single surface unit or intersecting surfaces, where several units are joined along valleys and ridges.

Later in this section it is shown that dome shells carry uniform vertical loads as radial arches in compression. These arches, in turn, are tied together by horizontal rings, which are in compression in the upper portion of the dome and in tension in the lower portion. In other words, shallow domes are in full compression and cannot provide any resistance to outward thrust due to arch action, in contrast to steep domes, where part or all of the tension is resisted within the dome shell by ring action.

Most early modern domes are of nearly hemispherical shape, thus not causing any lateral thrust; keep in mind, however, that horizontal force action appears for ribbed domes, which are not true surface structures, as discussed later. Hemispherical domes use more surface area than shallow domes to cover the same space. On the other hand, flat domes require a vertical support structure to increase the headroom, as well as large tension rings or external buttressing, because of the extensive lateral thrust exerted by the domes (i.e., horizontal component of meridional force at support). Tension rings and other support systems are identified in Fig. 8.24. A transition curvature

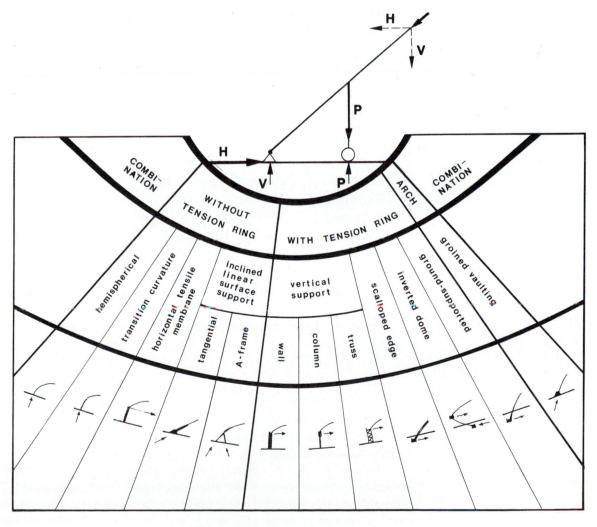

Figure 8.24 Dome support structures.

could be employed in order to smoothly adjust the dome curvature so that it is perpendicular to the base, such as for the synagogue of the Hebrew University in Jerusalem (see Fig. 8.27e), where the entire transition zone acts as a tension ring. Or the meridional arch forces could be directly transferred to the tension ring in the ground by an inclined support structure in a tangential fashion. The plan form of the dome is often closely related to the number, location, and type of supports, as the various cases in Fig. 8.27 reflect. Domes are generally organized according to the nature of their surface structure and according to the layout of bracing for skeletal domes. One generally distinguishes among the following:

- *Ribless thin-shell domes* (e.g., reinforced concrete, plastics)
- *Ribbed thin-shell domes*
- *Skeletal* or *framed, single-layer domes* (e.g., steel, aluminum, wood)
- *Skeletal domes* of the single-layer type at the perimeter and double-layer type at the center where special equipment loads are high
- *Trussed double-layer domes* for large spans (e.g., steel, aluminum, wood)
- *Stressed-skin domes* where the skin acts compositely with the framing (e.g., aluminum)
- *Formed surface domes* where the surface is bent or folded to form a spatial surface (e.g., reinforced concrete, metal, plastics)
- *Fabric domes* (air domes and cable domes); for discussion, see Chapter 9

Figure 8.25 attempts to indicate the infinite wealth of possibilities for braced dome patterns. From a conceptual point of view, the major framed domes are more clearly identified in Fig. 8.26, where nearly all the systems are shown triangulated so that they can be more easily compared. The primary dome structure types are as follows:

- *Braced domes* (with regular configuration)
    Ribbed domes, including polygonal domes
    Schwedler domes
    Polyhedral domes
    Lamella domes, including lattice domes (curved lamellas) and parallel lamella domes
    Network domes, including two- and three-way grid domes
    Geodesic domes, including single-layer framing, double-layer space trusses, and stressed-skin construction
    Tensegrity domes, including cable domes (see Section 9.7)
- *Thin shell domes*
    Ribless shell domes
    Ribbed shell domes
    Polygonal domes —
    Corrugated shell domes
    Double (cellular) shell domes
- *Air domes* (see Section 9.8)

These various dome types are now discussed, together with the cases in Fig. 8.27.
Thin-shell domes supported uniformly along their round bases can be designed for preliminary design purposes according to the conventional membrane theory, as demonstrated later in this section. But domes on polygonal bases form *segmental sur-*

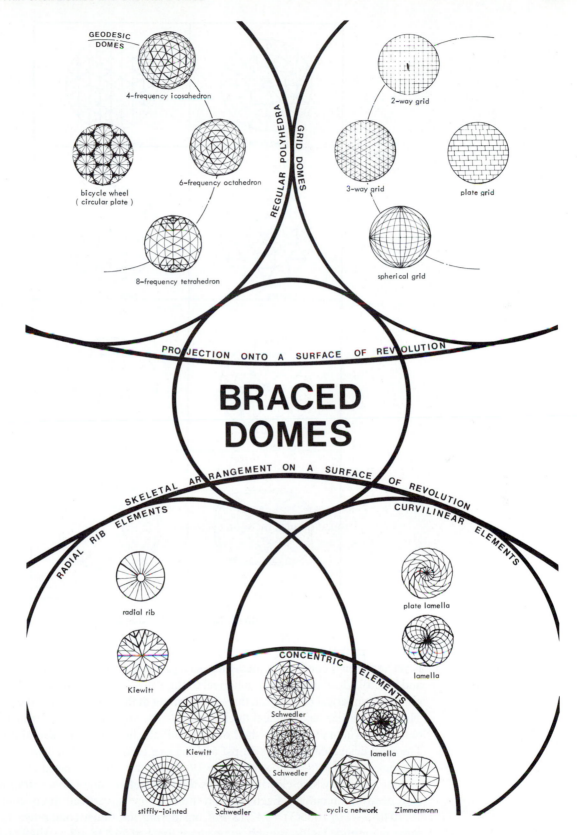

Figure 8.25 Braced dome types.

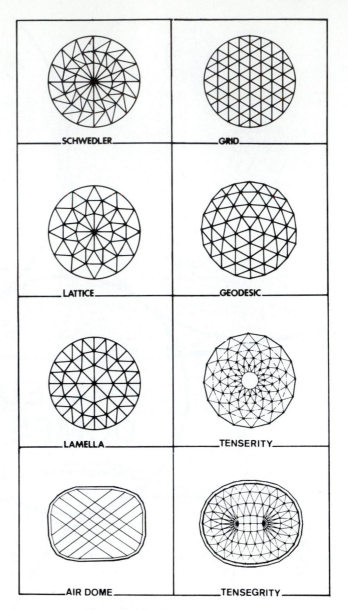

SCHWEDLER

GRID

LATTICE

GEODESIC

LAMELLA

TENSERITY

AIR DOME

TENSEGRITY

Figure 8.26 Major dome systems.

faces where the ordinary membrane force flow is very much altered. An example is Eero Saarinen's concrete dome of the Kresge Auditorium at MIT (1955, Figs. 8.27f and 8.30d), which comprises a spherical triangle that is one-eighth of a sphere. It is supported by spherical bearings on three pendentives at the intersection of the great circles. The shell surface has a typical thickness of $3\frac{1}{2}$ in., to which 2 in. of insulation and an outer nonstructural concrete shell of 2 in. are attached to reduce sound transmission. The total design dead load is 83 psf and the live load for wind and snow is 30 psf. A shell area of $4\pi R^2/8 = \pi(2 \times 112)^2/8 = 19{,}704$ ft^2 must be supported on less than 1 ft^2 of bearing area, which clearly indicates the tremendous stress concentration and special detailing required. Each support must carry a vertical force component of $19{,}704(83 + 30)/1000(3) = 742.18$ k. Considering the thrust component to be of the same magnitude yields, roughly, a resultant force of $742.18\sqrt{2} = 1049.60$ k, causing an approximate stress of $1049.60/0.33(144) \cong 22$ ksi, which requires pure steel for resistance at the point of investigation. Some of the window mullions act as columns

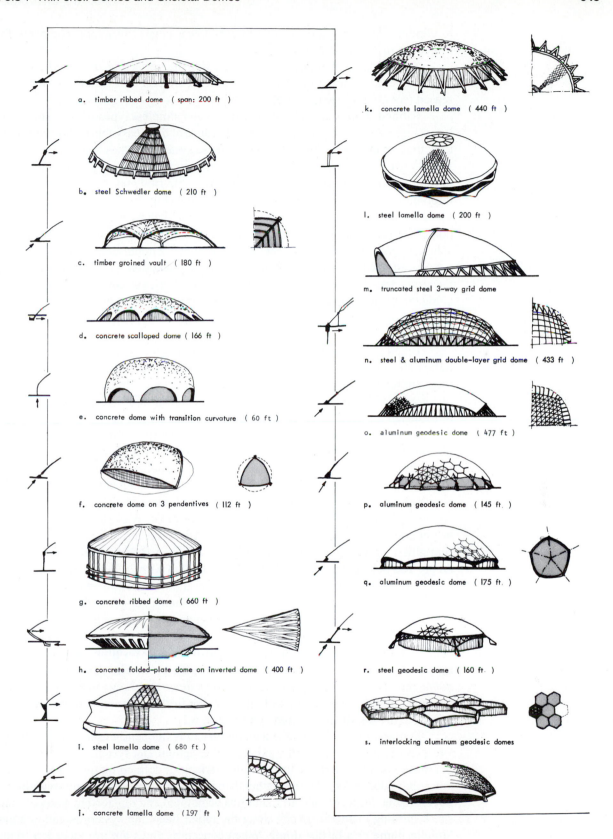

a. timber ribbed dome (span: 200 ft)

b. steel Schwedler dome (210 ft)

c. timber groined vault (180 ft)

d. concrete scalloped dome (166 ft)

e. concrete dome with transition curvature (60 ft)

f. concrete dome on 3 pendentives (112 ft)

g. concrete ribbed dome (660 ft)

h. concrete folded-plate dome on inverted dome (400 ft)

i. steel lamella dome (680 ft)

j. concrete lamella dome (197 ft)

k. concrete lamella dome (440 ft)

l. steel lamella dome (200 ft)

m. truncated steel 3-way grid dome

n. steel & aluminum double-layer grid dome (433 ft)

o. aluminum geodesic dome (477 ft)

p. aluminum geodesic dome (145 ft)

q. aluminum geodesic dome (175 ft)

r. steel geodesic dome (160 ft)

s. interlocking aluminum geodesic domes

Figure 8.27 Dome structure cases.

to eliminate twisting (torsion) due to curvilinear edges. Structural engineers for the project were Ammann & Whitney.

As compared to monolithic concrete domes, *segmental thin-shell construction* offers greater forming economy since segments are cast individually, so forms can be reused (see Seattle's Kingdome).

Ribbed domes are among the earliest and simplest types of bent surface structures (Fig. 8.27a). They are very popular because of their easy fabrication and construction; their span range is around 150 ft. In these systems, arches are arranged in a radial fashion and laterally supported at the top by a compression ring and at the base by a tension ring. For preliminary design purposes, the ribs may be treated as individual three-hinge arches with triangular gravity loading (Fig. 5.43).

We may conclude that, due to the individual behavior of their radial arches, rib domes are primarily direct thrust structures, which do not take advantage of the spatial spherical geometry of the surface. However, when circumferential or hoop rings are introduced, these ribs and rings form an interactive system. The rings at the base and the crown for ribbed domes are now replaced by a series of concentric rings positioned along the surface, thereby resisting the lateral thrust of the arches in a continuous manner. It is apparent that, as the members (stiffness) become more evenly distributed, true membrane action is approached.

Bracing the radial ribs with concentric rings is the basic principle of the popular **Schwedler dome**. For the condition of a quadrangular grid pattern, where the ribs and rings are hinged to each other, the roof skin may provide the necessary shear stiffness under asymmetrical loading. Diagonals may be introduced to form a bent surface truss or to stiffen the shell, or rings and ribs may be moment connected to form a rigid bent frame structure. There are many different layout patterns for the arrangement of the diagonal members. Pin-connected joints may be assumed for preliminary design purposes so that Schwedler domes can generally be treated as statically determinate.

The 328-ft Schwedler dome of the Convocation Center at Ohio University, Ohio, is one of the lightest steel dome structures in the United States; the roof steel framing weighs only 8.9 psf. The structure consists of 24 radial W24 steel ribs and three hoop beams supporting the 24-in.-deep steel joists. The dome thrust is resisted by a concrete ring girder as well as a spider web of radial cables extending horizontally across the dome. The Schwedler dome in Fig. 8.27b is framed with 24 radial steel arches and six concentric ring beams, which support the 5-in. lightweight concrete slab.

Polyhedral domes can be derived, as the name suggests, from polyhedra. The most simple dome forms may also be generated by joining horizontal (or spatial) polygonal rings (which may have different shapes at the various layers) with meridional members to form linear or spatial radial arches; the resulting rectangular surfaces are braced with diagonals. Braced surfaces like the *Schwedler dome,* the *Zimmermann dome,* the *plate-type dome,* the *cyclic network dome* (as shown in Fig. 8.25), and infinite other domes can be generated this way. For further discussion of polyhedral types of braced domes, the reader may refer to Fig. 6.1.

For larger-span domes, the lamella system is frequently used because of its even stress distribution and primarily axial member action. The system is a derivation of the Schwedler dome, where the major surface triangles defined by the radial ribs are further subdivided by parallel ribs, rather than radial ones, resulting in the **straight** or **parallel lamella dome.** Should all radial ribs be omitted and replaced by pairs of diagonal struts forming diamond shapes along the radial lines, the system is called a **curved lamella dome** or a **lattice dome**. When concentric rings are introduced, a triangular network is generated; often, however, the roof decking takes the place of the ring members, leaving the lozenge-shaped member pattern intact.

The lamella roof was developed in 1908 by Zollinger in Dessau, Germany, and introduced to the United States by G. R. Kiewitt in 1925. Numerous lamella structures, especially in wood, have been built since then. Possibly a high point of lamella construction was reached with Nervi's magnificent Small Sports Palace in Rome (1957) and Kiewitt's Houston Astrodome (1964).

Nervi's famous Little Sports Palace (Fig. 8.27j) for the 1960 Olympic Games in Rome, Italy, uses a 197-ft-span curved lamella dome consisting of diamond-shaped precast concrete panels of 13 different sizes, which acted as formwork for the cast-in-place concrete topping. The undulating dome edge is not a tension ring; it only collects the curved lamella ribs and redirects them to the column supports. Thus, the dome surface is continuous with the 36 inclined forked buttresses with vertical legs. These buttresses are, in turn, carried by the tension ring below the ground, which also functions as their foundation. The dead weight of the dome surface is about 80 psf (i.e., about 60 psf, taking shell and ribs together, and 20 psf for insulation and roofing). The dome must support the cupola, besides the snow and wind loading.

Houston's Astrodome with a span of 642 ft is also of the parallel lamella type. It consists of twelve 5-ft-deep main radial rib trusses, six hoop trusses, and diagonal lamella trusses. The roof was designed for a dead load of 30 psf, a live load of 15 psf, and an uplift pressure of 60 psf (see Problem 8.14).

The Louisiana Superdome in New Orleans (1975, Fig. 8.27i), with a span of 680 ft, is currently the world's largest steel dome; it was designed by Sverdrup and Parcel Structural Engineering. The roof is a straight lamella structure (Fig. 8.26) constructed with 7-ft 3-in.-deep welded steel trusses. The principal trussed framing consists of 12 main radial ribs connected by 5 concentric rings, about 56 ft apart, and the diagonal lamella members forming the diamond pattern. The roof is supported by an 8-ft 10-in.-deep, wide-flange, tension ring truss along its perimeter, which is mounted on rocker columns to allow for movement due to temperature changes. The dome must also support, at the center, a gondola for television screens with a total weight of 150 k. The roof system required 26 psf of structural steel.

Network domes are very popular. Intersecting ribs form a network; most common are two- and three-way grids. Network domes have been built all over the world and are usually based on some patented joint system (see Fig. 6.14), such as the popular three-way, single-layer Triodetic aluminum domes and the double-layer MERO domes. Grid domes have been produced not only over round but also over rectangular and hexagonal areas. In 1922, Walter Bauersfeld introduced in Jena, Germany, a triangular steel reinforcement network similar to a three-way grid dome that functioned both as formwork and reinforcement for the thin-shell domes.

The roof of the Sports Palace in Mexico City designed by Felix Candela (Fig. 8.27n) for the 1968 Olympics consists of 22 intersecting trussed steel arches spanning 433 ft. These 16-ft-deep arches, each supported by buttresses, form a two-way frame on a nearly square grid ranging in size from 33 to 43 ft. The square spaces, in turn, are covered with a triangular mesh of aluminum tubes that form a hyperbolic paraboloid surface, which is covered with two layers of 5/8 in. plywood and a copper membrane. The dead weight of the roof is 19 psf, while the structural steel weighs 12.3 psf. The dome rests on vertical concrete walls and columns, as well as on inclined V-shaped struts. A tension ring at the base of the inclined columns resists the outward thrust due to the arches.

The 502-ft-diameter roof for the Northern Arizona University stadium, built in 1977, is the first large-scale wood dome in the United States. The dome consists of a three-way grid of intermeshing great circle ribs with steel tension rings at the base. The curved, glue-laminated ribs are connected to hexagonal welded steel hubs to form triangles. The main members vary from 19.33 to 61.75 ft in length and $8\frac{3}{4} \times 27$ in. to

12¼ × 27 in. in section. Each triangle contains simple span purlins spaced at 8 ft on center. The system is known as the *Varax dome* and was developed by Western Wood Structures of Portland, Oregon. The roof was designed for a dead load of 17.4 psf, a snow load of 40 psf, and a 37.5-psf wind load, as well as hanging loads at the center. The largest wood dome in the United States is currently the 533-ft-span stadium dome at Northern Michigan University in Marquette (1991).

The year of 1954, when Buckminster Fuller obtained his first patent for **geodesic domes**, is often considered to be the beginning of the revival and the popularity of skeletal domes for the next 20 years. In the early years, Fuller was concerned with the mathematics defining the geometry; stress analysis occurred much later when computers became available.

In appearance, geodesic domes are three-way grid structures where the primary radial ribs of the upper dome portion (see Fig. 8.34) lie on the great circles of a sphere. While in three-way grid domes only triangular surfaces along individual rings may be equal, in geodesic domes the sphere is subdivided into 20 identical, equilateral, spherical triangles, which are then further subdivided. The resulting member network is nearly regular; that is, the members are virtually identical in length. These conditions are ideal with respect to ease of fabrication and construction, since most members and connectors are interchangeable. Other geodesic surface patterns, such as the rhombic and hexagonal ones, can be derived from the basic triangular network by process of elimination, or other patterns can be generated by projection of other regular polyhedra onto the spherical surface. The geometry of geodesic domes is discussed in more detail later in this section (Fig. 8.34).

Buckminster Fuller's U.S. Pavilion at Expo '67, Montreal, Canada (see Fig. 8.34), is the most famous example of a geodesic dome structure. It consists of a three-quarter sphere with a 250-ft diameter and a height of 200 ft. The shell is a 3-ft 6-in.-thick *double-layer space frame* with the tubular members slotted at their ends so that they can be pin connected to cast-steel spider connectors, which have central hubs and 12 radial arms for the outer layer (six exterior members and six web members), or six arms for the inner layer. The outer layer is triangulated, while the geometry of the inner layer is hexagonal. The varying lengths of the 3½-in. tubes of the exterior skin do not exceed 10 ft, nor do the 2⅞-in.-diameter inner chord and web members exceed 6 ft. The wall thicknesses of the tubes vary, being greater in the exterior layer below the equator. The inner layer supports the 27-in.-high, 1/4-in.-thick Plexiglas domes, most of them measuring 10 × 12 ft, thereby causing the members to be also subject to bending besides their axial action.

The double-layer geodesic dome, Poliedro de Caracas, Caracas, Venezuela (1975, Fig. 8.27o), with a 477-ft span, is one of the largest aluminum domes in the world. Its two layers consist of a triangular exterior grid and a hexagonal interior one; structural engineers were Synergetics, Inc.

The 175-ft-diameter Climatron (1960, Fig. 8.27q) in St. Louis is another exciting example of employing Fuller's geodesic dome principle; it consists of two layers of hexagonal grids on top of each other, about 30 in. apart. The two layers of aluminum tubing are connected diagonally by aluminum rods to develop spatial truss action. A third inner skin layer consisting of a triangular framework, which contains Plexiglas panels, is suspended from the double-layer aluminum dome structure.

In stressed-skin geodesic domes, skin and framing act together as a composite unit. In the mid-1950s, this system was introduced by Don L. Richter, who studied under Buckminster Fuller in 1949 at the Institute of Design in Chicago. He employed the system for the first time in 1957 when he worked for Kaiser Aluminum. Several years later in 1964 he formed his own firm, *Temcor,* in Torrance, California, where he continued designing the Temcor aluminum stressed-skin geodesic domes. For example, at the Placer County Offices (1968, Fig. 8.27s) in Auburn, California, five geodesic domes with spans of 82 ft each have been clustered. The dome surface structure consists of stressed-skin, diamond-shaped aluminum panels. The panels are braced with tubular struts so that each panel with its strut forms a tetrahedron; the struts, in turn, form a hexagonal grid pattern that is derived from the basic triangular geodesic network, where six triangles produce a hexagon.

The two following cases represent examples of **formed surface domes.** The 400-ft-span folded plate concrete dome of the University of Illinois Assembly Hall (1962, Fig. 8.27h) sits on an inverted, domelike structure consisting of 48 radial buttresses, which, in turn, rest on a continuous ring footing. Here, the posttensioned ring girder along the periphery must resist the horizontal thrust of both dome structures. The pattern of the branching roof folds resists buckling and absorbs the potential bending stresses efficiently. The maximum depth of the folds is 7.5 ft, with a typical depth of about 3.5 ft; the ribs at the valleys and ridges are connected by a $3\frac{1}{2}$-in.-thick concrete web. The stadium was designed by Harrison & Abramovitz, and Ammann & Whitney structural engineers.

The King County Stadium, Seattle, Washington, designed by J. Christiansen, is currently the world's largest concrete dome, with a span of 660 ft (1976, Fig. 8.27g). The roof consists of 40 identical, wedge-shaped, thin-shell segments, which are doubly curved hyperbolic paraboloids of 5 in. thickness. The sections are stiffened by 2- × 6-ft ribs along their edges and are 55 ft wide at their base. The lateral thrust of the dome is carried by a 24-ft-wide and 12-in.-thick posttensioned concrete ring beam, which, in turn, is supported by channel-shaped columns.

Fabric domes are discussed in Chapter 9. Currently, the 722-ft-span Pontiac Stadium in Michigan (1975, Fig. 9.24e) is the largest span *air dome* in the world, whereas the 770-ft-span Georgia dome (1992, Figs. 9.21e and 8.26) is the world's largest *cable dome.*

Membrane Forces in Spherical Dome Shells

Since many domes are spherical and since the spherical membrane has the most simple geometry of the domes of revolution, it is investigated here to develop a general understanding for the behavior of dome shells under typical loading.

The spherical membrane is a special case of the group of axisymmetrical forms, its surface being defined by a constant radius of curvature.

$$R_1 = R_2 = R$$

The surface geometry of the sphere is defined by the following equation if its center is the origin of the coordinate system.

$$x^2 + y^2 + z^2 = R^2 \tag{8.13}$$

Under uniform radial and gravity load action, only two membrane forces occur (i.e., shear forces are zero) in the shell. They are the following:

- Meridional forces, N_ϕ, in arch action
- Circumferential or hoop forces, N_θ, in ring action

For the preliminary design of dome shells, only membrane stresses are considered as generated by uniform radial force action and uniform gravity loading, that is, dead load along the shell surface and live load on the horizontal projection of the dome; asymmetrical loading is ignored at this early design stage.

In the following discussion of the membrane forces, the positive sign for the forces denotes compression and the negative sign tension.

Uniform Force Action Normal to the Surface.   The *meridional forces,* according to Eq. (8.3), due to normal pressure are

$$N_\phi = pR/2 \tag{8.3}$$

Substituting this expression into Eq. (8.4) yields the hoop or circumferential forces

$$N_\theta = pR\left(1 - \frac{R}{2R}\right) = pR/2 \tag{8.14}$$

or

$$N_\phi = N_\theta = pR/2$$

Hence, the axial forces in a spherical membrane, as caused by uniform force action normal to the surface, are constant at any point on the surface (e.g., a balloon).

The response of the membrane to forces acting normal to its surface constitutes one specific loading condition, as encountered, for instance, by the internal pressure stabilizing pneumatic structures or shallow surfaces under gravity loading. Roof structures, however, have to carry other types of loads, such as snow and the weight of the surface itself. Wind forces are considered to be of secondary importance and are neglected in this introductory discussion.

For the derivation of the membrane forces in the following section, refer to Schueller, 1983.

Uniform Loading on the Horizontal Projection.   First, the force flow along the membrane as caused by distributed uniform loading q acting on the horizontal projection of the roof (as provided by codes and other sources for snow and live loads) is given.

Visualize a spherical membrane (Fig. 8.28a, c) cut horizontally in order to be investigated for the range $\phi \le 90°$. For this range (Fig. 8.28c), vertical equilibrium necessitates that the external load acting on the projected area must be balanced by the vertical components of the meridional membrane forces acting along the circumference. Therefore, the meridional forces N_ϕ are equal to

$$N_\phi = qR/2, \qquad (0° \le \phi \le 90°) \tag{8.15}$$

From Fig. 8.28e, one may conclude that, for a spherical segment smaller or equal to a hemisphere, the *meridional membrane forces* are constant under horizontally projected loading. However, as the spherical segment increases beyond the hemisphere, the meridional forces increase, approaching infinity as ϕ nears 180°.

The *hoop* or *circumferential forces,* N_θ, for this loading case can be derived from the membrane equation [Eq.(8.2)], by letting $p = q\cos^2\phi$ (see Fig. 8.28b).

$$p = \frac{N_\phi}{R_1} + \frac{N_\theta}{R_2}, \quad \text{but} \quad R_1 = R_2 = R \tag{8.16}$$

$$N_\theta = pR - N_\phi = (qR/2)\cos 2\phi, \quad (0° \le \phi \le 90°)$$

Notice that when

$$\phi = 0, \quad N_\theta = qR/2 \tag{8.17}$$
$$\phi = 45°, \quad N_\theta = 0$$

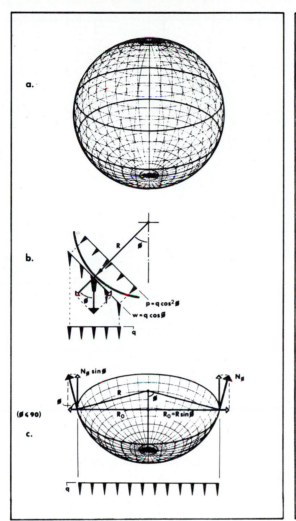

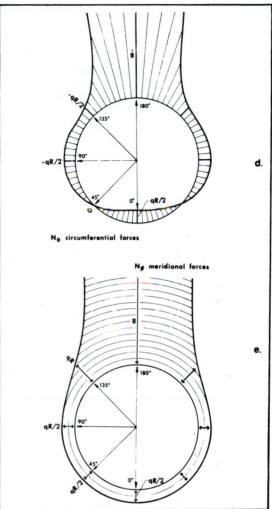

Figure 8.28   Membrane forces in a spherical dome shell due to live load *q*.

$$\phi = 90°, \quad N_\theta = -qR/2, \text{ for a hemisphere}$$

The distribution of the hoop forces is shown in Fig. 8.28d. It can be shown from these equations that the hoop forces are in compression when $\phi < 45°$, changing at that point into tension that approaches infinity as ϕ increases toward 180°.

   Uniform Loads along the Shell Surface.   The membrane forces caused by uniform loads *w* acting along the membrane surface, such as the weight of the membrane of constant thickness, are now presented. The derivation of the membrane forces is done similarly to the case of uniform loads on the horizontal roof projection.

   Vertical equilibrium (Fig. 8.29c) requires that the uniform load along the spherical surface be balanced by the vertical components of the meridional forces along the circumference. Therefore, the meridional forces N_ϕ are equal to

$$N_\phi = \frac{wR}{1 + \cos\phi}, \qquad (0° \le \phi \le 180°) \tag{8.18}$$

Notice that when

$$\phi = 0, \qquad N_\phi = wR/2 \tag{8.19}$$
$$\phi = 90, \qquad N_\phi = wR, \quad \text{for a hemisphere}$$

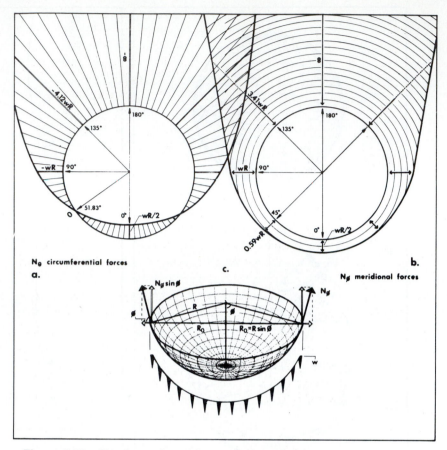

Figure 8.29   Membrane forces in a spherical dome shell due to self-weight *w*.

The distribution of the meridional forces is shown in Fig. 8.29b. With the increase of the spherical segment, the maximum meridional forces increase from $wR/2$ to wR for a hemisphere and approach infinity as ϕ nears $180°$.

The *hoop forces*, N_θ, are derived from the membrane equation. The derivation is similar to the one used to obtain Eq. (8.16). In this case, however, the load w is already projected onto the surface, and its component normal to the surface is $p = w \cos \phi$

$$N_\theta = pR - N_\phi = wR\cos\phi - N_\phi \tag{8.20}$$

$$= wR\left(\cos\phi - \frac{1}{1 + \cos\phi}\right), \quad (0° \le \phi \le 180°)$$

Notice that when
$$\begin{aligned}
\phi &= 0, & N_\theta &= wR/2 \\
\phi &= 51.83°, & N_\theta &= 0 \\
\phi &= 90°, & N_\theta &= -wR
\end{aligned} \tag{8.21}$$

The distribution of the hoop forces is shown in Fig. 8.29a. The hoop forces are in compression in the range $0° \le \phi < 51.83°$; after that point, they change into tension. Notice that the maximum forces increase rapidly as the spherical segment becomes larger than a hemisphere.

Structural Behavior of Dome Shells

In the previous discussion of membrane forces it was shown how a spherical thin shell responds to the gravity loads q and w. The meridian or arch forces N_ϕ along the lines of longitude are in compression, while the circumferential or hoop forces N_θ along the parallels of latitude are in compression for shallow domes, or compression and tension depending on the height-to-span ratio of the dome. They are in compression if the angle ϕ, which defines the shell boundaries, does not exceed a value between 45° and 51.83°. For the specific case where ϕ is equal to 51.83°, the dome edge is neither in compression nor tension. This plane of zero hoop stresses is called the *neutral plane*, similar to the neutral axis for beams.

The membrane behavior of a shell dome under uniform symmetrical loading can be visualized as the response of arches arranged in a continuous radial fashion but prevented from bending by the lateral support of the closed ring or hoop bands. As the arches tend to displace, the rings prevent movement by resisting the forces in an axial manner.

Although the membrane forces were developed for uniform symmetrical loading, the shell is still in the direct stress state under unsymmetrical uniform load action as caused, for example, by snow and wind, because of the capability of the surface to distribute forces spatially. For this case, the shear capacity of the rigid shell enters to resist the portion of the loads that the direct membrane forces cannot carry, similarly to a suspended membrane where the circumferential and meridional network is stiffened by diagonal members to resist shear.

The shell thickness for typical concrete domes is rarely selected based on force action; construction considerations are, in general, the determinant of thickness. Because of this, the membrane stresses are low, hence, additional loads such as those caused at the crown due to electrical and mechanical equipment, lanterns, or skylights can easily be supported. Keep in mind, however, that the surface loads must be continuous for the membrane theory to be valid. A concentrated load will cause a *kink* in the shell, in other words, local bending. By way of analogy it can be shown that it does not take much effort to pierce an eggshell with a needle, but considerable force is required to break it by squeezing it lengthwise between the palms of one's hands, where uniform load action causes the shell to respond in membrane action. The *collar load* at the crown of a dome, due to such conditions as mentioned before, is not considered a concentrated point load if it is uniformly supported along a circle concentric with the dome's axis.

Similarly, when openings interrupt the natural force flow along the shell, tension and compression members, which also may act as bending members, have to be provided around the opening to allow the transfer of hoop and meridional forces.

For the membrane theory to be truly applicable, domes should be continuously supported along their edges; in other words, they should not be incomplete domes. Where the dome is carried at discrete points, the flow pattern of the membrane is significantly influenced since the forces must collect at points of reaction. The *isocontour lines* in Fig. 8.5, top right, for continuous and concentrated reactive conditions clearly indicate the impact of the number of supports on the force flow pattern. The cases shown in Fig. 8.30 constitute segments of a surface of revolution with isolated (point) support conditions. For the *scalloped shell* (Fig. 8.30a), only the upper portion behaves according to the membrane theory, while in the lower portion the meridian forces must be carried by transverse transition arches. The concentration of forces at discrete points results in an increase of bending due to lateral thrust and the necessity of thickening the shell, as discussed later. In general, the membrane forces in the central crown portion of dome segments do not vary significantly from the ones for the full dome from which they were cut; however, they do become much larger as they approach the point supports and as the number of supports decreases.

The membrane theory constitutes a true representation of shell behavior if the continuous flow of membrane forces across the shell boundaries is not disturbed or interrupted. The support forces along the shell edge should be collinear and equal in magnitude to the arch forces, and the circumferential shell displacement should be

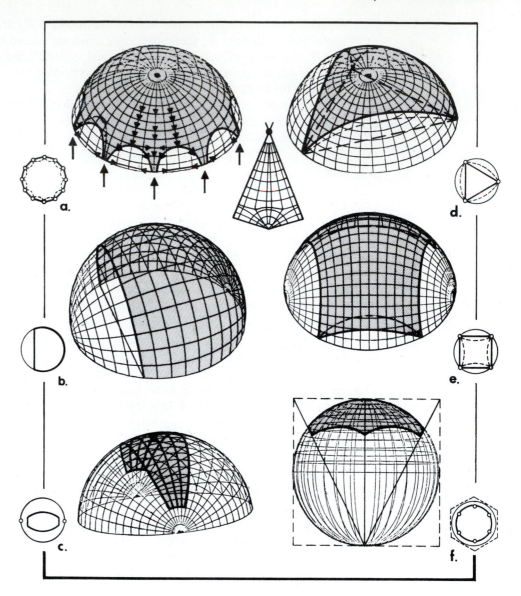

Figure 8.30 Dome shells on polygonal base.

compatible with the displacement of the support structure. This compatibility at the junction of the thin shell and rigid support structure is rarely achieved. For it to occur, the shell should sit on frictionless rollers so that it is not prevented from free movement. For the special case of a shallow dome with its edge located on the neutral plane, the shell may be pinned to the base, since loading will rarely affect a change in dome diameter. The geometry of a hemispherical dome (Fig. 8.31a) only requires its base structure to react vertically since there are no primary thrust forces at the edge of the shell; there are secondary ones, however. Under external force action, shrinkage, creep, and temperature change, the dome tends to displace laterally (Fig. 8.31a), but is not permitted to do so because of its fixed attachment to the boundary support structure. This incompatibility causes bending in the thin shell. The effect of continuity between shell and beam for preliminary design purposes, however, may be assumed to hardly influence the shell design, as the meridional moments along the edges are relatively small. A similar situation exists for shallow domes of cycloidal, elliptical, or spherical shapes that use transition curves, where the tangent to the shell curvature along the edges is vertical; also, here no tension ring is required since the arch forces

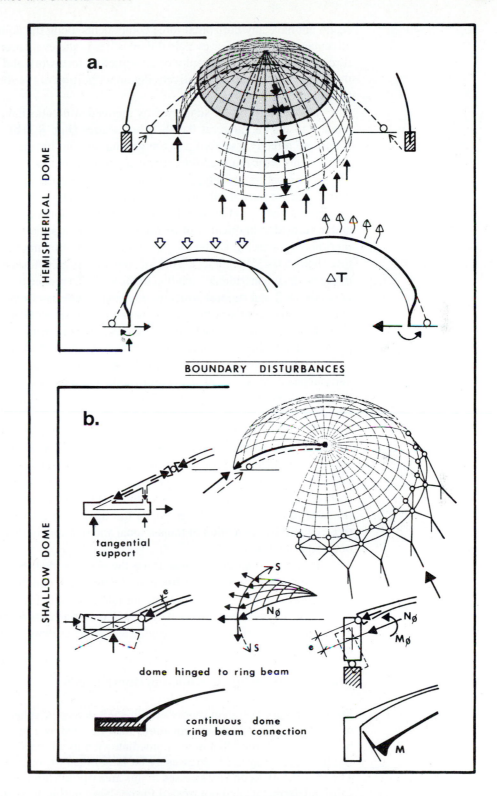

Figure 8.31 Junction of dome shell and support structure.

N_ϕ do not have any horizontal force components along the dome boundaries. The disadvantage of elliptical or cycloidal domes lies in the evaluation of the rather complex force distribution, the difficulty of constructing formwork and casting concrete along the steep portion, and the relatively flat upper portion of the shell, which is vulnerable to bending and buckling.

Domes can also be supported by inclined structures having a slope equal to the tangent to the curvature at the shell boundaries (Fig. 8.31b). For this case, the arch forces are directly transferred to the ground without having to be resolved into force components, as is required for vertical support structures.

The most common types of dome structures currently built are shallow spherical domes that are carried only by vertical support structures. They consist of large tension ring beams that resist the lateral thrust or horizontal force component of the arch forces and a vertical cylindrical wall or frame that supports the vertical components of the meridional forces N_ϕ. Under the thrust, the ring beam tends to expand, while the shallow shell ($\phi < 45°$) contracts. This incompatibility in deformation will cause twisting of the beam and meridional bending in the shell. To reduce the difference in movement between shell and support structure, the ring beam may be prestressed so that it is in compression rather than tension; but the main purpose of the prestressing is the prevention of any excessive expansion in the concrete ring beam. The proper placement of prestress bars reduces the tension due to meridional and vertical bending. The incompatibility between thin shell and rigid beam can be further reduced by integrating the vertical beam into the shell, using transition curvatures and smooth changes of material thickness along the meridian, thereby nearly eliminating the concentration of the lateral thrust force.

The shape of the ring beam and its location with respect to the shell determines the magnitude of the boundary disturbances (Fig. 8.31b). Visualize the shell to be hinged to the beam and hence free to rotate and slide. If the arch force N_ϕ does not pass through the centroid of the beam cross section, rotation is generated. Since the shell, in reality, is not free to move because of the stiff ring beam to which it is fixed, this rotation due to the eccentric placement of the ring beam causes moments in addition to those caused by prevention of translational and rotational movements, that is, the fixed boundary conditions.

In general, the disturbance along the shallow shell boundaries initiated by the incompatibility between the stiffness of the thin shell and the beam–wall are confined to the edge zone and do not penetrate far into the shell. The meridional displacements at the edge are quickly dampened by the circumferential shell rings, which tend to prevent expansion or contraction. The edge disturbances dissipate faster if the shell is thin in comparison to its overall dimensions.

Approximate Design of Concrete Dome Shells

For preliminary design purposes of thin-shell domes of spherical shape with continuous supports, the membrane equations may be used as a reasonable first approximation, keeping in mind that in the immediate vicinity of the supports the bending of the shallow shells may cause stresses by far larger than the ones due to membrane action. The analysis of secondary stresses in this area of transition from shell to beam is statically indeterminate and not treated further here; in this area, additional reinforcement must be provided and the shell must be thickened. The effect of openings, such as for a lantern at the dome crown, introduces boundary disturbances similar to the ones along the base periphery.

As already mentioned, the concrete shell thickness rarely depends on the magnitude of the membrane forces, but on the amount of concrete necessary for adequate coverage of two layers of reinforcement bars. Typical shell thicknesses range from 3 to 4½ in. for concrete domes having a span range of 100 to 200 ft, with an increase of shell thickness of approximately 50% to 70% near the periphery. For larger spans, however, the shell must be relatively thick, stiffened by concrete ribs, or designed as a double shell, in order to prevent buckling under compression. Nondevelopable, double-curvature surfaces are generally rather strong with respect to buckling. For hemispherical domes, the critical buckling load p_{cr} is nearly twice as high as for shallow domes. As a first approximation for shell thickness, in which buckling problems are greatly reduced, the following expression is often used (see Problem 8.3):

$$t \geq R/600 \qquad (8.22)$$

The loading condition (i.e., uniform gravity loading) for which the membrane forces were derived are a good approximation for preliminary shell design. In Fig. 8.32, the *stress trajectories,* due to wind loading perpendicular to the dome, express convincingly the compressive response of the inclined arches on the windward side and the tensile reaction of the cables on the leeward side. For preliminary design purposes, wind force action may be ignored.

EXAMPLE 8.4

A shallow concrete dome with a span of 200 ft is investigated. A concrete strength of 4000 psi and Grade 60 steel are used. The dome has the shape of a spherical segment, and for a typical height-to-span ratio for shallow domes of $h/L = 1/8$, a height of $h = 200/8 = 25$ ft is selected.

The radius of this dome, according to Eq. (9.64), is

$$R = (L^2 + 4h^2)/8h = (200^2 + 4(25)^2/8(25) = 212.5 \text{ ft}$$

The following geometrical relationship defines the location of the shell edge.

$$\sin\phi_o = \frac{L/2}{R} = 100/212.5 = 0.471, \quad \text{or} \quad \phi_o = 28.07° \leq 45°$$

Since the angle ϕ defining the shell edge is less than 45°, this shallow dome will be fully in compression according to the membrane theory [Eq.(8.17)]; there will be no tension in the shell! The horizontal components of the meridional (or arch) forces at the base must be resisted by a ring beam in tension. Also keep in mind, as previously discussed, that the incompatibility between the contraction of the

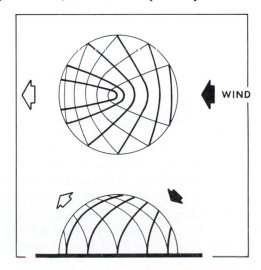

Figure 8.32 Dome under wind loading.

shell in compression and the expansion of the ring beam causes bending of the shell near the ring beam, which requires thickening of the shell and additional steel reinforcement.

The minimum shell thickness should not be less than about $R/600 = 212.5(12)/600 = 4.25$ in. so that the shell does not buckle. For this dome a thickness of $4\frac{1}{2}$ in. is selected. This thickness will be increased in the boundary area close to the ring beam.

The shell weight is

$$w = 150(4.5)/12 = 56.25 \text{ psf}$$

Adding the weight of roofing, interior finish, etc., results in a total area load of 60 psf. The assumed snow load of 30 psf is considered to act on the horizontal roof projection.

(a) The maximum compressive stresses occur along the meridians at the shell base (Figs. 8.28 and 8.29). They are larger than the circumferential stresses at the crown.

$$N_\phi = \frac{wR}{1 + \cos\phi} + \frac{qR}{2} \tag{8.18/8.15}$$

$$= \frac{60(212.5)}{1 + 0.882} + \frac{30(212.5)}{2} = 6775 + 3188 = 9963 \text{ lb/ft}$$

The critical compressive concrete stress is

$$f_c = N_\phi/A_c = 9963/4.5(12) = 185 \text{ psi} \le 0.45f_c' = 0.45(4000) = 1800 \text{ psi}$$

The membrane stresses are very small. It is apparent that the shell thickness is more than adequate from a strength point of view to support the design loads, although buckling must be checked in the final design.

Since there is no tension in the shell, minimum reinforcing must be provided along both principal directions.

$$A_{smin} = 0.0018A_g = 0.0018(12)4.5 = 0.0972 \text{ in.}^2/\text{ft} \tag{3.69}$$

The maximum bar spacing is $5t = 5(4.5) = 22.5$ in. ≤ 18 in. Try #3 bars at 13 in. both ways, $A_s = 0.102$ in.^{2}/ ft, or use welded wire fabric.

The horizontal thrust components of the arch forces $N_\phi(\cos\phi)$ in the shell are resisted by tension in the concrete ring beam.

$$H = N_\phi(\cos\phi_o) = 9963(\cos 28.07) = 8791 \text{ lb/ ft}$$

This horizontal thrust acts in a uniform radial fashion and thus causes the circular ring to respond in pure tension. It is shown in Eq. (9.57) that the constant tension in a circular cable under uniform radial pressure p is

$$T = pR = H(L/2) = 8.79(100) = 879 \text{ k}$$

The following reinforcement is required to resist this force:

$$A_s = T/F_t = 879/24 = 36.63 \text{ in.}^2$$

Try thirty #10 bars, $A_s = 38.1$ in.^{2}. When the ring beam is supported on columns, it must also be designed for beam action and possibly for torsion.

(b) The critical arch forces for this dome can also be quickly determined by simple statics without using the membrane equations.

The plan area of the dome is $A = \pi(L/2)^2 = \pi(100)^2 = 31,416$ ft^2. The surface area of the spherical segment (see Schueller, 1983), is

$$A = 2\pi R^2(1 - \cos\theta) = 2\pi Rh \qquad (8.23)$$
$$= 2\pi(212.5)25 = 33{,}379 \text{ ft}^2$$

Hence, the total gravity force is equal to

$$P = [60(33{,}379) + 30(31{,}416)]/1000 = 2003 + 942 = 2945 \text{ k}$$

or

$$V = P/2\pi R = 2945/\pi(200) = 4.69 \text{ k/ft}$$

The radial, horizontal thrust force is

$$H = V/\tan\phi_o = 4.69/\tan 28.07 = 8.79 \text{ k/ft}$$

The axial arch forces at the shell base are

$$N_\phi = V/\sin\phi_o = 4.69/0.471 = 9.97 \text{ k/ft}$$

For a steeper dome, the critical tensile hoop forces at the base can now be found from the membrane equation [Eq.(8.2)].

(c) Rather than using ordinary steel reinforcement to resist the side thrust of the shell, the edge beam may be designed as a prestressed element so that the compression induced by the tensioned cables will be at least as large as the tension induced by the external loading. Usually, the edge beam expands under gravity loading and temperature increase, while the diameter of the shell along the edges contracts. Posttensioning of the edge beam eliminates the expansion of the beam and overcomes some of the incompatibility between the beam and the shell. It also increases the bending capacity of the beam so that moments due to asymmetrical loading can be more effectively resisted.

It is assumed that the ring beam will expand about 25% more due to a temperature increase of 50° as compared to gravity loading (see Problem 8.7). Hence, the prestress force must resist the lateral thrust due to all the vertical loads increased by 25% for temperature effects. Therefore, the required prestress tendon area according to Eq. (3.74) is

$$A_s = P/0.56 f_{pu} = 1.25(879)/0.56(270) = 7.27 \text{ in.}^2 \qquad (3.74)$$

Try 34, 0.6-in. -diameter Grade 270 strands, $A_s = 7.34$ in.^{2} (Table A.5).

The concrete cross-sectional area required to resist the axial load due to posttensioning is based on the initial prestressing force and a relatively low compressive stress of concrete, say $f_c = 0.2 f'_c = 800$ psi. Hence, as a first estimate of the ring beam size (ignoring the effect of gravity bending) and taking into account that the shell weight counteracts the prestress force (i.e., it is assumed that the ring is prestressed in stages), that is, considering only the superimposed dead load and live load, we obtain [Eq.(3.75)] approximately

$$A_c = P_i/f_c \cong [879\,(34/90)/0.8]/0.80 = 519 \text{ in.}^2$$

Try a ring beam depth by width of 18×30 in. $A_c = 540$ in.^{2}

EXAMPLE 8.5

A hemispherical concrete dome is used to span the 200 ft of Example 8.4 with the same loads and materials. It should be kept in mind, however, that the shell thickness could be reduced to 3 in. because of the higher buckling capacity of the dome, $t_{min} = R/600 = 100(12)/600 = 2$ in.

In contrast to the shallow dome, for which a ring beam or abutments are needed to resist the lateral thrust forces, in a hemispherical dome no outward reaction forces occur; in other words, the sole reaction forces are vertical, tangential to the dome surface. In the hemispherical dome, the hoop forces below the neutral plane resist entirely by themselves the lateral displacement of the meridional or arch action, so no horizontal reactions are required under gravity loading, in contrast to semicircular arches.

The maximum compression along the meridian at the base, according to Eqs. (8.19) and (8.15), is

$$N_\phi = R(w + q/2) = 100(60 + 30/2)/1000 = 7.5 \text{ k/ft}$$

The corresponding compressive meridional stress is

$$f_c = N_\phi/A_c = 7.5/4.5(12) = 0.139 \text{ ksi} \le 0.45(4) = 1.8 \text{ ksi}$$

The compressive force action is no problem for the shell's allowable stress.

According to Eqs. (8.21) and (8.17), the maximum tension along the hoop at the base of the shell is equal in magnitude to the arch forces at this level (Figs. 8.28 and 8.29).

$$f_t = N_\theta/A_c = 139 \text{ psi} \le 1.6 \sqrt{f'_c} = 1.6 \sqrt{4000} = 101 \text{ psi}$$

Notice the maximum tensile stress is barely above the estimated allowable stress for concrete, indicating that the concrete has not even cracked, since the uniaxial tensile strength of approximately $6.5 \sqrt{f'_c}$ has not yet been reached. All the tension, however, must be carried by the reinforcement according to the ACI Code.

$$A_s = N_\theta/F_t = 7.5/24 = 0.313 \text{ in.}^2$$
$$A_{s\min} = 0.0035 A_g = 0.0035(4.5)12 = 0.189 \text{ in.}^2$$

Try #4 bars at 7.5 in. o.c., $A_s = 0.32$ in.^{2} at the base, which change to a maximum spacing of 12 in. at the neutral plane. The minimum steel reinforcement of #3 bars at 13 in. o.c., $A_s = 0.102$ in.^{2}/ ft (Example 8.4), is applicable above the neutral plane in both ways, as well as along the lines of longitude below the neutral plane.

Approximate Design of Skeletal Domes

The analysis of skeletal domes is complex and time consuming, but may be simplified by treating the skeletal dome as an equivalent or reticulated shell for preliminary design purposes. The main difference between these two dome systems lies in the fact that the thin shell responds primarily in membrane action, while the comparatively rigid members of the skeletal dome are also subject to bending. As the number of equal-sized members increases, the dome approaches the continuous material density of a shell surface. From a force flow point of view, forces applied uniformly to the surface can be distributed evenly within the surface of a continuous shell, but may follow only specific pathways along the linear members in the framework of a skeletal dome, even as the number of elements approaches infinity; in other words, the force flow depends on the configuration of the member arrangement. The material density of the framing should be uniform, in analogy to the constant thickness of a shell. In their geometry, triangulated grid and geodesic domes reflect a uniform member distribution, in contrast to radially framed domes, which have a higher member density at their centers than at their peripheries. This disparity of density in radially framed domes can be corrected by using smaller member sizes with less strength in order to obtain a more uniform strength distribution. The geodesic dome comes closest to the uniform force distribution of shell behavior and responds primarily in axial (membrane) action, rather than with combined axial and bending forces. It is assumed in the following investigation that the skeletal domes are of regular configuration, with all sectors identical, so that the membrane theory can be applied.

In the design of dome shells with continuous supports, it was found that one of the primary preliminary design determinants is the uniform gravity-loading case. For this condition, the membrane forces follow exactly the pattern of a Schwedler dome without diagonals, where the radially arranged arches are tied together by horizontal rings; here the arch forces N_ϕ are carried directly by the meridional rafters and the hoop forces N_θ by the ring purlins. The joints at the intersections of rings and arches must be rigidly connected to reduce distortion of the curvature. Under symmetrical gravity loading, this type of framed structure is very rigid, and the *shell analogy* as a preliminary design approach is reasonable if we assume an appropriate density of members. However, under asymmetrical loading, large deformations may appear, which can be reduced by diagonal bracing, such as trussing or tension tie-rod bracing and/or composite action with the roof skin. In this way the shear is not carried by bending of the ribs and rings, but by axial forces in the diagonals; hence, the framed surface may be considered stiff enough to justify the use of the shell analogy for preliminary design purposes. Furthermore, frame members should be selected so that the material density is uniform across the surface, as in a shell of constant thickness.

For the preliminary design of a ribbed dome, refer to Fig. 5.43.

EXAMPLE 8.6

A 40-ft-high Schwedler dome, with a span of 240 ft, is investigated (Fig. 8.33). It is composed of 36 radial ribs and 20 rows of concentric rings made of A36 steel. Diagonal bracing is added to carry the panel shear caused by asymmetrical loading. A dead load of 18 psf and a snow load of 20 psf are assumed. Because of the relatively flat profile of the dome, the entire roof is considered to be under wind suction; this loading condition will not influence the preliminary design of the members, only the design of the attachment of the skin to the framework.

The radius of the spherical cap, according to Eq. (9.64), is

$$R = (L^2 + 4h^2)/8h = (240^2 + 4(40)^2)/8(40) = 200 \text{ ft}$$

The angle ϕ, defining the dome base, is

$$\sin\phi = 120/200 = 0.6, \qquad \cos\phi = 0.8, \qquad \phi = 36.87° \le 45°$$

This shallow dome will be in full compression according to the membrane theory.

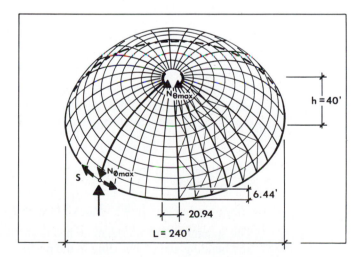

Figure 8.33 Schwedler dome (Example 8.6).

The spacing of the 36 arches at the base is

$$2\pi 120/36 = 20.94 \text{ ft}$$

The length of one-half radial arc [Eq.(9.59)] is

$$\frac{\pi R}{180}\phi = \frac{\pi 200}{180}36.87 = 128.70 \text{ ft}$$

The spacing of the 20 ring purlins is equal to

$$128.70/20 = 6.44 \text{ ft}$$

The spacing should be sufficient for the support of wood decking. The maximum compression forces along the meridians appear at the dome base and are found according to the membrane theory as follows:

$$N_\phi = \frac{wR}{1 + \cos\phi} + \frac{qR}{2} \qquad (8.18/8.15)$$

$$= \frac{18(200)}{1 + 0.8} + \frac{20(200)}{2} = 2000 + 2000 = 4000 \text{ lb/ft}$$

The rafters are spaced 20.94 ft apart at the base and must carry a maximum axial force of

$$N_{\phi r} = 4.00(20.94) = 83.76 \text{ k}$$

The deck acts in one-way action parallel to the arches; we may assume for this condition that the rafter does not have to resist any bending. However, for the case of two-way slab action or where the arch must also act as a joist (which spans between the ring purlins), the rafter must also be designed for local bending; it is conservative to consider it as simply supported between the points of intersection of the parallel ring members.

After several attempts, try, a W 10×22, $A = 6.49$ in.^{2} $r_y = 1.33$ in., as the preliminary maximum arch member size. The critical slenderness ratio for the section with the corresponding allowable axial stress is

$$\left(\frac{Kl}{r}\right)_y = \frac{1(6.44)12}{1.33} \cong 58, \text{ from Table A.12: } F_a = 17.62 \text{ ksi}$$

The actual stress is

$$f = N_{\phi r}/A = 83.76/6.49 = 12.91 \text{ ksi} < 17.62 \text{ ksi}$$

The trial section is stronger than necessary, but the next lower W10 section will not be sufficient.

Since $\phi = 36.87° \leq 45°$, the entire dome is in compression. The maximum hoop forces appear in the ring closest to the crown; the membrane forces at the crown are conservatively applied to the first ring.

$$N_\theta = (q + w)R/2 = (20 + 18)200/2 = 3800 \text{ lb/ft} \qquad (8.17/8.21)$$

For a typical ring spacing of 6.44 ft, the axial force that the purlin must resist is approximately

$$N_{\theta p} = 3.80(6.44) = 24.47 \text{ k}$$

The spacing of the meridian arches is so close at the crown that there is hardly any bending and no buckling in the purlin.

$$A = N_{\theta p}/0.6F_y = 24.47/0.6(36) = 1.13 \text{ in.}^2$$

Try $W6 \times 9$, $A = 2.68$ in.^{2}.

The axial compressive forces in the rings decrease as they approach the base; however, the effect of local bending increases because the length of the purlins increases. Again, we may conservatively consider the purlin as a simply supported beam spanning between the intersection of rings and arches, carrying the surface loads of the deck. Considering the snow to act together with the weight along the surface and using conservatively the span at the base and the total structure weight results in the following maximum moment: $M = 6.44(20 + 18)(20.94)^2/8(1000) = 13.41$ ft-k. This moment can be resisted by a $W8 \times 10$, ignoring the small axial force and possible biaxial bending and torsion, which is considered to be controlled by the roof skin. Comparing the ring member sizes at the crown and base, we notice that they do not vary greatly.

The horizontal thrust components of the arch forces at the base are resisted by a tension ring. The tensile force the ring must support, according to Eq. (9.57), is

$$T = pR = H(L/2) = (N_\phi \cos \phi)L/2 = 4.00(0.8)240/2 = 384 \text{ k}$$

The approximate size of the base ring, based on tension alone (ignoring any effects due to the boundary conditions and asymmetrical loading that results in lateral bending due to unsymmetrical radial forces, as well as vertical bending due to gravity loads), assuming yielding of the section, is

$$A = T/0.6F_y = 384/0.6(36) = 17.78 \text{ in.}^2 \qquad (3.82a)$$

Try $W 18 \times 65$, $A = 19.1$ in.^{2}.

Keep in mind that the tension ring may also be the beam portion of the vertical frame structure. In this case the vertical force components will cause torsion, which produces additional secondary bending on the W-section, but which may be ignored for closed, rectangular, beam cross sections. Hence, curved beams of rectangular shape may be treated as straight beams with respect to bending; the torsional stresses may be ignored for preliminary design purposes because of the slight beam curvature and the relative close spacing of the columns.

For the braced dome just discussed, the frame members are in line with the membrane forces, and a typical member must support the membrane forces in a shell width that is equivalent to the center-to-center distance of member spacing at the point of investigation. For the case of grid domes, the framing does not coincide with the membrane force flow of the shell. Here, the membrane forces must be resolved into components according to the surface geometry and relative member sizes.

Triangular grids can be generated in many different ways, forming various patterns such as the geodesic, grid, lamella, and lattice domes as Fig. 8.26 indicates. The three-way geodesic grid has the advantage that the variations in member lengths are rather small, resulting in a more or less uniform network of nearly equilateral triangles.

Three-way geodesic grids may be developed from the regular *Platonic solids* having triangular faces such as the tetrahedron, octahedron, icosahedron and their duals, or they may be generated from quasiregular polyhedra.

Buckminster Fuller, the famous inventor of geodesic domes, considered the triangulated icosahedron as the most efficient fundamental volume-controlling device of nature because it provides the most volume with the least surface; although not being the strongest (which the tetrahedron is), it is a stable structure. Visualize this icosahedron projected onto a spherical surface (Fig. 8.34a), thereby subdividing the surface into 20 equilateral spherical triangles, which constitute the maximum number of equal subdivisions possible. The extension of the icosa edges (Fig. 8.34b) form the 15 great circles, which are defined by a plane passing through the centroid of the sphere and cut-

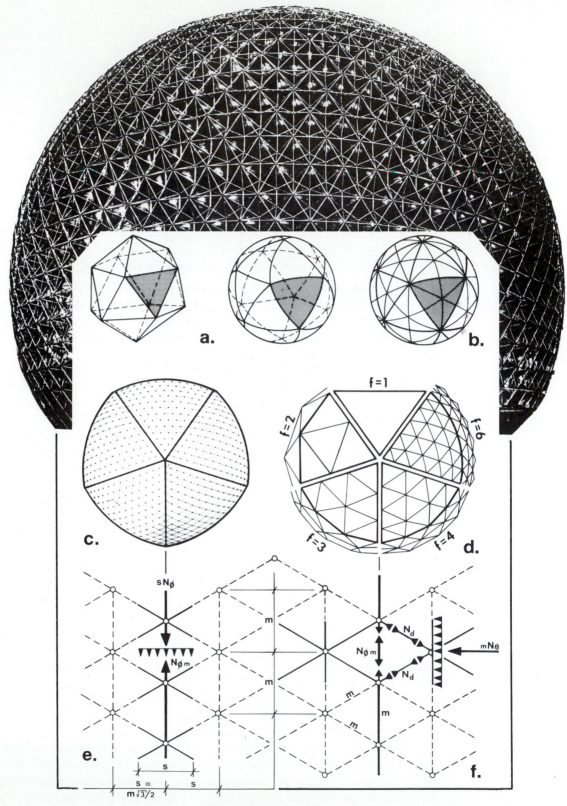

Figure 8.34 Geodesic dome.

ting the sphere exactly in half. Buckminster Fuller showed that these 15 great circles are regularly arranged and subdivide each of the 20 equilateral triangles of the icosahedron into 6 equal right-angled triangles, thus producing 120 identical right-angled triangles. This is the maximum number of identical triangles into which a spherical surface can be subdivided. Although this discovery is important, it does not solve the problem of equal member length. This can be approximately achieved by reducing the basic polyhedral face, in this case the 20 triangular icosahedral faces, into a larger number of components by further subdividing equally each of the equilateral triangles along their edges. The number of segments into which each of the principal sides is segmented is called the *frequency f,* as exemplified in Fig. 8.34d. The triangles formed by this subdivision are nearly equilateral. As the frequency or the number of components increases, the dome changes from a polyhedron, an icosahedron in this case, into a true spherical surface.

Approximate Design of Other Dome Shapes

Domes on polygonal bases (Fig. 8.30) may form segmental surfaces with straight or curvilinear edges. To cover a rectangular plan, for instance, either a surface of revolution (e.g., paraboloid of revolution or spherical dome) is cut by vertical or inclined planes, or a translational surface (e.g., elliptic paraboloid or section of a torus) is used.

    Elliptic paraboloids, also called *elpars,* can be generated in the same fashion as hyperbolic paraboloids by letting a convex, principal parabola slide parallel to itself along another convex parabola perpendicular to it. The double curvature surface is synclastic and not ruled; it does not have any straight-line generators. Vertical sections yield parabolas, while horizontal sections give ellipses (Fig. 8.35). The surface of the elliptic paraboloid on a rectangular plan is defined by Eq. (8.26).

    The natural distribution of the internal forces is very much disturbed when sections of the shells of revolution are cut off to fit other plan forms. The force flow must

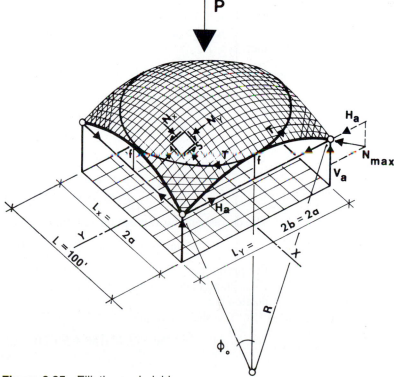

Figure 8.35 Elliptic paraboloid.

be redirected to the supports at the corners, resulting in large stress concentrations, which is particularly true for a shell segment forming a spherical triangle. We can visualize the crowns of these segmental surfaces to act as shells of revolution, which, in turn, are supported by arches that guide the loads to the supports. For example, a *square plan* dome can be treated as a pendentive dome, in other words, as a dome of revolution supported by four pendentives (e.g., spherical triangles) that form arches along the sides.

Usually, the crowns of segmental shells represent shallow domes where the difference in geometry between the various types of curves is insignificant, especially since the membrane stresses are negligible. Furthermore, the loads may be assumed to act normal to the shallow surface. We may conclude that for a square plan dome the paraboloid of revolution, the spherical segment, and the translational surface (all with the same arch rises at edges and crown) can be treated in the same fashion. The surface equation for the parabolic dome (i.e., paraboloid of revolution) requires the least calculations. It has the following form, with the origin of the coordinate system at the dome vertex, and can be derived from Eq. (8.26) for $R_x = R_y = R$ as

$$z = (x^2 + y^2)/2R \tag{8.24}$$

where R is the radius of curvature at the shell crown and, according to Eq. (9.17), is equal to

$$R = L^2/8f \tag{9.17}$$

EXAMPLE 8.7

Do a preliminary investigation of the force flow in a 100-ft square plan dome with a crown height of 20 ft and 10-ft-high arches along the sides. The 3-in.-thick concrete shell of 4000-psi concrete is supported at its four corners and must carry a roofing load of 5 psf and a live load of 20 psf (Fig. 8.35).

The vertical radius of the shallow side arches, assuming a spherical surface, is

$$R = (L^2 + 4h^2)/8h = (100^2 + 4(10)^2)/8(10) = 130 \text{ ft} \tag{9.64}$$

The corresponding half-central angle is

$$\sin \phi_o = \frac{L/2}{R} = \frac{50}{130} = 0.385, \quad \text{or} \quad \phi_o = 22.62°$$

Or, treating the surface as a parabolic dome, the radius of curvature at the crown is

$$R = L^2/8f = 100^2/8(10) = 125 \text{ft} \tag{9.17}$$

The corresponding slope at the reactions, according to Eq. (9.15), is

$$\tan \phi_o = 4f/L = 4(10)/100 = 0.4 \quad \text{or} \quad \phi_o = 21.8°$$

For preliminary design purposes, the results for the two types of geometries are close enough, considering how insignificant the membrane stresses in the crown portion are.

The shell loads are

$$w = (3/12)150 + 5 + 20 = 43 + 20 = 63 \text{ psf}$$

Since the crown is so shallow ($h/L = 10/100 = 1/10$), all the loads are assumed to act on the horizontal roof projection. Hence, the total weight of the crown acting at its supporting base (the circumferential tension ring) is

$$P = \pi(100/2)^2(43 + 20)/1000 = 337.72 + 157.08 = 494.80 \text{ k}$$

The vertical ring reactions are

$$V = 494.80/2\pi50 = 1.58 \text{ k/ft}$$

The radial horizontal thrust forces restrained by the tension ring are

$$H = V/\tan \phi_o = 1.58/\tan 22.62 = 3.79 \text{ k/ft}$$

The axial membrane arch forces at the base are

$$N_\phi = V/\sin \phi_o = 1.58/0.385 = 4.10 \text{ k/ft}$$

The membrane forces for shallow surfaces, however, could easily have been determined immediately by assuming the loads to act normal to the surface. For a spherical surface, according to Eq.(8.3), letting $p = w$,

$$N_\phi = (w/2)R = (63/2)130/1000 = 4.01 \text{ k/ft}$$

Or, by treating the curves like parabolas [see Eq.(9.17)],

$$N_\phi = (w/2)R = (w/2)L^2/8f = (63/2)100^2/8(10)/1000 = 3.94 \text{ k/ft}$$

The tensile force in the circular ring at the crown base is

$$T = pR = H(L/2) = 3.79(100/2) = 189.5 \text{ k}$$

This force is distributed across a wide band within the shell.

The total load for the entire square plan shallow dome is roughly estimated (assuming all the loads to act on the horizontal roof projection) as

$$P_{tot} = P_D + P_L = (43 + 20)100^2/1000 = 430 + 200 = 630 \text{ k}$$

The equivalent uniform loads along the perimeter, that is, the loads the side arches must carry, are approximately

$$w = w_D + w_L = (430 + 200)/4(100) = 1.08 + 0.5 = 1.58 \text{ k/ft}$$

The lateral thrust forces at the base of the arches, according to Eq. (5.84), are

$$H_a \cong wL^2/8h = 1.58(100)^2/8(10) = 197.5 \text{ k}$$

The vertical arch reactions are

$$V_a = wL/2 = 1.58(100/2) = 79 \text{ k}$$

The resultant reactions are assumed equal to the maximum compressive forces in the shell.

$$N_{max} = \sqrt{197.5^2 + 79^2} = 212.71 \text{ k}$$

The shallow side arches must also be designed for moments according to Eq. (5.80), as derived for one-half live loading.

$$M_{max} = w_L L^2/64 = 0.5(100)^2/64 = 78.13 \text{ ft-k}$$

Each foundation must carry the following vertical load:

$$P = P_{tot}/4 = V_a(2) = 630/4 = 79(2) = 158 \text{ k}$$

The tie rods (or abutments) must resist $H_a = 197.5 \text{ k}$

Refer also to the problems for the preliminary design of a parabolic dome (8.8), an elliptical dome (8.9), a conical dome (8.10), and a conoidal dome (8.12).

8.4 THE HYPERBOLIC PARABOLOID

The hyperbolic paraboloid (or hypar) is an invention of this century. It has opened a new era for the building of continuous, adaptive surface forms; it has added a new dimension to the traditional rotational dome shapes by introducing the endless formal potential of translational surface generation and the relatively easy building of double-curvature anticlastic surfaces using straight-line elements. Although in 1669 Christopher Wren discussed the hyperboloid from a mathematical point of view, and Antonio Gaudi, around the turn of this century, used the hyperbolic paraboloid as part of his complex, warped, thin-tile surface vaulting, Bernard Lafaille in 1933 was the first to actually build the shape as a reinforced concrete shell. In 1935 the French engineer F. Aimond developed the membrane theory for this new surface geometry, and Giorgio Baroni applied the hypar principle to the roof of the Alfa Romeo factory in Milan, Italy, in 1934. Finally, in 1935, Eduardo Torroja built the famous Zarzuela Hippodrome Grandstand near Madrid, Spain. The shallow cantilevering hyperboloidal sections are saddle-shaped surfaces that approximate a hypar. The real development, however, was initiated after World War II by Felix Candela in Mexico, who fully explored the visual richness and potential of the hyperbolic paraboloid. Through Candela, the hypar became part of the architectural language, as demonstrated by his numerous buildings, such as the magnificent church of La Virgen Milagrosa in Mexico City of 1955 (Fig. 8.38H). The church is composed entirely of deformed, inclined, umbrella-shaped units. Through this combination of the straight-edged hypar kits, none more than 1½-in. thick, the architect Felix Candela achieved a powerful interior space of impressive quality. Candela's restaurant at Xochimilco of 1958 is a 98-ft-span octagonal groined shell that consists of the intersection of four segments with anticlastic curvature. Since the forces accumulate along the groin lines to be brought to the eight supports, the shell edges can be thin and articulate lightness of form (similar to Fig. 8.42).

Many other architects experimented with hypar geometry. Le Corbusier's Philips Pavilion (Fig. 8.38F) for the Brussels World's Fair, 1958, is much more of a free-form, organic structure, rather similar to the computer drawing shown adjacent to it (Fig. 8.38E). The individual hypar units have straight edges above ground, whereas the edges along the ground are curvilinear.

Kenzo Tange, together with the engineer Y. Tsuboi, built the famous St. Mary's Basilica in Tokyo (1964, Fig. 8.38A), which is composed of eight hypar surfaces of three different shapes that sit directly on the ground and form a cruciform. The lateral thrust developed in the valleys along the intersection of the shells (i.e., A-frame action) is resisted at the base by tie beams that span diagonally across the building.

Experiments with hypar geometry were done not only in reinforced concrete; Lev Zetlin designed folded hypar shells made of light-gage steel decking in stressed-skin construction for the Los Angeles and San Francisco American Airlines hangars (1971, Fig. 5.56f), which reached unbelievable 230-ft cantilever spans on each side of the central core.

Hypar Types

The primary geometrical characteristics of the hypar are identified in this section. It is shown that the general shape resembles a horse saddle or mountain pass, where the hyperbolas in plan view look like a contour map. Keep in mind that in this section only special conditions are considered, such as the vertical position of the z axis, which

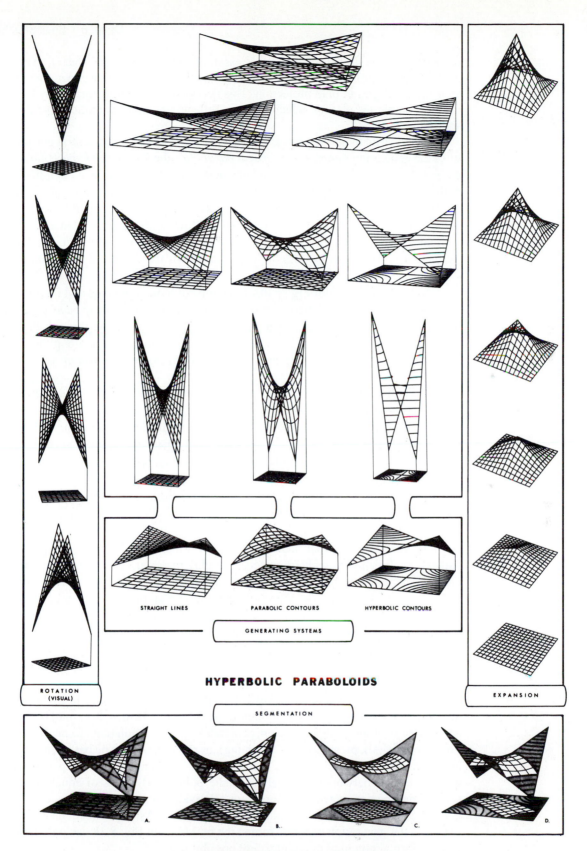

ROTATION
(VISUAL)

STRAIGHT LINES   PARABOLIC CONTOURS   HYPERBOLIC CONTOURS

GENERATING SYSTEMS

EXPANSION

HYPERBOLIC PARABOLOIDS

SEGMENTATION

Figure 8.36   Hyperbolic paraboliod.

obviously can be tilted, or the hypar is assumed to be rectangular, that is, to be made up of identical, principal parabolic generators with the straight-line generators forming an orthogonal grid in plan projection. This network may be transformed by flattening into a rhombic tesselation, which, in turn, reflects unequal parabolic generators or a skewed hypar.

Some basic hypar features are shown in Fig. 8.36. Here, the central portion identifies the generating systems: the *vertical parabolas*, the *horizontal hyperbolas,* and the *diagonal straight lines.* An important characteristic of the hypar is its **height-to-span ratio**, which in the drawing is shown as a variable for a given rectangular plan. The progression of a typical unit from its flat to its steep stage is clearly presented. Similar is the transition from a horizontal slab to a shallow, umbrella-type element to a steep domelike shape, as shown in the right portion of the drawing. In the left part, the basic hypar unit is rotated to express the visual power of the form and the relation of the straight lines to the curve at different positions.

In the bottom part of Fig. 8.36A to D, typical hypar elements and their locations within the saddle are identified. It is apparent that these elements can have any geometrical configuration and can be cut from any location. A section through the saddle always generates either a parabola, a hyperbola, or a straight line along the cut line. Some common solutions, using only linear sections, are the following:

- To cut the hypar diagonally along the straight-line generators to form straight edges (case A)
- To cut the hypar diagonally but skewed with respect to the straight-line generators, resulting in parabolic boundaries of concave or convex shape (case A)
- To cut the hypar vertically along the parabolas parallel to the principal curvatures to form parabolic boundaries (case C)
- To cut the hypar horizontally to form hyperbolic boundaries (partially case D)
- The curvilinear section in case B results in hyperbolic boundaries.

The shape of a hypar element depends on many architectural design determinants. It is a function of scale (span versus height), material, construction (in situ versus precast), expertise or function, and geometry of building plan, to name a few of the technical considerations. The shape of the building plan may be regular or irregular or it may be curvilinear (e.g., round, elliptic), lending itself to the use of a single hypar unit. For a *polygonal plan* (e.g., triangular, pentagon, rectangular, diamond, starlike), an assemblage of units may be considered, although this obviously is not necessary because a single hypar surface with linear or curvilinear edges can cover a rectangular space, for instance as shown in Fig. 8.37E. From a purely geometrical point of view, there exists an infinite number of ways of combining hypar elements to enclose a given building space. The process of combining surface kits, for instance, may be based on addition (e.g., growth, superposition, penetration), subtraction (e.g., truncation, dissection, subdivision), pressure (e.g., flattening), or rotation; it may be based on patterns of arrangement such as linear (e.g., radial, orbital, nuclear, branching), planar (regular and nonregular networks), or spatial (saddle polyhedra). Units of equal or different shapes may be connected in a smooth or relatively continuous fashion, or they may give the appearance of a folded-plate type of structure. As can be seen, there is no limitation to the generation of hypar forms; the buildings in Fig. 8.38 can only hint at the endless formal potential of architectural possibilities.

Different solutions for assembling warped rectangular surfaces to cover a square building enclosure are shown in Fig. 8.37. While cases B, D, and E. employ *single saddle surfaces* with straight or curvilinear edges, the other solutions consist of a combi-

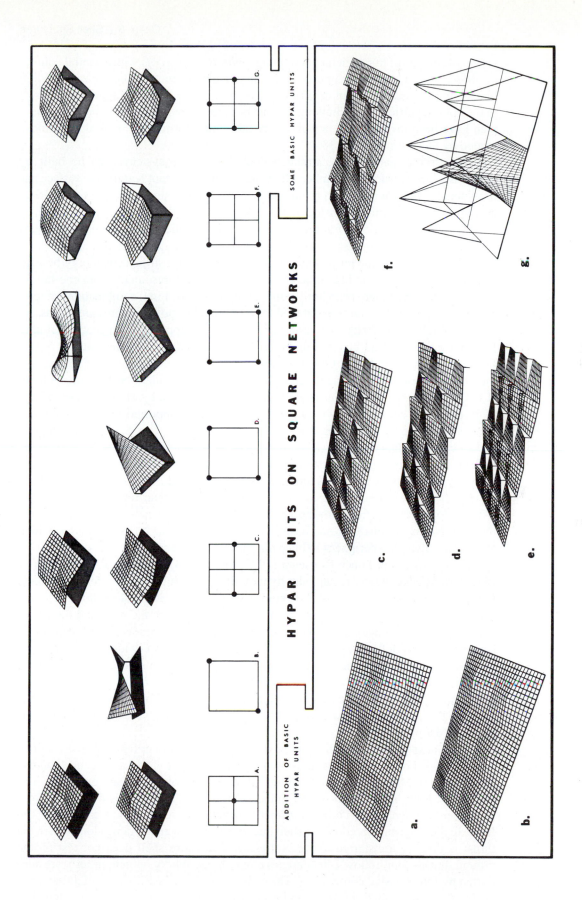

HYPAR UNITS ON SQUARE NETWORKS

ADDITION OF BASIC HYPAR UNITS

SOME BASIC HYPAR UNITS

Figure 8.37 Hypar units on square grids.

669

nation of four equal hypar elements forming *umbrella-type roofs* with a central column or *gabled roof types* with supports along the outside perimeter. These basic building blocks, in turn, can be added in various ways to form the multibay aggregates shown in the bottom portion of the drawing. It must be emphasized that only hypars on square plan have been investigated; however, any other plan forms can be used, as demonstrated by the cases in Fig. 8.38.

Hyperbolic paraboloids, although double-curvature systems, can be built by using only linear structural members. Because of this important feature, the surface has been constructed not only with reinforced concrete and ceramics (where formwork is an important economical consideration), but also in steel, aluminum, and wood by employing multidirectional framing systems and, for smaller scale, two or more layers of decking placed in opposite directions to act as two-way span systems.

Some of the abundant potential forms for anticlastic surfaces can only be suggested by the examples in Fig. 8.38. The scale of the structures ranges from one-family houses to stadiums; there is no limit to the wealth of forms. First, combinations of rectangular warped surface units assembled to relatively flat roof structures are discussed.

Umbrella-type shells consist of four hypar units supported by a central column. Examples are the tilted inverted umbrellas (60×116 ft) of the grandstand at Sciota Down Raceway, Columbus, Ohio (J) with a typical concrete thickness of $4\frac{1}{2}$ in., and the 3-in. concrete shells for the Berenplaat Water Treatment Plant, Rotterdam, Holland (L). The roof of the Saier House near Deauville, France (C), by Marcel Breuer after a 1958 proposal, consists of two square hypar surfaces supported on three columns, while the Centre Le Corbusier, Zürich, Switzerland (1967, Fig. 8.38B) has a hipped and an inverted hipped roof, each composed of four square steel panels. The roof of the athletic facility for the Pratt Institute, Brooklyn, New York (1975, Fig. 8.38 O), is a multibay aggregate of 12 hipped roof units designed by D. F. Tully. The principal bays are 130-ft square and are covered by 16 wooden square hypar units (32.5×32.5 ft), with the four central panels forming a pyramid. The warped panels are supported along the fold lines by laminated wood beams. The entire roof rests on 16 buttresses, which are located along the periphery in line with the A-frames and at the four corners.

The roof of the Ponce Coliseum (1971, Fig. 8.38D) in Puerto Rico consists of four straight-edge, nearly rectangular hypar units and is supported by four abutments located at the center of each of the exterior sides. The typically 4-in.-thick concrete shell units rest on the interior gable beams that span between the piers and the 138-ft inclined cantilever edge beams. These enormous cantilever beams had to be posttensioned to control deflection and stresses; the depth of the 30-in.-wide edge beams varies from 18 in. at the corner to 94 in. at the pier. The shell membrane is prestressed along the straight-line generators. The lateral thrust on the abutments, due to asymmetrical loading as transferred by the interior beams, is balanced by prestressed tie beams between the abutments below ground. T. Y. Lin International were the structural engineers for the project.

The next group of buildings to be discussed uses thin vertical shells as inclined walls arranged in such a way as to form pyramidal profiles. The La Virgen Milagrosa Church (H) in Mexico City and St. Mary's Basilica in Tokyo (A), are examples. The roof of St. Mary's Cathedral in San Francisco (1971, Fig. 8.38G) is developed from eight identical hypar shells joined together to form a square base and a cross at the top level. In the diagonal direction, the shell intersections form an A-frame along which the thrust is directly transferred to the buttresses. The entire shell does not rest on the ground, but on four hollow 140-ft arches along the periphery, which, in turn, are supported by four massive corner piers. The concrete shell surface consists of a triangular grid with a $5\frac{1}{2}$-in. shell thickness and 8-in.-wide ribs projecting $10\frac{1}{2}$-in. Design consultants for the project were Pietro Belluschi and Pier Luigi Nervi.

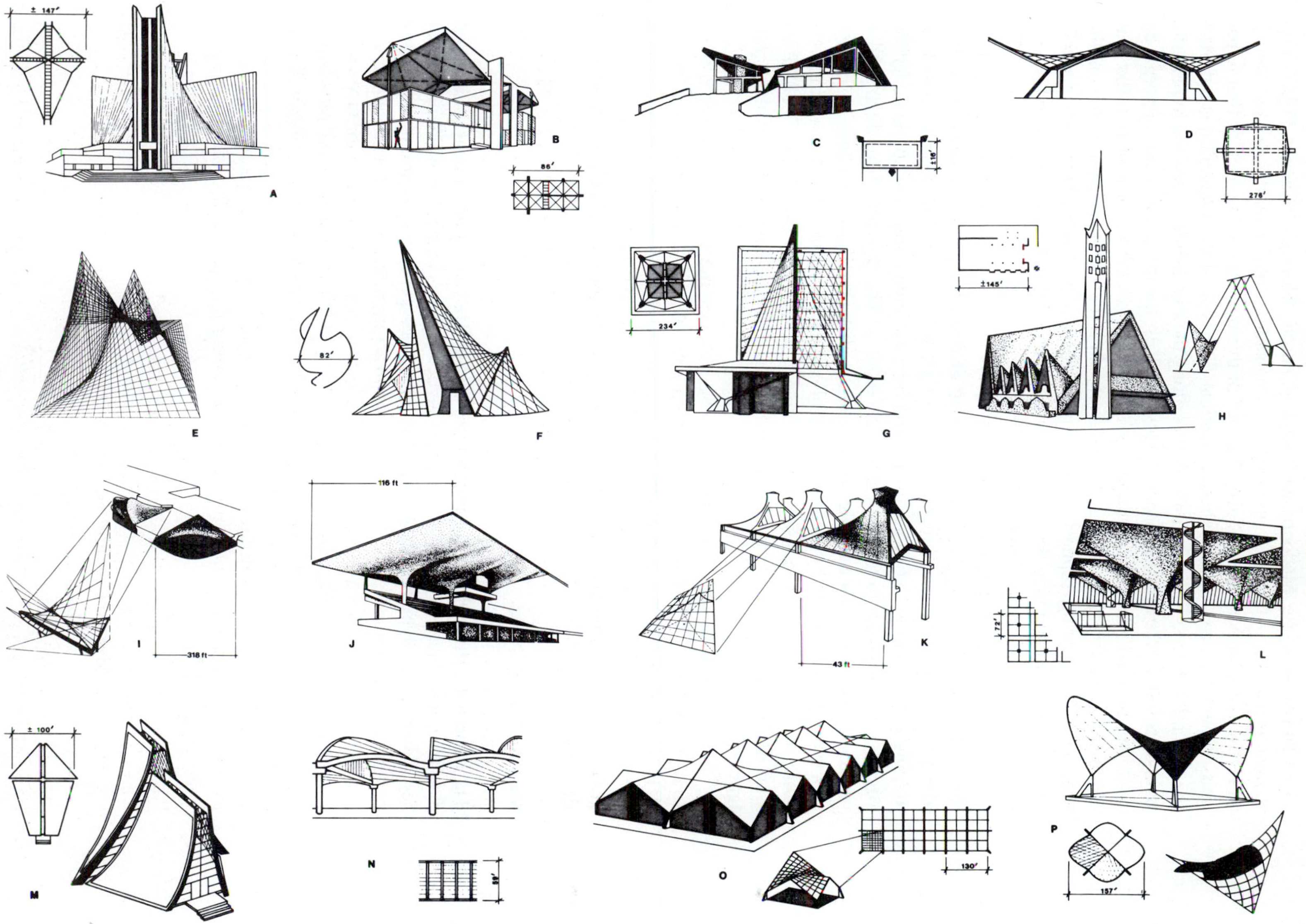

Figure 8.38 Case study of hypar roofs.

671

The roof of the Madonna di Pompei Church, Montreal, Canada (1967, Fig. 8.38P), is part of the larger group of intersecting saddle shapes similar to cross vaults, as is discussed further for Fig. 8.42. The roof of this church is composed of four hypar units, where each pair of opposites is identical. The typical shell quadrant is cut from the general saddle to form parabolic curves along the two inside edges and hyperbolic boundaries along the outside. Along the inside edges, the hypar units are supported by diagonal, nearly three hinge, parabolic groin arches, which, in turn, rest on the four abutments that carry the building. The concrete shell thickness varies from 6 in. at the crown to 10 in. near the supports. The primary structural behavior of the shell quadrant under gravity loading may be visualized as consisting of arch action between the supports in one direction and cantilever action in the other direction.

The huge hypar concrete shells of the TWA hangars of the Kansas City International Airport (1974, Fig. 8.38I) are made up of two principal shell surfaces, which, along their line of intersection, the central ridge, form a stiff 318-ft-span arch. This arching shell is supported by abutments and by additional column supports under the edge beams in the rear to resist asymmetrical loading. The typically 3-in.-thick shell is posttensioned to eliminate tensile stresses and is stiffened along the edges by hollow triangular beams defined by their own warped surface curvatures. Each of the intricate domical roof units for a foundry at Lohr-am-Main, Germany (1960, Fig. 8.38K), by Curt Siegel, is composed of two hypar surfaces at opposite sides and a circular ventilating shaft.

Although *conoidal shells* are anticlastic surfaces and can be constructed from straight lines (Fig. 8.4), they have rarely been built. A typical application is as north-light shells (Fig. 8.38N), similar to the half-cylindrical shells in Fig. 8.23. The thin, vertical, conoidal shell walls on the diamond-shaped plan of I. M. Pei's Memorial Chapel at Formosa's Tunghai University (1964, Fig. 8.38M) express the sculptural quality of the building.

Membrane Analysis

The hyperbolic paraboloid, also called *hypar* or *h-p* (HP), is a translational surface, not a surface of revolution. Cutting the surface vertically gives parabolas, while horizontal sections result in hyperbolas (see Fig. 8.36). It can be generated by sliding a concave parabola or suspended cable (called a *generatrix)* parallel to itself along a convex parabola or arch (called a *directrix),* which is located perpendicular to it, or vice versa. The parabolas follow the principal curvature lines and are called *principal parabolas.* They establish a rectangular coordinate system at the saddle point, as shown in Fig. 8.39f.

The parabolic generatrices can be defined at the origin of the coordinate system, as shown for parabolic cables in Section 9.2, letting R_0 equal to $R_{x'}$ and $R_{y'}$, respectively [Eq. (9.18)].

$$y' = 0: \quad z = \frac{(x')^2}{2R_{x'}} \tag{d}$$

$$x' = 0: \quad z = \frac{-(y')^2}{2R_{y'}} \tag{e}$$

The equation defining the surface of the hyperbolic paraboloid is

$$z = \frac{(x')^2}{2R_{x'}} - \frac{(y')^2}{2R_{y'}} \tag{8.25}$$

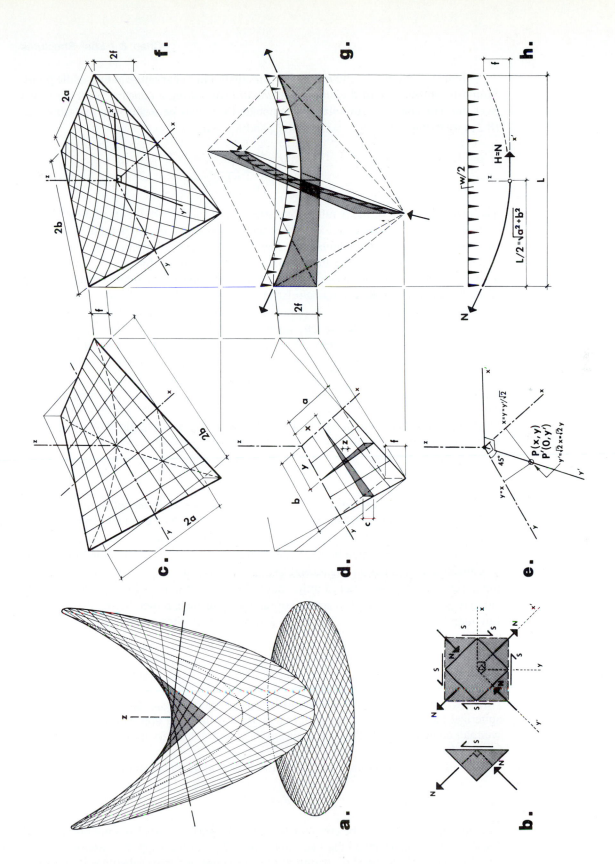

Figure 8.39 Membrane forces in a basic hypar unit.

In this equation, $R_{x'}$ and $R_{y'}$ are the principal radii of curvature at the saddle point. Similarly, the surface of the *elliptic paraboloid* (by letting a concave parabola move over another concave parabola perpendicular to it) can be easily derived from Eq. (8.25) by letting: $-y' = y$, $x' = x$, $R_{x'} = R_x$, and $R_{y'} = R_y$.

$$z = \frac{x^2}{2R_x} + \frac{y^2}{2R_y} \tag{8.26}$$

The plan projection of the parabolic generatrices of the hyperbolic paraboloid form a rectangular grid. For the condition where the convex and concave parabolas are identical or $R_{x'} = R_{y'} = R$, the generatrices form a *square grid* in plan projection and the hyperbolas are equilateral. For this special case, the membrane forces as caused by load action normal to the surface can be derived from the general membrane equation, Eq. (8.1), by substituting the following relationships:

$$R_x = -R_y = R, \;\; N_x = -N_y = N$$

$$p = \frac{N_y}{R_y} + \frac{N_x}{R_x} = \frac{2N}{R} \tag{f}$$

The membrane forces N are equal in magnitude but opposite in direction; they are in tension along the suspended parabolas and in compression along the parabolic arches. Hence, according to Eq. (f), the membrane forces are equal to

$$\pm N = pR/2 \tag{g}$$

Visualize a *rectangular* hypar element of size $2a \times 2b$ to be cut from the saddle surface of the membrane as shown in Fig. 8.39a. This surface element can be considered shallow if the rise f is less than the ratio of the shorter side to 5 (**$f \leq a/5$ if $a \leq b$**). For these very shallow membranes the external uniform loads w and q do not have to be considered separate; both dead and live loads can be treated as one load w to act on the horizontal roof projection. Furthermore, the uniform gravity loads w may be considered as equal to the loads p acting normal to the surface, remembering that for shallow membranes the radius of curvature R changes very little. Hence, Eq. (g) may be expressed as

$$\pm N = wR/2 \tag{8.27}$$

Note the similarity of this equation to Eq. (8.15) defining the axial force flow in a spherical membrane with a constant radius of curvature. For approximation purposes, we may consider the shallow parabolas to have a circular curvature since their radius hardly changes.

The radius of curvature for the parabola at its apex has been derived as

$$R_0 = L^2/8f \tag{9.17}$$

Considering this radius to be constant along the flat surface and also remembering that for shallow cables (arches) the maximum force T_{max} at the support is approximately equal to the cable force H occurring at the low point, we may substitute R_0 into Eq. (8.27).

$$\pm N = wR_0/2 = wL^2/16f \tag{8.28}$$

This expression can also be derived using a different approach, based on Fig. 8.39 *f, g,* and *h,* by considering the load to be shared equally by the arched membrane in compression and the suspended membrane in tension. Keep in mind that the parabola is the funicular shape of uniform loading on the horizontal roof projection only; the effect of dead load located along the surface geometry can only be neglected for flat surfaces!

As based on Fig. 8.39h, rotational equilibrium yields

$$N(f) = \frac{(w/2)L^2}{8} \quad \text{or} \quad N = \frac{wL^2}{16f}$$

The hyperbolic paraboloid is a *doubly ruled surface* ; it can be defined by two families of intersecting straight lines that form in plan projection a rhombic grid if related to the principal parabolas as surface generators (generatrices), in other words, they are the asymptotes of the hyperbolas. Here, the particular case is investigated where straight lines, perpendicular to each other, analogous to identical parabolas, are the generators. These straight lines form a rectangular grid as seen in plan view; the surfaces are called *rectangular hyperbolic paraboloids.* In Fig. 8.39d, the *warped surface* may be generated by dropping one corner below the plane described by the remaining three or by a straight line slid along two other straight lines slightly skewed with respect to each other, or vice versa. The following geometrical relationship is derived for a rectangular, warped surface based on Fig. 8.39d.

$$\frac{a}{x} = \frac{f}{c}, \text{ or } c = \frac{fx}{a}, \quad \text{and} \quad \frac{b}{y} = \frac{c}{z}, \quad \text{or } z = \frac{cy}{b} \qquad \text{(h and i)}$$

Substituting Eq. (h) into (i) gives the equation for the warped surface.

$$z = (f/ab)xy = kxy \qquad (8.29)$$

Hence, any point on the *h-p surface* can be defined by the linear equation just derived. In this expression, $k = f/ab$ is the *twist factor* representing the warping or sloping of the surface; it also defines the curvature along the y' and x' directions as explored next.

It can be shown that the two surface-generating systems are related to each other such that the straight-line generators in plan projection bisect the rectangular grid formed by the parabolic generators. Hence, the axes of the coordinate system xy are simply transformed by rotating the system through 45° in the case of a rectangular hypar to form the new coordinate system $x'y'$.

It is shown in Fig. 8.39e that any point P on the y' axis is defined as

$$x = y = y'/\sqrt{2} \qquad (j)$$

Substituting Eq. (j) into Eq. 8.29 yields

$$z = k(y')^2/2 \qquad (k)$$

The slope of the parabola, by using differentiation, is

$$dz/dy' = ky' \qquad (l)$$

But the slope is also equal to y'/R_0 according to the derivative of Eq. (9.18). Therefore, the radius of curvature of the parabolas at the saddle point is equal to

$$R_0 = 1/k = ab/f \qquad (8.30)$$

This expression shows that k not only defines the curvature of the parabola but also the geometry of the warped surface. Substituting R_0 into Eq. (8.27) yields the membrane forces for shallow rectangular hypars.

$$\pm N = wR_0/2 = w/2k = wab/2f \tag{8.31}$$

For the special condition of $z = f$, and $y' = x' = L/2$, and $a = b$, that is, for a single *square* HP saddle shape, Eq. (k) becomes

$$f = kL^2/8 \quad \text{or} \quad R_0 = 1/k = L^2/8f = a^2/f \tag{8.32}$$

Substituting this equation into Eq. (8.28) yields

$$\pm N = \frac{wR_0}{2} = \frac{wL^2}{16f} = \frac{w}{2k} = \frac{wa^2}{2f} \tag{8.33}$$

This is the membrane equation for the particular case of a *shallow, square* hyperbolic paraboloid saddle shape supporting uniform gravity loading, where the terms are as defined in Fig. 8.39. The forces are in compression along the arched parabolas and in tension along the suspended parabolas. **The equation reflects the important characteristics of shallow membranes in that it shows that the axial forces as constant along the entire surface.** Remember, there is no shear in the direction of the parabolas because they are funicular for the assumed symmetrical, uniform loading condition.

The principal axial forces, in turn, can be resolved into pure shear forces at 45° to them as shown in Fig. 8.39b. Hence, we can visualize the uniform loading to be balanced either by the principal axial forces in tension and compression or by their equivalent, the pure shear along the straight-line generators; this is the reason why hypar surfaces are often called *shear systems*. The diagonal tension or compression, which acts on a typical surface element 1×1 (as shown in Fig. 8.39b), is balanced by the shear forces along the straight-line generators without the help of any axial forces along these directions. The shear force $\overline{S}$ bar that acts along the length $2(1) \sin 45°$, replaces the equal normal forces in the diagonal direction along the parabolic generators, which act along the unit sides.

$$\overline{S} = 2N\sin 45°, \qquad N_x = N_y = 0$$

The stress due to the shear force is

$$f_s = \frac{\overline{S}}{A_s} = \frac{2N\sin 45°}{A(2\sin 45°)} = \frac{N}{A}$$

This expression shows that the shear stresses in the membrane are equal to the normal stresses, or the shear forces are equal to the axial forces.

$$S = N = \pm\frac{wab}{2f} \tag{8.34}$$

Along the straight-line generators there are no axial forces, the shallow membrane is in the state of pure shear. **We may conclude that the surface of a shallow hyperbolic paraboloid responds to uniform gravity loading by developing uniform stresses over its entire surface,** clearly reflecting its perfect efficiency from a force intensity point of view.

Supporting Structure Systems

The typical hypar roofs in Fig. 8.37, top, consisting either of single saddle shells or a combination of four warped surfaces, as used for the popular inverted umbrella and

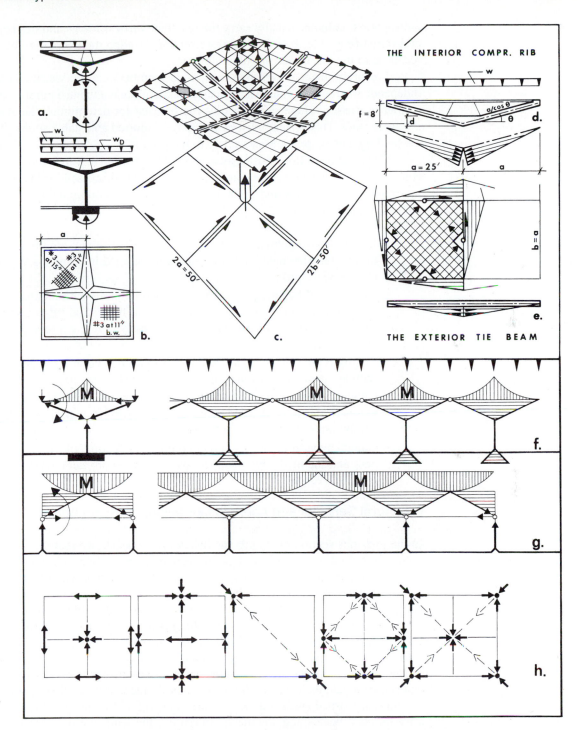

Figure 8.40 Some hypar characteristics.

gabled shells, all have exterior and possibly interior beam ribs. These members can be visualized to form an independent spatial frame that supports the hypar units.

Edge members are required along hypar units. When they are placed parallel to the straight-line generators of shallow hypars under uniform loading on the horizontal projection, they act only as axially loaded columns resisting the sum of the tangential shear along the shell edges; in other words, they serve as collectors of the edge shears. The axial forces increase linearly along the inclined or horizontal members and must balance at joints where there is no external support (e.g., Fig. 8.40e). At points of external supports, columns usually carry the resultant vertical forces and ties resist the lateral thrust (e.g., Fig. 8.40g). The cross section of edge members may be tapered in response to the linear increase of axial forces.

For preliminary design purposes, edge members can be visualized to form an independent structural system supporting the membrane units in a purely axial manner similar to trusses, where simple statics can be used for determining the magnitude of the force flow. This system must be properly supported and stable so that it can resist the loading and guide the forces safely to the ground.

The axial force flow in edge members is investigated in Fig. 8.40h for some of the common shell types on square plan in Fig. 8.37 and is discussed further in the following numerical examples and problems. The following conditions are shown in Fig. 8.40h from left to right:

- Inverted, symmetrical umbrella hypar roof on single column at center with horizontal edge members in tension and inclined interior ribs in compression (Fig. 8.37A, top).
- Symmetrical hypar cantilever roof on two columns with inclined interior ribs (of A-frame configuration) in compression, horizontal ridge member in tension, and horizontal and inclined edge members in compression (Fig. 8.37C, bottom).
- Single, symmetrical hypar cantilever roof on two columns at diagonal corners with inclined edge members in compression (Fig. 8.37B)
- Gabled hypar cantilever roof on four columns at midpoint of each side with inclined edge members and inclined interior ribs (of A-frame configuration) in compression (Fig. 8.37G, bottom).
- Gabled hypar roof on four corner columns with horizontal interior ridge members and inclined edge members (of A-frame configuration) in compression (Fig. 8.37F, top)

Structural Behavior and Approximate Design of Hypar Shells

In the previous section on membrane behavior, it was shown that the shallow hyperbolic paraboloid under uniform loading responds primarily as a shear system, where the shear forces, in turn, cause diagonal tension and compression. The behavior of the surface can be visualized as the interplay of thin compression arches in one direction and tension cables perpendicular to them. Here the spreading of the arches is prevented by the hanging cables, or vice versa, resulting in a full internal equilibrium without the help of shear and bending along the principal curvatures. These axial forces cause pure shear with no normal forces in the diagonal direction. It was found that the stresses caused by axial and shear action were constant over the entire surface, thus reflecting an optimal situation. This condition, however, is ideal and based on a rather shallow, rectangular warped surface under uniform loading and undisturbed along its boundaries by edge beams and abutments. In reality, the loads are not uniform, nor is the membrane hinged to its boundaries or the stiffness of the edge members in the plane of the shell negligible, nor is the edge beam (possibly of variable cross section) necessarily concentric with respect to the shell. This incompatibility between the thin shell and stiffer boundary members must cause additional shear and bending in the shell along the vicinity of the edges (Fig. 8.31). At corners, where beams intersect, the shell has hardly any double curvature and acts more like a slab. The *span-to-height* ratio of most hypars is in the range of $f > L/10$, or the *slope* of edge beams ($\tan\theta = f/a$) for rectangular units varies between **1:5 and 1:3**. The membrane approach is reasonable for these typical conditions; however, for steeper cases the shell weight cannot be considered anymore to act on the horizontal roof projection, but must be placed along the surface, which complicates the mathematical interpretation. Now axial forces parallel to the straight-line generators must be taken into account (which are zero for shallow hypars) and must be considered for the design of the edge beams, which resist the normal forces in beam action if they are linear or in arch action if they are curvilinear. The advantage of curved edge beams for steep hyperbolic paraboloids is apparent. The edge members for shallow hypars, if placed parallel to the straight-line generators, only act as axially loaded columns resisting the sum of the tangential shear along the shell edges, as discussed previously. At the intersection of shell segments, ribs may often be needed. For example, in the cross vault formed by single-curvature cylindrical shells, arches are generated along the diagonal intersection of the shells (Fig. 8.22). Here, free edges may be used along the outside perimeter, since surface action will be primarily in compression; that is, arch action parallel to the free edges transfers the membrane loads directly to the diagonal ribs along the intersection of the shells (see Fig. 8.42).

The membrane approach assumes uniform load action, which is only true for the weight of rather shallow membranes; live loads can take any position and edge beams exert a load concentration. The edge beams were found not to influence the overall membrane force flow along the shallow warped surface and hence, for preliminary design purposes, their weight can be evenly distributed across the horizontal projection of the roof; they behave as a natural extension and coherent part of the membrane if they are placed with their larger dimension parallel to the shell surface. As the weight or size of the boundary members increases, with increase in span of the roof, they cannot be treated as part of the membrane anymore; to the contrary, they seem to hang from the membrane and tend to stretch it, which may be beneficial to a soft surface because of the prestressing effect. For large-scale rigid shell structures, however, the edge beams may have to be supported directly by walls or frames.

It should be apparent that the membrane approach presented here can only be considered as helpful for preliminary design purposes and for the developing of a first understanding of hypar shells. Although hyperbolic paraboloids all have common geometrical characteristics, their structural behavior is dependent on how the hypar elements are combined and where and how they are supported. It must also be kept in mind that a shell cannot be too shallow since slab bending may start to control the design, rather than membrane action.

The shell thickness in reinforced-concrete construction is usually not based on strength, but on construction process (spraying, casting), the necessary cover for the steel reinforcement and leakage considerations, as has already been discussed. Keep in mind, however, that it does not have to be a minimum of 3 in. thick. This is clearly exemplified by the 5/8-in. shell of the Cosmic Ray Pavilion at the University of Mexico, which spans about 39 ft and was built by Felix Candela in 1951. Buckling may determine the shell thickness of hyperbolic paraboloids, but only for the flat portions or large spans. In general, stability considerations for anticlastic surfaces are not as critical, since the tensile curvatures tend to stabilize the arched curvatures. The shell thickness for the typical application of a single concrete saddle unit is approximately 3 in. up to a projection (maximum cable span) of 100 ft and is 4 in. up to 200 ft.

In the following examples and problems the most popular straight-edge hypar shell roofs are investigated, which are the following::

- HP saddle roof (Example 8.9)
- Inverted umbrella hypar roof (Example 8.8)
- Gabled or hipped hypar roof (Problem 8.17)

Usually, inverted umbrella hypars are used for multispan roofs (Fig. 8.40f), where each unit is designed as a free standing double-cantilever structure. Under uniform vertical loading, the perimeter beams act as tension flanges and the valley beams as compression flanges in response to the negative cantilever moment (Fig. 8.40f, left). Notice that when the umbrella roof is built continuously (i.e., no hinge at midspan) it becomes a gabled roof. Often, because of the economy of reuse of formwork, multi-span shells are built as individual units hinged together, such as for the gabled hypar roof in Fig. 8.40g, thereby forming simple span beams. Here, ridge members function as compression members, and ties that brace the columns function as tension flanges in response to the positive moment along the simple span gabled hypar (Fig. 8.40g and h).

EXAMPLE 8.8

An inverted, reinforced-concrete hypar umbrella shell of a multibay roof is investigated by treating it as free standing and not receiving any support from adjacent units. The unit is 50×50 ft in plan view and has a vertical rise of 8 ft (Fig. 8.40). A minimum shell thickness of 3 in. is selected. Grade 60 steel and 4000-psi concrete are used. The roof must support a snow load of 30 psf and an equivalent uniform dead load for edge beams and roofing of 5 psf. Wind does not have any effect on the preliminary design of the umbrella hypar. The column is fixed to the ground since it has to resist a maximum moment at the base due to lateral and asymmetrical vertical loading. The shell perimeter and valley beams will also have to carry some rotation, which will be ignored for this preliminary investigation. The cantilever deflections are generally controlled when $t/R = tf/ab > 0.003$. In this case, the slenderness is $(3/12) 8/25^2 = 0.0032$, which should be satisfactory.

The hypar shell must support the following surface load:

$$w = (3/12)150 + 5 + 30 = 73 \text{ psf}$$

Uniform loading conditions are assumed; therefore, the membrane equations can be used. Unbalanced live loading has only a small effect on the overall design of the shell, generating insignificant bending stresses, although it must be considered for the design of the pedestal. The membrane forces in tension, compression, and shear are based on one of the four warped surfaces.

$$S = N = \pm \frac{wab}{2f} = \frac{73(25)25}{2(8)} = 2852 \text{ lb/ft} \qquad (8.31)$$

The same result is obtained by basing the approach on arch and suspension action in the diagonal directions with spans of $L = a\sqrt{2} = b\sqrt{2}$ and a sag and rise according to Eq. (8.29) equal to $z = (f/ab)xy = f/4$, at $x = a/2$ and $y = b/2$. Therefore, the membrane forces according to Eq. (8.28) are

$$N = \pm w \frac{(a\sqrt{2})^2}{16(f/4)} = \frac{wa^2}{2f} = 2852 \text{ lb/ft}$$

The shear, compressive, and tensile stresses are

$$f_c = N/A_c = 2852/3(12) = 79 \text{ psi}$$

For design purposes, the membrane is treated as an axial system rather than a shear system. The allowable compressive stress is $F_c = 0.45f'_c = 0.45(4000) = 1800$ psi, clearly indicating that the shell thickness is not dependent on strength. The following steel reinforcement is required in the diagonal direction along the principal suspended parabolas, ignoring the tensile strength of the concrete

$$A_s = N/F_t = 2852/24{,}000 = 0.119 \text{ in.}^2/\text{ft}$$

Try #3 bars at 11 in. o.c., $A_s = 0.12$ in.^{2}/ ft.

The shrinkage (temperature) reinforcing is placed perpendicular to the main bars along the parabolic arches.

$$A_{smin} = 0.0018A_g = 0.0018(3 \times 12) = 0.065 \text{ in.}^2/\text{ft} \qquad (3.69)$$

The maximum bar spacing is $5t = 5(3) = 15$ in. ≤ 18 in. Select #3 bars at 15 in. o.c., $A_s = 0.09$ in.^{2}/ft.

According to the ACI Code, the shell reinforcement shall be provided at the middle surface of the shell and placed either parallel to the lines of principal tensile stress, as shown previously, or in two or three component directions. For hypar shells it may be more convenient to lay the reinforcement parallel to the straight-line generators, rather than having to cut each bar in order to fit them along the principal parabolic curves. Hence, #3 bars at 11 in. on center may be placed along each of the straight-line generators (Fig. 8.40b).

The shear forces or the resultant of the principal parabolic forces cause axial action in the beams along the perimeter edges (Fig. 8.40c). Any eccentricity between edge beam and shell and the effect of dead load (because $f/a = 8/25 = 1/3.13 \leq 1/3$) causing bending is neglected. The exterior perimeter beam is in *tension,* representing a funicular cable along the circumference, in response to equal diagonal compression forces at the corners of a square shape, by visualizing an inverted umbrella foundation supporting a column, for instance. The maximum axial force appears at the intersection with the interior rib. Since a typical beam is parallel to the straight-line generators, the maximum force is obtained by just summing up the tangential shear along the shell panel (Fig. 8.40e).

$$T_{B\max} = S(a) = wa^2b/2f = 2852(25)/1000 = 71.3 \text{ k}$$

The required steel reinforcement at the center of the edge member at the intersection point with the rib is

$$A_s = T/F_t = 71.3/24 = 2.97 \text{ in.}^2$$

Try six #7 bars, $A_s = 3.61$ in.^{2} Placing three bars in the upper and lower portions, respectively, yields an exterior member of about 12×12 in. The centroid of the reinforcement should coincide with the central axis of the membrane to avoid eccentric force action.

The interior inclined ribs are in compression. The maximum axial force at the column intersection is equal to the sum of all the shear along the edges of the adjacent warped panels accumulating linearly as they approach the support (Fig. 8.40d).

$$C_{R\max} = 2\,(Sa') = 2\,(Sa/\cos\theta) = \frac{wba^2}{f\cos\theta}$$

where $\tan\theta = f/a = 8/25,$ or $\cos\theta = 0.952.$
Hence, the maximum compression force in the rib is

$$C_R = (2(2852)25/0.952)/1000 = 149.79 \text{ k}$$

The approximate required concrete area is based on axial action only by considering the rib as a short column fully laterally supported by the shell [see Eq. (3.116)], realizing that the valley beams at the columns will have to resist some bending.

$$A = P/0.25f'_c = 149.79/0.25(4) = 149.79 \text{ in.}^2$$

The rib cross section is approximated as triangular (Fig. 8.40d).

$$A = d(2d\,\text{ctg}\theta)/2 = ad^2/f$$
$$149.79 = d^2(25/8) \quad \text{or} \quad d = 6.92 \text{ in.}$$

A larger beam depth $d = 9$ in. is selected so that the rib can resist the moment as caused by the cantilevering unsymmetrical live loading. This moment is larger than the one due to minimum eccentricity of axial forces as used in the preceding formula. The effective shell width, which helps to transfer the axial forces, is conservatively neglected.

The column must support the maximum axial load of

$$P_C = w(2a)(2b) = 4wab = 73(50)^2/1000 = 182.5 \text{ k}$$

The governing loading condition, however, is caused by unsymmetrical live load together with dead load (Fig. 8.40a). The live load is usually taken as one-half of the full live load, $30/2 = 15$ psf.

The column must support a moment, if the rotation is not distributed to adjacent umbrella units similar to three-hinged portal frame action, of

$$M_C = w_L(2a)b(b/2) = w_L ab^2$$

Hence, the live load moment is equal to

$$M_C = 0.015(25 \times 50)25/2 = 234.38 \text{ ft-k}$$

The corresponding axial dead and live loads are

$$P_C = [43(50)^2 + 15(50 \times 25)]/1000 = 126.25 \text{ k}$$

The equivalent eccentricity of the axial force is

$$e = M/P = 234.38(12)/126.25 = 22.28 \text{ in.}$$

The column size can be estimated from Eq. (3.121) using an average load factor of 1.5.

EXAMPLE 8.9

A single, straight-edged HP saddle shell, 100×100 ft in plan projection, is 15 ft high at the center (Fig. 8.41), so the slope of the edge member $f/a = 15/50 = 1/3.33$ is flat enough and the membrane theory can still be considered a reasonable first approximation. The reinforced concrete shell is 3¼ in. thick, the concrete's strength is 4000 psi, and the yield stress of the reinforcing bars is 60 ksi. The roof must support a snow load of 20 psf and 5 psf for roofing and the equivalent dead load for edge beams; wind forces are not critical for this preliminary design of the shell. The membrane is assumed hinged to its supports and hence can only be stable under symmetrical loading. To prevent the roof from tilting due to unsymmetrical gravity loading (in this case due to live load) and/or wind, the shell should be tied to the ground by tension or compression columns along the perimeter or only by single members at the high points, similar to a tent structure. On the other hand, the cantilever shell may also be fixed to its support structure so that it can resist rotation, but this may only be reasonable for small-scale buildings. This type of solution influences the membrane behavior since the edge beams will bend as cantilevers to resist unsymmetrical loading, thereby also bending the shell.

The total dead and live load is

$$w = (3.25/12)150 + 5 + 20 = 46 + 20 = 66 \text{ psf}$$

The force flow along the surface is composed of arch action straight across from support to support, with the span of the compression parabola equal to L (Fig. 8.41b) and tension cables acting in the perpendicular direction; the arches act diagonally with respect to the plan view. Based on this interpretation of stress flow, the membrane forces may be determined as

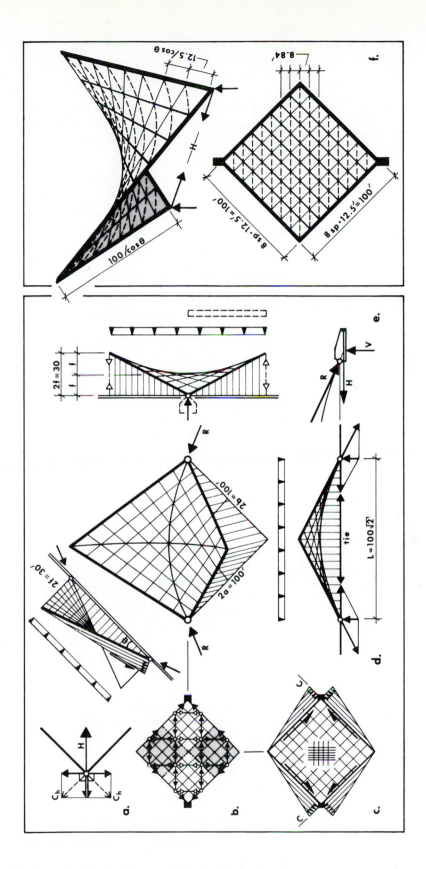

Figure 8.41 Examples 8.9 and 8.10.

$$\pm N = \frac{wL^2}{16f} = \frac{66\left(100\sqrt{2}\right)^2}{16\,(15)\,1000} = 5.5 \text{ k/ft} \tag{8.33}$$

The single hypar shell may also be visualized as consisting of four equal rectangular warped surfaces based on the surface subdivision by straight-line generators (Fig. 8.39c and d). From this interpretation, the membrane forces may also be obtained as

$$S = N = \pm\frac{wab}{2f} = \frac{66(50)\,50}{2\,(15)\,1000} = 5.5 \text{ k/ft} \tag{8.31}$$

Obviously, both approaches must give the same solution. The compressive stress is

$$f_c = N/A_c = 5.5/3.25(12) = 0.14 \text{ ksi} \le 0.45(4) = 1.8 \text{ ksi}$$

This stress is very low; the thickness of the shell is clearly not dependent on strength. The steel reinforcement along the principal parabolas is

$$A_s = N/F_t = 5.5/24 = 0.229 \text{ in.}^2/\text{ ft}$$

Select #4 bars at 10 in. o.c., $A_s = 0.24$ in.^{2}. The temperature (shrinkage) steel bars perpendicular to the main reinforcement are

$$A_{s\,\min} = 0.0018A_g = 0.0018(3.25)12 = 0.07 \text{ in.}^2/\text{ft}$$

The maximum spacing of the bars is $5t = 5(3.25) = 16.25$ in. ≤ 18 in. Try #3 bars at 16 in. o.c., $A_s = 0.083$ in.$^2/$ ft.

The diagonal parabolas in tension and compression or the shear along the four warped surface elements (Fig. 8.41b, c, and d) cause axial force flow along the edge beams, being zero at the highest points and increasing linearly to a maximum compression at the support.

The maximum axial force is equal to the sum of the shear along the shell edge.

$$C_{\max} = 2\,(Sa') = 2\,(Sa/\cos\theta), \text{ where } \tan\theta = \frac{f}{a} = \frac{15}{50}, \text{ or } \cos\theta = 0.958$$

$$= 5.5\,(100)\,/0.958 = 574.11\text{k}$$

There will also be some bending in the edge member, due to self-weight, that increases toward the abutments, which may be conservatively based on cantilever action: $M_D = w_D\,(2a)^2/2$. Equation (3.116) will be used for estimating the preliminary member size of this short column.

$$A_g = P/0.25f_c' = 574.11/0.25(4) = 574.11 \text{ in.}^2$$

Try a 24-in. square edge beam size at the abutment, $A_g = 576$ in.^{2}. The beam depth is decreased toward the high point.

The horizontal thrust component of the maximum axial edge beam force is equal to $C_h = S(2a)$. The thrust force H to be resisted at the base by a tie rod or by abutments (Fig. 8.41a) is

$$H = C_h\sqrt{2} = S(2a)\sqrt{2} = 5.5\,(100)\sqrt{2} = 777.82 \text{ k}$$

The required cross-sectional area of the tie rods according to Eq. (3.87) is

$$A_D = P/0.33\,F_u = 777.82/0.33(58) = 40.64 \text{ in.}^2$$

Try six 3-in.-diameter tie rods, $A = 6(7.07) = 42.41$ in.^{2}.

The total vertical force acting on each of the foundations is

$$V = [66\,(100 \times 100)/2]/1000 = 330 \text{ k}$$

EXAMPLE 8.10

The roof of Example 8.9 is designed in steel by using a triangular surface network (Fig. 8.41f). W-sections are placed in two layers along the straight-line generators to form a square grid that is 12.5×12.5 ft in plan view. This square grid must be stabilized against shear distortions by triangulating it with tie rods following the curvature of the concave tension parabolas. Consider dead and live load each equal to 20 psf along the horizontal roof projection. Use A36 steel.

The shell analogy is applied for this approximate design; in other words, the member grid is considered to behave as a solid surface (see also Approximate Design of Braced Domes). The membrane forces are equal to

$$N = S = \pm \frac{wab}{2f} = \frac{(20 + 20)\,50\,(50)}{2(15)\,1000} = 3.33 \text{ k/ft} \qquad (8.31)$$

Each parallel linear member must carry the load per foot multiplied by the spacing of the grid, which, in turn, varies depending on the slope or location of the member. Here, the maximum slope of the edge member is conservatively used.

$$P = 3.33 \frac{12.5}{\cos\theta} = 3.33 \frac{12.5}{0.958} = 43.45 \text{ k}$$

The bottom member acts as a simple column laterally supported at the grid intersection points. For the given slenderness ratio, a section can be selected from the column tables of the AISC Manual.

$$(Kl)_y = 1(12.5/\cos\theta) = 1(12.5/0.958) = 13.05$$

Try W 6×15.

The perpendicular members in the top layer not only carry the axial membrane forces, but also must support in bending the purlins, the corrugated steel deck, and roofing. The maximum moment for the continuous beam is approximated as (see also discussion of inclined beams),

$$M_{\max} = wl^2/10 = 0.04(12.5)12.5^2/10 = 7.81 \text{ ft-k}$$

This moment is transformed by the bending factor into a fictitious axial load. A bending factor of 0.46 is assumed for a W6 section.
The total axial load the top member must approximately carry is

$$P + P' = P + B_x M_x \qquad (3.99)$$
$$= 43.45 + 0.46(7.81)12 = 43.45 + 43.11 = 86.56 \text{ k}$$

The roof skin is considered stiff enough to prevent buckling of the member about its weak axis. Since the effect of buckling about the strong axis is relatively small, the member may be assumed to fail in yielding for preliminary design purposes. Hence, the required cross-sectional area is

$$A = P/0.6F_y = 86.56/0.6(36) = 4.01 \text{ in.}^2$$

Try W6 $\times$ 15, $A = 4.43$ in.^{2} and $B_x = 0.456$.

The bending factor approach, in general, overestimates the member size so that not the section required, but the next lower one, is selected as a trial section. But, in this case, that was not done because it is unconservative to ignore the buckling effect.

Diagonal members prevent the shear distortions of the rectangular grid under asymmetrical loading. For their design, it is conservatively assumed that the entire uniform load is carried not by the straight W-sections, but by the diagonal bars. The membrane forces along the diagonal directions are also $N = 3.33$

k/ ft; in this case, however, no arched parabolas are provided to support one-half of the load; hence, the suspended cables must carry double as much. Taking into account the tie spacing and considering the roof to be shallow (i.e., maximum cable slope is neglected) yields the following axial force (Fig. 8.41f):

$$T = 2[3.33(8.84)] = 58.87 \text{ k}$$

The required cross-sectional area according to Eq. (3.87) is

$$A_D = P/0.33F_u = 58.87/0.33(58) = 3.08 \text{ in.}^2$$

Select a flat bar $2\frac{1}{2} \times 1\frac{1}{4}$ in., $A = 3.13$ in.^{2}

The maximum compression in the edge beam at the support is equal to the sum of the shear along the membrane.

$$C_{max} = 3.33 \frac{100}{\cos \theta} = 347.60 \text{k}$$

The member must also resist one-half of the moment that a typical interior straight member of the top layer carries.

$$M_{max} = 7.81/2 = 3.91 \text{ ft-k}$$

This moment causes such small stresses as compared to the axial force that it can be ignored. The edge beam can be assumed to fail in compression yielding, since it is restrained from buckling by the roof framing. The required cross-sectional area is

$$A = C/0.6F_y = 347.60/0.6(36) = 16.09 \text{ in.}^2$$

Try W12 × 58, $A = 17.0$ in.^{2}

The tie member has to resist the following force:

$$H = 3.33(100) \sqrt{2} = 470.93 \text{ k}$$

The approximate cross-sectional area as based on yielding of the gross section [Eq.(3.82a)] is

$$A = H/F_t = 470.93/0.6(36) = 21.80 \text{ in.}^2$$

Try W14 × 74, $A = 21.8$ in.^{2}.

The magnitude of the vertical reaction forces is

$$V = 0.04(100 \times 100)/2 = 200 \text{ k}$$

8.5 OTHER SHELL FORMS

Of the infinite number of bent surface systems generated from the common basic shapes that were just discussed, the group of *intersecting shells* seems to appear frequently. The typical intersecting forms are single-curvature cylinders and double-curvature elliptic and hyperbolic paraboloids. The crossing of two similar shells results in a rectangular plan, while three shells produce a hexagonal plan; hence, the number of intersecting shells determines the plan form. The Romanesque *cross vault* is the most familiar example of intersecting cylindrical surfaces. Also familiar is the *cloister vault* constructed from four cylindrical sectors, yielding the polygonal dome (Fig. 8.22).

To develop a feeling for the behavior of these relatively simple forms, let two identical saddle surfaces intersect at right angles to cover a square plan (Fig. 8.42). First, visualize the four independent vault segments to be supported by diagonal arches. Each vault, in turn, consists of a series of adjacent parabolic arches parallel to

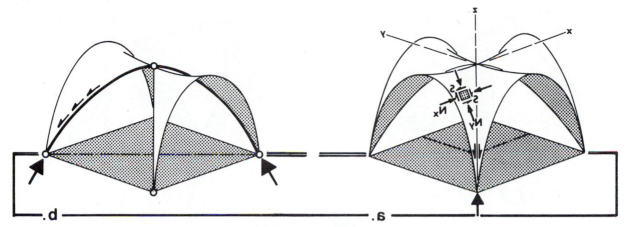

Figure 8.42 Groined shell.

the free edge, which transfer a uniform load, nearly in pure compression (i.e., no bending), to the diagonal ribs, or barrel intersections, which, in turn, carry the resultant forces to the abutments. In reality, however, the structure is not a groined vault but a *groined shell*, which does not need diagonal arches as support, but possibly only stiffening edge members.

The primary shell action will be along the arches (N_y), since little resistance can be provided perpendicular to the free edges (i.e., $N_x \cong 0$). While the arches in a vault are subject to bending, the arches in this thin shell cannot do so; here bending must be transformed into shear. Thus, no bending will be generated along the diagonal ribs, which will carry the resultant shell forces in a purely axial manner to the abutments.

Some typical examples of building structures using the principle of intersecting shells are shown in Fig. 8.43. Nervi's Good Hope Center at Cape Town, South Africa (1977, Fig. 8.43a), is derived from two intersecting cylindrical shells, forming a *cross vault* on a 280-ft square base supported on four buttresses. The shell is composed of nearly equilateral tringular precast components varying in length from 11.5 to 13 ft, and in depth from 1.5 to 2.5 ft. The three groin vaults of three intersecting, circular, cylindrical shell units for the Air Terminal at St. Louis (1954, Fig. 8.43d) are based on the same geometrical principles, although each shell cantilevers slightly beyond the 120-ft square base to form an octagonal plan projection. The typical concrete shell thickness is $4\frac{1}{2}$ in. and increases to 8 in. at the edges. The air terminal was designed by Minoru Yamasaki and the structural engineer Anton Tedesko.

The CNIT Exposition Hall, Paris, France (1958, Fig. 8.43f), designed by the structural engineer N. Esquillan, is the world's largest thin-shell concrete roof. Because of the enormous span of 720 ft and to avoid buckling, the roof had to be built as a *double shell* 6 ft apart at the crown, increasing to 10 ft at the abutments. The shells are tied together by a system of $2\frac{1}{2}$-in.-thick longitudinal and transverse diaphragms. Each shell layer is corrugated and $2\frac{1}{2}$-in. thick, changing to about 12 in. at the supports. From a geometrical point of view, the roof is generated by three 720-ft-wide, 152-ft-high parabolic cylinders that intersect along lines that radiate from the abutments, thereby covering an equilateral triangular plan of about 5.5 acres! From a structural point of view, we may visualize the roof as three wide arches meeting at the crown line, where each, in turn, consists of a series of adjacent arches that fan out from the supports. Because of the asymmetrical shape of the arch forms as seen in plan (that is, the inclination of the adjacent arches), lateral thrust is generated along the crown line, which is balanced along the horizontal starlike rib system. These ribs divide the roof into three sections, in which corrugations fan out from the abutments toward the ribs at the crown, reflecting the intensity of force flow along the arches. The horizontal

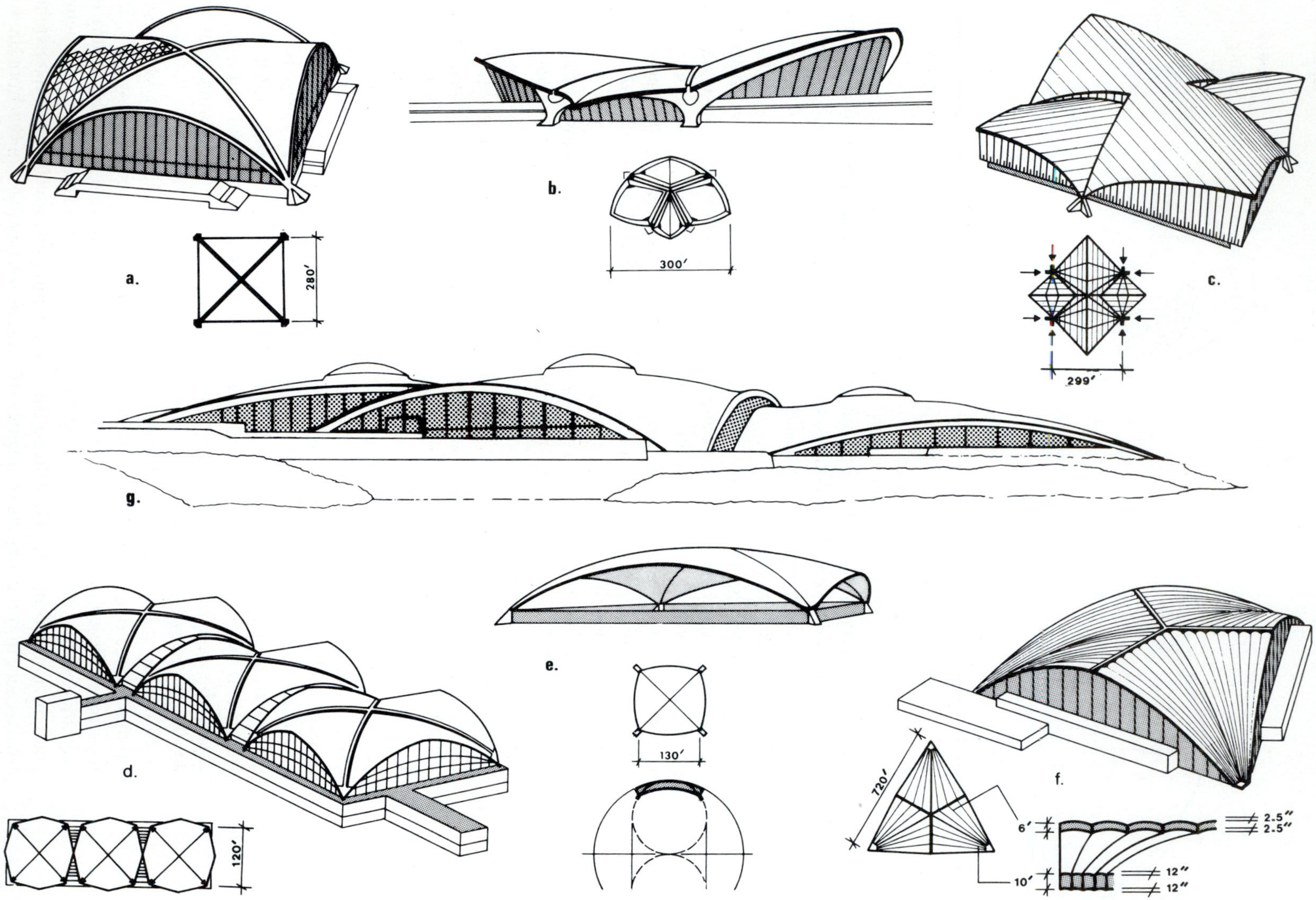

Figure 8.43 Intersecting shells.

building thrust is resisted at the base by cables that tie the abutments together along the building edges.

The Olympic Ice Stadium roof at Grenoble, France (1967, Fig. 8.43c), designed by N. Esquillan, consists of two orthogonally intersecting cylindrical double shells with spans of 312 and 213 ft, respectively. The large shell sits on four supports, while each of the two smaller ones is supported by only two abutments. A portal frame links the two small shells along the crown line across the large shell so that, under asymmetrical loading, they can stabilize each other. The thickness of the double shell is 4.26 ft, where the upper and lower shells are each typically 2.4 in. thick. The concrete surfaces are tied together with webs, which fan out from the abutments and also run parallel to the linear form generators. The roof structure can be visualized, similar to the CNIT Exposition Hall, as four arched shells, each of which consists of a series of adjacent arches where the thrust caused by the inclined arches is balanced along the prestressed rib at the crown line. In addition to this arch action, however, there is cantilever action. The roof thrust is balanced partly by ties that connect the four abutments and partly by inclined foundation piles.

Eero Saarinen had in mind sculptural considerations, rather than emphasis on engineering criteria, as the basis for the design of the roof for the TWA Terminal Building at Kennedy Airport, New York (1962, Fig. 8.43b). It consists of four independent, spherical, lightweight concrete shell segments; each of these cantilevering, winglike shells is bounded by edge beams and supported by two Y-shaped buttresses that are inclined along the direction of the resultant force action; the clear spans are 300 and 220 ft, respectively. Since each shell unit only rests on two abutments, the units must stabilize each other. This is achieved by connecting the interior edge beams to a common plate at the intersection of the four units, located near the center of the roof. Continuous skylights span the 3-ft gap between the shells along the interior edge beams. Structural engineers were Amman & Whitney.

An impressive shell structure is the sports center in Chamonix (1973, Fig. 8.43g) at the foot of Mont Blanc in the French Alps, designed by R. Taillibert and H. Isler. It consists of intersecting shallow equilateral spherical triangles with heavy peripheral arched edge beams; the span range of the shells is from 98 to 196 ft for the largest one.

The wealth of other bent surface forms can only be suggested by the examples in Fig. 8.44. The shells range from forms derived from basic geometries (p, o, n, s) to *catenary shells* that reflect the funicular shape of the shells' dead weight (r). They range from scalloped forms (e) and corrugated shells, such as beams or arches formed from a segment of the rotational hyperboloid (Figs. 8.4 and 8.5), to a habitable sculpture (q) or complex free forms (b), which may be obtained by combining surfaces that have the parabola as a common geometrical generator. For instance, the elliptic paraboloid or parabolic dome is often used in combination with the circular conoid, which looks like a truncated, inverted parabolic dome, with the hyperbolic paraboloid as a transition system between the other shapes; the slopes between the various basic surfaces must be perfectly matched along their edges.

The shell forms may evolve out of considerations of minimum weight or construction process. They may be derived as based on purely functional, economic, and engineering aspects, or the image of the building and its spiritual expression may be the essential issue.

Dynamic forms that seem to be fluid and in motion, such as the helix spiraling upward (g, f, i), the sprawling amoeba (r), the opening-up of the saillike triangular shells (c), the flying wing shapes (h), and the unfolding of fan-shaped surfaces (k) stand quite in contrast to more static forms, such as the egg sitting on a stem (s) or a container with its opening at the top (m).

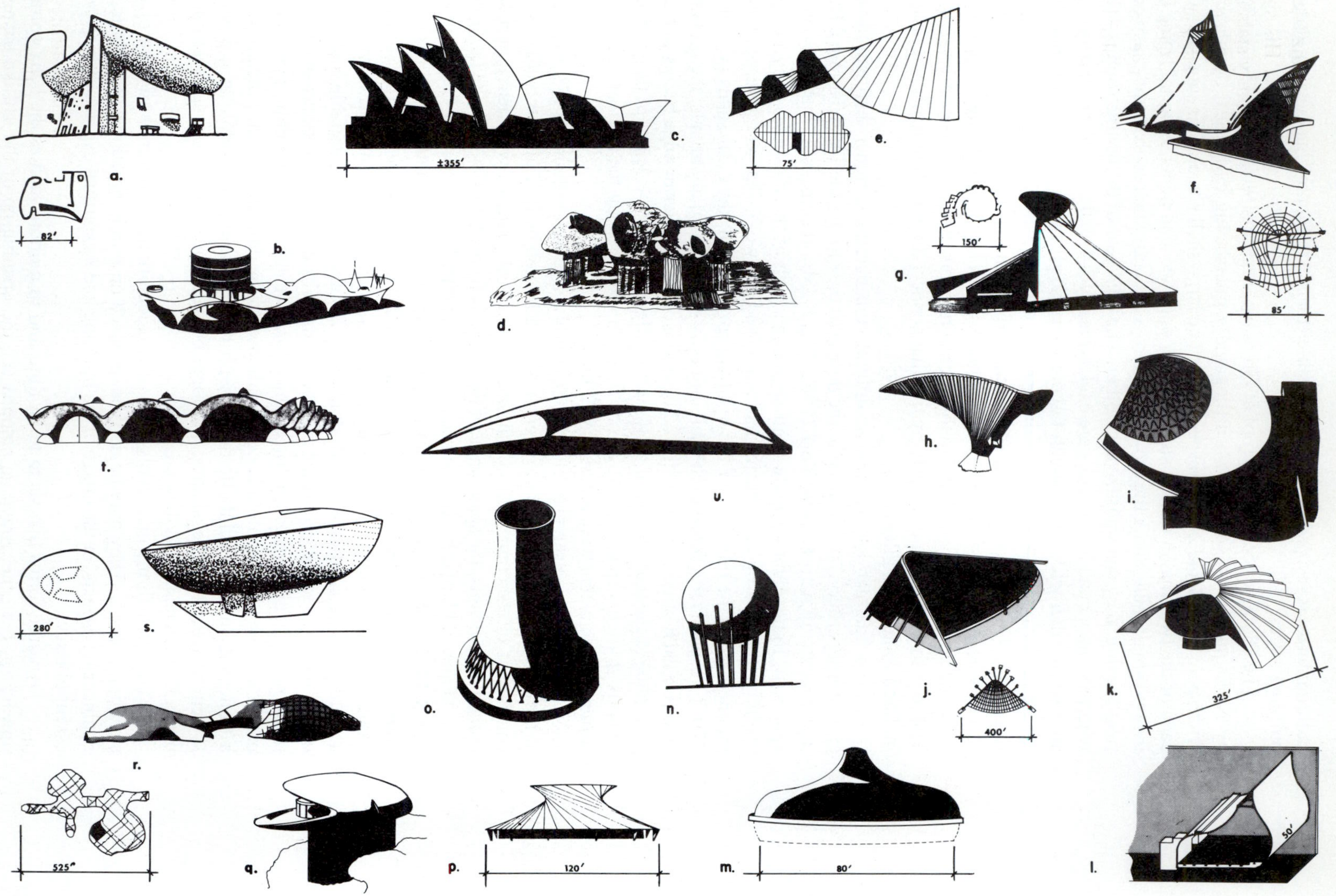

Figure 8.44 Other surface structures.

With respect to the development of freely shaped concrete shells, Heinz Isler of Switzerland must be considered as the most creative designer currently. His shells express such an elegant lightness, with their edges free of beams often only stabilized by countercurvatures; they seem to float in such a perfectly natural manner, clearly expressing their derivation from funicular considerations rather than mathematical geometries. For example, Isler's slender, graceful shell for the Garden Center in Camorino, Switzerland (1971, Fig. 8.44u), is $3\frac{1}{8}$ in. thick, with the free edges in tension, and is supported at four spreading points on a square grid of 89×89 ft. Isler determines the shell shapes primarily through experiments by reversing hanging membranes, by using pneumatic membranes as well as flowing forms. For shallow pneumatic bubble shapes, the air pressure perpendicular to the tensile membrane is nearly equal to the uniform gravity loading, thus causing only compression in the corresponding rigid shells. In his *flow method,* Isler uses the varied advancing velocity of viscous fluids inside a tube, where it is zero along the wall but maximum at the center, as the model for producing impressive shapes.

The shell forms range from surfaces derived from existing geometries, as well as mathematical and behavioral formulations, to the opposite, the antistructure. Frederick Kiesler called his endless house of 1959 (d) *continuous tension.* Here the free-modeled shell shape responds directly to the life of the interior of the organism; it is not derived from geometry. Le Corbusier's famous chapel at Ronchamp (1955, Fig. 8.44a) is not subject anymore to the laws of reasoning, such as structural interpretations or mathematical definitions. The suspended concrete double-shell roof, a conoid in shape, is resting on the tapered heavy walls; it creates a mysterious interplay of masses and surfaces, exposing a deeply spiritual and timeless expression. The chapel does not seem to be a building in the usual sense; its structure is resolved in its form, reflecting poetry in sculptural art. Its perfect correctness cannot be interpreted anymore simply with the language used in this context. The concrete roof structure consists of the main parallel flat beams connected by precast joists and covered with a $2\frac{1}{2}$-in.-thick slab.

Next, we discuss some of the buildings in Fig. 8.44, keeping in mind that several of the cases only look like shells but, in reality, are conventionally framed surface structures. The bent surfaces of Jørn Utzon's Sydney Opera House (c), built from 1959 to 1973, are segments of equivalent spheres ($R = 246$ ft), which allowed a maximum use of repetitive elements. The largest shell rises 220 ft. The shell surface is built of a fan of adjacent pointed arches that are comprised of precast concrete segments posttensioned together. The radiating, similar ribs lie on the great circles; they are hollow, wide-open Y-tubes at the upper surface portion and change to narrow, solid T-sections as they approach the pedestal. The behavior of the shell-like surface is that of individual inclined arches supported at the base and ridge. The ridge beam in turn is tangentially supported by the side shell that acts similarly to a tripod. The spherical surfaces of the Sydney Opera only give the appearance of the shell, but are not true shells! Structural engineers for the project were Ove Arup.

The wooden grid shell of the Mannheim Exhibition, Germany (1975, Fig. 8.44r), developed by Frei Otto and others, is of enormous proportions, covering an area of 80,000 ft square. The surface geometry of the catenary, two-way, double-layered lattice structure is derived by inverting a hanging chain model to a standing position (i.e., mirror image) and thus is primarily curved synclastically. The out-of-plane stiffness of the shell, necessary for buckling resistance, is obtained mainly by the double-layer laths, while the in-plane stiffness of the two-way grid is increased by cross cables connecting the joints. The western hemlock laths, nearly 2×2 in. in section, are bolted together to form an even 20×20 in mesh along the horizontal crown portion, while, along the sides, the grid is an uneven, diamond-shaped mesh. The wood lattice is covered with translucent PVC skin reinforced with woven Trevira net.

The saddle shell roof of 1963 in case (e) is constructed from 2- × 6-in. rafters, 18 ft long, placed side by side to form a solid surface. The roof shape is generated by translating the 18-ft members along the straight-line axis and the undulating outer edges. The complex, spirallike roof of the United Church of Rowayton, Connecticut (1962, Fig. 8.44f), designed by Joseph Salerno, consists of intricately shaped, glued-laminated arches tied together by purlins, with the sheathing forming a stressed-skin structure. Similarly, the spiral roof of 1969, case (g), is developed from inclined, glued-laminated rafters that are supported centrally by a compression ring at the top and by brick walls along the periphery.

It is apparent from the examples that there are no limits to form giving.

PROBLEMS

8.1   A single, long cylindrical shell with edge beams (Fig. 8.21c) spans 100 ft and has the following properties: shell width = 30 ft, shell height = 5 ft, shell thickness = 3 in., edge beam height = 5 ft, and edge beam width = 10 in. The shell must support a snow load of 30 psf. Use 4000-psi concrete and Grade 60 steel. Do a preliminary investigation of the shell and determine the steel reinforcement.

8.2   A typical interior, 3-in.-thick long barrel shell of a multiple system with edge beams (Fig. 8.21c) is 50 ft wide, 15 ft high, and spans 150 ft. Because the geometry of the shallow shell has not yet been determined at this preliminary design stage, treat the weight of the shell as if it were on the horizontal projection. Add an additional 5 psf for the edge beams and the effect of shell curvature and 5 psf for roofing and insulation. The snow load is 20 psf. Find the critical tensile reinforcement at the maximum moment location, the diagonal shear reinforcing, and the minimum reinforcing along the compression zone. Use Grade 60 steel and 4000-psi concrete.

8.3   Derive Eq. (8.22).

8.4   A $3\frac{1}{2}$-in. thick spherical concrete dome shell spans 150 ft and is 20 ft high. Investigate the shell for critical compression and tension if 4000-psi concrete and 60-ksi steel will be used. Assume a snow load of 25 psf and 6 psf for roofing. Determine the typical steel reinforcement.

8.5   The dome segment in Problem 8.4 is replaced by a 3-in. hemispherical concrete dome with the same material properties. Determine the critical tension stresses and select the necessary reinforcement.

8.6   A concrete dome shell spans 230 ft and is 70 ft high. The thickness varies from 4 in. on top to 8 in. at the footing. The dome must carry a live load of 30 psf plus a ceiling load of 15 psf and an added floor load of 35 psf. Check the critical compressive stresses in the concrete and determine the preliminary reinforcing steel in the shell and ring footing. Use 4000-psi concrete and Grade 60 steel.

8.7   Determine the change in radius at the dome base in Example 8.4 as caused by (a) expansion of the ring beam due to gravity loading, (b) temperature increase of 50°F and $\alpha = 0.0000065$ in./ in./°F, and (c) hoop shortening due to gravity loads. Refer to Eqs. (9.61 a and b).

8.8   Investigate the membrane forces due to uniform gravity load action in a parabolic dome shell so that the shell can be designed for fast approximation purposes.

8.9   Investigate a 2-in.-thick elliptical concrete dome shell spanning 64 × 50 ft that has a rise of 8 ft (see Example. 9.10 for further explanations). Assume a snow load of 20 psf and 5 psf for roofing. Use 4000-psi concrete and Grade 60 steel.

8.10 Determine the critical stresses in a 3.5-in.-thick conical concrete shell that spans 50 ft and is 10 ft high. Assume a snow load of 30 psf and 6 psf for roofing. For fast approximation purposes, treat the live load as dead load. Use 4000 psi concrete and Grade 60 steel.

8.11 A spherical dome shell has a lantern opening of radius R_0 at the crown. A stiffening ring beam must be provided to carry the loads since the thin shell cannot support the concentrated collar loads due to the lantern weight and due to other loads, such as lighting and ventilating equipment. Consider the resultant of the uniformly distributed collar load around the opening to be W_0. Determine the ring forces at the crown and at the base, and the meridional and hoop forces at the shell edge along the base as caused by the shell weight and lantern load.

8.12 Determine the hoop and arch forces at the base of a conoidal dome shell (pointed dome) composed of spherical segments carrying a single load at its vertex. Also, find the horizontal and vertical components of the arch forces at the base.

8.13 Design the prestressed concrete ring beam for a 150-ft-span spherical concrete dome of 30 ft height that carries a total gravity load of 1200 k. Use 4000-psi concrete and Grade 270 strands.

8.14 Houston's Astrodome has a span of 642 ft and a rise of 93 ft. The principal framing consists of twelve 5-ft-deep trussed ribs arranged radially at 30° intervals, spanning from the crown to the tension ring at the base, and six 5-ft-deep trussed loops. The sections are braced by two sets of lamella members each parallel to one of the main ribs. Check the stresses in the tension ring, which has a cross-sectional area of 201-in.^{2}, and check the stresses in the radial ribs at the base with an area of 279-in.^{2} for A36 steel. Use a dead load of 30 psf, and a snow load of 15 psf, which may be treated as dead load for this fast check.

8.15 A Schwedler dome made of A36 steel spans 225 ft and is 40 ft high. It consists of 36 meridian rafters and 18 rows of concentric ring purlins. The dead and live loads are each assumed as 20 psf. Determine the preliminary critical member sizes for the rafter, purlin, and tension ring.

8.16 Determine the membrane stresses in one of the 4-in.-thick saddle shells of the Ponce Coliseum (Fig. 8.38D); a typical quadrant is 116×138 ft and has a rise of $f = 20$ ft. Check the stresses for the loading case of a wind pressure of 56 psf upward or downward, together with the shell dead weight and an equivalent weight of edge beams of 16 psf. Then consider that the membrane was actually posttensioned by 10 to 20 k/ft to minimize crack formation. Use 4000-psi concrete.

8.17 A reinforced-concrete hipped roof (Fig. 8.37F, top), consisting of four warped rectangular surfaces with horizontal ridge beams, is supported by four corner columns. The roof covers an area of 120×100 ft and has a rise of 15 ft. It must support a live load of 20 psf and an equivalent uniform dead load for shell finish and edge beams of 5 psf. Use a minimum shell thickness of 3 in. and 4000-psi concrete together with Grade 60 steel. Determine the shell reinforcement and the approximate sizes of beams and tie rods.

8.18 Discuss the force flow conditions when the ridge beams for the gabled hypar in Problem 8.17 are folded downward and are supported in addition by an interior column.

8.19 If the inverted hypar umbrella shell in Example 8.8 is turned upside down to form a true umbrella with the same dimensions, what changes with respect to the force flow in the shell units and edge beams.

8.20 A 3-in.-thick hypar concrete shell consists of two single, straight-edged-hypar units, each 88×88 ft in plan projection, with a height of $f = 11$ ft; in other words, an identical unit is added to the one shown in Fig. 8.41. The roof is supported by three piers, at the corners on one side and at midspan on the opposite side. The shell must support a 30-psf snow load and an equivalent dead load for shell finish and edge beams of 5 psf. Use 4000-psi concrete and Grade 60 steel. For this preliminary investigation, determine the shell reinforcement, the approximate sizes of edge members, and the tie rod at the base, as well as the vertical reaction forces.

8.21 Design the roof of Problem 8.20 as a three-way surface grid structure similar to Example 8.10. The main straight lines are spaced at 11 ft in both directions. Consider the average uniform dead load to be 20 psf. Use A36 steel. Determine the preliminary member sizes.

8.22 An identical roof assemblage is added to the structure of Problem 8.21 to form a new roof consisting of four, single, straight-edged hypar units sitting on four abutments located at the middle of each outer face. Determine the critical forces in the membrane and edge members and the reaction forces.

8.23 Discuss the force flow conditions for the hypar roof in Problem 8.22 if (a) the ridge beams are placed horizontally and (b) the edge beams are placed horizontally.

8.24   Investigate the force action along the shell elements and beams in Fig. 8.37 C, bottom. Show the position of the tie rod if needed for resistance of lateral thrust.

8.25   Investigate an inverted HP concrete umbrella, 40 × 40 ft in plan, with a vertical rise of 5.5 ft; a thickness of 3 in. is selected. The live load is 20 psf, and 5 psf is used for roofing and to account for the weight of the edge beams. Design approximately the shell and edge members using 4000-psi concrete and A36 steel.

chapter 9

Tension Roof Structures

The principle of employing cables, nets, fabrics, and animal skins as tension structures has been known for thousands of years. Cables have been used as inclined stays to support beams and masts; nets and fabrics are to be found as fishing nets, hunting nets, umbrellas, awnings, sails, kites, tents, and for many other purposes.

Throughout the world, fabric roofs have been used as shade structures for streets, marketplaces, and courtyards to provide protection from the sun. For example, the Romans used sun awnings made of cloth to cover their amphitheaters. Also, umbrellas may be considered forerunners of tensile membrane structures, having been known already by Egyptians and Assyrians. The Romans used inflated bags of animal skins to support bridges and rafts. In 1783, the Montgolfier brothers in France reinvented the hot-air balloon made of paper and linen.

The development of today's lightweight tensile structures surely has been influenced directly by the suspension bridges and tent structures of the past. The Chinese and Tibetan cane and bamboo cable bridges go far back in time. The Quan-Xian bamboo cable bridge was mentioned as early as the third century B.C. The Chinese Emperor Ming supposedly built a 200-ft iron chain cable bridge in A.D. 65. Suspension bridges were quite common for a long time in China, India, Africa, and South America; vines, bamboo, hemp, birch, cane, twisted lianas, and similar materials were used for construction. The first major cable suspension bridge in the western world is credited to James Finley (Jacob's Creek Bridge, Pennsylvania, 1801). John A. Roebling is considered by many the inventor of the modern long-span suspension bridge, as demonstrated by the 1595-ft Brooklyn Bridge, where he employed a secondary cable system to counteract flutter. The bridge was finished in 1883 by his son Washington Roebling.

Tents are among human's oldest structural forms. They were made of animal skins, wool, cotton, silk, linen etc. They have been used by nomads and military for thousands of years. To them, the lightness, flexibility, portability, and ease of construction are all features necessary to their living style. The tents of the nomads can be found in extremely diverse climates, extending from the deserts of Saudi Arabia to the high altitudes of the Himalayan mountains, from the steppe of Mongolia and the taiga forest to the arctic tundra. Kings used often pavilion tents that looked like palaces.

Among the earliest applications of cables and fabrics as large roof structures that could be extended and retracted was over the 620- 513-ft Roman Colosseum in A.D. 70.

Covered circus tents were introduced last century, in the early 1820s. The Ringling Brothers' tent of the 1940s could seat an audience of 13,000! The early development of the tensile membrane in modern architecture may be traced back to V. G. Shookov's daring steel sheet tents for the Nijni–Novgorod Industrial Fair of 1896 (the circular tent had a diameter of 223 ft); the construction principle for the pavilions was far ahead of its time. In 1909, a masted, cable-supported, tent airship hangar was built in Frankfurt, Germany.

The influence of the concern for tension on the modern movement can be seen in the work of Italian futurist Antonio Sant' Elia who, before World War I, experimented with spatial tension and the dynamism of tension to express movement. At the same time, Antoine Pevsner and his younger brother Naum Gabo began to experiment with spatial tensile sculptures; they started the constructivist movement in Russia during the first two decades of this century. Since the early 1930s, Alexander Calder's kinetic sculptures have fascinated the public. His mobiles are tensile art; they express movement, balance, lightness, and antigravity. Along the same tradition are Kenneth Snelson's tensegrity sculptures of the 1960s as influenced by Buckminster Fuller's thinking. Bill Moss, with the invention of the pop tent in 1955, became the most famous fabric structure artist. Since then, he has been treating his tents as pieces of art and canopies and shade structures as sculptures; he perceives the tensioned fabric as space articulation. One of the first real, lasting tensile structures of the early years of development, which revolutionized furniture design, is Marcel Breuer's famous *Wassily chair* of 1925, in which canvas was stretched in tension between a bent tubular steel frame. About 10 years later, the b*utterfly chairs* (Hardoy chair) were produced consisting of twisted steel rods and canvas or leather to form a hammocklike seat. These chairs became the symbol of the avant-garde in the 1950s.

The real evolution of tensile structures in architecture was initiated in 1953 with Mathew Nowicki's State Fair Arena at Raleigh, North Carolina (Fig. 9.2k). The saddle roof clearly responds to the nature of surface tension. Its spirit influenced Frei Otto, one of the most prominent pioneers of tensile architecture, who started his research in the early 1950s. He first explored the potential of fabric membrane and then cable net for the large-scale tent, climaxing in the roofs for the Olympic Games at Munich, Germany, in 1972. Development of pneumatic structures began in the United States with Walter Bird's first radome, which he built in 1946 for the military to house radar for defense and space communications. The U.S. Pavilion at Expo '70 in Osaka, Japan (see Fig. 9.9a), designed by architects Davis Brody Associates, and engineer David Geiger, together with other pneumatic buildings at the exhibition, started the enthusiasm for and made the public aware of the formal and large-scale potential of the pneumatic principle. The first permanent fabric structure in the United States is the University of LaVerne student center, completed in 1973.

The development of cable-supported structures has been influenced by the simplicity and minimal proportions of the cable-stayed Rhine bridges of the 1960s. Renzo Piano experimented with the composite action of space frames, tensile membranes, and cables in his Italian Industry Pavilion at the 1970 Osaka Expo. The celebration of hinged lightweight structures as assembly systems that had to be stabilized with tensile columns and bracing is most convincingly demonstrated in the Pompidou Center in Paris (1977) designed by Piano and Rogers. It reflects some of the spirit of the 1960s as represented by the work of Archigram.

Finally, a new generation of tensile structures arrived in the late 1980s, integrating complex spatial networks with the fabric membrane. Examples are the Louvre Pyramid and the "cloud" at the base of LaGrande Arche in Paris, on the one hand, and the tensegrity domes for the Florida Suncoast and Georgia stadiums in the United States, on the other. Also mentioned must be the fabric skylight which are competing with the traditional glass skylights.

Among some of the important structural engineering pioneers who made the transition of tensile structures into architecture possible are Walter Bird, Fred Severud, Lev Zetlin, David Geiger, and Horst Berger of the United States; Ulrich Finsterwalder, Fritz Leonhard, and Frei Otto of Germany; Peter Rice, Tony Hunt, and Ted Happold of the United Kingdom; Riccardo Morandi of Italy, Rene Sarger of France, Robert LeRicolais of France and the United States, and Yoshikatsu Tsuboi of Japan.

Tensile structures are of light weight, which is especially true for fabric structures, contrary to the conventional rigid structures, where weight resists instability caused by lateral force action. Thin tensile surfaces are supported by air pressure, flexible members, or rigid members such as masts, arches, or frames. In traditional gravity-type buildings, the inherent massiveness of material transmits a feeling of stability and protection. Free-standing tensile structures, on the other hand, seem to be weightless and to float in the air; their stability is dependent on induced tension and on an intricate, curved, three-dimensional geometry in which the skin is prestretched. These antigravity structures require a new esthetics; now the curve, rather than the straight line, is the generator of space. This esthetics is closely related to biological structures, such as spider webs, the wings of flies, bird and bat wings, and liquid skins like water drops, soap bubbles, and bubble conglomerations in sea foam. This close relationship of the tensile membranes to natural forms makes it possible to truly integrate them with the landscape.

Although there is no real historical precedent for the complex forms of membrane structures, they remind one of the exuberance of architecture spaces of the Baroque and Rococo periods or the architecture of expressionism as represented by Antonio Gaudi's organic forms around the turn of this century. The use of complex spatial geometry reminds us also of Gothic cathedrals, which represent *forms in compression*. Here, the master builders responded with geometry together with weight to overcome the limits of stone. Fabric structures, in contrast, are *forms in tension;* as nearly weightless structures, they are pure, essential, and minimal. Spatial, curved geometry, together with induced tension, is necessary for structural integrity.

9.1 GENERAL PRINCIPLES

Tensile structures are often broadly organized as cable-supported structures, cable structures, and membrane structures. In other words, structures employing isolated individual tensile members as supports are called *cable-supported structures,* such as cable-stayed roofs or bridges, guyed towers, cable-supported beams, or any other cable-supported structure, keeping in mind that the tension members do not necessarily have to be cables, but can be more rigid members.

In *cable and membrane structures,* a system of cables or other types of tensile members form a surface by directly supporting the roof skin, as for hanging roofs, or pretensioned roofs often called *soft shells*. Familiar examples of pretensioned roofs are **saddle domes, tents, and pneumatic structures.** Three-dimensional tensile surface structures may be generated by intersecting cable beams or by the multidirectional tensegrity principle, where floating individual compression members are not touching each other and are held in place by the continuous tension cables; the principle is applied to the **cable domes.**

The following classification of tension structures should assist as an introduction to the organization of this chapter.

- **Cable-supported structures:** Cable-stayed roofs, suspended cable-supported roofs, cable-supported beams, intersecting cable-supported beams.
- **Cable and membrane structures**

—Simply suspended (hanging), single-layer cable roofs of single curvature or synclastic shape: cylindrical roofs with parallel cable arrangement, polygonal dishes with radial cables or cable nets.

—Simple double-layer cable systems: pretensioned cable beams and trusses, including air-inflated members.

—Prestressed membranes and cable nets

Edge-supported saddle surface structures

Arch-supported saddle surface structures

Mast-supported conical surface structures (tents)

Conbinations of the above

Air-supported (pneumatic) membrane structures

—Pretensioned double-layer nets (possibly formed by intersecting cable beams) or tensegrity surface structures

Little or no false work is necessary for the erection of suspended cable roofs. Since a long-span roof is about as easily erected as a short-span roof, the economy of construction increases for longer spans by offsetting the larger costs of fittings and abutment systems.

Stability Considerations

Cables and thin tensile membranes are inherently flexible and *form active.* A single cable must adjust its suspended form to the respective loading condition so that it can respond in tension. Because of the absence of bending and buckling and due to the high strength of cable material, cable roofs are extremely light.

Due to the lack of stiffness, a cable roof must be stabilized against formal instability so that it does not adjust its shape every time the loads change and thus can also absorb the uplift caused by the wind. Furthermore, the flexibility and lightness of the roof make it vulnerable to gusty winds, which tend to cause it to flutter or cause tensioned cables to vibrate like guitar strings, therefore requiring special stabilization so that the movement is damped.

Various methods for stabilizing a cable roof against formal instability (but not necessarily against dynamic load action, as is discussed later in this chapter), as shown in Fig. 9.1, are the following:

* *Dead weight* sitting on top or suspended below the roof, where the permanent dead load should be substantially larger than the asymmetrical live loads to prevent asymmetrical deformations
* *Rigid members* acting as suspended arches that provide bending stiffness
* *Rigid surfaces* behaving as inverted thin shells or ribbed tension shells with bending stiffness, usually prestressed concrete shells
* *Pneumatic pressure:* domical and cylindrical shapes
* *Mechanical prestress:* secondary cables prestressing the main cables so that the cables always remain in tension
   —The main cable is guyed to other structures.
   —The secondary cable is in plane with the main cable to form *cable beams*.
   —The secondary cable is placed perpendicularly/diagonally to the primary cable to form an *anticlastic net surface.*
   —Multiple cable systems, such as parallel suspended ridge and arched valley cables, are tied together by perpendicular cables or membranes (i.e., inclined web ties), as well as tensegrity structures including tensegric nets.

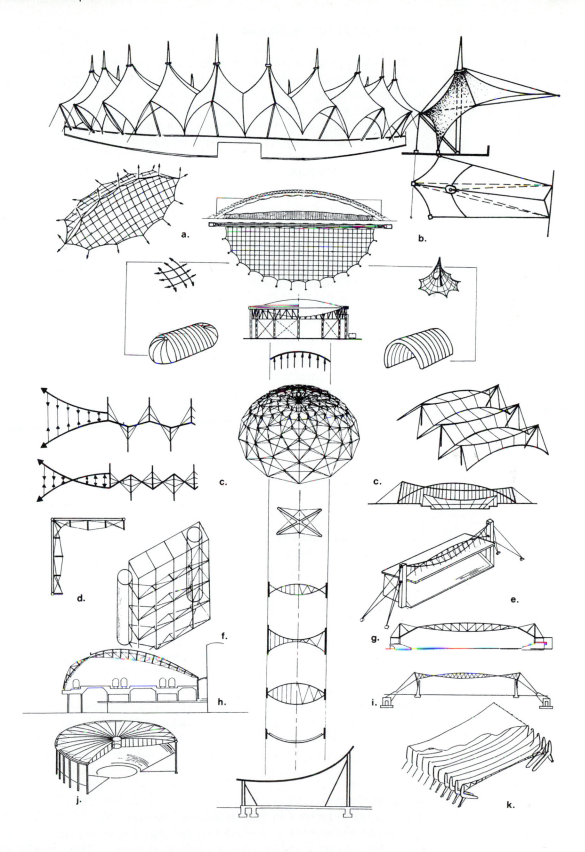

Figure 9.1 Methods for stabilizing cable structures.

Anchorage of Tensile Forces

A challenge to the designer is the anchorage of the tensile forces at the boundaries. These forces may be guided either directly to the ground and foundations, or they may be connected to flexible or rigid boundary structures. These support structures may be of the open type (e.g., frames, arches, cables), or they may be of the closed type where the internal forces balance each other (e.g., horizontal or spatial ring beams). Some typical solutions shown in Fig. 9.2 are the following:

- Simply suspended, single-curvature roofs with parallel arrangement of cables
    —Guyed masts or pendulum-type columns (a, b, c) with cables anchored to foundations (e.g., gravity anchors, plate anchors, tension piles, or self-balancing ties, struts, and rings)
    —Cantilever columns (f, g)
    —Legged columns (d, e)
    —Beams (i), e.g., frames, beams plus walls
    —Inclined arches or other funicular shapes (j)
    —Inclined, vertical, cylindrically curved walls
    —Form responsive boundary shapes (h)
- Dish-shaped, synclastic, hanging roofs
    —Individual buttresses for radial cable arrangement
    Circular compression edge rings for radial cable arrangement (with central tension ring) or two-way cable net arrangement in response to spherical surface forms (l)
    —Elliptical ring beams in response to elliptic paraboloids (m)
    —Combinations
- Anticlastic, pretensioned cable net roof forms
    —Open boundary arches with edge cables anchored to rigid boundaries (o, Fig. 9.1a)
    —Two intersecting inclined arches (k)
    —Space ring beams (n)
    —Ring beam together with buttresses
- For the complex boundary conditions of edge cables for tent structures (p) refer to Fig. 9.19.

By inclining the guyed masts (a, b), the magnitude of the forces in the columns and the diagonal guys is reduced (a.1), which is especially important because the tensile forces that the ground anchors must resist are decreased. When the pendulum-type columns are placed vertically, they do not resist any lateral thrust and leave this entirely to the guy cables requiring, in turn, an extensive tensile foundation (see Fig. 9.33). Furthermore, the magnitude of the tension in the guys decreases by making them less steep (a.2). Refer to Section 9.10 for a discussion of tensile foundations.

For the case where no diagonal guys are used, the columns must act as vertical cantilevers to resist the lateral thrust (f, g); the columns are now primarily bending members and require large foundations to resist the overturning moment. The lateral thrust must be carried directly by the soil or may be balanced by a horizontal compression strut (g), as for the City Hall in Bremen, Germany (h).

No tensile anchors are needed if closed edge ring arches are used that resist the lateral thrust forces internally in a self-balancing manner. These ring beams usually represent furnicular shapes under uniform gravity loading, so beam moments are minimized and are only generated by live loads. For example, circular compression rings

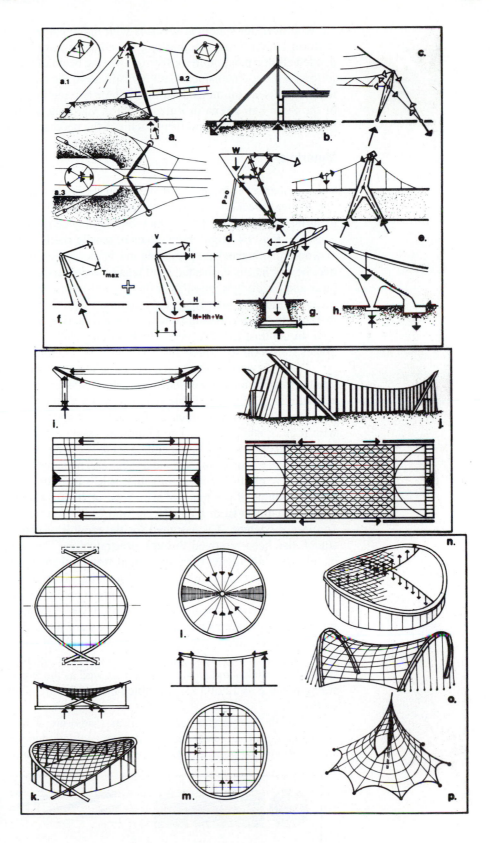

Figure 9.2 Anchorage of tensile forces.

are employed for spherical surfaces with radial cable or cable net layouts, causing a uniform lateral thrust action along the edges (l), or elliptical rings are used for suspended elliptic paraboloid roofs (m). The ring beam of the saddle dome (n), in contrast, follows a three-dimensional path in response to the nature of the surface so that bending moments are minimized. Notice that the two inclined, intersecting parabolic arches of the State Fair Arena in Raleigh, North Carolina (k), are still discontinuous, since they need the underground ties to achieve the necessary balance of forces.

Materials

The member types found in tensile structures range from linear members made generally of steel (Section 3.3) and occasionally of wood, plastics, or prestressed concrete, to surface membranes. Tensile membranes may be flexible, semirigid, or rigid, as discussed in Fig. 9.4. Rigid membranes or tensile shells (e.g., prestressed concrete, steel or wood membranes) are presented in Chapter 8. Semirigid (nonfoldable) membranes may be, for instance, sheeted steel tension members as used for large-span roof structures and for bridges, possibly reflecting stress-skin systems, or they may be thin, ferrocement ribbon roofs. Here, only common flexible membranes are introduced, such as films, meshes, and laminated and coated fabrics that are foldable, as well as cable nets. In contrast to cable nets which require cladding, fabric and metal sheet membranes form skin structures that serve both as structure and cladding.

Next the various materials, cables, cable nets, films and fabrics, used for tensile structures are briefly discussed. The currently common types of **cables** are steel strand and wire rope (Fig. 9.3, Table A.13). Wires are laid helically around a center wire to produce a strand, while ropes are formed by strands laid helically around a core (e.g., wire rope or steel strand). The minimum ultimate tensile strength of cables is in the range from 200 to 220 ksi, depending on the coating class. The strand has more metallic area than a rope of the same diameter and hence is stronger and stiffer. Notice that the effective elastic modulus for cables is less than for other structural steels. The minimum modulus of elasticity of wire rope is 20,000 ksi and 24,000 ksi for strands of nominal diameter from 1/2 to $2\frac{9}{16}$ in. and 23,000 ksi for the larger diameters, as based on class A coating. It is assumed for the moduli of elasticity that the constructional stretch has been removed through prestretching. The capacity of cables is approximately six times higher than for normal structural steels, while the material costs may only be about twice as high, resulting in savings if the structure is large enough so that the other additional costs are reduced. Three coating weights are available to meet the range of corrosion resistance requirements. Class A represents the least-weight coating for less corrosive atmospheres. Since ropes are more flexible than strands, they are easier to handle and can be used for smaller radii of curvature. Besides the common steel cables, among other cables are the Kevlar 49 rope with an ultimate tensile strength of 400 ksi and $E = 19,000$ ksi, as well as the S-glass rope with 450-ksi tensile strength and $E = 12,500$ ksi. In cable nets, cables are clamped together at their intersections.

Tensile membranes made of **films** (foils) are thin, flexible sheets of homogeneous material (e.g., clear vinyl, polyester, polyethylene) with the same properties in all directions that generally lack in strength, stiffness, and durability. Therefore, they

STRAND   WIRE ROPE

Figure 9.3 Common cable types.

are usually not used as primary architectural structures unless they are reinforced with webbing or cables and are protected with top coating should they be exposed to the environment. Currently, however, some new films are being developed that have properties required for more permanent structures.

The performance of *fabrics* depends primarily on the type of yarn, weave geometry, and coating material. Fabrics are produced by first spinning filament fibers into yarn, which is then woven or knitted into a fabric. There is a wide range of natural, synthetic, and mineral fibers from which textile membranes can be produced, as indicated in Fig. 9.4. Typical are cotton, aramid (Kevlar), nylon, polyester, rayon, and fiberglass. Architectural fabrics may be divided into meshes, laminated fabrics, and coated fabrics.

Meshes are porous fabrics that generally occur as polyester (or polyester–cotton blend) weaves lightly coated with acrylic or vinyl, or that occur as closely knitted fabric using polyethylene, polypropylene, or acrylic yarns, often occurring as acrylic painted or vinyl-coated cotton duck. Meshes are used only as secondary membrane structures for temporary purposes such as sun and wind shields or as for awnings with a life span of 5 to 8 years.

Laminated fabrics consist of a loosely woven or knitted fabric called a *scrim* (usually of polyester or nylon yarns) that is sandwiched between two sheets of plastic films. The open-weave scrim makes it possible for the outer sheets to bond together at the openings of the meshes. The sandwich is fused under pressure and heat; the structural behavior of the laminate clearly depends on the mechanical adhesion between the films. The common laminated fabrics are vinyl-laminated nylon fabric and vinyl-laminated polyester fabric.

Compared with nylon fabrics, polyester (although less strong and durable) offers greater dimensional stability (i.e., greater stiffness and less creep; see also Table 2.4), and because of its lower costs it is generally found in tensile membrane construction. Laminated fabrics are most economical and are therefore used for general applications such as canopies, awnings, and tents with a life span of 5 to 8 years. Although considered as temporary structures, laminated polyester fabrics, when protected by top coatings against exposure to the environment, are treated as more permanent structures (e.g., air-supported membranes of smaller scale) with a life span of up to about 15 years.

Coated fabrics are used for more permanent structures. They offer a higher strength, stiffness, and durability, but are also more expensive than similar laminated fabrics. They are constructed of a tightly woven or knitted base fabric coated usually on both sides with a bondable substance of high strength. The coatings are applied to the fabrics and then passed through a drying–curing oven, resulting in chemical adhesion to the yarn fibers.

There are several kinds of weaving methods, as indicated in Fig. 9.4B. The common *plain-weave fabrics* consist of sets of twisted yarns interlaced at right angles. The yarns running longitudinally down the loom are called *warp yarns*, and the ones running crosswise are called *filling yarns*, weft yarns, or woof yarns. The tensile strength of the fabric is a function of the material, the number of filaments in the twisted yarn, the number of yarns per inch of fabric, and the type of weaving pattern. The typical woven fabric consists of the straight *warp yarn* and the undulating *filling yarn* (Fig. 9.4). It is apparent that the warp direction is generally the stronger one and that the springlike filler yarn elongates more than the straight lengthwise yarn. From a structural point of view, the weave pattern may be visualized as a very *fine meshed cable network* of a rectangular grid, where openings clearly indicate the lack of shear stiffness. The fact of the different behavioral characteristics along the warp and filling, as well as in the diagonal directions, makes the membrane *anisotropic*. However, when the woven fabric is laminated or coated, the rectangular meshes are filled, thus effectively reducing the difference in behavior along the orthogonal yarns so that the fabric

TENSILE   MEMBRANES

ISOTROPIC   MEMBRANES

A.   HOMOGENEOUS SKINS

Flexible

- rubber membranes
- plastic sheets (films):
    polyester
    polyethylene
    polyvinyl chloride (PVC)
- metal membranes
    etc.

Semirigid

- metal membranes
    steel/aluminum sheets
- wood membranes
    plywood/sheathing
- ferrocement
- plastic foam

Rigid

- concrete membrane
- synthetic membrane (foam)

D.   CHOPPED OR CONTINUOUS STRAND MATS

- non-woven mats of randomly
  orientated fibers: such as
  glass fibers bonded with
  PVC or polyester

E.   COATED FABRICS

- fiber reinforced plastic sheets:
    vinyl (PVC)-coated nylon, fiber-
    glass, or polyester,
    Teflon-coated fiberglass
    Hypalon-coated polyester
    Neoprene-coated polyester

The following skins are isotropic
only with triangular nets and
where there is structural interaction of decking with net.

F.   FINE-MESH NETS AND DECKING

- welded wire mesh
- woven wire mesh
- welded wire fabric

G.   CABLE-REINFORCED FABRIC SKINS

- composite action of reinforcing with coated fabrics

H.   WIDE-MESH NETWORKS AND DECKING

- cable nets
- bar lattices
for decking see "A. Homogeneous Skins"

ANISOTROPIC   MEMBRANES

B.   WOVEN FABRIC

Weave Patterns
- twill
- satin
- basket
- plain

Natural Canvas

- from natural fibers:
    cellulose: flax, jute,
    cotton, etc.
    animal: hair, wool, silk
    metal: steel, aluminum, etc.
    mineral: asbestos, graphite,
             glass, ceramics, etc.

Synthetic Fabrics

- from regenerated organic/
  inorganic raw materials
- from synthetic fibers:
    polyester (Dacron, Diolene,
               Trevira, etc.)
    polyamides (Nylon, Perlon)
    polyacrylnitrilene (Dralon,
                        Orlon)
    etc.

C.   GRIDDED FABRIC

- course woven fabric

two
way

three
way

radial

geodesic

geometry:
some regular
patterns

net

member
material

- steel wires/cables/ropes
- steel bars/chains
- organic/synthetic fiber ropes
- wood lath

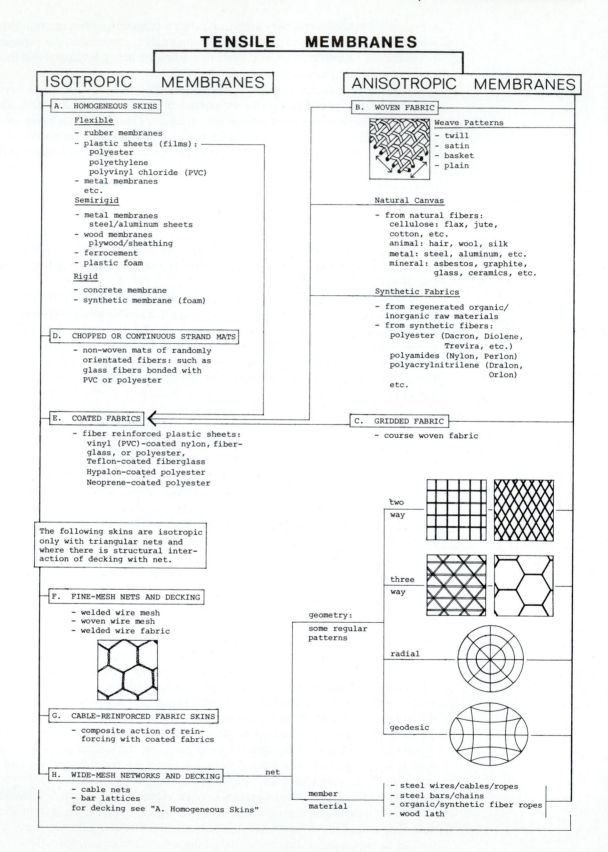

Figure 9.4   Classification of tensile membranes.

may then be considered *isotropic* for preliminary design purposes, similar to the cable network with triangular meshes, plastic films, metal skins, and other hanging semirigid or rigid shells. Keep in mind, however, that coated fabrics are complex structures for which the shear stiffness of the composite skin very much determines the degree of membrane action.

There is a great variety of coatings, each having its own characteristics. The common coated fabrics are either polyester based or glass based. They are as follows:

- PVC-coated polyester (polyester coated with vinyl) with an average life span of 15 years
- Tedlar clad, vinyl-coated polyester with an average life span of 20 years
- Vinyl-coated fiberglass with an average life span of 15 years
- Tedlar-clad, vinyl-coated fiberglass with an average life span of 20 years
- PTFE-coated fiberglass (better known by its commercial name, Teflon-coated fiberglass) and silicone-coated fiberglass with an average life span of 25 years.

Teflon and Tedlar are registered trademarks of the Du Pont Corporation. PVC-coated polyester is the most widely used material, whereas PTFE-coated fiberglass is applied for large-scale, permanent structures, such as domes or other sports or retail facilities that require fire ratings and a long fabric life.

The scale of the structure (from a structural point of view) determines the selection of the tensile membrane type, which is apparent from the previous discussion. The approximate tensile strengths of the most common coated fabrics may be taken as follows for preliminary design purposes:

Vinyl-coated nylon fabric	200–400 pli (lb/in.)
Vinyl-coated polyester fabric	300–700 pli
Vinyl-coated fiberglass fabric	300–800 pli
Teflon-coated fiberglass fabric	300–1000 pli

For example, the Owens–Corning Structo-Fab 450 fiberglass fabric coated with Teflon is 0.038 in. thick and weighs 45 oz/yd^2 (0.31 psf). Its tensile strip strength along the dry warp is 800 pli and 700 pli along the dry fill.

The strength of vinyl-coated cotton fabric (as used for small-scale temporary structures) is about 160 pli and, without coating, 120 pli. The common synthetic fabrics have a similar weight and strength to those of cotton fabrics. Plastic films have a relatively low tensile strength, such as 16 pli for a 20-mil vinyl film at low room temperature, which decreases with an increase in temperature. In contrast, vinyl-coated Kevlar fabrics reach a strip tensile strength of 3000 pli and urethane-rubber-coated Kevlar fabrics can resist even 4000 pli. It may be concluded that ordinary fabric membrane strengths up to 1000 pli are available on the market.

The joining of fabric strips is critical in textile membrane construction so that not only the necessary forces can be transferred, but also so that they are waterproof. In addition, since joints are visible, their layout becomes an important design consideration. Joints are made either in the shop and/or on site. The common connection systems are the following:

- *Welding*: Here the seams are bonded by heat and pressure; in other words, the coatings are fused. The heat sealing of joints is most common and is used for the joining of vinyl-coated polyester and Teflon-coated fiberglass fabrics.
- *Gluing*: Silicone-coated fiberglass joints are glued using special silicone adhesives.

- *Sewing*: Sewed seams have a lower strength than the membranes they are joining; they also tend not to be completely waterproof. Vinyl-coated Kevlar is sewn since it is impossible for the joints to develop the strength of the coated fabric.
- *Mechanical joints*: For site joints and temporary, low-strength fabric structures, common methods of joining are clamping (e.g., clamp plates), double-luff grooves, bolting, and sleeves to contain cables.

Most critical are the attachment points where the stress concentrations occur. The design of the details must allow smooth transition of forces and avoid wrinkling of the membrane by permitting it to move and adjust under load action. As the scale of the structure increases, the fabric must be reinforced so that the skin is relieved of stresses, which it may not be able to resist by itself; these forces are transferred and carried instead by high-strength polyester webbing or steel cables. The cables may be sewn into the edges of the prefab strips, which are then clamped together at the site, or they may be attached to the fabrics by being covered with fabric strips; they may be attached within heat-sealed sleeves directly into and across the fabric envelope for pneumatic structures. For larger tent structures, the fabric is attached to the cable network.

The stress relief systems for larger structures (i.e., webbings and cable reinforcing) may only appear at points of stress concentration, that is, at specific locations, or may form a regular pattern in one or two directions. Where the fabric strips are attached to each other, the overlapping and doubling of the material result in a reinforcing of the membrane that is equivalent to webbings. For very large scale structures, cable nets form the primary surface structure that supports the roof cladding, as is further discussed in Section 9.5.

Fabrics are coated not just from a stress point of view, but also to make them airtight and waterproof and weatherproof (against effects of sunlight, acids, gas, etc.) and to make them resistant to ripping and abrasion. In other words, besides strength and stiffness, energy efficiency, noise reduction, weathering properties, durability, and flammability and smoke toxicity may have to be considered. The performance and appearance of coated membranes are primarily influenced by ultraviolet (UV) light, together with heat causing loss of strength and stiffness, as well as by dirt and smog causing staining and discoloration of the coated membrane. From a fire safety point of view, the performance of fabric structures is better than that of all-wood structures. Since fabric structures generally enclose large volumes of space, smoke can easily disperse and rise away from occupants; fire will cause the fabric to melt (and does not add fuel to fire) and open up so that the smoke can escape. Although fabric structures are one of the lowest fire hazard construction types, the problem lies in the protection of the content, since overhead sprinkling systems are not compatible with the nature of the fabric structures.

Each of the coated fabric membranes has certain advantages and disadvantages. Vinyl-coated polyester is the most used architectural fabric because of its low cost, strength, and durability. But it must be regularly cleaned or painted unless a protective top coating (e.g., Tedlar) is laminated to it to prevent soiling and to diminish sunlight damage. Furthermore, creep must be considered in the design, which makes it difficult to predict its behavior. Vinyl-coated polyester fabrics rarely meet code fire-resistance test requirements for permanent conventional building materials.

As mentioned previously, Teflon-coated fiberglass fabric is employed for permanent noncombustible structures of larger scale. Although it is much more expensive than plain vinyl-coated polyester, it is much stronger, stiffer, more permanent, incombustible, and more reliable; its average life span is 25 years. In contrast to vinyl-coated polyester fabrics, Teflon-coated fiberglass fabric is much less susceptible to permanent elongations and is therefore more predictable. Fiberglass fabrics have a high melting temperature and are essentially noncombustible and do not contribute any fuel or smoke to a fire. They are self-cleaning, that is, dirt does not stick to them, and hence need no regular maintenance. The brittle character of the fiberglass fabric, however, must be considered in the design and construction of a structure; prefabricated sheets are handled in rolls rather than folded, as is the much more flexible polyester fabric.

Furthermore, Teflon-coated glassfiber fabrics have a high reflectivity and low heat absorption as well as a translucency ranging from 8% to 20%. In contrast, silicone-coated fiberglass fabrics offer a higher translucency of up to 60% besides being less expensive, but they have a high heat absorption, show more brittle stress–strain characteristics, and turn gray with age. Vinyl-coated polyester fabrics with Tedlar top coating have many of the same characteristics of Teflon-coated glass-fiber fabrics.

From an energy efficiency point of view, coated fabrics provide a highly reflective surface together with a low thermal mass. For example, some coated fabrics may reflect between 30% and 75% of the sunlight and, at the same time, allow up to 80% light transmission, thereby reducing the need for artificial lighting. They are energy savers in warm climates by keeping spaces cool; in other words, coated fabrics reflect enough light to eliminate or reduce air-conditioning costs and allow enough light through to eliminate daytime lighting costs. White tents are usually recommended, although dark ones give a more subdued daylight; but they also allow a high solar heat gain.

In moderate climates, insulation is provided with two layers of translucent coated fabrics (liners) with a dead air space of usually 4 in. between them. Here the elimination of daytime lighting costs must be weighed against the extra heat loss through the translucent area. In cooler climates, the two fabric layers may enclose translucent insulation of, say, 8 in., thereby, however, reducing the roof membrane's translucency, but still allowing adequate lighting.

Often it is claimed that savings on lighting and cooling costs in summer can outweigh the higher heating costs due to greater heat loss in winter. Energy efficiency can be optimized by controlling heat and light intake during the seasons through a *thermally active roof*, similar to responsive glazing, as part of the intelligent products of the future. Acoustical problems in large spaces, due to the geometry of the curved roof and lack of sound-absorbing material, are overcome by inner acoustical liners that absorb and reduce reverberations or by vertical banners that break up the continuity of the membrane surface. In contrast to air and tensile structures, which can be insulated and are much better sealed because they are fixed in place, tents need up to four times more heat to maintain a given temperature as compared to conventional structures.

In conclusion, it must be emphasized that the construction of fabric structures involves true team interaction, for example, among the following companies of the early period of tensile structure development: the yarn *producer* (e.g., Owens–Corning Fiberglas Corp. for Beta fiberglass filament), the *weaver* (e.g., Chemfab for fiberglass fabric), the *coater* (e.g., Du Pont Corp. for Teflon), the *fabricator/contractor* (e.g., Birdair Structures, Inc., for fabric structures coated with Teflon), and the *designers*. For further information on fabric structures, refer to the publications catalog of the Industrial Fabrics Association International (IFAI).

Loads

Tensile structures are generally of light weight. The magnitude of the roof weight is a function of the roof skin and the type of stabilization used. The typical weights of textile fabrics used for awnings range from about 10 to 18 oz/yd^2 (0.07 to 0.13 psf), and the weights of common coated polyester fabrics are in the range of approximately 24 to 32 oz/yd^2 (0.17 to 0.22 psf). A fabric membrane on a cable net may weigh up to approximately 1.5 psf, while plastics or corrugated metal sheets on cables may cause a load of 5 to 15 psf and wood or lightweight concrete between 15 and 30 psf. In other words, the roof weight ranges from about 0.20 psf for a coated fabric membrane to about 30 psf for concrete on cables. A typical fabric roof is about one-thirtieth the weight of a steel truss roof. The prestress required to maintain stability of the fabric membrane, depending on material and loading, is usually in the range of 25 to 50 lb/in.

The lightweight nature of membrane roofs is clearly expressed by the air-supported dome of the 722-ft-span Pontiac Stadium in Michigan, weighing only 1 psf, or by the 688-ft-span Suncoast cable dome in St. Petersburg, Florida, with 5 psf. For the weights of other roof structures, refer to the discussion of cases, which also include typical roof cladding systems that may consist of roofing, deck, insulation, and possibly framing, depending on the capacity of the deck.

Since the weight of typical *pretensioned* roofs is relatively insignificant, the stresses due to the superimposed primary loads of wind, snow, and temperature change tend to control the design. These loads may be treated as uniform loads for preliminary design purposes and the structure weight can be ignored. The typical loads to be considered are snow loads, wind uplift, dynamic loads (wind, earthquake), prestress loads, erection loads, creep and shrinkage loads, movement of supports, temperature loads (uniform temperature changes and temperature differential between faces), and concentrated loads, for example, due to a catwalk system and scoreboard for a stadium. Refer to Sections 2.1, 4.1, and 9.8 for a more detailed discussion of loads.

The load combinations that must be considered for the selection of cables can be based on the *Manual for Structural Applications of Steel Cables for Buildings* (AISI, 1973). According to this reference, the effective design breaking strength of cables in protected interior environments should not be less than the largest value produced by the typical cable tension conditions presented. In this context, for preliminary design purposes, a load factor $\gamma = 2.2$ is used as based on the cases of dead load plus prestress $(D + P)$, dead load plus live load plus prestress $(D + L + P)$, and $\gamma = 2.0$ is used when wind acts together with other loads, such as $(D + P + W)$ or $(D + L + P + W)$.

Therefore, the effective design breaking strength of the cable ϕT_n must be equal to or greater than the tension force multiplied by the respective load factor.

$$\phi T_n \geq T_u = \gamma T \tag{9.1}$$

With respect to the preliminary design of coated fabric membranes, a **minimum safety factor of four** is generally used as applied against the dry-strip tensile strength of the material under normal temperature ranges, as is the case in the United States. In other words, coated fabrics should have an ultimate strip tensile strength of not less than four times the maximum fabric stress in the warp and filling directions. For some materials such as polyester, this high safety factor takes into account the type of loading and reductions in fabric strength resulting from weathering (i.e., deterioration of fabric with time), biaxial loading, and handling. Keep in mind, however, that this safety factor may not be conservative since fabrics may fail by *tearing* as initiated at stress concentrations, at punctures (e.g., punched holes for joints), and at fabric stresses far below the strip tensile strength; it also may not be conservative for long-term loading of certain materials.

9.2 THE SINGLE CABLE

The cable is inherently flexible and form active since it can carry only tensile forces. Because of the absence of bending and buckling, as well as due to the high strength of cable material, the cables are rather light.

A single cable must adjust its suspended form to the respective loading condition so that it can respond in tension. For example, under single-load action it takes the shape of a string or *funicular polygon*, whereas under uniform load action the polygon changes to a curve and, depending on the type of loading, takes familiar geometrical forms, such as parabolic, elliptical, and circular shapes (Fig. 9.5), called *funicular curves*. These funicular tension lines are discussed in more detail in the following sections.

The form active nature of the cable can be illustrated by allowing a single load to move along a cable from midspan to the supports (Fig. 9.6a). The change of the string polygons clearly shows that the sag decreases as the load approaches the supports; the corresponding force diagrams indicate that the cable forces and hence cable sizes increase. In other words, the vertical reaction forces keep on increasing, while the magnitude of the horizontal reaction forces H decreases with decrease of beam moments and only slightly increases in proportion to the decrease in cable sag f.

When the cable is transversely loaded by a concentrated load P at midspan, the following observations can be made with respect to cable length (or cable sag at midspan) and force flow (Fig. 9.6b):

- The force polygon demonstrates that with decrease of cable sag the cable forces and horizontal reactions increase, whereas the vertical reaction forces do not change. The forces are infinitely large when the cable is in horizontal position, which obviously cannot be the case unless the cable is prestretched or replaced by a (stiff) bending member.
- With increase of cable length, the cable forces or cable sizes decrease, but the cable weight increases.

As a general guide, a sag-to-span ratio n in the range of 1:10 to 1:25 for buildings and 1:8 to 1:12 for suspension bridges is considered optimum. For a sag-to-span ratio of larger than 1:5, high boundary support structures have to be designed, which may not be economical. When the sag-to-span ratio becomes too low (i.e., for a very flat cable profile), large sizes are required for cables and boundary members.

For preliminary design purposes, the cable may be treated as inextensible, as based on the small deflection theory, so elementary statics can be used. In reality, the

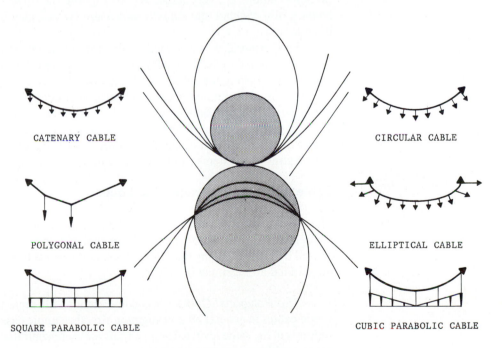

CATENARY CABLE

CIRCULAR CABLE

POLYGONAL CABLE

ELLIPTICAL CABLE

SQUARE PARABOLIC CABLE

CUBIC PARABOLIC CABLE

Figure 9.5 Funicular tension lines.

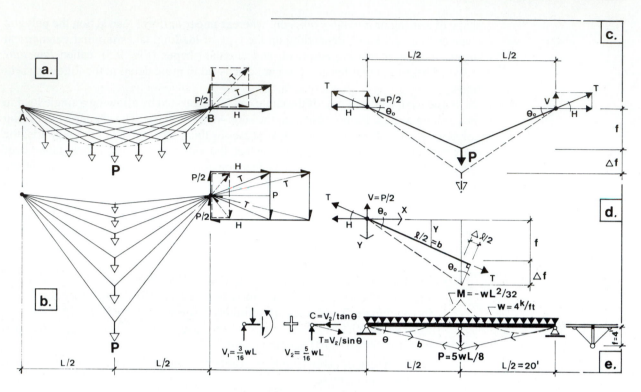

Figure 9.6 Polygonal cable.

problem is indeterminate since the flexible nature of the cable causes large deflections, which, in turn, influence the magnitude of the force flow. The assumption that the force flow is derived from the initial geometry of the structure, as based on the given cable length, may be reasonable for first estimation purposes as long as the suspended cable profile is not too shallow, in which case the axial forces and corresponding elastic deformations are very large. We also ignore the effects of cable extension due to *temperature changes* and *support movement* (which may become critical for flexible boundaries).

Furthermore, it is assumed in the following investigations that the transverse loads are applied to the horizontal projection of the cable (i.e., act over the horizontal span) and not along the cable, which includes the cable weight.

To develop an understanding of tension structures, first single cables under various load actions are studied. Single-cable action can be found, for instance, in simply suspended roofs where cables are not prestressed, cable-stayed roofs and bridges, and guyed towers. The loading conditions needed for the preliminary design of common tensile structure systems can be grouped in transverse and radial loads.

Cable Action under Transverse Loads

The typical transverse loads that will be discussed include single-load action at cable midspan, uniformly distributed loads on cables with horizontal and inclined chords as caused by parallel arrangement of cables, and various triangular loading cases reflecting a radial arrangement of cables.

The Polygonal Cable   It is convenient to illustrate the basic characteristics of single cables in general by investigating first the simple case of a suspended cable with supports at the same level carrying a transverse concentrated load at midspan and treating the cable weight as negligible (Fig. 9.6c).

Due to symmetry of loading, the vertical reaction forces are each equal to

$$A_v = B_v = V = P/2$$

The horizontal thrust forces H at the reactions are obtained by taking moments about the concentrated load at midspan (Fig. 9.6d).

$$\Sigma M = 0 = (P/2)L/2 - H(f) = M_{beam} - M_{cable\ profile}$$
$$H = (PL/4)/f = M_{max}/f$$
$$M_{beam} = H(f), \quad \text{or} \quad M_x \propto y, \qquad \text{since } H \text{ is constant} \tag{a}$$

Hence, in general,

$$A_h = B_h = H = M_{max}/f = M_x/y \tag{9.2}$$

The specific relationship, Eq. (a), can be generalized for any type of transverse loading on a simple cable by concluding that the cable sag f_x (which is equal to the cable shape y) at any point is directly proportional to the moment diagram of an equivalent beam on the horizontal projection carrying the same load, since the horizontal thrust force H is constant throughout. In other words, the beam moment M_x at any point along the cable is resisted by a force couple formed by the H forces separated by the cable sag $f_x = y$; at the largest sag, $f_x = f$ is the location of the maximum beam moment, M_{max}. In a rigid beam the moments are resisted by *bending stiffness*, while a flexible cable uses its *geometry* to carry the rotation in pure tension.

It is extremely helpful to visualize the deflected shape of the cable (i.e., cable profile) as the shape of the moment diagram of an equivalent, simply supported beam carrying the same loads as the cable. The *moment analogy method* is useful since the magnitude of moments can often be readily obtained from handbooks.

The maximum cable force occurs generally at the reaction where the angle of inclination is maximum. The slope of the cable curvature at the reaction is proportional to the reaction forces.

$$\tan \theta_0 = V/H = y/x = f/(L/2) = 2n \tag{9.3}$$
$$\cos \theta_0 = H/T_{max}, \qquad \sin \theta_0 = V/T_{max}$$

where $f/L = n$ = sag-to-span ratio.

The maximum *cable force* can be determined according to Pythagoras' theorem or as based on the slope of the cable curvature at the critical reaction.

$$T_{max} = H/\cos \theta_0 = V/\sin \theta_0 = \sqrt{V^2 + H^2} = H\sqrt{1 + \tan^2 \theta_0} \tag{9.4}$$

For this specific case for $\tan \theta_0 = 2n$,

$$T_{max} = H\sqrt{1 + 4n^2} \tag{9.5}$$

For the condition where the sag-to-span ratio is less than about 1:10, the following simplified version of the equation may be used.

$$\left(1 + \frac{4}{2}n^2\right)^2 = 1 + 4n^2 + \left(\frac{4}{2}n^2\right)^2, \quad \text{or} \quad 1 + 2n^2 \cong \sqrt{1 + 4n^2}$$

where the last term can be treated as equal to zero, so that the maximum cable force is equal to

$$T_{max} = H\sqrt{1 + 4n^2} \cong H(1 + 2n^2) \tag{9.6}$$

By placing the origin of the coordinate system at the reaction (Fig. 9.6d), the *cable profile* can be defined as

$$f/(L/2) = y/x, \quad \text{or} \quad y = 2nx \tag{9.7}$$

The *cable length* is obtained by simply adding up the inclined portions.

$$l = 2(l/2) = 2\sqrt{(L/2)^2 + f^2} = L\sqrt{1 + 4n^2}$$

When the sag-to-span ratio is less than about 1:10, this expression can be simplified as has been done for T_{max}.

$$l = L\sqrt{1 + 4n^2} \cong L(1 + 2n^2) \tag{9.8}$$

This equation can be transformed into a general form reflecting the loading condition of equally spaced transverse single loads of equal magnitude.

$$l \cong L(1 + kn^2) \tag{9.9}$$

where   number of P – loads: 1   2   3   4   5   6 P/ft
        with corresponding k: 2   3   2.5   2.8   2.6   2.7   2.67

To determine the *cable elongation* for the single-loading case, the two cable portions are assumed to act as independent segments, where each deforms under its axial load, assuming the material to behave linearly elastic. Therefore, according to Eq. (2.54),

$$\Delta l = 2(\Delta l/2) = 2(Tl/2)/AE = Tl/AE \quad \text{or}$$

$$\Delta l = \frac{TL}{AE} \cong \frac{TL}{AE}(1 + 2n^2) \tag{9.10}$$

The variation in cable length due to change in temperature, according to Eq. (2.8), is

$$\Delta l_t = \alpha l \, \Delta T \cong \alpha \, \Delta T L (1 + 2n^2) \tag{9.11}$$

To find the *increase in cable sag,* the linearly deformed cables are reassembled (Fig. 9.6c). The elongated end points intersect with the new end point at a distance Δf below the original location f. Because of the small scale of the sag increase due to the shallow profile, the inclined portion is assumed to be rotated along a straight line; this assumption simplifies the calculations considerably (see Fig. 9.6d).

$$\sin\theta = \frac{\Delta l/2}{\Delta f} = \frac{f}{L/2} = 2n, \quad \text{or} \quad \Delta f \cong \Delta l/4n \tag{9.12}$$

With respect to dynamic behavior, this cable system may be treated as a single-degree-of-freedom system where all the mass is assumed to be concentrated at midspan, and only vertical movement is allowed. The effect of dynamic behavior is discussed later in this section.

EXAMPLE 9.1

A cable-supported, single-strut steel beam (also called king-post truss) is investigated. This composite structure consists of a beam that is supported from below at midspan by a strut resting on a cable, as shown in Fig. 9.6e. This structure is externally determinate and internally indeterminate. It is generally conservative to treat the elastic strut support as not settling so that the magnitude of the maximum beam moment of $wL^2/8$ at the strut support can be looked up in the beam formulas tables in the AISC Manual or Table A.15. In the exact solution, the support moment at midspan and the axial forces would decrease, while the field moments would increase, as discussed for Fig. 9.14s, u; a slight settlement is beneficial. The depth of cable-supported beams is usually taken as $f \geq L/12$.

The maximum moment for the continuous two-span beam occurs at the midspan support and is equal to $M_{max} = wL^2/8$. But in this case the beam spans L are treated as $L/2$ with respect to the entire cable-supported beam spanning $L = 40$ ft. Therefore, the maximum beam moment at the strut due to 4 k/ft, is

$$M_{max} = w(L/2)^2/8 = wL^2/32 = 4(40)^2/32 = 200 \text{ ft-k}$$

The strut represents the interior reaction of the beam (see Table A.15) and carries the following axial compression force:

$$P = 2(5w(L/2)/8) = 5wL/8 = 5(4)40/8 = 100 \text{ k}$$

The exterior vertical reaction due to the strut load (or vertical component of the diagonal cable force) is

$$V_2 = P/2 = (5wL/8)/2 = 5wL/16 = 100/2 = 50 \text{ k}$$

Summing up the vertical forces at the exterior reaction joint yields the tension force T in the diagonals based on

$$\tan \theta = f/(L/2) = 4/20 = 0.2, \qquad \sin \theta = f/b = V_2/T$$
$$T = V_2(b/f) = V_2/\sin \theta = (5wL/16)/\sin \theta = 50/\sin \theta = 255 \text{ k}$$

Summing up the horizontal forces at the reaction yields the compression in the beam as caused by the diagonal tension.

$$\tan \theta = f/(L/2) = V_2/C$$
$$C = V_2(L/2f) = V_2/\tan \theta = 5wL^2/32f = 50(40)/2(4) = 250 \text{ k}$$

The horizontal thrust forces can also be found according to the moment analogy method using $P = 5wL/8$.

$$H = C = M_{max}/f = (PL/4)/f = (5wL/8)L/4/f = 5wL^2/32f$$
$$= (100(40)/4)/4 = 250 \text{ k}$$

The cable thrust force H is resisted by the beam in compression.

The beam must resist a maximum moment of 200 ft-k and an axial force of 250 k; A36 steel and a W14 section are assumed for this first trial. Since the beam is fully laterally supported by the floor structure, the effect of slenderness can be ignored and the allowable axial stresses may be assumed to be controlled by yielding. Hence, the required cross-sectional area of the beam can be estimated based on the bending factor approach (using $B_x = 0.18$ for W14) according to Eq. (3.100).

$$A_s = (P + B_x M_x)/0.6 F_y = (250 + 0.18(200 \times 12))/0.6(36) = 31.57 \text{ in.}^2$$

Because of the conservative nature of this approximation, not the section required, but the next lower one, will be tried.

$$\text{W14} \times 99, \qquad A_s = 29.1 \text{ in.}^2, \qquad B_x = 0.185$$

The cable must resist an ultimate tensile force, according to Eq. (9.1), of

$$T_u = \gamma\, T = 2.2(255) = 561 \text{ k}$$

Try a $2\frac{3}{16}$-in.-diameter galvanized structural strand (Table A.13) with a nominal breaking strength of 586 k.

The strut is assumed to act as a short column laterally braced at its ends for preliminary design purposes. The required cross-sectional area of a steel pipe with $F_y = 36$ ksi to carry 100 k is

$$A = P/F_a \cong P/0.6 F_y = 100/0.6(36) = 4.63 \text{ in.}^2$$

Try PIPE 6 STD, $A = 5.58 \text{ in.}^2$. This section can be verified by using the *column tables* in the AISC Manual for $KL \cong 1(4) = 4$ ft.

The Parabolic Cable   Under a uniform load w_x the simple cable takes a curved shape, a situation that is most typical for tensile roof structures. When this distributed transverse load is constant (w) and acting on the horizontal projection of the cable, the cable responds with the shape of a *parabola,* as will be shown in the following discussion. This situation occurs in suspension bridges where the suspended cables carry the roadway by ignoring the effect of the weight of the cables, and it also exists for shallow suspended roofs where the cables are arranged in a parallel fashion directly supporting the roof.

In reality, however, a cable carrying a uniform load along its length, such as its own weight, assumes the shape of a *catenary* or a hyperbolic cosine (cosh) curve. It can be shown that for small sag-to-span ratios ($n \le 1{:}10$) the geometry of a catenary and a parabola are practically the same. Since the equations defining a parabola are much simpler than the ones for a catenary, parabolic equations are generally used; that is, the loads are assumed to act uniformly along the span, rather than along the cable.

For the simple suspended cable with supports at the same level (Fig. 9.7h), the vertical reactions are equal to each other because of symmetry of loading.

$$A_v = B_v = V = wL/2 = W/2 \tag{9.13}$$

The horizontal reactions (or minimum cable force at midspan where the cable shape is zero) are found by using the moment analogy method and Eq. (9.2).

$$A_h = B_h = H = M_{max}/f = (wL^2/8)/f \tag{9.14}$$

The maximum cable slope occurs at the reactions and, according to Eq. (9.3), is equal to

$$\tan\theta_0 = V/H = 4f/L = 4n \tag{9.15}$$

The maximum cable force occurs generally at the reactions where the cable slope is steepest and, according to Eq. (9.4), is equal to

$$T_{max} = H/\cos\theta_0 = V/\sin\theta_0 = H\sqrt{1 + \tan^2\theta_0} = H\sqrt{1 + 16n^2} \tag{9.16}$$

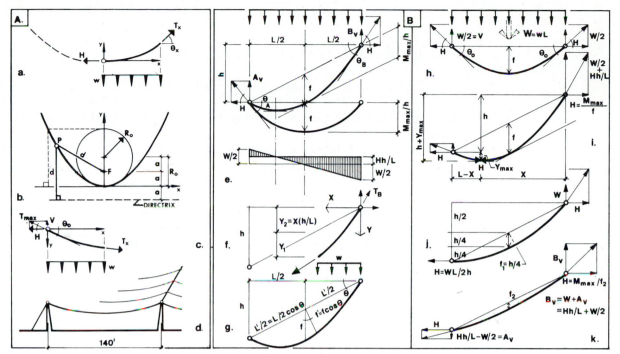

Figure 9.7 Parabolic cable.

The cable geometry is derived from the free body in Fig. 9.7a. The origin of the coordinate system is placed at the lowest point of the cable, where the cable slope is zero. Rotational equilibrium about T_x yields

$$\Sigma M_x = 0 = H(y) - wx(x/2), \quad \text{letting } H = wL^2/8f \qquad (a)$$
$$y = wx^2/2H = (4f/L^2)x^2$$

This is the equation for the parabola with its apex at the intersection of the coordinate axes. The general mathematical equation for this condition is

$$y = (1/4a)x^2 \qquad (b)$$

Mathematically, a parabola is defined as a locus of a point moving in a plane so that its distance d from a fixed line called the directrix is always equal to its distance from a fixed point F, called the focus, not on the line. In Eq. (b) the coefficient a (Fig. 9.7b) determines the shape, that is, the flatness of the parabola. For the uniform loading case the coefficient a can be derived from Eqs. (a) and (b) as follows:

$$1/4a = w/2H = 4f/L^2, \quad \text{or} \quad R_0 = 1/k_0 = 2a = H/w = L^2/8f \qquad (9.17)$$

where R_0 is the radius of curvature at the apex and k_0 is the *curvature* at that point (Fig. 9.7b). Hence, the general equation for the uniform loading with the coordinate system at the cable low point is

$$y = \frac{x^2}{2R_0} = \frac{k_0 x^2}{2} = \frac{wx^2}{2H} = 4f\left(\frac{x}{L}\right)^2 \qquad (9.18)$$

The slope of the curvature $\tan\theta_x$ is easily found by taking the derivative of y with respect to x.

$$\tan\theta_x = \frac{dy}{dx} = \frac{x}{R_0} = xk_0 = \frac{wx}{H} = \frac{8fx}{L^2}$$

At the supports, where $x = \pm L/2$ and $y = f$, the slope is

$$\tan\theta_0 = \pm 4f/L = \pm 4n = V/H \tag{9.15}$$

This result was already derived by using the ratio of vertical to horizontal reaction forces.

Sometimes it is helpful to express the geometry of the parabola by placing the origin of the coordinate system at one of the reactions (Fig. 9.7c). Rotational equilibrium about a point located at a distance x from the support yields

$$\Sigma M_x = 0 = \frac{wL(x)}{2} - \frac{wL^2(y)}{8f} - wx\frac{x}{2}$$

The equation of the parabola is

$$y = \frac{4f}{L^2}(Lx - x^2) = 4f\left[\frac{x}{L} - \left(\frac{x}{L}\right)^2\right] \tag{9.19}$$

The slope is

$$\tan\theta_x = \frac{dy}{dx} = \frac{4f(l-2x)}{L^2} \tag{9.20}$$

The cable length is obtained by integrating the differential length dl along the cable (see Schueller, 1983). Its initial length can be approximated for preliminary design purposes, according to Eq. (9.9), where f is much less than $L/4$, as

$$l \cong L\left(1 + \frac{8}{3}n^2\right) \tag{9.21}$$

The elongation of a linear cable of length L and cross-sectional area A due to elastic stretch in response to a constant force T is

$$\Delta L = TL/AE \tag{2.54}$$

Along the parabolic cable, however, the tensile force T_x varies. The total change in length of the cable is equal to the summation of the elongations of the differential length dl as caused by T_x (see Schueller, 1983). The *increase* in *cable length* due to elastic stretch can be approximated as

$$\Delta l \cong \frac{HL}{AE}\left(1 + \frac{16}{3}n^2\right) \tag{9.22}$$

The increase (or decrease) in cable length due to change in temperature is

$$\Delta l_t \cong \alpha l \Delta T \cong \alpha \Delta TL\left(1 + \frac{8}{3}n^2\right) \tag{9.23}$$

The approximate *increase in cable sag* may be obtained by taking the derivative of the cable length l with respect to cable sag f_x

$$\frac{dl}{df} = \left(\frac{d}{df}\right)L\left(1 + \frac{8}{3}n^2\right) = \frac{d}{df}\left(L + \frac{8f^2}{3L}\right) = \frac{16f}{3L} = \frac{16n}{3}$$

Considering dl/df as a finite quantity yields an approximate increase of cable sag equal to

$$\Delta f \cong \frac{3\Delta l}{16n} \tag{9.24}$$

For the more general case where the supports are not at the same level (Fig. 9.7B), the transverse uniform loading causes the same horizontal reaction forces as for the previous case where the supports are at the same level. It has been shown in the section on simply supported inclined beams under vertical loads that the moment flow is independent of beam inclination, so the maximum moment can be determined from the equivalent beam on the horizontal projection (Fig. 5.27). Therefore, we may conclude that the moment analogy method can also be applied to a cable with supports at different levels (Fig. 9.7e).

$$A_h = B_h = H = M_{\max}/f = (wL^2/8)/f \tag{9.14}$$

The vertical reactions are obtained by taking moments about the supports A and B, respectively, and letting $wL = W$.

$$A_v = A_v' - Hh/L = W/2 - Hh/L \tag{9.25a}$$
$$B_v = B_v' + Hh/L = W/2 + Hh/L$$

where A_v' and B_v' are the reactions for a horizontally supported cable carrying the same loads as the inclined cable. For a short, suspended, inclined cable with a small sag (Fig. 9.7k), the vertical reaction at support A is acting downward and will be equal to

$$A_v = Hh/L - A_v', \qquad B_v = Hh/L + B_v' \tag{9.25b}$$

The slope of the cable curvature at the reactions is proportional to the reaction forces.

$$\tan\theta_A = A_v/H, \qquad \tan\theta_B = B_v/H \tag{9.26}$$

The maximum cable force occurs at support B where the cable slope is maximum.

$$\tan\theta_B = B_v/H = wL/2H + h/L = 4n + h/L \tag{9.27}$$

The maximum cable force can be determined according to Pythagoras' theorem or as based on the cable curvature at the reaction.

$$T_B = T_{\max} = H/\cos\theta_B = B_v/\sin\theta_B \tag{9.28}$$
$$\doteq \sqrt{B_v^2 + H^2} = H\sqrt{1 + \tan^2\theta_B} = H\sqrt{1 + (4n + (h)/L)}$$

The cable geometry is determined by locating the origin of the coordinate system at the upper support (Fig. 9.7f). The vertical distance to the cable at a horizontal distance x is

$$y = y_1 + y_2$$

The vertical displacement y_1 is equal to the sag at that location for a cable with supports at the same level. The vertical distance y_2 is linearly related to the height h.

$$y_2/h = x/L \quad \text{or} \quad y_2 = hx/L$$

Hence, the equation for the parabola, by substituting Eq. (9.19), is

$$y = y_1 + y_2 = \frac{4f}{L^2}(Lx - x^2) + hx/L \tag{9.29}$$

The cable slope is found by taking the derivative of y with respect to x.

$$\tan\theta_x = \frac{dy}{dx} = \frac{4f}{L^2}(L - 2x) + \frac{h}{L} \tag{9.30}$$

The slope of the cable curvature at the supports is

$$x = 0: \quad \tan\theta_B = \frac{h}{L} + \frac{4f}{L} = \frac{h}{L} + 4n$$

$$\tag{9.31}$$

$$x = L: \quad \tan\theta_A = \frac{h}{L} - \frac{4f}{L} = \frac{h}{L} - 4n$$

The slope of the curvature at midspan is

$$x = \frac{L}{2}: \quad \tan\theta_{L/2} = 4f\left(\frac{1}{L} - \frac{1}{L}\right) + \frac{h}{L} = \frac{h}{L} \tag{9.32}$$

The slope of the curvature at midspan is parallel to the chord connecting the supports (Fig. 9.7e). It has been discussed already that for the inclined cable structure the maximum cable sag f is the same as for the projected horizontal cable system (Fig. 9.7e). The low points of the two structures, however, do not occur at the same position. Should the low point y_{max} be given for the inclined cable for the purpose of clearance requirements, its location must be found. The cable sag has its low point where the slope of curvature is zero. Hence, according to Eq. (9.30), the location x of the low point (Fig. 9.7i) is at

$$\tan\theta_x = \frac{dy}{dx} = 0 = \frac{4f}{L^2}(L - 2x) + \frac{h}{L}, \quad \text{or} \quad x = \frac{L}{2}\left(1 + \frac{h}{4f}\right) \tag{9.33}$$

For the special case where the curvature at the lower support is zero (Fig. 9.7j), the following relationship between maximum cable sag and difference in support elevation according to Eq. (9.31) or (9.33) for $x = L$ is

$$\tan\theta_A = 0 = \frac{h}{L} - 4n, \quad \text{or} \quad f = \frac{h}{4} \tag{9.34}$$

Should the lower support be at the ground, the maximum cable sag should not be larger than a quarter of the height h in order for the cable to clear the surface.

T_x at any point along the cable is given by the following familiar expression, although only the maximum cable force at one of the reactions is of interest since it is one of the determinants for the design of the cable.

$$T_x = \frac{H}{\cos\theta_x} = \frac{V_x}{\sin\theta_x} = H\sqrt{1 + \tan^2\theta_x} \tag{9.35}$$

For small sags it is considered accurate enough to derive the *length of the inclined cable* from the cable length for a parabolic cable with supports at the same level. According to Eq. (9.21), the cable length is

$$l \cong L'\left[1 + \frac{8}{3}(n')^2\right]$$

where (see Fig. (9.7g)) $n' = \dfrac{f'}{L'} = \dfrac{f\cos\theta}{L/\cos\theta} = n\cos^2\theta$

Hence, the approximate cable length for the inclined cable is

$$l \cong \frac{L}{\cos\theta}\left(1 + \frac{8}{3}n^2\cos^4\theta\right) \tag{9.36}$$

The cable elongation is obtained in a manner similar to that used for the cable with supports at an equal level (see Schueller, 1983). The increase in cable length as caused by the tensile forces is

$$\Delta l = \frac{HL}{AE}\left[1 + \frac{16n^2}{3} + \left(\frac{h}{L}\right)^2\right] \tag{9.37a}$$

The increase in cable length as caused by temperature increase is

$$\Delta l_t = \frac{\alpha\Delta T}{\cos\theta}\left(1 + \frac{8}{3}n^2\cos^4\theta\right) \tag{9.37b}$$

EXAMPLE 9.2

A typical cable of a single-layer suspension roof (Fig. 9.7d) is investigated for preliminary design purposes. The cables are spaced on 6-ft centers and span 140 ft. The maximum sag-to-span ratio is 1:15. Dead and live loads are 20 and 30 psf, respectively; they are assumed to act on the horizontal roof projection. Temperature change is 50°F.

The maximum cable sag is

$$n = f/L = 1/15 \quad \text{or} \quad f = 140/15 = 9.33 \text{ ft}$$

The uniform load that a typical interior cable must support is

$$w = 6(0.020 + 0.030) = 0.3 \text{ k/ft}$$

The minimum horizontal cable force, according to Eq. (9.14), is

$$H = \frac{wL^2}{8f} = \frac{0.3(140)^2}{8(9.33)} = 78.78 \text{ k}$$

The maximum cable slope, according to Eq. (9.15), is

$$\tan\theta_A = 4n = 4/15 = 0.267$$

The maximum cable force, according to Eq. (9.16), is

$$T_{max} = H\sqrt{1 + 16n^2} = 78.78\sqrt{1 + 16(1/15)^2} = 81.53 \text{ k}$$

or

$$T_{max} = H/\cos\theta = 78.78/0.966 = 81.53 \text{ k}$$

Notice that there is only about 3.5% difference between the largest (T_{max})and smallest (H) tensile force; the difference decreases as the cable profile becomes flatter. The ultimate tensile strength required is

$$T_u = \gamma T = 2.2(81.53) = 179.37 \text{ k}$$

Try $1\frac{1}{4}$-in.-diameter galvanized structural strand (see Table A.13).

$$A = 0.938 \text{ in.}^2, \qquad T_u = 192 \text{ k}, \qquad E = 24{,}000 \text{ ksi}$$

The equation defining the cable shape according to Eq. (9.18) is

$$y = \frac{wx^2}{2H} = \frac{0.3x^2}{2(78.78)} = \frac{x^2}{525.2}$$

For instance, for the cable displacement 30 ft from the support or 40 ft from mid-span,

$$x = 40 \text{ ft}, \qquad y = 40^2/525.2 = 3.05 \text{ ft}$$

or $9.33 - 3.05 = 6.28$ ft, measured from the horizontal chord connecting the two supports. The same result is obtained by using Eq. (9.19).

$$y = 4f\left[\frac{x}{L} - \left(\frac{x}{L}\right)^2\right] = 4(9.33)\left[\frac{30}{140} - \left(\frac{30}{140}\right)^2\right] = 6.28 \text{ ft}$$

The approximate cable length for one span from Eq. (9.21) is

$$l \cong L\left(1 + \frac{8n^2}{3}\right) = 140\left[1 + \frac{8}{3}\left(\frac{1}{15}\right)^2\right] = 141.66 \text{ ft}$$

Note that there is only about 1% difference between the cable length and its span. The cable elongation due to the tensile action from Eq. (9.22) is

$$\Delta l \cong \frac{HL}{AE}\left(1 + \frac{16n^2}{3}\right) = \frac{78.78(140)}{0.938\,(24{,}000)}\left[1 + \frac{16}{3}\left(\frac{1}{15}\right)^2\right] = 0.50 \text{ ft}$$

The cable elongation due to temperature increase from Eq. (9.23) is

$$\Delta l_t = \alpha\,\Delta Tl = 6.5 \times 10^{-6}(50)(141.66) = 0.05 \text{ ft}$$

Note that the influence of temperature at this scale is relatively small. Should there be a decrease in temperature, the cable will shorten and reduce the sag, thus increasing the maximum cable force. The approximate increase in cable sag due to elongation from Eq. (9.24) is

$$\Delta f \cong \frac{3\Delta l}{16n} = \frac{3}{16}(0.5 + 0.05)15 = 1.55 \text{ ft}$$

This increase in sag is about $(1.55/9.33)(100\%) = 16.61\%$ of the maximum sag, which clearly shows the large deflection character of the cable structure.

The increase in sag due to the superimposed live loads, neglecting the correction for temperature, is

$$\Delta f_{LL} = 1.55(30/50) = 0.93 \text{ ft} \quad \text{or} \quad \Delta f_{LL}/L = 0.93/140 = 1/151$$

which is quite high in comparison to $L/360$! Taking into account the increase in sag and assuming, for this approximate approach, that the sag is proportional to the maximum tensile force yields the corrected maximum tensile force.

$$T_{\max} = 81.53\frac{9.33}{9.33 + 1.55} = 69.92 \text{k}$$

This force, however, causes less increase in sag and therefore a larger cable force!

The Cubic Parabolic Cable   A cable carrying uniformly distributed, tapered, transverse loads along its horizontal base, such as triangular-, or trapezoidal-shaped loads, takes the funicular form of a cubic parabola. Situations like this usually occur where cables are arranged not in parallel but in a radial fashion. Three common cases will be investigated in the following discussion.

Case A: The cables of a circular suspension roof are arranged in a radial manner. Each cable (with horizontal chords) carries two symmetrically placed, triangularly distributed loads, as shown in Fig. 9.8A, that is, assuming a constant uniform load action on the horizontal projection of the roof. This assumption ignores the effect of the centrally suspended tension ring causing a concentrated load at midspan, as well as the weight being distributed uniformly along the length of the cable. The vertical reactions are equal to each other due to symmetry of loading (Fig. 9.8a).

$$A_V = B_V = V = (wL/2)/2 = wL/4 \qquad (9.38)$$

The horizontal reactions are found by using the moment analogy method (Table A.15).

$$A_h = B_h = H = M_{max}/f = (wL^2/24)/f \qquad (9.39)$$

The maximum cable slope at the reaction is

$$\tan\theta_0 = V/H = 6f/L = 6n \qquad (9.40)$$

The maximum cable force is

$$T_{max} = H/\cos\theta_0 = V/\sin\theta_0 = H\sqrt{1 + \tan^2\theta_0} = H\sqrt{1 + 36n^2} \qquad (9.41)$$

The cable geometry is derived from the free body in Fig. 9.8b. The origin of the coordinate system is located at the lowest point of the cable where the cable slope is zero. Rotational equilibrium about T_X yields

$$\Sigma M_x = 0 = H(y) - (w_x x/2)(x/3), \quad \text{where } w_x = 2w(x/2), \quad \text{or} \quad y = 8f(x/L)^3 \qquad (9.42)$$

The slope of the curvature at any distance x from the low point of the sag is

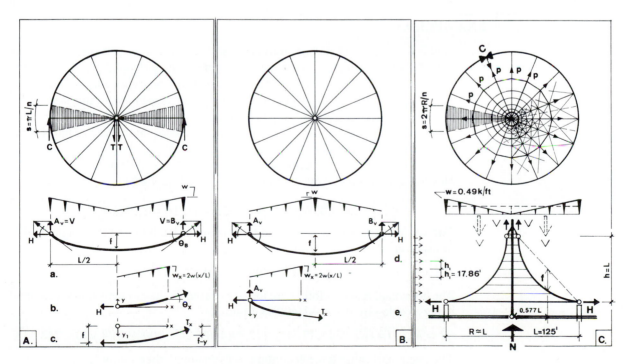

Figure 9.8 Cubic parabolic cable.

$$\tan\theta_x \ = \ \frac{dy}{dx} \ = \ \frac{24fx^2}{L^3} \ = \ 24n\left(\frac{x}{L}\right)^2 \tag{9.43}$$

The slope of the cable at the support at $x = L/2$ is

$$\tan\theta_B = 24n(1/2)^2 = 6n \tag{9.44}$$

The equation for the parabola may also be given by placing the origin of the coordinate system as shown in Fig. 9.8c as

$$y_1 \ = \ f - y \ = \ f\left[1 - 8\left(\frac{x}{L}\right)^3\right] \tag{9.45}$$

The cable length is derived in a similar manner as for the square parabolic cable.

$$l \cong L\left(1 + \frac{18}{5}n^2\right) \tag{9.46}$$

The change in cable length as caused by elongation due to loading is

$$\Delta l \cong \frac{HL}{AE}\left(1 + \frac{36}{5}n^2\right) \tag{9.47}$$

For the derivation of the formulas for elastic stretching and cable length, refer to Schueller, 1983.

The change in cable length due to change in temperature is

$$\Delta l_t \ = \ \alpha l\,\Delta T \cong \alpha L\,\Delta T\left(1 + \frac{18}{5}n^2\right) \tag{9.48}$$

For the approximate increase of cable sag refer to Problem 9.20.

EXAMPLE 9.3

A 400-ft-span, round stadium roof is supported by eight radially arranged cables (Fig. 9.8A). A uniform gravity load of 38 psf on the horizontal projection of the roof is assumed together with a single load of 100 k on each cable at midspan to take the mechanical equipment into account. The sag-to-span ratio is taken as $n = 0.1$. The sizes of the cables are estimated.

The magnitude of the triangular loading depends on the size of the circular sector that a typical cable must support. The perimeter length of the sector is obtained by dividing the perimeter length of the roof by the number of sectors N. Hence, the unit force at the exterior support (see Fig. 9.8A) is

$$w = 0.038(\pi L/N) = 0.038\pi(400)/16 = 2.99 \text{ k/ft}$$

The vertical reaction forces due to the triangular loading and the single load at midspan are

$$V = 2.99(200)/2 + 100/2 = 349 \text{ k}$$

The horizontal reaction forces are found as based on the moment analogy method using a cable sag of $f = 0.1(400) = 40$ ft.

$$H = M_{max}/f = (wL^2/24 + PL/4)/f = (2.99(400)^2/24 + 100(400)/4\ )/40 = 748.33 \text{ k}$$

The maximum cable force according to Pythagoras' theorem is

$$T_{max} = \sqrt{V^2 + H^2} \ = \ \sqrt{349^2 + 748.33^2} \ = \ 825.71 \text{ k}$$

The ultimate tensile strength required is

$$T_u = \gamma \, T = 2.2(825.71) = 1816.56 \text{ k}$$

Try 4-in.-diameter, single-strand, class A coating (see Table A.13) with a breaking strength of 1850 k.

Case B: Assume the superimposed loads such as snow and mechanical and electrical equipment to be concentrated at the central roof portion of the round suspension roof of case A, as shown in Fig. 9.8B. The reactions are (see Table A.15)

$$A_V = B_V = V = wL/4, \qquad H = M_{max}/f = (wL^2/12)/f \tag{9.49}$$

The maximum slope at the reactions is

$$\tan\theta_0 = V/H = 3n \tag{9.50}$$

The maximum cable force is

$$T_{max} = H/\cos\theta_0 = V/\sin\theta_0 = H\sqrt{1 + \tan^2\theta_0} = H\sqrt{1 + 9n^2} \tag{9.51}$$

The equation for the cable slope is derived based on the location of the origin of the coordinate system, as shown in Fig. 9.8e.

$$\Sigma M_x = 0 = \frac{wL}{4}(x) - H(y) - \frac{w_x(x)}{2}\left(\frac{x}{3}\right)$$

or

$$y = f\left[3\frac{x}{L} - 4\left(\frac{x}{L}\right)^3\right] \tag{9.52}$$

As a rough first approximation for the cable length (of any type of loading), the equation for the cable length of the square parabola [Eq. (9.21)] may be used.

Case C: The cables for a tentlike structure are radially arranged and supported by a central column (Fig. 9.8C). The tent has to be prestressed so that its surface does not flutter under the varied types of loading. By prestretching the radial cables, tension forces are induced in the ring cables, thus putting the whole cable net in tension. The magnitude of the prestress forces depends on, among other considerations, the superimposed loads, which should not cause the cables to slack. The design of a typical suspended cable is a function of the prestress force and the additional tension caused by the other loading cases.

The vertical reaction forces for the triangular gravity-loading case in Fig. 9.8C, but using the general cable condition in Fig. 9.7k, are found by taking moments about the supports.

$$\Sigma M_B = 0 = A_v(L) + \frac{wL}{2}\frac{2L}{3} - H(h), \quad A_v = \frac{Hh}{L} - \frac{wL}{3} \tag{9.53}$$

$$\Sigma M_A = 0 = B_v(L) + \frac{wL}{2}\frac{L}{3} - H(h), \quad B_v = \frac{Hh}{L} + \frac{wL}{6} \tag{9.54}$$

The horizontal reaction or minimum cable force is (see Table A.15)

$$H = M_{max}/f = wL^2/15.6f \tag{9.55}$$

Now the maximum cable force can be found by checking the resultant forces at both supports. The approximate cable length for any type of loading may be found with the following equation:

$$l \cong L\left[1 + \frac{8}{3}\,n^2 + \left(\frac{h}{L}\right)^2\right] \tag{9.56}$$

EXAMPLE 9.4

The round, point-supported tent structure in Fig. 9.8C is investigated for preliminary design purposes. The initial geometry of the typical radial cable is simplified so that it only resists vertical forces at the top and only horizontal forces at the perimeter. A uniform load of 10 psf is assumed on the horizontal projection of the roof.

The central mast must carry the entire weight of the roof, since the radial cables only provide thrust resistance at the perimeter.

$$N = 0.010[\pi(125)^2] = 490.87 \text{ k}$$

Each of the 16 cables resists this load at the top, which also constitutes the maximum tension force in the cables.

$$V = T_{\max} = 490.87/16 = 30.68 \text{ k}$$

The ultimate tensile strength required is

$$T_u = \gamma\,T = 2.2(30.68) = 67.50 \text{ k}$$

Try 3/4-in.-diameter, galvanized structural strand (Table A.13) with a breaking capacity of 68 k.

To estimate the magnitude of the initial prestress forces in the ring cables, it is assumed that the radial cables are of circular shape and that $H = V$ as caused by uniform radial load action. Therefore, to maintain the circular shape, the vertical forces would not have to be triangular but rectangular, as indicated by the dashed lines, and the H-force would have to be uniformly distributed over the height. Hence, a typical ring cable carries the following single loads P, as shown in Fig. 9.8C, top:

$$P = (H/h)h_1 = (V/L)h_1 = (30.68/125)17.86 = 4.38 \text{ k}$$

The ring cable takes the shape of a closed polygon under the loads. It is shown in the next section that the tension force in the cable, according to Eq. (9.70) for $\theta = 360/16 = 22.5°$, is

$$T = P/(2\sin\theta/2) = P/(2\sin 22.5/2) = 2.56P = 2.56(4.38) = 11.21 \text{ k}$$

The ultimate tensile strength required is

$$T_u = \gamma\,T = 2.2(11.21) = 24.66 \text{ k}$$

Try 1/2-in.-diameter, single-strand, class A coating (Table A.13) with a breaking capacity of 30 k. The compression force in the concrete ring at the base is approximately

$$C = 2.56P = 2.56(30.68) = 78.54 \text{ k}$$

The municipal refuse-recycling building in Vienna, Austria, completed in 1982, uses a layout similar to that just discussed. This circular, 561-ft-diameter building may very well be the world's longest-span hanging timber roof. The roof is supported by 48 suspended glulam timber ribs spanning about 269 ft horizontally from the central, 221-ft-high concrete tower to the radial perimeter concrete walls. This concept of construction was developed by the well-known Swiss structural engineer Julius K. Natterer.

Cable Action under Radial Loads

The funicular response of a cable under a constant uniform radial pressure is the *circular line*. The radial forces cause cable forces of constant magnitude that are proportional to the radius of curvature, as will be shown later. When these radial forces, however, are not constant and increase uniformly from a minimum at the center to a maximum at the edge, the cable takes the shape of an *elliptical line*.

The Circular Cable   Visualize a circular cable to resist a radial uniform pressure p as caused by the action of a fluid such as the air pressure of a pneumatic structure. The circular cable is cut to form a semicircle, where the lateral thrust forces balance, as shown in Fig. 8.7d. Now the uniform pressure p also acts perpendicular to the imaginary base line of the semicircle (Fig. 9.9d). This pressure may be considered to be a resultant or balancing pressure equal to $2Rp$. In reality, the uniform base pressure is nonexistent, but is provided by the resisting forces in the cable (Fig. 9.9e).

$$2T = 2Rp, \quad \text{or} \quad T = pR \tag{9.57}$$

Thus, for constant radial pressure p, the cable forces T are proportional to the radius of curvature of the cable. Since the radius of a circle is constant and the loads are constant, the cable forces must be constant. The circular cable length l is derived from Fig. 9.9b as follows:

$$\sin\theta_0 = \frac{L/2}{R}, \quad \text{or} \quad \theta_0 = \arcsin\frac{L}{2R} \tag{9.58}$$

$$\frac{2\theta_0}{360} = \frac{l}{2\pi R}, \quad \text{or} \quad l = \pi R \frac{\theta^\circ}{90^\circ} \tag{9.59}$$

The elongation of the cable with cross-sectional area A and modulus of elasticity E due to the constant axial force T, according to Eq. (2.54) for $\theta_0 = \theta$, is

$$\Delta l = \frac{Tl}{AE} = \frac{pRl}{AE} = \frac{p\pi R^2}{AE}\frac{\theta^\circ}{90^\circ} \tag{9.60a}$$

The increase in the radius of curvature may be approximated from Eq. (9.59) as

$$\Delta l \cong \pi\Delta R \frac{\theta^\circ}{90^\circ}, \quad \text{or} \quad \Delta R \cong \frac{\Delta l}{\pi}\frac{\theta^\circ}{90^\circ} \tag{9.60b}$$

and substituting Eq. (9.60a) yields the radial expansion due to thrust or ring forces.

$$\Delta R \cong pR^2/AE \cong TR/AE \tag{9.61a}$$

The increase or decrease in cable length due to change in temperature is

$$\Delta l = \alpha(\Delta T)l \tag{2.7}$$

Substituting Eq. (9.60b) and Eq. (9.59), yields the change in radius due to temperature change.

$$\Delta R \cong \alpha(\Delta T)R \tag{9.61b}$$

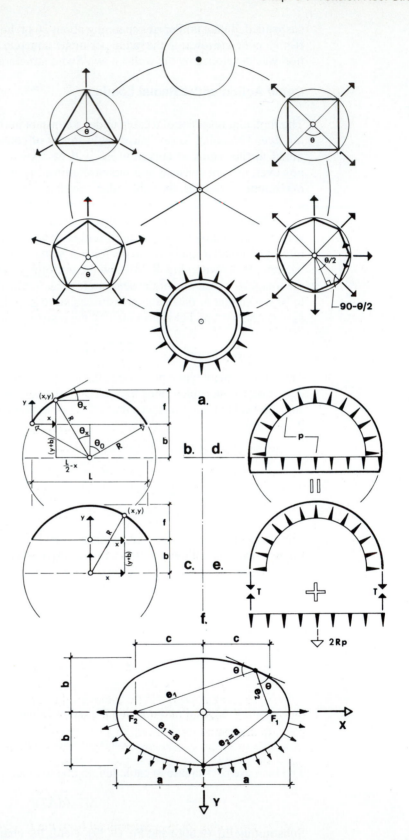

Figure 9.9 Cable action under radial loads.

By placing the origin of the coordinate system at the cable support, the shape of the circle and the slope may be defined as follows (Fig. 9.9b):

$$R^2 = (y+b)^2 + \left(\frac{L}{2}-x\right)^2 \quad \text{or} \quad y = -b + \sqrt{R^2 - \left(\frac{L}{2}-x\right)^2} \tag{9.62}$$

$$\tan\theta_x = \frac{dy}{dx} = \frac{L/2 - x}{\sqrt{R^2 - (L/2-x)^2}} \tag{9.63}$$

Hence, the slope at the reactions for $x = 0$, is $\tan\theta_0 = (L/2)/(R-f) = L/2b$.

For a given span L and sag f, the radius of curvature R is

$$R^2 = (L/2)^2 + b^2, \quad \text{where} \quad b = R-f$$

$$R = \frac{L^2 + 4f^2}{8f} \tag{9.64}$$

For a semicircle with $b = 0$, $L/2 = R$, and $f = R$, the shape of the circle is defined, according to Eq. (9.62), as

$$y = \sqrt{2Rx - x^2} \tag{9.65}$$

The slope of the curvature for the semicircular cable is

$$\tan\theta_x = \frac{dy}{dx} = \frac{R-x}{\sqrt{2Rx - x^2}} = \frac{R-x}{y} \tag{9.66}$$

Sometimes it may be more convenient to place the origin of the coordinate system at midspan (Fig. 9.9c). Now the shape of the circle is defined as

$$R^2 = (y+b)^2 + x^2, \quad \text{or} \quad y = -b + \sqrt{R^2 - x^2} \tag{9.67}$$

For a semicircle with $b = 0$ or when the origin of the coordinate system is the center of the circle, the shape of the circle is defined as

$$\text{Shape:} \quad y = \sqrt{R^2 - x^2}, \quad \text{or} \quad y^2 + x^2 = R^2 \tag{9.68}$$

$$\text{Slope:} \quad \tan\theta_x = \frac{dy}{dx} = \frac{-x}{\sqrt{R^2 - x^2}} = -\frac{x}{y} \tag{9.69}$$

In situations where single forces of the same magnitude act in a radial manner, the funicular shape of the cable is a regular, closed polygon (Fig. 9.9a). Familiar examples are the central tension ring of a round, dish-shaped suspension roof or the outside tension ring of a ribbed dome structure. The radial forces P are spaced at regular intervals; that is, the roof is subdivided into N parts

$$\theta = 360/N$$

The constant cable force is determined from the free body in Fig. 9.9a by summing up the horizontal forces.

$$\Sigma F_x = 0 = P - 2\left[T\cos\left(90 - \frac{\theta}{2}\right)\right], \quad \text{or} \quad T = \frac{P}{2\sin\theta/2} \tag{9.70}$$

EXAMPLE 9.5

The compression forces in the perimeter concrete ring for the suspended stadium roof in Example 9.3 are estimated.

The ring must resist uniform horizontal radial thrust forces of $H = 748.33$ k, which are spaced along the arch at

$$s = \pi L/N = 400\pi/16 = 78.54 \text{ ft}$$

The compression forces in the ring can be approximated according to Eq. (9.57) by treating the single loads as uniform loads.

$$C = pR \cong (H/s)R = (748.33/78.54)200 = 1905.60 \text{ k}$$

Or the compression forces can be found by the more precise approach according to Eq. (9.70) for a cable spacing at intervals of $\theta = 360/16 = 22.5°$.

$$C = P/(2\sin \theta/2) = 748.33/(2\sin 22.5/2) = 1917.91 \text{ k}$$

Since the forces must balance in the horizontal direction, the tension forces in the central ring must be equal to the axial forces in the perimeter compression ring ($C = T$), as indicated in Fig. 9.8A.

The Elliptical Cable   An ellipse can take many shapes (depending on the direction and magnitude of forces), varying from nearly round to narrow. An ellipse has two axes of symmetry (a major and minor axis) that are perpendicular to each other. It is defined as the graph of a point that moves in a plane so that the sum of its distances from two fixed points, called *foci* (F_1, F_2) is a constant (i.e., $e_1 + e_2 = 2a$), as shown in Fig. 9.9f. The equation of the ellipse with the origin at the center, where the major axis is along the x axis, is

$$(x/a)^2 + (y/b)^2 = 1, \qquad \text{where } a^2 > b^2 \tag{9.71a}$$

The location of the foci (F_1, F_2) is

$$a^2 = b^2 + c^2, \quad \text{or} \quad c = \pm \sqrt{a^2 - b^2} \tag{9.71a}$$

Equation (9.71a) can also be represented in the following form by letting $L_x = 2a$ and $L_y = 2b$.

$$(2x/L_x)^2 + (2y/L_y)^2 = 1 \tag{9.71b}$$

The circle can be treated as a special case of the many possible ellipses, where the two foci coincide:

$$c = 0, \quad \text{or} \quad a^2 = b^2 = R^2, \quad \text{or} \quad x^2 + y^2 = R^2 \tag{9.68}$$

This is the equation for the circle that was derived previously. The circle is the simplest and most symmetrical of all curves, whose shape never changes. Notice that the circle seen at an angle appears as an ellipse.

An elliptical cable responds to a specific load action (i.e., funicular loading). It is easy to visualize from Fig. 9.9f that the largest loads p_{max} must occur at the major axis with the minimum radius of curvature as to stretch the shape, whereas the smallest loads, p_{min}, occur at the minor axis with the maximum radius of curvature. It can be shown that the magnitude of the loads p is reciprocally proportional to the radii of curvature at each point along the elliptical line (i.e., $p_x/R_x = p_y/R_y$), in contrast to the circle where p/R is constant. The following relationship of loading and cable geometry must hold true for a funicular elliptical cable.

$$\frac{p_x}{p_y} = \frac{L_x^2}{L_y^2} = \frac{a^2}{b^2}, \quad \text{or} \quad p_x = p_y (a/b)^2 \quad \text{where} \quad p_x > p_y \tag{9.72}$$

Since the loads p_x and p_y are proportional to constants, each load is constant, and p_x is always larger than p_y (Fig. 9.19d). The maximum and minimum cable forces at the ends of the minor and major axes, respectively, can be easily obtained from the free body of half the ellipse (Fig. 9.19d), where the cable forces at the ends of the major axis must balance the uniform load p_y.

$$2T_y = p_y(2a) \quad \text{or} \quad T_y = T_{min} = p_y a = p_y L_x/2 \tag{9.73a}$$

Similarly, for the minor axis

$$T_x = T_{max} = p_x b = p_x L_y/2 \tag{9.73b}$$

The circumference of an ellipse is approximately

$$l \cong 2\pi \sqrt{(a^2 + b^2)/2} \tag{9.74}$$

The Prestretched Cable

Prestressed trampolinelike cable structures may be considered for multibay roof structures with bay spans up to about 50 ft. They have only a very small sag, which is due to loading and is limited to about $L/50$ or $n = f/L \le 0.02$. To achieve this small sag, very high pretensioning forces T_0 must be applied, which not only control the deflection but also stiffen the roof structure against fluttering. The selection of the cables depends on the control of the sag, rather than stress considerations.

The horizontal thrust forces are determined according to the moment analogy method [Eq. (9.2)], as shown in Fig. 9.10a.

$$(H_0 + H_1)(f + \Delta f) = M \tag{9.75a}$$

Letting the initial geometrical sag $f = 0$ yields

$$(H_0 + H_1)\Delta f = M \tag{9.75b}$$

For the condition where the tension caused by the applied loads $T_1 \cong H_1$ is small as compared to the initial prestress force H_0, we may, for preliminary design purposes, ignore the increase in tension due to gravity loading so that

$$T \cong H_0 \cong M/\Delta f \tag{9.76a}$$

For the situation where the cables are arranged in a parallel fashion Eq. (9.73) becomes

$$H_0 \cong (wL^2/8)/\Delta f \tag{9.76b}$$

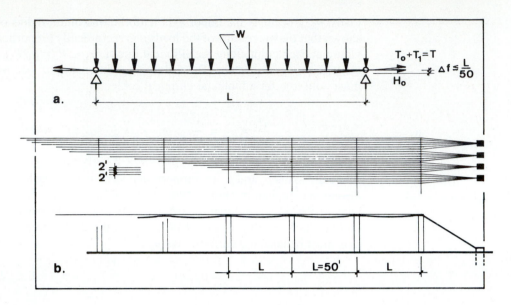

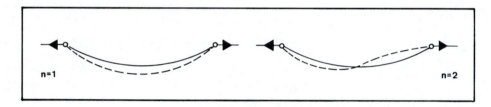

Figure 9.10 (a) Trampolinelike roof structure. (b) Cable vibration.

Here Δf can be replaced by the permissible deflection. The vertical displacement Δf caused by the loads, according to Eq. (9.24), is

$$\Delta f \cong 3\Delta l/16n \quad \text{or} \quad \Delta l \cong 16n\,\Delta f/3$$

The increase in cable length Δl, in turn, can be equated to Eq. (2.54) and then can be further simplified by letting $l \cong L$ and $n = \Delta f/L$.

$$\Delta l \cong \frac{16n\,\Delta f}{3} \cong \frac{HL}{AE}, \quad \text{where} \quad H = \frac{wL^2}{8\Delta f}$$

Therefore, the required cross-sectional area of the cables for preliminary design purposes is

$$A \cong \frac{3wL}{128n^3 E} \tag{9.77a}$$

Or a single-curvature membrane (or cable) can support the following load, w, for a given deflection, Δf.

$$w = 128AE(\Delta f)^3/3L^4 \tag{9.77b}$$

EXAMPLE 9.6

A prestretched cable roof for a rectangular multibay structure is investigated (Fig. 9.10a). It is assumed that pairs of cables are spaced 2 ft apart and span the 50-ft bays. For this preliminary study, only gravity loads are considered; the dead load is taken as 10 psf and the live load as 20 psf. A maximum deflection of $\Delta f = 0.5$ ft or a sag-to-span ratio of $n = 0.01$ is assumed.

Each cable must carry the following load.

$$w = 2(0.010 + 0.020)/2 = 0.03 \text{ k/ft}$$

Therefore, the required cross-sectional area of the cables according to Eq. (9.77a) is

$$A \cong \frac{3wL}{182n^3E} = \frac{3(0.03)\,50}{128\,(0.01)^3\,24{,}000} = 1.47 \text{ in.}^2$$

Try $1\tfrac{9}{16}$-in.-diameter galvanized structural strand (Table A.13).

$$A = 1.47 \text{ in.}^2, \qquad P_u = 2(150) = 300 \text{ k}$$

The maximum prestress force that can be applied to the cable is $300/2.2 = 136.36$ k. The actual thrust force due to gravity loading according to Eq. (9.76b) is

$$H = (wL^2/8)/\Delta f = (0.03(50)^2/8)/0.5 = 18.75 \text{ k}$$

Hence, this force is about 14% of the prestress force. A new cycle of analysis must take this force into account as based on Eq. (9.75b).

Dynamic Behavior

The lack of flexural stiffness in suspended roof structures makes them vulnerable to aerodynamic instability; in other words, wind gusts may cause the roof to flutter. Dynamic loads may initiate vibrations and loads larger than the comparative static ones. Furthermore, when the period of the wind approaches the natural period of the resisting structure, *resonant loads* are set up, which may cause collapse of the structure when it is not properly damped. The collapse of the too flexible Tacoma Narrows Suspension Bridge (Washington) in 1940 is a classic example of the lack of proper aerodynamic design. Wind turbulence is extremely variable in size and magnitude; it contains a wide range of periods so that the structure is only excited by a small portion of it. For preliminary design purposes of tensile roof structures, it is often assumed that the critical wind periods occur above 4 to 5 sec, so that the natural period of the roof structure should be below approximately 3 sec.

To determine the natural period of a multidegree-of-freedom system, such as a roof structure, is very complex and beyond the scope of this discussion. Some understanding, however, can be developed for the factors influencing the dynamic behavior by looking at the following simple case of free vibration [see also Eq. (4.4)]. It can be shown that the dynamic response, or natural periods of vibration P_n, of a taut single *undamped* cable carrying a uniformly distributed load along its horizontal projection can be approximated as

$$P_n = \frac{2\pi}{\omega_n} \cong \frac{2L}{n}\sqrt{\frac{m}{T}} = \frac{2L}{n}\sqrt{\frac{w/g}{T}} \quad \text{(sec)} \tag{9.78}$$

where ω_n = natural circular frequencies

m = w/g = uniformly distributed mass per unit length

w = weight of structure per unit length (lb/ft)

g = 32.2 ft/sec^2 = acceleration due to gravity

n = number of nth harmonic mode of vibration = 1, 2, 3, … (Fig. 9.10b)

L = cable span (ft)

T = tension force in cable (lb) $\cong H$

This equation can be simplified further for the common, relatively shallow sag-to-span ratios used in practice by letting $T = H$, where H is the horizontal component of the cable force T. When the uniform load is constant, as is the case for the dead load, $H = wL^2/8f$ according to Eq. (9.14). Therefore, Eq. (9.78) takes the simple form of

$$P_n = (2/n)\sqrt{8f/g} \qquad \text{(sec)} \qquad (9.79)$$

where f = maximum cable sag (ft).

We may conclude that the period P_n is independent of the loading; in other words, the increase in weight cannot stabilize the roof aerodynamically. Only the mode of vibration and the cable sag influence the period, which should not be in the vicinity of the period of the wind to prevent resonant loading, which obviously is very difficult to predict. Oscillations can be most effectively damped by using secondary cables to stabilize the primary cables, that is, through the use of prestressed double-cable systems, as discussed further in the next section. The reason that heavy cable-suspended roofs have performed well is due to the damping effect provided by the higher cable stiffness as a result of the larger loads, thereby keeping the vibrational deflection low.

EXAMPLE 9.7

The natural period for an undamped single cable of the roof structure in Example 9.2 is determined. The fundamental or first mode of vibration ($n = 1$) with the longest period, according to Eq. 9.79 is

$$P_n = (2/n)\sqrt{8f/g}, \quad \text{or} \quad P_1 = (2/1)\sqrt{8\,(9.33)/32.2} = 3.05 \ \text{sec}$$

This period is below the critical wind period of above 4 to 5 sec. Furthermore, keep in mind that damping through the roof structure has been conservatively ignored.

9.3 CABLE BEAMS AND CABLE TRUSSES

Cable beams and trusses represent the most simple case of the family of pretensioned cable systems that includes cable nets and tensegrity structures. Visualize a single suspended (concave) cable, the primary cable, to be stabilized by a secondary arched (convex) cable or prestressing cable. This secondary cable can be placed on top of the primary cable by employing compression struts, thus forming a lens-shaped beam (Fig. 9.11d), or it can be located below the primary cable (either by touching or being separated at center) by connecting the two cables with tension ties or diagonals (Fig. 9.11c). A combination of the two cable configurations yields a convex–concave cable beam (Fig. 9.11b).

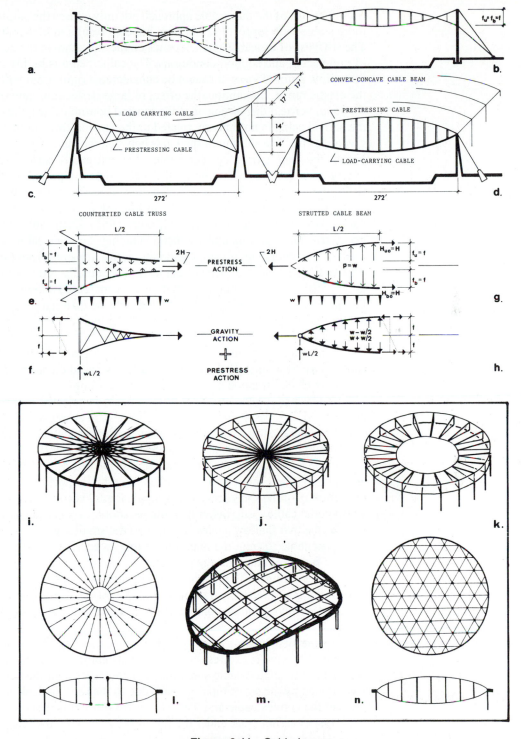

Figure 9.11 Cable beams.

The sags for the primary cables are designated by f_b and for the secondary cables by f_u. Their magnitude can be estimated as a function of beam span; they are in the range of about $f_u = f_b = L/15$ to $L/25$.

Cable beams can form simple span or multispan structures; they can also be cantilevers. They can be arranged in a parallel or radial fashion for various roof forms, or they can be used as single beams for any other application; they can also be arranged in a rectangular or triangular gridwork (Fig. 9.11). For further discussion, see Section 9.5.

The use of the dual-cable approach not only causes the single flexible cable to be more stable with respect to fluttering, but also results in higher strength and stiffness. The stiffness of the cable beam depends on the curvature of the cables, cable size, level of pretension, and support conditions. The cable beam is highly indeterminate from a force flow point of view; it cannot be considered a rigid beam with a linear behavior in the elastic range. In addition, the effect of large deflections cannot be neglected. Even for equal vertical displacement of the top and bottom cables, the compressive force in the arched cable is not linearly related to the tension force in the suspended cable at a given section. The sharing of the loads between the cables, that is, finding the proportion of the load carried by each cable, is an extremely difficult problem. It is surely overly conservative to assume all the loads to be supported by the suspended cable, while the secondary cable's only function is to damp the vibration of the primary cable. Lev Zetlin has shown that the oscillating of a single cable can be successfully damped by using a secondary cable (Zetlin, 1965). Each cable always takes a different geometric configuration under dynamic loading; thus one cable counteracts the movement of the other (Fig. 9.11a). Furthermore, it is apparent from Eq. (9.78) that the fundamental period in the primary cable is substantially reduced due to the additional large prestress force, while the weight is only increased slightly. It is apparent that the fundamental periods of the two cables should be different from each other, which is assured by always having a higher stress in the load-carrying cable as compared to the stress in the damping cable.

For rough first preliminary design purposes, a linearly related behavior of the cables may be assumed by sizing the cables as based on strength only and neglecting the effect of deformation as well as dynamic loading.

First, a trial geometry of the cable beam is selected and preliminary loads are determined. It is assumed that the cable beams are arranged in a parallel fashion; for a radial arrangement on round plan forms, a similar approach would be used. In this investigation a symmetrical, lens-shaped beam ($f_b = f_u = f$) is approximately designed to develop some feeling for the behavior of cable beams. Furthermore, a uniform force action along the cable beam is assumed, thereby neglecting the effect of the concentrated action of the spreaders causing single loads. These single loads actually generate a polygonal cable form, rather than the parabolic curvature as is assumed in this analysis for the two shallow cables. Because of the small sag-to-span ratio of cable beams, it is reasonable to treat the maximum cable force T as equal to the horizontal thrust force H for preliminary design purposes.

Under full gravity loading the suspended or load-carrying cable is stressed to a maximum, while the arched or prestressing cable is stressed to a minimum, requiring enough tension so that no compression can arise under any loading and the cable cannot slack.

In the first loading stage, prestress forces are induced into the beam structure. The initial tension (i.e., prestress force minus compression due to cable and spreader weight) in the arched cable should always be larger than the compression forces that are induced by the superimposed loads due to the roofing deck and live load; this is to prevent the convex cable and the web ties from becoming slack.

Let us assume that under the full loading stage all the loads w are carried by the suspended cables and that the forces in the arched cables are zero. Therefore, when the superimposed loads are removed, equivalent minimum prestress loads of $w/2$ are required to satisfy the assumed condition, which, in turn, is based on equal cross-sectional areas of the cables ($A_u = A_b$) and equal cable sags ($f_u = f_b$) so that the suspended and arched cables each carry the same loads. Naturally, the *equivalent prestress load* cannot be zero under maximum loading conditions since its magnitude is not just a function of strength as based on static loading and initial cable geometry, but also of dynamic loading including damping (i.e., natural period), stiffness, and possible considerations of the erection process. The determination of the prestress forces requires a complex process of analysis, which is beyond the scope of this introductory discussion. It is assumed, in this context, for rough preliminary approximation purposes that the *final equivalent prestress loads* are equal to $w/2$ (often designers use final prestress loads at least equal to the live loads w_L).

The preliminary design of the primary or load-carrying cables may be conservatively based on the assumption that all the gravity loads w together with the final prestress load $w/2$ are carried by the suspension cables (Fig. 9.11h).

$$H_b = (w + w/2)L^2/8f_b = 3wL^2/16f_b \qquad (9.80)$$

The preliminary design of the secondary or arched cables depends on the minimum gravity loads (since any additional gravity loads would reduce the tensile stresses) and possibly wind uplift. Therefore, to obtain a final equivalent prestress load of $w/2$ requires an initial *prestress load* of $w/2 + w/2 = w$, as based on equal cable areas (Fig. 9.11g). Actually, the superimposed loads are distributed to the secondary cable in proportion to the cable areas when $f_u = f_b$. Assuming that the wind uplift p is distributed equally to the cables and that $w \geq p/2$, then the size of the prestressing cable can be estimated from

$$H_{uo} = (w + p/2)L^2/8f_u \qquad (9.81)$$

Or the cross-sectional area of the secondary cable is roughly, when wind suction is ignored,

$$A_u = A_b(w/1.5w) = 0.67A_b \qquad (9.82)$$

For the general case where the cable sags are not equal to each other, the following relationship under initial pretensioning holds approximately true.

$$H_{bo}f_b = H_{uo}f_u, \qquad \text{or} \qquad H_{bo} = H_{uo}(f_u/f_b) \qquad (9.83)$$

In other words, the cables share the prestress loads in proportion to the inverse of the cable sags.

EXAMPLE 9.8

A stadium is covered by cable beams spaced parallel at 17 ft on center; the beams span 272 ft and are either strutted or countertied (Fig. 9.11). Sag and rise of the cables are each 14 ft. The following typical loads are used: 9-psf dead load, 20-psf live load, and a uniform wind suction of 12 psf. The wind loads are obviously not acting as given. However, the assumed uniform action should make it possible to consider the wind as a possible design determinant for a preliminary investigation. A typical interior cable beam must support the following loads:

$$
\begin{aligned}
&\text{Dead load} \quad w_D = 0.009(17) = 0.15 \text{ k/ft} \\
&\text{Live load} \quad w_L = 0.020(17) = 0.34 \text{ k/ft} \\
&\text{Total} \qquad\quad w = 0.15 + 0.34 = 0.49 \text{ k/ft} \\
&\text{Wind load} \quad p = 0.012(17) = 0.20 \text{ k/ft}
\end{aligned}
$$

The maximum cable slope, according to Eq. (9.15), is

$$\tan \theta_0 = 4n = 4f/L = 4(14)/272 = 0.206$$

The horizontal component of the maximum tensile force for the primary cable, according to Eq. (9.80), is

$$H_b = 3wL^2/16f_b = 3(0.49)272^2/16(14) = 485.52 \text{ k}$$

The maximum tension force, according to Eq. (9.16), is

$$T_{max} = H\sqrt{1 + \tan^2\theta_0} = H\sqrt{1 + 0.206^2} = 1.02H = 495.72 \text{ k}$$

Notice that the small difference between the minimum cable force H and the maximum force T can be considered unimportant in comparison to the other approximations.

The ultimate tensile strength required is

$$T_u = \gamma T = 2.2(495.72) = 1090.57 \text{ k}$$

Try $3\frac{1}{8}$-in.-diameter galvanized steel strand with a breaking strength of 1168 k (see Table A.13). The maximum cable force in the secondary cable due to prestressing and wind suction, according to Eq. (9.81), is

$$H_{uo} = (w + p/2)L^2/8f_u = (0.49 + 0.20/2)272^2/8(14) = 389.74 \text{ k}$$
$$T_{max} = 1.02H = 1.02(389.74) = 397.53 \text{ k}$$
$$T_u = \gamma T = 2.0(397.53) = 795.06 \text{ k}$$

Try $2\frac{5}{8}$-in.-diameter galvanized steel strand with a breaking strength of 834 k.

9.4 CABLE-SUPPORTED ROOF STRUCTURES

The concept of assisting beams, columns, soft and hard surface structures, or other member types with inclined stays such as ropes, rods, or iron chains that hang from abutments, columns, or towers has been known for thousands of years. Today, the principle is applied to cranes, ships, television towers, bridges, roofs, and entire buildings. In cable-supported structures, cables, rods, or other member types act as tensile columns to support linear members, surfaces, and volumes from above or to brace buildings against lateral forces. The cable can also be an integral part of a structural system and can give support from below rather than from above as, for example, the single-strut or double-strut cable-supported beams (e.g., king-post truss, queen-post truss and original Fink truss), the bottom chords of the simple scissors or German truss, and the Palladian truss typical in the Renaissance. The internally prestressed column trees for the Stansted terminal in London (1991, Fig. 4.1, left bottom) also demonstrate the concept spatially.

Although engineers experimented with the application of inclined cables, particularly in cable-stayed bridge construction during the last century, it was not until the 1950s that the concept gained in popularity. This trend is reflected, for example, by the 1572-ft-high guyed Oklahoma City television tower of 1954, the Bavinger House of 1955 in Norman, Oklahoma, by Bruce Goff, the Strömsund Bridge of 1955 in Sweden, and the cable-supported, single-cantilever roof for the TWA hangar in Philadelphia and the double-cantilever roof for the Pan Am hangar in New York, both completed in 1956, as well as the huge umbrellalike oval Pan Am terminal (528×422 ft with 114-ft cantilevers) at New York's J.F. Kennedy International Airport of 1959, where cable-stayed double cantilevers are arranged radially and anchored to tension columns at the building center. Another early example is the 300-ft-span roof for the Blyth Arena in Squaw Valley, California, built for the 1960 Olympic Winter Games, which consists of parallel, symmetrical, 150-ft-span, single-cantilever portions that are cable supported from tapered steel masts arranged in parallel bents.

We should also mention Buckminster Fuller's hexagonal, mast-hung, cable-supported Dymaxion House, which he proposed as early as 1927, predicting his later experiments with tensegrity structures. The concept of stayed tree houses was applied by Craig Zeidler to five interconnected pavilions over Lake Ontario, Toronto, in 1972. Robert LeRicolais's experiments with spatial structures at the University of Pennsylvania during the 1950s and 1960s have had much effect on understanding the interaction between tension and compression.

The visual power of tension and compression is most powerfully expressed in Santiago Calatrava's abstract sculpture of 1981 (Fig. 9.13, center and right). Here, massive cubes that barely rest on each other transfer compression forces to cones, while a wire in tension resists the rotation caused by the asymmetrical arrangement of the cubes. Tension and compression are anchored in the base plate that provides the necessary equilibrium. The dialogue between the individual cubes and the continuity and strength of the wire is very forceful. Calatrava seems to play with mass and minimum support of cables, with tension and compression, with equilibrium and movement, and with stability and instability. He articulates a minimal equilibrium that is on the verge of failure, thereby exposing a dynamic balance of forces.

Single-strut and Multistrut Cable-supported Beams

Although the primary theme of this subchapter is the cable-stayed roof structure, where diagonal cables give support to the horizontal or inclined framework from above, in this section we will look at the many possibilities of supporting the framework from below. The conventional king-post and queen-post trusses, which represent single-strut and double-strut cable-supported beams, are familiar. These systems form composite trusslike structures with steel or wood compression members as top chords, steel tension cables as bottom chords, and compression struts as web members. King-post trusses or single-strut, cable-supported beams can also be overlapped in plane or spatially, an example of this concept is the decktruss patented by Albert Fink in 1851. For instance, a mesh of overlapping king-post trusses with the bottom chords turned 45° to the top chords may form a space truss. The composite spatial trusses for the Memmingen city hall in Germany (1984, Fig. 1.11b), with a maximum span of 112 ft, consist of wood top chords and three suspended, parabolic, tensile steel bottom chords.

Several examples of subtensioned structures are shown in Fig. 9.12, ranging from simple parallel to two-way and complex spatial systems. Notice the cast iron strut of the queen-post truss in (b) is beautifully designed, expressing the power of transition from the tensile cable to the compression element. The inclined, tapered cantilever arms of the three-story Y-shaped steel frame in (a) are supported by horizontal cables on top and almost appear like inverted king-post trusses. Since the butterfly frame is pinned at the column base, it is laterally stabilized by vertical cable ties at the ends.

The light pyramids at the center of the 37- × 37-ft bays (c) are supported by simple, two-way, queen-post trusses. The roof supporting system in (g) is more complex. Here, two-story king-post trusses divide the 64- × 64-ft bays into four 32- × 32-ft bays with light pyramids, which, in turn, are each carried by a two-way, diagonal, king-post system.

In cases (d) and (f), parallel transverse truss action is replaced by more complex spatial composite trusses, causing surface action and a more uniform density of structure. The 147-ft-span roof in (f) consists of inclined, single-strut, cable-supported folded beams that form a continuous structure and give the appearance of a folded structure from below. Far more complex is the 80-ft-span spatial composite truss system in (d) supported on columns spaced at 16 ft that carry the central 32-ft-wide gable structure.

The main support structure for the roof of the open theater in (e) consists of two primary longitudinal, 170-ft-span, 32-ft-deep king-post trusses that split into Y-shapes at the struts. The two composite trusses, in turn, support a third, 100-ft-span king-post truss. Another interesting example of cable support from below is the suspended domes in Fig. 9.21.

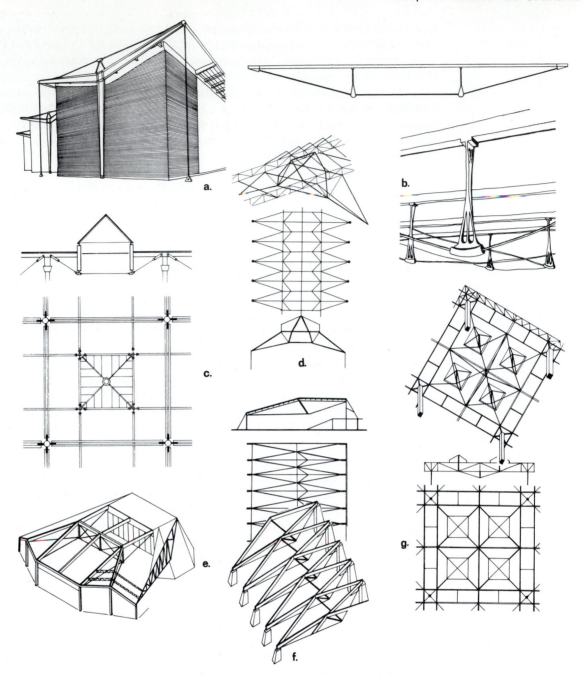

Figure 9.12 Single-strut and multi-strut cable-supported beams.

Cable-stayed Bridges

The modern development of cable-stayed bridges or cable cantiliver bridges has very much influenced architects, especially the simplicity and minimal proportions of the Rhine bridges in Germany during the 1960s.

The eminent German engineer Franz Dischinger rediscovered the cable-stayed bridge principle in 1938 and applied it to the 600-ft-span Strömsund Bridge in Sweden in 1955, the first bridge of this type. The first long-span, cable-stayed bridge in Germany was the 853-ft-span North Bridge over the Rhine at Düsseldorf with a harp-type cable arrangement designed by the architect Friedrich Tamms and completed in 1958. The 991-ft-span fan-type Severin Bridge in Cologne designed by the architect Gerd

Lohmer, was finished in 1959. Here, the cables spread from the top of the A-frame tower to the edges of the roadway deck in a spatial manner. A significant force in this early development of cable-stayed bridges in Germany was the renowned engineer Fritz Leonhard.

The first prestressed concrete, cable-stayed bridge was completed in 1966 across Lake Maracaibo in Venezuela (Fig. 9.13, bottom left). This 773-ft-span bridge was designed by the Italian Riccardo Morandi, one of the pioneers of prestressed concrete and one of the great concrete designers in the world.

A typical span consists of double-cantilever girders (Fig. 5.7) connected by 151-ft, simply supported, drop-in beams at center span to provide expansion joints. The hinged-beam system rests on inclined columns (145 ft apart) that form V-strut supports nestled between the towers. In addition, each of the enormous 310-ft-span, balanced cantilever, three-cell roadway box girders is supported by single pairs of cables suspended from each of the longitudinal 304-ft-high A-frame towers along the edge of the roadway, which are linked at the top by a transverse girder. This single-stay configuration requires the deck to carry loads in bending, in contrast to bridges with many stays in close spacing, where there is hardly any flexure and loads are resisted primarily axially in the deck.

Very different in appearance, although based on the same structural principle, is Christian Men's masterpiece, the 572-ft-span Ganter Bridge on the Simplon Road above Brig, Switzerland (1980, Fig. 9.13, top center). Here, the cable-stayed cantilever prestressed concrete bridge seems almost to change into a suspended bridge by letting the shallow prestressed cable stays be integrated into the thin triangular walls above the roadway. The elegance of the bridge profile lies in its purity and austere form, as well as in its lightness. The daring thinness of the slab between the cantilever towers (one column is 493 ft high) forms a magnificent contrast to the natural setting of the mountain landscape.

A fascinating design proposal (1978) is the Ruck-a-Chucky Bridge, Auburn, California (Fig. 9.13, center left) by Myron Goldsmith of SOM and the engineer T. Y. Lin. Here, the curved box girder roadway is supported by steel cables along the exterior edges. The stays are arranged in a hyperbolic paraboloid formation and anchored to the walls of the river canyon and to the neutral axis of the box girder to produce pure axial compression.

Meanwhile, cable-stayed bridges have become very popular in the United States, as demonstrated by the beautiful Sunshine Skyway Bridge across Tampa Bay (1987, Fig. 9.13, left) designed by Jean Muller. Currently, it is the world's longest precast concrete segmental bridge. Single plane stay cables of the fan-harp type support the 1200-ft central span roadway consisting of a single trapezoidal box section. The 1300-ft-span Broward Concrete Bridge at Dame Point, Jacksonville, Florida (1988), is currently the longest cable-stayed structure in the United States. The Normandy Bridge (1995) in France with a central span of 2808 ft is the world's longest cable-stayed bridge. An innovative steel-concrete longitudinal composite deck is used where the lighter steel deck in the central portion of the main span is mated to concrete cantilever spans supported by inverted Y-shaped concrete towers.

Cable-stayed Bridge Systems   Cable-stayed bridges consist of the towers, cable stays, and deck structure. The stays can give support to the deck structure only at a few points, using one, two, three, or four cables, or the stays can be closely spaced, thereby reducing the beam moments to a minimum and allowing much larger spans, possibly up to 6000 ft.

Multiple stays can be arranged in a *fan*-type fashion by letting them start all together at the top of the tower and then spread out (Fig. 9.14q). They can be arranged in a *harp*-type manner, where they are arranged parallel across the height of the tower

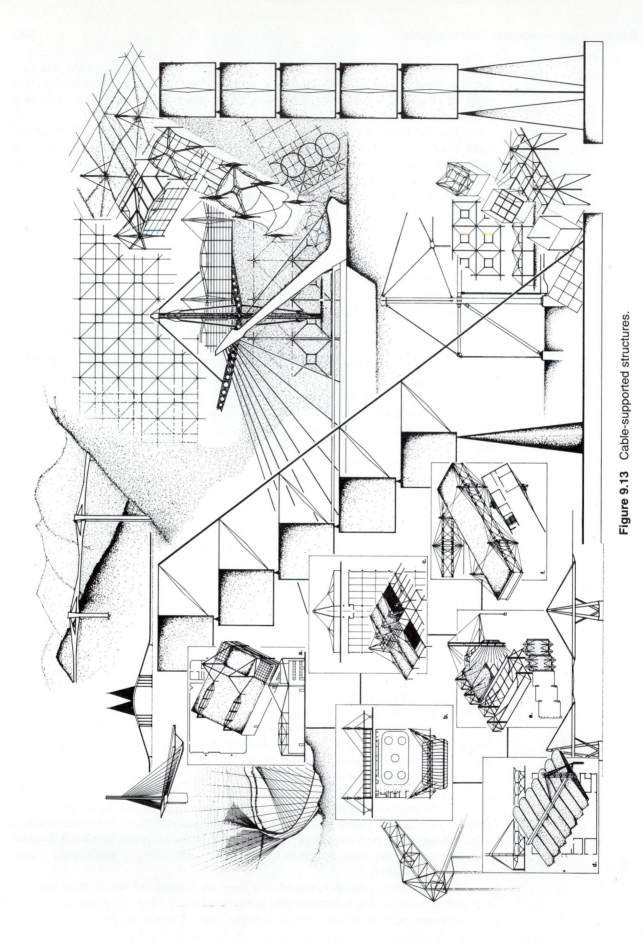

Figure 9.13 Cable-supported structures.

(Fig. 9.14r), or they can be arranged in the *star* configuration, where the stays are spaced along the tower and converge to a common point on the deck girder. The stay configuration may also fall between the *fan–harp* types. Furthermore, the stay configurations are not always symmetrical; they can also be asymmetrical.

Typical tower shapes are single or double free-standing pylons (vertical or inclined), double pylons connected by cross members, portal frame towers, braced frame towers, A-frame towers, and inverted Y-frame towers. The towers can be sloped rather than vertical, such as Calatrava's inclined sculptural tower for the asymmetrical cable-stayed bridge across the Segre River (proposal 1985, Fig. 9.13, center).

In the transverse direction, the stays can be arranged in one vertical plane at the center or off center, in two vertical planes along the edge of the roadway (i.e., edges of deck girders), in diagonal planes descending from a common point (e.g., top of A-frame) to the edge deck girders, or in a spatial manner by using, for instance, four inclined cable planes with pairs of opposite sloped cable planes descending to the edge deck girders. The purpose of the cable stays is to help the deck girder(s) to transfer the loads to the tower supports. The forces to be resisted are due to gravity loads, lateral loads, and asymmetrical loads causing torsion.

In a single, central, cable plane bridge, the cables only carry vertical loads. Torsion must be resisted entirely by a stiffened deck box girder, which also must resist the lateral loads in transverse bending. In a bridge with two vertical cable planes, the cables transfer gravity loads as well as torsional moments, so the deck girders must resist only lateral loads. In a bridge with two inclined cable planes forming an A-layout, the load transfer is the same as for the bridge with two vertical cable planes. In a bridge with a spatial cable stay arrangement, such as with four inclined cable planes, the vertical and horizontal loads together with the torsional moments are transferred solely by the cables without the help of the deck girders.

Cable-stayed Roof Structures

Examples of cable-stayed roof structures range from long-span structures for stadiums, hangars, and exhibition centers, to smaller scale buildings for shopping centers, and production or research facilities, to personal experiments with tension and compression, such as Shoei Yoh's bridgelike structure Wakita-Hi Tecs Head Office (1988, Fig. 9.13f), using a unique spatial suspension system.

All the general concepts of cable-stayed bridges can be transferred to the design of cable-stayed roof structures. They can be applied to simply supported beams, cantilever beams, frames, and planar framework using single or multiple stays in planar or spatial arrangement. The cable-stayed principle can be applied to grandstands, single-bay, multibay, or spinal structures.

Typical guyed roof structures, used either as planar or spatial stay systems, are the following:

- *Cable-stayed, double-cantilever roofs* for central spinal buildings requiring clear-span side wings, such as for hangars. In this case, the masts are located along a spine in the interior of the building. The cantilevers can also be end supported by columns to form continuous beams on elastic, intermediate supports.

- *Cable-stayed, single-cantilever roofs* as used for hangars and grandstands. Again, the cantilever can be end supported by a column to form a continuous beam on elastic, intermediate supports.

- *Cable-stayed beams,* where masts are located at each end of the beams; in other words, pairs of masts form exterior structures.

- Combinations of the preceding conditions.

- *Spatially guyed multidirectional composite roof structures* integrating concepts of cable-stayed beams or cable beams and cable trusses. These systems may be used for multibay buildings.

The concept of cable-stayed roofs is not solely associated with long-span structures. Particularly, architects like Richard Rogers, Norman Foster, Michael Hopkins, and Nick Grimshaw, together with the engineers Peter Rice and Tony Hunt, have experimented with a new architectural esthetics reflecting the spirit of the 1960s as represented by the work of Archigram, Buckminster Fuller, Konrad Wachsmann, and Jean Prouvé. Zeidler's suspended pavilions on Lake Ontario in Toronto (1971) and especially the Pompidou Center in Paris (1977), by Piano and Rogers, constitute a realization of that spirit.

This new language integrates inventive engineering technology and the process of construction into architectural design. It is concerned with the logic of structure and with exposing the structure components and expressing their structural responsibilities. Clearly identified are the parts of the assembly, that is, the rigid members, connections, or tension hangers and diagonals. Particularly important are the details; emphasis is on how the components come together. To articulate the lightweight character of structures and to minimize member sizes, the supporting frames are hinged (rather than rigid) or trussed and are laterally stabilized by diagonal bracing to express the play of tension and compression in the state of equilibrium. The site connections are generally pinned (e.g., cast-iron forks, single-eye plates) and occasionally bolted. The gerberettes in the Pompidou Center are in response to this world of thinking. They carry the floor loads in cantilever action and are stabilized by interior compression columns and exterior tension columns.

The following discussion of various historically important cases is based on this organization:

- Long-span, cable-stayed roof structures of the 1970s and some later work
- Spatially guyed, multibay roof structures
- Central spine buildings with clear-span wings
- Other imaginative applications of the cable-stay principle in roof construction

Among some of the important *long-span, cable-stayed roof structures* of the 1970s, surely influenced by the cable-stayed bridges of the 1960s in Germany, are the following buildings. A single, spinal cable-stayed two-hinge gable frame with cantilevers supports almost the entire roof of the ice rink in Braunlage, Germany (1974, Fig. 9.14t). The 270-ft twin box steel ridge beam is elastically supported by fan-type cables suspended from a pair of outward leaning, 161-ft-high masts at each end. Similarly, the supporting structure for the ice rink at Arosa, Switzerland (1978), consists of two parallel, longitudinal, cable-stayed beams of 245-ft span.

The 355- × 296-ft roof structure for the NEC Hall 7, Birmingham, U.K. (1980), consists of nearly equal space frame slabs of the Nodus joint system. It is supported by two steel box trusses in the 355-ft direction and by two box trusses in the other direction. Pairs of tubular steel stays are suspended from 118-ft-high masts to support the trusses at the four truss intersection points that occur approximately at the span third points.

For the powerful West Japan General Exhibition Center/Kitakyushu, Fukuoka Prefecture, Japan (1977, Fig. 1.2f), Arata Isozaki with the engineer M. Kawaguchi used a pair of 150-ft-high masts on top of the first floor, on each building side, to support four parallel 142-ft-span girders and six transverse beams with a spatial diagonal cable system of two groups of cables and 16 intersections, but four single-plane back-stay cables of the harp type extending 82 ft to the anchorage. The 362-ft-span roof of the magnificent flower market at Pescia, Italy, designed by Leonardo Savioli, was completed in 1981.

Among the current examples of cable-stayed, long-span structures is the Sydney Exhibition Center, Sydney, Australia (1988, Fig. 9.13e), which celebrates structural expressionism in a wonderful manner. Philip Cox & Partners are the architects and Ove Arup & Partners the structural engineers. The building consists of five staggered halls where each forms an independent structure of two pairs of masts. From the two opposite tall masts are suspended steel rods in a spatial manner to support three parallel, 303-ft-span, triangular, hinged steel trusses, as well as cross trusses of the same shape. Finally, the immense 500-ft-span roof of the beautiful Lufthansa Hangar at the Munich Airport, Germany (1992), is supported by diagonal cables suspended from 182-ft-tall concrete pylons.

Rogers' Fleetguard factory at Quimper and Foster's Renault Center in Swindon built during the early 1980s have had an important influence on other designers. The roof structures for these multibay buildings are spatially guyed. In the late 1980s, Rogers also used the same principle for a shopping center near Nantes in France.

In Rogers' buildings, the forest of slender, delicate masts and exposed tension rod rigging and the dynamics of roof suspension gives a light, airy, almost abstract diagrammatic appearance. In contrast, Foster does not expose all the tensile work; part of it penetrates into the enclosed space. He expresses strongly the force flow in the member assembly and emphasizes the crafting of the members in a playful way, thereby achieving a tactile experience and a more human scale.

The Fleetguard Factory, Quimper, France (1981, Fig. 9.13 bottom right), designed by Richard Rogers and the structural engineer Peter Rice of Ove Arup & Partners, is a multibay building where the columns are spaced on a 59×59-ft grid to form spatially guyed structures. Steel rods are suspended in a radial manner from the 56-ft-high pipe columns on a square grid of approximately 20 ft and are connected to tubular steel hangers, which penetrate the roof and carry the horizontal framework; in other words, only the tension elements are external. The complex spatial member network was developed to efficiently resist various loading conditions by activating only certain rod–strut sets. The uniform gravity loads are resisted by the suspended spatial cable system as, shown in Fig. 9.14l. Under wind uplift, inverted cable action is generated, as indicated in Fig. 9.14n. Asymmetrical loads are transmitted by a cable system linking the columns, which, in turn, activate inverse truss action in the adjacent bays, thereby not allowing deflections to be transmitted from the loaded bay (Fig. 9.14l and m). The lightweight nature of the tensile structure is apparent from the average steel weight of 9.62 psf, which is about 17% less than a steel structure of comparable bay size.

Similar structural principals were applied by Rogers and Rice to the 47-$\times 94$-ft bays of the shopping center near Nantes, France (1988). Here, 94-ft-high tubular masts, spaced at 47 ft, support 94-ft-span roof frameworks in a spatial fashion (Fig. 9.14o). As for the Fleetguard Factory, only certain combinations of the three-dimensional network of rods and struts are activated under various load actions, as indicated in Fig. 9.14D. Under wind uplift, the tensile rod–strut system forms an inverted V-shaped truss.

Also, the Renault Center, Swindon, U.K. (1983, Fig. 9.13, top right) designed by Norman Foster and Ove Arup & Partners as structural engineers, is a spatially guyed structure. Trusslike portal frames are placed not only along the 79-$\times 79$-ft square bays, but also along the diagonal directions. Rods are suspended from the top of the 53-ft-high tubular steel masts in the orthogonal and diagonal directions to support the tapered portal beams at their quarter-points. In the center portion the sloped beams are cable supported from below (i.e., strutted cable beams). In other words, we may visualize suspended cables to span from column to column using struts to support the central roof framework portion from below. The cable configuration follows the moment diagram of a multibay portal frame with hinged base under uniform gravity loads (Fig. 9.14i, j) by efficiently resolving the moment into compressive and tensile forces. Under wind uplift, the tapered portal beams act in tension and the prestressed ties in compression (Fig. 9.14k). The slender, tubular columns are laterally braced with four prestressed rods that are connected to the sloped beams, thereby providing a moment connection. In other words, lateral stability of the building is provided by portal frame action. The average steel weight of the structure is 12.7 psf.

The following two examples represent **linear, central, spinal buildings** with clear-span wings.

The Patscenter, Princeton, New Jersey (1984, Fig. 9.13c and Fig. 1.11d), designed by Richard Rogers and the structural engineer Peter Rice of Ove Arup & Partners, consists of parallel planar guyed structures. The central spine is divided into parallel 25-ft-wide portal frames set 30 ft on center that support on top 48-ft-high A-frames. A typical A-frame is formed by inclined pipe columns connected to a large ring plate from which are suspended steel rods to other ring plates on each side of the spine, from which, in turn, four hangers radiate down to provide support to the main 75-ft-span W14 roof beams. The two center hangers support the middle section of the beam against vertical loading, whereas inverted truss action is required for wind uplift where the outer hangers act as tensile chords (hence are rods) and the two central hangers act in compression (therefore are tubes). The exterior columns act either in compression or tension depending on loading. To control possible additional uplift, the roofing contains a crushed stone ballast. Platforms are suspended between the A-frames to carry the mechanical equipment and to stiffen the frames, besides providing longitudinal bracing together with the cross bracing between the frames. The critical wind loads are transferred from the roof level through the central portal frames to the ground. The average steel weight of the structure is 9.2 psf.

Similarly, the Inmos Factory, Newport, U.K. (1982, Fig. 9.14u), designed by Richard Rogers and the structural engineer Tony Hunt, consists of a 16-ft-wide central service spine along which are placed two 50-ft-high lattice masts spaced 43 ft on center. The masts are connected with compression cross bracing along the exposed upper part and form a lattice portal frame along the enclosed ground level. Tie rods are suspended from the top of the braced frames to give two additional supports at the one-third points of the 126-ft span spatial steel trusses (with double top chords and single bottom chord and nearly 5-ft depth), spanning from the central spine to the exterior columns. Secondary lattice trusses, spaced at 20 ft, span between the primary trusses to carry a tertiary beam grid. Upward wind loads are not critical since the roof dead load is larger. The average steel weight of the structure is 11.5 psf.

Among the architects using **other imaginative applications of the cable-stayed principle** in roof construction, instead of the repetitive cable-stayed beams of the previous cases, is the Englishman Nick Grimshaw. He articulates the concept in a very creative manner by selecting only one or two cable-supported primary beams (similar to Braunlage or Arosa) to support nearly an entire roof, as is demonstrated in the following examples. He celebrates the tall, highly visible, stayed masts as exciting images and landmarks, reminding us of the masts and rigging on sailing ships.

Most of the roof framing for the ice rink in Oxford, U.K. (1984, Fig. 9.13b and Fig. 1.2e), engineered by Ove Arup & Partners, is supported by a central cable-stayed beam in the longitudinal direction, which carries the perpendicular rib beams spaced 6 ft on center. The 237-ft-span spine beam (consisting of two steel box sections) is hung on four diagonal stainless steel rods from the 109-ft-high masts at each end of the building, thereby converting the spine beam into a five-span continuous beam. The spine beam is extended beyond the building to divert the loads efficiently to two groups of four vertically loaded tension piles. The masts consist of three independent, pin-jointed sections (two of them cable trussed), so bending stresses are nearly eliminated.

Similarly, for the Homebase project, Brentford, U.K. (1987, Fig. 9.13d), Grimshaw uses a 329-ft-long spinal box girder with cantilevers at each end and gives an additional support close to midspan with steel tension rods hung from one side of an 109-ft-high framed tower.

For the Ladkarn Haulage Headquarters, London, U.K. (1986, Fig. 9.13a), engineered by Ove Arup & Partners, Nick Grimshaw uses two parallel, exposed, 109-ft-long spatial cable trusses supported by strut and rod stays cantilevered from tall masts at one end and ordinary columns at the other ends. Lateral stability for the hinged steel framing is provided by the diagonally braced bays and the walls of the office building.

Structural Behavior   Most tensile structures are very flexible in comparison to conventional structures. This is particularly true for the current, fashionable, minimal structures where all the members want to be under axial forces. Here, repetitive members with pinned joints are tied together and stabilized by cables or rods. Not only the low stiffness of cables, but also the nature of hinged frame construction make them vulnerable to lateral and vertical movements. To acquire the necessary stiffness, special construction techniques have been developed, such as spatial networks, as well as the prestressing of tension members so that they remain in tension under any loading conditions.

Because of the lightweight and flexible character of cable-stayed roof structures, they may be particularly vulnerable with respect to vertical stiffness, wind uplift, lateral stability, and dynamic effects; redundancy must also be considered in case of tie failure. Temperature effects are critical when the structure is exposed to environmental changes. The movement in the exposed structure must be compatible with the enclosure. In the partially exposed structure, differential movement within the structure must be considered. Slotted connections may be used to relieve thermal movement.

Cable-stayed roof structures are lightweight structures with a live-to-dead load ratio of around four ($w_L/w_D \cong 4$), in contrast to bridges with a ratio of about 1 ($w_L/w_D \cong 1$). Since the dead load for cable-stayed roofs is generally less than the wind uplift, wind suction becomes a primary loading case.

In this section we briefly investigate the following common cable-stayed roof structures:

- Cable-stayed, double-cantilever roofs (Fig. 9.34b)
- Cable-stayed, single-cantilever roofs (Fig. 9.14h and 9.34a)
- Cable-stayed beams supported by masts from the outside (Fig. 9.14A)
- Spatially guyed, multidirectional composite roof structures

Cable-stayed cantilever roof structures allow column-free, long-span space as needed for hangars and grandstands. Whereas in symmetrical, double-cantilever construction (Fig. 9.34b) the dead loads of the two overhangs balance each other, in single-cantilever construction the dead load causes overturning. For this condition, the stay cable is generally anchored in the rear in heavy concrete bents, for example, which in turn may have to be supported by tension piles. The decking type for cable-stayed cantilever roof structures may have to be based on the minimum weight requirements that will resist uplift due to wind and reduce flutter. The horizontal thrust component of the cable force is carried by the roof beam in compression, which must also carry relatively high bending moments due to gravity loading. In case of extremely high winds, the beams may resist part of the uplift in cantilever bending, requiring the beams to be moment connected to the post. The reader may also want to study Problems 9.1 and 9.2, which deal with the preliminary design of cable-stayed cantilever roofs.

The structural behavior of the simple symmetrical cable-stayed roof in Fig. 9.14A is now investigated. For this preliminary approach we assume the following:

- The tensile stays are inextensible; in other words, the effect of elastic supports on the force flow in the continuous beam is disregarded.
- The roof girder is hinged to the mast.
- The mast bases are hinged.
- The masts carry only axial forces.

The response of the structure is studied with respect to symmetrical and asymmetrical gravity loading, wind uplift, lateral stability, and pretensioning.

In spatially guyed, multidirectional roof structures, composite versions of cable-stayed beams (Fig. 9.14B) or cable beams and trusses (Fig. 9.14C and D) are integrated. The exposed, complex, spatial cable–strut network in cases C and D in Fig. 9.14 is derived from efficiently resisting various loading conditions, where only certain member sets are activated for specific load actions.

A. Uniform Gravity Load Action: Because of symmetry, each cable in Fig. 9.14a transfers one-half of the concentrated load P acting at midspan to the masts. Therefore, the tension in the stays is

$$\Sigma F_y = 0 = T\sin\theta - P/2, \quad \text{or} \quad T = P/2\sin\theta$$

Pure axial compression is generated in the beam, assuming the cable is attached to the neutral axis of the section. The compression force C is equal to

$$C = (P/2)\text{ctg}\,\theta = T\cos\theta$$

There is no tension force in the central portion of the beam between the stays because of the balanced arrangement of the double-cantilever beams. Should the back-stay slope be different from the main stay slope (i.e., should the double cantilever not be self-balancing), tension forces occur in the center span portion between the stays. Furthermore, should the stays be arranged spatially (Fig. 9.14b), axial forces are also generated in the perpendicular direction, as shown. Notice that the principle of staying provides natural prestressing of beams! The axial compression forces in the masts can be easily obtained by balancing the forces at the top joint of mast and stays:

$$N = 2(P/2) = P$$

In Fig. 9.14d, the tensile ties are not anchored directly to the ground but are guided back diagonally to the supports so that they can resist also lateral forces. Therefore, the tension in the base stays is equal to

$$T = P/2\cos\phi$$

For this condition, horizontal reaction forces are generated by gravity action similar to portal frames. They are equal to

$$H = (P/2)\tan\phi = PL/8h, \quad \text{where} \quad \tan\phi = L/4h$$

Notice that there will be an axial compression force in the beam between the stays that is equal to H. The compression force in the beam along the cantilever portion is

$$C = (P/2)(\tan\phi + \text{ctg}\,\theta)$$

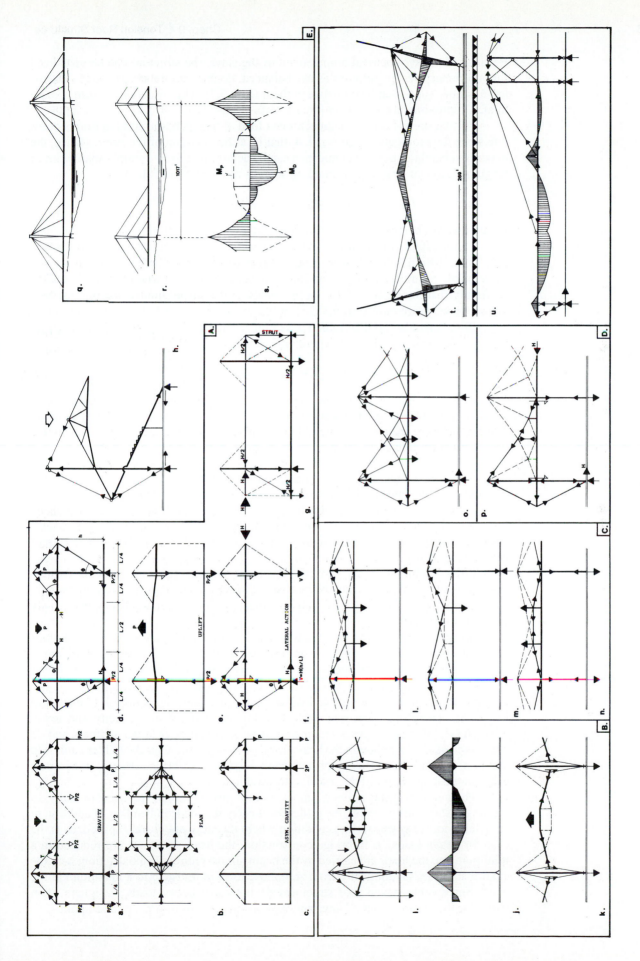

Figure 9.14   Force flow in cable-supported roofs.

Because of the symmetrical arrangement of the stays, the structure can be visualized for preliminary design purposes as two balanced, double-cantilever structures supporting a simply supported beam between them (Fig. 5.7). This situation, in turn, can be treated approximately as a beam fixed at both ends.

For the special case of a concentrated load at midspan, this assumption is accurate since the moments are zero at $L/4$, that is, at the location of the stays. Hence, the maximum bending moment at midspan can be determined from a simple span beam of length $L/2$, as can be the corresponding support moment M_s (Table A15).

$$M_{\max} = (P/2)L/4 = PL/8, \qquad M_s = PL/4 - PL/8 = PL/8$$

In other words, beam bending is only activated in the center portion between the stays, ignoring the effect of continuity. Should the cables continue below the roof beam and give support from below, load P is transferred directly to the columns, and the assistance of the beam in bending is no longer needed. However, when the single load is uniformly distributed ($w = P/L$), the moments at the suspension points are not zero unless the stays are moved from $0.25L$ to $0.21L$. We may conclude that the horizontal thrust force can also be determined based on a two-hinge portal frame, according to Eq. (5.6), using a hypothetical column moment M_s.

$$H = M_s/h = PL/8h$$

In Fig. 9.14B, the cable–arch configuration follows the moment diagram of a multibay portal frame with hinged bases by efficiently resolving the bending moment into compressive and tensile forces. Under wind uplift, the force action in the composite system is reversed.

In the previous discussion, it was assumed that the stays give a rigid support to the roof girder, which obviously cannot be true. Actually, the girder is a continuous beam on intermediate elastic supports, where the stays act as elastic springs. The stiffness of the stays is a function of the material, size, and length of members, as well as the support conditions. As the springlike supports settle, the support moments decrease, thereby increasing the field moments, as discussed for Fig. 2.27 and as indicated in Fig. 9.14s and u.

If the continuous beam were converted to a hinged beam with hinges at the suspension points, the beam would be insensitive to settlement. However, the increase in member size due to loss of continuity together with the larger flexibility may easily offset this approach.

We must also keep in mind that a shallow stay angle yields larger deflections than a steep stay. Usually, the stays are inclined at about 30° to 45° to the horizontal. Furthermore, to obtain the necessary vertical stiffness, solid rods are preferred over cables since they are three to four times stiffer than a cable of the same size.

Roofs are vulnerable to live load deflection because of their high live-to-dead load ratio. Therefore, steel rods with their higher modulus of elasticity and lower strength (yielding larger cross-sectional area or sections) offer larger stiffness and do not stretch as much, although the heavier rods will sag more than the lighter cables. In addition, the rods are easier to protect from corrosion. In bridge design, in contrast, cables are used because of the low live-to-dead load ratio.

The number of stays giving support to roof girders ranges from one to many. In the single-stay configuration, the girder must carry the loads in bending in addition to an axial load. However, for the condition where the stays are closely spaced, as is the case for heavy loads, as in bridge construction, the beam moments are reduced to a minimum and the loads are carried by the beam almost entirely in axial action; here the axial load varies parabolically from zero close to or at midspan to a maximum at the mast. Since cable-stayed roof structures do not carry heavy loads, designers often strive for the least number of stays and the least number of roof penetration points.

Comparing the fan-type arrangement (Fig. 9.14q) with the harp arrangement (Fig. 9.14r), notice that in the former configuration all tensile forces vary because of the different slopes, although each cable carries nearly the same load. The fan arrangement uses less steel since the lower cables are at a steeper slope and therefore produce less cable tension and less thrust in the roof girder. Since the upper cables are longer and shallower and hence carry larger forces, their extensions will be larger and thus cause larger beam settlement in the midspan region (see Fig. 9.14u). For the harp configuration, in contrast, the cable forces are the same, but the cable lengths vary, therefore causing different beam settlements, with the larger ones in the midspan portion. A disadvantage of the fan arrangement is the detailing at the top of the mast where all the cables come together; therefore, the modified layout of the fan–harp configuration is often used.

The design of the stayed girder in beam–column action is highly indeterminate because of the elastic cable or rod supports. The magnitude of the bending moments and deflections, however, can be controlled by adjusting the cable tensions, that is, by introducing prestress forces and/or connecting the stays eccentrically to the girder. In other words, the stays are not attached at the neutral axis of the beam but below to generate a counteracting bending moment (see Figs. 3.17 and 9.14s).

B. Other Loading Conditions: **Asymmetrical loads,** *P,* in one-bay structures can be easily transferred in single-cantilever action to the ground, as indicated in Fig. 9.14c. However, in continuous multibay structures these loads cause bending of the beams in adjacent bays and possibly bending of the columns. This situation can be controlled by stiffening the columns, for example, through guying (Fig. 9.14k) and by stiffening the roof framing with a spatial network of slender members, thereby developing frame action. For instance, the asymmetrical loads in Fig. 9.14m are transmitted by a cable system linking the columns and thereby activate inverse truss action in the adjacent bays so that deflections are not transferred from the loaded bay to other bays.

Since cable-stayed roof structures are of light weight in contrast to conventional buildings, **wind uplift** is a primary loading condition. The various methods for stabilizing cable roof structures are presented in Fig. 9.1. Among the design solutions for cable-stayed roofs, assuming an increase of dead load as not feasible, are as follows:

- Designing the roof beams for upward bending (Fig. 9.14e)
- Designing the roof beams for upward bending and the stays as compression members (e.g., using steel tubes)
- Developing a roof profile or other methods to reduce air suction
- To introduce tensile arch action (Fig. 9.14k) that is inverted suspended cable action or truss or cable beam action (Fig. 9.14C and D)

Lateral stability is a most important issue in hinged structures; it must be considered in the transverse and longitudinal directions of a building. The reader may want to refer to the response of buildings to lateral force action in general as is discussed in Section 4.2. Usually, in hinged structures the vertical bays or members are braced and the tension diagonals are activated as the roof plane transmits the horizontal force, as is the case in Fig. 9.14g, assuming that the exterior vertical ties are struts and can also act in compression. For the single-bay structure in Fig. 9.14f, the horizontal force is resisted at the side, where the tension diagonal is anchored (disregarding the pretensioning due to dead load in the diagonal on the right), similar to a portal frame with one leg pinned and the other supported on a roller (Fig. 5.8d). In multibay structures, the lateral forces may be resisted by the action of continuous portal frames pinned at their base (Fig. 5.20b) by directly guying the columns and linking them to the cable-stayed roof beams (Fig. 9.14B). In Fig. 9.14p, the lateral force is resisted in the end bay similar to a single-bay frame (Fig. 5.8f).

Although the principle of staying provides natural prestressing of roof beams, further **prestressing** may be required to efficiently counteract other loading conditions, as is the case for the stayed, two-hinge gable frame in Fig. 9.14t, where additional tension was introduced in the cables. Similarly, for the harp-type, cable-stayed bridge in Fig. 9.14s, the change of tension in the stays and the corresponding eccentric compression in the beams (i.e., bending) alter and control the moment distribution along the beams due to gravity loading. Hence, this relationship has to be optimized as to allow minimal member sizes.

EXAMPLE 9.9

The cable-stayed roof in Fig. 9.14a is investigated for preliminary design purposes as based on a uniform gravity load, using a dead load of 12 psf and a snow load of 25 psf. The masts are spaced 25 ft apart, and the beams span 100 ft from mast to mast. The 45° stays are assumed to support the beam at the fifth points, $L/5 = 100/5 = 20$ ft.

The beam carries a uniform load of

$$w = 25(0.012 + 0.025) = 0.93 \text{ klf}$$

Because of its layout, the structure can be visualized as two balanced, double-cantilever structures carrying a simply supported beam of 60-ft span at the center. This assumption is reasonable, since for a beam fixed at both ends (i.e., the symmetrical, 20-ft cantilevers act like a fixed end) carrying a uniform load, the moments are zero at $0.21L$ (see Fig. 2.10) which is close enough to $0.2L = 20$ ft. Therefore, the maximum moment at midspan, which is the longest moment in the stayed beam, can be quickly determined as

$$M_{max} = w(0.6L)^2/8 = 0.93(60)^2/8 = 418.50 \text{ ft-k}$$

The stays are assumed to support the following load:

$$V = 0.93(60 + 20)/2 = 37.20 \text{ k}$$

Therefore, the tension in the cables is

$$T = V/\sin \theta = 37.2/\sin 45° = 52.61 \text{ k}$$

The axial compression in the beam portion of the cantilever structure is

$$C = V/\tan \theta = 37.2/\tan 45° = 37.20 \text{ k}$$

Notice, because of conditions of symmetry, that there is no axial tensile force in the midspan portion between the stays.

The exterior anchor must resist a tensile force of 37.2 k, which is obtained by inspection from the symmetry of the cantilevers (Fig. 9.14a). The column must carry the two cantilever loads and the rest of the uniform load between stay support and column.

$$N = 2(37.2) + 10(0.93) = 83.70 \text{ k}$$

Assuming the columns to be laterally braced about both axes with an effective length $KL = 1(20) = 20$ ft, we obtain from the column tables in the steel manual as based on axial action only, for $F_y = 46$ ksi, the following structural tubing:

$$\text{TS } 6 \times 6 \times 5/16, \qquad P_a = 93 \text{ k}$$

The required section modulus for the beam [Eq. (3.4)], as based on A36 steel, is

$$S_x = M_x/F_b = 418.50(12)/0.66 (36) = 211.36 \text{ in.}^3$$

Try W27 × 84, $S_x = 213$ in.^{3} (Table A.11). Should the beam also act as a column, use Eq. (3.100) for design.

The required cross-sectional area of the diagonal rod, using A36 steel, according to Eq. (3.87), is

$$A_D = P/0.33F_u = 52.61/0.33(58) = 2.75 \text{ in.}^2$$
$$A_D = \pi d^2/4 \quad \text{or} \quad d = 1.87 \text{ in.}$$

Try 2-in.-diameter rod, $A = 3.14$ in.^{2}. Notice, however, that the sizing of the rods may depend on their stiffness rather than their strength.
The rod extension according to Eq. (2.54) is

$$\Delta L = PL/AE = 52.61 \, (20 \sqrt{2}) \, 12/3.14(29,000) = 0.20 \text{ in.}$$

Or the vertical support settlement, disregarding the effect of beam stiffness, is

$$\Delta_v = \Delta L \sin \theta = 0.20 \sin 45° = 0.14 \text{ in.}$$

Should there be a moment in the beam at the stay, it will be decreased by the settlement, but at the same time the maximum field moment will be increased. It is apparent that the elastic settlement must be controlled by selecting the appropriate stay stiffness.
At this stage, other stay arrangements and more stay supports should be investigated.

9.5 SIMPLY SUSPENDED ROOFS

Simply suspended or hanging roofs include cable roofs of single curvature and synclastic shape, that is, cylindrical roofs with parallel cable arrangement, and polygonal dishes with radial cable patterns or cable nets. The simply suspended cables may be of the single-plane, double-flange, or double-layer type. The concept of simply suspended roofs has surely been influenced by suspension bridge construction. Among the great long-span suspension bridges are the following:

- Roebling's 1595-ft-span Brooklyn Bridge (1883)
- Strauss's 4120-ft-span Golden Gate Bridge in San Francisco (1937)
- Ammann's 4260-ft.-span Verrazano Narrows Bridge in New York (1965)
- Wex's 4626-ft-span Humber Bridge in England (1981)

The longest suspension bridge in the world is currently Japan's Akashi-Kaikyo Bridge, 6800 ft in length. The principle of suspension bridges is directly applied to the **suspended, cable-supported roof systems** as expressed so powerfully by Nervi's Burgo Paper Mill (Fig. 9.15b), or the system may be applied using a hidden approach, as for Madison Square Garden in New York, where the suspended cables carry a two-story building that forms the roof.
The most common approach, however, is the **simply suspended roof,** where the suspended cables (or other member types) are integrated with the roof structure. Here, we distinguish whether single- or double-layer cable systems are used. One of the first modern simply suspended roofs is Bernard LaFaille's 110-ft-span, dish-shaped French Pavilion at the Zagreb Fair in 1935.
Simple, single-layer, suspended cable roofs must be stabilized by heavyweight or rigid members, as discussed for Fig. 9.1. Sometimes, prestressed suspended concrete shells are used where, during erection (and under ultimate loading), they act as simple suspended cable systems, while in the final state they behave like inverted prestressed concrete shells. In simple, double-layer cable structures, such as the typical *bicycle wheel* roof (Fig. 9.1j), stability is achieved by secondary cables prestressing the main suspended cables (Fig. 9.11).

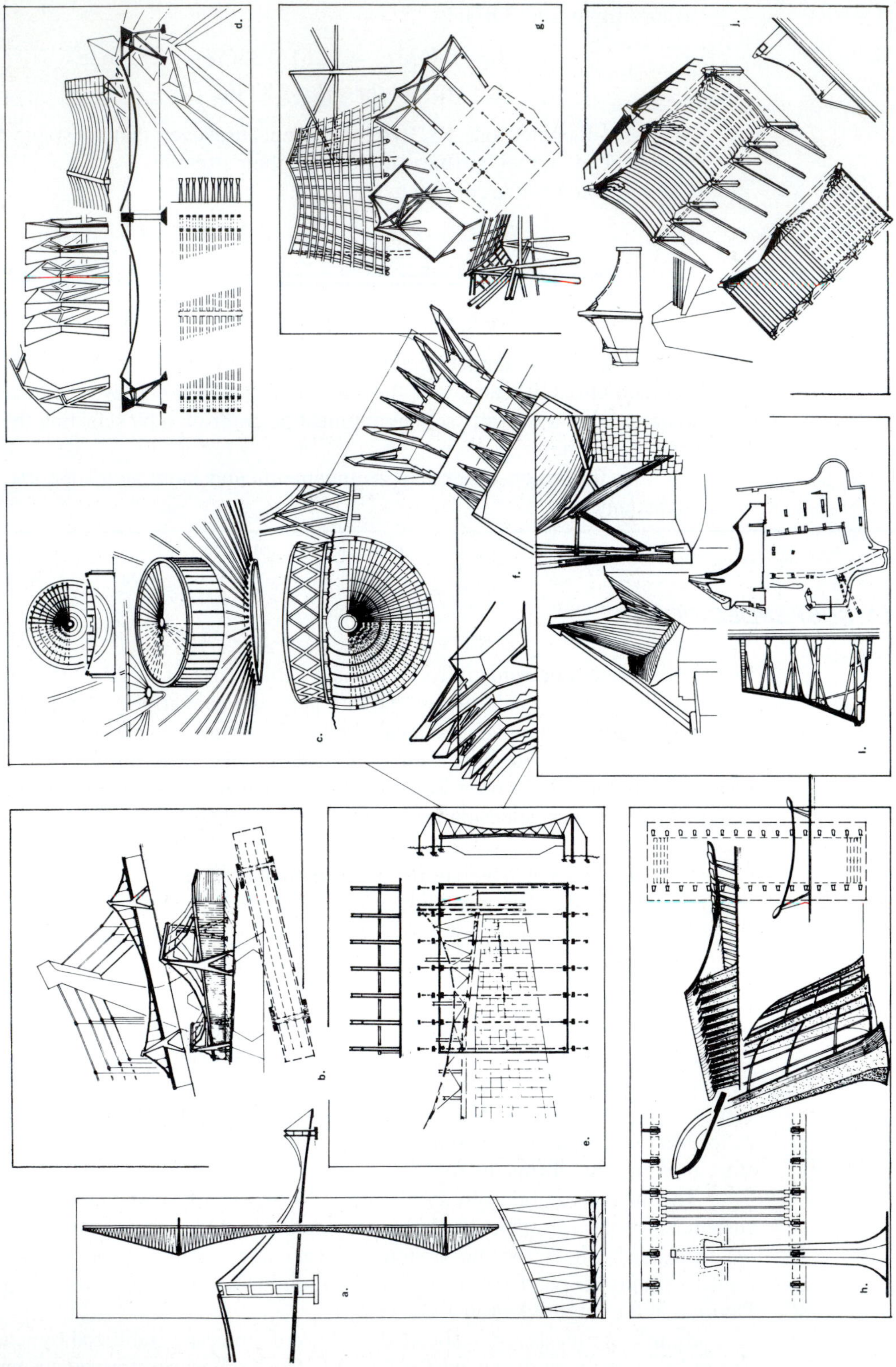

Figure 9-15 Simply suspended structures.

Most buildings using the suspended roof concept are either rectangular or round; in other words, the cable arrangement is either parallel or radial. For these typical conditions, the action of single cables under transverse loads is discussed in Section 9.2. However, in free-form buildings, the cable layout is irregular and the roof geometry is not a simple inverted cylinder or dish. The complexity of form is expressed by Mechelucci's wonderful Church of the Autostrada (Fig. 9.15i). Occasionally, structures are of hybrid nature, such as the timber A-frames supporting ridge cables from which are hung inclined tensile roof wood grids (Fig. 9.15g).

The drainage of cylindrical surfaces is easily achieved by sloping the roof in the longitudinal direction, that is, perpendicular to the curvature. Bowl-shaped roofs, however, collect the water at the center in tanks from which the water must be pumped back up to drain lines at the periphery wall. In the event of a catastrophic storm, the overflow from the pooling area is dumped into the arena. Since cables are subject to large deformations due to loading and temperature change, watertightness becomes an important design consideration.

Next the single-layer cases in Fig. 9.15, with spans ranging from about 150 to over 500 ft, are briefly discussed. First, some typical, suspended, cable-supported roof systems are presented.

For the Burgo Paper Mill in Mantua, Italy (1963, Fig. 9.15b), Pier Luigi Nervi applied the suspended bridge concept to cover the 817- × 98-ft column-free space. The central bay is 535 ft with 141-ft cantilevers on each end. The elegance and dynamics of the suspended cable structure from which the roof framing hangs is articulated by the sculptural qualities of the supporting legged concrete piers. The pier's straight long leg is supported by the short leg acting as a strut (see Fig. 9.2e). The roof structure for the ice rink in Memmingen, Germany (1987, Fig. 9.15e), is based on similar construction principles. Here, the timber roof framing is hung from the 184-ft-span suspended cables using diagonal ties, thereby giving the appearance of cable trusses.

The 404-ft-span circular roof of Madison Square Garden in New York (1967, Fig. 9.15c, top), designed by Charles Luckmann and Fred Severud, consists of radially arranged suspended 3 $\frac{3}{4}$-in.-diameter cables. The cables are attached to a central steel tension ring and an exterior steel compression ring and support a two-story, steel-framed roof structure, which houses the cooling tower and mechanical equipment (plumbing, heating, electrical, and air conditioning). The main structural elements of the conventionally framed roof structure are five concentric ring trusses that distribute the loads equally to adjacent cables and provide lateral stability. Since the cables are loaded with the heavy penthouse structure, sufficient stiffness is provided and the roof is prevented from vibrating.

In the discussion of the following cases, the principal of single-layer, simply suspended roofs is addressed.

The Oakland–Almeda County Coliseum (1966, Fig. 9.15c, bottom) designed by Myron Goldsmith of SOM, has a dished, single-layer, cable-supported, 420-ft-span roof. The cables are suspended from a concrete compression ring that rests on concrete cross columns. The cables are stabilized by precast concrete I-sections, which have a longitudinal slot in the bottom flange to fit over the cable and match the drape. In addition, the suspended roof is weighed down by the 260-ft-diameter steel-frame penthouse containing the mechanical and electrical equipment. The cross columns along the perimeter form a cylindrical trussed wall to effectively resist wind and seismic forces.

Eero Saarinen achieved a true integration of structure and architecture in the Terminal Building of Dulles Airport, Washington, D.C.; it is one of the great landmarks of modern architecture (1962, Fig. 9.15h); the building structure was engineered by Fred Severud. The powerful pull of the 161-ft suspended roof is perfectly counteracted by the giant outward sloping sculptured concrete columns (Fig. 9.2g). The inclined, suspended roof cables are supported by concrete edge beams, which, in turn, transfer the forces to the giant inclined columns carrying the forces in compression and bending. The roof was constructed by letting the cables carry precast, lightweight concrete panels. But when the spaces between the roof panel bands were filled with reinforced cast-in-place lightweight concrete, the stiff ribs forced the tensile roof to transform into a stiff tensile vault. This solution of a rigid inverted arched surface was considered necessary to eliminate vertical oscillations of the single-layer cable system.

The powerful shapes of the huge sculptured concrete piers of the City Center in Bremen, Germany (1964, Fig. 9.15f), designed by the architect R. Rainer and the engineer U. Finsterwalder, respond dramatically to the 328-ft-span, suspended, prestressed concrete tension members (see also Fig. 9.2h). The 250-ft-span inclined, concave, prestressed concrete vaults of Riccardo Morandi's Hangars 4 and 5 at the International Airport of Fiumicino in Rome, Italy (1969, Fig. 9.15j), are supported by cable-stayed beams along each face.

The Hangar V at Frankfurt airport in Germany (1969, Fig. 9.15d), designed by the architects Becker and Beckert and the construction company Dyckerhoff & Widmann, expresses elegantly the distribution of force flow. The enormous 440-ft spans of the suspended, prestressed, lightweight concrete strips are balanced dramatically at each end by the sculptured trussed concrete abutments with their massive counterweights on top (see also Fig. 9.2d).

The roof of the main oval stadium for the 1980 Olympic Games in Moscow has the shape of an elliptic paraboloid with the enormous dimensions of 735×610 ft along its main axes and is supported by a suspended steel membrane consisting of the steel skin and the hanging radial trusses.

The infinite freedom of form giving in contrast to the minimal design approach can only be suggested by the following cases.

One of the most exciting early modern tensile structures is Giovanni Mechelucci's Church of the Autostrada de Sole near Florence, Italy (1964, Fig. 9.15i), where he translated the suspended fabric and the pole supports of tent construction into the plasticity of concrete. The tensile shell roof is formed by steel cables suspended from bony forked columns. In the interior space, the sloping roof shell expresses the play of enormous tension and dramatizes the force flow to the supports through the articulation of the boniness of the branching columns and the marking of the joints.

The sculptural expression of the Mecca Auditorium roof (1974, Fig. 9.17C), designed by Frei Otto and Rolf Gutbrod together with the engineer Edmund Happold, represents an early example of a new direction in suspended roof design. Although the roof has the appearance of a tent structure, it consists of simply suspended steel cables approximately 1 in. in diameter supporting a heavyweight decking. The heavyweight cable roof is hanging from a steel frame at one end and edge cables spanning between guyed masts at the other end.

The deformed wheellike roof of the main temple of Sho-Hondo at Taiseki-ji (1972, Fig. 9.16e), designed by Kimio Yokoyama, rests on fan-shaped, inclined column clusters of various configurations that form pyramidlike space truss supports. The roof is not a flexible, lightweight, prestressed cable-net surface, but a heavyweight, semirigid, tensile structure where the radial hanging members are not cables but steel beams that provide the necessary rigidity against wind and other loads. These curved beams are suspended (363 ft at the major axis) between the inner circular tension and the outer elliptical compression rings; they are laterally braced by concentric ring beams. This beam grid, in turn, provides the support for the precast, lightweight concrete panels.

For preliminary methematical investigation of simply suspended cable roof structures, refer to Examples 9.2 and 9.3.

Instead of cables or metal sheets, suspended, prestressed, reinforced, lightweight concrete strips could be used. For example, thin slabs could first be cast horizontally on the ground, prestressed, and then lifted vertically and suspended from their supports.

The required prestress tendon area A_s can be determined according to Eq. (3.74) as

$$A_s = \frac{H}{0.56f_{pu}} = \frac{M/f}{0.56f_{pu}} = \frac{wL^2/8f}{0.56f_{pu}}$$

For a shallow suspended cylindrical surface, the maximum bending stresses f_b due to bending of the horizontal slab can be derived from Eq. (2.76) and be approximated as

$$M = 8\Delta EI/L^2, \quad \text{let } \Delta = f \tag{a}$$

or

$$\pm f_b = \frac{Md}{2I} = \frac{4fdE}{L^2} = 4n(d/L)E$$

where d = thickness of suspended concrete strip, and $n = f/L$.

The preliminary slab thickness can be estimated from Eq. (3.75).

$$f_c = f_b + f_a = M/S + H/A_g \le 0.45f_c'$$

$$\tag{9.84}$$

$$= 4n(d/L)E + \frac{wL^2/8f}{12d} \le 0.45f_c'$$

Notice that the bending stresses are small in comparison to the axial stresses due to prestressing and therefore can be ignored at the initial design stage.

Double-layer Simply Suspended Roofs

Double-surface structures may be formed by individual cables (e.g., "bicycle wheel" roof), planar cable beams, double-flange cable beams, or double-layer networks. Double-layer cable structures are stressed structures and are more stable than single-layer suspended roofs, as discussed in Section 9.3. Typical examples of double-layer cable layouts for round roofs are shown at the bottom of Fig. 9.11.

The double-surface bicycle wheel with independent cables was first used on a large-scale for the U.S. Pavilion at the Brussels 1958 Fair, designed by Edward D. Stone. From an exterior compression ring, cables were tied to tension rings at top and bottom of a central cylindrical drum; the upper cables were pretensioned but not linked to the lower cables (Fig. 9.11i).

Lev Zetlin developed the double-surface bicycle wheel with pretensioned linked cables, that is, cable beams. As for the bicycle wheel, where the radial tension of the spokes causes compression in the outer rim of the wheel and tension in the inner ring, the prestressed cables cause compression along the exterior compression ring and tension in the central hub. Lev Zetlin used this system for the first time for the 240-ft-span circular roof of the Utica Civic Auditorium, Utica, New York (1959, Fig. 9.1 l). Here, the prestressed, strutted, dual-cable system is anchored to a concrete compression ring; the cable beams are arranged at 5° intervals. The space between the two layers of cables is occupied by mechanical and air-conditioning equipment. When the tied radial cable beams in Fig. 9.11j are not continuous and pulled back, they form cantilever beams, as shown in Fig. 9.11k. Here, the upper loadbearing cables and the lower prestressed wind cables are held together by vertical ties. The radial cantilever cable beams are prestressed by the inner tension ring and held in place by the outer compression ring.

In the 1950s Robert LeRicolais developed a trigrid lenticular surface structure to achieve a more spatial distribution of unbalanced forces and to eliminate the central tension drum used for radial systems (Fig. 9.11n). Similarly, in the late 1980s Horst Berger proposed a two-way, double-layer saddle dome structure where the cables follow the warped compression ring (Fig. 9.11m).

In 1959, David Jawerth, a Swedish engineer, patented parallel cable beam construction to cover rectangular areas. The 273-ft-span cable trusses for the ice stadium at Stockholm–Johanneshov of 1965 represent his finest achievement.

Various types of cable beams and cable trusses arranged in a parallel manner are shown in Fig. 9.1e, g, i, and k. Complex spatial network structures are shown in Fig. 9.21 and discussed in Section 9.7. For the discussion of simple cable beams and cable trusses, refer to Section 9.3.

9.6 PRESTRESSED MEMBRANES AND CABLE NETS

In contrast to traditional buildings, tensile structures lack stiffness and weight. Whereas the basic building blocks of conventional *hard* structures are the rectangle and triangle, for *soft* structures (due to absence of stiffness) they are curvilinear geometry and built-in tension (prestress). The surface geometry is anticlastic, where the two opposing curvatures balance each other. In other words, the prestress in the membrane along one curvature stabilizes the primary load-bearing action of the membrane along the opposite curvature. The induced tension provides stability of form (see Fig. 9.1), while space geometry, together with prestress, provides strength and stiffness.

Since flexible membranes can resist loads only in pure tension, their geometry must reflect and mirror the force flow; surface geometry is identical with force flow. The geometry (forces) must be in equilibrium so that under superimposed loads it guarantees stability and safety. Membranes must have sufficient curvature and tension throughout the surface to achieve the desired stiffness and strength under any loading condition. In contrast to traditional structures, where stresses result from loading, in anticlastic structures, prestresses must be specified initially so that the resulting membrane shape can be determined.

The evolution of prestressed tensile surface structures is generally associated with the State Fair Arena at Raleigh, North Carolina, designed by the architect Mathew Nowicki and the structural engineer Fred Severud, which was completed in 1953. But it is the eminent architect Frei Otto who fully developed the formal potential of architectural tensile surface structures through his experiments and research, primarily at the Institute of Lightweight Structures (IL) at Stuttgart University, Germany. He made the transition possible from the small-scale fabric structure to the large-scale cable net structure, as demonstrated by the stadia of the 1972 Munich Olympics. Frei Otto became one of the significant innovators of the era of modern architecture.

Tensile membranes can be classified either according to their surface form or to support conditions. Basic surface forms are either *saddle shaped* and stretched between their boundaries or *conical-shaped* and center supported at high or low points; in other words, they represent either orthogonal or radial anticlastic surfaces. The combination of these basic surface forms yields an infinite number of new forms. The membrane supports may be rigid or flexible; they may be point or line supports located either in the interior or along the exterior edges. The following organization is often used based on support conditions:

- *Edge-supported* saddle surface structures
- *Arch-supported* saddle surface structures
- *Mast-supported* conical (including point-hung) membrane structures (tents)
- *Hybrid* structures, including tensegric nets (see Section 9.7)

The layout of the support types, in turn, results in a limitless number of new forms, such as the following:

- Ring-supported saddle roofs (Figs. 9.1 and 9.2)
- Parallel and crossed arches as support systems
- Parallel and radial folded point-supported surfaces
- Multiple tents on rectangular grids

Edge-supported Saddle Roof Structures

Various support types have already been introduced in the bottom portion of Fig. 9.2 and discussed for Fig. 5.37 with respect to exterior boundary forms. They range from exterior warped rings and inclined arches to open perimeter rings together with interior arches, as demonstrated by the cases along the outer portion of Fig. 9.16.

When the lightweight, long-span, anticlastic cable net roof structure of the State Fair Arena at Raleigh, North Carolina, was completed in 1953, it was the first structure of this type. At nearly the same time the warped, oval, tensile roof structure (151 ft $\times$ 234 ft along the primary axes) for the Schwarzwaldhalle in Karlsruhe, Germany, was designed by Ulrich Finsterwalder based on the same structural principles, but using prestressed concrete for the first time for this type of construction. The source of inspiration the Raleigh Arena had on many designers is expressed convincingly by Eero Saarinen's Hockey Rink at Yale University (1956) and Kenzo Tange's two magnificent Olympic stadia in Tokyo (1964).

The North Carolina State Fair Building in Raleigh, designed by Mathew Nowicki and Fred Severud (1953, Fig. 9.16j and Fig. 9.2k) has a nearly elliptical shape in plan view. Two inclined parabolic arches support in funicular action the uniform loading of the only slightly prestressed saddle-shaped cable roof, which spans about 300 ft in both directions. The vertical arch loads are carried by columns around the perimeter. The steel strands form a net pattern of 6- $\times$ 6-ft squares that supports the corrugated metal sheets (Fig. 9.31A). The primary suspended cable sizes vary from 3/4 to $1^5/_{16}$ in., while the arched cables range from 1/2- to 3/4 in.-diameter. Actually, the roof is more of a one-way system, since the tension in the bracing direction is by far less. Therefore, guy wires are attached to the primary cables and tied down to the perimeter columns to reduce fluttering.

Eero Saarinen's Hockey Rink at Yale University is probably the first true tensile surface architecture (1956, Fig. 9.16g). In this sculpturelike building the dynamism of the surface is fully absorbed into the total architecture. The anticlastic cable roof (324 $\times$ 183 ft) is strung between a vertical, spinal, parabolic arch 70 ft high and two curved horizontal edge walls that act as vertical cantilevers to resist the lateral thrust. Concrete struts under the base of the rink prevent the sliding of the wall foundations. The central parabolic arch spans 240 ft and then cantilevers with reverse curvature at both ends to support the roof overhangs at the entrances. The main transverse suspended cables are 15/16-in. in diameter and are spaced 6 ft apart. Nine lengthwise arched cables, also of 15/16-in. diameter, are located on each side of the spine and stretch between four steel trusses at each end of the building. Additional direct transverse cables on the outside brace the vertical spinal arch against unbalanced loading. Since the lower ends of the transverse cables are almost horizontal, most of the vertical loads are carried by the arch! A detail of the roof decking is shown in Fig. 9.31B. The structural engineer for the project was Fred Severud.

The sculptural shapes of the two stadia for the 1964 Olympics in Tokyo, Japan (Fig. 9.16d, f) are without question masterpieces of modern architecture; they were designed by Kenzo Tange and engineered by Y. Tsuboi. The roof of the larger gymnasium (394 $\times$ 702 ft) is supported by heavy steel cables stretched between two towering concrete masts and tied down to anchorage blocks. We may visualize the main sus-

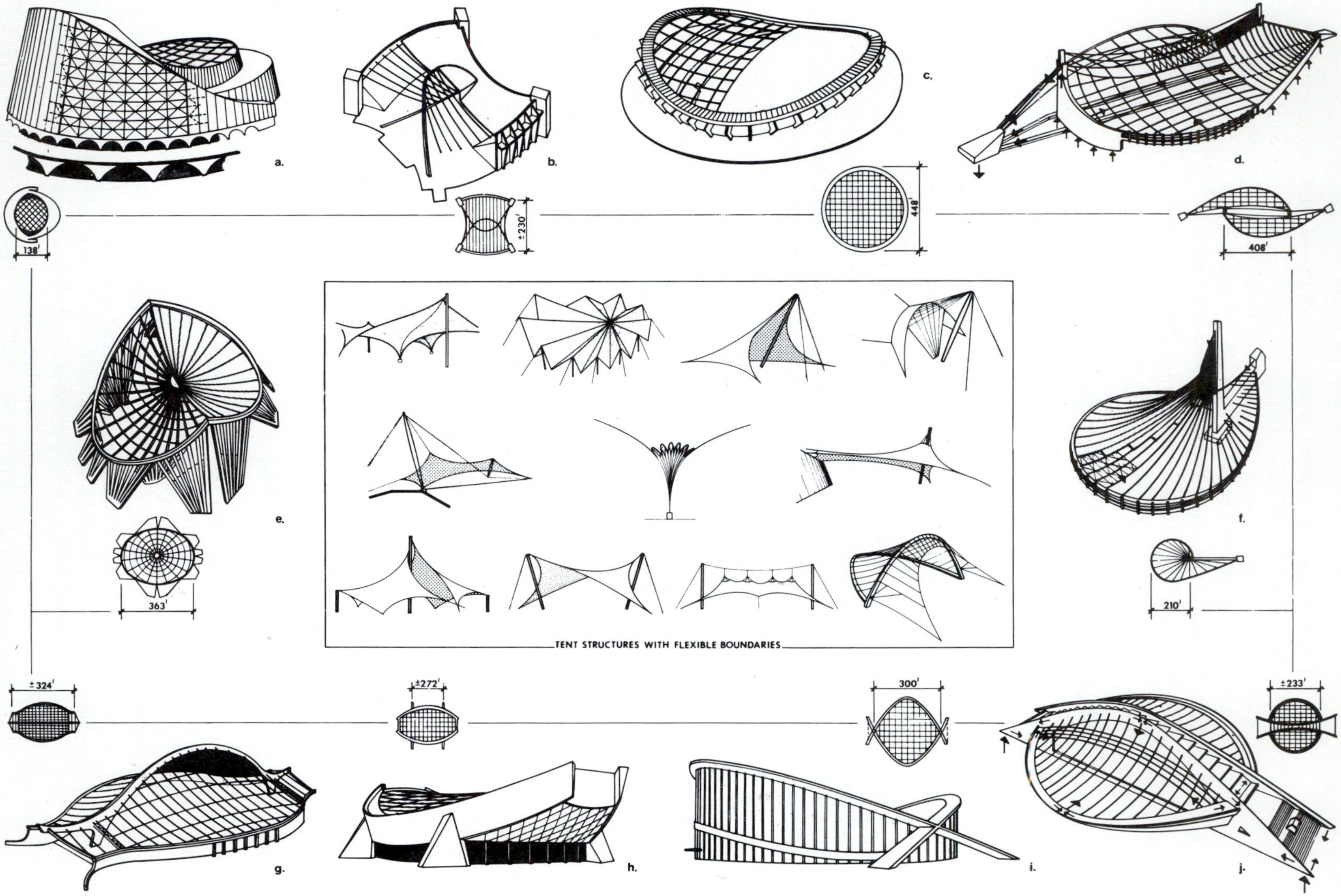

Figure 9.16 Tensile membrane structures.

758

pended roof as a net structure with hanging members of varying bending stiffness and the arched bracing members with zero stiffness. The hanging I-beams are hinged at one or two intermediate points along their spans; they provide the necessary stiffness because of the lack of curvature in the opposite direction and are supported by the heavy central steel cables and boundary arches.

The smaller gymnasium (Fig. 9.16f) employs a very different structural concept for its asymmetrical configuration. The radially arranged suspended members are not cables, but trusses with some bending capacity. Since the flat portion of the roof did not allow any prestressing, the roof had to be treated as a one-way suspended structure requiring a minimum rigidity. The hanging trusses are supported along the circumference by columns that are part of the grandstand structure, which forms a rigid closed cone (213-ft diameter), and at the off-center by a hanging steel pipe. This suspended pipe forms a spatial spiral curve between the top of the concrete mast and the anchor block.

The tensile roof for the ice rink at the Olympia Park in Munich, Germany (1982, Fig. 9.1a), was designed by the architect Kurt Ackermann and the structural engineer Jorg Schlaich. It consists of the central triangular, three-chord, trussed steel arch spanning about 342 ft, from which are hung on each side, in a nearly symmetrical fashion, anticlastic cable nets that are anchored along the perimeter to inclined columns.

The nearly elliptical Krylatskoe cycle track, built for the 1980 Olympic Games in Moscow, (551 × 453 ft) consists of two hyperbolic paraboloids joined along the center axis by two inclined parabolic arches similar in appearance to case j (Fig. 9.16). The anticlastic roof membranes of the bicycle track stadium are of particular interest since they are constructed not from cable nets, but from suspended steel sheets connected and stiffened in the transverse direction by continuous purlins.

One of the latest advances in ring-supported tensile roof structures is Calgary's 450-ft-span saddledome stadium (Fig. 9.20d, bottom) built for the 1988 Winter Olympics and designed by the engineer Jan Bobrowski. The doubly curved, hyperbolic, paraboloid roof was constructed by hanging precast concrete roof panels on the stressed two-way cable net, thus requiring no shoring. Then the cables were stressed again, and unbonded tendons were placed in all the joints between the panels, which were then filled with lightweight concrete. Finally, the unbonded tendons were stressed, thereby concluding the change from a net structure to a prestressed concrete shell.

For the preliminary design of an anticlastic tensile roof, refer to Example 9.10.

Arch-supported Saddle Surface Structures and Mast-supported Conical Surface Structures

Various arch- and mast-supported anticlastic surface structures are shown in Fig. 9.17. The combination of these two basic surface types results in an infinite number of possible shapes that may express a rich complexity of form and dynamic spatial movement, as can only be suggested by the cases in the center portion of Fig. 9.16.

Tensile membrane structures derive their strength from their shape and the amount of built-in tension, as has been discussed in the introduction to this section. Their overall configuration depends on the support conditions shown in Fig. 9.2. Ideally, the surface tension is transferred in pure compression by the supporting structure elements (e.g., columns, arches) directly to the ground, or the compression is self-balancing (rings).

Tensile surface structures are three-dimensional structures and therefore need at least four supports, in contrast to conventional planar structures that require at least three reacting force conditions. Hence, the most basic unit is the four-point structure; complex fabric structures consist of a combination of the basic units.

In traditional tent construction there are many forms of tents, ranging from conical, dome, vaulted, box, and wedge, to pole tents or any other combination. Familiar to the reader may be the *black tents* of the Bedouins, Berbers, Kurds, and many other tribes. These tents are generally woven from goats' hair and are primarily prestressed tensile structures, having only a few compression members as support. On the other hand, the *tepee* of the North American plains Indians employs straight poles to form a conical frame covered with buffalo skin. In contrast to the tents of vernacular architecture, which are primarily of single curvature and only lightly stressed, modern fabric structures are prestressed, double-curvature systems representing complex new forms.

The new breed of fabric structures or soft shells can be either temporary or permanent, open sided or closed, or free standing or integrated into hard architecture; they can be exterior or interior fabric structures, or they can be fixed or convertible. The world's largest convertible membrane structure, the retractable Kevlar fabric roof of the Montreal Olympic stadium (1987, Fig. 1.8, top left), is suspended from a curved, 55-story tower leaning at 45° from the vertical and covers the enormous 400 ft × 600 ft ellipse. Fabric structures can resist extreme weather in zones ranging from Saudi Arabia to Alaska. They provide relatively low cost shelters and can be prefabricated, easily transported anywhere, and quickly erected because they are lightweight. Membrane structures are also safe due to their lightness; nobody gets hurt if there is structural failure. And rarely will a local failure result in the collapse of the entire membrane because of its continuous character.

Soft shells range from small-scale tent structures of less than 60-ft span, PVC-coated polyester aluminum frame structures for temporary use (or PTFE-coated fiberglass for permanent fabric structures), to large-span, cable-reinforced membranes of possibly more than 300 ft. Small-scale fabric structures are used typically as roofs, ceilings, awnings, canopies, space dividers to diffuse light and control sound, shade structures, and space articulators. Fabric awnings and canopies not only give weather protection but also may be sculptural pieces that make an architectural statement. They consist of lightweight skeleton framing (using pipes or tubing with slip fittings or tube-to-tube welding) over which a fabric cover is attached. Awnings can be fixed, retractable, or movable; they are supported by the exterior wall of a building. Canopies are ground supported as well as being anchored to a building.

Large-scale fabric structures are used as disaster and construction shelters, as reservoir covers, as enclosures for sports and events facilities and for shopping malls, warehouses, airplane hangars, and manufacturing facilities.

Since the weight of flexible membrane structures is negligible compared to external loads, they are economical for large spans. Furthermore, since tensile membranes are flexible, they invite experimentation with respect to sculptural qualities and formal acrobatics. The design of spatial structures requires three-dimensional thinking, so stretch fabric models (e.g., nylon stocking material) must be used at the start of the exploration after preliminary sketches. Once an initial basic shape has been guessed using physical models, the computer is used for *form finding* (i.e., equilibrium shape generation). After the actual prestressed shape has been determined, *stress analysis* under superimposed loading is performed taking into account flexibility (i.e., nonlinear analysis). Finally, the *fabrication patterning* is done where the cutting patterns take into account the physical properties of the membrane in the warp and weft directions.

The full development of tensile membrane structures was made possible by Frei Otto. He was able to achieve large free spans by separating the cable net and cladding, in contrast to tent structures, where the fabric serves both structural and weatherproofing functions. With Frei Otto's German Pavilion at Expo '67 in Montreal, Canada (Fig. 9.17A), largescale tent construction reached a certain level of maturity and credibility; its form and structure established direction and identity. The complexity of the surfaces, reminding us of hills and valleys of various configurations, introduced an

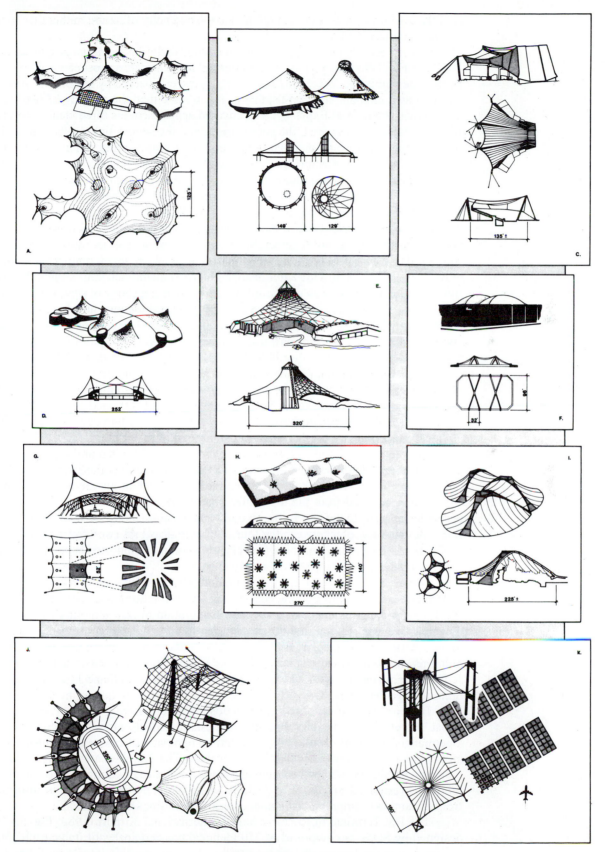

Figure 9-17 Tent architecture.

entirely new formal language that referred to the geometry of nature, rather than to traditional architectural forms.

But it was not until the construction of the stadia for the Munich Olympics in
1972 that Frei Otto became the great master of tensile structures respected all over the
world. The tentlike roofs for the Munich Olympics, engineered by Leonhardt–Andrae,
cover an area of about 808,000 ft^2, that is, much more than the roughly 87,000 ft^2 for
the German Pavilion at Expo '67. The covered space includes three stadia, the main
stadium, the swimming area, the sports arena, and the spaces linking the stadia. The
scale and method of construction of these roofs are a major achievement for Frei Otto;
he proved that tensile membranes can be permanent architecture. The western grandstand of the main outdoor stadium is covered by nine saddle-shaped, prestressed membrane units (Fig. 9.17J), which are supported along the outside by stayed cables
hanging from masts (an average 230 ft high) and along the inside by a huge edge cable
that runs along the inner edge of the grandstands. This main cable is 1440 ft long and
not only ties the entire roof together, but also makes the span of the roof possible. It is
composed of a bundle of ten 4.7-in.-diameter cables, which together must resist forces
of up to 10,000 k. To provide additional height, each membrane unit is supported by a
flying guyed mast. The cable net roof of 30-in. square meshes is composed of twin
steel cables of 1-in. diameter spaced at 2 in. and clamped together at cable intersection
points. The net, in turn, is connected to the edge cables. The roof covering is placed
above the network. It consists of translucent acrylic glass (Plexiglass) panels approximately 9½ ft square, which are bolted to the network at each of the clamped cable
intersections (Fig. 9.31C).

It was the development of the Teflon-coated fiberglass fabric and its first application to the student center of La Verne College (Fig. 9.17D), in 1973, that made this
building the first permanent fabric structure in the United States; it was designed by
architect John Shaver and the roof structure engineered by Horst Berger. The four
overlapping conical tent units enclose a space of 68,000 ft^2. Each unit is supported by
an inclined central tubular steel column. The Teflon-coated fiberglass fabric, weighing
only 45 oz/yd^2, is strengthened by a web of thirty-two 1-in.-diameter radial cables,
which are suspended from the top of the mast to the compression ring on top of the
perimeter wall and to the two 1⅞-in.-diameter main cables along the intersection of the
four tent units that arch between the four 24-ft cylindrical stairwell towers. The fabric
roof is insulated with a 2-in.-thick fiberglass blanket and supported by supplementary
3/16-in.-diameter polypropylene-coated cables.

In the 1970s, the engineer–designer Horst Berger became the driving force in the
United States in promoting a new generation of tent structures as permanent buildings.
By developing computer programs for the exploration of new structural shapes suitable for free spans, he replaced the previous design approach based on small-scale wire
models with soap bubbles, nets, or elastic membranes. Berger's studies, especially of
radial tension shapes, eventually lead to the design of the Haj Terminal of the New
Jeddah International Airport (1983) and the roof for the King Fahd International Stadium in Riyadh (1985).

Bullock's department store in San Jose, California (Fig. 9.17F), designed by
Horst Berger, became in 1978 the first retail establishment in the United States to use
a permanent membrane roof. It is supported by two pairs of crossed, laminated wood
arches hinged at the top, as well as by edge cables. The Teflon-coated fiberglass membrane covers the 96- × 160-ft space without the assistance of cable reinforcement; the
fabric is prestressed to a level of 35 lb/in. A second inner fabric layer is suspended
from the arches to improve insulation, acoustics, and fire safety.

The Haj Terminal of the New Jeddah International Airport (1983, Fig. 9.17K),
designed by SOM, is comprised of 210 identical quasiconical roof units, each 150-ft
square, that cover an area of 105 acres and make it the world's largest roof structure.
The Teflon-coated fiberglass fabric of a typical unit is strengthened and pretensioned

by a system of cables arranged in a radial pattern. The radial cables are supported by a 30-ft-diameter steel tension ring at the top, and at the bottom by ridge cables and edge cables that span between steel pylons. The tension ring at the top is suspended by four pairs of $1\frac{5}{8}$-in.-diameter cables from the 148-ft-high steel pylons at the four corners. The tents are under a prestress force of 66 lb/in. and are designed to resist a wind pressure upward and downward of 35 psf.

Horst Berger's major achievement is the roof structure of the new Fahd International Stadium in Riyadh, Saudi Arabia (1985, Fig. 9.1b), which is the largest stadium roof cover in the world at almost 1000 ft in diameter. The roof system consists of 24 tentlike modules arranged in a circle of 810 ft and forms radial, folded, point-supported surfaces. Each module is a cantilever unit supported by a 197-ft-high mast, a staying system, and a 440-ft-diameter inside circular ring cable, which supports a pair of suspension and stabilizing cables.

The more extensive involvement of architects in exploring the design potential of tensile membrane structures during the later part of the 1980s distinguish the latest generation from the more engineered solutions of the 1960s. Among the more recent outstanding tensile membrane structures is the sail-shaped roof of Vancouver's Canada Place for Expo '86 designed by architect Eberhard Zeidler and engineer Horst Berger.

One of the latest great tent structures is the Teflon-coated fiberglass roof of San Diego's Convention Center (1988, Fig. 5.1, top left and 9.18a). The tent spans 300 ft and is supported by flying poles, which are carried by an intricate web of cables strung between concrete abutments. Arthur Erickson conceived the structure as a great sailing ship beside the San Diego Bay. The structural engineer for the roof was Horst Berger, who also designed the tensile roof covering the Denver International Airport's Landside Terminal complex, which is the largest cable-supported fabric roof in the United States (1993). The translucent Teflon-coated fiberglass membrane roof covers the terminal's Great Hall of 220×900 ft. It consists of 17 pairs of tentlike conical units spaced 60 ft apart, which are supported by two rows of masts spaced at 150 ft.

The sculptural qualities together with complex spatial geometries of fabric structures can only be suggested by the tent for Schlumsberger's Paris headquarters (Fig. 9.18e and Fig. 1.2c) developed by Renzo Piano, Peter Rice, and others in 1985. The fabric roof for the Marylebone Cricket Club Mound Stand, London, U.K. (1987), consisting of umbrellalike units designed by Michael Hopkins in collaboration with structural engineers Peter Rice and John Thornton of Ove Arup, represents another much spoken about example.

With respect to the creative formal design of fabric structures, we must also mention, among others, Kent Hubbel (previously of Chrysalis), as well as Todd Dalland and Nick Goldsmith of FTL Happold in New York. They are among the tireless advocates in the United States attempting to integrate fabric structures into architecture. They are concerned with sculptural qualities and visual possibilities, as well as with the search for expression and meaning. They articulate the lightness of forms and its weightlessness and reinforce the spatial experience beneath tensile structures as emotional and sensual. Dalland and Goldsmith see the beauty of fabric structures in the connecting and unifying of inner and outer worlds, in working together with nature instead of subduing it, which they perceive as relaxation through tension.

For further discussion of membrane structures, refer to Section 9.7.

Approximate Design of Anticlastic Prestressed Membranes

Soft membranes such as fabric, film, and network lack rigidity; they are flexible and inherently unstable; they must adjust their shape to the loading so that they can respond in tension. In the section introducing linear cable systems we discuss several ways in which stability can be achieved (Fig. 9.1). For membranes, one obvious and most natural method is to use anticlastic surface geometry, where two opposing curvatures balance each other. Under gravity loads, the main (convex, suspended, lower, load-bearing, etc.) cable is prevented from moving by the secondary (concave, arched, upper, bracing, etc.) cable, which is prestressed and pulls the suspended layer down, thus stabilizing it. This interdependence of the two opposing systems is rather similar to the dual-cable beam, where the damping cable stabilizes the primary cable. Visualize the initial surface tension analogous to the one caused by internal air pressure in pneumatic structures. The prestress force must be large enough to keep the surface in tension under any loading condition, preventing any portion of the skin or any member to slack because of the compression being larger than the stored tension. In addition, the magnitude of the initial tension should be high enough to provide the necessary stiffness so that the membrane deflection is kept to a minimum. The type of response the skin takes to external loading depends on the magnitude of the prestress force. However, the amount of pretensioning not only is a function of superimposed loading but also is directly related to the roof shape and the boundary support conditions. An open, tentlike structure with flexible cables as edge supports is obviously more vulnerable to large deformations than a membrane attached to a closed rigid compression ring, which in comparison hardly deforms.

Small-scale temporary structures such as fabric awnings, canopies, and tents are given much more freedom to change their shape to resist loads; they are designed by experience as based on rules of thumb. Large-scale tensile membranes, in contrast, are engineered structures that are not allowed to fold or wrinkle. These structures are introduced in the following sections.

It has already been discussed that cables and fabric do not behave in a linear manner, but resist loads by going through large deformations and causing the magnitude of the membrane forces to depend on the final position in space. Other nonlinear aspects are due to creep of the surface material (canvas, coated fabric, cables) and to slippage at clamps and fittings allowing a change of net geometry. Temperature change and shrinkage also influence linear behavior. Membranes are vulnerable to dynamic loading causing material fatigue, which, in turn, results in reduction of strength with time. This nonlinear feature makes the design of membranes extremely difficult, requiring structural engineers specialized in this rather sophisticated field.

The determination of the roof shape for given building boundaries is not a simple undertaking. While the architect presents the surface geometry as based on static form of mass and gravity, thinking in terms of solid, rigid material as the decisive factors, the structural engineer must correct this shape. The engineer has to take into account the nonmaterial, weightless flexible character of a membrane surface—its relation to external loading and to the type and arrangement of the supports as the primary form determinants, as is discussed in the introduction to this section.

The surface geometry is directly dependent on the magnitude of the prestressing, which in turn makes it possible for the surface to be in equilibrium with varied external loading combinations. For preliminary design purposes, the surface geometry can be derived from its initial loading stage of *uniform, constant surface tension*. For this condition, the form is a function of the given boundary configuration, as is discussed for minimal surfaces (Fig. 8.5). They can be formed experimentally by dipping a closed spatial frame, having the shape of the roof boundaries, into a soap solution. The resulting liquid film surface is anticlastic and, if shallow enough, can be approximated by the hyperbolic paraboloid. Keep in mind that the actual surface material must respond to environmental changes and hence is very different from a soap film being supported by surface tension only and well-protected from any outside influence. The membrane is never really under uniform tension; it is constantly subject to other external forces. Furthermore, boundaries are not necessarily predetermined; there is an optimal solution between support form and surface shape, considering also the wind load distribution as a function of the roof form. We may conclude that the minimal surface, unfortunately, can only be considered as a first approximation of the real membrane shape. Study models using stretch fabric are generally necessary at the early design stage of generating the shape of the tensile membrane so that dimensions and curvatures together with slopes can be measured.

From a structural design point of view, prestressed anticlastic membrane surfaces may be organized in three main groups:

1. *Saddle surfaces* supported along exterior boundaries and possibly, in addition, supported by interior line supports (arches and cables), generating undulating and corrugating surface forms
2. *Anticlastic surfaces point supported* in the interior (stress concentrations)
3. Combinations

While saddle surfaces, in general, use nets with regular meshes, mostly triangular and rectangular, or fabric consisting of long narrow strips sewn or glued together acting along the overlapping edges similarly to small suspended cables, point-supported surfaces use radial net or fabric strip patterns that are generally radial cables and ring cables (and/or a fabric membrane). They have the appearance of cones and derive their net pattern from the radial and circumferential stress flow, resulting in a regular radial- or ring-type mesh, similar to a spider web if seen in plain view as in Fig. 9.18b for tents with rectangular bases. The cutting pattern of the fabric strips follows the radial stress flow or the suspended cables. Possible combinations of the two surface systems are shown in Fig. 9.18d where radial nets merge into an orthogonal network. In the point-supported tent, all the tension forces are collected at the column head, causing a stress intensity too high for the membrane. To overcome this problem, either the membrane is reinforced with additional radial cables or relief is provided by one of the methods shown in Fig. 9.18c.

The preliminary design of *point-supported membranes* is very complex. The design of the upper cone-shaped portion is usually controlled by wind pressure and wind suction. The ring cables form closed stress lines, which is obviously advantageous. They may be assumed circular and prestressed by a constant force, thereby determining the shape of the load-bearing radial cables, that is, the tent. The design of the lower valley regions between the peaks may be governed by the snow that accumulates there; these portions may be treated as saddle surfaces for preliminary design purposes. Refer to Example 9.4 for the first investigation of a mast-supported tent structure (Fig. 9.8C). The design of open-sided, free-standing fabric structures is complex since the wind acts upward from below.

In the following discussion, simple *saddle surfaces* supported along their edges by rigid or flexible members are further investigated (Fig. 9.19). The stress state of the membrane for the ideal condition of *pretension* only, ignoring any other influences, can be derived from the general membrane equation [Eq. (8.1)] for zero external loading conditions.

$$p = \frac{T_y}{R_y} + \frac{T_x}{R_x} = 0 \tag{8.1}$$

$$-\frac{T_y}{R_y} = \frac{T_x}{R_x} \quad \text{or} \quad \frac{T_y}{T_x} = \frac{-R_y}{R_x} = \frac{f_x}{f_y}\left(\frac{L_y}{L_x}\right)^2 \tag{9.85}$$

where $R = L^2/8f$ according to Eq. (9.17)

As can be seen, the tensile forces along the principal curvatures must be proportional to their respective radii (or their sag and span ratios), independent of cable stiff-

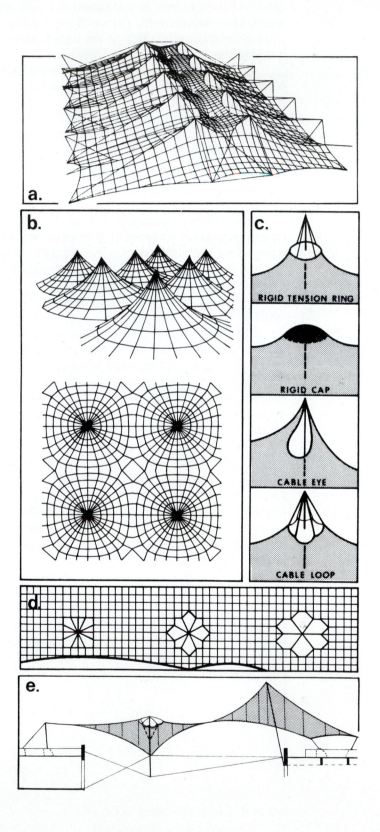

Figure 9.18 Point-supported tents.

ness, so that they are in balance along the surface. Furthermore, the surface must be anticlastic to make equilibrium possible, as the negative sign indicates. Here, for preliminary design purposes, the special case of identical shallow parabolas ($T \cong H$, $l \cong L$) or equal radii of curvature is investigated (i.e., $f_x = f_y = f$ and $L_x = L_y = L$, so the sag and span ratios are equal to 1). For this condition the pretension forces must be equal to each other.

$$T_{y0} = T_{x0} = T_0 \qquad (9.86)$$

The membrane is in a state of *uniform tension*; any effect of roof weight, temperature change, material creep and shrinkage, and flexible boundary supports and the fact that fabric and net membranes are not truly isotropic, all of which will influence the ideal uniform stress distribution, are neglected. The stress state is obtained by assuming the cable network or netlike coated fabric to follow the principal parabolas, forming a square grid in plan view; for this condition the mesh sizes vary, increasing from a minimum in the flat portion to a maximum in the steeper areas (Fig. 8.36). Prefabricated nets, on the other hand, will have an equal square grid, where the deviation of its net pattern from the one based on the principal stress flow is only small for *shallow surfaces*. Hence, it is reasonable, for approximation purposes, to assume for shallow surfaces that the square mesh projects an equal square on the horizontal plan. However, keep in mind that for steeper surfaces the square mesh along the inclined surface portions wants to adjust its shape to coincide with the principal stress flow, thereby causing grid distortions and shearing forces at the cable cross points. In general, cables are clamped together at the points of intersection in order to get shear resistance and to prevent them from sliding. For a square mesh, it can be shown from the equilibrium of forces at the cable intersection (Fig. 9.20b) that the forces must be equal to each other, as was concluded before based on the membrane equation. Furthermore, the surface is considered truly continuous and not to consist of discrete points, as is actually the case for network surfaces.

In the following discussion, the square mesh net is not placed along the straight-line generators of the hyperbolic paraboloid. The lack of curvature along the linear cables causes large deformations of the surface, which can only be overcome by increasing the prestress forces similar to the strings of a tennis racket. Furthermore, it is assumed that the surface is not too flat so that deflection does not control.

The initial tension may be induced directly to the cables or by tensioning the boundary cables using prestress equipment; the tension can also be induced by an initial external overload p. Even if this approach should not be used, the load p is helpful as a concept and can be seen as an imaginary or equivalent external load to express the prestress force.

According to the membrane equation, for shallow hyperbolic paraboloids under uniform vertical load action, letting $w = p$, the membrane forces per unit width along each principal direction can be approximated according to Eq. (8.28) for the assumed conditions of symmetry and $T \cong H$ as

$$T_{y0} = T_{x0} = T_0 = \frac{pR}{2} = \frac{pL^2}{16f} \qquad (9.87a)$$

Or the magnitude of the equivalent prestress load p for a surface under constant tension T is a function of the curvature

$$p = 2T/R \qquad (9.88)$$

Note that with a decrease of surface curvature (or increase of radius of curvature) less prestress force is needed! Surfaces with more curvature are stiffer than the ones with less curvature for the same amount of tension.

It may be concluded from Eqs. (9.87a) and (9.85) that for conditions of asymmetry, such as for different span ratios and sag ratios, the membrane forces in each direction under the equivalent uniform prestress load p are

$$T_{x0} = pL_x^2/16f_x, \quad T_{y0} = pL_y^2/16f_y, \quad T_{x0} = \frac{f_y}{f_x}\left(\frac{L_x}{L_y}\right)^2 T_{y0} \qquad (9.87b)$$

The following relationship, important for determining the basic membrane shape, can be derived from Eq. (9.87b) for a surface under constant tension, $T_{x0} = T_{y0} = T_0$.

$$f_y/f_x = L_y^2/L_x^2 \qquad (9.89)$$

The magnitude of the prestress forces is not easy to determine; it depends on strength, stiffness, dynamic considerations, and for smaller structures often simply on experience. The pretension force should be large enough so that the membrane is always in tension and does not slack because of the compression caused by external loading. The critical stress stage for a suspended cable or an imaginary fabric strip is, in general, under full gravity loading. Yet how are these loads distributed to the hanging and arched membrane portion? Is there any difference in response between the continuous fabric and the noncontinuous network, where the cable intersection points may be considered hinged? In the discussion of the approximate load flow along a cable truss (Example 9.8), it is shown that a much higher portion of the load is carried by the suspended cable compared to the damping cable. The true distribution of the loads is highly indeterminate. However, to develop an appreciation for prestressed membranes and some feeling for the importance of geometry, fabric and orthogonal nets are both treated in the same fashion, as membranes. They are assumed to behave linearly, neglecting the effect of larger deformation in the flatter surface areas. Rigid boundaries are assumed, while the effect of flexible edge members on the membrane behavior is ignored, thus allowing some slack. Furthermore, conditions of perfect symmetry of loading, geometry, and material are assumed for first preliminary design purposes. Not only are the boundary conditions symmetrical, but also the cables in each principal direction are considered identical in size and spacing so that the stiffness ratio $(EA)_x/(EA)_y = 1$. The membrane is positioned horizontally; for if it were inclined and wall-like, horizontal wind forces become primary design determinants. Cable sags in the range of 4% and 6% of the span generally give satisfactory results with respect to structural behavior.

With respect to superimposed loading, the membrane action of a rectangular hyperbolic paraboloid, as applied for the design of thin shells, is assumed to reflect also the behavior of the flexible skin, ignoring the effect of prestressing. For this condition of symmetry, it is reasonable to consider the external loads to be equally shared by the suspended and arched cables. Keep in mind that perfect symmetry never exists; even for material symmetry, environmental changes will cause asymmetry, developing a force flow that is very different from the one assumed.

Under *full gravity loading*, the lower hanging cable or yarns (i.e., fabric strip), T_L, are stressed to a maximum, while the arched membrane portion T_T is stressed to a minimum, requiring only enough tension so that it does not slack. Hence, each cable system, taken as a membrane per unit width, carries an equal pretensioning force and an equal share of the superimposed loading.

$$T_L = T_0 + \frac{wL^2}{16f} \qquad (b)$$

$$T_T = T_0 - \frac{wL^2}{16f} \geq 0 \qquad (c)$$

For first-approximation purposes (under normal wind conditions where wind suction is less than snow loading), let the tensile force in the stabilizing cable or yarns, Eq. (c), be equal to zero ($T_T = 0$) by not considering any safety factor. The result is a *prestress force* generated by an imaginary load equivalent to one-half of the maximum superimposed loading, $p = w/2$.

$$T_0 = wL^2/16f \tag{d}$$

The magnitude of the prestress force T_0 is only preliminary; it has to be changed to take into account the dead-to-live load ratio, different cable sizes (membrane flexibility), surface flutter, rigidity of boundaries, etc. Substituting T_0 into Eq. (b) yields the following maximum membrane force per unit width.

$$T_T = T_{max} = wR = wL^2/8f \tag{9.90a}$$

The cable force is obtained by multiplying the unit force T_{max} by the cable spacing. The terms in the equation have already been defined in the discussion of the membrane behavior of the hyperbolic paraboloid [Eq. (8.28)]. We may conclude that, for the preliminary design of shallow membranes and cable nets, all external loads, such as snow, are carried by the suspended portion of the surface, similarly to a single curvature system when the arched portion has lost its prestress and goes slack (Fig. 9.20e). Also notice that at least one-half of the permitted tension in the cables or yarns is consumed by the initial stored tension.

The design of the arched cable system for lightweight roof structures is derived, in general, from the loading condition where maximum wind suction w_u causes uplift and increases the stored prestress tension, which is considered here equal to one-half of the full gravity loading, minus the relatively small effect of membrane weight. In other words, under upward loading, the maximum forces occur in the arched cables and can be approximated according to Eq. (9.90a) as $w_uL^2/8f$. For most cases it is conservative to consider for preliminary design purposes the cable sizes in the arched direction as equal to those in the suspended direction. We may conclude that an anticlastic surface with sufficient curvature throughout is required in order to resist the snow loads in one direction and wind in the opposite direction.

For conditions of asymmetry, it may be concluded from Eq. (9.90a) for preliminary design purposes that the superimposed downward loads w are resisted entirely by the suspended cables (or yarns) when the arched cables go slack.

$$T_x = wL_x^2/8f_x \tag{9.90b}$$

Similarly, the uplift wind forces w_u, which usually control the design of lightweight tensile membrane structures, are resisted by the arched cables or fabric strips when the suspended portion has lost its prestress and goes slack.

$$T_y = w_uL_y^2/8f_y \tag{9.90c}$$

Sometimes it may be advantageous for the investigation of shallow membranes, at the very early design stage, to treat the radius of curvature of the parabola $L^2/8f$ as a constant that can be scaled off the drawing or model (i.e., replacing the parabola by a circle).

$$T = wR \tag{9.90a}$$

We also should keep in mind, however, that under load action the membrane stretches and decreases its radii of curvature, which, in turn, results in smaller stresses (see Example 9.2). The effect of edge details causing stress concentrations where the thin membrane meets the stiffer edge cable is a critical design consideration, which, however, is beyond the scope of this discussion.

Approximate Design of Edge Members

Tensile membranes may be attached to rigid or flexible boundary members that act as transition structures to guide the membrane forces to other support structures or possibly directly to the foundations. Flexible boundary members are generally steel cables or fabric belts for small-scale tensile surfaces. Typically, the edge cables are heat sealed into the fabric, whereas for large-scale structures they are external with the fabric clamped to them (see also discussion of Fig. 9.31).

The transfer of membrane forces acting perpendicular to the edge cables is less difficult to deal with than the tangential or *sliding forces* along unclamped cables, keeping in mind that only small tangential forces can be transmitted by friction. In this context, only force action perpendicular to the edge cables is briefly investigated. In other words, constant membrane stresses in the suspended and arched directions are assumed so that the tangential forces along the edge cables are zero.

For the special condition of a *square mesh* net under uniform tension, a boundary cable will respond with the funicular shape of a circular segment [Eq. (9.57)]. The cable forces cause radial resultants of equal magnitude along the edges, as is shown conceptually in Fig. 9.19g. Hence, a membrane under uniform tension causes its edge cables to take the shape of a circular segment (Fig. 9.19c and h) with a constant tension force Sn_0 of

$$S_{n0} = T_0 r_n = pRr_n/2 \qquad (9.91)$$

With the decrease in the radius of curvature along the edges, the cable force decreases, yielding smaller cable sizes (Fig. 9.19h).

For the case where the cables are spaced differently in each principal direction, that is, where a *rectangular mesh* net is being used, the radial force action on the edge cable will no longer be constant as for the square grid; it may force the edge member to take the shape of an elliptical segment, for example. For preliminary design purposes, edge cables with a shallow sag s may be assumed to take a parabolic configuration and to be loaded only by perpendicular membrane forces of constant magnitude (Fig. 9.19h) and are treated as hanging cables. The maximum horizontal component of the cable force in the horizontal plane (since the loads are on the horizontal plane), according to Eq. (9.90a), is

$$S_h = T_y L_x^2/8s \qquad (9.92)$$

The maximum vertical component of the cable force in the horizontal plane is roughly

$$S_v = T_y L_x/2 \qquad (9.93)$$

Hence, the maximum cable force is approximately

$$S_{max} = \sqrt{S_v^2 + S_h^2} \qquad (9.94)$$

Rigid edge members may have a variety of forms. They may be planar, such as horizontal circular rings, or spatiallike warped rings, or inclined intersecting arches, which may be straight or curved (Fig. 9.2). While straight members have certain advantages from a constructional and functional point of view, they must also carry the membrane forces in bending, requiring large beam sections. The behavior of arched

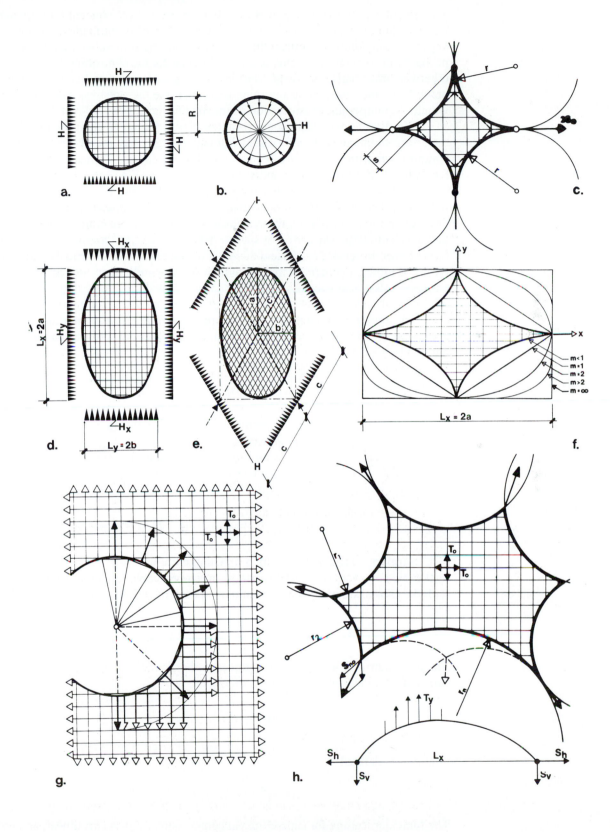

Figure 9.19 Edge supports for cable nets.

edge members may be interpreted as being the inverse of edge cables, although the flexible cable will adjust its shape to respond to the different loads (actual change is kept to a minimum by prestress forces), which the rigid member cannot. Arches act primarily in compression, being possibly funicular for the controlling loading case, but otherwise must carry moments. The ring arch is stiffened significantly as the membrane loads increase and, at the same time, increase the prestress effect on the ring, which is an important consideration in concrete design.

We briefly investigate some common closed funicular compression rings with their corresponding constant loading. Funicular rings carry membrane forces in pure compression without bending moments and shear forces. Typical funicular shapes have already been introduced under radial load action (Fig. 9.9a). Common funicular compression rings for tension structures are circular, elliptical, superelliptical, and polygonal. They can be planar or spatial. Usually, they are planar for dish-shaped roofs or air-supported domes (i.e., for synclastic surfaces), but are warped for saddle roofs that have anticlastic surfaces. For shallow roofs, the vertical components of the membrane forces are small ($T \cong H$) and therefore do not cause much vertical bending on the ring. For roofs, in general, it may be assumed at the preliminary design stage that the vertical loads are transferred directly to columns or walls and hence do not cause bending.

The constant compression force N_R in a funicular *circular ring* due to the horizontal components H of the constant radial cable forces (Fig. 9.19b) has already been defined by Eq. (9.57) as

$$N_R = H(R) \tag{9.95}$$

The same result holds true for a uniform square cable net (Fig. 9.19a) as based on Fig. 9.19g; here, H is constant for all cables.

For an *elliptical compression ring* to be funicular, the rectangular cable net (Fig. 9.19d) must exert constant horizontal cable force components H_x and H_y in each direction, respectively, according to Eq. (9.72).

$$H_x/H_y = L_x^2/L_y^2 = a^2/b^2, \quad \text{or} \quad H_x = H_y(a/b)^2 \tag{9.96}$$

Similarly, it can be shown for a funicular elliptical compression ring with a diagonal cable net layout (Fig. 9.19e) that the horizontal cable force components H, in line with the cables, are of constant magnitude in each direction.

$$H_x/a^2 = H_y/b^2 = H/c^2, \quad \text{or} \quad H = H_x(c/a)^2 = H_y(c/b)^2 \tag{9.97}$$

The maximum and minimum axial forces in the funicular elliptical compression ring can be determined, according to Eq. (9.73), as

$$N_{\max} = H_x b, \qquad N_{\min} = H_y a \tag{9.98}$$

Often *superelliptical compression rings* are used (Fig. 9.19f), where the family of curves is defined as

$$(x/a)^m + (y/b)^m = 1, \qquad m > 2 \tag{9.99}$$

The funicular loading for superelliptical compression rings is curvilinear, and not uniform as for the ellipse with parabolic cables, and an exponent $m = 2$ (see Eq. 9.71a). Refer to Fig. 9.19f for the definition of various boundary conditions as a function of the exponent m.

Asymmetrical live loads (i.e., nonfunicular loads) cause significant horizontal bending of the ring beam. Allowing the ring to move, rather than restraining it, results in a redistribution of membrane stresses, which, in turn, may cause smaller bending moments in the ring. To keep the size of the ring beam within acceptable limits may require additional lateral support by buttresses such as A-frames, thereby reducing the arch action and bending and twisting of the ring beam.

EXAMPLE 9.10

An elliptical, paraboloid, dish-type membrane roof covers an elliptical plan of 300×400 ft (Fig. 9.20d). The prestressed cable net forms an approximately 10- $\times$ 10-ft grid that supports a metal deck. For the closed elliptical ring to support the saddle-shaped roof, it must vary in elevation. The selection of the right surface shape becomes most critical; it is a function of the shape of the ring beam.

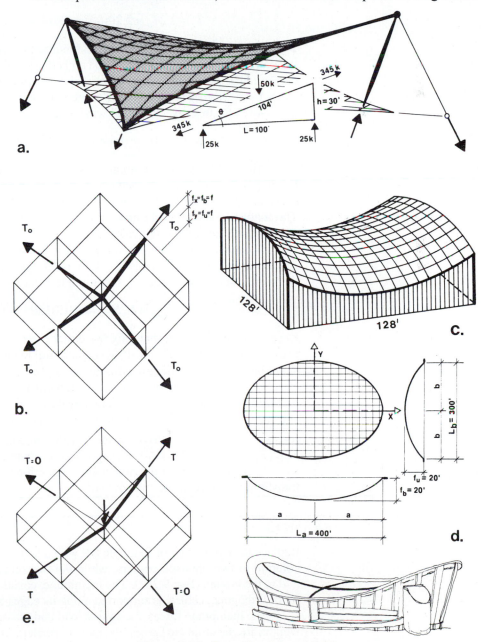

Figure 9.20 Examples 9.9 and 9.10.

An efficient saddle shape may require the ring beam to be considerably warped, thereby causing large bending moments on the ring following the three-dimensional path. On the other hand, a shallow membrane will have high stresses, but reduce the warp of the ring beam and the corresponding bending moments. Here, a cable rise and sag along the central axes of $f_u = f_b = f = 20$ ft is selected as a first trial using a sag–span ratio of 1:20. Dead and live loads of 20 psf each are assumed for this preliminary investigation.

It is assumed for this initial approach that each cable along the critical major and minor axes carries one-half of the tributary load. Therefore, it may be concluded from Eq. (9.90b) that the maximum force $(T_a \cong H_a)$ in the suspended load-bearing cable along the major axis (as based on the maximum loading condition) is roughly equal to

$$T_a = wL_a^2/8f = (0.040 \times 10)400^2/8(20) = 400 \text{ k}$$

$$T_u = 2.2T_a = 2.2(400) = 880 \text{ k} = 440 \text{ t}$$

Try a $2\frac{3}{4}$-in.-diameter steel structural strand with class A coating (Table A.13). In contrast, the maximum force on the arched cable along the minor axis is based on the initial prestress case, possibly together with wind uplift. Since the prestress force was based on one-half of the maximum gravity loading, which may be too high for this case, potential wind effects may have been taken care of.

$$T_b = wL_b^2/16f = (0.040 \times 10)300^2/16(20) = 112.5 \text{ k}$$

Although shallow membranes with high stresses cause less horizontal bending moments on the warped ring beam, they are very difficult to determine, particularly since the loads at various stages of construction must be considered. The analysis of the warped ring is approximated by its projection on the horizontal plane as a funicular elliptical ring.

The maximum axial compressive force in the ring beam (by assuming conservatively constant uniform load action) occurs at the ends of the minor axis and therefore can be approximated, according to Eq. (9.98), as

$$N_{max} = H_x b = (400/10)150 = 6000 \text{ k}$$

Keep in mind, however, that only the initial pretension forces cause a funicular response of the elliptical ring. That is, under full loading, the ring is not funicular any more since the loading conditions do not respond to Eq. (9.96). To take the effect of horizontal bending due to nonfunicular loads into account, for rough preliminary design purposes, the ring may be visualized as cut along its long axis and treated as a fixed-ended beam carrying the uniform horizontal membrane forces T_a due to live loads with maximum end moments of $M = (T_a/s)L_a^2/12 = (T_a/s)a^2/3$; this approach is obviously too conservative.

Vertical bending of the ring beam is not critical when the vertical loads are transferred directly to columns or walls.

EXAMPLE 9.11

Replace the hypar shell of Example 8.9 with a tent structure that is supported by columns at two opposite corners, while the two remaining corners are directly tied to the ground (Fig. 9.20a). The membrane, consisting either of a coated fabric or orthogonal cables, is supported along its edges by diagonal cables.

The membrane forces due to live load and prestressing, ignoring the small weight, are, according to Eq. (9.90a) and a corrected version of Eq. (8.33), equal to

$$T_{max} = \frac{wL^2}{8f} = \frac{20(100\sqrt{2})^2}{8(15)} = 3333 \text{ lb/ft}$$

or

$$T_{max} = wa^2/f = 20(50)^2/15 = 3333 \text{ lb/ft}$$

The minimum required strip tensile strength for the coated fabric is

$$T_U = LF(T) = 4(3.33) = 13.32 \text{ k/ft} = 1.11 \text{ k/in.}$$

Since the maximum dry-strip tensile strength of the strongest ordinary fabric, which at this time is Teflon-coated glass fabric, is about 1000 lb/in., a cable net must be selected.

For a mesh size of 4×4 ft, the cable force is

$$T_{max} = 4(3.33) = 13.33 \text{ k}$$
$$T_U = LF(T) = 2.2(13.33) = 29.33 \text{ k}$$

From Table A.13, try a 1/2-in.-diameter single-strand cable. The arched cables must be designed for the prestress force and wind from below.

The length of the edge cable is $l \cong L/\cos\theta$, where $\tan\theta = h/L = 30/100$, or $\cos\theta = 0.958$; therefore, $l \cong 100/0.958 = 104$ ft. Since the loads are assumed on the horizontal roof projection, the resisting membrane forces are transformed into loads along the member (see Fig. 5.26) or $T\cos\theta = 3.33(0.958) = 3.19$ k/ft. Realize, however, that under full loading only the suspended cable portion of the network applies a force to the edge member, while the forces in the secondary cables are zero according to the previous assumptions. Hence, the edge cable is only loaded diagonally from one direction, thereby causing sliding forces along the cable. The evaluation of the maximum force for this loading condition is very complex. To still be able to get a preliminary cable size, the shallow membrane is assumed, under full loading, to cause only perpendicular force action so that the shallow edge cable responds approximately as a parabola. Now the maximum cable force may be roughly approximated for a cable sag-to-span ratio = 0.12, or $s = 104(0.12) = 12.5$ ft, similar to a hanging cable, according to Eq. (9.92), as

$$S \cong \frac{TL^2}{8s} = \frac{3.19(104)^2}{8(12.5)} = 345 \text{ k}$$

The vertical displacement of each of the four edge cables due to

$$V = 0.020(100 \times 100)/4 = 50 \text{ k}$$

causing a vertical reaction of 25 k of the cables, is negligible in comparison to the 345-k thrust force. Therefore, the edge cable must resist an ultimate tensile force of

$$S_U = LF(S) = 2.2(345) = 759 \text{ k}$$

From Table A.13, try a $2\frac{9}{16}$-in.-diameter galvanized single-strand cable.

9.7 HYBRID TENSILE SURFACE STRUCTURES

There is a new generation of tensile structures appearing that ranges from the large scale of cable domes for the Florida Suncoast and Georgia stadiums to the complex spatial networks holding up the glass skin of the Louvre Pyramid in Paris. The inventive minds of designers in their search to resist loads primarily in tension, in turn, requiring reduction of weight to a minimum, have created this new breed of lightweight structures that take into account the construction process. These structures consist of self-stressing spatial networks and floating compression acting together with thin skins. On the one hand, they have been generated in response to particular situations related to esthetical and engineering requirements, such as the spatially guyed multidirectional composite roof structures, integrating concepts of cable-stayed beams and cable beams designed by Rogers and Rice (Fig. 9.13). Earlier, Renzo Piano suggested a new language with the Italian Industry Pavilion at Expo '70 in Osaka, Japan, by employing an exterior membrane on a prestressed steel frame. On the other hand, these structures have their origin in the self-supporting, self-tensioning polyhedric networks, such as single- and double-curvature, double-layer tensegrity structures derived from the work of Buckminster Fuller in the United States and David Emmerich in France, particularly their patents related to self-stressing of the early 1960s.

When Buckminster Fuller introduced at the Black Mountain College in North Carolina (1947–1948) the concept of *tensegrity* or tensional integrity, explaining the continuous tension throughout space with compression elements becoming small islands in a sea of tension, it was Kenneth Snelson, one of his students, who gave birth to the geometric tensegrity structure. He explored the spatial interaction of tension and compression in his numerous sculptures. His tensegrity mast, the Needle Tower (1968), a tapered tensile cable-tube sculpture at the Hirschorn Museum in Washington, D.C., is now famous. Various building cases (in Fig. 9.21 and others) are now briefly discussed to develop appreciation for this complex world of network structures.

The spirit of Michael Hopkin's Schlumberger Research Center, Cambridge, U.K. (1985, Figs. 9.21g and 1.11a), with its shiplike masts and rigging as well as its high-level technology and detailing, reminds one of Rogers's Fleetguard Factory (Fig. 9.13), although the structures are very different. Whereas the masts and aerial rod network at Fleetguard support the horizontal rigid roof framework, at Schlumberger it stresses the spatial, domelike, undulating tensile fabric membranes.

The central portion of the building is subdivided by four parallel, exposed, portal steel frames into three bays, each 79×59 ft in size. The typical hinged portal frame is laterally stabilized by bracing. It consists of horizontal, 79-ft-span, triangulated truss girders (about 5 ft deep) and nearly 8-ft-wide vertical trusses, which, in turn, support two pairs of upper and lower booms. The two inclined upper tubular masts are supported by tie rods, which are braced by lower masts (struts). Cables are suspended from the 35-ft-long masts to give support to two parallel ridge cables at certain pickup points. The translucent, Teflon-coated, double-curvature, fiberglass membrane, in turn, is clamped and stretched between the ridge cables and steelwork. The aerial cables incorporate adjustable pinned joints in order to allow rotation when live loads change. The design of the tensile structure takes into account possible failure of a member by duplicating all critical structural members. Tony Hunt and Brian Foster of Ove Arup, for the membrane and cables, were the structural engineers for the project.

The intricate web of cables required to support the 300-ft-span roof fabric membrane of San Diego's Convention Center (1988, Fig. 9.18a) is another example of the new generation of complex tensile structures.

The BOP Air France pallet-handling building at Charles de Gaulle airport, Roissy, France (1989, Fig. 9.21h), is one of Europe's largest textile roof structures. This multibay building, designed by Marc Malinowski of Arcora, powerfully expresses the lightness of the taut fabric skin and clearly articulates the structural principles. The steel frame that supports the textile membrane consists of tubular columns linked by pairs of 62-ft span, inclined stayed arches in one direction, and in the other direction box beams with spans between 36 to 62 ft. In other words, the tubular polygonal arches are supported by diagonal tie rods suspended from small vertical masts 46 ft off the ground. The double-curvature, translucent, PVC-coated polyester fabric membrane is attached to the arches and beams by double lacing. Along certain bays, in

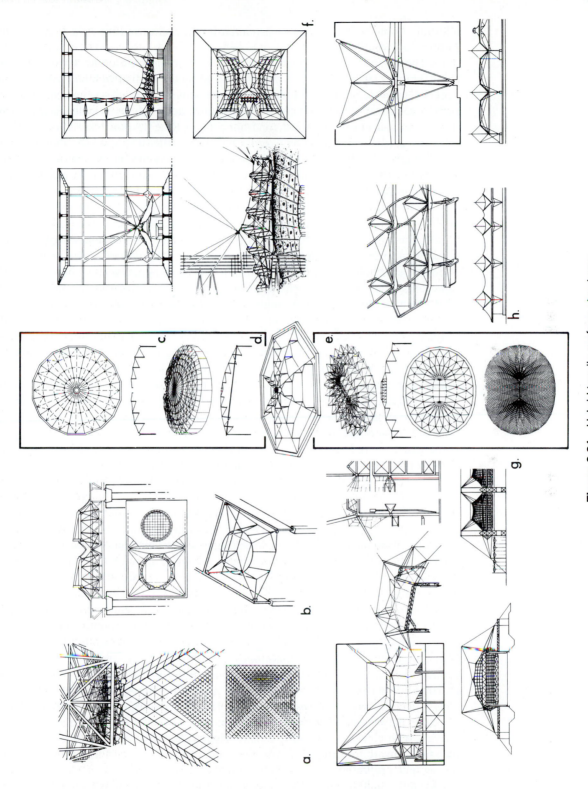

Figure 9.21 Hybrid tensile surface structures.

addition, framed pyramids are suspended from the arches and columns to allow natural lighting and ventilation.

The floating, tensile, textile membrane over the base of La Grande Arche in Paris, France (1989, Fig. 9.21f), was visualized by the Danish architect Johan Otto von Spreckelsen as reflecting the lightness and spontaneity of the cloud and contrasting the perfect geometry of the giant cube, thereby introducing a human scale, besides providing shelter and improving wind conditions. Peter Rice designed the complex "cloud" structure consisting of diagonally cross-braced, parallel lens-shaped cable beams prestressed against edge cables. The translucent fabric membrane is stressed against the underside of the cable beams. This composite prestressed structure is supported by suspension cables that are anchored to the walls of the cube.

An extraordinarily complex spatial steel framework supports the glass skin of the 71-ft-high, 115- × 115-ft Pyramid, the entrance to the Louvre in Paris, above the underground galleries (1989, Fig. 9.21a). The architect I. M. Pei and the Quebec structural engineer Roger Nicolet, in their search for visual lightness, developed a difficult geometrical layout of structure, which cannot be understood immediately. In a way, it represents a celebration of structural complexity, still achieving the goal of transparency and an almost immaterial lightness with its thin member fabric.

Stainless steel bowstring trusses form a two-way diagrid structure on each plane of the pyramid; in other words, 16 crossed beams of different lengths are placed parallel to the edges. The flat, diamond-shaped top compression chord bars (which are welded to nodes) give support to the laminated glass skin by extending the truss struts to the intersection of the aluminum mullion frame. The curvilinear, very light bottom chord rods that form the tensile net reach their largest depth near the center span portion of the triangular faces where the loads cause the largest stresses. To prevent the outward thrust of the pyramid and to stabilize and stiffen the shape, the four faces are tied together by 16 horizontal countercables (similar to belts) in a third structural layer, thereby bracing and stressing the diamond-shaped network.

The two natural lighting domes above the entrance hall of the City of Science and Industry, Parc de La Villette, Paris, France (1986, Fig. 9.21b), engineered by Rice, Francis, and Ritchie, look from the outside like sliced off cones. The braced cone framing is suspended by a tensile network from the building frame. The transition from the round, 56-ft-diameter dome lights to the rectangular bay support structure required a special cable structure layout. Its geometry had to go through the transition from the rectangle to the octagon and polygon to the circle. The cones are covered with a flexible, translucent fabric membrane.

David Geiger invented a new generation of low-profile domes after his air domes, which he called *cable domes*. He derived the concept from Buckminster Fuller's *aspension* (ascending suspension) *tensegrity domes* (Fig. 8.26), which are triangle based and consist of discontinuous radial trusses tied together by ascending concentric tension rings; but the roof was not conceived as made of fabric. Geiger's prestressed domes, in contrast, appear in plan like simple, radial Schwedler domes with concentric tension hoops (Fig. 9.21c). His domes consist of a radioconcentric spatial cable network and vertical compression struts. In other words, radial cable trusses interact with concentric floating tension rings (attached to the bottoms of the posts) that step upward toward the crown in accordance with Fuller's aspension effect. The trusses get progressively thinner toward the center, similar to a pair of cantilever trusses not touching each other; the heaviest members occur at the perimeter of the span. In section, the radial trusses appear as planar and the missing bottom chords give the feeling of instability, which, however, is not the case since they are replaced by the hoop cables that tie the trusses together.

The cable dome concept can also be perceived as ridge cables radiating from the central tension ring to the perimeter compression ring. They are held up by the short compression struts, which, in turn, are supported by the concentric hoop (or ring) cables and are braced by the intermediate tension diagonals, as well as by the radial cables. A typical diagonal cable is attached to the top of a post and to the bottom of the next post. The pie-shaped fabric panels span from ridge cable to ridge cable and then are tensioned by the valley cables, thus being shaped into anticlastic surfaces; they contribute to the overall stiffness of the dome. The maximum radial cable spacing is limited by the strength of the fabric and detail considerations. The number of tension hoops is a function of the dome span. The sequence of erection of the roof network, which is done without any scaffolding, is critical, that is, the stressing sequence of the posttensioned roof cables to pull the dome up into place.

The first tensegrity domes built were the gymnastic and fencing stadiums for the 1988 Summer Olympics in Seoul, South Korea. The 393-ft-span dome for the gymnastic stadium (Fig. 9.21c) required three tension hoops and has a structural weight of merely 2 psf. The 688-ft-span Florida Suncoast Dome in St. Petersburg (1989, Fig. 9.21d) is one of the largest cable domes in the world. The dome is a four-hoop structure with 24 cable trusses and has a structural weight of only 5 psf. The dome weight is 8 psf, which includes the steel cables, posts, center tension ring, the catwalks supported by the hoop cables, lighting, and fabric panels. The translucent fabric consists of the outer Teflon-coated fiberglass membrane, the inner vinyl-coated polyester fabric, and an 8-in.-thick layer of fiberglass insulation sandwiched between them. The dome has a 6° tilt and rests on an all-precast, prestressed concrete stadium structure.

The world's largest cable dome is currently Atlanta's Georgia Dome (1992, Fig. 9.21e), designed by engineer Mattys Levy of Weidlinger Associates. Levy developed for this enormous 770- × 610-ft oval roof the *hypar tensegrity dome,* which required three concentric tension hoops. He used this name because the triangular-shaped roof panels form diamonds that are saddle shaped. In contrast to Geiger's radial configuration primarily for round cable domes, Levy uses triangular geometry, which works well for noncircular structures and offers more redundancy, but also results in a more complex design and erection process. An elliptical roof differs from a circular one in that the tension along the hoops is not constant under uniform gravity load action. Furthermore, while in radial cable domes, the unbalanced loads are resisted first by the radial trusses and then distributed through deflection of the network, in triangulated tensegrity domes, loads are distributed more directly.

The oval plan configuration of the roof consists of two radial circular segments at the ends, with a planar, 184-ft-long tension cable truss at the long axis that pulls the roof's two foci together. The reinforced-concrete compression ring beam is a hollow box girder 26 ft wide and 5 ft deep that rests on Teflon bearing pads on top of the concrete columns to accommodate movements. The Teflon-coated fiberglass membrane, consisting of the fused diamond-shaped fabric panels approximately 1/16 in. thick, is supported by the cable network but works independently of it (i.e., filler panels); it acts solely as a roof membrane but does contribute to the dome stiffness. The total dead load of the roof is 8 psf. The roof erection, using a simultaneous lift of the entire giant roof network from the stadium floor to a height of 250 ft, was an impressive achievement of Birdair, Inc.

9.8   PNEUMATIC STRUCTURES

Pneumatic objects, possibly internally braced with tensile membranes (e.g., chambered pneus) filled with air or liquids, are effective structures. They are not only used as air buildings but also as inflatable dams, as pillow-style tanks for storage and transportation, as lifting cushions, as cushions for sealing leaks in pipelines and tanks, and for many other purposes. Helium-filled balloons are used for construction purposes and unloading of cargo ships. Pneumatic bubbles have been proposed as ideal lunar and Martian habitats.

In nature, pneumatic forms are found as liquid skins like water drops and bubble conglomerations in sea foam, as soap bubbles, and organic flexible cells stabilized by fluids. The hydrostatic skeleton of vermiform animals and the lack of stiff components, typical for some worms, clearly respond to the flexible nature of pneumatic structures.

The Romans employed inflated bags of animal skin to support bridges and rafts. The modern development of pneumatic structures goes back to 1783, when the Montgolfier brothers in France reinvented the hot-air balloon made of paper and linen and when Ferdinand von Zeppelin in 1900 launched the first successful dirigible airship. The English engineer Frederick William Lanchester proposed in 1917 an air-supported building for a portable field hospital. The real development of pneumatic structures is generally associated with Walter Bird of Buffalo, New York, who built his first inflatable radar dome (50 ft in diameter and 40 ft high) for the military in 1946. In 1956, Walter Bird and several associates founded Birdair Structures, Inc., to promote the design and manufacture of pneumatic structures. The high point was reached in 1970 with the large-span, shallow cable dome structure for the U.S. Pavillion in Osaka, Japan, designed by Davis Brody Associates and the engineer David Geiger. But the other pneumatic buildings at the exhibition also initiated enthusiasm and made the public aware of the formal and large-scale potential of the pneumatic principle. Furthermore, the research at the Institute of Lightweight Structures (IL) at Stuttgart University, Germany, for which Frei Otto became director in 1964, has had much effect on the development of pneumatic structures.

Pneumatic structures can be classified into three major groups (Fig. 9.22) as follows:

1. *Air-supported structures* are single-membrane structures, such as the basic domical and cylindrical forms; they are often called *low-pressure systems* because only a small pressure is needed to hold the skin up. This pressure can be positive, causing a convex response of the tensile membrane, or it can be negative, resulting in a concave shape. The basic shapes can be combined in infinitely many ways and can be partitioned by interior tensile columns or membranes to form chambered pneus.

2. *Air-inflated structures* or simply *air members*, such as high-pressure tubes and lower-pressure cellular mats. These air members may act as columns, arches, beams, frames, and so on; they need a much higher internal pressure than air-supported membranes.

3. *Hybrid air structures* are formed by a combination of the preceding two systems (e.g., air cushion) or when one or both of the pneumatic systems are combined with any kind of rigid support structure (e.g., arch supported). In *double-walled air structures*, the internal pressure of the main space supports the inner skin and must be larger than the pressure between the skins, which, in turn, must be large enough to withstand the wind loads. This type of construction allows better insulation, does not show the deformed state of the outer membrane, and has a higher safety against deflation. It provides rigidity to the structure and eliminates the need for an increase of pressure inside the building.

Air-supported Structures

The tensile membrane floats like a curtain on top of the enclosed air, whose pressure exceeds that of the atmosphere; only a small air pressure differential is needed. The pneumatic shell must have a proper geometry and is prestressed by internal air pressure so that it can support the external loads, which act in compression on the tensile skin. The membrane must be anchored along its periphery to prevent its being lifted

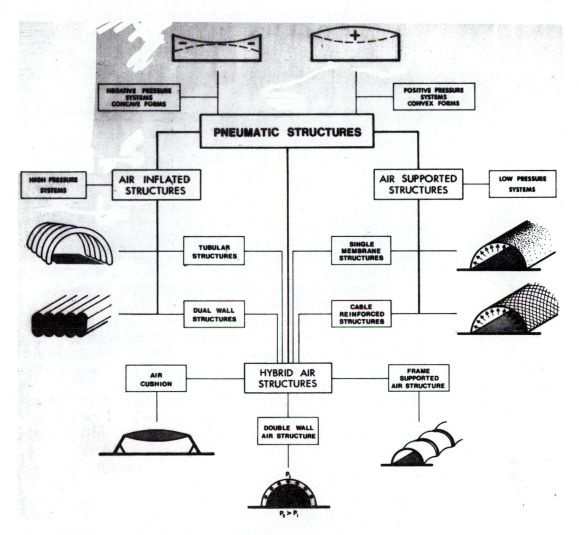

Figure 9.22 Classification of pneumatic structures.

up; it is an antigravity structure, in contrast to the heavy traditional building construction that we are so used to. Air-supported skins cannot be completely *airtight* and therefore need a continuous air supply to replace the air lost through leakage, thus maintaining internal pressure. People movement in and out is handled by revolving doors, which help to maintain an airtight inner space; airlocks accommodate vehicular movement. At least two blowers are required, a primary blower and an auxiliary one as a backup, which, in case of an emergency, is activated automatically when the air pressure drops to an unsafe level. Each fan should be capable of supporting and prestressing the membrane to the desired value. The range in low-pressure pneumatic systems is roughly from 3.5 to 20 psf above atmospheric pressure. The magnitude of the design pressure depends on wind conditions, possibly snow loads, size of structure, and its tightness (i.e., leakage through skin and along its base, type and number of doors and windows, as well as size and number of vents and air venting requirements). The typical normal operating pressure for air-supported membranes in the United States is in the range of 4.5 to 8 psf or roughly 1.0 to 1.5 in. of water as read from a water-pressure gauge.

  To develop a feeling for the magnitude of this pressure, remember that the atmospheric pressure decreases with altitude. Near sea level or the bottom of an ocean of air, the atmospheric pressure is caused by the weight of all the air above; a column of air with a cross section of 1 in.[2] extending through the atmosphere weighs 14.7 lb. The local effect of changing pressure due to moving air currents and altitude variations is in the range of 1 psi. The standard atmospheric pressure (atm) of 14.7 psi will support a mercury column of 29.92 in. at 0°C as measured by a barometer. The following relationships hold true:

$$1 \text{ atm} = 14.7 \text{ psi} = 2116.22 \text{ psf} = 29.92 \text{ in. Hg} \cong 406.8 \text{ in. } H_2O \text{ at } 4°C$$

Near sea level the air pressure decreases roughly 1 in. of mercury or 13.60 in. of water for each 1000 ft of altitude. Hence, we may conclude that 1 in. of H_2O or 5.2 psf is approximately equivalent to 1/400 atm or 1000/13.60 = 74-ft height difference, which corresponds to the pressure difference between the seventh floor of a building and ground level. This small pressure variation is hardly noticeable; it is even caused inside a house when the wind blows.

Air-supported structures may be organized as *ground-mounted air structures,* usually of high profile (Fig. 9.23), and berm- or wall-mounted, *low-profile roof membranes* (Fig. 9.24).

Ground-mounted air structures generally have a cylindrical shape; typical conventional profiles are shown in the upper portion of Fig. 9.23. They are employed as enclosures for recreational and industrial use, such as for tennis courts, swimming pools and ice-skating rinks, exposition pavilions, and warehouses. The various cylindrical shapes can be distinguished by their different cutting patterns and by the layout of the reinforcement, as well as by the type of ends (e.g., round or nearly rectangular). Usually, as pure fabric structures without cable reinforcing, they are moderate-span structures up to 120 ft (as for tennis courts or gymnasiums), although they are commercially available as reinforced membrane structures up to 250-ft width, 600-ft length, and 70-ft height. Typical height-to-width ratios range from 0.3 to 0.5. Semicircular

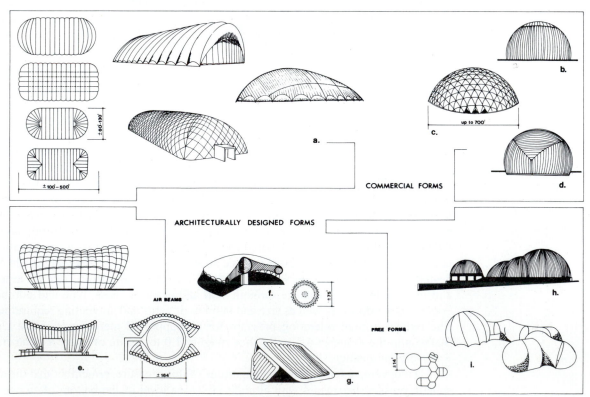

Figure 9.23 Pneumatic structures.

cylindrical shapes ($h/L = 0.5$) have the advantage that the membrane tension is resisted directly by gravity. Depending on the span, the fabric must be reinforced with webbing or steel cables. The reinforcement may form parallel cable systems or a two-way bias net in which the cables crisscross each other at 45° angles. A disadvantage of the crisscross bias net system from a structural point of view is that the membrane stresses in the cross direction are not the same as in the longitudinal direction; they are twice as high and therefore the steel is not used effectively. The cables at the ends, which have a double-curvature shape, either follow a radial pattern, a layout perpendicular to the main reinforcement, or the layout of the main cables in an armadillolike pattern. At points of stress, concentrating around openings (e.g., doors, windows, vent pipes, ducts), or where the membrane is attached to rigid structural elements (e.g., columns, arches, foundations), special reinforcement of the fabrics or isolation of the loads (e.g., transition shroud for entrance) is required. The joints (seams) are developed so that they equal the capacity of the fabrics. The structural design of ground-mounted air structures is generally controlled by wind pressure, as discussed later in this section.

Although basic air structure shapes are cylindrical or spherical, as typically commercially available, there are no limits to custom-made variations, as some of the cases at the bottom portion of Fig. 9.23 suggest.

Berm or wall-mounted air domes usually are large-scale, shallow roof structures attached to a rigid base. They are permanent structures, therefore requiring the roof membrane to be noncombustible and to have a minimum distance of about 25 ft from the occupied space. Common plan configurations are round, elliptical, superelliptical, and polygonal, ranging in span from 200 to over 700 ft, as indicated in Fig. 9.24. David Geiger determined in the mid-1980s that for roofs with spans exceeding 800 ft air structures are the best choice. Air domes are generally supported by two-way cable grids, although occasionally one-way cable systems are found. The cable layouts (that restrain the fabric roof) are directly related to the funicular compression rings, as is discussed for Fig. 9.19; they range from dense to sparse grids. The minimum number of cables results in the most economical roof structures since the number of attachments and anchorages is reduced. Keep in mind, however, that there are limits to the maximum cable spacing due to fabric strength and transportation considerations.

The typical rise-to-span ratio of air domes is in the range of 0.1 to 0.2. Because of the low profile, winds will only cause suction (uplift) on the roof, whereas the wind pressure is resisted by the supporting rigid perimeter structure. An important feature of the air dome is that it cannot collapse; it can only deflate, as may be the case in emergencies, when the membrane will still hang free of the occupied space, not harming anybody.

Air-supported roofs are *active* structural systems that need mechanical support. They are maintained and managed structures in contrast to the conventional *passive* roof structures. Automated mechanical operating procedures are required to deal with different situations of weather and/or roof behavior, particularly snow loading. Although large-scale air-supported roofs are relatively inexpensive to construct, the weakness of the system as dependent on mechanical devices becomes apparent from the number of deflations at several major U.S. air domes.

Most important for the development of large-scale air domes was David H. Geiger; he is considered the pioneer of the field. Nearly all the cases discussed now were designed by him. Geiger's fame began with the U.S. Pavilion at Expo '70 in Osaka, Japan (Fig. 9.24f), the first air-supported, clear-span cable structure ever built of such large scale. The membrane is a translucent, vinyl-coated, fiberglass fabric about 3/32 in. thick with a strip strength of about 500 lb/in. It spans between the cables, which form a diamond grid; the steel cables are 20 ft apart and vary in size from $1\frac{1}{2}$ to $2\frac{1}{4}$ in. in diameter (Fig. 9.31F). The cable network is anchored to the concrete compression ring, which encloses the superelliptical shape of the pavilion (Fig. 9.33C). The ring rests on top of the shallow earth berm, which greatly deflects the wind

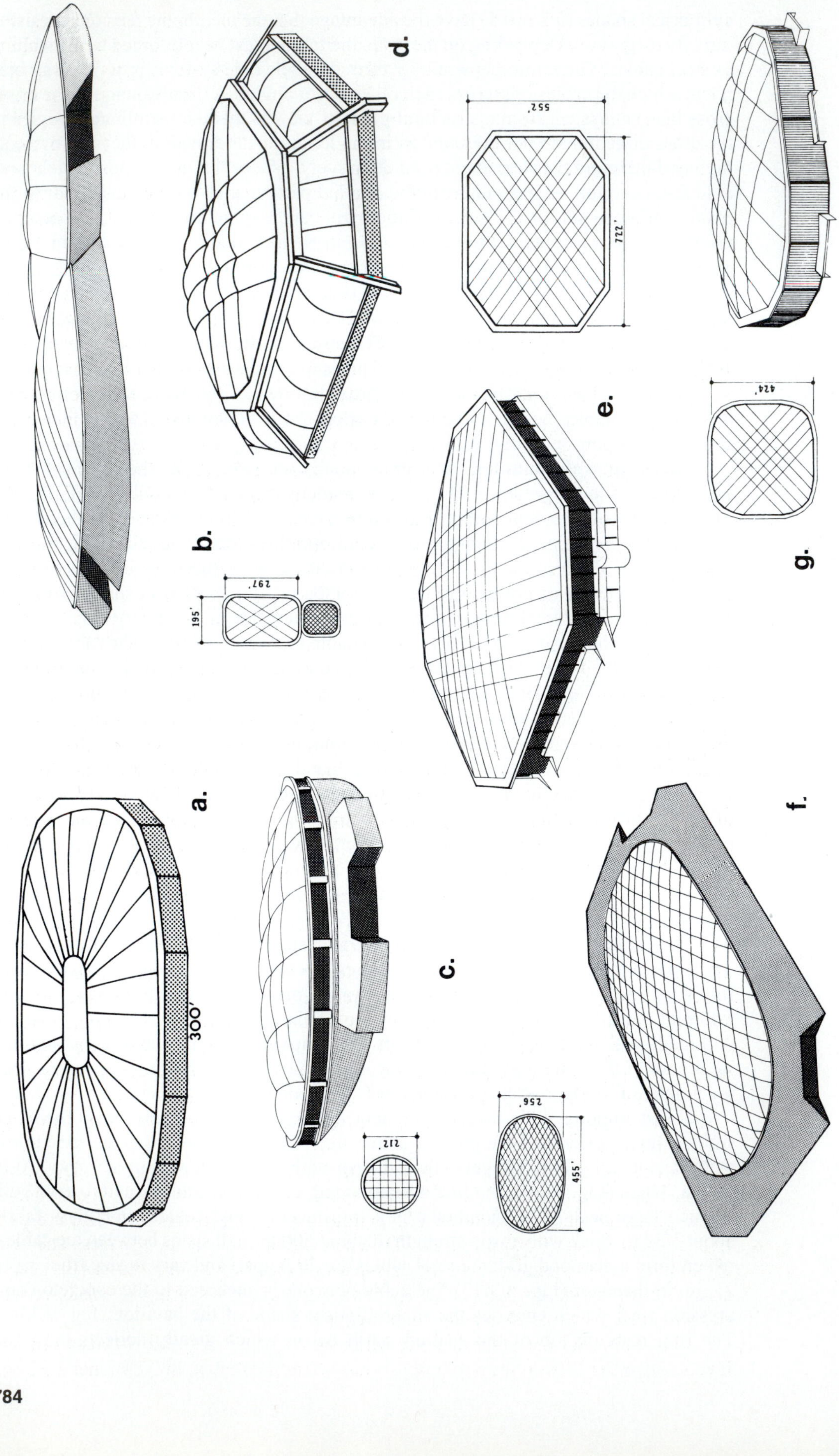

Figure 9.24 Low-profile, long-span pneumatic roof structures.

a.

b.
195'
297'

c.
212'
256'
455'

d.

e.
255'
722'

f.

g.
424'

300'

upward to create suction on the flat dome, which has a rise of only 23 ft. The skewed cable layout causes a funicular response of the ring under normal uniform loading; bending is only generated by unbalanced cable forces due primarily to wind action. The roof weighs 1.25 psf and is supported by an operating pressure of 5 psf; the membrane is designed to resist a wind suction of 30 psf.

Some other early examples of air domes are presented in Fig. 9.24 and are now briefly discussed. The roof of the Silverdome at Pontiac, Michigan (1975, case e) covers about 10 acres, 1/2 acre more than the Superdome in New Orleans, and covers about four times more area than the U.S. Pavilion in Osaka. Its span of 722 ft makes it currently the largest-span dome structure in the world, although the Korakuen Dome in Tokyo, completed in 1988, measuring over 660 ft in both directions, is the largest air-supported dome structure. The octagonal dome of the Pontiac Stadium is covered with 100 Teflon-coated fiberglass panels (32 nearly rectangular, 4 large triangular, and 64 diamond-shaped). The panels, with their edges bordered by $1\frac{1}{2}$-in.-diameter nylon ropes, are clamped to a network consisting of eighteen 3-in.-diameter galvanized structural steel strands spaced at 41.63-ft intervals (Fig. 9.31E). The roof weighs only 1 psf; it is designed for a wind suction of 15 psf and a snow load of 12 psf by increasing the internal air pressure. The compression ring is a rectangle with the corners truncated parallel to the diagonals. This octagonal ring is not a funicular shape, although the moments are greatly reduced, and it is wide enough to keep the pressure line within the prestressed concrete plate girder under uniform loading (Fig. 9.33D). There are 29 air blowers, but only 2 of them are used to support the roof with about a 3.5-psf normal operating pressure when the stadium is not occupied. The high fan capacity is needed to provide ventilation and heating; the number of fans in operation depends on the type of event, the outside temperature, and the magnitude of snow load.

The roof of the UNI-Dome at the University of Northern Iowa (1975, case g) is similar in principle to the Pontiac Stadium but smaller in scale. It only needs 12 cables (2.875-in. diameter) spaced 42.43 ft apart to support the Teflon-coated, fiberglass fabric roof. The roof weighs 1.0 psf and is designed for a wind suction of 15 psf and a snow load of 30 psf. The operating air pressure required to sustain the dome is approximately 5 psf. Sixty percent of the roof area is double layered to allow for moving warm air so that heavy snow can be melted. The double layer also serves as an insulator and sound absorber. The compression ring is supported by the exterior wall constructed from precast double tees. The Leavey Center at the University of Santa Clara (1974, case b) consists of two air-supported roof structures formed of superellipses. The larger of the two is covered with a Teflon-coated fiberglass fabric and the smaller one with a vinyl-coated polyester fabric that is retractable. The larger roof skin is supported by six $1\frac{7}{8}$-in.-diameter steel cables spaced at 40-ft intervals and anchored to the compression ring cast atop earth berm walls (Fig. 9.33B). The air pressure required to sustain the air structure is 1.5 to 5 psf. The 212-ft span roof of the field house of Milligan College (1974, case c) is an insulated membrane consisting of 4-in.-thick blankets of fiberglass insulation sandwiched between two layers of coated fiberglass fabric. Lighting, with an equivalent weight of 0.5 psf, is hung from the roof cables. The cables ($1\frac{1}{2}$- to $1\frac{1}{16}$-in. diameter) are spaced at 30 ft (Fig. 9.31G). The roof supports a wind suction of 15 psf and a snow load of 20 psf through increased air pressure.

The Stephen C. O'Connell Center at the University of Florida (1980, Fig. 9.24d), designed by Caudill Rowlett Scott, integrates imaginatively conventional construction with lightweight fabric structures. The architecture expresses how the hard concrete framing restrains the outward thrust of the soft but taut tensile membranes. The 300-200-ft main arena is covered with an air-supported membrane roof, whereas the spaces around the arena for smaller-scale activities are enclosed by short-span tensioned fabrics stretched over precast concrete half-arches.

The fabric for the Metrodome in Minneapolis (1981, Fig. 8.26) consists of two layers, the outer 1/16-in.-thick, Teflon-coated fiberglass layer and the inner 1/32-in.-

thick acoustical fabric layer. The dead air space between acts as an insulation layer. During winter, warm air will be blown between the layers to melt the snow that may have accumulated. Other notable early large-scale air domes besides the Silverdome and Metrodome are the Hoosier Dome in Indianapolis (1983) and B.C. Place in Vancouver (1982).

The world's first air-supported stainless steel roof (case a) was built in the early 1980s at Dalhousie University in Halifax, Nova Scotia, Canada; the concept was developed by Donald A. Sinoski. The low-profile roof consists of twenty-four 1/16-in.-thick trapezoidal steel sheet segments arranged radially around an oval center piece. The giant, pie-shaped, steel sheet slices are connected to special wave-shaped contraction joints that run between the sheet sections. The internal air pressure of 7.2 psf supports the roof weight of 2.6 psf and lights and other fixtures of 0.4 psf, as well as 27-psf snow load.

The elliptical, air-inflated cushion roof for the Roman arena in Nîmes, France, engineered by Jörg Schlaich, indicates another trend in the application of large-scale pneumatic structures. The roof, built in 1989, measures about 200×300 ft and can be taken down in summer for open-air events.

Effect of Internal Air Pressure on Geometry   The stress condition for a *spherical pneumatic structure* loaded only by an inflation pressure p has already been discussed in the section concerning the general treatment of a spherical membrane as it responds to a uniform load action normal to its surface [Eq. (8.14)]. There, it was shown that the axial tension forces (T) along the membrane are optimal, as they are constant at any point. The forces are equal to

$$N = T = p_i R/2 \qquad (8.14)$$

The stresses do not vary for different spherical segments with the same curvature and internal pressure. The surface forces are independent of the profile of the pneumatic structure (Fig. 9.25a). Each case in this figure, however, covers a different ground surface area. For the condition where the same base area is to be enclosed by spherical membranes of varying radius (Fig. 9.25b), the decrease of height necessitates an increase of the radius of curvature and thus an increase of the membrane forces if the internal pressure is to remain constant. A decrease in membrane forces may also be obtained by reducing the curvature of the membrane through the use of cable ribs (line supports) or tension columns (point supports). These systems relieve the membrane, similar to a beam or column added to support a floor structure (Fig. 9.25b, bottom).

In contrast to spherical membranes, the radial pressure in the single-curvature *circular cylindrical tubes* causes membrane forces twice as large. The circumferential or hoop forces according to Eq. (8.8) are

$$T_\theta = pR$$

The longitudinal membrane forces, according to Eq. (9.115), are

$$T_x = T_\theta/2$$

The spherical membrane represents a *minimal surface* under the action of radial pressure, since not only stresses and mean curvature are constant at any point on the surface, but also because the sphere by definition represents the smallest surface for the given volume. Some examples found readily in nature are the sea foam, soap bubbles floating on a surface forming hemispherical shapes, and flying soap bubbles. The small deviation from the spherical form, due to distortion caused by the soap film weight, will be ignored for this introductory approach.

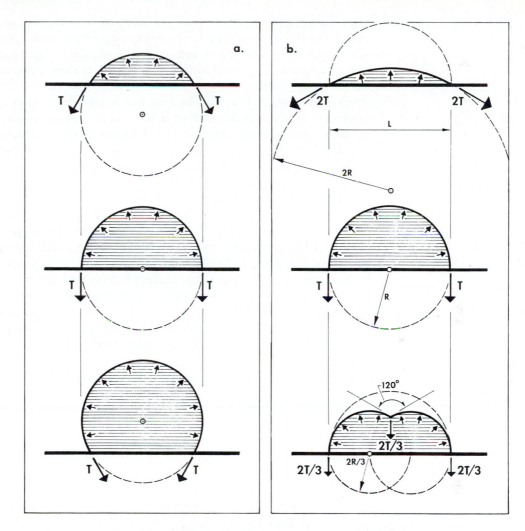

Figure 9.25 Effect of internal air pressure on geometry.

The soap bubble is the result of an equilibrium reached between constant surface tension and the enclosed air pressure; the surface cannot withstand any shearing forces. Any decrease of internal pressure must increase the size of the bubble, because its surface is always under the same state of stress; that is, it cannot respond to force action in any other way and must always be stretched with a constant force independent of the bubble size.

$$T = p_1 R_1/2 = p_2 R_2/2 = \cdots = p_n R_n/2 \qquad (9.100)$$

This equation shows that the internal pressure p is inversely proportional to the radius and directly proportional to the curvature of the bubble when the surface tension is constant.

$$p = 2T/R \propto 1/R \qquad (9.101)$$

Large bubbles have a small internal pressure, while small bubbles have a large internal pressure. This phenomenon is exemplified when different-sized bubbles are in contact with each other (Fig. 9.26), and a smaller bubble always tries to inflate the larger one.

To develop some understanding of the shapes of pneumatic structures, it is helpful to study the formal response of soap bubbles as they come in contact with each other. However, keep in mind that the shape of soap bubble conglomerations is not necessarily the most efficient for pneumatic structures, which must respond to many other types of loading (snow and wind in particular). The spherical membrane no longer represents a minimal surface with respect to these loads, as the stresses along the surface are varying. Only for the one loading condition of inflation pressure does it represent a minimal structure. Since the live loads constantly change, it is impossible to develop a truly minimal structure to respond to them, although for given conditions an optimal geometry may be derived. For a more accurate determination of pneumatic forms, rubber models are better suited to experiments. They respond to any type of loading and are not limited, as soap bubbles are, to one type of force action or constant surface tension. For preliminary studies, however, it is reasonable to use soap bubbles as a first step for finding pneumatic forms.

Typical combinations of bubbles are shown in Fig. 9.26. No more than three ever come in contact with each other in two-dimensional packing. The three soap film edges meeting at one point must form tangent angles of 120° so that the tensile forces balance at that point. In three-dimensional packing, not more than four bubbles will have a common contact point. The four edges meet each vertex at an angle of 109° 28', called the *Maraldi angle*, which appears as an angle of 120° in an orthogonal view.

When a free soap bubble falls on a flat surface, it forms a hemispherical floating shape; that is, a sphere is transformed into a hemisphere without any change of contained volume, since the soap film is stretched to its limits

$$\frac{4}{3}\pi R_s^3 = \frac{1}{2}\left(\frac{4}{3}\pi R_h^3\right), \quad \text{or} \quad R_h = 1.26R_s$$

The radius of the floating bubble (R_h) must be 26% larger than the radius of the flying bubble (R_s) in order to enclose the same volume. With a change of radius, the internal pressure must change since the surface tension is always constant in a soap film.

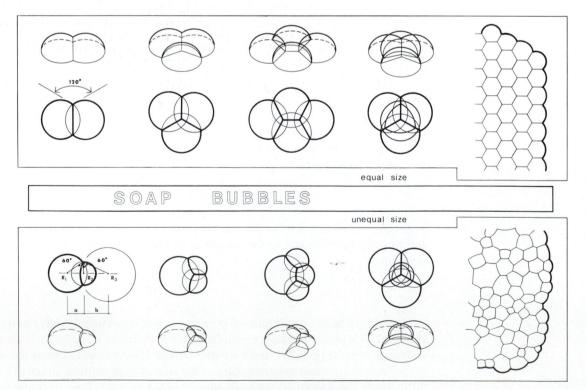

Figure 9.26   Soap bubbles.

$$T = p_s R_s/2 = p_h R_h/2 = p_h(1.26 R_s)/2 \qquad (9.102)$$

or
$$p_h = 0.79 p_s, \qquad p_s = 1.26 p_h$$

The pressure is proportional to the curvature ($1/R$), or the pressure in the sphere is 26% larger than in the hemisphere.

When two bubbles of equal size meet, they must connect at an angle of 120° so that the equal tensile forces at the point of contact are in equilibrium (Fig. 9.26).

$$\Sigma\, V = 0 = T - 2(T \sin 30°)$$
$$\Sigma\, H = 0 = T \cos 30° - T \cos 30°$$

Since the radii of both bubbles are the same, the internal pressure for both is equal as well, causing a straight intermediate partition.

This partition is curved for bubbles of different size, due to the fact that it must balance the internal pressure difference between the two or the larger pressure of the smaller bubble. The relationship of the radii for a twin bubble formed by two unequal soap bubbles can be derived as follows by considering that the internal pressures are inversely proportional to their radii (Fig. 9.26).

$$T = p_2 R_2/2 = p_1 R_1/2 = (p_2 - p_1)R_3/2$$

hence,
$$\frac{R_3}{R_1} = \frac{p_1}{p_2 - p_1} = \frac{1}{p_2/p_1 - 1}; \quad \text{but} \quad \frac{p_2}{p_1} = \frac{R_1}{R_2}$$

or,
$$R_3 = \frac{R_1 R_2}{R_1 - R_2} \qquad (9.103)$$

$$\frac{1}{R_2} = \frac{1}{R_1} + \frac{1}{R_3} \qquad (9.104)$$

The geometrical relationship between the three surfaces is clearly governed by pressure adjustment, as explained in the mathematical rules. Other arrangements of soap bubbles, like those shown in Fig. 9.26, are subject to the geometrical laws just described.

Bubbles arrange themselves in different ways, mostly in a radial manner forming round (closed) shapes rather than linear (open) ones. As the spherical bubbles attach to each other, the interior enclosed ones change into polyhedral forms. Equal-sized bubbles form a hexagonal network in plain view. Bubbles of unequal size can form many different combinations, creating a great variety of polygonal arrangements as seen in plan.

Effect of Membrane Weight   For typical commercial pneumatic structures, the weight of the fabric can be neglected in structural calculations because of its insignificance in comparison to the superimposed snow and wind loads. The membrane weight ranges roughly from about 0.1 psf for canvas to about 0.2 psf for coated fabric to 1 psf for a large stadium roof, such as that of Pontiac Stadium in Michigan.

However, for air-supported forming used in concrete shell construction, load action along a tensile membrane is a design determinant. Adjustable air forming is used not only for concrete shell homes or storage spaces, but also for the construction of arched bridges, pipes, box culverts, storm drains, and manholes. Air-formed concrete shells may be constructed according to one of the following more popular methods:

- *Interior concreted method*: Here, the interior is coated with 4- to 6-in. urethane foam. Then reinforcing steel is placed and shot-creted in multiple thin layers. The foam is necessary to stiffen the air form.
- *Externally concreted method*: This technique requires a higher air pressure so that the skin does not deform under the wet concrete weight.

Rather than using a reusable air form and steel reinforcing bars, the fabric can be used as shell reinforcing. Furthermore, keep in mind that the design of concrete foundations may be controlled by the inflation stage where uplift and horizontal thrust must be resisted, rather than by the final stage of supporting the finished shell.

EXAMPLE 9.12

A 2-in.-thick spherical segment of $L = 50$-ft span and $h = 15$-ft height is constructed by spraying lightweight concrete (100 pcf) on the outside of a pneumatic skin that supports the steel mesh reinforcing. This mesh also prevents the wet concrete from sliding off the pneumatic form. Inflation pressure and material strength of the membrane are to be determined.

The radius of curvature for the spherical membrane (Fig. 9.9b) is

$$R^2 = (L/2)^2 + (R - h)^2 = 25^2 + (R - 15)^2$$
$$= 25^2 + R^2 + 15^2 - 30R, \quad \text{or} \quad R = 28.33 \text{ ft}$$

Or, according to Eq. (9.64),

$$R = (L^2 + 4f^2)/8f = (50^2 + 4(15)^2)/8(15) = 28.33 \text{ ft}$$

The cosine of the semicentral angle ϕ (Fig. 8.29c) is

$$\cos\phi = \frac{R - h}{R} = \frac{28.33 - 15}{28.33} = 0.471$$

The concrete shell weighs

$$w = 2(100/12) = 16.67 \text{ psf}$$

The tension in the skin as caused by air pressure p_i must be at least as large as the maximum compression due to the concrete weight so as to preclude any folding of the skin. For this loading condition, the maximum compression appears along the meridian at the base of the membrane [Eq. (8.18)].

$$\frac{p_i R}{2} = \frac{wR}{1 + \cos\phi}, \quad \text{or} \quad p_i = \frac{2w}{1 + \cos\phi} \tag{9.105a}$$

For a hemispherical dome ($\phi = 90°$), the inflation pressure is

$$p_i = 2w \tag{9.105b}$$

The inflation pressure is independent of the curvature; however, it is a function of the profile: it increases rapidly as the spherical segment approaches the full sphere. The minimum air pressure for this example is

$$p_i = \frac{2(16.67)}{1 + 0.471} = 22.67 \text{ psf}$$

or a gauge pressure of $22.67(0.192) = 4.35$ in. of water, but keeping in mind that this air pressure may not be sufficient to prevent the deformation of the air form. The maximum tensile force acts along the circumference at the base of the membrane (Fig. 8.29a), as caused by air pressure and shell weight [Eq. (8.20)].

$$T_{\theta max} = wR\left(\cos\phi - \frac{1}{1 + \cos\phi}\right) - \frac{p_i R}{2}$$

$$= 16.67(28.33)\left(0.471 - \frac{1}{1 + 0.471}\right) - \frac{22.67(28.33)}{2}$$

$$= -98.61 - 321.12 = -419.73\,\text{lb/ft} = -34.98\ \text{lb/in.}$$

For the given conditions, the safety factor can be neglected since nobody is working underneath the shell during its construction, so for example, a polyethylene or vinyl film may be used as air form.

For a hemispherical dome ($\phi = 90°$), the maximum tensile force is

$$T_{\theta max} = -wR - p_i R/2 = -2wR = -p_i R \tag{9.106}$$

It is apparent that the preceding approach can only be applicable up to a certain span.

For example, for a 200-ft-span, 4-in.-thick hemispherical concrete shell, a membrane capacity of 100 pli = 1200 plf is used. According to Eq. (9.106), this membrane can support an internal air pressure of

$$p_i = T_\theta/R = 1200/100 = 12 \text{ psf}$$

According to Eq. (9.105b), this air pressure can support a load of

$$w = p_i/2 = 12/2 = 6 \text{ psf}$$

But 4 in. of concrete weighs 50 psf, far more than the form can support! In this case, the dome has to be built layer by layer from the bottom up so that the partially finished shell can carry some of the load.

Effect of Snow Loading   Typical standard air-supported shapes, such as cylindrical and spherical structures of relatively *high profile*, are designed for wind loading only; snow is not considered critical at the preliminary design stage. Not only does the wind blow the snow off the roof, but the heat loss through the skin melts the snow so that it easily slides off the smooth surface. Possible adhesion of snow to the skin, together with accumulation due to flexibility of the skin at the flat crown portion, can be broken by alternating changes of inflation pressure. The snow loading must be controlled in this manner since, from a static resistance point of view, it may require a higher inflation pressure than for wind. In cases where the enclosed space is not heated or where the skin has a high thermal insulation, the membrane may have to be heated directly. In order for the snow not to cause any extensive skin deformations and thus allow for further snow accumulations, high inflation pressure should be kept to maintain the original shape. Minimum operating air pressures for high-profile structures are 5 psf for $h/L = 0.3$ and 6.7 psf for $h/L = 0.75$.

The effect of snow on *low-profile roofs* is more critical; because of their flexibility, they are also susceptible to ponding, particularly when the air pressure is low. They can support additional loads up to a certain amount by increasing the internal air pressure, which will not increase the fabric tension if the snow loads are uniformly distributed ($p_i = q$, $\phi \leq 90°$). However, in reality, the snow is not evenly distributed, resulting in an increase in membrane tension together with an increase of inflation pressure. Hence, it is clear that melting systems have to be employed. In case of failure, the roof will slowly deflate and eventually act as a freely suspended membrane, where the loads will be released through some new openings in the skin. For complex pneumatic forms designed by combining different shapes, snow can no longer be dispersed that easily and may gather at certain locations. For such cases, snow must be considered in the structural analysis. The effect of snow loading on structural design is investigated in the following example of an air-inflated cushion roof.

EXAMPLE 9.13

A lens-shaped pneumatic bubble structure, 200 ft in diameter, is being used as a roof enclosure (Fig. 9.27). It consists of an upper and lower spherical membrane spaced 30 ft apart at center span and stabilized by internal air pressure, causing equal tension in both surfaces. Not much air pressure is required under ordinary loading conditions. When the bubble is under heavy snow or rain, the following two design approaches may be considered:

1. The normal, low air pressure will automatically be increased with decrease of volume as the heavy loads act on the top membrane, thereby forcing the lower membrane to carry the additional superimposed loads.

2. The superimposed heavy loads are carried by a higher internal air pressure, allowing less change in volume. The membranes are not connected to each other by any webbing; hence, the bottom membrane only carries the inflation pressure, while the top membrane behaves like a prestressed dome by resisting the internal air pressure and the superimposed loads.

The actual design approach falls somewhere between these two design positions. Furthermore, since temperature affects the internal air pressure, either it must be kept constant or the air pressure must be adjusted. The building is closed, hence the wind will only cause suction on the upper membrane, which will be ignored for this investigation. The air cushion is held in place along the perimeter by a

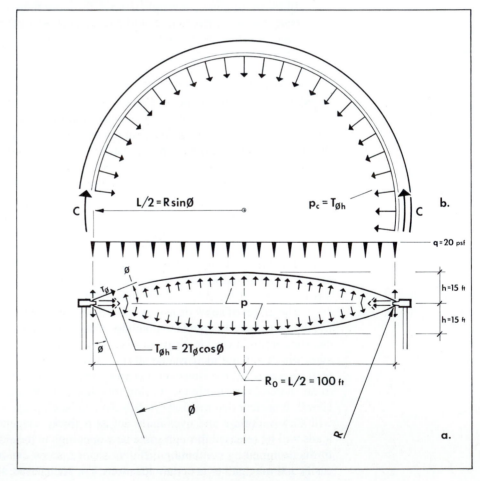

Figure 9.27 Example 9.12.

concrete compression ring (4000 psi concrete, Grade 60 steel), which in turn is carried as a continuous beam by columns. For this example, a snow load of 20 psf on the horizontal roof projection is assumed to control the preliminary design of the membrane. Conservatively, the ground snow load is used, it is not converted to a roof load. The weight of the membrane is considered negligible. The radius of curvature for the upper and lower spherical membranes (Fig. 9.27a), according to Eq. (9.64), is

$$R = (L^2 + 4f^2)/8f = (200^2 + 4(15)^2)/8(15) = 340.83 \text{ ft}$$

The semicentral angle is defined as

$$\cos\phi = \frac{R-15}{R} = \frac{340.83 - 15}{340.83} = 0.956$$

$$\sin\phi = 0.293, \quad \phi = 17.06°$$

Here, the prestressed dome approach with the high internal air pressure is used because of the simple statically determinate analysis. To preclude any folding of the skin, the maximum compression in the upper membrane due to snow loading must be balanced by the tension force caused by internal air pressure, which is at least equal in magnitude. The critical compression is generated by the constant meridional forces [Eq. (8.15)] or the circumferential forces at the crown [Eq. (8.16)] which, however, does not control.

$$p_i R/2 = qR/2, \quad \text{or} \quad p_i = q \tag{9.107}$$

Note that for spherical and cylindrical structures ($\phi \leq 90°$) the inflation pressure is only a function of snow loading and is independent of the profile and curvature. For this case, the required inflation pressure is

$$p_i = q = 20 \text{ psf}$$

or a gauge pressure of $0.192(20) = 3.84$ in. of water. The air pressure, however, may have to be larger than the snow load to prevent excessive deformation of the skin. The maximum tensile stresses appear in the top and bottom membrane due to inflation pressure only, since there will be no circumferential tension due to snow in the upper membrane because $\phi = 17.06° < 45°$ (Fig. 8.28d). Keep in mind that potential wind uplift forces, which have been considered as secondary in this example, may potentially increase the tensile forces in the top membrane, although less air pressure is required when there is no maximum snow load.

$$T_{\phi max} = p_i R/2 = 20(340.83)/2 = 3408.30 \text{ lb/ft} = 284.03 \text{ lb/in.}$$

The minimum required strip tensile strength is

$$T_u = SF(T_{\phi max}) = 4(284.03) = 1136.12 \text{ lb/in.}$$

The required tensile capacity is too high for common coated fabrics. The membrane has to be reinforced so that the fabric is relieved from carrying all the loads. Stressed stainless steel membranes are much stronger than reinforced fabrics. For instance, a stainless steel skin 1/16 in. thick with an allowable tensile stress of 20 ksi can support the following load:

$$P = F_t A = 20(1/16)1 = 1.25 \text{ k/in.}$$

This load is by far larger than the 284.03 lb/in. to be covered in this example.

The maximum compressive forces in the concrete ring are caused by the maximum tensile membrane forces along the perimeter. For this case, the critical loading is due to inflation pressure only. Notice that the ring is funicular under the uniform loads, causing equal radial forces p_c (Fig. 9.27b). The horizontal inward-acting thrust components $T_{\phi h}$ of the membrane forces T_ϕ along the perimeter are

$$T_{\phi h} = 2(T_\phi \cos\phi) = p_c \tag{a}$$

This uniform radial pressure p_c compresses the concrete ring with a constant axial force C [see Eq. (9.57)].

$$C = p_c R_0 = 2(T_\phi \cos\phi)L/2 = T_\phi L \cos\phi \tag{9.108}$$
$$= 3408.30(200)0.956/1000 = 651.67 \text{ k}$$

The ultimate compression ring load, considering the constant inflation pressure as a dead load, is

$$C_u = SF(C) = 1.4(651.67) = 912.34 \text{ k}$$

The required concrete cross section to resist this axial force is estimated as

$$A_G = P_u/0.55f_c' = 912.34/0.55(4) = 414.7 \text{ in.}^2 \tag{3.115}$$

Try a cross section of 13×32 in., $A_G = 416$ in.^{2}.

Additional reinforcement will be necessary, besides the assumed 1% of the ring cross section, to carry the moments caused by asymmetrical roof loading. Also, the spacing of the columns determines the design of the ring as a beam with respect to the vertical plane.

Effect of Wind Loading   In general, air-supported *high-profile structures* are designed for wind loading, the condition in which they are subject to the greatest stresses. The wind pressure distribution, in contrast to the loading due to inflation or weight of the skin, is variable and complex. However, the critical wind forces must be known in order to determine the membrane stresses and the anchorage uplift forces. In this preliminary study, it is assumed that the wind response of rigid bodies is also applicable to pneumatic structures. This assumption constitutes only an approximation and is not conservative; it would only be true if the air structure under lateral pressure maintains its original shape, which is not the case for soft-shell structures; as they deform, they further influence the wind pressure distribution. This deformation causes some benefit while, as the volume decreases, the internal pressure increases, which in turn increases the structure's rigidity. The building rigidity depends on the ratio of inflation pressure to wind velocity pressure p_i/q; as this ratio increases, the rigidity of the building increases. For fast approximation purposes, the static wind pressure formula [Eq. (4.1)] is used and simplified to

$$p = C_e C_q q_s I \cong C_q q_s = 0.00256 C_q V^2 \tag{9.109}$$

Here, only low-rise buildings are investigated for exposure B; hence the increase of velocity with height can be neglected. The minimum pressure coefficients for the following ordinary building forms, disregarding the effect of internal pressure on high-profile structures, are defined as follows:

- Cylindrical shapes (Fig. 4.2)
     Maximum pressure: $C_q = 1.4r = 1.4h/L$
     Maximum suction: $C_q = -0.7 - r = -0.7 - h/L$
- Spherical dome shapes (Fig. 9.28)

Maximum pressure:		Maximum suction:	
Hemisphere	$C_q = 0.9$	*Hemisphere*	$C_q = -1.0$
Three-quarter sphere	$C_q = 1.0$	*Three-quarter sphere*	$C_q = -1.25$

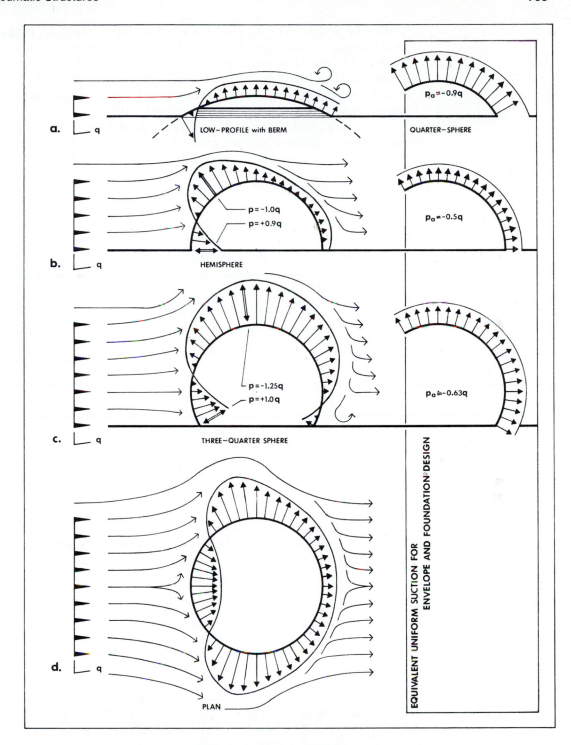

Figure 9.28 Effect of wind loading on spherical membrane shapes.

The difference in the distribution of the equivalent static wind pressure p for structures of various profiles is apparent from Fig. 9.28.

The membrane of a *low-profile structure* is only stressed by suction or uplift forces. The compression at the base is carried by a rigid structure (e.g., earth berm, perimeter framing or walls). For this case, only a minimum inflation pressure is needed to carry the membrane weight, a possible snow load, and to stabilize the membrane against gusty winds. *Minimum air pressures* range from 3.5 psf for $h/L = 0.1$ to 5 psf for $h/L = 0.3$. Under calm weather conditions, the membrane weight will, theoretically, already be supported by the *buoyancy effect* caused by the small temperature difference between outside temperatures and the warmer air generated within the enclosure. High-profile structures, on the other hand, must resist large wind forces, as previously discussed. High wind pressure necessitates high inflation pressure to minimize deformation and folding of the skin. *Minimum air pressures* for high-profile structures range from 5 psf for $h/L = 0.5$ to 6.7 psf for $h/L = 0.75$.

We may conclude that with increase of height the external wind forces increase, causing higher inflation pressure and higher membrane forces, while at the same time the radius of curvature decreases, resulting in smaller membrane forces. These seemingly contradictory statements indicate that there must be some optimum solution with respect to material economy. According to Geiger (see *Zodiac 21*, 1972), the optimum rise of the dome is about 20% of the span if no rigid base structure is used and about 6% of the span where a solid base structure resists the positive wind pressure.

Next, three air-supported buildings are investigated in order to study the effect of wind on the preliminary structural design of tensile membranes.

EXAMPLE 9.14

The pneumatic bubble roof of Example 9.13 is replaced with a shallow, air-supported spherical membrane; that is, only the upper membrane of the cushion structure is left in place. The roof is to be designed for a uniform wind suction of 14 psf. For snow emergencies, the minimum air pressure of 3.5 psf is increased to 12 psf, or the roof has to be deflated so that it can act as a suspended membrane. The minimum internal air pressure easily balances the roof dead load of, say, 1 psf.

The maximum tensile membrane stresses due to inflation pressure and wind uplift are

$$T_\phi = (p_i + p)R/2 = (3.5 + 14)340.83/2 = 2982.26 \text{ lb/ft} = 248.52 \text{ lb/in.}$$

Notice that the uniform loads (due to the minimum air pressure and wind suction of 17.5 psf) are larger than the maximum air pressure of 12 psf for snow emergencies (assuming that snow is not yet on the roof); therefore, control the preliminary design of the membrane. However, keep in mind that the membrane must still be checked in the deflated state under maximum uniform snow loads.

The minimum required strip tensile strength is

$$T_u = SF(T_\phi) = 4(248.52) = 994.09 \text{ lb/in.}$$

This capacity can be provided by ordinary coated fabrics.

EXAMPLE 9.15

Design an air-supported hemispherical dome of 100-ft diameter to withstand a wind velocity of 80 mph. The structure is anchored to the ground by a continuous concrete grade beam. The small weight of the membrane is ignored.

The wind velocity pressure is

$$q = 0.00256V^2 = 0.00256(80)^2 = 16.38 \text{ psf}$$

The approximate maximum equivalent static wind pressures perpendicular to the surface, according to Eq. (9.109), are

Suction: $p = q_sC_q = 16.38(-1.0) = -16.38 \text{ psf}$

Pressure: $p = q_sC_q = 16.38(0.9) = 14.74 \text{ psf}$

To maintain the shape and to prevent any folding of the envelope, the inflation pressure must be at least equal to the maximum positive pressure at the windward face at the base.

$$p_i = p = 14.74 \text{ psf} = 2.83 \text{ in. of water}$$

Some designers allow local folding of the membrane because of the short loading conditions by using an internal pressure of 60% to 70% of the maximum equivalent wind load pressure, but not less than a pressure of 1 in. of water (5 psf). The maximum local membrane tension appears close to the crown as caused by wind suction and air pressure. It would be overconservative to design the fabric based on this critical stress location only and to ignore the effect of the redistribution of forces along the skin. Therefore, an average or equivalent uniform suction p_a equal to one-half of the maximum value (Fig. 9.28a, b, and c) is assumed. Hence, the maximum tension due to inflation and aerodynamic lift is

$$T_{\phi max} = (p_i + p_a)R/2 \tag{9.110}$$
$$= (14.74 + 16.38/2)50/2 = 573.25 \text{ lb/ft} = 47.77 \text{ lb/in.}$$

The minimum required strip tensile strength is

$$T_u = SF(T_{\phi max}) = 4(47.77) = 191.08 \text{ lb/in.}$$

The foundations anchor the envelope to the ground. They resist primarily uplift forces resulting from wind suction and inflation pressure, that is, the average maximum membrane tension along the base perimeter. In this case, the self-weight of a concrete grade beam is employed as the anchorage system. The membrane is attached to the footing continuously, thus reducing stress concentrations in the fabric (Fig. 9.33), while at the same time providing an efficient air seal and drainage control. Since the membrane forces for the hemisphere along the base are vertical, they must be fully resisted by the weight of the foundation. The necessary self-weight to balance the uplift for a footing width of 1 ft and a concrete weight of $w_c = 150$ pcf with V_c equal to the concrete volume, and with a safety factor SF = 1.1 (see *Design and Standards Manual*, ASI-77) is

$$\Sigma V = 0 = T_{\phi max} - w_c(V_c) = 1.1(573.25) - 150[1.0(d)1.0], \quad \text{or} \quad d = 4.20 \text{ ft}$$

Select a 1-ft × 4-ft 3-in. concrete footing. The possible increase in air pressure because of emergencies should be taken into account, as should the fact that the wind suction does not act as a uniform force across the entire building, as was conservatively assumed.

There are no lateral thrust forces acting on the ring foundation under the investigated loading conditions since the dome is hemispherical ($\phi = 90°$). However, should the dome be shallow ($\phi < 90°$), the ring acts as an arch in compression, or for the condition where $\phi > 90°$, the foundation behaves like a tension ring.

EXAMPLE 9.16

A ground-mounted, air-supported cylindrical membrane 70 ft wide and 30 ft high is investigated. The structure must resist a wind of 70 mph.

The radius of curvature according to Eq. (9.64) is

$$R = (L^2 + 4f^2)/8f = (70^2 + 4(30)^2)/8(30) = 35.42 \text{ ft}$$

The semicentral angle is defined as based on Fig. 8.19, top, as

$$\cos\theta = \frac{R-h}{R} = \frac{35.42-30}{35.42} = 0.153, \quad \text{or} \quad \sin\theta = 0.988, \quad \theta = 81.20°$$

The wind velocity pressure is

$$q = 0.00256V^2 = 0.00256(70)^2 = 12.54 \text{ psf}$$

The approximate maximum equivalent static wind pressures perpendicular to the membrane surface are

Pressure: $p = q_s C_q = q_s(1.4r) = 12.54(30/70)1.4 = 7.52 \text{ psf}$
Suction: $p = q_s C_q = q_s(-0.7-r) = 12.54(-0.7-30/70) = -14.15 \text{ psf}$

The inflation pressure p_i must balance the positive pressure p.

$$p_i = p = 7.52 \text{ psf} = 1.44 \text{ in. of water}$$

The maximum circumferential membrane forces due to the internal air pressure p_i and the assumed average suction pressure p_a are

$$T_{\theta\max} = (p_i + p_a)R \tag{9.111}$$
$$T_{\theta\max} = (7.52 + 14.15/2)35.42 = 516.96 \text{ lb/ft} = 43.08 \text{ lb/in.}$$

The required minimum strip tensile strength is

$$T_u = SF(T_{\theta\max}) = 4(43.08) = 172.32 \text{ lb/in.}$$

The longitudinal membrane forces are zero for the condition where rigid end walls carry the wind independent of the air structure (Fig. 8.15b). However, should the end structure be part of the membrane and act together with the cylindrical portion as a total unit, then longitudinal membrane forces T_x are generated. For this example, it is assumed that the cylindrical unit is capped at each end by spherical elements. Some structural designers approximate the magnitude of the longitudinal forces by considering the same load that is carried by the cylinder in one direction to be shared by the spherical end segment in two directions as based on uniform radial force action. The circumferential forces and, in particular, the meridional forces parallel to the cylinder must be supported by the longitudinal membrane forces in the cylinder (Fig. 9.29a), which may then be approximated as

$$T_x = T_{\theta\max}/2 \tag{9.112}$$

Laboratory tests, however, have obtained longitudinal forces by far larger than those suggested by the preceding approximation. Critical wind action against the end of the structure may generate longitudinal forces T_x larger than the circumferential forces T_θ. Special considerations have to be given to the junction of cylinder and end piece where incompatibility of forces causes differential movement, since the circumferential or hoop forces in the cylinder are twice as large as the ones in the sphere. The difference in stresses can be corrected by providing reinforcement at the intersection of the two shapes. The anchoring system for this structure will be provided at intermittent points along the perimeter. ASI (see *Design and Standards Manual*, ASI-77) suggests a safety factor of 3 for earth anchors and 1.5 for concrete pylons. The capacity of the anchorage system depends on the shear capacity of the soil, as well as other soil characteristics. The uplift force to be resisted by each anchor, regularly spaced at a distance $s = 5$ ft apart, is due to the vertical component of the maximum membrane tension, increased by a safety factor of 1.5 for concrete piers (Fig. 9.29c).

$$V = SF(T_{\theta\max})(\sin\theta)s \tag{9.113}$$
$$= 1.5(516.96)(0.988)5/1000 = 3.83 \text{ k}$$

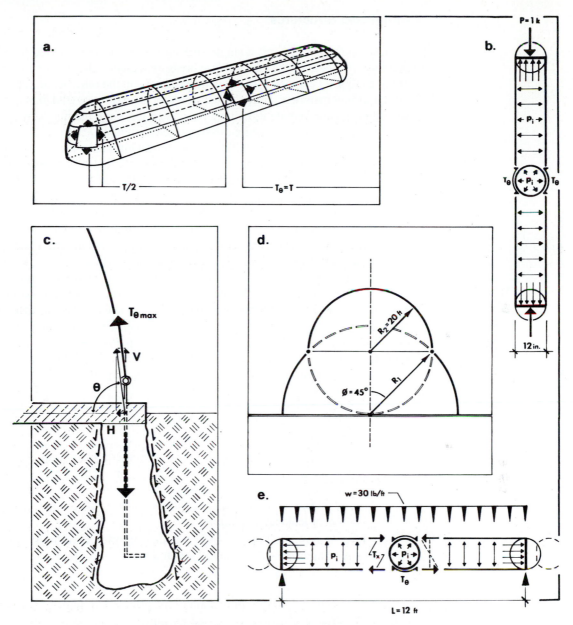

Figure 9.29 Air-inflated members and Example 9.14.

This uplift force must be resisted by each pier in friction with the soil and its self-weight, similarly to tension piles.

The horizontal thrust component of the membrane force is resisted by the ground or, if too large, by the concrete floor acting as a compression strut. For this example, the thrust force acting at each pier (without safety factor) is equal to

$$H = T_{\theta max}(\cos\theta)s \tag{9.114}$$
$$= 516.96(0.153)5 = 395.47 \text{ lb}$$

This thrust force is easily resisted by a 12-in.-wide strip of a 4-in.-thick concrete slab of 3000-psi concrete [Eq. (3.116)].

$$f_c = P/A = 395.47/4(12) = 8.24 \text{ psi} \le 0.22 f_c' = 0.22(3000) = 660 \text{ psi}$$

Air-inflated Structures

In air-supported structures, a single structural membrane is carried by a low internal pressure in excess of the atmospheric pressure, causing the membrane to float on top of the enclosed air space. Air-inflated structures, on the other hand, replace traditional rigid building members. Here, only the members are pressurized, and not the enclosed space. Air-inflated elements behave very much like rigid members, as they are subject to bending and buckling, while air-supported structures act truly as thin membranes. Because of their relatively small cross section, air-inflated structures are high-pressure systems with an inflation pressure in the range of 2 to 100 psi, or a difference in air pressure of about 0.2 to 7.0 atm. This pressure differential is about 100 to 1000 times higher than for an air-supported structure. A car tire is a typical example of an air-inflated element with an inflation pressure of about 30 psi or about 2 atm.

High-air-pressure structures need high-strength materials. The performance of these materials is much more critical than for single-membrane systems, since they must contain the pressure over a greater time period, hence requiring higher safety factors against failure. In general, air tubes are made of coated fabric with some type of inner lining for airtightness. While air-supported structures need continuous air supply, typical air-inflated members do not, although leakage and temperature variation cause a change in pressure and thus require adjustments from time to time. In structures with large air volumes, such as the Fuji Group Pavilion (Fig. 9.23e), air compressors are required to maintain a continuous flow of air.

Air-inflated structures can be organized in a way similar to rigid structures.

- *Tubular systems* (line elements) have a strong curvature in one direction and much less or no curvature in the other direction: columns, beams, arches, and combinations such as grid structures.
- *Dual-wall systems* or airmats (surface elements) are double-membrane envelopes that are joined to form walls, slabs, or shells. The webbing is made of threads or diaphragms. They can be arranged in many different ways, forming many different types of cellular patterns. These cell compartments allow better control of leakage and failure.
- *Hybrid systems*: For example, the air-supported membrane spheres of some radar domes are stiffened with inflated membrane ribs, thus representing composite structures.

There is no limit to the application of the principle of high-pressure systems with respect to intricate forms as expressed by uniquely shaped pneumatic sculptures or shade structures. However, from a practical point of view, they are in an early stage of development with respect to broad application in the building field. Some typical examples are shown in Fig. 9.23. The Fuji Group Pavilion at Expo '70 in Osaka (case e) consists of air-beam arches 13 ft in diameter and inflated to about 16.2 psi. They are fastened to steel cylinders embedded in a concrete ring foundation. The arches are strapped together to cover a 164-ft-diameter circular base area.

The Open-Air Restaurant, also at Expo '70 (case f), is composed of a high-pressure tubular perimeter ring about 10 ft in diameter and a central high-pressure sphere about 20 ft in diameter. Both are supporting a negative low-pressure, double-membrane roof. Along the periphery, wire ropes anchor the roof to the foundations. The Events Structure for the Three Rivers Festival in Pittsburgh, Pennsylvania (1976, case g) is composed of air mats along the perimeter enclosing about 2000 ft^2 of area; the structure is 35 ft high. Here, the architects of *Chrysalis* were primarily concerned with the sculptural qualities of tensile architecture by blending the art and technology of fabric structures.

Many other experiments have been done with high-pressure structures. For instance, nylon-reinforced plastic air tubes have been connected to multidirectional air ball joints to form space frames or grid shells. The deflated nets can easily be shipped to the site and inflated; the principle is applicable particularly as form work for the construction of shells.

The structural behavior of high-pressure tubes is very complex. Since there is no established design procedure, linear behavior is assumed for the very approximate design of the following simple structural elements so that some basic understanding may be developed.

EXAMPLE 9.17

The membrane material and inflation pressure are to be determined for an air-inflated, 1-ft-diameter tube closed at its ends with rigid plates. The behavior of the tube is first investigated as a column and then as a beam. The equation for the membrane forces in the circular cylinder due to inflation pressure p_i only have already been determined. The circumferential membrane forces are

$$T_\theta = p_i R \qquad (8.8)$$

The longitudinal membrane forces are derived from the equilibrium of the internal pressure against the end plates and the longitudinal membrane forces along the circumference.

$$\Sigma F = 0 = p_i(\pi R^2) - T_x(2\pi R), \quad \text{or} \quad T_x = p_i R/2 = T_\theta/2 \qquad (9.115)$$

This expression was already given as an approximation for the longitudinal forces in an air-supported cylindrical unit capped at each end with spherical segments [Eq. (9.112)].

(a) The high-pressure tube will first be designed as a column to carry an axial load of $P = 1000$ lb (Fig. 9.29b). Visualize the rigid end plates to be separated from the membrane boundaries and to float on the internal air pressure. Vertical equilibrium yields

$$P = p_i \pi R^2$$

Hence, the required minimum inflation pressure is

$$p_i = P/\pi R^2 = 1000/\pi(6)^2 = 8.84 \text{ psi} \qquad (9.116)$$

We could have also reasoned that the compression exerted in the membrane by the external load must be balanced by the tension due to the air pressure.

$$p_i R/2 = P/2\pi R \quad \text{or} \quad p_i = P/\pi R^2 \qquad (9.116)$$

Although both approaches give the same result, each system shows distinctly different behavior. The separation of the rigid end plates from the membrane similar to a piston, assuming this approach to be practically possible, makes it impossible for the skin to fold or buckle. Since the external load is supported solely by the fluid, which, in turn, is supported by the circumferential membrane forces only, there will be no longitudinal forces in the membrane, as is assumed for the other approach. The circumferential membrane forces are

$$T_\theta = p_i R = 8.84(6) = 53.04 \text{ lb/in.}$$

For the selection of membrane material, a safety factor of 6 is used. In general, they are larger for air-inflated structures than for air-supported skins, because of the sudden explosivelike nature of their failure. The required minimum tensile strength is

$$T_u = SF(T_{\theta max}) = 6(53.04) = 318.24 \text{ lb/in.}$$

The capacity of the membrane in the longitudinal direction should be at least one-half of that in the circumferential direction as based on inflation pressure only.

(b) The high-pressure tube, tied together by internal webs, is assumed to act as a simply supported beam spanning 12 ft and supporting a uniform snow load of 30 lb/ft. The weight of the fabric is neglected (Fig. 9.29e). The beam must support a maximum moment of

$$M_{max} = wl^2/8 = 30(12)^2/8 = 540 \text{ lb-ft}$$

Small deflection theory is assumed for this very approximate investigation, keeping in mind that the load–deflection relationship is not linear due to local deformations and creep of the fabric. Hence, the bending stresses can be defined as

$$f = \pm M/S, \qquad \text{where } S_0 = \pi R^2 t \quad \text{(Table A.14)}$$

The axial forces at top and bottom face due to bending are

$$N = \pm tf = M/\pi R^2 \tag{9.117}$$

The prestress force T_x in the longitudinal direction due to inflation pressure must be at least as large as the critical compression force at midspan in order to prevent any folding of the membrane.

$$p_i R/2 = M/\pi R^2$$

$$p_i = \frac{2M}{\pi R^3} = \frac{2(540)\,12}{\pi(6)^2} = 19.10 \text{ psi} \tag{9.118}$$

A minimum air pressure of 19.10 psi is necessary for equilibrium, although it was found in experiments that local folding on the compression face does not cause collapse of the beam! The maximum tensile forces in the longitudinal direction appear at the bottom face at midspan due to air pressure and load action.

$$T_{xmax} = \frac{p_i R}{2} + \frac{M}{\pi R^2} = \frac{19.10(6)}{2} + \frac{540\,(12)}{\pi\,(6)^2} = 114.60 \text{ lb/in.} \tag{9.119}$$

The maximum tensile forces in the circumferential direction are

$$T_{\theta max} = p_i R = 19.10(6) = 114.60 \text{ lb/in.}$$

The critical tensile stresses in the longitudinal and circumferential directions are equal to each other. Should the internal air pressure be increased by a safety factor, the circumferential membrane forces will have a larger controlling value. The ultimate tensile strength of the membrane must be at least equal to

$$T_{u\,max} = SF(T_{\theta max}) = 6(114.60) = 687.60 \text{ lb/in.}$$

Shear must also be checked, since the fabric along the sides of the beam close to the supports may wrinkle before the top face at midspan folds due to bending.

Air-inflated elements are extremely flexible. The deformation of air beams is not controlled by bending, as for rigid beams, but by shear, because of the extremely low shear rigidity of the inflation pressure, which takes the place of web material in rigid members. How the internal pressure affects the stiffness of the air beam is beyond the scope of this discussion.

Some Other Soft Shell Structures

Rather than using air as the primary support medium for columns and walls, water or sand may replace the gas and be contained in a tensile membrane. The liquid or soil carries the compression, while the enclosing membrane prevents its lateral displacement in tensile action.

The principle of the sail or parachute with natural wind or a mechanically generated airflow produces a dynamic airborne roof structure. Additional support for the canvas may be provided by helium-filled balloons, rings, or other shapes anchored to the ground.

The ultimate minimum weight structure may be the massless air curtain enclosure (Fig. 9.30), where the static material response is replaced by a dynamic energy system by utilizing, for this invisible architecture, rapidly moving jets of air to protect the enclosed space from rain, snow, and wind.

9.9 TYPICAL MEMBRANE ROOF DETAILS

For very large scale structures, cable nets form the primary surface that supports the roof cladding, as is further discussed in the following pages. Some typical network layouts are shown in Fig. 9.4; naturally, many more net forms are possible. The nets with meshes of the same size and member length offer the advantage of repetitive infill panels, although we must keep in mind that under load the regular pattern will be distorted. For hanging polygonal or synclastic dish-shaped roofs, as well as for flat anticlastic surfaces, semirigid or rigid tensile members or surfaces may have to be used since the missing proper countercurvature cannot provide the necessary stability.

Typical roof details for tensile membranes are shown in Fig. 9.31. While simply suspended roofs need stiffness and/or weight as provided by concrete shells and continuous rigid members (Fig. 9.15), prestressed membranes should have light and flexible cladding so that the necessary pretensioning forces can be reduced to a minimum, since larger prestress forces require larger cable sizes, but also larger boundary structures. Furthermore, the roof cladding should be flexible to allow for the large deflections of the cable network. Typical lightweight decking employed includes wooden panels and corrugated sheeting made from steel, aluminum, plastics, asbestos-cement, and reinforced gypsum; insulation may be placed on top and covered with several layers of roofing felt. For large-scale pneumatic and tensioned roofs, fabric panels (e.g., Teflon-coated fiberglass) are clamped to the cable lattice.

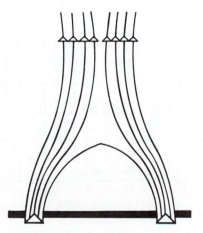

Figure 9.30 Air curtain enclosure.

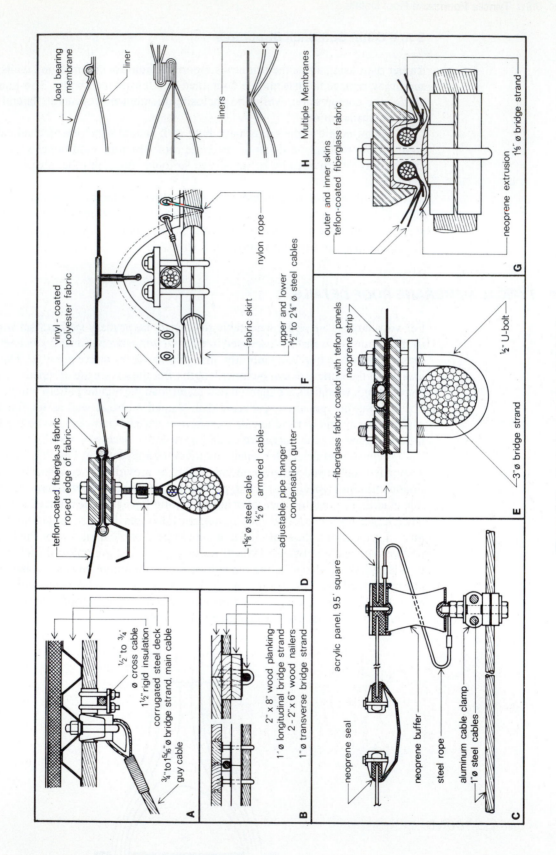

Figure 9.31 Typical membrane roof details.

804

The intersecting cables of the State Fair Arena in Raleigh (Fig. 9.31A) are fastened to each other at points of intersection and form a network. The corrugated metal sheets, in turn, are bolt-clipped to the cables (see also Fig. 9.2k). The 2- × 8-in. wood plank roof deck of the Ingalls Hockey Rink at Yale University (B) is laid across and nailed to double 2- × 6-in. wood nailers, which are bolted to the transverse cables spaced at 6 ft (see Fig. 9.16g).

The air-supported, low-profile cable roofs of larger scale (see Fig. 9.24) consist of a cable network of relatively large spacing to which the edges of the fabric panels are clamped. The fabric spans the distance between the cables. Typical connection details are shown for cases D to G of Fig. 9.31.

For the anticlastic membrane roofs of the Krylatskoe cycle track of the 1980 Moscow Olympic Games, not cable nets but 0.158-in.-thick steel bands about 13 ft wide were used and connected in the transverse direction by continuous steel purlins.

The cables of the roof of the U.S. Pavilion at Osaka (see Fig. 9.24f) are clamped together at the intersection points and the skirts of the vinyl-coated fiberglass fabric are attached to the cables below (Fig. 9.31F). The joining of fabric to fabric was done by heat sealing with the same vinyl that covers the fabric.

The Pontiac Stadium roof (see Fig. 9.24e) consists of nine 3-in.-diameter steel cables spanning in each of the diagonal directions. The edges of the Teflon-coated fiberglass panels are bordered with $\frac{1}{2}$-in.-diameter nylon ropes. The adjoining fabric panels are covered with neoprene and then clamped between aluminum strips (case E). The overlapping neoprene provides a waterproof seal for the bolted joint.

Using a second or multiple inner membranes called liners below the main roof skin (H) changes the ceiling geometry from concave to convex. A stagnant air space is created that, together with the coating, acts as a thermal insulation and reduces heat loss, as well as improving acoustics and fire safety; in winter, hot air can be passed through the air space to melt snow.

Many tent-type structures use arches and frames as major supports, which usually directly support the fabric. Tents using masts as primary supports generally employ a radial cable network to carry and pretension the fabric membrane. The Teflon-coated fiberglass fabric of La Verne College (see Fig. 9.17D) is factory fabricated; it was unrolled at the site and placed over the central mast to which a system of radial cables was already attached. This cable network supports and pretensions the membrane. Also, the roof skin of the umbrellalike tent structures for the Jeddah International Airport in Saudi Arabia (see Fig. 9.17K) consists of a Teflon-coated fiberglass fabric stretched between the tension ring at the crown and the cables between the four steel pylons along the base. Radial cables were laced into sleeves in the fabric to stretch and strengthen the membrane.

Large-scale tent structures, such as the Munich Olympic Stadium (Fig. 9.31C and Fig. 9.17J), require special membrane detailing. The primary structure of the cable-net supports the separated secondary structure of acrylic panels on top of it. The panels are bolted to the network at cable intersection points. Neoprene buffers separate the panels from the steel cables. They act as shock absorbers to prevent the breakage of the plastic panels since the differential movement of the net cannot be transferred to the panels and since the temperature difference causes extensive change of panel sizes. A steel rope connecting the panel to the cable-net is provided as an additional safety precaution in case of buffer failure.

9.10 TENSILE FOUNDATIONS

Tensile foundations must withstand the vertical uplift as well as the horizontal thrust forces due to external and prestress loading. The lift forces can be resisted by counteracting weight as well as by anchorage to the ground or to other building structures. The different ways in which foundations or ground anchors can support tension are shown in Fig. 9.32. Vertical reactions can be provided by gravity, such as the weight of the foundation, ballast, and earth loads, as well as by friction of the anchorage along its contact surfaces, and any combination of these. The horizontal thrust forces are either balanced by struts and/or slabs, which act as braces from one side to the other, or by compression rings; or they are resisted by passive earth pressure and/or skin friction. Some other considerations related to tension anchor design are identified in Fig. 9.32. Tensile membrane forces can be transferred to the foundations in a continuous uniform manner by employing, for example, strip footings, or they can be transferred with the help of edge cables to point supports, such as anchors or concrete piers. Tensile forces of large magnitude, as in guy cables, may be carried directly by one foundation or they may be broken down and resisted by a group of anchors.

The selection of the anchorage system depends, among other criteria, on the magnitude of the force action, the capacity of the soil, the strength of the membrane, the capacity and spacing of the anchors, the construction of the membrane along its base, the nature of the structure (e.g., permanence), and available expertise of labor, as well as costs.

Various tension foundations are identified in Fig. 9.33 and may be classified as follows:

- Earth anchors
- Gravity anchors
- Tension piles
- Rock anchors
- Self-balancing structures (e.g., struts, ties, slabs, compression or tension rings)
- Special foundations (e.g., inclined L-shaped retaining wall anchors)

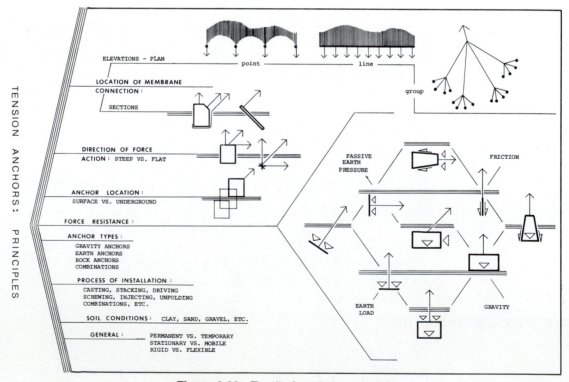

Figure 9.32 Tensile foundation principles.

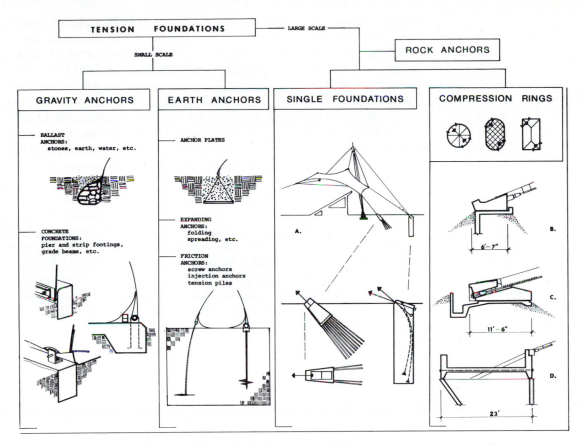

Figure 9.33 Tension foundations.

Common earth anchors are *plate* and *mushroom anchors,* where the weight of the anchors and the weight of the soil counteract the tension forces. Certain soil anchors, such as *screw anchors* (for lighter soils) and *spread anchors* can be easily transported, installed, and removed. Their capacity depends on the strength of both the anchor itself and the soil. There are many types of soil anchors; for more precise information, the reader must refer to the literature provided by the producers.

Since the soil capacity may possibly be reduced due to the process of installation, that is, due to the ramming or screwing of the anchor, a safety factor of 3.0 is generally applied to the maximum calculated loads for the selection of the soil anchors. The soil capacity around the anchor may be increased by injecting cement grout.

Gravity anchors prevent the vertical uplift simply by having enough counteracting weight, which can be provided by some type of ballast (e.g., sand bags, water bags or tanks) on top of or in the ground, by more permanent cast-in-place or precast concrete foundations, or by the soil burden above the anchor plates. Concrete foundations may be of the continuous strip type (i.e., grade beams) or they may be individual piers. A safety factor of 1.5 is generally applied to the design loads for concrete piers. While the vertical uplift component of the tension force is resisted by the weight of the foundation, ignoring any frictional resistance, the horizontal force component must be supported by the soil, slab, struts, or rings (see Fig. 9.29c).

In the case of curvilinear structures, closed ring beams replace the linear struts. Where membranes are relatively flat, such as pneumatic roofs for stadiums, the horizontal thrust forces are extremely large. The thrust is carried by compression rings (Fig. 9.33 B to D) while the vertical uplift component due to internal air pressure and wind suction is resisted either by the ring alone or by the ring together with the walls or frames along the perimeter. For instance, the concrete ring for the U.S. Pavilion at Expo '70, Osaka, Japan (C) is 4 ft high and 11.5 ft wide. It is in pure compression under the uniform horizontal thrust caused by the internal air pressure and its own weight; however, it will bend under asymmetrical wind action. The ring alone also balances the upward lift. To prevent the transfer of horizontal forces to the supporting earth berm, the ring is separated from the berm by a slip joint, which still provides enough frictional resistance to prevent the roof from sliding off its base.

Unlike the air domes shown, the Hubert H. Humphrey Metrodome in Minneapolis (1981) has a sloping compression ring tangential to the roof skin, thus much more effectively resisting the thrust forces. Furthermore, the ring is not supported by the facade walls or earth berms, but directly on the seating structure to which it is hinged so that it is allowed to float as an independent structure.

Individual or groups of *tension piles* (e.g., reinforced concrete piles) may support a guy cable to transfer large forces to the ground. These piles are inclined so that the forces are transferred directly in axial action; they resist the tensile forces primarily by their mantle friction and possibly by their self-weight.

Special tensile foundations were designed for the Munich Olympics (Fig. 9.33A). The ground anchor foundation directly resists the tension by the earth weight filled onto the foundation, as well as by its tension piles. The other anchorage system shown is of the gravity type.

PROBLEMS

9.1   The single-cantilevered roof for a hangar is cable supported. Masts are spaced 30 ft apart and carry pairs of steel strands that support steel trusses upon which the purlins rest. The weight of the roof is 14 psf, just enough to counteract the uplift due to wind; the snow load is 30 psf. Only uniform gravity loading is considered for this preliminary design. For the building geometry, refer to Fig. 9.34a.

    (a)   Determine the cable size by considering only the front roof portion; verify that the steeper cable for the back will be larger. Neglect the cable extension.

    (b)   Determine the cable extension and the vertical displacement of the beam support.

    (c)   Estimate the W12 chord sections for the truss (A36), assuming a truss depth of $d/L = 1/12$.

    (d)   What is the major problem in a single-cantilever structure? Compare it to the double-cantilever structure in Problem 9.2.

9.2   For the cable-supported, double-cantilevered hangar (Fig. 9.34b), determine the critical cable size. The cables support the steel girders set at 25.5-ft intervals. Assume a roof dead load of 15 psf and a live load of 40 psf. Also estimate the size of the wide flange beam sections (50 ksi steel) using $d/L = 1/20$.

9.3   The 156-ft-span suspended roof of a high school gymnasium consists of steel plates welded together and 3-in. lightweight concrete on top of it. Assume an average weight for the steel blanket of 13 psf and 25 psf for the concrete; the snow load is 40 psf. Determine the plate size using a sag-to-span ratio of 9:156.

9.4   Investigate the George Washington Bridge (1931, O. H. Ammann) spanning the Hudson River from New York City to New Jersey. The approximate dimensions for the bridge are given in Fig. 9.34d. The road deck is supported by two huge cables on each side, which, in turn, are supported by the towers. For this preliminary investigation, consider only a gravity deck load of 39 k/ft for dead loading and 8 k/ft for live loading. Determine the number of 4-in.-diameter cables required. Do not forget to consider the side span in your investigation. Find the displacement of the cable at the quarter-point of the main span.

9.5   Do a preliminary investigation of the queen-post truss (Fig. 9.34e).

    (a)   Select the steel beam W-section (A36), ignoring the effect of elastic cable deformations.

    (b)   Determine the approximate cable size.

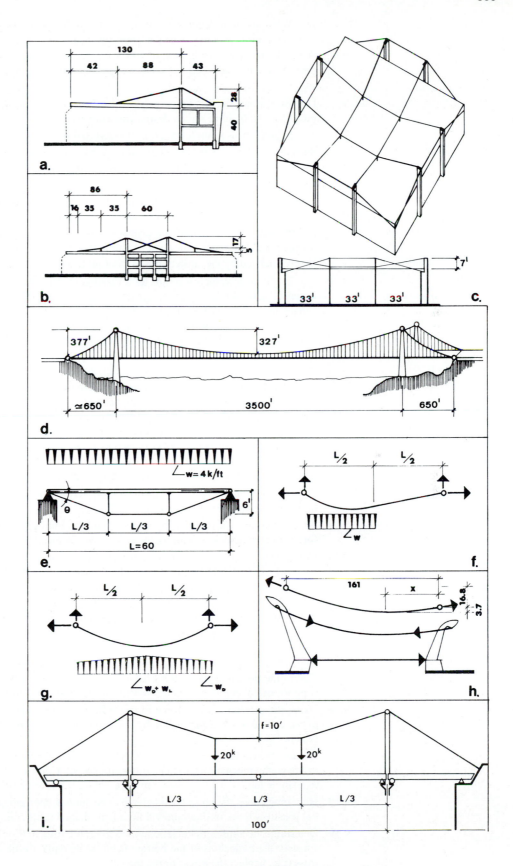

Figure 9.34 Exercises for tensile roof structures.

(c)   What are your conclusions if the elastic deformations (i.e., settlement of posts) are taken into account? Does the beam size change? Reason!

9.6   A 100-ft-span cable with a sag of 10 ft carries a uniform load of 100 lb/ft. Determine the size, geometry, sag at 20 ft from the support, slope, length, elongation, and increase in sag.

9.7   Decrease the sag in Problem 9.6 to $f = 1$ ft. What impact does this change of sag-to-span ratio to $n = 0.01$ have on the design and analysis?

9.8   The length of the cable in Problem 9.6 was determined for a temperature of 50°F. Find the change in the unstressed length of the cable and the change in sag at a temperature of 110°F. The coefficient of linear expansion is $\alpha = 0.0000065$ in./in./°F. Draw your conclusions.

9.9   A prestretched or trampolinelike cable roof is used rather than the suspended one of Problem 9.6. The cables are pretensioned each to 350 k to stiffen the roof against fluttering and to reduce the deflection due to gravity loading. Determine the approximate displacement of the cable at center span. Assume that the magnitude of the prestress force is hardly changed by the superimposed external loads.

9.10   The single-layer suspension roof for a rectangular building with parallel cables spaced 5 ft apart spans 240 ft and has a maximum sag-to-span ratio of $n = 1/12$. The roof must support a dead load of 20 psf and a live load of 30 psf assumed to act on the horizontal roof projection. Determine the cable size, cable geometry, cable sag at $L/4$, cable slope at $L/4$, cable length, increase in cable length and cable sag; assume a temperature difference of 50°F. Ignore dynamic considerations.

9.11   For the roof structure shown in Fig. 9.34h, determine the size of a typical interior cable spaced at 10 ft on center. The cable size is based on the construction stage, where the roof acts as a true suspension structure supporting only the dead load of 33 psf due to the precast concrete panels and concrete ribs and ceiling. In the final stage, the roof does not act as a true suspension system since the concrete ribs along the cables together with the concrete slab form a suspended ribbed shell that, due to its stiffness, controls deflection and the danger of flutter.

9.12   A dome spans 138 ft. It is supported by eight ribs that exert a lateral thrust of 1706 k at the base (see the dome of S. Maria Del Fiore in Florence, Italy). Determine the size and number of cables to resist this thrust.

9.13   For Fig. 9.34f, determine the location of the low point of the cable, the maximum cable force, and the equation defining the geometry of the cable.

9.14   Consider the snow load in Problem 9.10 to be zero at the edges but twice as much at the center (i.e., 60 psf). Does this new loading condition change the design of the cable?

9.15   A round stadium roof with a span of 420 ft is supported by radially arranged cables. The single layer of cables is weighed by precast I-sections to stabilize the roof against flutter. The cables are spaced at 3.75° intervals, framing outward from a 45-ft-diameter tension ring in the center to a compression ring that rests on 32 pairs of cross columns. Assume a total load of 50 psf to act on the horizontal projection of the roof. For this approximate solution, neglect the 45-ft center opening (i.e., line load) and continue the surface loading. Determine the cable size, cable shape, vertical displacement of cable at 100 ft from the support, cable length, and cable elongation. Assume a sag-to-span ratio of $n = 0.1$.

9.16   Using the horizontal cable force found in Problem 9.15, determine the approximate size of the central 45-ft tension cable by assuming the cable forces to act as a uniform radial load p.

9.17   Assume the cable roof of Problem 9.15 has to carry an additional steel-framed penthouse to house the mechanical and electrical equipment. Consider the weight to be equivalent to a single load of 10 k per cable to act at the center span. Find the cable size.

9.18   How much does the bottom cable have to be spread, approximately, to introduce the initial prestress force in Example 9.8 for a lens-shaped cable truss?

9.19   Determine the approximate deflection of the cable truss in Problem 9.18. Assume conservatively the deflection of the bottom cable to be equal to the deformation of the whole truss (i.e., neglect the effect of the top cable).

9.20 A prestretched, flat, horizontal roof is designed for very high prestress forces so as to allow only a small sag and to reduce fluttering. Investigate this circular roof of 120-ft diameter to be prestressed for 600 lb/ft and permit a maximum sag of 3 ft. Neglect the small amount of additional tension due to the superimposed loading and ignore any cable extensions. First, assume the flat roof to be built of radially arranged cables and then of a surface membrane. Determine the loads that can be carried.

9.21 A flat roof consists of a series of parallel cable pairs arranged 2 ft on centers horizontally and $1\frac{1}{2}$ ft apart vertically. The cables are pretensioned and strung between concrete abutments 200 ft apart. A maximum deformation of only 3 ft is permitted for the 15-psf dead load and the 20-psf live load. Determine the approximate cable sizes.

9.22 The behavior of a suspension cable with a sag ratio $n = 1:10$ is investigated for a simple, symmetrical loading condition. Assume the cable to carry two single loads of 20 k each. For the geometry of the structure, refer to Fig. 9.34i.

9.23 Do a rough, preliminary design of the membrane, first by using a coated fabric and then a network $4\frac{1}{4} \times 4\frac{1}{4}$ ft and find also the edge cable size for the tent shown in Fig. 9.20a. The tent covers a square base of 80 ft and has a maximum height at the masts of 20 ft; it must support a live load of 15 psf. The sag of the edge cables is 8 ft.

9.24 Do a quick preliminary design for one of the nine saddlelike net surfaces of the Olympic Stadium in Munich (Fig. 9.17J). The membrane consists of a square mesh of about $2\frac{1}{2} \times 2\frac{1}{2}$ ft. The radii of curvature of the hanging cables vary between 164 and 230 ft. Assume the radius of curvature for the typical edge cable as 80 ft. The membrane must support its own weight of 4 psf and a snow load of 26 psf.

9.25 A saddle-type canvas roof with negligible weight must support a snow load of 15 psf and a wind suction of 20 psf. If the mean radius of curvature of the membrane surface is 80 ft, what must be the capacity of the canvas?

9.26 Determine the approximate cable size for a prestressed square mesh network 8×8 ft covering a square plan of 128×128 ft (Fig. 9.20c). The saddle-shaped roof has a typical sag-to-span ratio of 1:12, resulting in a sag and rise of 10.67 ft. The dead and live loads are 8 and 20 psf, respectively.

9.27 Determine the increase in circumference and radius for a semicircular cylindrical structure as caused by inflation pressure only.

9.28 For the combined spherical membranes (Fig. 9.29d), determine the membrane forces for both domes as caused by an internal pressure of 5 psf. Find also the circumferential tension force for the ring at the junction of the two domes.

9.29 Determine the initial inflation pressure and the material strength to support a pneumatically stretched skin upon which is sprayed, from the inside, a heat insulating structural plastic that when stiffened, will support itself as a dome; at this first stage the shell weighs 5 psf. Once the plastic shell is rigid, further materials will be glued onto the surface to form a sandwich shell. Consider a spherical dome of 100-ft span and 30-ft height.

9.30 An air cushion of 60-ft height spans a stadium of 700 ft. It consists of equal spherical membranes made of nickel stainless steel with an allowable stress of 60 ksi. The roof membrane must carry a uniform 10-psf wind suction perpendicular to the surface, 25-psf snow load on the horizontal roof projection, and 2-psf roof weight. First, determine the internal air pressure and then the thickness of the membranes by neglecting their weight.

9.31 An open summer theater is covered with a lens-shaped roof that consists of an upper and lower spherical membrane of equal size, stabilized by internal air pressure producing tension in both surfaces. The roof has a diameter of 145 ft and a depth of 20 ft at the center. Both membrane layers only weigh about 1/2 psf and will be neglected. Other live loads due to gravity are insignificant because the structure is only used during summer. Assume a uniform wind action of 3 psf perpendicular to the surface for this preliminary design. Determine the inflation pressure and the tensile capacity of the membrane.

9.32 Investigate the compression ring in Problem 9.31 by using A36 steel.

9.33 Do a preliminary investigation for a cylindrical membrane with quarter-spheres as end pieces. The structure has a floor area 500 ft in length and 90 ft in width and a maximum height of 30 ft. Assume as the critical loading a wind velocity of 90 mph. Determine the internal air pressure, the tensile capacity of the membrane, and the depth of the 1.5-ft-wide continuous concrete footing (150 pcf). Can the horizontal force components be carried by the 4-in.-thick concrete slab of 4000-psi strength?

9.34 Investigate an air-supported, three-quarter spherical dome structure with a base diameter of 100 ft. Design the dome for a wind velocity of approximately 75 mph. Determine the internal air pressure, the tensile capacity of the membrane, and the anchorage capacity of the foundation.

9.35 An air-inflated, 15-in. cylindrical column carries an axial load of 5 k. Determine the internal air pressure and the capacity, as well as the type of membrane to be selected.

9.36 Replace the gas medium air in Problem 9.35 with the liquid, water ($w = 62.4$ pcf). Assume a column height of 12 ft. What material capacity do you select?

9.37 A high-pressure cylindrical air beam spans a distance of 45 ft. If a working pressure of 100 psi is to be maintained, what minimum-size tube must be selected to support 75 lb/ft? Determine also the tensile capacity of the membrane.

9.38 A 10-in. pressurized tube of parabolic shape (parabolic air arch) spans 40 ft and has a height of 20 ft. It must support an equivalent load of 100 lb/ft on the horizontal roof projection. Determine the internal air pressure and the capacity of the membrane material.

9.39 Determine the inflation pressure and material strength for a truncated conical membrane that has a height of 30 ft and a diameter of 50 ft at the base and 12.5 ft at the top.

(a) Neglect material weight and assume a 20-psf snow load to act on the flat top portion.

(b) Consider the snow load to act on the horizontal projection of the cone.

High Rise Building Structures

The building shape and building function very much determine the nature of the support structure, aside from the design attitude toward structure. Vertical structures range from massive building blocks to slender towers; they may occur as isolated objects within the urban fabric, or they may form urban megastructures. They encompass types from simple symmetrical to complex asymmetrical forms, from boxes to terraced buildings, and from ordinary bearing wall, skeleton, and core construction to bridge buildings, cellular clusters, suspension buildings, tubes, superframes, and the new breed of compound hybrid skyscraper forms. From an appearance point of view, buildings may disclose themselves as organisms by exposing how they were constructed, how they stand up, and how they function. These buildings may celebrate technology, experimentation, and innovation as art. On the other hand, buildings may represent sculptures or other formal objects that are inspired by symbolism, ornamentation, and the preservation of past values. In this case, the building image is controlled by external considerations as may be expressed by complex spatial geometry, massing, and imposed surface composition, where the facade wall forms an independent decorative enclosure articulating values possibly reminiscent of the Art Deco style. Whatever the position of structure in architecture, the designer must understand structure in order to either control it as a builder or to articulate its spirit as an artist.

The early development of the skyscraper occurred in Chicago from about 1880 to 1900 where block- and slablike building forms reached merely 20 stories. Then the soaring towers of New York introduced the true skyscraper, the symbol of American cities.

The birth of the skyscraper is usually associated with the construction of the 10-story Home Life Insurance Company Building in Chicago, designed by William LeBaron Jenney, and completed in 1885 (the building was demolished in 1931). There were no bearing walls; the entire building weight was carried solely by the metal skeleton, consisting of round cast-iron columns filled with cement mortar, with wrought-iron I-beams for the first six floors and steel beams for the remaining floors. The typical beams were on 5-ft centers and supported the flat tile floor arches. The masonry facade, that is, the piers and spandrels, were hung from the frame like a curtain; they were carried on shelf angles fastened to the spandrel beams. The building facade did not express the skeleton-skin concept, but rather that of the traditional load-bearing masonry piers; it was organized in horizontal layers in a vaguely Romanesque revival style. The significance of the building lied in the technology rather than in its esthetics.

Finally, a high point (but also a conclusion) of the Chicago frame was reached with Louis Sullivan's masterpiece, the 12 story Carson Pirie Scott Department Store of 1904. The building exposes the static, neutral nature of the frame, clad with white terra cotta. The spans and heights of the skeleton bays were determined by functional requirements alone, and not by abstract rules of regularity and symmetry. The building is an impressive statement of modern architecture and underscores Sullivan's famous declaration that "form follows function."

The concern of the Chicago School in expressing the function of the building was not shared by New York designers. They strongly believed in a separate ornamental system within the order of past styles (as represented primarily by the Classic and Gothic periods), attached to the steel frame and disguising it. While Sullivan tried to make ornament and building inseparable, the designers of the New York School composed the ornament through skilled academic exercises directly from eclectic forms by concentrating on the overall image of the building. When buildings continued to grow taller, the palazzo model became outmoded, and designers were forced to search for new historical precedents. Tall towers had replaced the relatively low building blocks of the Chicago period. They thrust into the sky above other buildings and became symbols defining the skyline and the silhouette of the American city: the true skyscraper was born, connecting the sky to the earth. But the symbolic role of the skyscraper was much broader; it was also supposed to be beautiful and reflect an image of the power of American corporations.

The Gothic cathedral style was successfully introduced as a more appropriate model for skyscraper design with the 57-story, 792-ft-high Woolworth Building in 1913 by Cass Gilbert, often called the Cathedral of Commerce. As the world's tallest building of its time, it was praised as the first true skyscraper, ideally expressing verticality and upward thrust, as well as the power of commerce in American life.

The Art Deco style is highlighted by William Van Alen's 77-story Chrysler Building in New York, and in particular in its stainless steel spire (1930). Its facade architecture uses modern form language with a composition reminding us of industrial design concepts, perhaps symbolizing the building as a machine; a dressing that seems to reflect the flamboyant jazz age of Manhattan. The composition of the setback massing (consisting of a large base, enclosing slabs, tower, and spire, together with the conventional vertical piers and recessed spandrels at the center portion of the tower, but horizontally banded around the corners) truly built vertically into the sky, proving that it was, with 1046 ft, the tallest building in the world for a short time. The Chrysler Building is surely an advertisement and represents one of the most powerful skyscraper images in America.

In contrast, the 102-story Empire State Building by Shreve, Lamb, and Harmon (1931), uses purer forms. Soaring verticality, as achieved by the setback idiom and the continuous piers with recessed Art Deco aluminum spandrels, is clearly dominant. The upward thrust culminates in the spectacular top as a transition into the sky, celebrating the achievement of height as the tallest building in the world for many years. The building, however, does not externally express the complexity and effort of the organism in making this incredible height of 1250 ft possible, a feature that the Modern Movement is very much concerned with.

Toward the end of the 1920s, the more austere, linear geometric patterns of the modernists became predominant. They expressed function, efficiency, and economy as ornament to glorify materialism and business, or the building as advertisement, thereby creating the basis for the great streamlined modernist skyscrapers of the early 1930s. Raymond M. Hood was the most inventive and brilliant representative of this period, due to his sense for composition, together with a clear understanding of pragmatic requirements.

Hood was largely responsible for the conceptual design of the 70-story RCA Building of Rockefeller Center in Manhattan (1933), possibly his greatest achievement. Here, he molded the vertical slab masterfully with thin setbacks to transcend its character by resolving the flat mass into a series of floating slender slabs. The stepping is based not only on esthetic but also functional considerations as it occurs at floor levels where elevator shaftways terminate. The verticality of the facade composition is expressed by the uniform arrangement of the piers, where the 27-ft spacing of the load-bearing columns along the long faces is clearly articulated by the wider piers alternating with two narrower ones. The upward thrust is somehow softened by the horizontality of the cast aluminum spandrel panels, which are nearly flush with the piers.

All the early tall skyscrapers were constructed with steel skeletons. Reinforced concrete was generally used only for lower buildings and was treated merely as a substitute for steel. In 1903, the world's first reinforced concrete skyscraper was completed, the daring 16-story, 210-ft-high Ingalls Building by Elzner & Anderson in Cincinnati. At the same time, Auguste Perret and engineer François Hennebique were the first to employ the reinforced concrete skeleton concept in high-rise construction in Europe. In the eight-story Franklin Apartment Building in Paris (1904), Perret exposed the concrete skeleton to unify structure and architecture. But it was not until the 1950s that reinforced concrete, as a material and structure, established its own identity in high-rise construction. With the Johnson Wax Tower (1950) at Racine, Wisconsin, Frank Lloyd Wright became one of the first designers to break away from the traditional skeleton concept in high-rise construction. Although the 15-story building cannot be considered advanced in terms of height, Wright innovatively integrated the nature of concrete into the building design. He used the tree concept, in his urge toward the organic, by letting the mushroom-type floor slabs cantilever from the central core, which is rooted deep in the ground. With the 16-story Price Tower (1956) in Bartlesville, Oklahoma, Wright freely used the plastic quality of concrete and helped to even further identify the potential of the material. Finally, in 1962 the 60-story Marina Towers in Chicago by Bertrand Goldberg reached 588 ft. These may be considered the first true reinforced-concrete skyscrapers expressing the character of the material. The tallest concrete building in the United States is currently the 71-story 311 South Wacker Building in Chicago (1990) with 969 ft, designed by Kohn Pedersen Fox Associates.

The high-rise buildings of the modern architecture period, after World War II, in a way continue where the Chicago frame left off. They have generally been pure and simple shapes with flat tops, reaching a high point with the 1368-ft-high World Trade Center Towers in New York (1972) and the 1454-ft high Sears Tower in Chicago, which has been the world's tallest building until the mid-1990s.

In contrast to the high-rise buildings of the modern period, the ones of the postmodern era, starting in the late 1970s, seem to have complete freedom of formgiving, using the early New York skyscrapers as a source of inspiration. They often represent compound hybrid forms where the mass of the building is broken up vertically and horizontally to reduce the scale of the building and to project a desired image. The tops of these buildings usually have spectacular crowns to establish a distinctive silhouette and form a skyline. There is no formal limit to these articulated caps or peaks, as demonstrated in Fig. 10.1. They range from sloped, gabled, stepped, and arched tops to pyramids, domes, cylinders, spirals, assembly of turrets, and any combination of shapes.

Figure 10.1 High-rise building crowns.

Today, a new generation of skyscrapers is rising in Asia. The 1476-ft high twin Petronas Towers in the Malaysian capital of Kuala Lumpur will be 22 feet taller than Chicago's Sears Tower when finished in 1996; Cesar Pelli is the architect and Thornton-Tomasetti are the structural engineers. The tallest reinforced-concrete building structure in the world is currently the 1228-ft-high, 78-story Central Plaza office tower with a triangular-shaped plan in Hong Kong (1992); architects were Ng Chun Man and Associates and consulting engineers Ove Arup Partners.

10.1 GENERAL INTRODUCTION TO HIGH-RISE STRUCTURE SYSTEMS

A building structure can be visualized as consisting of horizontal planes or floor framing and the supporting vertical planes of walls and/or frames. The horizontal planes tie the vertical planes together to achieve a box effect and a certain degree of compactness. It is obvious that a slender, tall tower building must be a compact, three-dimensional closed structure where the entire system acts as a unit. The tubular, core-interactive, and staggered truss buildings are typical examples of three-dimensional structures. On the other hand, a massive building block only needs some stiff, stabilizing elements that give lateral support to the rest of the building. In this sense, the building structure represents an open system where separate vertical planar structure systems, such as solid walls, rigid frames, and braced frames, are located at various places and form stand-alone systems that provide lateral stability.

Every building consists of the load-bearing structure and the nonload-bearing portion. The main load-bearing structure, in turn, is subdivided into the *gravity structure,* which carries only the gravity loads, and the *lateral force-resisting structure,* which supports gravity forces but also must provide stability to the building. The condition where the lateral bracing only resists horizontal forces but does not carry gravity loads, with the exception of its own weight, is considered a *secondary structure* . Failure of secondary members is not as critical as that for main members, where an immediate collapse of a building portion may occur depending on the redundancy of the structure. The nonload-bearing structural building elements include wind bracing as well as the membranes and skins, that is, the curtains, ceilings, and partitions that cover the structure and subdivide space.

The lateral force-resisting structure in a building tower may be concentrated entirely in the central core, for instance when an optimal view and thus a light perimeter structure is required. Conversely, rather then hiding the lateral force-resisting structure in the interior, it may be exposed and form the perimeter structure, as for tubes.

The structure represents an assembly system that consists of components and their linkages. The basic elements are lines (columns, beams), grids (floor framework, frames), surfaces (slabs, walls, plates), spatial units (cells, tubes), and any combination of these. The interaction or degree of continuity between these elements depends on the type of linkage (hinged, semirigid, or rigid). These basic components can be combined in an endless variety to form a building.

Before discussing fundamental concepts of structure behavior, typical structure systems are introduced, but purely from a geometrical point of view. Although buildings are three dimensional, their support structures can often be treated from a behavioral point of view as an assembly of two-dimensional vertical planar elements in each major direction of the building. In other words, structures can usually be subdivided into a few simpler assemblies, since structural elements are rarely placed randomly in plan. The most common high-rise structure systems are identified in Fig. 10.2. They are shown simply as planar, two-dimensional structures, although they may act in combination with each other and, in context of the building, may form spatial

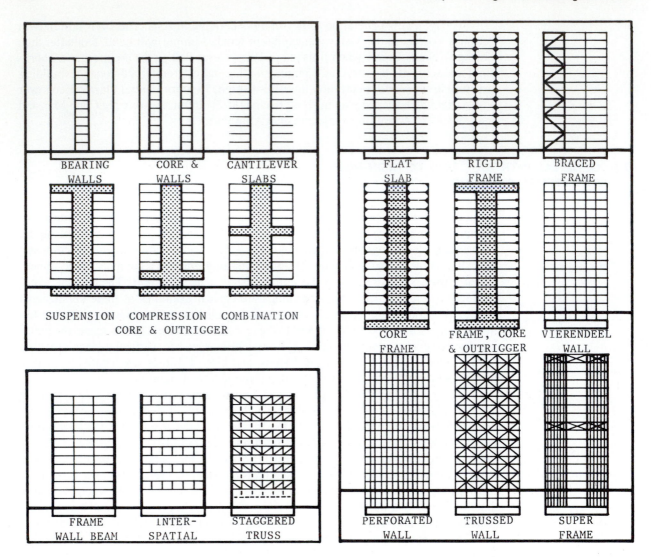

Figure 10.2 High-rise structure systems.

structures. They range from pure structure systems, such as skeleton and wall construction, and systems requiring transfer structures to composite systems and megastructures. As the buildings increase in height, different structure systems are needed for reasons of efficiency (i.e., least weight). The following classification of the various systems is roughly in accordance with these efficiency considerations, as discussed in the next section.

- *Two-dimensional structures*

    Bearing wall structures: combinations of single walls and connected walls, cross walls, long walls, two-way walls, stacked boxes

    Light framing construction (e.g., wood platform framing for three- to four-story buildings)

    Skeleton (frame) structures: rigid frame, braced frame, braced rigid frame, truss, flat slab, Vierendeel wall beam (interspatial, bridge type)

    Connected walls and frames

    Core structures: they may be considered three-dimensional from a structural point of view, but do not necessarily integrate the entire building shape: cantilevered slab, bridge structures (multicore), cores with outriggers on top (suspension), at the bottom, and at intermediate levels

Combinations of these systems

- *Three-dimensional structures*

Staggered wall beams

Cores plus outriggers plus belt trusses: single-, double-, and multiple-outrigger systems

Tubes: Vierendeel tube, deep-spandrel tube, perforated wall/shell tube, trussed tube, tube with belt trusses and head, etc.

Megastructure: superframe, superdiagonals

Hybrid structures

Typical combinations of structure systems are the following:

Walls + core(s)

Frames + core(s) and/or walls

Tube + frame(s) or wall(s)

Tube + core (tube-in-tube)

Tube + tube (bundled tubes)

Other combinations include the following:

Vertical stacking of structures: connected towers of the bridge type

Series of superframes

Internally braced structures

Cellular structures

Stayed structures

Other mixed systems

The selection of a structure system is not a simple undertaking. Among other criteria, it depends on the overall geometry, the vertical profile, height restrictions, the slenderness (that is, the building height-to-width ratio), and the plan configuration (depth-to-width ratio, degree of regularity, etc.) and is a function of strength, stiffness, and possibly ductility demands in response to loading conditions. Selection also depends on building base conditions, site conditions, and construction coordination, including preconstruction and construction time. In order not to give the impression that the preceding pure structure systems are imposed upon the architecture and do not allow any flexibility in the form-giving process, various building cases are presented in Fig. 10.3. They demonstrate some of the endless possible combinations of the structure systems for low- and mid-rise buildings, realizing that the smaller-scale buildings allow more freedom than large-scale towers. It is shown that the structures respond to setbacks, cavities, changing spans, varying story heights, altering bay proportions, sudden changes of stiffness, sloping site conditions, space inclinations, and so on. Most structures are treated as planar, with the exception of the central core-type buildings with diagonal outriggers connected to the corner columns or that are stabilized by a tensile network along the perimeter.

To gain a better understanding of the structure systems in Fig. 10.2, they must be seen within the building space; hence their location must be known. For this reason, solid surface elements have been placed into the uniform beam–column grid of the various plans in Fig. 1.12. They represent the lateral force-resisting structure systems of walls, cores, frames, tubes, or any other combination; they may form either planar or spatial assemblies.

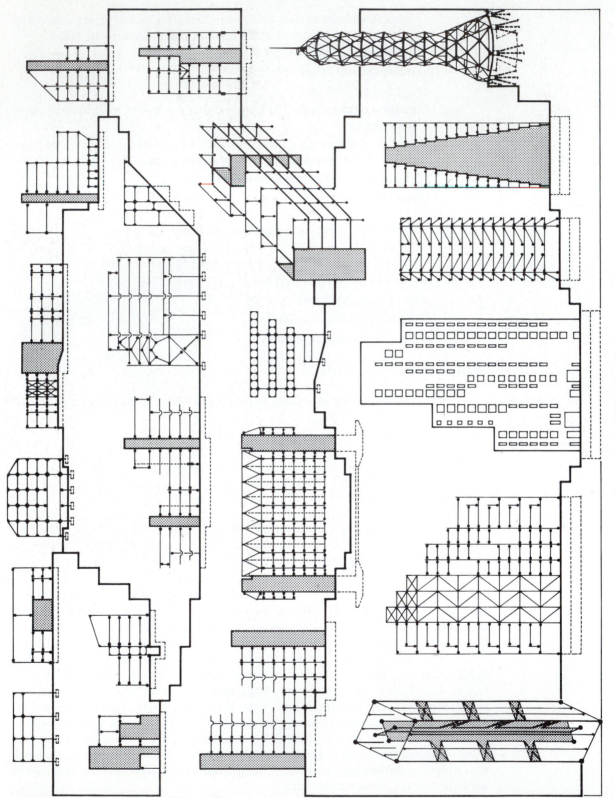

Figure 10.3 High-rise structures.

Considerations of Efficiency

A building must resist the primary loads of gravity and lateral force action. With respect to gravity loads, the weight of the structure increases almost linearly with the number of stories. In this context, it is the weight of the vertical structural elements (columns, walls) that increases roughly linearly with height, since their weight is proportional to the axial stresses that determine the vertical member sizes, while the weight for each floor remains constant. It is also interesting to note that the floor weight, and hence floor cost, constitutes more than one-half of the cost of the entire structure for buildings not higher than the 30- to 40-story range. However, with an increase in building height and slenderness, the importance of lateral force action rises in a much faster nonlinear fashion as compared to the gravity loads and becomes dominant. Therefore, the cantilever action becomes more critical than the column action, or the rotation M/S is more important than the axial action N/A. The section modulus S (or the moment of inertia I), rather than the cross-sectional area A, controls the stress and becomes the determinant of form. Hence, the material needed for the resistance of lateral forces increases as the square of the height, that is, at a drastically accelerating rate.

For typical *medium-rise* structures in the 20- to 30-story range, the vertical load resistance nearly offsets the effect of the lateral forces; only about 10% of the total structure material is needed for lateral force resistance. The late eminent structural engineer Fazlur Khan of SOM has shown that it may be economical to select (for a tall building) a structure system in which the bending stresses, due to lateral force action, do not exceed one-third of the axial gravity stresses so that the effect of the lateral forces can be ignored. At a certain height, however, the lateral sway of a building becomes critical, so that considerations of stiffness, rather than the strength of the structural material, control the design. The degree of stiffness depends on the building shape and the spatial organization of the structure.

It has been a challenge to building designers to minimize the effect of the horizontal forces by developing optimum lateral force-resisting structure systems. The efficiency of a particular system is directly related to the quantity of material used, at least for steel structures. It is measured as the weight per square foot (psf), that is, the weight of the total building structure divided by the total square footage of the gross floor area. Therefore, optimization of a structure for given spatial requirements should yield the maximum strength and stiffness with the least weight. This results in innovative structure systems applicable to certain height ranges. Naturally, it must be kept in mind that not only material costs, but also fabrication costs and erection time, must be considered.

Fazlur Khan argued, in the mid-1960s (and was supported by the development of computer simulations), that the rigid frame that had dominated high-rise building construction was not the only system associated with tall buildings. Due to a better understanding of the mechanics of materials and how materials and members interact, the structure could now be treated as a whole or the building form as a three-dimensional unit. Khan later proposed a range of structure systems for office buildings of ordinary proportions and shapes that are appropriate for certain heights. He showed that the weight of steel structure systems range from about 30 psf for a 100-story building to 6 psf for a 10-story structure (Table 10.1). This effect of scale is known from nature, where animal skeletons become much bulkier with an increase in size, since the weight increases with the cube, while the supporting area only increases with the square. For example, the bones of a mouse make up only approximately 8% of the total mass, in contrast to about 18% for the human body.

Furthermore, Frank Lloyd Wright's experiments, as early as the 1920s, with slender, treelike, concrete cantilever structures that were opposite in nature to the traditional skeleton construction, should not be forgotten. This approach has recently lead

TABLE 10.1
Typical Weight of Historically Important High-Rise Steel Structures

Building	Year	Stories	Height/Width	psf	Structure System
Empire State Building, New York	1931	102	9.3	42.2	Braced rigid frame
John Hancock Center, Chicago	1968	100	7.9	29.7	Trussed tube
World Trade Center, New York	1972	110	6.9	37.0	Framed tube
Sears Tower, Chicago	1974	109	6.4	33.0	Bundled tubes
Chase Manhattan, New York	1963	60	7.3	55.2	Braced rigid frame
U.S. Steel Building, Pittsburgh	1971	64	6.3	30.0	Shear walls + outriggers
I.D.S. Center, Minneapolis	1971	57	6.1	17.9	+ belt trusses
Boston Company Building, Boston	1970	41	4.1	21.0	K-braced tube
Alcoa Building, San Francisco	1969	26	4.0	26.0	Latticed tube
Housing, Brockton, Massachusetts	1971	10	5.1	6.3	

to buildings more than 50 stories high, where large concrete cores alone provide lateral force resistance. Many of the concrete structures of the 1960s exposed the cores in order to articulate the strength of the three-dimensional support structure (often of the bridge type) and to express the servicing as clearly separated from the served spaces. In this case, the design philosophy is very different from most of the steel skeleton structures of the same period. The efficiency of a concrete structure is evaluated (to a great extent) in terms of the process of construction, in addition to the quantities of materials used (roughly between 0.5 to 1.0 ft^3/ft^2 concrete and reinforcing steel of 2 to 4 psf), in contrast to steel, which considers only the quantity of material used.

Many of the tall buildings today no longer represent the pure shapes of the 1960s and early 1970s. Compound, hybrid building shapes have become fashionable. With the aid of computers, in response to complex geometric shapes, a wealth of new structure layouts has been made possible, which basically consist of combinations of the fundamental structure systems. In addition, wind tunnel studies have become very accurate in evaluating the response of a building to wind flow. Despite this increased sophistication of structural analysis and design, however, the fundamental fact should not be overlooked that the material and layout of the structure should not provide the stiffness solely by themselves; the *form* of the structure must also be searched for with the help of computers so as to efficiently reduce the use of materials.

The new structure systems reflect optimum solutions for given complex building shapes, which include composite structures and the mixed construction of concrete and steel. Large buildings are broken down into smaller zones; megaframes give support to the supertall buildings of the future. It is hoped that the architect will use the potential richness of structure to express its power and its purpose as support, rather than just letting the engineer plug the structure into a form that was derived independently of its nature.

10.2   FORCE FLOW IN HIGH-RISE BUILDING STRUCTURES

The horizontal and vertical structural building planes must disperse the external and internal forces to the ground. Some basic concepts of vertical and lateral load transmission for various structure systems are discussed in a simplified fashion in Figs. 10.4 and 4.12.

Visualize a gravity load acting on the slab and transferred by the floor framing in bending (Fig. 10.4, top left) to one of the vertical structure building planes, which may transmit the load axially directly to the ground. The type and pattern of force flow depend on the arrangement of the vertical structural planes as indicated at the top of Fig. 10.4 for two-dimensional structures. The columns may be vertical or inclined, continuous or staggered; they may be evenly distributed or concentrated in the center or along the periphery, possibly to form cores. The path of the force flow may be continuous along the columns or may be suddenly interrupted and transferred horizontally to another vertical line. The transmission of the loads may be short and direct or long and indirect with a detour, such as for a suspension building. From an efficiency point of view, the vertical loads should be carried along the shortest path possible to the foundations.

When columns are inclined, gravity will cause lateral thrust, which increases as the column moves away from the vertical supporting condition. The cases at the bottom of Fig. 10.4 indicate that the horizontal floor beams at the top act as ties in tension when the columns lean outward but as struts in compression at the bottom. For a symmetrical structure, the thrust due to the dead load will self-balance, but the horizontal forces due to asymmetrical live loads must still be resisted, as for an asymmetrical building, where the weight also causes thrust. Hence not only wind and earthquake, besides the centrifugal outward effect of cars upon curved bridges, but also gravity together with the respective geometry may cause lateral force action on a building.

Optimum, free ground-level space with a minimum of columns is often required for high-rise buildings. Examples range from grand entrances and wide lobby spaces, loading docks, and parking aisles to open public plazas. For these conditions the upper building mass must be linked to the ground by using a different structure system. The geometrical pattern of the building structure cannot extend to the foundation walls; it becomes discontinuous and is replaced by another structure system. For preliminary design purposes, the upper and lower structure portions can be analyzed separately. Various transition types are shown in the central part of Fig. 10.4. They range from suspension buildings and lifting an entire building up on frames or stilts, to changing the column spacing to a wider pattern by using transfer systems within the framed tube grid. The latter may be accomplished, for example, by increasing the spandrel beam sizes toward the main columns or by changing the column sizes in a treelike fashion, thereby generating a natural, archlike, gradual transition of the loads. The load transition can also be achieved by heavy transfer systems, such as girders, trusses, wall beams, arches (direct or indirect action), or V- and Y-shaped tree columns (two-, and three-forked columns) to collect the columns above. These V-columns are effective in resisting wind and earthquake forces, but unfortunately respond to asymmetrical gravity loading with horizontal thrust, as previously discussed. The reader may want to study the behavior of the various inclined column cases in Fig. 10.4.

The horizontal forces are transmitted along the floor and roof planes, which act as deep, flat beams spanning between the vertical lateral force-resisting structures, as described in Fig. 4.12. Once the lateral forces are distributed to the resisting vertical structure planes, these systems must act as vertical cantilevers to carry the forces down to the ground. The various two- and three-dimensional structure systems in Fig. 4.13 demonstrate how the overturning moment due to wind action is resisted at the base by different cantilever types, consisting of solid walls or skeleton construction. The skeletons may act either as *flexural systems* when they are rigid frames (where the members react in bending) or they may be predominately *axial systems*, as for braced frames where the forces are effectively resisted directly in tension and compression.

Conflicts of force flow are generated when plan forms or structure systems change, possibly at locations of setbacks often found at the base, top, or intermediate levels of buildings. For example, when a triangular plan changes to an L-shaped base or when a perimeter structure such as a tube cannot be continued to the base, an extensive horizontal *transfer structure* is necessary not only to redirect the vertical forces, but also to act as a diaphragm to transfer the horizontal forces (see also Fig. 10.6).

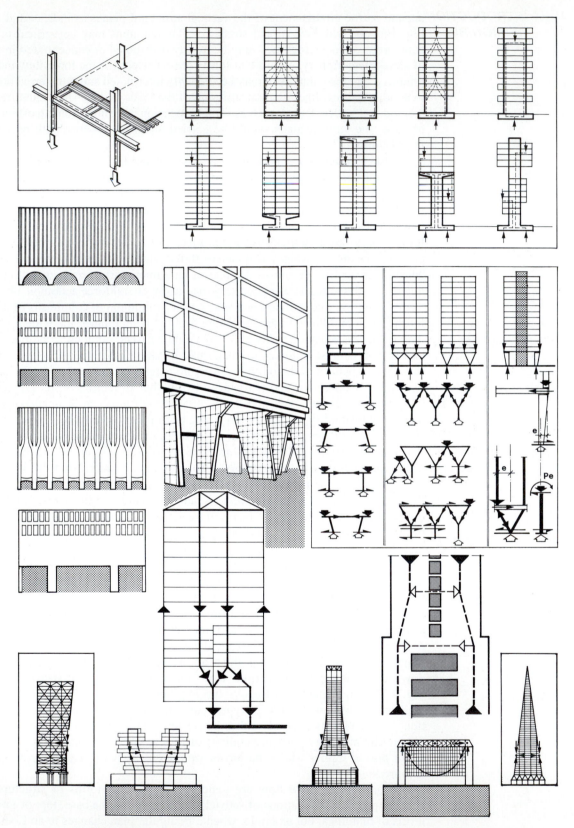

Figure 10.4 Vertical force flow (Reproduced with permission from *The Vertical Building Structure,* Wolfgang Schueller, copyright © 1990 by Van Nostrand Reinhold.)

10.3 INTRODUCTION TO BASIC BEHAVIOR OF HIGH-RISE BUILDING STRUCTURES

Strength and stiffness are the primary characteristics activated when the building structure responds to the load action described in Fig. 10.5; in areas of strong seismic activity, ductility also becomes an important criterion. Considerations of stability and other load-related effects have already been briefly introduced elsewhere. In structural analysis, the real structure is replaced with an idealized model. Its response to loading is based on analytical theory derived from material properties and member behavior.

First, the reaction of the structure to lateral loading is conceptually investigated. It is shown in Fig. 4.12 that, with respect to horizontal force action, the floor structure acts as a rigid diaphragm supported on elastic vertical elements. These lateral force-resisting elements in Fig. 10.5 are interior cores, core plus walls (Fig. 10.5j), and a perimeter tube (Fig. 10.5m). The cores are of an open or closed type and may be located at the centroid of the building, or their location may be eccentric. The building

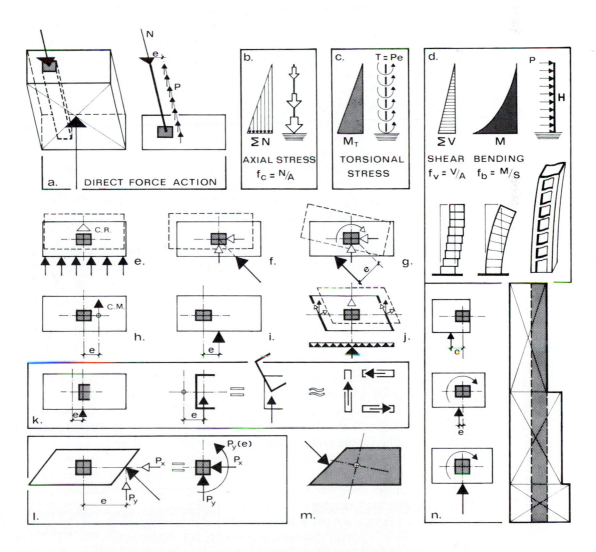

Figure 10.5   The building response to load action. (Reproduced with permission from *The Vertical Building Structure*. Wolfgang Schueller, copyright ©1990 by Van Nostrand Reinhold.)

form determines where the resultant wind pressure acts, and the arrangement of the building masses defines the location of the resultant seismic force, which the core must resist. Since lateral stiffness requirements reduce in the upper portion of the building, the core walls can be dropped off or stepped back at the termination of the low- and mid-rise elevator banks.

The tall buildings in Fig. 4.16 respond to lateral forces primarily as *flexural cantilevers* if the resisting structure consists of shear walls or braced frames. The behavior of these systems is controlled by rotation rather than shear; they have a high shear stiffness provided by the solid wall material or axial capacity of the diagonals, so the shear deformations can be neglected. Tall buildings act as *shear cantilevers* when the resisting elements are rigid frames since the shear can only be resisted by the girders and columns in bending. In this context, the effect of rotation (i.e., axial shortening and lengthening of columns) is secondary and may be ignored for preliminary design purposes. The combined action of different structure systems, such as rigid frames together with a braced core (depending on the relative stiffness of each system) may have the appearance of a flat S-curve with a shear-type frame building sitting on top of a flexural cantilever-type structure.

In the preceding discussion, it has been assumed that the structure systems were for tall buildings and were of the same height. It is apparent that when the shear wall or braced frame is no longer shallow and slender, as for the extreme case of a horizontal panel in a low-rise building, such systems do behave like shear cantilevers and not flexural cantilevers.

As known from basic mechanics of materials (Section 2.6), flexural resistance to lateral loads is expressed by the axial bending action M/S and an average shear action of V/A for certain conditions of symmetry (Fig. 10.5d). For the given uniform lateral loading case, the shear increases linearly toward the base ($V \propto H$), while the moment grows much faster following, a second-degree parabola ($M \propto H^2$). Since this special condition of simple bending due to symmetry is often not present because of asymmetry of the resisting structure and/or the eccentric action of the resultant lateral force, some general concepts of structural behavior of bending members are briefly reviewed first. Furthermore, thin-walled beam behavior, as for tubular structures, is ignored in this context.

In general, to determine the stresses due to pure bending of an unsymmetrical section with no *axes of symmetry* requires complex calculations. First, the *principal axes,* which are always mutually perpendicular and about which the moments of inertia are maximum and minimum, respectively, must be located, then the direction of the *neutral axis* has to be found. All these axes together with the *centroidal axes* pass through the centroid of the cross section. For this general condition, the simple bending formula $f_b = Mc/I = M/S,$ which applies only to symmetrical bending, cannot be used!

In addition, the loads must act through the shear center or center of twist, which is located at the intersection of the *shear axes,* in order to not generate torsion in addition to unsymmetrical bending. Therefore, this shear center must be located; it does not coincide with the centroid of the cross section. We may conclude that, when the load is applied at the centroid, the member will twist as it bends. Lack of symmetry results in eccentric loads, unsymmetrical bending, and torsion! Fortunately, cross sections usually have a certain degree of symmetry (i.e., one axis of symmetry), which simplifies understanding of the behavior and the stress calculations, remembering that an axis of symmetry is always a principal axis.

Forces are not always transferred in a straightforward fashion as by the pure structure systems discussed previously. For example, the 500-ft-high concrete structure of the Metro Dade Center in Miami, Florida (1983), designed by LeMessurier, is a hybrid structure that consists of frame construction with two shear walls at each end and huge 60-ft-deep spandrel girders near the base, as shown in Fig. 10.6. Here, the gravity loads are carried by the columns and end walls. The exterior columns along the broad faces, in turn, transfer the loads to the spandrel girders that bridge the space between the end walls.

   The wind against the broad building face is transferred by the floor slabs to the two coupled shear walls at the building ends, which act as cantilevers above the spandrel girders. Below this level the walls act similarly to end cores or open tubes, where the tension due to rotation is suppressed by the gravity loads. In other words, gravity loads are used as a stabilizing agent.

   The lateral force action on the short face is resisted in the upper building portion by the combined action of the cantilevering end shear walls and the exterior rigid frames and at the base by the huge portal frames, as shown in Fig. 10.6.

10.4 BRIEF INVESTIGATION OF COMMON HIGH-RISE STRUCTURES

The most common structure types, identified conceptually in Fig. 10.2, are briefly discussed in this section.

Bearing Wall Structures: The bearing wall was the primary support structure for high-rise buildings before the steel skeleton and the curtain wall were introduced in the 1880s in Chicago. The traditional tall masonry buildings were massive gravity structures where the walls were perceived to act independently; their action was not seen as part of the entire three-dimensional building form. It was not until after World War II that engineered thin-walled masonry construction was introduced in Europe. Bearing wall construction is used mostly for building types that require frequent subdivision of space, such as for residential application. The bearing walls may either be closely spaced, e.g., 12 to 18 ft and directly define the rooms, or they may be spaced, for instance, 30 ft apart and use long-span floor systems that support the partition walls subdividing the space. Bearing wall buildings of 15 stories or more in brick, concrete block, precast large-panel concrete, or cast-in-place reinforced concrete are common-

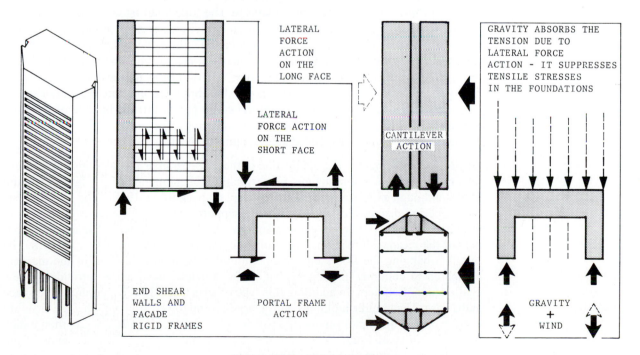

Figure 10.6 Hybrid structure.

place today; they have been built up to the 26-story range. The bearing wall principle is adaptable to a variety of building forms and layouts. Plan forms range from slab-type buildings and towers of various shapes to any combination. The wall arrangements can take many different forms, such as the cross-wall, long-wall, double cross-wall, tubular, cellular, and radial systems. An endless variety of hybrid systems is possible by combining these cases. The walls may be continuous in nature and in line with each other, or they may be staggered; they may intersect or they may function as separate elements to form individual wall columns. Bearing walls usually carry both gravity and lateral forces.

Core Structures: Many multicore buildings with their exposed service shafts have been influenced by the thinking of the Metabolists of the 1960s, who clearly separated the vertical circulation along cores and the served spaces. According to Kenzo Tange, "buildings grow like organisms in a metabolic way." Their urban clusters consisted of vertical service towers linked by multilevel bridges, which, in turn, contained the cellular subdivisions. Other examples of urban-type megastructures can be found in hospital planning of the 1970s. The linear bearing wall structure works well for residential buildings where functions are fixed and energy supply can be easily distributed vertically. In contrast, office and commercial buildings require maximum flexibility in layout, calling for large open spaces subdivided by movable partitions. Here, the vertical circulation and the distribution of other services must be gathered and contained in shafts and then channeled horizontally at every floor level. These vertical cores may also act as lateral stabilizers for the building. There is an unlimited variety of possibilities related to the shape, number, arrangement, and location of cores. They range from single-core structures (core with cantilevered floor framing, core with massive base cantilever, core with large top and/or intermediate cantilevers, core with other structure systems) to multiple core structures.

Bridge Structures: The idea of the bridge structure was vitalized by the designers of the 1960s, who were concerned with large-scale urban architecture and wanted to separate the ground and services from social activities. These megastructures or urban structures were proposed by the Metabolists in Japan, Archigram in England, and designers such as Yona Friedman in France and Eckhard Schulze-Fielitz in Germany, who used horizontal space-frame structures. The long span from vertical support to vertical support can be achieved through an endless number of possibilities, as has been expressed in architecture. Closely related to the bridge concept is the core structure, where many of the buildings formed megaframes to support, in bridgelike fashion, secondary building packages. Similarly, several of the suspension structures are based on the bridge principle, as are supertall buildings that use megaframes or super-diagonals to gather the building weight to certain points for the purpose of stability. Space can be bridged by using one of the following structural concepts: Vierendeel trusses, trusses, arches, suspended arches, and wall beams.

Suspension Buildings: The application of the suspension principle to high-rise buildings rather than to roof structures is essentially a phenomenon of the late 1950s and 1960s, although experiments with the concept go back to the 1920s. The structuralists of this period discovered a wealth of new support structure systems in the search to minimize the material and to express antigravity, that is, lightness of space and openness of the facade, allowing no visual obstruction with heavy structural members. The fact that hanging the floors on cables required only about one-sixth of the material compared to columns in compression, as in skeleton construction, provided a new challenge to designers. In addition, this type of structure allowed a column-free space at the base. The treelike buildings with a large central tower, from which giant arms are cantilevered at the top to support the tensile columns at their ends, are common today. The floors or spatial units (e.g., capsules, entire building blocks) are suspended from the support sturctures by using either vertical or diagonal tensile members. Typical

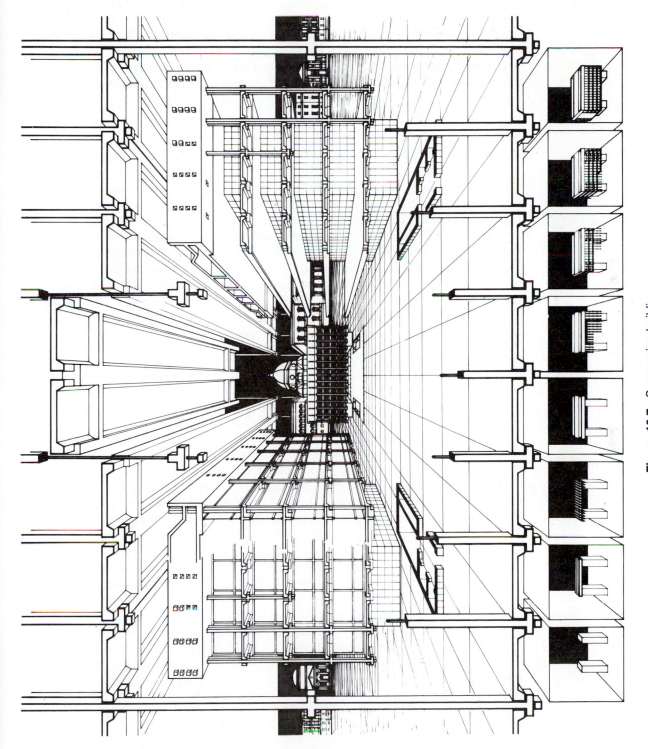

Figure 10.7 Suspension building.

suspension systems use the rigid core principle (single or multiple cores with outriggers or beams, megaframes, treelike frames, etc.), the guyed mast principle, and the tensegrity or space-net principle.

For example, the architect Alberto Galardi uses the suspension principle for the Olivetti building in Florence, Italy (1973, Fig. 10.7) to form a bridge and thus allow open space at the base. The exterior, prestressed concrete hangars supporting the four floors are hung from the roof structure, which, in turn, is carried by two towerlike concrete cores.

Staggered Wall Beam Structures: In this innovative structure system, developed in the mid-1960s by a team of architects and engineers at M.I.T., story-high wall beams span the full width of the building on alternate floors of a given bay and are supported by columns along the exterior walls; there are no interior columns. The wall beams are usually steel trusses, but can also be pierced reinforced-concrete members. The steel trusses are concealed within the room walls. In the interstitial system, wall beams are used at every other floor to allow for uninterrupted free flow in the floor space between, while in the staggered wall beam system, the wall beams are used at every floor level, but arranged in a staggered fashion between adjacent floors. The arrangement of the story-high members depends on the layout of the functional units. We can visualize apartment units to be contained between the wall beams and to be vertically stacked to resemble masonry bond patterns. As the unit sizes change, the spacing of the wall beams may be adjusted or additional openings may be provided. The most common system of organization is the running bond or checkerboard pattern.

Skeleton Structures: When William Le Baron Jenney, in the 10-story Home Life Insurance Building in Chicago (1885), used iron framing for the first time as the sole support structure carrying the masonry facade walls, the all-skeleton construction was born. The tradition of the Chicago Frame was revived after World War II when the skeleton again became a central theme of the modern movement in its search for merging technology and architecture. Famous landmarks became SOM's Lever House in New York (1952) and Mies van der Rohe's two 860–880 Lake Shore Drive Apartment Buildings (1951, Fig. 10.8). These landmarks have been most influential to the subsequent generation of designers; they symbolized at the time, with their simplicity of expression, the new spirit of structure and glass. Although the pure, boxy shapes of the 1960s are closely associated with skeleton construction, as derived from Miesian minimalism, other high-rise building skeletal forms, based on different design philosophies, have been built, for example, the unusual hammer-shaped Velasca Tower in Milan, Italy (1957). Today, there seems to be no limit to the variety of building shapes; the skeleton as an organizing element for this new generation of hybrid forms has been extensively experimented with. Odd-shaped towers, possibly with tapered frames, reflect the change of irregular plan forms with height; skeleton buildings may be stepped at various floor levels where large setback terraces may be fully landscaped. In the Lloyd's of London Building (1986) by Richard Rogers, the braced perimeter concrete frame is surrounded by six satellite service towers, while the internal perimeter columns carry the elaborate central atrium structure. Kisho Kurokawa articulated the regularity of the three-dimensional grid and its adaptability to growth and change by constructing the Takara Beautillion for Osaka's Expo '70 from single six-pointed spatial cross units. Facade framing ranges from long-span, deep girder systems and Vierendeel frames to perforated walls. The open, airy skeleton is contrasted with the framed tubular wall. Frames may be organized as continuous rigid frames, hinged frames, and any combination. The behavior of moment-resisting frames (i.e., rigid frames), which resist both gravity and lateral forces, is investigated briefly in Fig. 5.23.

Flat Slab Building Structures: Flat slab buildings, developed during the mid-1940s in New York, consist of horizontal planar concrete slabs directly supported on columns, thus eliminating the need for floor framing. This results in a minimum story

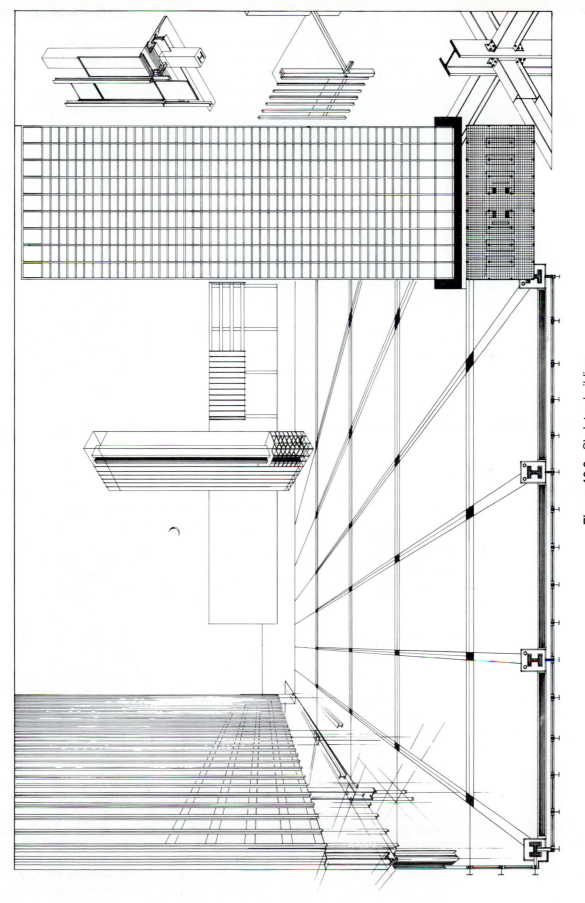

Figure 10.8 Skeleton building.

height, an obvious economic benefit that is especially advantageous for apartment buildings. Drop panels and/or column capitals are frequently used because of high shear concentrations around the columns. Slabs without drop panels are commonly called *flat plates*. This system is adaptable to an irregular support layout. From a behavioral point of view, flat slabs are highly complex structures. The intricacy of the force flow along an isotropic plate, in response to uniform gravity action, is reflected by the principal moment contours (Figs. 1.4, 2.1). Here, the main moments around the column support are negative and have circular and radial directions, while the positive field moments basically connect the columns linearly. The patterns remind one of organic structures, such as the branching grids of leaves, the delicate network of insect wings, radial spider webs, and the contour lines of conical tents, realizing a similar relationship between cable response and loading as well as the corresponding moment diagram. Pier Luigi Nervi, for the Gatti Wool Factory (1953) in Rome, Italy, actually followed the principal bending moments with the layout of the floor ribs. Centuries earlier, however, the late medieval master builders had already intuitively developed patterns for ribbed vaulting predicting these tensile trajectories; the fan vaults of the Tudor period in England are a convincing example.

Braced Frame Structures: The concept of resisting lateral forces through bracing is the most common construction method; it is applied to all types of buildings, ranging from low-rise structures to skyscrapers. At a certain height, depending on the building proportions and the density of frame layout, the rigid frame structure becomes too *mushy* and may be uneconomical, so it must be stiffened by, for example, steel bracing or concrete shear walls. The basic bracing types for frames are single diagonal bracing, cross-bracing, K-bracing, lattice bracing, eccentric bracing (single diagonal or rhombic pattern), knee bracing, and combinations. When the diagonal members are kinked for the placement of openings, they must be stabilized by additional members.

The architects Burnham and Root developed the concept of vertical shear wall (or the vertical truss principle) in the 20-story Masonic Temple Building (1892) in Chicago. The spirit of a braced frame structure is investigated conceptually in Fig. 10.9. In braced frames, the frames carry the gravity loads, whereas the bracing resists the lateral loads. In contrast, in braced rigid frames the frame not only carries the gravity loads but also, together with the bracing, resists lateral loads.

Trussed Frame Structures: Trusses not only constitute support structures hidden within the building, but may also be revealed on the exterior. One of the earliest examples of braced skeleton buildings is the Chocolate Factory at Noisiel-sur-Marne near Paris by Jules Saulnier (1872), where the walls consist of exposed trussed iron framework. This method of construction was surely inspired by trussed bridge construction, as well as by the timber framing, that first occurred in Europe during the Middle Ages. Here, each region developed its own distinct pattern of braced wall heavy timber framing, with space between the timber members infilled with masonry or other material mixtures. An early example of high-rise braced frame construction is Gustave Eiffel's interior-braced iron skeleton for the 151-ft-high Statue of Liberty (1886) in New York. He also designed the braced skeleton wrought iron structure of the Eiffel Tower (1889) in Paris, at almost 1000 ft the tallest building of its time; this first modern tower became a symbol for a new era with its daring lightness of construction. In contrast to braced frames, trussed frames are bearing wall structures that carry both gravity and lateral loads. In other words, the diagonal members also carry gravity loads.

During the early part of this century, the elaborate tops of skyscrapers required complex bracing systems. For example, a high spire structure with a needlelike termination was designed to surmount the dome of the Chrysler Building (1930, Fig. 10.1). Currently, postmodern building tops with their spires and pinacles revive ornamentation and the architectural styles of the past. Intricate braced frames are required for the various roof shapes, such as pyramids, domes, spirals and gabled, stepped, folded, or

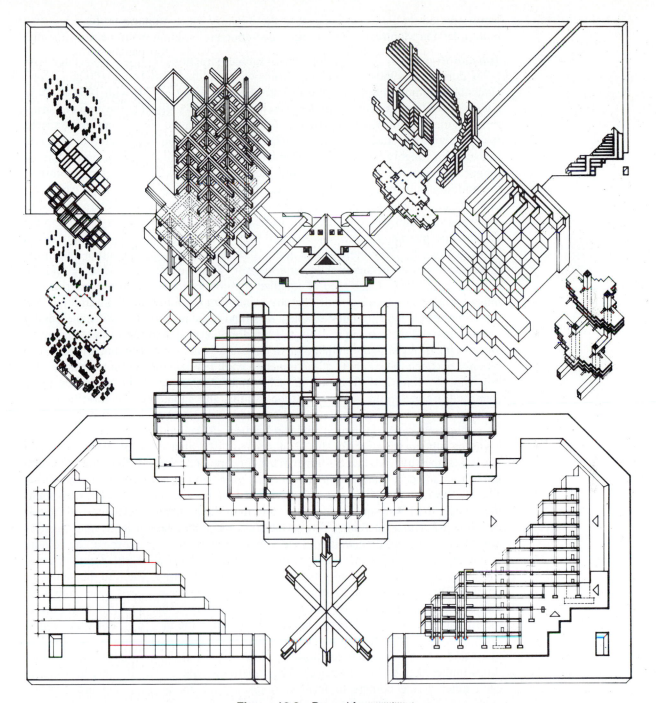

Figure 10.9 Braced frame structure.

arched forms. These structural complexities are not only found in the roof spires, but also in lobby entrances and atria of high-rise buildings.

Shear Walls with Outriggers: At a certain height the braced frame will become uneconomical, particularly when the shear core is too slender to resist excessive drift. Here, the efficiency of the building structure can be greatly improved by using story-high or deeper outrigger arms that cantilever from the core at one or several levels and tie the perimeter structure to the core by either connecting directly to individual columns or to a belt truss. This interaction activates the participation of the perimeter columns as struts and ties, thus redistributing the stresses and eccentric loading. Pier Luigi

Nervi applied the outrigger concept to the 47-story Place Victoria (1964, Fig. 10.3, bottom left) in Montreal, the first reinforced-concrete building to utilize the principle.

Tubular Structures: The development of tubular structures is closely associated with SOM during the 1960s; the 38-story Brunswick Building (1964), the 100-story John Hancock Building (1968), and the 110-story Sears Tower (1974) with 1454 ft, all in Chicago, are famous early examples. Much credit must be given to the eminent structural engineer Fazlur Khan, a partner of SOM, who invented the concept in the search for optimizing structures with the use of computers. As the building increases in height in excess of roughly 60 stories, the slender interior core and the planar frames are no longer sufficient to effectively resist the lateral forces. Now the perimeter structure of the building must be activated to provide this task by behaving as a huge cantilever tube. Here, the outer shell may act as a three-dimensional hollow structure, that is, as a closed box beam where the exterior walls are monolithically connected around the corners and internally braced by the rigid horizontal floor diaphragms. The concept evolved from the three-dimensional action of structure as found in nature and in the monocoque design of automobiles and aircraft. The dense column spacing and the deep spandrell beams also tend to equalize the gravity loads on all the exterior columns, similar to a bearing wall, thereby minimizing column sizes. In addition, the closed perimeter tube provides excellent torsional resistance. In the 1960s, the tubular concept revived the bearing wall for tall building construction, but in steel, concrete, and composite construction rather than in masonry. Now, window lights can be placed directly between the columns of the punched wall; hence the need for a separate curtain wall is eliminated. The pure tubular concepts include single perimeter tubes (punched, framed, or trussed walls), tube-in-tube, and bundled tubes. Modified tubes include interior braced tubes, partial tubes, and hybrid tubes.

The well-known structural engineer Leslie E. Robertson of New York developed a unique tubular structure for the 72-story Bank of China Building (1988) in Hong Kong designed by I. M. Pei, as shown in the conceptual drawing of Fig. 10.10, consisting of four adjacent triangular prisms of different heights rising out of the square base. The 1209-ft high tower is a *space-frame braced tube* organized in 13-story truss modules, where the 170-ft square plan at the bottom of the building is divided by diagonals into four triangular quadrants. The space truss resists the lateral loads and transfers almost the entire building weight to the four supercolumns at the corners; the column at the center of the four quadrants is discontinued at the twenty-fifth floor, where it transmits the loads to the top of the tetrahedron, which carries them to the supercolumns. Midway through the 13-story truss modules, transverse trusses wrap around the building to transfer the gravity loads from the internal columns to the supercolumns at the corners; the horizontal trusses are not expressed in the facade. The loading conditions in Hong Kong, in contrast to the United States, are much more severe: the live loads and wind loads are twice those in New York, and the earthquake load is four times higher than in San Francisco. The superdiagonals are not directly attached to each other at the corners to form complex spatial connections, but are, instead, anchored in the massive concrete columns, thereby forcing the concrete to behave as a shear transfer mechanism. The mixed construction of the primary structure consists of the separate steel columns at the corners (to which the diagonals are connected), which are encased and bonded together by the massive concrete columns. The giant diagonal truss members are steel box columns filled with concrete. The open space at the base of the building did not allow the diagonals to continue to the bottom; at the fourth level, a specially reinforced floor diaphragm was required to transfer the lateral shear to steel-plated core walls, which were designed as three-cell shear tubes.

Composite and Mixed Steel–Concrete Buildings: The integral interaction of reinforced concrete and steel can be seen not only in the popular composite metal deck and floor framing system, but also on a much larger scale. It is not the composite action of

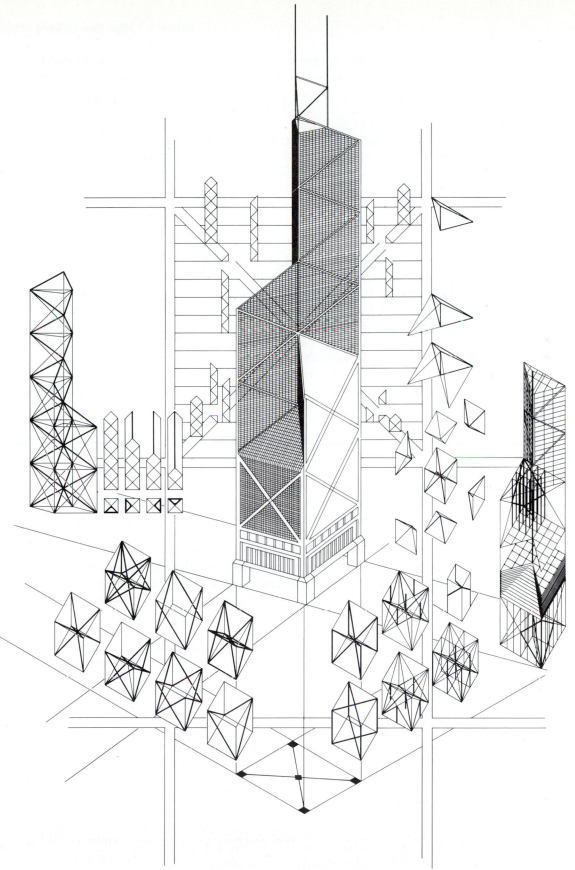

Figure 10.10 Space-frame braced tube.

the structure members—the slabs, beams, and columns—that are of interest here, but rather the combination or interaction of these members that are blended into a single structure system. Typical *composite building types*, which have developed over the last decade or so, are composite framed tubes, composite steel frames, composite panel-braced steel frames, composite interior core-braced systems, composite mega-frames, and hybrid composite structures. Recently, *mixed steel–concrete buildings* have also become popular; the combining of major structure components of concrete, steel, or composite buildings is a relatively new development. For example, it may now be economical to place a steel building on top of a concrete building, or vice versa; alternatively, a central concrete core may be slip-formed to a predetermined height and then the steel frame built around it.

Megastructures: In this context, the term megastructure does not refer to the vision-ary concepts of the 1960s, expressing the comprehensive planning of a community or even an entire city, but solely to the support structure of a building. However, this megastructure is still formulated on the basic concept of a primary structure that sup-ports and services secondary structures or smaller individual building blocks. In the early 1970s, Fazlur Khan of SOM proposed to replace the multicolumn concept by the *four massive corner column supporting superframe* by using supertransfer trusses at every 20 floors or so on the interior and exterior of the building, thereby allowing all gravity loads to flow to the four supercolumns. The principle can be traced to Khan's studies of superframes for multiuse urban skyscrapers, with the John Hancock Center in Chicago representing the forerunner of this idea. Surely, one of the most important first examples for the new breed of megastructures is the 59-story Citicorp Center in New York (1977). In this building the renowned structural engineer William J. LeMes-surier introduced a unique structure and a new way of thinking about structure. This is also reflected by LeMessurier's ingenious support structure, which is not, however, integrated in articulating the building form for Helmut Jahn's Bank of the Southwest proposal (1982, Fig. 10.11) in Houston, an obelisklike, 82-story square tower with chamfered corners. The slender, 1220-ft-high structure tapers from a 165-ft square base to a 135-ft square plan at the top. The entire building is supported by eight super-columns, two on each side, which reduce in size from 10×15 ft at the bottom to 5×5 ft at the top of the building. Interior steel superdiagonals straddling the core cross the plan to connect the massive perimeter concrete columns on the opposite sides. Similar to a Greek cross configuration, they gather and then transfer the gravity loads at the base of each module, as well as act as the web with respect to wind shear. The chevron configuration of the primary interior bracing is organized in nine-story modules.

Hybrid Structures: The current trend away from pure building shapes toward irreg-ular complex ones, that is, hybrid solutions, as expressed in geometry, material, struc-ture layout, and building use, is apparent. In the search for more efficient structural solutions, especially for very tall buildings, a new generation of systems has developed with the aid of computers, which, in turn, have an exciting potential for architectural expression. These new structures do not necessarily follow the traditional classifica-tion of the previous sections. Now, the selection of a structure system as based on the primary variables of material and the type and location of structure is no longer a sim-ple choice between a limited number of possibilities. Mathematical modeling with computers has made mixed construction possible, which may vary with building height, thus allowing nearly endless possibilities that could not have been imagined only a few years ago. The computer simulates the effectiveness of a support system so that the structure layout can be optimized and nonessential members can be eliminated to obtain the stiffest structure with a minimum amount of material. Naturally, other design considerations besides structure will have to be included, but the design con-cepts can be tested quickly and efficiently by the computer.

The New Generation of Tall Building Structures

The new generation of high-rise buildings is not necessarily solely based on economic, functional, or stylistic considerations, that is, traditional design attitudes. Tall buildings may not just represent the power of cooperations, but express the life of vertical cities. In other words, single-use buildings may be replaced by multiuse ones where the urban context at the base extends into the building. This design position, in turn, may be articulated by stacking various building shapes on top of each other or connecting buildings horizontally with each other or by placing buildings within buildings.

Figure 10.11   Megastructure.

The new building types may have an irregular profile and a complex spatial organization requiring a complex support structure.

The mixed-use concept may raise concerns for environmental issues and quality of life; it may introduce the *green building*. Here energy conservation, natural light, and natural ventilation become primary issues. Vertical interior light shafts or atria, together with horizontal courtyards linked to the exterior walls, act as ducts for providing ventilation and natural light to the interior spaces. Exterior walls have adjustable openings, possibly recessed for purposes of shading. They are not seen anymore as sealed surfaces but permeable membranes or filters.

Communal interaction vertically as a natural extension of the urban fabric at the base, in the spirit of the vertical city, together with other design positions that are often very personal and even irrational in nature all support the tendency of breaking up the tall building mass internally (e.g., multistory atria, vertical landscaping) and externally by connecting different buildings vertically and/or horizontally. An example is the 775-ft First Bank Place in Minneapolis by James Ingo Freed (1992, Fig. 10.13, center right).

The 574-ft-high, 43-story Hongkong Bank (1985, Fig. 10.12) in Hong Kong by Norman Foster and Ove Arup structural engineers, is surely a unique structure and may be considered as having introduced the new generation of high-rise building structures. The supporting bridgelike structure, which allows opening up the central space, is articulated by placing the structural tower and service cores along the sides of the building, opposite in approach to the central core idea of a conventional high-rise building. The Hongkong Bank building is supported by a cluster of eight towers, where pairs of towers form four parallel megaframes. Structurally independent buildings stacks of varying height, separated by double-story spaces, are suspended from the megastructure, reflecting the five-level vertical zoning of the building. The primary structure is made up of the four parallel, fire-protected, aluminum-clad steel megaframes, each consisting of two towers connected by two double-story, pin-connected cantilever suspension trusses, which span 110 ft between the masts and cantilever 35 ft beyond. Three vertical steel tube hangers are suspended from the bridgelike trusses, one at midspan and one at each end of the cantilevers, to support the primary floor girders of as many as nine decks. For architectural reasons, the top chords of the suspension trusses between the masts are omitted on the facades. The efficiency of the gravity flow is convincingly manifested by the minimum of tensile material needed to bridge the (for high-rise buildings) unusually large span from tower to tower. The megaframe clearly expresses the character of stayed bridges vertically stacked on top of each other. The building, with rectangular plan dimensions of about 180×236 ft, is divided horizontally by the four parallel megaframes into vertical sections of 30, 37, and 43 stories. The megaframes also form the 16.7-ft-wide circulation and edge zones and support the 36-ft-span composite floor beams, spaced at about 8 ft in the three general zones. The towers are linked at the truss levels by cross bracing located on the inner line of the tubular tower columns. The lateral loads are resisted by the masts (together with the trusses and cross bracing) in frame action, for the two principal wind directions, indicating that shear wracking of the columns in the multistory portal frames is predominant. The towers are composed of four-corner supercolumns rigidly connected at every floor by haunched beams to form a Vierendeel box truss with plan dimensions of 15.8×16.7 ft, center-to-center. The columns are circular steel tubes that taper from 4.6-ft diameter, with a wall thickness of about 3.9 in. at the base, to 2.6-ft diameter and 1.6-in. wall thickness at the top.

Typical examples of the new generation of high-rise buildings are shown in Fig. 10.13. They range from innovative structural concepts of the 947-ft Collserola Broadcast Tower in Barcelona, Spain (1992), the impressive portallike 40-story Umeda Sky Building in Osaka, Japan (1993, Fig. 10.13 center) with an almost 500-ft high open atrium space between two building towers crowned by a sky garden bridge, to Eisen-

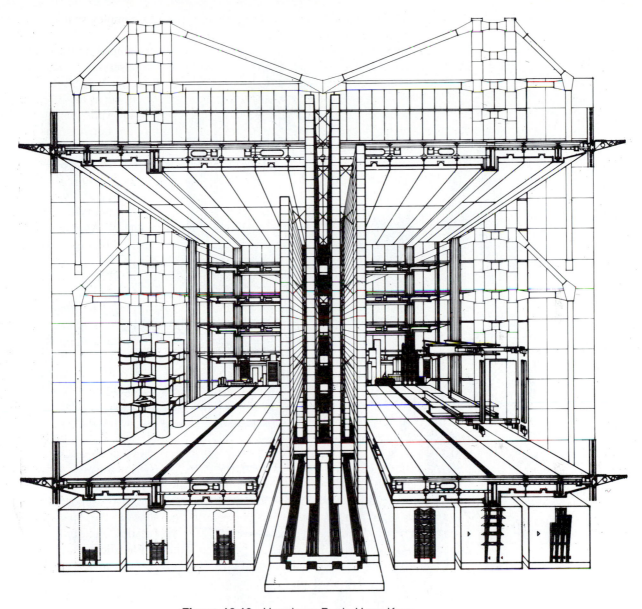

Figure 10.12   Hongkong Bank, Hong Kong.

man's almost Expressionist 10-story Alteka Office Building in Tokyo (1991, Fig. 10.13, middle right) derived from geometric operations of infolding, unfolding, and enveloping of the basic L-shape.

The 947-ft Collserola cable-stayed communication tower by Foster & Arup is of a curvilinear equilateral triangle plan along its main portion (Fig. 10.13, bottom left). It consists of the tapering central, circular, guyed concrete shaft, nearly 15 ft in diameter at the base changing to a steel mast at the upper portion of the tower. The concrete core is braced by three vertical steel trusses set 120° to each other in plan and supports around it a 12 story steel structure, with the floors hanging from it by the three main trusses. The lateral stiffness of the tower is achieved through prestressing of the principal guys.

The tapered, 1000-ft-high, 73-story Landmark Tower in Yokohama, Japan (Fig. 10.13, top center left), designed by Hugh A. Stubbins, architect, and LeMessurier, structural engineering consultants, uses a tube-in-tube structure system and a semi-active mass damper below the roof to control vibrations due to wind and earthquakes.

Figure 10.13 New generation of high-rise building structures.

The exposed perimeter steel frame of the 45-story Hotel de las Artes tower in Barcelona, Spain, by SOM (1992, Fig. 10.13, center bottom) consists of cross-braced, L-shaped frames at each corner linked horizontally by cross-braced frames at various locations in the center bays to allow for tubular behavior of the structure.

With the sloping Domino's Farms Tower proposal for Ann Arbor, Michigan (1989, Fig. 10.13, top right), Gunnar Birkerts tried to free the building from the vertical to make it independent from functional restraints and traditional design attitudes. Keep in mind that in leaning high-rise buildings lateral forces are not only generated directly by wind and earthquakes, but also by geometry, that is, eccentricity of gravity loads. A triangulated structure provides an efficient solution, as proposed by Leslie Robertson and others. For the case shown, the structure is in balance according to the seesaw principle. The weight of the lower building block is hanging at its apex and balances the weight of the upper cantilevering portion.

The design of Foster's green office tower proposal for Frankfurt, Germany (1991, Fig. 10.13, bottom right) is based on living in harmony with nature and energy conservation. Here, the typical interior elevator cores are replaced by a 608-ft-high atrium space, allowing the interior office spaces to be lighted and ventilated naturally. The draughts in this light chimney are drawn by convection due to stack effect, assisted by large fans, and converted into energy. The shaft, located at the center of the triangular plan, is connected to three-story gardens at various levels arranged in a staggered fashion as indicated in Fig. 10.13. The floor plans are free of columns, and the loads are carried by perimeter Vierendeel trusses.

We may conclude from the cases in Fig. 10.13 that there seem to be no limits in the form-giving process and that the imagination and ingenuity of designers know no boundaries.

appendix a

Tables for Structural Design

TABLE A.1
Total Areas for Various Numbers of Reinforcing Bars (in.^{2})

Bar Size No.	Nominal Diameter (in.)	Weight (lb/ft)	Number of Bars									
			1	2	3	4	5	6	7	8	9	10
3	0.375	0.376	0.11	0.22	0.33	0.44	0.55	0.66	0.77	0.88	0.99	1.10
4	0.500	0.668	0.20	0.40	0.60	0.80	1.00	1.20	1.40	1.60	1.80	2.00
5	0.625	1.043	0.31	0.62	0.93	1.24	1.55	1.86	2.17	2.48	2.79	3.10
6	0.750	1.502	0.44	0.88	1.32	1.76	2.20	2.64	3.08	3.52	3.96	4.40
7	0.875	2.044	0.60	1.20	1.80	2.40	3.00	3.60	4.20	4.80	5.40	6.00
8	1.000	2.670	0.79	1.58	2.37	3.16	3.95	4.74	5.53	6.32	7.11	7.90
9	1.128	3.400	1.00	2.00	3.00	4.00	5.00	6.00	7.00	8.00	9.00	10.00
10	1.270	4.303	1.27	2.54	3.81	5.08	6.35	7.62	8.89	10.16	11.43	12.70
11	1.410	5.313	1.56	3.12	4.68	6.24	7.80	9.36	10.92	12.48	14.04	15.60
14	1.693	7.650	2.25	4.50	6.75	9.00	11.25	13.50	15.75	18.00	20.25	22.50
18	2.257	13.600	4.00	8.00	12.00	16.00	20.00	24.00	28.00	32.00	36.00	40.00

TABLE A.2
Average Area per Foot of Width Provided by Various Bar Spacings (in.^{2}/ft)

Bar Size No.	Nominal Diameter (in.)	Spacing of Bar in Inches													
		2	2½	3	3½	4	4½	5	5½	6	7	8	9	10	12
3	0.375	0.66	0.53	0.44	0.38	0.33	0.29	0.27	0.24	0.22	0.19	0.17	0.15	0.13	0.11
4	0.500	1.18	0.94	0.79	0.67	0.59	0.52	0.47	0.43	0.39	0.34	0.29	0.26	0.24	0.20
5	0.625	1.84	1.47	1.23	1.05	0.92	0.82	0.74	0.67	0.61	0.53	0.46	0.41	0.37	0.31
6	0.750	2.65	2.12	1.77	1.51	1.33	1.18	1.06	0.96	0.88	0.76	0.66	0.59	0.53	0.44
7	0.875	3.61	2.89	2.41	2.06	1.80	1.60	1.44	1.31	1.20	1.03	0.90	0.80	0.72	0.60
8	1.000	4.71	3.77	3.14	2.69	2.36	2.09	1.88	1.71	1.57	1.35	1.18	1.05	0.94	0.79
9	1.128		4.80	4.00	3.43	3.00	2.66	2.40	2.18	2.00	1.71	1.50	1.33	1.20	1.00
10	1.270			5.07	4.34	3.80	3.38	3.04	2.76	2.53	2.17	1.90	1.69	1.52	1.27
11	1.410			6.25	5.35	4.68	4.16	3.75	3.41	3.12	2.68	2.34	2.08	1.87	1.56

TABLE A.3
Maximum Number of Bars as a Single Layer in Beam Stems at Interior Exposures with #3 Stirrups and Maximum Aggregate Size ¾ in.

Bar Size No.	Beam Width, b_w (in.)								
	8	10	12	14	16	18	20	22	24
4	3	4	6	7	8	10	—	—	—
5	3	4	5	6	8	9	10	—	—
6	3	4	5	6	7	8	9	10	—
7	2	3	4	6	7	8	9	10	—
8	2	3	4	5	6	7	8	9	10
9	2	3	4	5	5	6	7	8	9
10	2	2	3	4	5	6	6	7	8
11	2	2	3	4	4	5	6	6	7

TABLE A.4
Maximum Reinforcement Ratio for Singly Reinforced Rectangular Beams: $\rho_{max} = 0.75\rho_b$

f_y (psi)	f_c' (psi)				
	3000 ($\beta_1 = 0.85$)	3500 ($\beta_1 = 0.85$)	4000 ($\beta_1 = 0.85$)	5000 ($\beta_1 = 0.80$)	6000 ($\beta_1 = 0.75$)
40,000	0.0278	0.0325	0.0371	0.0437	0.0491
50,000	0.0206	0.0241	0.0275	0.0324	0.0364
60,000	0.0160	0.0187	0.0214	0.0252	0.0283

TABLE A.5
Typical Characterisitics of Stress-Relieved 7-Wire Strands (ASTM A416)

ASTM Type or Grade	Nominal Diameter		Nominal Area		Minimum Tensile Strength, f_{pu}	
	in.	mm	in.^{2}	mm^2	ksi	MPa
Grade 250	0.25	6.35	0.036	23.22		
	0.313	7.94	0.058	37.42		
	0.375	9.53	0.080	51.61		
	0.438	11.11	0.108	69.68	250	1725
	0.500	12.54	0.144	92.90		
	0.600	15.24	0.216	139.35		
Grade 270	0.375	9.53	0.085	54.84		
	0.438	11.11	0.115	74.19		
	0.500	12.54	0.153	98.71	270	1860
	0.563	14.29	0.192	123.87		
	0.600	15.24	0.216	139.35		

TABLE A.6
Deformed Reinforcing Steel Bars for Concrete and Masonry

Type of Steel	Bar Size No.	Grade/Yield Strength, ksi	Allowable Tensile Stress, ksi	Minimum Tensile Strength, ksi
Billet steel (ASTM A615)	3–6	40	20	70
	3–11	60	24	90
	14, 18			
Rail steel (ASTM A616)	3–11	50	20	80
	3–11	60	24	90
Axle steel (ASTM A617)	3–11	40	20	70
	3–11	60	24	90
Low-alloy steel (ASTM A706)	3–11	60	24	80
	14, 18			

TABLE A.7
Moment Coefficients for Two-Way Slabs on Beams (ACI, Method 2)[a]

	Short Span						Long Span, All Values of m
	Values of m						
Moments	1.0	0.9	0.8	0.7	0.6	0.5 and Less	
Case 1: Interior panels							
Negative moment at:							
Continuous edge	0.033	0.040	0.048	0.055	0.063	0.083	0.033
Discontinuous edge	—	—	—	—	—	—	—
Positive moment at midspan	0.025	0.030	0.036	0.041	0.047	0.062	0.025
Case 2: One edge discontinuous							
Negative moment at:							
Continuous edge	0.041	0.048	0.055	0.062	0.069	0.085	0.041
Discontinuous edge	0.021	0.024	0.027	0.031	0.035	0.042	0.021
Positive moment at midspan	0.031	0.036	0.041	0.047	0.052	0.064	0.031
Case 3: Two edges discontinuous							
Negative moment at:							
Continuous edge	0.049	0.057	0.064	0.071	0.078	0.090	0.049
Discontinuous edge	0.025	0.028	0.032	0.036	0.039	0.045	0.025
Positive moment at midspan	0.037	0.043	0.048	0.054	0.059	0.068	0.037
Case 4: Three edges discontinuous							
Negative moment at:							
Continuous edge	0.058	0.066	0.074	0.082	0.090	0.098	0.058
Discontinuous edge	0.029	0.033	0.037	0.041	0.045	0.049	0.029
Positive moment at midspan	0.044	0.050	0.056	0.062	0.068	0.074	0.044
Case 5: Four edges discontinuous							
Negative moment at:							
Continuous edge	—	—	—	—	—	—	—
Discontinuous edge	0.033	0.038	0.043	0.047	0.053	0.055	0.033
Positive moment at midspan	0.050	0.057	0.064	0.072	0.080	0.083	0.050

[a] Reproduced from *Building Code Requirements for Reinforced Concrete (ACI 318-63)*, courtesy of ACI

TABLE A.8
Some Typical Approximate Design Value Ranges for Common Dimension Lumber and Timbers Under Normal Duration Loading and Dry Service Conditions (to be used for preliminary design purposes only)

Single-member bending: $F_b \cong 1800$ psi to 800 psi	Repetitive-member bending: $F_b \cong 2000$ psi to 900 psi
Tension parallel to grain: $F_t \cong 1000$ psi to 500 psi	Shear parallel to grain: $F_v \cong 100$ psi to 85 psi
Compression parallel to grain: $F_c \cong 1800$ psi to 600 psi	Compression perpendicular to grain: $F_{c\perp} \cong 700$ psi to 500 psi
Modulus of Elasticity: $E \cong 1,800,000$ psi to $1,300,000$ psi	

TABLE A.9
Partial List of Properties of Sawn-Lumber Sections[a]

Nominal Size, b(inches)d	Standard Dressed Size (S4S), b(inches)d	Area of Section, $A(in.^2)$	Moment of Inertia, $I(in.^4)$	Section Modulus, $S(in.^3)$	Nominal Size, b(inches)d	Standard Dressed Size (S4S), b(inches)d	Area of Section, $A(in.^2)$	Moment of Inertia, $I(in.^4)$	Section Modulus, $S(in.^3)$
1 × 3	¾ × 2½	1.875	0.977	0.781	6 × 2	5½ × 1½	8.250	1.547	2.063
1 × 4	¾ × 3½	2.625	2.680	1.531	6 × 3	5½ × 2½	13.750	7.161	5.729
1 × 6	¾ × 5½	4.125	10.398	3.781	6 × 4	5½ × 3½	19.250	19.651	11.229
1 × 8	¾ × 7¼	5.438	23.817	6.570	6 × 6	5½ × 5½	30.250	76.255	27.729
1 × 10	¾ × 9¼	6.938	49.466	10.695	6 × 8	5½ × 7½	41.250	193.359	51.563
1 × 12	¾ × 11¼	8.438	88.989	15.820	6 × 10	5½ × 9½	52.250	392.963	82.729
					6 × 12	5½ × 11½	63.250	697.068	121.229
2 × 3	1½ × 2½	3.750	1.953	1.563					
2 × 4	1½ × 3½	5.250	5.359	3.063	6 × 14	5½ × 13½	74.250	1127.672	167.063
2 × 5	1½ × 4½	6.750	11.391	5.063	6 × 16	5½ × 15½	85.250	1706.776	220.229
2 × 6	1½ × 5½	8.250	20.797	7.563	6 × 18	5½ × 17½	96.250	2456.380	280.729
2 × 8	1½ × 7¼	10.875	47.635	13.141	6 × 20	5½ × 19½	107.250	3398.484	348.563
2 × 10	1½ × 9¼	13.875	98.932	21.391	6 × 22	5½ × 21½	118.250	4555.086	423.729
2 × 12	1½ × 11¼	16.875	177.979	31.641	6 × 24	5½ × 23½	129.250	5498.191	506.229
2 × 14	1½ × 13¼	19.875	290.775	43.891					
					8 × 1	7¼ × ¾	5.438	0.255	0.680
3 × 1	2½ × ¾	1.875	0.088	0.234	8 × 2	7¼ × 1½	10.875	2.039	2.719
3 × 2	2½ × 1½	3.750	0.703	0.938	8 × 3	7¼ × 2½	18.125	9.440	7.552
3 × 4	2½ × 3½	8.750	8.932	5.104	8 × 4	7¼ × 3½	25.375	25.904	14.803
3 × 5	2½ × 4½	11.250	18.984	8.438	8 × 6	7½ × 5½	41.250	103.984	37.813
3 × 6	2½ × 5½	13.750	34.661	12.604	8 × 8	7½ × 7½	56.250	263.672	70.313
3 × 8	2½ × 7¼	18.125	79.391	21.901	8 × 10	7½ × 9½	71.250	535.859	112.813
3 × 10	2½ × 9¼	23.125	164.886	35.651	8 × 12	7½ × 11½	86.250	950.547	165.313
3 × 12	2½ × 11¼	28.125	296.631	52.734	8 × 14	7½ × 13½	101.250	1537.734	227.813
3 × 14	2½ × 13¼	33.125	484.625	73.151	8 × 16	7½ × 15½	116.250	2327.422	300.313
3 × 16	2½ × 15¼	38.125	738.870	96.901	8 × 18	7½ × 17½	131.250	3349.609	382.813
					8 × 20	7½ × 19½	146.250	4634.297	475.313
4 × 1	3½ × ¾	2.625	0.123	0.328	8 × 22	7½ × 21½	161.250	6211.484	577.813
4 × 2	3½ × 1½	5.250	0.984	1.313	8 × 24	7½ × 23½	176.250	8111.172	690.313
4 × 3	3½ × 2½	8.750	4.557	3.646					
4 × 4	3½ × 3½	12.250	12.505	7.146	10 × 1	9¼ × ¾	6.938	0.325	0.867
4 × 5	3½ × 4½	15.750	26.578	11.813	10 × 2	9¼ × 1½	13.875	2.602	3.469
4 × 6	3½ × 5½	19.250	48.526	17.646	10 × 3	9¼ × 2½	23.125	12.044	9.635
4 × 8	3½ × 7¼	25.375	111.148	30.661	10 × 4	9¼ × 3½	32.375	33.049	18.885
4 × 10	3½ × 9¼	32.375	230.840	49.911	10 × 6	9½ × 5½	52.250	131.714	47.896
4 × 12	3½ × 11¼	39.375	415.283	73.828	10 × 8	9½ × 7½	71.250	333.984	89.063
4 × 14	3½ × 13¼	46.375	678.475	102.411	10 × 10	9½ × 9½	90.250	678.755	142.896
4 × 16	3½ × 15¼	53.375	1034.418	135.66	10 × 12	9½ × 11½	109.250	1204.026	209.396
					10 × 14	9½ × 13½	128.250	1947.797	288.563
5 × 2	4½ × 1½	6.750	1.266	1.688	10 × 16	9½ × 15½	147.250	2948.068	380.396
5 × 3	4½ × 2½	11.250	5.859	4.688	10 × 18	9½ × 17½	166.250	4242.836	484.896
5 × 4	4½ × 3½	15.750	16.078	9.188	10 × 20	9½ × 19½	185.250	5870.109	602.063
5 × 5	4½ × 4½	20.250	34.172	15.188	10 × 22	9½ × 21½	204.250	7867.879	731.896
					10 × 24	9½ × 23½	223.250	10274.148	874.396
6 × 1	5½ × ¾	4.125	0.193	0.516					

[a] Reproduced from the *National Design Specification® for Wood Construction–NDS® Supplement*, courtesy of the American Forest & Paper Association, Washington, D.C.

TABLE A.10
Partial List of Approximate Section Properties of Structural Glued Laminated Timber (to be used for preliminary design purposes only)

Depth, d(in.)	Area, A(in.²)	Modified Section Modulus, SC_f (in.³)	Moment of Inertia, I(in.⁴)
3⅛" Width			
6.0	18.8	18.8	56
7.5	23.4	29.3	110
9.0	28.1	42.2	190
10.5	32.8	57.4	302
12.0	37.5	75.0	450
13.5	42.2	93.7	641
15.0	46.9	114.3	879
16.5	51.6	136.9	1,170
18.0	56.3	161.3	1,519
19.5	60.9	187.6	1,931
21.0	65.6	215.8	2,412
22.5	70.3	245.9	2,966
24.0	75.0	277.8	3,600
5⅛" Width			
7.5	38.4	48.0	180
9.0	46.1	69.2	311
10.5	53.8	94.2	494
12.0	61.5	123.0	738
13.5	69.2	153.6	1,051
15.0	76.9	187.5	1,441
16.5	84.6	224.5	1,919
18.0	92.3	264.6	2,491
19.5	99.9	307.7	3,167
21.0	107.6	354.0	3,955
22.5	115.3	403.2	4,865
24.0	123.0	455.5	5,904
25.5	130.7	510.8	7,082
27.0	138.4	569.0	8,406
28.5	146.1	630.2	9,887
30.0	153.8	694.3	11,531
31.5	161.4	761.4	13,349
33.0	169.1	831.3	15,348
34.5	176.8	904.1	17,538
36.0	184.5	979.8	19,926
6¾" Width			
12.0	81.0	162.0	972
13.5	91.1	202.4	1,384
15.0	101.3	246.9	1,898
16.5	111.4	295.6	2,527
18.0	121.5	348.4	3,280
19.5	131.6	405.3	4,171
21.0	141.8	466.2	5,209

Depth, d(in.)	Area, A(in.²)	Modified Section Modulus, SC_f (in.³)	Moment of Inertia, I(in.⁴)
22.5	151.9	531.1	6,407
24.0	162.0	600.0	7,776
25.5	172.1	672.8	9,327
27.0	182.3	749.5	11,072
28.5	192.4	830.0	13,021
30.0	202.5	914.5	15,188
31.5	212.6	1,002.8	17,581
33.0	222.8	1,094.9	20,215
34.5	232.9	1,190.8	23,098
36.0	243.0	1,290.5	26,244
37.5	253.1	1,393.9	29,663
39.0	263.3	1,501.1	33,367
40.5	273.4	1,612.0	37,367
42.0	283.5	1,726.6	41,674
43.5	293.6	1,845.0	46,301
45.0	303.8	1,967.0	51,258
46.5	313.9	2,092.6	56,556
48.0	324.0	2,222.0	62,208
8¾" Width			
12.0	105.0	210.0	1,260
13.5	118.1	262.3	1,794
15.0	131.3	320.1	2,461
16.5	144.4	383.2	3,276
18.0	157.5	451.7	4,252
19.5	170.6	525.4	5,407
21.0	183.8	604.4	6,753
22.5	196.9	688.5	8,306
24.0	210.0	777.7	10,080
25.5	223.1	872.1	12,091
27.0	236.3	971.5	14,352
28.5	249.4	1,076.0	16,880
30.0	262.5	1,185.5	19,688
31.5	275.6	1,299.9	22,791
33.0	288.8	1,419.3	26,204
34.5	301.9	1,543.6	29,942
36.0	315.0	1,672.8	34,020
37.5	328.1	1,806.9	38,452
39.0	341.3	1,945.9	43,253
40.5	354.4	2,089.6	48,439
42.0	367.5	2,238.2	54,022
43.5	380.6	2,391.6	60,020
45.0	393.8	2,549.8	66,445
46.5	406.9	2,712.7	73,314

TABLE A.10 (continued)

Depth, d(in.)	Area, A(in.²)	Modified Section Modulus, SCf (in.³)	Moment of Inertia, I(in.⁴)	Depth, d(in.)	Area, A(in.²)	Modified Section Modulus, SCf (in.³)	Moment of Inertia, I(in.⁴)
48.0	420.0	2,880.3	80,640	37.5	403.1	2,219.9	47,241
49.5	433.1	3,052.7	88,439	39.0	419.3	2,390.6	53,140
51.0	446.3	3,229.8	96,725	40.5	435.4	2,567.3	59,510
52.5	459.4	3,411.6	105,513	42.0	451.5	2,749.8	66,370
54.0	472.5	3,598	114,818	43.5	467.6	2,938.3	73,739
55.5	485.6	3,789.1	124,654	45.0	483.8	3,132.6	81,633
57.0	498.8	3,984.9	135,037	46.5	499.9	3,332.7	90,071
58.5	511.9	4,185.3	145,980	48.0	516.0	3,538.7	99,072
60.0	525.0	4,390.3	157,500	49.5	532.1	3,750.5	108,653
10¾" Width				51.0	548.3	3,968.0	118,833
15.0	161.3	393.3	3,023	52.5	564.4	4,191.4	129,630
16.5	177.4	470.8	4,024	54.0	580.5	4,420.4	141,062
18.0	193.5	554.9	5,224	55.5	596.6	4,655.2	153,146
19.5	209.6	645.5	6,642	57.0	612.8	4,895.7	165,902
21.0	225.8	742.5	8,296	58.5	628.9	5,141.9	179,347
22.5	241.9	845.8	10,204	60.0	645.0	5,398.8	193,500
24.0	258.0	955.5	12,384	61.5	661.1	5,651.4	208,379
25.5	274.1	1,071.4	14,854	63.0	677.3	5,914.5	224,000
27.0	290.3	1,193.6	17,633	64.5	693.4	6,183.3	240,384
28.5	306.4	1,321.9	20,738	66.0	709.5	6,457.8	257,548
30.0	322.5	1,456.4	24,188	67.5	725.6	6,737.8	275,511
31.5	338.6	1,597.0	28,000	69.0	741.8	7,023.4	294,289
33.0	354.8	1,743.7	32,194	70.5	757.9	7,314.6	313,902
34.5	370.9	1,896.4	36,786	72.0	774.0	7,611.3	334,368
36.0	387.0	2,055.2	41,796	73.5	790.1	7,913.6	355,704

TABLE A.11
Allowable Stress Design Selection Table (a partial list) for Shapes Used as Beams[a]

S_x in.³	Shape	F'_y ksi	F_y = 36 ksi — L_c ft	F_y = 36 ksi — L_u ft		S_x in.³	Shape	F'_y ksi	F_y = 36 ksi — L_c ft	F_y = 36 ksi — L_u ft
3170	W 36 × 848	—	19.1	89.0		310	W 18 × 158	—	11.9	38.3
2980	W 36 × 798	—	19.0	85.7		**299**	**W 30 × 108**	—	**11.1**	**12.3**
2690	W 36 × 720	—	18.8	78.5		299	W 27 × 114	—	10.6	15.9
2590	W 40 × 655	—	17.8	63.4		295	W 21 × 132	—	13.1	27.2
2420	W 36 × 650	—	18.6	71.2		291	W 24 × 117	—	13.5	20.8
2340	W 40 × 593	—	17.6	57.9		282	W 18 × 143	—	11.8	35.1
2180	W 36 × 588	—	18.4	65.2		273	W 21 × 122	—	13.1	25.4
2090	W 40 × 531	—	17.4	52.6		**269**	**W 30 × 99**	—	**10.9**	**11.4**
1950	W 36 × 527	—	18.2	59.4		267	W 27 × 102	—	10.6	14.2
1890	W 40 × 480	—	17.3	47.7		258	W 24 × 104	58.5	13.5	18.4
1710	W 40 × 436	—	17.1	43.7		256	W 18 × 130	—	11.8	32.2
1560	W 40 × 397	—	17.0	40.3		249	W 21 × 111	—	13.0	23.3
1450	W 36 × 393	—	17.8	45.4		**245**	**W 30 × 90**	58.1	**10.0**	**11.4**
1420	W 40 × 362	—	16.9	36.7		245	W 24 × 103	—	9.5	6.7
1340	W 40 × 328	—	18.9	35.9		243	W 27 × 94	—	10.5	12.8
1280	W 40 × 324	—	16.8	33.1		231	W 18 × 119	—	11.9	29.1
1220	W 40 × 298	—	18.8	32.8		227	W 21 × 101	—	13.0	21.3
1170	W 40 × 297	—	16.7	30.3		222	W 24 × 94	—	9.6	15.1
1120	W 44 × 285	—	12.5	22.0		**213**	**W 27 × 84**	—	**10.5**	**11.0**
1100	W 40 × 277	—	16.7	29.1		204	W 18 × 106	—	11.8	26.0
1090	W 40 × 268	—	18.7	29.5		**196**	**W 24 × 84**	—	**9.5**	**13.3**
992	W 40 × 249	—	16.6	26.3		192	W 21 × 93	—	8.9	16.8
983	W 40 × 244	—	18.7	26.5		190	W 14 × 120	—	15.5	44.1
889	W 44 × 224	—	12.5	17.9		188	W 18 × 97	—	11.8	24.1
858	W 40 × 215	—	16.6	22.8		**176**	**W 24 × 76**	—	**9.5**	**11.8**
776	W 44 × 198	—	12.5	15.5		175	W 16 × 100	—	11.0	28.1
708	W 40 × 192	37.1	17.8	19.7		173	W 14 × 109	58.6	15.4	40.6
682	W 40 × 183	—	12.5	17.1		171	W 21 × 83	—	8.8	15.1
623	W 36 × 182	—	12.7	18.2		166	W 18 × 86	—	11.7	21.5
599	W 40 × 167	—	12.5	14.5		157	W 14 × 99	48.5	15.4	37.0
542	W 36 × 160	—	12.7	15.7		155	W 16 × 89	—	10.9	25.0
512	W 40 × 149	—	11.9	12.6		**154**	**W 24 × 68**	—	**9.5**	**10.2**
448	W 33 × 141	—	12.2	15.4		151	W 21 × 73	—	8.8	13.4
439	**W 36 × 135**	—	**12.3**	**13.0**		146	W 18 × 76	64.2	11.6	19.1
436	W 30 × 148	—	11.1	18.7		143	W 14 × 90	40.4	15.3	34.0
406	**W 33 × 130**	—	**12.1**	**13.8**		**140**	**W 21 × 68**	—	**8.7**	**12.4**
359	**W 33 × 118**	—	**12.0**	**12.6**		134	W 16 × 77	—	10.9	21.9
355	W 30 × 124	—	11.1	15.0		**131**	**W 24 × 62**	—	**7.4**	**8.1**
345	W 27 × 129	—	10.6	18.4		**127**	**W 21 × 62**	—	**8.7**	**11.2**
344	W 18 × 175	—	12.0	41.7		127	W 18 × 71	—	8.1	15.5
329	**W 30 × 116**	—	**11.1**	**13.8**		123	W 14 × 82	—	10.7	28.1
329	W 24 × 131	—	13.6	23.4		118	W 12 × 87	—	12.8	36.2
329	W 21 × 147	—	13.2	30.3		117	W 18 × 65	—	8.0	14.4

[a] Reproduced from the *Manual of Steel Construction*, 9th ed., courtesy of American Institute of Steel Construction, Inc.

TABLE A.11(continued)

S_x in.³	Shape	F'_y ksi	L_c ft	L_u ft
117	W 16 × 67	—	10.8	19.3
114	**W 24 × 55**	—	**7.0**	**7.5**
112	W 14 × 74	—	10.6	25.9
111	W 21 × 57	—	6.9	9.4
108	W 18 × 60	—	8.0	13.3
107	W 12 × 79	62.6	12.8	33.3
103	W 14 × 68	—	10.6	23.9
98.3	**W 18 × 55**	—	**7.9**	**12.1**
97.4	W 12 × 72	52.3	12.7	30.5
94.5	**W 21 × 50**	—	**6.9**	**7.8**
92.2	W 16 × 57	—	7.5	14.3
92.2	W 14 × 61	—	10.6	21.5
88.9	**W 18 × 50**	—	**7.9**	**11.0**
87.9	W 12 × 65	43	12.7	27.7
81.6	**W 21 × 44**	—	**6.6**	**7.0**
81.0	W 16 × 50	—	7.5	12.7
78.8	W 18 × 46	—	6.4	9.4
78.0	W 12 × 58	—	10.6	24.4
77.8	W 14 × 53	—	8.5	17.7
72.7	W 16 × 45	—	7.4	11.4
70.6	W 12 × 53	55.9	10.6	22.0
70.3	W 14 × 48	—	8.5	16.0
68.4	**W 18 × 40**	—	**6.3**	**8.2**
66.7	W 10 × 60	—	10.6	31.1
64.7	**W 16 × 40**	—	**7.4**	**10.2**
64.7	W 12 × 50	—	8.5	19.6
62.7	W 14 × 43	—	8.4	14.4
60.0	W 10 × 54	63.5	10.6	28.2
58.1	W 12 × 45	—	8.5	17.7
57.6	**W 18 × 35**	—	**5.3**	**6.7**
56.5	W 16 × 36	64.0	7.4	8.8
54.6	W 14 × 38	—	7.1	11.5
54.6	W 10 × 49	53.0	10.6	26.0
51.9	W 12 × 40	—	8.4	16.0
49.1	W 10 × 45	—	8.5	22.8
48.6	**W 14 × 34**	—	**7.1**	**10.2**
47.2	**W 16 × 31**	—	**5.8**	**7.1**
45.6	W 12 × 35	—	6.9	12.6
42.1	W 10 × 39	—	8.4	19.8
42.0	**W14 × 30**	55.3	**7.1**	**8.7**
38.6	**W 12 × 30**	—	**6.9**	**10.8**
38.4	**W 16 × 26**	—	**5.6**	**6.0**
35.3	**W 14 × 26**	—	**5.3**	**7.0**

S_x in.³	Shape	F'_y ksi	L_c ft	L_u ft
35.0	W 10 × 33	50.5	8.4	16.5
33.4	**W 12 x 26**	57.9	**6.9**	**9.4**
32.4	W 10 × 30	—	6.1	13.1
31.2	W 8 × 35	64.4	8.5	22.6
29.0	**W 14 × 22**	—	**5.3**	**5.6**
27.9	W 10 × 26	—	6.1	11.4
27.5	W 8 × 31	50.0	8.4	20.1
25.4	**W 12 × 22**	—	**4.3**	**6.4**
24.3	W 8 × 28	—	6.9	17.5
23.2	**W 10 × 22**	—	**6.1**	**9.4**
21.3	**W 12 × 19**	—	**4.2**	**5.3**
21.1	**M 14 × 18**	—	**3.6**	**4.0**
20.9	W 8 × 24	64.1	6.9	15.2
18.8	W 10 × 19	—	4.2	7.2
18.2	W 8 × 21	—	5.6	11.8
17.1	**W 12 × 16**	—	**4.1**	**4.3**
16.7	W 6 × 25	—	6.4	20.0
16.2	W 10 × 17	—	4.2	6.1
15.2	W 8 × 18	—	5.5	9.9
14.9	**W 12 × 14**	54.3	**3.5**	**4.2**
13.8	W 10 × 15	—	4.2	5.0
13.4	W 6 × 20	62.1	6.4	16.4
13.0	M 6 × 20	—	6.3	17.4
12.0	**M 12 × 11.8**	—	**2.7**	**3.0**
11.8	W 8 × 15	—	4.2	7.2
10.9	W 10 × 12	47.5	3.9	4.3
10.9	**M 12 × 10.8**	—	**2.5**	**3.1**
10.3	**M 12 × 10**	—	**2.3**	**3.3**
10.2	W 6 × 16	—	4.3	12.0
10.2	W 5 × 19	—	5.3	19.5
9.91	W 8 × 13	—	4.2	5.9
9.72	W 6 × 15	31.8	6.3	12.0
9.63	M 5 × 18.9	—	5.3	19.3
8.51	W 5 × 16	—	5.3	16.7
7.81	**W 8 × 10**	45.8	**4.2**	**4.7**
7.76	**M 10 × 9**	—	**2.6**	**2.7**
7.31	W 6 × 12	—	4.2	8.6
6.94	**M 10 × 8**	—	**2.3**	**2.7**
6.57	**M 10 × 7.5**	—	**2.2**	**2.7**
5.56	W 6 × 9	50.3	4.2	6.7
5.46	W 4 × 13	—	4.3	15.6
4.62	**M 8 × 6.5**	—	**2.4**	**2.5**
2.40	**M 6 × 4.4**	—	**1.9**	**2.4**

TABLE A.12
Allowable Stress for Compression Members of A36 Steel[a,b]

$\frac{Kl}{r}$	F_a (ksi)	$\frac{Kl}{r}$	F_a (ksi)	$\frac{Kl}{r}$	F_a (ksi)	$\frac{Kl}{r}$	F_a (ksi)	$\frac{Kl}{r}$	F_a (ksi)
1	21.56	41	19.11	81	15.24	121	10.14	161	5.76
2	21.52	42	19.03	82	15.13	122	9.99	162	5.69
3	21.48	43	18.95	83	15.02	123	9.85	163	5.62
4	21.44	44	18.86	84	14.90	124	9.70	164	5.55
5	21.39	45	18.78	85	14.79	125	9.55	165	5.49
6	21.35	46	18.70	86	14.67	126	9.41	166	5.42
7	21.30	47	18.61	87	14.56	127	9.26	167	5.35
8	21.25	48	18.53	88	14.44	128	9.11	168	5.29
9	21.21	49	18.44	89	14.32	129	8.97	169	5.23
10	21.16	50	18.35	90	14.20	130	8.84	170	5.17
11	21.10	51	18.26	91	14.09	131	8.70	171	5.11
12	21.05	52	18.17	92	13.97	132	8.57	172	5.05
13	21.00	53	18.08	93	13.84	133	8.44	173	4.99
14	20.95	54	17.99	94	13.72	134	8.32	174	4.93
15	20.89	55	17.90	95	13.60	135	8.19	175	4.88
16	20.83	56	17.81	96	13.48	136	8.07	176	4.82
17	20.78	57	17.71	97	13.35	137	7.96	177	4.77
18	20.72	58	17.62	98	13.23	138	7.84	178	4.71
19	20.66	59	17.53	99	13.10	139	7.73	179	4.66
20	20.60	60	17.43	100	12.98	140	7.62	180	4.61
21	20.54	61	17.33	101	12.85	141	7.51	181	4.56
22	20.48	62	17.24	102	12.72	142	7.41	182	4.51
23	20.41	63	17.14	103	12.59	143	7.30	183	4.46
24	20.35	64	17.04	104	12.47	144	7.20	184	4.41
25	20.28	65	16.94	105	12.33	145	7.10	185	4.36
26	20.22	66	16.84	106	12.20	146	7.01	186	4.32
27	20.15	67	16.74	107	12.07	147	6.91	187	4.27
28	20.08	68	16.64	108	11.94	148	6.82	188	4.23
29	20.01	69	16.53	109	11.81	149	6.73	189	4.18
30	19.94	70	16.43	110	11.67	150	6.64	190	4.14
31	19.87	71	16.33	111	11.54	151	6.55	191	4.09
32	19.80	72	16.22	112	11.40	152	6.46	192	4.05
33	19.73	73	16.12	113	11.26	153	6.38	193	4.01
34	19.65	74	16.01	114	11.13	154	6.30	194	3.97
35	19.58	75	15.90	115	10.99	155	6.22	195	3.93
36	19.50	76	15.79	116	10.85	156	6.14	196	3.89
37	19.42	77	15.69	117	10.71	157	6.06	197	3.85
38	19.35	78	15.58	118	10.57	158	5.98	198	3.81
39	19.27	79	15.47	119	10.43	159	5.91	199	3.77
40	19.19	80	15.36	120	10.28	160	5.83	200	3.73

[a]When element width-to-thickness ratio exceeds noncompact section limits of Sect. B5.1, see Appendix B5. (Manual of Steel Construction)
Note: $C_c = 126.1$.
[b] Reproduced from the *Manual of Steel Construction,* 9th ed., courtesy of American Institute of Steel Construction, Inc.

TABLE A.13
Cable Design Tables[a]

Properties of Zinc-coated (Galvanized) Steel Structural Strand ASTM A 586-86, Class A Coating Throughout

Nominal Diameter, in.	Approximate Weight, lb/ft	Approximate Gross Metallic Area, in.^{2}	Minimum Breaking Strength, tons
1/2	0.52	0.150	15.0
9/16	0.66	0.190	19.0
5/8	0.82	0.234	24.0
11/16	0.99	0.284	29.0
3/4	1.18	0.338	34.0
13/16	1.39	0.396	40.0
7/8	1.61	0.459	46.0
15/16	1.85	0.527	54.0
1	2.10	0.600	61.0
1 1/16	2.37	0.677	69.0
1 1/8	2.66	0.759	78.0
1 3/16	2.96	0.846	86.0
1 1/4	3.28	0.938	96.0
1 5/16	3.62	1.03	106.0
1 3/8	3.97	1.13	116.0
1 7/16	4.34	1.24	126.0
1 1/2	4.73	1.35	138.0
1 9/16	5.13	1.47	150.0
1 5/8	5.55	1.59	162.0
1 11/16	5.98	1.71	176.0
1 3/4	6.43	1.84	188.0
1 13/16	6.90	1.97	202.0
1 7/8	7.39	2.11	216.0
1 15/16	7.89	2.25	230.0
2	8.40	2.40	245.0
2 1/16	8.94	2.55	261.0
2 1/8	9.49	2.71	277.0
2 3/16	10.05	2.87	293.0
2 1/4	10.64	3.04	310.0
2 5/16	11.24	3.21	327.0
2 3/8	11.85	3.38	344.0
2 7/16	12.48	3.57	360.0
2 1/2	13.13	3.75	376.0
2 9/16	13.80	3.94	392.0
2 5/8	14.47	4.13	417.0
2 11/16	15.16	4.33	432.0
2 3/4	15.88	4.54	452.0
2 7/8	17.36	4.96	494.0
3	18.90	5.40	538.0
3 1/8	20.51	5.86	584.0
3 1/4	22.18	6.34	625.0
3 3/8	23.92	6.83	673.0
3 1/2	25.73	7.35	724.0
3 5/8	27.60	7.88	768.0
3 3/4	29.50	8.43	822.0
3 7/8	31.50	9.00	878.0
4	33.60	9.60	925.0

Properties of Multiclass Zinc-coated (Galvanized) Steel Structural Wire Rope ASTM A603-88, Class A Coating Throughout

Nominal Diameter, in.	Approximate Weight, lb/ft	Minimum Breaking Strength, tons
3/8	0.24	6.5
7/16	0.32	8.8
1/2	0.42	11.5
9/16	0.53	14.5
5/8	0.65	18.0
11/16	0.79	21.5
3/4	0.95	26.0
13/16	1.10	30.0
7/8	1.28	35.0
15/16	1.47	40.0
1	1.67	45.7
1 1/8	2.11	57.8
1 1/4	2.64	72.2
1 3/8	3.21	87.8
1 1/2	3.82	104.0
1 5/8	4.51	123.0
1 3/4	5.24	143.0
1 7/8	6.03	164.0
2	6.85	186.0
2 1/8	7.73	210.0
2 1/4	8.66	235.0
2 3/8	9.61	261.0
2 1/2	10.60	288.0
2 5/8	11.62	317.0
2 3/4	12.74	347.0
2 7/8	13.90	379.0
3	15.11	412.0
3 1/4	18.00	475.0
3 1/2	21.00	555.0
3 3/4	24.00	640.0
4	27.00	730.0

Minimum modulus of elasticity of prestretched structural wire rope: 20,000 ksi.

Minimum modulus of elasticity of prestretched structural strand:

1/2 to 2 9/16:	24,000 ksi
2 5/8 and larger	23,000 ksi

[a] Reproduced from ASTM Standards in Building Codes, 29th ed. 1992, courtesy of ASTM.

TABLE A.14
Geometrical Properties of Simple Areas

	RECTANGLE		TRIANGLE
X_o — X_o, $d/2$, $d/2$, a — a	$A = bd$ $I_{xo} = bd^3/12$ $I_a = bd^3/3$ $S_{xo} = bd^2/6$ $r_{xo} = 0.289d$	d, X_o, a — a, $d/3$	$A = bd/2$ $I_{xo} = bd^3/36$ $r_{xo} = 0.236d$ $I_a = bd^3/12$
d, $0.707d$	**SQUARE ON DIA.** $A = d^2$ $I = d^4/12$ $S = 0.118d^3$ $r = 0.289d$	R, R	**CIRCLE** $A = \pi R^2$ $I = \pi R^4/4$ $S = \pi R^3/4$ $r = R/2$
t, t, $d/2$, $d/2$, 2	**THIN-WALLED TUBE** $(t \ll d)$ $A = 4dt$ $I = 2td^3/3$ $S = 4td^2/3$ $r = 0.408d$	R, $0.42R$	**SEMICIRCLE** $A = \pi R^2/2$ $I = 0.11R^4$ $S = 0.19R^3$ $r = 0.264R$
$3b/8$, a, 1 — 1, $0.4a$, 2, b	**HALF PARABOLA** $A = 2ab/3$ $I_1 = 0.0457ba^3$ $I_2 = 0.0396ab^3$	d, R, R, t	**CIRCULAR RING** $(t \ll d)$ $A = 2\pi Rt = \pi dt$ $I = \pi R^3 t$ $S = \pi R^2 t$ $r = 0.707R$
a, $0.7a$, $0.75b$	**COMPLEMENT OF HALF PARABOLA** $A = ab/3$ $I_1 = 0.0176ba^3$ $I_2 = ab^3/80$	t, R, $0.64R$, t, d, $\approx \frac{2}{3}d$	**SEMICIRCULAR RING** $(t \ll R)$ $A = \pi Rt$ $I = 0.3R^3 t$ $r = 0.309R$ **SHALLOW ARCH**

TABLE A.15
Maximum Bending Moments and Deflections for Common Beams

SINGLE SPAN BEAMS	MAXIMUM BENDING MOMENT: M	MAXIMUM BENDING DEFLECTION: △
$wL/2$ $wL/2$	$wL^2/8$	$\dfrac{5wL^4}{384EI}$
wL	$-wL^2/2$	$\dfrac{wL^4}{8EI}$
$5wL/8$ $3wL/8$	$-wL^2/8$	$\dfrac{wL^4}{185EI}$
$wL/2$ $wL/2$	$-wL^2/12$	$\dfrac{wL^4}{384EI}$
$P/2$ $P/2$	$PL/4$	$\dfrac{PL^3}{48EI}$
P	$-PL$	$\dfrac{PL^3}{3EI}$
$11P/16$ $5P/16$	$-3PL/16$	$\dfrac{PL^3}{107.3EI}$
$P/2$ $P/2$	$\pm PL/8$	$\dfrac{PL^3}{192EI}$
$wL/3$ $wL/6$	$wL^2/15.6$	$\dfrac{wL^4}{153.4EI}$
$wL/2$	$-wL^2/6$	$\dfrac{wL^4}{30EI}$
$wL/2$	$-wL^2/3$	$\dfrac{11wL^4}{120EI}$
$2wL/5$ $wL/10$	$-wL^2/15$	$\dfrac{wL^4}{419.3EI}$
$7wL/20$ $3wL/20$	$-wL^2/20$	$\dfrac{wL^4}{964EI}$
$wL/4$ $wL/4$	$wL^2/12$	$\dfrac{wL^4}{120EI}$
$wL/4$ $wL/4$	$wL^2/24$	$\dfrac{3wL^4}{640EI}$
M/L M/L M	$3EI\triangle/L^2$	$\dfrac{ML^2}{3EI}$
$2M/L$ $2M/L$ M	$\pm 6EI\triangle/L^2$	$\dfrac{ML^2}{6EI}$
M/L M/L M θ	$3EI\theta/L$	$\dfrac{ML^2}{15.6EI}$
$1.5M/L$ $M/2$ $1.5M/L$ M θ	$4EI\theta/L$	$\dfrac{ML^2}{27EI}$

FOR CONTINUOUS BEAMS REFER TO FIG. 2.27

TABLE A.16
Metric Conversion Table: U.S. Customary Units to SI Metric Units or Vice Versa

Length	1 m = 100 cm = 1000 mm = 1.09 yd = 3.28 ft = 39.37 in. 1 km = 1000 m = 0.62 mi, 1 mm = 0.039 in. 1 ft = 12 in. = 0.305 m = 0.3 m, 1 in. = 2.54 cm = 25.4 mm 1 mi = 1.61 km ≅ 1.6 km, 1 yd = 0.914 m ≅ 0.9 m	Force	1 MN = 10^3 kN = 10^6 N = 225 k 1 N = 1 kgm/s^2 = 0.225 lbf 1 kN = 1000 N = 225 lbf = 0.225 k 1 lbf = 4.45 N ≅ 4.5 N, 1 k ≅ 4.45 kN
Length/time, velocity	1 m/s = 3.28 ft/s, 1 km/h = 0.62 mi/h 1 ft/s = 0.305 m/s, 1 mi/h = 1.61 km/h	Line load	1 kN/m = 10 N/cm = 1 N/mm = 68.52 lbf/ft = 0.069 k/ft 1 k/ft = 14.59 kN/m, 1 lbf/ft = 14.59 N/m
Acceleration	1 m/s^2 = 100 cm/s^2 = 3.28 ft/s^2, 1 ft/s^2 = 0.305 m/s^2	Surface load	1 kN/m^2 (kPa) = 20.89 lbf/ft^2 1 k/ft^2 = 47.88 kN/m^2, 1 lbf/ft^2 = 47.88 N/m^2 ≅ 48 N/m^2 (Pa)
Area	1 m^2 = 10^4 cm^2 = 10^6 mm^2 = 10.76 ft^2 ≅ 1.2 yd^2 1 cm^2 = 10^2 mm^2 = 0.155 $in.^2$ 1 ha = 2.47 acres 1 $in.^2$ = 6.45 cm^2 = 645 mm^2, 1 acre = 0.405 ha ≅ 0.4 ha 1 ft^2 = 0.093 m^2 ≅ 0.09m^2, 1 sq mi ≅ 2.6 km^2	Unit weight	1 kN/m^3 = 6.36 lbf/ft^3 1 k/ft^3 = 157 kN/m^3, 1 lbf/ft^3 = 157 N/m^3
		Bending moment	1 kNm = 738 lbf-ft = 0.738 k-ft, 1 Nm = 0.738 lbf-ft 1 k-ft = 1.356 kNm, 1 lbf-ft = 1.356 Nm

Volume, section
modulus
	1 m^3 = 10^6 cm^3 = 10^9 mm^3 = 1.31 yd^3 = 35.31 ft^3
	1 cm^3 = 10^3 mm^3 = 0.061 $in.^3$, 1 L = 0.22 gal
	1 $in.^3$ = 16.39 cm^3 = 16387 mm^3, 1 gal = 4.55 L
	1 ft^3 = 0.028 m^3 = 0.03 m^3, 1 yd^3 = 0.765 m^3
		Stress	1 MPa = 1 MN/m^2 = 0.1 kN/cm^2 = 1 N/mm^2
			     = 145 psi = 0.145 ksi
			1 kPa = 1 kN/m^2 = 20.89 psf = 0.145 psi
			1 Pa = 1 N/m^2 = 1 kg/ms^2 = 0.000145 psi
			1 ksi = 6.895 MPa (MN/m^2 = N/mm^2)
			     = 0.69 kN/cm^2
			1 psi = 6.895 kN/m^2 (kPa) ≅ 6.9 kPa

Moment of inertia	1 cm^4 = 0.024 $in.^4$, 1 ft^4 = 0.0086 m^4 1 $in.^4$ = 41.62 cm^4 = 0.4162$(10)^6$ mm^4		
Mass	1 kg = 1000 g = 2.205 lbm = 35.27 oz 1 t = 1000 kg = 2205 lbm = 1.102 tons 1 lbm = 16 oz = 0.454 kg, 1 oz = 28.35 g	Energy Temperature Linear expansion coefficient	1 J = 1 Nm = 0.738 lbf-ft, 1 lbf-ft = 1.356 Nm (J) °C = (°F − 32)/1.8, °F = 32 + 1.8°C 1°C^{-1} = 0.556°F^{-1}, 1°F^{-1} = 1.8°C^{-1}
Mass/volume, density	1 kg/m^3 = 0.062 lbm/ft^3 1 lbm/ft^3 = 16.109 kg/m^3, 1 lbm/yd^3 = 0.593 kg/m^3		

TABLE A.17
Typical Equivalent Metric Material Properties and Other Values for Preliminary Design Purposes

Material Properties	Steel 36.26 ksi = 250 N/mm^2 (MPa)	Reinforced Concrete 3.6 ksi = 25 N/mm^2 (MPa)
Compressive stresses	$0.6F_y$ ≅ 22 ksi ≅ 150 MPa	$0.25f'_c$ = 900 psi ≅ 6 MPa
Tensile stresses	$0.6F_y$ ≅ 22 ksi ≅ 150 MPa	Steel: 22 ksi ≅ 150 MPa
Flexural stresses	$0.66F_y$ ≅ 24 ksi ≅ 165 MPa	$0.45f'_c$ = 1620 psi ≅ 11 MPa
Shear stresses	$0.4F_y$ ≅ 14.5 ksi ≅ 100 MPa	$1.1\sqrt{f'_c}$ = 66 psi ≅ 0.5 MPa
Bearing stresses	$0.66F_y$ ≅ 24 ksi ≅ 165 MPa	$0.3f'_c$ = 1080 psi ≅ 7 MPa
Elastic modulus	29,000 ksi = 200 kN/mm^2	3420 ksi ≅ 25 kN/mm^2
Unit weight	490 pcf = 77 kN/m^3	150 pcf = 24 kN/m^3
Linear expansion coefficient	6.5×10^{-6} /°F = 12×10^{-6}/°C	5.5×10^{-6}/°F = 10×10^{-6}/°C

Material Properties	Brick Masonry 2 ksi ≅ 14 N/mm^2 (MPa)	Wood
Compressive stresses	$0.2f'_m$ = 400 psi ≅ 2.8 MPa	1000 psi ≅ 7 MPa
Tensile stresses	28 psi ≅ 0.2 MPa	600 psi ≅ 4 MPa
Flexural stresses	$0.33f'_m$ = 660 psi ≅ 4.6 MPa	1200 psi ≅ 8 MPa
Shear stresses	22 psi ≅ 0.15 MPa	87 psi ≅ 0.6 MPa
Bearing stresses	$0.25f'_m$ = 500 psi ≅ 3.4 MPa	385 psi ≅ 2.7 MPa
Elastic modulus	2400 ksi ≅ 17 kN/mm^2	1600 ksi ≅ 11 kN/mm^2
Unit weight	120 pcf = 19 kN/m^3	35 pcf ≅ 6 kN/m^3
Linear expansion coefficient	3.6×10^{-6}/°F ≅ 6×10^{-6}/°C	2.1×10^{-6}/°F ≅ 4×10^{-6}/°C

Soil bearing pressure	5200 psf ≅ 250 k N/m^2(kPa)
Live load	40 psf ≅ 2 kN/m^2 (kPa)
Water weight	62.4 pcf = 9.81 kN/m^3 ≅ 10 kN/m^3
Member span	20 ft ≅ 6 m, 100 ft ≅ 30 m, 500 ft ≅ 150 m
Member spacing	16 in. ≅ 400 mm, 24 in. ≅ 600 mm, 4 ft ≅ 1.20 m
Member thickness	1/4 in. ≅ 6 mm, 1/2 in. ≅ 13 mm, 2 in. ≅ 50 mm

Answers to Selected Problems

Chapter 2

(2.2) 14 pcf; (2.5) P_1 = 1029 k, P_{13} = 211 k; (2.6) 36 psf, 33 psf (UBC); (2.7) 15.08 ksi; (2.8) 13.20 ksi; (2.9) 88.28 ft; (2.15) −23.3°F; (2.21)$\Delta_{st} = \Delta_{AL} / 3$; (2.26) 75 times; (2.30) 28'9" × 5'3".

Chapter 3

(3.1) (a) 108 ft-k, (b) 28.13 ft-k, (d) 108 ft-k, (e) 96 ft-k, (f) 108 ft-k, (k) 16.31 ft-k, (m) 11.63 ft-k, (n) 5.53 ft-k; (3.6) W18 × 35; (3.7) M12 × 10; (3.8) (a) W12 × 14, (b) W16 × 40; (3.12) W16 × 57; (3.14) W33 × 118; (3.17) 2 × 12 in.; (3.18) 8¾ × 52 ½ in.; (3.20) w = 1.8 klf; (3.22) (a) M_u = 262 ft-k, (b) M_u = 280 ft-k; (3.23) 4 #8; (3.24) (a) 14 × 28 in., (b) 5 #8; (3.26) (a) 10 × 23 in., (b) 3 #8; (3.27) (a) 10 × 20 in., (b) 3 #8 top, 2 #8 bottom; (3.30) #3 at 6" and #3 at 12"; (3.32) (a) 4-in. slab, (b) #4 at 12 in. in the field, (c) #4 at 8.5 in. at the supports, (d) #3 at 13.5 in. temperature steel; (3.35) 138.62 k; (3.36) 158.76 k; (3.37) 2L 6 × 3 ½ × ⅜; (3.39) 2-in. rod; (3.42) W12 × 120; (3.43) W14 × 193; (3.44) W14 × 120; (3.46) TS 5 × 0.258 pipe; (3.48) W14 × 99; (3.50) W10 × 100; (3.52) section is satisfactory; (3.53) W14 × 159; (3.54) W12 × 106; (3.55) 4 × 8 in.; (3.56) 16 in. o.c.; (3.57) 6 × 6 in.; (3.58) column is satisfactory; (3.63) 4 #9, #3 ties at 12 in.; (3.64) (a) 16 × 16 in., (b) 8 #8, (c) #3 ties at 16 in.; (3.66) (a) 16 × 22 in., (b) 18 × 24 in.; (3.67) (a) 16 × 16 in., (b) 4 #10 and 4 #8, (c) #3 ties at 16 in.; (3.69) 0.213 in.; (3.70) (a) 10 × 10 in., (b) 4 #5, (c) #3 ties at 10 in.; (3.72) (a) 15 × 15 in., (b) 4 #8, (c) #3 ties at 15 in.; (3.73) 14 × 14 in.; (3.76) 28 × 28 in.; (3.77) W16 × 40; (3.78) 6 #9; (3.81) (a) 4.33 × 1 ft footing, (b) #4 at 9 in.; (3.82) 17 in. thick; (3.83) (a) 8'-10" square footing, (b) 26 in. thick, (c) 12 #6 e.w.; (3.86) PL 1½ × 10 × 20 in.; (3.87) PL ½ × 11 × 13 in.; (3.88) (a) 10 bolts on each side, (b) 6 bolts on each side.

Chapter 4

(4.2) $V = 0.103W$; (4.5) $V = 0.059W$; (4.7) $M_s = 360$ ft-lb, $M_f = 540$ ft-lb; (4.10) $P_1 = 0.25P$, $P_2 = 0.75P$; (4.18) 2.54 ft; (4.19) 38.88 ft-k/ft; (4.23)$M_{max} = 1.57$ ft-k/ft; (4.24)$M_{max} = 0.38$ ft-k.

Chapter 5

(5.7) (a) W30 × 90 (beam), W10 × 33 (column), (c) W30 × 90 (beam), W27 × 114 (column), (e) W21 × 62 (beam), W18 × 71 (column); (5.10) 4.71 ksi; (5.11) 32.13 ft-k; (5.18) W30 × 90 (beam), W16 × 77 (ext. column), W10 × 39 (int. column); (5.20) (a) 2 × 10 in., 4 × 10 in., (b) 6 × 12 in.; (5.21) 2 × 6 in.; (5.22) 2 × 6 in.; (5.25) 3⅛ × 12 in., 1⅛ - in.-diameter rod; (5.26) W10 × 12; (5.27) W30 × 90; (5.28) W24 × 84; (5.29) W33 × 118; (5.34) W10 × 22; (5.35) 3⅛ × 9 in. (top chord), 3⅛ × 7½ (bottom chord).

Chapter 6

(6.11) (a) $wL^2/9$, (b) $wL^2/6$; (6.12) WT8 × 13 (bottom chord), W6 × 15 (top chord),2L 3 × 3 × ⅜ (diagonal); (6.13) TS 5 OD × 0.258 (top chord), TS 6 OD × 0.280 (bottom chord), TS 3 ½ OD × 0.226 (diagonal); (6.14) TS 3 OD × 0.216 (top chord), TS 20 D × 0.154 (bottom chord),TS 3½ OD × 0.226 (diagonal); (6.16) TS 4 OD × 0.237.

Chapter 7

(7.2) 1.17 ksi; (7.3) L 8 × 8 × 9/16 (chords), W4 × 13 (end post), W4 × 13 (diagonal), L 3½ × 3 × ¼ (end wall tie); (7.6) 9 #8, main reinforcing; (7.7) 16 #11, main reinforcing.

Chapter 8

(8.1) 13 #9 main steel bars; (8.2) 25 #10 main steel bars; (8.4) #3 at 17 in. b.w.; (8.5) #4 at 13.5-in. hoop reinforcement along base area; (8.6) 23 #10 reinforcing bars in ring footing; (8.13) twelve 0.438-in. diameter Grade 270 strands, 16- × 20-in. ring beam; (8.15) W6 × 9 (rafter), W8 × 10 (purlin), W14 × 53 (tension ring); (8.17) #4 at 9 in. (along tension lines), #3 at 15 in. (along compression lines); (8.20) #4 at 8.5 in. (along tension lines), #3 at 15 in. (along compression lines); (8.21) W6 × 15 grid members, 3¼ × 1-in. bars in diagonal direction; (8.25) #3 at 13.5 in. (along tension lines), #3 at 15 in. (along compression lines);

Chapter 9

(9.1) (a) 2⅞-in.-diameter strand, (b) 1.23 vertical displacement, (c) W12 × 58; (9.3) ¼ in.; (9.4) 312 cables; (9.5) (a) W14 × 99, (b) 2 ⅜-in.-diameter strand; (9.6) ½ -in diameter strand, 6.4-ft sag at 20 ft, 102.67 ft long, 0.366-ft elongation, 0.686-ft increase in cable sag; (9.9) 0.36 ft; (9.10) 1⅜-in.-diameter strand, 15-ft cable sag at $L/4$, 9.46° cable slope at $L/4$, 244.44 ft long, 0.826-ft elongation, 0.079-ft elongation due to temperature, 2.04-ft increase in cable sag; (9.12) three- 3¾-in.-diameter strands; (9.14) 1½ -in.-diameter strand; (9.15) 1⅝-in.-diameter strand; (9.16) three-3½-in.-diameter strands; (9.17) 1¼-in.-diameter strand; (9.18) 1.92 ft; (9.19) 2.51 ft; (9.20) 2 psf; (9.21) 2³⁄₁₆ - in.-diameter strand; (9.22) 1⅛-in.-diameter strand; (9.23) ½-in.-diameter rope (net), 2⅛-in.-diameter strand (edge cable); (9.25) 0.53 k/in.; (9.26) 1-in.-diameter strand; (9.29) 6.80 psf, 84 lb/in.; (9.30) 27.02 psf, 1/16 in.; (9.31) 3 psf, 268 lb/in.; (9.33) 9.68 psf, 331 pli; (9.34) 14.4 psf, 276 lb/in.; (9.37) 2-ft diameter, 7.2 k/in.; (9.38) 28.47 psi, 854 lb/in.

Bibliography and References

For names of designers and buildings used in the drawings refer to List of Buildings in Figures which can be obtained from the author.

General

Ackermann, Kurt. *Grundlagen für das Entwerfen and Konstruieren,* Karl Krämer Verlag, Stuttgart, Germany, 1983.

————. *Industriebau,* Deutsche Verlags-Anstalt (DVA), Stuttgart, Germany, 1984.

————. *Tragwerke in der konstruktiven Architektur*, Deutsche Verlags-Anstalt (DVA), Stuttgart, Germany, 1988.

Acland, James H. *Medieval Structure. The Gothic Vault,* University of Toronto Press, Toronto, 1972.

American Iron and Steel Institute, *Long-span Steel Roof Structures,* The Institute, Washington, D.C., 1978.

Architects and Earthquakes, AIA Research Corporation, Washington, D.C., NSF/RA-770156, 1977.

Arnold, C., and Reitherman, R. *Building Configuration and Seismic Design,* Wiley, New York, 1982.

ASCE Standard. *Minimum Design Loads for Buildings and Other Structures,* ASCE 7-88, American Society of Civil Engineers, New York, 1990.

Barratt, Krome. *Logic and Design,* Eastview Editions, Inc., Westfield, N.J., 1980.

Beckett, Derrick. *Bridges,* Paul Hamlyn, London, 1969.

Bill, Max. *Robert Maillart,* Verlag für Architektur AG, Erlenbach-Zurich, Switzerland, 1949.

Billington, P. David. *The Tower and the Bridge,* Basic Books, Inc., New York, 1983.

———— and Mark, Robert. *Structures and the Urban Environment,* Civil Engineering Department, Princeton University, Princeton, N.J., 1983.

Blaser, Werner, ed. *Myron Goldsmith, Buildings and Concepts,* Rizzoli, New York, 1987.

———— ed. *Santiago Calatrava,* Birkhäuser Verlag, Basel, Switzerland, 1989.

Breuer, György. *Gyakorlati Szerkezettervezés*, Vols. 1 and 2, Müszaki Könyvkiadó, Budapest, Hungary, 1973.

Brookes, Alan J., and Grech, Chris. *The Building Envelope,* Butterworth Architecture, London, 1990.

Building Arts Forum/New York. *Bridging the Gap,* Van Nostrand Reinhold, New York, 1991.

Buildings at Risk: Seismic Design Basics for Practicing Architects, AIA/ACSA Council on Architectural Research, Washington, D.C., 1992.

Burt, Michael. Spatial Arrangement and Polyhedra with Curved Surfaces and Their Architectural Applications, M.S. Thesis, Technion, Haifa, Israel, 1966.

Büttner, Oskar, and Hampe, Erhard. *Bauwerk, Tragwerk, Tragstruktur,* Vol. 1, Verlag Gerd Hatje, Stuttgart, Germany, 1976; Vol. 2, Ernst & Sohn, Berlin, Germany, 1985.

Cavallari-Murat, Augusto. Static Intuition and Formal Imagination in the Space Lattices of Ribbed Gothic Vaults, *Student Publications of the School of Design,* Vol. 11, No. 2, North Carolina State College, Raleigh, 1963.

Cole, Campbell B., and Rogers, Elias R., eds. *Richard Rogers + Architects,* St. Martin's Press, New York, 1985.

Cook, Peter. *Experimental Architecture,* Universe Books, New York, 1970.

Cowan, Henry J. *An Historical Outline of Architectural Science,* 2nd ed., Elsevier, New York, 1977.

———. *The Master Builders,* Wiley, New York, 1977.

———. *Science and Building,* Wiley, New York, 1978.

———. and Wilson, Forrest. *Structural Systems,* Van Nostrand Reinhold, New York, 1981.

Dahinden, Justus. *Urban Structures for the Future,* Praeger, New York, 1972.

Degenkolb, Henry J. Earthquake, Booklet 2717A, Bethlehem Steel, 1977.

Dietz, Albert G. H. *Plastics for Architects and Builders,* MIT Press, Cambridge, Mass., 1969.

Eekhout, Mick. *Product Development in Glass Structures,* Uitgeverij 010 Publishers, Rotterdam, The Netherlands, 1990.

Elliott, Cecil D. *Technics and Architecture,* MIT Press, Cambridge, Mass., 1992.

Faegre, Torvald. *Tents, Architecture of the Nomads,* Anchor Press/Doubleday, Garden City, N.Y., 1979.

Fehn, Sverre. *The Thought of Construction,* Rizzoli, New York, 1984.

Fonatti, Franco. *Basic Principles of Architectural Design,* Akademie der bildenden Künste, Vienna, Austria, 1982.

Forest Products Laboratory Forest Service, U.S. Department of Agriculture. *Wood Handbook,* U.S. Government Printing Office, Washington, D.C., 1974.

Foster, Jack Stroud, and Harrington, Raymond. *Structure and Fabric,* Part 2, B. T. Batsford Ltd., London, 1976.

Francis, A. J. *Introducing Structures,* Ellis Horwood Ltd., Chichester, England, 1989.

Gordon, J. E. *The New Science of Strong Materials,* Walker and Company, New York, 1968.

Gordon, J. E. *The Science of Structures and Materials,* Scientific American Library, New York, 1988.

———. *Structures,* Plenum, New York, 1978.

Götz, Karl-Heinz, Hoor, Dieter, Möhler, Karl, and Natterer, Julius. *Timber Design and Construction Sourcebook,* McGraw-Hill, New York, 1989.

Graefe, Rainer, ed. *Zur Geschichte des Konstruierens, Deutsche Verlags-Anstalt,* Stuttgart, Germany, 1989.

Graver, Jack E., and Baglivo, Jenny A. *Incidence and Symmetry in Design and Architecture,* Cambridge University. Press, New York, 1984.

Green, Norman B. *Earthquake Resistant Building Design and Construction,* Van Nostrand Reinhold, New York, 1978.

Grillo, Paul Jacques. *Form Function and Design,* Dover, New York, 1975.

Grube, Oswald W. *Industrial Buildings and Factories,* Praeger, New York, 1971.

Grube, Oswald W., Pran, Peter C., and Schulze, Franz. *100 Years of Architecture in Chicago,* Follett, Chicago, 1977.

Günschel, Günter. *Grosze Konstrukteure I,* Ullstein, Berlin, Germany, 1966.

Hancocks, David. *Master Builders of the Animal World,* Harper & Row, New York, 1973.

Hart, Franz. *Kunst und Technik der Wölbung,* Verlag Georg D. W. Callwey, Munich, Germany, 1965.

Hawkes, Nigel. *Structures,* Macmillan Publishing Co. New York, 1990.

Hayden, Martin. *The Book of Bridges,* Galahad Books, New York, 1976.

Henn, Walter. *Buildings for Industry,* London Iliffe Books Ltd., London, 1965.

Hildebrandt, S., and Tromba A. *Mathematics and Optimal Form,* Scientific American Library, New York, 1985.

Hilson, Barry. *Basic Structural Behavior,* Thomas Telford, London, 1993.

Hodgkinson, Allan, ed. *AJ Handbook of Building Structure,* Architectural Press, London, 1974.

Hollaway, Leonard. *Glass Reinforced Plastics in Construction,* Wiley, New York, 1978.

Howard, Seymour H., Jr. *Structure: An Architect's Approach,* McGraw-Hill, New York, 1966.

Huxtable, Ada Louise. *Pier Luigi Nervi,* George Braziller, New York, 1960.

Jacobs, Harold R. *Mathematics. A Human Endeavor,* W. H. Freeman, San Francisco, 1970.

Kepes, Gyorgy, ed. *Module, Symmetry, Proportion,* Studio Vista, London, 1966.

———, ed. *Structure in Art and in Science,* Studio Vista, London, 1965.

Klotz, Heinrich, ed. *Vision der Moderne—Das Prinzip Konstruktion,* Prestel-Verlag, Munich, Germany, 1986.

Koncz, Tihamér. *Manual of Precast Concrete Construction,* Vols. 1–3, Bauverlag, Wiesbaden and Berlin, Germany, 1968.

Lawlor, R. *Sacred Geometry,* Crossroads, New York, 1982.

Lebedew, J. S. *Architektur und Bionik,* VEB Verlag, Berlin, 1983.

Leonhardt, Fritz. *Brücken/Bridges,* MIT Press, Cambridge, Mass., 1984.

Levy, Matthys, and Salvadori, Mario. *Why Buildings Fall Down,* W. W. Norton, New York, 1992.

Lotus International 45. *Engineering in Architecture,* Milan, Italy, 1985.

Mainstone, Rowland. *Developments in Structural Form,* MIT Press, Cambridge, Mass., 1975.

March, Lionel, and Steadman, Philip. *The Geometry of Environment,* MIT Press, Cambridge, Mass., 1974.

Mark, R. *Experiments in Gothic Structure,* MIT Press, Cambridge, Mass., 1982.

———. *Light, Wind, and Structure,* McGraw-Hill, New York, 1990.

———. The Structural Analysis of Gothic Cathedrals, *Scientific American,* November 1972; Structural Experimentation in Gothic Architecture, *American Scientist,* 10/11, 1978.

McMahon, Thomas A., and Bonner, John T. *On Size and Life,* Scientific American Library, New York, 1983.

Metal Building Manufacturers Association, *Metal Building Systems Manual,* The Association, Cleveland, Ohio, 1981.

Michaels, Leonard. *Contemporary Structure in Architecture,* Reinhold Publishing Co., New York, 1950.

Michailenko, W. I., and Kaschtschenko, A. W. *Natur-Geometrie-Architektur,* VEB Verlag, Berlin, 1986.

Monasa, Frank F., ed. *Approximate Methods and Verification Procedures of Structural Analysis and Design,* American Society of Civil Engineers, New York, 1991.

Nachtigall, Werner. *Biotechnik,* Quelle & Meyer, Heidelberg, Germany, 1971.

Nervi, Pier Luigi. *Aesthetics and Technology in Building,* Harvard University Press, Cambridge, Mass., 1966.

Nervi, Pier Luigi. *Buildings, Projects, Structures, 1953–1963,* Praeger, New York, 1963.

Nicholas Grimshaw & Partners. *Book 1 Product, Book 2 Process,* NGP, London, 1988.

Orton, Andrew. *The Way We Build Now,* Van Nostrand Reinhold (UK), Co. Ltd, Wokingham Berkshire, England, 1988.

Quarmby, Arthur. *The Plastics Architect,* Pall Mall Press, London, 1974.

Peters, Tom F. *Transitions in Engineering,* Birkhäuser Verlag, Boston, 1987.

Petroski, Henry. *To Engineer Is Human. The Role of Failure in Successful Design,* St. Martin's Press, New York, 1985.

Report of Summer Seismic Institute for Architectural Faculty, Stanford University, August 1977, AIA Research Corporation, Washington, D.C., 1977.

Rickenstorf, Günther. *Tragwerke für Hochbauten,* BSB B. G. Teubner Verlagsgesellschaft, Leipzig, German Democratic Republic, 1972.

Rickey, George. *Constructivism,* George Braziller, New York, 1967.

Ruske, W. *Holzskelettbau,* Deutsche Verlags-Anstalt (DVA), Stuttgart, Germany, 1980.

Sandaker, Bjørn N., and Eggen, Arne P. *The Structural Basis of Architecture,* Whitney Library of Design, New York, 1992.

Sandori, Paul. *The Logic of Machines and Structures,* Wiley, New York, 1982.

Salvadori, Mario. *Why Buildings Stand Up,* W. W. Norton, New York, 1980.

Schmitt, Heinrich. *Hochbau Konstruktion,* 6th ed., Vieweg & Sohn, Braunschweig, Germany, 1977.

Schodek, Daniel L.. *Structure in Sculpture,* MIT Press, Cambridge, Mass., 1993.

Schueller, Wolfgang. *Horizontal-span Building Structures,* Wiley, New York, 1983.

Schulitz, Helmut C., Schüsseler, Jan, and Sprysch, Michael. *Aircraft Hangars,* Quadrato Verlag, Braunschweig, Germany, 1989.

Shawcroft, Brian. Building Skeletons, *Student Publications of the School of Design,* Vol. 17, No. 1, North Carolina State University, Raleigh, 1967.

Siegel, Kurt. *Structure and Form in Modern Architecture,* Reinhold, New York, 1962.

Sontag, H., Hart, F., and Henn, W. *Multi-Storey Buildings in Steel,* Wiley, New York, 1978.

Space Forms in Steel, *AISC Engineering Journal,* December 1965, American Institute of Steel Construction, New York.

Stevens, Garry. *The Reasoning Architect,* McGraw-Hill, New York, 1990.

Stevens, Peter S. *Patterns in Nature,* Little, Brown and Co., Boston, 1974.

Thompson, D'Arcy. *On Growth and Form,* abridged ed., John Tyler Bonner, ed., Cambridge University Press, New York, 1979.

Torroja, Eduardo. *Philosophy of Structures,* University of California Press, Berkeley and Los Angeles, 1958.

———. *The Structures of Eduardo Torroja,* F. W. Dodge Corp., New York, 1958.

Underground Space Center, University of Minnesota. *Earth Sheltered Housing Design,* Van Nostrand Reinhold, New York, 1979.

Von Büren, Charles. *Function & Form,* Birkhäuser Verlag, Basel, Switzerland, 1958.

Von Frisch, Karl. *Animal Architecture,* Harcourt Brace Jovanovich, New York, 1974.

Wachsmann, Konrad. *The Turning Point of Building—Structure and Design,* Reinhold, New York, 1961.

Walker, Derek. *Great Engineers,* Academy Editions, London, 1987.

Weidlinger, Paul. Visualizing the Effect of Earthquakes on the Behavior of Building Structures, *Architectural Record,* May 1977.

Whyte, Lancelot Law. *Aspects of Form,* Indiana University Press, Bloomington, 1961.

Wilkinson, Chris. *Supersheds,* Butterworth Architecture, London, 1991.

Wilson, Forrest. *Emerging Form in Architecture. Conversations with Lev Zetlin,* Cahners Books, Boston, 1975.

Wunderlich, K., and Gloede, W. *Natur als Konstrukteur,* Edition Leipzig, Germany, 1977.

Yanev, Peter. *Peace of Mind in Earthquake Country,* Chronicle Books, San Francisco, 1974.

Zamos, A. *Form and Structure in Architecture,* Van Nostrand Reinhold, New York, 1987.

Zuk, William, and Clark, Roger H. *Kinetic Architecture,* Van Nostrand Reinhold, New York, 1970.

Structural Design and Analysis

American Institute of Timber Construction. *Timber Construction Manual,* 3rd ed., Wiley, New York, 1986.

Andersen, Paul, and Nordby, Gene, M. *Introduction to Structural Mechanics,* Ronald Press, New York, 1960.

Benjamin, B. S. *Structures for Architects,* 2nd ed., Van Nostrand Reinhold, New York, 1984.

Breyer, D. E. *Design of Wood Structures,* 3rd ed., McGraw-Hill, 1993.

Building Code Requirements for Reinforced Concrete (ACI 318-89), American Concrete Institute, Detroit, December 1989.

Coleman, R. A. *Structural Systems Design,* Prentice Hall, Englewood Cliffs, N.J., 1983.

Cowan, Henry J. *Design of Reinforced Concrete Structures,* 2nd ed., Prentice Hall, Englewood Cliffs, N.J., 1989.

Crawley, S. W., and Dillon, R. M. *Steel Buildings, Analysis and Design,* 4th ed., Wiley, New York, 1993.

Fanella, David A., and Ghosh, S. K., eds. *Simplified Design, Reinforced Concrete Buildings of Moderate Size and Height,* 2nd ed., Portland Cement Association (PCA), Skokie, Ill, 1993.

Fling, R. S. *Practical Design of Reinforced Concrete,* Wiley, New York, 1987.

Gaylord, Edwin H., and Gaylord, Charles N., eds. *Structural Engineering Handbook,* 3rd ed., McGraw-Hill, New York, 1989.

Gurfinkel, German. *Wood Engineering,* Southern Forest Products Association, New Orleans, La., 1973.

Lauer, K. R. *Structural Engineering for Architects,* McGraw-Hill, New York, 1981.

Lin, T. Y., and Stotesbury, S. D., *Structural Concepts and Systems for Architects and Engineers,* 2nd ed., Van Nostrand Reinhold, New York, 1988.

Lisborg, Niels. *Principles of Structural Design,* 2nd ed., B. T. Batsford Ltd., London, 1967.

MacGregor, J. G. *Reinforced Concrete,* Prentice Hall, Englewood Cliffs, N.J., 1988.

Manual of Steel Construction (ASD), 9th ed., American Institute of Steel Construction, Chicago, 1989.

McCormac, J. C. *Structural Analysis,* 2nd ed., International Textbook, Scranton, Pa., 1967.

———. *Structural Steel Design,* 3rd ed., Harper & Row, New York, 1981.

Nash, Alec. *Structural Design for Architects,* Nichols Publishing Co., New York, 1990.

Nilson, Arthur H. *Design of Prestressed Concrete,* 2nd ed., Wiley, New York, 1987.

Norris, Charles H., and Wilbur, John B. *Elementary Structural Analysis,* 2nd ed., McGraw-Hill, New York, 1960.

Salvadori, Mario. *Statics and Strength of Structures,* Prentice Hall, Englewood Cliffs, N.J., 1971.

———, and Levy, Matthys. *Structural Design in Architecture,* 2nd ed., Prentice Hall, Englewood Cliffs, N.J., 1981.

Schodek, Daniel L. *Structures,* Prentice Hall, Englewood Cliffs, N.J., 1980.

Shaeffer, R. E. *Reinforced Concrete,* McGraw-Hill, New York, 1992.

Spiegel, Leonard, and Limbrunner, G. F. *Applied Structural Steel Design,* 2nd ed., Prentice Hall, Englewood Cliffs, N.J., 1993.

———, and ———. *Reinforced Concrete Design,* 3rd ed., Prentice Hall, Englewood Cliffs, N.J., 1992.

Stalnaker, J. J., and Harris, E. C. *Structural Design in Wood,* Van Nostrand Reinhold, New York, 1989.

Teng, Wayne C. *Foundation Design,* Prentice Hall, Englewood Cliffs, N.J., 1962.

White, R. N., and Salmon, G. G., eds. *Building Structural Design Handbook,* Wiley, New York, 1987.

——— Gergely, P., and Sexsmith, R. G. *Structural Engineering,* Vol. 1, Wiley, New York, 1972.

Winter, George, and Nilson, Arthur H., et al. *Design of Concrete Structures,* 8th ed., McGraw-Hill, New York, 1972.

Wood Construction, 1986, ed. National Design Specification, National Forest Products Association, Washington, D.C.

Space Frame and Surface Grid Structures

Baer, Steve. *Zome Primer,* Zomeworks Corp., Albuquerque, N.M., 1970.

Barr, Stephen. *Experiments in Topology,* Thomas Y. Cromwell, New York, 1964.

Benjamin, B. S. *The Analysis of Braced Domes,* Asia Publishing House, Bombay, India, 1963.

Borrego, John. *Space Grid Structures,* MIT Press, Cambridge, Mass., 1968.

Büttner, Oskar, and Stenker, Horst. *Metalleichtbauten, Deutsche Verlags-Anstalt,* Stuttgart, Germany, 1970.

Critchlow, Keith. *Order in Space,* Viking, New York, 1970.

Cuoco, Daniel A. Today's Space Frame Structures, *Architectural Record,* June 1982.

Davies, R. M., ed. *Space Structures,* Wiley, New York, 1967.

Domebook 2, Pacific Domes, Bolinas, Calif., 1971.

Eekhout, Mick. *Architecture in Space Structures,* Uitgeverij 010 Publishers, Rotterdam, The Netherlands, 1989.

Emmerich, David G. *Constructive Geometry,* Department of Architecture, U. of Washington, Seattle, Washington, 1970.

Engel, Heinrich. *Structure Systems,* Praeger, New York, 1968.

Fuller, Buckminster R. *Synergetics,* Macmillan, New York, 1975.

———, and Marks, Robert. *The Dymaxion World of Buckminster Fuller,* Anchor Press/Doubleday, Garden City, New York, 1973.

Gheorghiu, Adrian, and Dragomir, Virgil. *Geometry of Structural Forms,* Applied Science Publishers Ltd., London, 1978.

Ghyka, Matila. *The Geometry of Art and Life,* Dover, New York, 1977.

Gugliotta, Paul. Architecs' Guide to Space Frame Design, *Architectural Record,* mid-August, 1980.

Kappraff, Jay. *Connections,* McGraw-Hill, New York, 1991.

Kenner, Hugh. *Geodesic Math,* University of California Press, Berkeley, 1976.

Lee, Hung-gum, and Makowski, Stanislaw. Study of Factors Affecting Stress Distribution in Double-layer Grids of the Square and Diagonal Type, *Architectural Science Review,* December 1977.

Loeb, Arthur L. *Space Structures,* Addison-Wesley, Reading, Mass., 1976.

Makowski, Zygmunt S., ed. *Analysis, Design, and Construction of Double-layer Grids,* Applied Science Publishers, London, 1981.

———. *Steel Space Structures,* Michael Joseph, London, 1965.

Mengeringhausen, M. *Raumfachwerke aus Stäben und Knoten,* Bauverlag GmBH, Wiesbaden, Germany, 1975.

Mimram, Marc. *Structures et Formes,* Dunod Presses, Paris, 1983.

Miyazaki, Koji. *Form of Space,* Asakura Publishing Co., Tokyo, 1983.

Naslund, Kenneth C. Design Considerations for Horizontal Space Frames, *Architectural Record,* August 1964.

Pearce, Peter, and Pearce, Susan. *Experiments in Form,* Van Nostrand Reinhold, New York, 1980.

Pearce, Peter. *Structure in Nature Is a Strategy for Design,* MIT Press, Cambridge, Mass., 1978.

Popko, Edward. *Geodesics,* School of Architecture, University of Detroit, Detroit, Michigan, 1968.

Pugh, Anthony. *Polyhedra,* University of California Press, Berkeley, 1976.

———, *Tensegrity,* University of California Press, Berkeley, 1976.

Richter, Don L. Geodesic Domes, *Forum,* January/February 1972.

Rühle, Hermann. *Räumliche Dachtragwerke Konstruktion und Ausführung,* Vol. 2, Verlagsgesellschaft Rudolf Müller, Köln, Germany, 1970.

Second International Conference on Space Structures, Department of Civil Engineering, University of Surrey, Guildford, England, September 1975.

Sedlak, Vinzenz. Paper Shelters, *Architectural Design,* December 1973.

Space Forms in Steel, American Institute of Steel Construction, New York, 1965.

Stevens, David E., and Odom, Gerald S. The Steel Framed Dome, *AISC Engineering Journal,* July 1964.

Subramanian, N. *Principles of Space Structures,* Wheeler & Co., Allahabad, India, 1983.

Wendel, Wendel R. *Spaceframe Basics,* 4th ed., Starnet Structures, Inc., West Babylon, N.Y., 1988.

Williams, Robert. *Natural Structure,* Eudaemon Press, Moorpark, Calif., 1972.

Wilson, Forrest. Of Space Frames, Time, and Architecture, *Architecture,* August 1987.

Shell and Folded Plate Structures

Abercrombie, Stanley. *Ferrocement,* Schocken Books, New York, 1977.

Angerer, Fred. *Surface Structures in Building,* Reinhold, 1961.

Billington, David P. *Thin Shell Concrete Structures,* McGraw-Hill, New York, 1965.

Carney, J. M. Plywood Folded Plates, Laboratory Report 212, American Plywood Association, Tacoma, Wash., 1971.

Catalano, Eduardo F. Structures of Warped Surfaces, *Student Publication of the School of Design,* Vol. 10, No. 1, North Carolina State College, Raleigh, 1960.

Christiansen, Jack, ed. *Hyperbolic Paraboloid Shells—State of the Art,* SP 110 ACI, American Concrete Institute, Detroit, Mich., 1988.

Chronowicz, Albin. *The Design of Shells,* 3rd ed., Crosby Lockwood & Son Ltd., London, 1968.

Concrete Thin Shells, Publication SP-28, American Concrete Institute, Detroit, Mich., 1971.

Design Examples. Space Forms in Steel, American Institute of Steel Construction, 1966.

Design of Circular Domes, Portland Cement Association, ST-55, Chicago.

Dome Structures, *Consulting Engineer,* December 1959.

Elementary Analysis of Hyperbolic Paraboloid Shells, Portland Cement Association, ST-85, Chicago, 1960.

Engel, Heinrich. *Structure Systems,* Praeger, New York, 1968.

Faber, Colin. *Candela, The Shell Builder,* Reinhold, 1963.

Francis, A. J. Domes, *Architectural Science Review,* November 1962.

Frostick, Peter. Antiprism Based Form Possibilities for Folded Surface Structures, *Architectural Science Review,* September 1978.

Haas, A. M. *Design of Thin Concrete Shells,* Vols. 1 and 2, Wiley, New York, 1962, 1967.

Heinz Isler as Structural Artist, the Art Museum, Princeton University, Princeton, N.J. 1980.

Hyperbolic Paraboloid Shells, Western Wood Products Association, Portland, Ore., Technical Guide TG-4.

Iffland, Jerome S. B. Folded Plate Structures, *Journal of the Structural Division,* American Society of Civil Engineers, Vol. 105, No. ST1, Proc. Paper 14300, 1979.

Joedicke, Jürgen. *Shell Architecture,* Reinhold, New York, 1963.

Ketchum, Milo. Design of Shell Structures. Barrel Vaults, *Consulting Engineer,* September 1961.

Makowski, Z. S., ed. *Braced Domes,* Nichols Publishing Co., New York, 1984.

Melaragno, Michele. *Shell Structures,* Van Nostrand Reinhold, New York, 1991.

Nilson, Arthur H. Steel Shell Roof Structures, *AISC Engineering Journal,* January 1966.

Pflüger, Alf. *Elementary Statics of Shells,* 2nd ed., F. W. Dodge, New York, 1961.

Ramaswamy, G. S. *Design and Construction of Concrete Shell Roofs,* McGraw-Hill, New York, 1968.

Ramm, Ekkehard, and Schunck, Eberhard. *Heinz Isler,* Krämer Verlag, Stuttgart, 1986.

Rühle, Hermann. *Räumliche Dachtragwerke Konstruktion und Ausführung,* Vol. 1, Verlagsgesellschaft Rudolf Müller, Köln, Germany, 1969.

Sedlak, Vinzenz. Paper Shelters; Folded Structural Forms in Paperboard, *Arcitectural Design,* December 1973.

Smith, Baldwin E. *The Dome,* Princeton University Press, Princeton, N.J., 1971.

Torroja, Eduardo. *The Structures of Eduardo Torroja,* F. M. Dodge, New York, 1958.

Winter, George, and Nilson, Arthur H. *Design of Concrete Structures,* 8th ed., McGraw-Hill, New York, 1972.

Tensile Structures

Air Structures, *Engineering News-Record,* August 1974.

Air Structures, Proceedings of the International Conference on the Practical Application for Air Supported Structures, 1974, Canvas Products Association, St. Paul, Minn., 1976.

Berger, Horst. The Engineering Discipline of Tent Structures, *Architectural Record,* February 1975.

Boys, C. V. *Soap-Bubbles,* Dover, New York, 1959.

Bubner, Ewald. *Zum Problem der Formfindung vorgespannter Seilnetzflächen,* IGMA Dissertationen 2, University of Stuttgart, Karl Krämer Verlag Stuttgart/Bern, Germany, 1972.

Cable Construction in Contemporary Architecture, Booklet 2264A, Bethlehem Steel Corp., Bethlehem, Pa., 1966.

Cable Roof Structures, Booklet 2318A, Bethlehem Steel Corp., Bethlehem, Pa., 1968.

Dent, Roger N. *Principles of Pneumatic Architecture,* Halsted Press, New York, 1972.

Design and Standards Manual, Publication ASI-77, Air Structures Institute (ASI), St. Paul, Minn., 1977.

Drew, Philip. *Frei Otto, Form and Structure,* Westview Press, Boulder, Colo., 1976.

———. *Tensile Architecture,* Granada Publications, London, 1980.

Engel, Heinrich. *Structure Systems,* Praeger, New York, 1968.

Era of Swoops and Billows, The, Fabric Structures, *Progressive Architecture,* June 1980.

Fabric Structures Grow Up, *Architecture Plus*, October 1973.

Faegre, Torvald. *Tents. The Architecture of Nomads,* Anchor Books, Doubleday, Garden City, New York, 1979.

Geiger, David H. Pneumatic Structures, *Progressive Architecture*, August, 1972.

Hanging Roofs, Booklet 2319, Bethlehem Steel Corp., Bethlehem, Pa., 1967.

Hatton, E. M. *The Tent Book*, Houghton Mifflin, Boston, 1979.

Herzog, Thomas. *Pneumatic Structures*, Oxford University Press, New York, 1976.

Howard, Seymour H., Jr. Suspended Structures Concepts, United States Steel, Pittsburgh, Pa., 1966.

Institute for Lightweight Structures (IL), University of Stuttgart, Germany: numerous publications, e.g., IL 1 to IL 36.

International Conference on Tension Roof Structures, Polytechnic of Central London, 1974.

Joseph, Marjory L. *Introductory Textile Science*, 2nd ed., Holt, Rinehart and Winston, New York, 1972.

Krishna, Prem. *Cable-suspended Roofs,* McGraw-Hill, New York, 1978.

Light Structures, *Zodiac 21*, Milan, Italy, 1972.

Manual for Structural Applications of Steel Cables for Buildings, 1973 ed., American Iron and Steel Institute, Washington, D.C.

Nicholas Grimshaw & Partners. *Book 1 Product, Book 2 Process,* NGP, London, 1988.

Otto, Frei, ed. *Tensile Structures*, MIT Press, Cambridge, Mass., 1973.

Proceedings of the First International Colloquium on Pneumatic Structures, IASS, University of Stuttgart, Germany, 1967.

Proceedings of the IASS Pacific Symposium—Part 2 on Tension Structures and Space Frames, Tokyo and Kyoto, 1971, Architectural Institute of Japan, Tokyo, 1972.

Proceedings of the IASS World Congress on Space Enclosures, Vol. 2, Concordia University, Montreal, Canada, 1976.

Proceedings of the International Symposium on Architectural Fabric Structures. The Design Process, Vol. 1, Orlando, 1984, IFAI, St. Paul, Minn.

Roland, Conrad, and Otto, Frei. *Tension Structures*, Praeger, New York, 1970.

Rühle, Herrmann. *Räumliche Dachtragwerke Konstruktion und Ausführung,* Vol. 2, Verlagsgesellschaft Rudolf Müller, Köln, Germany, 1970.

Schierle, Gotthilf Goetz, ed. *Lightweight Tension Structures*, Department of Architecture, University of California, Berkeley, 1968.

Stainless Steel Membrane Roof, Committee of Stainless Steel Producers, American Iron and Steel Institute, Washington, D.C., Booklet AISI SS 902-480-25M-GP, April 1980.

Tension Structures. Their Theory and Practice, *The Architects' Journal*, May 1973.

Thornton, J. A. *The Design and Construction of Cable-stayed Roofs*, Structural Engineer, September 1984.

Zetlin, Lev. Steel Cable Creates Novel Structural Space Systems, Space Forms in Steel, *AISC Engineering Journal*, 1965.

High-rise Buildings

Council on Tall Buildings and Urban Habitat, Lehigh University, Bethlehem, Pa., numerous publications.

Guise, David. *Design and Technology in Architecture,* rev. ed., Van Nostrand Reinhold, New York, 1991.

Hart, F., Henn, W., and Sontag, H. *Multi-Storey Buildings in Steel,* Granada Publications, London, 1978.

Heinle, Erwin, and Leonhardt, Fritz. *Towers,* Rizzoli, New York, 1989.

Jahn, Helmut. *Genesis of a Tower,* Architectural Technology, American Institute of Architecture (AIA), Fall 1983.

Sabbagh, Karl. *Skyscraper,* Viking Penguin, New York, 1990.

Schueller, Wolfgang. *Exercise Manual to the Vertical Building Structure,* Virginia Tech, Blacksburg, Virginia, College of Architecture and Urban Studies, 1990a.

———. *High-Rise Building Structures,* Wiley, New York 1977.

———. *The Vertical Building Structure,* Van Nostrand Reinhold, New York, 1990b.

Taranath, Bungale S. *Structural Analysis and Design of Tall Buildings,* McGraw-Hill, New York, 1988.

Index

A-frame, 459–460
Abnormal loads, 88
Abutment. *See* Buttress
Acceleration of gravity, 19, 90, 332
Accidental torsion, 339
Adjustment factors, 183
Admixtures, concrete, 187
Aesthetics, 7
Air
  curtain, 803
  domes, 640, 642
  inflated members, 799
  inflated structure, 780, 800–803
  pressure, 318, 793, 795
  structures
    ground mounted, 782
    wall mounted, 783
  supported structure, 780–799
Allowable stress approach, 190, 227, 279, 346, 608
Aluminum, 10, 125, 556
Anchorage, 355, 700–702
Angle of twist, 151
Anisotropic material, 179, 703
Anticlastic prestressed membranes, 763–775
Antigravity, 59
Antiprism, 536
Arch, 24, 395, 467–491
  axes: centroidal, 114
    neutral, 133
    principal, 144
    shear, 144
    symmetrical, 144
  circular, 481–484
  fixed, 484
  funicular, 396, 452, 478
  length, 488
  parabolic, 478–481
  trussed box, 527

Archimedean,
  grids, 535
  solids, 540
Archimedes principle, 16, 384
Architectural integrity, 356
Architectural theory, 2–4, 7
Area of reinforcing bars, 189
Aspension, 778
Assembly, building, 344
Asymmetry, effect of, 144, 205, 374
Axial
  deformations, 120, 122
  forces. *See* Forces
  systems, 351, 823

Backfill, 387
Balanced cantilever beam concept, 406, 739
Balloon wood frame, 36, 62
Barrel shells,
  short, 625, 629
  long, 624
Barrel vault, 468
Basement walls, 382–384
Base shear formula, 335
Beam,
  action, 570
  bearing plates, 298
  behavior, 145
  boundary conditions, 146
  building, 514–517
  column, 255, 264, 266, 273, 274, 282
  column loading, 260
  connections, 287
  deflection, 138–143
  span, 48, 508
Beam design, 191–248
  built-up wood beams, 219–221
  composite beams, 246–248

concrete beams, 224–239
  glulam beams, 216–219
  prestressed concrete beams, 239–246
  steel beams, 200–206
  steel reinforced beams, 221–223
  wood beams, 209–223
Beams, 49, 143–148
Beam types, 51, 105–108, 147
  bent, 406
  cable, 733
  cantilever, 105
  continuous, 108
  curved, 145, 408
  deep, 143
  fixed, 108
  grid, 543
  hinge-connected cantilever beams, 107
  inclined, 448–450
  overhanging, 107
  ring, 146
  shallow, 49, 143
  shell, 143, 148
  simple, 105
  trussed, 50, 146
  two-way grid, 557
  wall, 143
Bearing,
  capacity of soil, 164, 167
  connection, 289
  plates, 295–299
  stiffeners, 202
  stress, 131, 202, 212
  wall, 60, 336, 827
Belt trusses, 833
Bending, 143–148
  biaxial, 144, 204, 210
  factor, 267, 269
  members, 143–148, 191
  moment, 103–104

stress, 132
  symmetrical, 144
  unsymmetrical, 144, 204
Bicycle wheel roof, 751, 755
Blast loads, 319
Blocking, 363
Bolted connections, 288–292
Bolts,
  bearing capacity, 290
  shear capacity, 290
Boundary conditions, 105–108
Box girders, 206, 220
Braced,
  buildings, 258, 532
  dome, 591, 640
  frame, 336, 357
Brick
  veneers, 58
  wall, 60
Bridge buildings, 514, 528
Bridge structures, 47
Brittle, 83, 122
Buckling,
  elastic, 257
  lateral, 353
Building,
  braced, 258, 336
  codes, 70
  flexible, 333
  foundations, 162
  irregular, 334, 378
  mass, 19, 332
  materials, 124–128
  nonbraced, 258
  response to load action, 825
  shapes, 3, 32
  structure, 33–36
  symmetrical, 365
  use, 8. *See* Occupancy
  weight, 75
Built-up members, 219
Buoyancy, 16, 381, 384

Butterfly shell, 637
Buttress, 396, 468, 486, 589, 639

Cable,
    beam, 50, 732, 733
    bracing, 66
    catenary, 714
    circular, 725
    column, 154, 157
    dome, 778–779
    elliptical, 728
    elongation, 712, 716
    geometry, 709
    inclined, 718
    length, 716
    material, 702
    nets, 702, 703
    parabolic, 714, 715
    polygonal, 710
    prestretched, 729
    profile, 241
    properties, 702
    sag, 712
    truss, 732
    vibration, 731
Cable-stayed,
    bridges, 738–741
    roofs, 741–751
Cable-supported structures, 50, 697, 736
Caissons, 166
Camber, 153, 218, 502
Cantilever, 407
    beams, 105, 406
    columns, 439
    effect of, 406
    flexural, 355
    footings, 169
    pendulum, 332
    retaining walls, 386
    shear, 355
    structure, 508–514
    tower, 32
Capacity reduction factor, 191
Carbon steel, 177
Cast iron, 470
Castings, 178
Catenary, 474, 592
Cavity walls, 61
Cement, 186
Center of,
    gravity, 113, 249
    force action, 113
    mass, 113, 374
    rigidity, 374
Centrifugal forces, 87, 318
Centroid,
    cross section, 113
    force action, 98, 100
    location, 114–115
Chemical prestressing, 156
Chicago frame, 814, 830
Chord deflection, 503
Circle, 726
Circular bending, 140
Cladding systems, 56, 58
Closepacking polyhedra, 540, 541
Coated fabrics, 703

Codes, 70
Coefficient of,
    active earth pressure, 380
    expansion, 85, 129
    friction, 389
Collapse, 69, 398, 400, 498
Collar,
    frames, 457
    load, 651
    strut, 457
    tie, 457
Column base plates, 296
Column design, 255–283
    composite, 280
    concrete, 275–283
    steel, 261–270
    wood, 270–275
Columns, 53–55
    buckling, 257
    cable-trussed, 154
    cantilever, 439
    effective length, 257
    failure, 261
    foundation, 305
    long, 257, 260, 261, 271, 276
    sections, 256
    shapes, 256
    short, 257, 261, 271, 276
    slenderness, 257
    steel ratio, 278
    straight-line formula, 262
Combustible, 10
Compact sections, 195
Composite beams, 219, 222, 246
Composite structure, 834
Composite members,
    beams, 246–248
    columns, 280
    skins, 58
    walls, 61
Compression ring, 771
Concrete,
    cover, 11, 189
    material, 10, 186
    walls, 61
Concurrency, 91, 93, 95
Cone, 692
Conic sections, 467
Connections, 284–301
    behavior, 286
    reinforced concrete, 301
    steel, 286–299
    tension members, 252
    types, 285
    wood, 299–301
Connector types, 284
Conoidal shells, 672
Construction, 12–15
    conventional, 13
    industrialized, 14
    lift-up method, 548
    loads, 82
    materials, 124–128
    noncombustible, 10
    scaffold method, 548
    segmental, 158
    special, 15
Constructivists, 595, 696
Continuous beams, 108

Contrafleture, points. See
    Inflection points
Corbeled arch, 468
Corbeled domes, 587
Core buildings, 828
Corrosion-resistant steels, 178
Costs, construction, 13
Creep, 188
Cribbed domes, 587
Cubic parabolic cable, 720
Curtain walls, 56
Curvature factor, 216
Curvatures, 24, 599
Cyclic loading, 332
Cycloid, 467, 618, 628
Cylindrical shells, 617–637

Damping, 4, 86, 88
Dead loads, 72, 74
Deconstruction, 4
Deep beam behavior, 49
Deflection,
    lateral, 357–364, 427–430
    permissible, 142, 197, 198, 212, 224
    thermal, 85
    vertical, 138–143, 197, 198, 212
Deformation, axial, 122
Degree of freedom, 332, 334
Density, 19
Depth-to-span ratio, 132, 224, 544
Design determinants, 6–15
Determinacy, 398–400, 498
Determinate, statically, 91, 398, 498
Diagonals. See Bracing, Trussing.
Diagonal tension, 630
Diaphragm,
    action, 350–355
    chord, 354
    flexible, 366
    floor, 351
    rigid, 366
    roof, 354
Differential movement, 58
Dihedral angle, 537
Dimension lumber, 182
Dimensions, 22, 27
Domes, 588, 637–665
    cable, 778
    ribbed, 481, 592
    skeletal, 640, 658
    thin-shell, 640
    tensegrity, 778
    types, 638
Double-layer,
    domes, 590
    grids, 523
    space frames, 538, 539, 646
    suspension roofs, 755
Drainage, 387
Drift, 58, 356
Drop panels, 832
Dual-cable systems, 734
Dual systems, 337
Ductile, 83, 122
Ductility, 336

Dynamic,
    loads, 86
    properties, 18
    response, 731
    lateral force approach, 334
Dynamics, 15

Earth pressure,
    active, 381
    passive, 382
Earthquake action, 332–345
Earth-sheltered buildings, 382–383
Eccentrically braced frames, 337, 348
Eccentric force action, 249, 278
Economy, 12
Effective column length, 257
Effective column length factor, 257–259
Effective depth, 224, 227–228
Effective flange width, 247, 571
Edge members, 770–772
Elastic design, 608
Elastic range, 122
Elongation, tension, 122
Ellipse, 618, 628
Elliptic paraboloid, 663
Emperical design, 16, 49
Environmental context, 6
Epicenter, 329, 330
Equilibrium-form systems, 33
Euler's formula, 257
Expansion, 129, 429
Exponential profile, 326
Eye-bars, 253

Fabrication patterning, 760
Fabric
    domes, 597, 647
    skylights, 696
Fabrics, 703–707
    teflon-coated fiberglass 705
    vinyl-coated fiberglass 705
    vinyl-coated polyester 705
Fasteners. See Connector types
Faults, 330
Ferrocement, 39, 596, 609
Fibonacci series, 28–29
Filler beams, 194
Fillet weld, 293, 294
Films, 702
Fire, 10
    fireproofing, 10–12
    flame shielding, 12
    insulation, 11
    intumescent materials, 11
    isolation, 12
    loads, 9
    protection, 10–12
    resistance, 9, 226
    safety, 8–12
    subliming materials, 11
    wall, 9
First moment of area, 114, 118
Fixed beams, 107
Flat plate (slab) buildings, 830
Flexible connections, 288
Flexural systems, 34, 351, 823

Flexural stiffness, 140
Flexure. *See* Bending
Flitch beams, 221–223
Floating foundations, 165
Floor
    beams, 191–248
    concrete, 224–239
    diaphragms. *See*
        Diaphragms, floors
    framing, 111, 193, 194
    steel, 193–209
    structure, 235
    wood, 209–222
Flying buttresses, 590
Foam shells, 616
Focus of earthquake, 329, 330
Fold,
    arrangement, 565
    member cross-section, 565
    parallel, 572–581
    triangular, 581, 582
    type, 565
Folded plate structures,
    564–583, 566, 568
Footings
    column, 302–309
    combined, 165
    isolated, 165
    plain concrete, 306
    spread, 165
    strap (cantilever), 165
    strip, 165
    wall, 302–309
Force, 91–101
    bending. *See* Bending
    buckling, 257
    buoyant, 381. *See also*
        Buoyancy
    circumferential (hoop). 604
    couple, 97
    cyclic, 86, 87
    drag, 320
    dynamic, 86–88
    eccentric, 249
    flow, 42–45, 351, 400, 401,
        747, 822, 824
    gravity, 69
    hydrostatic, 381
    impact, 87
    inertial, 19, 332
    internal, 102–112
    lateral, 347–390
    lift, 320
    locked-in, 84–86
    meridional (arch), 604
    moment, 103–104
    normal, 320
    overturning. *See* Overturning
    parallelogram, 90
    polygon, 90, 94, 95, 709
    ponding, 79
    prestressing, 157
    properties, 89
    resonant. *See* Resonant
        Loading
    seismic, 328–345
    shear, 103–104
    suction, 328
    surcharge, 381

    thermal, 85
    torsion. *See* Torsion
    transverse, 320
    twisting. *See* Twisting
    uplift, 328, 381
    whiplash, 339
    wracking. *See* Wracking,
        Lateral
Forgings, 178
Form-resistant structures, 34
Foundations, 159, 165–172
    column, 302–309
    deep, 166
    systems, 162, 165
    tensile, 167, 805–808
    wall, 302–309
Frame, 395
    braced, 317, 348
    characteristics, 412
    cantilevers, 414
    footings, 302
    gable, 345
    lateral deflection, 357
    pitched, 445–466
    portal, 403–446
    profile, 451
    types, 408, 415
Framing factor. *See* Building
    type factor.
Free body, 91, 400, 409
Frequency of vibration, 87, 332
Friction, 389
Friction piles 166
Function, 7, 8
Fundamental period of
    vibration, 335, 357
Funicular cable analogy,
    109–110, 709
Funicular shapes, 33, 50, 98,
    138, 397, 410, 452, 473,
    628, 709

Gable
    frame, 460–466
    roof types, 452, 455, 670
Gambrel roof, 452, 455
Geodesic dome, 646, 662,
Geometry
    patterns, 23
    three-dimensional, 522
Gerber beams. *See* Hinge-
    connected cantilever
    beams
Gerberettes, 68
Girders, 206, 220
Glass curtain walls, 59
Glass fiber reinforced plastics,
    616
Glued laminated timber,
    184–185
Glulam beams, 216–219
Glulam frames, 464
Gothic rib patterns, 636
Grade beam, 166
Grandstands, 512–514
Gravity
    anchors, 807
    flow, 31, 45
    loads, 72

    structures, 31, 817
    towers, 32
Green building, 838
Grid, 23
    dimensional, 27
    double-layer, 523
    geodesic, 661
    planning, 26
    space, 536
    stuctural, 26
    triangular, 642, 661
Grillage, 542, 553
Ground
    acceleration, 333
    failure, 330
    movement, 330
    shaking, 330
Groundwater pressure, 167
Guyed mast systems, 50. *See
    also* Stayed structures

Hangar, 510–511, 594, 696
Hardwood, 179
Harmonic vibration, 332
Height-to-span, 474, 571
Height-to-width ratio of
    building, 362
Helix, 149, 375
Hidden loads, 84
High-rise
    building structures, 813–841
    pressure systems, 800–802
    structure systems, 817–820
High-strength bolts, 289
Hinge-connected cantilever
    beams, 107, 432
Hinge location, 411
Hip members, 457
Holes in beams, 197
Hooke's Law, 20, 132, 133, 143
Hoop forces, 622
Horizontal force action,
    318–345
Horizontal force factor. *See*
    Building type factor
Horizontal force flow, 351
Horizontal-span structure
    systems, 35
House construction, 36, 40
Hurricanes, 319
Hybrid building shapes/
    structures, 836
Hydration, 186
Hydrostatic pressure, 319
Hypartensegrity dome, 779
Hyperbola, 618
Hyperbolic paraboloid, 512,
    613, 666–686
Hyperstatic, 91, 398
Hypocenter. *See* Focus of
    earthquake

Icosahedron, 531, 661
Impact forces, 318
Importance factor, 80, 323, 335
Inclined beams, 448–450
Indeterminacy, degree of
    structure, 398–400, 484

Indeterminate, statically, 91,
    398, 414. 484, 498
Industrial architecture, 430
Industrialized construction, 14
Inelastic behavior, 122
Inertial forces, 19, 69, 83, 332
Inflection points, 257, 406, 418
Insulation, 11
Instability. *See* Stability failure
Intelligent structure, 4, 33
Interaction diagram, beam-
    column, 266
Interaction equation, beam-
    column, 267
Internal forces, 102–112, 401
Internal couple method, 634
Irregular buildings, 334
Intumescent materials, 11
Isocontour lines, 651
Isotropic plate behavior, 68
Iron, 177
Isolation, 11
Isolators, 6

Joint method, 95
Joints, 62–65
    construction, 63
    control, 63, 84
    expansion, 63
    movement, 63, 65, 84
    seismic, 63
    structural, 63
    type, 63, 92
Joist floors/roofs, 193, 214,
    452, 455
Jack arches, 489, 491

K-bracing, 832
Kern of member, 170, 378, 397
Kinematics, 15
Kinetics, 15
King-post truss, 153, 710, 713,
    737
Knee-braced portal frame, 421

L-frame, 409
Lamella roofs, 591, 644
Laminated fabrics, 703
Lateral
    bracing, 317
    deflection, 355–364,
        427–430
    force distribution, 365–378,
        367
    force resisting structures, 31,
        38, 348–349, 817
    forces, 318–347
    stability. *See* Stability
    systems, 317, 347–352
Lattice structure, 644
Lightweight concrete, 187
Line diagram, 21–29, 402
Liquid-filled columns, 12
Live load
    arrangement, 432
    reduction, 77, 78
Live loads, 76
Live load-to-dead load ratio,
    78, 745

Load
    action on foundations, 169
    arrangement, 146, 403
    balancing method, 241
    duration, 183
Load bearing structure, 31
Load factors, 345
Load combination probability
        factor, 88, 345
Load flow, 45, 351, 824
Loading patterns, 108–110
Loads, 71, 707
    abnormal, 88
    combination, 88, 345–347
    construction, 82
    cyclic, 322
    dead, 72,74
    dynamic, 86
    earth, 381, 383
    fabrics, 708
    fire, 9
    gravity, 72
    hidden, 84
    hydrostatic, 381
    impact, 86
    inertial, 83
    live, 76
    occupancy, 76
    ponding, 556
    prestress, 735
    radial, 725
    reactive, 84
    resonant, 87, 322, 731
    roof, 79
    seismic, 82, 328–345, 339
    snow, 79, 80
    thermal, 11
    vibrational, 86
    water, 79
    wind, 82, 322, 319–328
Load types, 108–110
Locked-in forces, 84
Logarithmic profile, 323
Long columns, 260
Long-span structures, 394,
        508–515
Long shell. See Shell beam
Low-pressure system, 780
Lumber sizes, 182
Lumped-mass system, 334

Mansard roof, 452, 455
Maraldi angle, 788
Masonry, 10
Masonry arches, minor,
        489–491
Mass, 19, 86, 429
    building, 83
    center of, 83
    dampers, 4
    distribution, 83
    production, 14, 394
Mat foundation, 165
Material properties, 120–129,
        702–707
    brittleness, 123
    ductility, 123
    density, 19

mechanical, 120–124, 126,
        127
    weights, 73, 128
Mechanics, 15
Mechanics of materials, 15
Megastructures, 8, 514, 836,
        837
Membrane,
    action, 603, 647
    cylindrical, 622, 624
    equation, 604
    forces, 603, 604, 622,
        647–650
    hoop forces, 622, 647
    hyperbolic paraboloid,
        672–676
    material, 702–707
    meridional forces, 604, 647
    roof details, 803–805
    spherical, 647–650
    tensile, 603
Membrane skin structures, 697
Mero joint, 523, 546, 645
Meshes, 703
Method of joints, 95, 492, 500
Method of sections, 101, 492,
        500
Mixed construction, 834, 836
Mobile homes, 14, 39
Model codes, 70
Modified Mercally Intensity
        Scale (MMI), 331
Modular coordination, 14,
        21–29, 394
Modular ratio, 28
Modulus, shearing, 124
Modulus of elasticity, 122, 128,
        188
Modulus-to-weight ratio, 129
Moment, 90
    analogy, 711
    bending, 103–104
    biaxial, 144
    coefficients, 233
    connections, 287
    diagrams, 106, 109
    magnification, 264, 266, 273
    two-way slabs, 551–555
Moisture content, 180
Moments of inertia, 116–119
Morphology, 4
Mortar, 61
Movement, 58, 84
Mullions, 59
Multibay structure, 39, 41, 431,
        433
Multi-degree of freedom
        system, 334
Multilayer space frames, 535–
        563
Multistory rigid frames, 441–
        444
Multistory structures. See
        Highrise structures

Nail slip, 364
Natural period of building, 86,
        87, 332, 338, 429, 731
Net section, 250, 255

Network domes, 645
Neutral plane, 651
Newton, 19, 90
Newton's three laws of motion,
        18, 83, 91, 120
Nonbraced buildings, 258
North light shell, 637
Normal force method, 325

Occupancy, 8, 76
Occupancy loads, 76
Occupant comfort, 356
Octahedron, 531
Ogee arches, 470
One-way span, 110
Open-web steel joists, 193
Origami, 564
Oscillations, 86
Overturning, 352, 378–379,
        387, 388
Overhanging beams, 107

P-A effect, 264, 266, 435
Parabolic
    arch, 478–481
    cubic, 720, 721
    funicular, 410
    quadratic, 715–716
    tendon, 242
Partition walls, 60
Passive soil pressure, 382
Pendentives, 589, 639, 664
Pendulum column, 69
Performance, 13
Period of vibration, 86, 332
Permeability, 385
Piece-by-piece method, 215
Pier foundations, 166
Piles, 166, 170
    batter, 166, 171
    bearing, 166
    friction, 166
    tension, 166
    types, 166
Pitched frames, 445
Pitched roofs, 452, 453
Photoelasticity, 68, 138
Plain concrete footings, 306
Planar structures, 393
Planning grid, 26
Plastic
    design, 346
    deformation, 123
    shells, 613–617
Plastics, 125, 127
Plate analogy, 551
Plate girders, 206
Plate tectonics, 328
Platform framing, 61, 62
Platonic solids, 522, 537, 540,
        661
Plywood
    beams, 219
    sheathing, 215
    shells, 610, 620
    stressed-skin panels, 221
Pneumatic
    classification, 781
    dual-wall systems, 800

hybrid systems, 800
    lens-shaped structure, 792
    structure, 779–803
    tubular member systems, 800
Pointed arch, 468, 589
Poisson's ratio, 123, 124
Polar moment of inertia, 117,
        150
Pole construction, 39
Polyester, 125, 705
Polygonal cable, 710
Polygons, 25, 658
Polyhedra, 524, 537, 540, 541
Polyhedral roofs, 525, 526, 565,
        644
Ponding, 556
Portal frames, 405, 408
Portal method, 437, 442
Portland cement, 186
Post and beam structures, 404,
        406, 409, 432, 454
Posts and timbers, 182
Post-tensioning, 153, 239, 606
Pratt truss, 206
Prefabrication, 14
Pressure
    air, 320
    bearing, 388
    contact, 388
    earth/water, 388
    gage, 793
    ground, 388
    external, 325
    hydrostatic, 387
    internal, 325
    stagnation, 320
    uplift, 321, 385
    wind, 322
    velocity, 320
Prestress loss, 243
Prestressed concrete, 155–161,
        239–246, 633
Prestressed membranes, 698,
        756–779
Prestressing, 153–161, 239,
        606, 633, 654
Prestretched cable, 729
Pretensioning. See Prestressed
        concrete
Principal
    axes, 144
    parabolas, 672
    radii of curvature, 604
    stress flow, 138
Product of inertia, 117
Progressive collapse, 69, 319
Projected area method, 325
Proportions, 28
Punching shear, 304
PVC-coated polyester, 705
Pyramid, 532
Pyramid, truncated, 534

Quasi-static approach, 322, 334
Queen-post truss, 491, 737

Radial stresses, 465, 485, 486
Radiolaria, 522, 585
Radius of gyration, 117

Raft foundation, 165
Rafter
  jack, 458
  roof, 552, 458
Rankine theory, 380
Reduction of live loads. *See*
  Live load reduction
Redundancy, 34, 398–400, 498,
  530, 745
Regular polygons, 25
Reinforced concrete, 186–191
  beams, 224–239
  columns, 275–283
  footings, 302–309
  shells, 607–609
  slabs, 232–234
Reinforcing steel, 189
  cover, 11, 189
  cut off and bent points, 233
  properties, 189
  stirrups, 229–230
  temperature, 232
  ties, 277
  wire fabric, 189
Resistance factor, 345
Resonant loading, 87
Response spectrum, 333
Resultant load action, 93
Retaining walls, 385–390
Reticulated shell, 658
Ribbed domes, 481, 592
Richter scale, 331
Ridge beam, 457
Rigid connections, 288, 435
Rigid floor diaphragms, 351
Rigidity. *See* Stiffness
Ring beams, 655, 773
Roof
  slope, 52, 80
  shapes, 430, 452
  structures, 5, 50
Roof pressure coefficients, 327
Roof slope factor, 81
Ropes, 254, 702
Rules of thumb, 47–49
Rupture length, 327

Saddle roof structures, 757–763
Safety factors, 190, 708
Sandwich
  panels, 221, 585
  shells, 612, 616
Sawn Lumber, 179
Scale, effect of, 45–49, 75
Schwedler dome, 523, 644, 659
Second moment of area. *See*
  Moment of inertia
Section, method of, 101
Section properties, 113–119
Seismic
  action, 332–335
  coefficient, 332
  loads, 328–345
  zone factor, 335
  zones, 335
Seismograph, 330
Sequence, 26
  arithmetic, 28
  geometric, 28

harmonic, 28
Shaft, thin-walled, 375
  closed, 375
  open, 377
Shallow beams, 49
Shear, 103–104
  allowable stresses, 197, 229
  cantilevers, 356
  center, 149
  connections, 286
  connectors, 246–248
  deflections, 212
  failure, concrete beams, 228
  lag, 249
  punching, 304
  reinforcement, 228
  stirrups, 229–230
  strain, 124
  stress, 150, 197, 211
  system, 676
  walls, 359–365
  walls with outriggers, 833
  wracking, 440
Shearing modulus, 124
Sheet piling, 386
Shed roof, 452, 455
Shell, 584
  analogy, 659
  arch, 607
  beam, 148, 607
  butterfly, 637
  catenary, 689
  conoidal, 672
  intersecting, 688
  material, 607–617
  northlight, 637
  plastic, 613–617
  reinforced concrete, 607–609
  reticulated, 584
  segmental, 606
  steel, 612–613
  stressed skin, 512
  tubes, 834
  wood, 610–612
Shells
  compressive, 584
  tensile, 584
Short shell. *See* Shell arch
Shotcrete, 608
Shrinkage, 85, 188, 430
Sidesway, frames, 403–408
Single-degree of freedom, 332,
  334, 342
Single-layer space frames, 524–
  535
Site coefficient, 335
Skeleton
  buildings, 830
  domes, 591
  shells, 612
Skyscraper, 814
Slab
  action, 570
  analogy, 552, 555
  design, 232, 551
  flat, 554
  one-way, 232–234
  two-way, 554
  waffle, 553

Slenderness ratio, 250, 257, 262
Sliding, 387, 389
Slip-critical connection, 289
Snow, 79
Snow load map, 80
Snow load reduction factor, 81
Soap bubbles, 788
Softwood, 179
Soil, 161–165
  bearing capacity, 164, 167
  classification, 163, 164
  cohesionless, 163, 380
  cohesive, 163, 380
  contact pressure, 388
  expansive, 165
  pressure, 303
  properties, 163–165
  settlement, 164
  weight, 380
Space frame, 521–563
  double layer, 538, 539, 646
  grids, 536
  joints, 545–548
  multilayer, 535–563
  single-layer, 524
Space statics, 527–532
Span
  one-way, 110
  two-way, 110
Span-to-depth ratio, 49, 224
Span-to-width ratio, 572
Specifications, 13
Specific gravity, 180
Specific elasticity, 129
Specific strength, 129
Spread footings, 165, 166
St. Venant's torsion, 150, 151,
  375, 377
Stability, 379, 398, 498, 698,
  749
  building, 316–392
  collapse. *See* Collapse.
  elastic buckling, 257
  inelastic buckling, 260
  material failure, 260
  structure, 1, 30, 356, 378
Staggered hole arrangement,
  250
Staggered trusses, 830
Stagnation pressure, 320
Statics, 15
  planar, 89–101
  spatial, 527–529
Static pressure approach (force
  procedure), 322, 334
Stayed structures, 50
Stand-alone structures, 350
Steel
  allowable stresses, 196
  beams, 193–208
  coefficient of expansion, 126
  columns, 261–270
  connections, 286–299
  flexure, 195
  fire safety, 10–12
  floor framing, 194
  frames, 394
  material, 176–179
  modulus of elasticity, 126
  plate girders, 206–208

prestress, 240
properties, 176–179
ratio, 226, 230
reinforcing, 608
shells, 612–613
strength, 177–178
Stiffeners, girder, 207
Stiffness, 86
  absolute, 371
  building, lateral, 355
  flexural, 140
  frame, 417, 420
  relative, 370, 372
  structure, 1, 30
  torsional, 151
Stirrup spacing, 229
Story drift, 356
Strain, 120
Strain diagram, 132
Strain hardening range, 123
Strands, 254, 702
Strength, 1, 30, 128
Strength design method, 190,
  345
Strength-to-weight ratio, 129
Strength-reduction factors, 191
Stress
  axial, 130
  bearing, 131, 202, 212
  bending, 130, 132–134
  combined axial and bending,
    134
  combined shear and bending,
    137
  direct, 130
  radial, 465, 485, 486
  shear, 130, 134–137, 150,
    197, 229
  simple, 130–132
  thermal, 130
  torsional, 130, 150
  trajectories, 132, 138
Stress-strain curve, 120–122
Stressed-skin systems, 512,
  572, 612, 613, 646
Strip beam method, 185
Structuralism, 2
Structural stability, 356
Structure
  bearing, 31
  braced, 258
  horizontal-span, 33, 37
  planar, 36, 818
  lateral force resisting, 31, 38,
    348–349
  layouts, 43
  nonbraced, 258
  secondary, 31, 350, 817
  systems, 348–349
  vertical, 36
Stud walls, 61
Subliming materials, 11
Superelliptical, 772
Support
  conditions, 49, 92, 529, 552,
    556, 639
  structure, 29–49, 676–678
  types, 531, 532
Surcharge loads, 381, 387

Surface
  anticlastic, 599
  classification, 598–603
  complex, 602
  developable, 600
  free form, 602
  funicular, 602
  hybrid structures, 775–779
  minimal, 602
  natural, 586
  optimal, 602
  rotational, 599
  ruled, 599
  saddle, 599
  sculptural, 602
  synclastic, 599
  translational, 599
  warped, 675
Suspension
  building, 828
  cables, 50
  hanger, concrete, 159
  roof, 719, 752
Suspended
  cables, 50
  curtains, 59
  glazing, 59
  roofs, 751–756
Symbolism, 2
Symmetry, degree of, 144
Synthetic materials, 125, 127
Systems,
  building, 14
  structure, 32

T-beams, 224
Tectonics, 4, 25, 55
Teflon-coated fiberglass fabric,
  705
Temperature
  difference, 85
  loads, 657
  reinforcement, 232
Tendons, prestressed, 157, 239
Tensegrity, 523, 642, 778–779
Tension field action, 206, 207
Tension
  anchors, 807
  cables, 254
  connections, 252
  design, 248–255
  effective net area, 250
  elongation, 122
  foundations, 805–808
  hangers, 253
  members, 248–255
  membranes, 697, 698
  piles, 808
  prestressed concrete, 239
  ring, 481, 639, 661
  rods and bars, 253
  slenderness ratio, 250
  trajectories, 132, 159
Tent structure, 724
Tension roof structures, 695–
  812
Tetrahedron, 530
Theorem of parallel axes, 118
Thermal

loads, 11
  movement, 59, 441
Thermoplastics, 125, 127
Thermosets, 125, 127
Thin-shell structures, 592
Three-dimensional structures,
  33
Three-hinge frame, 409, 462
Three-moment theorem, 416
Thrust, lateral, 30
Thrust lines, 450
Tied columns, 275
Tie plates, 249
Tie rods, 254, 474
Ties, 478
Timber, 181
Tolerances, 63
Topography, 6
Tornadoes, 319
Torsion, 148–153, 230, 374
  bending, 152, 377
  ring, 396
  warping, 152, 377
Torsional shear stress, 150, 377
Torus, 600
Towers, 31
Trajectories of principle
  stresses, 138, 655
Trampoline like roof structure,
  730
Transfer structure, 161, 823
Transformed area method, 222,
  246
Translucency, 707
Transmissibility, 90
Tree columns, 53
Triangle method. See Force
  polygon.
Tributary area, 77
Tripod, 530
Truss, 159, 395, 493, 496,
  491–507
  analogy, 152
  beam, 50
  belt, 833
  bowstring, 494, 500
  box arches, 527
  cable, 732
  complex, 492
  compound, 492
  design, 500–507
  determinacy, 498
  geometry, 493
  Fink, 492, 495
  Funicular, 500
  Howe, 492, 495
  inclined, 575
  k-bracing, 832
  king-post, 491
  knee-bracing, 421
  Latticed, 502
  outrigger, 833
  Pratt, 492, 495
  Queen-post, 491, 737
  Scissor, 495
  simple, 492
  space, 542, 543
  staggered, 830
  two-way, 542, 543, 557
  types, 493

Warren, 495, 550
  x-bracing, 502
Trussed frame structures, 832
Tsunamis, 330
Tubular buildings, 834
Turbulence, 320
Twist factor, 675
Twisting, 149, 230
Two-dimensional structures,
  393
Two-hinge frame, 419, 463
Two-way horizontal-span
  systems, 33
  span ranges, 47

U-factor, 249
Ultimate strength, 120
Ultimate strength design. See
  Strength design method
Umbrella-type roofs, 670
Unbraced length, 196
Underreinforced concrete, 227
Unstable structure, 400
Unsymmetrical bending, 204–
  206
Uplift, 385, 749
  Use, 8. See also occupancy

Varignon's theorem, 98, 113
Vaults, 584, 617, 619
  cloister, 587, 686
  cross, 589, 686
  groined, 587, 686
  iron-glass, 635
  Lamella, 595, 618, 635
  ribbed, 472, 835
  thin tile, 592
  timbrel, 592
Velocity pressure, 320
Velocity profile, 323
Veneers, 58
Vertical force flow, 824
Vertical building structure, 36,
  818
Vibration, 86
Vierendeel members, 66, 403,
  443, 444, 548

W-folds, 580
Waffle slabs, 542
Wall
  beams, 49
  footings, 162, 306
  structures, 60, 827, 834
Walls, 55–62
  basement, 384
  bearing, 60, 827, 834
  buttress, 387
  cantilever, 385
  cavity, 61
  counterfort, 387
  curtain, 56
  gravity, 385
  lateral deflection, 359–362
  partition, 60
  masonry, 60
  reinforced concrete, 61
  retaining, 386
  sheathed, 362
Warped surfaces, 668, 675

Warping torsion, 152, 377
Water and earth pressure loads,
  380–385
Water-cement ratio, 187
Water table, ground, 380
Web buckling, 206
Web crippling, 202, 208, 298
Web deflection, 358, 503
Web reinforcement. See Stirrup
  spacing.
Weep holes, 387
Weight of
  building, 73, 75
  materials, 73, 128
Welded connections, 293–295
Welded wire fabric, 189
Weld types, 293
Whiplash effect, 338, 339
Width-to-thickness ratio, 195
Wind, 82, 319
  action, 321
  character, 319–328
  exposure categories, 323
  hurricanes, 319
  loads, 322
  map, 324
  periods, 86, 87
  pressure, 321, 325
  pressure coefficients, 324
  suction, 328
  tornadoes, 319
  tunnel testing, 320
  turbulence, 320
  uplift, 328, 749
  velocity, 322
  velocity profile, 323
Wire rope, 254
Wood
  beam design, 209–223
  column design, 270–275
  connections, 299–301
  frame design, 460, 464
  material, 179–186
  shells, 610–612
  tension member design,
    254–255
Wood-frame shear walls.
  362–364
Working-stress design. See
  Allowable stress approach
Wracking, lateral, 419, 440
Wrought iron arches, 470

X-bracing, 317

Y-shaped tree columns, 53, 824
Yarn
  filling (weft, woof), 703
  warp, 703
Yield stress, 123, 126
Young's modulus. See Modulus
  of elasticity.

Zeiss-Dywidag System, 609
Zones
  seismic, 331
  snow, 80
  wind, 324